S0-DTY-988

Biochemistry
AND
Human Metabolism

Biochemistry AND *Human Metabolism*

BY

BURNHAM S. WALKER, M.D., Ph.D.

Associate Pathologist in Chemistry
Burbank Hospital, Fitchburg, Mass.
Formerly Professor of Biochemistry
Boston University School of Medicine

WILLIAM C. BOYD, Ph.D.

Professor of Immunochemistry
Boston University School of Medicine

ISAAC ASIMOV, Ph.D.

Associate Professor of Biochemistry
Boston University School of Medicine

THIRD EDITION

BALTIMORE

THE WILLIAMS & WILKINS COMPANY

1957

First Edition, 1952
Second Edition, 1954
Third Edition, 1957

Made in United States of America

Library of Congress
Catalog Card Number
57-8453

Composed and Printed by
THE WAVERLY PRESS, INC.
Baltimore 2, Md., U.S.A.

Preface to the Third Edition

The field of biochemistry has been expanding at a rate which is gratifying and, to authors of textbooks on the subject, somewhat appalling. An interval of three years since an earlier edition is sufficient time to cause a text to seem almost archaic in spots.

The general structure of this book has been altered only in minor ways. The chapter on "Tissue Chemistry" has grown to unwieldy size and has undergone fission into two chapters, the first of which is entitled "Carbohydrates and Lipids", the second remaining "Tissue Chemistry". The chapter on "Enzymes" has gone through a similar metamorphosis, and now there are chapters entitled "Enzymes and Coenzymes" and "Enzymes and Enzyme Systems" in its place. The information contained in two of the appendices of the earlier edition has been distributed through the text. In particular, an enlarged and, in our opinion, a useful section on thermodynamics serves to introduce the chapter on "Enzymes and Coenzymes".

The main effort in preparing this edition has been, of course, to keep the text as nearly abreast of the field as is humanly possible. As examples, we have introduced new material on:

optical configuration of the phosphoglycerides
meromyosins
the high-energy acyl-mercaptan bond in metabolism
abnormal hemoglobins
blood group substances
the role of chelation in enzyme activity
the function of metallo-flavo-enzymes
corticotrophin A structure
melanocyte-stimulating hormone
aldosterone
the sodium pump theory of membrane potentials
nucleic acid fine structure and theories of autoreproduction
purine antagonists in cancer chemotherapy
the carbon cycle in photosynthesis
interpretations of obesity
the role of coenzyme A in the tricarboxylic acid cycle

carbamyl phosphate and the urea cycle
manganous ion and enzyme action
the structure of vitamin B_{12}
properdin

This is only a partial list chosen at random. We have not hesitated, also, to expand or rewrite older material when experience indicated that it could be made clearer to students. As examples we may mention the section on blood clotting and the fatty acid oxidation cycle.

The authors are grateful for the six years that have now been spent on the three editions of the book. This opportunity has enabled them to continue to follow the growth of a vigorous science and to appreciate anew its beauties. It is to these and the other, nearly all intangible, rewards of authorship that this book is dedicated.

BSW
WCB
IA

Contents

PART I

Structure

CHAPTER 1

Proteins and Amino Acids

Just as it is necessary when beginning the study of medicine to learn first of all the gross anatomy of the human body, so in beginning the study of the chemistry of life processes it is necessary to study first of all the structural chemistry of living matter. The characteristic material of living cells is called *protein*. The term was first applied to complex nitrogenous substances found in plant and animal tissues by Mulder in 1838. The word comes from Greek roots meaning "preeminence". If we compare a living cell with a machine, then we can compare the protein with the steel and brass out of which the engine is constructed.

All living cells contain protein, carbohydrate, lipids, water, and inorganic ions, linked together in various ways to form compounds and complexes. Most of our chemical study of these materials comes only after the cells have been taken apart and the individual constituents isolated. The method of taking the cell apart often influences the result. For instance, if we represent the various constituents of a cell by the letters, A, B, C, D . . . , and assume that by rough treatment of some sort we split the cell up into fragments which can be dealt with chemically, it is easy to see that we might come out with complexes such as AB, BC, ABC, CD, CDE . . . , depending on the methods used and the severity of the treatment. This does in fact happen and it is always difficult to make sure that what has been isolated is a reproducible material which actually represents a constituent of the cell, for if at one time we obtain AB from the cell and another time BC or ABC, our chemical analyses in the two instances will be somewhat contradictory, or at least inconsistent. Indeed, since there is every reason to suspect that the protein in the cell is rarely if ever free, the idea of pure protein is in itself an idealization and represents something which, probably, is not to be found in the average tissue. But, like many other abstract ideas, it is a very useful one; and we shall find it very valuable to examine the properties of purified proteins and to try to apply this knowledge to the behavior of the living cell.

That the concept of protein is not entirely an imaginary one has been demonstrated by the isolation of proteins with reproducible properties from various types of tissue. It is true that the situation is not as simple as it was thought to be by Mulder, who believed that there was only one protein which combined with sulfur or phosphorus or both to give various compounds, and that it was these which were present in the tissues of plants and animals. Indeed, we know that there are a very large number of proteins although we have not obtained very many of them in a pure state. It is not usually to be expected that even a single organ will contain but one sort of protein, for diversity of chemical structure is required by the manifold functions performed by most of the organs of the human body. We can not state, for instance, that there is such a thing as a "kidney protein" because the kidney contains proteins of various types. So do most of the tissues. However, there are proteins more or less characteristic of certain types of tissues and our ability to isolate them in more or less pure form justifies our naming them for the tissue from which they come or for some special property or activity which they exhibit.

DEFINITION OF PROTEIN

It is at once difficult and easy accurately to define a protein. It is easy in the sense that it is not difficult to give the reader a fairly clear idea of what we mean by the word. It is difficult in the sense that we find difficulty in being absolutely precise about our definition. Suppose we simply say for the time being that proteins are large molecules, of molecular weight of the order of several thousand to several million, occurring in the tissues of plants and animals and containing carbon, hydrogen, oxygen, nitrogen, and sometimes other elements, and constructed largely from *amino acids.* This definition serves to differentiate proteins from all the other compounds we shall study in this book.

AMINO ACIDS

The fundamental structural units of proteins are amino acids which are held together by the *peptide linkage.* That is, the carboxyl group of one amino acid has combined with the amino group of another, water is split off, and the —CONH— link is formed.

$$R\text{—}COOH + R'\text{—}NH_2 \rightarrow R\text{—}CONH\text{—}R' + H_2O$$

All the amino acids naturally occurring in proteins are α-amino acids—in other words, have an amino group attached to the carbon next to the carboxyl group. Some have also an amino group elsewhere. Also, certain amino acids have two carboxyl groups. It is mainly the presence of these extra amino and carboxyl groups, which seem to take no part in the peptide linkage holding the protein together, which explains the amphoteric properties of proteins and their behavior as dipolar ions. At suitable pH a free carboxyl group can ionize, giving R—COO⁻ plus a proton, and under other conditions an amino group can acquire a proton, giving us $R'—NH_3^+$. Some amino acids contain also other groups capable of ionizing such as hydroxyl, imidazole, and guanidine.

The amino acids which are of general occurrence in proteins may be divided into the following classes. Each class contains amino acids with a certain type of group which confers characteristic properties on the molecule.

1. Monoamino-monocarboxy-α-amino acids whose side chains differ only in the length of the hydrocarbon chain and its degree of branching.

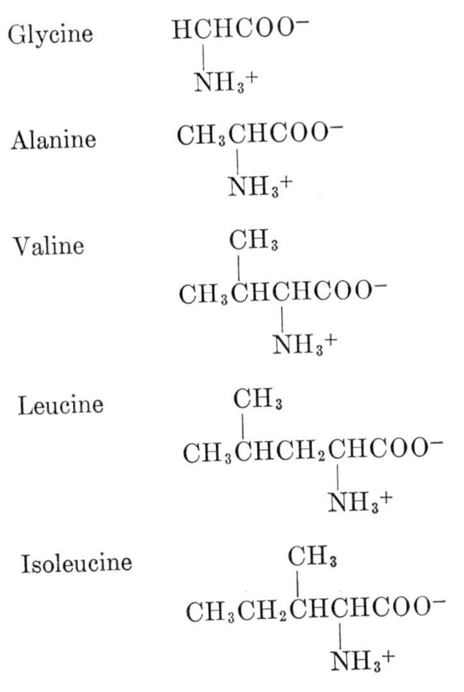

2. Amino acids containing pyrrolidine rings rather than paraffin side chains, and which are imino acids rather than amino acids.

Proline $CH_2CH_2CHCOO^-$ / $CH_2 — NH_2^+$

3. Amino acids containing benzene or indole rings in the side chains.

Phenylalanine

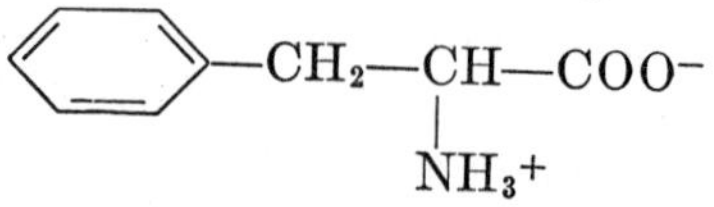

$$\text{—}CH_2\text{—}\underset{\displaystyle NH_3^+}{\underset{|}{CH}}\text{—}COO^-$$

Tyrosine

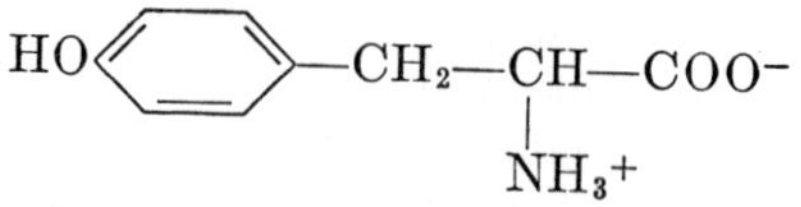

$$HO\text{—}\quad\text{—}CH_2\text{—}\underset{\displaystyle NH_3^+}{\underset{|}{CH}}\text{—}COO^-$$

Tryptophane

$$\text{—}CH_2\text{—}\underset{\displaystyle NH_3^+}{\underset{|}{CH}}\text{—}COO^-$$

N
H

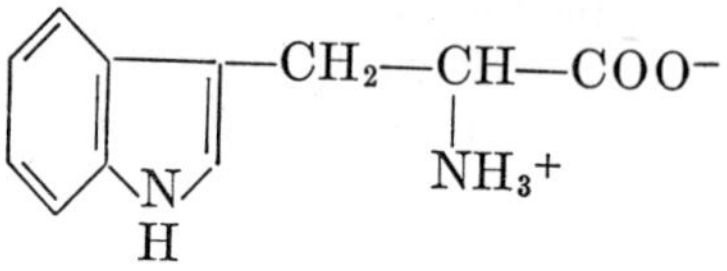

4. Amino acids containing hydroxyl groups whose acidity is so feeble that it makes no contribution to the amphoteric properties of the protein.

Serine

$$HO\text{—}CH_2\text{—}\underset{\displaystyle NH_3^+}{\underset{|}{CH}}\text{—}COO^-$$

Threonine

$$CH_3\text{—}\overset{\displaystyle OH}{\overset{|}{CH}}\text{—}\underset{\displaystyle NH_3^+}{\underset{|}{CH}}\text{—}COO^-$$

5. Dicarboxy-amino acids and their amides.

Aspartic acid (aspartate ion)

$$\overset{\displaystyle COO^-}{\overset{|}{CH_2}}\underset{\displaystyle NH_3^+}{\underset{|}{CH}}COO^-$$

Asparagine

$$\overset{\displaystyle CONH_2}{\overset{|}{CH_2}}\underset{\displaystyle NH_3^+}{\underset{|}{CH}}COO^-$$

Glutamic acid (glutamate ion)

$$\overset{\displaystyle COO^-}{\overset{|}{CH_2}}CH_2\text{—}\underset{\displaystyle NH_3^+}{\underset{|}{CH}}COO^-$$

Glutamine

$$\overset{\displaystyle CONH_2}{\overset{|}{CH_2}}CH_2\underset{\displaystyle NH_3^+}{\underset{|}{CH}}COO^-$$

6. Diamino and other basic amino acids.

Lysine (lysinium ion)

$$^{+}H_3N{-}CH_2{-}CH_2{-}CH_2{-}CH_2{-}\underset{\displaystyle NH_3^{+}}{\underset{|}{CH}}{-}COO^{-}$$

Histidine

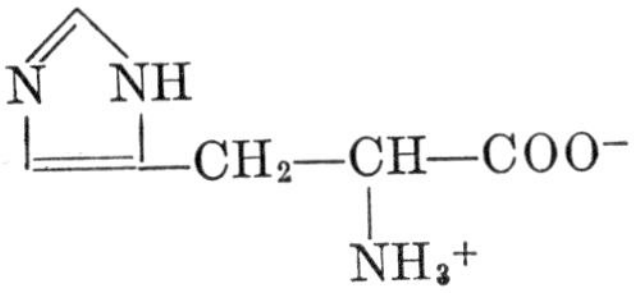

$$\text{(imidazole ring: N, NH)}{-}CH_2{-}\underset{\displaystyle NH_3^{+}}{\underset{|}{CH}}{-}COO^{-}$$

Arginine (argininium ion)

$$\begin{array}{l} NH_2 \\ | \\ C{=}NH_2^{+} \\ | \\ NHCH_2{-}CH_2{-}CH_2{-}\underset{\displaystyle NH_3^{+}}{\underset{|}{CH}}{-}COO^{-} \end{array}$$

7. Sulfur-containing amino acids.

Cysteine $HSCH_2\underset{\displaystyle NH_3^{+}}{\underset{|}{C}}HCOO^{-}$

Cystine

$$\begin{array}{l} SCH_2\underset{\displaystyle NH_3^{+}}{\underset{|}{C}}HCOO^{-} \\ | \\ SCH_2\underset{\displaystyle NH_3^{+}}{\underset{|}{C}}HCOO^{-} \end{array}$$

Methionine $CH_3SCH_2CH_2\underset{\displaystyle NH_3^{+}}{\underset{|}{C}}HCOO^{-}$

The 21 amino acids listed above include all those generally found in proteins. Several other amino acids occur only in a few proteins. In collagen, the chief protein of connective tissue, two hydroxyl-containing amino acids occur, *hydroxyproline* and *hydroxylysine* (formula I).

In thyroglobulin, a protein formed in the thyroid gland, iodine-containing amino acids occur. The more important of these are derivatives of *thyronine* (formula II), an amino acid with a diphenyl ether group in the side chain. The iodothyronine derivative longest known is 3,5,3′,5′ tetra-iodothyronine or *thyroxine* (formula III). More recently, *3,5,3′ tri-iodothyronine* (formula IV) has been isolated. *Monoiodotyrosine* and *di-iodotyrosine* (formula V) also occur in thyroglobulin.

Still another group of amino acids include those which do not form a significant part of the structure of any proteins but which are formed from one or another of the structural amino acids in the normal chemical changes they undergo in the body. Thus, arginine gives rise in the body to *citrulline* and *ornithine* (formula VI). Aspartic acid and glutamic acid give rise respectively to *beta-alanine* and *gamma-aminobutyric acid* (formula VII).

A number of amino acids have been reported in addition to those listed. These, however, either await confirmation or else have been reported only in non-mammalian proteins.

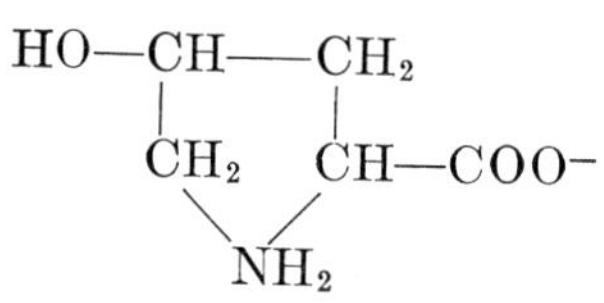

HO—CH——CH_2
CH_2 CH—COO^-
NH_2

Hydroxyproline

CH_2—NH_3^+
CH—OH
CH_2
CH_2
CH—NH_3^+
COO^-

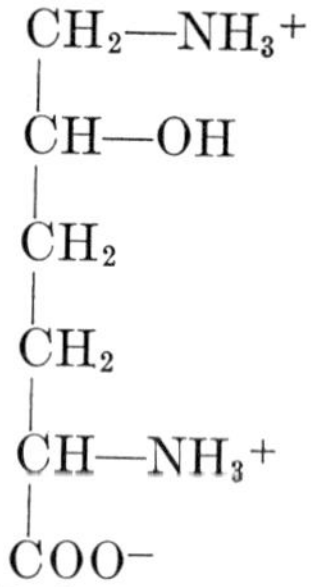

Hydroxylysine
(hydroxylysinium ion)

I

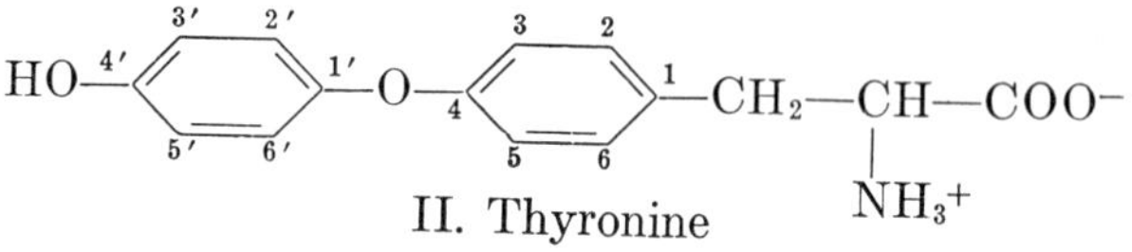

3′ 2′ 3 2
HO—4′ 1′—O—4 1—CH_2—CH—COO^-
5′ 6′ 5 6 NH_3^+

II. Thyronine

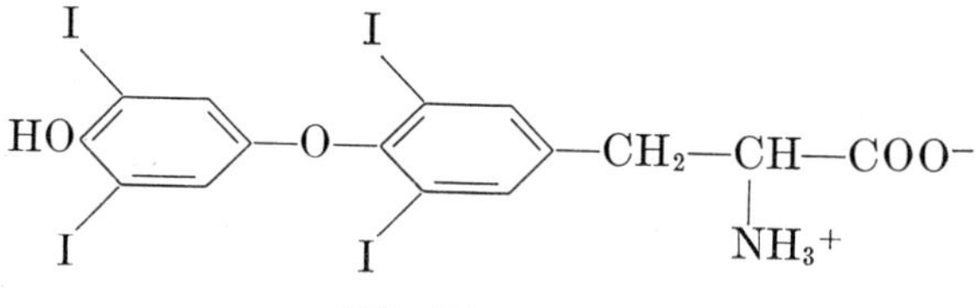

I I
HO —O— —CH_2—CH—COO^-
I I NH_3^+

III. Thyroxine

I I
HO— —O— —CH_2—CH—COO^-
I NH_3^+

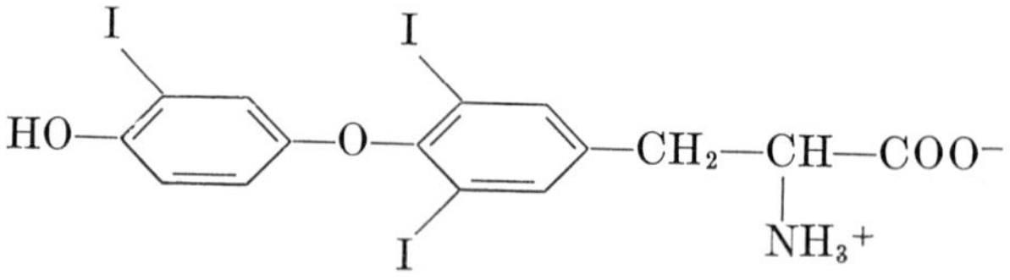

IV. Tri-iodothyronine

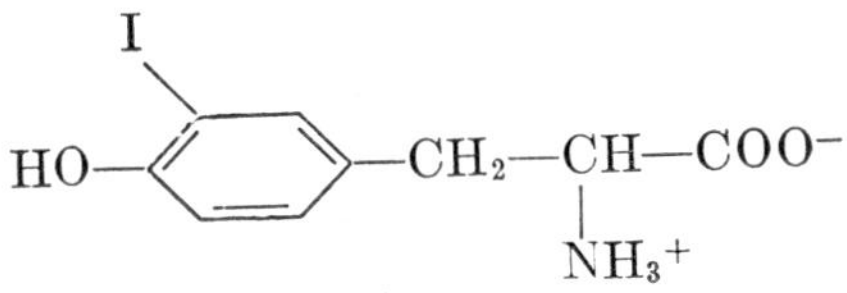

$$\text{HO}-\text{C}_6\text{H}_3(\text{I})-\text{CH}_2-\underset{\displaystyle |\atop \displaystyle \text{NH}_3^+}{\text{CH}}-\text{COO}^-$$

Monoiodotyrosine

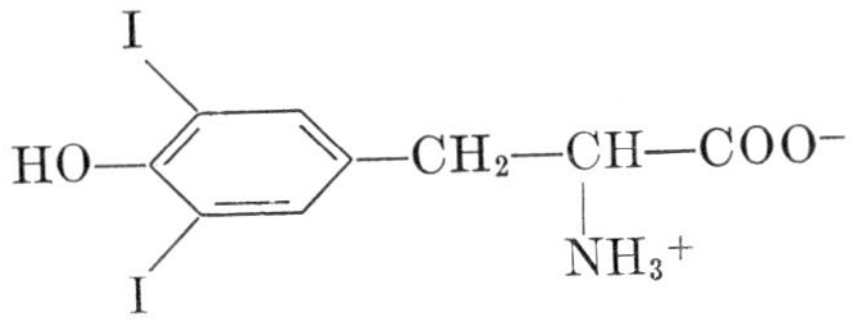

$$\text{HO}-\text{C}_6\text{H}_2(\text{I})_2-\text{CH}_2-\underset{\displaystyle |\atop \displaystyle \text{NH}_3^+}{\text{CH}}-\text{COO}^-$$

Di-iodotyrosine

V

$$\begin{array}{l} \text{NH}_2 \\ | \\ \text{C}=\text{O} \\ | \\ \text{NH} \\ | \\ \text{CH}_2 \\ | \\ \text{CH}_2 \\ | \\ \text{CH}_2 \\ | \\ \text{CH}-\text{NH}_3^+ \\ | \\ \text{COO}^- \end{array} \qquad \begin{array}{l} \text{NH}_3^+ \\ | \\ \text{CH}_2 \\ | \\ \text{CH}_2 \\ | \\ \text{CH}_2 \\ | \\ \text{CH}-\text{NH}_3^+ \\ | \\ \text{COO}^- \end{array}$$

Citrulline

Ornithine (ornithinium ion)

VI

$$\begin{array}{l} \text{CH}_2-\text{COO}^- \\ | \\ \text{CH}_2-\text{NH}_3^+ \end{array} \qquad \begin{array}{l} \text{CH}_2-\text{COO}^- \\ | \\ \text{CH}_2 \\ | \\ \text{CH}_2-\text{NH}_3^+ \end{array}$$

Beta-alanine

Gamma-aminobutyric acid

VII

All of the amino acids have been written in the form called in German a "Zwitterion", which means that an internal salt has been formed by the loss of a proton from the carboxyl group and its acquisition by the amino group. It is in this form that amino acids commonly exist, and the form explains their amphoteric properties, dielectric constants, and electric dipole moments. The German word has been in use for some time, but there is a rapidly growing tendency to employ the equivalent English term "dipolar ion", and this will be done in the remainder of this book.

In order to understand the dipolar ion concept and the light it throws on protein behavior, it will be necessary to devote a little time to a consideration of modern concepts of acids and bases.

ACIDS AND BASES

Current usage defines an *acid* as a molecule or ion which has a tendency to lose a proton. A molecule or ion which has a tendency to accept a proton is a *base*. For each acid there is a corresponding *conjugate base*, formed when the acid loses a proton. An acid and its conjugate base form a *conjugate acid-base pair*; examples are shown in table I-1.

A strong acid is an acid with a strong tendency to lose a proton. The conjugate base of a strong acid is necessarily weak because such a base has little tendency to accept a proton. The conjugate base of a weak acid is strong.

Although it is often convenient, as we have done in table I-1, to define acids and their conjugate bases by the simple equation

$$\text{Acid} \rightleftharpoons H^+ + \text{Base} \qquad \text{Equation I-1}$$

it should be kept in mind that effective proton transfers (acid-base reactions) take place between two conjugate acid-base pairs,

$$\text{Acid}_1 + \text{Base}_2 \rightleftharpoons \text{Acid}_2 + \text{Base}_1 \qquad \text{Equation I-2}$$

The proton is transferred from the acid of one pair to the base of the other. Thus the dissociation of hydrogen chloride in water may be represented

$$HCl + H_2O \rightleftharpoons H_3O^+ + Cl^- \qquad \text{Equation I-3}$$

TABLE I-1

Acid				*Conjugate base*
HCl	$\rightleftharpoons$	H^+	+	Cl^-
H_2SO_4	$\rightleftharpoons$	H^+	+	HSO_4^-
HSO_4^-	$\rightleftharpoons$	H^+	+	$SO_4^=$
H_2CO_3	$\rightleftharpoons$	H^+	+	HCO_3^-
HCO_3^-	$\rightleftharpoons$	H^+	+	$CO_3^=$
H_3O^+	$\rightleftharpoons$	H^+	+	H_2O
H_2O	$\rightleftharpoons$	H^+	+	OH^-
NH_4^+	$\rightleftharpoons$	H^+	+	NH_3
NH_3	$\rightleftharpoons$	H^+	+	NH_2^-

Some molecules and ions are both acids and bases (H_2O, NH_3, HSO_4^- and HCO_3^- are examples in table I-1), and are called *ampholytes* or *amphoteric substances*. Water, an ampholyte present in all living organisms, dissociates into hydrated hydrogen ions and hydroxyl ions:

$$HOH + HOH \rightleftharpoons H_3O^+ + OH^- \qquad \text{Equation I-4}$$

Therefore in systems containing water,

$$[H_3O^+][OH^-] = Kw = 1.2 \times 10^{-14} \text{ at } 25 \text{ C} \qquad \text{Equation I-5}$$
$$= 3.13 \times 10^{-14} \text{ at } 37 \text{ C}$$

The H_3O^+ or hydrated hydrogen ion is also called the oxonium or hydronium ion. It proves convenient to ignore the hydration and write H^+. The molar concentration of $[H_3O^+]$ can be expressed as $[H^+]$ without changing the numerical value so it is possible to state approximately that at the temperatures of living organisms

$$[H^+][OH^-] = 10^{-14} \qquad \text{Equation I-6}$$

It is very convenient to define a quantity

$$pH = -\log[H^+] \qquad \text{Equation I-7}$$

and to use numerical values of pH to describe the acidity (pH less than 7) or alkalinity (pH greater than 7) of dilute aqueous solutions, in which pH plus the analogous quantity pOH must equal 14. In an aqueous solution 0.1 M in hydrogen ion activity, $[H^+] = 10^{-1}$ and pH $= 1$. This would be the situation in an 0.1 N solution of a strong acid such as HCl. Weak acids are incompletely dissociated; the extent of their individual dissociations is commonly expressed by the value of pK for the acid in question. The simplest definition of pK is the pH at which half of the molecules of the particular weak acid are ionized.

This definition of pK can be explained by recognizing that the law of mass action applies to equation I-2 and arranging the equation in the form of an equilibrium.

$$\frac{[Acid_2]\ [Base_1]}{[Acid_1]\ [Base_2]} = K' \qquad \text{Equation I-8}$$

In a water solution of a weak acid, such as acetic acid, if we recall that $[H_3O^+] = [H^+]$,

$$\frac{[H^+]\ [CH_3\ COO^-]}{[CH_3\ COOH]\ [H_2O]} = K' \qquad \text{Equation I-9}$$

If the solution is dilute, the value of $[H_2O]$ will not change significantly with changes in the other quantities, therefore $[H_2O]$ can be absorbed into the constant, giving

$$\frac{[H^+]\,[CH_3\,COO^-]}{[CH_3\,COOH]} = K'\,[H_2O] = K \qquad \text{Equation I-10}$$

or for any weak acid

$$\frac{[H^+]\,[\text{Base}]}{[\text{Acid}]} = K \qquad \text{Equation I-11}$$

Taking the negative logarithm of Equation I-11, we have:

$$-\log\,[H^+] = -\log K + \log \frac{[\text{Base}]}{[\text{Acid}]} \qquad \text{Equation I-12}$$

The negative logarithm of the hydrogen ion concentration is symbolized as pH as we have said above, and by analogy, the negative logarithm of the ionization constant is symbolized as pK. Equation I-12 becomes:

$$\text{pH} = \text{pK} + \log \frac{[\text{Base}]}{[\text{Acid}]} \qquad \text{Equation I-13}$$

Equation I-13 is known as the Henderson-Hasselbalch equation. When an acid is half-ionized, half its molecules are in the form of the base, half remain in the acid form. The ratio is 1, and the log of 1 is zero. At the point of half-ionization, therefore, pH equals pK.

The symbol pK can thus be defined in either of two ways: as the negative logarithm of the ionization constant of a particular acid, or as the pH at which a particular acid is half-ionized. The pK of acetic acid is 4.6.

Dissociation constants, and pK values, can be obtained for anion acids and cation acids as well as for uncharged acids, such as acetic acid. The dissociation constant of the anion acid $H_2PO_4^-$ is

$$\frac{[H^+]\,[HPO_4^{=}]}{H_2PO_4^-]} = K \qquad \text{Equation I-14}$$

and is often called the second dissociation constant of H_3PO_4. The dissociation constant

$$\frac{[H^+]\,[NH_3]}{[NH_4^+]} = K \qquad \text{Equation I-15}$$

represents the acid strength of the cation acid NH_4^+. In all instances, high values of K and low values of pK represent great acid strength with associated weakness of the conjugate base.

A weak acid which, in pure water, will ionize only to a minor extent can be made to ionize more by the addition of hydroxyl ion:

$$CH_3COO^-H^+ + OH^- \rightarrow CH_3COO^- + H_2O \quad \text{Equation I-16}$$

The weaker the acid, the higher the concentration of hydroxyl ion (*i.e.*, the higher the pH) required to bring about half-ionization, and therefore the higher the pK of the acid.

The connection of pK with *buffer action* can now be made plain. In any mixture of an acid and its conjugate base, the addition of a base such as hydroxyl ion will change the pH of the solution less than we might expect since some of the hydroxyl ion will react with the acid to form water as shown in equation I-16. The addition of hydrogen ion will also change the pH of the solution less than we might expect since some of the hydrogen ion will combine with the conjugate base to form the undissociated acid. This resistance to change in pH is what we mean when we speak of buffer action.

From Equation I-13, it can be seen that if the pK of the acid were independent of the concentrations of the acid and the ions present, the pH of a buffer would depend only upon the ratio of salt to acid in the mixture. If the total molality of salt plus acid is M, and a fraction y is in the form of the acid, the equation can be written

$$\text{pH} = \text{pK} + \log \frac{(1 - y)M}{yM} \quad \text{Equation I-16a}$$

or

$$\text{pH} = \text{pK} + \log \frac{1 - y}{y} \quad \text{Equation I-17}$$

The pK of the acid is, in fact, sufficiently constant for the purpose of rough calculations. The pK of acetic acid, for example, varies only by about 8 percent over a range of salt concentrations from 0 to 3 molal.

The change produced in the pH of a buffer mixture by the addition of a small amount of H^+ (or OH^-) ion is easily calculated. Suppose we have a mixture that is 0.10 molal in acetate ion and 0.10 molal in acetic acid. The pH at this point of half-ionization would then be equal to the pK, that is, to 4.73.

If to such a mixture we add 0.01 mol of H^+, we convert 0.01 mol of acetate ion to acetic acid, changing the ratio $(1 - y)/y$ from 0.10/0.10 to 0.09/0.11. The pH changes from 4.73 to (4.73 + log 0.09/0.11) or 4.64. This is a decrease of 0.09 pH units.

If we had started with a mixture of 0.16 molal in acetate ion and 0.04 molal in acetic acid, the initial pH would have been (4.73 + log 0.16/0.04) or 5.33. Adding 0.01 mol of H^+ would, calculating as before, reduce the pH to 5.21, a decline of 0.12 pH units.

Again, if we had started with a mixture 0.04 molal in acetate ion and 0.16 molal in acetic acid, the pH would be 4.13, and the addition of 0.01 mol of H^+ would reduce the pH to 3.98, a decline of 0.15 pH units.

It would appear that buffer action is greatest, therefore, at the point of half-ionization, where both y and $1 - y$ are equal, and where pH equals pK. It is easily shown that this is so.

From Equation I-17, we find that the instantaneous rate of change in pH with change in y (which is obtained by differentiating the equation with respect to y) is:

$$\frac{d\,(\mathrm{pH})}{dy} = -\frac{0.4343}{1 - y} - \frac{0.4343}{y} \qquad \text{Equation I-18}$$

The value of y for which the rate of change in pH with change in y is a minimum (that is, the point at which buffer action is a maximum) is found by differentiating Equation I-18 a second time with respect to y and setting this second derivative equal to 0. We find that

$$\frac{d^2(\mathrm{pH})}{dy^2} = -\frac{0.4343}{(1 - y)^2} + \frac{0.4343}{y^2} \qquad \text{Equation I-19}$$

If Equation I-19 is set equal to 0 and solved for y, it turns out that y is equal to 0.5 and $1 - y$ is therefore also equal to 0.5. The ratio $(1 - y)/y$ is therefore unity and it is at such a ratio that buffer action is greatest.

The *buffer capacity* of a buffer mixture is the total amount of H^+ or OH^- required to bring about a specified change in pH in a specified volume of buffer. This depends not only upon the ratio of salt to acid but also upon the total molality, M, of acid and salt. The buffer capacity at any point during the process of adding H^+ or OH^- is thus directly proportional to M and inversely proportional to d(pH)/dy. If the buffer capacity is symbolized as x', the relationship is

$$x' = \frac{M}{d(\mathrm{pH})/dy} = \frac{y(1 - y)}{0.4343} M \qquad \text{Equation I-20}$$

It can be seen from Equation I-20 how the buffer capacity falls off as y changes. For y equal to 0.5, buffer capacity is at a maximum of $0.576M$. For y equal to 0.333 or 0.667, buffer capacity is 0.512 (89 per cent of maximum). For y equal to 0.1 or 0.9, buffer capacity is $0.207M$ (36 percent of maximum).

Although a buffer system is most efficient at the pH corresponding to the pK of its acid, buffers of specified pH within one or two units of the pK value can be prepared by deliberately varying the proportions of acid and base. The pH of any buffer solution is given by equation I-13, the Henderson-Hasselbalch equation. In the practical use of buffers, the acid and conjugate base are added in high concentration relative to other reactants, in order to control the pH of the mixture and the equilibria of other acid-base pairs. Acid-base *indicators* also follow the Henderson-Hasselbalch equation, but these acid-base pairs should be added in the smallest useful concentration so as not to disturb the pH of the solution under study.

TITRATION CURVES

The end point of the titration of a strong alkali such as sodium hydroxide with a strong acid such as hydrochloric acid is the complete conversion of the two into the salt, in this case sodium chloride, and water. A solution of sodium chloride is neutral, and it follows that the end point we are trying to reach in such a titration is neutrality, where all the hydroxyl ions of the alkali have been combined with hydrogen ions from the acid, and the solution contains hydroxyl ions and hydrogen ions in the same concentrations (to a close approximation) as plain water. The theoretical indicator for such a titration would be one which changes color at pH 7.0. Of the common indicators, bromthymol blue most nearly fills the bill.

Nevertheless, in titrating a strong acid with a strong alkali, it is much commoner to use some other indicator. One which is readily available and which has a striking color change is phenolphthalein. Its color change occurs at a pH well above 7.0, but it is perfectly satisfactory for such a titration. This fact depends ultimately on the shape of the titration curve of a strong alkali and a strong acid. It will be seen from the curve (figure I-1) that as alkali is added to acid the pH changes, at first slowly, but as neutrality is reached the pH changes quite abruptly to pH 7.0, and as soon as a fraction of a drop of alkali is added in excess, rises to considerably higher values. In other words, the slope of the titration curve is very steep at the neutral point. Consequently, it makes little difference if we use phenolphthalein as an indicator instead of bromthymol blue, for the amount of alkali which must be added to bring the pH of the titration mixture from pH 7.0 to 9.7 (the turning point of phenolphthalein) is too small to measure in ordinary titrations. The situation is different if we titrate a weak acid like acetic acid. In this case the pH of the mixture changes at first with considerable rapidity, rising to values of 3 to 4 when only one-tenth of the equivalent amount of alkali has been added. But then the curve changes slope, and there is a point of inflection. At this point, where the acid is just half neutralized, the slope of the curve is slight, and the addition of further alkali, within limits, makes but little difference in the pH. For instance, if we have just half neutralized a liter of normal acetic acid by adding 500 ml. of normal sodium

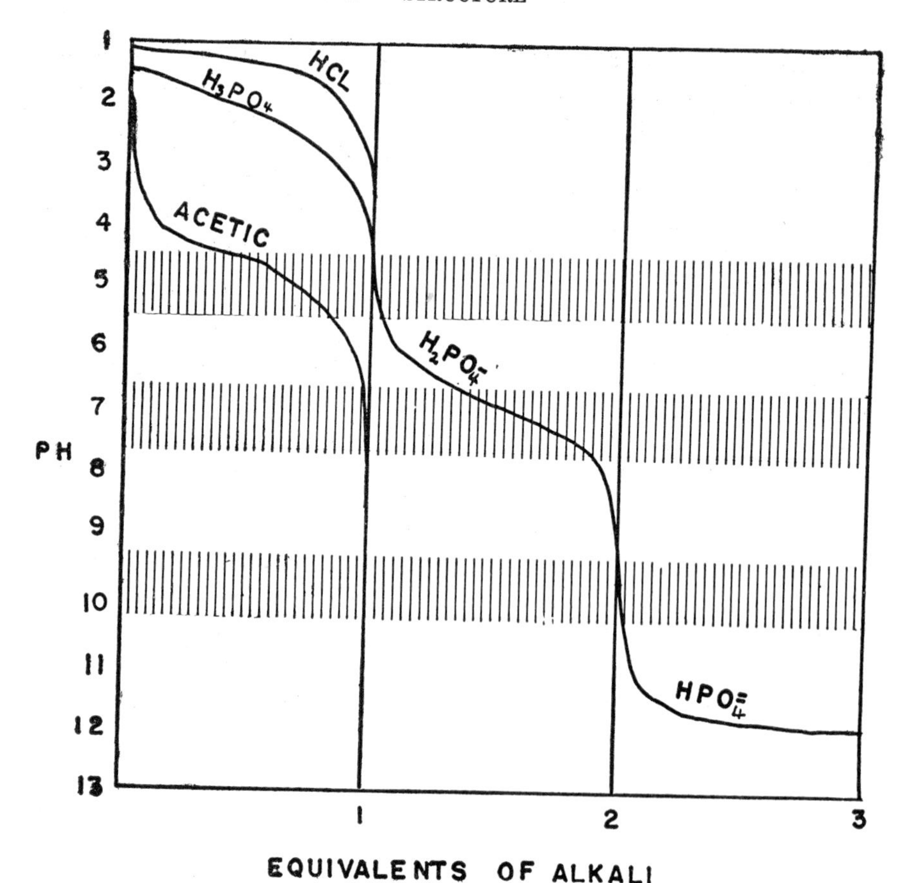

FIG. I-1 Titration curves of hydrochloric, phosphoric, and acetic acids. Upper shaded band, color change of methyl red; middle shaded band, color change of bromthymol blue; lower shaded band, color change of phenolphthalein.

hydroxide solution, then the further addition of 50 ml. of normal sodium hydroxide solution will change the pH only 0.05 pH unit. Such a mixture evidently does not have its pH affected much by the addition of alkali.

Neither would the addition of acid to such a mixture make any greater difference. If we put 50 ml. of normal hydrochloric acid into the above mixture, the result is simply the formation of water (from the hydrogen ions of the acid and the hydroxyl ions of the alkali). This leaves the pH at the value we had before the last 50 ml. of sodium hydroxide solution was added, and this was found to be only 0.05 pH unit less.

As the amount of alkali added to the acetic acid approaches one equivalent, the pH begins to change more rapidly again, and the rate of change is quite fast at the point where one equivalent has been added.

Now we observe another difference. A solution of sodium acetate is not neutral, for sodium acetate, being the salt of a strong alkali and a weak acid, forms some hydroxyl ions in solution by salt hydrolysis. Consequently our end point, where one chemical equivalent of alkali has been added to the acetic acid, is not neutrality, but a pH on the alkaline side of neutrality. Phenolphthalein is a suitable indicator in this also, but this time precisely because its turning point is well above pH 7.0. (In titrating a weak base such as ammonia an indicator which turns at a pH on the acid side, such as methyl red, would be desirable.)

Polyvalent Acids

One gram molecular weight of phosphoric acid contains three equivalents of acid (see Chapter 17), since it possesses three replaceable hydrogens. If we titrate phosphoric acid, we find that the titration curve has three distinct parts, each corresponding to one of the ionization constants, and each resembling the titration curve of an ordinary monobasic weak acid. Each shows the rise in pH, the point of inflection at the pK for the particular hydrogen involved, and the steep portion in which the pH changes rapidly. These steep portions occur at the points where one and two equivalents of phosphoric acid have been added. The three points of inflection occur at the points of half neutralization of the three ionizable hydrogens (at about pH 2, 7, and 12, respectively).

If we titrate an amino acid such as glycine (see formula VIII), we find the titration curve consists of two portions of the sort we have come to associate

$$\begin{array}{ccccc} COOH & & COO^{(-)} & & COO^{(-)} \\ | & & | & & | \\ CH_2 & \xrightarrow[\text{NaOH}]{\text{1 equiv.}} & CH_2 & \xrightarrow[\text{NaOH}]{\text{1 equiv.}} & CH_2 \\ | & & | & & | \\ NH_3^{(+)} & & NH_3^{(+)} & & NH_2 \end{array}$$

Glycine (acid ion)	Glycine (dipolar ion) VIII	Glycine (basic ion)

with the titration of a weak acid, connected by a steep portion which corresponds to the addition of one equivalent of alkali. In the case of glycine the points of inflection occur at pH 2.3 and 9.6. The first corresponds to the pK of the —COOH group and the second, to the $-NH_3^+$ group.

It is in connection with titrations of amino acids that we can appreciate particularly the advantages of the newer way of looking at acids and bases where, at the neutral point, the amino acid exists in the dipolar form. According to the older point of view in which the amino acid was an uncharged molecule, the point of inflection at pH 9.6 had to correspond to the COOH group, since that was the only group that seemed to be titratable with base.

A little thought will show that it is not likely that the COOH group of glycine should have a pK value differing so much from that of the closely related acetic acid (pK = 4.7). That it does not is shown by the results of titrating glycine to which formaldehyde (HCHO) has been added. Formaldehyde combines readily with amino groups to form methylol derivatives. When we titrate glycine, to which formaldehyde has been added, with sodium hydroxide, the curve shown by the dotted line in figure I-2 is obtained. This new curve has a different point of inflection and the end point is about pH 9 instead of pH 12 as before. This suggests that in the titration from pH 6 (the pH of a solution of glycine) to pH 12, the amino group, the one affected by the formaldehyde, was acting as an acid and must therefore have been in the form of $-NH_3^+$.

This conclusion is confirmed by the observation that the titration curve of glycine with hydrochloric acid is not affected by the addition of formaldehyde, a point of inflection corresponding to pH 2.3 being obtained in both cases. Evidently this pK value is that of the $-COOH$ group, which is not modified by the introduction of the formaldehyde. Above pH 2.3, the carboxyl group exists chiefly as the COO^- form, below pH 2.3, chiefly as the COOH form.

This effect of formaldehyde is utilized in the Sørensen "formol titration" for the estimation of amino groups (and amino acids). The mixture to be titrated is neutralized to phenolphthalein (pK 9.7) and an excess of neutralized formaldehyde is added. Then the mixture is again titrated to neutrality with phenolphthalein, using standard base. The new hydrogen ions

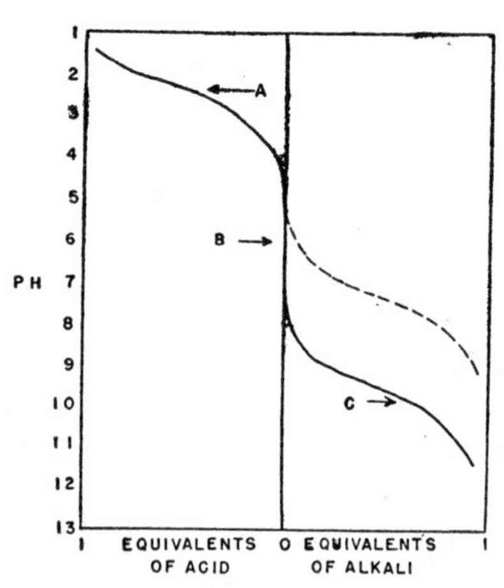

Fig. I-2. Titration curves for glycine. A, first point of inflection (pK = 2.3); B, isoelectric point; C, second point of inflection (pK = 9.6). Solid curve, titration in water; dotted curve, titration in presence of formaldehyde.

which appear in the neutral mixture when the formaldehyde is added, and which are then titrated, come from the NH_3^+ groups of the amino acids, which partly still retained their hydrogen ions even at pH 9.7. The formaldehyde, by converting these NH_3^+ groups to neutral methylol groups, releases the hydrogen ions and converts the procedure to one of bringing the pH of an organic acid to pH 9.7. Since the fresh hydrogen ions are released from the amino groups, the titration actually measures the amino groups present, and not the carboxyl groups, as was formerly erroneously thought.

Since the pK of the COOH group in glycine is 2.3 and that of the NH_3^+ group is 9.6, the molecule is chiefly in the dipolar form at all intermediate pH values. Most significantly, it is in the dipolar form at the pH of the internal environment of the body, which is 7.4.

As a pH of 9.6 is approached, a growing fraction of the glycine molecules is in the basic form which carries a net negative charge. As the pH sinks toward 2.3, a growing fraction is in the acidic form which carries a net positive charge. At some intermediate pH value the net charge of all the glycine molecules in solution is neither positive nor negative, but is zero. This intermediate pH value is called the *isoionic point* and for glycine, its value is 5.97.

The pK of a COOH group or an NH_3^+ group varies somewhat with the structure of the molecule in which it occurs. For instance, the pK of the COOH group in alanine is 2.34 and that in methionine is 2.28. The pK of the respective NH_3^+ groups are 9.69 and 9.21. For this reason, the isoionic point varies somewhat from amino acid to amino acid even when one COOH group and one NH_3^+ group are the only two ionizing groups present in the molecule.

Nevertheless for the neutral amino acids (i.e., those containing neither an acidic nor a basic group in the side chain), including asparagine and glutamine, the isoionic point does not vary greatly. The highest value is 6.30 for proline and the lowest 5.05 for cystine.

For the acidic amino acids (aspartic acid and glutamic acid) and the basic amino acids (histidine, lysine and arginine), the story is quite different. In the case of each of these, four and not three ionic forms are possible. The various ionic forms of aspartic acid and lysine are presented, by way of example, in formula IX.

Note that in the case of aspartic acid, the ionic form with a net charge of zero is predominant in the pH range from 1.9 to 3.6. The isoionic point of aspartic acid is 2.77. (For glutamic acid, the second carboxyl group of which has a somewhat higher pK value than that of aspartic acid, the isoionic point is 3.22). Note also that at the physiological pH of 7.4, the aspartic acid molecule with a net negative charge of 1 is predominant. The amino

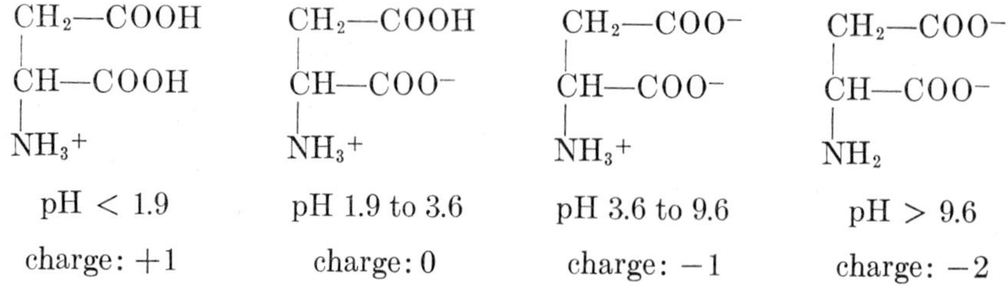

CH_2—COOH	CH_2—COOH	CH_2—COO^-	CH_2—COO^-
|	|	|	|
CH—COOH	CH—COO^-	CH—COO^-	CH—COO^-
|	|	|	|
NH_3^+	NH_3^+	NH_3^+	NH_2
pH < 1.9	pH 1.9 to 3.6	pH 3.6 to 9.6	pH > 9.6
charge: +1	charge: 0	charge: −1	charge: −2

Ionic forms of aspartic acid

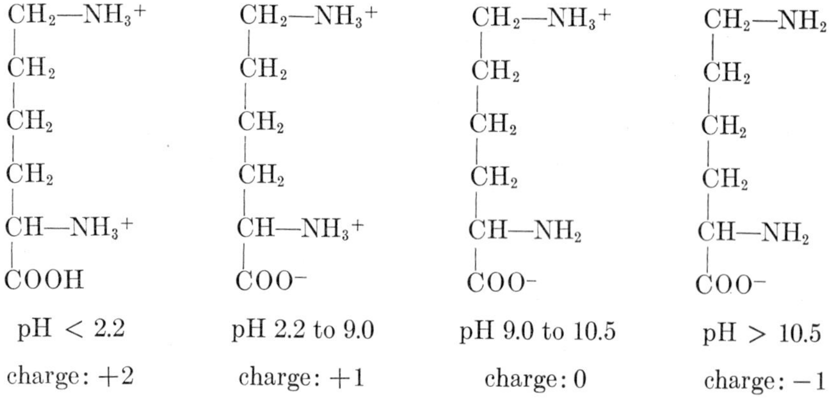

CH_2—NH_3^+	CH_2—NH_3^+	CH_2—NH_3^+	CH_2—NH_2
|	|	|	|
CH_2	CH_2	CH_2	CH_2
|	|	|	|
CH_2	CH_2	CH_2	CH_2
|	|	|	|
CH_2	CH_2	CH_2	CH_2
|	|	|	|
CH—NH_3^+	CH—NH_3^+	CH—NH_2	CH—NH_2
|	|	|	|
COOH	COO^-	COO^-	COO^-
pH < 2.2	pH 2.2 to 9.0	pH 9.0 to 10.5	pH > 10.5
charge: +2	charge: +1	charge: 0	charge: −1

Ionic forms of lysine

IX

acids listed at the beginning of this chapter are shown there in the ionic form predominantly present at physiological pH.

In the case of lysine, the ionic form with a net charge of zero is on the basic side of neutrality, in the pH range from 9.0 to 10.5, and the isoionic point of lysine is 9.74. The imidazolium group in the histidine side chain is a weaker base than the NH_3^+ group in the lysine side chain, and the isoionic point of histidine is 7.59. The guanidinium group in the arginine side chain is a stronger base, and the isoionic point of arginine is 10.76.

The pH at which a dipolar ion has a net charge of zero is determined experimentally by finding the pH at which that dipolar ion remains motionless in an electric field. This pH is called the *isoelectric point.* The isoelectric point is not necessarily equal to the isoionic point. Isoelectric points are determined in systems including buffers, where, in addition to H^+, other anions are present. This results in minor differences, although the terms are frequently used interchangeably.

HYDROGEN ION CONCENTRATION AND THE MEASUREMENT OF pH

In addition to knowing that we have present in solution a substance which tends to give off hydrogen ions, and in addition to knowing the concentration of this substance, we need, in order to be able to estimate the actual acidity properly, another piece of information, which is the extent to which our acid substance is actually dissociated into hydrogen ions and the conjugate base. For it is the concentration of hydrogen ions present, and not the concentration of substances potentially capable of yielding hydrogen ions, which immediately matters for biological systems. We customarily express the hydrogen ion concentration by giving the pH of a solution.

The hydrogen ion concentration, or actual acidity of a solution, which we symbolize by pH, should be carefully distinguished from the titratable acidity which we would obtain by titrating the solution with standard NaOH solution. The latter may be much greater in the case of systems containing weak acids, for such acids are not fully dissociated into hydrogen ions (protons) and anions (or in terms of modern physical chemistry, the activity of the hydrogen ions is much less than that of a strong acid in the same concentration). As a weak acid is titrated, and H^+ ions are combined with OH^- ions to form water, more H^+ ions are released from the acid, until ultimately all the dissociable hydrogen of the acid has come into play.

The way in which hydrogen ion concentration may be measured can be understood from rather elementary electrochemical considerations. If a strip of copper is immersed in water or salt solution, a small amount of the copper dissolves and some copper ions are produced until the concentration of copper ion is sufficient to counteract the dissolving tendency of the copper. At this point we have equilibrium, and as many copper ions come out of solution and deposit on the copper as go in solution in any given time. In other words, we have a dynamic equilibrium.

If a copper strip is immersed in a solution containing Cu^{++}, such as a solution of copper sulfate, there will be a tendency for copper ions to deposit on the copper strip, forming, when discharged, metallic copper. Each ion which does so, however, leaves two positive charges on the metallic strip (or takes away two electrons), and before long the charge on the metal is sufficient to repel the Cu^{++} ions and prevent further deposition of ions so that the action ceases. Suppose, however, we have two solutions of copper sulfate at different concentrations and in each of these we immerse a copper strip. If we connect the two pieces of copper by an electrical conductor such as a piece of wire, and connect the two solutions by a tube containing ions so as to permit ion transfer from one solution to the other, we will observe the following action: in the more concentrated solution copper is deposited from solution on to the metallic strip, leaving a positive charge on the strip. In the less concentrated solution, copper dissolves into the solution, forming metallic ions and this strip is left with a negative charge. Since by convention current flows from the positive to the negative pole, we shall have an electric current flowing from the positively charged strip to that with the negative charge. However, since the convention of positive and negative which was adopted at the suggestion of Benjamin Franklin proves to have been just the reverse of the truth, it is unfortunately true that the flow of electrons is in the opposite direction—that is, from the negative to the positive pole. This, however, need not confuse us if we always speak of the electric current as passing from the positive to the negative pole. This flow of current will continue with the above set-up until the concentration of copper ion has become equal in the two halves of the cell. An electric cell of this type is called a concentration cell because the electric potential produced is due to differences in concentration of the same substance.

It is not necessary that the cell be made up of solutions of the same material, however; in fact it is commoner to use cells which are made up of different material. This can be illustrated by the following principle. If a copper wire is dipped in a tube of water or salt solution, some copper ion will dissolve until the back pressure of copper ion tending to deposit out equals the tendency of the copper to dissolve and equilibrium is attained. Similarly, if a strip of zinc is dipped in a test tube of water or salt solution, some zinc ions will dissolve but somewhat more since zinc is more soluble than copper in aqueous solution. If now we connect the two strips of zinc and copper by conducting materials such as a piece of wire and connect the two solutions by glass tubing containing salt solution so that ions may pass from one to the other, we will find that an electric current flows through the wire from the copper to the zinc; in other words, in this cell the copper is the positive pole.

At a certain stage of telegraphy not too remote to be remembered by two of the authors of the present book this principle was made use of in the type of battery constructed somewhat as follows: in a glass jar was placed a copper electrode covered with a layer of copper sulfate solution. On top of this was a layer of zinc sulfate solution of somewhat lower density so it did not sink into the copper sulfate solution. Suspended in this solution was a zinc electrode. The two were connected by a wire and a current was found to flow from the copper to the zinc. (A salt bridge was not necessary in this case as the two solutions were in direct contact.) The zinc was gradually eaten away while copper deposited out of solution onto the copper electrode. As long as the battery was kept connected the electric action itself prevented the two solutions from mixing by diffusion. So-called dry cells

box, lose their characteristic properties even more rapidly than albumin.

er to isolate proteins with as little damage as possible the tem- must be kept low—lower than body temperature whenever con- ermit; otherwise random thermal agitation of the atoms will cause aturation. Also, enzymes which are originally present in the impure ion may begin to act on the protein before they are separated from pH should be maintained as near that of the environment of the protein as possible. The dielectric constant of the medium should t as high as possible, which means that other things being equal, eous medium is best. Organic solvents such as alcohol should be used ith great caution and preferably at low temperatures. They appear age certain proteins only slightly but may rapidly damage others. se of high salt concentrations to precipitate proteins, though tradi- , should be avoided when possible. In some cases this seems to result ry slight damage but ultimately damage always does result. Some ins are damaged immediately beyond repair by this procedure.

bility

me proteins are insoluble in all ordinary solvents; others are more or soluble in various mixtures of water and other compounds. Pure water dissolve some proteins; in other cases it acts as a better solvent if s are present. Some proteins will dissolve best in mixtures of water l less polar solvents such as alcohol. In general, the more polar the vent the greater its power of dissolving proteins, which suggests that e protein is itself polar in nature. If sufficient salt, especially a salt such ammonium sulfate, is dissolved in a protein solution, the protein be- mes less soluble and most proteins are completely precipitated. There e certain general rules about the solubility of proteins which may now e stated.

1. A protein is least soluble in the neighborhood of its isoelectric point see below). The pH of minimum solubility varies with the nature and con- entration of the salt which is used. Unless the salt concentration is very dilute, the pH of minimum solubility is generally found to be somewhat different from the true isoelectric point of the protein.

2. Solubility of proteins in water without salt varies a great deal. Serum albumin dissolves in water readily and it seems to be miscible with water in all proportions. Other proteins are soluble only if salt is present; some, like edestin from hempseed, require concentrations of neutral salt of the order of 5 per cent to get them into solution.

3. The solubility of a protein in water or other solvents depends upon the nature of the amino acids of which it is composed. Thus proteins rich

containing similar materials in the form of pastes are still in use today for flashlights and electric bells.

If we now go back to the concentration cell, we see that it seems reason- able to expect that there will be a relationship between the concentration of ions in the two sides of the cell and the electromotive force developed be- tween the electrodes. This is in fact true and the equation for the electro- motive force is well known. This means that if we knew the concentration of ions in one side of the cell and measured the electromotive force, we could calculate from the equation what the concentration of ions was in the other side of the cell. On this principle is based the determination of hydrogen ions.

To determine hydrogen ions one would want a set-up which would con- tain two cells, one with a known hydrogen ion concentration, the other with an unknown hydrogen ion concentration which we desired to deter- mine. If we could immerse electrodes of solid hydrogen in these and get electrical action when the two electrodes were connected by a wire and the solutions connected by a salt bridge, such measurements could be made. Unfortunately, the freezing point of hydrogen is far too low for us to make use of such a simple device, not to mention other difficulties. However, it has been found possible to use *gaseous* hydrogen, if it is kept in the small pores and crevices on the surface of some non-reactive con- ducting surface. One of the so-called noble metals such as platinum, gold, or palladium is satisfactory for this purpose. Then if we have an electrode of such material and keep the pores filled with gaseous hydrogen by bub- bling hydrogen over the electrodes constantly, we have what is the equiva- lent of a hydrogen electrode. If we put a known hydrogen ion concentration on one side then we measure the ion concentration in an unknown solution on the other side of the cell.

In practice it has not proved convenient to make up known hydrogen ion concentrations for use in such determinations, because they do not remain constant in concentrations, but tend to change by evaporation and other effects. But from what has been said above about the copper- zinc cell it will be apparent that if we replace the hydrogen half-cell on one side of the hydrogen cell just described by some other half-cell, we will still get an electric potential and it is known that this will differ by a con- stant from the potential which we would have obtained with a hydrogen electrode of known hydrogen ion concentration. The so-called calomel electrode has been found most satisfactory for this purpose.

We now see that if we have a calomel half-cell, or calomel electrode as it is usually called, and a half-cell containing a hydrogen electrode and a hydrogen ion concentration which we wish to determine, we may do so by connecting the two by a suitable salt bridge and measuring the electromotive force between the two electrodes with a potentiometer. When this is done, we may calculate from an equation the hydrogen ion concentration in the unknown solution.

The hydrogen electrode as just described is the basis of all accurate hydrogen ion determinations. For practical purposes, however, it has been superseded by other more convenient types of apparatus. For example, it was discovered that if a thin glass membrane was brought into contact with a solution containing hydrogen ions, a potential was developed across the membrane which varied with the hydrogen ion concentration on the two sides. By making use of this observation, it is possible to replace the hydrogen electrode by a so-called glass electrode, which consists essentially of a thin glass membrane which is brought in contact on the outside with the unknown solution and on the inside with a solution of fixed hydrogen ion concentration, so arranged that this known H^+ concentration can be sealed in and thus will not change. Such a glass electrode in connection with a calomel electrode together form a cell capable of measuring hydrogen ion concentration. It is possible to calibrate the instrument so that it will read directly in pH units. These are the most accurate and simplest of the common laboratory instruments for the measurement.

A brief mention should be made of other methods of determining hydrogen ions. Many chemical compounds change color at certain pH (hydrogen ion concentrations)—one color when alkaline to this pH and another color,

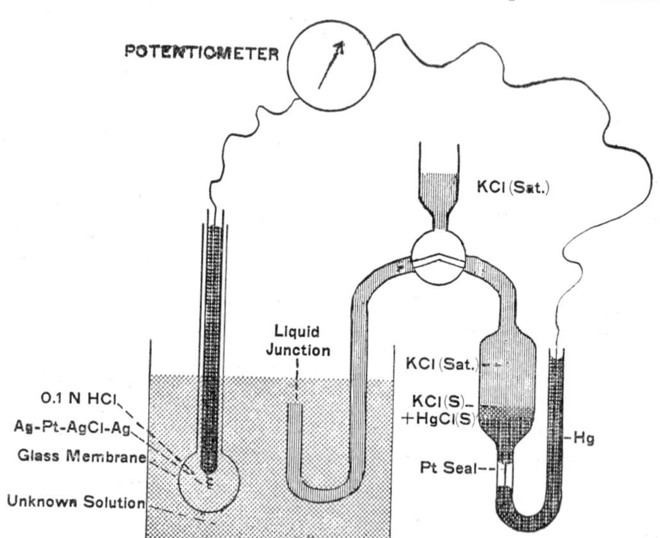

FIG. I-3. Diagram of glass electrode pH meter. Glass electrode at left, calomel halfcell at right. The potentiometer consists of a galvanometer plus batteries, a Wheatstone bridge circuit, and a vacuum tube amplifying circuit, not shown here.

or colorless, when acid to this pH. Suc
and their characteristic color change m
pH. During the range in which part of th
and part in a form having a different colc
results, giving a range of different shades o
these with the color obtained in solutions o
known solution may be determined. The bes
litmus, which can be deposited on paper, and
will have used litmus paper in his early study
now known which cover the whole range of pH
gate. Their results are not as accurate as those
they are quicker to use and often convenient.

PROPERTIES OF PROTI

In discussing the properties of the proteins we
thing about their behavior *in vivo*—that is, in the in
that is, as they behave in the laboratory after we

Lability

One of the most striking characteristics of the prot
extremely sensitive to change and it is difficult to kee
during chemical manipulation. Probably one of the fir
changes which takes place in a cell when the organism
part has died is alteration in the constituent proteins. T
reach the stage where they are irreversible and the cell is
As an example of this lability we may consider the prop
albumin which forms the principal protein of the white
This protein can be isolated from the egg and obtained in c
but as a rule only if the egg is less than 24 hours old. If t
than this, great difficulty is experienced in trying to crystalli
The egg is still viable; it will still hatch, but nevertheless s
happened to the albumin during the first twenty-four hours
been outside the body of the hen. And keeping the egg at bo
ture, instead of arresting this process, merely accelerates it. A
bumin has been crystallized, it will keep in the form of cryst
with mother liquor for a considerable time, but nevertheless
temperature or even at icebox temperatures slow changes take
that the solubility of the crystals in water gradually becomes less,
a few years they may become completely insoluble, showing that ev
the protein is in the form of crystals and relatively pure, someth
happened to it. Crystals of the blood protein hemoglobin, even whe

in the ice
does egg
In ord
perature
ditions p
some de
prepara
it. The
native
be kep
an aqu
only w
to dar
The u
tional
in ve
prote

Solu

S
less
will
salt
and
sol
th
as
co
a
b

in non-polar groups such as paraffin side chains, benzene rings, or pyrrolidine rings tend to dissolve better in alcohol-water mixtures than in water; whereas those poor in non-polar groups but rich in polar (electrically charged) groups tend to be precipitated even by small amounts of alcohol or acetone.

4. Proteins which are insoluble in water but have large numbers of charged groups become more soluble in the presence of neutral salts or other dipolar ions.

5. Proteins are usually more soluble when combined with acids or bases than in the neutral state. This will be discussed immediately below.

6. Formation of salts between proteins and another protein or between a protein and ion may result in compounds which are more or less soluble. Thus, protamines form a compound with insulin which is less soluble than either protamine or insulin. Protamines also form insoluble salts with casein.

Amphoteric Behavior of Proteins

Most proteins can behave either as acids or bases and are thus called amphoteric. Consequently it is possible to dissolve these proteins either in dilute acid or dilute alkali, forming a salt in either case. This amphoteric behavior of proteins is of great importance to understanding other properties in general. For instance, whether a protein combines with an anion or a cation depends upon the pH of a solution or in other words, on which side of the isoelectric point of the protein we find ourselves. This was well shown in the classical experiment of Jacques Loeb, described on page 29.

The amphoteric behavior of proteins is due to the presence of acidic and basic groups in their molecules. Some of these groups are ordinarily charged, positively or negatively as the case may be, and thus account for the presence on the surface of the molecule of fixed charges. Except at the isoelectric points, the positive and negative charges do not usually balance exactly, so the molecule has an over-all net charge which is positive or negative, depending on pH. The distributions of the two sorts of charges are seldom the same, so negative charges will predominate in one part of the molecule and positive charges in another. The effect of this is the same as the localization of a positive charge of varying magnitude on one part of the molecule and a negative charge on another part. This causes the protein molecule, even at the isoelectric point, to behave like an electric dipole. Therefore, protein molecules in solution will orient themselves in an electric field.

A dipolar ion will tend to orient itself in an electric field, with the end which is predominantly negative pointing towards the positive pole and the positive end towards the negative pole.

The dielectric constant of a solution can be interpreted as being almost entirely a measure of the number of molecules oriented by the electric field. This orientation is hindered by frictional forces which vary with the size and shape of the molecule. Hence, a study of the dielectric constant of a protein solution subjected to alternating electric fields of various frequencies gives information about the physical characteristics of the protein molecules (18).

Isoelectric point. The isoelectric point of a protein has been defined as a pH value such that the net charge (the algebraic sum of all charges, positive and negative, on the molecule) on the amphoteric molecule is zero and it will not move towards either the positive or the negative electric pole. The isoelectric points of a few typical proteins are given in table I-2. It will be observed that a good many proteins have isoelectric points very near together so that this method alone would not distinguish these various proteins very well.

Since a good many aspects of the behavior of proteins are due to their possession of electrical charges, it is not surprising that many of these effects are larger when the protein is not at its isoelectric point and are at a

TABLE I-2

Isoelectric points of proteins

PROTEIN	SOURCE	ISOELECTRIC POINT
Actomyosin	Muscle	6.2–6.6
Bushy stunt virus	Infected tomato plant	4.11
Casein	Cow's milk	4.6
Catalase	Beef liver	5.7
Chymotrypsin	Beef pancreas	5.4
Cytochrome C	Beef heart	9.7
Fibrinogen	Human blood	ca. 5
Gamma globulin	Human blood	ca. 7
Gelatin	Skin	4.7–4.85
Hemoglobin	Horse blood	6.79–6.83
Insulin	Pig's pancreas	5.30–5.35
β-Lactoglobulin	Cow's milk	5.18
Myogen	Muscle	6.2–6.4
Old yellow enzyme	Yeast	5.2
Ovalbumin	Hen's egg	4.84–4.90
Papain	Papaya	9
Pepsin	Pig's stomach	2.75–3.0
Salmine	Salmon sperm	12.0–12.4
Serum albumin	Human blood	4.9
Trypsin	Beef pancreas	10.8
Urease	Jack bean	5.0–5.1

minimum at or near the isoelectric point. Such properties are osmotic pressure, electrophoretic mobility, and swelling of proteins in the solid phase.

The importance of the isoelectric point may be illustrated by the fact that if the pH of a protein is greater than the isoelectric point, it can only combine with cations, forming as a rule a metal protein salt; whereas if the pH is less than the isoelectric point, the protein combines with anions forming a salt of protein and some ions such as chloride ions and so forth. This was proved thirty years ago by Jacques Loeb by experiments on gelatin. Gelatin was brought to different pH by treatment with varied amounts of nitric acid. After washing, the samples were all treated with a solution of silver nitrate of a certain concentration. After being washed again, the various samples of gelatin were put in test tubes, melted, and exposed to the light at room temperature, after the pH of the sample had been determined on a sample from each tube. All of the gelatin solutions with pH greater than 4.7, which is the isoelectric point of gelatin, became opaque and then brown or black due to the reduction of the protein-silver salt to metallic silver (as in a photographic film), while the solutions with a pH less than 4.7 remained transparent even when exposed to light for months or years.

On the acid side of the isoelectric point, that is, when the pH is less than 4.7, gelatin is in combination with the anion of the salt used. Loeb demonstrated this by bringing different samples of powdered gelatin to different pH, then treating them for one hour with a dilute solution of potassium ferrocyanide. After this treatment the gelatin was washed thoroughly with cold water and 1 per cent solutions of these different samples made by dissolving in hot water, whereupon it was found that when the pH was less than 4.7 the gelatin solution turned blue after a few days due to the formation of ferric ferrocyanide, some of the iron present having been oxidized to the ferric state. All the solutions of gelatin with a pH of 4.7 or above remained permanently colorless.

Complex Formation

Proteins, having a fair number of charged groups which may be of either sign, can and do form complexes of various sorts. Some of them form insoluble salts with anions or with cations. For instance, the vegetable protein, edestin, is not only relatively insoluble itself but forms a relatively insoluble hydrochloride. Some proteins which form soluble sodium or potassium salts form insoluble calcium salts. Others form insoluble zinc salts. A large number of substances form insoluble compounds with nearly all proteins and are thus used as tests for proteins. Among these we may list tannic acid, picric acid, phosphotungstic acid, and trichloracetic acid. These will be mentioned under tests for proteins. The process of tanning leather seems to be a process of formation of insoluble complexes with various agents such as tannic acid, chromic acid, and so forth. Proteins may also form insoluble complexes with other proteins. A very well known example is the complex of insulin and the simple protein, protamine. The combination of insulin and protamine is used clinically because the insolubility of the complex results in its liberating insulin into the circulation more slowly than if the hormone were injected by itself.

EXTRACTION AND PURIFICATION OF PROTEINS

In order to make chemical studies of proteins, it is necessary to extract them from the tissues in which they occur and follow this by purification procedures. In the case of blood, the proteins of the plasma are already in solution and it is only necessary to add something to prevent the clotting of the blood, if we wish to study also the fibrinogen, or to take the serum which separates from the clot in case we are not interested in the fibrinogen. But blood is a rather special tissue, although one of great importance, and it is somewhat easier to study than most of the others. If we have tissue consisting of cells, connective tissue, fat, and so forth, and wish to study one of the proteins contained therein, one of our first problems is how to break up the cells so as to get out the proteins. To break up the cells of the tissue, several methods are available. Alternate freezing and thawing will usually accomplish it; or grinding with sand in some sort of mill, or the action of intense sound waves of high frequency ("ultrasonic vibration"). In a given case one of these may have some advantage over the other. It is probable that none is quite ideal. In some cases simple mincing of the tissue, followed by extraction with dilute salt solution will extract a large amount of protein present. This is perhaps the mildest of the methods of extraction.

Concept of purification. It is somewhat more difficult to establish the purity of a protein than in the case of simpler chemical compounds, although even in these cases it is not always as easy as might be supposed. In the case of simple organic compounds which can be crystallized readily, it is usual to repeat crystallization until one or several of the properties of the substance remain unchanged after further crystallization. For instance, it is customary to follow the melting point and crystallize until the product no longer changes its melting point after further crystallizations. Elementary analyses (C, H, N, and so forth) also help one to follow the degree of purity attained. But some compounds such as cholesterol, for instance, even after repeated crystallization, still contain impurities which are very difficult to remove. To understand how this may occur consider the distillation of ethyl alcohol from beer, wine, and other products of alcoholic fermentation. Distillation to constant boiling point is usually a good method of fractionating liquids and obtaining pure components. But the reader will recall that if this is attempted in the case of mixtures of ethyl alcohol and water, the constant boiling mixture as finally obtained contains between 95 and 96 per cent alcohol and no further distillation will improve the purity because a *mixture* with a constant boiling point has been obtained. Other methods must be relied upon to remove the rest of the water. A similar situation can arise in crystallization if compounds are formed which separate out in the form of crystals and the solubility of the compound is less than that of either component. There is good reason to think that this does often happen with protein materials. The usual definition of "pure" for an organic substance is a substance which consists of a single molecular species—that is, all of its molecules are exactly alike. This concept, though satisfactory enough for most organic compounds, especially those dissolved in non-polar solvents, is not at all satisfactory when applied to proteins—even ideally. In fact there is very good evidence in many cases that it is not so. Repeatedly crystallized material can be shown to be non-homogeneous. In the case of proteins we are inclined to rely upon physical and relatively crude criteria such as solubility, the sedimentation constant as determined in the ultracentrifuge (see below), and the electrophoretic mobility as determined in the Tiselius electrophoresis apparatus. It is desirable that supposedly pure protein should conform to the phase rule of solubility, that is, if the amount which goes into solution is plotted against the amount added, a straight line of definite slope should be obtained up to a certain point after which this line abruptly becomes horizontal and does not change its slope thereafter. The point at which the line changes its slope, of course, represents the point at which saturation is obtained. (See figure I-4.)

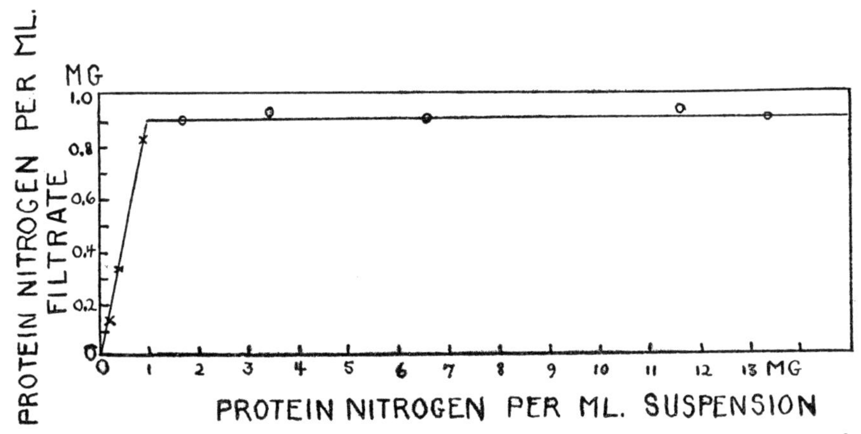

FIG. I-4. Solubility of crystalline chymotrypsinogen in one-quarter saturated ammonium sulfate at 10 C in the presence of increasing quantities of chymotrypsinogen (12).

It is natural to suppose that all the molecules of a protein which fulfills a single function in a single tissue are exactly alike. But this is not necessarily so. Human serum albumin, after careful separation and repeated crystallization, is homogeneous electrophoretically, ultracentrifugally, and immunologically. But amino acid analysis suggests that there is only half as much tryptophane present as would be needed for one tryptophane residue to be present in each molecule.

Even when there is no difference in structure between the different molecules of a purified protein, it is unlikely that all of them are at any one time in the same state. Thus Cohn and Edsall (2) calculate that in the case of the hemoglobin studied by them at pH 6.4, the isoelectric point, only 22.4 per cent of the molecules possessed zero net charge. They estimated that 21.2 per cent possessed a net charge of −1, and 0.03 per cent, a net charge of −7.

Salting out. After the protein has been obtained in the form of a solution from the cells, the next procedure is to purify it. A number of different possibilities present themselves, some better than others. One of the older methods consists in what is called "salting out"—that is, the addition of a sufficient concentration of a neutral salt such as sodium chloride, magnesium chloride, or ammonium sulfate, the latter being the one most commonly used, to precipitate the protein and throw it out of solution. We may consider that salting out depends on the monopolization of so much of the

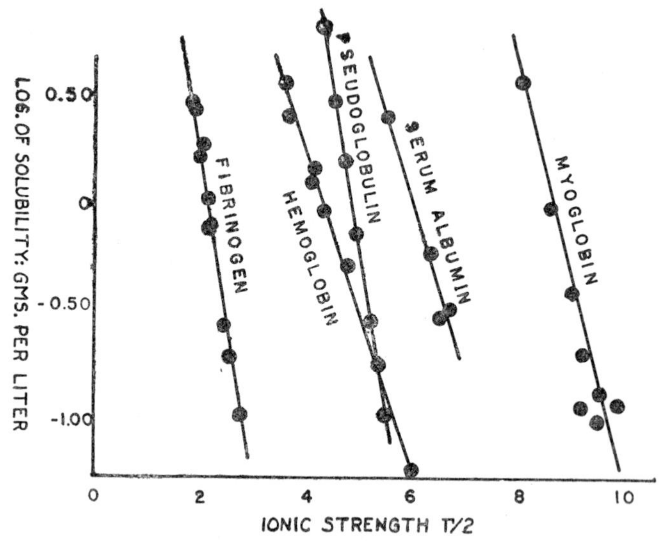

FIG. I-5. Solubility of five different proteins in ammonium sulfate solutions of various ionic strengths. Γ/2 = ionic strength. Note linear relation between log of solubility and ionic strength.

water by the more polar salt that not enough is free to keep the protein in solution. If the precipitate is then filtered off, it may be redissolved and the process repeated. This will often result in considerable purification because small amounts of other proteins, particularly if they are relatively soluble, will be left behind in the filtrate so that eventually a relatively pure protein preparation will result. Crystallization can be achieved in many cases by bringing salt concentrations to the right concentration and adjusting the pH to a value in the vicinity of the isoelectric point. Such methods have allowed the crystallization of egg albumin, serum albumin, and various other proteins. However, the addition of high concentrations of salts to proteins is not without danger and in many cases, without doubt, results in denaturation, which may be mild or in some cases pronounced.

It has been shown that the salting out of a protein depends upon the character of the protein and also upon the particular neutral salt which is used. The characteristics of salting out with ammonium sulfate for a number of proteins are shown in figure I-5 taken from Cohn (2) and for a number of salts in figure I-6.

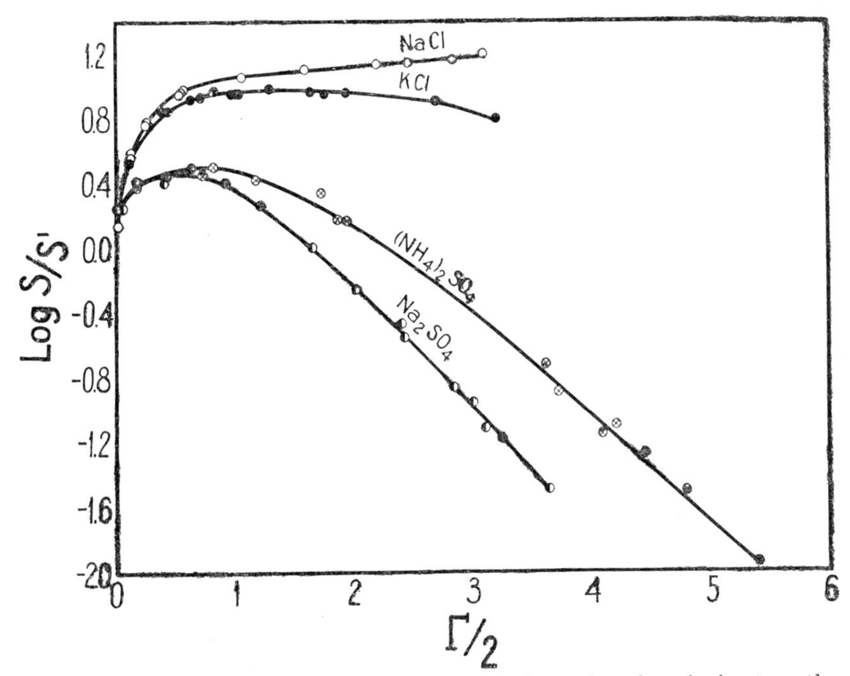

FIG. I-6. Solubility of hemoglobin in salt solutions of various ionic strengths. S/S′ stands for the ratio of the solubility at the indicated ionic strength to the solubility in distilled water. Note initial increase in solubility (salting in), followed by subsequent larger decrease (salting out).

The solubility of proteins in salt solutions depends not only on the total concentration of ions present but also on the ion charge. A quantity which takes both of these into account is the ionic strength ($\Gamma/2$). To calculate the ionic strength of a solution, multiply the concentration in mols of each ion (c_i) by the square of the number of positive or negative charges on that ion (z_i) and divide the sum of all these products by 2. This is expressed in symbols as

$$\Gamma/2 = \frac{1}{2} \sum c_i z_i^2 \qquad \text{Equation I-21}$$

Crystallization. It has been possible to purify many proteins by crystallization. The most successful of the early procedures were based on the principle of lowering the solubility of the protein by adding an amount of a polar salt such as ammonium sulfate which brings the protein to the verge of precipitation (impurities of lower solubilities are removed by a preliminary step). Then the pH is brought down by the addition of buffer or dilute acetic acid to the vicinity of the isoelectric point, or until a slight cloudiness appears. This cloudiness, under favorable conditions, develops into a deposit of protein crystals in the course of 24 hours or so. Sometimes it is better to adjust the pH and add saturated ammonium sulfate solution until cloudiness begins to develop. By modifications of these methods many enzymes have been crystallized (12).

The use of these high salt concentrations is not without danger to the protein. Many proteins may be crystallized from mixtures of water with organic liquids such as methanol, ethanol, acetone or dioxane. The presence of chloroform, or, especially, aliphatic alcohols such as decanol makes crystallization easier. In some cases the use of decanol or some such substance is absolutely essential, and makes possible the crystallization of proteins which have never been crystallized from solutions of salts such as ammonium sulfate. The damage to the protein in this method seems to be very slight, provided temperatures are kept low.

Proteins from different sources vary in the ease with which they crystallize. The albumin from the white of the hen's fresh egg is relatively easy to crystallize, but that from duck's egg is very difficult. Crystallized proteins show evidence of being purer than amorphous preparations, but are not always completely homogeneous, even after repeated crystallization (10, 14).

Dialysis. Another means of purifying proteins depends upon the fact that they will not diffuse through fine membranes which possess pores large enough to admit the passage of smaller molecules, ordinary ions, or even some of the larger organic compounds. This process of allowing the impurities of a protein to pass into water through a membrane such as cellophane is called dialysis, and has been extensively employed as a method of purifying proteins. It is not usually a good method of separating proteins unless one is interested in separating two proteins, one of which is soluble in pure water and the other only in salt solutions. In that case dialysis may be continued to the point where the salt concentration is so low that the second protein is precipitated whereas the water-soluble protein remains in solution. Filtration then allows their separation. This method of purification has been effective, for instance, with the hemocyanins (see page 53), and Dhéré succeeded in crystallizing a number of hemocyanins by slow dialysis.

In order to free proteins completely from all impurities except ions with which they are in chemical combination, it is usually necessary to employ electrodialysis, in which case the movement of the ions through the membrane is speeded up and facilitated by the use of an electric potential between the inside and outside of the membrane. Here, too, there is danger of denaturation and the process must be used with caution, special attention being directed to avoiding local heating effects.

Adsorption. Another method which has been employed to purify proteins is the method of adsorption. The protein is adsorbed on a suitable material such as diatomaceous earth or powdered aluminum hydroxide and after this material has been washed free of non-adsorbed or loosely adsorbed impurities, the protein is removed (eluted) by suitable means. This method was used in the early days of protein chemistry in attempts to purify enzymes and antibodies. It gave very small yields, however, and in many cases seems to have produced some denaturation. Nevertheless, ion-exchange resins (high molecular weight synthetic molecules containing reactive groupings) have proved useful in protein work, and have made possible, for example, the separation of adult and fetal hemoglobin and also hemoglobin from methemoglobin (1).

Ultracentrifugation. Another method of separating proteins depends upon the use of the ultracentrifuge which is discussed immediately below. Proteins which differ to any great extent in molecular weight can be separated by this method, the larger molecules being deposited first on the bottom of the centrifuge tube. However, only relatively small amounts can be handled by this technique and it is primarily an analytical and not a manipulative tool.

The ultracentrifuge, as its name implies, is a centrifuge which attains much higher centrifugal force than the ordinary laboratory centrifuge. It was invented by Svedberg and a full account of it may be found in the book by Svedberg and Pedersen (15). It consists essentially of a small turbine, driven by oil or compressed air, or an electric motor, which in turn drives a metal chamber called the rotor, in which there is a place to put a cell containing the protein solution to be studied. This cell has quartz windows so that the sedimentation of the protein under the influence of the centrifugal field may be observed visually or photographically. By use of the ultracentrifuge an intense gravitational field may be produced varying from several thousand to as much as a million times gravity. In these intense gravitational fields, protein molecules which otherwise would be kept in solution by forces of diffusion are caused to sediment because the forces of diffusion are no longer able to counteract the tendency of the molecules to move away from the center of rotation. If a colored protein such as hemoglobin is employed, visible light may be used to record the sedimentation of the protein. A boundary is formed between clear solvent and the protein solution beyond it, and the movement of this boundary can be followed by successive photographs taken at known intervals. Knowing the speed of the centrifuge and its dimensions, the centrifugal field may be calculated and thus the *sedimentation constant* of the protein becomes known. If the protein were spherical, it would be possible from this alone to calculate its molecular weight. Since most protein molecules are not spherical, however, it becomes necessary to know one other thing, the *diffusion constant*, in order that the molecular weight may be determined. However, even if this is not known, a calculation of the molecular weight on the assumption that the molecule is spherical gives a minimal value for the molecular weight and we know that the true molecular weight can not be below this figure. It may, however be considerably above it. Table I-3 shows some typical results for important proteins.

If the protein is colorless, as so many are, visible light will not enable sedimentation to be followed by direct observation. Instead, it is necessary

TABLE I-3

Molecular weights of certain proteins from ultracentrifuge studies

PROTEIN	SEDIMENTATION CONSTANT S_{20}	PARTIAL SPECIFIC VOLUME V_1	FRICTIONAL RATIO f/f_0	MOLECULAR WEIGHT	
				Assuming spherical shape	Corrected for shape
Bacillus phlei protein	1.8	0.748	1.22	13,000	17,000
Tubercle bacillus (human) protein	3.3	0.70	1.25	23,000	32,000
Human serum albumin	4.6	0.733	1.28	47,000	69,000
Diphtheria toxin	4.6	0.736	1.30	49,000	72,000
Rabbit antibody	7.0			95,000	157,000
Human serum gamma globulin	7.2	0.739	1.38	96,000	156,000
Urease (jack bean)	18.6	0.73	1.19	370,000	480,000
Thyroglobulin (pig)	19.2	0.72	1.43	370,000	630,000
Antipneumococcus antibody (horse)	19.3	0.715	1.86	360,000	910,000

to use ultraviolet light which proteins absorb. Or by taking advantage of the fact that the refractive index of the protein solution and the solvent will be different, suitable optical equipment makes it possible to photograph with visible light the boundary and obtain a photograph in which the boundary is represented by a peak, and the sharper the boundary the sharper and more abrupt will be the peak. The student should consult some other source, such as Svedberg and Pedersen (15), for details of operation and theory. Svedberg's high speed ultracentrifuge is driven by a turbine fed by oil under pressure, and the rotor runs in an atmosphere of hydrogen at a pressure of a few millimeters of mercury. The hydrogen is necessary to conduct away heat which develops in the rotor during the run. A somewhat simpler type of centrifuge driven by compressed air has been developed by various workers in this country. A blast of compressed air turns the air turbine which forms the upper portion of the instrument from which the rotor is suspended by a piece of piano wire. The piano wire is completely surrounded by a packing which fits very closely and just leaves it free to turn, the very small gap between the wire and the casing being filled by oil. The rotor is completely encased and runs in a high vacuum so that the heating effects are extremely small. This type of ultracentrifuge is useful not only for molecular weight determinations but for the concentration of viruses and other very large protein molecules, and in general in protein preparations when large amounts are not required. The maximum centrifugal force, and thus the resolving power, which can be obtained with this type of instrument is somewhat less than with Svedberg's oil centrifuge, but it is satisfactory for all except the smallest protein molecules and is less expensive to construct and operate.

Electrophoresis. Another method of separating and characterizing proteins depends upon their characteristic mobility in electric fields. Since proteins are amphoteric and can behave either as weak acids or weak bases, they can form salts with strong acids or strong bases and these salts are very highly dissociated—dissociating into a charged protein anion or cation and a metallic cation or anion as the case may be. The farther from the isoelectric point we bring the pH, the larger will be the charge on the protein molecule, for then more charged groups will be present. Since the isoelectric points of most proteins are somewhat on the acid side of neutrality, it is customary to use an alkaline buffer. Plasma proteins are ordinarily run at a pH of about 8. The protein solution is placed in a U-shaped tube of square cross-section which is kept at a constant temperature by a thermostated bath. Low temperatures are used, not far from the temperature of the maximum density of water, for here the disturbing effects of heating due to passage of electric current are least. It can be shown theoretically and it is found experimentally, that once a boundary between solution and proteins has been formed, the electric current will tend to maintain this boundary even though the protein molecules are in motion and subjected to the random forces of Brownian motion. On the alkaline side of the isoelectric point the protein ion has a negative charge; consequently it tends to move towards the positive pole of the apparatus. An electric potential of the order of 10 volts per centimeter is ordinarily used. A schematic diagram of the apparatus is shown in figure I-7, which gives some idea of the apparatus as developed by Tiselius. The boundary between solvent and protein solution is formed mechanically as indicated and the movement of the boundary between solvent and protein solution is followed optically by a method very similar to that used in following the boundary in the ultracentrifuge. It is unnecessary to give technical details here, but, again, photographs are obtained which show a peak corresponding to the boundary. If more than one protein with different electrophoretic mobilities is present, more than one peak will be obtained, since the molecules will migrate at a different rate. (The rate of movement is practically independent of the size of molecule.) In that case more than one peak may be obtained. A characteristic photograph is shown in figure I-8 which shows results obtained with human plasma. Several components will be recognized. The designation "A" represents serum albumin; the Greek letter φ stands for fibrinogen; and three species of globulins are found designated by the Greek letters, α, β, and γ. The gamma globulins are probably a mixture of proteins of slightly different electrophoretic mobilities. More will be said about this when we discuss the components of the blood.

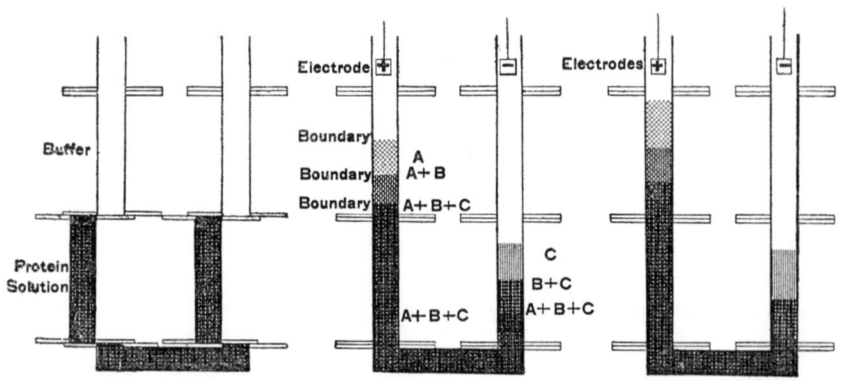

Fig. I-7. Separation of proteins in electrophoresis. At left a solution of three different proteins is placed in the apparatus. Buffer is placed in the upper tubes, and the lower tubes containing protein solution slid into contact, giving an uninterrupted path for the electric current (right) and making a sharp boundary. In the two diagrams at the right, protein A moves fastest, protein C slowest. The various boundaries, when photographed, yield characteristic peaks (figure I-8) corresponding to variations in refractive index of the solutions.

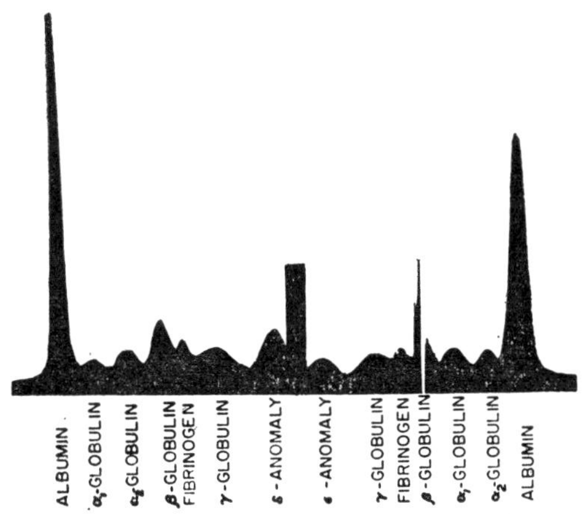

FIG. I-8. Electrophoretic schlieren (shadow) diagrams of normal human plasma (courtesy of Dr. J. L. Oncley). Note that in this instance the symbol φ is not used for the fibrinogen.

There seems to be no simple exact correspondence between protein fractions as identified by electrophoresis and those obtained by salting out procedures. However, Cohn *et al.* (3) and Svensson (16) reported a general parallelism between solubility and electrophoretic mobility, so that if relatively low salt concentrations were used, the slower moving component (gamma globulin) was precipitated before the faster components. Working with serum, Popper *et al.* (13) found that using Wolfson's technique a good correlation between the fractions obtained by salting out and by electro phoresis, except with the alpha and beta globulins, was achieved.

Electrophoretic separations may also be carried out on a strip of filter paper moistened with a suitable buffer and connected to electrodes to establish a potential difference of a few hundred volts. If a solution containing a mixture of proteins is placed on the strip, the components move at rates depending on their molecular charge. *Paper electrophoresis* is cheaper and simpler than ordinary electrophoresis, but not quite as accurate. Some proteins, moreover, are so strongly adsorbed to paper that they do not separate cleanly and at best can only be partially eluted.

CHARACTERIZATION OF PROTEINS

Elementary Analysis

When we have isolated a protein from tissues and wish to characterize it and differentiate it from other proteins there are a number of ways in which we may attempt to do this. One of the most obvious by analogy with procedures in organic chemistry would be to analyze the protein for its constituent amino acids and other substances. An elementary analysis of the protein for carbon, hydrogen, nitrogen, and so forth is not likely to give such decisive information because the large size of the molecule means that proteins which differ enormously from each other may have very much the same elementary composition. Also it is rather difficult to obtain proteins entirely free from contamination so that one can seldom be certain that the elementary analysis represents the composition of the protein and nothing else.

However, one analysis which is usually performed, and on which estimation of the amount of protein present is often based, is the determination of nitrogen. Proteins vary somewhat in the percentage of nitrogen they contain but it has long been assumed that a fair average is 16 per cent. This gives us a conversion factor of 6.25 by which we should multiply the weight of nitrogen in order to obtain the weight of protein present. The nitrogen analysis is usually done by the Kjeldahl method, which involves digesting the protein with sulfuric acid and certain catalysts. This converts the nitrogen of the protein into ammonium sulfate. The mixture is then cooled and made alkaline, and the ammonia liberated is distilled into standard acid. Back titration enables the amount of ammonia, and thus of nitrogen, to be estimated.

Analysis of a protein into its constituent amino acids is an important method of characterization. It is described in Chapter 2.

Solubility

The solubilities of proteins enable us to characterize them as water-soluble, salt-soluble, alcohol-soluble, insoluble, and so forth. This alone allows only a rough characterization to be made, but before modern methods developed it was one of the chief methods, and still has a place in protein chemistry. In particular, separating proteins into albumins, which are in general soluble even when the solution is half-saturated with ammonium sulfate, and globulins, which are precipitated under such conditions, has proved useful.

Another use of solubility is in testing the purity of a protein preparation (see page 31).

Molecular Weight

One important method of characterizing proteins depends upon a determination of their molecular weight. Protein molecules range in size from about thirteen thousand to several million, and a knowledge of the molecular weight is of value in characterizing the protein. However, the number of distinguishable classes of molecular weight is not nearly as great as the number of different proteins; so the molecular weight simply enables us to arrange the protein roughly into a molecular-weight class. It looks as if proteins of practically any molecular weight over 12,000 may occur.

Osmotic pressure. The classical method of determining the molecular weight of proteins depended upon measurement of the osmotic pressure. The concept of osmotic pressure, though explained in most books on elementary chemistry and in all books on physical chemistry, is not always clearly understood by the student. The osmotic pressure may be defined as the hydrostatic pressure which should be applied to the solution to produce equilibrium if the solution is separated from pure solvent by a membrane which is freely permeable to the solvent but impermeable to the solute. In the case of proteins, because of their large molecular size, such membranes are not very difficult to procure. Note that the pressure which we have defined as the osmotic pressure has to be exerted *on* the solution. The reason for this is that when the solution is in contact with the solvent through this membrane, solvent tends to pass more rapidly in the direction from solvent to solution than in the opposite direction, so that solvent is gradually drawn into the solution. To counteract this, pressure must be exerted on the solution until this counterpressure is sufficient to counteract the tendency of the solvent to move from one part of the membrane to the other. It is only in this sense that osmotic pressure is a pressure; otherwise it would be more accurate to describe it as a kind of suction, for it is the solution which sucks water into itself. The reason osmotic pressure is important biologically is because solutions have the power of drawing water into themselves through a semipermeable membrane. The body contains many such membranes.

If we represent the osmotic pressure by the Greek letter π, its value is given for dilute solutions by the expression

$$\pi = \frac{RTx}{V_o} \qquad \text{Equation I-22}$$

where R is the gas constant, T the absolute temperature, x the mol fraction of the solute, and V_o the volume occupied by one mol of solvent. Or, clearing of fractions, we get

$$\pi V_o = xRT. \qquad \text{Equation I-23}$$

An occasional source of confusion is the emphasis which has been laid in the past upon the formal similarity between these equations for osmotic pressure and the perfect gas law.[1] It is true that in very dilute solutions

[1] The perfect gas law is an expression giving the behavior of a "perfect gas", as idealized from the behavior of actual gases, all of which are "imperfect", although not always in the same way. It is given by

$$PV = NRT$$

where N is the number of mols of gas in the container, P is equal to pressure, and the other symbols have the same significance as above. The perfect gas law is a law approached as a limit by actual gases as the pressures studied become less and less.

the osmotic pressure is the same as the pressure which would be exerted at the same temperatures by gas molecules in the same volume concentration as the solute molecules. There, however, the similarity ends. The mechanisms for these two phenomena are entirely different. Gas pressure is due to the impact of the molecules on the walls of the container; osmotic pressure is due to the decrease of thermodynamic activity of the solvent due to the presence of the solute. Also, in the thermodynamic derivation of the expression for osmotic pressure, it will be found that the way in which V, R, and T come into the expression is quite different from that in the gas law.

The osmotic pressure of a solution depends on the number of particles of solute contained in unit volume. From this it follows that if one gram of protein is dissolved in 100 ml. of water, the osmotic pressure will be much less than that produced by dissolving one gram of sugar, and this in turn is less than that produced by dissolving one gram of sodium chloride—the reason being, of course, that the number of particles is larger the smaller the molecular weight and also, in the case of sodium chloride, dissociation increases the number still further.

Sedimentation constant. The ultracentrifuge has been discussed previously and it was pointed out that from the sedimentation constant a value of the molecular weight could be deduced if it were shown that the molecule were spherical. If the molecule is not spherical, it is necessary to know the diffusion constant of the protein, and diffusion constants are in general known less accurately than sedimentation constants. Therefore, it is customary to give the sedimentation constant of a protein as one of its characteristics without attempting an estimated molecular weight unless a reliable value of the diffusion constant is available. The sedimentation constant is now expressed in units of svedbergs, named after the inventor of the ultracentrifuge. Values of sedimentation constant, usually symbolized S_{20}, of the order of 2, 3, 3.5, and so on, have been observed up to a value of 280 for a virus preparation. This latter preparation was considered to have a molecular weight about 47 million.

Light-scattering methods. When light is passed through a protein solution, some of it is scattered in various directions by the molecules of proteins. Quantitative measurements of this phenomenon enable us to estimate (a) the size of the protein molecules. This depends upon the fact that the apparent turbidity varies with the wave length of the light used and also depends on the size of the molecules. The dependence of the turbidity on the wave length has been shown to make an estimate of the particle size, or molecular weight, possible. (b) The way in which the light is scattered in different directions has been shown to depend on the shape of the particles, and a study of this with light of different wave lengths enables the shape of the particles to be calculated (5).

Titration Curves

The amino acids of which proteins are composed contain several different types of acid and basic groups. Therefore, one characteristic of any given protein is the amount of acid or base which must be added to bring it from one given pH to another; or if we start at the extreme acid or base range, we obtain a curve of the equivalents of acid or base added plotted against pH. This is called the titration curve of the protein.

If the titration is begun in a strongly acid solution so that the protein molecule carries its maximum positive charge, the addition of hydroxyl ion brings about the removal of protons from the acidic group of the protein. If the proton is removed from an uncharged acid group, a negatively charged group is produced. Thus, the total charge on the protein is increased by one unit while the net charge is decreased by one unit. If a proton is removed from a cationic acid group, both the net charge and the total charge decrease by one unit. Thus, if the protein contains +n cationic acid groups the net charge and also the total charge in strongly acid solutions is +n. When protons are removed from the protein ions by the addition of base, the net charge on the protein molecule falls from +n to zero, although the total charge on the protein in this condition may be, in fact usually is, greater than on the cationic protein. After the protons have been removed by the addition of base, the pH reached is called the isoionic point. If only protons are involved, this may correspond to the isoelectric point of the protein as determined by electrophoresis. In this book we shall make no distinction between the two.

It is clear that the increase in net charge on the protein molecule between the isoionic point and the region of maximum acid-binding capacity equals the number n—that is, the number of cationic groups in the protein. By similar reasoning it is clear that the maximum base-binding capacity of the protein should be equal to the number of uncharged acid groups of molecules. Thus, from titration curves of the protein we are able to draw some conclusions as to the number of basic and acidic groups present in the molecule. The titration curve of human serum albumin is shown in figure I-9. It will be noted that the curve shows two steep drops. The first, and steepest, is roughly between pH values of 3.5 and 4.5. The second is roughly from pH 9 to 12. It is probable that in the first region the carboxyl groups of the dicarboxylic amino acids are being titrated, and in the second the amino or other basic groups, lysine and arginine (or hydroxyl groups of tyrosine) residues. In the flatter region in between, the histidine groups probably come into play (2).

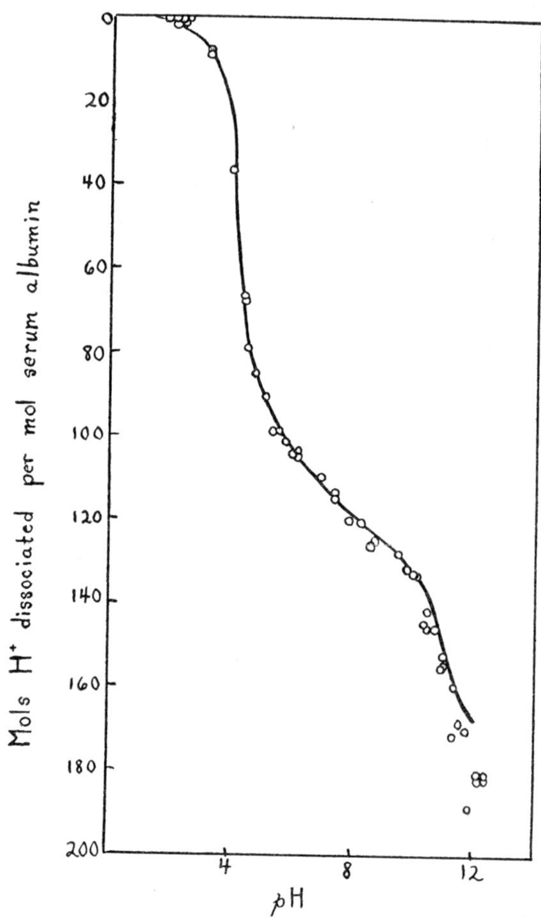

FIG. I-9. Titration curve of normal human serum albumin (Tanford). Mols of hydrogen ion dissociated per mol of serum albumin plotted against pH.

Immunological Characterization

When a foreign protein is injected into an animal, the animal frequently responds by the production of substances capable of combining with this protein, sometimes in a visible way so as to produce a precipitate. The substances produced in response to the injection are called antibodies and they are relatively specific for the protein in response to which they are produced. For instance, if we inject the albumin from the hen's egg into a rabbit, we usually obtain antibodies capable of precipitating the albumin from the hen's egg. This will not react with most other proteins of the egg or chicken tissues, but will react, although somewhat less satisfactorily, with the protein from the egg of the duck and other birds. The specificity is not complete but tends to follow taxonomic lines; in other words, when species are closely related, their corresponding proteins are likely to be similar immunologically. The immunological method affords the best way of characterizing proteins sharply and enables us to distinguish proteins which may seem identical from the point of view of molecular weight, electrophoretic mobility, and so forth. By quantitative studies of this precipitation reaction it is possible to estimate the degree to which the proteins are related to each other.

However, these immunochemical, or serological methods as they are sometimes called, have certain drawbacks as methods of characterizing proteins. Individual animals do not always respond in the same way to the injection of foreign proteins, and the antibodies produced by one animal may indicate more similarity between proteins than do the antibodies produced by another. Also, the antibodies do not keep indefinitely after the animal has been bled so that the antiserum is not a reagent of constant properties—this introduces some complications into its routine use to characterize and identify proteins. However, the method has had valuable applications.

COLLOIDAL BEHAVIOR OF PROTEINS

The distinction between crystalloids and colloids was proposed by Graham in 1861; crystalloids were characterized by a tendency to form crystals when separating from watery solutions, and colloids by a tendency to separate out in the form of gelatinous or amorphous masses. Graham found that these two groups of substances differed also in two other respects: first, in their diffusive mobility; and second, in their peculiar physical aggregation. The crystalloids diffuse readily through various membranes, for example, the wall of the pig's bladder; but the colloids are able to diffuse not at all or only very slowly through these membranes. The second peculiarity was the tendency of the colloids to form aggregates when in solution; while this property was lacking or less pronounced in crystalloids.

Later workers came to the conclusion that there was no such fundamental distinction between colloidal and crystalloidal substances as Graham had thought, and distinguished instead between colloidal and crystalloidal states of matter. For instance, Krafft observed that the alkaline salts of the higher fatty acids, stearate, palmitate, oleate, dissolve in alcohol as crystalloids of normal molecular weight, but in water they are true colloids. The reverse is true of sodium chloride. In water, of course, sodium chloride is a typical crystalloid, but Paal found that the latter gave a colloidal solution in benzene.

Later, Loeb showed that it was possible to explain the practically all of characteristics of colloidal behavior merely by the difference in diffusibility between colloids and crystalloids. The tendency to form aggregates is of slight importance in explaining the colloidal behavior of the proteins, although some protein molecules will aggregate under suitable conditions. It is also true that the protein molecule as we ordinarily know it can often be dissociated into smaller units under more extreme conditions. But these units are, nevertheless, of the order of the size of protein molecules and are still colloidal in their behavior. Therefore the formation of aggregates is not the secret of the colloidal behavior of proteins; it is their large size, their amphoteric properties, and their inability to pass through membranes which allow ordinary electrolytes to pass that give this behavior.

Proteins pass either in very small quantities or not at all through membranes such as cellophane which allow electrolytes to pass freely. As already mentioned, this difference in diffusibility can be taken advantage of in attempts to purify proteins. Also, this non-diffusibility of proteins is of great importance to living matter, serving to keep the structural units in their proper place.

Donnan Membrane Equilibrium

The British chemist, Donnan, proved in 1911 (4) that when a membrane separates two solutions of electrolytes, one of which contains an ion which can not diffuse through the membrane while the other ions can diffuse through the membrane, the result will be an unequal distribution of the diffusible ions on the opposite sides of the membrane. At equilibrium the products of the concentration of each pair of oppositely charged diffusible ions are the same on the opposite sides of the membrane. This unequal concentration of the crystalloidal ions gives rise to potential differences and osmotic forces, and it was shown by Loeb that these forces furnish the explanation for the colloidal behavior of proteins.

Donnan derived his results from thermodynamic considerations, and students familiar with these methods may find this derivation either in the original paper (4) or quoted in Loeb's book (11). Actually, Willard Gibbs (8) had derived the general equations for this and similar equilibria, although he did not cite any specific examples. Indeed, at the time Gibbs wrote, none was known. Modern writers familiar with Gibbs' work often refer to the result as the Gibbs-Donnan equilibrium.

To derive the Donnan equilibrium let us consider a system of two compartments each containing an aqueous solution, separated by a semipermeable membrane such that all but one of the ions present can penetrate it. This non-diffusible ion must perforce remain in the compartment in which it is first placed. Water and the diffusible ions are, however, able to penetrate the membrane and will eventually be distributed, as a result of diffusion, between the two compartments in such a way that we have an equilibrium. If the electrolytes present are hydrochloric acid (HCl) and protein chloride, which we may represent by RCl, the membrane being impermeable to ion R^+, then the ions H^+ and Cl^- can penetrate the membrane, but they must be considered as going through in pairs, not separately, because of the electrostatic attraction which one exerts on the other. Electrostatic forces are so great that in a solution no large number of positive charges is ever separated from a large number of negative charges. Now the probability that a single ion arrives at a given point of the membrane during a given time interval is obviously proportional to the concentration of that ion in the solution on that side of the membrane. The probability that another ion will hit a point on the membrane is proportional to *its* concentration in the solution. The probability that two independent events will happen simultaneously is equal to the product of the individual probabilities. Therefore, the probability that an H^+ ion and a Cl^- ion will simultaneously arrive at a given point at the membrane is equal to the product of their concentrations. Thus the rate of diffusion of HCl through the membrane in one direction is proportional to the product of the concentrations of the H^+ ion and Cl^- ion in the side from which it is diffusing and the same is true for its diffusion through the membrane from the opposite side. At equilibrium the two rates of diffusion must be equal. The diagram represents the equilibrium state.

(z)	R^+	|		
(y)	H^+	|	H^+	(x)
(y + z)	Cl^-	|	Cl^-	(x)

In the diagram the vertical line represents the membrane and the chemical symbols the various ions. We let the letter x represent the concentration of H^+ (or Cl^-) ions, which we remember must be equal because of the electrostatic forces between them. Let y represent the concentration of H ions on the other side of the membrane; and z the concentration of R^+ ions. Then, $y + z$ = concentration of Cl^- ions on that side of the membrane, for here, too, electroneutrality must be maintained. The Donnan equilibrium, which as we have seen implies that the products of the concentrations on the two sides must be equal for the two ions H^+ and Cl^-, gives us the relation $x^2 = y(y + z)$. From this equation we see that x is the geometric mean of y and $y + z$ and thus lies between y and $y + z$ in magnitude. Therefore we see that, although the ions H^+ and Cl^- can pass freely through the membrane, their concentrations on the two sides at equilibrium will not be equal if a non-diffusible ion is present on one side. This unequal distribution of ions is the outstanding feature of the Donnan equilibrium and accounts for the influence of electrolytes on various properties of colloid systems.

One consequence of the difference in concentrations of the diffusible ions is that there will be a potential difference between the opposite sides of the membrane. The measurement and significance of such potentials in the body form a part of biophysics. Such potentials are involved in neuromuscular activity and in glandular secretion.

For the operation of the Gibbs-Donnan membrane effect it is not necessary to have the protein enclosed in a membrane such as collodion or cellophane. If the protein is in the solid state or forms part of a structure such as a cell which prevents it from moving through the solution, it is just as effectively immobilized as if it were confined within a semipermeable membrane. Under such conditions the Gibbs-Donnan equilibrium is observed. For instance, if small particles of gelatin are suspended in cold water or salt solution at various pH, it will be found that the salt concentration and the pH inside and outside differ in accordance with the membrane equilibrium as defined above. It is also found that the sign of the charge inside the particles of gelatin depends upon the pH, just as would be predicted. The same considerations apply to a suspension of any amphoteric compound capable of combining with the electrolytes of the medium.

The Donnan membrane equilibrium also enables us to account for the effect of salts on colloidal systems. If we have an acid gelatin solution, or if particles of gelatin are suspended in an aqueous acid medium, the potential difference between the positive gelatin ions and the solution with which it is in contact is determined by the values in the Donnan equation for the anions. The values of the cations do not enter into the expression on the acid side of the isoelectric point. Loeb (11) carried out experiments which demonstrated quite clearly that this prediction holds in practice and was able to depress the observed potential difference, by roughly the expected amounts, by adding calculated amounts of various salts. He was also able to show that the effect depended on the anions added and was independent of the concentration and valency of the cations. Loeb noted that this explained the precipitating effect of the salt on otherwise stable colloidal suspensions and thus corrected an error which had long appeared in the literature of colloidal chemistry, to the effect that neutral salts precipitated colloidal particles by bringing them to their isoelectric point. What happens is that the addition of neutral salts diminishes the potential difference between the suspended particles and the liquid; and if enough salt is added it completely wipes out this potential difference. It is true that if the particles, gelatin granules, for example, are brought to their isoelectric point by a change in hydrogen-ion concentration, the potential difference between the particles and the surrounding liquid also becomes zero; but this is for a different reason, mainly because the net charge on the gelatin molecule is zero at this point.

The Gibbs-Donnan equilibrium also explains the effect of salts on the osmotic pressure of proteins. It has been known for a long time that the osmotic pressure of a protein solution as measured at pH values not at the isoelectric point may be considerably larger than would be expected. Errors of considerable magnitude have resulted from making osmotic pressure measurements not at the isoelectric point. The measured osmotic pressure of a protein separated from water by a semipermeable membrane is the sum of the pressure due to the colloidal particles and the pressure caused by the difference in the number of dialyzable ions inside and outside the membrane. This difference can be calculated and measurements have been made on colloidal systems which agree well with theory. The great difficulty of obtaining proteins absolutely free from electrolytes and exactly at their isoelectric points accounts for the inaccuracy of the osmotic pressure measurements and the resulting molecular weight determinations with the larger protein molecules. It is for this reason that other methods, such as the ultracentrifuge, often give more reliable results.

SHAPE OF PROTEIN MOLECULES

It was stated above that there was evidence that most protein molecules are not spherical. Measurements of the viscosity of protein solutions indicate that protein molecules are either fairly elongated ellipsoids—something like a cigar—or else flat disks—something like a peppermint wafer. The diffusion and viscosity data could be interpreted on either one of these assumptions. These calculations are based, however, on the assumption that the protein molecule is not hydrated—in other words, that it does not bind any water but simply swims in a sea of solution. It is likely that water does combine in a relatively firm way with proteins in solution so that proteins are *hydrated*; and if this is taken into account, the ratio of the major and minor axes of the molecule and the viscosity are related to the hydration. The decision between the elongated ellipsoid and the flattened disk as the true shape of protein molecules in solution has to be made by determination of dielectric increment and dispersion and relaxation time. Such work has been published by Oncley (2) and he states that the agreement between the various methods for evaluation hydration and the ratio of the two axes yields strong support for the view that protein molecules appear to rotate as rigid units which can be mathematically described as if they were elongated ellipsoids of revolution with varying ratios of major-minor axes. In figure I-10 we give pictures of the probable appearance of certain well known protein molecules. Further information on this subject may be found in a review by Edsall (6).

Protein molecules are sufficiently large so that some of the larger ones can be seen directly (dehydrated and *in vacuo*, to be sure) by means of the electron microscope. The molecule of snail hemocyanin (the blue copper-containing protein of its blood) is made up of cubes which are in turn made up of four short rods packed closely together. Molecules of catalase and edestin, each with a molecular weight of about 250,000, are made up of clusters of subunits, as many as ten per molecule. Some protein molecules, like edestin, seem to be spherical. The edestin molecule, as seen under the electron microscope, has a diameter of 8 millimicrons. (A millimicron is 10^{-9} meters or 10^{-6} millimeters.) Some proteins are rodlike. Fibrinogen molecules are 70 millimicrons long and 4 millimicrons wide. The molecule of the tobacco mosaic virus is 280 millimicrons by 15.

DENATURATION AND COAGULATION OF PROTEINS

Proteins, as we have already stated, are very labile molecules. Any change which leaves a protein molecule in a state which we judge to be different from the state in which it is found in those tissues which are its natural source is called denaturation. Since we usually have to judge this by comparison with the sample of protein which we had extracted by what we think is a safe procedure, it is not always easy to make certain that a

Horse antipneumococcus antibody

Horse antitoxin	Rabbit antiovalbumin
Human antipneumococcus	Rabbit antipneumococcus
Human gamma globulin	Human serum albumin

FIG. I-10. Probable appearance of certain protein molecules (Models made and photographed by W. C. Boyd)

given sample of protein has not been denatured. It is easier to state in some cases that it definitely has been denatured. One of the first symptoms of denaturation is a loss of solubility; this loss may be partial or complete. When the loss is complete the protein is said to be coagulated; it becomes an insoluble solid. This may be effected by drastic treatment such as boiling or the action of nitric acid and other powerful reagents. Albumins or globulins are coagulated by heating their solutions to the boiling point. Some are coagulated at lower temperatures. One seems to be reversibly coagulated and redissolves at higher temperatures; this is the Bence-Jones protein which occurs in the urine in certain types of malignancy (see page 386). But not all proteins are coagulated by heating to the boiling point. Casein, for instance, may be boiled or even autoclaved, and still remain in solution. What happens during denaturation will be discussed in Chapter 2.

CLASSIFICATION OF THE PROTEINS

Proteins might be classified according to their molecular weight, electrophoretic mobilities, or in various other ways. Some workers rely mainly on such characteristics. If sufficient analytical data were available, they might be classified by their amino acid composition; but existing data are not yet adequate for this purpose. Some years ago, before much of the modern physicochemical data had been obtained, a classification of proteins which is still in common use was proposed and adopted. This classification is not very informative or very useful, but since the terms employed are still in common use, it is felt that it should be presented here. It is to a large extent based on rather obvious differences in physical properties, particularly in solubility.

Simple Proteins

The simple proteins on hydrolysis yield α-amino acids or their derivatives only.

1. *Albumins* are soluble in water and are coagulated by heat. They are obtained from animal and vegetable cells and body fluids. Good examples are egg albumin, serum albumin from blood, and perhaps leucosin from wheat. It is interesting that these definitions begin to break down even at the very beginning of the classification, for some of the so-called simple proteins—characteristic in other ways—contain other components in addition to amino acids; for instance, crystalline egg albumin, even from repeatedly crystallized preparations of the highest purity, contains a polysaccharide made up of mannose and glucosamine groups.

2. *Globulins*, found in the same sources as the albumins, are insoluble in pure water but are soluble in dilute salt solution and are coagulated

by heating. Like albumins, they are precipitated at higher salt concentrations, somewhat more readily than are albumins. The best examples of globulins are perhaps serum globulin, which has been prepared in quite pure although not crystalline form, myosin from muscle, and ovoglobulin from egg yolk. Also certain plant proteins are put in this class: edestin from hemp seed, amandin from almond, and excelsin from Brazil nuts. These proteins, however, require considerably higher salt concentrations, of the order of 5 per cent NaCl, to keep them in solution, and they are more readily crystallized.

Sørensen distinguished two kinds of globulins. The first kind, euglobulin, is insoluble in pure water but soluble in dilute salt solutions. It was reported to contain phosphorus. The second kind, pseudoglobulin, is soluble in pure water and was reported not to contain phosphorus. The pseudoglobulins were distinguished from the albumins by being salted out, like other globulins, by half-saturation with ammonium sulfate. In practice it is hard to distinguish euglobulin from pseudoglobulin largely because of the great difficulty of getting proteins entirely salt-free. Globulins, homogeneous by ultracentrifugal and electrophoretic test, sometimes prove to be partly soluble in pure water and partly insoluble, with no striking difference in other properties between the two fractions.

3. The *glutenins*, obtained from cereal seeds, are insoluble in all neutral solvents, but are soluble in dilute acid and dilute alkali. Examples are glutenin from wheat and oryzenin from rice.

4. Alcohol-soluble proteins (*prolamines*), also from seeds, are soluble in 70 to 80 per cent alcohol but insoluble in either water or absolute alcohol. Examples are gliadin from wheat, hordein from barley, and zein from Indian corn (*Zea mays*).

5. The *scleroproteins* or *albuminoids* are also insoluble in neutral solvents and practically insoluble in other solvents. Examples are collagen from hide, bone, and cartilage, and keratin from skin, horns, hair, and feathers. The keratins have a high content of the sulfur-containing amino acids.

6. The *histones*, from animal cells, are soluble in water but insoluble in dilute ammonia because of their basic properties. Solutions of many other proteins will precipitate with histones. Examples are globin, which is the protein part of hemoglobin, thymus histone, and scombrone from mackerel.

7. The *protamines* are found in ripe generative cells of some fish and are the simplest of the simple proteins. They have a low molecular weight for proteins; they are soluble in water and not coagulated by heating, they precipitate many other proteins from their aqueous solutions. They contain relatively few amino acids and these are largely basic in character. Therefore the protamines possess strong basic properties and form stable salts with the strong acids. Examples of protamines are salmine from

salmon, sturine from sturgeon, clupeine from herring, scombrine from mackerel, and cyprinine from carp.

Conjugated Proteins

Conjugated proteins are proteins combined with non-protein compounds other than salts. The non-protein portion is called the prosthetic group.

1. The *nucleoproteins* are combinations of proteins with nucleic acids. Nucleoproteins will be discussed more in detail in Chapter 9.

2. The *glycoproteins* are combinations of proteins with substances which contain a carbohydrate group. For further discussion and some examples, see page 143.

3. The *phosphoproteins* are combinations of proteins with phosphorus-containing substances other than nucleic acid or phospholipid. Examples are casein from milk, and possibly vitellin from egg yolk.

4. The *chromoproteins* are combinations of protein with a colored prosthetic group. The best known example is hemoglobin. Copper-containing chromoproteins called hemocyanins occur in the blood of a number of invertebrates such as lobsters and crabs.

5. The *lipoproteins* are combinations of proteins with lipids. They are found in cell nuclei, egg yolk, milk, and in the blood. They are widely distributed in blood and tissues, in some viruses, and in certain bacterial antigens.

Derived Proteins

As the name implies derived proteins are proteins which have been derived from other proteins by modification of one sort or another. They are divided into: 1) primary protein derivatives, which include proteans, metaproteins, and coagulated proteins; and 2) secondary protein derivatives, which include proteoses, peptones, and peptides. The first group has been changed relatively slightly from the original protein; the second group has been more extensively changed, and in the case of peptides, has been degraded to relatively small molecules, in many cases so small that they can be identified as definite chemical compounds of known composition. The best known derived protein, gelatin, is difficult to put snugly into either of these classes. It is prepared by the treatment of collagens (from bone and so forth) with superheated steam or boiling dilute acids, and is certainly radically altered from its parent proteins. In regard to size it is intermediate, consisting of a mixture of molecules of various sizes, from molecular weights of the order of the albumins down to relatively small molecules. The average molecular weight is so small that much of an ordinary gelatin preparation can pass through membranes such as the capillary walls. Another notable feature of gelatin is its non-antigenicity.

1. The *proteans* are insoluble products which result from the action for a short time of various reagents, such as dilute acids, or enzymes, or even water. Examples are myosan from myosin, and edestan from edestin.

2. The *metaproteins* are the result of still further action of acids and alkalis. They are soluble in dilute acids and alkalis but insoluble in solutions of neutral salts. Examples are acid albuminate and alkali albuminate.

3. The *coagulated proteins* are insoluble products resulting either from the action of heat or other denaturing agents, such as alcohol or formaldehyde.

4. The *proteoses* are soluble in water and are not coagulated by heat. They can be precipitated if their solutions are saturated with ammonium sulfate or zinc sulfate, typical protein precipitants.

5. The *peptones* are soluble in water and not coagulated by heat. They are not precipitated even if their solutions are saturated with ammonium sulfate. This is partly due to their lower molecular weight. Certain reagents which are known to precipitate alkaloids also precipitate peptones. An example is phosphotungstic acid.

6. The *peptides* are combinations of two or more amino acids joined together by the peptide link. Many of them have been definitely identified chemically. An interesting example is glutathione, which is γ-glutamylcysteinylglycine.

QUALITATIVE TESTS FOR PROTEINS

In our original definition of proteins we mentioned that they were made up chiefly of α-amino acids and that the separation of α-amino acids from the products of hydrolysis of a substance would establish its protein nature. Since this is difficult and time-consuming, it is convenient to have various color tests which indicate that a protein, or some other substance which happens to give the test, is present. If several of these tests give positive reactions it becomes virtually certain that proteins are present, for the different color reactions are due to different groups in the protein molecule and no substance except a protein will respond to any considerable number of them.

The first tests we consider are not color reactions, but precipitation reactions which are characteristic of proteins. We have already mentioned that proteins are precipitated by saturated solutions of ammonium sulfate, zinc sulfate, and other salts. These precipitates are usually soluble if the precipitate is filtered off and water or dilute salt solutions added. Proteins may also be precipitated by the salts of heavy metals such as copper sulfate, lead acetate, and mercuric nitrate. It is likely that in many cases these precipitates are the protein salts of the metal cations. Proteins are also precipitated by alkaloid reagents such as picric acid, phosphotungstic acid, and tannic acid.

The Biuret Reaction

If a protein solution is mixed with sodium hydroxide solution and a very weak solution of copper sulfate, a violet color, or in some cases a much more intense blue color than could be explained by the small amount of copper sulfate used, is obtained. This test probably depends upon one or more of the following groups of the protein molecule: two (—CONHR) groups, or one (—CONHR) plus one of the following: $—CSNH_2$, $—C(NH)NH_2$, $—CH_2NH_2$, $—CHRNH_2$, $—CHOHCH_2NH_2$, $—CHNH_2CH_2OH$, $—CHNH_2CHROH$.

Millon's Reaction

Millon's reagent is prepared by dissolving mercury in nitric acid and represents a solution of mercuric nitrite and mercuric nitrate in a mixture of nitric and nitrous acid. If a protein is heated with this reagent, a red color or precipitate is obtained, evên with solid proteins. The test is due to the presence in the protein molecule of the phenol group of tyrosine.

The Hopkins-Cole (Glyoxylic Acid) Reaction

If a protein solution is mixed with glyoxylic acid and concentrated sulfuric acid is poured in to form a layer on the bottom, a violet ring is obtained. This test is due to the tryptophane group which contains the indole nucleus. The discovery of the amino acid tryptophane was due to a systematic search for the substance which gave this color reaction. Gelatin, which has practically no tryptophane, fails to give the reaction.

Folin's Phenol Reagent

The phenol reagent of Folin and Ciocalteu (7) (a phosphotungstic-phosphomolybdic acid) gives a blue color with those proteins (the great majority) which contain tyrosine. This has been used to determine small amounts of protein quantitatively (9).

Sakaguchi Reaction

Proteins which contain arginine (most proteins) develop an intense red color when treated with α-naphthol and sodium hypochlorite (8a). This reaction is also given by other substances containing the guanidine group.

The Xanthoproteic Reaction

Xanthoproteic simply means "yellow protein" and refers to the fact that if a protein is heated with nitric acid a yellow color is obtained. This is evidently due to nitration of the benzene ring in phenylalanine, tyrosine, or tryptophane. Addition of ammonia after this heating causes the yellow color to become more intense and to shift towards the orange.

The Ninhydrin (Triketohydrindene Hydrate) Reaction

When proteins are boiled with ninhydrin

$$\text{(structure of ninhydrin: benzene ring fused to five-membered ring with two C=O groups and } C(OH)_2\text{)}$$

a blue color is obtained. This reaction is due to the α-amino acid groups present in the protein molecule. Naturally it is also obtained with the α-amino acids themselves.

When amino acids are heated with ninhydrin they are quantitatively deaminated. The keto acid formed is decomposed by heating into an aldehyde and carbon dioxide. The latter is evolved quantitatively in acid solution, and this forms the basis of a method for the determination of amino acids (17).

REFERENCES

1. Boardman, N. K., and Partridge, S. M. Separation of neutral proteins on ion-exchange resins. Biochem. J., **59:** 543–552, 1955.
2. Cohn, E. J., and Edsall, J. T. *Proteins, Amino Acids and Peptides*. New York, Reinhold Publishing Corp., 1943.
3. Cohn, E. J., *et al.* Chemical, clinical, and immunological studies on the products of human plasma fractionation. I. J. Clin. Investigation, **23:** 417–432, 1944.
4. Donnan, F. G. The theory of membrane equilibrium in the presence of a non-diffusible electrolyte. Zeitschr. Electrochem., **17:** 572, 1911.
5. Doty, P., and Edsall, J. T. Light scattering in protein solutions. Advances Protein Chem., **6:** 35–121, 1951.
6. Edsall, J. T. The size, shape and hydration of protein molecules. *in* Neurath, H., and Bailey, K. *The Proteins*. Volume I, Part B. New York, Academic Press, 1953.
7. Folin, O., and Ciocalteu, V. On tyrosine and tryptophane determinations in proteins. J. Biol. Chem., **73:** 627, 1927.
8. Gibbs, W. On the equilibrium of heterogeneous substances. Trans. Conn. Academy, **3:** 108–248, 343–524, 1875–1878. See *The Collected Works*, p. 55–371. New York, Longmans, Green and Co., 1928.

8a. Gilboe, D. D. and Williams, J. N. Jr. Evaluation of the Sakaguchi reaction for quantitative determination of arginine. Proc. Soc. Exptl. Biol. Med., **91:** 535–536, 1956.

9. Kabat, E. A., and Mayer, M. M. *Experimental Immunochemistry*. Springfield, Ill., Charles C Thomas, 1948.
10. Landsteiner, K., and van der Scheer, J. On cross reactions of egg albumin sera. J. Exptl. Med., **71:** 445–454, 1940.

11. LOEB, J. *Proteins and the Theory of Colloidal Behavior*. New York, McGraw-Hill Book Co., 1922.
12. NORTHROP, J. H. *Crystalline Enzymes*. 2nd edit. New York, Columbia Univ. Press, 1948.
13. POPPER, H., *et al*. Chemical versus electrophoretic separation of proteins. Am. J. Clin. Path., **20:** 530–538, 1950.
14. SHARP, D. G., *et al*. The electrophoretic properties of serum proteins. J. Biol. Chem., **144:** 139–147, 1942.
15. SVEDBERG, T., AND PEDERSEN, K. O. *The Ultracentrifuge*. Oxford, Clarendon Press, 1940.
16. SVENSSON, W. Q. Fractionation of serum with ammonium sulfate and water dialysis, studied by electrophoresis. J. Biol. Chem., **139:** 805–825, 1941.
17. VAN SLYKE, D. D., MACFAYDEN, D. A., AND HAMILTON, P. Determination of free amino acids by titration of the carbon dioxide formed in the reaction with ninhydrin. J. Biol. Chem., **141:** 671–680, 1941.
18. WYMAN, J. Studies on the dielectric constant of protein solutions. I. Zein. J. Biol. Chem., **90:** 443–476, 1931.

CHAPTER 2

Protein Structure

To stay alive, an organism must constantly adjust its internal structure and functions to changes in external environment. These external changes can be complex, and, even when important to the organism, slight on an absolute scale. It therefore seems fitting that important units of the chemical structure of the body, the protein molecules, are themselves unstable and delicate. This delicacy and instability are among the most characteristic features of protein molecules. A protein solution is often recognizably altered merely by standing at temperatures no higher than those of the human body, or by having a gentle current of air passed through it.

Proteins are also complicated molecules. This is partly merely the consequence of their large size. For instance, consider insulin, a pancreatic hormone, which has an empirical formula which is $C_{496}H_{730}O_{150}N_{122}S_{12}$. Here we have a molecule in which we can count over fifteen hundred individual atoms of five different kinds. The molecular weight is 11,187, which means that the molecule is over six hundred times as heavy as a water molecule. And yet insulin is a protein of comparatively simple structure. Certainly, its molecular weight is well below the average for proteins. Molecular weights in the hundreds of thousands are common and those in the millions are not unknown.

Yet the size of protein molecules is only part of the story. There are other types of molecules produced by living organisms which compare with proteins in molecular weight. There is cellulose, for instance, impressive in size, and yet used for nothing more in the living plant than to enclose the cell in a sturdy box. It is a huge molecule, yet so stable that we build houses out of it.

PROTEINS AS AMINO ACID CHAINS

The Nature of the Chain Unit

In considering protein structure, we might for the moment focus our attention upon a typical protein molecule as compared with a molecule of cellulose. Both are alike in that they are repetitive structures. That is, they are constructed of repetitions of comparatively simple units, somewhat in the manner that beads are strung together to form a necklace.

Both proteins and cellulose can be so treated as to be broken down into smaller chains or into the ultimate mixture of the units themselves. Cellulose, after strenuous hydrolysis with mineral acids, is broken down to molecules of glucose, a simple sugar with a molecular weight of 180. In order to build the cellulose molecule, as many as two thousand glucose molecules may be condensed (with loss of water). Under acid or alkali hydrolysis, a protein molecule is similarly degraded to its units, the amino acids discussed in Chapter 1, and here again thousands may be condensed with one another to build a protein. But there is an important difference to be noted. Whereas cellulose possesses a single kind of unit, glucose, proteins possess more than twenty different units, all amino acids to be sure, but differing among themselves otherwise.

This difference is characteristic of proteins. Other repetitive structures, either natural or synthetic, are built up of not more than one or two different units (nucleic acids are a partial exception—see Chapter 9), and here would seem to be one of the answers to the problem of protein complexity. Whereas in the case of cellulose, for instance, individual molecules can vary only in the number of glucose units in the chain; in the case of proteins, not only the number but the *type* and *arrangement* of the units becomes significant. The difference implied here is somewhat analogous to that existing between an artist compelled to paint a scene in a single color and one who is allowed twenty colors.

The first task in investigating protein structure is then not to count atoms but to investigate the manner in which amino acids are combined.

The Peptide Linkage

A series of amino acids may condense with the elimination of one molecule of water per amino acid in the manner indicated in formula I. That this is the actual linkage between the amino acids in the protein molecule is well established. The amide linkage (—CO—NH—), when it involves the alpha-carboxyl group of one amino acid and the alpha-amino group of another, is known as a *peptide bond.* Groups of amino acids combined in this manner are called *dipeptides*, *tripeptides*, and so forth, according to the number of amino acids in the chain. Where the chain is long and the number of amino acids composing it is indeterminate, the generic term *polypeptide* is applied. The amino acids in the chain being minus the elements of water are termed *amino acid residues.*

The lines of evidence pointing to the existence of the peptide linkage are many and conclusive. Some of the evidence may be indicated:

a. Proteins in their natural state contain far fewer free amino groups or free carboxyl groups than would correspond to the number of amino acid residues they contain, indicating these groups to be involved in bond forma-

$$
\cdots \; H_3\overset{(+)}{N}\text{—CHR—}C{\overset{\displaystyle O}{\|}}\text{—}O^{(-)} \quad H_3\overset{(+)}{N}\text{—CHR—}C{\overset{\displaystyle O}{\|}}\text{—}O^{(-)} \quad H_3\overset{(+)}{N}\text{—CHR—}C{\overset{\displaystyle O}{\|}}\text{—}O^{(-)} \quad H_3\overset{(+)}{N}\text{—CHR—}C{\overset{\displaystyle O}{\|}}\text{—}O^{(-)} \; \cdots
$$

$$
\xrightarrow{-4H_2O}
$$

$$
\text{—NH—CHR—}\underbrace{\overset{\displaystyle O}{\overset{\|}{C}}\text{—NH}}_{\text{Peptide linkage}}\text{—CHR—}\underbrace{\overset{\displaystyle O}{\overset{\|}{C}}\text{—NH}}_{\text{Peptide linkage}}\text{—CHR—}\underbrace{\overset{\displaystyle O}{\overset{\|}{C}}\text{—NH}}_{\text{Peptide linkage}}\text{—CHR—}\overset{\displaystyle O}{\overset{\|}{C}}\text{—}
$$

I. Polypeptide chain

tion. Such free amino groups as are detected approximate the number of lysine residues contained by the protein. This is to be expected since lysine possesses a second amino group not concerned in the peptide linkage. Similarly free carboxyl groups result from the presence of aspartic acid and glutamic acid in the molecule. Furthermore when proteins are hydrolyzed, amino groups and carboxyl groups are invariably liberated in equivalent quantities, exactly as would be expected if the hydrolysis consisted of the breaking of peptide bonds by the addition of the elements of water.

b. Protein fragments more complicated than amino acids can be isolated after incomplete hydrolysis. These frequently turn out to be identical chemically with known synthetic short-chain peptides. Thus, glycyl-tyrosine has been isolated from the incompletely hydrolyzed fibrous protein of silk (formula II).

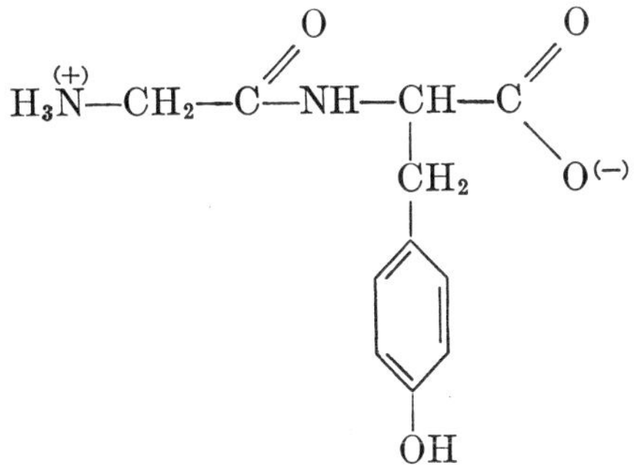

$$H_3\overset{(+)}{N}-CH_2-\overset{O}{\overset{\|}{C}}-NH-\underset{\underset{\underset{\underset{OH}{|}}{C_6H_4}}{|}}{\underset{CH_2}{\underset{|}{CH}}}-C\begin{matrix}\nearrow O \\ \searrow O^{(-)}\end{matrix}$$

II. Glycyl-tyrosine

c. Enzymes which catalyze the hydrolysis of proteins in the digestive tract act in similar fashion upon synthetic peptides. In view of the specificity of enzyme action (see Chapter 6), this is excellent evidence in favor of the peptide linkage.

d. The absorption spectrum produced by the passage of infra-red radiation through protein films has been found to show an appearance similar to the spectra of other substances known to contain the —CO—NH— linkage (26).

Naturally-occurring Peptides

The peptide link or links similar to it exist in certain naturally-occurring compounds with molecular weights between those of proteins and of single amino acids. Since the molecules of these intermediate compounds are smaller and more stable than those of proteins, it has proved simpler to establish their structure. The study of these compounds (11) has made it convenient to introduce new terms and redefine old ones. A *peptide*, for instance, is a substance with a molecular weight less than 10,000 which, on hydrolysis, yields amino acids. The figure 10,000 is quite arbitrary and is chosen because substances of smaller molecular weight can diffuse across a cellophane membrane. Similar substances with higher molecular weights are polypeptides or proteins.

A *homeomeric peptide* is a peptide which on hydrolysis yields amino acids plus, possibly, ammonia. A *heteromeric peptide* yields, on hydrolysis, substances other than amino acids in addition. An example of a heteromeric peptide which occurs in all living cells is pteroylglutamic acid, the formula of which is given on page 811. On hydrolysis, pteroylglutamic acid yields, in addition to glutamic acid, para-aminobenzoic acid and a pteridine derivative.

Of the homeomeric peptides, the most generally distributed is *glutathione*, which probably occurs in all living cells. It is a tripeptide with the structure shown in formula III. Notice that the bond between the cysteine and glycine residues is an ordinary peptide bond involving the alpha-carboxyl group of the cysteine and the alpha-amino group of the glycine. The bond between the glutamic acid residue and the cysteine residue is

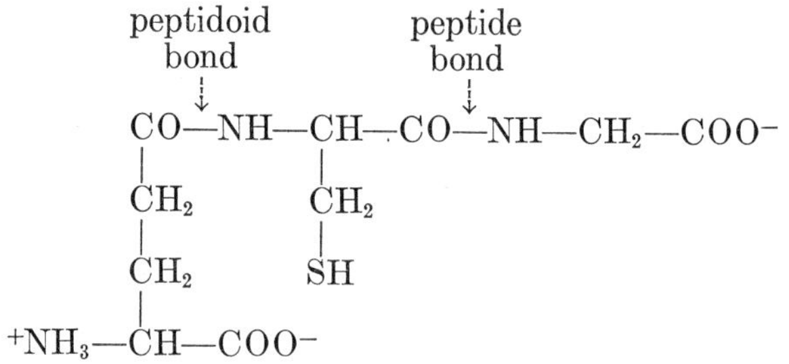

```
                 peptidoid        peptide
                   bond             bond
                     ↓                ↓
                CO—NH—CH—CO—NH—CH2—COO-
                |          |
                CH2        CH2
                |          |
                CH2        SH
                |
         +NH3—CH—COO-
```

III. Glutathione

unusual, however, in that the side chain carboxyl group, the one attached to the gamma-carbon, is involved. When two amino acid residues are linked through any groups other than the alpha-carboxyl group of one and the alpha-amino group of another, the two residues are said to be bound by a *peptidoid bond.*

Among the more complex homeomeric peptides of importance to the human body are oxytocin and vasopressin. These are octapeptides, the structures of which have now been completely worked out (see page 307).

X-Ray Diffraction

X-rays have proven themselves powerful tools in the investigation of the structure of some of the simpler protein molecules. Through their use it has been possible to estimate the distance between the various atoms of the polypeptide chain (5, 22).

It is not proposed here to go into the theory of x-ray diffraction in detail. It need only be said that when a parallel beam of x-rays impinges upon any substance, the constituent atoms of that substance become centers for the scattering of radiation. Where the atoms are randomly placed, as in amorphous material, scattering is likewise random, and if a photographic plate is placed behind the material, one will detect a dark central spot due to that portion of the x-ray beam which pierced the substance undeflected. Surrounding it will be a hazy, featureless area of darkening, fading as the distance from the central spot increases.

However, where an orderly array of atoms is presented to the x-ray beam as in the case of crystals, planes of atoms exist which reinforce each other's scattering properties, so that portions of the impinging beam are deflected as subsidiary beams. In this case, the central spot on the photographic plate would be surrounded not by a featureless haze, but by well defined spots located in beautifully symmetrical fashion. From the position of these spots it is possible to determine the relationship of the planes of similar atoms existing within the crystal, in terms of actual distance.

It is not necessary that material be crystalline in the ordinary sense for x-ray diffraction to be of use. It is only necessary that there be present periodically repeated chemical groups. Thus, such non-crystalline substances as silk, hair, and cotton are suitable for x-ray diffraction studies since they consist of protein or cellulose which are polymers made up of repeating units. If silk fibroin is the protein studied, the data obtained are consistent with the view that the molecule is a puckered polypeptide chain of the dimensions shown in formula IV. The chain, as thus pictured, is a fully

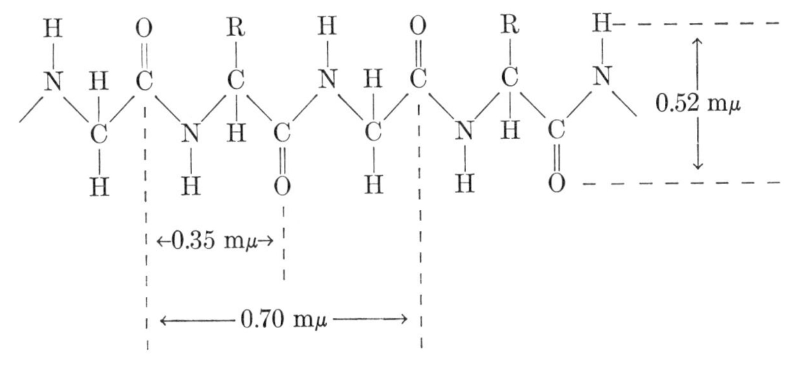

IV. Silk fibroin

extended one. It might seem as though there were room for further extension to allow the carbon and nitrogen peptide backbone to become completely linear. Because of the arrangement in space of the valence bonds about the carbon and nitrogen atoms, such an extension is impossible. X-ray diffraction studies show further that adjacent polypeptide chains of fibroin in silk fiber are separated by a distance of 0.35 mμ. Occasionally, when a bulky side chain, such as that of tyrosine is involved, the separation between chains is as high as 0.57 mμ (28).

Silk fibroin, which is an inelastic protein, maintains its configuration under all ordinary conditions. The protein of hair, keratin, which is elastic presents a somewhat more complicated case. When wetted, it can be stretched to as much as three times its original length. When fully extended in this fashion, x-ray diffraction patterns indicate a structure similar to that of silk fibroin. In its normal contracted state the polypeptide backbone of keratin forms sinuous loops, so that the molecule as a whole becomes shorter and thicker.

The fibers of hair and wool (keratin fibers) show distinctly different x-ray diffraction patterns before and after being stretched. The diagrams obtained before the fiber is stretched are called alpha-diagrams, and the pictures obtained from the stretched fibers, which are similar to those obtained from silk fibroin, are called beta-diagrams. It is apparent that in the alpha form the polypeptide chains of the protein are folded up. Infra-red spectra of proteins have also been used to determine the types of folding in peptide chains (1).

Artificial polypeptides have been synthesized with molecular weights nearly in the protein range. Such synthesis has involved chain reactions beginning with N-carboxyanhydride derivatives of the amino acids (formula V) which in water spontaneously polymerize (12). Since a single type of N-carboxyanhydride is usually used (or at most a mixture of two types), the resulting polypeptide consists of a chain of a single type (or at most two) of amino acid residues. Polylysine, polyarginine and similar compounds have been made in this fashion, as well as copolymers of two components such as glutamic acid and lysine.

```
HN—CHR—CO
|       |
OC———————O
```

V. N-Carboxyanhydride of amino acid

Such artificial polypeptides are far removed from the complex proteins that make up living tissue and represent only a small step toward the problem of artificial synthesis of physiologically active proteins. However, x-ray diffraction studies of such polypeptides have shown their structure to be similar to that of the natural fibrous proteins. Usually, they were found to exist in the beta form (maximum stretching) in which case they were insoluble. Under special conditions they could be prepared in soluble form, in which case they were in the folded or alpha form (2).

Quantitative Determinations of Protein Amino Acids

Having decided on the manner in which amino acids are combined in the protein molecule and having obtained some notions as to the size and shape of the polypeptide backbone in some of the less complex proteins, there remains the problem of determining exactly the numbers of the several amino acids in the molecule and, if possible, their arrangement as well. A number of techniques have been evolved for the purpose of investigating this problem of amino acid content of proteins.

Chemical methods. These are the oldest, and, currently the least used and least useful means of analyzing for amino acids. It is not necessary to outline the many tests that have been employed by various workers to determine the various components of the hydrolysis mixture of proteins. These can be found by the interested student in a monograph such as that of Schmidt (39). Many of these methods depend upon the formation of derivatives which may be fractionally distilled (as the ethyl esters of the amino acids) or fractionally precipitated (as in the case of uramino or hydantoin derivatives). Individual amino acids may be precipitated from the mixture (as, for instance, arginine and histidine in the form of silver salts) or determined by a more or less specific colorimetric reaction while still forming part of the mixture as tyrosine is, through use of a phosphomolybdate solution.

Spectrophotometric methods. Spectrophotometric analysis depends on the ability of chemical substances to absorb selectively specific wave lengths of light. The reason why certain wave lengths and not others are absorbed by a given compound lies in the field of quantum mechanics and is beyond the scope of this text. The fact of this selective absorption is, however, familiar to all of us from youth. Colored glass, for instance, is colored because it will transmit some wave lengths of light and absorb others. Inorganic ions such as cupric ion or chromate ion are colored for the same reason, as are the many organic coloring matters, both natural pigments and synthetic dyes.

Those substances which to us appear colorless also absorb light but at wave lengths in the ultra-violet or infra-red where the effects must be detected by photocells or by the photographic film. The spectrophotometer is an instrument designed to measure the amount of this light absorption. A beam of visible or ultra-violet monochromatic light falls upon a photocell after passing through a cell of standard dimensions containing the solvent in which the substance to be investigated will later be dissolved. The electric current produced is compared with that obtained when the light passes through an identical cell containing solvent *plus* the compound to be investigated. Spectrophotometers are usually so designed that the difference in current intensities may be read directly as per cent light transmitted or *percentage transmittance.*

From percentage transmittance, one can obtain *absorbency* usually called *optical density* or extinction, as follows:

$$\text{absorbency} = \log \frac{\text{percentage transmittance of solvent alone}}{\text{percentage transmittance of (solvent plus compound being investigated)}}$$

Since the percentage transmittance of the solvent is usually arbitrarily set at 100, the equation becomes:

$$\text{absorbency} = \log \frac{100}{\text{percentage transmittance}}$$

$$= 2 - \log \text{(percentage transmittance)}.$$

Many instruments are so devised as to allow absorbency to be read directly.

The advantage of absorbency over the apparently simpler and more straightforward percentage transmittance lies in the fact that in the case of most substances *absorbency varies directly with concentration*, all other factors being equal. This is a statement of *Beer's Law* which is so fundamental n analyses by optical methods and so useful that there is a tendency to forget that it does not always hold true. For Beer's Law to be valid there must be no change in the chemical nature of the substance as the concentration is changed, no association nor dissociation of molecules, no significant changes in per cent ionization, nor any reaction with the solvent. In photometric analyses it is advisable therefore to plot standard calibration curves first. This is done by measuring the absorbencies of *known* concentrations of the substance being studied at the wave length being used. Usually it is found that over some range of concentrations the relationship is virtually linear (i.e., Beer's Law holds). Outside that range, photometric measurements can be valid only when comparative readings are made with standard solutions closely approximating the concentrations of the unknowns.

In photometric analyses, another factor of great importance is the wave length of light used in measuring absorbencies. It seems obvious, upon reflection, that the most useful wave length is the one at which the absorbency varies most with concentrations, since here the method would be most sensitive. This point is usually (but not invariably) at a wave length where absorption is greater than it is at lower and higher wave lengths, that is, at absorption peaks.

Both the location of the absorption peak and its intensity per unit concentration are characteristic of the electronic configuration of the substance being investigated. While the subject, if treated rigorously, is most complex, certain generalizations can be made which are both simple and useful for beginners. Among organic compounds, it is a general rule that unsaturated compounds show absorption peaks at longer wave lengths than do saturated compounds. This is particularly true if the unsaturated bonds are conjugated (i.e., if single and double bonds alternate in the carbon chain). Thus, in the case of benzene there is an absorption peak at about 270 mμ, whereas in the case of hexane no absorption peaks of any significance occur until wave lengths well below 200 mμ are reached.

Of the amino acids contained in proteins, three contain the benzene ring. These are phenylalanine, tyrosine, and tryptophane. (To these, the specialized amino acid, thyroxine, may be added, but except in proteins of the thyroid gland, its possible presence can be neglected.) These amino acids show absorption peaks in the region 260–290 mμ, whereas all other amino acids are almost perfectly transparent there.

It is thus possible to determine the "aromatic amino acid content" of proteins without the necessity of any separatory procedures on the hydrolysis mixture (25). One can indeed determine it in the intact protein. In fact, the tyrosine-tryptophane-phenylalanine absorption band has been used for the determination of protein content of solutions (21).

While spectrophotometry has the virtue of simplicity and elegance, there are serious drawbacks as far as its use in amino acid analysis is concerned. In the first place, the method is restricted to the three amino acids mentioned. Secondly, it can not very well distinguish among the three. Although the absorption peak is slightly different in each case, it is to be remembered that a peak is never a sharp upsurge of absorbency at a given wave length but is rather merely the most pronounced point of a fairly broad region of absorption. In the case of tyrosine, tryptophane, and phenylalanine, the three regions of absorption overlap badly, so that it is difficult to determine how much of the observed absorption is contributed by each of the three without the use of subsidiary methods of analysis. Finally, there exist common cellular constituents (chiefly the purine and pyrimidine bases of the nucleic acids—see Chapter 9) in close association with proteins, which absorb in the same region of the ultraviolet spectrum with an intensity some ten times as great as that of the aromatic amino acids. The possibility of a contamination by even small quantities of such substances is therefore very serious.

Isotopic methods. Recent years have brought an extremely powerful tool to the aid of the biochemist in the form of isotopes (see Appendix). Isotopes of a given element are atomic species which differ among themselves only in the number of neutrons in their nuclei. The number of protons in the nucleus and consequently the number of planetary electrons in the neutral atom remain the same. The result is that the *chemical* properties of the various isotopes of a given element (which depend upon the number of electrons in the neutral atom) are virtually identical. On the other hand, isotopes of a given element can be distinguished one from the other by various physical methods. Particularly, where one isotope is radioactive it can be detected by radiation counting devices which have an almost fantastic sensitivity. In this manner it can be distinguished from its non-radioactive isotopic brother, with which it still shares in common all chemical properties. Furthermore, the radioactive properties of a given isotope, let us say C^{14}, are not in the least affected by the incorporation thereof into a molecule such as an amino acid.

If now, a known quantity of an amino acid such as arginine, containing C^{14} as part of its structure is added to a protein hydrolysate and the mixture stirred to homogeneity, the ratio of stable and radioactive arginine will remain the same at all stages of chemical isolation. Once separated in pure form (not necessarily quantitatively), the proportion of the mixture which is radioactive can be determined. Since the ratio is just what it was at the start and since the total quantity of radioactive arginine added is known, the quantity of arginine in the original hydrolysate is easily calculated.

Although this *isotopic dilution method* (40) is quite accurate and is applicable to any amino acid, it has certain practical drawbacks. There is always the necessity of isolating the given amino acid in great purity, and this is sometimes difficult. Furthermore, radioactive amino acids and equipment used in dealing with radioactivity are quite expensive and safety precautions that are not required in other analytical techniques must be taken in any procedures involving the use of radioactive substances.

Nevertheless, despite expense and the special precautionary measures involved, isotopic procedures are becoming more prominent yearly in biochemical research, particularly in the study of metabolism.

Enzymatic methods. The use of enzymes in analyzing for individual amino acids in a hydrolysate (3) is not very different in principle from the use of a specific chemical precipitant. Enzymes are proteins which catalyze certain reactions in a very specific way (see Chapter 6), and the advantage of their use lies in this specificity.

Thus an enzyme known as arginase will, under the proper conditions of pH, split urea from arginine in the manner shown in formula VI. Arginase will not affect any other amino acid in the slightest so that a measure of the rate of urea formation is an accurate representation of the amount of arginine originally present.

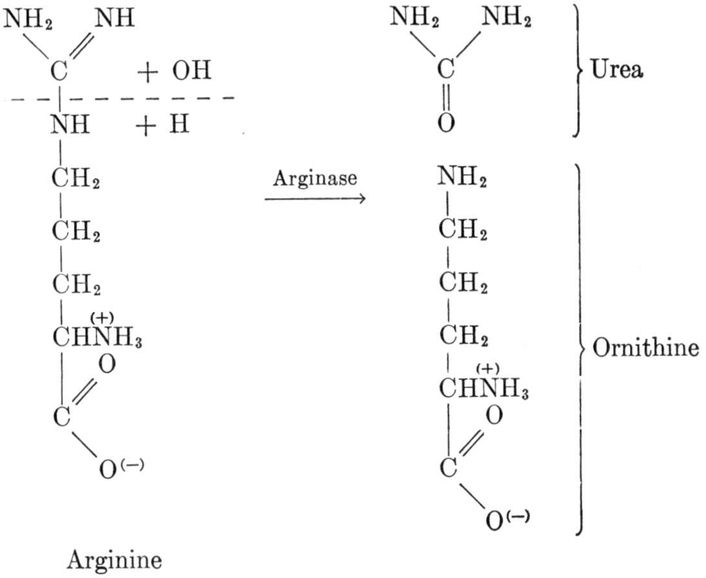

$$
\begin{array}{l}
\text{Arginine: } NH_2\text{–}C(=NH)\text{–}NH\text{–}CH_2\text{–}CH_2\text{–}CH_2\text{–}CH\overset{(+)}{N}H_3\text{–}C(=O)\text{–}O^{(-)} \;+\; OH \;+\; H \\
\xrightarrow{\text{Arginase}} \\
\text{Urea: } NH_2\text{–}C(=O)\text{–}NH_2 \\
\text{Ornithine: } NH_2\text{–}CH_2\text{–}CH_2\text{–}CH_2\text{–}CH\overset{(+)}{N}H_3\text{–}C(=O)\text{–}O^{(-)}
\end{array}
$$

Arginine

VI

Microbiological methods. The use of enzymes not as separate entities but as parts of living micro-organisms is gaining in popularity. The basic principle underlying their use in amino acid analysis is the fact that strains or varieties of certain micro-organisms can often be cultivated which require for normal growth the presence of a specific amino acid in their culture medium (41).

A typical micro-organism so used is *Neurospora*, a common mold. In its natural wild form it can grow normally upon a medium containing certain necessary minerals, water, and a small quantity of the organic food factor, biotin. From these plus carbon dioxide and oxygen, it can manufacture in quantities suitable for its needs all the amino acids. Exposure of *Neurospora* to ultraviolet or x-radiation, however, may cause mutations to arise through disturbances within the genetic factors of the cell (see Chapter 11). These mutations demonstrate themselves often in a disability to manufacture some compound vitally needed for metabolism. The compound required is usually a vitamin or an amino acid. As far as a given mutant strain is concerned, the compound it can not synthesize must be supplied in its culture medium, and over a certain range the rate of growth is proportional to the concentration of the needed factor in its food supply.

If, then, one had a supply of a strain of *Neurospora* which could not manufacture lysine from inorganic constituents (a so-called "lysine-less" strain), one need only grow it upon a medium containing in part the protein hydrolysate to be analyzed, and then measure its rate of growth to determine the concentration of lysine.

The difficulties here lie largely in the techniques required to grow the micro-organism and in forming, isolating, and identifying the useful mutant strains. Moreover, strains useful in analyzing each of the amino acids can not be obtained with equal ease. In the case of some amino acids, i.e., cystine, the appropriate mutant strain is rarely or never met. It should be stressed that the strains arise not through any purposeful action on the part of the experimenter but through the random and unpredictable actions of high-energy radiation.

Chromatographic methods. Chromatography, in its various phases, is now nearly half a century old. Originally, it was a technique by which closely related compounds were separated through their differing tendencies to be adsorbed on a material such as powdered aluminum oxide. Earliest experiments were made on such plant pigments as the chlorophylls and the carotenoids. When these (in solution) were passed through a column packed with aluminum oxide, the individual compounds of the mixture were adsorbed and held by the powder. Those compounds which were strongly adsorbed clustered at the very top of the column, while those with somewhat less tendency for adsorption percolated through the column a longer distance before surface forces held them. Further addition of pure solvent to the column tended to wash the mixture downward. Again, the component which was less strongly held by the alumina was more easily loosened and was washed down further than the component which was more strongly held. Eventually the bands representing the various components of the colored mixture were entirely separate and visible to the eye as varicolored regions in the column. Hence the original name "chromatographic adsorption". The components of the mixture, through adsorption, are "written in color". This technique is not restricted to visibly colored compounds. Actually, any mixture for which the appropriate solvents and adsorbents can be found can be chromatographed. Where the components can not be readily distinguished by eye, as in the case of amino acids, it is only necessary that more indirect chemical methods be used. Many natural and synthetic substances besides aluminum oxide are used in chromatography. Starch in particular has been used recently with considerable success as a chromatographic medium for amino acids (30). Its use does not vary in principle from that of aluminum oxide.

A great many synthetic polymers have since been devised for the specific purpose of facilitating certain separations. These polymers depend for their action, not upon the surface phenomenon of adsorption, with its secondary valence links, but upon reactions of known chemical groups of the polymer with chemical groups of the mixture passing through the column. Thus a given polymer may, at a pH of 7 for instance, contain many carboxylate ions. These will attract, by electrovalent forces, compounds sufficiently basic to be positively charged at pH 7. Furthermore, the place in the column at which the ester of a given component will tend to cluster will depend on how strongly basic it is. Of two components, one of which is slightly more basic than the other, the more basic one will be found higher in the column. The addition of stronger base, by replacing the compounds being investigated, washes or "elutes" them down through the column, again replacing the least basic most easily. If the liquid emerging from the column is collected in small fractions, it is found that the various fractions will contain portions of single components of the mixture. The polymer can then be regenerated for further use by passing buffer at pH 7 through it. Other polymers can be designed to possess basic groups which will separate acidic components of a mixture, or indeed to possess groups specifically intended to take advantage of almost any peculiar chemical properties of the substances being studied. These polymers are known as *ion-exchange resins* and such is their efficiency that they have been used to de-salt water, the water so obtained being as salt-free as ordinary distilled water.

The type of chromatography, however, which has been applied most fruitfully to the analysis of complex amino acid mixtures is known as *paper partition chromatography* (48). This has become an extremely popular technique for amino acid analysis and for other problems of similar nature. It is less successful with complex molecules, but successful separations of proteins are occasionally reported (31).

In paper partition chromatography advantage is taken of the different partition coefficients of two closely related substances. The partition coefficient may be defined as the ratio of the solubilities of a substance in two different mutually insoluble solvents. If for instance two amino acids in aqueous solution are shaken with butyl alcohol in a separatory funnel and the two phases allowed to separate, both amino acids will be found not only in the aqueous layer but in the butyl alcohol layer as well. The amount of each in each layer depends upon the solubilities of each amino acid in the two solvents. Let us suppose that amino acid A is more soluble in water than is amino acid B, and conversely, less soluble in butyl alcohol. Another way of saying this is that the water/butyl alcohol partition coefficient of amino acid A is greater than that of amino acid B. The result of all this is that when an aqueous mixture of the two has been shaken with butyl alcohol, the proportion of amino acid B in the butyl alcohol layer will be higher than it was in the original mixture, while in the aqueous layer the proportion of amino acid B will be lower.

One can well imagine that if the aqueous layer is extracted a second time with butyl alcohol, and a third and a fourth *ad infinitum*, eventually only amino acid A will be left in the water, and that if the butyl alcohol layer is extracted and re-extracted with water, only amino acid B will be left there. To be sure, only vanishingly small quantities of either will be left after such treatment, but nevertheless this principle has been put to use in what is known as the countercurrent technique (19). Here a series of separatory funnels or their equivalents are used in many successive extractions, so that each component of a mixture is spread throughout the funnels at varying rates and may thus be separated.

In amino acid separations (16), a strip of ordinary filter paper takes the place of the separatory funnels. A drop of the mixture to be separated is placed near one end of the filter paper strip and allowed to dry. That end is then dipped into butyl alcohol (which has been water-saturated) and the liquid allowed to travel through the filter paper by capillary action past the spot where the dried mixture has been placed. The components of the mixture are now under the influence of two tendencies. On the one hand, they have the tendency to stay where they are, remaining "dissolved" in the water film which is adsorbed on the surface of the filter paper. (This water film is present under all ordinary circumstances, and its removal would require prolonged desiccation of the filter paper.) On the other hand, there is the tendency to be dragged along by the advancing butyl alcohol. The extent to which a given amino acid follows each of these tendencies depends upon its comparative solubilities in water and in butyl alcohol. A highly water-soluble amino acid such as lysine would be bound comparatively strongly to the water film and would move but slightly with the butyl alcohol front. A comparatively water-insoluble amino acid such as phenylalanine, on the contrary, will be bound weakly to the water film and will move comparatively rapidly with the butyl alcohol. The tendency of the amino acids thus to separate is accentuated with time, and eventually the amino acids which were originally all superimposed upon the original drop of mixture will be strung out in various positions along the filter paper.

It is, of course, not impossible that two amino acids will travel at so nearly the same rate that even after hours of separation they will remain at least partially superimposed. A second separation is therefore usually performed. After allowing the filter paper to dry, it is rotated through ninety degrees, dipped into a second solvent (say, water-saturated phenol) in such a way that the line of amino acids now lies horizontally an inch or so above the surface of the phenol. As the phenol rises past each spot there is again a separation, and if anywhere along the line, two or three amino acids are superimposed, it is extremely unlikely that their phenol/water partition coefficients will also be equal as their butyl alcohol/water partition coefficients were. As a matter of fact, after such a two-dimensional separation, no two amino acids will be found occupying the same spot on the filter paper. A complete separation of all naturally-occurring amino acids can be performed thus.

The position of the spots is of course invisible to the naked eye. The spots can be visualized by allowing the filter paper to dry and then spraying it with ninhydrin solution and warming. Wherever an amino acid is located a visible spot will appear. By conducting experiments where each amino acid is treated singly in this manner, a map of spots can be developed so that through the R_f values (the ratio of the rate of movement of a given spot to that of the solvent boundary) one can, by noting the position of the spots, determine the identities of the amino acids present in the mixture.

Paper chromatography is simple, sensitive, and inexpensive. For most accurate results certain precautions must be taken. Thus of the two solvents used, one is usually water although this is not absolutely necessary, since by first washing the filter paper intensively with another solvent a "standing" solvent other than water can be used. The other must be a solvent which is not entirely miscible in water, and which is sufficiently similar chemically so that the compounds being investigated will be appreciably soluble in both. Butyl alcohol and water are an ideal such pair. The separation must be conducted in a closed container where the atmosphere can be saturated with both butyl alcohol and water in order that evaporation effects not affect the procedure. For similar reasons the butyl alcohol used must be water-saturated in order that it may not dissolve the adsorbed film of water on the filter paper and thus destroy the partition effect upon which the separation is based. For best results separations should be conducted at constant temperature. The filter paper itself plays a minor role and can be viewed merely as a support for the two liquid phases. Nevertheless it, too, must be chosen with care. It must be quite pure and quite homogeneous to prevent the production of annoying artifacts.

Paper chromatograms can be evaluated quantitatively by measuring the area and optical density of individual spots (densitometry), or the spots may be eluted and the substance measured by photometric or other type of analysis. The spots may also be compared with those resulting from the treatment of known quantities of the amino acid concerned. It should be mentioned that through paper chromatography, mixtures containing components present in quantities of only one to five micrograms (or even less in some cases) can be easily separated and the components accurately identified. Paper electrophoresis is also useful in separating mixtures of amino acids.

Amino Acid Contents of Proteins

All the techniques listed have been used in quantitative determinations of the amino acids in proteins. Brand and his group (10) reported the amino acid composition of the protein β-lactoglobulin. They also proposed a shorthand notation for the various amino acids involving, usually, the use of the first three letters of the common chemical name. The results are shown in table II-1.

Two points should be mentioned in connection with the table. The first concerns amino acid amides. There is good reason to think that both glutamine and asparagine form part of the structure of almost all proteins. The amide group in both of these compounds is, however, quite easily hydrolyzed. Any method commonly used to hydrolyze proteins to amino acids will also hydrolyze that amide group. The result is that glutamine and asparagine residues present in the original protein molecule appear in the final amino acid mixture as glutamic acid and aspartic acid and are reported as such. The amide groups appear as ammonia and are commonly reported as *amide nitrogen*. There is no way of telling after such hydrolyses

TABLE II-1

Amino acid composition of β-lactoglobulin

AMINO ACID	SYMBOL USED	NUMBER PER MOLECULE
Glycine	Gly	8
Alanine	Ala	29
Leucine	Leu	50
Isoleucine	Ileu	27
Tyrosine	Tyr	9
Serine	Ser	20
Aspartic acid	Asp	36
Glutamic acid	Glu	24
Glutamine	Glu·NH_2	32
Phenylalanine	Phe	9
Lysine	Lys	33
Arginine	Arg	7
Histidine	His	4
Tryptophane	Try	4
Proline	Pro	15
Valine	Val	21
Methionine	Met	9
Threonine	Thr	21
Cysteine	CySH·	4
½-Cystine	CyS—	8
Water	H_2O	4

how many glutamine residues and how many asparagine residues were present in the original protein since there is no way of knowing how to divide the ammonia molecules produced among them. In table II-1, all the ammonia molecules were arbitrarily assigned to glutamine. The presence of asparagine and glutamine can be demonstrated, rather indirectly, by treating the original protein with diazomethane to methylate the side chain carboxyl groups of aspartic acid and glutamic acid. Further treatment with lithium borohydride reduces the methyl esters to the respective alcohols. The amide groups of asparagine and glutamine remain unchanged. If the protein is now hydrolyzed, the original aspartic acid and glutamic acid residues appear in the final amino acid mixture as hydroxymethylalanine and hydroxyethylalanine, while the original asparagine and glutamine residues appear as aspartic acid and glutamic acid.

Secondly, there is the device of counting the number of *half*-cystine molecules in the protein. This is done because of the fact that cystine has in its molecule two amino groups and two carboxyl groups, thus giving the appearance of a double amino acid. Furthermore, cystine is easily converted to two molecules of cysteine (formula VII). This reaction is an important one in protein chemistry and it is convenient to avoid changing the number of residues present. Thus, two cysteine molecules can be considered as being oxidized to two half-cystine molecules.

```
         (+)                                          (+)
         NH3  O                                       NH3  O
        /    //                                      /    //
S—CH2—CH—C                              HS—CH2—CH—C
|          \                                          \
|           O(−)        + 2H                           O(−)
|                    ⇌                       +
|           O           − 2H                           O
|          //                                         //
S—CH2—CH—C                              HS—CH2—CH—C
        \   \                                        \   \
         NH3  O(−)                                    NH3  O(−)
         (+)                                          (+)

      Cystine                                    2 Cysteine
                              VII
```

Unfortunately, such catalogues of amino acids, while representing an enormous advance, are but a small part of the story. There is still the question of arrangement of the amino acids in the peptide chain. Methods for determining this experimentally are tedious, and not surprisingly so in view of the tremendous complexity of the problem.

Sanger introduced the use of 2,4-dinitrofluorobenzene (formula VIII) in this connection (36). This reagent will form colored compounds with *free* amino groups and will thus add to the epsilon-amino groups of any lysine residues present in the protein and to the free amino group of the N-terminal

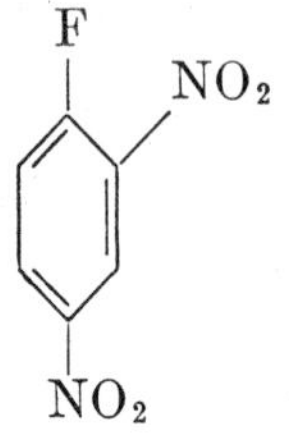

F
NO_2
NO_2

VIII. 2,4-Dinitrofluorobenzene
(Sanger's reagent)

amino acid of the peptide chain. (Every non-cyclic peptide chain has a free amino group at one end as part of the N-terminal amino acid, and a free carboxyl group at the other as part of the C-terminal amino acid.) If the protein treated with Sanger's reagent is now hydrolyzed, the presence of the yellow dinitrophenylated amino acids can be detected and identified by paper chromatography. If the presence of lysine is allowed for, the N-terminal amino acids on the chain can be determined. In this manner, it has been found that insulin, for instance, is a combination of four polypeptide chains, two of which end in glycine and two in phenylalanine.

Another such end-residue determination has been introduced by Edman (20). He found that by reacting a peptide with phenylthiourea, a condensation compound is formed which can be split off from the rest of the chain by a mild hydrolysis which would not affect ordinary peptide linkages. The fragments can be separated, identified chromatographically, and the process repeated on what remains of the chain. Presumably, amino acids could be identified one by one as they were split from the chain, given enough patience. The use of this method has shown that the N-terminal sequence in human serum albumin is aspartylalanine, while in bovine serum albumin, it is aspartylthreonine (43).

Methods have also been devised for the determination of C-terminal groups (23, 46). It has been reported, for instance, that the C-terminal amino acid of lysozyme is leucine, of ovalbumin, alanine, and of ovomucoid, phenylalanine.

To determine the arrangement of the amino acids along the whole polypeptide chain, a protein is hydrolyzed incompletely and the di- and tripeptides in the hydrolysis mixture are identified by a combination of end-group analysis and paper chromatography. Assuming that the order of amino acids in the resulting peptides is the same as in the original protein, the pieces can eventually be put together like a jigsaw puzzle.

The application of such methods has resulted in the complete elucidation of the structure of the protein hormone insulin. The two types of polypeptide chains in insulin are held together by —S—S— bridges. These can be separated by oxidizing the disulfide groups to $—SO_3H$ groups. The two kinds of chains are a more basic phenylalanyl chain and a more acidic glycyl chain (37, 38). The sequence of amino acids in the phenylalanyl chain has been found to be—

> Phe-Val-Asp($-NH_2$)-Glu($-NH_2$)-His-Leu-Cys-Gly-Ser-His-Leu-Val-Glu-Ala-Leu-Tyr-Leu-Val-CyS-Gly-Glu-Arg-Gly-Phe-Phe-Tyr-Thr-Pro-Lys-Ala

The glycyl chain is—

> Gly-Ileu-Val-Glu-Glu-CyS-CyS-Ala-Ser-Val-CyS-Ser-Leu-Tyr-Gly-Leu-Gly-Asp-Tyr-CyS-Asp

Minor differences exist among the insulin molecules from various species. The glycyl chain in beef insulin contains the -ala-ser-val- sequence as shown above. Sheep insulin contains -ala-gly-val- instead, and pig insulin contains -thr-ser-ileu- (13).

Optical Isomerism

In considering the ordinary chemical formula as written on the blackboard or on a sheet of paper a great and unavoidable flaw enters into the symbolism since an attempt is being made to represent an object which exists in three dimensions upon a flat surface which can only show two. In order to understand optical isomerism it is absolutely vital that the third dimension be taken into account. Since it is not always easy to visualize three dimensions from two-dimensional representations, however painstakingly drawn and described, it would be advisable for students to follow the discussion below with ball-and-rod atom models.

In space, the four valence bonds of carbon are *not* located in a single plane and directed toward the four corners of a square as usually represented. Rather, the valences are located in two planes (a pair in each) which are mutually perpendicular. As nearly as one can represent the three-dimensional situation on paper, the carbon atom may be depicted as in formula IX.

IX. Tetrahedral carbon atom

In version "a" the solid bonds are to be visualized as sticking out in front of the paper, and making an angle of nearly 36° with the plane of the paper. The dotted bonds, on the other hand, are sticking into and behind the paper again making a 36° angle with it. The angle between *any two bonds* is a little over 109°. In version "b" the two light solid bonds are in the plane of the paper. The bold-face solid bond sticks out from the paper at an angle of about 55° and the dotted bond sticks into and behind the paper at the same angle. (Actually if the two light solid bonds were in the plane of the paper, the other two bonds would be superimposed since they would lie in a plane perpendicular to that of the paper. Their position in version "b" has been slightly distorted so that both might be shown.) A little reflection, or better still, atom models, will prove to the student that the two versions, "a" and "b", are identical.

This spatial configuration of carbon leads to important results when each of the four valence bonds of a carbon atom is attached to a group of a different nature. Let us consider carbon atoms to which four such different groups, W, X, Y, and Z, are attached (formula X).

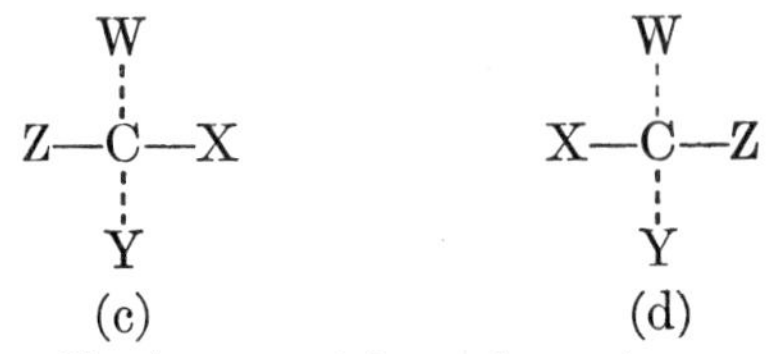

X. Asymmetric carbon atoms

The question now arises whether any significant difference arises from the manner of attaching the four groups to the carbon atom. In "c" and "d" for instance, we have added them first clockwise, then counterclockwise. Does that matter? At first sight to the student trained in the planar representation of the carbon atom this would seem a frivolous question. It would appear obvious to him that if one were to lift "d" up and out of the paper, turn it around so that the back is in front and put it back, he would have "c" and that therefore the two compounds are the same

Actually, this is not so where three-dimensional configurations of carbon are concerned. If we perform the process just described of lifting "d" out of the paper, reversing, and returning it, we end up with "e" (formula XI). The placing of the attached groups in "e" would seem superficially

```
    W
    |
Z---C---X
    |
    Y
```

(e)

XI. (d) turned through 180°

$$\begin{array}{c} H \\ | \\ {}^{(+)}NH_3—C—COO^{(-)} \\ | \\ H \end{array} \quad \text{Plane of symmetry}$$

Glycine

$$\begin{array}{c} R \\ | \\ {}^{(+)}NH_3—C—COO^{(-)} \\ | \\ H \end{array} \quad \text{No plane of symmetry}$$

Other amino acids

XII

atom which thus fulfills the criterion of asymmetry and is optically active. And of each pair of optically active amino acids the body uses but one.

Which one? This is a question almost impossible to answer in an absolute way; that is, by actually indicating the position of each of the four substituents in space. It can be answered relatively, however, by indicating how the various amino acids compare to one another. In order to explain this let us go back to the first investigations on optically active substances which were conducted on tartaric acid by Pasteur. He discovered that one isomer of tartaric acid rotated the plane of polarized light to the right (clockwise) and one to the left (counterclockwise). Without going into the details of formulas, which would at this point add nothing to the principles expounded above, it will suffice to say that he therefore called the two isomers, "dextro-tartaric acid" and "levo-tartaric acid". These are usually written as D-tartaric acid and L-tartaric acid. Similarly, other optically active substances were termed dextro or levo in accordance with the direction in which they rotated polarized light. (Two common sugars are sometimes called dextrose and levulose for their possession of those very properties.)

However, it was soon discovered that there is no simple way of predicting the direction of rotation from the spatial arrangement of the atoms. For instance, let us suppose that of the two model enantiomorphs presented on page 78, "c" is dextrorotatory and "d" levorotatory. If in each case, the group Z is replaced by another substituent V in a manner which is known not to affect the comparative position in space of the other three substituents (and this is by no means an easy thing to be sure of), compounds "f" and "g" are obtained (formula XIII).

Spatially, "f" is analogous to "c" and "g" to "d". It is therefore natural to assume that "f" like "c" is dextrorotatory, and "g" like "d" is levorotatory. *This is not necessarily the case.* One can not predict the direction of light rotation from the spatial configuration.

```
       W                W
       ¦                ¦
   V—C—X            X—C—V
       ¦                ¦
       Y                Y
      (f)              (g)
```

XIII. Spatial analogs of (c) and (d)

It has been necessary for chemists to decide whether it was better to classify organic compounds on the basis of light rotation without regard to spatial configuration, or vice versa. The second alternative was adopted since it was obvious that spatial configuration was the more fundamental property as far as structural problems were concerned. Glyceraldehyde was taken as the starting point (formula XIV). The levorotatory isomer

```
         CHO                  CHO
          ¦                    ¦
     HO—C—H                H—C—OH
          ¦                    ¦
         CH2OH                CH2OH
  L-Glyceraldehyde      D-Glyceraldehyde
                   XIV
```

was termed L-glyceraldehyde and the dextrorotatory isomer was termed D-glyceraldehyde. No decision as to which isomer had which spatial configuration was made or was necessary. Of the two possible configurations, one was arbitrarily assigned to D-glyceraldehyde and one to L-glyceraldehyde as shown in formula XIV. Evidence has since been presented that the arbitrary choice was actually the correct one (8). Thereafter all compounds with asymmetric carbons that have the same spatial configuration as L-glyceraldehyde (as decided by the various tools at the disposal of the synthetic organic chemist) are termed L-compounds, *regardless of the direction in which they rotate plane-polarized light.* The equivalent statement may be made for compounds spatially related to D-glyceraldehyde.

It is now common to distinguish between two D-compounds that rotate light in opposite directions by means of an added plus or minus sign. Thus, D-(+)-glucose, or D-(−)-levulose.

To return to amino acids, the question as to which of the optical pairs is utilized by the body may now be answered. In the case of each amino acid (except the inactive glycine, of course), the L-amino acid occurs in proteins, never (or almost never) the D-.

It should be noted that in carbohydrates D or L refers to the configuration of the highest numbered asymmetric center (carbohydrates are numbered away from the functional group). In amino acid nomenclature, D or L refers to the configuration of the alpha-carbon atom, i.e., the *lowest numbered* asymmetric center (47). Since the configurations of D- and L-serine (formula XV) (the standard amino acid for configuration) have been correlated

$$
\begin{array}{ccc}
\begin{array}{c}
\mathrm{COOH} \\
| \\
\mathrm{H_2N—C—H} \\
| \\
\mathrm{CH_2OH}
\end{array}
& \qquad &
\begin{array}{c}
\mathrm{COOH} \\
| \\
\mathrm{H—C—NH_2} \\
| \\
\mathrm{CH_2OH}
\end{array} \\
\text{L-serine} & & \text{D-serine}
\end{array}
$$

XV

with D- and L-glyceraldehyde, no confusion results from this difference in practice in the case of the amino acids.

Three of the common amino acids, cystine, threonine, and isoleucine, have more than one asymmetric carbon atom in the molecule. In the case of threonine and isoleucine, the second asymmetric carbon is in the side chain, and there it serves as a further focal point of optical isomerism. Thus, if we take the case of the naturally-ccurring L-threonine, we find the hydroxyl group in the side chain to have a definite orientation. An isomer of L-threonine can be prepared in which the side chain hydroxyl has the opposite or "unnatural" orientation. This isomer is said to belong to the "allo" series and is L-*allo*threonine. These two isomers are shown in formula XVI. The reader can see for himself that D-*allo*threonine is the enantio-

$$
\begin{array}{cccc}
\begin{array}{c}
\mathrm{COOH} \\
| \\
\mathrm{H_2N—C—H} \\
| \\
\mathrm{H—C—OH} \\
| \\
\mathrm{CH_3}
\end{array}
&
\begin{array}{c}
\mathrm{COOH} \\
| \\
\mathrm{H_2N—C—H} \\
| \\
\mathrm{HO—C—H} \\
| \\
\mathrm{CH_3}
\end{array}
&
\begin{array}{c}
\mathrm{COOH} \\
| \\
\mathrm{H—C—NH_2} \\
| \\
\mathrm{H—C—OH} \\
| \\
\mathrm{CH_3}
\end{array}
&
\begin{array}{c}
\mathrm{COOH} \\
| \\
\mathrm{H—C—NH_2} \\
| \\
\mathrm{HO—C—H} \\
| \\
\mathrm{CH_3}
\end{array} \\
\text{L-threonine} & \text{L-}\textit{allo}\text{threonine} & \text{D-threonine} & \text{D-}\textit{allo}\text{threonine}
\end{array}
$$

XVI

morph (or mirror-image) of L-threonine, while D-threonine is the enantiomorph of L-*allo*threonine. The situation with respect to isoleucine is quite analogous to that of threonine.

In the case of cystine, the two optically active carbons are the two alpha-carbons (remembering that cystine is a "double" amino acid). Here, too, there is now the possibility of four different compounds, all of which are drawn in formula XVII. L-Cystine is the isomer occurring naturally in proteins, and D-cystine is its enantiomorph. The remaining two configurations are identical, since either can be converted to the other by rotating it through 180° in the plane of the paper. The possession of a line of symmetry, as indicated in the figure, causes the compound to be optically

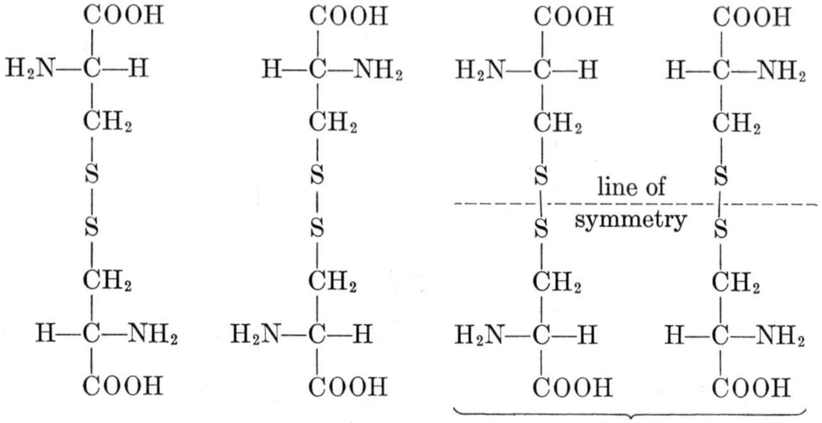

```
      COOH             COOH              COOH            COOH
       |                |                 |               |
  H2N—C—H           H—C—NH2         H2N—C—H          H—C—NH2
       |                |                 |               |
      CH2              CH2               CH2             CH2
       |                |                 |               |
       S                S                 S    line of    S
       |                |           ------|---------------|------
       S                S                 S   symmetry    S
       |                |                 |               |
      CH2              CH2               CH2             CH2
       |                |                 |               |
   H—C—NH2         H2N—C—H          H2N—C—H          H—C—NH2
       |                |                 |               |
      COOH             COOH              COOH            COOH
                                    \_____________  _____________/
                                                  \/
   L-cystine        D-cystine              meso-cystine
                              XVII
```

inactive despite the fact that it contains two asymmetric carbons. The action of one of them is exactly balanced by the equal and opposite action of the mirror-image other. Such a compound is called a *meso* compound, so that there are only three varieties of cystine altogether, L-cystine, D-cystine, and *meso*-cystine. We shall come across other meso compounds later on.

L-amino acids and protein structure. The significance of this universal occurrence of L-amino acids in connection with protein structure is not immediately obvious. The importance becomes apparent, however, when the spatial relationships of peptides are considered. Let us take the case of two tetrapeptides, one made up of four molecules of L-alanine and one made up of two molecules of L-alanine and two of D-alanine alternately placed. First, the case of the four L-alanines (formula XVIII). Despite appearances, the four molecules are all L-alanine. Although the second and fourth appear to be different from the first and third, it is only that they

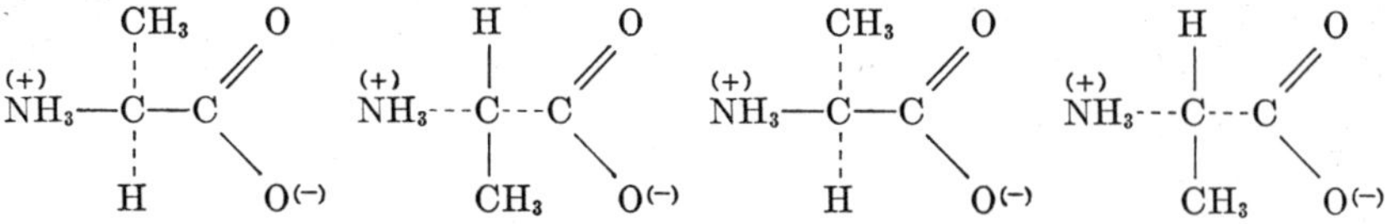

```
       CH3    O           H     O            CH3    O           H     O
 (+)    :    //     (+)   |    //      (+)    :    //     (+)   |    //
 NH3—C—C           NH3---C---C        NH3—C—C           NH3---C---C
        :    \             |    \              :    \             |    \
        H     O(-)        CH3    O(-)          H     O(-)        CH3    O(-)
   L-alanine          L-alanine           L-alanine          L-alanine
                                  XVIII
```

are viewed from the back and upside down. If this is not obvious to the student from a consideration of the diagrams above, models will make it so.

It might appear arbitrary to turn alternate molecules backwards and upside down, but again that is the fault of the inadequacy of the plane to picture solids. If models were used instead, it would become apparent that only so could we place amino groups and carboxyl groups together so that a natural unstrained tetrahedral bond can be formed between them. With three molecules of water split out the resulting peptide is as shown in formula XIX.

```
         CH3   O    H     O   CH3   O    H     O
(+)       |   //    |    //    |   //    |    //
NH3—C—C—NH—C—C—NH—C—C—NH—C—C
          |         |          |         |    \
          H        CH3         H        CH3    O(−)
```

XIX. L-alanyl-L-alanyl-L-alanyl-L-alanine

The formula is not an accurate picture even according to the symbolism we have ourselves adopted. Actually, the carbon and nitrogen atoms in the horizontal chain are alternately above and below the plane of the paper, while the hydrogen and methyl side chains are in a plane at a slight angle to the paper. It is unnecessary to introduce those complications, however, since at the moment the point which should be brought out is that the methyl side chains occur alternately above and below the peptide backbone.

In contrast, the situation with regard to peptide chains formed by alternate L- and D-alanine may be pictured as in formula XX. Again, the final result is not exactly as pictured. The four methyl groups are not all in the same plane. That of the second and fourth is slightly tilted to that in which the first and third are placed. Essentially, however, they may be looked upon as being on the same side of the peptide backbone.

Is this difference between the appearance of the two types of peptide

```
     CH3   O            CH3   O            CH3   O            CH3   O
(+)   :   //       (+)   |   //       (+)   :   //       (+)   |   //
NH3—C—C           NH3---C---C          NH3—C—C           NH3---C---C
      :    \              |    \             :    \              |    \
      H     O(−)          H     O(−)         H     O(−)          H     O(−)
  L-alanine           D-alanine          L-alanine           D-alanine

                                  |
                                  | −3H2O
                                  ↓

         CH3   O   CH3    O   CH3   O   CH3    O
(+)       |   //    |    //    |   //    |    //
NH3—C—C—NH—C—C—NH—C—C—NH—C—C
          |         |          |         |    \
          H         H          H         H     O(−)
```

XX. L-alanyl-D-alanyl-L-alanyl-D-alanine

chains presented here important? It is. If the two peptide chains here presented were built with models of atoms designed to represent actual atomic sizes proportionally (these are quite expensive) it would be seen that the D-L-peptide with the methyl groups on the same side of the backbone is comparatively crowded although the methyl group is the smallest side chain possessed by any amino acid other than glycine. Where amino acids such as tryptophane or tyrosine are considered, the extra space made available by the alternate placement of residues in the all L-peptide is most useful. It should be pointed out that very elaborate theories of protein structure, representing much labor and ingenuity, have eventually come to grief because of lack of room for side chains *even when placed alternately* (31).

There is, of course, no magic about the L-configuration. A peptide chain containing all D-molecules would be as roomy as its L sister. Perhaps proteins began as L-polymers through chance. Once established, however, there would be no choice but to continue on the path chosen.

D-Amino Acids. The D-amino acids are sometimes termed "unnatural" amino acids, but this is a rather unfortunate term, representing an unwarranted intrusion of emotionalism into science. It is like speaking of "improper fractions" and "imaginary numbers". Actually, D-amino acids *are* natural in the sense that they *do* occur, albeit rarely. For instance, D-phenylalanine, D-aspartic acid and D-glutamic acid occur in the proteins of various bacteria (42, 44). It is always important to remember that one optical isomer can often be partially converted to its sister form by various chemical means, one of which is heating with strong acid or alkali. This process of conversion of one enantiomorph to the other is known as *racemization.*

PROTEIN "FINE STRUCTURE"

Evidence for Protein Linkage Other Than Peptide

Thus far proteins have been considered simply as long chains of amino acids, yet it is easy to show that such chains do not in the least represent the structural situation as it exists in most proteins. It is not that it is wrong; it is simply quite incomplete.

If we were to consider first those proteins known to be composed actually of little more than such chains of amino acids, we would find that they fall into the class of "fibrous proteins". Typical examples are silk fibroin, keratin, collagen—which are the proteins of silk, hair or wool, and connective tissue, respectively. Fibrous proteins in general are characterized by extremely asymmetric molecules, insolubility in water, resistance to hydrolysis by enzymes, and comparative stability. In some cases, they are elastic. Certainly these are not the proteins referred to at the beginning of the chapter as being so unstable that the warmth of the hand or gentle bubbling could change their properties. In fact, the fibrous proteins have more of their physical properties in common with such a polymer as cellulose than with other non-fibrous proteins. Silk, which is protein in nature, has a greater tensile strength by far than cotton, which is cellulose.

As opposed to fibrous proteins, there are the "globular proteins" which are the typical proteins of the organism and which play roles other than the merely protective or structural ones characteristic of the fibrous proteins. The term "globular" is somewhat of a misnomer, since globular proteins are not to be looked on as globes. All or almost all proteins are asymmetric to a certain extent and should be viewed as ellipsoidal rather than spherical. The asymmetry of the average globular protein, however, is considerably less than that of the average fibrous protein. In the latter, the ratio of the long axis to either short one is not less than thirty. Globular proteins, moreover, are generally soluble in water or salt solutions, are easily digested by enzymes, and are sensitive to slight changes in the environment. Actomyosin, the protein of muscle, may be mentioned as an exception. Although a globular protein in its properties, it has the asymmetry of a fibrous protein. Actomyosin, however, has special properties which will be considered in Chapter 4.

The ease with which globular proteins are damaged by environmental stress as compared with the stability of the fibrous proteins would seem to make evident the existence of important differences in their structure. Fibrous proteins are composed of amino acids linked through the firm and stable peptide linkage. Globular proteins, on the other hand, must have subsidiary linkages of far lesser stability, which are essential to the maintenance of their properties. The nature of these subsidiary linkages has not yet been entirely determined, but important and illuminating advances have been made.

Subsidiary covalent links. Perhaps the first variation on the peptide-chain theme that might occur to the student is the possibility of branched chains. One amino acid, lysine, possesses a second amino group, and two, aspartic acid and glutamic acid, possess each an additional carboxyl group. It would seem reasonable that under certain conditions these additional

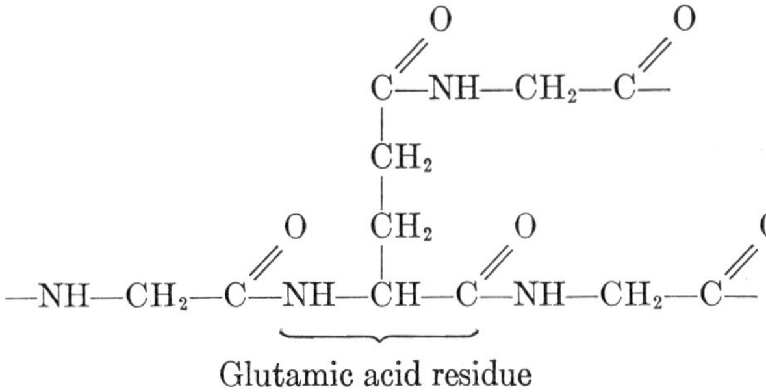

```
                          O           O
                         //          //
                        C—NH—CH2—C—
                        |
                        CH2
                        |
               O        CH2      O           O
              //        |       //          //
—NH—CH2—C—NH—CH—C—NH—CH2—C—
                 \_________/
```

Glutamic acid residue

XXI. Branched chain peptide

groups could form parts of new peptide chains branching off from the main backbone (formula XXI) by means of a peptidoid link such as that in glutathione (see page 62).

Peptidoid links involving the gamma-carboxyl group of glutamic acid have been found in a number of common proteins (24), while in the capsular polyglutamic acid of the anthrax bacillus, such links predominate (14). It should be noted, however, that the presence of gamma-carboxyl peptidoid links does not necessarily imply branching of the peptide. If the alpha-carboxyl group of glutamic acid remains unesterified, the peptide remains unbranched (formula XXII).

$$
\begin{array}{l}
CH_2—CO—NH—CHR—COO^- \\
| \\
CH_2 \\
| \\
^+H_3N—CH—COO^-
\end{array}
$$

XXII. Gamma-glutamyl peptide

Another form of branching and one with more important consequences arises as a result of the chemical nature of the amino acid cystine. Since it possesses a second amino group *and* a second carboxyl group, peptide chains can be conceived as branching off in two directions at once. The net result would be to have two peptide chains linked together by means of a disulfide bond (formula XXIII). This is quite different from the types

$$
\begin{array}{c}
—NH—CH_2—\overset{O}{\overset{\|}{C}}—NH—CH—\overset{O}{\overset{\|}{C}}—NH—CH_2—\overset{O}{\overset{\|}{C}}— \\
| \\
CH_2 \\
| \\
S \\
| \\
S \\
| \\
CH_2 \\
| \\
—NH—CH_2—\overset{O}{\overset{\|}{C}}—NH—CH—\overset{O}{\overset{\|}{C}}—NH—CH_2—\overset{O}{\overset{\|}{C}}—
\end{array}
$$

XXIII. Cystine-linked peptide chains

of branching discussed immediately above since the disulfide bond can be split by methods that would leave the peptide bond untouched. Thus, under the influence of reducing agents, it is split to form two sulfhydryl (or thiol) groups. By using oxidizing agents such as performic acid, the disulfide bond can be broken to form two *cysteic acid* residues (formula XXIV).

$$
\begin{array}{c}
\text{—NH—CH—CO—} \\
| \\
CH_2 \\
| \\
S \\
| \\
S \\
| \\
CH_2 \\
| \\
\text{—NH—CH—CO—} \\
\text{Cystine residue}
\end{array}
\xrightarrow{HC(=O)OOH}
\begin{array}{c}
\text{—NH—CH—CO—} \\
| \\
CH_2 \\
| \\
SO_3H \\
+ \\
SO_3H \\
| \\
CH_2 \\
| \\
\text{—NH—CH—CO—} \\
\text{2 Cysteic acid residues}
\end{array}
$$

XXIV

Proteins are indeed known in which adjacent peptide chains are held together by disulfide links in just the fashion described. The keratins of hair, wool, and feathers are rich in cystine residues (hence, their characteristic odor when burnt) which link chains together, forming a three-dimensional network that in great part accounts for the tensile strength of hair. The breaking of such interchain linkages under conditions of moist heat, the consequent sliding of peptide chains past one another, and their hardening into a new mold by re-formation of disulfide bonds upon cooling are the basis of the "permanent wave". Similarly, the four peptide chains of insulin, which have been discussed above, are held together in the intact molecule by six disulfide links.

A great many types of protein molecule, when denatured (subjected to heat or to other environmental conditions that cause them to lose their specific physiological properties) seem to possess sulfhydryl groups that could not be detected prior to denaturation. While the source of such sulfhydryl groups is still disputable, one hypothesis is that denaturation involves the breaking of disulfide links that served to hold the protein together in a specific fashion. Since denaturation of a protein does not usually involve a change in its molecular weight, one may visualize a single polypeptide chain, coiled and looped into an intricate design, held in place by the periodic cystine residues. Denaturation would then be a breaking of the disulfides and a consequent uncoiling. Unfortunately, while the picture drawn is attractive, it is oversimplified. On the one hand, many proteins do not form sulfhydryl groups on denaturation. On the other, denaturation can be accomplished by methods so mild that it is difficult to see how they would suffice to break disulfide linkages. Still other types of bondings have evidently yet to be considered.

Electrovalent links. The variations of structure among the twenty-odd commonly occurring amino acids are such as to allow considerable flexibility as far as possible types of interlinkages are concerned. So far only covalent bonds have been considered, but the possibilities of electrovalences can not be ignored. Of the various amino acids, the side chains of aspartic acid and glutamic acid are sufficiently acidic so that at body pH they exist as negatively charged ions. Similarly, the side chains of lysine and arginine are quite basic and at body pH would exist as positively charged ions (see page 7).

Two peptide chains or two regions of a single peptide chain can thus be conceived as being held together by the mutual attractions of these positively and negatively charged ions. Bonds so formed would be quite sensitive to changes in the pH and in the ionic strength of the medium in which proteins are dissolved. Indications are not lacking that in certain specialized instances some protein properties may be explained in just this manner. Insulin (to use it once again as an illustration) exists in solution as a complex of protein molecules. It is built up of three (sometimes four) identical sub-molecules with weights of 11,000. Each sub-molecule is composed, as has been already mentioned, of four peptide chains held together by disulfide bonds. The whole, apparently, is bound to the other such sub-molecules through electrovalent linkages since at least some dissociation takes place with mere shift of pH. Similarly, the ease with which such protein molecules as hemoglobin and hemocyanin are dissociated and re-associated under mild conditions without loss of physiological properties would seem to indicate the presence of such linkages between sub-molecules there as well.

Hydrogen bonds. Modern electronic theory of interatomic bonding has completely disposed of the older viewpoint that valence is a discontinuous phenomenon; that an atom can have only integral valences of one, two, three and so on. The current attitude is to view atoms as being held together by electronic interaction, the strength of which is dependent upon the make-up of the molecule as a whole.

Thus, the benzene molecule, which is ordinarily represented as possessing alternate single and double bonds, may more accurately be considered as having six carbon atoms held to one another by "one and a half" bonds. According to this view each carbon atom in benzene has three ordinary valence bonds and a fourth which can be looked upon as having been broken into two "half-bonds" and each half used for one of the neighboring carbons.

This modern concept of fractional valences is particularly illuminating when applied to the hydrogen atom. The properties of many hydrogen-containing compounds can be explained only on the assumption that hydrogen shares its single valence between two atoms, thus forming a link between those atoms known as "hydrogen bond".

Not all atoms can take part in hydrogen bonding. The only atoms, in fact, which can do so to any significant extent are those of fluorine, oxygen, and nitrogen in order of diminishing ability. Hydrogen-containing compounds of these elements are in some ways unique, as may be exemplified by the most important of all such compounds, water. Water is commonly looked upon as being simply H_2O. Actually, it is more complicated than that because of hydrogen bonding. Each water molecule "shares" its two hydrogens with two other molecules so that in the end each oxygen atom is surrounded, not by two but by *four* hydrogen atoms, much as a carbon atom is. This type of bonding, while not very strong as compared with ordinary covalent links, serves to hold neighboring water molecules together and to impart an abnormal stability to the liquid and solid states. (In steam, water molecules are too far apart and too energetic in their thermal motion to be able to form hydrogen bonds.)

Thus, ice is abnormally difficult to melt, and water abnormally difficult to boil since in passing first from solid to liquid and then from liquid to vapor, it is necessary to break these hydrogen bonds in addition to overcoming the normal cohesive forces among neutral atoms. This can best be shown by comparing the melting and boiling points of H_2O with those of H_2S. Ordinarily, among similar compounds, those with a higher molecular weight show higher melting and boiling points. However, although H_2S has a molecular weight of 34, as compared with 18 for H_2O, it has a melting point of $-82.9°C$. and a boiling point of $-59.6°C$., as compared with the well known 0°C. and 100°C. for the latter compound.

In similar fashion, ammonia (NH_3) has a melting and a boiling point higher than those of phosphine (PH_3), and hydrogen fluoride (HF) bears the same anomalous relationship to the heavier hydrogen chloride (HCl). Again, if one of the hydrogens of water is replaced by a methyl group, the increase in molecular weight is more than compensated for by the loss of half the hydrogen-bond forming capacity. The boiling point of methyl alcohol (CH_3OH) is, therefore, only 64.7°C. If the second hydrogen atom is also replaced by a methyl group, all possibilities for hydrogen bonding disappear and the boiling point of dimethyl ether (CH_3OCH_3) is $-23.7°C$.

In this manner, the hydrogen bond can be shown to have a very real existence and the question arises as to whether the folding of the protein molecule depends upon the formation of hydrogen bonds between various portions of the polypeptide chain (6). The following neutral amino acids contain oxygen-bound hydrogen in their side chains: serine, threonine, tyrosine and hydroxyproline. The following, tryptophane and histidine, contain nitrogen-bound hydrogen. These amino acids are theoretically capable of forming hydrogen bonds. In addition, the peptide-chain backbone itself is capable of forming hydrogen bonds with other such backbones as shown in formula XXV.

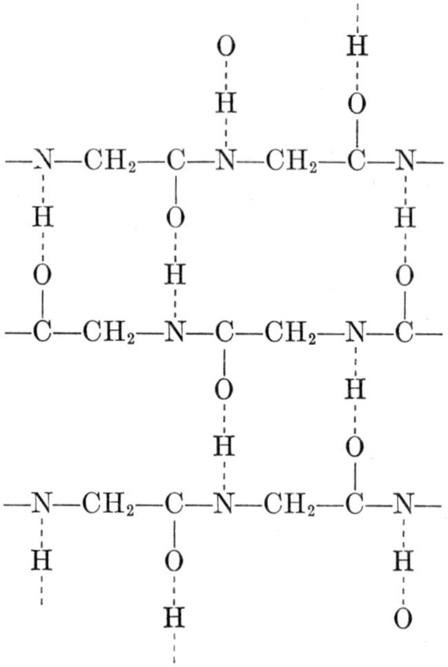

```
                                       :
                           O           H
                           :           :
                           H           O
                           :           |
 —N—CH2—C—N—CH2—C—N—
  :          |                        :
  H          O                        H
  :          :                        :
  O          H                        O
  |          :                        |
 —C—CH2—N—C—CH2—N—C—
                    |          :
                    O          H
                    :          :
                    H          O
                    :          |
 —N—CH2—C—N—CH2—C—N—
  :          |                        :
  H          O                        H
  :          :                        :
             H                        O
             :
```

XXV. Hydrogen bonds (---H---) and protein structure

Basing their picture on the results of modern data on the interatomic distances and bond angles in amino acids and polypeptides, Pauling and Corey (17, 33) have proposed a picture of the possible configurations of polypeptide chains which would be stable. One of these, a coiled structure called a helix, is thought to correspond to synthetic polypeptides and proteins which give x-ray diagrams of the alpha-keratin type. It is thought to occur also in some synthetic polypeptides and some globular proteins, including hemoglobin, myoglobin, and serum albumin. In this helical structure there are about 3.7 amino acid residues per turn, and each amide group is attached by hydrogen bonds to the third amide group from it, in either direction, along the helix. Pauling and Corey predict that this alpha-helix may be found to be the most common mode of folding of the polypeptide chain in proteins.

Other configurations, called by Pauling and Corey the "pleated sheet" configurations, are thought to characterize proteins which give x-ray diagrams of the beta-keratin type. These structures are not so easy to show in two dimensions. They also involve hydrogen bonds.

These concepts of protein internal structure account satisfactorily for many of the x-ray and crystallographic data on proteins and have been used in proposing various detailed structures for such proteins as collagen (35) and insulin (27).

Eight of the common amino acids are not involved in any of the subsidiary linkages. They are: glycine, alanine, valine, leucine, isoleucine, phenylalanine, proline, and methionine. All of these but methionine have side chains which are purely hydrocarbon in nature. Their particular functions with regard to protein structure are somewhat obscure. It may be that their presence in such chemical variety allows the varying sizes of their side chains partly to dictate the manner of folding of the polypeptide backbone. For instance, there has been a suggestion (32) that the presence of a series of glycine residues somewhere along the chain would result in a region where there would be a complete absence of side chains with a consequent increase in chain flexibility. These glycine areas would thus be natural folding points for the chain.

Again, proline and hydroxyproline present unusual features as far as amino acids are concerned. Actually, they contain not the $—NH_2$ group but an —NH— group. Consequently, as part of the polypeptide backbone, they alone among the amino acids possess a nitrogen atom to which is attached not a single hydrogen and which therefore is completely incapable of forming a hydrogen bond. One could speculate, therefore, that regions of the polypeptide backbone which were particularly rich in proline would be relatively "non-sticky", or, in other words, would have less tendency to form subsidiary linkages with other polypeptide backbones.

Denaturation

A denatured protein is, if the word is accepted literally, a protein that is no longer exactly as it originally existed in nature. The most apparent effect of extreme denaturation on protein is its loss of solubility in water and dilute salts and its tendency to coagulate irreversibly. It is this which happens when eggs are boiled. In that sense, practical knowledge of denaturation and its effects is as old as the art of cooking.

There are more subtle chemical changes which can not be detected by the eye. One of these is the uncovering of sulfhydryl groups. Their appearance may be due to the reductive cleavage of cystine in the process of denaturation. It has also been suggested that in native proteins, sulfhydryl groups form hydrogen bonds (S—H . . . N) with neighboring nitrogen atoms, and do not react to the usual tests. Denaturation breaks the bond and unmasks the group (7). Although hydrogen bonds involving sulfur are certainly not common, they have been postulated as occurring in glutathione, for instance, and accounting for the difference in behavior of the sulfhydryl group in that compound as compared with the same group in cysteine (15). When proteins are dissolved in urea or in guanidine solutions, a dissociation into smaller sub-molecules frequently takes place (notably in the case of hemoglobin). This process can be reversed by the removal of the urea or guanidine by dialysis.

There are biological changes in protein molecules which are symptomatic of denaturation and which may precede any detectable physical or chemical variations. There are proteins which, in extremely minute quantities, display very characteristic and detectable properties. Examples of these are viruses and enzymes. Even when every effort is made to preserve them in their native states, losses in their specific activity may be noted in the absence of any other detectable change.

The number of agencies which can effect denaturation is also indicative of the sensitivity and lability of proteins. These agents include heat, short-wave radiation, ultrasonic waves, pressure, various chemicals, high acidity, high alkalinity, electron bombardment, interface effects (as in bubbling or frothing), dilution with water, and freezing.

The protein molecule has a very complex and specific inner arrangement. If some of the bonds holding that arrangement firm are broken, the situation may be somewhat like that resulting from the careful removal of one or two blocks from the bottom of a house of blocks. The structure may remain standing, and the bonds may be reformed when the disruptive influence has been removed. Thus, there is evidence that mild heating or mild changes in pH, when not too prolonged, can cause changes in protein properties which can be reversed to normal upon cooling or upon neutralization (29).

If too many bonds are broken so that a whole loop of the protein falls out of place, the effect is that of the removal of one block too many from the house of blocks. The entire structure comes down rather suddenly and irreversibly (9). In such a case, the highly ordered protein molecule becomes disorganized. Instead of all the proteins being virtually identical, all have succeeded in falling apart in different ways (just as no two heaps of blocks will be exactly alike even though resulting from the disruption of two identical structures). Such an increase of disorder upon denaturation ought, if the above picture is a correct one, to result in an increase in the thermodynamic property known as entropy. The entropy of denaturation can be determined indirectly by measuring such things as the heat evolved in the process, and is indeed found to increase markedly.

Protein Isomerism

We have spoken of proteins as possessing great flexibility because of their size and complicated structure. There is a way of attempting to express this flexibility in numbers, and that is to calculate the various proteins that can be built up from the twenty-odd amino acids that act as monomers. It is obvious that every different arrangement of amino acids would represent a different protein with different properties. It is also obvious that the number of such proteins is limited and in view of the enormous chemical intricacies of even a simple organism, it is reasonable to wonder if the limit is large enough to allow for the complexities of life. If it is not, it would be a strong indication that all our theories of protein structure are wrong.

It might be well to start with the very simple protein salmine. Salmine is a protamine occurring in salmon sperm. It presents a certain problem to biochemists. It, together with a nucleic acid (see Chapter 9), forms a large part of the sperm and therefore within themselves they must carry at least some of the genetic factors that go to organize a particular salmon. And yet salmine is almost indecently simple for a protein. It contains only 72 amino acid residues, of which no less than 50 are the single amino acid arginine (18). It has a molecular weight of about 10,000. Can such a simple molecule, even in combination with nucleic acid, contain sufficient complexity within itself to possess the capacity for producing one particular salmon and not another; for certainly no two salmon are alike?

Let us consider first only the arrangement of the 72 amino acid residues along the chain. Any of the 72 residues can be the first in the chain. There are 72 possibilities there. For each of the 72 cases, any one of the remaining 71 amino acid residues can form the second in the chain. We now have 72×71 different possible chains. If we continue down the chain in this manner, we end up with a total number of possibilities equal to $72 \times 71 \times 70 \times 69$ and so on all the way down to $5 \times 4 \times 3 \times 2 \times 1$. A number of this sort is called in shorthand mathematical notation "factorial 72" and is symbolized as "72!"

But not all the amino acid residues are different. There are 50 arginines, for instance. In any given salmine chain, the 50 arginines can be shifted among themselves in any conceivable way and it would still be the same salmine chain. Since 50 items can be shuffled among themselves in any of 50! ways (the reasoning is the same as that in the preceding paragraph), the number of possible salmine chains now becomes 72!/50!. In addition to the 50 arginines, salmine contains 7 serines, 6 prolines, 4 glycines, 3 valines, 1 isoleucine and 1 alanine. Taking the remaining duplications into account the formula for the number of possible salmine chains becomes:

$$\frac{72!}{50! \times 7! \times 6! \times 4! \times 3!}$$

The above expression turns out to be, approximately, 4×10^{30}. That is certainly greater than the number of salmon in the world. It is undoubtedly greater than all the salmon that ever lived. Actually, the amino acids in protamines do not occur randomly. There are patterns in their arrangement that considerably reduce the number of permutations that exist (see page 366). However, there are other causes of variations among proteins besides amino acid arrangement, so the number of protamine isomers is still enormous.

Larger molecules such as hemoglobin have possibilities of isomerism that are incomparably greater. With over 500 individual amino acid residues of 20 different kinds, the number of arrangements possible would be of the order of 4×10^{619} (4). A number such as that is quite beyond comprehension. If the universe were a cube one billion light years to a side, (with each light year about six trillion miles in length) the number of neutrons (each 10^{-13} centimeters in diameter) which could be packed into it tightly would be something like 10^{120}.

We need go no further to decide that there is enough complexity in the protein molecule to provide different protein specimens for all the vast number of chemical reactions going on within all the millions of species of living organism in the world. There is enough complexity to make reasonable —indeed, almost inevitable—the fact that no two humans are ever exactly alike (not even identical twins), as are no two rabbits or salmon or amoebae. There is, in fact, enough complexity in the protein molecule to make it possible for every one of them to be different from every other one of them, taking into consideration not only those now existing, but all those that ever existed in the past or will exist in the probable future.

REFERENCES

1. Ambrose, E. J., and Elliot, A. Infrared spectra and structure of fibrous proteins. Proc. Roy. Soc. (London), A**206**: 206–219, 1951.
2. Ambrose, E. J., and Hanby, W. E. Evidence of chain folding in a synthetic polypeptide and in keratin. Nature, **163**: 483–484, 1949.
3. Archibald, R. M. Enzymatic methods in amino-acid analysis. Ann. Nwe York Acad. Sc., **47**: 181–185, 1946.
4. Asimov, I. Potentialities of protein isomerism. J. Chem. Ed., **31**: 125–127, 1954.
5. Astbury, W. T. X-ray studies of protein structure. Cold Spring Harbor Symp. Quant. Biol., **2**: 15–27, 1934.
6. Astbury, W. T. The hydrogen bond in protein structure. Tr. Faraday Soc., **36**: 871–880, 1940.
7. Benesch, R. E., and Benesch, R. A model for the configuration of sulfhydryl groups in proteins. J. Am. Chem. Soc., **75**: 4367–4369, 1953.
8. Bijvoet, J. M. Determination of the absolute configuration of optical antipodes. Endeavour, **14**: 71–77, 1955.
9. Boyd, G. A., and Eberl, J. J. Absolute reaction kinetics of tobacco mosaic virus and a proposed theory of denaturation. J. Physical & Colloid Chem., **52**: 1146–1153, 1948.
10. Brand, E. Amino-acid composition of simple proteins. Ann. New York Acad. Sc., **47**: 187–228, 1946.

11. Bricas, E., and Fromageot, C. Naturally-occurring peptides. Advances Protein Chem., **8:** 1–125, 1953.
12. Brown, C. J., *et al.* Further studies in synthetic polypeptides. Nature, **163:** 834–835, 1949; Kotchalski, E. Poly-α-amino acids. Advances Protein Chem. **6:** 123–185, 1951.
13. Brown, H., *et al.* Structure of pig and sheep insulins. Biochem. J., **60:** 556–565, 1955.
14. Bruckner, V., *et al.* Structure of poly-D-glutamic acid isolated from capsulated strains of Bacillus anthracis. Nature, **172:** 508, 1953.
15. Cecil, R. Quantitative reactions of thiols and disulfides with silver nitrate. Biochem. J., **47:** 572–584, 1950.
16. Consden, R., *et al.* Qualitative analysis of proteins: A partition chromatographic method using paper. Biochem. J., **38:** 224–232, 1944.
17. Corey, R. B., and Pauling, L. Fundamental dimensions of polypeptide chains. Proc. Roy. Soc. *B*, **141:** 10–20, 1953.
18. Corfield, M. C., and Robson, A. Amino acid composition of salmine. Biochem. J., **55:** 517–522, 1953.
19. Craig, L. C. Identification of small amounts of organic compounds by distribution studies. II. Separation by countercurrent distribution. J. Biol. Chem., **155:** 519–534, 1944.
20. Edman, P. A method for the determination of the amino-acid sequence in peptides. Arch. Biochem., **22:** 475–476, 1949.
21. Eiger, I. Z., and Dawson, C. R. Spectrophotometric method for the determination of the protein content of mushroom tyrosinase preparations. Arch. Biochem., **21:** 181–193, 1949.
22. Fankuchen, I. X-ray diffraction and protein structure. Advances Protein Chem., **2:** 15–27, 1945.
23. Fox, S. W., *et al.* A method for the quantitative determination of C-terminal amino acid residues. J. Am. Chem. Soc., **77:** 3119–3122, 1955.
24. Haurowitz, F., and Bursa, F. The linkage of glutamic acid in protein molecules. Biochem. J., **44:** 509–512, 1949.
25. Holiday, E. R. Spectrophotometry of proteins. Biochem. J., **30:** 1795–1803, 1936.
26. Klotz, I. M., and Griswold, P. Infrared spectra and the amide linkage in a native globular protein. Science, **109:** 309–310, 1949.
27. Lindley, H., and Rollett, J. S. Investigation of insulin structure by model-building techniques, Biochim. et Biophys. Acta, **18:** 183–193, 1955.
28. Marsh, R. E., *et al.* Structure of silk fibroin. Biochim. et Biophys. Acta, **16:** 1–34, 1955.
29. Mirsky, A. E. Protein denaturation. Cold Spring Harbor Symp., Quant. Biol., **6:** 150–163, 1938.
30. Moore, S., and Stein, W. H. Chromatography of amino acids on starch columns. J. Biol. Chem., **178:** 53–91, 1949.
31. Neurath, H. Intramolecular folding of polypeptide chains in relation to protein structure. J. Physical Chem., **44:** 296–305, 1940.
32. Neurath, H. The role of glycine in protein structure. J. Am. Chem. Soc., **65:** 2039–2041, 1943.
33. Pauling, L., and Corey, R. B. Stable configurations of polypeptide chains. Proc. Roy. Soc. *B*, **141:** 21–33, 1953.
34. Piantanida, M., *et al.* Paper-strip chromatography of proteins. Arch. Biochem. and Biophys., **57:** 334–339, 1955.

35. Ramachandran, G. N., and Kartha, G. Structure of collagen. Nature, **176:** 593–595, 1955.
36. Sanger, F. Some chemical investigations on the structure of insulin. Cold Spring Harbor Symp. Quant. Biol., **14:** 153–161, 1950.
37. Sanger, F., and Thompson, E. O. P. The amino-acid sequence in the glycyl chain of insulin. I. The identification of lower peptides from partial hydrolysis. Biochem. J., **53:** 353–366, 366–374, 1953.
38. Sanger, F., and Tuppy, H. The amino-acid sequence in the phenylalanyl chain of insulin. Biochem. J., **49:** 463–490, 1951.
39. Schmidt, C. L. A. *The Chemistry of Amino Acids and Proteins*, pp. 183–220. Springfield, Ill., Charles C Thomas, 1938.
40. Shemin, D., and Foster, G. L. The isotope dilution method of amino acid analysis. Ann. New York Acad. Sc., **47:** 119–134, 1946.
41. Snell, E. E. Microbiological methods in amino acid analysis. Ann. New York Acad. Sc., **47:** 161–179, 1946.
42. Stevens, C. M., *et al.* The cellular D-amino acids of Bacillus brevis. J. Biol. Chem., **212:** 461–467, 1955.
43. Thompson, E. D. P. The N-terminal sequence of serum albumins; observations on the thiohydantoin method. J. Biol. Chem., **208:** 565–572, 1954.
44. Thorne, C. B., *et al.* Production of glutamyl polypeptide by Bacillus subtilis. J. Bacteriol., **68:** 307–315, 1954.
45. Tristram, G. R. The amino acid composition of proteins; *in* Neurath, H., and Bailey, K. *The Proteins*. New York, Academic Press, 1953.
46. Turner, R. A., and Schmerzler, G. Identification of C-terminal residues in peptides and proteins through formation of thiohydantoins. Biochim. et Biophys. Acta, **13:** 553–559, 1954.
47. Vickery, H. B. On the use of the capital letter prefixes L and D. Science, **113:** 314–315, 1951.
48. Williams, R. T., and Synge, R. L. M. (eds.). Partition chromatography. Biochem. Soc. Symp., No. 3. Cambridge, Eng., Cambridge University Press, 1950.

CHAPTER 3

Carbohydrates and Lipids

In the two preceding chapters the chemistry of proteins, which form the fundamental living substratum of all tissues, has been discussed. We are now ready to take up those biochemical characteristics of individual tissues which are of significance in the study of human metabolism. Blood, a tissue which is unique both in its physical state since it is the only liquid tissue, and in many of its functions, will be dealt with in a separate chapter. Before proceeding to the individual tissues, however, it would be well to recall a few aspects of the general chemistry of carbohydrates and lipids. These two classes of compounds, like the proteins, are universally distributed among living cells. Their function is primarily, but not entirely, that of serving as fuel for the body.

CARBOHYDRATES

Carbohydrates have been defined as *aldehydic or ketonic derivatives of polyhydric alcohols*. The names of the carbohydrates usually have the generic ending *ose*, as for example, glucose. Carbohydrates can be divided into groups depending upon the length of the carbon chain involved. A carbohydrate containing six carbons is termed a *hexose*. By the same principle we have dioses, trioses, tetroses, pentoses, and so on. The simplest carbohydrates of metabolic significance are the trioses, and two such are *glyceric aldehyde* (formula I) (or glycerose) and *dihydroxyacetone* (formula II). Note that the former is aldehydic and the latter ketonic. For this

```
H     O
 \   //
   C                        CH2OH
   |                        |
  HCOH                      C=O
   |                        |
  CH2OH                     CH2OH
```

I. Glyceric aldehyde   II. Dihydroxyacetone

reason glyceric aldehyde is referred to as an *aldotriose*, and dihydroxyacetone as a *ketotriose*. Similar distinctions can be made with respect to carbohydrates with longer carbon chains. Two very important carbohydrates, glucose and fructose, are an aldohexose and ketohexose, respectively. One important ketoheptose, *sedoheptulose,* is formed in the course of carbohydrate utilization in the body. Carbohydrates containing aldehyde groups are spoken of collectively, regardless of the length of carbon chain, as *aldoses,* those with ketone groups as *ketoses.* Among the ketoses, the ketone group is invariably located on the second carbon. The numbering system of carbohydrates begins at the aldehyde end (or at the end nearest the ketone group).

Aside from the ketone or aldehyde group, each carbon atom in a carbohydrate is usually attached to a hydroxyl group. Carbohydrates exist, however, in which a hydroxyl group is missing along the carbon chain. These may be named in two ways according to whether the hydroxyl group is missing from a carbon at the end of the chain or one not at the end. An example of the former is rhamnose (formula III). Such a compound is

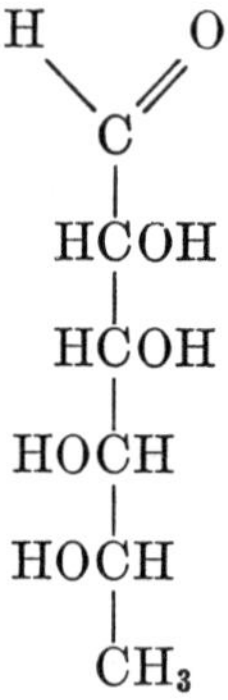

H O
C
HCOH
HCOH
HOCH
HOCH
CH_3

III. Rhamnose

usually referred to as a *methylpentose,* or more generally still, a *methylose.* An example where the missing hydroxyl group is not at the end may be taken from the two important pentoses which occur in nucleic acids (see Chapter 9). The fully hydroxylated pentose is named ribose (formula IV), while the other is named deoxyribose (formula V). A compound

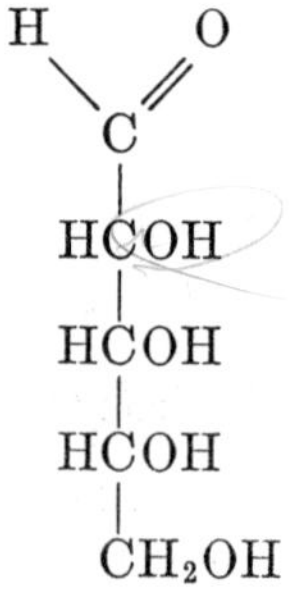

H O
C
HCOH
HCOH
HCOH
CH_2OH

IV. Ribose

H O
C
HCH
HCOH
HCOH
CH_2OH

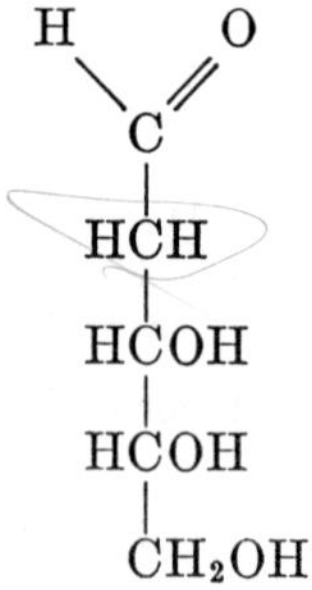

V. Deoxyribose

such as the latter may be considered a deoxypentose, or more generally still, a *deoxy-sugar*. Deoxy-sugars in which more than one hydroxyl group is missing occur among the components of the digitalis glycosides.

Carbohydrates which can be represented structurally as a single carbon chain (or as a single oxide ring) are termed *monosaccharides*. The carbohydrates so far referred to in this chapter all belong to this class. Carbohydrates which are composed of two monosaccharides linked by an oxygen bridge after removal of the elements of water are *disaccharides*. Still more complicated compounds, containing more than two such oxygen-linked units also occur. When the number of monosaccharide units is relatively few, such multiple compounds are *oligosaccharides*, and these include the disaccharides, trisaccharides, and so on. Where the number of units is indefinite, or relatively large, the compound is termed a *polysaccharide*. As the number of monosaccharide units becomes larger, the compounds gradually lose the characteristic monosaccharide properties of water-solubility and sweetness. Oligosaccharides retain these properties and are classified along with the monosaccharides under the general term, *sugars*.

Monosaccharides

Optical activity. Referring back to the formulas for glyceric aldehyde and dihydroxyacetone, earlier in this chapter, we notice that of the two it is glyceric aldehyde which possesses an asymmetric carbon. In other words the center carbon of the three in glyceric aldehyde is surrounded by four different groups: H, OH, CHO, and CH_2OH. For this reason glyceric aldehyde possesses optical activity, while dihydroxyacetone, which does not possess an asymmetric carbon atom, does not. In this respect it is glyceric aldehyde which is typical of the carbohydrates and dihydroxyacetone which is exceptional. Although the principles of optical isomerism have been discussed in the previous chapter, certain further complexities arise in the study of carbohydrates.

The typical aldohexose has four asymmetric carbons; carbons number 2, 3, 4, and 5. In the case of each carbon the exact spatial orientation of the hydrogen and the hydroxyl groups about the carbon is of significance in determining the optical properties of the compound. For each asymmetric carbon there are two and only two ways of arranging the four groups about it. If these two ways are denoted as *a* and *b*, then where two such carbons are found in a single compound four possibilities exist: *aa*, *ab*, *ba*, and *bb*; where three such are found, eight possibilities exist: *aaa*, *aab*, *aba*, *baa*, *abb*, *bab*, *bba*, and *bbb*. In general, where *n* asymmetric carbons exist in a compound, 2^n optical isomers may exist. In the case of an aldohexose such as glucose, with four asymmetric carbons, sixteen isomers may exist, of which only one is the sugar commonly termed glucose, or grape-sugar. And to confirm theory, sixteen aldohexoses have indeed been isolated or synthesized, and no more than sixteen.

The structure of glucose, spatially, is conventionally represented as shown in formula VI. Again, the exact spatial relationships are best followed by atom models since the molecule is not, as represented here, planar. However by restricting efforts at superimposition to the plane of the paper, optical relationships can be adequately presented. Thus, if glucose (properly termed D-glucose is compared with its mirror image, L-glucose (formula VII), we see that the two compounds can not be superimposed.

The differing extents to which closely related optical isomers occur in living tissue may be illustrated by these two glucoses. D-Glucose, either free or in the form of derivatives, is nearly as widespread in occurrence as a natural compound can be. L-Glucose, on the other hand, occurs in nature with extreme rarity. Its most notable appearance (in the form of a methylamino derivative) is in the antibiotic, streptomycin. This difference may be generalized. Among carbohydrates as a group, the D series occurs naturally, while the L does not. For this reason, simple references to carbohydrates without reference to the optical series imply the D form.

As a matter of fact, the majority of the 16 aldohexoses do not occur in nature to any significant extent and are merely laboratory curiosities whose main importance lies in bearing witness to the validity of the stereochemical theory. In addition to D-glucose, only two other aldohexoses, D-galactose (formula IX) and D-mannose (formula VIII), are found widespread in nature.

A distinction must here be drawn among three often confused terms—stereoisomers, optical isomers, and enantiomorphs. *Stereoisomers* are any group of compounds which resemble one another in being composed of identical chemical groups, and differ from one another in the spatial arrangement of those groups. *Optical isomers* are those stereoisomers which

VI	VII	VIII	IX
H—C=O	H—C=O	H—C=O	H—C=O
H—C—OH	HO—C—H	HO—C—H	H—C—OH
HO—C—H	H—C—OH	HO—C—H	HO—C—H
H—C—OH	HO—C—H	H—C—OH	HO—C—H
H—C—OH	HO—C—H	H—C—OH	H—C—OH
CH_2OH	CH_2OH	CH_2OH	CH_2OH
VI. D-Glucose	VII. L-Glucose	VIII. D-Mannose	IX. D-Galactose

involve arrangements about an asymmetric carbon atom. *Enantiomorphs* are those optical isomers which are mirror images of one another.

D-Galactose and D-glucose, for instance, are stereoisomers, and since they involve asymmetric carbon atoms, optical isomers as well.[1] However, they are *not* enantiomorphs since they are not mirror images of one another. Their optical rotations for that reason bear no necessary relationship to one another. D-Glucose and L-glucose are not only optical isomers, but enantiomorphs as well. For this reason the specific optical rotations of D-glucose and L-glucose are necessarily the same, but in opposite directions. D-Galactose and D-mannose both naturally have their own enantiomorphs in the compounds L-galactose and L-mannose, the formulas of which the student should be able to write without further help. The sixteen existing aldohexoses are, in fact, composed of eight pairs of enantiomorphs. It should once again be emphasized that the prefixes D and L apply to the spatial structure of the compound only and not to the direction of optical rotations.

Sugars which differ only in the spatial arrangement about one carbon atom are called *epimers*. D-Glucose and D-galactose are 4-epimers because they differ only in the configuration about carbon 4. D-Glucose and D-mannose are 2-epimers.

Passing to the ketohexoses, we see from the general formula (formula X) that they have only three asymmetric carbons (carbons number 3, 4, and 5), so that there are only eight ketohexoses. Of these eight, one only, D-fructose (formula XI), is of importance in the human body. Similarly, of the eight possible aldopentoses, only D-ribose is significant in human metabolism. Of all the many monosaccharides, the physician need be well acquainted with only six: glucose, fructose, galactose, mannose, ribose, and deoxyribose, all of the D series.

For further details on carbohydrate stereochemistry and on the chemical

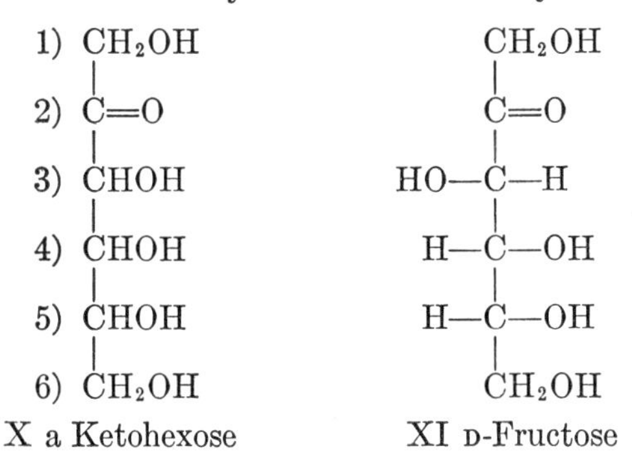

X a Ketohexose XI D-Fructose

[1] Examples of stereoisomers which are *not* optical isomers will be found later in the chapter in the discussion of sterols.

methods used to determined the structure of individual sugars, see Pigman and Goepp (11).

Monosaccharide ring structure. Until now we have been considering monosaccharides as straight-chain aldehydes or ketones. They do, indeed, possess several properties characteristic of simple aldehydes. For instance, they reduce Cu^{++} and Ag^{+} ions in alkaline solution, a property which forms the basis of important clinical tests. It is this reducing ability which puts monosaccharides as a group into the category of *reducing sugars.* However, monosaccharides do not share all properties of aldehydes. They do not form sulfite addition products, for instance. This, and other evidence, has resulted in the conclusion that monosaccharides exist only to a minor extent as straight-chain aldehydes and ketones. The straight-chain forms are in tautomeric equilibrium with less reactive ring systems in which an oxygen bridge is formed between the aldehyde or ketone group and one of the other carbons, hydrogen being transferred to the aldehyde or ketone oxygen in the process (formula XII).

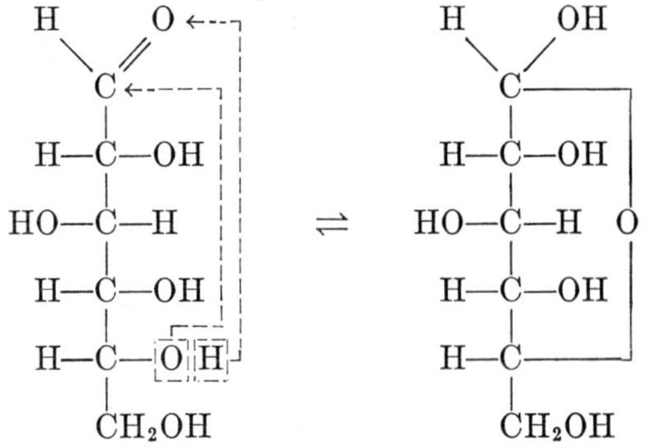

```
  H       O ←--┐              H       OH
   \     //    ¦               \     /
     C ←-----┐ ¦                 C ──────┐
     |       ¦ ¦                 |       │
  H—C—OH     ¦ ¦              H—C—OH     │
     |       ¦ ¦                 |       │
 HO—C—H      ¦ ¦     ⇌       HO—C—H     O
     |       ¦ ¦                 |       │
  H—C—OH     ¦ ¦              H—C—OH     │
     |    ┌--┘ ¦                 |       │
  H—C—[O][H]---┘              H—C──────┘
     |                           |
   CH₂OH                       CH₂OH
```

D-Glucose (straight-chain) D-Glucose (ring)

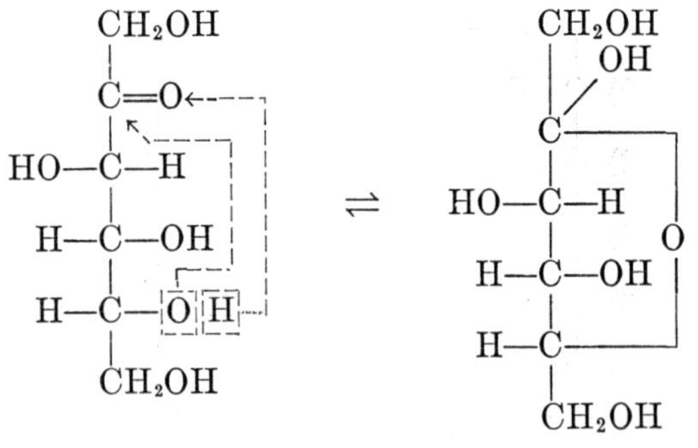

```
   CH₂OH                       CH₂OH
     |                           |  OH
     C=O←---┐                    | /
     | ↖---┐ ¦                    C ──────┐
 HO—C—H    ¦ ¦               HO—C—H      │
     |      ¦ ¦    ⇌             |        O
  H—C—OH    ¦ ¦               H—C—OH     │
     |   ┌--┘ ¦                  |        │
  H—C—[O][H]--┘               H—C───────┘
     |                           |
   CH₂OH                       CH₂OH
```

D-Fructose (straight-chain) D-Fructose (ring)

XII

Theoretically, such an oxygen bridge can be formed between the aldehyde carbon and any of the others. Actually, however, unions between carbons separated by two or three others, forming five and six membered rings, respectively, predominate. This is because the angles between carbon valences are such that in forming a carbon chain, which is usually made to look straight when represented on paper, loops are formed in which a given carbon finds itself lying quite close to another carbon separated from it by three or four others. This can be readily shown with the aid of atom models. It is for this reason that five and six-membered rings predominate among organic compounds. They are relatively "strain-free" since the valence bonds need not be greatly distorted from their normal positions in their formation.

The representation of the ring structure shown in formula XII, in which the linking oxygen is placed to one side of a straight chain of carbon atoms and is attached to two of them by long, bent "bonds" is quite unrealistic. Another method for presenting such cyclic carbohydrate structures was introduced in the 1920's by Haworth. A modification of this system will be used in the remainder of this book.

Monosaccharides in the form of a six-membered ring containing five carbons and one oxygen are spoken of as pyrane derivatives, since pyrane is the simplest compound containing such a ring system. Similarly, those consisting of five-membered rings containing four carbons and one oxygen are spoken of as furane derivatives. These facts are included in the names given the rings formed. Thus glucose exists largely in the form of *glucopyranose* (formula XIII), and fructose in the form of *fructofuranose* (formula XIV).

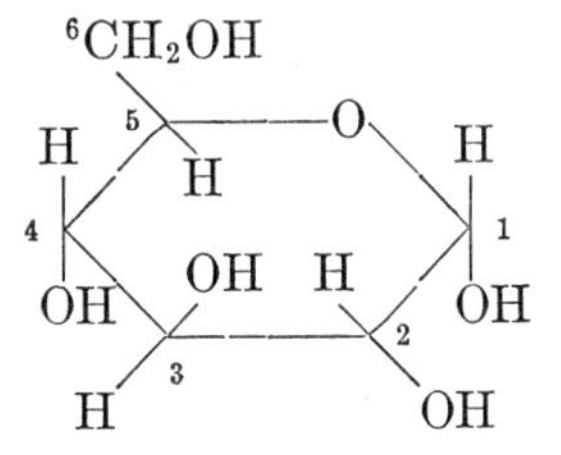

XIII. D-Glucopyranose

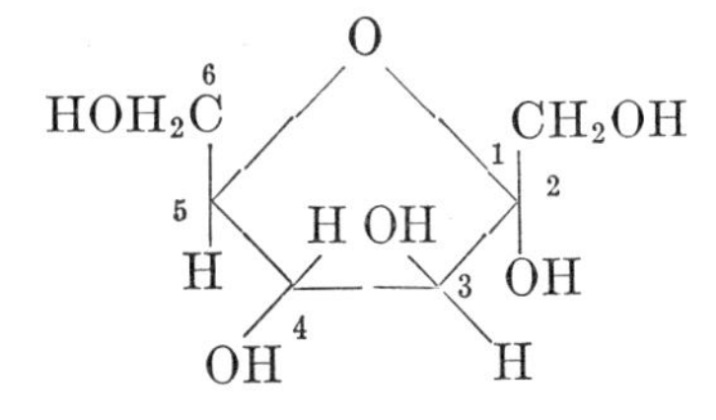

XIV. D-Fructofuranose

For reference, ring structures are given for galactose, mannose, ribose, and deoxyribose (formulas XV, XVI, XVII, and XVIII). Notice that in the formation of an oxygen bridge the aldehyde or ketone carbon becomes asymmetric. Two additional optical isomers of glucose (and of the other monosaccharides) may be expected and are found. They are termed alpha-D-glucopyranose and beta-D-glucopyranose, or more simply, alpha-glucose and beta-glucose. Each of these forms has a different specific optical rotation (note that they are *not* enantiomorphs). The structures shown in formulas XIII to XVIII inclusive are of monosaccharides in the alpha con-

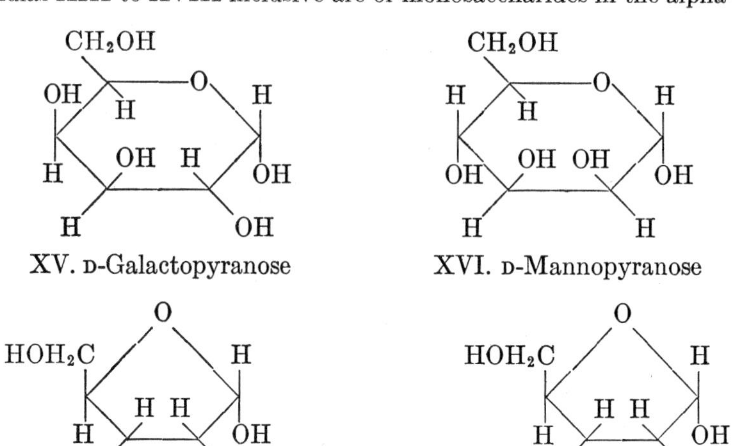

XV. D-Galactopyranose

XVI. D-Mannopyranose

XVII. D-Ribofuranose

XVIII. D-Deoxyribofuranose

figuration. A shift in orientation of the groups around carbon 1 takes place much more readily than is the case with any of the other carbons, probably through a process whereby the oxygen ring is alternately opened and closed. In solution, therefore, an equilibrium between the alpha and beta

Alpha-glucose

$\underset{-H_2O}{\overset{+H_2O}{\rightleftharpoons}}$

Hydrated intermediate

$\underset{+H_2O}{\overset{-H_2O}{\rightleftharpoons}}$

Beta-glucose

XIX. Mutarotation

forms of the sugar exists. If pure alpha-glucose or pure beta-glucose is dissolved in water and its optical activity observed, the specific rotation will be found to change steadily until in each case the same equilibrium figure is reached. This process is known as *mutarotation* (formula XIX).

The carbon atom about which mutarotation takes place (carbon 1 in aldoses and carbon 2 in ketoses) is called the *anomeric carbon*, and sugars differing only in the spatial relation about that carbon (as do alpha-glucose and beta-glucose) are *anomers*. Note that an anomer is a special variety of an epimer (see page 103).

Monosaccharide derivatives. The monosaccharides we have discussed rarely occur free in nature. Fructose and glucose are found in honey and in numerous fruit juices, and glucose is the sugar of blood and the body's extracellular fluids in general. Within the cell, on the other hand, the monosaccharides exist in the form of phosphate esters. These esters play important roles in the energy-releasing breakdown of carbohydrates and will be discussed in Chapter 14.

Amino sugars, such as *glucosamine* (2-aminoglucose or, more precisely, 2-aminodeoxyglucose) (formula XX) and *galactosamine* (2-aminogalactose), occur in the body as components of more complex substances such as the mucopolysaccharides.

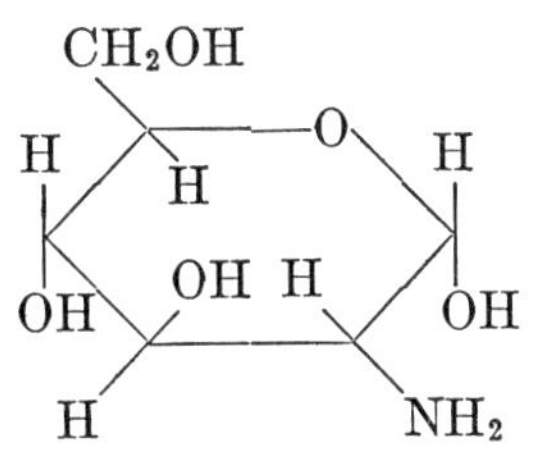

XX. 2-Aminoglucose

Oxidized monosaccharides. The *uronic acids* (formula XXI) are monosaccharide derivatives in which the primary alcohol group of carbon 6

Glucuronic acid Mannuronic acid Galacturonic acid

XXI. Uronic acids

is oxidized to a carboxyl. Individual members of this group are named after the parent monosaccharide. Uronic acids have, in common with other gamma- or delta-hydroxy acids, the property of forming cyclic esters, called *lactones*. Thus, glucuronic acid very easily forms *glucuronolactone* (formula XXII), the accepted formula of which is a double ring, one a furanose structure and the other a gamma-lactone.

Uronic acids occur, along with amino sugars, among the hydrolysis products of mucopolysaccharides. Glucuronic acid frequently acts in the body as a detoxifying agent, forming *glucuronides* which are excreted in the urine.

The stereochemical nomenclature of the uronic acids has been a source

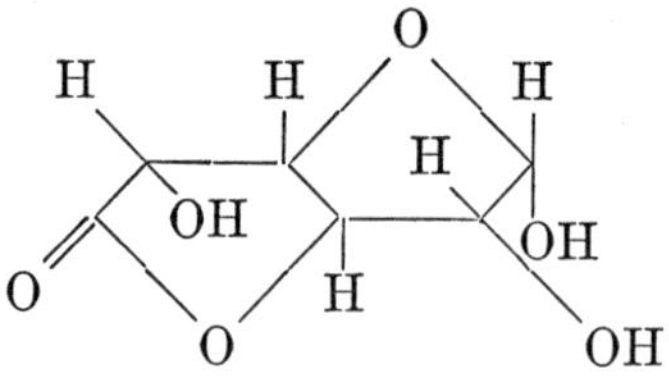

XXII. Glucuronolactone

of confusion. The tendency is to call the uronic acid obtained by the oxidation of D-glucose, D-glucuronic acid. Artz and Osman (1), however, point out that a sugar or sugar derivative is judged to belong to the D or L series by the configuration of the asymmetric carbon furthest from the functional group. In the case of the hexoses, this is carbon 5. In uronic acids, however, the carboxyl is the functional group which gives the compound its name and carbon 2 is the asymmetric carbon furthest from that. In glucuronic acid, carbon 2 happens to have the L-configuration as is easily seen if glucuronic acid is written in the straight-chain form with the functional group, COOH, at the top and the CHO group at the bottom (formula XXIII).

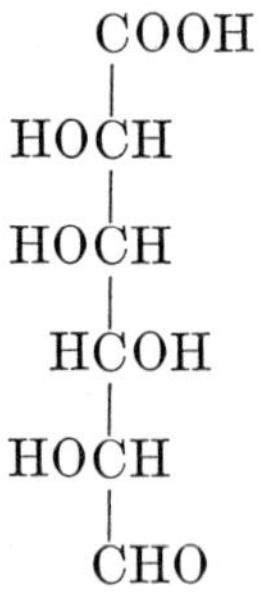

$$\begin{array}{c} COOH \\ | \\ HOCH \\ | \\ HOCH \\ | \\ HCOH \\ | \\ HOCH \\ | \\ CHO \end{array}$$

XXIII. L-Glucuronic acid

It is L-glucuronic acid, then, that is derived from D-glucose. By similar lines of reasoning, the uronic acids derived from D-galactose and D-mannose are L-galacturonic acid and D-mannuronic acid. An alternate method of nomenclature is to stress the aldehyde (or ketonic) group of the uronic acids by calling them the *onoses*, thus pointing up the relationship to the *oses* or sugars. In that case, the aldehyde group can still be considered as reference and the three natural compounds would be D-glucuronose, D-galacturonose, and D-mannuronose.

In the *glyconic acids*, also known as *aldonic acids*, it is the aldehyde carbon (carbon 1) which is oxidized to a carboxyl group. Again, individual aldonic acids derive their names from the parent compound, as gluconic acid. The aldonic acids do not form oxide rings as do the parent monosaccharides but form lactones easily, splitting out water (formula XXIV). The

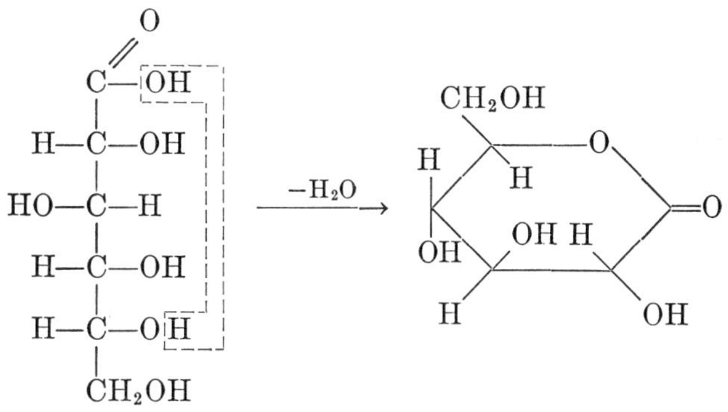

$$\begin{array}{c} O \\ // \\ C—OH \\ | \\ H—C—OH \\ | \\ HO—C—H \\ | \\ H—C—OH \\ | \\ H—C—OH \\ | \\ CH_2OH \end{array} \xrightarrow{-H_2O} \text{Gluconolactone}$$

Gluconic acid Gluconolactone

XXIV

aldonic acids are of little or no significance in body structure, but one of the important vitamins, L-ascorbic acid is the lactone of an aldonic acid (see page 820). Also of interest to the physician is the fact that in cases of calcium or iron deficiency, the metal in question is often administered as the salt of gluconic acid.

A still more complete oxidation of the monosaccharides results in derivatives in which both terminal carbons are oxidized to carboxyl groups. These are termed the *saccharic acids* (less frequently, aric acids). Saccharic acid itself (formula XXV) is derived from glucose, while *mucic acid* (formula XXVI) is derived from galactose.

Reduced monosaccharides. Reduction of the monosaccharides converts the aldehyde group to an alcohol, the resulting compound being a *sugar alcohol*. Individual names are not always derived from the parent compound. Thus the sugar alcohol of glucose is sorbitol (formula XXVII)

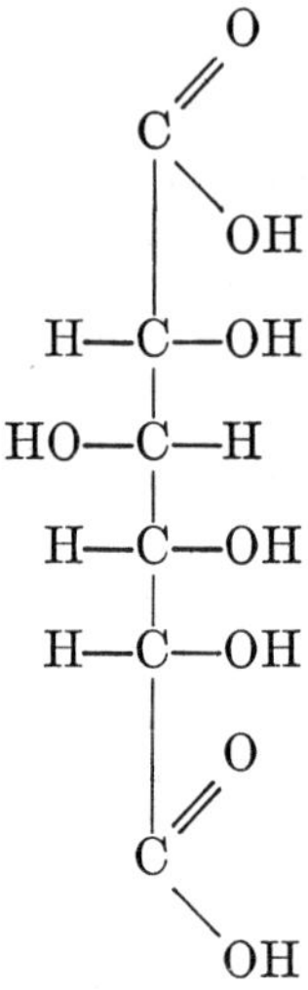

$$\begin{array}{c} \mathrm{O} \\ \| \\ \mathrm{C{-}OH} \\ | \\ \mathrm{H{-}C{-}OH} \\ | \\ \mathrm{HO{-}C{-}H} \\ | \\ \mathrm{H{-}C{-}OH} \\ | \\ \mathrm{H{-}C{-}OH} \\ | \\ \mathrm{O} \\ \| \\ \mathrm{C{-}OH} \end{array}$$

XXV. D-Saccharic acid (from glucose)

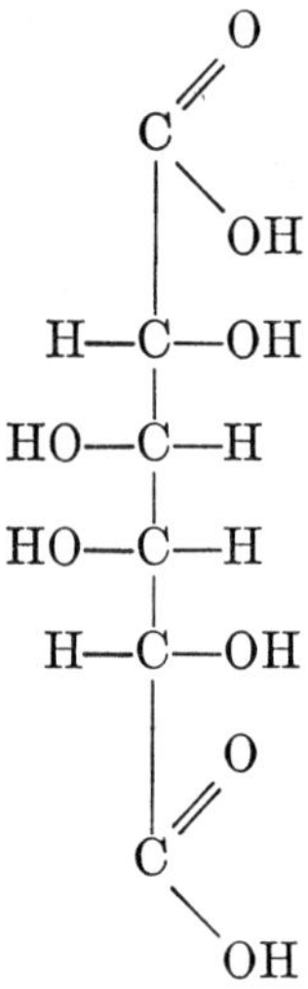

$$\begin{array}{c} \mathrm{O} \\ \| \\ \mathrm{C{-}OH} \\ | \\ \mathrm{H{-}C{-}OH} \\ | \\ \mathrm{HO{-}C{-}H} \\ | \\ \mathrm{HO{-}C{-}H} \\ | \\ \mathrm{H{-}C{-}OH} \\ | \\ \mathrm{O} \\ \| \\ \mathrm{C{-}OH} \end{array}$$

XXVI. Mucic acid (from galactose)

and that of galactose is dulcitol (formula XXVIII). That of mannose, on the other hand, is mannitol (formula XXIX). These sugar alcohols occur in various plants. A five-carbon sugar alcohol, ribitol (formula XXX), forms

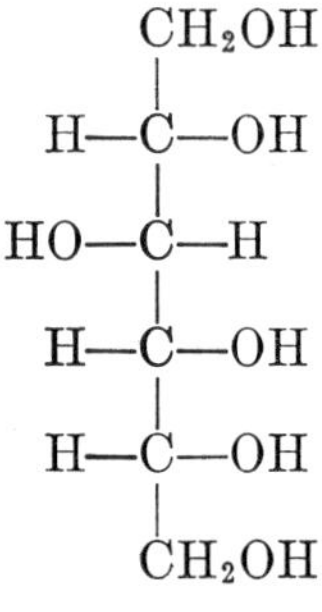

$$\begin{array}{c} \mathrm{CH_2OH} \\ | \\ \mathrm{H{-}C{-}OH} \\ | \\ \mathrm{HO{-}C{-}H} \\ | \\ \mathrm{H{-}C{-}OH} \\ | \\ \mathrm{H{-}C{-}OH} \\ | \\ \mathrm{CH_2OH} \end{array}$$

XXVII. D-Sorbitol

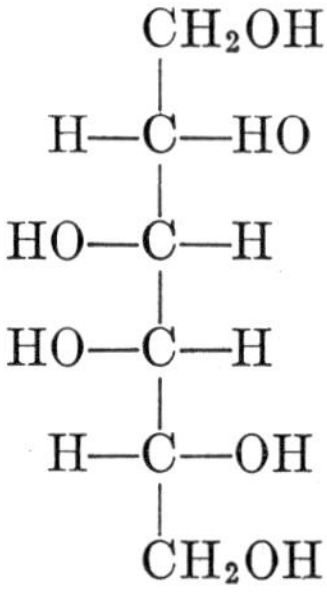

$$\begin{array}{c} \mathrm{CH_2OH} \\ | \\ \mathrm{H{-}C{-}HO} \\ | \\ \mathrm{HO{-}C{-}H} \\ | \\ \mathrm{HO{-}C{-}H} \\ | \\ \mathrm{H{-}C{-}OH} \\ | \\ \mathrm{CH_2OH} \end{array}$$

XXVIII. Dulcitol

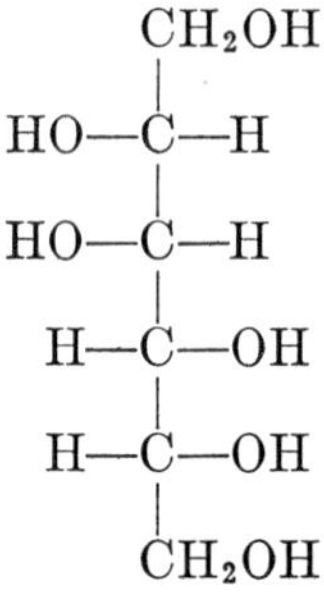

$$\begin{array}{c} \mathrm{CH_2OH} \\ | \\ \mathrm{HO{-}C{-}H} \\ | \\ \mathrm{HO{-}C{-}H} \\ | \\ \mathrm{H{-}C{-}OH} \\ | \\ \mathrm{H{-}C{-}OH} \\ | \\ \mathrm{CH_2OH} \end{array}$$

XXIX. D-Mannitol

$$\begin{array}{c} \mathrm{CH_2OH} \\ | \\ \mathrm{H{-}C{-}OH} \\ | \\ \mathrm{H{-}C{-}OH} \\ | \\ \mathrm{H{-}C{-}OH} \\ | \\ \mathrm{CH_2OH} \end{array}$$

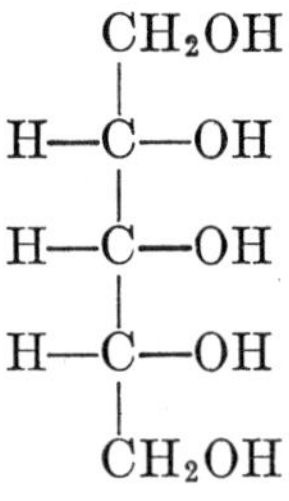

XXX. Ribitol

part of vitamin riboflavin. Dulcitol and ribitol are meso compounds (see page 84). If their formulas are inspected it will be seen that the former has four carbons and the latter two, which, considered individually, are asymmetric. However, the compounds taken as a whole are symmetrical (formula XXXI). This can be shown in two ways. If an imaginary line is drawn

Dulcitol	Ribitol
CH_2OH	
H—C—OH	CH_2OH
HO—C—H	H—C—OH
- - - - - - - - - - - - - -	H—C—OH - - - - plane of symmetry
HO—C—H	H—C—OH
H—C—OH	CH_2OH
CH_2OH	

XXXI

perpendicular to the long axis of the molecule through the bond connecting carbons 3 and 4 in dulcitol, or through the middle of carbon 3 in ribitol, the top half of the molecule in each case is the mirror image of the bottom half. The lines drawn therefore represent lines of symmetry. Secondly, if the mirror images of dulcitol or ribitol are constructed, they can be made to superimpose on the original molecule by a rotation of 180 degrees in the plane of the paper. Dulcitol and ribitol thus do not have enantiomorphs. Nor, for similar reasons, does mucic acid (see formula XXVI).

Inositols, sometimes called *cycloses*, are cyclic sugar alcohols. One such, *meso*-inositol (formula XXXII), so-called because a plane of symmetry can be passed through the molecule, is particularly important. Note that there is no heterocyclic oxygen atom in the inositol ring.

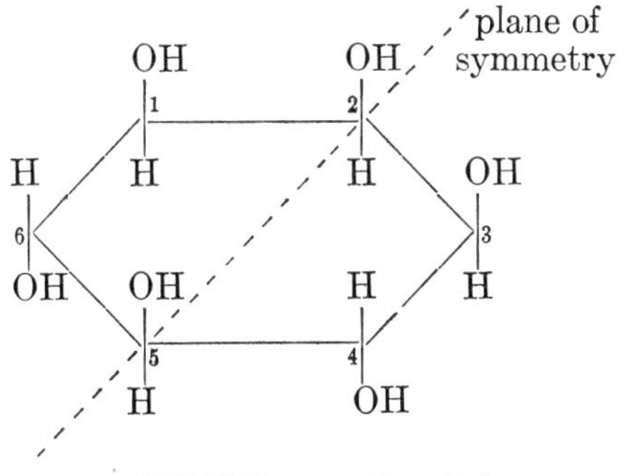

XXXII. *meso*-Inositol

Glycosides. When the hydroxyl upon an anomeric carbon is replaced by an —OR group, the compounds is called a *glycoside* and the sugar —O—R linkage a *glycoside link*. Glycosides derive their names, such as glucoside, fructoside, or galactoside, from the sugar.[2] Both alpha and beta glycosides exist (formula XXXIII). These are stable and do not exhibit

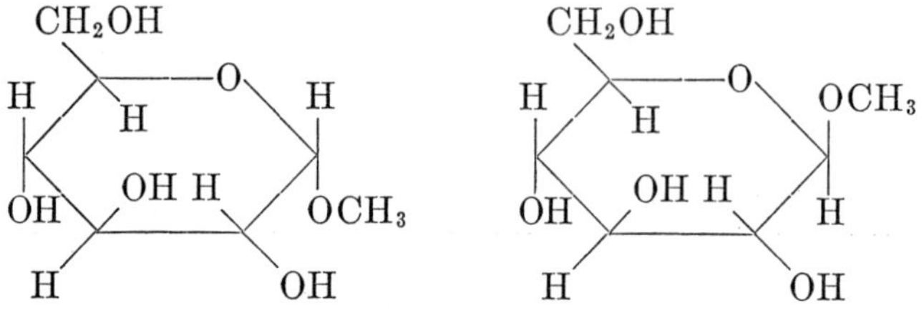

Alpha-methyl glucoside Beta-methyl glucoside

XXXIII. Glycosides

mutarotation since the glycosides contain a relatively complex and immobile group in the place of the mobile hydrogen of the parent alpha or beta monosaccharide.

If the glycoside-linked group is not itself a sugar, it is called an *aglycone*. In the glycosides of plants the sugar of most frequent occurrence is glucose. In many cases, however, specific glycosides occur in which the sugar component is an unusual one, often occurring nowhere else in nature. In the case of the digitalis glycosides, for instance, sugars with methoxy groups or lacking *two* hydroxyl groups occur. Naturally-occurring aglycones vary very widely in chemical nature, including phenols, alcohols, cyanide compounds, flavones, and steroids among others. Many glycosides are of great pharmacological importance. The digitalis group already mentioned is important in the treatment of cardiac disease. Phlorizin produces glycosuria on injection and phlorizinized dogs have been most useful in the study of carbohydrate metabolism. Streptomycin (see page 868) contains two glycoside links.

Disaccharides

Two monosaccharides connected by a glycoside link constitute a disaccharide. The sugar found in milk is lactose, a disaccharide consisting of one glucose unit and one galactose unit. Note that the glycoside link is between the carbon 4 of glucose and the carbon 1 of galactose (formula XXXIV). Since the anomeric carbon of galactose is involved in the glycoside link, the compound is a galactoside. Since glucose is also involved, but *not* by

[2] Note the distinction between "glycoside" and "glucoside". A glycoside is a general term involving any sugar. A glucoside is a glycoside in which the anomeric carbon of glucose is specifically involved.

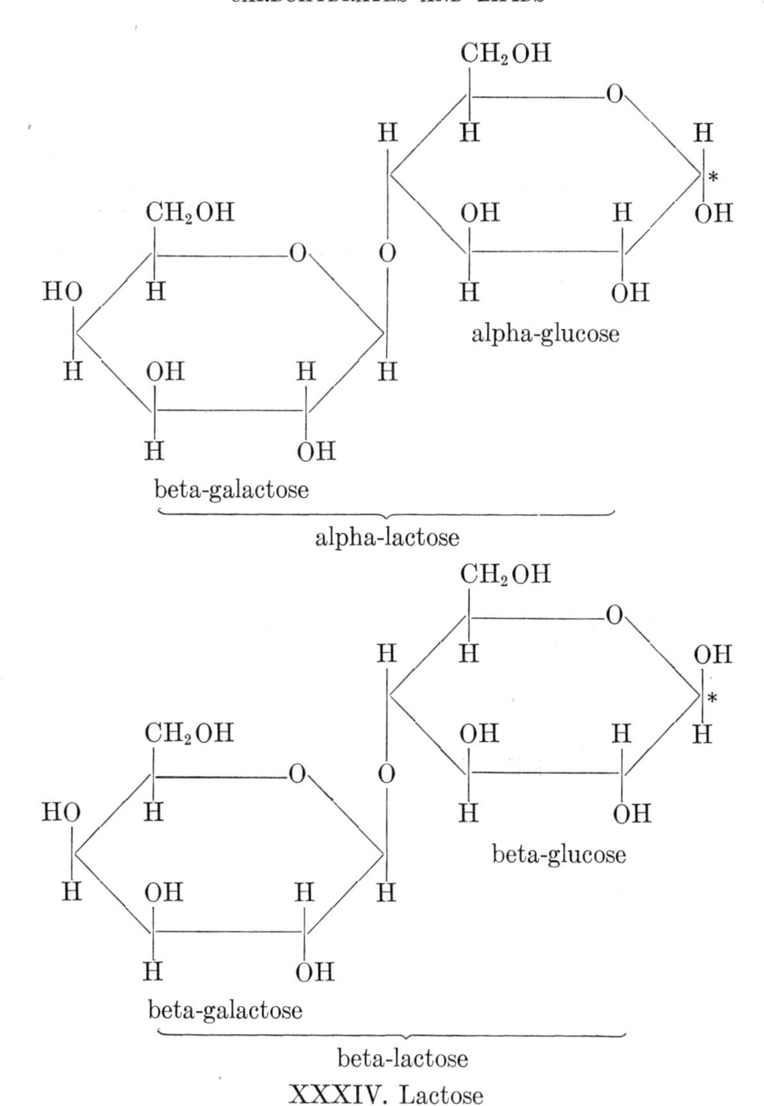

XXXIV. Lactose

Note: The alpha and beta lactoses are distinguished by the configuration about the free anomeric carbon (marked by an asterisk) of the glucose portion. Both alpha- and beta-lactose are beta-galactosides.

an anomeric carbon; lactose is a glucose galactoside. Furthermore, the glycoside link possesses the beta configuration, so that the full chemical name of lactose is D-glucopyranose-4-(beta-D-galactopyranoside). Note also that the anomeric carbon of the glucose portion of lactose remains free. This freedom leads to two results: (a) the anomeric CHOH has reducing properties because of its easy conversion to the open-chain aldehyde form by a shift in the hydrogen of the hydroxyl group and the breaking of the oxygen bridge. Lactose is thus a *reducing sugar* and exhibits mutarotation. And, (b) it is conceivable that the free anomeric CHOH can be utilized to form still another glycoside link with a third monosaccharide.

An example of a non-reducing sugar is *sucrose*, the common sugar of the dinner table (formula XXXV). It is a disaccharide of glucose and fructose

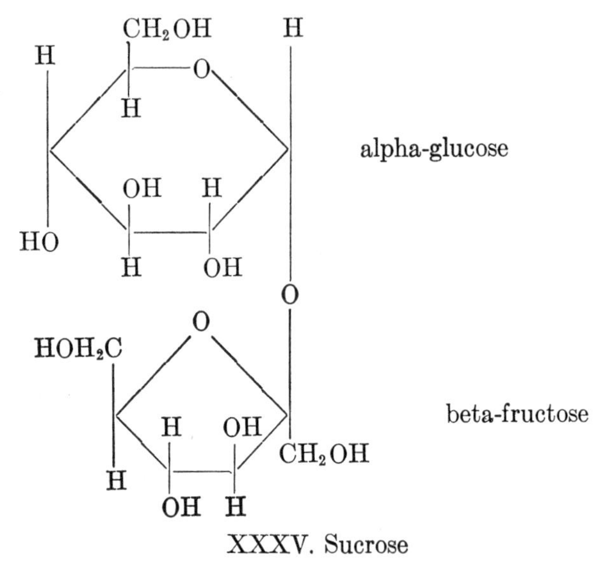

XXXV. Sucrose

and is notable in that the anomeric carbons of both sugars are involved. Sucrose can thus be looked upon either as a glucoside or a fructoside. Since the glucose is connected by an alpha glycoside link, and the fructose by a beta, the chemical name is alpha-D-glucopyranosyl-beta-D-fructofuranoside. The suffix -syl indicates that the anomeric carbon of glucose is also involved. Sucrose can equally well be called beta-D-fructofuranosyl-alpha-D-glucopyranoside. The glycoside-forming groups of both sugars are involved in the linkage so that no aldehyde can be formed through a break in either ring. Sucrose is therefore non-reducing and does not display mutarotation.

Sucrose is readily hydrolyzed through the action of acid or of a specific enzyme named *sucrase*. In either case, an equimolecular mixture of glucose and fructose results. This mixture is termed *invert sugar* because of the behavior of the plane of polarized light during the progress of hydrolysis. The specific rotation of sucrose is +66.5. The specific rotation of D-glucose is +52.7 and that of D-fructose, −92.4. The specific rotation of invert sugar is the average of that of its two components, or −20.3. In the process of hydrolysis, therefore, sucrose changes from a dextrorotatory substance to a levorotatory substance. This "inversion" of rotation gives the name to the final mixture and an alternate name to sucrase, which is also called *invertase*.

Although sucrose is the most familiar sweet substance, it is not the sweetest of the sugars. If the sweetness of sucrose is taken as 100, that of fructose is 173. Fructose is the sweetest of the common sugars, and although glucose is less sweet than sucrose (the rating of glucose is 74), the fructose portion is sufficient to make invert sugar sweeter than sucrose. Invert sugar is used in candies for that reason. It also occurs in honey. Some sugars are only slightly sweet; galactose has a rating of 32 and lactose of 16. On the other hand, the non-sugar, saccharin, has a rating, on the same scale, of 55,000.

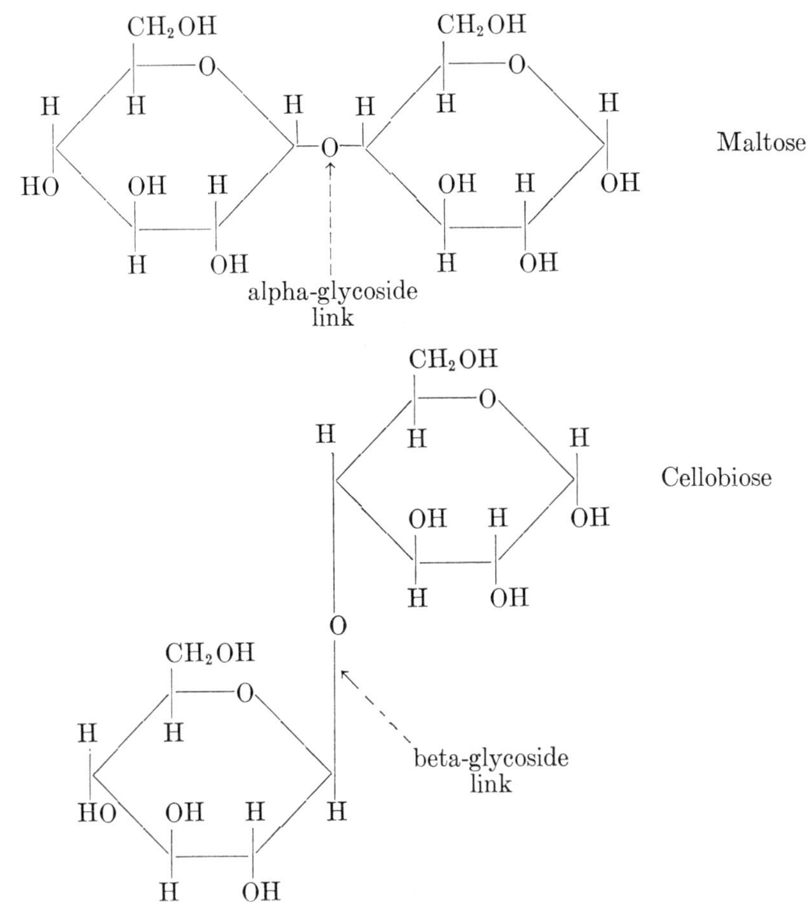

XXXVI

Two disaccharides, *maltose* and *cellobiose*, illustrate the importance of the nature of the glycoside link. Both consist of two molecules of glucose, in which a carbon 1 of one glucose is linked via an oxygen atom to the carbon 4 of the other. Maltose, however, is D-glucopyranose-4-(alpha-D-glucopyranoside), while cellobiose is D-glucopyranose-4-(beta-D-glucopyranoside) (formula XXXVI). The difference, as can be seen, lies in the orientation about the connecting oxygen atom. The two compounds can not be superimposed upon one another by turning one upside down or by twisting one ring through 180 degrees. Physiologically, the difference between the two is immense. Maltose can be converted to glucose in the gastro-intestinal tract through the action of the enzyme maltase which acts specifically on the alpha-glucoside linkage. There is no enzyme in the human body which can act upon a beta-glucoside linkage; thus cellobiose is nutritionally useless to man. (There are, however, enzymes capable of catalyzing the hydrolysis of beta-galactosides such as lactose or beta-fructosides such as sucrose.) Both maltose and cellobiose are reducing sugars capable of mutarotation and the formation of further glycoside links.

Other disaccharides are known to occur in nature as well as trisaccharides and even a tetrasaccharide, but none are nutritionally significant.

Polysaccharides

Starches. Multiple glycosides occur in which the number of monosaccharide units is large, ranging into the hundreds and even thousands. These are the polysaccharides. The starches are polysaccharides consisting exclusively of glucose units connected by alpha-glucoside linkages after the manner of maltose. Because the digestive tract contains enzymes capable of splitting the alpha-glucoside linkage, the starches are nutritionally useful to man and, in fact, such starch-containing foods as wheat, rice, potatoes, taro, and manioc form the great non-protein staples of the human diet.

Starch is by no means to be considered a well defined chemical species, any more than protein may be so considered. Starch molecules vary not only in the length of individual chains, or in average length among species, but also in the nature of the chain itself. One may visualize a starch molecule composed of a linear chain of glucose units. Such a linear chain is referred to as an *amylose* molecule, and in it the carbon 1 of each glucose unit is connected by a glucoside link to the carbon 4 of its neighbor. It is, however, possible to form a branched chain by connecting the carbon 1 of a glucose unit to the carbon 6 of another already within an amylose chain. Such a branched molecule is referred to as an *amylopectin*. Amylopectins can vary among themselves in the average straight-chain interval between branchings and in the degree of branching and sub-branching.

Amylose and amylopectin are affected differently by the *amylases* or starch-splitting enzymes. There are two broad classes of amylases, arbitrarily designated as alpha- and beta-amylases. (Both hydrolyze only alpha-glucoside links, however.) The alpha-amylases, found notably in saliva and pancreatic juice, hydrolyze glucoside links well inside the starch molecule, producing fragments generally larger than maltose. Beta-amylases, found notably in sprouted grains (malt), split off maltose moieties in systematic manner. An amylose molecule subjected to the action of beta-amylase is, in this way, converted entirely to maltose. The action of beta-amylase on amylopectin, however, is limited to the end pieces beyond a branching point and the 1-6 linkages are unaffected. Iodine will oxidize amylose to form a characteristic, intensely blue-black compound, while with amylopectin it yields a red product. Most naturally-occurring starches are mixtures of amylose and amylopectin. Amylopectins have molecular weights much greater than those of amyloses (14). In rare cases, as in certain varieties of maize, rice, barley, and other cereals, starch can be found which is entirely amylopectin in nature, but in most starches, the amylose/amylopectin ratio varies from 1:5 to 1:3.

Upon hydrolysis, either acid or enzymatic, the starch chain is gradually broken into smaller and smaller fragments called *dextrins*. As the average molecular weight grows smaller, the dextrins lose the typical starch properties of insolubility, coloration with iodine, and so on. The change is a continuous one. What is called in the laboratory *soluble starch* is actually a relatively high molecular weight dextrin produced by the partial acid hydrolysis of raw starch.

Glycogen, otherwise known as animal starch, is a comparatively high molecular weight polysaccharide, values as high as 1,500,000 being reported. Carbohydrates in the animal body are stored in this form. Chemically it resembles an amylopectin but is particularly highly branched, the interval between branches varying from as little as three glucose units to six or seven. In ordinary plant amylopectins the corresponding intervals may extend up to 25 or 30 glucose units.

Cellulose and other polysaccharides. Starch and other polysaccharides containing glucose residues may be referred to generally as *glucosans*. *Cellulose* which is characteristic of plant organisms, is a glucosan. The glucose units, however, are connected by beta-glycoside linkages after the manner of cellobiose. Cellulose, like cellobiose, is nutritionally useless to man except indirectly, since it can be utilized by herbivorous livestock through the action of the bacteria of their alimentary tracts.

Dextran is a glucosan produced by certain micro-organisms. It differs structurally from starch and cellulose in that the majority of the linkages are 1-6. Depending on the micro-organism which produces it, the 1-6 linkages in dextran make up 75 to 95 per cent of the whole. The remaining linkages are 1-4.

Glucose is not the only monosaccharide which can occur in polysaccharides. The Jerusalem artichoke stores carbohydrate in the form of *inulin*, which, being made up of fructose units only, is a *fructosan*. Inulin is used in medical laboratories to determine renal clearance (see Chapter 20). Mannosans and galactosans also exist in the plant kingdom. *Agar*, for instance, so useful to bacteriologists, contains a galactosan. Collectively, polysaccharides consisting of hexose units are *hexosans*.

Pentosans, such as *xylan* (consisting of units of a five-carbon sugar, xylose), occur in woody plants along with cellulose. Related substances containing more than one sugar (usually including glucose, galactose, xylose, and another five-carbon sugar, arabinose) are the *hemicelluloses*, which together with the celluloses they usually accompany, form a great part of the "bulk" in the human diet. Mucilages, gums, and pectic substances represent even more complicated plant polysaccharides, containing uronic acids as well as hexoses and pentoses.

Mucopolysaccharides. These are polysaccharides that contain a hexosamine as one component. Usually, though not invariably, they contain uronic acids as well. Thus, *hyaluronic acid*, a mucopolysaccharide occurring in connective tissue, vitreous humor, umbilical cord, skin and synovial fluid, is composed of equimolar proportions of glucuronic acid and N-acetylglucosamine. It is an unbranched polymer in which these two units are placed alternately, with 1-3 beta linkages throughout (9). A portion of such a chain is presented in formula XXXVII. The molecular

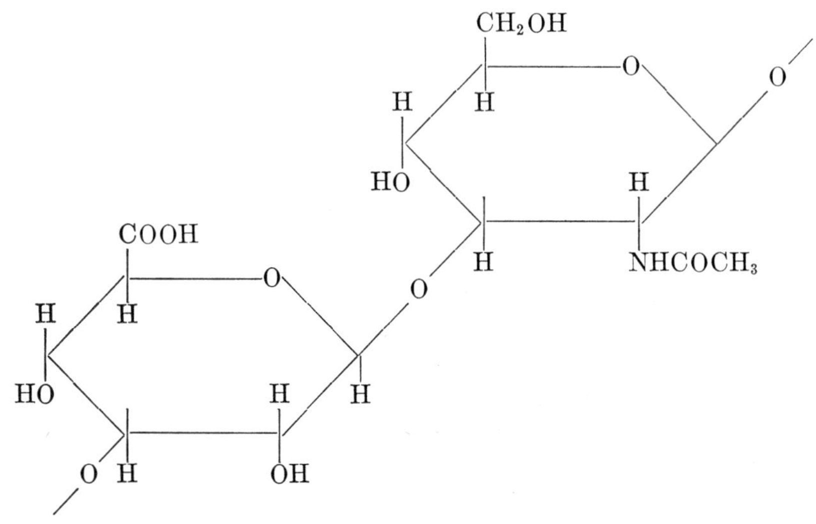

XXXVII. Portion of chain of hyaluronic acid

weight of hyaluronic acid from vitreous humor has been reported to be well over a million (11a).

The mucopolysaccharide, *heparin*, has a structure very similar to that of hyaluronic acid, differing mainly in the presence of five sulfate ester groups per tetrasaccharide unit. Heparin is found in liver, lung, and muscle. When added to blood, it acts as an anticoagulant and can be used in maintaining blood in a liquid state when this is desirable. The molecular weight of heparin is reported as 20,000 (13).

In addition to the free mucopolysaccharides mentioned above, others exist which are found bound to protein such as *mucoitin sulfuric acid*, occurring in saliva and gastric juice. Its formula resembles that of hyaluronic acid except that it contains a sulfate ester group on the carbon 6 of each N-acetylglucosamine unit. *Chondroitin sulfuric acid*, occurring in tendon, cartilage, and bone, differs from mucoitin sulfuric acid in that the N-acetylglucosamine sulfate units are replaced by N-acetylgalactosamine sulfate units. The molecular weight of these mucopolysaccharides has been placed between 200,000 and 400,000.

Other naturally-occurring substances of medical interest which have been found to be mucopolysaccharide, at least in part, include various pituitary hormones and the A, B and O blood group substances.

LIPIDS

Chemically speaking, lipids are far more diverse as a class than are the carbohydrates. Actually, any organic substance capable of being metabolized by a living organism may be classified as a lipid if it is insoluble in water and soluble in such "fat solvents" as ether, chloroform, benzene, carbon tetrachloride, or carbon disulfide. Three major subdivisions of lipids can be distinguished:

1. *Simple lipids* are esters which on hydrolysis yield only alcohols and aliphatic monocarboxylic acids. They contain only carbon, hydrogen, and oxygen.

2. *Compound lipids* are esters which on hydrolysis yield substances in addition to alcohols and aliphatic monocarboxylic acids. Such extraneous substances are most often nitrogen-containing bases and phosphoric acid, and in some cases sulfuric acid or monosaccharides.

3. *Associated lipids* are not esters but are found together with simple and compound lipids in the early stages of most isolation procedures, because of the similarity of their solubility properties.

Along with these, a group known as the *derived lipids* is frequently spoken of, these being the fragments resulting from the hydrolysis of simple or compound lipids. However, since many of these completely lack the typical lipid solubility properties, we prefer to discuss hydrolysis fragments along with the parent compounds, rather than to devote a separate section to them.

Simple Lipids

The simple lipids vary in the nature both of the alcohol and the carboxylic acids that make them up, but it is the identity of the alcohol which is used as the basis of a further subclassification. Thus, the simple lipids may be divided into the fats and oils on one hand, and the waxes on the other. *Fats* and *oils* are mixtures of triesters of fatty acids with the trihydroxy alcohol glycerol. *Waxes* are solid esters of fatty acids with any alcohol other than glycerol. The alcohols occurring in waxes are usually monohydroxy compounds. The distinction between fats and oils is a purely physical one, the former name being given to those naturally-occurring mixtures of glycerol esters which are solid under ordinary conditions, while the latter is applied to those which are liquid. This is a purely arbitrary distinction, born of common usage, and chemically it is more satisfactory to apply the name *glyceride* to all glycerol esters, regardless of their melting points.

Glycerides. The glycerides are the concentrated fuels of the body. Where carbohydrate, in the form of glycogen, is the immediately available fuel (less than one day's supply being present in liver and muscle), glycerides can be stored, as circus sideshows often demonstrate, practically indefinitely. The greater economy of the glycerides as a fuel store when compared with carbohydrates is obvious from the fact that the former have a larger percentage of carbon and hydrogen and a lower percentage of oxygen than the latter. The empirical formula of glycogen, for instance, is $(C_6H_{10}O_5)_n$, whereas that for tristearin, a typical glyceride, is $C_{57}H_{110}O_6$. A given weight of fat has over twice the fuel value of the same weight of carbohydrate.

Tristearin)formula XXXVIII) consists of glycerine esterified with three molecules of stearic acid. *Stearic acid* (formula XXXIX) belongs to the group of *fatty acids*, a name usually applied to the aliphatic monocarboxylic acids, because of their association with fats. A variety of fatty acids has been isolated from lipids, but of these some half a dozen are most wide-

$$
\begin{array}{l}
CH_3CH_2CH_2CH_2CH_2CH_2CH_2CH_2CH_2CH_2CH_2CH_2CH_2CH_2CH_2CH_2CH_2C(=O)\text{—}OCH_2 \\
CH_3CH_2CH_2CH_2CH_2CH_2CH_2CH_2CH_2CH_2CH_2CH_2CH_2CH_2CH_2CH_2CH_2C(=O)\text{—}OCH \\
CH_3CH_2CH_2CH_2CH_2CH_2CH_2CH_2CH_2CH_2CH_2CH_2CH_2CH_2CH_2CH_2CH_2C(=O)\text{—}OCH_2
\end{array}
$$

XXXVIII. Tristearin ($C_{57}H_{110}O_6$)

$$CH_3CH_2CH_2CH_2CH_2CH_2CH_2CH_2CH_2CH_2CH_2CH_2CH_2CH_2CH_2CH_2CH_2C\begin{matrix} \diagup\!\!\diagup O \\ \diagdown OH \end{matrix}$$

XXXIX. Stearic acid ($C_{17}H_{35}COOH$)

spread and important, stearic acid being one of them. Two closely related fatty acids of frequent occurrence are *palmitic acid* (formula XL) and *arachidic acid* (formula XLI). Note that these three acids are very similar,

$$CH_3CH_2CH_2CH_2CH_2CH_2CH_2CH_2CH_2CH_2CH_2CH_2CH_2CH_2CH_2C\begin{matrix} \diagup\!\!\diagup O \\ \diagdown OH \end{matrix}$$

XL. Palmitic acid ($C_{15}H_{31}COOH$)

$$CH_3CH_2CH_2CH_2CH_2CH_2CH_2CH_2CH_2CH_2CH_2CH_2CH_2CH_2CH_2CH_2CH_2CH_2CH_2C\begin{matrix} \diagup\!\!\diagup O \\ \diagdown OH \end{matrix}$$

XLI. Arachidic acid ($C_{19}H_{39}COOH$)

differing only in the length of the hydrocarbon chain. Palmitic acid contains sixteen carbons, stearic acid contains eighteen, and arachidic acid contains twenty.

Fatty acids of this type containing still longer chains than arachidic or still shorter than palmitic are also found in nature, but in smaller quantity. Particularly short-chain fatty acids, such as *butyric acid*, *caproic acid*, *caprylic acid*, and *capric acid*, which contain 4, 6, 8, and 10 carbons, are of interest since they occur in the milk fat of cows (hence in butter) and, to a lesser extent, in coconut and palm nut oils. These short-chain fatty acids are volatile and possess an offensive odor. The odor of rancid butter is the result of liberation of butyric acid by hydrolysis. Volatile fatty acids also contribute to the characteristic odors of certain cheeses and of perspiration.

It will be noticed that the fatty acids so far mentioned all possess an even number of carbon atoms. This is not because odd numbers of carbons render a fatty acid unstable. Odd-numbered fatty acids have been prepared synthetically and do not differ fundamentally from even-numbered fatty acids in their chemical properties. In the biosynthesis of fats, however, the fundamental building stone is a two-carbon fragment (see Chapter 15) which on being added to itself two or more times naturally yields even-numbered acids only.

In addition to the saturated fatty acids just mentioned, certain unsaturated fatty acids are also of frequent natural occurrence. Three such acids containing eighteen carbon atoms each and respectively, one, two, and three double bonds are *oleic acid* (formula XLII), *linoleic acid* (formula XLIII), and *linolenic acid* (formula XLIV). These differ from the saturated

$$CH_3CH_2CH_2CH_2CH_2CH_2CH_2CH_2CH{=}CHCH_2CH_2CH_2CH_2CH_2CH_2CH_2C\begin{array}{l}\diagup\!\!\!\diagup O\\ \diagdown OH\end{array}$$

XLII. Oleic acid ($C_{17}H_{33}COOH$)

$$CH_3CH_2CH_2CH_2CH_2CH{=}CHCH_2CH{=}CHCH_2CH_2CH_2CH_2CH_2CH_2CH_2C\begin{array}{l}\diagup\!\!\!\diagup O\\ \diagdown OH\end{array}$$

XLIII. Linoleic acid ($C_{17}H_{31}COOH$)

$$CH_3CH_2CH{=}CHCH_2CH{=}CHCH_2CH{=}CHCH_2CH_2CH_2CH_2CH_2CH_2CH_2C\begin{array}{l}\diagup\!\!\!\diagup O\\ \diagdown OH\end{array}$$

XLIV. Linolenic acid ($C_{17}H_{29}COOH$)

fatty acids most obviously in their lower melting points both as free acids and as glycerides. Those fats which are generally liquid at room temperature and which are therefore termed oils usually contain considerable oleic acid.

The unsaturated fatty acids are more reactive chemically than are the saturated fatty acids. They will add on halogen atoms quite readily at the double bond, a fact which is important in fat analysis. In addition they will autoxidize at the double bond, forming evil-smelling peroxides which contribute to the rancidity of glycerides. If two or more double bonds are located in the molecule, it can autoxidize, polymerizing in the process, into a tough, insoluble, amorphous material. Linoleic acid and linolenic acid are found in considerable quantities in linseed oil, which because of this last-mentioned polymerizing property is much used in paints.

Other fatty acids are of limited occurrence. Fish oils contain several fatty acids of very high degree of unsaturation. In castor oil there is an important hydroxyl-containing fatty acid, *ricinoleic acid* (formula XLV),

$$CH_3CH_2CH_2CH_2CH_2CH_2\underset{\displaystyle OH}{\underset{|}{C}}HCH_2CH{=}CHCH_2CH_2CH_2CH_2CH_2CH_2CH_2C\begin{array}{l}\diagup\!\!\!\diagup O\\ \diagdown OH\end{array}$$

XLV. Ricinoleic acid ($C_{17}H_{32}(OH)COOH$)

which because of the hydroxyl group, differs from other fatty acids of like carbon content in being alcohol-soluble. For a full discussion of these and even rarer fatty acids, the student is referred to Hilditch (8) or Deuel (3).

Of all the fatty acids mentioned oleic acid, palmitic acid, stearic acid, and linoleic acid are by far the most commonly occurring. Stearic acid occurs chiefly in animal glycerides, linoleic acid in the vegetable glycerides, oleic acid and palmitic acid in both. *Soaps* are the metallic salts of fatty acids and may be formed by alkaline hydrolysis of glycerides.

While fatty acids do occur free in nature, they are much more commonly found esterified as glycerides or other lipids. Fatty acids are examples of what we have referred to earlier as derived lipids, since they are obtained from the naturally-occurring glycerides by hydrolysis. The trihydroxy-alcohol, glycerol, is in this sense also a derived lipid. The glycerides are named according to the fatty acids that compose them. The generic suffix for glycerides is *in*. The glyceride containing three stearic acid groups has already been referred to as *tristearin*. It is sometimes called simply *stearin*. Similarly, we would have such glycerides as *tripalmitin*, *triolein* and *trilinolein*. Glycerides such as these which contain only one kind of fatty acid are *simple glycerides*, while those containing more than one kind of fatty acid are *mixed glycerides*. Examples of the latter (the names being self-explanatory) are *stearodiolein*, *oleodipalmitin*, and *oleopalmitostearin*. In the case of the mixed glycerides, isomerism is possible. Where two fatty acids are present in a glyceride, as in oleodipalmitin, two isomers exist; where three fatty acids are present, as in oleopalmitostearin, three arrangements exist.

Mixed glycerides are of far more frequent occurrence in nature than are simple glycerides, and the tendency among organisms seems to be to avoid the formation of simple glycerides unless there is such a preponderance of a given fatty acid that the process becomes unavoidable. In olive oil, for instance, 75 per cent of all the fatty acids is oleic acid and 50 per cent of the glyceride content is triolein.

The *chemical characterization of naturally-occurring fats and oils* is not easy in view of the fact that they usually consist of complex mixtures of glycerides which differ among themselves only slightly in their chemical properties. It has therefore long been the custom to apply such tests to fats and oils as will yield average values for certain of their characteristics, and in this way distinguish roughly among them.

As an example, the grams of iodine which can be taken up by 100 grams of fat, is known as the *iodine value*, and is a measure of the average content of unsaturated fatty acids present, since it is only at the double bonds of those acids that iodine is taken up. Thiocyanogen will add to only one double bond in linoleic acid and only two in linolenic acid. The *thiocyanogen value* (grams of thiocyanogen taken up by 100 grams of fat) combined with the iodine value, would thus give an idea of the oleic acid content as differentiated from the combined linoleic and linolenic acid content. The *acetyl value* is the milligrams of KOH required to neutralize the acetic acid from hydrolysis of one gram of acetylated fat. Since only hydroxylated fatty acids can be acetylated, the acetyl value is a measure of the average content of such acids. Such values are of great use in characterizing and identifying fats and oils. Linseed oil, for instance, would have a high iodine value, and castor oil, a high acetyl value.

The *saponification value* is the milligrams KOH required to hydrolyze one gram of fat. This is a measure of the average molecular weight of the fatty acids present. That this is so may not be obvious without reflection. Each glyceride to be hydrolyzed requires the action of three molecules of KOH, one for each ester linkage, regardless of the length of the fatty acid chains in the molecule. Where the average length of the fatty acid chain is comparatively small, a given weight of fat will contain more glyceride molecules and therefore have a higher saponification value. A comparatively low saponification value would indicate the presence of long-chain fatty acids. Specialized saponification values, such as the *Reichert-Meissl value* (milliliters N/10 KOH required to neutralize the steam-distillable acids from the hydrolysate of 5 grams of fat), are sometimes useful. Only fatty acids with less than ten carbons are steam-distillable, and these occur appreciably only in butter and, to a lesser extent, in a few plant oils. Butter thus has an appreciable Reichert-Meissl value and can in this manner be easily distinguished from margarine.

Waxes. Waxes have already been defined as solid esters of fatty acids with any alcohol other than glycerol. For the most part these alcohols are long-chain aliphatic monohydroxy compounds containing, usually, an even number of carbon atoms. Such waxes are generally of little importance in human metabolism, since they occur chiefly as insect secretions (beeswax, cochineal wax) or as protective coatings in plants (carnauba wax).

Waxes of greater importance to the human structure are esters of the higher fatty acids with cholesterol, a secondary alcohol belonging to the class of sterols (a group which will be taken up in greater detail below). Cholesteryl esters, sometimes known as *cholesterides*, are found in the blood and form the major portion of the secretions of the sebaceous glands. As sebum, cholesterides fulfill the function of maintaining the gloss and flexibility of hair, and less usefully accumulate within the ear as ear wax. *Lanolin*, used frequently in pharmaceutical and cosmetic preparations, is sheep sebum and owes its properties to its cholesteride content. Some of the fatty acids of lanolin belong to the so-called "lano" series, i.e., fatty acids with branched chains.

Compound Lipids

Phospholipids. Those compound lipids which yield phosphoric acid among their hydrolysis products are termed phospholipids or phosphatides. These can be further subdivided according to the nature of the alcohols present as phosphoglycerides, phosphosphingosides, and phosphoinositides.

Of these, the *phosphoglycerides* most closely resemble the simple lipids already discussed. Phosphoglycerides are fatty acid esters of glyceryl phosphate. Examples are the *phosphatidic acids* (formula XLVI). Phosphatidic acid minus a hydroxyl is the *phosphatidyl group*. Phosphatidic acid itself does not occur in the body, but phosphatidyl derivatives do. One such is *phosphatidyl choline*, an older and more common name for which is *lecithin* (formula XLVII).

$$\begin{array}{l} CH_2O\text{—fatty acid} \\ \quad | \\ \text{fatty acid—}OCH \qquad \overset{O}{\|} \\ \quad | \\ CH_2O\text{—}P\text{—}OH \\ \qquad\quad | \\ \qquad\quad OH \end{array}$$

XLVI. Phosphatidic acid

$$\begin{array}{l} CH_2O\text{—fatty acid} \\ \quad | \\ \text{fatty acid—}OCH \qquad \overset{O}{\|} \qquad\qquad\qquad CH_3 \\ \quad | \qquad\qquad\qquad\qquad\qquad\quad {}^{(+)}/ \\ CH_2O\text{—}P\text{—}OCH_2CH_2N\text{—}CH_3 \\ \qquad\quad | \qquad\qquad\qquad\qquad \backslash \\ \qquad\quad O \qquad\qquad\qquad\qquad\quad CH_3 \end{array}$$

XLVII. Phosphatidyl choline

Phosphatidyl choline is the commonest and most widespread of the compound lipids. It probably occurs in all cells. In addition to the phosphate group, it contains another structure in its molecule that is not present in simple lipids. That is *choline* (formula XLVIII). Choline contains a quater-

$$\begin{array}{l} CH_2\text{—}CH_2 \\ | \qquad\quad | \\ OH \qquad | \quad CH_3 \\ \qquad\quad | \;/ \\ \quad {}^{(+)}N\text{—}CH_3 \\ \qquad\quad \backslash \\ \qquad\quad CH_3 \end{array}$$

XLVIII. Choline (cholinium ion)

nary nitrogen and is therefore positively charged (see page 268). This positive charge must be neutralized by a negative ion. In aqueous solution, the negative ion most available is hydroxyl ion. The isolation of choline from solution is therefore usually in the form of choline hydroxide which, yielding hydroxyl ions when returned to solution, is a strong base.

A glycerol derivative in which a group is attached to one of the end carbons is said to be an alpha-derivative. A group attached to the middle carbon forms a beta-derivative. Formula XLVI, for instance, shows an alpha-phosphatidic acid and formula XLVII an alpha-phosphatidyl choline. The weight of the evidence now indicates that only the alpha-isomers occur naturally (2).

The naturally-occurring phosphatidyl cholines are a group of closely related compounds which differ among themselves in the nature of the fatty acid groups in the molecule. An investigation of phosphatidyl cholines obtained from the livers of cattle, rabbits, dogs, guinea pigs and rats has shown that the fatty acid in the alpha-position (end carbon) is always unsaturated, while that in the beta-position (mid-carbon) is always saturated (7).

It will be noted that the middle carbon of the glycerol portion of phosphatidic acid and its derivatives (and also of many triglycerides, for that matter) is asymmetric. There arises the question, therefore, whether the naturally-occurring phosphatidyl compounds are L or D in nature. The structure of the L and D isomers together with their relationship to the glyceraldehyde reference compounds (see page 82) is shown in formula XLIX. It seems quite definite now that the naturally-occurring phospha-

$$\begin{array}{c}
\mathrm{CHO} \\
| \\
\mathrm{HO{-}C{-}H} \\
| \\
\mathrm{CH_2OH}
\end{array}
\qquad
\begin{array}{l}
\qquad\quad \mathrm{CH_2OOCR} \\
\qquad\quad | \\
\mathrm{RCOO{-}C{-}H} \qquad \mathrm{O} \\
\qquad\quad | \qquad\quad \| \\
\qquad\quad \mathrm{CH_2OPOH} \\
\qquad\qquad\quad | \\
\qquad\qquad\quad \mathrm{OH}
\end{array}
\qquad
\begin{array}{c}
\mathrm{CHO} \\
| \\
\mathrm{H{-}C{-}OH} \\
| \\
\mathrm{CH_2OH}
\end{array}
\qquad
\begin{array}{l}
\qquad \mathrm{CH_2OOCR} \\
\qquad | \\
\mathrm{H{-}C{-}OOCR} \quad \mathrm{O} \\
\qquad | \qquad\quad \| \\
\qquad \mathrm{CH_2OPOH} \\
\qquad\qquad\quad | \\
\qquad\qquad\quad \mathrm{OH}
\end{array}$$

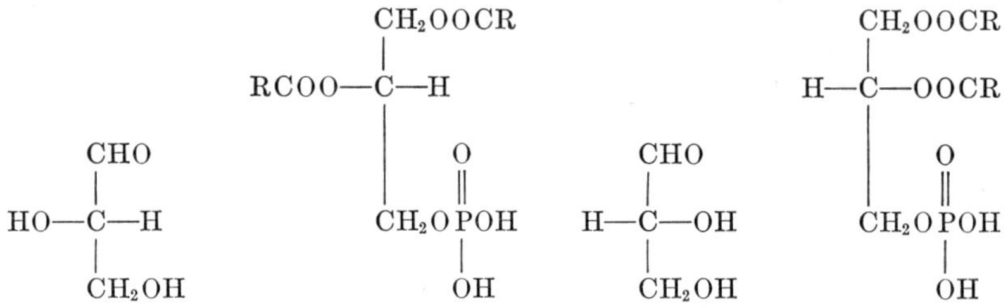

L-Glyceraldehyde | L-Phosphatidic acid | D-Glyceraldehyde | D-Phosphatidic acid

XLIX. Stereoisomerism among the phosphatidyl compounds

tidyl compounds, like the naturally-occurring amino acids and unlike the naturally-occurring monosaccharides, belong to the L series (10). The compound shown in figure XLVII is, in actual fact, L-alpha-phosphatidyl choline.

Phosphatidyl ethanolamine (formula L) and *phosphatidyl serine* (formula LI) have structures similar to the lecithins except that choline is replaced by ethanolamine (colamine) (formula LII) and serine, respectively. In such

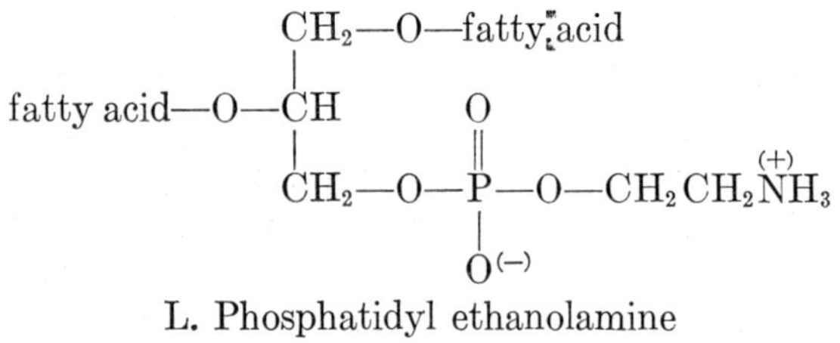

$$\begin{array}{l}
\qquad\qquad\qquad \mathrm{CH_2{-}O{-}fatty\ acid} \\
\qquad\qquad\qquad | \\
\mathrm{fatty\ acid{-}O{-}CH} \qquad\quad \mathrm{O} \\
\qquad\qquad\qquad | \qquad\qquad\quad \| \qquad\qquad\qquad\qquad (+) \\
\qquad\qquad\qquad \mathrm{CH_2{-}O{-}P{-}O{-}CH_2CH_2NH_3} \\
\qquad\qquad\qquad\qquad\qquad\quad | \\
\qquad\qquad\qquad\qquad\qquad\quad \mathrm{O^{(-)}}
\end{array}$$

L. Phosphatidyl ethanolamine

compounds, the two fatty acid residues may be replaced by one long carbon-chain aldehyde such as stearal or palmital, connected by an acetal link (formula LIII). Such compounds are called *acetal phosphatides* or *plasmalogens*.

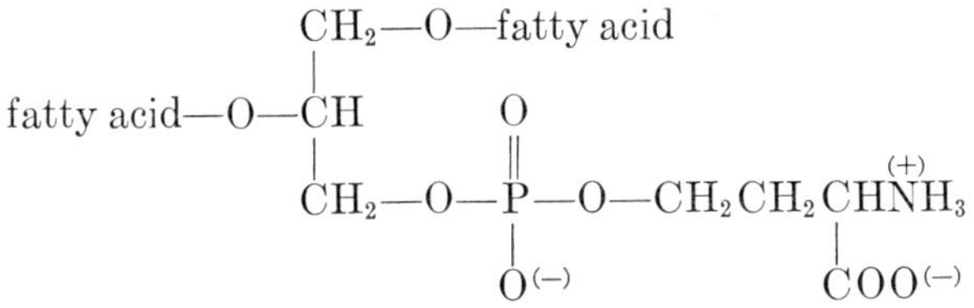

```
                    CH2—O—fatty acid
                    |
fatty acid—O—CH          O
                    |          ||                    (+)
                    CH2—O—P—O—CH2CH2CHNH3
                               |                  |
                               O(−)               COO(−)
```

LI. Phosphatidyl serine

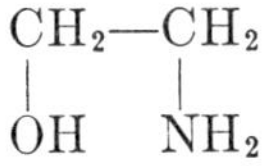

```
CH2—CH2
|        |
OH     NH2
```

LII. Ethanolamine

```
CH2—O
|        \
|         CHR
|        /
CH—O     O
|           ||              (+)
CH2—O—P—OCH2CH2NH3
            |
            O(−)
```

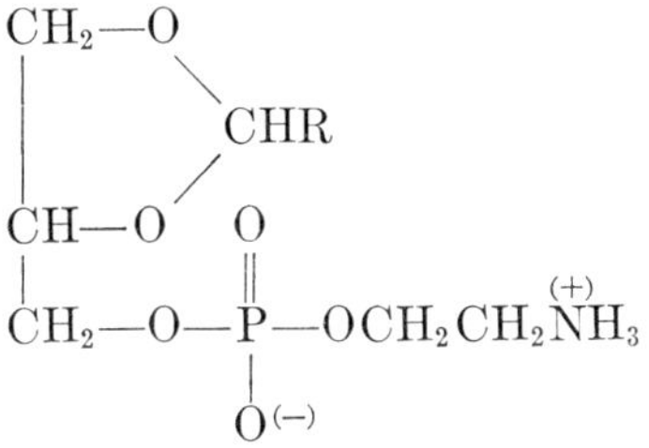

LIII. Plasmalogen

A lecithin in which one of the fatty acids has been hydrolyzed away is known as a *lysolecithin*. An enzyme, lecithinase A, is capable of hydrolyzing the phospholipids to their "lyso" forms removing the fatty acid attached to the alpha carbon. The lysolecithins have strong hemolytic properties and lecithinase A is at least one of the active principles of the venom of bees, scorpions, and some snakes.

The *phosphosphingosides*, in addition to choline, phosphoric acid, and fatty acid, contain the complex amino alcohol, *sphingol* (formula LIV), or

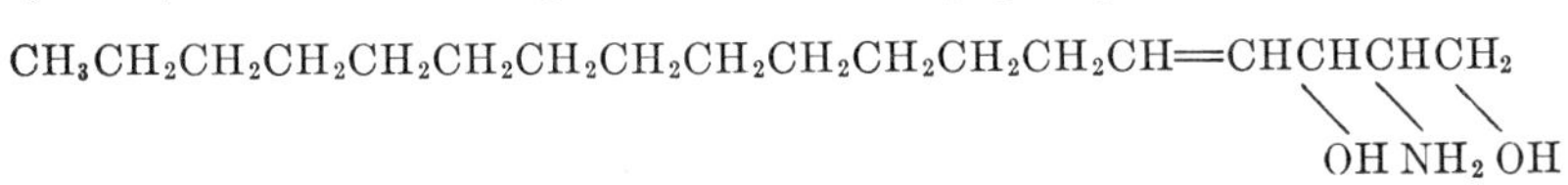

```
CH3CH2CH2CH2CH2CH2CH2CH2CH2CH2CH2CH2CH2CH=CHCHCHCH2
                                                  \    \    \
                                                  OH NH2 OH
```

LIV. Sphingol

sphingosine. The structure of these compounds is not as well established as is those of the phosphoglycerides. The structure of *sphingomyelin* (formula LV) is the best known. It may be roughly considered an analog of phophatidyl choline, with sphingol replacing glycerol. The stereochemical structure of sphingol is analogous to that of the four-carbon sugar, D-erythrose, with a *trans* configuration about the double bond (formula LVa) (2a)

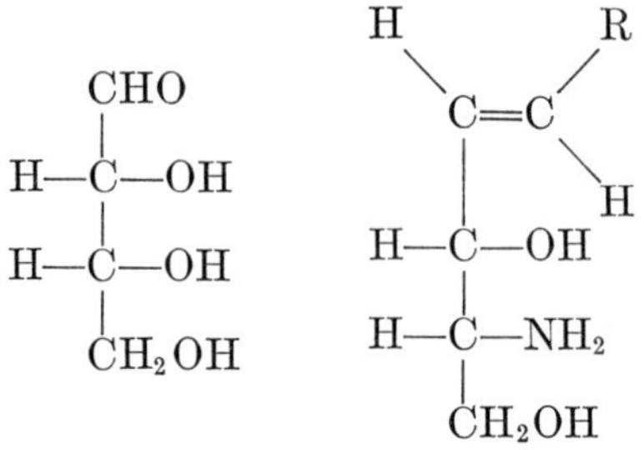

$$\begin{array}{ccc} & & \\ & \text{CHO} & \\ & | & \\ \text{H—} & \text{C} & \text{—OH} \\ & | & \\ \text{H—} & \text{C} & \text{—OH} \\ & | & \\ & \text{CH}_2\text{OH} & \end{array} \qquad \begin{array}{ccc} \text{H} & & \text{R} \\ & \diagdown \quad \diagup & \\ & \text{C=C} & \\ & | \quad \diagdown & \\ & | \qquad & \text{H} \\ \text{H—} & \text{C} & \text{—OH} \\ & | & \\ \text{H—} & \text{C} & \text{—NH}_2 \\ & | & \\ & \text{CH}_2\text{OH} & \end{array}$$

D-Erythrose Sphingol

LVa-Configuration of sphingol

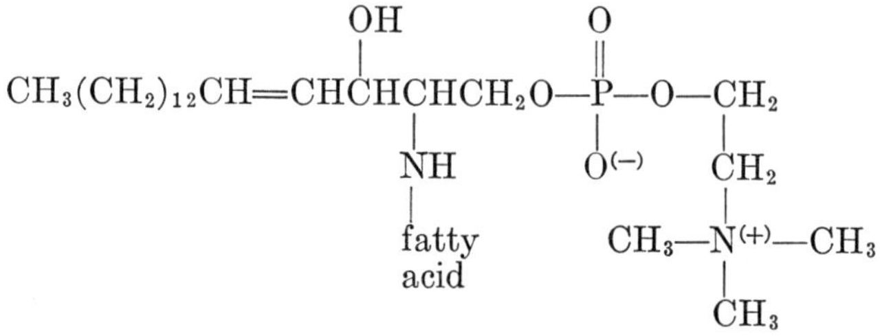

$$\begin{array}{l} \qquad\qquad\qquad\qquad\quad \text{OH} \qquad\qquad\quad \text{O} \\ \qquad\qquad\qquad\qquad\quad | \qquad\qquad\qquad \| \\ \text{CH}_3(\text{CH}_2)_{12}\text{CH=CHCHCHCH}_2\text{O—P—O—CH}_2 \\ \qquad\qquad\qquad\qquad\qquad\;\; | \qquad\qquad\quad | \qquad\quad | \\ \qquad\qquad\qquad\qquad\qquad \text{NH} \qquad\quad \text{O}^{(-)} \quad \text{CH}_2 \\ \qquad\qquad\qquad\qquad\qquad\;\; | \qquad\qquad\qquad\qquad | \\ \qquad\qquad\qquad\qquad\qquad \text{fatty} \qquad \text{CH}_3\text{—N}^{(+)}\text{—CH}_3 \\ \qquad\qquad\qquad\qquad\qquad \text{acid} \qquad\qquad\qquad | \\ \qquad\qquad\qquad\qquad\qquad\qquad\qquad\qquad\quad \text{CH}_3 \end{array}$$

LV. Sphingomyelin

Phosphoinositides are characterized by the presence of *meso*-inositol in the molecule. (4) The phosphoinositide of brain has inositol, phosphoric acid, glycerol and fatty acid present in the proportions of 1:2:1:1.

Glycolipids. The most important of the compound lipids which do not contain phosphoric acid are the glycolipids. These contain fatty acid, sphingol and a monosaccharide. The glycolipids isolated from brain contain galactose; those isolated from spleen contain glucose. An older name for the glycolipids, reflecting their occurrence in nerve tissue, is *cerebrosides.* Four fatty acids are known to occur in glycolipids. One is the normal 24-carbon lignoceric acid. The other three are cerebronic, nervonic and oxynervonic acids, which are also 24-carbon acids, differing from lignoceric acid in the presence of a 2-hydroxy group, a double bond between carbons 15 and 16, and both, respectively. The individual glycolipids contain fatty acid, sphingol and monosaccharide in equimolar proportions. They differ structurally from the phosphosphingosides in that the monosaccharide unit replaces the phosphorylated choline on the terminal carbon of sphingol. The *gangliosides,* which are closely related substances, contain in addition, neuraminic acid which behaves chemically like a condensation product of aminohexose with pyruvic acid. For a proposed formula, see (6).

Strandin, a brain component closely related to the gangliosides, has a molecular weight over 250,000. Its repeating unit contains fatty acid, carbohydrate, a substance resembling sphingol, and an unidentified substance (5).

One sulfur-containing lipid, *cerebron sulfuric acid*, has been isolated and found to be a sulfate ester of a glycolipid. Other lipids which contain sulfur are present in brain, but have not yet been identified.

Solubility characteristics of the compound lipids. The difference in the chemical structure of the compound lipids as compared with the simple lipids is reflected in their solubility behavior. The simple lipids, composed as they are of hydrocarbon chains primarily, are non-polar compounds (see Chapter 7) and therefore highly insoluble in water, or other polar solvents such as alcohol. They are, on the other hand, readily soluble in solvents which are themselves non-polar, such as ether, chloroform, or benzene. The compound lipids, however, while containing hydrocarbon chains in the form of fatty acids, sphingol, or both, contain in addition such highly polar fragments as phosphoric acid, choline, ethanolamine, inositol, galactose, or combinations thereof. These polar fragments, while insufficient to convert the molecule as a whole into a water-soluble substance, nevertheless exert a pronounced effect in that direction.

Since the compound lipid molecules contain both water-soluble groups (choline, ethanolamine, serine, inositol, monosaccharide) and oil-soluble groups (fatty acids, sphingol), they are surface active and tend to concentrate at cell surfaces and along cell membranes. It is significant that even in cases of starvation and extreme emaciation, the compound lipid of the body remains relatively untouched and is not used for energy production even though much of the body protein has been thrown into the furnace in order to maintain life.

Associated Lipids

Of those lipids which are not esters, the various steroids are the most important. A *steroid* may be defined as an oxygen-containing derivative of cyclopentanoperhydrophenanthrene (CPP) (formula LVI). Despite the apparent complexity of CPP, such is the medical importance of its

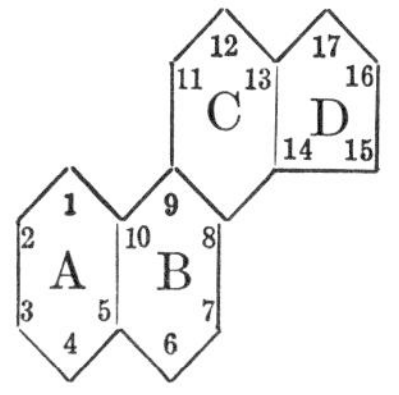

LVI. Cyclopentanoperhydrophenanthrene (CPP)

various derivatives that the student will do well to learn not only its structure, but also the numbering system used to identify the individual carbons.

Sterols. Sterols include those steroids which contain an OH group attached directly to the CPP nucleus and are otherwise hydrocarbon. One such, cholesterol (formula LVII), has already been mentioned in the

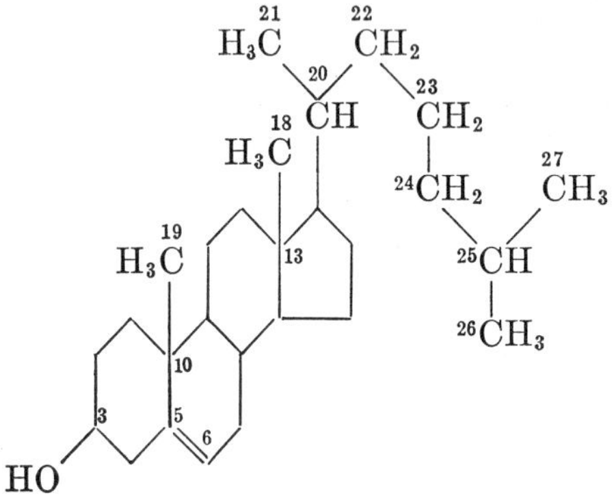

LVII. Cholesterol

section on waxes. Note here that the CPP nucleus is modified in three ways: (a) there is a double bond between carbons 5 and 6; (b) there is the characteristic hydroxyl group of sterols on carbon 3; and, (c) there are hydrocarbon side chains on carbons 10, 13, and 17. Of the side chains, those on carbons 10 and 13 are methyl groups, referred to because of their peculiar position on the angle between two rings as *angular methyls*. These are characteristic of steroids in general and are absent only in a few cases. The third side chain which is, in the case of cholesterol, an iso-octyl group, shows considerably more variability among the different steroids. The numbering system of the CPP nucleus is extended to the carbons of the side chains in the manner shown in the cholesterol formula.

Cholesterol is by far the most commonly-occurring sterol in the human body. Its occurrence in blood serum and in sebum, in ester form, has been mentioned. In addition, it occurs in the free state in high concentration in nerve tissue, liver and adrenal cortex. Bile also is rich in cholesterol. The presence of cholesterol in the skin is significant since it is the precursor of 7-dehydrocholesterol, which accompanies it in small quantities and is itself the precursor of vitamin D_3. It is accompanied, to the extent of 2 per cent, by its dihydro derivative, beta-cholestanol (formula LVIII). Another dihydro derivative, coprostanol (formula LIX), occurs in the feces. It will be noted that the formulas for these two compounds seem to differ only in that one bond is dotted in one case and solid in the other, representing respectively bonds which are directed below and above the plane of the paper. This form of isomerism has not been encountered previously in this book, and merits some discussion.

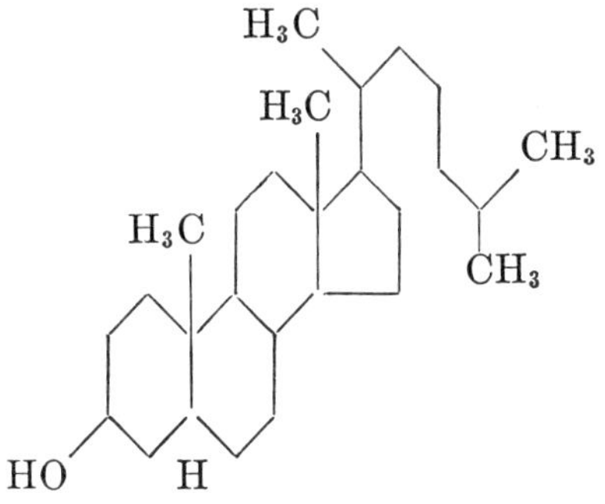

LVIII. Beta-Cholestanol

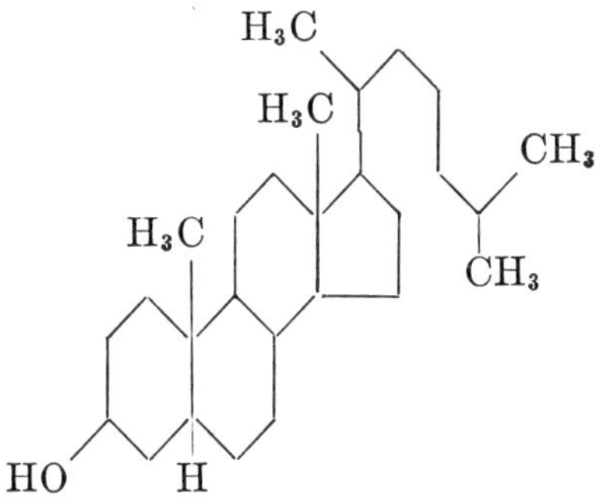

LIX. Coprostanol

Cis-trans isomerism. Mention has already been made of the fact that the four bonds of a carbon atom are not distributed in a single plane as would appear in the usual blackboard representation, but that two bonds are located in a plane perpendicular to that in which the other two bonds are found. Under these circumstances a molecule of ethane may be represented as shown in formula LX—where the carbon-carbon bond is in

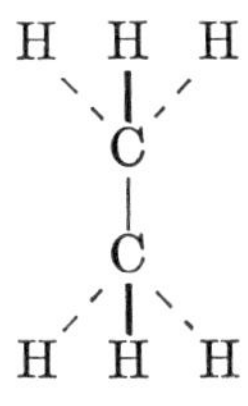

LX. Ethane

the plane of the paper, the bold-face bonds are visualized as projecting upwards from the paper and the dotted bonds projecting downwards into the paper. If, in such a model, one hydrogen on each carbon is replaced by a chlorine, so that ethylene dichloride is formed, it will be seen that each chlorine can be placed on any one of three bonds, making a total of nine combinations. Three of these are not superimposable if the molecule is viewed as a rigid unit (formula LXI), so that at least three varieties of

H Cl H — C — C — H Cl H H Cl H — C — C — Cl H H H Cl H — C — C — H H Cl

LXI. Three non-superimposable forms of ethylene dichloride if the molecule is viewed as a rigid unit

ethylene dichloride should be distinguishable. Actually only one kind of ethylene dichloride is known. The reason for this is that "free rotation" is possible about the carbon-carbon single bond. By free rotation is meant that the amount of energy required to impart rotation to a group about a particular bond is so low that the kinetic energy derived from the normal thermal motions of molecules at room temperature is sufficient. If rotation about the carbon-carbon single bond of ethylene dichloride is assumed, then each of the three "varieties" shown can be made to superimpose so that only one variety will in fact exist, which is in accord with experimental findings.

Where rotation about a bond is restricted, however, cases of isomerism due to different arrangements of substituents become possible. The energy required for rotation about a carbon-carbon double bond, for instance, is far more than can be acquired from thermal motions at room temperature. The molecule of ethylene can be pictured as lying entirely in the plane of the paper, with the two CH_2 groups incapable of rotating out of it. This rigidity means that two non-superimposable forms of dichloroethylene can exist. This form of isomerism is known as *cis-trans isomerism*, the *cis* form being that in which the two substituents in question are on the same side of the molecule, and the *trans* form that in which they are on opposite sides (formula LXII).

H Cl — C = C — H Cl H Cl — C = C — Cl H

cis-Dichloroethylene *trans*-Dichloroethylene

LXII. *Cis-trans* isomerism

Cis-trans isomerism differs from optical isomers in that *cis* and *trans* forms are not optically active, involve no asymmetric carbon atoms and are not mirror images of one another. Unlike optical isomers, *cis-trans* isomers usually differ considerably in their chemical and physical properties. The *trans* form is nearly always the more stable of the two, and usually has the higher melting and boiling point.

An example of *cis-trans* isomerism among substances of biochemical interest is found in the case of oleic acid (formula LXIII), which has the *cis*

```
                              H       H
                               \     /
                                C=C                              O
                               /     \                          //
CH3CH2CH2CH2CH2CH2CH2CH2              CH2CH2CH2CH2CH2CH2CH2C
                                                                 \
                                                                  OH
```

LXIII. Oleic acid

configuration. The equivalent molecule in *trans* configuration, *elaidic acid* (formula LXIV), can be formed from oleic acid by the action of nitrogen

```
                                                                 O
                                                                //
                              H       CH2CH2CH2CH2CH2CH2CH2C
                               \     /                          \
                                C=C                              OH
                               /     \
CH3CH2CH2CH2CH2CH2CH2CH2              H
```

LXIV. Elaidic acid

trioxide, N_2O_3, or by heating in the presence of selenium. The differences between the two compounds are quite marked. Whereas oleic acid melts at 14°C., elaidic acid has a melting point at 45°C. The *cis* configuration is considerably more common than the *trans* configuration in naturally-occurring unsaturated fatty acids.

Cis-trans isomerism is not restricted to compounds containing a double bond. Rotation about a carbon-carbon single bond may be restricted if that bond forms part of a ring. In the case of decahydronaphthalene (decalin), for instance, two forms exist depending on the arrangement of the bonds about the carbon-carbon single bond held in common by the two rings. In *cis*-decalin, the two bonds *a* and *b* both emerge on the same side of the plane of the paper, while in *trans*-decalin they emerge on opposite sides (formula LXV).

In beta-cholestanol, rings A and B of the CPP nucleus are fused in the manner of *trans*-decalin, while in coprostanol the fusion is in the manner of *cis*-decalin and that is the only difference between them. Cholesterol, itself, is neither *cis* nor *trans* in this respect since the double bond between carbons 5 and 6 removes the possibility of a substituent hydrogen at carbon 5. Steroids in which rings A and B are *cis* are said to belong to the normal

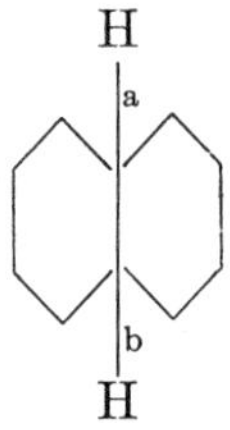

cis-Decalin

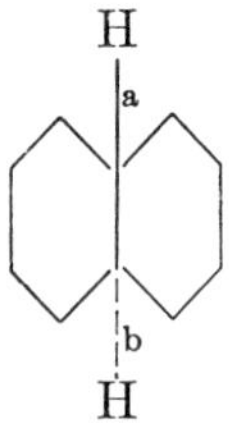

trans-Decalin

LXV. *Cis-trans* configuration in ring structures

series, while those in which rings A and B are *trans* belong to the *allo* series. An alternate name for beta-cholestanol would thus be *allo*-coprostanol. Similar isomerism can exist between other rings in the CPP nucleus, but in naturally-occurring compounds only A-B orientation is ever found to vary.

In the case of the hydroxyl on carbon 3, the group may project either below or above the plane of the paper, the two isomers being distinguished arbitrarily as alpha or beta hydroxy compounds. Where the hydroxyl bond is on the same side of the plane of the paper as the carbon 19 bond, it is arbitrarily assumed to lie above the plane and given the beta configuration. Cholesterol and all other natural sterols belong in this group. Sterols in which the hydroxyl and the carbon 19 bond are *trans* are said to have the hydroxyl extending below the plane of the paper in the alpha position. These latter belong to the *epi* series. Portions of cholesterol and *epi*-cholesterol are shown in figure LXVI. Digitonin, a compound used in the determination of cholesterol will not precipitate *epi*-compounds.

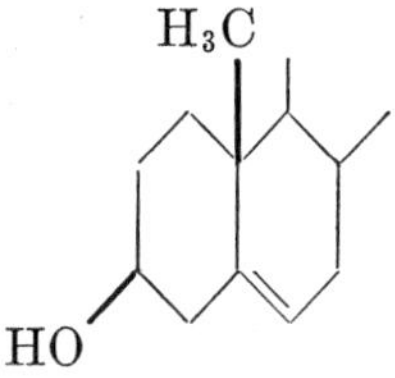

Cholesterol
(Beta hydroxyl)

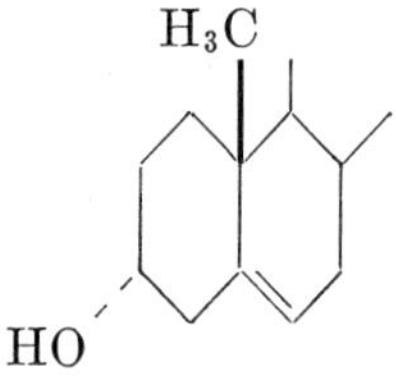

epi-Cholesterol
(Alpha hydroxyl)

LXVI. Alpha and beta configurations in steroids

The importance to the medical student of the phenomenon of *cis-trans* isomerism lies in the fact that the body can distinguish between such isomers and use one in preference to the other. Thus, while beta-cholestanol is found in tissues, its *cis*-isomer, coprostanol, is not. Again, in the steroid hormones (see Chapter 8), trace quantities of a given compound may have profound effects on body chemistry, while an isomer involving the geometric configuration of a single bond may be nearly or entirely devoid of action. For additional details on steroid isomerism, see a review by Shoppee (12).

Other steroids. Steroids other than sterols are best discussed in those chapters applicable to their functioning in the human body. The *bile acids*, discussed in Chapter 13, are carboxylated steroids. *Steroid hormones*, distinguished by the presence of two or more hydroxy or keto groups on the CPP nucleus, are taken under consideration in Chapter 8. The *D vitamins*, while not steroids themselves, since they do not possess an intact CPP ring system, are derived physiologically from sterol precursors, and are dealt with in Chapter 21. Bacteria, yeasts, and filamentous fungi accomplish numerous alterations in the structure of steroids. Such microbe activity has been utilized in the production of steroids of definite structures associated with desirable pharmacological effects (12a).

Certain steroids which occur only in plants in the form of glycosides are of interest to the physician because of their pharmacological properties. The chief of these occurs among the active principles of digitalis.

Isoprenoids. Among the non-steroid associated lipids are the isoprenoids, which are structurally related to isoprene (formula LXVII). An

$$\begin{array}{c} \quad CH_3 \\ \quad | \\ CH_2{=}C{-}CH{=}CH_2 \end{array}$$

LXVII. Isoprene

example is the hydrocarbon carotene ($C_{40}H_{56}$), which occurs as a mixture of two closely related substances—alpha- and beta-carotene (formulas LXVIII and LXIX), plus a small amount of still a third isomer, gamma-carotene (formula LXX). The relation ship of these compounds to isoprene is readily seen from a consideration of the formulas. Carotene of dietary origin is found in the human body, and because of its intense color (due to its possession of numerous conjugated double bonds) it contributes notably to the pigmentation of skin, corpus luteum and blood serum. The various double bonds of the isoprenoids each offers the possibility of *cis-trans* isomerism, a fact which is of significance in the chemistry of vision (see page 165).

Carotene is a precursor of vitamin A. The other fat-soluble vitamins, vitamins E and K, have isoprenoid side chains. For discussions of all of these, see Chapter 21.

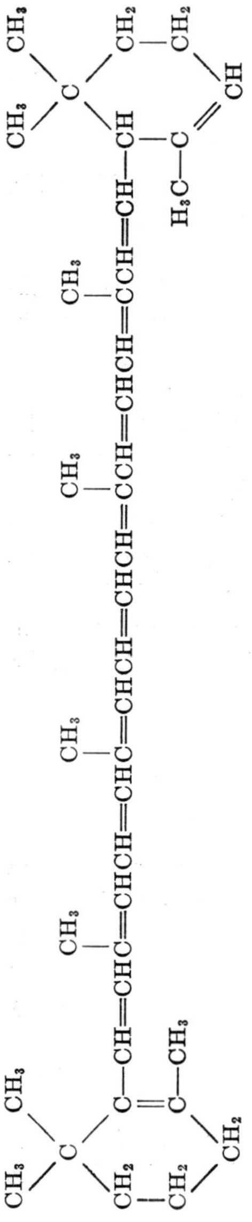

C—CH=CHC=CHCH=CHC=CHCH=CHCH=CCH=CHCH=CCH=CH—CH

LXVIII. α-Carotene

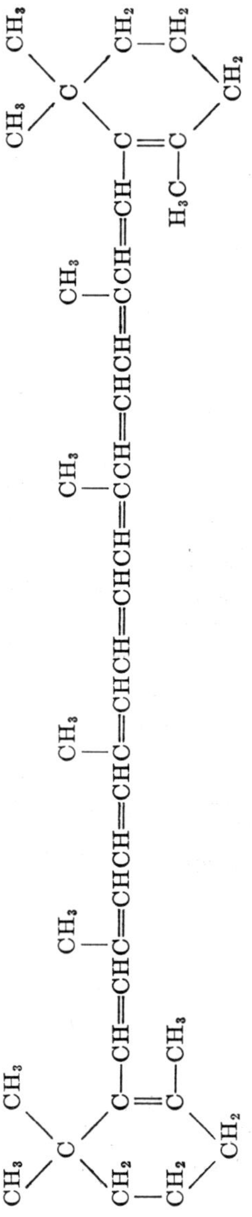

C—CH=CHC=CHCH=CHC=CHCH=CHCH=CCH=CHCH=CCH=CH—C

LXIX. β-Carotene

C—CH=CHC=CHCH=CHC=CHCH=CHCH=CCH=CHCH=CCH=CH—CH

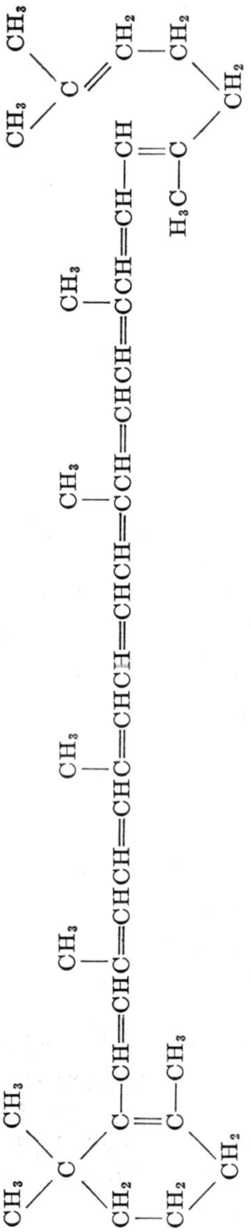

LXX. γ-Carotene

Squalene (formula LXXI) another isoprenoid hydrocarbon occurs in small amount in all fat deposits of the body and in sebum up to 8 per cent.

$$
\begin{array}{l}
\quad\;\; CH_3 \qquad\qquad\quad CH_3 \qquad\qquad\quad CH_3 \\
\quad\;\; | \qquad\qquad\qquad\;\; | \qquad\qquad\qquad\;\; | \\
CH_3C{=}CHCH_2CH_2C{=}CHCH_2CH_2C{=}CHCH_2 \\
\qquad\qquad\qquad\qquad\qquad\qquad\qquad\qquad\quad | \\
CH_3C{=}CHCH_2CH_2C{=}CHCH_2CH_2C{=}CHCH_2 \\
\quad\;\; | \qquad\qquad\qquad\;\; | \qquad\qquad\qquad\;\; | \\
\quad\;\; CH_3 \qquad\qquad\quad CH_3 \qquad\qquad\quad CH_3
\end{array}
$$

LXXI. Squalene ($C_{30}H_{50}$)

OTHER UNIVERSAL CELL CONSTITUENTS

In addition to proteins, carbohydrates, and lipids, one other group of substances may be found in considerable quantities in all living cells. These are the *nucleic acids*, usually found in conjunction with protein in both nucleus and cytoplasm. Such is the importance of these substances that it is thought wise to devote an entire chapter (Chapter 9) to them. They are phosphorus-containing compounds, the molecules of which rival in size and complexity of structure the proteins themselves.

Cells also contain a number of other substances in very low concentrations, yet indispensable to proper function. These include the B vitamins, for instance, which will be considered individually in Chapter 21. The mineral constituents of cells are considered in Chapter 17.

Lastly, as is perhaps needless to emphasize, there is water. In terms of sheer bulk it is the predominant constituent of the body.

REFERENCES

1. Artz, N. E., and Osman, E. M. *Biochemistry of Glucuronic Acid.* New York, Academic Press, 1950.
2. Baer, E., and Kates, M. Migration during hydrolysis of esters of glycerophosphoric acid. I. The chemical hydrolysis of L-α-glycerylphosphorylcholine. J. Biol. Chem., **175**: 79–88, 1948.

2a. Carter, H. E. *et al.* Chemistry of the sphingolipides. Can. J. Biochem. Physiol. **34**: 320–330, 1956.

3. Deuel, H. J. *The Lipids: Their Chemistry and Biochemistry.* New York, Interscience Publishers, 1951.
4. Folch, J. and LeBaron, F. N. Chemistry of the phosphoinositides. Can. J. Biochem. Physiol., **34**: 305–319, 1956.
5. Folch, J., *et al.* Isolation of brain strandin, a new type of large-molecule tissue component. J. Biol. Chem., **191**: 819–831, 1951.
6. Gottschalk, A. Structural relation between sialic acid, neuraminic acid and 2-carboxypyrrole. Nature, **176**: 881–882, 1955.
7. Hanahan, D. J. Positional asymmetry of fatty acids on lecithin. J. Biol. Chem., **211**: 313–319, 1954.
8. Hilditch, T. P. *The Chemical Constitution of Natural Fats.* 2nd edit. New York, John Wiley & Sons, Inc., 1947.

9. JEANLOZ, R. W., AND FORCHIELLI, E. Studies on hyaluronic acid and related substances. J. Biol. Chem., **190**: 537–546, 1951.
10. LONG, C., AND MAGUIRE, M. F. Structure of naturally-occurring phosphoglycerides. I. Evidence derived from alkalin-hydrolysis studies. Biochem. J., **54**: 612–617, 1953.
11. PIGMAN, W. W., AND GOEPP, R. M., JR. *Chemistry of the Carbohydrates*. New York, Academic Press, 1948.
11a. ROWEN, J. W. *et al.* Form and dimensions of isolated hyaluronic acid. Biochim et Biophys. Acta, **19**: 480–489, 1956.
12. SHOPPEE, C. W. Steroid configuration. Vitamins and Hormones, **8**: 255–308, 1950.
12a. SHULL, G. M. Transformations of steroids by molds. Trans. N. Y. Acad. Sci., Ser. II, **19**: 147–172, 1956.
13. WOLFROM, M. L., *et al.* The structure of heparin. J. Am. Chem. Soc., **72**: 5796–5797, 1950.
14. ZIMM, B. H., AND THURMOND, C. D. The molecular weight of amylopectin. J. Am. Chem. Soc., **74**: 1111–1112, 1952.

CHAPTER 4

Tissue Chemistry

A complete chemical analysis of a fresh adult human cadaver, well nourished, in the prime of life, and without significant disease or abnormality has been reported by Forbes *et al.* (12). The over-all results are presented in table IV-1, modified from their paper. Note that carbohydrate makes no sizable contribution to the body make-up.

The body protein content of rats, chicks and pigs has been analyzed for amino acid composition at various stages of growth by Williams *et al.* (55). The differences in relative concentrations of amino acids among the species tested were not very great (see table IV-2), and the over-all composition of human protein is doubtless similar. Serine, alanine, aspartic acid, proline and cysteine are not given for any of the species listed. The figures for glycine and glutamic acid are given for chicks and are 135 and 176, respectively.

Harrison (16) reports the composition of rat liver cells. The diameter of a liver cell in a male rat is given as 15 microns (that is, 0.015 mm.). Its volume is 1.9×10^{-9} ml. and its mass 2.03×10^{-3} micrograms. The nucleus makes up 8 per cent of the cell mass. The mass of the major constituents of such a cell is given in table IV-3. It should be noted that even the trace substances are atomically abundant. For instance, the mass of copper in a single liver cell is equivalent to one hundred million copper atoms. The one significant difference between liver cells of the male and female rat is that the quantity of iron in the liver cell of the female is three times that in the liver cell of the male. This may be related to the fact that the female mammal must supply its new-born infant with a store of iron sufficient to last it during that period of time when its nourishment is chiefly milk.

CONNECTIVE AND SUPPORTING TISSUE

Connective and supporting tissue performs its functions through the possession of three types of substances of widely divergent nature: (a) inorganic structures, (b) albuminoids, and (c) mucopolysaccharides and mucoproteins.

TABLE IV-1

Chemical composition of adult human body (12)

ORGAN	WATER	FAT	PROTEIN	ASH
	%	%	%	%
Skin	57.71	14.23	27.33	0.62
Skeleton	28.17	25.04	19.71	26.62
Striated muscle	70.09	6.60	21.94	1.01
Brain and nerve	75.09	12.35	11.50	1.37
Liver	71.58	3.11	22.24	1.35
Heart	62.95	16.58	17.48	0.61
Lungs	77.28	1.32	19.20	1.03
Kidneys	70.58	7.18	19.28	0.87
Alimentary tract	77.40	9.17	12.77	0.53
Adipose tissue	23.02	71.57	5.85	0.20
Whole Body	55.13	19.44	18.62	5.43

Inorganic Components

The distinctive hardness and rigidity of bones and teeth are due to their inorganic content. Although organic material is the continuous matrix of the bones and teeth, that matrix is reinforced by inorganic salts. The final inorganic content of bone at maturity is about 45 per cent. Mineralization proceeds even further in the dentine and cement of teeth, where about three-fourths of the total is inorganic, and reaches an extreme in enamel where some 98 per cent is inorganic. Ninety-nine per cent of the body's calcium, and 90 per cent of its total mineral content occur in the bones

TABLE IV-2

Relative proportions of amino acids present in whole body protein (55)

	RATS	CHICKS	PIGS
Lysine	100	100	100
Leucine	85	89	84
Arginine	77	90	83
Valine	72	90	70
Threonine	51	54	44
Phenylalanine	48	53	44
Isoleucine	46	55	45
Tyrosine	38	33	30
Histidine	28	26	31
Methionine	22	24	21
Cystine	20	24	12
Tryptophane	10	10	9

TABLE IV-3

Constituents of the liver cell of the male rat (16)

SUSTANCE	MASS (IN 10^{-4} MICROGRAMS)
water	12.5
protein	4.35
glycogen	1.17
neutral lipids	0.41
phospholipids	0.76
RNA	0.26
DNA	0.06
potassium	0.85
iron	0.0035
zinc	0.0011
copper	0.00011

and teeth. The inorganic content of bones and teeth consists essentially of positive calcium ions and negative phosphate and carbonate ions, the phosphate/carbonate molar ratio being about 2.25. The whole is arranged in a crystal lattice similar to that of inorganic minerals known as apatites. Details on the structure and quantitative ionic content of bones and teeth will be found in Chapter 17.

Albuminoids

As a class, albuminoids are highly inert, either to solution or to digestion, and possess the ability to form fibers of considerable toughness and tensile strength. Usually they are composed of fewer kinds of amino acids than are the typical globular proteins, and the simpler amino acids predominate. The most abundant such protein in connective tissue is *collagen.* Fully one third of the organic matter of tendons consists of this protein. It, or proteins similar to it, may be found in cartilage, bone, and ligament as well, and in those portions of the teeth other than enamel. Collagen extracted from rat-tail tendon by dilute acetic acid was found to have a molecular weight of about 700,000. If the molecules are assumed to be prolate spheroids, their dimensions are 500 by 1.8 millimicrons. This extraordinary asymmetry is typical of fibrous proteins (41).

Collagen is unusually high in the simple amino acids, glycine and alanine, and in the imino acids, proline and hydroxyproline. It is remarkable in its content of hydroxyproline, which is found in no proteins other than those in connective tissue. Hydroxylysine is another amino acid found only in collagen (to the extent of 1 per cent) and related proteins. Methionine is present in small quantities only, while tryptophane is completely absent. Since these amino acids belong to the group which is dietarily essential (see

Chapter 12), the edible protein derived from collagen (i.e., gelatin) has serious nutritional shortcomings.

The dentine protein of human teeth has an amino acid pattern similar to the collagen of connective tissue and it is, therefore, considered a collagen. On the basis of the solubility properties of protein derived from human bones as well as those of rabbits and oxen, it can be concluded that 89 to 97 per cent of bone protein is collagen (45).

Although collagen is an inert and tough material, it is converted in boiling water to *gelatin*. Gelatin is a derived protein with no albuminoid properties. It would seem that interpolypeptide linkages can be broken during boiling and that gelatin consists of single polypeptide chains, with an average molecular weight of 60,000 (7). This increased simplicity of physical structure as compared with collagen probably explains the greater solubility of gelatin and the ready digestibility of gelatin by pepsin or trypsin. Gelatir is not coagulated by heat, withstanding boiling water without loss of any of its characteristic properties. This is not an indication of any extraordinary resistance to denaturation, but rather a sign that gelatin is already as denatured as it can be.

Reticulin fibers have properties similar to collagen. Reticulin differs from collagen in having fatty acid (mostly myristic acid, a saturated 14-carbon acid) as an integral part of the molecule (56). The lipid portion of reticulin may explain its stability to boiling water and its resistance to acid hydrolysis. Reticulin is easily soluble in hot NaOH, which could be explained by the saponification of the lipid.

Elastin is another albuminoid of connective tissue, which resembles collagen in its properties. It differs from collagen physically in that it is not converted to gelatin or a gelatin-like substance on heating with boiling water. Chemically, the amino acid distribution varies considerably. Collagen is richer in amino acids with polar side chains such as arginine and lysine among the basic amino acids, and aspartic acid and glutamic acid among the acidic ones. Elastin is higher in such neutral acids with non-polar side chains as proline and valine (13). The distribution of these two albuminoids is by no means identical. Collagen, considerably the more common of the two in the body, is found in particularly high concentrations in tendons and in white connective tissue generally, while elastin is found in higher concentration in ligaments and yellow connective tissue. Thus, on a dry weight basis, the *tendo calcaneus* of man is 96 per cent collagen and 2 per cent elastin (31), while the *ligamentum nuchae* of the ox is 75 per cent elastin and 17 per cent collagen.

Mucopolysaccharides and Mucoproteins

The elastic fibrils of collagen and elastin are embedded in a matrix composed of mucoprotein—that is, protein containing as a prosthetic group a mucopolysaccharide. Meyer has suggested the following terminology (35):

1. *Mucopolysaccharides* are polysaccharides containing a hexosamine (usually acetylated) as one component. The hexosamine involved is almost always either 2-aminoglucose (glucosamine) or 2-aminogalactose (galactosamine or chondrosamine).

a. *Acid mucopolysaccharides* are those which, in addition to hexosamine, contain uronic acid groups, sulfate groups, or both. Hyaluronic acid is an example.

b. *Neutral mucopolysaccharides* do not contain acid groups. One such is *chitin*, which is the principal constituent of the exoskeletons of arthropods. On hydrolysis, chitin yields only N-acetylglucosamine (formula I).

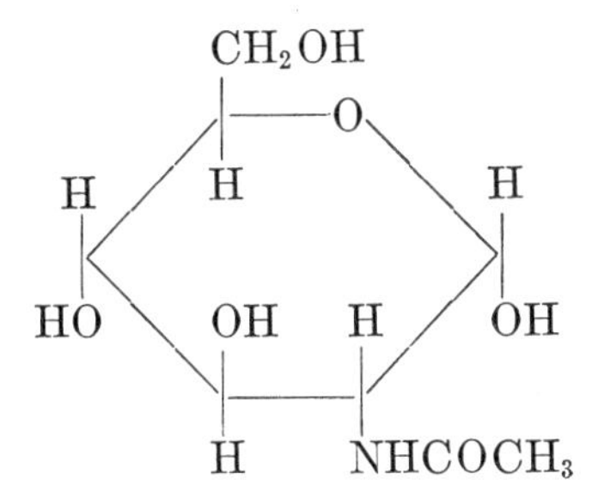

I. N-Acetylglucosamine

2. *Glycoproteins* are proteins whose molecules contain a mucopolysaccharide as a prosthetic group but with hexosamine making up less than 4 per cent of the molecule. According to this definition, glycoproteins would include many proteins usually considered to be "simple proteins", such as egg albumin, serum albumin and serum globulin.

3. *Mucoproteins* are proteins whose molecules contain an acid mucopolysaccharide as a prosthetic group and with hexosamine making up more than 4 per cent of the molecule. The prosthetic group of such proteins is easily split away by changes in salt concentration or pH.

4. *Mucoids* are proteins whose molecules contain a neutral mucopolysaccharide as a prosthetic group and with hexosamine making up more than 4 per cent of the molecule. The prosthetic group of such proteins is bound by firm covalent links and is not easily split off.

The mucoproteins that occur in the various connective and supporting tissues resemble one another closely. All possess the prosthetic group, *chondroitin sulfuric acid*. Individual collagen fibers are surrounded and held together by a layer of mucoprotein (24). There is evidence for chemical bonds between mucoprotein and collagen, suggesting that mucoprotein or a component thereof may have a place in the structure of collagen fibrils (18). Free condroitin sulfuric acid also forms linkages with collagen, estimated at 40 per cent electrovalent and 60 per cent hydrogen bonds, which add to the stability of the collagen fibril. The same may be true of elastin. The mucoprotein/collagen ratio is higher in weight-bearing than in non-weight-bearing cartilage (32).

Most mucoproteins are characterized by high solubility in water and relative resistance towards denaturation. The amount of polysaccharide content necessary to so protect a protein is the basis for the otherwise arbitrary distinction of Meyer's between mucoproteins and glycoproteins.

Mucopolysaccharides in the free state are also found in connective tissue. The chief of these is *hyaluronic acid* (see page 118). In connective tissue, its occurrence would seem to be established by the action of hyaluronidase and by histochemical staining techniques. Intercellular gels of hyaluronic acid act partly as a cement to hold the cells together and partly as viscous barriers in connective tissues, slowing the exchange of metabolites and water. The substance also occurs in the synovial fluid of the joints where it serves as a lubricant and protective agent for their internal surfaces.

Hyaluronidases are enzymes capable of catalyzing the depolymerization of hyaluronic acid, or in other words, of breaking the glycoside bonds and forming end products of relatively small molecular weight. It was originally called "spreading factor" because in its presence injected fluids seemed to permeate tissues at a more rapid rate. It does this by causing the depolymerization and consequent loss of viscosity of the hyaluronic acid between cells and within connective tissue, thus allowing substances to soak through more quickly. Studies of the end products of the enzymatic action of hyaluronidase have yielded further information concerning the structure of hyaluronic acid. Pneumococcal hyaluronidase hydrolyzes hyaluronic acid to a disaccharide consisting of N-acetylglucosamine and glucuronic acid. Testicular hyaluronidase, however, hydrolyzes hyaluronic acid to a tetrasaccharide only. This indicates that there are two different kinds of acetylglucoaminidic bonds in hyaluronic acid, occurring alternately, and that pneumococcal hyaluronidase can hydrolyze both kinds, while testicular hyaluronidase can only attack one (43). The difficulty of obtaining pure preparations of hyaluronidase, however, renders these conclusions doubtful (5). Hyaluronidase is thought to play a part in facilitating entrance of spermatozoa into ova during fertilization (see Chapter 11) and is one of the factors involved in bacterial invasiveness.

Lysozyme is an enzyme noted originally for its lytic effect upon certain bacteria. Some of its sources are nasal mucus, tears, leukocytes, and egg white. Meyer and Hahnel (36) have shown the enzyme to be a mucopolysaccharase similar in its properties to hyaluronidase. Presumably, it lyses bacteria through its hydrolytic action upon the bacterial capsule, which is usually a mucopolysaccharide. Its role as a protective agent against bacterial invasion is, however, uncertain. Most of the bacterial species on which its lytic activity has been demonstrated are non-pathogenic.

SKIN AND ITS APPENDAGES

Skin, like connective tissue, must be tough and flexible to fulfill its purpose. It is even more roughly treated since it withstands directly the buffeting of the outside environment. Like connective tissue it is rich in tough, inert albuminoids and in mucoproteins which act as an intercellular ground substance. The albuminoid which is characteristic of skin is *keratin.*

Keratin

Keratin possesses the properties of collagen to a greater extreme. Like collagen it is insoluble in any solvent that does not attack it chemically, but it lacks even the limited digestibility of collagen. Neither pepsin nor trypsin, both of which will slowly attack collagen, has any appreciable effect on keratin in its natural state. This is not because of any fundamental difference between keratin and other proteins in terms of its chemical make-up, but is rather due to the physical nature of the albuminoid. If keratin is finely ground in a ball mill, not only does it become somewhat digestible but it even becomes appreciably soluble.

Keratin is the characteristic protein not only of skin itself but of the various skin appendages, such as hair, wool, nails, hooves, horns, and feathers. What protein exists in the enamel of teeth is keratin, although that in the dentine and cement is collagen. A form of keratin exists in nerve tissue, a biochemical indication of ectodermal origin. Keratin differs chemically from the albuminoids of connective tissue chiefly in its unusually high content of cystine, human hair containing 15 to 20 per cent of that amino acid. It is the cystine content which imparts the "odor of burning feathers" to burning feathers, skin, hair, or wool. The presence of cystine helps explain the toughness and inertness of keratin. It forms disulfide bridges (see Chapter 2) between adjacent polypeptide backbones, forming a three-dimensional polymer, in which the separate fibrils mutually strengthen one another.

Keratin is also characterized by a relatively constant proportion of the basic amino acids histidine, lysine, and arginine. The ratio of these is 1:4:12 in the keratin of skin appendages. In skin itself, the ratio is somewhat different and the two types of keratin are sometimes referred to respectively as *eukeratin* and *pseudokeratin.* Pseudokeratins contain only about half the percentage of cystine that eukeratins do, which is reasonable if one compares the toughness required of hair or nails with that required of skin itself.

Skin Pigments

One of the most noticeable variations among human beings is that of their external coloring—that is, the extent of pigmentation of the hair, skin, and eyes. These variations in pigmentation are not due to the possession by one group of human beings of any pigment not possessed by others, but rather to differences in the proportions of several pigments, each of which is possessed by all normal humans. Of these, the most important by far is *melanin.*

Melanin. This is a dark brown pigment, insoluble in all ordinary reagents but alkali and only slowly soluble in that. It is a polyphenol polymer synthesized in the body from the amino acid tyrosine. Melanin is present in the epidermis of all men who are not albinos. It is partly in the form of complexes with protein which are referred to as *melanoproteins.* In the classical protein classification these come under the heading of chromoproteins.

The amount of melanin in the skin varies in a roughly direct manner with the average local intensity of solar ultraviolet radiation. Scandinavians are generally "fair", that is, have but little melanin in the skin, while the inhabitants of west and central Africa have skin rich in melanin and are consequently dark skinned. Inhabitants of the Mediterranean shores are intermediate. Melanin acts as a radiation absorber and protects the body from the bad effects of excessive exposure to the ultraviolet light of the sun. The positive survival value of a high melanin content of skin in hot, sunny climates is obvious. The difference between a blond who "burns" at the seashore and a brunette who "tans" is familiar to all.

Under stimulation of sunlight, production of melanin is increased and there is a general darkening of the skin, known as tanning. Where the natural capacity for melanin production is deficient, the darkening is often not uniform but occurs spottily in the form of freckles. There are local increases in melanin-formation under the influences of various normal and pathological changes. Examples are the darkening of the areolar area of the breast during pregnancy, and the increase in melanin in the malignant proliferation of pigment-producing cells. These last are known as *melanomas*, and ordinary moles are benign forms thereof.

Melanin is also partly responsible for hair and eye color. As its concentration is progressively diluted, hair may be black, brown, red, blond, and white with innumerable shades between. Eyes shade through the various intensities of brown to hazel, and eventually blue. The color of blue eyes is due not to the actual color of melanin, but to the fact that it is present in sparsely-distributed particles of the proper size to induce a Tyndall effect, whereby blue light is scattered and partially reflected to a greater extent than the light at the red end of the spectrum.

Individuals are occasionally born without the capacity to form melanin at all. This is probably due to the absence of the enzyme, *tyrosinase*, which catalyzes the first stage of the conversion of tyrosine to melanin. Such individuals are called *albinos* and they may occur in families whose members possess any degree of pigmentation. They are characterized by extremely fair skin, white hair, and unpigmented eyes.

The amount of melanin in the skin in any degree has no known biochemical connection with the presence or absence of any other human characteristics.

Other skin pigments. The role of *carotene*, which is present in small percentages in skin lipid fractions, is second in importance only to melanin. Its presence imparts a yellow coloration to the skin. Where melanin is present in large concentrations the effect is, of course, masked; but among the melanin-poor groups of humans, differences in carotene content may be markedly visible to the naked eye. Human groups in central and east Asia are relatively high in carotene pigmentation, for instance, while those of Europe are relatively low.

The various pigments present in the blood make their contribution to apparent skin color. Of these, hemoglobin and oxyhemoglobin are most important. They impart a ruddy color to the skin which is most noticeable in individuals poor in melanin and carotene. In albinos, the color due to blood itself is the only significant pigmentation effect, so that the eyes, for instance, being devoid of melanin, possess pink irides. Unusual contributions to skin color occur in various pathological states. There is the yellow-green color in jaundice which is due to increased circulating bile pigment, and also pigmentation as the result of intake of chemicals, accidental or otherwise, as in the use of atabrine, the chief antimalarial of World War II. For a discussion of skin pigmentation as an aid to diagnosis see Jeghers (25).

Hair

Hair is eukeratin fiber in an almost pure state. As such it has been much used in physical chemical investigation of protein structure. Hair, in its ordinary state, is a cluster or bundle of partially folded polypeptide chains bound together by disulfide links. Keratin in this state is known as *alpha-keratin.* When soaked in water, hair can be stretched out to about twice its original length and x-ray diffraction studies on stretched hair reveal that the polypeptide backbone has been extended to a simple zig-zag pattern similar to that of silk fibroin. In this state, the protein is *beta-keratin.* If hair is subjected to damp heat, or to alkaline solutions at room temperature, the disulfide links are broken and the polypeptide chain can then fold up to an even greater extent than it normally does. A "supercontracted" form thus makes its appearance, in which the hair shrinks to about 70 per cent of its original length.

It is the last phenomenon which is made use of in "permanent waving". When hair is subjected to damp heat so that disulfide links are broken, individual polypeptide chains hitherto bound together can slide past one another and take up new relative positions if the hair is artificially curled while in this state. The use of an alkaline agent such as ammonia hastens the procedure. If the hair is maintained in the curled position and allowed to dry, disulfide links are again formed between neighboring chains, holding them firmly in the new positions. Permanent waving can be performed without the use of heat if the disulfide groups are broken by reduction to sulfhydryl groups. So called "cold waves" make use of various reducing agents for this purpose, including mercaptans, thioglycolate, and other such compounds.

Ordinary baldness, while not usually a pathological state, is often considered a social handicap, and those suffering from it are frequently quite disturbed about it and seek relief either within or without the borders of medicine. Unfortunately, non-pathological baldness is not yet amenable to treatment and may never be. There are at least three factors involved in baldness. First, there is age. Hair generally decreases in quantity with age in both men and women. Secondly, there are genetic factors. In some men, the hair thins markedly during middle life and sometimes as early as in the twenties. This is apparently a hereditary trait which can not be altered by external treatment by lotions and salves any more than the pigmentation of the eye can be changed by the use of eyewash. Lastly, loss of hair is apparently associated with the presence of androgens. That is, even in the presence of a gene for baldness, hair will not fall out excessively if androgen concentration is comparatively low. Eunuchs, for instance (where castration has occurred before baldness has already appeared), practically never lose their hair. It is through the effect of increased androgen concentration that baldness occurs in adults rather than in children, and in males rather than in females.

Dietary correctives for undesirable hair characteristics have been established for animals. Thus, addition of cystine to the diet increases the weight of wool grown by sheep in a given time. Para-aminobenzoic acid achieved some notoriety as a corrective for premature graying of hair in rats. Neither substance, unfortunately, has as yet proved useful in human beings.

ADIPOSE TISSUE

The lipid content of the human body can be divided into two broad classes. There are first the lipids of tissue cells. These are, in the main, compound lipids and sterols, which are essential to the cell mechanism and are not appreciably consumed during periods of emaciation. They make up less than one per cent of the body weight (26). Secondly, there is *depot fat*, which is composed of fat cells which are in essence merely fat droplets surrounded by a thin shell of protoplasm scarcely more than a membrane. Such fat is almost entirely simple lipid in nature and its function is mainly that of serving as an energy store. Depot fat decreases in quantity during periods of caloric restriction to nearly nothing and may increase in obese individuals to half the total body weight or more. Mayer (33) defines obesity as a situation where more than 30 per cent of the body weight is fat.

About half of the depot fat is deposited in the subcutaneous regions. Considerable quantities are located in the mesentery and about the kidneys. In the latter case it serves to provide mechanical support and protection. Subcutaneous fat is poorly vascularized and is itself a poor heat conductor. It therefore serves as insulation against cold. The chemical nature of depot fat varies to a certain extent with the nature of the fat in the diet. The subcutaneous fat of a dog fed on linseed oil remains liquid at 0°C. as a result of the unusually large quantities of unsaturated acids incorporated therein. When a dog is fed on mutton fat, his adipose tissue becomes firmer and its melting point higher.

Human fat, under normal conditions, is liquid at blood temperature due to a considerable content of oleic acid, linoleic acid, and palmitoleic acid. The three together account for about 60 per cent of the fatty acid content of adipose tissue. The remaining fatty acid content is almost entirely saturated. Palmitic acid is the most common saturated acid, composing about 20 per cent of the total.

MUSCLE

About 40 per cent of the normal human body is muscle, which is thus the largest single tissue component. One quarter of the muscle weight is solid matter and of this, four-fifths is protein. It is the protein content, then, that will first be discussed.

Muscle Protein

Actomyosin. This composes 60 per cent of the protein in muscle. It is often called simply myosin. Actomyosin is the chief, and perhaps the only, protein of the myofibril itself, other muscle proteins occurring in the sarcoplasm. Furthermore, actomyosin is *the* contractile substance in muscle, so that studies of muscle contraction have centered upon it. Actomyosin is a unique protein, combining within itself the solubility characteristics of a typical globulin and the physical structure of an albuminoid. As far as the classical protein classification scheme is concerned, however, solubility properties take precedence and actomyosin is generally included under the globulins.

Solutions of actomyosin show marked birefringence of flow and high viscosity, properties characteristic of highly asymmetric molecules. X-ray diffraction patterns indicate structures similar to those of keratin, an alpha-keratin pattern in normal dried muscle and a beta-keratin pattern in stretched muscle. Contracted muscle is somewhat analogous to "supercontracted" keratin, except that where hair under supercontraction shrinks to 70 per cent of its normal length, muscle will contract to as little as 30 per cent of its normal length.

In the natural state, myofibrils contain actomyosin in the form of long polypeptide chains held together in parallel by subsidiary side chains. These are not the cystine disulfide links of keratin, and are undoubtedly more "fluid", allowing a certain degree of sliding. It is thought that this is due to the fact that the side chains are polar in nature and highly hydrated. In this connection it should be pointed out that actomyosin has a higher percentage of charged or polar side chains than any known protein except the protamines.

The actomyosin micelles within a myofibril have been studied by ultracentrifugation and their size calculated as 55 millimicrons in length and 5 millimicrons in thickness. This is large enough to contain about 40 actomyosin chains, if the molecular weight is taken to be 500,000 (27). Under the electron microscope, actomyosin fibrils show periodic striations at approximately 40 millimicron intervals and the thought arises that the ultimate contractile unit is not the actomyosin fibril as such but a smaller "monomer" out of which the fibril is built up.

Myosin, if exposed briefly to the action of trypsin, is broken up into two types of subunits: *L-meromyosin*, with a molecular weight of 96,000, and *H-meromyosin*, with a molecular weight of 232,000. The "L" and "H" stand for "light" and "heavy". The meromyosins behave like native proteins. Trypsin apparently merely breaks a few peptide links without disturbing inner organization. The myosin molecule may be made up of 4 L-meromyosin and 2 H-meromyosin subunits (28).

Tropomyosin is estimated to have a molecular weight of from 50,000 to 90,000 and can exist in the alpha, beta and supercontracted form. It is tempting to think that it represents the contractile unit of actomyosin and there seems to be evidence that this is so in the case of uterine muscle, at any rate (47). Both actomyosin and tropomyosin show very few terminal amino acids when tested with Sanger's reagent. They do, however, possess C-terminal amino acids (29). Myosin and tropomyosin, for instance, have a C-terminal isoleucine and actin a C-terminal phenylalanine.

The fact that C-terminal amino acids but rarely N-terminal amino acids are present indicates a *sigma-peptide*, one shaped like the Greek letter sigma (σ) which is circular but has one end (the C-terminal end) sticking out. The action of trypsin on myosin (see above) converts the sigma-peptide to open-chain highly asymmetric meromyosins. N-terminal amino acids have been detected and identified in these subunits (28).

Actomyosin is not a homogeneous protein. Straub separated it into two proteins called *actin* and *myosin*. Szent-Györgyi (49) has over many years investigated muscular contraction as a phenomenon related to a reversible complex formation of actin and myosin. The term myosin is thus used in the literature to cover two substances: (a) the contractile protein of muscle, and (b) one component of this contractile protein. Needless to say, a certain confusion has arisen from such practices. In this book the component of the actomyosin complex will be referred to as myosin, while the complex itself will be referred to as actomyosin. It should be pointed out, however, that, on the basis of light-scattering studies, Blum and Morales (2) maintain that the contractile protein of muscle does not dissociate in the process of muscle contraction but remains a single protein throughout. They prefer the older name, myosin, for the protein, therefore.

According to Szent-Györgyi, myosin consists of asymmetric rods 200 to 400 millimicrons long and 2.5 millimicrons wide. Myosin is not markedly fibrous. The viscosity of its solutions is intermediate between those of solutions of globular proteins and fibrous proteins. It is labile and readily denatured. Actin is more stable and possesses properties one would expect of a highly fibrous protein. Its solutions have extremely high viscosities and show pronounced birefringence of flow. One gram of actin can combine with 4.3 grams of myosin (50). This combination probably takes place through the H-meromyosin subunit since that, in isolation, will combine with actin, whereas L-meromyosin will not (51). Neither myosin nor actin by itself displays any contractile properties. The complex, however, is highly contractile.

Actin displays another remarkable property. It is capable of being transformed from its highly fibrous state into a normal globular protein of rather small molecular weight. Szent-Györgyi distinguishes between the two forms by naming them *F-actin* and *G-actin*, standing for the fibrous and globular forms, respectively. The molecular weight of G-actin is estimated to be 57,000. The conversion of G-actin to F-actin is probably not due to any "unwinding" of the globular arrangement, therefore, but to an association of the G-actin individuals in a unidimensional manner to form an F-actin fiber, much as beads may be strung into a necklace. The transition between G-actin and F-actin is reversible and can be repeated many times without damage to the molecule. The conversion of G-actin to F-actin is catalyzed by small concentrations of certain metallic ions. Monovalent ions will bring about the change at concentrations of 0.1 M, divalent ions at concentrations of 0.005 M. Magnesium ions seem to be quite specific in this respect, and in their entire absence polymerization of G-actin to F-actin seems not to take place at all. If salts are removed by dialysis, F-actin will depolymerize to G-actin. Myosin itself also catalyzes the conversion of G-actin to F-actin as will adenosine triphosphate.

Myoglobin. A second important protein constituent of muscle tissue is myoglobin, the concentration of which in some muscle fibers is as high as 0.8 per cent. It is an iron porphyrin protein that resembles in structure and function the hemoglobin of the blood. Its chemical properties are also similar to hemoglobin. Thus, it is oxygenated by molecular oxygen to form *oxymyoglobin* and, in addition, it autoxidizes *in vitro* to *metmyoglobin.* Myoglobin differs from hemoglobin in possessing but one-quarter its molecular weight. Hemoglobin has a molecular weight of about 67,000 and possesses four iron-porphyrin groups per molecule (see Chapter 5). Myoglobin has a molecular weight of about 17,500 and contains only one iron-porphyrin group per molecule. Myoglobin appears in the urine after crushing injury to the muscles.

It should not be assumed from the foregoing that myoglobin is nothing but a "quarter-hemoglobin" molecule. While it is true that hemoglobin may readily be split into semi-molecules in the presence of solutions of substances such as urea and guanidine and that each semi-molecule contains two iron-porphyrin groups, remains functional and can reform into the original hemoglobin on removal of the urea or guanidine by dialysis, a further reversible splitting of the semi-hemoglobin molecules into functional quarter-molecules has not yet been successfully achieved. Myoglobin, while closely resembling what one would expect a quarter-hemoglobin to be, had better therefore be considered an individual in its own right. The iron-porphyrin groups in hemoglobin and myoglobin are identical but the protein portions of the molecule (globin) differ in amino acid composition (46).

Functionally, myoglobin serves as an emergency source of readily available oxygen in the muscle and differs from hemoglobin in having a somewhat greater affinity for oxygen. At 40 mm. Hg oxygen partial pressure, hemoglobin is 38 per cent saturated, while myoglobin is 60 per cent saturated. The transportation of oxygen from the blood vessel to the contracting fibril is thus hastened by the intervention of myoglobin, which absorbs oxygen from the less stable oxyhemoglobin and gives it up in turn to the respiratory enzyme systems of the cells which have a still greater affinity for oxygen.

Muscle Carbohydrate

Glycogen. Although the liver is generally considered the organ in which glycogen is chiefly stored and although it is truly the organ in which that carbohydrate is present in the highest concentration (as high as 6 per cent of the organ), the *total* glycogen content of resting muscle is greater than that of liver, since the muscles of the body have a total weight some fifteen times that of the liver. The glycogen concentration of resting muscle is 0.5 to 1.0 per cent, but decreases during muscular activity and may approach the vanishing point when the activity is particularly intense and prolonged. In the complete absence of glycogen, normal muscular activity can not continue.

Carbohydrate metabolic products. The utilization of glycogen by muscle takes place via a series of phosphorylative and oxidative steps which, in the presence of adequate oxygen, is continued to the final conversion of the carbohydrate to carbon dioxide and water. During vigorous muscular exercise, the accelerated utilization of glycogen is more than the available supply of oxygen can take care of. Energy is then obtained by anaerobic glycolysis (see Chapter 14) in which case the end product is *lactic acid* (formula II). The accumulating lactic acid must eventually

$$CH_3{-}CH(OH){-}C(=O){-}OH$$

II. Lactic acid

be oxidized or reconverted to glycogen if muscular activity is to continue. Resting muscle contains about 0.02 per cent of lactic acid, while during exercise the concentration rises rapidly to about 0.25 per cent. Concentrations as high as 0.4 per cent have been detected in muscle artifically put into rigor by heat, injury, or chloroform, and allowed to remain so to death of the tissue, or by complete deprivation of oxygen to the point of tissue death (54). Small quantities of intermediates in carbohydrate metabolism, other than lactic acid, have also been detected in muscle.

Non-Protein Nitrogenous Constituents

Two water-soluble phosphorus-containing substances, *phosphocreatine* (also called creatine phosphate or phosphagen) and *adenosine triphosphate* (referred to for convenience's sake as ATP), occur in appreciable concentration in muscle. Conway and Hingerty (6) found 24.4 millimols of phosphocreatine per kilogram of rat muscle (0.5 per cent) and 19.5 millimols of ATP phosphorus or 6.5 millimols of ATP itself per kilogram of rat muscle (0.4 per cent). Both play a vital role in the chemistry of muscular contraction (see p. 557). Significant differences in ATP and phosphocreatine content of various types of muscle have been found. Mommaerts (38) states that cardiac muscle contains only one-third the concentration of ATP that skeletal muscle contains and only one-tenth the concentration of phosphocreatine. Uterine muscle contains only one-seventh the concentration of ATP found in skeletal muscle (8). Guanosine triphosphate (GTP) and uridine triphosphate (UTP), compounds very similar to ATP in structure, have also been isolated from mammalian muscle. Each is present at about one to two per cent the concentration of ATP (1). The functions of these and similar substances will be discussed on page 557.

Two dipeptides of unknown function occur in muscle. *Carnosine* (formula III) is a dipeptide of histidine and beta-alanine. The latter amino

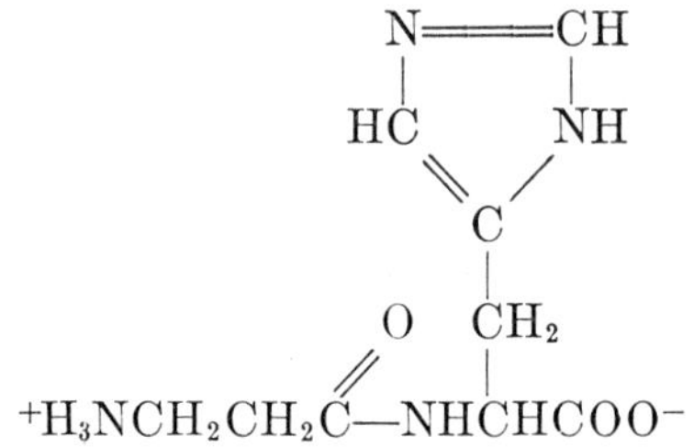

```
                   N═══CH
                   |    |
                  HC    NH
                    ⟍  /
                     C
                     |
                  O  CH2
                 //  |
+H3NCH2CH2C—NHCHCOO−
```

III. Carnosine

acid is an unusual example of a naturally-occurring amino acid with no amino group in the alpha position. It is further interesting to note that beta-alanine also occurs in the food factor, pantothenic acid, one of the B vitamins (see Chapter 21). *Anserine* (formula IV) is a methylcarnosine.

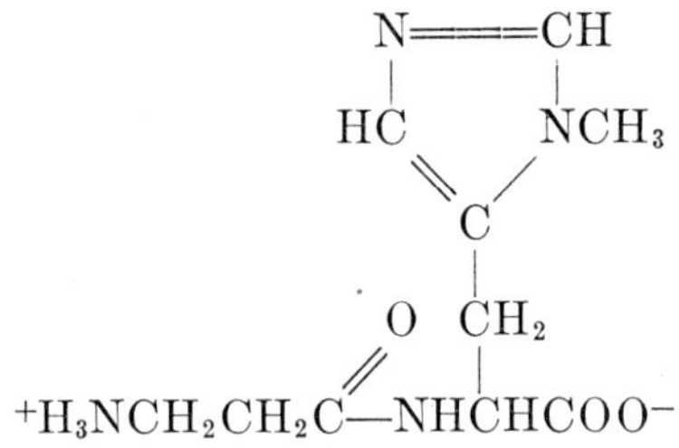

```
                   N═══CH
                   |    |
                  HC    NCH3
                    ⟍  /
                     C
                     |
                  O  CH2
                 //  |
+H3NCH2CH2C—NHCHCOO−
```

IV. Anserine

Their quantitative occurrence in rat muscle in millimols per kilogram of tissue is 1.4 and 20.3, which in percentage is 0.03 and 0.48 for carnosine and anserine, respectively. A methylated gamma amino acid, *carnitine* (see page 818), is also found in muscle.

NERVE TISSUE

Proteins

Two groups of proteins are characteristic of the nervous system. *Neurokeratin* is the fraction of brain protein which remains after hydrolysis by the protein-splitting enzymes, pepsin and trypsin. In the brain, and in the nervous system generally, neurokeratin fulfills a connective function, being the chief constituent of the fibrils (neuroglia) which are the supporting tissues of the brain. A neutral mucopolysaccharide ground substance can be demonstrated (17) surrounding nerve cells and neuroglia. *Proteolipids* of the nervous system differ from other known lipoproteins in that they are water-insoluble but soluble in chloroform-methanol mixtures.

Other protein components of neural tissue include several globulins, two of which coagulate at the unusually low temperature of 47°C. In addition nerve cytoplasm is unusually rich in *ribonucleoprotein* (see Chapter 9) which probably plays some role in nerve condition since it is reported to disappear in part on nerve stimulation (40). Monné (40) speculates that the high nucleoprotein content of nerve cytoplasm is necessary for the ready formation upon demand of a vast variety of highly specific proteins (perhaps the relatively unstable globulins already referred to) by means of which elusive processes as memory, conception, and imagination are made possible. Each new idea or experience might be "stored" in a particular protein molecule, the brain as a whole functioning in a manner similar to the cybernetic machines of modern mathematical technology. Instinct would be interpreted in such a scheme as the inheritance of specific nucleoproteins adapted to prepare certain pre-established proteins from generation to generation.

Nerve Lipids

Nerve tissue is remarkable for the wealth and variety of lipid material found within it. Of the solid matter of brain, about 40 per cent is protein and more than 50 per cent lipid, practically all of the latter being compound or associated lipids. Of the lipid content of the cerebrum, for instance, 52 per cent is phospholipid in nature, 19 per cent is cholesterol, and 13 per cent is galactolipid. The remainder is composed largely of incompletely characterized inositides and sulfur-containing lipids.

The myelin sheaths are particularly rich in lipids which occur in layers alternating with lipoprotein (11). White matter has four times as much cholesterol, phosphatidylserine and phosphosphingosides as gray matter and five times as much galactolipids. Phosphatidylcholine and phosphatidylethanolamine are more evenly divided between the two types of tissue (34).

The function of lipid in the myelin sheaths is probably one of insulation. The nerve impulse is accompanied by electrical phenomena and any device which will prevent loss of electrical potential by leakage will enable a smaller current to be effective and hence permit a thinner nerve fiber. Vertebrate axons, for instance, are only a few microns in diameter so that a two-inch length will weigh one hundredth of a milligram. A squid, on the other hand, the nerves of which are not surrounded by a lipid-filled myelin sheath, must prevent potential loss and maintain efficiency by lowering the resistance of the neural paths. This is done by thickening the axon to, in some cases, nearly a millimeter. A two-inch length of such an axon would weigh 10 to 20 milligrams. For this reason much chemical and physiological research concerns itself with the nervous system of the squid, a creature which is not at all closely related to man.

Monné (40) cites reasons for believing that the compound lipids of the cell have as their chief function that of insulating enzyme systems from one another, and, until necessary, individual enzymes from their substrates (see Chapter 7). Thus, the myriad reactions within a cell would not occur at random but would be moderated and organized by the presence or removal of the compound lipid barriers.

Electrical Potentials in Cells

There is normally in the resting nerve cell, as in all plant and animal cells, a difference in electrical potential, the *resting potential* measurable between the inside and the outside of the cell. In all cells, including resting nerve cells, this difference in potential is approximately 0.1 volt, and agrees with the voltage obtained from the Nernst equation

$$V = 0.059 \log (C_2/C_1)$$

where V is the difference in potential in volts, and C_2 and C_1 are the concentrations inside and outside the cell, respectively, of K^+. In the resting state, the concentration of K^+ inside a cell is typically about 40 times the concentration of K^+ outside. The situation is quite different for Na^+, the concentration of which within the typical cell is only about $\frac{1}{7}$ that of the surrounding fluid.

The resting potential is clearly related to the differences in concentration of K^+ inside and outside the cell. Such observations do not tell us which is cause and which is effect. Of numerous theories proposed, the most acceptable is that the difference in potential (the resting potential) causes the unequal distribution of K^+, and that the resting potential is the result of the existence within the cell of a sodium-extruding mechanism.

The sodium pump. The typical cell membrane, a structure about 10 millimicrons thick and composed of both lipids and proteins, is demonstrably permeable to both Na^+ and K^+ ions. The intact resting cell has a membrane less permeable than that of an injured cell, or of a nerve cell which has been stimulated. In all cells, however, it is apparent that Na^+ and K^+ can both traverse the membrane, and if no other mechanisms were operative would eventually reach the concentrations predicted by the Donnan equilibrium. The distribution of K^+ is in accord with such prediction, as modified by the presence of the membrane potential. The distribution of Na^+ is unrelated to such prediction and is in an opposite direction.

Although the facts as observed strongly support the hypothesis of a sodium-extruding mechanism or pump, no direct anatomical evidence of its existence has been pointed out. The most probable mechanism depends upon the constant formation within the cell of a substance *P* (for pump) which can replace the water of hydration of the Na^+ ion. When this happens, it is presumed that the complex ion *P* Na^+ is more soluble in the lipid and protein components of the cell membrane than either free *P* or hydrated Na^+, and can pass through more rapidly. It is further presumed that *P* Na^+ dissociates in the extracellular fluid and the substance *P* is promptly bound, destroyed, or otherwise inactivated. This prevents the rediffusion of *P* Na^+ back into the cell, leaving an excess of the relatively slowly diffusible (hydrated) Na^+ outside the cell. In brief, the sodium pump is considered to be a stream of *P* leaving the cell and carrying the Na^+ with it in the form of a coordination compound, *P* Na^+.

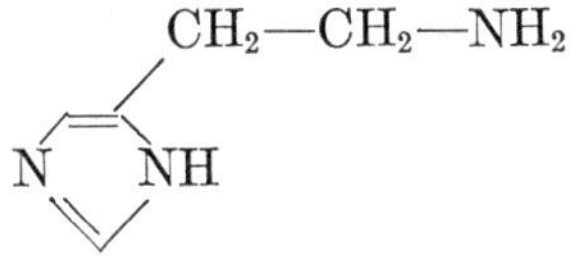

$CH_2—CH_2—NH_2$

N NH

V. Histamine

There is considerable indirect evidence (10) that the pump substance *P* is *histamine* (beta-imidazolylethylamine (formula V)) formed by the decarboxylation within cells of the amino acid histidine, which contains the imidazole ring. Histamine is a known component, in small amounts, of all mammalian cells and tissues. Even if histamine is eventually clearly identified as a pump substance, there remains the possibility that other pump substances exist, and that other ions than Na^+ may be pumped. The pumping of sodium ions remains the most satisfactory explanation for the highly unequal distribution of Na^+ between cells and surrounding fluid, and the resting potential of nerve and other cells is best explained by this constantly maintained inequality of Na^+ concentrations. Since the resting potential of the giant axon of the squid is insensitive to experimental changes in concentration of K^+ or Cl^- or both within the cell, it has been suggested that the pump mechanism may operate within the cell membrane itself (15). Increase of K^+ concentration outside the cell decreases the resting potential.

Conduction in Nerve. The membrane surrounding the resting nerve cell, including its axon, is *polarized* since the resting potential of approximately 0.1 volt exists between the inside and outside of the cell, as described in the previous section. The membrane is charged positively (electron deficit) on the outside and negatively (electron excess) on the inside. The polarization of the cell membrane results from the difference in Na^+ concentration on the two sides of the membrane, this difference in concentration being maintained by the activity of the sodium-extruding mechanism of the cell.

When a stimulus of threshold value or greater is applied to the resting cell, three changes may be observed at the stimulated area. An *action potential* replaces the resting potential; the membrane rapidly depolarizes, then a potential in the opposite direction builds up, falls back to zero, and the resting potential then returns—all in about half a millisecond. A *decrease in electrical resistance* of the cell membrane occurs, becoming maximal in half a millisecond, and slowly returning to the resting value over a period of ten milliseconds. At the stimulated area, *sodium ions enter* and *potassium ions leave* the cell. Sodium entry takes place during the development of the action potential; potassium exits during the return to resting potential. The action potential is believed to be the result of the sodium entry. In a medium containing no sodium ion, there is no action potential (19), except that certain quaternary ammonium ions such as tetraethyl ammonium can substitute for sodium (30, 57).

The disturbances produced locally by an adequate stimulus are propagated along the axon or nerve fiber at velocities varying from a few meters to over 100 meters per second, the higher velocities occurring in larger fibers. The location of the propagated impulse is indicated by an area of local electron excess (negative potential) as compared to adjacent points on the outside of the axon. Inside the axon the impulse moves as a wave of electron deficit or positive potential. These electrical disturbances are propagated across synapses as well as along axons, without loss of intensity of the action potentials.

The energy required for these electrical manifestations of the nerve impulse arises from the utilization by neurons of foodstuffs, chiefly glucose, by energy-liberating chemical processes which will make up the subject matter of several later chapters. The steps which liberate the greatest amount of energy require oxygen, and we find that the central nervous system has a high requirement for oxygen and for glucose. The requirement of the brain for oxygen is not diminished in sleep or increased by normal mental activity.

Acetylcholine. The electrochemical processes which add up to the propagated nerve impulse were studied as purely electrical phenomena in the early years of neurophysiological research. The chemical aspects have been studied intensively since 1920. It appears futile to attempt to characterize the propagated nerve impulse as either purely electrical or purely chemical. As an electrical disturbance, the nerve impulse can be followed along axons and across synapses by electrical measuring devices. *Acetylcholine* (ACh) is one of the important chemical compounds involved in the generation of electrical potentials within axons, and at synaptic junctions. Acetylcholine does not move along axons with the nerve impulse. It is the electrochemical disturbance, in many instances initiated or renewed by acetylcholine release, which travels.

Inspection of the structural formula of acetylcholine (formula VI) gives few if any clues to its astounding physiological potency. Acetylcholine is an ester in which the ethoxy hydroxyl of choline is linked to the carboxyl

$$CH_3—CO—O—CH_2—CH_2—\overset{+}{N}(CH_3)_3$$

VI. Acetylcholine

of acetic acid. The linkage is stable at pH 4, with the tendency to spontaneous hydrolysis remaining slight at pH 7.8 and at intermediate values.

Acetylcholine occurs in all vertebrates, in insects, crustacea, molluscs, annelids, planaria, and some few protozoa. In the mammalian nervous system, its highest concentrations are in more primitive regions such as autonomic ganglia and the myenteric plexus. The function of acetylcholine in neural transmission is characteristic of all preganglionic neurons in the autonomic nervous system, of all postganglionic neurons in the parasympathetic system, but in the sympathetic system only of postganglionic neurons leading to sweat glands and the *arrectores pilorum* muscles. The function of acetylcholine in the central nervous system is less clearly localized, with some evidence that acetylcholine is functional in alternate neurons. Sensory neurons contain small but measurable amounts of acetylcholine.

Acetylcholine is formed by the condensation of choline possibly through the intermediate formation of *phosphorylcholine* (1a) with acetyl groups carried by *coenzyme A* (the structure of which will be considered in detail later). The formation of acetylcholine is catalyzed by an enzyme, *choline acetylase*, and takes place within nerve cells which are *cholinergic*, meaning that they make use of acetylcholine in their functional activity. Acetylcholine may be formed in other types of cells, including the red cells of the blood (22).

In the resting or unstimulated nerve cell, acetylcholine is combined in an inactive form, probably to a mitochondrial storage protein. Stimuli (electron flow, possibly proton flow) release acetylcholine and permit its combination with a receptor substance, probably also a protein. Upon union with acetylcholine, the receptor substance undergoes a change in configuration which favors rapid Na^+ influx—the reverse of the action of the sodium pump and the trigger by which the action potential is generated.

Actually only a few molecules of acetylcholine are liberated with each nerve impulse. This makes it probable that the receptor for acetylcholine is an enzyme, requiring acetylcholine for its activation, and catalyzing the breakdown of phospholipoproteins in the neural membrane, thereby decreasing resistance of the membrane to the passage of Na^+ ions (21). Less probable is the hypothesis that acetylcholine actually replaces phospholipid, with similar results.

Cholinesterase. Indirect evidence concerning the functioning of ACh in nerve transmission is derived from studies of *cholinesterase*, an enzyme which catalyzes the hydrolysis of ACh to acetic acid and choline. The facts about cholinesterase which bear upon the problem may be summarized as follows:

1) Cholinesterase is localized at the neuronal surface where the bioelectric phenomena occur. This means that cholinesterase tends to be localized at the synaptic junction where the repeated division and subdivision of the axon results in finer fibers and a consequently greater surface-volume ratio.

2) Cholinesterase in nerve and muscle is much more specific than the average esterase (ester-hydrolyzing enzyme). Cholinesterase will catalyze the hydrolysis of ACh and one or two other very closely related choline esters. Another esterase exists which can split ACh, but this catalyzes the hydrolysis of many other esters. It is distinguished from cholinesterase by the name pseudocholinesterase. Presumably, after formation of ACh in the process which accompanies, or perhaps initiates, nerve impulse transmission, it must be quickly inactivated to render that portion of the nerve amenable to initiation of another impulse. Cholinesterase would be the agent employed in that process.

3) The high concentration of cholinesterase in nerve tissue makes possible the removal of ACh at a speed comparable to that of the electrical phenomena. Nachmonsohn, working with the frog's sartorius muscle, showed that 1.6×10^9 molecules of ACh could be hydrolyzed in one-thousandth of a second at a single motor end plate. This would correspond

to the ACh content of 100 to 250 square microns of neuronal surface. In mammalian brain, 10^{14} to 10^{15} molecules of ACh may be inactivated per gram of tissue, which would correspond to the ACh content of 10 to 100 square millimeters of neuronal surface.

4) Substances which inhibit cholinesterase alter, and in high concentrations, abolish the nerve action potential. Such *anticholinesterases* include many quaternary ammonium bases (but not betaine), physostigmine (an alkaloid of the calabar bean) and several related urethanes, alkyl fluorophosphates (such as di-isopropyl fluorophosphate, DFP), and alkyl polyphosphates (such as tetraethyl pyrophosphate, TEPP). Anticholinesterases have found application in medicine, in chemical warfare, and as insecticides. By permitting ACh to persist, these substances cause a continuous depolarization of nerve endings, and thus prevent the transmission of the nerve impulse.

5) A direct proportionality between voltage and cholinesterase activity has been established in the electric organ of *Electrophorus electricus*, the electric eel.

Catechol Amines

Just as those neurons which utilize acetylcholine in the process of impulse transmission are designated as cholinergic, other neurons and neuron chains which utilize the catechol amines, *nor-adrenalin* (L-arterenol) or to a less extent *adrenalin* (formula VII), in transmitting impulses are designated as *adrenergic*. All neurons belong in one or the other category.

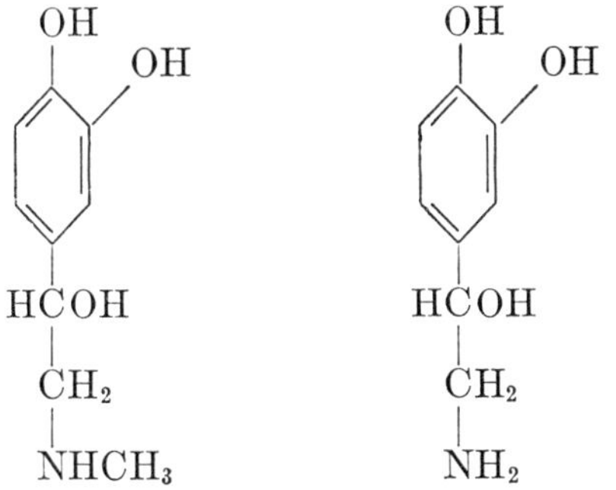

OH, OH (benzene ring)	OH, OH (benzene ring)
HCOH	HCOH
CH_2	CH_2
$NHCH_3$	NH_2
Adrenalin	*nor*-Adrenalin

VII

Nor-adrenalin is the principal catechol amine of adrenergic neurons. Although the distribution in individual cells is not known, the output in a stimulated organ is in the ratio of 90 per cent *nor*-adrenalin to 10 per cent adrenalin.

Adrenalin is the principal catechol amine of the medulla of the adrenal gland. The human adrenal contains about 0.5 mgm. of adrenalin and 0.1 mgm. of *nor*-adrenalin per gram of tissue. The secretion of the adrenal medulla is a *hormone*, distributed by way of the blood stream. The cells of the adrenal medulla can be considered as neurons which have acquired a specialized secretory function. All adrenal medullary cells can form *nor*-adrenalin. It is possible that methylation of this compound to form adrenalin is accomplished only by a limited number of adrenal medullary cells. Adrenalin secretion is stimulated by the sympathetic nervous system, and the hormonal action on muscle and blood vessels is sympathomimetic. Adrenalin causes liver and muscle glycogenolysis, with resulting hyperglycemia and possibly glycosuria, if there is glycogen in the liver (see page 527). It also increases oxygen consumption, its effect in this respect being much more rapid than that of thyroxine. Adrenalin is also considered to have a stimulating effect, mediated by pituitary ACTH, on adrenocortical activity. Adrenalin contracts the arterioles, increasing the blood pressure. The hormone is very active and the concentration in blood under normal conditions is one part in one or two billion. In periods of stress, as in fear, pain, or anger, there is hypersecretion of adrenalin into the blood stream with resultant blood pressure increase and glycogen breakdown enhancement in both liver and muscle so that the body is prepared better to meet the emergency, either by flight or by physical resistance.

The effect of a single injection of adrenalin is of short duration. It is quickly destroyed by oxidation in the capillary circulation through muscle and possibly other tissues, but not in the liver. Among the oxidation products have been identified *adrenochrome* (formula VIII) and *protocatechuic acid* (formula IX). The enzyme *amine oxidase* catalyzes the oxidative inactivation.

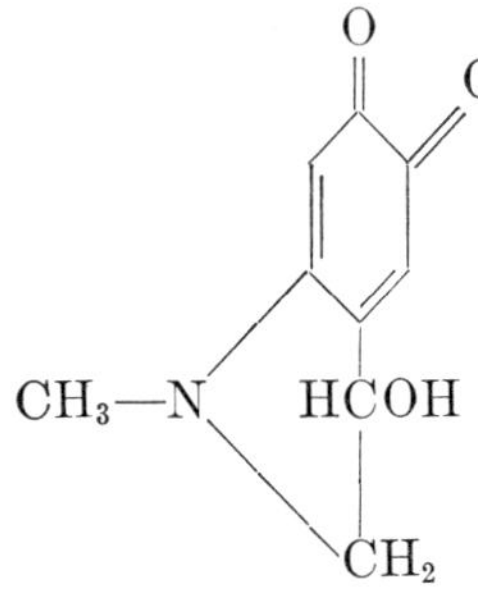

VIII. Adrenochrome

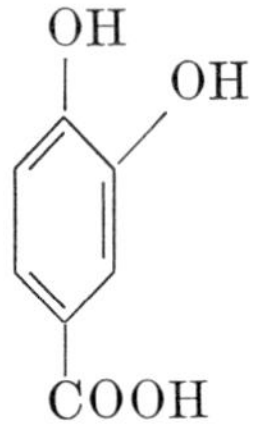

IX. Protocatechuic acid

In contrast to the emergency functions of adrenalin, *nor*-adrenalin is chiefly involved, by its vasoconstrictor effect, in maintaining a steady arterial blood pressure. Although both of the catechol amines are pressor substances, meaning that they elevate arterial blood pressure, the net effect of adrenalin is vasodilator in peripheral tissues, while that of *nor*-adrenalin is vasoconstrictor. The stimulating effects of adrenalin upon the central nervous system, upon oxygen consumption, and upon the conversion of liver glycogen to glucose, are all quantitatively greater than the similar actions of *nor*-adrenalin.

Neurohumors and Drug-induced Psychoses

Acetylcholine and the catechol amines, since they are essential reactants in the functions of the nervous system, are examples of a class of substances designated as neurohumors, neurosecretions, or neurohormones.

Another neurohumor, a pressor amine related to the catechol amines, is *5-hydroxytryptamine* (5HT) (formula X) which is also known as *serotonin*. It can be identified in the brain, particularly in the hypothalamus, and in the intestinal mucosa. In the latter location it was formerly designated as *enteramine*, with its presumptive origin the enterochromaffin cells. The same amine has been identified in blood platelets, and in this location given the name of *thrombotonin*.

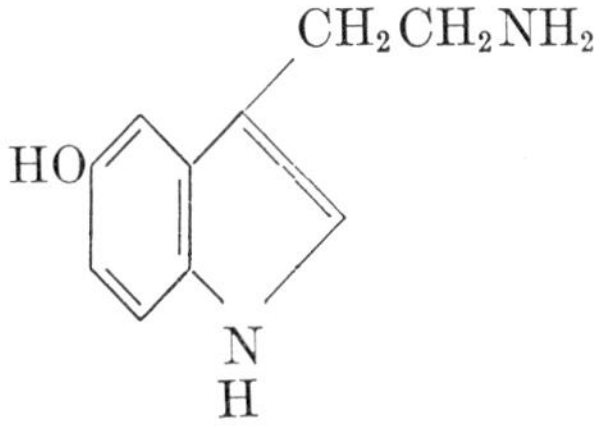

$CH_2CH_2NH_2$

HO

N
H

X. 5-Hydroxytryptamine
(Serotonin)

Abnormal amines, or excessive amounts of normally-occurring amines or their breakdown products, may have a part in the cause of mental diseases. Adrenochrome, mentioned above as an oxidation product of adrenalin, produces disturbances comparable to psychoses when injected, while the drug mescaline (3,4,5-trimethoxyphenylethylamine), long known to produce hallucinations, has a recognizable similarity to adrenalin in its structure (20). Mescaline inhibits oxidation of glucose in the brain (42), and adrenochrome has a similar inhibiting effect (37).

A number of amines, including 5-hydroxytryptamine and *nor*-adrenalin, may be considered as competing for amine oxidase in the brain. Unlike adrenalin, *nor*-adrenalin occurs in significant amounts in the brain (52) particularly in the hypothalamus.

One of the drugs most effective in the production of transitory psychoses in human subjects is lysergic acid diethylamide (LSD). LSD can increase the brain content of 5HT by blocking amine oxidase, but it also antagonizes the pharmacological actions of 5HT on smooth muscle, an antagonism shown also by 2-brom-D-lysergic acid diethylamide, which does not produce psychic disturbances (4). Serotonin (5HT) is also excreted in normal urine, 60 to 160 micrograms per 24 hours, and this is consistently decreased following administration of LSD (44). Reserpine, a tranquilizing drug, decreases the 5HT of brain and platelets and brings about increased urinary excretion of 5-hydroxyindolacetic acid, the chief product of 5HT in the body. *Bufotenine*, the N-dimethyl derivative of 5HT, is known to produce bizarre behavior in men, monkeys, and cats. In the form of cohaba snuff (from seeds of *Piptadenia peregrina*) it has been used for centuries by South American Indian priests and necromancers (48). When 5HT is injected, it does not enter the brain, but its precursor 5-hydroxytryptophane does so readily and is converted to 5HT by the aid of the enzyme 5-hydroxytryptophanase, in amounts sufficient to raise the 5HT of the brain to 20 times the normal content and show abnormalities of behavior like those induced by LSD or bufotenine.

Chemistry of Sense Perception

Odor and taste. Of the varieties of sense perception available to man, the sense of smell and the sense of taste are known to be primarily chemical in nature. Unfortunately, the nature of the reactions involved is as yet little understood. It has been suggested that the sensations of odor and taste result from the inhibition by various substances of one or more enzymes present in the taste buds and olfactory mucosa. High concentrations of phosphatase have been found in nasal mucosa of rabbits and high concentrations of both phosphatase and esterase in their taste buds. There is some evidence to the effect that substances such as vanillin inhibit the phosphatase activity in the taste buds, while quinine inhibits that of the esterase (9). If the mechanism which makes the sensations of taste and odor possible were indeed to involve enzyme inhibition, it would account both for the minute quantities of some substances capable of being smelled or tasted, and for the wide variety of chemicals yielding similar taste or odor, since enzymes can usually be inhibited by many different substances of widely varying nature. For information on the classification of tastes and odors and on the effects of individual substances, the student is referred to Moncrieff (39).

The quantity and quality of taste of a given substance can vary from individual to individual. Thus, phenylthiourea (formula XI) and certain related sulfur-containing substances, while bitter in varying degrees to some people, are completely tasteless to others (3). Such "taste-blindness" is apparently subject to Mendelian inheritance and has been consequently studied for its relationship to problems of anthropology.

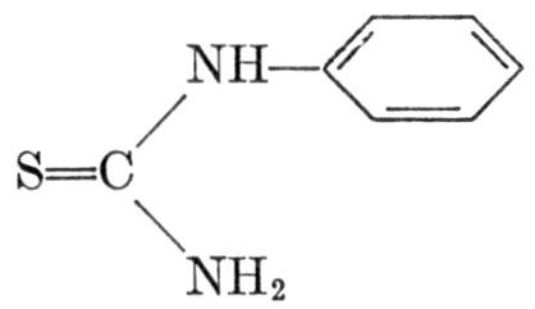

XI. Phenylthiourea

Sight. The retina contains chromoproteins, usually termed *visual pigments* because of their role in the phenomenon of sight. Of these, the best known is *rhodopsin,* also called erythropsin or "visual purple". Rhodopsin is a conjugated protein, its prosthetic group being the isoprenoid, *retinene,* which is the aldehyde of vitamin A. Vitamin A enters the retina from the blood stream and there it is kept in equilibrium with retinene through the catalytic action of retinene reductase. The equilibrium, considered by itself, is far in the direction of vitamin A, but the protein *opsin* (in the dark) takes up retinene to form rhodopsin, driving the reaction in the retinene direction. In the light, rhodopsin breaks down once more to retinene and opsin, through poorly-defined stages (23). The mixture of retinene and opsin is less intensely colored than is rhodopsin and is therefore called "visual yellow".

Vitamin A and retinene can exist in a number of *cis-trans* isomers (53) of which the two important ones, those shown in formula XII, may be termed, for convenience sake, *cis*-vitamin A (or retinene) and *trans*-vitamin A (or retinene.) In the rhodopsin cycle, it is *cis*-vitamin A which is first oxidized to *cis*-retinene then combined with opsin to form rhodopsin. Rho-

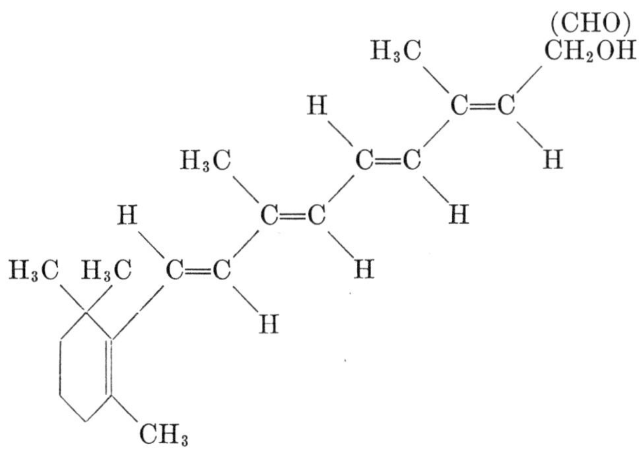

Trans-vitamin A (retinene)

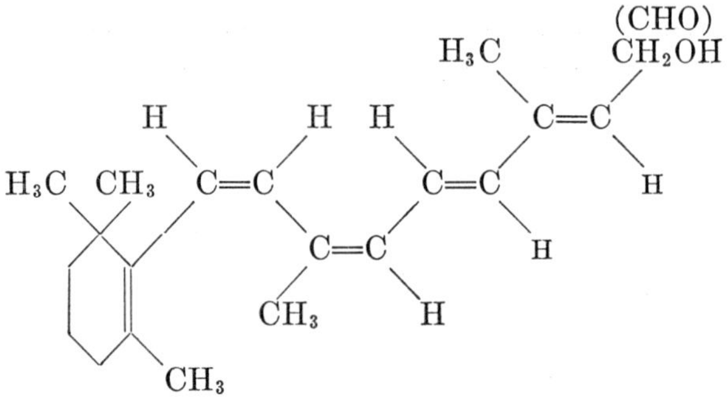

```
                                              (CHO)
                                    H3C       CH2OH
                                       \     /
              H          H  H           C=C
               \        /    \         /    \
H3C  CH3        C=C           C=C           H
   \ /   /         \         /   \
   / \  /           C=C           H
  |   ||           /   \
  |   ||        CH3     H
   \ /  \
         CH3
```

Cis-vitamin A (retinene)

XII. *Cis-trans* isomerism in vitamin A and retinene

dopsin, on breaking down, liberates *trans*-retinene which is in turn reduced to *trans*-vitamin A. The *cis* and *trans* forms of vitamin A and retinene are apparently interconvertible in the body. A continuous supply of vitamin A is apparently necessary, perhaps because some of the retinene is lost through decomposition during its free existence. Rhodopsin is concerned with vision under faint illumination, which is a function of the rods of the retina. It is for this reason that avitaminosis A leads to "night blindness".

The crystalline lens and the cornea are unique among body solids for their high degree of transparence to light. This is not due to any unusual tissue components. Both consist almost entirely of water and protein. The crystalline lens is unusually high in protein content which amounts to 35 per cent of its weight. Some 85 per cent of the protein are *crystallins*, which are chiefly remarkable in their low degree of specificity among varying species of animals. When an animal is sensitized to the crystallins of a particular species of mammal, antibodies are formed which will react with the crystallins of almost any other mammalian species but not with those from birds. The crystallins have been divided into two groups, alpha and beta, the former exhibiting high antigenic power and the latter, low. The remainder of the crystalline lens protein is albuminoid in nature. The cornea contains 18 per cent proteins of which four-fifths is collagen and the remainder mucoprotein.

The aqueous humor is formed from the blood plasma. It differs from blood plasma in being almost pure water. Only a little over one per cent of its weight is solid matter and of this some three-fifths is sodium chloride. The protein content, about 0.015 per cent, is composed of albumins and globulins similar in nature and relative proportions to those of plasma (58). The aqueous humor has in it traces of glucose, glutathione, and ascorbic acid, all of which play a part in the maintenance of the metabolic processes of the crystalline lens, which is, of course, free of any direct vascularization.

Beta-crystallin may play a role as oxidizing enzyme in this respect. The *vitreous humor*, the other intraocular fluid, contains *vitrosin*, a protein which, in amino acid composition, resembles the collagens (14).

Tears are similar to the humors in chemical composition, possessing in fact a somewhat larger percentage of solid matter (1.8 per cent in tears as compared with 1.1 per cent in the humors) including proteins and glucose as well as the commonly recognized sodium chloride. The most interesting component of tears is lysozyme, which, through its lytic action upon bacteria, gives to tears the qualities of a mild disinfectant.

REFERENCES

1. Bergkvist, R., and Deutsch, A. Guanosine triphosphate and uridine triphosphate from muscle. Acta Chem. Scand., **7:** 1307–1308, 1953.

1a. Berry, J. F. and Stotz, E. Role of phosphorylcholine in acetylcholine synthesis. J. Biol. Chem., **218:** 871–874, 1956.

2. Blum, J. J., and Morales, M. F. The interaction of myosin with adenosine triphosphate. Arch. Biochem. Biophys., **43:** 208–230, 1953.

3. Boyd, W. C. "Taste blindness" to phenylthiocarbamide and related compounds. Psychol. Bull., **48:** 71–74, 1951.

4. Cerletti, A., and Rothlin, E. Role of 5-hydroxytryptamine in mental diseases and its antagonism to lysergic acid derivatives. Nature, **176:** 785–786, 1955.

5. Chauncey, H., Lionetti, F., and Lisanti, V. Hydrolytic enzymes in hyaluronidase preparations. Science, **118:** 219–220, 1953.

6. Conway, E. J., and Hingerty, D. The influences of adrenalectomy on muscle constituents. Biochem. J., **40:** 561–568, 1946.

7. Courts, A. The N-terminal amino acid residues of gelatin. I. Intact gelatin. Biochem. J., **58:** 70–74, 1954.

8. Csapo, A., and Gergely, J. Energetics of uterine muscle contraction. Nature, **166:** 1078–1079, 1950.

9. El-Baradi, A. F., and Bourne, G. H. Theory of tastes and odors. Science, **113:** 660–661, 1951.

10. Eyring, H., and Dougherty, T. F. Molecular mechanisms in inflammation and stress. American Scientist, **43:** 457–467, 1955.

11. Finian, J. B. The structure of myelin. Exptl. Cell Research, **5:** 70–74, 1954.

12. Forbes, R. M., *et al.* The composition of the adult human body as determined by chemical analysis. J. Biol. Chem., **203:** 359–366, 1953.

13. Graham, C. E., *et al.* The amino acid content of some scleroproteins. J. Biol. Chem., **177:** 529–532, 1949.

14. Gross, J., *et al.* Vitrosin: a member of the collagen class. J. Biophys. Biochem. Cytol., **1:** 215–220, 1955.

15. Grundfest, H. The nature of the electrochemical potentials of bioelectric tissues; *in* Shedlovsky, T. *Electrochemistry in Biology and Medicine*, pp. 141–166. New York, John Wiley & Sons, Inc., 1955.

16. Harrison, M. F. Composition of the liver cell. Proc. Roy. Soc. (London), **B141:** 203–216, 1953.

17. Hess, A. The ground substance of the central nervous system revealed by histochemical staining. J. Comp. Neurol., **98:** 69–91, 1953.

18. Highberger, J. H., *et al.* Interaction of mucoprotein with soluble collagen; an electron-microscope study. Proc. Natl. Acad. Sci. U. S., **37:** 286–291, 1951.

19. Hodgkin, A. L., and Katz, B. The effect of sodium ions on the electrical activity of the giant axon of the squid. J. Physiol., **108:** 37–77, 1949.
20. Hoffer, A., *et al.* Schizophrenia: a new approach. II. J. Mental Sci., **100:** 29–45, 1954.
21. Hokin, M. R., and Hokin, L. E. Effects of acetylcholine on phospholipides in the pancreas. J. Biol. Chem., **209:** 549–558, 1954.
22. Holland, W. C., and Grieg, M. E. The synthesis of acetylcholine by human erythrocytes. Arch. Biochem. Biophys., **39:** 77–79, 1952.
23. Hubbard, R., and Wald, G. The mechanism of rhodopsin synthesis. Proc. Natl. Acad. Sci. U. S., **37:** 69–79, 1951.
24. Jackson, D. S. Chondroitinsulfuric acid as a factor in the stability of tendon. Biochem. J., **54:** 638–641, 1953; Nature of collagen-chrondroitinsulfate linkages in tendon. Biochem. J., **56:** 699–703, 1954.
25. Jeghers, H. Medical progress: Pigmentation of the skin. New England J. Med., **231:** 88–100, 122–137, 181–189, 1944.
26. Keys, A., and Brožek, J. Body fat in adult man. Physiol. Rev., **33:** 245–325, 1953.
27. Laki, K., and Carroll, W. R. Size of the myosin molecule. Nature, **175:** 389–390, 1955.
28. Lauffer, M. A., and Szent-Györgyi, A. G. Comments on the structure of myosin. Arch. Biochem. Biophys., **56:** 542–548, 1955.
29. Locker, R. H. C-terminal groups in myosin, tropomyosin and actin. Biochim. et Biophys. Acta, **14:** 533–542, 1954.
30. Lorente de Nó, R. On the effect of certain quaternary ammonium ions upon frog nerve. J. Cell. Comp. Physiol., **33:** supplement, 1949.
31. Lowry, O. H., *et al.* The determination of collagen and elastin in tissues, with results obtained in various normal tissues from different species. J. Biol. Chem., **139:** 795–804, 1941.
32. Matthews, B. F. Collagen/chondroitin sulfate ratio of human articular cartilage related to function. Brit. Med. J., **1952:** II, 1295.
33. Mayer, J. Genetic, traumatic and environmental factors in the etiology of obesity. Physiol. Rev., **33:** 472–508, 1953.
34. McIlwain, H. *Biochemistry and the Central Nervous System.* Boston, Little, Brown, and Co., 1955.
35. Meyer, K. Mucoproteins and mucoids. Ann. Conf. Protein Metabolism, **9:** 64–73, 1953.
36. Meyer, K., and Hahnel, E. The estimation of lysozyme by a viscosimetric method. J. Biol. Chem., **163:** 723–732, 1946.
37. Meyerhof, O., and Randall, L. O. Inhibitory effects of adrenochrome on cell metabolism. Arch. Biochem., **17:** 171–182, 1948.
38. Mommaerts, W. F. H. M. *Muscular Contraction.* New York, Interscience Publishers, 1950.
39. Moncrieff, R. W. *The Chemical Senses.* New York, John Wiley & Sons, Inc., 1946.
40. Monné, L. Functioning of the cytoplasm. Advances Enzymol., **8:** 1–69, 1948.
41. Noda, H. Physico-chemical studies on the soluble collagen of rat-tail tendon. Biochim. et Biophys. Acta, **17:** 92–98, 1955.
42. Quastel, J. H., and Wheatley, A. H. M. The effects of amines on oxidations of the brain. Biochem. J., **27:** 1609–1613, 1933.
43. Rapport, M. M., *et al.* The hydrolysis of hyaluronic acid by pneumococcal hyaluronidase. J. Biol. Chem., **192:** 283–291, 1951.

44. Rodnight, R., and McIlwain, H. Serotonin, lysergic acid diethylamide and mental status. Brit. Med. J., **1:** 108, 1956.
45. Rogers, H. J., *et al.* Skeletal tissues. II. Collagen content of bones from rabbits, oxen, and humans. Biochem. J., **50:** 537–542, 1952.
46. Rossi-Fanelli, A., *et al.* Amino-acid composition of human crystallized myoglobin and hemoglobin. Biochim. et Biophys. Acta, **17:** 377–381, 1955.
47. Snellman, O., and Tenow, M. A contractile element containing tropomyosin (actotropomyosin). Biochim. et Biophys. Acta, **13:** 199–208, 1954.
48. Stromberg, V. L. The isolation of bufotenine from *Piptadenia peregrina*. J. Am. Chem. Soc., **76:** 1707, 1954.
49. Szent-Györgyi, A. *Chemistry of Muscular Contraction.* 2nd edit. New York, Academic Press, 1951.
50. Szent-Györgyi, A. G. A new method for the preparation of actin. J. Biol. Chem., **192:** 361–369, 1951.
51. Szent-Györgyi, A. G. Meromyosins, the subunits of myosin. Arch. Biochem. Biophys., **42:** 305–320, 1953.
52. Vogt, M. The concentration of sympathin in different parts of the central nervous system under normal conditions and after the administration of drugs. J. Physiol., **123:** 451–481, 1954.
53. Wald, G., *et al.* Hindered cis isomers of vitamin A and retinene: the structure of the neo-b isomer. Proc. Natl. Acad. Sci. U. S., **41:** 438–451, 1955.
54. Wilhelmi, A. E. *in* Fulton, J. F. (ed.) *A Textbook of Physiology.* 17th edit. Philadelphia, W. B. Saunders Co., 1955.
55. Williams, H. H., *et al.* Estimation of growth requirements for amino acids by assay of the carcass. J. Biol. Chem., **208:** 277–286, 1954.
56. Windrum, G. M., *et al.* The constitution of human renal reticulin. Brit. J. Exptl. Pathol., **36:** 49–59, 1955.
57. Woolley, D. W. Biosynthesis and energy transport by enzymic reduction of "onium" salts. Nature, **171:** 323–326, 1953.
58. Wunderly, C., *et al.* Investigations of the aqueous humor of the human eye. Experientia, **10:** 432–433, 1954.

CHAPTER 5

Blood and the Anemias

Although it is a liquid, blood may be regarded as a tissue. If we wish to summarize briefly the usefulness of blood to the other tissues, we may say that it combines in itself the functions of transport and protection; that is, it carries substances to and from the various tissue cells, it protects the tissue cells from abrupt and undesirable changes in the interstitial fluid which is their immediate environment, and it protects them against invasion by foreign organisms.

The blood is the chief medium of distribution of nutrients such as monosaccharides and amino acids to the various tissues. It also brings to the tissues the oxygen without which energy-producing substances could not be utilized fully, takes away from the tissues waste products such as carbon dioxide, urea, uric acid, and conducts away the excess heat so that it can be radiated from the surface of the body. The blood also carries hormones from one tissue to another and thus mediates the control exercised upon physiological function by the various endocrine glands. It also conveys antibodies to fight infection, and helps bring phagocytes to the site of an invasion so that they may devour the micro-organisms attempting to establish a foothold.

In addition to transport, the blood has the function of protecting the various tissues. It does this in various ways. In addition to the transport of antibodies and phagocytes, it supplies to all of the tissues a fluid of suitable osmotic pressure for correct function. By means of its buffer power it protects the tissues against changes in hydrogen ion concentration which might impair their efficiency. In order that the volume of blood may be maintained relatively adequate, there are two necessary additional functions which the blood must have. The first of these involves the clotting mechanism, which restricts loss of blood. A second exists, whereby water can be drawn into the circulation so that the plasma is kept suitably dilute and the total blood volume at a suitable level. This osmotic function also enables blood to mobilize water from tissues in case of emergency.

THE CLOTTING OF THE BLOOD

When blood escapes from the blood vessels, it ordinarily soon solidifies into a gelatinous mass called the clot. Clotting is one of the mechanisms involved in hemostasis which occurs in the following stages:

1. Vessels are severed by injury, and become dilated by the histamine liberated as the result of tissue damage. Blood therefore flows from the wound.

2. After a short time the vessels contract from diminished intravascular pressure and from their elastic retraction. Blood ceases to flow and now has time to clot. Blood platelets agglutinate (26) and yield an activator of thromboplastin (see below) which initiates the clotting process.

3. After the period of contraction (0.5 to 2 hours), the vessels again dilate, a first step toward wound repair. Recurrence of bleeding is prevented by the already formed clot, and the adhesion of endothelial cells.

The Stages of Clotting

The clotting of blood takes place in three stages, which are distinct and separate from the three stages of hemostasis described above. We shall take them up in reverse order.

Stage 3. *Fibrinogen to fibrin.* The framework of the clot consists of *fibrin*, a glycoprotein (41) which has a fibrous structure which accounts for its name. A network mass of fibrin traps the cells of the blood, and if the blood is normal and the injury not too extensive, bleeding is checked and tissue repair eventually begins on the basis of the blood clot. The fibrin itself accounts for less than one per cent of the entire clot.

Fibrin is formed from its precursor, *fibrinogen* (Factor I), a globulin which occurs in blood plasma in a concentration of approximately 0.3 per cent. Fibrinogen solutions are highly viscous and show high birefringence of flow. The molecules are elongated ellipsoids with an axial ratio of the order of 20 to 1 and a molecular weight of 330,000 (39). Fibrinogen is formed exclusively in the liver. A normal animal "defibrinogenated" by first removing and then replacing blood after the fibrinogen has been removed by whipping will restore the fibrinogen within a few hours; replacement of fibrinogen does not occur in the hepatectomized animal. In poisoning with substances such as chloroform or arsenic which damage the liver there is often fibrinogen deficiency. In the normal body, fibrinogen is present well in excess of normal requirements for clotting, and may be increased further during acute infections, suppurative processes and after hemorrhage.

The conversion of soluble fibrinogen to insoluble fibrin is catalyzed by the protein, *thrombin*. Thrombin is a protein-splitting enzyme (see page

243) of unusually high specificity. It will apparently attack only two or three peptide bonds out of the 3,800 present in the fibrinogen molecule. The bonds attacked are those between the carboxyl group of arginine and the amino group of glycine (2a, 38). Fibrinogen, under the influence of thrombin splits off a peptide portion called *fibrino-peptide* which consists of two separate peptides, each with a molecular weight of about 3,000 (3a). What is left of the molecule spontaneously polymerizes, by way of hydrogen bonding, forming polymers which, eventually, become the fibrin clot.

Stage 2. *Prothrombin to thrombin.* Thrombin does not, of course, occur in circulating blood as such, since that would cause coagulation within the blood vessels. The circulating precursor of thrombin is *prothrombin* (Factor II), a protein occurring in the globulin fraction of plasma and containing 4 per cent carbohydrate. Prothrombin has a molecular weight of 40,000 and precipitates at pH 5.3. It has been estimated that normal human plasma contains less than 20 mgm. of prothrombin per 100 ml. of blood but nevertheless this low concentration exerts a powerful action. Plasma can be diluted three hundred times and still, when the prothrombin is fully activated, show a clotting effect upon fibrinogen solutions.

Prothrombin deficiency may result from liver disease, since the liver is the exclusive site of prothrombin formation, or from vitamin K deficiency which is most commonly a defect of absorption except in the newborn, in whom vitamin K production by intestinal bacteria has not yet started. In the absence of the K vitamins, the liver is unable to make prothrombin.

The conversion of prothrombin to thrombin involves several substances. The most important of these are *thromboplastin* (Factor III) and *calcium ion* (Factor IV). Thromboplastin may possibly act by catalyzing the splitting off of a peptide fraction from prothrombin. This possibility is indicated by the fact that trypsin, the only known function of which is to split certain peptide bonds, will convert prothrombin to thrombin (24). The reaction between prothrombin and thromboplastin appears to be stoichiometric, since if small amounts of prothrombin are used in the presence of adequate amounts of thromboplastin, doubling the amount of prothrombin doubles the yield of thrombin.

The removal of calcium ion by precipitation with oxalate, by complex-formation with citrate, or by adsorption upon ion-exchange resins, will prevent the conversion of prothrombin to thrombin even in the presence of thromboplastin.

There are other substances involved in the conversion of prothrombin to thrombin. One of these is *plasma accelerator globulin* (AcG or Factor V). (Thrombin, once formed, converts plasma accelerator globulin to *serum accelerator globulin*, which is more active and less stable.) Another is *serum prothrombin conversion accelerator* (SPCA) of which *proserum prothrombin conversion accelerator* (PPCA or Factor VII) is the precursor. More commonly, they are called *convertin* and *proconvertin*, respectively.

Stage 1. *Thromboplastin precursors to thromboplastin.* Thromboplastin is not a single substance, but rather a group of tissue components (protein, lipid and mucopolysaccharide) which are liberated as a result of injury and initiate the clotting mechanism. There are also thromboplastin precursors in plasma. One of these is Factor VIII, which is *antihemophilic globulin*, or *antihemophilic factor* (AHF), sometimes referred to as *thromboplastinogen*. Another is *plasma thromboplastic component* (PTC or Factor IX). PTC occurs in the beta$_2$-globulin fraction of plasma to the extent of 1 mg./100 ml. plasma (1). Still another precursor is *plasma thromboplastic antecedent* (PTA). These precursors are essential for production of thromboplastic activity by interaction with material released by the lysis of blood platelets when blood is shed and this, too, initiates the clotting mechanism since, as mentioned above, thromboplastin catalyzes the conversion of prothrombin to thrombin. The importance of platelets is shown by the severe hemorrhages which occur in patients with platelet deficiency (thrombocytopenia) and the improvement brought about in such patients by transfusion of platelets (40). Factors V and VII, involved in the second stage of clotting, also take part in the first stage. For example, brain extract which is not thromboplastic becomes so after reaction with Factors V and VII. These are also necessary for the formation of thromboplastic substances in blood (4). In turn, Factor VII is activated by thromboplastin and calcium, a fact which may account for the autocatalytic course of blood clotting.

Summary. The clotting mechanism is initiated by tissue injury which liberates thromboplastic substances into the blood, or by bleeding, which liberates activators from platelets which in turn convert plasma thromboplastin precursors to active thromboplastin. Thromboplastin and calcium ion plus other factors catalyze the formation of thrombin from prothrombin. Thrombin catalyzes the formation of fibrin from fibrinogen. Fibrin forms the network of the clot. The different clotting factors are listed in Table V-1. (Clotting can be deliberately brought about by the use of thrombin preparations of human, beef, or rabbit origin. These can be used by surgeons in local hemostasis.)

Clotting Defects

Congenital. The best known of the congenital diseases characterized by a lifelong tendency to prolonged hemorrhage after even slight injury is *hemophilia*. This disease, which is sex-linked in its inheritance (see page 426), occurs almost exclusively in males, though females can, in rare cases

TABLE V-1

Clotting factors

FACTOR	NAMES
I	Fibrinogen
II	Prothrombin
III	Thromboplastin
IV	Calcium ion
V	Plasma accelerator globulin, AcG, proaccelerin
VI	(see note)
VII	Proserum prothrombin conversion accelerator, PPCA, proconvertin
VIII	Antihemophilic globulin, antihemophilic factor, AHF, thromboplastinogen
IX	Plasma thromboplastic component, PTC, Christmas factor
X	(see note)

Note: Factor VI is a hypothetical substance which appears during coagulation and coagulates oxalated plasma in the presence of low concentrations of calcium ion. Factor X has been demonstrated, shown to be necessary for thromboplastin formation, and distinguished from other factors (10). It has not yet been named.

(35), be affected. Hemophilia is the result of an inherited inability on the part of the body to form antihemophilic globulin (6).

Lack of other essential clotting factors, either inherited or acquired, can produce diseases with hemophilia-like symptoms. Ordinary hemophilia or "classical hemophilia" is sometimes called "hemophilia A", while bleeding due to the hereditary deficiency of plasma thromboplastic component is called "hemophilia B". Because plasma thromboplastic component is also called "Christmas factor" from the name of a child suffering from its lack, "hemophilia B" is also called "Christmas disease".

The subtypes of hemophilia are deficiencies of antihemophilic globulin, plasma thromboplastin component, plasma thromboplastin antecedent and of a fourth thromboplastin component. Deficiency of a fifth thromboplastin precursor yields a blood with notably defective coagulation as tested in the laboratory, but is not associated with hemorrhagic symptoms (13). Clotting disorders due to *hypofibrinogenemia* and *hypoprothrombinemia* may be congenital or acquired.

Acquired. As has been stated, liver disease may result in the deficiency of fibrinogen or prothrombin, both of which are synthesized in the liver. Deficiency of proconvertin may also be acquired in this manner. Vitamin K deficiency may interfere with prothrombin production. All these conditions interfere with clotting.

In addition, there are substances which act as inhibitors or antagonists of one or another of the factors involved in clotting. *Hirudin* is an antithrombin from the leech. (The importance of hirudin to the leech, a blood-sucking animal, is obvious.) A cattle disease, resulting from ingestion of improperly cured sweet clover, is caused by *dicumarol* (see page 838). It is an antagonist to vitamin K and therefore interferes with prothrombin production. Dicumarol is used to prevent thrombosis (intravascular clotting) in man, as in patients following operation, but its action is delayed 24 hours or more and it is therefore useless in emergencies.

A substance which does act immediately is *heparin.* It is a mucopolysaccharide, containing mucoitin sulfuric acid and is distributed throughout the body, being most concentrated in the liver and lungs. Heparin plus a cofactor which comes from the plasma can act as an anticoagulant since it has properties of antithromboplastin, antiprothrombin, and antithrombin. Purified preparations of heparin are used to render the blood promptly incoagulable *in vivo.*

The simplest way to keep blood from coagulating when collected for storage is to collect it in a vessel containing a substance such as citrate or oxalate which will serve to bind the calcium ion necessary for clotting. Alternatively, fibrinogen may be removed by whipping.

Fibrinolysis

If clotted blood is allowed to stand, the clot first shrinks (syneresis) and eventually dissolves (lysis). The lysis of blood clots, and possibly also the syneresis, can be explained by the action of an enzyme in plasma. Enzymes are proteins with specific catalytic actions; the next chapter in this book is given over to the subject of enzymes. The particular enzyme most concerned in clot lysis is *plasmin,* which catalyzes the hydrolysis of certain peptide bonds in proteins, including fibrin and fibrinogen, yielding products of lower molecular weight and greater solubility. From the action of plasmin on synthetic peptides, it is concluded that the enzyme splits those peptide bonds involving residues of arginine or lysine (43).

The catalytic activity of plasmin in circulating blood is small for two reasons: most of it is present as a precursor, *plasminogen,* which has no catalytic activity, and there are substances in plasma which inhibit the catalytic activity of plasmin itself. Conversion of inactive plasminogen to active plasmin is brought about by several *kinases* (enzyme activators) of varying origin. One kinase is present in plasma, and presumably accounts for the small activity of plasmin normally observed, which increases on standing. Another kinase can be identified in tissues. Other kinases, which act more rapidly, can be obtained from cultures of certain pathogenic bacteria. Examples of these are *streptokinase* and *staphylokinase.* Activation with bacterial kinases shows much variation in different animal species;

in the human, streptokinase is definitely effective, but staphylokinase very slightly. Conversion of plasminogen to plasmin can also be moderately accelerated *in vitro* by chloroform; this is usually explained as the result of alteration of a substance in plasma which normally blocks the conversion. Patients who have had streptococcal infections show a decrease in proteolytic activity of serum as activated by streptokinase. Severe injuries have been shown to cause an increase in spontaneous proteolytic activity, decreasing with recovery from traumatic shock (7).

BLOOD PLASMA

Osmotic Pressure

The hydrostatic pressure in the capillaries, which is produced by the contractions of the heart, tends to expel fluids from the blood, and this is opposed by the osmotic pressure of the plasma proteins which has the effect of drawing water into the blood from the tissue spaces. The walls of the capillaries are permeable to small ions such as Na^+, Cl^-, and HCO_3^-, so there is no differential osmotic effect due to these ions. It is the larger molecules, particularly the proteins, of the blood which are important in maintaining the volume of the circulation. The osmotic pressure due to these molecules is often called the colloid osmotic pressure. This reminds us again that the osmotic pressure of a solution can have more than one meaning. In one sense it is a thermodynamic function calculated from the concentration of dissolved particles present. In another sense it is the pressure which must be exerted on the solution to prevent the entrance of additional solvent through a semi-permeable membrane with which the solution is in contact. Thus, the colloid osmotic pressure acts to draw water into the blood and to maintain a proper volume of circulating fluid.

As we shall see later, the proteins of the blood plasma may be roughly divided into two classes—albumin and the globulins. Serum albumin has a molecular weight of about 70,000. Most of the serum globulins have molecular weights of the order of 160,000. It will be recalled that the osmotic pressure depends on the number of dissolved particles per unit volume. It will therefore be appreciated that a one per cent solution of serum albumin will have more than twice the effect on the colloid osmotic pressure than will a one per cent solution of serum globulin. There is about twice as much albumin present as globulin, and as far as our present knowledge goes, the osmotic activity of serum albumin is its chief function.

Another factor which increases the importance of serum albumin in maintaining the osmotic pressure of the blood is based on the fact that it has a lower isoelectric point than globulins. Consequently, at physiological pH it carries a higher negative charge, and thus is associated with a larger number of cations such as Na^+. These cations function also as particles in determining osmotic pressure and although the walls of the capillaries are permeable to such ions, enough cations must remain in the blood to neutralize the negative charges on the protein molecules. It has been found that one gram of serum albumin holds about 18 ml. of water in the blood. This figure is arrived at both from clinical experiments following the injection of concentrated solutions and laboratory experiments on the osmotic pressure of purified albumin.

Although some, perhaps all, of the globulins have functions of importance other than maintenance of osmotic pressure, they also act as negatively charged colloid ions and contribute about 20 per cent of the colloid osmotic pressure of the blood.

The importance of the osmotic function of the plasma proteins is clearly shown when there has been an acute loss of blood. Even before the amount of blood lost is sufficient to lower the oxygen and carbon dioxide transporting power seriously, grave results are observed, and in severe cases the the patient goes into shock (see Chapter 17).

In conditions of shock resulting from an acute loss of plasma proteins, the introduction of human plasma has been extensively employed and is of great value. The plasma was originally obtained by centrifuging off the formed elements of blood to which citrate had been added as an anticoagulant. Later it was demonstrated that this plasma could be dried from the frozen state and after being redissolved in the proper amount of distilled water, would still carry out its osmotic function.

Since the albumin of plasma accounts for about 80 per cent of the colloid osmotic pressure, it would be expected that human serum albumin would be effective in combating shock. This was found to be true. The unit of serum albumin was established as 100 ml. of a 25 per cent solution, and it may be computed from the preceding figures that this represents the osmotic equivalent of about 500 ml. of citrated plasma.

It was observed that patients usually responded almost immediately to the injection of concentrated albumin with a fall in hemoglobin concentration and hematocrit readings and a subsequent decrease in serum protein concentration, indicating that extravascular water had been transferred into the circulation. In cases in which the patient was markedly dehydrated, it was found best also to administer saline solution to provide the necessary fluid.

Serum albumin has been found useful in combating edema, which is due, at least in some cases, to loss of plasma protein particularly of the serum albumin which is osmotically the most effective. It is an old clinical observation that when a great deal of albumin has been lost by albuminuria and the plasma proteins are low, proportional degrees of edema are observed. The most effective preparations of serum albumin for this purpose are those containing a relatively low concentration of salts.

Buffer Action

Since most of the plasma proteins have isoelectric points distinctly acid to ordinary physiological pH, most of these proteins are behaving under physiological conditions as anions. The blood proteins and their salts are mixtures of weak acids and their conjugate bases and are therefore typical buffer systems (see Chapter 1). In addition to the protein buffer systems in the blood, we have other systems such as the dissolved carbonic acid and bicarbonate ion and mixtures of phosphates.

Plasma Proteins

Blood plasma contains a very complicated mixture of proteins which, out of necessity, we deal with in groups, as serum albumin (or plasma albumin) and various kinds of globulins. In addition, there are proteins which are treated more nearly as individuals. Some have already been mentioned; for instance, fibrinogen and the other protein-clotting factors.

Of the proteins, fibrinogen, prothrombin and serum albumin are formed in the liver exclusively. Hepatectomized rats show a steep decline in serum albumin levels and so do patients with cirrhotic livers. At least a portion of the serum globulins, notably the gamma-globulins, which are the antibody fraction of plasma proteins, seem to be synthesized in tissues other than the liver, particularly in lymphatic tissues (8).

Protein synthesis by the liver is quite rapid. This may be demonstrated by *plasmapheresis*, a technique in which an animal is bled and the washed blood cell fraction, suspended in physiological saline, is restored to the animal. In this way, plasma protein concentration is reduced to a minimum. Within 24 hours thereafter, however, a considerable percentage of the plasma protein has been restored. There is a daily turnover of up to 25 per cent of the total circulating plasma protein. A dog weighing 11 kilograms can produce 13 grams of plasma proteins per week or about 170 mg./kilogram body weight/day.

Paraproteins are abnormal plasma proteins and are unusually homogeneous, as shown by sharp peaks in the electrophoretic patterns (19). Among these abnormal proteins may be listed: *macroglobulins*, which have molecular weights of approximately 1,000,000 and sedimentation constants in the 19 to 20 S range; *cryoglobulins*, which form gels at low temperatures, and which vary widely in molecular weight and sedimentation constant; and the *myeloma proteins*, which are usually of molecular weight less than 300,000. All of these paraproteins have amino acid compositions not strikingly different from that of the normal gamma-globulins. The presence of paraproteins in the blood plasma is associated with diseases involving the reticuloendothelial system, where gamma-globulins are produced.

Salt fractionation of plasma proteins. The traditional method of separation of plasma proteins, as with other proteins, was fractional salting out with ammonium sulfate. It was found that if plasma was about 20 to 25 per cent saturated with ammonium sulfate, the fibrinogen was precipitated. One-third saturation (33.3 per cent) separated out a fraction which was called the euglobulin fraction because it would not redissolve in distilled water and thus behaved as a true globulin. Further increase in the concentration of ammonium sulfate to half-saturation (50 per cent) removed the pseudoglobulin fractions which were soluble in distilled water. The remaining protein was called the albumin, and was nearly completely precipitated if the supernatant containing it was brought to full saturation with ammonium sulfate. The fractions so obtained were not pure and were capable of further separation.

In 1921, Howe (16) published a method which has been widely used clinically. Howe used sodium sulfate instead of ammonium sulfate, since in this way his precipitate could be directly digested in the Kjeldahl procedure for nitrogen determination; whereas with precipitates from ammonium sulfate, dialysis was necessary to remove the NH_4^+, and in some cases doubt remained that it had all been removed. The majority of the data available on plasma protein changes in states of disease were until very recently obtained by the method of Howe.

In spite of the wide use of the method, it was early apparent that Howe's separation into components was exceedingly crude, and the precipitates overlapped in their content of individual proteins. This has since been shown by electrophoretic analysis of precipitates obtained by the Howe technique. A further difficulty involved in the Howe method was the uncertainty of the factor for the conversion of the nitrogen to protein. The nitrogen figures were multiplied by the conventional 6.25 which, as has been stated previously, was based on the assumption that the average nitrogen percentage of protein is exactly 16. The average nitrogen factor found experimentally in dried proteins from pooled normal human plasma was 6.3. The conversion factor of the individual proteins, however, varied from 6.1 to 8.4.

In spite of the deficiencies of the Howe method, results of definite clinical usefulness were obtained. The total serum protein according to the Howe technique varied from 6.5 to 7.5 grams per 100 ml. of blood, with a mean of 7.2 grams. The mean for the albumin value was 5.2 grams, and for the total globulins, 2.0 grams. An average A/G ratio therefore was about 2.6. It is obvious that either lowering of the serum albumin content or an increase of the serum globulin content, the first of which might occur in nephrosis for example and the latter in many types of infection, would give a decreased A/G ratio. However, such a change could result from a combined decline in albumin and rise in globulin such as is found in hepatic cirrhosis. Obviously, the single ratio was not adequate and it is preferable to give the absolute values of the serum albumin and serum globulin.

Alcohol fractionation of plasma proteins. Ammonium and sodium sulfates are, of course, by no means the only protein precipitants, and studies with various others were reported from time to time. In particular, it had long been known that ethyl alcohol, although at room temperature and ordinary concentrations a protein denaturant, might be used to precipitate plasma proteins if the addition were made carefully at low temperatures. Felton had made use of this procedure to separate the antibodies from horse serum.

During World War II, Cohn and collaborators developed large scale procedures for the low temperature separation of plasma fractions by the addition of ethyl alcohol. They pointed out that ethyl alcohol offers some advantages over salting out by the older methods. In the first place, since the alcohol is non-ionized, one may control the ionic strength of the mixture by the addition of suitable amounts of salts. Also, by the addition of suitable buffers, the pH, a very important factor in protein fractionation, may be controlled. The presence of alcohol permitted the use of temperatures below the freezing point of water.

The details of the process and results of the low temperature, low salt ethanol fractionation have been reviewed by Edsall (11). We may summarize them briefly here. The fractions were designated by Roman numbers. Originally two of the separate fractions were distinguished as II and III, but as these were later found to be very similar chemically, this procedure was abandoned and a combined II + III was removed.

Fraction I was obtained by precipitation with 8 to 10 per cent of ethanol by volume at pH 7.2—in other words, the usual pH of pooled citrated plasma—ionic strength 0.14 at a temperature of −3°C. which was very close to the freezing point of this mixture. This fraction contained about 60 to 65 per cent fibrinogen. In combination with thrombin preparations, it found considerable use as a hemostatic agent. Another component of this fraction is antihemophilic globulin.

Fraction II + III was obtained from the supernatant of Fraction I by raising the alcohol concentration to 25 per cent, changing the pH to 6.8, the ionic strength to 0.09, and the temperature to −5°C. This contained virtually all of the gamma-globulins, prothrombin, and most of the isoagglutinins (see p. 199). It also contains plasminogen, (the inactive precursor of plasmin, the proteolytic enzyme of plasma), the midpiece of complement (C'_1) (see page 852), a large proportion of the beta-globulins including β_1-lipoprotein which has been called x-protein by Macfarlane, and two other distinct β_1-globulins, a β_2-globulin, and some fibrinogen and antihemophilic globulin not removed in Fraction I. Phospholipids and sterols are consistently found in plasma coupled with protein as lipoprotein, and not as free lipids (44). Pathological increases in these lipids are usually associated with the β_1-lipoprotein fraction.

Fraction IV, which was usually further subdivided, appeared as Fraction IV-1 which was obtained from the supernatant of II + III by lowering the ethanol concentration to 18 per cent and the pH to 5.2. It contained a large proportion of the alpha-globulin, mostly as an α_1-lipoprotein containing about 35 per cent of lipids, and in addition, a blue-green pigment and other components. Fraction IV-4, obtained from the supernatant of IV-1 by increasing the ethanol concentration to 40 per cent by volume and the pH to 5.8, contained nearly lipid-free alpha and beta-globulins and some albumin. It contained an esterase and a β_1-globulin of relatively low molecular weight.

Fraction V was obtained from the supernatant of Fraction IV-4 by lowering the pH to 4.8 with acetate buffer so as to give an ionic strength of 0.11 while maintaining the ethanol concentration at 40 per cent. It was separated at a temperature of -5°C., and was composed almost entirely of albumin. The supernatant of Fraction V contained less than 2 per cent of the total protein and was made up mostly of albumin and alpha-globulin.

Methods of ethyl alcohol fractionation did not produce pure components of plasma, although by re-working some of the fractions substantially pure proteins could be obtained—notably in the case of serum albumin and gamma globulin. Some success was achieved in purifying the anti-A and anti-B isoagglutinins.

Pillemer and Hutchinson (33) revived the use of methyl alcohol in the fractionation of serum proteins. They found that at a concentration of 42.5 per cent methanol, at pH 6.7 to 6.9, ionic strength about 0.03 and temperature about 0°C., the globulins are almost quantitatively precipitated while almost all the albumin remains in solution. The agreement was within 5 per cent with electrophoretic data for normal sera; and for abnormal sera, within 5 to 10 per cent. The albumin values obtained were consistently and considerably lower than those derived by the older Howe method, being around 70 to 80 per cent of the Howe albumin levels.

The protein components of normal human plasma which have thus far been obtained in a relatively homogeneous state are set forth in table V-2 (31). The plasma also contains enzymes in addition to those concerned with coagulation and fibrinolysis. They make up only a small fraction of the plasma proteins, less than 0.1 per cent (44). They include amylase, beta-glucuronidase, esterases, phosphatases, and deoxyribonuclease.

TABLE V-2

Protein components of human plasma thus far obtained in relatively homogeneous state

ELECTROPHORETIC CLASS	APPROX. CONC. IN PLASMA	APPROX. MOLECULAR WEIGHT
	grams/100 ml.	
Albumin	3.2	69,000
α_1-globulin	0.2	200,000
α_2-globulin	0.1	300,000
β_1-globulin	0.2	90,000
β_1-globulin	0.2	150,000
β_1-globulin	0.1	500,000–1,000,000
β_1-globulin	0.2	1,300,000
β_2-globulin	0.2	150,000
γ-globulin	0.5	156,000
γ-globulin	0.1	300,000
Fibrinogen	0.3	330,000

Proteins (except for fibrinogen) arranged in order of decreasing electrophoretic mobility at pH 8. Thus, albumin moves faster than α-globulin, α-globulin moves faster than β-globulin, etc. Not all plasma proteins have been obtained in as pure a form as these listed here, consequently the values given here do not add up to the total value for proteins in plasma.

An unusual plasma enzyme is *ceruloplasmin*, a blue, copper-containing protein, with a copper content of 0.34 per cent. Since the enzyme has a molecular weight of 150,000, this means 8 copper atoms per molecule. Ceruloplasmin catalyzes the oxidation of amines, phenols and ascorbic acid. This enzyme is deficient in patients with Wilson's disease, a hereditary, progressive and fatal condition involving cirrhosis of the liver, degeneration of the basal ganglia, and abnormalities in copper and amino acid metabolism (36).

The ethanol method for fractionation of plasma, applied on a large scale by Cohn and his colleagues, has been applied to the measurement of gamma-globulin in small samples of plasma (22). This method gives a mean value for gamma-globulin of 0.61 ± 0.18 grams per 100 ml. of plasma or serum.

Electrophoresis of plasma proteins. Electrophoresis has been used to characterize plasma proteins. It has the advantage of giving relatively sharp values but is limited by the fact that it depends essentially on a single property of the protein, the mobility in an electric field at the pH used. We know that some plasma proteins, although very different in molecular size, shape, chemical constitution, and biological significance, happen to have the same or very similar mobilities and thus appear together as an apparently homogeneous peak. It is for this reason that with few exceptions the numerous antibodies contained in human plasma have not been separated from the other gamma-globulins. When the electrophoretic method was first introduced it was hoped that different and specific electrophoretic spectra would characterize different diseases. We now know that this is not in general the case, but the changes follow certain common patterns much as in the case of sodium sulfate fractions. For instance, a decrease in serum albumin is associated with many kinds of disease, especially those accompanied by wasting or malnutrition or an acute febrile illness. This decrease is especially pronounced when albumin is being lost in the urine as in nephrosis, or by exudation through denuded surfaces as in burns, or in disturbances in albumin formation such as cirrhosis of the liver. A summary of plasma protein changes in various diseases will be found in the reviews by Gutman (15) and Fisher (12).

Non-protein Constituents

Remembering that one of the chief functions of blood is transport, it is not surprising to find that most of the non-protein constituents are either substances going on their way to the tissues from the point at which they were absorbed, traveling from one tissue to another, or going from the tissues as waste products to be excreted. Many of these substances will be considered in later chapters. Some of them are so significant for diagnostic purposes that physicians must be familiar with their normal ranges. Table V-3 expresses the normal range (51) for each substance by four figures: the interval between the extreme values includes 98 per cent of the findings in normal subjects; the interval between the middle figures includes 80 per cent of normal findings. An alternative method of expressing the range is by the mean and standard deviations, which are given in the last two columns of the table. The range included between ($M + 2\sigma$) and ($M - 2\sigma$) contains 95 per cent of the normal values. This notation is not precise unless the frequency distribution of the observed values follows a normal curve. In many instances (51) this is not the case, but a normal frequency distribution curve can be obtained by plotting the logarithms of the observed values against their frequencies. This type of distribution is designated as lognormal.

Values outside the normal range are indicative, and values at the extremes of the normal range are suggestive of nutritional, metabolic, or excretory disturbance. The mechanisms of some such disturbances will be presented in later chapters. Blood analysis as related to the diagnosis and management of particular diseases is part of the subject matter of pathology and medicine, and is treated in the literature of these disciplines.

The total lipids of blood plasma were found to be 507 ± 73 (on a series of 21 subjects) (18). The components of the blood lipids will be discussed in Chapter 15. Hormones, vitamins, organic acids, and phenols are present

TABLE V-3

Normal values for diagnostically significant blood constituents

SUBSTANCE	RANGE				FREQUENCY DISTRIBUTION*	MEAN† M	STANDARD DEVIATION† σ
	Lower 1%	Lower 10%	Upper 10%	Upper 1%			
In whole blood: (mgm. per 100 ml.)							
Glucose	55	68	96	109	Normal	91.4	6.2
Non-protein nitrogen	25	29	43	51	Lognormal	33.8	2.75
Urea nitrogen	12	16	35	47	Lognormal	15.0	1.6
Uric acid	0.6	1.6	3.9	4.9	Normal	3.2	0.5
Inorganic P	2.0	2.4	3.5	3.9	Normal	3.5	0.38
In serum or plasma: (mgm. per 100 ml.)							
Cholesterol	123	153	260	324	Lognormal	180	28
(mEq. per liter)							
Sodium	133	137	148	152	Normal	144‡	3.6‡
Potassium	3.5	3.9	5.0	5.6	Lognormal	4.52‡	0.45‡
Calcium	4.5	4.8	5.4	5.7	Normal	5.25	0.15
Chloride	99	101	106	108	Normal		

* From data of Wooton and King (51).

† From data of Jellinek and Looney (18), except for sodium and potassium values.

‡ From data of Marinis *et al.* (28).

in the blood in small amounts. In diabetes and starvation appreciable amounts of acetone and acetoacetic acid may occur. The color of the plasma is chiefly due to bilirubin, with a minor contribution of carotenoids.

The *amino acid nitrogen* of the various amino acids on their way to and from the tissues occurs to the extent of 5 to 8 mgm. of nitrogen per 100 ml. All the amino acids are represented, glutamic acid, lysine and alanine being present in highest concentration (46). Amino acid concentration may be increased in leukemia, diseases involving damage to the liver and in severe nephritis. *Ammonia nitrogen* does not occur in the blood in significant amounts. The ammonium ion is rapidly converted to urea in the liver (see Chapter 16). By this means the concentration of the quite toxic ammonia is kept at very low levels in the blood. In hepatic disease, the concentration of ammonia in the blood may increase. At concentrations above two micrograms per ml., ammonia is taken up by the brain. Hepatic coma, a terminal event in cirrhosis of the liver, is associated with increased blood ammonia.

THE FORMED ELEMENTS OF THE BLOOD

Within the blood plasma float numerous cells which have a definite structure and a definite form, although some of them are plastic and can admit to a considerable amount of deformation. These are known as the formed elements of the blood; they amount to about 42 to 50 per cent by volume of the normal blood of the male, 40 to 48 per cent for the female.

The formed elements include the red discs or erythrocytes, several types of white cells or leukocytes, and the platelets or thrombocytes. The erythrocytes far surpass the other formed elements in number, and are responsible for the sex difference in total volume of formed elements. The thrombocytes are involved in the clotting process, as already described. The white cells are in many ways similar to tissue cells in chemical composition and physiological activity, but have certain distinguishing characteristics.

The White Cells (Leukocytes)

Some types of leukocytes are motile and are capable of phagocytosis, an important function of the blood and the chief defense against microbial invasion.

All types of leukocytes, like other tissue cells, possess the catalytic proteins (enzymes) and associated organic compounds (coenzymes) necessary for the conversion of glucose to glycogen, and for the utilization of these substances as energy sources, with CO_2 and H_2O as the end products. The glycogen stored in granulocytic leukocytes greatly exceeds the amount stored in other types of blood cells. The basophils contain more histamine and heparin than other leukocytes. White cells in general contain sulfhydryl compounds, including glutathione, in concentrations comparable to those in tissue cells. The sulfhydryl content of white cells is about seventy times that of red cells. Variations in disease are not clearly significant.

The enzyme *glyoxalase*, which is present in red cells, is in even greater amount in white cells. The ratio of activities in white and red cells is over 700 to one. The reaction catalyzed by glyoxalase is the conversion of pyruvic aldehyde (also called methyl glyoxal) to lactic acid by the addition of one molecule of water (formula I). The presence of this enzyme is somewhat of a puzzle since there is no known definite metabolic pathway for the formation of methyl glyoxal in the body, and this substance must be added when testing for the enzyme *in vitro*. Glutathione is a necessary co-

$$\underset{\text{Pyruvic aldehyde}}{CH_3\overset{\overset{O}{\|}}{C}CHO} + H_2O \xrightarrow[\text{+ glutathione}]{\text{glyoxalase}} \underset{\text{Lactic acid}}{CH_3\overset{\overset{OH}{|}}{C}HCOOH}$$

I. Hydration of pyruvic aldehyde

enzyme for glyoxalase, but the difference in activity between white and red blood cells does not depend upon the difference in glutathione content (29).

The mean life span of a white cell has been estimated to be about 13 days. The life of lymphocytes is longer, and of granulocytes shorter.

The Red Cells (Erythrocytes)

The red cells are the most numerous cells in the blood and their function is well understood. Mature erythrocytes are composed almost entirely of hemoglobin, water, and stroma. The stroma is often spoken of as merely the cell membrane, but there is some evidence that it permeates the entire cell. This is supported by observation. For instance, erythrocytes can be partially hemolyzed, that is, when immersed in mildly hypotonic solutions such that the osmotic pressure allows water to enter and rupture the cell membrane, not all the hemoglobin is lost. If restored to an isotonic solution a large number of gradations between ordinary cells and completely hemolyzed cells, commonly referred to as ghosts, may be obtained. The stroma may be obtained after hemolyzing the corpuscles with distilled water. It is found to be rich in phospholipid and cholesterol. These constitute about 20 per cent or more of the dry weight of the stroma. There are also inorganic materials, and proteins peculiar to the erythrocyte envelope. The stroma also contains the numerous blood group substances which will be discussed later in this chapter. The red corpuscles of man and most animals tend to accumulate K^+ preferentially to Na^+, so that there is a characteristic difference in the concentration of the two ions within and without the cell.

Sulfhydryl groups appear to be necessary for the integrity of the red cell surface. Compounds of silver and mercury (including some organic mercurials) and of other metals which form insoluble mercaptides will cause lysis of red cells. Such lysis can be inhibited by sulfhydryl compounds such as reduced glutathione.

Glutathione and ergothioneine (another sulfhydryl compound, see page 643) are present in red cells, but not significantly in plasma. Glutathione is present in red cells at a mean concentration of 55 mgm. per 100 ml.; ergothioneine is present in white adults at a mean concentration of 11 mgm. per 100 ml., and in Negro adults at 15 mgm. per 100 ml. Values for infants are lower, but there is still a significant difference between Negro and white (42). This may be connected with the greater susceptibility of white infants to poisoning by well-water contaminated with nitrate.

The average life span of a circulating red cell, as measured by tagging with radioactive Cr^{51}, is about 110 days (30).

As compared with white cells, red cells are much less active in the utilization of oxygen and of simple foodstuffs such as glucose. The enzyme systems involved in these processes are present in red cells in lower concentrations, or may be incomplete or lacking. The enzyme concentration of red cells falls off as the cells increase in age (2).

Hemoglobin

The main protein contained in the red cells of the blood is hemoglobin. This protein is present in such high concentration that individual molecules are within 0.8 millimicrons of touching one another (34). Certain abnormal hemoglobins, less soluble than normal hemoglobin, actually do precipitate under low partial pressure of oxygen (see page 197). Hemoglobin molecules are not arranged entirely at random as they would be if they existed in solution within the erythrocyte. Electron microscopy shows micelles of hemoglobin in loose, thread-like arrangement.

Hemoglobin is a conjugated protein made up of a porphyrin derivative, heme, as the prosthetic group of the basic protein, globin, usually classified as a histone. The iron-containing prosthetic group has been synthesized and successfully coupled to the protein portion, globin. Hemoglobin contains 0.335 per cent iron. If we assume one atom of iron (atomic weight 55.9) per molecule, this percentage leads to a minimum molecular weight of about 16,000. As in many other examples of calculation of minimal molecular weight, it turns out that this value is too small. Osmotic pressure and ultracentrifugal measurements show that in fact the molecular weight of hemoglobin is about 68,000, which shows there must be four atoms of iron per molecule. Hemoglobin has the property of combining reversibly with oxygen, and it is this property which makes it possible for oxygen to be transferred in adequate amounts from the lungs to the tissues. If the absorption spectrum of *oxyhemoglobin*, that is, hemoglobin combined with oxygen, is examined, it shows two sharp absorption bands in the green, leaving the transmitted light which we see a bright red. Dilute solutions of oxyhemoglobin have a yellowish tinge. Examination of red cells with high magnification under the microscope demonstrates the same yellow color.

Porphyrins. The iron-containing portion of hemoglobin is a porphyrin derivative. Porphyrins are compounds containing the porphin ring (formula II), which is made up of four pyrrole rings connected by four methene (—CH=) bridges. The porphin ring contains eleven conjugated double bonds so that like the benzene ring, it is stabilized by resonance. The position of the double bonds, if shown as in formula II, is therefore strictly arbitrary, and in actuality all the bonds in the ring have a certain amount of double-bond character.

For convenience, the porphin ring is usually shown in various conventionally simplified forms. The one we shall use is that suggested by Lemberg and Legge (25), according to which the formula for porphin would be as

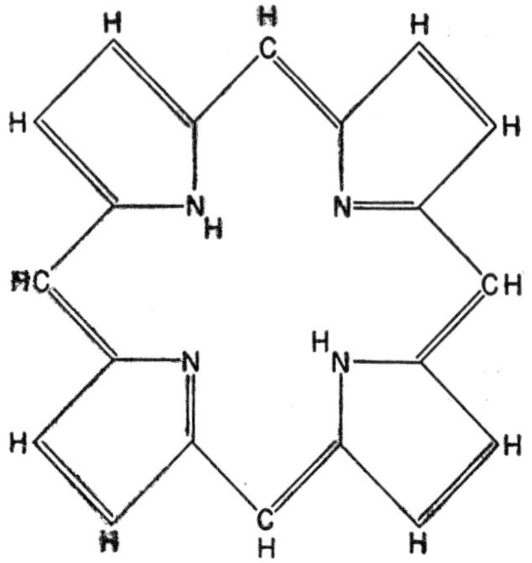

II. Porphin

in formula III, where the inward pointing angles signify nitrogens and the rest of the pyrrole rings are not represented, and where the outward pointing angles are the methene bridges between the pyrrole rings. The free corners of the four pyrrole rings, indicated by the dotted lines in the Lemberg-Legge representation are numbered from 1 to 8 and the four methene bridges are marked by the Greek letters alpha to delta.

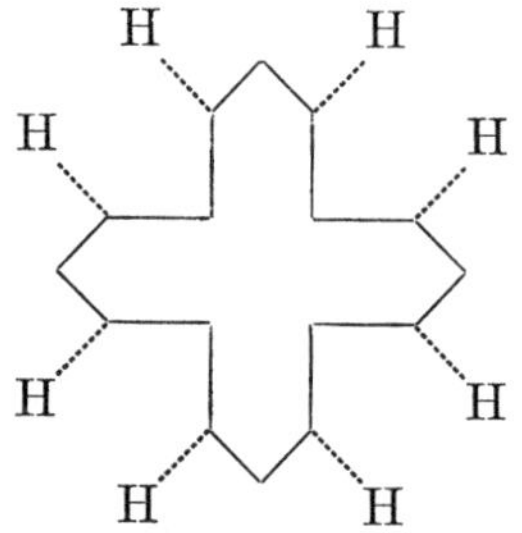

III. Porphin according to Lemberg's notation (25)

The naturally-occurring porphyrins all contain side chains on each of the eight free corners of the pyrrole rings. The naturally-occurring side chains are methyl, ethyl, hydroxyethyl, formyl, vinyl, carboxymethyl and carboxyethyl groups. When natural porphyrins were first isolated, the side chains were converted by reduction or decarboxylation to methyl and ethyl groups only. It was found then that each pyrrole ring contained one methyl side chain and one ethyl side chain, making four of each altogether. Such a porphyrin is an *etioporphyrin*.

There are four possible etioporphyrins, which differ among themselves only in the positions of the 4 methyl groups and 4 ethyl groups (where each pyrrole ring has but one of each). These four isomers are numbered from

I to IV and they are shown in formula IV, where M stands for methyl group and E stands for ethyl group.

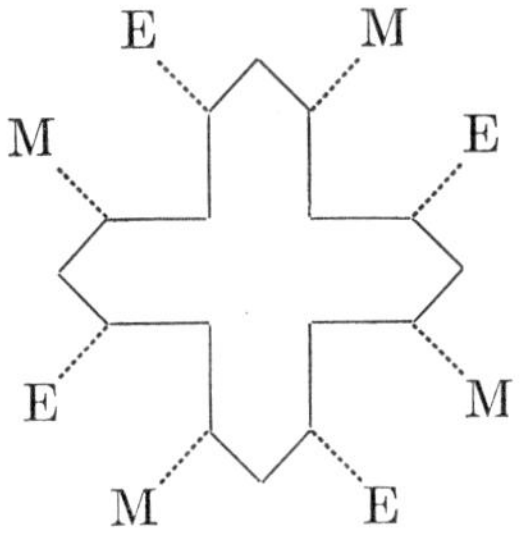

IV. Isomeric forms of etioporphyrin

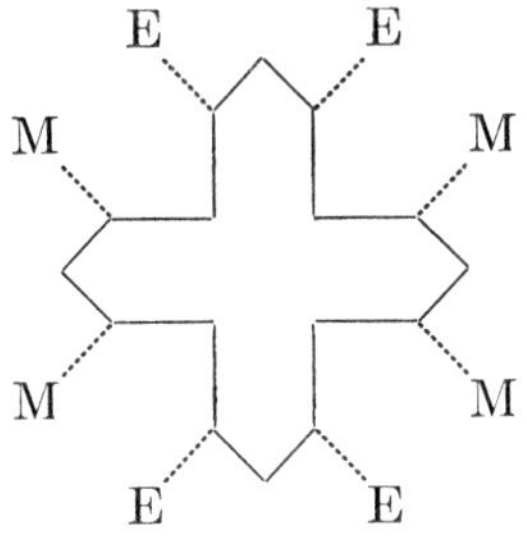

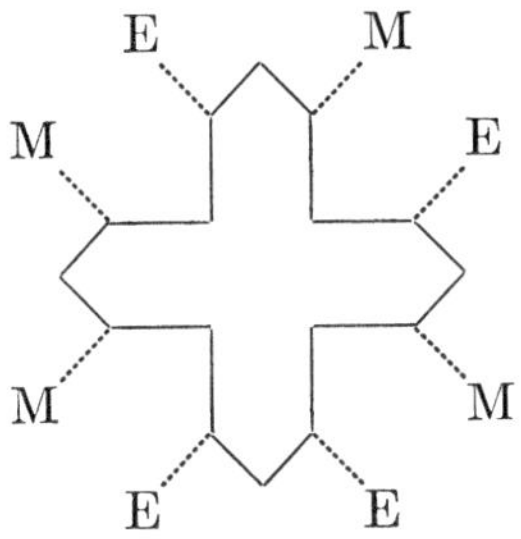

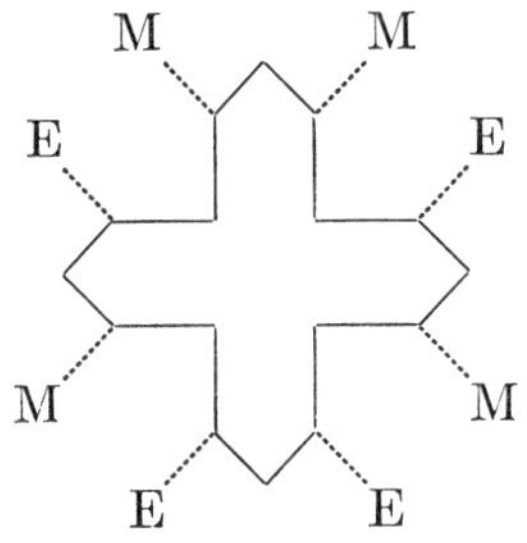

All naturally-occurring porphyrins can be considered as belonging to one of these series of etioporphyrin isomers and may be numbered accordingly. The etioporphyrin III series occurs most commonly in nature, though the etioporphyrin I series is also represented. For instance *coproporphyrin III* (formula V) is a normal constituent of human feces and, in very low concentration, of human urine. The coproporphyrins in general contain 4 methyl groups and 4 carboxyethyl (propionic acid) groups, each pyrrole ring bearing one of each. If the carboxyethyl groups are decarboxylated to ethyl groups in the case of the coproporphyrin shown in formula V, it becomes etioprophyrin III.

A *uroporphyrin* is one which contains four carboxymethyl (acetic acid) groups and four carboxyethyl groups on the porphin ring, each pyrrole ring bearing one of each. One is shown in formula VI and this occurs in the urine of people with the disease, congenital porphyria (see page 775). If the eight side chains of the uroporphyrin shown in formula VI are decarboxylated to methyl and ethyl groups, etioporphyrin I is formed. The original compound is therefore named *uroporphyrin I*.

This numbering system falls down in the case of the *protoporphyrins*,

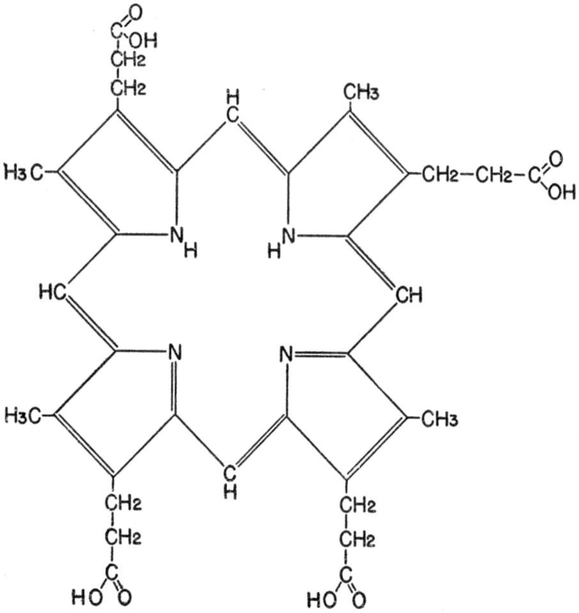

V. Coproporphyrin III

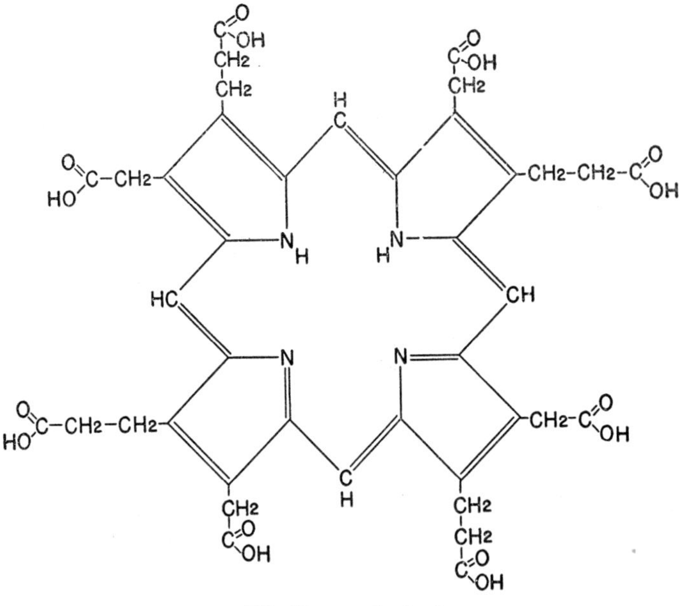

VI. Uroporphyrin I

which are porphyrins containing 4 methyl groups, 2 vinyl groups and 2 carboxyethyl groups, each pyrrole ring bearing one methyl group and one of the others. With three different kinds of side chains, 15 isomers are possible. One such protoporphyrin occurs in hemoglobin and in order to determine which it was, Hans Fischer numbered the possible isomers from 1 to 15 and supervised the synthesis of each. The isomer found to be identical with the natural substance happened to be the one numbered 9, so that the natural porphyrin of hemoglobin is called *protoporphyrin IX* (formula VII).

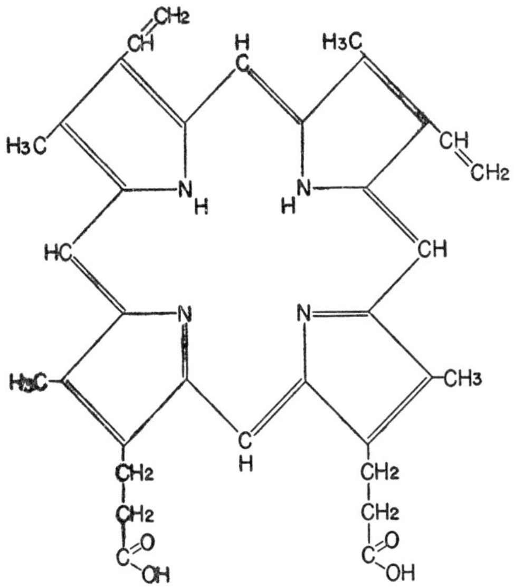

VII. Protoporphyrin IX

If the vinyl groups of protoporphyrin IX are reduced to ethyl groups and the carboxyethyl groups decarboxylated to ethyl groups, etioporphyrin III results. Protoporphyrin IX therefore belongs to the etioporphyrin III series but it must never, on that account, be referred to as protoporphyrin III, as it sometimes is (inaccurately) in the literature. It is possible to interchange the vinyl and carboxyethyl groups of protoporphyrin IX in such a way that a different protoporphyrin results, yet this different one is also reducible to etioporphyrin III and is also a member of that series.

The porphyrins, in general, absorb various wave lengths of visible light, just as the carotenoids do, and for the same reason, the presence of conjugated double bonds. In fact, the word, porphyrin, comes from the Greek word for "purple". It is the porphyrin of hemoglobin that is responsible for the red coloring of blood. In this case, the color has no known relationship to function, but in the case of chlorophyll, which also contains a porphyrin ring, the absorption of visible light is of the very essence of its function.

Heme. The porphyrins that occur in nature are usually associated with a metallic ion. In hemoglobin and in certain enzymes, the metal is iron. In chlorophyll, it is magnesium, and in the cobalamins (see page 816), a porphyrin-like ring system is associated with cobalt. Copper and zinc are also involved in some cases.

The combination of iron with protoporphyrin IX is *heme*. Of the six coordination valences of the ferrous ion, four are linked to the four pyrrole nitrogens of the porphyrin ring. The remaining two are available for the formation of other links. In the heme of hemoglobin the iron is in the divalent ferrous state (Fe^{++}) and the charge is balanced by a single negative charge on each of two pyrrole nitrogens which have lost their nitrogens (compare formulas VIII and VII). As a result of resonance, the charge is distributed over all the nitrogens and so all the Fe—N bonds are equivalent.

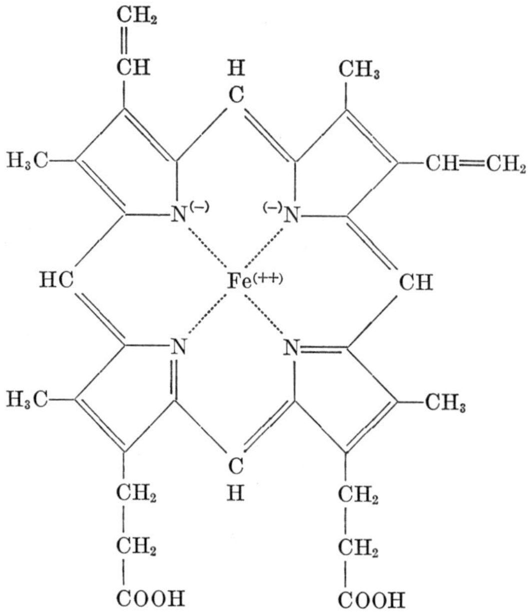

VIII. Heme

The iron atom in hemoglobin can, under abnormal conditions, be oxidized to the trivalent ferric state (Fe^{+++}), and in some enzymes it occurs normally in that condition at least part of the time. This oxidized heme is *ferriheme*, often called *hemin*. Hemin has a net positive charge of 1 and must be balanced by a negative ion, which, when the substance is isolated, usually turns out to be Cl^-. Ferriheme can be isolated with its charge balanced by OH^-, in which case the two carboxyethyl groups ionize to COO^- and must be balanced by two positive ions, usually Na^+. This form of ferriheme is called *hematin*.

Hemoglobin structure. The hemoglobin molecule is made up of two equivalent halves which can be reversibly split apart in urea solutions. The molecule as a whole is 6.5 millimicrons long, 5.5 millimicrons high and 5.5 millimicrons wide (14). There are two hemes at each end of the hemoglobin molecule, making four in all. The heme group is 1.4 by 1.7 millimicrons.

We have said that four of the coordination valences of the ferrous ion of heme are bound to the four pyrrole nitrogens of the porphyrin. Of the remaining two, one is bound to the nitrogen of the imidazole side chain of a histidine residue of the protein portion of hemoglobin. The sixth and last is bound to the oxygen of a water molecule (23). (When the hemoglobin is oxygenated, forming *oxyhemoglobin*, this last bond is attached to an oxygen molecule instead.)

Several of the side chains of the heme molecule also serve to help bind the heme to the protein. The carboxyethyl side chains, in the carboxylate ion form at physiological pH, form electrovalent links with the positively-charged side chains of lysine or arginine residues. The vinyl side chains probably link up with the sulfhydryl side chains of cysteine. This certainly happens in some heme-containing proteins (see page 278), and globin itself has sulfhydryl groups occurring in clusters that may fit the vinyl positions in heme neatly (17). Each heme is thus bound to protein by five linkages. The methyl side chains of heme fulfill no obvious function.

Although the rate of hemoglobin synthesis in the body can vary significantly, the rate of heme biosynthesis is found to be consistently equivalent to the rate of globin biosynthesis. This suggests that hemoglobin is synthesized complete, rather than by the union of preformed heme and globin (24a).

In the body, hemoglobin molecules are eventually degraded, the iron removed and stored for further use, the porphyrin ring broken and then detached from the globin to form *bile pigments* which are excreted chiefly through the liver bile. For details of this process, see page 508.

If the iron of hemoglobin is oxidized to the ferric state, we then have the compound called by Lemberg hem*i*globin, and by others *methemoglobin*, which no longer has the power of transporting oxygen. The oxidation of hemoglobin to methemoglobin will occur in circulating red cells as a result of poisoning with nitrates or nitrites. Hemoglobin removed from the red cells and kept in solution above pH 4 and in the presence of oxygen will slowly change to methemoglobin. Hemoglobin in solutions below pH 4 becomes denatured, with separation of heme from the denatured globin. This denaturation of globin is to some degree reversible on restoring the pH to 7.3.

While man and other warm blooded animals make use of hemoglobin as the oxygen-transporting compound, certain lower forms have red heme-containing proteins of larger molecular weight possessing iron. Other lower organisms possess a large protein molecule which contains copper but not porphyrins. These compounds are called *hemocyanins*. Some snails have copper-containing hemocyanin, others have iron-containing compounds related to hemoglobin. Some of the smaller snails have no detectable respiratory pigment.

Fetal hemoglobin. Maternal blood does not circulate through the veins and arteries of the unborn child, although popular superstition would often have it so. The fetal circulation is distinct, separated in the placenta from the maternal only by a thin membrane. Fetal hemoglobin, or hemoglobin F, in most respects resembles adult hemoglobin. The molecular weights are the same and there are no differences in the heme portion. Such differences as exist must therefore be attributed to the globin fraction. Fetal hemoglobin is less soluble than adult hemoglobin in phosphate buffer solutions, and is more resistant to denaturation with alkali. The two hemoglobins have different crystal forms, and can be differentiated by immunological methods (5), by electrophoresis, and by ultraviolet absorption spectrometry. The production of fetal hemoglobin is stimulated by low partial pressures of oxygen such as prevail in fetal tissues. Fetal hemoglobin will combine with more oxygen at a given low partial pressure than will adult hemoglobin and therefore works more efficiently for the fetus (45). The change from fetal to adult hemoglobin begins during intra-uterine life: at 20 weeks of pregnancy, hemoglobin in the fetus contains about 6 per cent adult hemoglobin (3); at birth, up to 35 per cent; and at 4 months after birth, 80 to 90 per cent. Small quantities of fetal hemoglobin (0.3 to 0.4 per cent of total hemoglobin) persist in normal adults (16a).

Transport

The blood fulfills its function as a transporting medium in a variety of ways. Objects of cellular or near cellular size, such as the leukocytes, erythrocytes, and thrombocytes are carried along in suspension by the force of the current. Some of the leukocytes also have the faculty of independent motion. Soluble substances such as glucose, amino acids, urea and the plasma proteins are transported in the plasma. Material may also be transported in association with proteins of the plasma or of the erythrocytes. Oxygen is carried chiefly by hemoglobin and carbon dioxide in a variety of forms (see page 726) including compound formation with hemoglobin and plasma proteins.

Lipids, such as the fat-soluble vitamins, steroids and carotenoids, which are insoluble in water, and hence in the blood plasma, are probably transported by means of the lipoproteins of the plasma. These are about three-quarters lipid (consisting mostly of phosphoglycerides and cholesterides) and therefore behave as fat solvents. Plasma albumins and globulins transport less abundant substances such as hormones. Some gamma-globulins (antibodies) transport antigens.

Some of the heavier metallic ions, notably iron, are carried by reversible combination with plasma proteins. The $beta_1$-globulin, *siderophilin*, can associate reversibly with up to 1.25 micrograms of iron per milligram of protein. This means that each molecule can bind two atoms of iron. Siderophilin has a molecular weight of 90,000 and contains about 1.8 per cent carbohydrate. Siderophilin will also bind copper and zinc, but not as strongly.

The Anemias

The term *anemia* is generally used in clinical medicine to refer to a reduction below normal in the number of red corpuscles per cubic millimeter, the concentration of hemoglobin, the volume of packed red cells obtained by centrifugation. Conversely, polycythemia designates a condition where the number of red cells is increased.

There are a very large number of types of anemias; they are usually classified by laboratory examination of the blood. If there is a proportionate decrease in the number of corpuscles, the quantity of the hemoglobin, and the volumes of the packed red cells, indicating that the average content of hemoglobin in the cell and the average size of the cell has not been changed, then this is called a *normocytic anemia.* In some cases there is a greater decrease in the number of cells than in the volume of packed cells due to the fact that the majority of such red cells as are produced is larger than normal; this is called a *macrocytic anemia.* More commonly, the reverse of this is found where the majority of the corpuscles is smaller than normal; this is called a *microcytic anemia.* The word anemia, as can be seen from its derivation, implies a lack of blood. One of the most striking symptoms of anemic patients is pallor, indicating less hemoglobin underneath the skin.

Macrocytic anemias. Pernicious anemia has been conclusively demonstrated by Minot and Murphy to be due to deficiency of factors concerned in red cell maturation. It was shown not to be due to deficiency of iron.

Much later, an antipernicious anemia factor, cyanocobalamine (page 816), was isolated. It is not absorbed by patients affected with pernicious anemia. Normally, cyanocobalamine is stored in the liver; and liver feeding was the first effective treatment of pernicious anemia, being later replaced by the use of concentrated liver extracts. Other macrocytic anemias arise as a result of a deficiency of other substances necessary in the formation of red cells, such as folic acid (page 810). A progressive macrocytic anemia is one of the numerous congenital disorders seen in the Fanconi syndrome.

Microcytic anemia. Microcytic anemia results from lack of iron. Since the time of Hippocrates iron salts have been used for anemia by physicians. It has been stated that the origin of this therapy dates back to sympathetic magic, because the weak patient hoped to acquire the strength of steel by drinking water in which a sword had rusted. Three centuries ago the use of iron in the form of filings which had been steeped in cold Rhenish wine was introduced by Sydenham into clinical medicine in treating a type of microcytic anemia, chlorosis. No doubt the acetic acid of the wine dissolved enough of the iron to benefit the patient. In 1832, Pierre Blaud emphasized the specific value of iron in the treatment of chlorosis, and described his now famous pills which contain ferrous carbonate.

Post-hemorrhagic anemia results from loss of blood, which of course involves a loss of iron. It may be treated by blood transfusion, or if it is not too severe, iron plus protein therapy may be used.

The normocytic anemias. The anemia associated with the majority of chronic infections is sometimes called simple chronic anemia. There is no very radical alteration from the normal red cell morphology. The beginning is insidious and recovery is often very slow. Myelophthisic anemia is associated with destructive processes in the bone marrow. High protein diets have been effective in the treatment of simple chronic anemia. The pathological changes leading to myelophthisic anemia, on the other hand, are frequently severe and irreversible so that no form of therapy has any prolonged effect.

Hemolytic anemias include a large number of conditions which vary according to their cause and severity and have one feature in common which is excessive blood destruction. The acute and subacute hemolytic anemias include anemias due to malaria, bacterial toxins, and chemical hemolytic agents. Anemias due to blood destruction following mismatched blood transfusions or erythroblastosis fetalis, otherwise known as hemolytic disease of the newborn, or the action of cold agglutinins or the peculiar disease called paroxysmal hemoglobinuria, which is also dependent upon the action of cold, all fall in this class. Paroxysmal hemoglobinuria is characterized by sudden passage in urine of hemoglobin following local or general exposure to the cold. It was shown by Donath and Landsteiner in 1904 to be due to a sudden hemolysis of the blood by the action of a hemolytic agent contained in the patient's own blood which, however, was active only at low temperatures.

Under the chronic hemolytic anemias we include familial or congenital hemolytic jaundice, sickle cell anemia (sicklemia), chronic hemolytic anemia with nocturnal hemoglobinuria, the so-called acquired hemolytic jaundice, and Cooley's anemia, otherwise known as thalassemia.

Abnormal hemoglobins (molecular anemias). Normal adult hemoglobin, also called *hemoglobin A*, is virtually the only hemoglobin in the blood of most humans 30 months after birth and throughout their subsequent life. Hemoglobin A is present in prenatal and early life along with fetal hemoglobin, *hemoglobin F*. Fetal hemoglobin persists into adult life in significant quantities in many hereditary anemias, and may appear in some acquired anemias.

Thalassemia is a form of hereditary anemia in which the production of hemoglobin A is deficient and, under the stimulus of hemoglobin deficiency, hemoglobin F may be produced. When an individual is heterozygous (see page 412) for the thalassemia gene, the result is *thalassemia minor*, characterized by moderate microcytic anemia. An individual homozygous for the thalassemia gene will be the victim of *thalassemia major*, a severe microcytic anemia in which hemoglobin F makes up 12 to 100 per cent of the total hemoglobin. The effect of the gene is to cut down the production of hemoglobin A, thus producing the microcytic anemia. The severity of the anemia, which may be fatal early in life, is not related to the percentage of hemoglobin F.

In addition to normal adult hemoglobin A and fetal hemoglobin F, a number of other hemoglobins have been identified, some in cases of hereditary anemias. In these abnormal hemoglobins, the heme does not vary, but the globins can be shown to be different (47), with sometimes startling differences in such properties as solubility and electrophoretic mobility.

Drepanocytosis, more commonly known as *sickle-cell anemia*, or the hybrid word, *sicklemia*, is a hereditary trait (when heterozygous) or disease (when homozygous) characterized by the presence of an abnormal *hemoglobin S*, which has three more net positive charges on its molecule than normal hemoglobin A. When uncombined with oxygen, S is much less soluble than A or F, and may form solid aggregates or a gel. This insolubility of the hemoglobin produces the bizarre angular and sometimes sickle-shaped cells observed in preparation of blood, under conditions of low oxygen, taken from patients with this abnormality. The hemoglobin in sickled cells has aggregated and the cell membrane has been distorted (32).

The sickle-cell trait is found in about nine per cent of American Negroes. Much higher percentages are observed in some of the tribes of African

Negroes, and in the Veddoids of southern Arabia and India. In affected individuals, who are presumably heterozygous for the S-forming gene, 25 to 45 per cent of the hemoglobin is S, and sickling can be demonstrated in red cell suspensions examined under a low partial pressure of oxygen. There is no anemia and no other characteristic abnormality. There appears to be a rather complex relationship between the sickle-cell trait and resistance to the severe form of malaria caused by *Plasmodium falciparum*, suggesting that the mutation which leads to the presence of the S-forming gene may have survival value in malarial regions.

Sickle-cell anemia results from a double dose of the gene responsible for formation of hemoglobin S; both parents must have carried the gene, and shown either the trait or the disease. Hemoglobin F is commonly found in affected individuals, in amounts up to 40 per cent of the total hemoglobin, the rest being hemoglobin S. These patients are anemic from excessive destruction of circulating sickled cells, and are particularly susceptible to clot formation within the blood vessels. Sickle-cell anemia occurs in about 0.2 per cent of American Negroes. *Microdrepanocytosis* or sickle-thalassemia is a rare combination of the sickling and the thallassemia genes, and the blood shows both the microcytic anemia of thalassemia and the sickling phenomenon.

Other abnormal hemoglobins C, D, E, G, H, I, J, and K have been identified. The C-trait is present in about 3 per cent of American Negroes and is accompanied by an increased percentage of target cells. *Hemoglobin C* is more electropositive than S, and more soluble than A. In the C-trait, about one-third of the hemoglobin is C, the rest A. In the disease which occurs when an individual is homozygous for the C gene, hemoglobin C occurs alone, or with a small amount of hemoglobin F. The genes for the formation of S and of C are probably alleles (see page 420) and an individual may obtain one from each parent, giving him the C disease plus the sickle-cell trait. If the C and thalassemia genes occur together, there will be some hemoglobin A along with C and F. *Hemoglobin D* trait or disease is characterized by slow sickling. Hemoglobin E is common in Southeast Asia. About 13 per cent of the population of Thailand and 6 per cent of the population of Jakarta have been reported to possess it. The abnormal hemoglobins have so far been characterized chiefly by differences in electrophoretic mobility and their mode of inheritance.

Abnormal genes, in general, determine the type of hemoglobin formed, but abnormal hemoglobins are produced at lower rates than normal hemoglobin A. The production of fetal hemoglobin F is not controlled by the alleles responsible for production of abnormal hemoglobins, but is rather a response to the stimulus of chronic anemia. All abnormal hemoglobins are broken down in the body more rapidly than hemoglobin A.

For a review of the biochemical and clinical aspects of the molecular anemias, see (6a).

BLOOD GROUP SUBSTANCES

A, B, O Factors

It was discovered by Landsteiner and his pupils about 1900 that all human blood could be classified into one of four groups; these are now designated as *O*, *A*, *B*, and *AB*. Landsteiner, in his original papers, pointed out the possible importance of this for blood transfusion, but it was not until World War I that wide recognition of this importance prevailed, and from time to time transfusions of unmatched blood were attempted—sometimes successfully and sometimes with fatal results (5, 37, 48).

The phenomenon of isohemagglutination depends upon the fact that in the stroma of the red blood cells there are mucopolysaccharides which we designate as A and B, called the blood group agglutinogens. These may occur, singly, together, or neither may occur, giving the four classical blood groups as shown in table V-4. This, in itself, would not complicate transfusions. However, in the liquid part of the blood there almost invariably occur complementary agglutinating substances, globulins in nature, according to a rule which may be formulated thus: In your blood you will always find agglutinins for the people unlike yourself—never for your own blood. When red blood cells containing agglutinogen A, for example, are mixed with plasma containing anti-A agglutinin, the red cells stick together, or agglutinate. This leads to such dangerous consequences as blockage of small blood vessels, hemolysis, and release of histamine. Shock and death often result.

It will also be seen that if group B blood is transfused into a person of group A, the anti-B agglutinin in the plasma of the group A individual will react with the introduced red cells, agglutinating them, and if complement is present, as it usually is, causing them to dissolve and producing a serious or fatal reaction. Transfusion reactions are not always fatal, but are sufficiently dangerous to make it absolutely essential to determine the blood

TABLE V-4

Classification of blood groups

SUBSTANCE IN CELLS	AGGLUTININ IN SERUM	BLOOD GROUP
—	anti-A and anti-B	O
A	anti-B	A
B	anti-A	B
A + B	—	AB

groups of recipient and donor before attempting a transfusion. The agglutinin in the blood of the donor may also have bad effects, but is not so important since it is considerably diluted—not more than 0.5 liters of blood being usually administered—to restore a circulating volume of some 7 to 8 liters. Also, dissolved blood group antigens corresponding to the patient's own agglutinogens are found in his various tissues and body fluids, and these help neutralize the introduced agglutinins. These facts permit us to set up the possibilities of transfusion shown in formula IX.

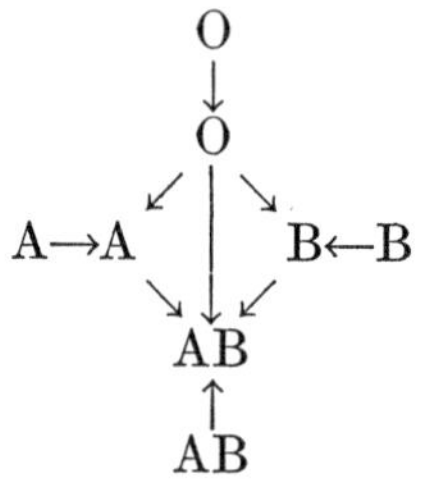

$$\begin{array}{ccccc} & & O & & \\ & & \downarrow & & \\ & & O & & \\ & \swarrow & | & \searrow & \\ A \rightarrow A & & | & & B \leftarrow B \\ & \searrow & \downarrow & \swarrow & \\ & & AB & & \\ & & \uparrow & & \\ & & AB & & \end{array}$$

IX

Nevertheless, it is better not to follow this scheme unless it is absolutely impossible to obtain a person of exactly the same blood group as the patient. Persons of group O are often called universal donors, and it is true their blood has often been used in transfusions into people of other groups. However, if their anti-A and anti-B agglutinins have a high titer and are unusually avid, they may react with the red blood cells of the recipient. Many cases are on record in which this has occurred. In a case observed by one of the authors (27), 25 per cent of the red blood cells of the recipient had been destroyed by the introduced agglutinins from the donor. It is obvious that in such cases no good, but actual harm, is done by the transfusion. During World War II, when fresh refrigerated blood was being flown from this country to the theaters of operation, only group O blood was sent, to avoid the necessity of doing a grouping before performing the transfusion As a routine precaution, before blood was flown across, a 1:80 dilution of the plasma was tested against known sensitive A and B red blood corpuscles. This was a purely arbitrary procedure, but it was found by experience that it excluded the so-called dangerous universal donors—that is, persons of group O who have an unusually high titer of anti-A and anti-B agglutinins. Their blood could still be used for conversion to dried plasma.

The A, B, and O blood group substances are all mucopolysaccharides. The sugars or amino-sugars that have been isolated from them include three which we have mentioned before, D-galactose, D-glucosamine and D-galactosamine. (Both glucosamine and galactosamine are present in the blood

TABLE V-5

Composition of the A, B and O blood group substances

Constituent	Occurrence in blood group substance (in per cent)		
	A	B	O
Nitrogen	5.7	5.8	5.3
Reducing sugar	54	50	54
Glucosamine	11.8	12.5	11.5
Galactosamine	13.2	4.5	10.4
Acetyl	9	7	8.7
Fucose	18	18	14
Amino acid N (as per cent of total N)	38	50	41

group molecule as the N-acetyl derivatives.) In addition, there is a fourth sugar, the methylpentose, L-*fucose* (formula X), which, stereochemically, belongs to the "unnatural" series of sugars. As usually isolated, the A, B, O blood group substances contain a number of amino acids which, however, play no role in the specificity. Some quantitative data concerning the structure of these substances are given in table V-5 (adapted from (20)).

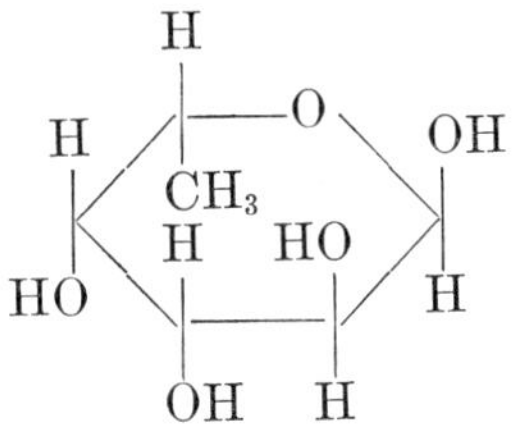

```
              H
              |
      H     / |----O    OH
      |   /   |        \ |
      | <    CH3        >|
      |   \   H    HO  / |
     HO     \ |_____| /  H
              |     |
              OH    H
```

X. L-Fucose

Substances resembling the A, B and O blood group substances in serological behavior occur in a variety of tissues and secretions (gastric mucosa, saliva, ovarian cysts, gastric juice and so on) and in a number of animals (hogs, horses, cows, apes) as well. The student should not think of them as individual compounds, but rather as a spectrum of compounds related to one another by their similar behavior in the presence of a particular antiserum.

Relatively small portions of the polysaccharide determine the specificity of the particular blood group substances. N-acetyl-D-galactosamine plays an important role in the specificity of A substance, and galactose in the specificity of B substance. As a result of inhibition studies, Kabat and Leskowitz (21) have concluded that these sugars are the terminal units of the oligosaccharide chains responsible for the specificities. In the case of

B blood group substance, they identify two end units as the disaccharide, galactosido-1,6-alpha-N-acetylglucosamine (formula XI).

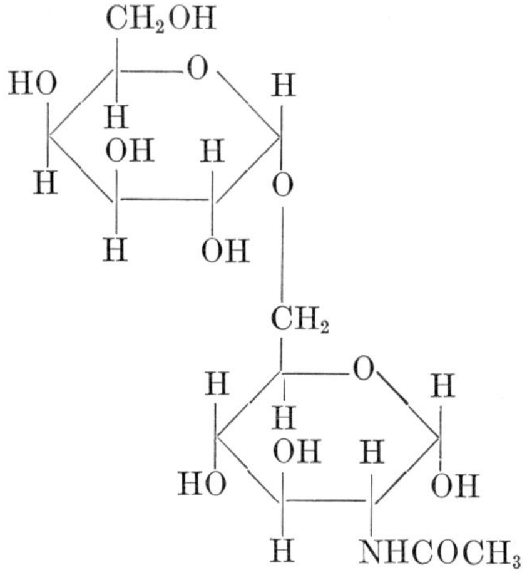

```
          CH2OH
           |
 HO     /|----O\     H
  |    / H       \   |
  |   <  OH   H   >  |
  H    \ |----|  /   O
           |    |        |
           H    OH       |
                         |
                       CH2
                         |
                  H     /|----O\     H
                  |    / H       \   |
                  |   <  OH   H   >  |
                  HO   \ |----|  /   OH
                         |    |
                         H    NHCOCH3
```

XI. Galactosido-1,6-alpha-N-acetylglucosamine

Antigenic differences among normal human bloods are not confined to the OAB systems of antigens, although these are by far the most important for transfusions. In 1927 Landsteiner and Levine discovered three new factors, M, N, and P. Agglutinins for these agglutinogens are not usually found in human plasma, however, and M and N are detected by the reactions of the absorbed sera of rabbits which have been injected with washed red cells of these types. It might have been supposed that it would be necessary to pay close attention to the differences in M, N, and P for transfusion as well as A and B, and this was in fact suggested. Actually, blood of type N could be transfused into patients of type M repeatedly without causing the formation of dangerous antibodies. The same apparently was true for P and other combinations of these types.

Rh Factors

The M antigen is found to exist in the anthropoids as indeed do the O, A, and B, but in the case of M there seems to be a marked difference because some absorbed sera which are very satisfactory for detecting M in human beings are quite unsatisfactory in the case of certain anthropoids. While studying this phenomenon, Landsteiner and Wiener in 1940 discovered a new blood factor by injecting the blood of Rhesus monkeys into rabbits and absorbing to remove the undesired agglutinins. From the initial letters of the word Rhesus we get the symbol *Rh*. Rh might have met the same fate of being practically forgotten, which fate befell numerous other blood group antigens discovered after O, A, and B, had it not been that Wiener (49) was able to demonstrate that certain transfusion reactions, especially those in which the recipient had been repeatedly transfused, produced reactions involving the Rh factor. Soon after this Levine and co-workers discovered that Rh was in fact responsible for most cases of a rare disease of infants called erythroblastosis fetalis or hemolytic disease of the newborn, which had been described by Diamond (9) previously.

If a woman becomes pregnant with a fetus which carries the Rh positive factor and she herself is Rh negative, the antigen from the fetus may diffuse across the placenta either as whole cells or simply as dissolved antigen, and cause the production in her circulation of anti-Rh antibodies (50). In this respect Rh seems to be a more potent antigen than the other minor M, N, P, and so forth. If a mother is so sensitized (although fortunately this happens very rarely—only about once in 400 births), the agglutinins which she forms may diffuse across the placenta into the circulation of the fetus and cause damage to the red cells of the fetus. The disease which appears in the newborn is called erythroblastosis, since in stained preparations of the blood young red blood cells called erythroblasts are seen. Other symptoms such as liver damage, edema, and jaundice are often involved. In subsequent work it has been shown that the Rh factor is actually extremely complex (see page 436).

Since the discovery of the Rh factor a number of other blood group antigens have been discovered, mostly by British workers, by the technique of observing incompatibilities in people who seem to be otherwise of the same blood group. These have only limited clinical importance, but are of interest to students of genetics and anthropology.

IDENTIFICATION OF BLOOD

Blood stains are at times found at the scene of violent crime and it is often important to be able to say positively that these stains are blood, since many other substances, such as dried chocolate syrup, resemble old blood stains closely. It is often desired to detect small amounts of blood in gastric samples, exudates, urine, and feces, all as evidence of internal bleeding.

One of the traditional tests depends upon the liberation of the prosthetic group heme from hemoglobin. The blood stain is extracted and the extract dried on a microscope slide, and then treated with glacial acetic acid and a little sodium chloride and heated until the acetic acid boils. It is then cooled slowly. Chlorohemin, which is generally referred to as *hemin*, crystallizes in chocolate colored, rhombic plates which are very characteristic. Spectroscopic examination of blood to detect the characteristic absorption spectra of hemoglobin or its derivatives has also been used to identify extracts as containing blood. Other methods of detecting blood which are not specific depend on the fact that hemoglobin acts as a peroxidase. One of the tests is the guaiac test. The extract is treated with a few drops of an alcoholic solution of gum guaiac and then with hydrogen peroxide. If blood is present, a blue color due to the oxidation of the guaiac to guaiaconic acid is obtained.

Another somewhat more sensitive test is the benzidine test. The suspected solution is treated with a saturated solution of benzidine in glacial acetic acid. Hydrogen peroxide is added, and in the presence of blood, a brilliant greenish-blue or blue oxidation product of benzidine forms. This test will detect blood in a dilution of about 1 in 1,000,000.

After blood has been detected, it is necessary in medico-legal cases to determine whether or not it is human blood, for the presence of animal blood does not indicate murder. This may be done by the precipitin reaction (5).

REFERENCES

1. Aggeler, P. M., *et al.* Purification of plasma thromboplastin factor B (plasma thromboplastin component) and its identification as a $beta_2$-globulin. Science, **119:** 806–807, 1954.
2. Allison, A. C., and Burn, G. P. Enzyme activity as a function of age in the human erythrocyte. Brit. J. Haematol., **1:** 291–303, 1955.
2a. Bailey, K. and Bettelheim, F. R. Clotting of fibrinogen. I. Liberation of peptide material. Biochem. et Biophys. Acta, **18:** 495–503, 1955.
3. Beaven, G. H., *et al.* The hemoglobins of the human foetus and infant. Electrophoretic and spectroscopic differentiation of adult and foetal types. Biochem. J., **49:** 374–381, 1951.
3a. Bettelheim, F. R. Clotting of fibrinogen. II. Fractionation of peptide material liberated. Biochim. et Biophys. Acta, **19:** 121–130, 1956.
4. Biggs, R., *et al.* The initial stages of blood coagulation. J. Physiol., **122:** 538–553, 1953; The action of thromboplastic substances. *ibid*: 554–569.
5. Boyd, W. C. *Fundamentals of Immunology.* 3rd edition New York. Interscience Publishers, 1956.
6. Brinkhous, K. M., and Graham, J. B. Hemophilia and the hemophilioid states. Blood, **9:** 254–257, 1954.
6a. Chernoff, A. I. The human hemoglobins in health and disease. New Eng. J. Med., **253:** 322–331, 365–374, 416–423, 1955.
7. Cliffton, E. E. Variations in the proteolytic and the antiproteolytic reactions of serum: effect of disease, trauma, x-ray, anaphylactic shock, ACTH, and cortisone. J. Lab. & Clin. Med., **39:** 105–121, 1952.
8. Coons, A. H. Labelled antigens and antibodies. Ann. Rev. Microbiol., **8:** 333–352, 1954.
9. Diamond, L. K., *et al.* Erythroblastosis fetalis and its association with universal edema of fetus, icterus gravis neonatorum, and anemia of newborn. J. Pediat., **1:** 269–309, 1932.
10. Duckert, F., *et al.* Clotting factor X, its physiologic and physicochemical properties. Proc. Soc. Exptl. Biol. Med., **90:** 17–22, 1955.
11. Edsall, J. T. The plasma proteins and their fractionation. Advances Protein Chem., **3:** 384–479, 1947.

12. FISHER, B. Recent contributions of electrophoresis to clinical pathology. Am. J. Clin. Path., **23:** 246–262, 1953.
13. FRICK, P. G., AND HAGEN, P. S. Severe coagulation defect without hemorrhagic symptoms caused by a deficiency of the fifth plasma thromboplastin precursor. J. Lab. Clin. Med., **47:** 592–601, 1956.
14. GRANICK, S. Anatomy of hemoglobin and some functions of its parts. Annual Conference on Protein Metabolism, **9:** 2–18, 1953.
15. GUTMAN, A. B. The plasma proteins in disease. Advances Protein Chem., **4:** 155–250, 1948.
16. HOWE, P. E. The use of sodium sulfate as the globulin precipitant in the determination of proteins in blood. J. Biol. Chem., **49:** 93–107, 1921.
16a. HUISMAN, T. H. J. *et al.* Is foetal haemoglobin present in the blood of normal human adults? Biochim. et Biophys. Acta, **18:** 576–577, 1955.
17. INGRAM, V. M. Sulfhydryl groups in hemoglobin. Biochem. J., **59:** 653–661, 1955.
18. JELLINEK, E. M., AND LOONEY, J. M. Statistics of some biochemical variables on healthy men in the age range of twenty to forty-five years. J. Biol. Chem., **128:** 621–630, 1939.
19. JIM, R. J. S., AND STEINKAMP, R. C. Macroglobulinemia and its relationship to other paraproteins. J. Lab. & Clin. Med., **47:** 540–561, 1956.
20. KABAT, E. A. *Blood Group Substances.* New York, Academic Press, Inc., 1956.
21. KABAT, E. A., AND LESKOWITZ, S. Immunochemical studies on blood groups. XVII. Structural units involved in blood group A and B specificity. J. Am. Chem. Soc., **77:** 5159–5164, 1955.
22. KAITZ, E. H., *et al.* The estimation of gamma-globulins in normal and myelomatous plasmas by chemical fractionation. J. Lab. & Clin. Med., **41:** 248–257, 1953.
23. KEILIN, D. Position of hems in the hemoglobin molecule. Nature, **171:** 922–925, 1953.
24. KLEINFELD, G., AND HABIF, D. V. Effect of trypsin on prothrombin. Proc. Soc. Exptl. Biol. Med., **84:** 432–437, 1953.
24a. KRUH, J. AND BORSOOK, H. Hemoglobin synthesis in rabbit reticulocytes in vitro. J. Biol. Chem., **220:** 905–915, 1956.
25. LEMBERG, R., AND LEGGE, J. W. *Hematin Compounds and Bile Pigments.* New York, Interscience Publishers, 1949.
26. LUTZ, B. R. Intravascular agglutination of the formed elements of blood. Physiol. Rev., **31:** 107–130, 1951.
27. MALKIEL, S., AND BOYD, W. C. A transfusion reaction due to a dangerous universal donor. J. A. M. A., **129:** 344, 1945.
28. MARINIS, T. P., *et al.* Sodium and potassium determinations in health and disease. J. Lab. & Clin. Med., **32:** 1208–1216, 1947.
29. MCKINNEY, G. R. Glyoxalase activity in human leucocytes. Arch. Biochem. Biophys., **46:** 246–248, 1953.
30. MOLLISON, P. L., AND VEALL, N. Use of the isotope Cr^{51} as a label for red cells. Brit. J. Haematol., **1:** 62–74, 1955.
31. ONCLEY, J. L., *et al.* Physicochemical characteristics of certain of the proteins of normal human plasma. J. Phys. & Colloid Chem., **51:** 184–198, 1947.
32. PAULING, L., *et al.* Sickle cell anemia, a molecular disease. Science, **110:** 543–548, 1949.
33. PILLEMER, L., AND HUTCHINSON, M. C. The determination of the albumin and globulin contents of human serum by methanol precipitation. J. Biol. Chem., **158:** 299–301, 1945.

34. PONDER, E. Present concepts of the structure of the mammalian red cell. Blood, **9:** 227–235, 1954.
35. QUICK, A. J., AND HUSSEY, C. V. Hemophilic conditions in a girl. A. M. A. Am. J. Dis. Children, **85:** 698–705, 1953.
36. SCHEINBERG, H. Clinical implications of plasma fractionation. Bull. N. Y. Acad. Med., **30:** 735–749, 1954.
37. SCHIFF, F., AND BOYD, W. C. *Blood Grouping Technic.* New York, Interscience Publishers, 1942.
38. SHERRY, S., AND TROLL, W. The action of thrombin on synthetic substrates. J. Biol. Chem., **208:** 95–105, 1954.
39. SHULMAN, S. The size and shape of bovine fibrinogen. Studies of sedimentation, diffusion and viscosity. J. Am. Chem. Soc., **75:** 5846–5852, 1953.
40. STEFANINI, M., AND DAMESHEK, W. Collection, preservation, and transfusion of platelets, with special reference to the factors effecting the "survival rate" and the clinical effectiveness of transfused platelets. New England J. Med., **248:** 797–802, 1953.
41. SZARA, S., AND BAGDY, D. On the polysaccharide of fibrinogen and fibrin. Biochim. et Biophys. Acta, **11:** 313–314, 1953.
42. TOUSTER, O., AND YARBRO, M. C. The ergotheioneine content of human erythrocytes; the effect of age, race, malignancy and pregnancy. J. Lab. & Clin. Med., **39:** 720–724, 1952.
43. TROLL, W., *et al.* The action of plasmin on synthetic substrates. J. Biol. Chem., **208:** 85–93, 1954.
44. TULLIS, J. L. *Blood Cells and Plasma Proteins.* New York, Academic Press, 1953.
45. WALKER, J. Fetal and adult hemoglobin in the blood of the human fetus, a preliminary communication. Cold Spring Harbor Symposia Quant. Biol., **19:** 141–142, 1954.
46. WESTFALL, B. B., *et al.* The amino-acid content of the ultrafiltrate from horse serum. J. Natl. Cancer Inst., **15:** 27–35, 1954.
47. WHITE, J. C., AND BEAVEN, G. H. A review of the varieties of human haemoglobin in health and disease. J. Clin. Path., **7:** 175–200, 1954.
48. WIENER, A. S. *Blood Groups and Transfusion.* 3rd edit. Springfield, Ill., Charles C Thomas, 1943.
49. WIENER, A. S., AND PETERS, H. R. Hemolytic reactions following transfusions of blood of homologous groups, with three cases in which same agglutinogen was responsible. Ann. Int. Med., **13:** 2306–2322, 1940.
50. WIENER, A. S., AND WEXLER, I. B. Transfusion therapy of acute hemolytic anemia of newborn. Am. J Clin Path., **13:** 393–401, 1943.
51. WOOTTON, I. D. P., AND KING, E. J. Normal values for blood constituents. Lancet, **264:** 470–471, 1953.

PART II

Control

CHAPTER 6

Enzymes and Coenzymes

In Section I we have discussed the chemical nature of the materials composing the human body. This means thus far that the body has been considered as a static phenomenon. In reality it is a system in which the various components are in a state of rapid flux, and are related to and dependent upon one another in complex fashion. For instance, the dynamic state of the protein constituents of the body was first proved by Schoenheimer and his group, using N^{15} to label introduced amino acids. They found that all the complex molecules of the body, including proteins and their components, amino acids and nucleic acids, were constantly involved in more or less rapid chemical reactions, notably the constant exchange of amino groups. Fats and carbohydrates are likewise in a state of flux. Ester, peptide, and other linkages open, fragments are liberated and merge with those derived from other large molecules, and these form a metabolic pool of components which can no longer be distinguished in regard to origin.

These liberated molecules are again subject to numerous processes; fatty acids are dehydrogenated, hydrogenated, degraded or elongated and in general continually converted. While some individual molecules of fatty acids are completely degraded, others are formed, notably from carbohydrate.

Similar activity goes on among the split products of proteins. The free amino acids are deaminated and the liberated ammonia transferred to other previously deaminated molecules to form new amino acids. Intermediate products of carbohydrate oxidation may shift forward in the direction of complete oxidation, backward toward resynthesis of glycogen or sideways toward conversion into fat or amino acid.

This chemical flux is most rapid in the soft tissues; muscular, glandular and epithelial; and is comparatively slow in bones and connective tissue, but nothing in the body remains completely static. The remainder of the book will therefore deal with the human body as a dynamic phenomenon.

The many chemical reactions involved in birth, growth, and maturation, and in the conversion of food into living tissue and energy, are obviously not random or unorganized. If they were, they would continue unaltered by death, and it is the very essence of life and death that they do not. Even a relatively slight change in the nature of only one of the chemical reactions proceeding in the living body may result in serious illness or death. Consider the extraordinarily small quantity of cyanide which is required to bring the entire mechanism of life to an abrupt halt. Conversely, all maladjustments of the human organism involve—originally or eventually—the non-function or malfunction of one or more of these reaction-supervising factors. The nature of these factors and their normal functioning becomes therefore a matter of vital interest.

What are these factors? We will begin by simply supplying them with a name: *enzymes* (derived from the Greek words meaning "in yeast" because it was in yeast that enzymatic reactions were first systematically studied). Enzymes were recognized and their catalytic activity extensively studied long before any reliable data were available as to their chemical natures.

For an understanding of enzymes, and of what they can and cannot do, a knowledge of basic thermodynamic principles is helpful.

THERMODYNAMICS

There are sometimes said to be three laws of thermodynamics, but we shall need only the first two here.

The First Law

The first law is so well known and universally accepted that it will hardly cause any difficulty. It is simply the familiar law of the conservation of energy, and states that in the process of converting heat into work, chemical energy into work, or, generally, in any energy conversion, energy can neither be created nor destroyed. If a handful of coal, or a cheese sandwich, is considered, we know that it contains just so much chemical energy, and that no chemical machine or heat engine, however ingenious, will enable us to get more than the mechanical equivalent of that energy out of it. (Of course we live in an age in which certain elements of matter can be transformed, to a very limited degree, into energy, and such operations are excepted from the law as we have stated it. For matter-energy transformations there is a more general statement of the law which includes both matter and energy; this more general law will not concern us here.)

From the first law it follows that no perpetual motion machine "of the first type"—that is, one getting energy from nowhere, or producing more energy than is fed into it—can ever be constructed.

The total energy of a system is symbolized in English language textbooks as E, and is usually expressed in heat units such as calories.

The pioneers in thermodynamics were interested primarily in converting heat into work, and this preoccupation is partly responsible for the form which the subject took under their hands. We may state the first law in the form

$$\Delta E = \Delta Q - \Delta W \qquad \text{Equation VI-1}$$

where ΔE is the increase in the total energy of the system, while ΔQ is the heat absorbed, and ΔW is the work done by the system. Or, if we consider a change so small that we may replace ΔE by the differential dE, we may write (ignoring some questions of mathematical rigor)

$$dE = dQ - dW \qquad \text{Equation VI-2}$$

We must now consider carefully what such an equation means. It looks as if it means that when we measure the infinitesimal increase in the total energy of a system, we find experimentally that it equals the experimentally determined infinitesimal absorption of heat, minus the experimentally determined infinitesimal amount of work done. But this is not the meaning of the equation, for we do not have a "total energy meter" which we can attach to the system and thereby measure the change in total energy. The only way we have of measuring dE is by measuring dQ and dW and taking the difference. It might seem, therefore, that equation VI-2 is a trivial tautology. It is not, however, because it turns out that there is an essential difference between dE, on the one hand, and dQ and dW, on the other. The difference is that the differential dE is exact, whereas dQ and dW are not exact (11).

The exact differential has an important consequence for us, which is this: if a differential, dX, is exact, the values of X at two different points, X_1 and X_2, depend solely upon the initial and final values of the independent variable of which X is a function, whereas if the differential is inexact, the values of X depend upon the particular route by which we go from X_1 to X_2. The differential, dW, is not exact, for it is shown in elementary physics that the amount of work done by a system depends upon how we go about extracting the work, that is, on the nature of the route going from the initial to the final state. The same is true of the heat absorbed. In equation VI-2, the value of E, but not of Q and W, depends only upon the initial and final states of the system, and that is what makes the equation more than simply a definition. Specifically, in figure VI-1, this means that as we go from point 1 to point 2, the change in E is completely determined by the initial and final values of P (pressure) and V (volume), whether we follow path A, path B, or any other path, but the changes in Q and W will in general be different for different paths. Furthermore, if we go from point 1 to point 2, then back to point 1 (a reversible, cyclic process), ΔE must equal zero, while ΔQ and ΔW may equal any value. All of this is really only the first law of thermodynamics.

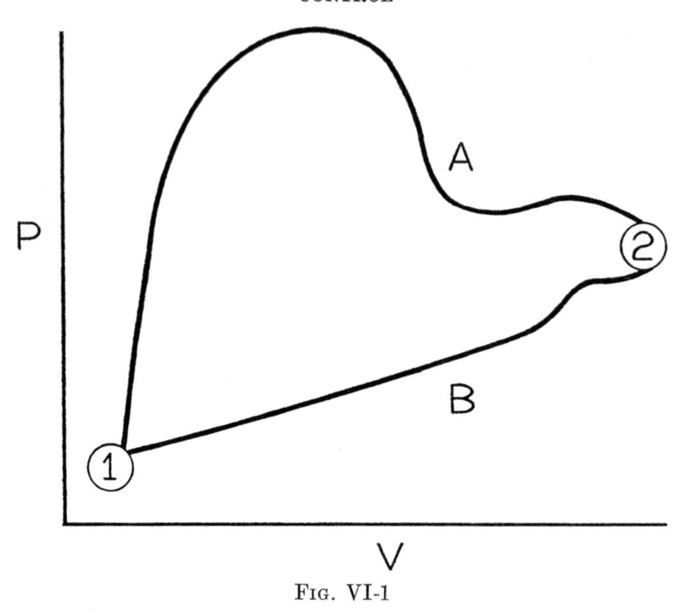

FIG. VI-1

A thermodynamic quantity which depends only upon the values of the independent variables (which in heat engines can be conveniently taken to be P and V) is called a *thermodynamic function*. Knowing that E is a thermodynamic function, we can define others, for instance

$$H = E + PV \qquad \text{Equation VI-3}$$

H is called the total heat or *enthalpy*.

It is obvious that this is a thermodynamic function, for we have just seen that the first law ensures that E depends solely upon the values of P and V, and obviously the product PV also depends only upon these variables. Therefore dH is an exact differential.

The reason for defining H in this manner is that it proves a convenient function to use when P is constant. It can be shown that

$$(\Delta H)_P = (\Delta Q)_P \qquad \text{Equation VI-4}$$

where the subscript, P, indicates that pressure is constant.

Since most chemical reactions are carried out at atmospheric pressure, this relation applies to them, and states that the change in H equals the heat absorbed, or in other words, the increase in enthalpy is equal to the negative heat of reaction. (Since thermodynamics was developed by engineers who were interested in what could be done with the heat absorbed, a positive sign for ΔQ means heat is absorbed and a negative sign that heat is given off. If thermodynamics had been developed by chemists the opposite convention might have been adopted.)

Now the heat of reaction is obviously an important index of the energy of this reaction and of the tendency of the reaction to go. In fact, for many years it was believed that ΔH was an exact measure of the spontaneity of a reaction. If this had been true, it might not have been necessary to introduce any further thermodynamic notions into this discussion. Unfortunately, it is not quite true. But before we can define a function which does measure exactly the tendency of a reaction to go, we must discuss the second law of thermodynamics.

The Second Law

Going back to equation VI-2, we may write it as

$$dQ = dE + dW \qquad \text{Equation VI-5}$$

If P and V are the variables, then it can be shown that, P being constant, $dW = PdV$. Then

$$dQ = dE + PdV \qquad \text{Equation VI-6}$$

Now we have seen that E is a function of P and V, and from elementary calculus, we know that the differential dE is defined, in terms of the independent variables P and V, as

$$dE = (\partial E/\partial V)dV + (\partial E/\partial P)dP \qquad \text{Equation VI-7}$$

where ∂ indicates partial differentiation. Substituting in Equation VI-6, we obtain

$$dQ = (\partial E/\partial V + P)dV + (\partial E/\partial P)dP \qquad \text{Equation VI-8}$$

We have already seen that dQ is not an exact differential, and consequently equation VI-8 cannot be integrated as it stands. It is shown, however, in the calculus (17) that whenever you have an equation of the form

$$dQ = XdV + YdP \qquad \text{Equation VI-9}$$

where X and Y are functions of the variables V and P, there is always an integrating factor $B = f(P, V)$, in fact a number of such factors, functions of P and V, which, when both sides of equation VI-9 are multiplied by B, makes the product BdQ an exact differential. It is not always easy to find the function which does this, but in the case of equation VI-7 the problem is easily solved. Since the values of P and V determine the state of the system, they determine the value of the absolute temperature, T. It turns out that the simplest function of P and V which is an integrating form for equation VI-7 is $1/T$. Multiplying through by this factor, we obtain

$$dQ/T = \frac{(\partial E/\partial V + P}{T} dV + \frac{(\partial E/\partial P)}{T} dP \qquad \text{Equation VI-10}$$

and the expression dQ/T is now an exact differential.

That dQ/T is an exact differential is proved in thermodynamics by showing that (a) dQ/T is an exact differential for an ideal gas carried through a certain sequence of reversible changes called a Carnot cycle, (b) dQ/T is an exact differential for any substance carried through a Carnot cycle, (c) dQ/T is an exact differential for any substance carried through any reversible cycle.

The changes in the Carnot cycle are deliberately made simple and symmetrical. As a consequence it is easy to show that

$$W/Q_2 = (T_2 - T_1)/T_2 \qquad \text{Equation VI-11}$$

where W is the work done by the cycle, Q_2 is the heat taken in at the higher temperature, T_2, and T_1 is the lower temperature. The fraction W/Q_2 is called the efficiency of the cycle. It is shown in thermodynamics that (a) the efficiency of a real substance carried through a Carnot cycle cannot exceed that of an ideal gas and cannot be less, (b) the efficiency of any substance carried through any reversible cycle is the same as that of an ideal gas carried through a Carnot cycle. Consequently equation VI-11 gives the maximum efficiency which any heat engine, taking in its heat at temperature T_2 and discharging it at temperature T_1, can ever attain. This proves that the efficiency of an actual heat engine can never exceed that of the hypothetical heat engine which carries an ideal gas through a Carnot cycle.

Since dQ/T is an exact differential, we know that it may be integrated and that the value of the function so obtained will depend only upon the values of P and V, and will thus be a new thermodynamic function. If we have a new thermodynamic function, we can give it a name, and the name given is *entropy*, with the symbol S, and

$$dS = dQ/T \qquad \text{Equation VI-12}$$

It therefore follows that the value of ΔS, in going from point 1 to point 2 (figure VI-1), depends only upon the initial and final values of P and V, and that for any cyclic change which leaves P and V at their original values, $\Delta S = 0$.

The units in which entropy is expressed are not independent of the units of temperature used, for we see from equation VI-12 that the product TS must come out in energy units, in order to be directly comparable with Q. The usual units for S are calories per degree. In biochemistry we are more likely to use kilocalories (Kcal.) instead of the calorie. One Kcal. equals 1000 calories.

The discovery that the integral of dQ/T defines a new thermodynamic function constitutes a statement of the second law of thermodynamics, and serves to define the otherwise mysterious concept of entropy, which is not, like the concepts of temperature, pressure, heat content, and so on, an obvious generalization of earlier concepts already more or less familiar to the ordinary man, but represents a subtle and powerful new concept.

The second law is harder to state non-mathematically in a way which sounds immediately obvious, although its validity is well established. One way of stating it is that no work can be obtained from a heat engine which has no temperature difference between the highest temperature at which its working substance operates and the lowest temperature available to it. For instance, the ocean has a mean temperature many degrees above the absolute zero, and we know that temperature is an expression of the kinetic energy of the molecules of the substance whose temperature we measure. Consequently the ocean contains enormous stores of energy. Nevertheless nothing is more certain than the fact that you can not devise a heat engine for an ocean liner or battleship which will drive the ship by utilizing the kinetic energy of the molecules of water in the ocean. The trouble is that you do not have a "sink" at a still lower temperature into which to dump your ocean water after it has gone through your engine.

The reason this "sink" is necessary is not clear to all students. But consider the following analogy. If you have in your possession a cylinder of compressed gas, say at about 80 pounds per square inch, you can discharge this gas through a little turbine and make it do some work. In interplanetary space you could use the stored energy directly as a reaction motor which operates on the same principle as a rocket. But suppose you and your little cylinder are on a planet where the atmospheric pressure is 80 pounds per square inch—the same as in your cylinder. If you then open the valve of your cylinder, nothing happens. The compressed gas is still there with all its stored energy, but you no longer have a "sink" of lower pressure. No pressure gradient, no work. Similarly with heat engines; no temperature gradient, no work.

In the tropics, where the surface temperature of the ocean may be much higher than that of the depths below, a temperature gradient is available; and utilization of this has enabled work to be produced, although no machine capable of doing this has yet proved worth the financial investment involved.

It should be noted that a machine which drew heat from a single reservoir at one temperature and converted the heat to work (and of course as a result cooled the reservoir without the benefit of a "sink"), although a perpetual motion machine, would not violate the first law of thermodynamics, for energy would not be obtained from nothing. Such a machine, combining in one apparatus the best features of a heat engine and a refrigerating machine (18), would be marvellous to have if it could be built. The second law of thermodynamics, which states that such a machine, called a perpetual motion machine of the second class, cannot be constructed, is a generalization of the universal experience of inventors and engineers that such machines are impossible.

Free Energy

The significance of the second law for us is that it furnishes a criterion by which we may decide whether a process that occurs in nature is reversible or irreversible. After an irreversible process has taken place, the entropy of the system has changed; after a reversible process has taken place, the entropy of the system is unchanged.

In the case of a heat engine, where we are interested in getting work out of the system by letting P and T vary, we take in heat at one temperature and let this cause the expansion of a gas (the "working substance"), thereby doing work. Some of the heat is discharged into the "sink" at a lower temperature, and the final energy and volume of the working substance return to their initial values. This constitutes a cycle, and if the cycle is reversible, we know that

$$\Delta S_{V,E} = 0 \qquad \text{Equation VI-13}$$

where the subscripts mean that V and E are held constant.

The value of ΔS is a measure of the tendency of a reaction to go, V and E being constant, and the larger the value of ΔS the more spontaneous a process and the greater the tendency for it to take place.

Now the use of the size and sign of ΔS as a criterion of spontaneity, though fine in the study of heat engines, is not satisfactory in the study of chemical reactions, which is what interests us here, partly because the energy of the system does not remain the same. It proves possible to remedy this situation, however, by defining a new thermodynamic function by the equation

$$F = H - TS \qquad \text{Equation VI-14}$$

where H is the enthalpy, T the absolute temperature, and S the entropy. Under laboratory conditions most useful in biochemistry, T and P are constant and V and S vary. It is easy to show that the new function F is determined solely by the values of V and S and is thus a thermodynamic function. The proof is that since $H = E + PV$, then

$$F = E + PV - TS \qquad \text{Equation VI-15}$$

We saw above that $E = f(P, V)$. If P and T are constant, this means that $E = f(V)$, $PV = f(V)$, and $TS = f(S)$, so that $F = f(V, S)$.

This new function is called the *Gibbs free energy*, and is represented by G by some writers.

For processes occurring at constant temperature and pressure the sign and magnitude of ΔF provides the index of spontaneity which ΔS provided under conditions of constant E and V. In a system at equilibrium (and therefore reversible) at fixed pressure and temperature, with only mechanical work possible,

$$\begin{aligned} \Delta F_{P,T} &= \Delta H - \Delta(TS) \\ &= \Delta E + P\Delta V - T\Delta S \\ &= \Delta Q - P\Delta V + P\Delta V - T\Delta S \end{aligned} \qquad \text{Equation VI-16}$$

Since $\Delta Q/T = \Delta S$, the right hand expression vanishes and we have

$$\Delta F_{P,T} = 0 \qquad \text{Equation VI-17}$$

as the criterion of a reversible reaction (or of equilibrium).

If a system is doing no work except mechanical work, $dW = PdV$, we may substitute from the equation stating the first law of thermodynamics $dE = dQ - PdV$ and obtain the differential equation corresponding to equation VI-16, as follows:

$$dF = dQ - PdV - TdS - SdT \qquad \text{Equation VI-18}$$

When the change in F is due to some reversible transformation, we have from the definition of entropy, $dQ = TdS$, and equation VI-18 becomes

$$dF = -PdV - SdT \qquad \text{Equation VI-19}$$

If the temperature is constant this reduces to

$$dF = -PdV \qquad \text{Equation VI-20}$$

For an ideal gas,

$$P = nRT/V \qquad \text{Equation VI-21}$$

so that

$$dF = -nRTdV/V \qquad \text{Equation VI-22}$$

Integrating, we obtain

$$\Delta F = nRT \ln (V_1/V_2) \qquad \text{Equation VI-23}$$

From the gas law, equation VI-21, we see that at constant temperature $V_1/V_2 = P_2/P_1$, so that

$$\Delta F = nRT \ln (P_2/P_1) \qquad \text{Equation VI-24}$$

Strictly, this equation applies only to an ideal gas, but it applies without serious error to many real gases, and if we replace P_2 and P_1 by the activities, which for dilute solutions may not differ appreciably from the concentrations, we may apply it to substances in solution. As an example of the power of the simple condition that at equilibrium $dF = 0$, let us apply equation VI-24 to the Donnan membrane equilibrium, already discussed in less rigorous fashion on page 46. We can now give a treatment which is practically identical with that given by Donnan in his original paper.

The diagram on page 47 represents the equilibrium state, so the free energy change caused by an infinitesimal change in either direction away from equilibrium ought to be zero. If we represent the free energy change caused by the transport of an infinitesimal fraction of a mol of hydrochloric acid from the left side to the right side by δF, we have, from equation VI-24

$$\delta F = \delta nRT \ln (x/y) + \delta nRT \ln [(x/(y + z)] = 0 \quad \text{Equation VI-25}$$

where the first term on the right represents the free energy change produced by the transport of δn mols of H^+ from the left to the right side, and the second term on the right is the free energy change due to the transport of δn mols of Cl^-. Since the sum of the first and second terms on the right must be zero, it is evident that

$$\ln (x/y) = -\ln [x/(y + z)] \quad \text{Equation VI-26}$$

or

$$x/y = (y + z)/x \quad \text{Equation VI-27}$$

or

$$x^2 = y (y + z) \quad \text{Equation VI-28}$$

which is the same relation as that derived previously.

Spontaneity of a Reaction

The relation between the size of ΔF and the degree of spontaneity of a reaction are found by considering the relation between ΔF and the equilibrium constant.

It is obvious that ΔF depends upon the state of the reactants and products. In order to be able to tabulate standard values of ΔF it is necessary to agree upon certain definitions of *standard states* of these substances. The standard state of a solid is its most stable form at atmospheric pressure and the specified temperature, the standard state of a gas is at 0°C. and atmospheric pressure, and the standard state of a dissolved substance is unit activity. Activity is a thermodynamic concept which is defined below; it is sufficient for our present purposes to know that the activity of very dilute substances, such as we often deal with in biochemistry, can usually be equated without serious error to their concentration. Since we are interested merely in the meaning of ΔF, we need not consider the practical difficulties involved in calculating it.

Suppose we have a chemical reaction between two perfect gases (A and B) to give two products (C and D) which are also perfect gases. Then if the lower case letters (a, b, c, and d) represent numbers of mols of the respective gases, P_A and P_B the initial pressure of the reactants, and P_C and P_D the final pressures of the products, we have

$$aA(P_A) + bB(P_B) \rightarrow cC(P_C) + dD(P_D) + \Delta F \qquad \text{Equation VI-29}$$

This reaction is accompanied by a free energy change, ΔF, as shown. If we wish to calculate the standard free energy change, $\Delta F°$, which is the free energy change resulting when both reactants and the products are under standard conditions (in this case, one atmosphere pressure), we can do so by systematically adding suitable equations to equation VI-29, in each of the new equations carrying one of the gases from the standard pressure of one atmosphere to its partial pressure P_A, and P_B, etc., adding each time the free energy change produced by such an alteration. For instance, the first equation we add is

$$aA\ (P_A = 1) \rightarrow aA\ (P_A = P_A),\ \Delta F = a\ RT \ln (P_A/1) \qquad \text{Equation VI-30}$$

When we perform all these additions and combine logarithmic terms, we obtain

$$\Delta F° = \Delta F + RT \ln \frac{(P_A)^a (P_B)^b}{(P_C)^c (P_D)^d} \qquad \text{Equation VI-31}$$

or

$$\Delta F° = \Delta F - RT \ln \frac{(P_C)^c (P_D)^d}{(P_A)^a (P_B)^b} \qquad \text{Equation VI-32}$$

Since we are by definition dealing with equilibrium conditions, $\Delta F = 0$, and the last term in equation VI-32 is $-RT \ln K$, where K is the equilibrium constant as usually defined, then we see that

$$\Delta F° = -RT \ln K \qquad \text{Equation VI-33}$$

an expression of the important relation between the standard free energy change and the equilibrium constant. From its form it is seen at once that when $\Delta F°$ is large and negative, the reaction has a strong tendency to run to the right, and when $\Delta F°$ is large and positive, the reaction has a strong tendency to run to the left.

It can be seen that if the equilibrium constant of a reaction is known (and thus $\Delta F°$ is known) and $\Delta H°$ has been determined by calorimetric measurements, then $\Delta S°$ can be determined from equation VI-14. If $\Delta H°$ is not known, it may be estimated from Van't Hoff's equation

$$d(\ln K)/dT = \Delta H°/RT^2 \qquad \text{Equation VI-34}$$

The great importance of F stems from the fact that it is a measure of the driving force of a chemical reaction at constant temperature and pressure. If a reaction as we write it involves a decrease of free energy, that is, the combined free energies of the products are less than those of the reactants, thermodynamics predicts that this reaction will go spontaneously, although thermodynamics can not predict the rate, and in the absence of the proper catalysts the reaction may go so slowly that we are quite unable to measure its rate. If the reaction as written involves an increase in F, then we know it will never go, unless external energy is applied and it thus becomes part of a more complex total reaction in which there is an over-all net decrease in free energy. This is a law of great generality and power, and no exceptions to it have ever been observed.

An important feature of free energy values is that they are additive. If we know the free energies of the starting materials of a reaction, and the free energies of the final products, it does not matter through how many intermediate stages, or by what steps, the reaction proceeds, for the net free energy change will still be the same. This, of course, has to be true of a thermodynamic function. Consider the following hypothetical reactions

$$A + B \rightarrow C + D, \Delta F_1 = a$$
$$C + D \rightarrow E + F, \Delta F_2 = b$$
$$\cdots\cdots\cdots\cdots\cdots\cdots$$
$$W + X \rightarrow Y + Z, \Delta F_n = z$$

then the sum

$$(a + b + \ldots. + z) = (\Delta F_1 + \Delta F_2 + \ldots. + \Delta F_n) = \Delta F \qquad \text{Equation VI-35}$$

is the free energy change of the over-all reaction

$$A + B \rightarrow Y + Z$$

We may illustrate the use of free energy by two examples. We have the reactions

$$C \text{ (graphite)} + O_2\text{(gas)} \rightleftarrows CO_2\text{(gas)}, \Delta F = -94.26 \text{ Kcal.} \qquad \text{Equation VI-36}$$

$$C \text{ (graphite)} \rightleftarrows C \text{ (diamond)}, \Delta F = 0.685 \qquad \text{Equation VI-37}$$

In equation VI-36, the free energy change is negative, that is, the product, carbon dioxide, has less free energy than the reactants, graphite and oxygen. Therefore the reaction will go spontaneously and energetically once it is started. It is true that graphite may be exposed to the air indefinitely without getting oxidized, but that is because, under ordinary conditions, the requisite energy of activation (see page 296) is lacking. Once the energy of activation is supplied by sufficient local heating, the oxidation will proceed energetically.

The situation is otherwise with equation VI-37, which has long been of great interest to those who would get rich quickly. Graphite is a cheap and plentiful substance, and both graphite and diamonds are only forms of carbon. Can we make diamonds out of graphite? It is obvious from equation VI-37 that we can not, at least not under the usual conditions of temperature and pressure, for ΔF is positive, which shows that the reaction as written will not go to any appreciable extent. Under ordinary conditions graphite is the stable and diamond the unstable form.

It will be noted that in equation VI-37 ΔF is not large, which suggests that under other conditions of temperature and pressure it might be reduced in value to the point of becoming negative. This is evidently true, for the General Electric Laboratories have recently announced the artificial production of diamonds under high temperatures and pressures.

The important conclusions about equilibrium reactions which can be drawn from equation VI-33 will illustrate the power of the concept of free energy. Suppose we have a set of reactions

$$A \rightleftarrows B \qquad A' \rightleftarrows B' \qquad A'' \rightleftarrows B''$$

and we find that the reactant and product in each case exist in the reaction mixture in equilibrium with each other. If we assume that all the B present came originally from the conversion of A and arbitrarily set the original amount of A equal to 100, so that at any later time $A + B = 100$ (4), we may express the equilibrium constant of these reactions in terms of the per cent composition of the mixture with respect to B. If $B = 50$ for instance, then $A = 50$ and $K = 50/50 = 1$. If $B' = 80$, then $A' = 20$ and $K' = 4$. Substituting these values in equation VI-33, we can obtain the standard free energy change (ΔF°) of the various reactions in terms of the per cent composition of the reaction mixtures at equilibrium. This relation is shown in figure VI-2.

From the graph it is evident that if the reactant and product are both found in appreciable amounts at equilibrium, the standard free energy change of the reaction can not be very large. When A and B are found in equal concentrations at equilibrium the free energy change is zero. Even if the reaction goes so far that 95 per cent of A is converted into B, the free energy change is only about 1.75 Kcal. Conversely, it can be seen that if the free energy change is at all large, say of the order of 10 Kcal., the amount of the reactant remaining (if ΔF° is negative) or product formed (if ΔF°

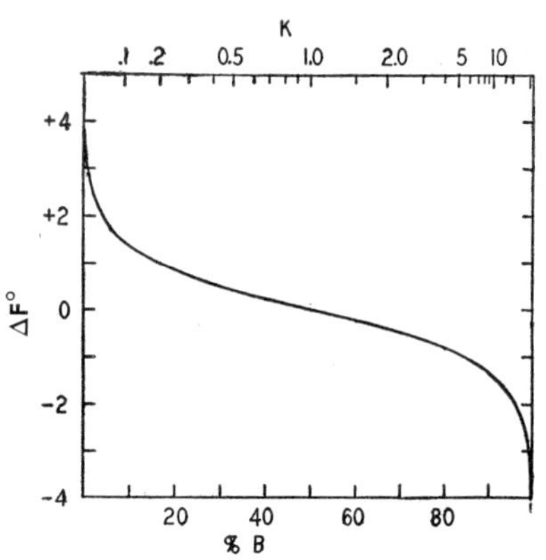

FIG. VI-2. Standard free energy change (Δ F°) of the reaction A ⇄ B as a function of the per cent composition of the equilibrium mixtures in respect to B at 25°C. (Slightly modified from H. B. Bull. *Physical Biochemistry*, 2nd edit. New York, Wiley, 1951. By permission.)

is positive) at equilibrium will be too small for easy detection (about one part in 20 million).

Chemical Potential

In discussing the effects of changes in the concentration of reagents on equilibrium it is useful to have the concept of partial molar free energy

$$[\partial F/\partial n_i]_{T,P,n_j} = \mu_i \qquad \text{Equation VI-38}$$

where n_i represents the molar concentration of the i-th constituent of the mixture, T and P are temperature and pressure, and n_j stands for the concentration of each of all the other constituents. It can be seen that μ_i measures the effect on the free energy of the system of adding a small amount of constituent i, keeping temperature, pressure, and other concentrations constant. Gibbs called the partial molar free energy the *chemical potential*, and introduced the symbol μ for it.

The quantity μ_i is a measure of the force which constituent i is exerting to drive a reaction in a particular direction. If for instance we have a reaction in solution

$$mM + nN \rightleftarrows qQ + rR \qquad \text{Equation VI-39}$$

where M, N, Q and R represent different chemical substances, and m, n, q and r are the relative numbers of molecules involved, then at equilibrium we must have

$$m\mu_M + n\mu_N = q\mu_Q + r\mu_R \qquad \text{Equation VI-40}$$

It should be noted that chemical potential can not be identified with concentration. It is a necessary condition of heterogeneous equilibrium that the chemical potential of any component must be the same in each phase. For example, if we dissolve iodine in carbon tetrachloride and shake this solution up with water, at equilibrium the chemical potential of the iodine in the water and in the carbon tetrachloride will be the same, although the actual concentrations will be very different.

Chemical potential is closely related to the *activity* of a substance. In fact, the chemical potential of any component in a solution is

$$\mu = \mu_o + RT \ln A \qquad \text{Equation VI-41}$$

where A is the activity of the component, μ_o is its chemical potential in the standard state, R is the gas constant, T the absolute temperature, and ln signifies logarithm to the base e.

Since the free energy change of a reaction is zero when the reactants and products are present in equilibrium concentrations, it follows that the work required at such concentrations to form a small amount of the products, or conversely to change a small amount of the products back to the reactants, is zero. In other words, such a reaction can easily be forced to go in either direction up to a point. The existence of such reactions is important to the body. For instance, since both glucose-1-phosphate and glycogen are present in appreciable amounts at equilibrium, it can be seen that it will take very little free energy to form more glycogen from glucose-1-phosphate or more glucose-1-phosphate from glycogen.

Free energy considerations also help us understand the way in which one reaction can be driven by another in the body. Suppose we have a reaction, $A \rightleftarrows B$, which has a $\Delta F°$ at 37°C. of +2.85 Kcal. The fact that $\Delta F°$ is positive means that this reaction will go only slightly by itself. At equilibrium the concentration of B is less than 1 per cent of A.

Suppose, however, there is also another reaction, $B \rightleftarrows C$, with a $\Delta F°$ at 37 C of −4.27 Kcal. This reaction will go far to the right. Consequently, as B is formed from A it will be converted into C. We shall have the over-all reaction, $A \rightleftarrows C$, with its own $\Delta F°$ which is the algebraic sum of those of the two component reactions, or $\Delta F° = +2.85 - 4.27 = -1.42$ Kcal. This over-all reaction will go on until at equilibrium the concentration of C is 10 times that of A. The larger negative free energy change of reaction $B \rightleftarrows C$ has been employed to drive reaction $A \rightleftarrows B$ in a forward direction. This type of coupling between reactions seems to be the only one known to occur in the body (4).

For an electrochemical cell ΔF may be calculated from the voltage and the number of equivalents of electricity transferred per mol. The equation is

$$\Delta F = -n_e \mathbf{F} V \qquad \text{Equation VI-42}$$

where F represents the free energy, n_e the number of equivalents of electricity per mol, **F** the value of the Faraday, and V the voltage of the cell.

We are so accustomed to heat engines in daily life that we are likely to regard other methods of converting energy into work, such as the battery-powered motor or the animal body, as exceptional and to feel surprised if they compete in mechanical efficiency with heat engines. Actually, there is no reason to think that burning a fuel and then converting as much of the resulting heat as possible into work in a heat engine make the best use of the contained energy. It sometimes does not. To pursue this argument a little further, let us take some actual figures (7). The heat of oxidation of metallic zinc in contact with a concentrated solution of zinc sulfate is about 55.2 Kcal. per mol of zinc consumed. The electrical energy we can get from a cell employing this as one electrode is somewhat less (50.5 Kcal. per mol). Nevertheless, we see that if we assume the heat utilized in one of our best heat engines, with an efficiency of 40 per cent, and the electrical energy utilized by an electric motor with an efficiency of 80 per cent, we can get, by utilizing the heat of reaction, at the most 22.0 Kcal. of work per mol of zinc, whereas our battery motor combination will give us 40.4 Kcal., or nearly twice as much.

High-energy Bonds

We have just spoken of coupled reactions, that is, of a reaction with a positive free energy change being driven forward by a second and related reaction with a negative free energy change sufficiently large to make a net negative free energy change for the over-all reaction. In the body the reactions used as driving forces for otherwise "uphill" or non-spontaneous reactions involve certain molecular groups which, when hydrolyzed, are responsible for a particularly large negative free energy change. Chemical bonds which involve unusual amounts of negative free energy change when broken (and require the input of the same unusual amounts when formed) are termed high-energy bonds, and of these, the most familiar are the *high-energy phosphate bonds.*

Phosphate bonds fall into two large groups differing widely in their energy content. One group, known as "low-energy phosphate bonds", is more stable, easier to form, harder to hydrolyze and for those very reasons has a lower energy content. This group of compounds includes the *phosphate esters of alcohols and phenols.* Typical examples of compounds containing low-energy phosphate bonds are glyceryl phosphate and phenyl phosphate (formula I). The various glucose and fructose phosphates which are of importance in carbohydrate metabolism (see Chapter 14) also contain low-energy bonds.

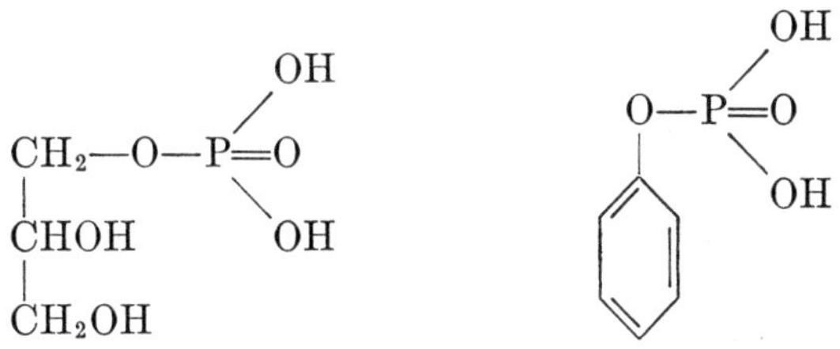

Glyceryl phosphate　　　　Phenyl phosphate

I. Low-energy phosphate bonds

A second group, the "high-energy phosphate bonds", is less stable, harder to form, and easier to hydrolyze, and for those very reasons has a higher energy content. This group of compounds includes *acid anhydrides of phosphoric acid and those compounds containing a phosphorus-nitrogen bond*. It has been suggested (29) that phosphorus-sulfur bonds are also included. The group containing high-energy bonds may be subdivided further and in view of the great importance of these compounds in metabolism, they will be taken up in some detail.

1. Acid anhydride formation between two molecules of phosphoric acid to form a *pyrophosphate link*. Compounds containing such links may be typified by ATP and *adenosine diphosphate* (ADP) (formulas II and III). It is conventional in writing these and other such formulas to use a wavy line (∼) to indicate a high-energy phosphate bond. The phosphate group

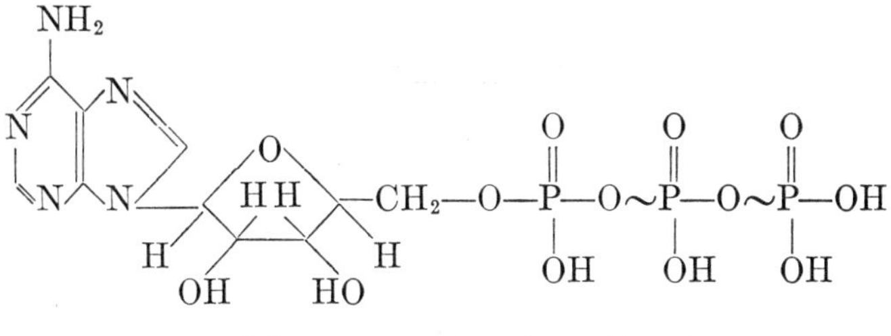

II. Adenosine triphosphate (ATP)

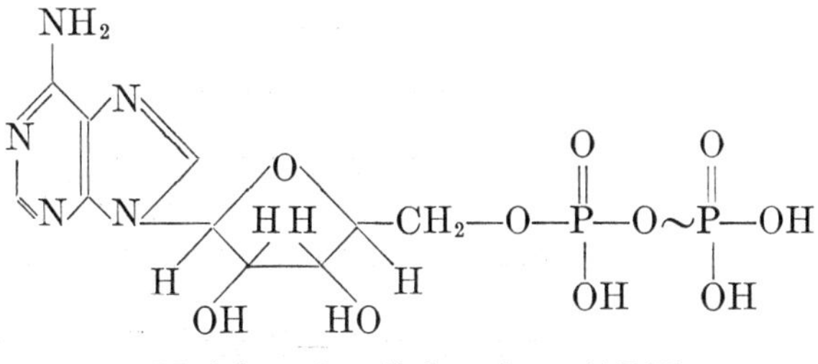

III. Adenosine diphosphate (ADP)

itself is sometimes abbreviated as ph or Ⓟ. Note that ATP has two pyrophosphate links. In breaking such a link, as in the hydrolysis of ATP to ADP or ADP to *adenylic acid* (formula IV) (also called adenosine mono-

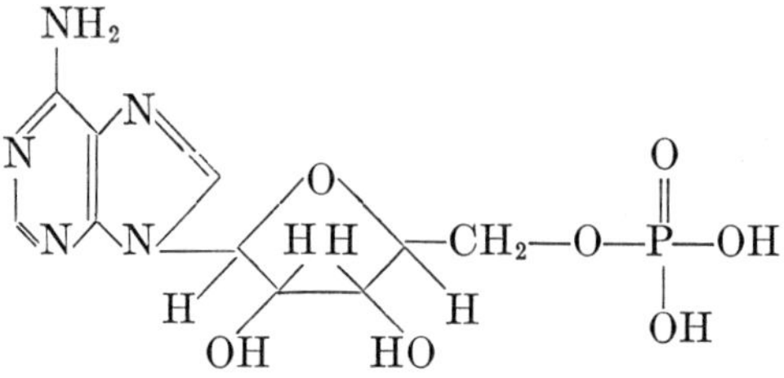

```
      NH2
       |
      / \  N
    N//  \/ \\
    |    ||   >     O                     O
    \\N/\N/      /     \                   ||
                <   H H   >—CH2—O—P—OH
             H/   \___/   \H              |
                 OH     HO               OH
```

IV. Adenylic acid (AMP)

phosphate, or AMP), comparatively large quantities of energy are released. All this holds true for guanosine triphosphate (GTP) and uridine triphosphate (UTP), which differ from ATP only in the heterocyclic ring structure. The number of consecutive pyrophosphate links in a single molecule is not limited to two. Adenosine tetraphosphate (with three high-energy bonds) has been identified in horse muscle (13).

2. Acid anhydride formation between a molecule of phosphoric acid and one of a carboxylic acid. Such compounds are acyl phosphates and an example is *1,3-diphosphoglyceric acid* (formula V) which is an intermediate in the glycolytic breakdown of glycogen (see Chapter 14).

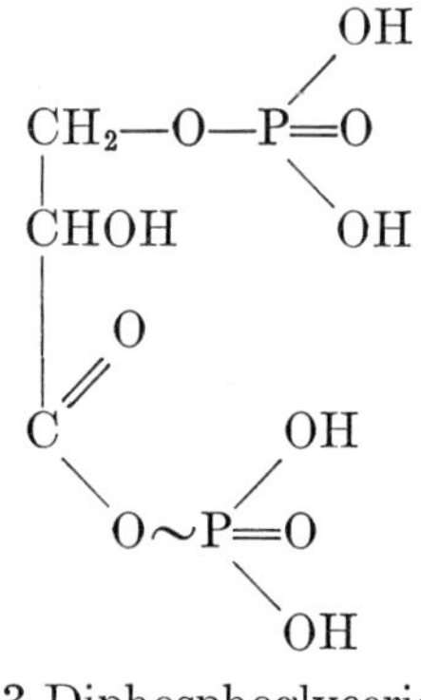

```
                   OH
                  /
CH2—O—P=O
|                 \
CHOH               OH
|
|   O
|  //
C          OH
 \        /
  O~P=O
         \
          OH
```

V. 1,3-Diphosphoglyceric acid

3. Acid anhydride formation between a molecule of phosphoric acid and an enol form of a keto-acid (formula VI). An example is *phosphoenolpyruvic acid* (formula VII), which is a step in the glycolytic breakdown of glycogen.

4. A compound containing a direct N—P bond. The most important such compound in the vertebrate body is phosphocreatine (formula VIII), already mentioned as a muscle component. Note that the phos-

phate group is attached to the nitrogen of a guanidino group. Arginine contains such a guanidino group and, in fact, in invertebrate muscle, phosphocreatine does not exist and *phosphoarginine* (formula IX), or arginine phosphate as it is sometimes called, is found instead.

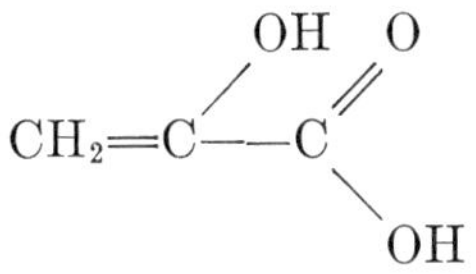

```
                                                  OH
                                                 /
                                           O~P=O
                                           |    \
                                           |     OH
        OH   O                             |
       /    //                             |      O
CH2=C---C                           CH2=C—C  //
            \                                   \
             OH                                  OH
```

VI.
Pyruvic acid (enol form)

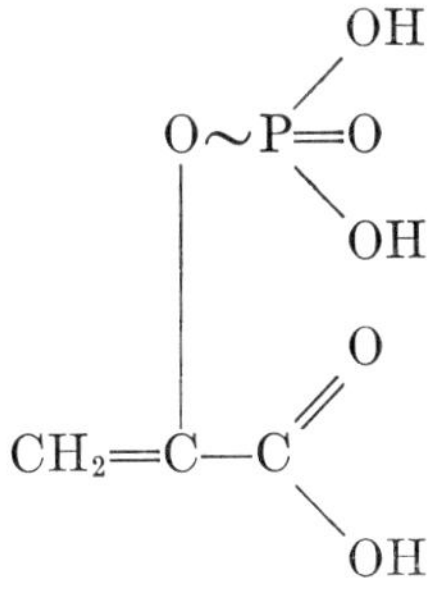

VII.
Phosphoenolpyruvic aci

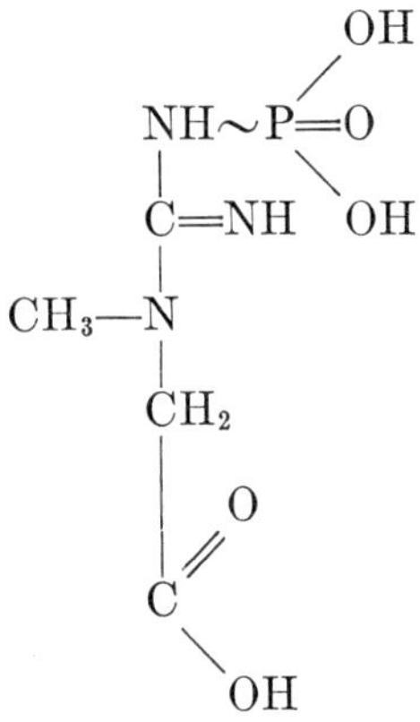

```
                                       OH
                                      /
                                NH~P=O
                                |  (+) \
                                C=NH2   OH
                                |
                               HN
         OH                     |
        /                       CH2
  NH~P=O                        |
  |     \                       CH2
  C=NH   OH                     |
  |                             CH2
CH3—N                           |
  |                             CHNH2
  CH2                           |
  |   O                         |   O
  |  //                         |  //
  C                             C
   \                             \ (−)
    OH                            O
```

VIII.
Phosphocreatine

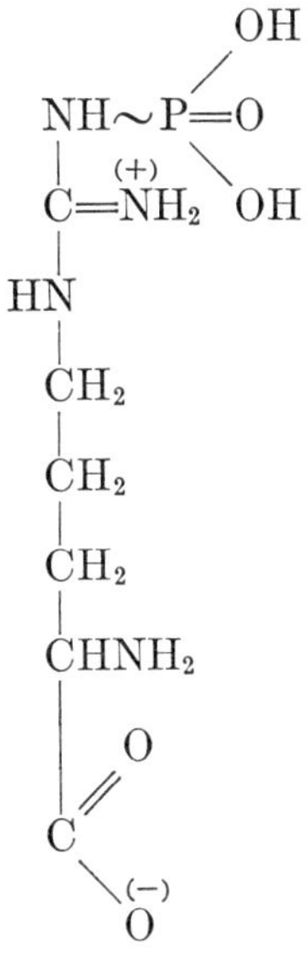

IX.
Phosphoarginine

A logical question at this point would be why an acyl phosphate, for instance, should be high-energy, while an alkyl phosphate is low-energy. In the case of the acyl phosphate, the carboxyl group of the product of hydrolysis is stabilized by resonance. Such stabilization tends to occur in those compounds which possess double bonds and a certain amount of symmetry. The carboxyl, phosphate, and guanidinium ions (formula

X) fulfill both these requirements. The addition to any of these of a phosphate group decreases the symmetry and diminishes resonance stabili-

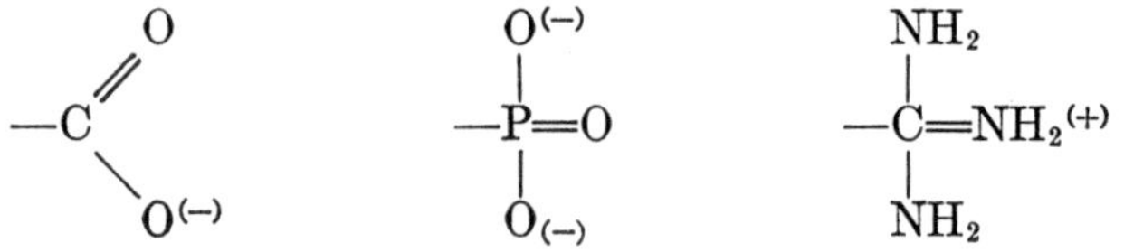

$$-C\begin{matrix}\nearrow O \\ \searrow O^{(-)}\end{matrix} \qquad -\overset{O^{(-)}}{\underset{O_{(-)}}{P}}=O \qquad -\overset{NH_2}{\underset{NH_2}{C}}=NH_2^{(+)}$$

Carboxyl ion Phosphate ion Guanidinium ion

X. Ions stabilized by resonance

zation. There is thus a particularly large increase in chemical energy in passing from these compounds to their respective phosphate esters or, to look at it in another way, an abnormally high tendency for the phosphate ester to hydrolyze back to the more symmetrical compound.

In the case of *enol* phosphates such as phospho*enol*pyruvic acid, resonance is not the answer. At least part of the answer, however, is the fact that the presence of the phosphate group "freezes" the molecule in the less stable *enol* configuration, so that to the natural energy content of the phosphate bond is added the energy required to maintain the *enol* form. For more on the physical chemistry of the phosphate bond, see (10).

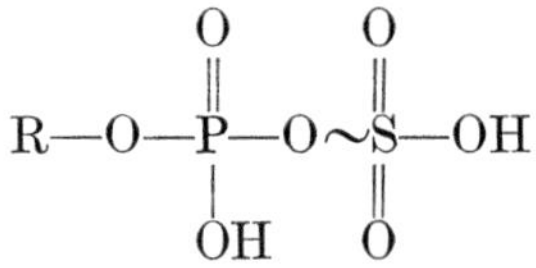

$$R-O-\overset{O}{\underset{OH}{\overset{\|}{\underset{|}{P}}}}-O\sim\overset{O}{\underset{O}{\overset{\|}{\underset{\|}{S}}}}-OH$$

XI. Active sulfate group

It has been suggested that an active sulfate bond (possibly high-energy) can exist when the sulfate group is bound to a phosphate group (formula XI) and that such compounds take part in enzymatically-catalyzed sulfate group transfers (21).

The function of the high-energy phosphate bond in living tissue is to act as a driving force in coupled reactions. If ATP, for instance, is hydrolyzed to ADP, energy is released which can be utilized to force a second reaction, requiring energy uptake, to completion. An example of such a two-reaction system is the formation of glucose-6-phosphate from glucose in the presence of the appropriate enzyme when ATP is added (formula XII). The reaction does not take place in the absence of ATP. Note that the phosphate bond formed on the glucose molecule is low-energy. The heat liberated by the hydrolysis of ATP to ADP, in the presence of myosin as catalyst, has been measured and found to be 4.7 ± 0.7 Kcal./mol (19). The formation of a low-energy phosphate bond such as that in glucose-6-phosphate requires only 3.1 Kcal. per mol. It is obvious then that a low-energy bond can be formed at the expense of a high-energy bond. The reverse of such a reaction (that is, the formation of ATP from ADP by hydrolysis of glucose-6-phosphate to glucose) can not take place unless the difference in energy is supplied.

Some way must exist, therefore, of forming high-energy phosphate bonds without the utilization of such a bond, as otherwise the body's limited supply of such bonds would be quickly consumed with fatal results.

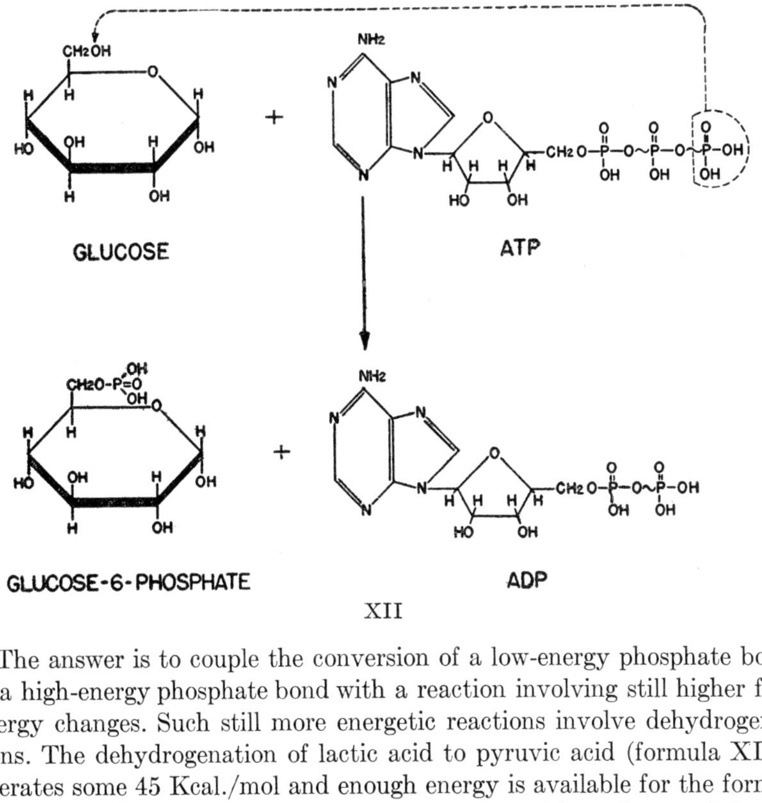

GLUCOSE + ATP

GLUCOSE-6-PHOSPHATE + ADP

XII

The answer is to couple the conversion of a low-energy phosphate bond to a high-energy phosphate bond with a reaction involving still higher free energy changes. Such still more energetic reactions involve dehydrogenations. The dehydrogenation of lactic acid to pyruvic acid (formula XIII) liberates some 45 Kcal./mol and enough energy is available for the formation of many high-energy phosphate bonds. In actual fact, three such bonds are formed (see Chapter 14). It is possible that in the process of forming these bonds, high-energy acyl-mercaptan bonds play a role. These are formed by the condensation of sulfhydryl groups with carboxyl groups (formula XIV).

The free energy liberated by the hydrolysis of an acyl-mercaptan bond is 8.2 Kcal./mol (5) and this allows energy and to spare for the formation of a high-energy phosphate bond.

The formation of energy-rich substances, such as carbohydrates, the gradual dehydrogenation of which is used to start the whole chain of coupled reactions, supplying energy for all, is the result, fundamentally, of the direct utilization of solar energy through photosynthesis by the green plant. On this, the whole elaborate system of energetics in all heterotrophs (including man) depends.

$$CH_3\overset{\overset{OH}{|}}{C}HCOOH \xrightarrow{-2H} CH_3\overset{\overset{O}{\|}}{C}COOH$$

XIII. Dehydrogenation of lactic acid

$$R\overset{\overset{O}{\|}}{C}—OH + HSR' \rightarrow R\overset{\overset{O}{\|}}{C}—O \sim SR' + H_2O$$

XIV. Formation of acyl-mercaptan

THE NATURE OF ENZYMES

Enzymes as Catalysts

An *ideal catalyst* is a substance which when present in small quantities will alter the rate of a chemical reaction without itself being altered in the process. An ideal catalyst would be expected, therefore, to exert its rate-changing effect over indefinite periods of time if sufficient quantities of the reagents concerned are present. Actually, no real catalyst is ideal, and all are, in the course of time as the catalyzed reaction proceeds, rendered inert or "poisoned".

Well known inorganic catalysts affecting many reactions include many metals such as platinum, palladium, or nickel, and such compounds as vanadium pentoxide and copper chromite. Water itself is perhaps the most versatile and important inorganic catalyst known, since a surprisingly wide array of reactions exist (the chemical union of hydrogen and oxygen to form water being the best known) in which the presence or absence of a trace of water effects a tremendous change in rate. Within the body there exist catalysts (enzymes) which, by contrast, are complex organic compounds. However, as far as the basic principles of catalysis are concerned no distinction can be made between catalysts derived from living or non-living sources. Thermodynamically, the enzyme is no more esoteric than platinum or water.

Let us consider several of the properties that all catalysts of whatever nature share. A catalyst has already been defined as a substance which alters the rate of a reaction. It is important to realize how limiting such a definition is. It says nothing about affecting the nature of a reaction, its direction, or its extent; it affects only the rate. The implication is that any reaction taking place under the influence of an enzyme (or any other catalyst) would also take place, albeit at a different rate, in the absence of an enzyme (or any other catalyst). This is, indeed, true. As a case in point let us consider hydrogen peroxide.

When hydrogen peroxide solution is placed upon an open cut, there is a rapid effervescence which can be determined chemically to be due to the decomposition of the compound into water and oxygen (formula XV). This

$$2H_2O_2 \xrightarrow[\text{[catalase]}]{} 2H_2O + O_2\uparrow$$

XV

decomposition of hydrogen peroxide is extraordinarily rapid, and takes place with similar rapidity in the presence of almost any type of tissue. This is attributed to the widely distributed presence of an enzyme catalyzing the reaction. This enzyme has been given the name *catalase.* This catalytic property is not confined to catalase; manganese dioxide will act similarly, but less effectively. However, hydrogen peroxide in the absence of any catalyst will slowly decompose into water and oxygen. The effect of the enzyme, catalase, is not to change this reaction, but merely to accelerate it many thousandfold.

The case of hydrogen peroxide solution was chosen purposely because its decomposition proceeds in the absence of catalysis of any sort at a rate that is easily measurable. In the case of many enzymatic processes, the corresponding reactions can not be shown to occur at all in the absence of the enzyme under comparable conditions. Such would be the case, for instance, in the hydrolysis of maltose to glucose under the catalytic influence of the enzyme, maltase (formula XVI). Here, maltose is a stable organic compound which in aqueous solution at neutral pH and body temperature will persist unchanged for extended periods. Is this a case where the presence of an enzyme has done more than simply change the rate of reaction? Has it acted to initiate a reaction? The view generally held by enzyme chemists is that these questions must be answered in the

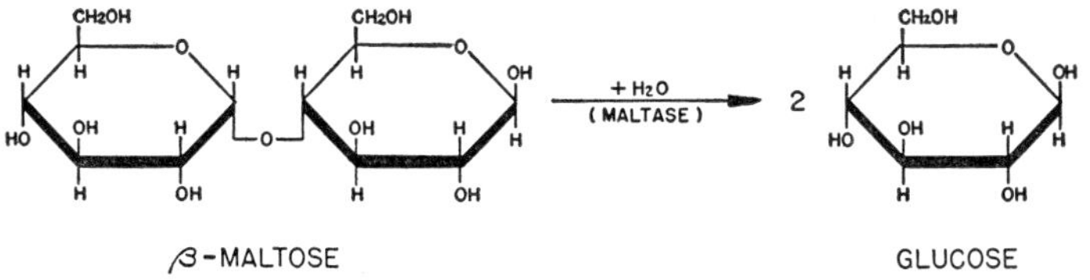

β-MALTOSE GLUCOSE

XVI

negative. Even here, maltose is supposed to hydrolyze spontaneously to form glucose, but at a rate too slow to be measured.

To summarize, then, a catalyst (whether an enzyme, a metallic ion or anything else) can not alter the thermodynamic characteristics of a reaction. It can not alter the free energy relationships and therefore does not alter the direction of a reaction or the position of the equilibrium point. It affects only the rate of the reaction, concerning which thermodynamics makes no predictions. A reaction may be spontaneous and yet almost infinitely slow because of a high energy of activation (see page 296). It is upon this energy of activation that the enzyme exerts its effect, not upon the free energy change of the over-all reaction.

Enzymes, through their ability to accelerate thermodynamically possible reactions, are capable of determining the path of a reaction where several are possible. Let us take the case of pyruvic acid, an important intermediary in carbohydrate metabolism. In the human organism, pyruvic acid may undergo many chemical changes two of which are reduction to lactic acid and transamination to alanine (formula XVII). All the reactions, including these two can take place spontaneously; all involve loss of free energy at least in the first stages of the reaction. Each reaction is catalyzed by a separate enzyme or group of enzymes. However, all these changes do not take place in the body in any random, catch-as-catch-can way. At a given time in a given tissue only certain of these reactions will take place; perhaps even only one of them. One way of choosing among these multiple reactions is by controlling the chemical environment. In the case of the two reactions presented in formula XVII, a source of hydrogen atoms such

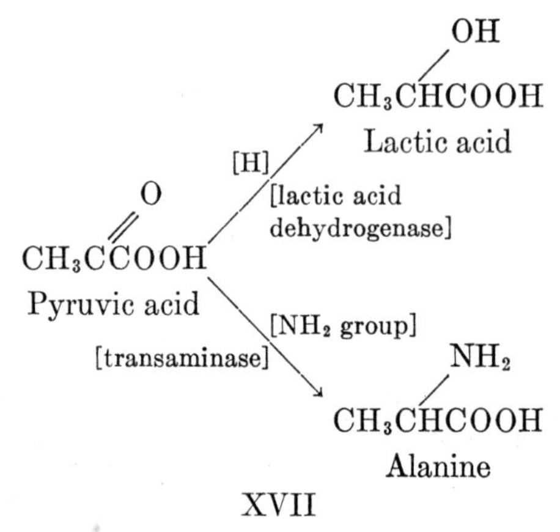

XVII

as reduced coenzyme I (see page 276) must be present for pyruvic acid to be reduced to lactic acid, while a source of amino groups, such as glutamic acid, must be present for it to be transaminated to alanine. If all such substances are present, the relative rates of the reactions are controlled by the relative concentration of specific enzymes. The reduction of pyruvic acid is catalyzed by the enzyme, lactic acid dehydrogenase, the transamination by transaminase. Thus if only the enzyme transaminase were present, only the reaction it catalyzed would be accelerated in rate. All the other reactions would still proceed, provided all necessary substances were present, but at their natural uncatalyzed rates. The catalyzed transamination would so far outstrip the rest that all the pyruvic acid would be converted to alanine before significant quantities of the acid had reacted in any other way.

Enzymes as Proteins

Until now we have discussed those properties that enzymes hold in common with all catalysts; it is now time to discuss those which differentiate them from other catalysts. The differentiating property may be stated simply: *Enzymes are proteins*. Enzymes may therefore be completely defined as catalytic proteins. The protein nature of enzymes was long unrecognized because of the very fact of their intense activity. Extremely small concentrations of enzyme suffice to bring about very marked acceleration of the reactions they catalyze. It was thus possible a quarter century or more ago to obtain purified enzyme extracts which, although very active, yielded negative results to the protein tests of that day—even the most sensitive. That enzymes are proteins is indicated nevertheless by several lines of evidence:

1. Enzymes exhibit properties entirely similar to those of proteins. They are precipitated by protein precipitants, such as phosphotungstic acid or concentrated ammonium sulfate solutions; they will dissolve, as will proteins, in water or in dilute salt solutions; they will not pass through a dialyzing membrane; and they will lose their activity when exposed to any of the environmental factors known to denature proteins.

2. When solutions of enzymes are allowed to stand at ordinary temperatures, their activity declines with time. This decline in activity is hastened by increasing the temperature. If this decline is viewed as the result of some chemical action, such a change in rate with increasing temperature is not unusual since the rate of all chemical reactions increases as the temperature is raised. Whereas in the case of almost all ordinary chemical reactions the rate is only doubled or tripled for every ten-degree rise in temperature, in the case of this decline in enzymatic activity with time the rate is increased many hundredfold for each ten-degree rise in temperature. There is only one other reaction known which shows a similar startlingly steep increase in rate with rise in temperature, and that is protein denaturation. This seems strong evidence that enzymes are proteins and that their inactivation is due to the denaturing effects of, in this case, heat.

3. In 1926 Sumner isolated an enzyme in crystalline form for the first time in history (27)—an achievement for which he later received the Nobel

Prize. The enzyme was urease, which catalyzes the hydrolysis of urea to carbon dioxide and ammonia. Since then a number of other enzymes have been crystallized (16). In the case of every enzyme so crystallized examination showed it to be protein.

Polynucleotides (see Chapter 9) which may be as complex in their structure as proteins, could conceivably act as enzymes. Binkley (3) offers evidence that cellular enzymes catalyzing the formation or rupture of a peptide bond are polynucleotides.

THE MULTIPLICITY OF ENZYMES

Terminology

The number of known enzymes is immense and as methods of investigation are refined it increases constantly. It is useful, therefore, to learn the systematic terminology for the individual enzymes. The suffix *ase* is accepted as indicating an enzyme. The rest of the name is usually chosen in one of the following ways:

(1) From the name of the chief *substrate* (the substrate of an enzyme being the substance or substances affected by the reaction catalyzed). Thus urease is an enzyme which catalyzes the hydrolysis of urea.

(2) From the name of the chemical process involved. A transphosphorylase, for instance, is an enzyme catalyzing the transfer of a phosphate group from one compound to another.

These two systems of nomenclature can be combined as in ascorbic acid oxidase where both the substrate and the chemical process involved are named. General groups of enzymes are named similarly. A proteinase (or protease) would be any one of many different enzymes which catalyze the hydrolysis of proteins. An oxidase (or a reductase) is any of many different enzymes which catalyze an oxidation-reduction reaction. Closely related enzymes can be differentiated by including a reference to the nature of the environment in which they are active as in the case of acid phosphatase as opposed to alkaline phosphatase; or by naming the source from which the enzyme is derived as in the case of kidney phosphatase as opposed to intestinal phosphatase.

Exceptions to these general rules occur among enzymes identified and studied during the infancy of enzymology before a systematic nomenclature had been developed, e.g., catalase, where the substrate is hydrogen peroxide and the chemical process involves a dismutation. Some of the digestive enzymes have names which even lack the almost universal *ase* suffix. Pepsin and trypsin are the best known of these.

Pepsin and trypsin, as secreted by the gastric mucosa and pancreas, respectively, are inactive. These inactive forms are termed *pepsinogen* and *trypsinogen* and differ from the active forms in possessing a polypeptide addendum which, apparently, shields certain reactive portions of the enzyme surface. Such inactive precursors of enzymes are called *zymogens* or *proenzymes*. Trypsinogen is activated by several proteolytic enzymes, including the enzyme, *enterokinase*, which occurs in the duodenal mucosa and whose action is to hydrolyze the shielding polypeptide off the enzyme surface. Such activators of zymogens are called *kinases*. Pepsinogen is not activated by a specific kinase, but rather through the proteolytic action of pepsin itself. In other words, as soon as a molecule of pepsinogen spontaneously changes to form pepsin, the pepsin formed catalyzes the change of more pepsinogen into pepsin, which thus adds to the supply of catalyst and further increases the rate of reaction. A system such as this where the product formed in a reaction is itself the catalyst of the reaction is an *autocatalytic system*. Since trypsinogen is similarly activated by trypsin, as well as by enterokinase, its activation is also autocatalytic. Not all enzymes pass through a zymogen stage. When, as in the case of pancreatic lipase, they are initially secreted in the active state, they are termed *preformed enzymes*.

Classification

There are many different ways of classifying enzymes and no one way is the "right" way. On the basis of the type of chemical action catalyzed, enzymes can be divided into five groups: (a) splitting enzymes, (b) hydrolyzing enzymes, (c) transferring enzymes, (d) isomerizing enzymes, and (e) oxidizing enzymes. This classification is, in the main, that proposed by Baldwin (1).

Splitting enzymes. These catalyze reactions which form two or more molecules out of one. The reactions usually involve the separation of such simple groups as water, carbon dioxide, or ammonia. Examples of such enzymes are fumarase, tyrosine decarboxylase, and aspartase, which catalyze reactions as shown in formula XVIII. The enzyme carbonic anhydrase, which catalyzes the decomposition of carbonic acid to carbon dioxide and water, also belongs to this group (formula XIX).

In some cases, larger groups are split off. Aldolase, for instance, an enzyme of the glycolytic chain (see Chapter 14) splits fructose-1,6-diphosphate into two three-carbon fragments, 3-phosphoglyceraldehyde and phosphodihydroxyacetone (formula XX).

Enzymes catalyze reversible reactions in both directions. It follows that what we call splitting enzyme could equally well be termed *adding enzymes*. This latter term is used by Baldwin (1).

Hydrolyzing enzymes. This group of enzymes, also called *hydrolases*, as the name indicates, catalyze reactions that split compounds with the addition of the elements of water. This subgroup, a large one, can be di-

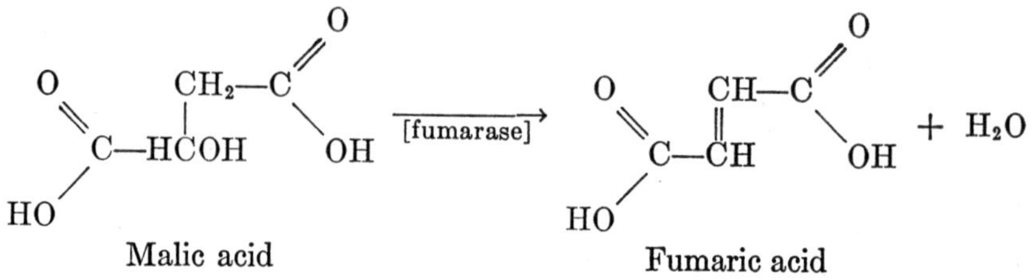

$HOOC{-}CH(OH){-}CH_2{-}COOH \xrightarrow{\text{[fumarase]}} HOOC{-}CH{=}CH{-}COOH + H_2O$

Malic acid → Fumaric acid

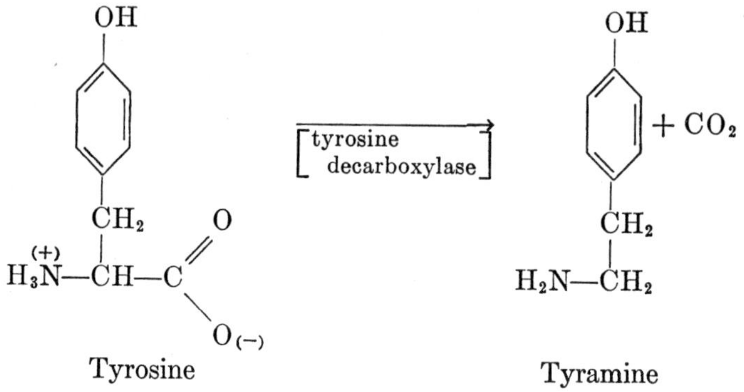

$HO{-}C_6H_4{-}CH_2{-}CH(\overset{(+)}{N}H_3){-}COO^{(-)} \xrightarrow{\text{[tyrosine decarboxylase]}} HO{-}C_6H_4{-}CH_2{-}CH_2{-}NH_2 + CO_2$

Tyrosine → Tyramine

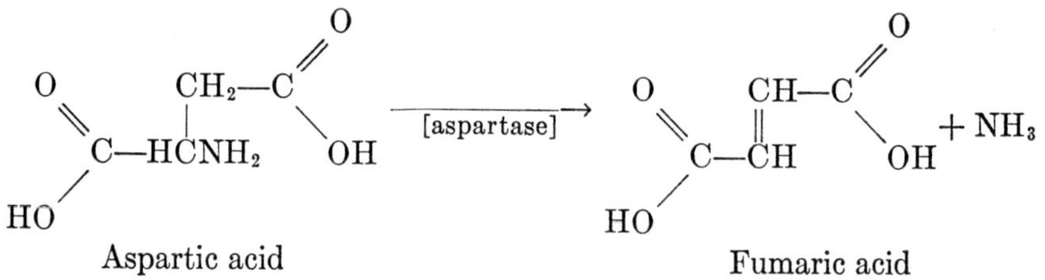

$HOOC{-}CH(NH_2){-}CH_2{-}COOH \xrightarrow{\text{[aspartase]}} HOOC{-}CH{=}CH{-}COOH + NH_3$

Aspartic acid → Fumaric acid

XVIII

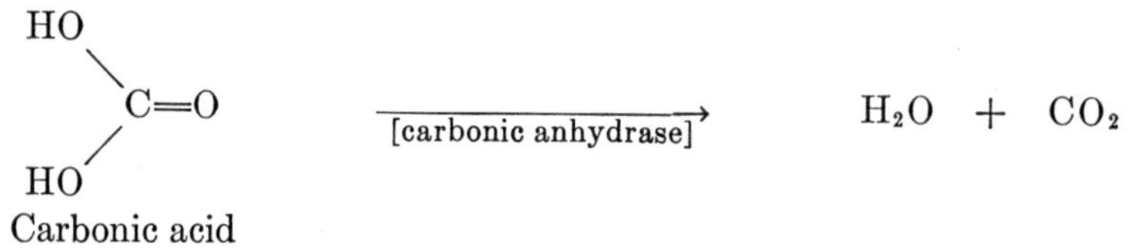

$(HO)_2C{=}O \xrightarrow{\text{[carbonic anhydrase]}} H_2O + CO_2$

Carbonic acid

XIX

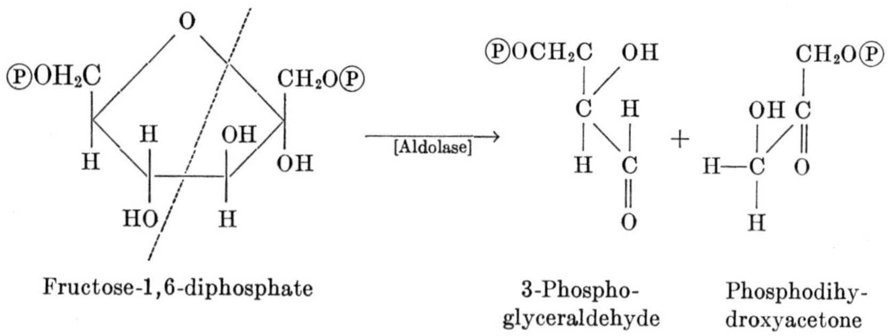

Fructose-1,6-diphosphate $\xrightarrow{\text{[Aldolase]}}$ 3-Phosphoglyceraldehyde ($ⓅOCH_2{-}CH(OH){-}CHO$) + Phosphodihydroxyacetone ($CH_2OⓅ{-}CO{-}CH_2OH$)

XX

vided and subdivided on the basis of the nature of the bond split. In general, it is sufficient at this point to state that one of four types of linkages is usually involved: (a) —C—N—; (b) —C—O—; (c) —O—P—; (d) —N—P—; and (e) —O—S—. As examples of enzymatic reactions involving each type, see formula XXI. The most familiar enzymes of all, the digestive enzymes, are all hydrolases. The anylases (carbohydrate-digesting) and lipases (fat-digesting) hydrolyze —C—O— bonds, whereas proteinases hydrolyze —C—N— bonds.

Transferring enzymes. These catalyze reactions in which a chemical group is transferred from one compound to another. Actually, such a definition is exceedingly broad. Hydrolases might be considered transferring enzymes in which water is the acceptor molecule. Thus a phosphatase might be said to transfer a phosphate group from glycerol to water in catalyzing the hydrolysis of glyceryl phosphate. Again, the large group of oxidizing enzymes (see page 275) may be said to catalyze the transfer of two

$$\underset{\text{peptide}}{RC(=O)\vdots NHR'} + H_2O \xrightarrow{\text{peptidase}} \underset{\text{amino acid}}{RC(=O){-}OH} + \underset{\text{amino acid}}{NH_2R'}$$

$$\underset{\text{ester}}{RC(=O)\vdots OR'} + H_2O \xrightarrow{\text{esterase}} \underset{\text{fatty acid}}{RC(=O){-}OH} + \underset{\text{alcohol}}{HOR'}$$

$$\underset{\text{phosphate}}{RO\vdots Ⓟ} + H_2O \xrightarrow{\text{phosphatase}} \underset{\text{alcohol}}{ROH} + \underset{\text{phosphoric acid}}{HOⓅ}$$

$$\underset{\text{N-phosphate}}{RNH\vdots Ⓟ} + H_2O \xrightarrow{\text{phosphamidase}} \underset{\text{amine}}{RNH_2} + \underset{\text{phosphoric acid}}{HOⓅ}$$

$$\underset{\text{sulfate}}{RO\vdots SO_3H} + H_2O \xrightarrow{\text{sulfatase}} \underset{\text{alcohol}}{ROH} + \underset{\text{sulfuric acid}}{HOSO_3H}$$

XXI. Action of hydrolyzing enzymes

hydrogen atoms from one molecule to another. Because of the importance of these two groups of enzymes, it is useful to consider them separately and confine the classification of transferase to enzymes catalyzing transfer reactions in which hydrogen atoms are not the group transferred and in which water does not act as an acceptor molecule.

Examples of transferring enzymes are the transaminases, such as that which catalyzes the transfer of an amino group from glutamic acid to pyruvic acid (formula XXII). The transfer of a phosphate group from one molecule of ADP to another by the catalytic action of myokinase (see page 558) is an example of transphosphorylation. Enzymes mediating such reactions are called transphosphorylases or, sometimes, phosphokinases. Other important transferring reactions are transpeptidation, transmethylation, transacylation and transformylation, all of which are important in the reactions taking place within living tissue and which will be taken up at appropriate points later in the book, (see Chapters 16 and 21).

Phosphorylases are a group of enzymes which it has been customary to view as catalyzing the splitting of a —C—O— bond with the addition of the elements of phosphoric acid. By analogy to the process of hydrolysis this reaction is termed a phosphorolysis and phosphorylases are generally discussed along with the hydrolases they seem to resemble so closely. The substrates of phosphorylases are generally carbohydrate in nature. Thus, sucrose phosphorylase phosphorylyzes sucrose into fructose and glucose-1-phosphate (formula XXIII) while amylophosphorylase splits glucose-1-phosphate off the end of starch and glycogen chains.

However, it now seems that the characteristic feature of this type of enzymatic reaction is not the phosphorolysis but the transfer of the glucose from the sucrose molecule to, in this case, the phosphoric acid molecule. That this is the important process can be shown by the fact that "sucrose phosphorylase" will catalyze the transfer of the glucose from sucrose to another sugar without involving phosphoric acid at all. In other words it is the transfer that holds constant, while the nature of the acceptor molecule

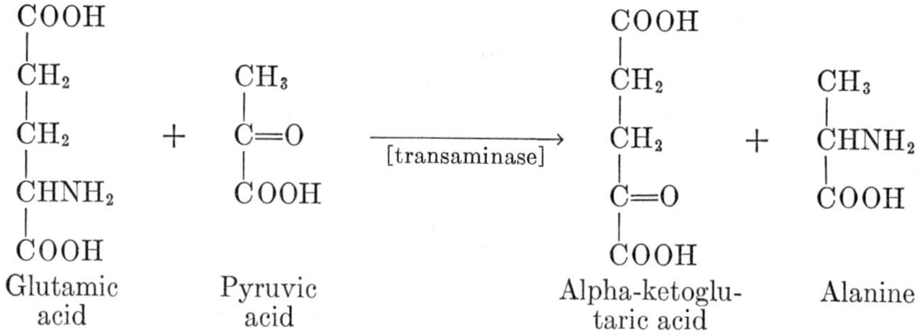

$$\begin{array}{c} COOH \\ | \\ CH_2 \\ | \\ CH_2 \\ | \\ CHNH_2 \\ | \\ COOH \end{array} + \begin{array}{c} CH_3 \\ | \\ C{=}O \\ | \\ COOH \end{array} \xrightarrow[\text{[transaminase]}]{} \begin{array}{c} COOH \\ | \\ CH_2 \\ | \\ CH_2 \\ | \\ C{=}O \\ | \\ COOH \end{array} + \begin{array}{c} CH_3 \\ | \\ CHNH_2 \\ | \\ COOH \end{array}$$

Glutamic acid — Pyruvic acid — Alpha-ketoglutaric acid — Alanine

XXII. Transamination

can vary. For this reason, it is now suggested that phosphorylases be termed transglycosidases (9).

Phosphatases, on the other hand, which hydrolyze phosphate groups, are actually transferring those groups to the water molecule. In such reactions, various alcohols may also act as acceptor molecules so that phosphatases can behave as transphosphorylases (6). This is but one example of the difficulties involved in any system of arbitrary enzyme classification.

Isomerizing enzymes. These enzymes catalyze the transfer of chemical groups from one position in a molecule to another position in the same molecule. Examples of such enzymes are phosphotriose isomerase, and phosphoglyceromutase (formula XXIV). The change in position of the group affected may result in nothing more than a stereochemical alteration. Thus, there are enzymes which can convert a D-amino acid or D-hydroxyacid to the L-enantiomorph and vice versa. Such enzymes are called *racemases*. Another enzyme catalyzes the racemization of alpha-glucose to beta-glucose and is therefore named *mutarotase*. It will also catalyze the mutarotation of galactose but not of mannose. *Phosphogalactoisomerase* (see page 530) racemizes the hydroxyl group about carbon 4 of a sugar molecule, catalyzing the interconversion of galactose-1-phosphate and glucose-1-phosphate.

Oxidizing enzymes. For purposes of convenience, consideration of this important group of enzymes will be deferred to a later point in the chapter (see page 275).

Specificity

Implicit in the notion of enzyme multiplicity is that of enzyme specificity. Since, except for the few enzymes whose crystalline forms can be characterized, enzymes can be differentiated only by the reactions they catalyze, it must follow that each enzyme is fairly specific in its action; otherwise differentiation and classification of enzymes would be meaningless. That such specificity exists is easily demonstrated experimentally and is one of the most characteristic properties of enzymes. In comparison, such non-enzymatic catalysts as platinum black, water, or light radiation are quite unspecific and catalyze all sorts of not necessarily closely related reactions.

Absolute specificity. Some enzymes, in fact, may catalyze one reaction and one reaction only. Such enzymes show absolute specificity. Examples are barley maltase, and succinic acid dehydrogenase (formula XXV). Catalase is almost but not quite absolutely specific. It has been shown to catalyze the decomposition of ethyl hydrogen peroxide to ethyl alcohol and oxygen (formula XXVI).

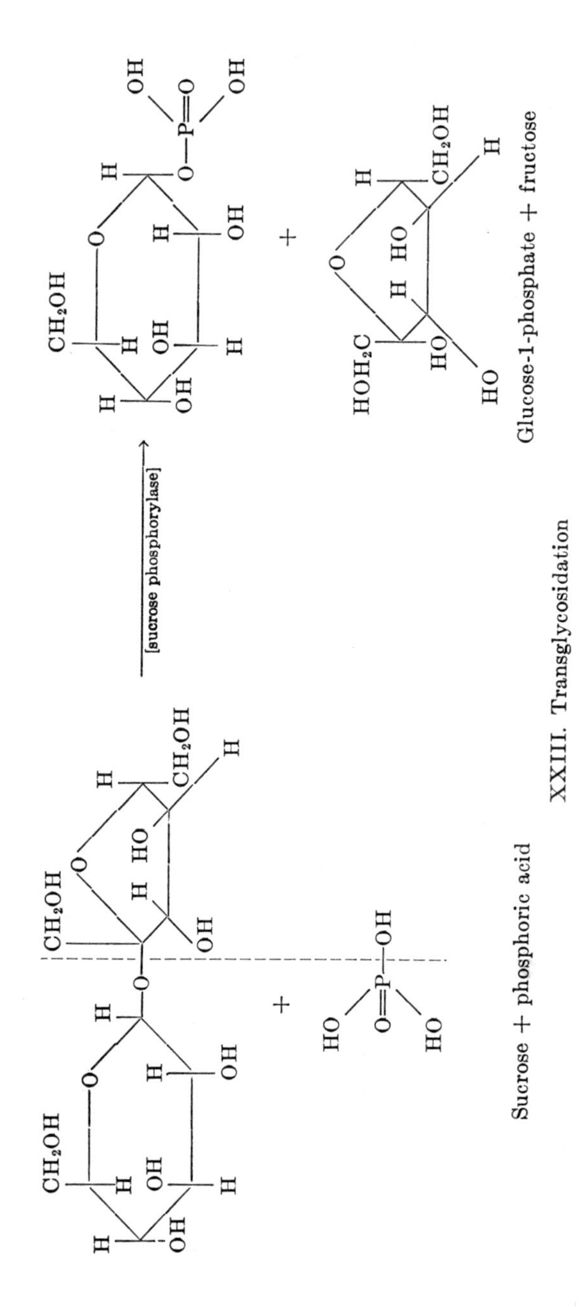

XXIII. Transglycosidation

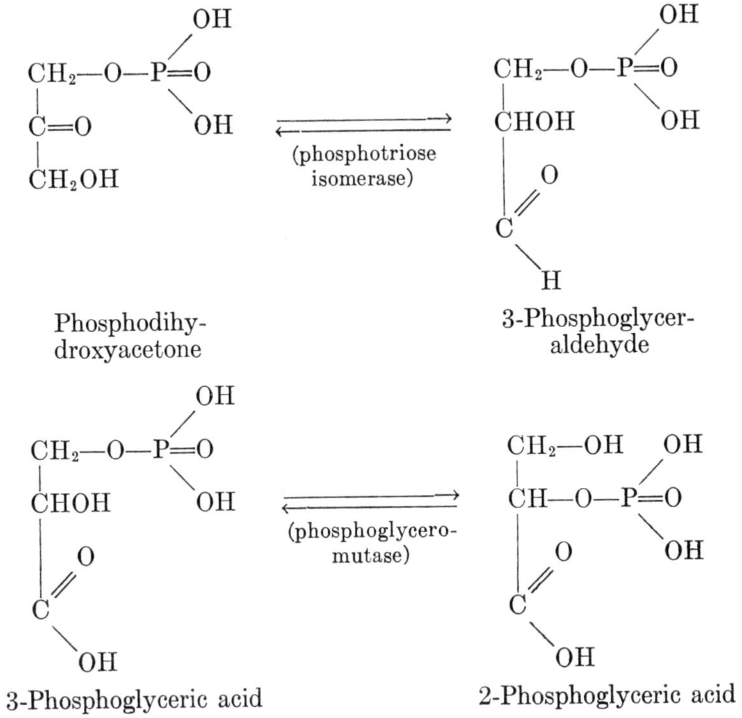

$$\begin{array}{l} CH_2{-}O{-}P(=O)(OH)_2 \\ | \\ C{=}O \\ | \\ CH_2OH \end{array} \quad \underset{\text{(phosphotriose isomerase)}}{\rightleftharpoons} \quad \begin{array}{l} CH_2{-}O{-}P(=O)(OH)_2 \\ | \\ CHOH \\ | \\ C(=O)H \end{array}$$

Phosphodihydroxyacetone — 3-Phosphoglyceraldehyde

$$\begin{array}{l} CH_2{-}O{-}P(=O)(OH)_2 \\ | \\ CHOH \\ | \\ C(=O)OH \end{array} \quad \underset{\text{(phosphoglyceromutase)}}{\rightleftharpoons} \quad \begin{array}{l} CH_2{-}OH \\ | \\ CH{-}O{-}P(=O)(OH)_2 \\ | \\ C(=O)OH \end{array}$$

3-Phosphoglyceric acid — 2-Phosphoglyceric acid

XXIV. Isomerizing enzymes

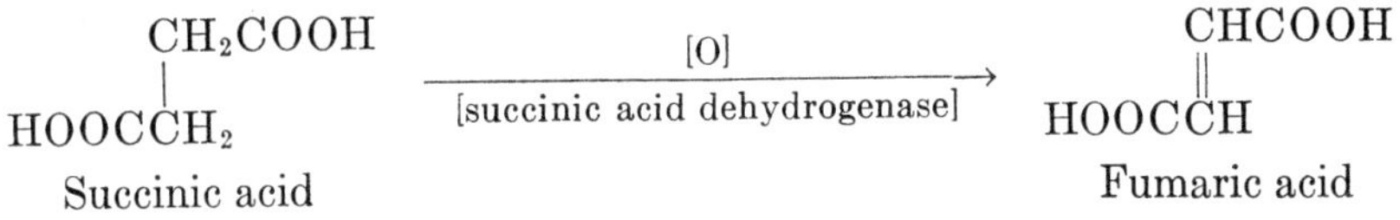

$$\begin{array}{l} CH_2COOH \\ | \\ HOOCCH_2 \end{array} \quad \xrightarrow[\text{[succinic acid dehydrogenase]}]{[O]} \quad \begin{array}{l} CHCOOH \\ \| \\ HOOCCH \end{array}$$

Succinic acid — Fumaric acid

XXV

$$2CH_3CH_2{-}O{-}O{-}H \xrightarrow{\text{[catalase]}} 2CH_3CH_2OH + O_2$$

Ethyl hydrogen peroxide — Ethyl alcohol

XXVI

Group specificity. Other enzymes display group specificity which indicates that provided a certain chemical grouping is present in the molecule, the nature of the remainder of the molecule is a matter of relative indifference. This concept may be conveniently exemplified by a consideration of the group of enzymes known as *phosphatases*, which in the previous section have been listed as those hydrolases which split an —O—P— linkage. The phosphatases can be divided into several subgroups depending upon the type of —O—P— linkage concerned. Three of such subgroups are:

1. *Phosphomonoesterases*, the substrates for which must contain a singly esterified phosphoric acid group, e.g., glyceryl phosphate (formula XXVII).

2. *Phosphodiesterases*, the substrates for which must contain a doubly esterified phosphoric acid group, e.g., diglyceryl phosphate (formula XXVIII).

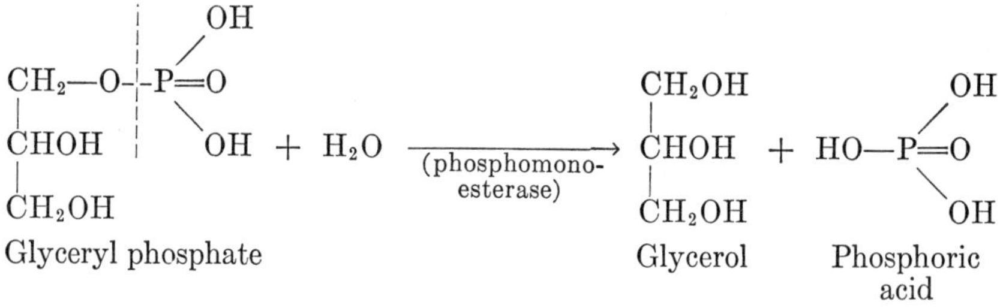

$$\underset{\text{Glyceryl phosphate}}{\begin{array}{l} CH_2{-}O{-}P(=O)(OH)_2 \\ | \\ CHOH \\ | \\ CH_2OH \end{array}} + H_2O \xrightarrow[\text{(phosphomono-esterase)}]{} \underset{\text{Glycerol}}{\begin{array}{l} CH_2OH \\ | \\ CHOH \\ | \\ CH_2OH \end{array}} + \underset{\text{Phosphoric acid}}{HO{-}P(=O)(OH)_2}$$

XXVII

$$\underset{\text{Diglyceryl phosphate}}{\begin{array}{l} CH_2{-}O{-}P(=O)(OH){-}O{-}CH_2 \\ | \qquad\qquad\qquad\qquad | \\ CHOH \qquad\qquad\quad CHOH \\ | \qquad\qquad\qquad\qquad | \\ CH_2OH \qquad\qquad\; CH_2OH \end{array}} + H_2O \xrightarrow{\text{phosphodiesterase}}$$

$$\underset{\text{Glycerol}}{\begin{array}{l} CH_2OH \\ | \\ CHOH \\ | \\ CH_2OH \end{array}} + \underset{\text{Glyceryl phosphate}}{\begin{array}{l} HO{-}P(=O)(OH){-}O{-}CH_2 \\ \qquad\qquad\qquad\quad | \\ \qquad\qquad\qquad CHOH \\ \qquad\qquad\qquad\quad | \\ \qquad\qquad\qquad CH_2OH \end{array}}$$

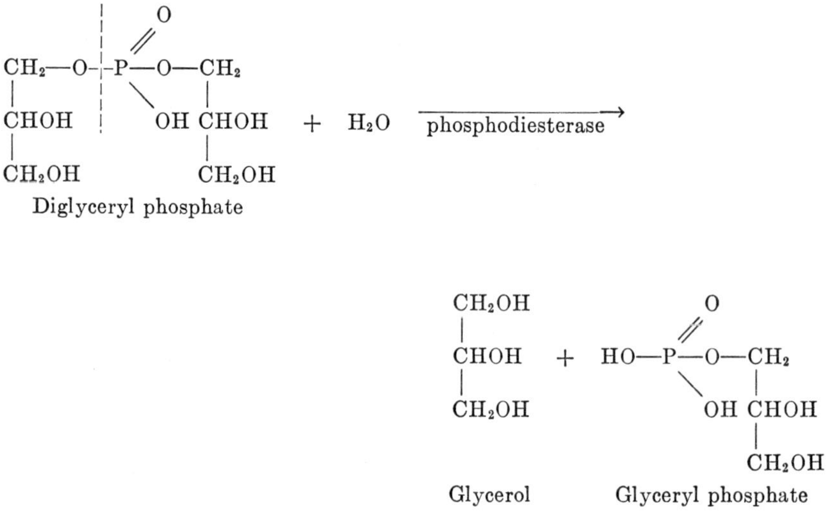

XXVIII

3. *Diphosphatases*, the substrates for which must contain two singly esterified phosphoric acid groups, e.g., glyceryl diphosphate (formula XXIX).

As far as the nature of the phosphate linkage is concerned the enzymes within each subgroup listed show absolute specificity. As far as the nature of the organic radical with which the phosphoric acid is esterified, however, the enzymes are relatively unspecific. Thus a phosphomonoesterase will hydrolyze phosphoric acid from such diverse compounds as glyceryl phosphate, phenyl phosphate, phenolphthalein phosphate, and adenylic acid, and is therefore only group specific. Even here, however, it should be noted

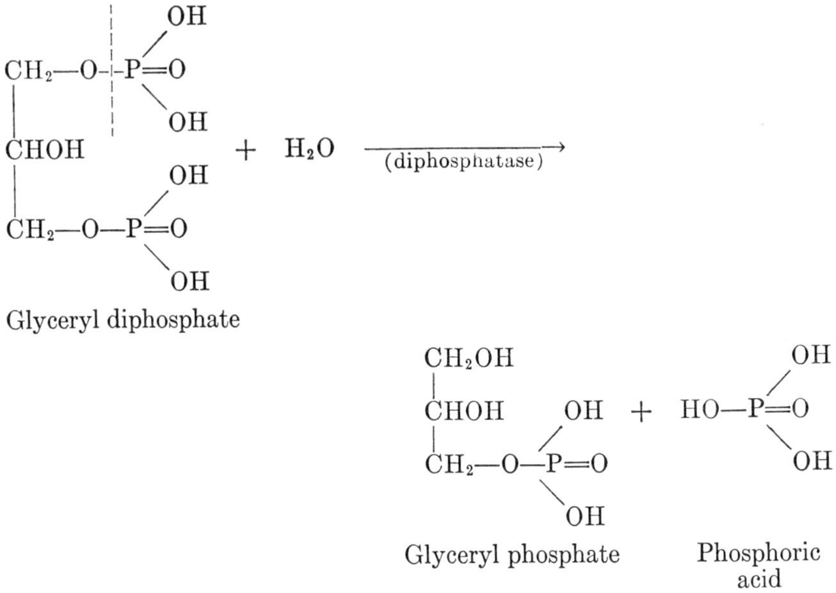

```
         ¦  OH
         ¦ /
CH2—O-¦-P=O
|        ¦ \
|        ¦  OH
CHOH              +  H2O  ——————————————→
|           OH            (diphosphatase)
|          /
CH2—O—P=O
           \
            OH
Glyceryl diphosphate

                                  CH2OH                       OH
                                  |                          /
                                  CHOH      OH   +   HO—P=O
                                  |        /                 \
                                  CH2—O—P=O                   OH
                                           \
                                            OH
                                Glyceryl phosphate       Phosphoric
                                                            acid
```

XXIX

that not all such compounds are hydrolyzed at equal rates, and the relative speed with which various mono-esterified phosphoric acids are hydrolyzed varies with phosphomonoesterases isolated from various species. Phosphatases also exist with narrower specificities. Adenosine triphosphatase (see page 557) is an example of a phosphatase which, as far as we know, is specific for ATP and a few closely related substances.

Enantiomorphic specificity. Enzymes catalyzing reactions involving such optically active substances as sugars or amino acids frequently act primarily or exclusively upon one of the enantiomorphs. Arginase will act upon L-arginine but not upon D-arginine. D-Amino acid oxidase, as its name indicates, will oxidize only D-amino acids. Furthermore, enzymes can form optically active compounds from inactive precur ors. Thus, lactic acid dehydrogenase, which catalyzes the interconversion of lactic acid and pyruvic acid, will oxidize only L-lactic acid, and will fo m only L-lactic acid from the optically inactive pyruvic acid (formula XXX).

Proteinase specificity. The protein-hydrolyzing enzymes of the digestive juices present a special problem with regard to questions of specificity because of the complexity of their substrates. From the fact that the human digestive system can hydrolyze most proteins one would suppose the enzymes involved to be relatively unspecific, or at most to display group specificity for the peptide linkage. However, the specificity involved is far more delicate than had been thought. As examples we can take pepsin and

$$\underset{\text{Pyruvic acid}}{CH_3-C(=O)-C(=O)OH} \;\underset{-2H}{\overset{+2H}{\underset{\text{(lactic acid dehydrogenase)}}{\rightleftharpoons}}}\; \underset{\text{L-Lactic acid}}{CH_3-HOCH-C(=O)OH}$$

XXX

trypsin. Both will hydrolyze proteins or polypeptides of high molecular weight, forming as the end product not amino acids but rather peptides of moderate molecular weight. The hydrolyzed products resulting from prolonged action of either enzyme will, however, yield readily to further hydrolysis by the other. This would indicate that both pepsin and trypsin will hydrolyze only certain peptide linkages and that each will hydrolyze various linkages left untouched by the other. By studies on artificial peptides of known composition Bergmann (2) showed that pepsin will hydrolyze a peptide bond which is on the amino side of a residue of phenylalanine or tyrosine, and on the carboxyl side of a residue of glutamic acid or aspartic acid, provided that a molecule of lysine or arginine is not bound nearby on the polypeptide chain, whereas trypsin will act upon the carboxyl side of a residue of lysine or arginine provided a residue of glutamic acid or aspartic acid is not too close. Chymotrypsin will act upon the carboxyl side of a residue of phenylalanine, tyrosine, tryptophane or methionine, provided it is not also the amino side of a residue of glutamic acid or aspartic acid. Formula XXXI summarizes the situation.

$$-C(=O)-NH-\underset{\underset{COOH}{|}}{\underset{CH_2}{\overset{}{CH}}}-C(=O)\overset{\text{attacked by pepsin}}{\overset{HO\,|\,H}{\vdots}}NH-\overset{\overset{C_6H_5}{|}}{\underset{}{\overset{CH_2}{CH}}}-C(=O)\overset{\text{attacked by chymotrypsin}}{\overset{HO\,|\,H}{\vdots}}NH-\underset{CH_3}{CH}-C(=O)-NH-\overset{CH_2NH_2-CH_2-CH_2-CH_2}{CH}-C(=O)\overset{\text{attacked by trypsin}}{\overset{HO\,|\,H}{\vdots}}NH-\underset{CH_3}{CH}-C(=O)-$$

XXXI. Peptide chain under enzymatic hydrolysis

The peptide fragments remaining from the action of pepsin and trypsin are substrates of various peptidases. Here again there is a division of function. There are carboxypeptidases which split off the amino acid residue from that end of the peptide chain containing a free carboxyl ion. The other end of the chain would analogously be attacked by an aminopeptidase. There are specialized peptidases which will hydrolyze off only a particular residue, as for instance prolinase, which will act only when proline is at the end of a chain. In addition there are numerous dipeptidases which catalyze the final hydrolysis of dipeptides to amino acids. Peptidases are sometimes distinguished from one another by the position of the peptide linkage catalytically attacked. Thus pepsin, under whose influence nonterminal peptide linkages can be directly hydrolyzed, is termed an *endopeptidase* whereas carboxypeptidase, which can catalyze the hydrolysis of a terminal amino acid residue is called an *exopeptidase*.

It would thus seem characteristic of living tissue that groups of necessary chemical reactions are catalyzed by many specific enzymes rather than by a few unspecific ones. The assembly-line technique in industry has been anticipated (by the manner in which proteins are digested, for instance) by several hundreds of millions of years in the living cell.

THE STRUCTURE OF ENZYMES

Essential Protein Groupings

It is reasonable at this point to ask whether an enzyme is a typical protein or whether it possesses certain structural peculiarities that set it apart from those proteins which do not possess catalytic properties. As far as we know now the former is the case. In molecular weight, in amino acid content, in general structure, enzymes can not be differentiated as a class from other proteins. Efforts have been directed at determining whether any particular portion of the enzyme molecule is responsible for its activity. Thus pepsin acetylated to varying degrees has been tested for activity (16). It was found that acetylation first took place on the free amino groups of the three lysine residues of the pepsin molecule. The activity of this triacetyl pepsin remained unchanged. When further acetylation took place, activity declined by stages. Since this further acetylation took place upon the phenolic side chains of the tyrosine residues, there would appear to be a close relationship between the tyrosine of pepsin and the proteolytic activity of the protein molecule. Similar results were obtained when pepsin was iodinated and the tyrosine residues converted to diiodotyrosine.

The sulfhydryl group of the cysteine side chain is essential to the function of many enzymes. Heavy metal ions which form unionized complexes with —SH inhibit many enzymes. Sulfhydryl groups, in some instances, may form acyl-mercaptan bonds in the enzyme-substrate complex. Such high-energy bonds may be part of the mechanism which forms and transfers high-energy phosphate groups. Creatine phosphokinase is an example of an enzyme containing essential —SH groups.

In any case, while some groups may serve as the immediate reactive center of the protein molecule, the molecule as a whole contributes to enzyme specificity. In fact, any significant disruption of the enzyme structure, as in the case of all forms of denaturation, will decrease or erase enzyme activity. The capacity for catalytic activity is therefore obviously a function of the enzyme as a whole and not of any particular part alone.

Prosthetic Groups

Not all enzymes are simple proteins. Cytochrome oxidase and catalase, which are of universal occurrence in animal cells, are heme proteins containing as a prosthetic group the same iron porphyrin compound which occurs in hemoglobin (see Chapter 5). Hemocyanin and certain plant oxidases contain copper closely associated with the protein molecule, but the nature of the prosthetic group of which it may form a part is not yet known.

Activators

Many enzymes are without catalytic effect upon their substrates in the absence of definite amounts of certain ions, which do not appear to be so closely bound to the protein of the enzyme as to justify their consideration as parts of prosthetic groups. The exact function of the ions is in many cases not known, and biochemists have been content to refer to them simply as activators.

The enzyme-ion relationship is often not very specific. In some cases it would seem that almost any bivalent metallic ion would do. Thus, the enzyme deoxyribonuclease is strongly activated by Mg^{++}, Mn^{++}, Co^{++}, and Fe^{++}, and less strongly by Ca^{++}, Ba^{++}, Sr^{++}, Ni^{++}, Cd^{++} and Zn^{++} (15). Enzymes are inactivated if the necessary ionic activators are physically removed from solution, as by dialysis, or if the ions are removed chemically by complex formation. Fluoride ion is usually a good inhibitor for those enzymes which require bivalent metallic ions for their functioning.

It should be mentioned that activators need not necessarily be metallic or even cationic in nature. Salivary amylase requires chloride ion for any activity whatever and the action of pancreatic lipase is accelerated in the presence of bile salts.

Peptidases, in contrast to the proteinases such as pepsin, trypsin, and chymotrypsin, are often metal-activated enzymes, the activators being usually bivalent ions. Smith (24) advances the theory that the metal ion acts as a link between the peptidase and the peptide substrate, governing, at least in part, the specificity of the enzyme through the distribution of its co-ordinate valences. Thus, the specific leucine aminopeptidase requires

Mn^{++} as an activator. The Mn^{++} ion possesses four co-ordinate valences, two of which are bound to specific polar groups in the enzyme and two with the amino groups of a substrate such as alanylleucine (formula XXXII).

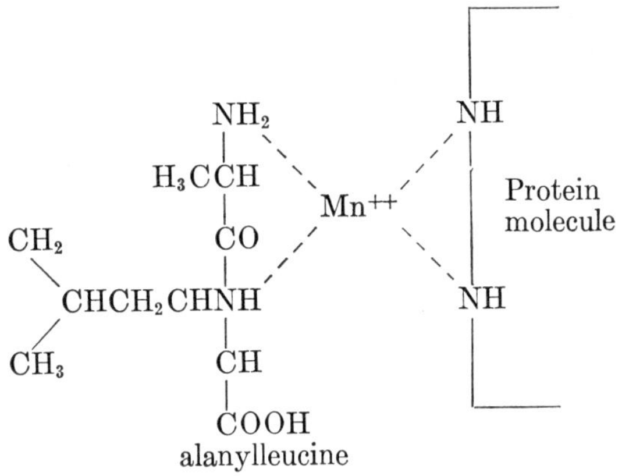

```
                                            ┌──────
                                            │
                  NH2                       NH
                   |  \                   / |
              H3CCH    \                /   |   Protein
                   |     Mn++         /     |   molecule
   CH2             CO   /      \            |
      \            |  /          \          |
       CHCH2CHNH                   NH
      /            |                        |
   CH3             CH                       |
                   |                        └──────
                   COOH
              alanylleucine
```

XXXII. Chelation

This form of bond formation between a metal ion and two or more polar groupings of a single molecule is known as *chelation* (30). In this case Mn^{++} has been chelated by both enzyme and substrate.

A chelating agent is a compound containing more than one polar group, so situated as to allow two or more of the polar groups to be attached to a single ion. Proteins in general are chelating agents and so are simpler compounds, such as ethylenediamine tetraacetate (EDTA) (formula XXXIII). Similarly, the Co^{++} activator of glycylglycine dipeptidase, with

$$(^{-}OOCH_2C)_2NCH_2CH_2N(CH_2COO^{-})_2$$

XXXIII. Ethylenediaminetetraacetate(EDTA)

its six co-ordinate links, forms a complex with glycylglycine. Apparently the metal ions will by themselves form complexes with various dipeptides and tripeptides, exhibiting considerable specificity in doing so, but the presence of the enzyme is necessary for rapid hydrolysis.

Coenzymes

If an enzyme such as carboxylase is dialyzed, it is found to lose its activity. If the inactive dialysate is added to the inactive dialyzed enzyme, activity is restored. Dialysis removes a substance necessary to enzyme activity, a *coenzyme*, of relatively small molecular weight. The coenzyme, unlike the protein itself, is thermostable, i.e., is not inactivated by heating. It is organic and not simply an inorganic ionic activator of the type we have already discussed, since if the dialysate is dried and ashed, activity after the re-addition of water is lost. A coenzyme is distinguished from a prosthetic group in that it is more loosely bound to the enzyme and can be removed by simple dialysis while a prosthetic group can not. (The distinction among the three terms, activator, coenzyme, and prosthetic group grows dimmer with the advance of knowledge and perhaps the time is coming when the one general term, coenzyme, will do for any non-protein adjunct of an enzyme.)

In carboxylase, the coenzyme has been identified as *diphosphothiamine* (formula XXXIV), otherwise known as *cocarboxylase*. The structure of a number of other coenzymes has been elucidated. These include *coenzyme I* and *coenzyme II*, each of which is the coenzyme for various oxidizing enzymes known as dehydrogenases (see page 275). From a consideration of its chemical structure, coenzyme I is more properly termed *diphosphopyridine nucleotide* (DPN) (formula XXXV), while coenzyme II is, analogously, *triphosphopyridine nucleotide* (TPN) (formula XXXVI). (A nucleotide is a compound in which a nitrogen-containing ring is bound to a phosphorylated sugar, see page 351.) A third member of the group, *coenzyme III* (formula XXXVII), aids in the catalysis of one of the steps in cysteine oxidation (see page 628). Coenzymes I, II, and III are grouped together as *pyridine nucleotide coenzymes* since the nicotinamide portion of these compounds is a pyridine derivative and it is the nicotinamide portion which is most significant nutritionally (see page 801).

In most cases the pyridine nucleotides will not substitute for one another. Usually, a particular enzyme will require a particular one of the group and

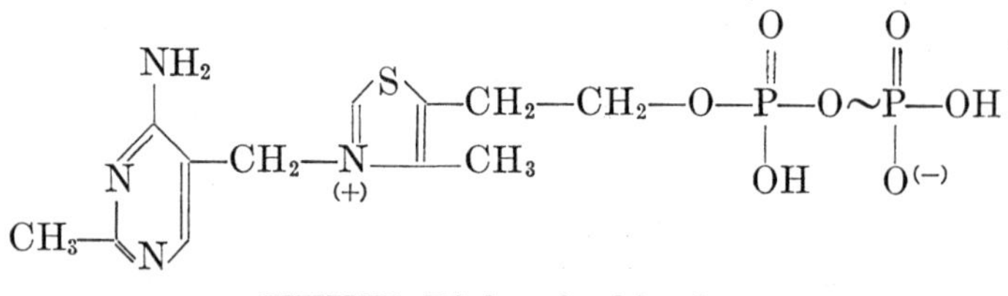

XXXIV. Diphosphothiamine

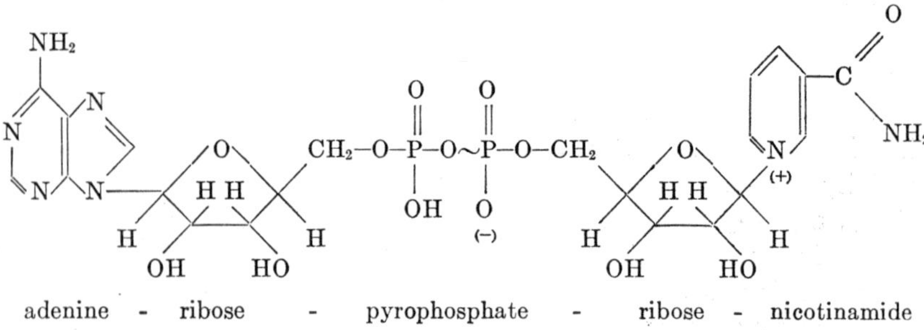

adenine - ribose - pyrophosphate - ribose - nicotinamide

XXXV. Diphosphopyridine nucleotide (coenzyme I or DPN)

will not be active in the presence of another of the group. The pyridine nucleotides seem to be linked to certain specific points in the apoenzyme, the number of such spots being at least 3 in triosephosphate dehydrogenase and 4 in alcohol dehydrogenase (20). Sulfhydryl groups may be involved in the binding.

Flavin mononucleotide (FM) (formula XXXVIII) and *flavin adenine dinucleotide* (FAD) (formula XXXIX) are coenzymes of such enzymes as cytochrome reductase and diaphorase and are sometimes referred to as the isoalloxazine coenzymes. Note that in both FM and FAD the five-carbon

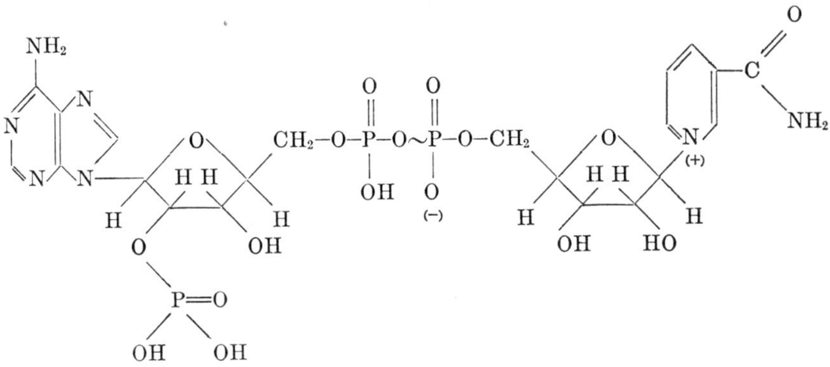

adenine-ribosephosphate - pyrophosphate - ribose - nicotinamide

XXXVI. Triphosphopyridine nucleotide (coenzyme II or TPN)

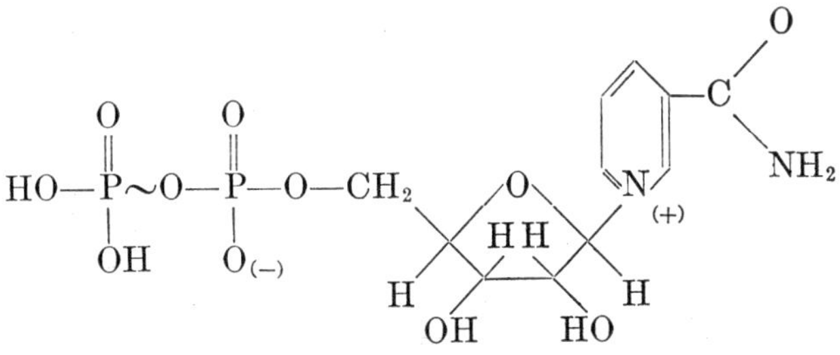

Pyrophosphate - ribose - nicotinamide

XXXVII. Coenzyme III

group attached to the isoalloxazine ring system is not a ribose residue, but a residue of the alcohol derived from ribose, ribitol. FM and FAD are often referred to as the *flavin coenzymes*.

Enzymes for which DPN and TPN are coenzymes are called *pyridinoenzymes*, while those containing FM or FAD are the *flavoenzymes*.

Pyridoxal phosphate (formula XL), or *codecarboxylase*, is the coenzyme for various enzymes involved in transamination, decarboxylation, thioether cleavage, and for the deamination of such amino acids as serine and threonine.

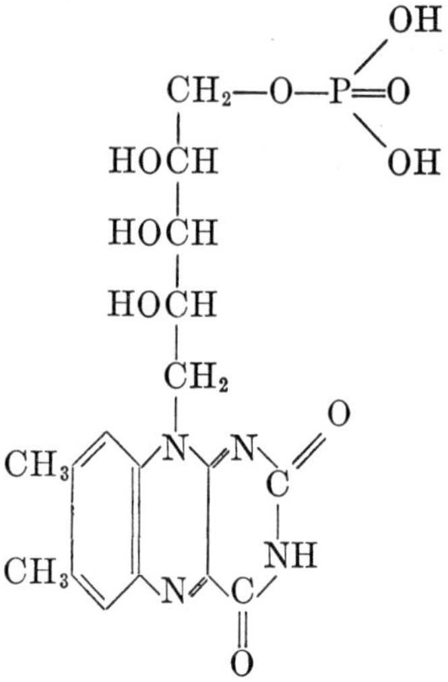

OH
CH_2—O—P=O
HOCH OH
HOCH
HOCH
CH_2
O
CH_3 N N C
NH
CH_3 N C
O

XXXVIII. Flavin mononucleotide (FM)

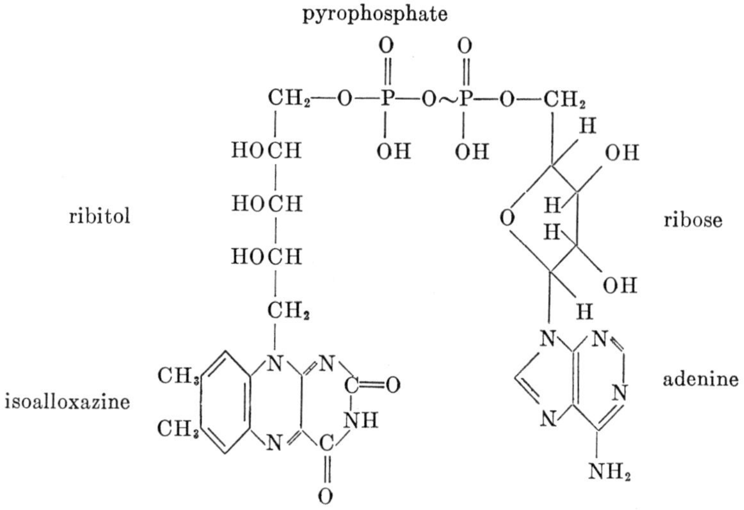

pyrophosphate
O O
CH_2—O—P—O~P—O—CH_2
H
HOCH OH OH OH
ribitol HOCH H O ribose
H
HOCH OH
CH_2 H
N N
CH_3 N N C=O N adenine
isoalloxazine NH
CH_3 N C N
O NH_2

XXXIX. Flavin adenine dinucleotide (FAD)

Uridinediphosphoglucose (UDPG) (formula XLI) is associated with phosphogalactoisomerase (also called galacto-Waldenase), an enzyme catalyzing the conversion of galactose-1-phosphate to glucose-1-phosphate and with the formation, in the body, of glucuronic acid from glucose.

Coenzyme A (see page 804) is essential for various acetylation reactions in the body and leucovorin (see page 811) for various transformylation reactions. Leucovorin is not itself an enzyme but is part of a larger molecule, probably including phosphate groups, which is the enzyme.

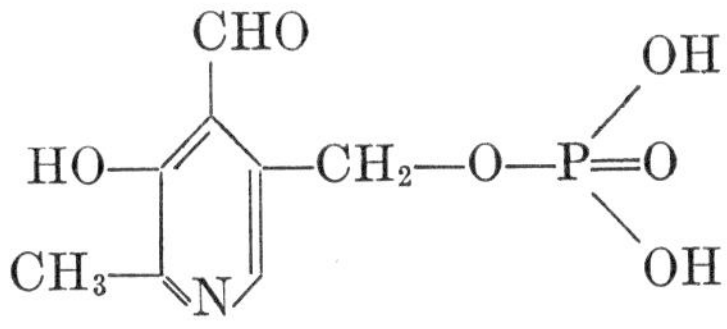

XL. Pyridoxal phosphate

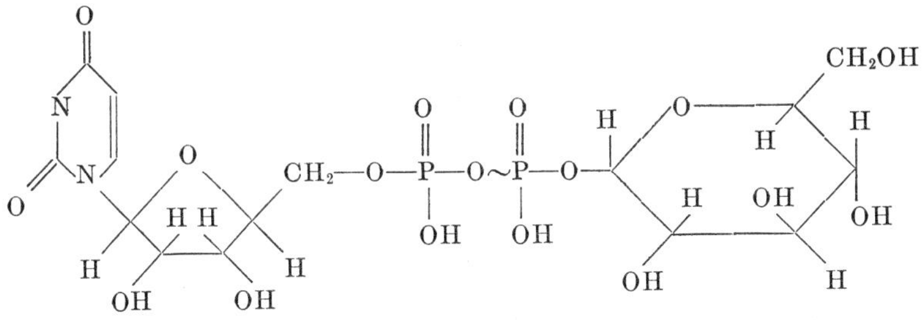

uracil - ribose - pyrophosphate - glucose

XLI. Uridinediphosphoglucose (UDPG)

If the coenzyme structures here presented are viewed as a group, three points can be made.

1) There is a close relationship between certain of the B vitamins (see Chapter 21) and various coenzymes. Thus, nicotinamide is part of DPN and TPN, riboflavin is part of FM and FAD, thiamine is part of cocarboxylase, pyridoxal is part of codecarboxylase, pantothenic acid is part of coenzyme A and leucovorin is a folic acid derivative.

2) All the coenzymes listed here contain the phosphate group, and all but FM and pyridoxal phosphate contain a pyrophosphate group. The student is reminded that the pyrophosphate link is high-energy.

3) FAD, DPN, TPN, UDPG, and coenzyme A all contain a nucleotide structure.

At least one coenzyme has been definitely established which is characterized by none of the above and which does not even contain phosphorus. It is *glutathione* (see page 62), which is a coenzyme for glyceraldehyde-3-phosphate dehydrogenase (12) and, along with DPN, is a coenzyme for formaldehyde dehydrogenase. Here, glutathione probably performs its function by way of the intermediate formation of high-energy acyl-mercaptan bonds (26).

The manner in which at least some of these coenzymes function will be discussed later in the section on oxidizing enzymes. Certain of their characteristics, however, may be pointed out here. Coenzymes do not generally display the sharp specificities of the protein portions of the enzyme. (The protein portion is often referred to as the *apoenzyme* and the enzyme and coenzyme together as the *holoenzyme*.) As an example, DPN will act as coenzyme for alcohol dehydrogenase, lactic acid dehydrogenase, malic acid dehydrogenase, and others. In each case the apoenzyme itself is specific with respect to both the substrate and the coenzyme. Enzymes which can make use of either of two coenzymes are not common. Certain dehydrogenases are active in the presence of either DPN or TPN (23). To summarize, while a single apoenzyme can almost never make use of more than one coenzyme, a single coenzyme can often be used by several apoenzymes. The situation is here similar to that whereby such proteins as hemoglobins of various species of animals, catalase, peroxidase, cytochrome c, and cytochrome oxidase (discussed later), differing widely in function, may all possess the same prosthetic group, heme.

MEASUREMENT OF ENZYME ACTIVITY

Methods

The relative concentrations of the enzymes in slices, suspensions, or extracts of tissues may be estimated by measuring the rates at which the catalyzed reaction is proceeding. In a reaction in which substance A is enzymatically converted to substance B, it is often possible to determine with great accuracy the concentration of either A or B at various times after the start of the reaction. For example, where catalase is catalyzing the decomposition of hydrogen peroxide, aliquots of the reaction mixture may be removed at given times, added quickly to a solution of molar sulfuric acid (which, by denaturing catalase protein, stops the enzyme reaction from proceeding further) and then titrated with potassium permanganate. From the rate at which hydrogen peroxide disappears with time, one can estimate the concentration of catalase. Similarly when alkaline phosphatase catalyzes the hydrolysis of phenyl phosphate, the rate at which phenol and inorganic phosphate are formed can be measured. In some cases it is possible to determine directly the rate at which some component of the reaction system varies without its being necessary to remove aliquots. This is possible in the hydrolysis of phenolphthalein phosphate in the presence of alkaline phosphatase. As free phenolphthalein is formed under the alkaline conditions of the experiment, it of course turns red. If the reaction is conducted in a cell of some colorimetric instrument, the increase in intensity of color with time can be measured continuously.

Where the enzymatically catalyzed reaction involves either the utilization or the formation of a gas, measurements can be made with great ease and refinement by *manometric methods*. In these methods the enzyme reaction is allowed to proceed in a vessel which is attached to a manometer, a device which measures changes in gas pressure by the rise and fall of a column of liquid. By using specially designed micromanometers, changes in gas volume as small as 0.05 cubic millimeters (0.00005 ml.) can be detected (25). Manometric methods are particularly adapted to the study of respiration of tissue slices, since this involves the interchange of oxygen and carbon dioxide. It can also be used to measure the rate of oxygen formation during the decomposition of hydrogen peroxide in the presence of catalase and the rate of carbon dioxide formation resulting from the decomposition of various amino acids by the decarboxylases present in bacterial suspensions. Many variations have been rung on this theme, many refinements and adaptations evolved to meet specific problems, so that manometry has become a large and highly specialized branch of enzymatic analysis. For further information concerning it, the student is referred to the monograph by Umbreit *et al.* (28).

The concentration of enzyme determined by rate measurements is not usually expressed absolutely as weight of enzyme per volume of solution. It is almost impossible to determine activity in terms of weight of enzyme, unless one first purifies and crystallizes it and then determines the activity of a known weight of the crystals. This would also involve the assumption that the enzyme crystals were unchanged in efficiency as compared with the enzyme originally present in the tissues. It is much more convenient to express enzyme concentration relatively in terms of the reaction rates themselves. Concentrations are then expressed simply as *units*. Thus a unit of catecholase has been defined as that concentration of enzyme which under certain specified environmental conditions will oxidize catechol to quinone at a rate involving the uptake of ten cubic millimeters of oxygen per minute. The number of cubic millimeters (or microliters) taken up per milligram of tissue or specified tissue component per hour is usually symbolized as Q_{O_2}.

It is not necessary to use well defined chemical reactions in measuring enzyme activity. Amylase activity, for instance, can be measured in terms of the time required to hydrolyze a given weight of starch under standardized conditions to the point where it will no longer give the characteristic blue color with iodine or with potassium triiodide. In many cases, the chemical reaction involved in the enzyme action may be ignored completely and a very accurate measure of activity obtained by observing changes in a selected physical property. Invertase, as an example, will hydrolyze sucrose to glucose and fructose. The optical rotation of sucrose is clockwise, whereas that of the equimolar mixture of glucose and fructose (due to the intense levo-rotatory property of fructose) is counterclockwise. If the enzymatic hydrolysis is conducted in a polarimeter, the activity of invertase (hence its relative concentration) can be measured by the rate of change of optical rotation of the solution. Invertase, in fact, obtains its name from the fact that it changes or "inverts" the sign of the optical rotation. Nucleases, through their action in depolymerizing the long asymmetric molecules of nucleic acids (see Chapter 9) cause solutions of those substances to become less viscous and to lose their anisotropic properties. The rate of decline of either viscosity or anisotropy can thus be used as a measure of nuclease activity.

Two precautions must be emphasized and re-emphasized in the use of the indirect method of determining enzyme concentration by enzyme activ ity. In the first place, *measurements of enzyme activity must always be con ducted under rigidly controlled environmental conditions*, or they are meaningless. The nature and importance of some of the environmental conditions concerned will be discussed immediately below. Secondly, it often happens that different workers investigating a particular enzyme will define varying units. This may happen because they will

1. Use a different chemical reaction as a measure of enzyme activity —one may determine phosphatase activity by its catalytic effect upon phenyl phosphate and another by its effect upon glyceryl phosphate;
2. Use different aspects of the same reaction—one may determine phosphatase activity by using phenyl phosphate and measuring the free phosphate formed and another will measure the free phenol formed;
3. Using the same aspect of the same reaction but under varying environmental conditions—one may determine phosphatase activity at 25°C. and another at 37°C.

In doing any work on enzymes it is always advisable to make use of units already established in the literature, rather than to invent new ones however convenient the latter alternative might seem. Again, in comparing one set of data in the literature with another, or either with your own, particular attention must always be paid to the exact manner in which the units of enzyme activity were determined.

Since the study of enzyme reactions involves primarily the interpretation of rates of reaction, that branch of physical chemistry known as *reaction kinetics* is of particular interest to the enzymologist. Some elementary principles of kinetics are discussed in the next chapter.

Environmental Effects

Nature of the enzyme. Since enzymes are proteins it should not be surprising, in view of the complexities of protein structure, that different samples of what would appear to be the "same" enzyme would vary in their properties. Many enzymes not only vary from species to species but also from tissue to tissue within a given animal. In many cases, the properties of a given enzyme may depend upon the particular method used in isolating it. It is therefore important to maintain the source and the method of isolation of enzymes as constant as possible.

Concentration of the enzyme. It is a fundamental assumption in the measurement of enzyme activity that the rate at which a catalyzed reaction proceeds is directly proportional to the concentration of enzyme. That this is generally so can be shown from the fact that doubling the amount of enzyme preparation used usually results in doubling the rate of the reaction. This is so, however, only as long as the enzyme concentration is comparatively small. As more and more enzyme is used, there comes a point where the reaction rate is no longer increased proportionately but begins to lag behind. The reason for this should not be difficult to see. As the concentration of enzyme increases, the reaction becomes so rapid that the supply of reactants is insufficient to feed it at maximum. In the case of the manometric measurement of ascorbic acid oxidase, for instance, the use of excessive quantities of enzyme leads to a depletion of oxygen in the solution, since it is consumed faster than it can be replaced from the air-filled portion of the vessel. If the manometer is shaken more rapidly, or if an oxygen atmosphere replaces the air in the vessel, oxygen is dissolved at a greater rate and the concentration limit of enzyme that can be efficiently used is raised. From this we see that in any experiment designed to measure enzyme concentration through reaction rates, it must be determined that *under the conditions of the experiment* a direct relationship between the two does indeed exist.

Concentration of the substrate. The question of substrate concentration is merely the inverse of what has been discussed immediately above. If we return to the case of ascorbic acid oxidase, it is obvious that if ascorbic acid is completely absent there will be no oxygen uptake regardless of the quantity of enzyme present. As ascorbic acid is increased in concentration the reaction rate, as measured by oxygen uptake, rises. For a short time it will rise in direct proportion with the substrate concentration and then it will begin to lag until a point is reached where further addition of ascorbic acid causes little or no increase in the rate of oxygen uptake. Again this results from the fact that at small substrate concentrations, not enough exists in solution to keep the enzyme molecules functioning at maximum rate. Once the substrate concentration is high enough to induce full enzyme activity, the addition of still more substrate will obviously not increase activity further. The enzyme, in other words, is then working as quickly as it can and additional substrate must simply wait its turn. It is desirable to measure enzyme activity in the region of this substrate plateau so that the decrease in substrate concentration during the course of the reaction does not affect the reaction rate. Indeed, it often happens that where still more substrate is added the reaction rate begins to decline again for reasons that are not as yet entirely clear. In that case a definite *substrate optimum* exists, at which it is desirable to maintain substrate concentration during the course of activity measurements.

Temperature. Chemical reaction rates are invariably affected by temperature change. As has been stated before, most reaction rates increase by a factor of two to three for every ten-degree rise in temperature. This increase is symbolized as Q_{10}. This rule applies to enzymatically catalyzed reactions as well. Another factor peculiar to such reactions, however, must also be considered. As the temperature rises, the rate of enzyme inactivation through heat-induced alterations in its protein structure increases by a factor of hundreds in that ten-degree interval. Small variations in temperature which would not be expected to alter the rate of a non-enzymatic reaction significantly may very easily render enzyme rate studies meaningless. (Enzyme action is also affected by pressure, but not significantly by the pressure changes ordinarily encountered in biochemical systems.)

pH. Proteins in general are multivalent, amphoteric acid-base compounds and their properties are changed with change in pH. This holds true specifically for enzyme activity. If enzyme measurements are made in a series of buffered media under conditions in which only the pH is varied from experiment to experiment, the rate of the catalyzed reaction will vary considerably. In general, the reaction will proceed most rapidly at a certain pH, and this is the *pH optimum.* On either side of this pH optimum there is a decline in activity and often activity ceases at distances of as little as two pH units from the optimum. Provided the departure from pH optimum is not too extreme or too prolonged, activity can be fully restored by a return to optimum. This would indicate that the decline in activity is due not to permanent changes in protein structure but to changes in the ionized state of the enzyme, i.e., in the distribution of charge on the enzyme surface.

Enzyme experiments are usually conducted in buffered solutions maintained at the optimum pH for that enzyme. It should be noted that each different enzyme has its characteristic pH optimum and that known optima vary from a pH of 1.5 for pepsin to one of 9.7 for arginase. Optima at either extreme of the pH range are rare, however, and the majority of enzymes possess optima nearer the neutral point. The optimum may vary slightly with the substrate, the buffer used, or with other factors in the reaction system. Enzyme reactions will vary in rate with *ionic strength* (see page 34) so that in preparing a buffer, the nature and concentration of the salts used must be taken into account as well as the final pH.

Enzyme Inhibition

The term, enzyme inhibition, is usually restricted to such loss of enzyme activity as results from the presence of small quantities of certain chemical substances. Such *inhibitors* may be classified into three groups, according to the method whereby the inhibiting process takes place.

Protein precipitants. Of the three the least significant are those substances which attack the protein moiety of the enzyme specifically. Trichloracetic acid, for instance, which is an excellent protein precipitant is in small concentrations an enzyme inhibitor. Other compounds which would act to precipitate proteins or to denature them in any way, such as by the oxidation or acetylation of essential groups, would also by their presence inhibit enzyme activity. Such inhibition is usually not very specific since obviously all enzymes would be vulnerable to any protein-damaging agent.

Activator inhibition. A more specific and more medically significant form of inhibition includes the effects of ions which are capable of forming complexes with metals whose presence is essential for enzyme activities. Usually a trace metal in order to function as an enzyme activator must be unencumbered by chemical linkages with any groups other than those with which it is physiologically concerned, the enzyme and the substrate or substrates. In the case of an iron enzyme a small concentration of cyanide ion would form a very stable ferro- or ferricyanide complex which, by occupying the available co-ordination linkages of iron in a more or less permanent fashion, makes it impossible for the iron further to fulfill its function. Since the cellular respiratory system in man includes an iron enzyme, cytochrome oxidase, cyanide acts as a virulent poison.

Competitive inhibition. Competitive inhibition is due to the lack of absolute specificity in complex formation on the part of the enzyme. Succinic acid dehydrogenase, for instance, has already been described as possessing absolute specificity. This is true insofar as it will catalyze the oxidation of succinic acid only. However, if to a system containing the enzyme succinic acid (formula XLII), and oxygen, is added also malonic acid (formula XLIII) the rate of oxygen uptake will decline. If sufficient malonic acid is added, it will virtually cease. This effect is not due to any deleterious action of malonic acid upon the protein. It will be noticed, however, that malonic acid is very much like succinic acid—differing in fact only in that there is one less methylene group between the two carboxyls. It would seem then that in the formation of the intermediate enzyme-substrate complex, the enzyme is unable to distinguish completely between succinic acid and malonic acid. In this respect the specificity is faulty. The enzyme-malonate complex can not be oxidized by oxygen and

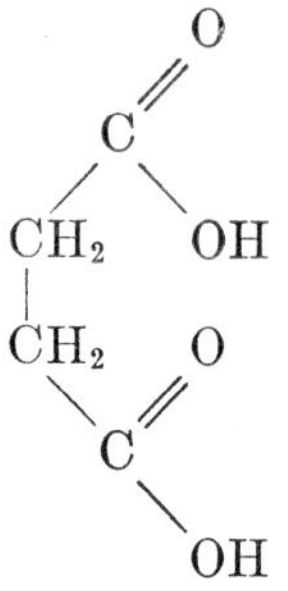

```
           O                      O
          //                     //
         C                      C
        / \                    / \
     CH2   OH                 /   OH
      |                    CH2
     CH2   O                 \    O
       \  //                  \  //
         C                      C
          \                      \
           OH                     OH
```

XLII. Succinic acid 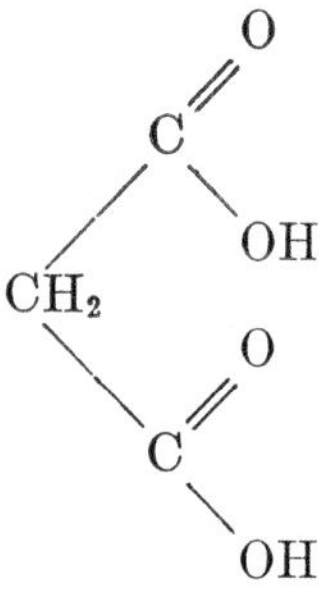 XLIII. Malonic acid

in that respect the enzyme is specific. However, the presence of the inert malonic acid upon the enzyme surface prevents those portions from combining with succinic acid. In this manner enzyme activity is inhibited. Since the formation of an enzyme-substrate complex depends upon the rate at which the substrate under the influence of thermal agitation collides with the enzyme surface, which in turn depends upon the concentration of substrate, the extent to which malonic acid will inactivate succinic acid dehydrogenase depends upon the relative concentrations of succinic acid and malonic acid in the system. The exact relationship of the degree of inhibition to relative concentrations of substrate and inhibitor varies with the system under consideration. A ratio of malonic acid to succinic acid equal to 1:50 results in a 50 per cent reaction inhibition.

Competition need not be between substrates alone. Substances similar to but not identical with coenzymes or portions of coenzymes will similarly compete for enzyme surface. Much of chemotherapy is based upon this principle (see Chapter 22).

An unusual type of competitive inhibition has been noted in the presence of *enzymoids*, which are enzymes with the structural groups involved in catalytic activity altered, while the substrate-binding groups remain. Enzymoids can form a complex with the substrate, but there is no formation of products. An example is lysozyme methyl ester (8). The inhibitory effect of enzymoids is most apparent when substrate concentrations are low in proportion to enzyme concentrations.

Still another form of competitive inhibition of medical interest is that existing among ionic activators. As an example we may consider cases of beryllium poisoning which followed wounds received from broken fluorescent bulbs made with beryllium-containing phosphors or from inhalation of the resulting dust. Non-healing granulomata in lungs or on skin are formed. The Be^{++} competes with Mg^{++} which is an activator for a wide variety of enzymes, notably those catalyzing reactions involving ATP. The competition is very one-sided in favor of beryllium. Mg^{++} added in concentrations 40,000 times as great as Be^{++} serves to reduce inhibition only from 80 per cent to 30 per cent. This is attributed to the greater stability of the beryllium-protein complex. Beryllium inhibition can be countered by the addition of a chelating agent such as aurintricarboxylic acid (ATA) (formula XLIV) which will compete successfully with the protein for the beryllium (22).

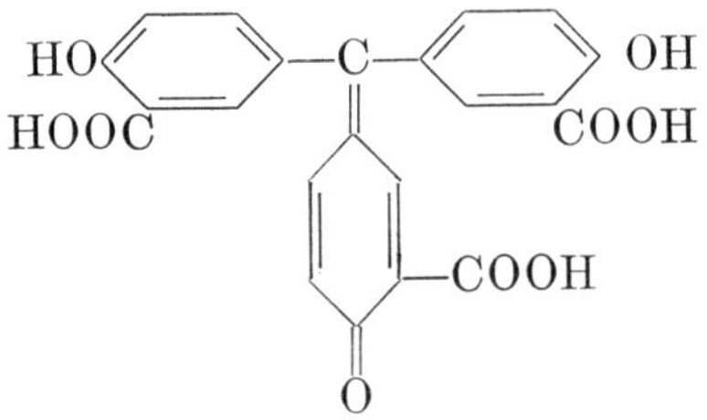

XLIV. Aurintricarboxylic acid (ATA)

Enzyme Reaction Rates

Even where enzyme systems are carefully standardized with regard to temperature, pH, and the other factors discussed above, the reaction rates we observe are seldom constant. The velocities of reactions catalyzed by enzymes may decrease with time for any of the following reasons: (a) the cumulative denaturing effect of temperature and gas-liquid interface upon the enzyme; (b) the decline in concentration of substrate as enzyme activity proceeds; and, (c) the possible enzyme-inhibiting nature of the products of the reaction. *In vitro*, the third item is the most difficult to deal with, since the first can be minimized by conducting enzyme reactions at low temperatures and the second by use of an excess of substrate. In the case of many oxidizing enzymes, for instance, the oxygen used to oxidize the substrates is itself reduced to hydrogen peroxide, a powerful protein denaturant. Unless measures are taken to remove peroxide as quickly as it is formed, e.g., by the addition of a small quantity of catalase, enzyme activity rapidly declines.

It is therefore advisable to make measurements at intervals, graph the results, and choose that portion of the curve which is linear for calculations of enzyme activity. Sometimes changes in enzyme activity are so rapid that no portion of the curve is linear. When this occurs, mathematical devices may be used to determine the rate of *enzyme activity at zero time* (14).

Cases where velocity increases with time also occur. Such periods of increasing reaction rates at the start of an enzyme reaction are known as *induction periods*. An example of such a case is found in the action of the enzyme tyrosinase upon *p*-cresol. The first step of the oxidation is the conversion of *p*-cresol to the corresponding catechol with oxygen being consumed (formula XLV). When the reaction is followed manometrically it is found that, initially, oxygen uptake is very slow but increases until after several minutes it is proceeding at a linear rate. When a small quantity of catechol is added to the system to begin with, however, oxygen uptake is linear from the very start. Where catechol is excluded from the mixture, the enzyme can not perform its catalytic function until some has been formed by the non-catalytic oxidation of the *p*-cresol. This would account for the initially slow but rapidly increasing rate of the reaction. Such induction periods can be detected and allowed for only by making periodic measurements during the course of the enzyme reaction.

$$\text{OH-}C_6H_4\text{-}CH_3 + \tfrac{1}{2}O_2 \xrightarrow[\text{[tyrosinase]}]{} (OH)_2C_6H_3\text{-}CH_3$$

p-Cresol → 4-Methyl catechol

XLV

An absolute value for enzyme reaction rates can be obtained if a known weight of enzyme of known molecular weight is present in the system. In such a case it is possible to calculate the number of molecules of substrate which react in a given time under standardized conditions for each molecule of enzyme present. This quantity is known as the *turnover number*. Some turnover numbers are extremely high. One molecule of catalase will decompose five million molecules of hydrogen peroxide in one minute at 0°C. while cholinesterase is reported to have a turnover number as high as twenty million. Turnover numbers in the thousands are much more common, and the value for Warburg's yellow enzyme is only fifty.

REFERENCES

1. Baldwin, E. *Dynamic Aspects of Biochemistry.* 2nd edit. Cambridge, England, Cambridge University Press, 1952.
2. Bergmann, M. A classification of proteolytic enzymes. Advances Enzymol., **2:** 49–68, 1942.
3. Binkley, F. Organization of enzymes in the synthesis of peptides. Proc. Royal Soc. (London) Series B, **142:** 170–174, 1954.
4. Bull, H. B. *Physical Biochemistry.* 2nd edit. New York, John Wiley & Sons, 1951.
5. Burton, A. K. Free energy change associated with the hydrolysis of the thiol ester bond of acetyl coenzyme A. Biochem. J., **59:** 44–46, 1955.
6. Davison-Reynolds, M. M., *et al.* Transphosphorylating action of human prostatic extracts. Enzymologia, **17:** 145–151, 1954.

7. Eucken, A., *et al. Fundamentals of Physical Chemistry.* New York, McGraw-Hill, 1925.
8. Frieden, E. H. "Enzymoid" properties of lysozyme methyl ester. J. Am. Chem. Soc., **78:** 961–965, 1956.
9. Hehre, E. J. Enzymic synthesis of polysaccharides. Advances Enzymol., **11:** 297–337, 1951.
10. Hill, T. L., and Morales, M. F. "High energy phosphate bonds" of biochemical interest. J. Am. Chem. Soc., **73:** 1656–1660, 1951.
11. Klotz, I. M. *Chemical Thermodynamics.* New York, Prentice-Hall, 1950.
12. Krimsky, I., and Racker, E. Glutathione, a prosthetic group of glyceraldehyde-3-phosphate dehydrogenase. J. Biol. Chem., **198:** 721–729, 1952.
13. Lieberman, I. Identification of adenosinetetraphosphate from horse muscle. J. Am. Chem. Soc., **77:** 3373–3375, 1955.
14. Miller, W. H., *et al.* A new method for the measurement of tyrosinase activity. II. Catecholase activity based on the initial reaction velocity. J. Am. Chem. Soc., **66:** 514–519, 1944.
15. Miyaji, T., and Greenstein, J. P. Cation activation of desoxyribonuclease. Arch. Biochem. Biophys., **32:** 414–423, 1951.
16. Northrop, J. H., *et al. Crystalline Enzymes.* 2nd edit. New York, Columbia University Press, 1948.
17. Osgood, W. F. *Advanced Calculus.* New York, Macmillan & Co., 1925.
18. Planck, M. *Theory of Heat.* Translated by H. L. Brose. London, Macmillan & Co., 1932.
19. Podolsky, R. J., and Morales, M. F. The enthalpy change of adenosine triphosphate hydrolysis. J. Biol. Chem., **218:** 945–959, 1956.
20. Racker, E. Mechanism of action and properties of pyridine nucleotide-linked enzymes. Physiol. Revs., **35:** 1–56, 1955.
21. Robbins, P. W., and Lipmann, F. Identification of enzymatically active sulfate as adenosine-3′-phosphate-5′-phosphosulfate. J. Am. Chem. Soc., **78:** 2652–2653, 1956.
22. Schubert, J., and Lindenbaum, A. L. The mechanism of protection by aurintricarboxylic acid in beryllium poisoning. J. Biol. Chem., **208:** 359–368, 1954.
23. Singer, J. P., and Kearney, E. B. Chemistry, metabolism, and scope of action of the pyridine nucleotide coenzymes. Advances Enzymol., **15:** 79–139, 1954.
24. Smith, E. L. The specificity of certain peptidases. Advances Enzymol., **12:** 191–257, 1951.
25. Stern, H., and Kirk, P. L. A versatile microrespirometer for routine use. J. Gen. Physiol., **31:** 239–242, 1948; Entner, N., and Kirk, P. L. Study of some enzyme systems by means of the differential microrespirometer. J. Gen. Physiol., **34:** 431–437, 1951.
26. Strittmatter, P., and Ball, E. G. Formaldehyde dehydrogenase, a glutathione-dependent enzyme system, J. Biol. Chem., **213:** 445–461, 1955.
27. Sumner, J. B. The isolation and crystallization of the enzyme urease. Preliminary paper. J. Biol. Chem., **69:** 435–441, 1926.
28. Umbreit, W. W., *et al. Manometric Techniques and Tissue Metabolism.* 2nd edit. Minneapolis, Burgess Publishing Co., 1949.
29. Walsh, E. O'F. Thiophosphates as possible intermediates in phosphate transfer. Nature, **169:** 546, 1952.
30. Warner, R. C. The metal chelate compounds of proteins. Trans. N. Y. Acad. Sc., **16:** 182–188, 1954.

CHAPTER 7

Enzymes and Enzyme Systems

In passing on next to that class of enzymes earlier classified as oxidizing enzymes, it is necessary first to take up, in some detail, the general subject of oxidation.

BIOLOGICAL OXIDATIONS

Electron Chemistry

All of life, except that of viruses and anaerobic bacteria, is made possible by the fact that molecular oxygen can be made to combine with the carbon and hydrogen of foodstuffs to yield carbon dioxide, water, and *energy*. It is this chemical fact that underlies the necessity for food and air, for eating and breathing. Oxygen, on the one hand, is a powerful oxidizing agent, capable under certain conditions of reacting with foodstuffs in order to yield comparatively large amounts of useful energy. On the other hand, it is sluggish in its actions, and will not spontaneously oxidize most foodstuffs. Oxygen may thus be visualized as a powerful tool which remains quietly in its case until needed and then, under the influence of the oxidizing enzymes of the body, is capable of great feats. To gain insight into this curious phenomenon and the life processes that depend upon it, a brief excursion into electron chemistry is necessary. To begin with, some knowledge of the electronic configuration of the biochemically significant elements is required.

Electronic configuration of the elements. Beginning with the simplest elements, hydrogen possesses a single electron and helium, two. They can be symbolized as $H\cdot$ and $He\colon$. The higher elements all contain an inner shell of two electrons as in helium, and begin adding to an outer shell. Since only the outermost shell of electrons in any given element is concerned in chemical reactions, it is conventional to omit consideration of all but those and to symbolize lithium, the third element, as $Li\cdot$—the two inner electrons being understood but not expressed. The process continues, an electron being added for each additional element, until a complete shell (containing eight electrons this time) is obtained in element number 10, neon, which would be expressed as $:\overset{\cdot\cdot}{\underset{\cdot\cdot}{Ne}}:$. Once again, a new shell is begun, and element 11, sodium, is $Na\cdot$. This new shell is completed with element 18, argon, which is $:\overset{\cdot\cdot}{\underset{\cdot\cdot}{A}}:$.

Electronic structures become considerably more complicated as atomic number increases. All but the innermost electron shell are composed of four or more subshells, the electrons of which differ in energy content. The energy spread among subshells becomes greater with each successive shell until, by the time the third shell is reached, overlapping with the next higher shell begins. The result is, for instance, that potassium and calcium represent atoms which have begun adding one and two electrons, respectively, to the innermost subshell of the fourth electron shell, while the outermost subshells of the third electron shell are as yet unfilled. The next ten atoms retain two electrons in the fourth shell and proceed to complete the electron content of the third shell. Such overlapping grows continually worse with increasing atomic number, and results in the phenomenon of varying valence, since the overlapping subshells are closely spaced (energically speaking). The heavier atoms may therefore avail themselves of only the outermost electrons or, on the other hand, may use electrons from an inner subshell as well in forming valence bonds. Thus iron and cobalt with two electrons in their outermost subshells form many compounds in which they exhibit a valence of two. Each, however, may and frequently does utilize a third electron from an inner subshell to form compounds in which they possess a valence of three. Similarly, copper with one electron in its outermost shell may draw one more from an inner shell, while manganese with two electrons in its outermost shell may utilize as many as five electrons from an inner shell. These, however, are the only elements of biochemical significance (see table VII-1) which exhibit this phenomenon. All others, from hydrogen to iodine, behave much more uniformly from an electronic standpoint.

Electronic concept of valence. In the formation of chemical compounds from the elements we can usually detect a tendency on the part of the elements concerned (particularly in the case of those of low atomic weight) to gain, lose, or share electrons in such a way as to end up with a complete electron shell, a state of maximum electronic stability. The formation of simple compounds such as methane, water, ammonia, hydrogen fluoride, carbon dioxide, and carbon tetrachloride, follows these electronic principles which thus account for the usual valence bond configurations. The same principles apply, with rare exceptions, to organic compounds of greater complexity. In the electronic formulas presented here (formulas I–VI), the electrons are pictured as little dots occupying particular places

TABLE VII-1

Electronic configuration of elements of biochemical interest

ELEC-TRON SHELLS	I	II	III	IV	V	VI	VII	VIII			O
1	1) H·										2)
2	3)	4)	5)	6) ·C·	7) :N·	8) :O·	9) :F·				10)
3	11) Na·	12) Mg:	13)	14)	15) :P·	16) :S·	17) :Cl·				18)
4	19) K·	20) Ca:	21)	22)	23)	24)	25) Mn:	26) Fe:	27) Co:	28)	
	29) Cu·	30) Zn:	31)	32)	33)	34)	35)				36)
5	37)	38)	39)	40)	41)	42)	43)	44)	45)	46)	
	47)	48)	49)	50)	51)	52)	53) :I·				54)
6	55) etc.										

between two atoms. Actually the picture is much more complicated, and for the understanding of many problems in chemistry it is necessary to assume the electronic charges to be distributed among various portions of the molecule and to be spread about like jam, rather than placed like

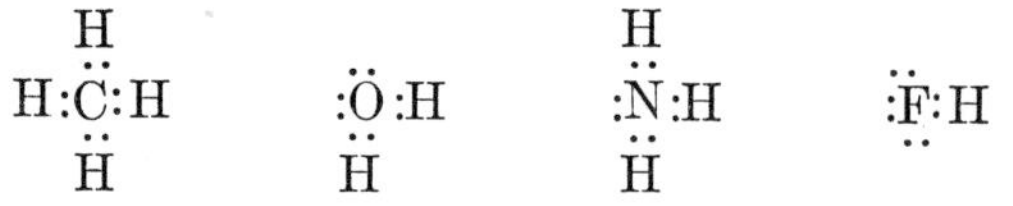

H
H:C:H
H

I.
Methane

:O:H
H

II.
Water

H
:N:H
H

III.
Ammonia

:F:H

IV.
Hydrogen
fluoride

O::C::O

V.
Carbon
dioxide

:Cl:
:Cl:C:Cl:
:Cl:

VI.
Carbon
tetrachloride

raisins. Such concepts, usually referred to by the term *resonance*, are however beyond the scope of this book and for our purposes the simpler images here presented will be sufficient.

Two major types of valence bonds may be formed, depending upon the electronic structure of the elements involved. *Electrovalent bonds* are usually formed between elements of widely different electronic structure. Sodium, for instance, having only one electron in its outer shell need only lose that one to gain a stable configuration, and therefore has a great tendency to lose it. Chlorine, on the other hand, has seven electrons in its outer shell and has therefore a similarly great tendency to gain the single electron that it requires for maximum stability. The reaction between them is one of complete transfer (formula VII). Sodium, with the loss of one electron is left with a net unit positive charge, while chlorine, having gained one, possesses a net unit negative charge. They are no longer neutral elements but are now charged ions, which attract one another according to the laws of electrostatics. Elements such as sodium which have a tendency to lose electrons are termed *electropositive*; while those like chlorine which have a tendency to gain electrons are termed *electronegative*. Methods exist for measuring quantitatively the relative tendencies of elements to gain or lose electrons and scales of electronegativity and electropositivity have

$$\underset{\text{Sodium atom}}{\mathrm{Na}^{\times}} + \underset{\text{Chlorine atom}}{\cdot\ddot{\underset{..}{\mathrm{Cl}}}:} \rightarrow \underset{\text{Sodium ion}}{\mathrm{Na}^{(+)}} + \underset{\text{Chloride ion}}{{}_{\times}^{\cdot}\ddot{\underset{..}{\mathrm{Cl}}}:^{(-)}}$$

VII[1]

been prepared. Of the elements of biochemical significance, sodium, potassium, and calcium are among the electropositive elements, while oxygen, nitrogen, and chlorine are electronegative. Carbon and hydrogen are intermediate in character.

Where elements are of similar electronic structure, *covalent bonds* are formed. Here the electron transfer is incomplete, and the electron is more or less shared between the two and can be considered as forming part of the electron shell of each atom. This is especially so in the case where the two atoms concerned are identical, as in chlorine gas, Cl_2 (formula VIII), and the two electrons between the atoms are equally shared by both. Since carbon and hydrogen are about equally electronegative (each having exactly half a complete shell of outer electrons), they also form compounds

[1] In this and similar formulas the electrons of various atoms are distinguished as dots and crosses. This is for ease of picturing the phenomena described and does *not* imply any difference in the nature of the electrons. Actually the electrons of all elements are identical in nature.

in which electrons are equally shared, as in butane (formula IX). Where such equal sharing of electrons exists, compounds are termed *non-polar*, since there is no net separation of electrical charge.

Where elements possess moderately different electronic structures, a bond is formed which is intermediate in nature and is neither electrovalent nor yet entirely covalent. In other words, electrons are neither transferred

$$\underset{\text{atom}}{\underset{\text{Chlorine}}{\overset{\times\times}{\underset{\times\times}{{}_{\times}^{\times}\mathrm{Cl}\times}}}} \;+\; \underset{\text{atom}}{\underset{\text{Chlorine}}{\overset{\cdot\cdot}{\underset{\cdot\cdot}{\cdot\mathrm{Cl}:}}}} \;\rightarrow\; \underset{\text{molecule}}{\underset{\text{Chlorine}}{\overset{\times\times\;\cdot\cdot}{\underset{\times\times\;\cdot\cdot}{{}_{\times}^{\times}\mathrm{Cl}{}_{\times}^{\cdot}\mathrm{Cl}:}}}}$$

VIII

$$(4\cdot\dot{\underset{\cdot}{\mathrm{C}}}\cdot,\ 10\mathrm{H}\times)$$

$$\downarrow$$

$$\begin{array}{ccccccc} & \mathrm{H} & \mathrm{H} & \mathrm{H} & \mathrm{H} & \\ & \cdot\times & \cdot\times & \cdot\times & \cdot\times & \\ \mathrm{H}\overset{\times}{\cdot} & \mathrm{C}: & \mathrm{C}: & \mathrm{C}: & \mathrm{C} & \overset{\times}{\cdot}\mathrm{H} \\ & \times\cdot & \times\cdot & \times\cdot & \cdot\times & \\ & \mathrm{H} & \mathrm{H} & \mathrm{H} & \mathrm{H} & \end{array}$$

IX. Butane

entirely, nor are they shared equally. They are shared unequally. As examples, we can consider the electronic structure of water (formula X) and of acetone (formula XI). In each case the electrons binding oxygen and hydrogen, or oxygen and carbon, are held more tightly by the oxygen since that element is more electronegative than either carbon or hydrogen.

$$\begin{array}{ccc} \delta^{(-)}\;\cdot\cdot & & \delta^{(+)} \\ :\dot{\mathrm{O}}\times & & \mathrm{H} \\ \cdot\times & & \\ \delta^{(+)} & & \\ \mathrm{H} & & \end{array} \qquad\qquad \begin{array}{ccccc} & & \delta^{(-)} & & \\ & & :\mathrm{O}: & & \\ & & \cdot\times & & \\ & \mathrm{H} & \cdot\times & \mathrm{H} & \\ & \times\cdot & & \cdot\times & \\ \mathrm{H}\underset{\times}{\cdot} & \mathrm{C}: & \mathrm{C}: & \mathrm{C} & \underset{\times}{\cdot}\mathrm{H} \\ & \cdot\times & \delta^{(+)} & \times\cdot & \\ & \mathrm{H} & & \mathrm{H} & \end{array}$$

X. Water[2] XI. Acetone

The result is that there is an accumulation of some negative charge at the oxygen end of the molecule and of positive charge at the carbon or hydrogen end of the molecule. Molecules in which separation of charge exists are termed *polar*.

Water and ammonia are typical polar compounds, while the various

[2] The symbols $\delta^{(-)}$ and $\delta^{(+)}$ are used to express fractional charges due to unequal electron sharing. The charges so represented are usually less than unit charges which are symbolized in the usual manner as (+) or (−).

hydrocarbons and such compounds as carbon tetrachloride are non-polar. Polar compounds tend to dissolve in polar solvents and non-polar compounds in non-polar solvents. Organic compounds containing oxygen or nitrogen tend to be polar, and if enough of these elements exist in the molecules they are water-soluble. For this reason the sugars and several amino acids are water-soluble. Where an organic compound contains a large hydrocarbon component, as is the case in lipids and in such amino acids as phenylalanine, it is sparingly soluble in water and relatively soluble in non-polar liquids. Non-polar liquids are often referred to collectively as fat solvents.

Double and triple bonds result from the sharing of four and six electrons respectively. The formation of a double bond from a single bond or of a triple from a double involves the loss of electrons and is therefore an oxidative process. This can be exemplified in the conversion of ethane to ethylene and then to acetylene (formula XII).

Covalent bonds are not necessarily formed by the contribution of equal numbers of electrons by two different elements. If we consider ammonia, for instance, we see that the nitrogen is surrounded by four pairs of electrons, three of which are shared with hydrogens. The fourth and last pair remains unshared. It is possible for such an unshared pair of electrons to

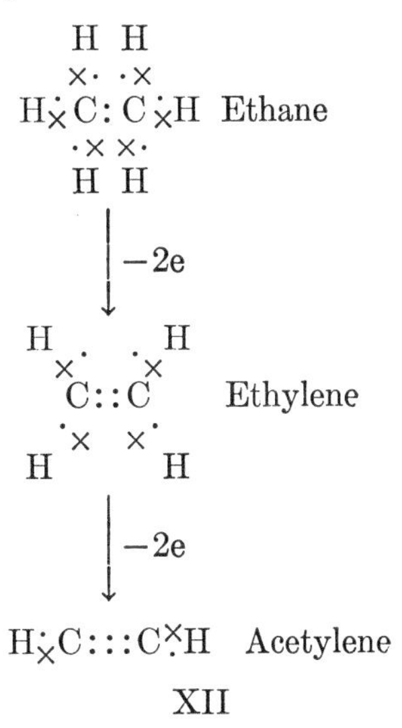

XII

be used as a means of attaching a hydrogen ion, as shown in formula XIII, so that nitrogen would be surrounded by four hydrogens, and would possess a positive charge, becoming, in fact, ammonium ion. Such a bond in which one atom contributes both shared electrons is known as a *co-ordinate* or *semi-polar bond*. Note that it is *not* a hydrogen *atom* which is attached to the nitrogen, since it would possess an electron of its own for which there would be no place in the octet.

$$\begin{array}{c} \text{H} \\ \times\cdot \\ \text{H}\overset{\cdot}{\times}\text{N}: \\ \cdot\times \\ \text{H} \end{array} + \text{H}^{(+)} \rightarrow \left[\begin{array}{c} \text{H} \\ \times\cdot \\ \text{H}\overset{\cdot}{\times}\text{N}:\text{H} \\ \cdot\times \\ \text{H} \end{array}\right]^{(+)}$$

Ammonia Ammonium ion

XIII

The positively charged ammonium ion exists in solution only in balance with an equivalent number of negatively charged ions, as is true of any other positive ion. In fact, the charge and ionic size of the ammonium ion are similar to those of the potassium ion so that ammonium compounds have many properties similar to those of the analogous potassium compounds.

One might expect that just as ammonium chloride is similar to potassium chloride, so ammonium hydroxide ought to be similar to potassium hydroxide. Here, however, another factor interferes. If NH_4OH existed and ionized to yield ammonium ion and hydroxyl ion, it would behave as a strong base in solution, just as KOH and NaOH do, by virtue of the formation of OH^-. However, if NH_4 and OH^- are present simultaneously in solution, they react to form ammonia and water:

$$NH_4^+ + OH^- \rightarrow NH_3 + H_2O$$

with the equilibrium far to the right. If an attempt is made to form ammonium hydroxide by adding sodium hydroxide to a solution of ammonium chloride, it can be seen that the product is only ammonia, which is a weak base.

The same is true of those organic derivatives of the ammonium ion which allow at least one hydrogen to remain bound to nitrogen. Trimethylammonium ion will react with hydroxyl ion as follows:

$$(CH_3)_3NH^+ + OH^- \rightarrow (CH_3)_3N + H_2O$$

trimethyl-ammonium ion trimethylamine

A *quaternary ammonium ion*, that is, one in which the nitrogen is bound to four organic radicals, as in tetramethylammonium ion, $(CH_3)_4N^+$, does not react with OH^- in this manner. Whereas less substituted ammonium ions are positively charged only in acid solution, the quaternary ammonium ion is positively charged in basic solutions as well. A quaternary ammonium ion can therefore be isolated in solid form together with OH^- as a *quaternary ammonium hydroxide*. Such a solid, upon solution, will liberate OH^- and will thereforefore behave as a "base" in the same way that NaOH and KOH will.

Several substances of biochemical interest are quaternary ammonium compounds. An example is *choline* (formula XIV) which is trimethyl-beta-hydroxyethylammonium ion. Choline is a positively charged ion which must be balanced by a negative ion. In the case of choline as isolated, this is usually an OH^- or a Cl^-, depending on the method of preparation. Choline hydroxide behaves as a strong base in solution due to the OH^- liberated. The choline ion itself is neither acidic nor basic in the Brønsted sense, see page 10. When choline forms part of a larger molecule, containing an acidic group capable of forming a negatively charged ion, a dipolar ion results as in the case of phosphatidylcholine (see page 125.) DPN (see page 248) and thiamine (see page 795) are additional examples of important quaternary ammonium compounds.

Electronegative elements, such as fluorine, oxygen, and nitrogen, accumulate a sufficiently large negative charge in compounds to possess considerable attraction for hydrogens which have accumulated positive charges through being attached to oxygen, nitrogen, or fluorine. These are the *hydrogen bonds* so important with reference to protein structure.

Electronic concept of oxidation-reduction. Oxidation has already been defined as a chemical process involving the loss of electrons. Such a process is easy to visualize in simple ionic reactions as when ferrous ion is oxidized to ferric ion or chloride ion is oxidized to chlorine gas, in each case through the loss of one electron (formula XV). The conversion of ferric ion to ferrous ion or chlorine gas to chloride ion would be the corresponding reductions, each reaction involving the gain of one electron. It is not to be thought, however, that either oxidations or reductions can exist independently. Where ordinary chemical reactions are concerned, there are at no time measurable concentrations of free electrons present in solutions. What does take place, therefore, is the coupling of an oxidation and reduction involving the transfer of electrons from the substance being oxidized to the substance being reduced. An example of such a complete oxidation-reduction reaction (or as it is sometimes called, *redox reaction*) can be obtained by combining the two partial reactions discussed immediately above. Molecular chlorine will oxidize solutions of ferrous salts to ferric salts, itself being reduced to chloride ion (formula XVI).

In the case of the non-ionic reactions which are so important in biochemistry similar principles can be applied. Thus when carbon burns in the presence of oxygen the process can be represented electronically as shown in formula XVII. Since ions are not involved, there is no complete transfer of electrons. However, it will be recalled that oxygen is far more

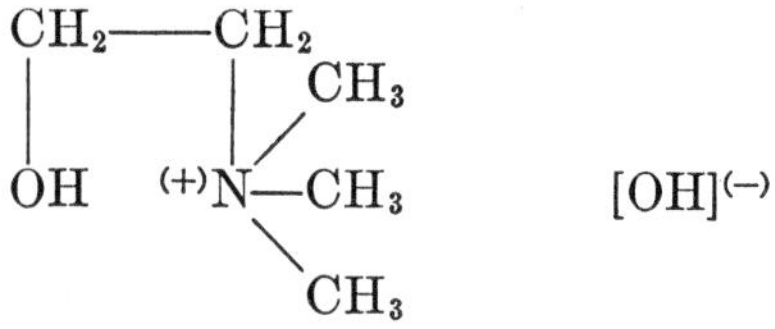

$$\begin{array}{l} CH_2\text{——}CH_2 \\ \ \ | \qquad\quad |\ \ \ CH_3 \\ \ \ | \qquad\quad |\,/ \\ OH \quad ^{(+)}N\text{—}CH_3 \qquad\qquad [OH]^{(-)} \\ \qquad\qquad\ \ \backslash \\ \qquad\qquad\quad CH_3 \end{array}$$

XIV. Choline

$$Fe^{++} \xrightarrow{-1e} Fe^{+++}$$

$$Cl^{-} \xrightarrow{-1e} Cl^{0}$$

XV[3]

1e transfer

$$Cl^{0} + Fe^{++} \rightarrow Cl^{-} + Fe^{+++}$$

XVI

electronegative than carbon, and has a stronger attraction for the shared electrons than has carbon. The process may be considered on that account to be a partial transfer of electrons from carbon to oxygen, so that in the process carbon has been oxidized and oxygen reduced. When atoms share electrons equally, as in the formation of a chlorine molecule from two chlorine atoms there is naturally neither oxidation nor reduction.

A special type of oxidation of great importance in biochemistry is that known as *dehydrogenation*. Its electronic significance is exemplified by the oxidation of hydroquinone to *p*-quinone (formula XVIII). In step 1, two electrons are lost, one from each pair shared by the hydrogens and oxygens of the phenol groups. *This is the oxidation step.* (It occurs only in the

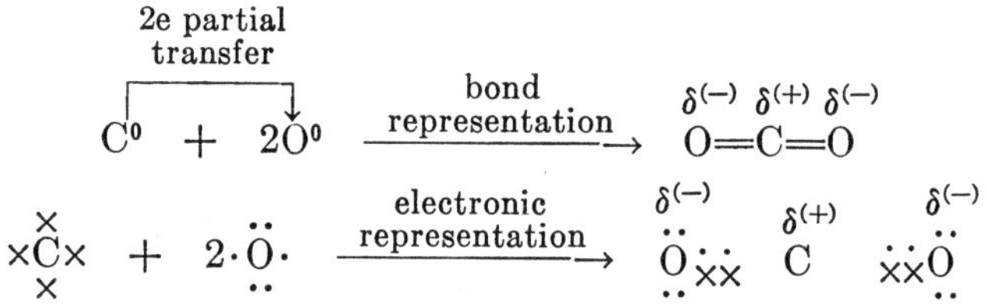

2e partial transfer

$$C^{0} + 2O^{0} \xrightarrow{\text{bond representation}} \overset{\delta(-)}{O}=\overset{\delta(+)}{C}=\overset{\delta(-)}{O}$$

$$\times\overset{\times}{\underset{\times}{C}}\times + 2\cdot\overset{\cdot\cdot}{\underset{\cdot\cdot}{O}}\cdot \xrightarrow{\text{electronic representation}} \overset{\delta(-)}{\ddot{O}}{\times\times}\ \overset{\delta(+)}{C}\ {\times\times}\overset{\delta(-)}{\ddot{O}}$$

XVII

presence of an electron acceptor such as oxygen, which is reduced in the process.) With the indicated electrons lost, only a single electron is shared by the hydrogens and oxygens of the phenol groups. Since *two* shared electrons are required for stable bonds, the hydrogen is now very weakly

[3] The superscript ^{0} as in Cl^{0} indicates the uncharged atom. It does not necessarily represent the uncharged molecule, which in the case of chlorine is Cl_2. It is simpler for our purposes to deal here with single atoms.

held and in step 2 spontaneously leaves the molecule as a hydrogen ion. *This is the dehydrogenation step.* The loss of hydrogen by a molecule is thus not itself oxidation, but is rather the consequence of oxidation. Step 3 represents only a rearrangement of electrons into a more stable configuration.

It is important to distinguish between dehydrogenation and ionization. In the case of hydroquinone, dehydrogenation is, as has been indicated, the loss of two electrons followed by the loss of two hydrogen ions, while ionization would be the loss of hydrogen ions without loss of electrons, leaving a charged negative ion (formula XIX). Ionization is thus not an oxidation-reduction process though the extent to which it is present may influence the ease of oxidation. In the case of hydroquinone, the preliminary dissociation of hydrogen ion encourages oxidation since electrons are more easily removed from the negatively charged quinonate ion than from the

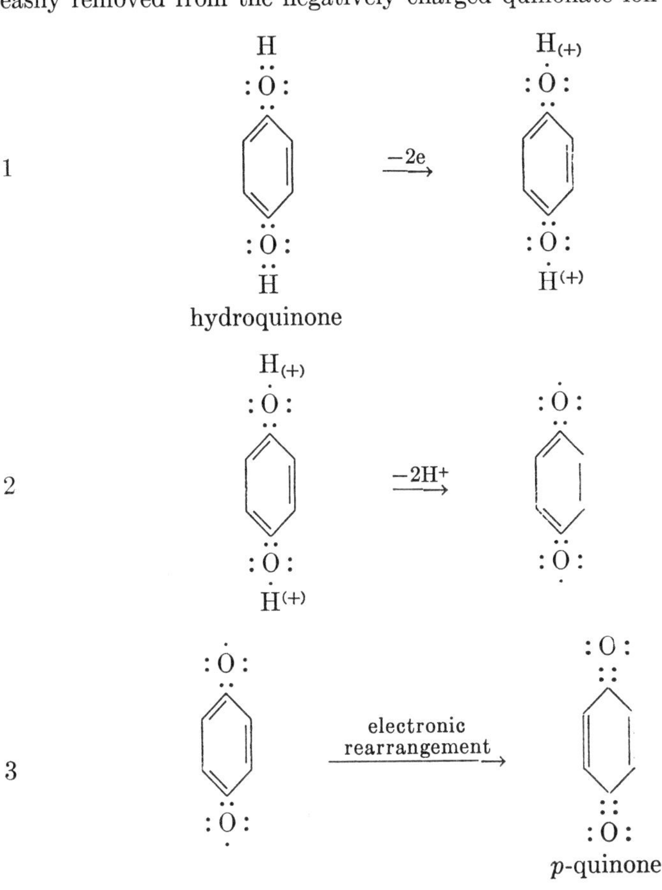

XVIII

neutral hydroquinone itself. It is for this reason that substances such as hydroquinone are particularly liable to oxidation at alkaline pH.

In the oxidation of ferrous ion by chlorine, electrons move from the ferrous ion to the chlorine but not vice versa. Similarly, carbon can be completely oxidized by oxygen to carbon dioxide, and hydroquinone to *p*-quinone, the reactions being not measurably reversible. This is so because of the considerably greater attraction for electrons on the part of chlorine and oxygen as compared with the other substances mentioned. It is possible

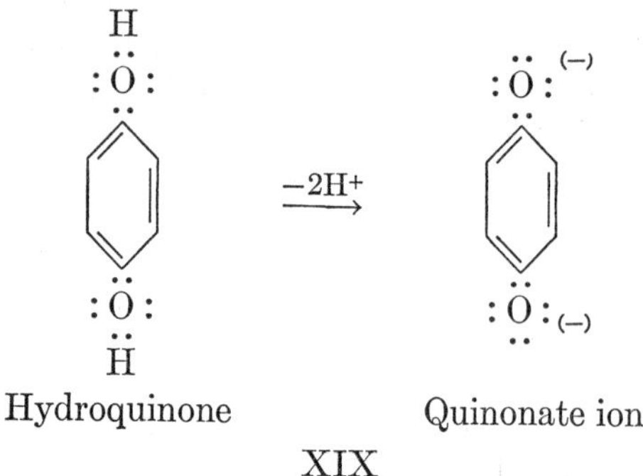

XIX

to measure by electrochemical methods the relative attraction various substances have for electrons (14) and this measure is termed the *oxidation potential.*

The oxidation potential is always given for a system which includes the reduced and oxidized form of a single substance, the reduced form being conventionally placed first. Since the absolute values of oxidation potentials can not be measured, the oxidation potential of the hydrogen/hydrogen ion system is arbitrarily set at zero, and other potentials are related to it. Systems in which the reduced form has a greater tendency to lose electrons than has hydrogen have negative oxidation potentials, while those in which the reduced form has a lesser tendency to lose electrons have positive oxidation potentials. It follows, then, that as one progresses up the scale of oxidation potential in the direction of decreasing negative and increasing positive values, systems are reached which show a progressive tendency to retain electrons and remain in the reduced forms. The oxidized forms of such systems, with their great tendency to gain electrons and be reduced, are therefore *oxidizing substances.* At the other end of the scale are located the *reducing substances.* It is important to realize that both these terms are relative. The hydrogen/hydrogen ion system is an oxidizing substance with respect to potassium (explosively so), but is a reducing substance with respect to oxygen. Oxygen, in turn, usually considered an oxidizing substance (it gave its very name to the process), is a reducing substance with respect to fluorine. In general, if two redox systems are mixed, electrons will flow in the direction of the higher positive or smaller negative potential. In other words, a given redox system will oxidize, and be reduced by, any other system with a less positive or more negative potential.

In the cases cited in the earlier part of this section, for instance, the chloride ion/chlorine system has an oxidation potential of 1.3583 volts, while that of the ferrous ion/ferric ion system is 0.771. Again, the oxidation potential of water/oxygen is 0.815, while that of hydroquinone/*p*-quinone is 0.6994. The oxidation potentials of many systems of biochemical interest have been determined, and in Lardy (1) one can find a listing of 231 of these. In most of these systems the oxidized form is colored and the reduced form is colorless (e.g. 2,6-dichlorophenolindophenol and methylene blue). Such systems are very useful as oxidation-reduction indicators if their oxidation potential is intermediate between those of the two systems whose interaction is being studied. The principle is quite analogous to the use of pH indicators. In using oxidation potentials, it is always important to remember that they vary with the temperature and pH of the system.

Although many oxidation-reduction reactions go virtually to completion, pairs of systems which have oxidation potentials close to one another yield equilibria in which measurable quantities of both the oxidized and reduced forms of the systems concerned may exist. Such systems can respond readily to the addition of either oxidizing or reducing systems and are referred to as being *poised,* just as certain acid-base systems are *buffered.* The living cell is such a poised system, and much of its behavior can be explained on the basis of that fact.

Semiquinone theory of oxidation. The oxidizing reactions with which physicians are primarily concerned are those of oxygen on foodstuffs. If we consider the oxidizing properties of molecular oxygen in this connection, we are brought face to face with an apparent contradiction. Oxygen is the second most electronegative element known, yielding first place only to fluorine (26). It is more electronegative than chlorine as is evidenced by the fact that oxygen in combination with carbon or hydrogen can attract the shared electrons sufficiently to accumulate enough charge to form hydrogen bonds, whereas chlorine attached to carbon or hydrogen can not so successfully compete for electrons. (This ability on the part of oxygen to form hydrogen bonds while chlorine can not is not entirely due to the greater electronegativity of the former atom. Oxygen is the smaller atom and can approach closer to the hydrogen so that there is a stronger attraction between the two unlike charges.) One would assume from this that oxygen would be the stronger oxidizing agent since that is what is meant by this greater tendency to gain electrons. Yet in actual practice the reverse seems to be true. Chlorine (in the presence of light) will react readily with hydrogen or hydrocarbons at room temperature, while oxygen requires elevated temperatures. What is the answer to this apparent contradiction?

In the previous section we have considered oxidations involving oxygen and one involving chlorine, and one difference between them—electronically speaking—stands out. In the oxidation of ferrous ion by chlorine, one electron is transferred, since chlorine requires only one to make up its octet. In the oxidation of hydroquinone by oxygen, *two* electrons are transferred since oxygen requires that many for octet formation. There is evidence (5, 19) that the two-electron transfer involved in oxidations by oxygen is accomplished one electron at a time. This means that in the case of hydroquinone there must be an intermediate stage which is neither hydroquinone nor *p*-quinone, but something in between. This intermediate is known as a *semiquinone* and shown in formula XX. Note that the semiquinone has lost both an electron and a hydrogen ion, and is therefore not charged. It must not be confused with the quinonate ion existing in strongly alkaline solutions in which only a hydrogen ion has been lost and which has a unit negative charge in consequence.

All hydrocarbon derivatives, when oxidized by oxygen, pass through such a one-electron-loss semiquinone stage. The name is applied even to intermediates of compounds which do not form quinones on oxidation. The effect was first studied in quinones and their reduced derivatives, in certain of which it is even possible to demonstrate visually the presence of such intermediates. The significance of this stage in oxidation by oxygen

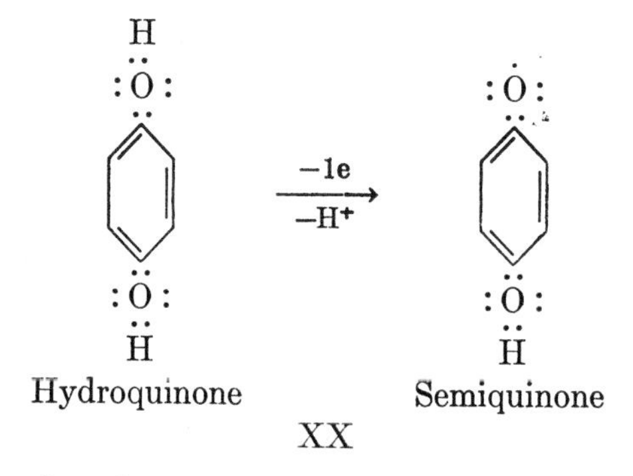

XX

rests on the fact that the semiquinone is an extremely unstable structure, the formation of which requires an input of energy. In other words, semiquinone formation involves an increase of free energy and is therefore not itself a spontaneous reaction. Once formed, as through the application of heat energy, the semiquinone readily loses the second electron, the loss of free energy in this second step more than making up for the gain in the first so that the over-all reaction is spontaneous. For this reason a candle *once lit* will continue to burn freely. Similarly, oxygen and hydrogen combine at room temperature, but at infinitesimal rates since only the rare molecules which collide with sufficient energy to form the free radical stage can proceed to the final oxidation. (It is evident from the foregoing discussion that nitrogen which requires a *three*-electron shift to form its octet will behave in practice almost like an inert gas, despite the fact that it is about as electronegative as chlorine, and since it is smaller can form hydrogen bonds.)

Oxidizing Enzymes

Oxidizing enzymes catalyze the transfer of electrons from one substance to another. Since biological oxidations in general usually involve the C:H and the O:H bond, the transfer of electrons involves the transfer of hydrogen as well.

A reversible oxidation-reduction system, or redox system as it is sometimes termed, requires the presence of both the oxidized and reduced forms of a substance in equilibrium with one another. If we consider two redox systems, AH_2/A and BH_2/B, of which the latter has the higher oxidation potential, then the general reaction catalyzed by an oxidizing enzyme would be

$$AH_2 + B \xrightarrow[\text{enzyme}]{} A + BH_2$$

AH_2 would represent the *metabolite* or the *hydrogen donor*, while B is the *hydrogen acceptor*. It is possible to classify the various oxidizing enzymes according to the nature and specificity of the hydrogen acceptors involved. Three categories can be listed:

1. Enzymes which can utilize organic compounds as hydrogen acceptors. These are the *dehydrogenases.*
2. Enzymes which specifically require molecular oxygen as the hydrogen acceptor. These are the *oxidases.*
3. Enzymes which specifically require hydrogen peroxide as the hydrogen acceptor. These are the *hydroperoxidases.*

Dehydrogenases. In the presence of certain tissue extracts a variety of organic acids, aldehydes, and alcohols can be oxidized in the absence of atmospheric oxygen if an appropriate hydrogen acceptor is also included in the system. Pure solutions of lactic acid are stable in the presence of either methylene blue or homogenized liver tissue. In the presence of both methylene blue and tissue, however, lactic acid is oxidized to pyruvic acid. This can be demonstrated by the fact that in the process methylene blue is reduced to a colorless compound. If methylene blue is mixed with lactic acid, none of the blue color is lost even over extended periods. If tissue extract is added as well, decolorization takes place rapidly. Again, if methylene blue and tissue extract are mixed, no decolorization takes place until lactic acid is added. The lactic acid dehydrogenase which catalyzes this reaction will not itself function except in the presence of DPN as coenzyme. Apparently the reaction can thus be divided into two steps, each of which involves the transfer of two hydrogen atoms:

1. Lactic acid is oxidized to pyruvic acid in the presence of the specific apoenzyme by DPN which is reduced to $DPN \cdot H_2$;
2. $DPN \cdot H_2$ is re-oxidized to DPN by methylene blue which is itself reduced to the colorless leucomethylene blue.

Substances such as DPN which in a process of this sort first gain hydrogen and then lose it are termed *hydrogen carriers*. Such hydrogen carriers shuttle repeatedly between their oxidized and reduced forms so that one molecule can handle indefinite numbers of metabolite molecules. Such a substance need be present only in small quantities in order to perform its function. Among the coenzymes, the pyridine and isoalloxazine nucleotides act in this fashion (see formulas XXI and XXII).

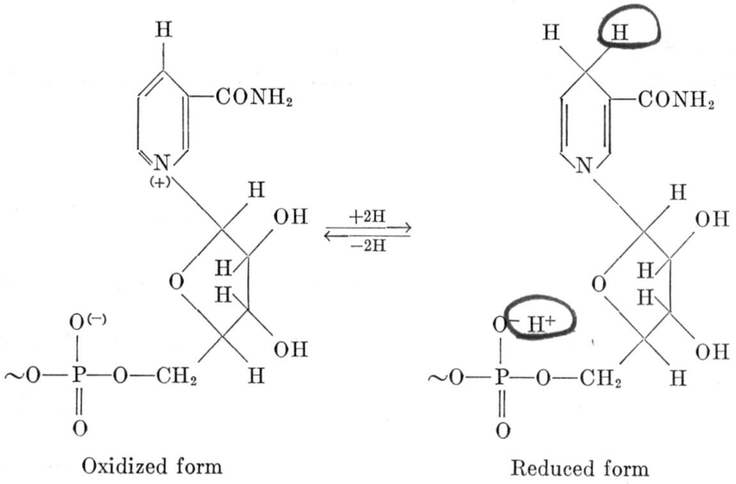

$$\xrightleftharpoons[-2H]{+2H}$$

Oxidized form — Reduced form

Note: One of the hydrogens in reduced DPN is added covalently at position 4 of the pyridine ring (27), the other is added as H^+ in equilibrium with the phosphate ion. Reduced DPN may therefore be represented by the symbol $DPNH \cdot H^+$ perhaps more appropriately than as $DPN \cdot H_2$.

XXI. Pyridine nucleotide coenzymes as redox system (see page 248 for complete formulas of the coenzymes).

Of the two steps listed in the enzymatic oxidation of lactic acid, there is a lack of specificity in the second which is characteristic of the dehydrogenases. With few exceptions any redox system with a suitable oxidation potential can substitute for methylene blue as a hydrogen acceptor. Oxygen

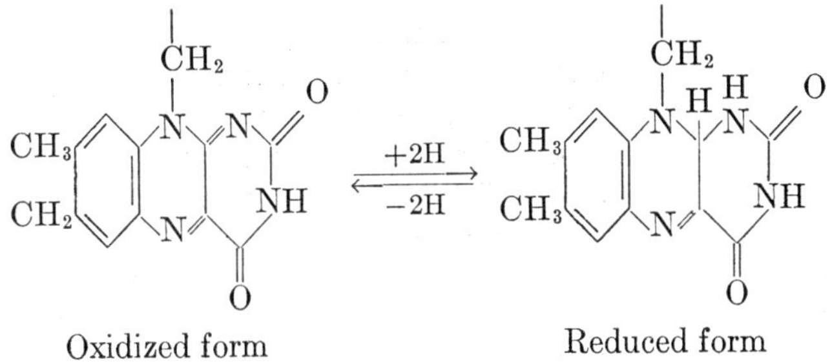

$$\text{Oxidized form} \underset{-2H}{\overset{+2H}{\rightleftharpoons}} \text{Reduced form}$$

XXII. Isoalloxazine nucleotide coenzymes as redox system (see page 250 for complete formulas of the coenzymes).

itself is the most important exception. It can not accept the hydrogen from DPN directly, presumably because of the sluggishness involved in the semiquinone step which was discussed in the previous section. Some dehydrogenases show even less specificity with respect to hydrogen acceptors. In the case of xanthine dehydrogenase, in which FAD is the coenzyme, oxidation of xanthine to uric acid can take place not only in the presence of such oxidizing agents as indophenol, alloxan, chlorates, iodine, nitrates, nitrobenzene, permanganate, quinone, methylene blue, and hydrogen peroxide—but in the presence of molecular oxygen as well. Dehydrogenases can be divided into aerobic and anaerobic according to whether they can or can not use molecular oxygen as a possible hydrogen acceptor.

Aerobic dehydrogenases can be distinguished from true oxidases not only by the fact that the former can utilize substances other than molecular oxygen as the hydrogen acceptor, while the latter is restricted to oxygen, but also by the fact that the use of oxygen by the former results in its reduction to hydrogen peroxide while by the latter it is reduced to water. The difficulty of attempting to make any enzyme classification can be exemplified by uricase, also known as uric acid dehydrogenase and uricooxidase. This enzyme catalyzes the oxidation of uric acid only in the presence of molecular oxygen as hydrogen acceptor. It can not, however, be considered a true oxidase, since the oxygen is reduced to hydrogen peroxide.

In general, the anaerobic dehydrogenases are pyridinoenzymes while the aerobic dehydrogenases are flavoenzymes. There may be some connection between this and the fact that a number of flavoenzymes contain metallic ions more or less tightly bound to the enzyme. Xanthine dehydrogenase contains molybdenum; succinic acid dehydrogenase contains iron (15); and butyryl CoA dehydrogenase contains copper.

Oxidases. Oxidases require molecular oxygen as the hydrogen acceptor but, as we have just said, differ from aerobic dehydrogenases in that the oxygen is reduced to water and not to hydrogen peroxide. All oxidases contain a metal ion. The ability to use oxygen as a hydrogen acceptor appears to be restricted to metal-containing enzymes. Two types of oxidases may be recognized, depending upon the nature of the metal involved. The copper oxidases include such enzymes as tyrosinase, catecholase, and ascorbic acid oxidase. They are of more frequent occurrence in plants than in animals. Much more generally distributed are the iron oxidases, the chief of which belong to the *cytochrome system.*

The cytochrome system was first detected by spectroscopic studies. On the basis of differences in the positions of the absorption bands, three types of cytochromes were distinguished: cytochrome *a*, cytochrome *b*, and cytochrome *c*. Still finer spectroscopic distinctions split cytochrome *a* into four subgroups, a, a_1, a_2, *and* a_3, and cytochrome *b into* b_1 *and* b_2. Cytochrome *c* may also be made up of two components (16). In the living cell, the cytochromes are bound firmly to subcellular particles such as mitochondria; so firmly, indeed, that for a long time only cytochrome *c* could be obtained in aqueous solution. Even there, there is strong reason to think that in extracting cytochrome *c*, bonds are broken whose disruption causes marked changes in its enzymatic properties (37). A hemochromogen, thought to be cytochrome b_1, has been isolated from liver microsomes (35).

Because cytochrome *c* can be extracted from cell particles, more is known about its structure than that of the other components of the cytochrome system. Cytochrome *c* is a small protein molecule with a molecular weight of about 15,000 and containing one heme group per molecule. It is a basic molecule with an isoelectric point at 10 and is particularly rich in lysine, of which there are 22 residues in a molecule containing 104 amino acid residues altogether.

Cytochrome *c* is not autoxidizable and the reason for that is thought to lie in the manner in which the iron atom is attached to the protein portion of the molecule. Four of the co-ordinate bonds of iron are connected to the four nitrogens of the pyrrol rings of the porphyrin molecule (as in hemoglobin) and both the remaining two (not one only as in the case of hemoglobin) are connected to the protein via two of its three histidine residues. The iron, fully bound, is thus not able to combine with oxygen. The heme of cytochrome *c* is also bound to the protein via its vinyl groups, which are connected to cysteine residues. In fact, a porphyrin, called *porphyrin c*, has been isolated from cytochrome *c*, in which the cysteine residues yet remain (formula XXIII). The partial hydrolysis of cytochrome *c* from horses, pigs and cattle resulted in the isolation of a heme-containing peptide that was the same in each case. Paper chromatography (38) showed the peptide structure to be as in formula XXIIIa:

Reduced cytochrome *c* contains iron in its ferrous state (Fe^{++}) and the oxidized form contains it in its ferric state (Fe^{+++}). The shuttle between the oxidized and reduced forms thus involves a one-electron transfer at each

```
      S—CH2CHCOOH
      |       \
      CHCH3    NH2
      |          CH3        S—CH2CHCOOH
      |     CH               /       \
CH3<     \//   \      /    —CH        NH2
     \   |      |   //        \
      ==N    HN—           CH3
     /          \
   CH            HC
   \\            //
     —NH     N—
CH3<    |    ||    >CH3
        \\  CH  /
     |              |
     CH2            CH2
     |              |
     CH2            CH2
     |              |
     COOH           COOH
```

XXIII. Porphyrin *c*

```
lys—CyS—ala—gluNH2—CyS—his—thr—val—glu—lys
      \                /
       \              /
        \heme/
```

XXIIIa. Heme-containing peptides

heme prosthetic group. The same can be said of other iron enzymes and of copper enzymes as well, since in the latter, the shuttle is between cuprous ions (Cu^{+}) and cupric ions (Cu^{++}).

Comparatively little is known of cytochrome *b* and its subdivisions, except that its prosthetic group is heme and that, unlike cytochrome *c*, it is autoxidizable. Cytochrome b_2, as prepared from yeast, contains FM as well as heme and, in fact, 8 atoms of iron not bound as heme for every one that is, so that it may be looked upon as a metallo-flavo-enzyme, rather than as a heme-enzyme like cytochrome *c* (6).

Of the subdivisions of cytochrome *a*, cytochromes a_1 and a_2 have been found only in bacteria. In animal cells, only cytochrome *a* and a_3 are significant. They are probably present in equimolecular quantities (22). Whether the prosthetic group of cytochrome *a* is heme or some closely related iron-porphyrin is still undecided. Cytochrome a_3 is generally thought to be identical with *cytochrome oxidase*, the terminal enzyme of the cytochrome system and the one which utilizes moledular oxygen as the hydrogen acceptor.

Hydroperoxidases. All enzymes which utilize hydrogen peroxide as a hydrogen acceptor are classified by Theorell as hydroperoxidases. In the past these enzymes have been separated into two groups: *catalases*, in which hydrogen peroxide is the substrate (or hydrogen donor) as well as the hydrogen acceptor, and *peroxidases* for which numerous easily oxidized organic compounds (notably the polyphenols) act as substrate (or hydrogen donor), see formula XXIV. Catalase may react with other substrates, either directly as when it uses ethyl peroxide as a substrate, or indirectly by means of coupled reactions. The end result is then similar to peroxidase action. In the case of all the hydroperoxidases, the mechanism of catalysis is the formation of a reactive complex between enzyme and hydrogen peroxide.

Peroxidases are found in both plant and animal tissues. The most important animal peroxidase is hemoglobin, which acts enzymatically in addition to its function in oxygen transport. Specific peroxidases have been identified in leukocytes and in milk. These two peroxidases are not identical. Catalases have been crystallized from liver, kidney, and red blood cells, as well as from many species of bacteria. Heme is the prosthetic group in all catalases, in horse-radish peroxidase, and probably in other peroxidases.

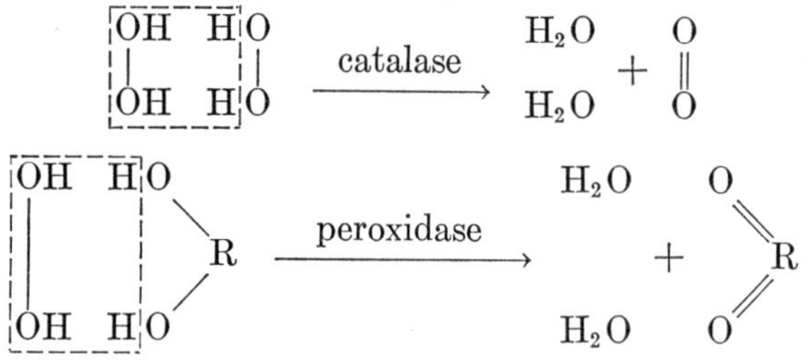

$$\begin{array}{c} OH\ \ H\,O \\ |\qquad\ | \\ OH\ \ H\,O \end{array} \xrightarrow{\text{catalase}} \begin{array}{c} H_2O \\ H_2O \end{array} + \begin{array}{c} O \\ \| \\ O \end{array}$$

$$\begin{array}{c} OH\ \ H\,O \\ |\qquad\ \ \diagdown \\ \qquad\qquad R \\ |\qquad\ \ \diagup \\ OH\ \ H\,O \end{array} \xrightarrow{\text{peroxidase}} \begin{array}{c} H_2O \\ \\ H_2O \end{array} + \begin{array}{c} O \\ \diagdown\!\!\diagdown \\ R \\ \diagup\!\!\diagup \\ O \end{array}$$

XXIV. Hydroperoxidase-catalyzed reactions

MULTIPLE ENZYME SYSTEMS

Up to this point we have discussed enzymes as single entities, and in so doing we have indulged in an oversimplification. It can not be too often stressed that the living cell is a unit and that within it no one enzyme acts independently of the others.

The Cellular Respiratory Chain

As an example we may take the oxidizing enzymes discussed in the previous section. We have mentioned the dehydrogenases which require the presence of a hydrogen acceptor which may not be oxygen and which, indeed, sometimes can not be. On the other hand, there are the oxidases which require oxygen and oxygen only as the hydrogen acceptor. Two questions arise. Why does the cell require both kinds of oxidizing enzymes? And, since substances such as methylene blue and ferricyanide ion do not occur in the cells, what is the natural hydrogen acceptor for the dehydrogenases?

The second question may be answered at once if we consider that the two enzyme systems are complementary. The natural hydrogen acceptor of a dehydrogenase is the cytochrome oxidase system, either directly or indirectly through other dehydrogenases. For example, a known three-component system in cellular respiration would be:

1. A pyridino-dehydrogenase such as isocitric acid dehydrogenase or lactic acid dehydrogenase.
2. A metallo-flavo-dehydrogenase such as cytochrome reductase.
3. The cytochrome system with its heme prosthetic groups.

What actually occurs in this system is that a "bucket brigade" for hydrogen atoms (or more fundamentally, for electrons) is set up. The metabolite, lactic acid, for instance, is oxidized in the presence of the specific dehydrogenase, passing two atoms of hydrogen to DPN, which is reduced in the process. The reduced DPN passes the hydrogen to the FM coenzyme of a flavo-dehydrogenase and is thus re-oxidized while the latter is reduced. The reduced FM passes the hydrogens to the heme of one of the cytochromes and is re-oxidized while the latter is reduced. The hydrogens then pass from cytochrome to cytochrome. The exact part taken by each cytochrome is not completely clear, largely because of our limited information concerning cytochromes *a* and *b*. It is certain that cytochrome a_3 (cytochrome oxidase) is last in line and that its reduced heme passes its hydrogen (finally) to molecular oxygen and is itself re-oxidized while the oxygen is reduced to water. It therefore follows that of the cytochrome system, only cytochrome a_3, strictly speaking, is a true oxidase. It is also certain that cytochrome *c* lies between cytochrome a_3 and the metalloflavo-dehydrogenases. For reasons that will be explained below, cytochrome *b* may belong between the metalloflavo-dehydrogenases and cytochrome *c* while cytochrome *a* may belong between cytochrome *c* and cytochrome a_3.

The presence of two coenzymic factors in metallo-flavo-dehydrogenases, a flavin and a metal ion, is perhaps significant here. In the redox shuttle of the flavin group, two electrons are alternately taken up and given off. The flavin can thus work along with the pyridino-dehydrogenases in which the same situation exists. The metal adjunct of the metallo-flavo-dehydrogenase, which is Fe^{++} in the case of cytochrome reductase (17), alternately takes up and gives off a single electron during its redox shuttle, and this fits in well with the situation among the cytochromes, in which a similar 1-electron shuttle exists. The "two-headedness" of the metallo-flavo-dehydrogenases thus suits them for their intermediate position in the respiratory chain (6, 18).

The energy released in the oxidation of a compound such as isocitric acid by means of an enzyme chain such as that described above is stored in the form of high-energy phosphate. For every two hydrogens transferred from isocitric acid to oxygen, 3 high-energy phosphate groups are formed (see page 224).

The various coenzymes and prosthetic groups are not used up in the respiratory enzyme chain. The only substances that are used up are the isocitric acid (or other such metabolite) at one end of the chain and oxygen at the other. From this it can be seen that extremely small quantities of the coenzymes are sufficient for the purpose and that similarly small quantities of coenzyme poisons will end the respiratory process and with it life itself. Such substances as HCN through the formation of stable iron complexes interfere with the oxidation of reduced cytochrome *c*, and block that particular step and the cell, or organism, dies. The step can be just as effectively and fatally blocked through interferences with the reduction of oxidized cytochrome *c* as is accomplished by many of the hypnotics, such as the barbiturates. Nor are the other components of the system less vulnerable. The action of atabrine (formula XXV) on the malaria parasite is thought to be the result of its competitive inhibition of FM, which it resembles somewhat.

We have not yet answered the question why oxidations must be conducted in this stepwise fashion. The answer is that in each individual step the difference in oxidation potential is comparatively small. The oxidation

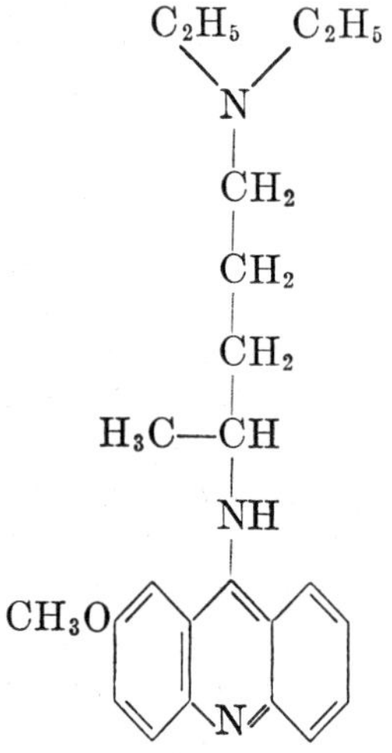

$C_2H_5 \quad C_2H_5$

N

CH_2

CH_2

CH_2

$H_3C—CH$

NH

CH_3O

N

XXV. Atabrine

potentials of the various components of the oxidative chain are as listed in table VII-2. (It is on the basis of such determinations that the order of

TABLE VII-2

Oxidation potentials (in volts) of components of the oxidative chain (2, 16a, 28, 39)

Water/oxygen	+0.81
Reduced cytochrome a_3/oxidized cytochrome a_3	+0.29
Reduced cytochrome *a*/oxidized cytochrome *a*	+0.29
Reduced cytochrome *c*/oxidized cytochrome *c*	+0.25
Reduced cytochrome *b*/oxidized cytochrome *b*	−0.05
Reduced flavoenzyme/oxidized flavoenzyme	−0.22
Reduced pyridinoenzyme/oxidized pyridinoenzyme	−0.32

the cytochromes in the chain is thought to be *b*, *c*, *a*, a_3 though it should be remembered that the potentials of the enzymes as part of an intact enzyme system may differ somewhat from their potentials as isolated *in vitro*). Because of the small potential differences, each step is more or less easily reversible and the whole represents a well poised system, the action of which can be more delicately controlled by the concentration of nutrient metabolites and the partial pressure of oxygen in the cells. The difference between bucket-brigade oxidation and one-step oxidation may therefore be compared with that between sliding downhill and falling off a precipice.

Phosphate-Linked Chains

Phosphate transfer to and from ADP and ATP (see page 225) is not only a means of transferring and storing chemical energy, but also, by the cyclic nature of the process, links different but related enzyme reactions. Phosphate-linked enzyme systems are of particular importance in carbohydrate metabolism (Chapter 14). Some inspection of the principles involved will be useful at this point, however.

Glucose, in the process of absorption into the intestinal mucosal cells, is converted to glucose-6-phosphate at the expense of ATP, which loses a high-energy phosphate bond and becomes ADP. This reaction is catalyzed by the enzyme glucokinase. Under physiological conditions, this reaction is irreversible since to reverse it directly would require an input of energy equivalent to that released by the original conversion of a high-energy phosphate bond to one of low energy. In order to regain free glucose from the glucose-6-phosphate, the mucosal cell hydrolyzes it in the presence of a phosphatase to form glucose and phosphoric acid. These two reactions can be represented by the Baldwin system as shown in formula XXVI.

Such a representation shows that in the two reactions described glucose and glucose-6-phosphate are not consumed. The over-all reaction is:

$$ATP + H_2O \rightarrow ADP + H_3PO_4$$

yet phosphoric acid does not accumulate, nor does ATP entirely disappear. They take part in other reactions, both in the mucosal cells and elsewhere in the body. Let us take up phosphoric acid first. In the presence of phosphorylase, it will convert glycogen to glucose-1-phosphate, which under the influence of phosphoglucomutase can be converted by phosphate transfer within the molecule to glucose-6-phosphate. Adding these reactions to the system, the Baldwin representation would be as shown in formula XXVII. It is not necessary at this point to concern ourselves with the renewal of glycogen and water, since both are obtained directly or indirectly from foodstuffs in ample quantity. What is important, however, is the

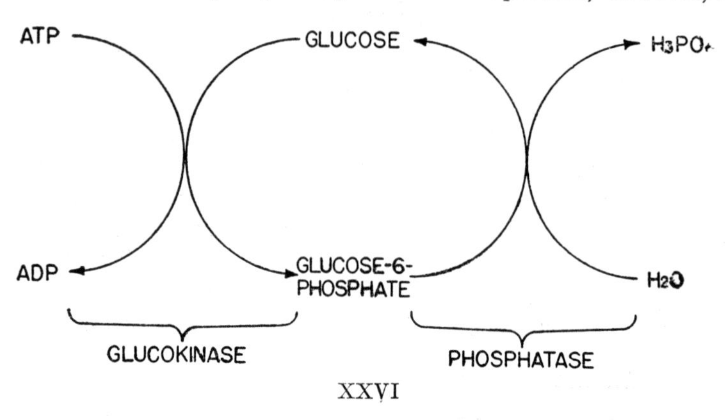

XXVI

manner in which the high-energy phosphate bond is recreated and ADP converted to ATP.

There are several physiological reactions involving oxidation, by which

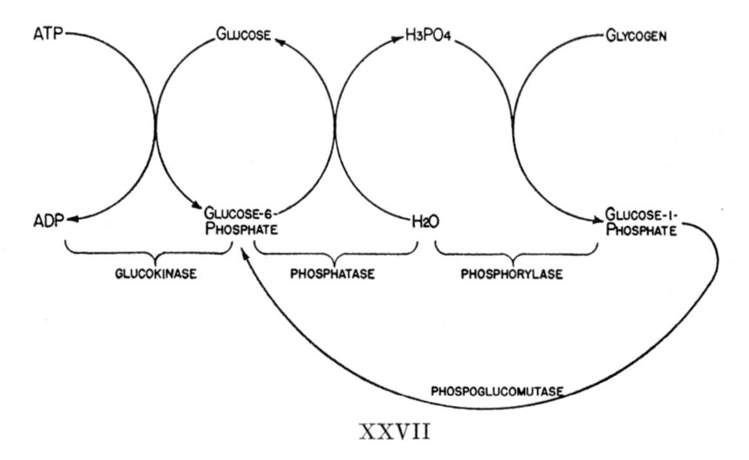

XXVII

a low-energy phosphate bond is converted to one of high-energy. One such is the oxidation of 1,3-diphosphoglyceraldehyde to 1,3-diphosphoglyceric acid, under the influence of triosephosphate dehydrogenase, an enzyme utilizing DPN as coenzyme. In the process a low-energy ester phosphate is converted to a high-energy acyl phosphate, the free energy gain being more than made up for by the free energy loss involved in the oxidation. 1,3-Diphosphoglyceric acid can now transfer its high-energy phosphate to ADP under the influence of a phosphokinase, forming ATP and itself being hydrolyzed to 3-phosphoglyceric acid. These last two reactions can be represented as shown in formula XXVIII. This system, involving an oxi-

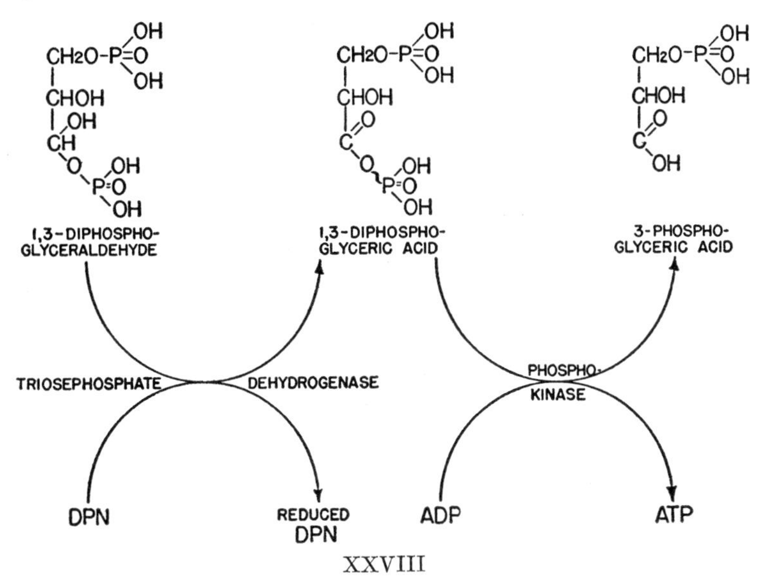

XXVIII

dation, can now be combined with the previous system involving phosphorylation through the ATP/ADP shuttle as shown in formula XXIX.

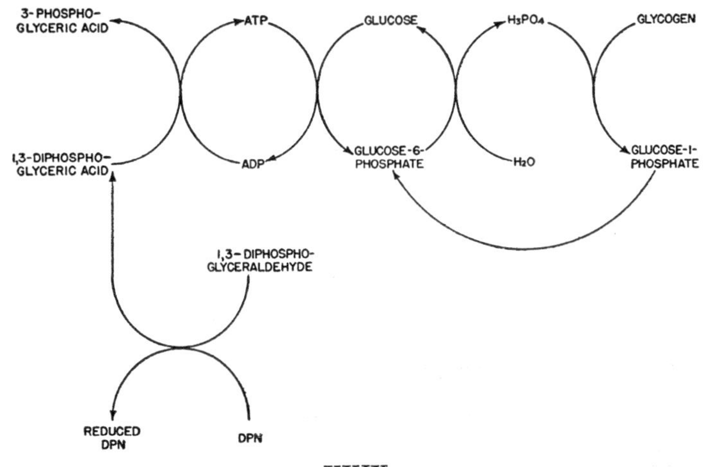

XXIX

By proceeding in this manner still further it is possible to link up enzyme systems in ever more complicated fashion so that it is not too difficult to imagine the entire cell as consisting of a single multi-enzyme system, intricately interrelated and, in health at least, working with smooth precision. Naturally, it is not a closed system. Not all the components are renewed. Some materials are used up and not reformed, while others are formed and not re-used. In the system just described, glycogen is the source of 1,3-diphosphoglyceraldehyde and is itself formed from foodstuffs by processes to be described in Chapter 14. The system, moreover, is linked with the respiratory chain and, therefore, with the ultimate use of oxygen as an energy source through the DPN/DPN.H_2 step (formula XXX). The over-all change is thus:

$$\text{carbon and hydrogen of foodstuffs} + \text{oxygen} \rightarrow \text{carbon dioxide} + \text{water} + \text{usable energy}$$

If green plants did not exist, this over-all reaction would be irreversible. Fortunately, the photosynthetic process can reverse the reaction through the utilization of solar energy (formula XXXI).

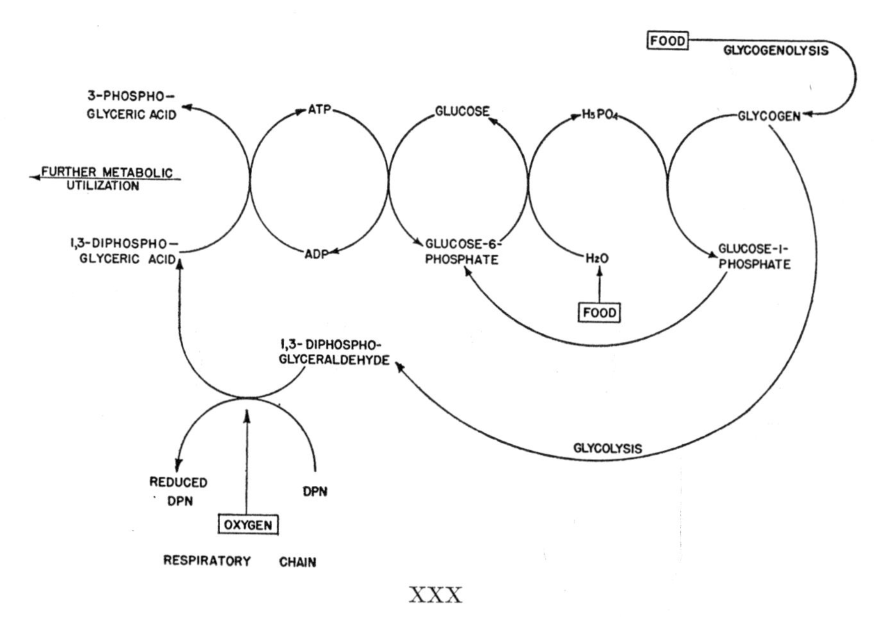

XXX

The over-all reaction now becomes:

$$\text{solar energy} \rightarrow \text{available energy} + \text{unusable heat,}$$

a relationship which is, as far as we know, unqualifiedly irreversible.

However, the total supply of solar energy is as yet sufficient to last the human race for about ten to twenty billions of years.

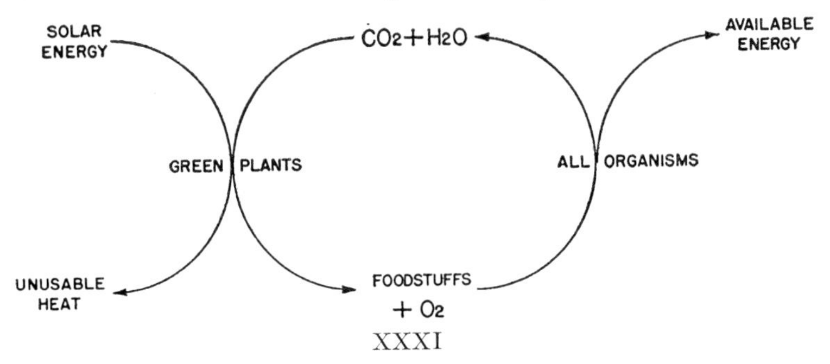

XXXI

LOCATION OF ENZYMES IN CELLS

Understanding of enzymes and enzyme systems in the cellular economy can be enhanced if these are studied not merely as substances indiscriminately present in tissue or its extracts, but as chemical entities organized in a definite manner within the cell. Many techniques have been developed to locate concentrations of specific enzymes within the cell.

Cytochemistry

The general procedures and precautions to be observed in the cytochemical location of enzymes have been described (11). The following examples are representative.

Esterases. The localization of esterases within cells may be demonstrated by adding indoxyl acetate or indoxyl butyrate in an appropriate buffer to fixed tissue slices (3). The indoxyl acetate or butyrate during incubation is hydrolyzed by the action of esterase to indoxyl which autoxidizes rapidly to the insoluble and highly colored indigo blue (formula XXXII). The indigo precipitates, presumably, in just those spots within the cell where the esterases were localized. Liver, kidney and duodenum slices were perceptibly blue in two minutes, showing numerous small particles of precipitate within the cell. Nerve tissue required ten minutes of incubation before appearing grossly colored, and showed only a few large particles per cell. Indoxyl acetate is, apparently, hydrolyzed by lipases, cholinesterase, and other esterases. Indoxyl butyrate is not hydrolyzed by cholinesterase. The existence of weak indigo staining in spinal section slices using that ester as substrate thus indicates the presence of small quantities of esterases other than cholinesterase.

Dehydrogenases. Tetrazolium salts exist in reduced and oxidized form and are unusual in that the oxidized form is colorless and the reduced form is colored. (The reverse is true in the majority of redox systems.) This property of tetrazolium salts is useful in the intracellular detection of dehydrogenases (9, 32) since tissue slices incubated with properly buffered tetrazolium salt will be colored with a precipitate of reduced tetrazolium in those regions where dehydrogenase is active, the coenzyme of the dehydrogenase being oxidized in the process.

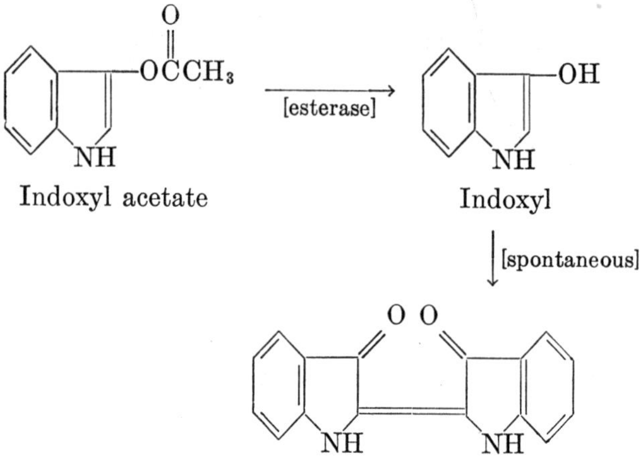

Indoxyl acetate → [esterase] → Indoxyl

↓ [spontaneous]

XXXII. Indigo blue

The best known of the tetrazolium salts is 2,3,5-triphenyl tetrazolium chloride (TTC) (formula XXXIII). TTC is red in the reduced form. Newer tetrazolium derivatives are even more intensely colored, some being blue in the reduced form (31) and some being nearly black. Since a cell which is free of dehydrogenase activity is a dead cell, tetrazolium indicators can be used to test the viability of such cells as seeds and sperm and thus prove of considerable economic importance to farmers and stock-raisers.

Phosphatases. In the Gomori method for phosphatase, tissue sections are incubated with a substrate such as glyceryl phosphate in the presence

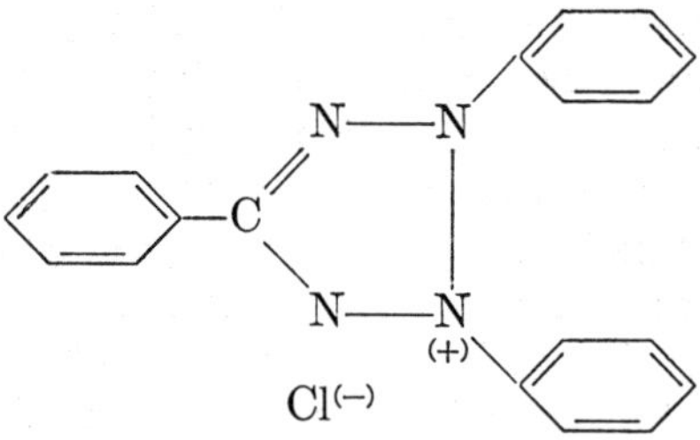

XXXIII. 2,3,5-Triphenyl tetrazolium chloride (TTC)

of small quantities of an ion, such as Ca^{++}, capable of forming an insoluble phosphate. The glyceryl phosphate is hydrolyzed by the phosphatase present in the tissue section, and the free phosphate ion thus formed reacts with the calcium present to form the insoluble calcium phosphate. The sections are then removed from the incubating medium, washed with distilled water, and placed in solutions of cobalt acetate or lead acetate. The calcium phosphate is converted to cobalt or lead phosphate. Finally, the sections are placed in a solution of ammonium sulfide, so that the phosphate is converted into the black cobalt sulfide or lead sulfide. Under the microscope, the cells are seen to contain dark brown or black patches.

The assumption that these patches indicate the original location of the phosphatase present may, however, be erroneous. Novikoff (23) and Gomori (12) point out that calcium phosphate is diffusible and tends to be absorbed at the nuclear membrane so that eventual staining of nuclear regions is not necessarily indicative of nuclear concentration of phosphatase.

Separation of Cell Fragments

Another increasingly prominent technique for determining the intracellular location of enzymes may be summarized as the subdivision of cells into fractions containing specific organelles. Thus, Dounce (8) has devised methods whereby cell nuclei can be isolated free of any significant cytoplasmic admixture. This is done, generally speaking, by homogenizing tissue at low temperature in the presence of citric acid solution. By this procedure, the cell membrane is disrupted while the nuclear membrane remains intact. The nuclei can then be separated from the cytoplasmic debris by filtration. Such nuclei can be tested for enzyme activity. Thus, arginase and catalase are present in liver nuclei but not in kidney nuclei (34) although both tissues are rich sources of both enzymes. Presumably kidney arginase and kidney catalase occur in the cytoplasm. At least two substances have been located exclusively in nuclei. These are deoxyribonucleic acid (see Chapter 9) and an enzyme which catalyzes the synthesis of DPN (13) and which may actually be located in the nucleolus. On the other hand, nuclei lack flavoenzymes and the cytochromes and cannot make use of molecular oxygen (33). They contain pyridinoenzymes, however, so that anaerobic oxidation can take place within the nucleus.

A still finer separation of cellular components can be brought about by *differential centrifugation* (29). In this method, cellular homogenates or extracts are centrifuged in several stages in which the centrifugal force is increased each time. At each stage, the sediment is collected. In this way four major fractions (beginning with the most easily sedimented) can be collected: (a) a *nuclear fraction* containing about 15 per cent of the original total nitrogen and including all the nuclei and such intact cells as may have survived homogenization; (b) a *mitochondrial fraction*, containing about 25 per cent of the nitrogen and consisting largely of mitochondria, particles the diameters of which range from 1 to 3 microns; (c) a *microsomal fraction*, containing 20 to 25 per cent of the nitrogen and consisting of microsomes, particles the diameters of which are from 0.03 to 0.15 microns, and (d) a *supernatant fraction*, containing 35 to 40 per cent of the nitrogen and made up of the soluble material of the cell plus very small lipid droplets, secretory granules and so on. The cytoplasmic fractions are heterogeneous and may be further subdivided (24, 25).

The mitochondria and microsomes are together referred to as *cytoplasmic particulates*. They are composed largely of nucleoprotein and phospholipid. Mitochondria are surrounded by semi-permeable membranes. Disruption of the membrane reveals half the protein content to be water-soluble and half bound to the mitochondrial structure itself. In general, the mitochondria contain those enzymes and coenzymes required for aerobic oxidation; that is, the metallo-flavo-dehydrogenases and the cytochromes.

Enzymes controlling glycolytic and hydrolytic reactions exist in solution in the supernatant fraction, but even here association with particulates is suspected in some cases. Acid phosphatase, for instance, a very soluble enzyme, has been found to be associated with mitochondria in inactive form (4). It can be irreversibly liberated into the cell sap where it is highly active.

THEORY OF ENZYME ACTION

Kinetics

In general, the rate at which a reaction proceeds (all environmental factors being held constant) is proportional to the number of molecules of the substances involved in the reaction; that is, it is proportional to the molar concentration. Where only one species of molecule is involved in the reaction,

$$A \rightarrow B$$

the rate of the reaction is expressed as $-\,d[A]/dt$; i.e., the rate at which the molar concentration of substance A (the brackets representing "molar concentration") decreases with time, t, as it is converted to B. This rate is directly proportional to molar concentration of substance A at any given time, so that we have the equation:

$$-\frac{d[A]}{dt} = k[A] \qquad \text{Equation VII-1}$$

Since the increase in concentration of B is equal to the decrease in concentration of A:

$$+\frac{d[B]}{dt} = k[A] \qquad \text{Equation VII-2}$$

Reactions whose rates obey the relationship expressed in equation VII-1 are called *reactions of the first order*, and the proportionality constant, k, is the *rate constant*. Radioactive decay (see Appendix) is the ideal case of a first-order reaction.

Where the reaction involves the interaction of two different molecules,

$$A + B \rightarrow C$$

then the rate of the reaction is directly proportional to the product of the molar concentrations of both molecules involved.

$$-\frac{d[A]}{dt} = -\frac{d[B]}{dt} = k[A][B] \qquad \text{Equation VII-3}$$

Reactions whose rates obey the relationship expressed in equation VII-3 are *reactions of the second order*.

It is also possible to have *reactions of zero order* when a reaction rate is independent of the concentration of a particular substance. Thus, in *in vitro* enzyme experiments where substrate is present in large excess, the addition of still more substrate does not, generally, affect the reaction rate. The reaction is thus zero order with respect to substrate.

In the case of a reversible reaction such as:

$$A \rightleftarrows B$$

where the conversion of A to B and of B to A are both first-order reactions (with, usually, different rate constants), it is important to remember that at equilibrium, the rates at which the two opposing reactions are taking place are equal. In other words, $-\, d[A]/dt$ is equal to $-\, d[B]/dt$, or

$$k[B] = k'[A] \quad \text{(at equilibrium)} \qquad \text{Equation VII-4}$$

From this, it follows that at equilibrium

$$\frac{[A]}{[B]} = \frac{k}{k'} = K \qquad \text{Equation VII-5}$$

Similarly, in reversible reactions such as:

$$A \rightleftarrows B + C$$

or:

$$A + B \rightleftarrows C + D$$

the concentrations of the various reactants at equilibrium can be related to one another as, respectively

$$\frac{[B][C]}{[A]} = K' \qquad \text{Equation VII-6}$$

$$\frac{[C][D]}{[A][B]} = K'' \qquad \text{Equation VII-7}$$

Now we can restate one of the principles of enzyme action that we presented at the beginning of the chapter. In the catalysis of one direction of a reversible reaction, an enzyme can change the value of the rate constant, k, usually to a very marked extent. However, it invariably changes the value of the rate constant of the reverse reaction, k′, by an exactly proportionate amount, so that the ratio k/k′ (i.e., K, usually called the *equilibrium constant*), is not changed. This is another way of saying that an enzyme may affect the rate of a reaction but does not affect the equilibrium point.

If, during the course of a reaction, additional quantities of one of the involved substances are added to the reaction mixture, the equilibrium concentration of all substances involved will change since K must remain constant. In general the addition of more of either C or D to the reaction:

$$A + B \rightleftarrows C + D$$

must result in a higher concentration of A and B at equilibrium. In other words, adding either C or D (or removing A or B) drives the reaction to the left. Similarly, addition of either A or B (or removal of C or D) drives the reaction to the right. Carried to extremes, such a process can prevent a reversible reaction from ever reaching equilibrium, a fact which can be useful to the body. Thus, if a product of hydrolysis diffuses out of a cell as quickly as it is formed, the parent substance (undiffusible) is continuously hydrolyzed despite the fact that the equilibrium point in a closed system may be heavily on the side of the unhydrolyzed substance. This is not an enzymatic phenomenon but a consequence of the laws of chemical equilibrium.

Enzyme-Substrate Complexes

The first step in the action of an enzyme upon a substrate is a combination of the two to form an enzyme-substrate complex. Such a complex is an unstable and short-lived step in the over-all process and therefore a difficult one to detect by direct observation. Long before such a complex was directly observed, however, kinetic calculations were made on the assumption of its existence and useful results were obtained.

Free enzyme, E, combines reversibly with substrate, S, to form an enzyme-substrate complex, ES, and this complex, under the appropriate conditions (e.g., combination with water in the case of a hydrolase), breaks down to enzyme and the reaction product, P. The situation may be summarized as follows:

$$E + S \rightleftarrows ES \rightarrow E + P$$

The enzyme, substrate, and enzyme-substrate complex, reacting reversibly can be interrelated as:

$$K_s = \frac{[E][S]}{[ES]} \qquad \text{Equation VII-8}$$

The constant, K_s, is often termed the *Michaelis-Menten constant* (20). Since it relates the concentration of the enzyme-substrate complex to that of the component parts of the complex, it is a measure of the rate of breakdown of the complex and is therefore also its *dissociation constant.*

The concentration of free enzyme (i.e., the enzyme which is not bound in an enzyme-substrate complex, and which we have been symbolizing as E) is very difficult to measure at any given time in an enzyme reaction system. The total enzyme, free and bound, (E_0) is, however, usually known, since it is the quantity originally added. The free enzyme is obviously the total enzyme minus the bound enzyme. The molar concentration of bound enzyme is equal to that of the enzyme-substrate complex, since an equimolar complex is assumed. Therefore:

$$[E] = [E_0] - [ES] \qquad \text{Equation VII-9}$$

If we combine equations VII-8 and VII-9, we have:

$$K_s = \frac{([E_0] - [ES])[S]}{[ES]} \qquad \text{Equation VII-10}$$

Solving equation VII-10 for ES, we obtain:

$$[ES] = \frac{[E_0][S]}{K_s + [S]} \qquad \text{Equation VII-11}$$

Now the reaction which actually represents the completion of the enzymatic process is the breakdown of ES to E and P. That process depends solely upon the concentration of ES. The velocity, v, of the enzyme reaction may be written therefore, as:

$$v = k[ES] \qquad \text{Equation VII-12}$$

Or, combining equations VII-11 and VII-12, we see that:

$$v = \frac{k[E_0][S]}{K_s + [S]} \qquad \text{Equation VII-13}$$

For a given value of E_0, the maximum velocity, V, will occur when all the enzyme is in the form of enzyme-substrate complex. The molar concentration of ES will then be equal to that of E_0, and equation VII-12 will become:

$$V = k[E_0] \qquad \text{Equation VII-14}$$

Combining equations VII-13 and VII-14, we have:

$$v = \frac{V[S]}{K_s + [S]} \qquad \text{Equation VII-15}$$

Equation VII-15 is sometimes known as the *Michaelis-Menten equation.* Graphically, the equation is represented by a curve such as that shown in figure VII-1, in which the velocity of the enzyme reaction increases as the concentration of substrate increases, but more slowly with each additional increment of substrate. Eventually, a "plateau" is reached, in which region, the reaction becomes zero order with respect to substrate. Ideally, the form of the curve is that of a rectangular hyperbola, with V as one of its asymptotes.

It is possible to calculate the value of K_s from equation VII-15 since the other variables are either known or can be measured. A time-honored method of doing so consists of plotting the reciprocal of the velocity of the enzyme reaction against the reciprocal of the substrate concentration. The reason for this can be seen if we take the reciprocal of equation VII-15,

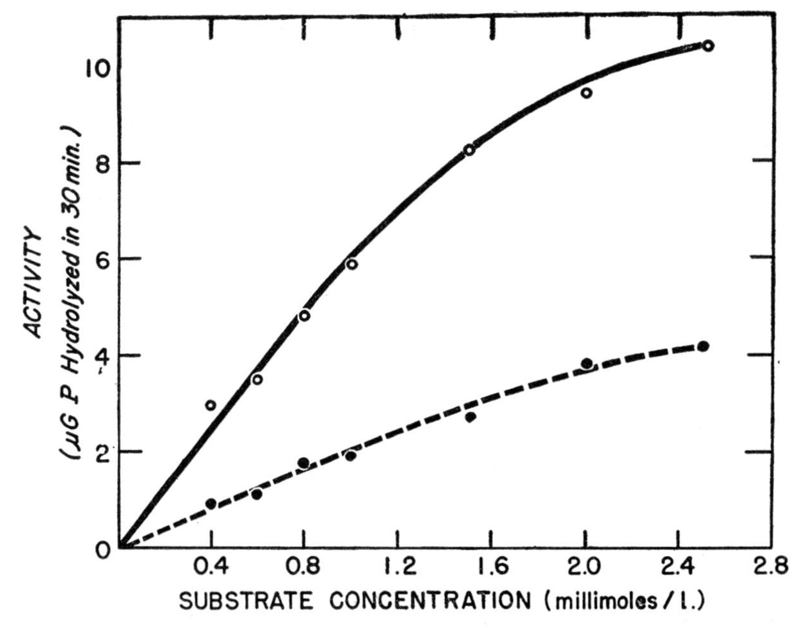

FIG. VII-1. (Schwartz *et al.* (30))

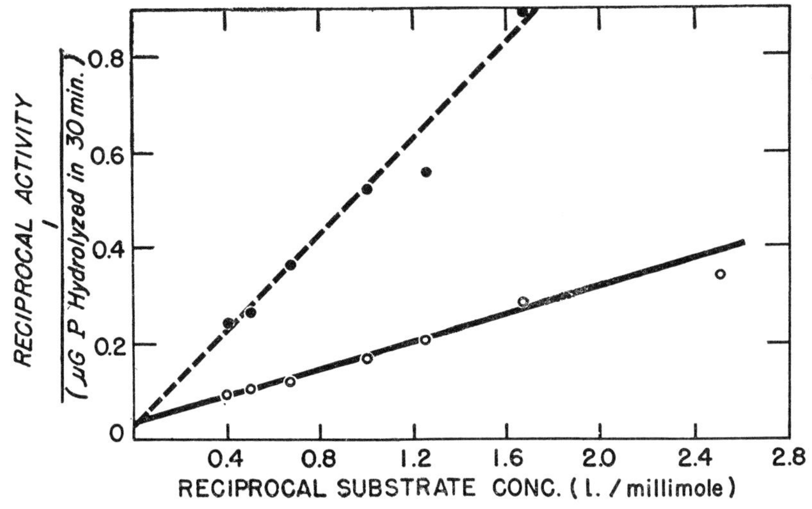

FIG. VII-2. (Schwartz *et al.* (30))

as follows:

$$\frac{1}{v} = \frac{K_s + [S]}{V[S]} = \frac{k_s}{V}\left(\frac{1}{[S]}\right) + \frac{1}{V} \qquad \text{Equation VII-16}$$

The relationship of 1/v to 1/S is a straight line (see figure VII-2) in which the slope is equal to K_s/V and the y-intercept to 1/V. Dividing the slope by the y-intercept (both can be easily measured), we arrived at $K_s/V \div 1/V$, which is equal to K_s, which is the value we are after.

Other methods for determining K_s are occasionally used, an example being that of Thorn (36).

The Michaelis-Menton relationship can be further extended to include the effects of inhibitors on enzyme reaction rates and to predict the effect of competitive inhibition (40). It can also be extended to include possible interaction of enzyme with the products of the enzymatically-catalyzed reaction (10) and cases of non-competitive inhibition (21).

The fact that kinetic equations based on the assumption of the existence of intermediate enzyme-substrate complexes often fit observed data (30) is a point, albeit an indirect one, in favor of the actual existence of such complexes. Direct evidence for the existence of enzyme-substrate complexes, particularly in the case of the heme enzymes, has been described, however (7).

In the case of many enzyme reactions, the Michaelis-Menton equation does not hold. Even where enzyme reaction rates are determined, by extrapolation or otherwise, at zero time to avoid the disturbing effects of possible enzyme inactivation in the course of the reaction, increasing substrate concentration may result in a curve which, instead of approaching and remaining in the plateau region, passes through a peak, i.e., a *substrate optimum,* after which reaction rates once more begin to decline. One explanation is the possible formation at high concentrations of substrate, of complexes containing additional molecules of substrate, ES_2 or ES_3, which unlike the active ES, are inert.

Energy of Activation

Granted the existence of an enzyme-substrate complex, it appears that a substrate is converted to a product much more quickly when it forms part of a complex, than when it exists as a separate entity. That is the essence of enzymatic catalysis. In order to get an insight into why this should be so, we must first define the energy of activation of a reaction.

Even where a reaction, under given conditions, is thermodynamically possible, there is always an intermediate stage during which a molecule has begun to react but has not yet completed its reaction and at that stage its energy content is higher than at either end of the reaction. This energy "hump" varies in height, and is equal to the energy input required to initiate the reaction, i.e., the energy of activation. When the energy of activation is low, as in the reaction of metallic sodium and ethyl alcohol, the kinetic energies of molecules at room temperature are sufficient to supply it and the reaction proceeds rapidly under ordinary conditions. Where the energy of activation is high, as in the reaction of gasoline and molecular oxygen, ordinary thermal energies of the molecules involved are not sufficient to get over the hump, and gasoline may therefore stand indefinitely in contact with air without perceptible oxidation. If, however, enough energy is supplied the gasoline-air mixture (in the form of a lighted match or an electric spark) to allow a relatively small number of the gasoline molecules to gain enough energy to pass the hump and be oxidized, enough energy is liberated to allow neighboring molecules to do the same. The result is a reaction which can take the form of a steady flame as in a cigarette lighter or an explosion as in the cylinder of an internal combustion engine.

We may say, then, that the function of an enzyme, catalyzing a given reaction, is to reduce the energy of activation of that reaction, thus allowing, at a given temperature, a greater number of substrate molecules to pass over the lowered hump per unit time, and down the free-energy chute to the thermodynamic equilibrium point.

How the enzyme reduces this energy of activation is still a matter of speculation. It may increase the energy of the substrate molecule when it is part of the enzyme-complex so that the energy excess of the reaction intermediate (i.e., its energy of activation) is smaller. It can do this by so orienting the molecule as to decrease its entropy and increasing its free energy. (It must be remembered that a system in which molecules are uniformly oriented has a lower entropy than one in which they are randomly oriented.) Physically, one might imagine a substrate molecule fitting into its niche on the enzyme surface in such a fashion that a given bond is, say, particularly exposed to bombardment by water molecules or the action of a coenzyme (8a).

An alternative explanation is that the substrate is not made less stable (i.e., more energetic), but that its reaction intermediate is made more stable (i.e., less energetic) as part of the enzyme complex. This would also reduce the energy of activation. The reaction intermediate is made more stable since as a part of an enzyme complex, it has more opportunity for resonance stabilization (i.e., the distribution of molecular strains and charges over a number of symmetrically placed bonds and groups) over the entire protein-complex surface than it has by itself. Analogously, a man who would ordinarily sink into a snow-drift, would be borne neatly on its surface were he to distribute his weight more widely by donning snow-shoes.

Enzyme Specificity

In thinking of enzyme specificity, it is useful to think of enzyme-substrate complexes as being the result of a "lock-and-key" arrangement.

Here, the enzyme surface is thought to be constructed in such a manner as to allow a substrate or a coenzyme to fit snugly as a key would fit into a lock. The phenomenon of competitive inhibition would be easily explained in this manner. A compound similar to the one for which the enzyme surface is adapted might be able to squeeze in, yet not snugly enough to allow proper orientation, just as a slightly wrong key might be forced into a lock and remain motionless (or, worse, break off) when an attempt is made to turn it.

The simplest view of such a lock-and-key mechanism would consider the atoms themselves as being so arranged as to form a depression in the surface into which the substrate molecule or a portion thereof will just fit. Another view would consider the arrangement of surface charge on the enzyme to be such that the electrostatic attraction between it and the substrate molecule in a certain orientation is greater than it would be for any other orientation or any other molecule.

REFERENCES

1. Anderson, L., and Plaut, G. W. E. Table of oxidation-reduction potentials; *in* Lardy, H. A., ed. *Respiratory Enzymes*, rev. edit., pp. 71–84. Minneapolis, Minn., Burgess Publishing Co., 1949.

2. Ball, E. G. Energy relation of the oxidative enzymes. Ann. N. Y. Acad. Sci., **45:** 363–375, 1944.
3. Barnett, R. J., and Seligman, A. M. Histochemical demonstration of esterases by production of indigo. Science, **114:** 579–582, 1951.
4. Berthet, J., and De Duve, C. Tissue fractionation studies. I. The existence of a mitochondria-linked, enzymically inactive form of acid phosphatase in rat-liver tissue. Biochem. J., **50:** 174–181, 1951.
5. Blois, S. Free-radical formation in biologically occurring quinones. Biochim. et Biophys. Acta, **18:** 165, 1955.
6. Boeri, E., and Tosi, L. Properties of cytochrome b_2 from yeast. Arch. Biochem. Biophys., **60:** 463–475, 1956.
7. Chance, B. Enzyme-substrate compounds. Advances Enzymol., **12:** 153–190, 1951.
8. Dounce, A. L. Enzyme studies on isolated cell nuclei of rat liver. J. Biol. Chem., **147:** 685–698, 1943.
8a. Dounce, A. L., and Fisher, N. E. Function of the protein component in enzymic catalysis. Enzymologia, **17:** 182–192, 1955.
9. Fahmy, A. R., and Walsh, E. O'F. The quantitative determination of dehydrogenase activity in cell suspensions. Biochem. J., **51:** 55–56, 1952.
10. Foster, R. J., and Niemann, C. The evaluation of the kinetic constants of enzyme-catalyzed reactions. Proc. Natl. Acad. Sci. U. S., **39:** 999–1003, 1953.
11. Glick, D. *Techniques of Histo- and Cytochemistry.* New York, Interscience Publishers, 1949.
12. Gomori, G. Alkaline phosphatase of cell nuclei. J. Lab. Clin. Med., **37:** 526–531, 1951.
13. Hogeboom, G. H., and Kuff, E. P. Relation between cell structure and cell chemistry. Federation Proc., **14:** 633–638, 1955.
14. Johnson, M. J. Oxidation-reduction potentials; *in* Lardy, H. A., ed. *Respiratory Enzymes*, rev. edit., pp. 58–70, Minneapolis, Minn., Burgess Publishing Co., 1949.
15. Kearney, E. B., and Singer, T. P. Prosthetic group of succinic dehydrogenase. Biochim. et Biophys. Acta, **17:** 596–597, 1955.
16. Keilin, D., and Hartree, E. F. Relation between certain components of the cytochrome system. Nature, **176:** 200–206, 1955.
16a. Lowe, H. J., and Clark, W. M. Oxidation-reduction. XXIV. Oxidation-reduction potentials of flavine adenine dinucleotide. J. Biol. Chem., **221:** 983–992, 1956.
17. Mahler, H. R., and Elowe, D. Metalloflavoproteins. II. The role of iron in diphosphopyridine nucleotide cytochrome c reductase. J. Biol. Chem., **210:** 165–179, 1954.
18. Mahler, H. R., and Green, D. E. Metalloflavoproteins and electron transport. Science, **120:** 7–12, 1954.
19. Michaelis, L. Fundamentals of oxidation and reduction; *in* Green, D. E., ed. *Currents in Biochemical Research*, pp. 207–228. New York, Interscience Publishers, 1946.
20. Michaelis, L., and Menten, M. L. Die Kinetik der Invertinwirkung. Biochem. Ztschr. **49:** 333–369, 1913.
21. Morales, M. F. If an enzyme-substrate modifier system exhibits non-competitive interaction, then, in general, its Michaelis constant is an equilibrium constant. J. Am. Chem. Soc., **77:** 4169–4170, 1955.

22. Morrison, M., and Stotz, E. Partition chromatography of hemins. Separation of the prosthetic groups of cytochromes a and a_3. J. Biol. Chem., **213:** 373–378, 1955.
23. Novikoff, A. B. Validity of histochemical phosphatase methods on the intracellular level. Science, **113:** 320–325, 1951.
24. Novikoff, A. B., *et al.* Biochemical heterogeneity of the cytoplasmic particles isolated from rat-liver homogenate. J. Histochem. Cytochem., **1:** 8–26, 1953.
25. Paigen, K. The occurrence of several biochemically distinct types of mitochondria in rat liver. J. Biol. Chem., **206:** 945–957, 1954.
26. Pauling, L. *The Nature of the Chemical Bond*, p. 64. Ithaca, New York, Cornell University Press, 1939.
27. Racker, E. Mechanism of action and properties of pyridine nucleotide-linked enzymes. Physiol. Revs., **35:** 1–56, 1955.
28. Rodkey, F. L. Oxidation-reduction potentials of the diphosphopyridine nucleotide system. J. Biol. Chem., **213:** 777–786, 1955.
29. Schneider, W. C., and Hogeboom, G. H. Cytochemical studies of mammalian tissues: The isolation of cell components by differential centrifugation: A review. Cancer Research, **11:** 1–22, 1951.
30. Schwartz, M. K., *et al.* The action of acid phosphatase from cancerous and noncancerous human prostate on various substrates. Cancer, **6:** 924–929, 1953.
31. Seligman, A. M., and Rutenburg, A. M. Histochemical demonstration of succinic dehydrogenase. Science, **113:** 317–320, 1951.
32. Smith, F. E. Tetrazolium salt. Science, **113:** 751–754, 1951.
33. Stern, H. The intranuclear environment. Science, **121:** 144–146, 1955.
34. Stern, H., *et al.* Some enzymes of isolated nuclei. J. Gen. Physiol., **35:** 559–578, 1952.
35. Strittmatter, C. F., and Ball, E. G. Hemochromogen component of liver microsomes. Proc. Natl. Acad. Sci. U. S., **38:** 19–25, 1952.
36. Thorn, M. B. Method for determining the ratio of the Michaelis constants of an enzyme with respect to two substrates. Nature, **164:** 27–29, 1949.
37. Tsou, C. L. Exogenous and endogenous cytochrome c. Biochem. J., **50:** 493–499, 1952.
38. Tuppy, H., and Bodo, G. Cytochrom c I. Über die der prosthetischen G. Gruppe benachbarten Amino-säurereste. Monatsh., **85:** 807–821, 1954.
39. Wainio, W. W. Reduction of cytochrome oxidase with ferrocytochrome c. J. Biol. Chem., **216:** 593–599, 1955.
40. Wilson, P. W. Kinetics and mechanisms of enzyme reactions; *in* Lardy, H. A., ed. *Respiratory Enzymes*, rev. edit., pp. 16–57. Minneapolis, Burgess Publishing Co., 1949.

CHAPTER 8

Hormones

In addition to enzymes, another group of chemical substances known as *hormones* acts to regulate the reactions continually proceeding in living tissue (19). It is important to differentiate clearly between these two types of chemical mediators.

(1) While enzymes are biosynthesized in all cells, hormones are generally produced by specialized glands within the body, such as the pituitary, the thyroid, and the adrenals.

(2) Enzymes exert their function in the cells where they are formed or extracellularly after leaving the parent cell. Hormones are carried by the bloodstream from the glands where they are formed to the cells within which they act. Because the hormone-producing glands secrete their products directly into the blood, they are often termed the *ductless glands* to differentiate them from such organs as the liver or salivary glands which deliver their secretions by way of a duct into regions other than the bloodstream. They are also called *endocrine glands*, because they secrete their products into the bloodstream which is within the body, whereas the glands with ducts secrete their products into spaces which are topologically outside the body. For the same reason, hormones (derived from a Greek term meaning "I arouse to activity") are sometimes called *internal secretions*.

(3) Enzymes are invariably proteins, whereas hormones although often proteins are not necessarily so.

The precise chemical functions of the various hormones have so far proven to be more elusive than those of the various enzymes. Whereas the reaction catalyzed by a given enzyme is usually well known, and indeed it is by means of the discovery of the reaction that the enzyme is discovered, the manner in which hormones influence reactions is not certainly known for even a single hormone. It is logical to suppose, however, that since they are active in exceedingly small concentrations they must act through enzyme systems, either as activators or inhibitors or, in the case of the protein hormones, conceivably as enzymes in their own right.

The effect of hormones on the body is usually marked, even startling. It is through their presence that body growth is regulated, that secondary sexual characteristics appear at puberty, that the body is physiologically mobilized to meet the stresses of fear and anger, and so on. The hormones of the human body may be divided into two large chemical classes, the protein hormones (including polypeptide hormones) and the steroid hormones, plus a few hormones which are simple amino acid derivatives.

THE PROTEIN HORMONES

The Pituitary Hormones

If there is a master organ in the body, one which might poetically be considered the seat of life, it is not the liver (as the ancient Greeks thought) or either the heart or brain (as popular modern thought might have it) but the obscure little gland known as the *pituitary* or the *hypophysis*. In man, it weighs just over half a gram and is situated beneath the brain in the hypophyseal fossa, and is attached to the base of the brain by a thin stalk. It is a well hidden, well protected, centrally placed, difficult-to-reach organ as though there had proved to be positive survival value in keeping the pituitary as safe and secure as possible. Many of its hormones have as their chief function that of controlling the secretion of hormones produced by the other endocrines, so that the pituitary would seem to be the master mediator of the body. The pituitary consists of an anterior lobe, a *pars intermedia*, and a posterior lobe, each of which produces its own characteristic hormone or hormones, so that functionally the pituitary is really a triple gland. The *pars intermedia* has no known significant function in the human. All pituitary hormones that have been identified are found to be proteins or peptides.

In carrying out its function as a master mediator, the pituitary is subject to controls and restraints. Master it may be, but no autocrat. It responds to neurohumoral stimuli from the nearby and closely connected hypothalamus, one of the more primitive centers in the nervous system. There is also a feed-back type of control exerted by the hormones produced by other endocrine glands under pituitary stimulation. Deficiency of gonadal hormone production, for example, leads to increased production of gonadotropic hormone by the pituitary, while excess hormone production by the gonads causes an inhibition of the output of the corresponding pituitary hormone.

The anatomical relationships of the hypothalamus to the pituitary, and the effects upon pituitary-controlled functions following injection of certain hypothalamic extracts (29), have raised the question of the possible endocrine activity of the hypothalamus. The evidence for the existence of a substance produced by the hypothalamus and activating the pituitary has been derived from physiological studies and sheds no light upon the chemical nature or action of such a substance.

Anterior pituitary hormones. What has been said about the pituitary gland applies particularly to the anterior lobe, from which about 25 different hormonally active fractions have been reported in the literature. Of these, six hormones have been well characterized and will be discussed here. Two of these may be listed as *gonadotropic hormones*, that is, hormones which stimulate the growth of the gonads and their production of sex hormones. These are *follicle-stimulating hormone* (FSH) and *interstitial-cell-stimulating hormone* (ICSH). FSH stimulates ovarian follicles and male germ cells, while ICSH (also known as luteinizing hormone or LH) similarly stimulates the interstitial cells of the testes or the formation of the corpus luteum. A third hormone affecting tissues associated with the reproductive process is *lactogenic hormone* (also known as prolactin), which initiates and maintains lactation. *Luteotropin*, which maintains the corpus luteum and promotes the formation of progesterone (see page 322), is probably identical with lactogenic hormone.

The other three hormones may be considered metabolic hormones, since they control the rate of metabolism, either indirectly through their stimulation of the thyroid or adrenal cortex in the case of *thyroid-stimulating hormone* (TSH) or *adrenocorticotropic hormone* (ACTH), respectively, or through some as yet unknown mechanism in the case of *growth hormone* (also called *somatotropic hormone* or STH).

These hormones are all produced by the chromophil cells of the anterior lobe. The acidophilic cells are associated with the secretion of the growth hormone, and certain special forms of the basophils with the secretion of the glycoprotein hormones (TSH, FSH, and ICSH) (50). The cellular source, acidophil or basophil, of the other hormones is not clearly ascertained. The chromophobe or neutrophilic cells of the anterior lobe show no particular evidence of secretory activity.

ICSH. Among the easily available animal sources of pituitary gland tissue, sheep pituitary is the richest in ICSH, possessing some five times the concentration of that in swine pituitary and about twelve times that in beef pituitary. Human pituitary is rather low in ICSH, being rather comparable to cattle in this respect. Pure preparations of ICSH have been prepared both from sheep and from swine, the proteins in each case behaving homogeneously when subjected to electrophoresis, the ultracentrifuge, and solubility tests. The hormone preparations from these two animals are quite distinct, differing markedly in molecular weight (sheep ICSH has a 40,000 MW, while that of swine is 100,000) and in isoelectric point (that of sheep ICSH is 4.6, while that of swine is 7.45). Both types of ICSH are glycoproteins containing both mannose and hexosamine. Sheep ICSH is about twice as rich as is swine ICSH in both these components, containing 4.5 per cent of mannose and 5.8 per cent of hexosamine as compared to the corresponding figures of 2.8 per cent and 2.2 per cent in swine. The activity of ICSH is destroyed when it is treated with ketene and by sufficiently vigorous treatment with cysteine. This is interpreted as indicating the essentiality of free amino groups and certain difficultly-reducible disulfide groups to the activity of the hormone.

FSH. Purified FSH has a molecular weight of 67,000 and an isoelectric point at pH 4.5; it is a glycoprotein containing both mannose and hexosamine. As in the case of ICSH, the pituitaries of swine and sheep are the best source, and the two gonadotropins can be separated by differential precipitation with ammonium sulfate. FSH is soluble in 50 per cent saturated ammonium sulfate solution, which is sufficient to precipitate ICSH. FSH, in fact, is the only anterior pituitary hormone which is not precipitated by half-saturation with ammonium sulfate. FSH is inactivated by cysteine as is ICSH. This would indicate that intact cystine links (S—S) are necessary to the hormone activity in each case. Unlike ICSH, however, FSH is not inactivated by ketene. FSH is more resistant than is ICSH to the action of trypsin (6). Human pituitary is rich in FSH.

A gonadotropin of non-pituitary origin appears in the blood and urine of pregnant women. Originally, it was thought to be identical with the pituitary agents, but it is now known to be formed in the chorion and is for that reason known as *human chorionic gonadotropin* (HCG). HCG has been crystallized in a form which is electrophoretically homogeneous (7). Earlier analyses of purified preparations indicated that HCG, like the pituitary gonadotropins, was a glycoprotein. It differs from ICSH and FSH in that it is richer in carbohydrate, since it contains 10 to 12 per cent hexose and 5 to 6 per cent hexosamine. It further differs in that the hexose is not mannose as in the case of the pituitary gonadotropins but is galactose. Its molecular weight is 60,000 to 80,000 and its isoelectric point is about 3.2. The clinical importance of HCG lies in the fact that during the second and third months of pregnancy sufficient hormone passes into the urine for that urine to show pronounced physiological effects upon suitable experimental animals. Thus, injection of urine of pregnancy into immature female mice or rats will result in the formation of corpora lutea or "blood points" in their ovaries within 96 hours. This was the first useful biological test for pregnancy, and it and later modifications are termed Aschheim-Zondek tests (A-Z tests) after the discoverers. Later modifications of the A-Z test stressed increased quickness of response. Male frogs will discharge spermatozoa into the cloaca as a specific response to HCG two hours after an injection of urine of pregnancy. HCG stimulates growth and endocrine function of the immature human testis, and is used in the treatment of undescended testicle and in some types of eunuchoidism (40). Normal adult interstitial cells respond to HCG by secreting increased amounts of androgens (page 325) and estrogens (page 318).

Lactogenic hormone. Again, in the case of this anterior pituitary hormone, sheep pituitary is a particularly rich source. Lactogenic hormone has been prepared in pure crystalline form (75). It is a protein of comparatively low molecular weight, various determinations yielding values between 22,000 and 32,000, the lower value probably being more nearly correct. Its isoelectric point is about 5.7. Its solubility properties are rather unusual, since it is extremely insoluble (0.01 per cent at 8°C.) in distilled water but soluble in methyl or ethyl alcohol in the presence of a small amount of acid. As far as is now known, lactogenic hormone is a simple protein and amino acids are its only hydrolysis products.

The reaction of lactogenic hormone with iodine or ketene results in the loss of activity indicating that tyrosine and free amino groups are essential to its hormone action. The hormone is comparatively resistant to heat. Activity of dilute solutions is maintained at a temperature of 60°C. over periods of an hour and at 100°C. for fifteen minutes. The hormone is somewhat more stable in acid than in alkaline solution.

Prolactin is concerned with the secretion of the mammary glands, but is not apparently in any way concerned with their development, which is under control of ovarian secretion, estrogens and progesterone. The stimulus for the release of prolactin seems to be of a duplex nature. One of the factors concerned is the fall in estrogen level which occurs at the time of childbirth and another seems to be a reflex stimulation from the act of nursing the baby. The commercial possibilities of prolactin applied to dairy cows were thought of quite a while ago. It does prolong the period of milk secretion. It will not prolong it indefinitely. Mammary glands must occasionally still be reconditioned either by a pregnancy or by properly spaced injection of estrogens and progesterone. The therapeutic use of prolactin in increasing milk production in the human is by no means uniformly successful. We still have to learn more about its necessary conditions of action.

TSH. The best practical source of TSH, also known as thyrotropic hormone, is the pituitary of cattle or swine. It has not yet been prepared in purified form so that its chemical properties are not very well known. It is an even smaller molecule than is lactogenic hormone, its molecular weight being estimated as about 10,000. It is a glycoprotein containing 2.5 per cent glucosamine. TSH can definitely be shown to bring about increase in the size of the thyroid, hypertrophy of its cells, and the signs of hyperthyroidism. The absence of TSH results in the direct opposite. The lowering of the basal metabolic rate (see page 754) of an animal or of a patient following complete loss of anterior lobe function (by hypophysectomy or otherwise) is identical numerically with the fall in metabolic rate that takes place following total thyroidectomy.

ACTH. The pituitaries of swine and sheep are used as sources for ACTH. ACTH extracts, purified by adsorption on oxycellulose (i.e., cellulose which has been oxidized by NO_2 so as to convert most or all of the glucose residues to glucuronic acid residues), are termed *corticotropin A*. Corticotropin A can be further separated by countercurrent distribution into at least eight active peptide fractions. Of these the most common is *beta-corticotropin*, which makes up a third of the total ACTH.

Beta-corticotropin is a polypeptide made up of 40 amino acids (32). The sequence is, beginning with the N-terminal: ser-tyr-ser-met-glu-his-phe-arg-try-gly-lys-pro-val-gly-lys-lys-arg-arg-pro-val-lys-val-tyr-pro-ala-asp-gly-ala-glu-asp-glu. NH_2-leu-ala-glu-ala-phe-pro-leu-glu-phe. It has been suggested that the sequence, -lys-lys-arg-arg-, which introduces a local area of highly basic side chains is what makes it possible to adsorb ACTH on the acidic oxycellulose or on acidic ion-exchange resins.

Pepsin will hydrolyze portions of the beta-corticotropin molecule at the C-terminal end. As many as eleven amino acids can be broken off without seriously reducing ACTH activity, which is thus localized to the 29-amino acid peptide at the N-terminal end (11).

An important mechanism by which the adrenal cortex is stimulated is through the secretion of pituitary ACTH, which acts to increase the activity and size of the adrenocortical cells. In hypophysectomized rats, the adrenal cortex atrophies, but can be restored both in structure and function by injections of ACTH. Growth hormone (see below) is also active in adrenal maintenance, performing this task by increasing the mitotic rate. The two hormones together act synergistically. That is, the increase in adrenal weight following an injection of both is greater than the sum of the increases following the injection of each singly. Some of the adrenal secretions tend to inhibit ACTH production by the pituitary so that a feedback mechanism is set up. Adrenal overactivity depresses ACTH production, which cuts down adrenal activity. Adrenal underactivity decreases the inhibition effect, which increases ACTH production, which increases adrenal activity. This type of stabilizing interaction between various hormones is quite common.

Growth hormone. Of all the hormones of the pituitary, and perhaps of the entire organism, growth hormone is the most spectacular in its effects when present in greater or less than the normal amount. Growth can be briefly described in chemical terms as the conversion of amino acids into proteins and the orderly storage of such proteins. Congenital deficiency of growth hormone results in dwarfism. The state of *pituitary infantilism* presumably occurs from the absence of growth and gonadotropic hormones.

In some cases, although not in all, the pathologist is able to demonstrate atrophy or destruction of the anterior lobe. In the classical picture of pituitary infantilism, the body is properly proportioned, although small.

Conversely, *gigantism* occurs either from hyperplasia or adenoma of the acidophil cells. In addition to the oversized body, one usually notes increased basal rate, hyperglycemia, glycosuria, and low sugar tolerance. All of these are brought about by the usually associated excess production of other tropic hormones aside from the growth or somatotropic hormone. A late stage in this disease may be a degeneration of the anterior pituitary and the reversal of the metabolic findings. Hyperproduction of growth hormone in adults who have previously been normal in this respect stimulates the growth of those parts of the skeleton not yet so mineralized as to be incapable of further extension. The main enlargements take place in the feet, hands, and lower jaw, and the condition is known as *acromegaly*. With the progress of the disease, the chemical findings are frequently reversed, as in gigantism. Growth hormone is particularly concerned with skeletal growth. Uptake of radioactive Ca^{++} by bone in hypophysectomized rats is doubled when they are regularly injected with growth hormone (71).

Growth hormone is one of the anterior pituitary hormones to be prepared in pure form (38), the protein proving homogeneous by electrophoresis and solubility criteria. Its molecular weight is high for an anterior pituitary hormone, about 45,000. Its isoelectric point has been reported as 6.85. It contains no carbohydrate. Amino acid analysis indicates the molecule to contain 396 residues but, as in the case of ACTH, not all are necessary for activity. Considerable hydrolysis by chymotrypsin leaves peptide portions with full activity (37a, 38a).

The growth hormone is still present in the glands of normal adult animals who have stopped growing. So the presence of some sort of inhibitor of the growth hormone which becomes effective at maturity is logical. One immediately thinks of the sex hormones as such a possible inhibitor and indeed, experimentally, the sex hormones do have an inhibiting effect on the growth hormone of the anterior lobe. But this again leads us into a logical trap because castrated animals or castrated humans, although a little overgrowth may be observed, certainly do not consistently, or for any great period of time, show continued increase in size. So this is one of the unsolved problems. What is it in the normal adult man or animal that, at the proper time, and in almost all instances, holds in check the activity of the growth hormone which can be demonstrated with very little trouble in the adult pituitary gland? The size of almost any animal appears to bear a close relationship to the size of his pituitary gland and whether we deal with species now living or whether we examine skeletons of fossilized vertebrates, the relationship between the size of the sella and the size of the animal remains mathematically proportional (18).

In pan-hypopituitarism or *Simmonds' disease* we have the unmistakable signs of extreme and often complete functional failure and often structural absence or structural destruction of the anterior lobe. These patients are characterized by their extreme emaciation, and extreme or complete regression of the gonads. All metabolic activities are depressed, and there is hypoglycemia. Like rats that have had their anterior lobe removed, they develop a profound anemia. There are rather few really verified cases where the pathologist has actually found destruction of the anterior lobe. Emaciation and loss of practically all body functions certainly can often arise and probably oftener arise from causes other than pan-hypopituitarism.

Posterior pituitary hormones. Extracts of the posterior pituitary lobe have been found to exhibit at least three effects in mammals: (a) excitation of uterine muscle contractions; (b) increase of the blood pressure; and (c) antidiuresis. Two hormones have been isolated. One, *oxytocin*, is the posterior pituitary principle to which uterine contractions can be attributed, while the other, *vasopressin*, is held responsible for the increase in blood pressure. In the purification of these substances, antidiuretic activity remains with the vasopressin fraction, indicating that the antidiuretic and pressor effects are both produced by vasopressin (48).

It is in accord with the available evidence (58) to consider the posterior pituitary as the depot for storage and release of these hormones. Their site of production may be in cells of the supraoptic and paraventricular nuclei, and they may migrate to the posterior pituitary along the axons of the supraopticohypophyseal tract.

Oxytocin is an octapeptide with the structure shown in formula I. The structure has been proved by synthesis (17) and the synthetic product shown to be identical chemically and in hormone effect with the oxytocin isolated from beef or hog pituitaries.

Vasopressin has been shown, by similar methods, to have a structure like that of oxytocin, but with phenylalanine substituted for isoleucine, and arginine (in beef) or lysine (in hog) for leucine. Purified or synthetic oxytocin has no pressor or antidiuretic effects, but purified vasopressin possesses oxytocic activity varying, according to the assay method used, from 0.05 to 0.2 of that of oxytocin (48). Both varieties of vasopressin, when administered intravenously to human subjects cause a significant increase in the plasma 17-hydroxycorticosteroids (40a).

Melanocyte-stimulating hormone (MSH) has been isolated from hog posterior pituitary (36). It is a peptide made up of 34 amino acids, and with a molecular weight of 4,500. Its most pronounced effect is that of causing an expansion of the melanophores of many chordates. In man (37), it causes

```
                 leucine

                  CH3
                   |
                  CH—CH3      glycinamide
                   |
                  CH3
                   |
CO—NH—CH—CO—NH—CH2—CO—NH2
|
CH—CH2
|      \
|       CH2
|      /
N——CH2
|
CO                         asparagine
|
CH—NH—CO—CH—NH—CO
|         |     |
CH2       CH2   CH—CH2—CH2—CO—NH2        glutamine
|         |     |
S         CO    NH
|         |     |
S         NH2   CO      CH3
|               |      /
CH2             CH—CH
|               |      \
NH2—CH—CO—NH—CH—CO—NH    CH2—CH3          isoleucine
              |
             CH2
              |
          (C6H4)
              |
             OH

          tyrosine
```

I. Oxytocin

darkening of the skin. An earlier name for the hormone is *intermedin* because of its location in the *pars intermedia* of the pituitary of various cold-blooded chordates. In none of the hormones of the posterior lobe or pars intermedia are any hydrolysis products other than the amino acids or ammonia found.

The amino acid sequence of a peptide with MSH activity has been found to be: asp-glu-gly-pro-tyr-lys-met*-glu*-his*-phe*-arg*-try*-gly*-ser-pro-

pro-lys-asp (27a). The portion of this sequence indicated by asterisks also occurs in corticotropin (see above) and this may account for the fact that even pure samples of corticotropin show some MSH activity (11a).

The Thyroid Hormone

The rate at which oxidative metabolic reactions in general proceed in the body is determined by the endocrine activity of the thyroid gland. Following experimental thyroidectomy, the rate of oxygen consumption by the resting, fasting animal (more succinctly, the basal metabolic rate as explained in Chapter 19) is notably decreased below the rate observed in a normal or euthyroid animal. The same observation has been repeatedly made in human patients who had been subjected to total or partial thyroidectomy. Feeding of thyroid substance to such animals or patients leads to a predictable and measurable increase in basal metabolic rate. Feeding of adequate amounts of thyroid substance to normal animals or men will increase the metabolic rate above normal. Measurement of oxygen consumption of animal tissues *in vitro* confirms the observations made on the intact animal. Tissues of thyroid-deficient or *hypothyroid* animals show diminished rates; and those from animals rendered *hyperthyroid* by thyroid feeding show increased rates as compared with euthyroid controls. Human patients may develop spontaneous hypothyroidism, which responds favorably to the feeding of thyroid, or they may develop spontaneous hyperthyroidism, which may be treated by thyroid-blocking drugs or by partial removal or destruction of the thyroid by physical means. The chemical mode of action of the thyroid hormone, which is not clearly understood, is probably related to the increase of coenzyme A in tissues of thyroidectomized animals after administration of thyroid hormone (68) and the uncoupling, observed *in vitro*, of oxidation and phosphorylation by thyroid hormone (30). It has also been suggested that the primary site of action of thyroid hormone is on the TPN-cytochrome *c* reductase of liver tissue (44a).

Inadequacy of thyroid hormone permits certain disorganizations of structure to occur in special tissues, in addition to the general lowering of metabolic rate. *Myxedema* is the elaboration and deposit of an abnormal protein in intercellular spaces. *Epiphyseal dysgenesis* is a change from the orderly calcification which occurs at the epiphyses of normally-growing bone, to a diffuse and irregular distribution of areas of calcification. These disturbances can not be corrected by raising the metabolic rate by nonhormonal drugs. Because *thyroxine* as well as certain other iodinated amino acids will act to relieve the conditions arising from hypofunction of the thyroid, a certain confusion arises as to exactly what chemical entity may be referred to as the hormone. Salter (56) suggested that "thyroid hormone" be considered a generic term to indicate any substance that will relieve human myxedema when properly administered. Included in such a term would therefore be *thyroglobulin* (the iodinated protein biosynthesized by the thyroid gland), various thyroglobulin degradation products, artificially iodinated proteins, thyroxine and related amino acids, and the proteins which act as carriers for hormone activity through the bloodstream and into the cell.

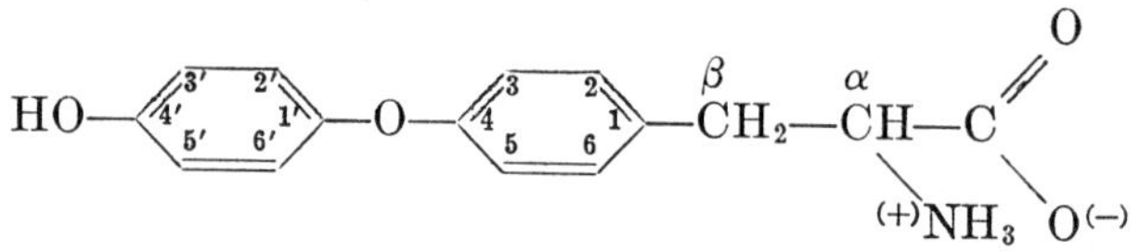

$$HO-\langle 3',2',1',6',5',4' \rangle-O-\langle 3,2,1,6,5,4 \rangle-\overset{\beta}{CH_2}-\overset{\alpha}{CH}(^{(+)}NH_3)-C(=O)O^{(-)}$$

II. Thyronine

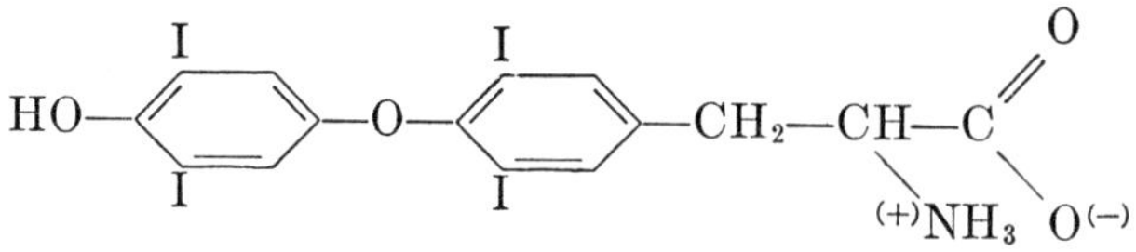

$$HO-\langle 3'\text{-}I,5'\text{-}I \rangle-O-\langle 3\text{-}I,5\text{-}I \rangle-CH_2-CH(^{(+)}NH_3)-C(=O)O^{(-)}$$

III. Thyroxine

$$HO-\langle 3'\text{-}I \rangle-O-\langle 3\text{-}I,5\text{-}I \rangle-CH_2-CH(^{(+)}NH_3)-C(=O)O^{(-)}$$

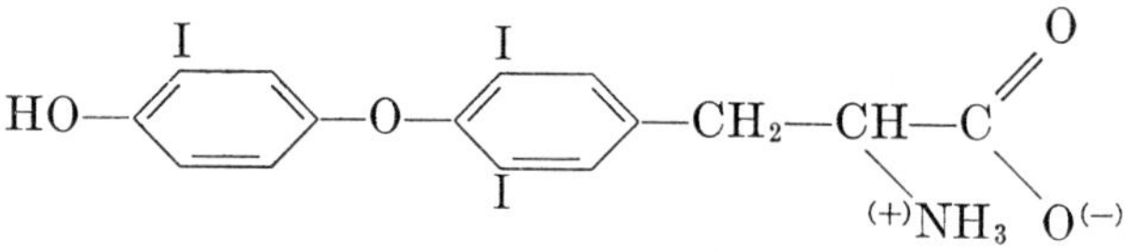

IV. 3,5,3′-Triiodothyronine

Iodine-containing amino acids. The amino acids which display significant thyroid hormone activity are halogen derivatives of thyronine (formula II). The best known of these is L-thyroxine, or 3,5,3′,5′-tetraiodo-L-thyronine (formula III), which was isolated and named by Kendall in 1915. For many years thyroxine was the most active known thyroid hormonal substance. A more potent compound is 3,5,3′-triiodo-L-thyronine (TITh) (formula IV), which has 3 to 8 times the activity of L-thyroxine when assayed by various methods. Both of these iodothyronines can be demonstrated in alkaline or enzymatic hydrolysates of animal thyroid, indicating that they are amino acid components of the iodine-containing protein, thyroglobulin. Their relative proportions in the thyroid protein are about one triiodothyronine residue to 20 thyroxine residues. These two compounds account for nearly all the hormonal activity of thyroid substance, in which they are bound as components of thyroglobulin, and for the hormonal activity of circulating blood, in which they are bound to plasma protein.

Even in the absence of the thyroid gland, thyroxine can be converted to triiodothyronine in tissues other than the thyroid (47). Both thyroxine and triiodothyronine are excreted in the bile, previous to which they are in part deiodinated and in part converted to glucuronide conjugates.

The hormonal potency of synthetic D-thyroxine and D-triiodothyronine is insignificant compared to the natural L-isomers. The activity of 3,5-diiodo-L-thyronine is 4 per cent of that of L-thyroxine. If the four iodines of L-thyroxine are replaced by four bromines, or if the ether oxygen binding the two aromatic rings is replaced by a sulfur, traces of hormone activity (as judged by the ability to relieve human myxedema) remain. If the two benzene rings are each attached directly to the beta-carbon, or if three benzene rings (rather than two) are strung in a row, all activity is lost. If the phenol OH is transferred to the 6′ position, 4 per cent of the activity remains, but if it is transferred to the 5′ position while the two neighboring iodine atoms are shifted to 4′ and 6′, all activity is lost.

In experiments on tadpoles (whose metamorphosis into frogs can be hastened by thyroid hormone), hormone activity was detected in compounds not containing the thyronine skeleton, as for instance in dibromotyrosine and diiodotyrosine (formula V). The amphibian test is, however, less spe-

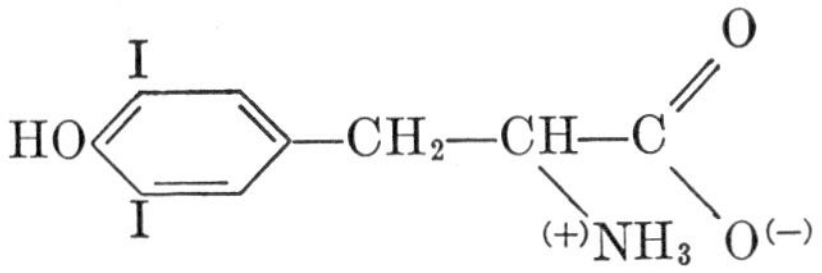

$$HO-C_6H_2I_2-CH_2-CH(NH_3^{(+)})-C(=O)O^{(-)}$$

V. Diiodotyrosine (DIT)

cific than is that of the relief of human myxedema, and by the former criterion even potassium iodide shows 10 per cent of the hormone activity of thyroxine. It seems probable that compounds other than thyroxine and triiodothyronine show hormone activity in a manner related to the efficiency with which the body can convert them to these compounds.

The steps by which thyroid hormone is formed may be listed as follows (27):

1. The concentration of iodide ion in the thyroid;
2. The oxidation of iodide ion to iodine;
3. The iodination of tyrosine to monoiodotyrosine (MIT) and to diiodotyrosine (DIT);
4. The coupling of two molecules of DIT to form thyroxine.
5. The deiodination of thyroxine to TITh.

No free iodine has been located in the thyroid and a plausible explanation for that is that iodine is incorporated into the tyrosine molecule at a rate potentially considerably greater than its formation from iodide ion. The iodination of tyrosine is catalyzed by the enzyme, *tyrosine iodinase*, which occurs in the thyroid. It occurs also in the salivary glands (21) where it may catalyze the reverse deiodinating reaction.

In a hypophysectomized animal, the binding of iodide ion is decreased but not abolished. In the coupling of two molecules of DIT to form thyroxine a dehydroalanine molecule is split out, which may either be hydrolyzed to pyruvic acid and ammonia or reduced to alanine (formula VI). Such coupling proceeds even more readily when the DIT is part of a peptide chain.

There are two possibilities for the mechanism of formation of TITh: a) that TITh is formed by the coupling of DIT with MIT and b) that TITh

$$\text{DIT} + \text{DIT} \xrightarrow{-2H} \text{Thyroxine} + \text{Dehydroalanine } (CH_2{=}C(NH_2){-}COOH)$$

DIT DIT Thyroxine Dehydroalanine

VI. Coupling of DIT to produce thyroxine

is formed by the deiodination of thyroxine. The weight of the evidence seems now to be in favor of the latter suggestion (63), since deiodinating systems have been demonstrated in liver and other tissues of the rat which are capable of deiodinating thyroxine to TITh and of deiodinating TITh further.

Thyroglobulin. The thyroid gland contains iodothyronines and iodotyrosines as part of a large protein molecule, thyroglobulin, which apparently serves as a storage place for hormone activity. The isoelectric point of the protein is at a pH of about 5. Thyroglobulin contains 0.6 per cent iodine, and since its molecular weight is 700,000, each molecule must have about 32 iodine atoms. A technique has been worked out whereby the hydrolyzed thyroid gland of animals who have been treated with radioactive iodine is fractionated by paper chromatography and the iodine fractions visualized by the effect of the radioactive radiation upon a photographic plate (69). In this way, it was determined that 15 per cent of the iodine in rat thyroid existed as monoiodotyrosine (formula VII), and 30 per

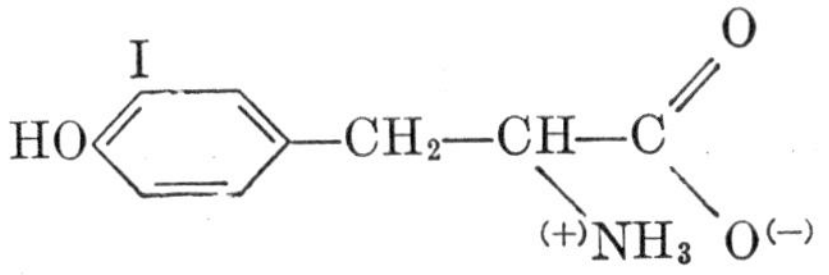

$$HO\langle\text{(I)-ring}\rangle—CH_2—CH(NH_3^{(+)})—C(=O)O^{(-)}$$

VII. Monoiodotyrosine (MIT)

cent as diiodotyrosine. The method was not entirely satisfactory for the determination of thyroxine, but it was estimated that at least 20 per cent and probably much more of the iodine was in that form. About 4 per cent of the thyroxine was later shown to be in the form of 3,5,3′-triiodothyronine. No 3,5-diiodothyronine (that is, thyroxine minus the two iodines on the outermost ring) was located. Iodotyrosines and iodothyronines exist free in the thyroid only to the extent of about one per cent of the total iodine. Free iodide ion makes up another one per cent. The remaining 98 per cent is in the form of thyroglobulin. Monoiodohistidine is detectable in small amounts in the enzymic hydrolysate of rat and dog thyroglobulin, with no known function as a hormone or hormone precursor (52). In addition to these iodinated amino acids, analysis of thyroglobulin indicates the presence in each molecule of 120 cystine residues, 110 tyrosine residues, and 60 residues apiece of tryptophane and methionine. It is particularly high in arginine which makes up 12.4 per cent of the total by weight, and this amounts to 550 residues per molecule (3). Thyroglobulin is a glycoprotein containing 8.3 per cent carbohydrate. Of the total molecule, 4.0 per cent is glucosamine, 2.7 per cent is mannose, 1.2 per cent galactose, and 0.4 per cent fucose (25a).

The non-iodinated portions of thyroglobulin are not at all essential to thyroid hormone activity. Proteins such as egg albumin, serum albumin, and casein can be treated with iodine and converted into iodinated derivatives which exhibit thyroid hormone activity. The reason for this is that the added iodine will first enter the tyrosine residues of the protein molecule to form diiodotyrosine, and then by some oxidative process not yet completely elucidated will form thyroxine directly in the polypeptide chain. Proteins containing as much as 4 per cent thyroxine have been produced through iodination (51); and it has been shown that diiodotyrosine could to a small extent be converted to thyroxine by alkaline iodination in the test tube.

Salter (57) has suggested that the globulin portion of thyroglobulin is synthesized in the thyroid gland independently of iodine metabolism. Such an iodine-free protein would be a specially designed "iodine trap" which could rapidly fix iodine absorbed from the foodstuffs. The thyroid gland will also collect astatine (27), the unstable halogen below iodine in the periodic table, but not the smaller halogens. Thyroglobulin itself, because of its large molecular weight, can not be expected to be secreted as such into the bloodstream. Secretion is preceded by proteolysis (10), by an enzyme controlled by pituitary TSH, into polypeptides small enough in molecular weight to diffuse readily through the cell walls into the bloodstream.

Thyroid-blocking agents. The existence of chemical agents which block the action of the thyroid gland is of interest to the physician in that it offers a method for countering hyperfunction of the gland without resort to surgery. Pitt-Rivers (46) divides such agents into three groups.

(1) Thiocyanate ion, which by its resemblance in size and charge to iodide ion apparently interferes with the mechanism of iodide collection by the thyroid through competitive inhibition. Thiocyanate is metabolized by thyroid tissue and fixed in organic combination, whereas other tissues do not fix thiocyanate sulfur. In large doses it will even cause the release of some iodide already concentrated in the thyroid.

(2) Radioactive iodine, I^{130} or I^{131}, which by destroying thyroid tissue mechanically reduces the ability of the gland to manufacture hormone.

(3) A variety of compounds, most of which contain sulfur and which directly or indirectly inhibit hormone synthesis. Since such inhibition tends to induce goiter (an increase in the size of the thyroid), as though the gland attempts by this means to restore its hormone production to normal, these thyroid-blocking agents are sometimes referred to as *goitrogens*.

The most important goitrogens are derivatives of thiourea (also called thiocarbamide), such as *thiourea* itself (formula VIII), *2-thiouracil* (formula IX), or *2-mercapto-imidazole* (formula X). An example of a sulfur-containing goitrogen which is not a thiourea derivative is *2-aminothiazole* (formula

```
        NH2                       NH—C(=O)
       /                         /       \
  S=C                       S=C           CH
       \                         \       //
        NH2                       NH—CH
```

VIII. Thiourea

IX. 2-Thiouracil

```
        NH—CH                  CH——S
       /    ||                 ||    |
  S=C       ||                 CH    C—NH2
       \    ||                   \  //
        NH—CH                      N
```

X. 2-Mercapto-imidazole

XI. 2-Aminothiazole

XI). Various sulfonamides and even such a non-sulfur-containing compound as p-aminobenzoic acid have been shown to possess goitrogenic activity.

It is thought that in the case of the thiourea goitrogens, at least, interference occurs in the step involving oxidation of iodide to iodine (see page 311) (1) because of the reducing action of the sulfhydryl group. Thiourea, for instance, will reduce iodine rapidly to form formamidine disulfide hydroiodide (formula XII). Similar reactions taking place in the body would obviously reduce the amount of iodine available for the iodination of tyrosine.

Goitrogens have been found to occur naturally in certain foods, notably rutabaga and turnips. The active goitrogenic principle in turnips has been isolated and characterized (26) as a sulfur-containing compound which is not, however, a thiourea derivative. It is 5-vinyl-2-thio-oxazolidone (formula XIII). The recognized cause of most cases of simple goiter is iodine deficiency. The use of foods containing natural goitrogens is an additional cause which may produce or increase goiter even with adequate iodine intake.

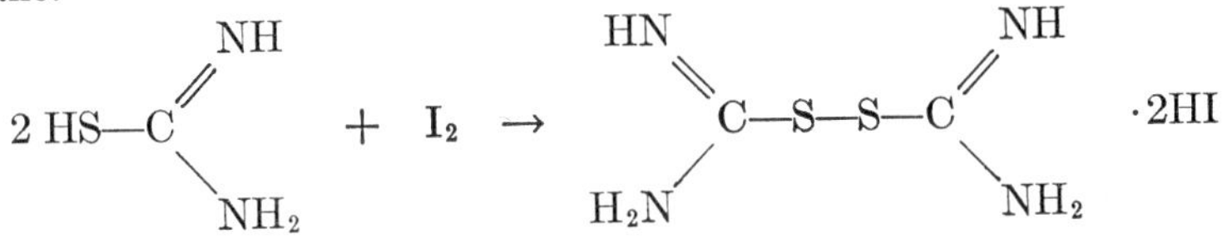

$$2\,HS{-}C(=NH){-}NH_2 + I_2 \rightarrow (HN{=})(H_2N)C{-}S{-}S{-}C(=NH)(NH_2) \cdot 2HI$$

Thiourea
(Isothiourea)

XII

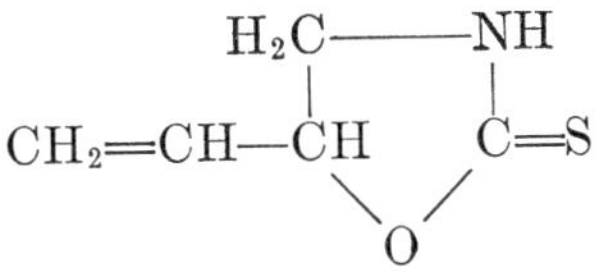

$$CH_2{=}CH{-}\underset{\text{ring: } CH{-}CH_2{-}NH{-}C(=S){-}O}{CH}$$

XIII. 5-Vinyl-2-thio-oxazolidone

The Pancreatic Hormones

While the pancreas is primarily a ducted gland (second only to the liver in size) discharging a secretion containing a number of digestive enzymes into the duodenum, the groups of specialized cells within the pancreas known as the *islets of Langerhans* elaborate two hormones, *insulin* and *glucagon*, and secrete them directly into the blood stream.The islets constitute from 0.9 per cent to 2.7 per cent of normal pancreatic tissue, and in diabetics the percentage is usually below 0.9 per cent. From 250,000 to 2,500,000 individual islets exist in the human pancreas.

Insulin. The function of insulin is to promote the uptake of glucose by cells and the metabolism of glucose within cells. In this way it acts to decrease the amount and concentration of glucose in the circulating blood. The assay of insulin preparations is on the basis of the lowering of blood sugar in rabbits, and is expressed in units, defined as the lowering of blood sugar produced by 0.125 mgm. of the international standard insulin preparation. The known facts and the theories of its action will be explained in later portions of this book. Insulin is secreted by the beta cells of the islets of Langerhans and is a protein having its isoelectric point at a pH of 5.3. Its molecular weight as judged from ultracentrifugal sedimentation is 46,000 (41), but the molecule has been found to consist of conglomerates of sub-molecules of 12,000 molecular weight which are held together by electrovalent bonds. The sub-molecules consist of four polypeptide chains bound together by disulfide linkages (see page 88). There is some evidence that the smallest active sub-molecule of insulin is of 6,000 molecular weight and composed of two polypeptide chains (61).

The activity of insulin is destroyed by alteration of the free amino groups, as by reaction with formaldehyde, or by the reduction of the disulfide linkages through the action of such chemicals as cysteine, thioglycolic acid, or leucomethylene blue. It is inactivated by pepsin and chymotrypsin, but is resistant to tryptic digestion.

Zinc has been considered to form an integral portion of the insulin molecule, although neither the exact manner in which it is bound nor the manner in which it fulfills a function is known certainly. Utilizing radioactive zinc, the zinc content of crystalline insulin prepared in various ways has been shown to vary from 0.3 per cent to 0.6 per cent depending upon the pH of crystallization (8), although a crystalline insulin containing only 0.15 per cent has been reported (55). The zinc content of human pancreas ranges from 18.5 to 30.4 mgm. per kgm. of fresh gland. Insulin will also crystallize as a salt of other bivalent ions such as those of cobalt or cadmium, and will form a complex with protamine which is of clinical importance since protamine-insulin maintains hormonal activity after injection longer than insulin itself (see Chapter 14).

Insulin is comparatively resistant to denaturation by organic solvents, dilute acids, or by film formation. If it is heated in weakly acid solution, inactive fibrils are formed which aggregate into insoluble spherites (74). This change is reversible and fully active insulin can be regained upon gentle treatment with alkali.

Glucagon. This hormone, secreted by the alpha cells of the islets of Langerhans, has an action reciprocal to that of insulin. Glucagon secretion is stimulated by growth hormone (23) and this in turn may account for the insulin-blocking or diabetogenic effect of growth hormone which some have reported. By increasing the amount of the active form of a liver enzyme, *phosphorylase*, glucagon increases the rate of breakdown of hepatic glycogen, causing the glucose of the circulating blood to increase. The increase in blood glucose induced by injections of glucagon is associated with a prompt suppression of gastric hunger contractions and a decrease in the feeling of hunger (66). This same effect may be augmented by other actions (9); inhibition of glycogen formation and direct antagonism to insulin. It had long been known that preparations of insulin usually carried a "contaminant" which transiently elevated the blood sugar, but the crystallization of glucagon, also called the hyperglycemic-glycogenolytic factor of the pancreas (HGF), was not reported until 1953 (65). The crystallized material is a protein or polypeptide. The amino acid sequence of glucagon is reported to be: his-ser-glu·NH_2-gly-thr-phe-thr-ser-asp-tyr-ser-lys-tyr-leu-asp-ser-arg-arg-ala-glu·NH_2-asp-phe-val-glu·NH_2-try-leu-met-asp·NH_2-thr (3a).

The Parathyroid Hormone

The parathyroid glands are small glands, two to four in number, usually situated adjacent to the dorsal surface of the thyroids. Their total weight is only 100 to 200 mgm. The secretion of the parathyroids stimulates proliferation of the osteoclasts of cancellous bone, increasing the liberation of calcium and phosphate ions from bone (62); there is also stimulation of urinary excretion of phosphate to such an extent that serum phosphate actually falls in hyperparathyroidism. High urinary calcium output is characteristic of hyperparathyroidism. The removal of the parathyroids from the body results in hypocalcemic tetany.

The nature of parathyroid hormone (PTH) is less well understood than that of any of the hormones thus far discussed. Investigation has scarcely reached beyond the point of demonstrating the hormone to be protein in nature. That it is a protein is indicated by the fact that hormone activity is lost when active parathyroid extracts are exposed to peptic or tryptic digestion. Furthermore, such protein denaturants as acid and alkali destroy activity, while protein precipitants remove activity from solution. The hormone is not a glycoprotein, since active extracts react negatively to carbohydrate-detecting reagents.

PTH is less stable in neutral or alkaline media than in slightly acid media. Exposure of hormone activity to various reagents indicates that the hormone is stable to reducing agents and unstable to oxidizing agents, that free amino groups are essential to its activity, and that disulfide linkages are not present.

The Gastrointestinal Hormones

The chemical substances described in Chapter 13 as gastrointestinal hormones are somewhat atypical of hormones as a class, since they are not elaborated by special glands. The gastrointestinal mucosa in general seems their region of origin. Nevertheless, in one important respect they are hormone-like in their action, since they are secreted directly into the blood by means of which they are carried to the particular region of the digestive tract where they are effective. They therefore can be considered internal secretions. They are all polypeptide in structure with relatively low molecular weights—secretin, for example, has a molecular weight of about 5,000.

If we now pause to view the protein hormones as a group, one important property they seem to hold in common is that of low molecular weight. Many seem to be polypeptides rather than proteins in the more common sense of that word, and few have molecular weights of more than 20,000. Where larger molecular weights exist, there is the possibility that the isolated molecules represent the "storage" form existing within the hormone-synthesizing cells and that the active hormone activity resides in polypeptide components of the protein that fragment off and enter the bloodstream. This is certainly the case in thyroid hormone and posterior pituitary hormone, and probably the case in ACTH and parathyroid hormone. This small size would seem to be a necessity, of course, if hormones are to be proteins or polypeptides and yet be able to diffuse through the cell walls of the gland and blood vessels into the bloodstream and then from the bloodstream into the receptor cells.

THE STEROID HORMONES

Almost all hormones which are not proteins or polypeptides are steroids. The endocrine glands which secrete steroid hormones are the ovaries, the testes, and the adrenal cortices. The methods for bioassay of the steroid hormones have been critically reviewed by Dorfman (15).

The Ovaries

The reproductive function of the human being is not a vital part of the individual's life, regardless of how reluctant the individual may be to have it ignored completely, and is not a matter with which the body is continually preoccupied, as it is with respiration and excretion. A woman, for instance, is capable of bearing young for only about thirty years of a life of which the average expectancy in the United States is nearly seventy. Even within that period a complex four-week cycle exists, during which she may be fertile for only a day or two.

The onset of the menstrual cycle in adolescence, its cessation in middle age, the details of change within each cycle, and the interruption of the cycle by the advent of pregnancy and parturition all involve profound physiological and even anatomical changes. To induce such changes in individual parts of the body without at any time allowing the organism as a whole to have its metabolic machinery seriously interfered with is an important function of those chemical messengers and regulators, the hormones. The ovaries and testes in addition to their reproductive function also elaborate a variety of important hormones.

The ovarian hormones fall into two classes, the estrogens and progesterone. The *estrogens* are biosynthesized by the maturing ovarian follicle, and by the corpus luteum cells which arise from the matured follicle. *Progesterone* is formed in the corpus luteum, and therefore in significant amounts only during the luteal phase of the menstrual cycle. The production of these ovarian hormones is regulated by the gonadotropins of the anterior pituitary.

Estrogens. Three closely related steroids with estrogenic activity are formed in the human ovary, differing only in the number of hydroxy or keto groups present on the steroid nucleus (CPP). They are *estradiol* (formula XIV), *estrone* (formula XV), and *estriol* (formula XVI). If we compare the three we see that estradiol and estrone differ only in that the 17-hydroxyl of the former is oxidized to a 17-ketone in the latter. Estriol differs from estradiol in the possession of a third hydroxyl group on carbon 16. As a group, the three estrogens differ from most other steroids in that ring A is fully aromatic so that these steroids lack the usual angular methyl on carbon 10. This aromaticity is carried even further in *equilin* (formula XVII) and *equilenin* (formula XVIII), related estrogens which occur in the horse but not in the human. Estrone and estradiol occur in all species studied, but estriol seems characteristic of the human (43). Estradiol-17β has been identified in human ovaries, estradiol-17α in mare's urine. The estrogens may be regarded as derivatives of the hypothetical saturated hydrocarbon *estrane* (formula XIX).

The estrogens control the development of secondary sexual characteristics in the female and are responsible, as well, for the maintenance of the menstrual cycle. After ovariectomy the cycle is interrupted and can be restored by injections of estrogens, which will also reverse the consequent atrophy of organs essential or accessory to reproduction—as the uterus, vagina, oviducts, and breasts. Circulating estrogens are also believed to be responsible for rebuilding the endometrium after menstruation and for sensitizing the pregnant uterus to the action of oxytocin. Estrogens inhibit pituitary gonadotropic hormone production and in high concentrations prevent ovulation. Estradiol is the most active of the estrogens, with estrone and estriol following in that order.

The estrogens are not formed in the ovaries alone. Estrogen excretion rises considerably during the course of pregnancy, even in women subjected to bilateral ovariectomy during pregnancy, and drops very rapidly after parturition, indicating that the placenta also is involved with estrogen

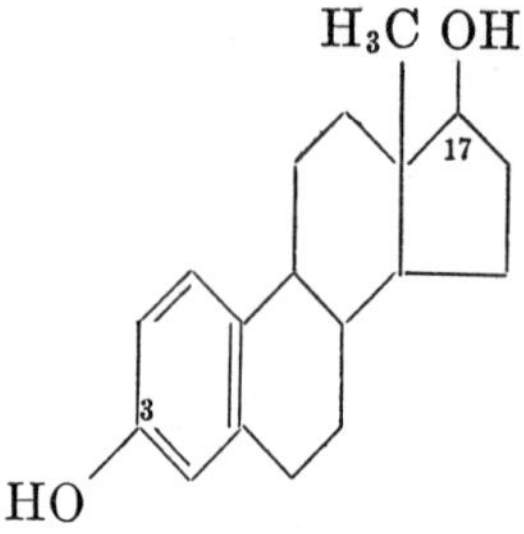

XIV. Estradiol

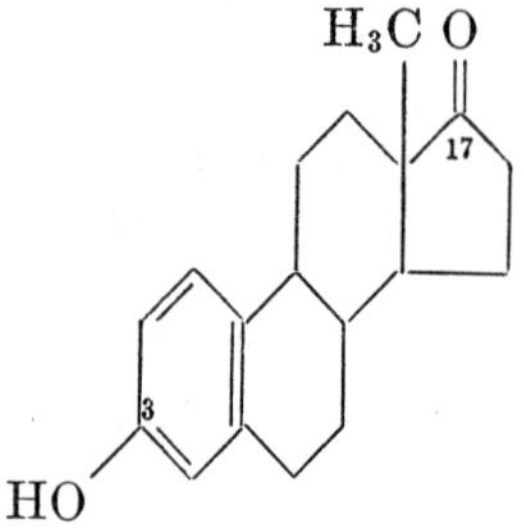

XV. Estrone

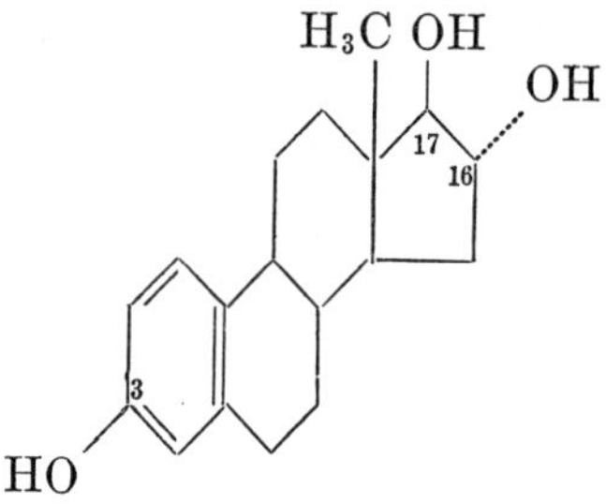

XVI. Estriol

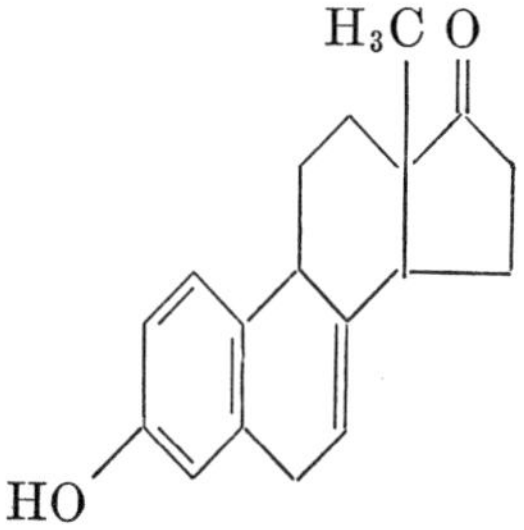

XVII. Equilin

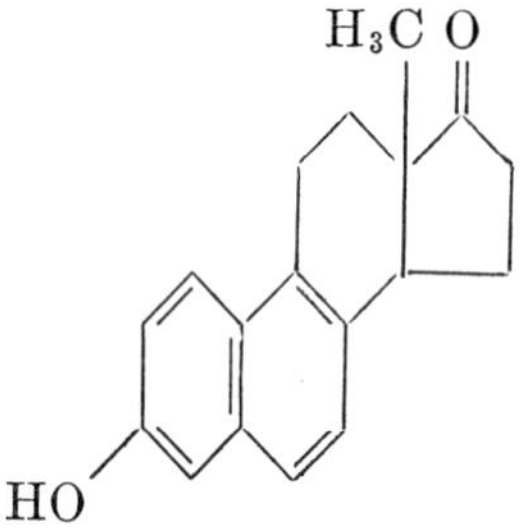

XVIII. Equilenin

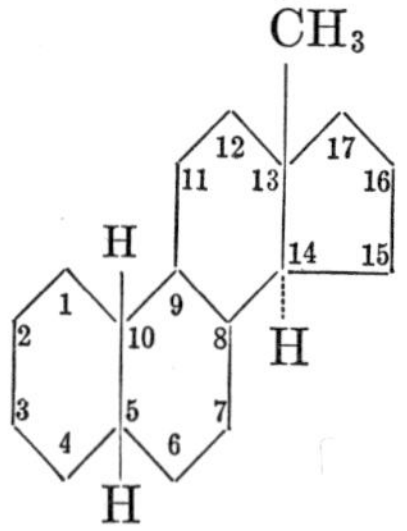

XIX. Estrane

TABLE VIII-1

Estrogen values in urine of men and non-pregnant women (5)

In International Units/24 hours

	RANGE	AVERAGE
Female (days of menstrual cycle)		
7	65–160	110
14 (mid-cycle peak)	160–660	330
21 (corpus luteum peak)	160–660	500
28	30–110	65
Menopausal woman	10– 65	under 65
Male	25–100	50

One International Unit (I.U.) is equivalent in estrogenic activity to that of 0.1 μg. of crystalline estrone.

synthesis. Placental tissue contains a high concentration of estrogens. In addition, estrogens have been isolated from the other two organs known to synthesize steroid hormones, the adrenal cortex and the testes. The last may seem somewhat unexpected. For some reason the horse testis is an extremely active estrogen producer and the richest known natural source of estrogens is the urine of stallions. The Leydig cells of the human testis respond to stimulation with chorionic gonadotropin by increased production of estrogens as well as of androgens (40).

The urine of pregnant women is rich in estrogen, its excretion rising to 8,000 mouse units per day just before parturition (44). Twelve days after parturition urinary excretion has dropped to virtually indetectable values and remains so for two weeks. In non-pregnant women and in men, urinary estrogen values are low (see table VIII-1).

Urinary estrogens occur as conjugates, either of glucuronic acid as in the case of estriol or of sulfuric acid as in the case of estrone. Estradiol-17β, the most active of these hormones, is apparently the parent compound of human estrogens. It is in equilibrium with estrone, to which it may be reversibly oxidized in the body. Estrone may be converted to estriol in an irreversible fashion. This has been determined by feeding individual estrogens and then analyzing the urine for excreted products. When estrone or estradiol is fed, all three forms may be found in the urine; but when estriol is fed, only it is excreted. However, some 80 to 90 per cent of ingested estrogen is converted to non-hormonal metabolic products.

Considerable evidence exists to the effect that liver is a major site of estrogen catabolism. Estrogens have been found to lose activity when incubated with liver slices or certain liver extracts (20). The inactivation can be quite rapid and is attributed by some to the presence of an estrinase in the liver cell. *In vivo* studies confirm this liver role. Animals with damaged or partially removed livers proved to be more sensitive to estrogen administration, indicating a slowdown in the rate of estrogen inactivation (59). In male patients with liver disease, increased concentrations of estrogens probably are present in blood and tissues, since such patients have shown atrophy of testicles and enlargement of breasts. Many human tissues other than liver have the ability to degrade estrone and estradiol to unidentified inactive products (54). The liver is not the only, or necessarily the chief, organ of estrogen catabolism.

The end products of estrogen inactivation are not known. The reaction catalyzed by the postulated estrinase must be more than a "detoxication" by such usual mechanisms as glucuronide formation since the estrogen glucuronides display feeble estrogenic activity, whereas the liver-inactivated products are completely inactive. It may be that the *in vivo* inactivation of estrogens by the liver is far from as rapid as *in vitro* studies have indicated, and that it may not keep up with the supply reaching it via the blood. This is indicated by the fact that the bile in dogs and in humans has been found to be high in estrogens in the case of both endogenous secretion and exogenous administration (4). Excretion by the bile would thus be another mechanism whereby surplus estrogens are disposed of by the liver.

There are several synthetic estrogens. The most important of these is *diethylstilbestrol* (formula XX), also known simply as stilbestrol. It is of great importance clinically, since it is not only cheaper than the natural estrogens, but is more potent than is estrone.

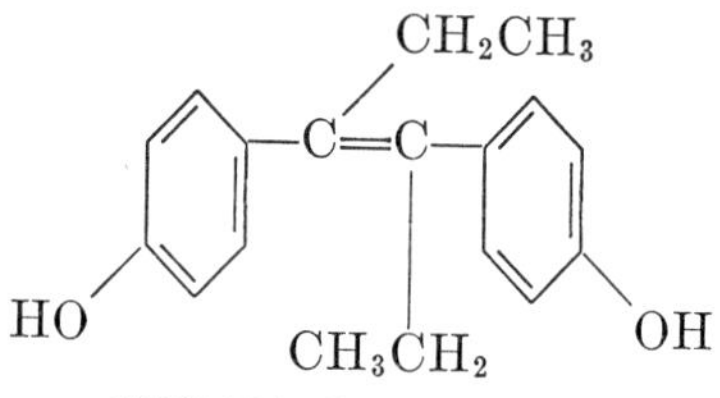

CH_2CH_3

HO — C=C — OH

CH_3CH_2

XX. Diethylstilbestrol

Progesterone. Progesterone is synthesized by the corpus luteum, which is formed from the ovarian follicle after ovulation. The secretion continues during the second half of the menstrual cycle, stopping a few days before the onset of menstruation. Its effects thus follow the earlier action of the estrogens. Progesterone causes an increase in vascularity and secretory activity of the endometrium, preparing it for the implantation of the fertilized ovum. Uterine motility is inhibited by progesterone. Menstruation follows cessation of progesterone secretion, but should pregnancy intervene progesterone continues to be elaborated by the persistent corpus luteum under the influence of the pituitary gonadotropins, and the vascularity of the endometrium continues to increase. In the later stages of pregnancy, the placenta becomes an important site for the formation of progesterone. The biosynthesis of progesterone in the adrenal cortex has been demonstrated from acetate, from cholesterol, and from pregnenolone.

Progesterone, like the other ovarian hormones, is a steroid. It differs from the estrogens in the following respects: (a) ring A is not aromatic but has one double bond between carbons 4 and 5. This does not involve carbon 10, and progesterone (formula XXI) has an angular methyl at that position—unlike the estrogens. (b) The oxygen at position 3 is ketonic rather than phenolic. And, (c) there is a $-COCH_3$ at position 17 rather than $-OH$ or $=O$.

Synthetic compounds similar to progesterone also possess hormone activity, as, for instance, those in which the C-17 side chain is $-COCH_2CH_3$ or $-CHO$, or where the angular methyl at C-10 is absent. This last compound, *10-norprogesterone* (formula XXII), is as active as and possibly more

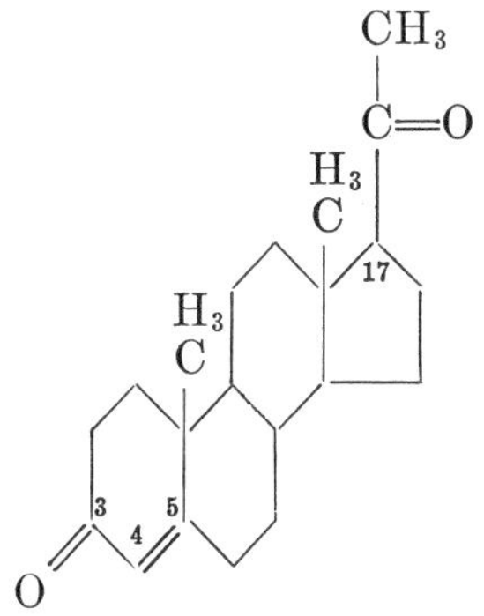

XXI. Progesterone

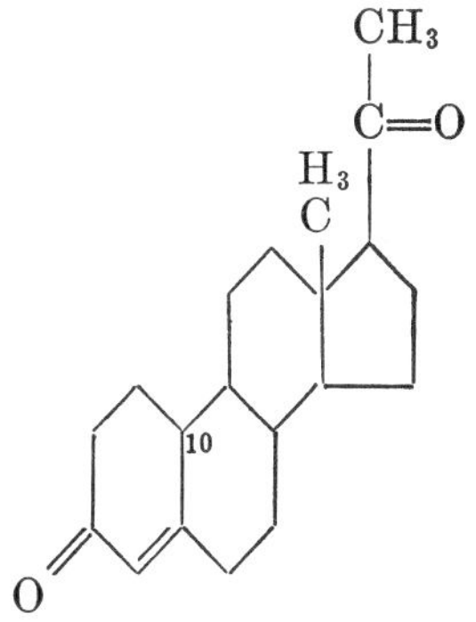

XXII. 10-Norprogesterone

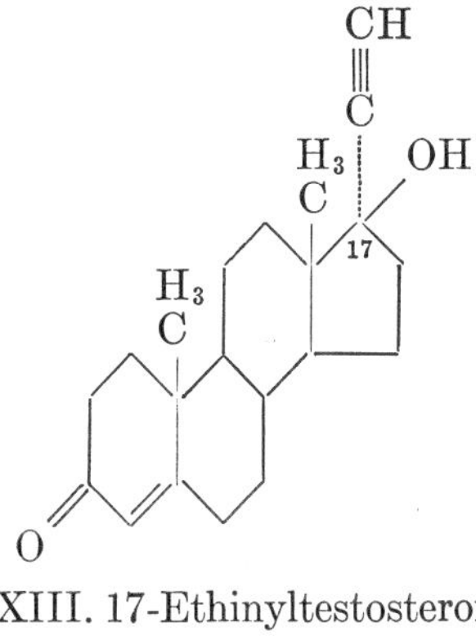

XXIII. 17-Ethinyltestosterone

active than progesterone itself. Still another compound, *17-ethinyltestoster-*

one (formula XXIII), which differs from progesterone in that there are two substituents on the C-17, a hydroxy group and a —C≡CH (ethinyl) group, not only displays about one-third the progestational activity of progesterone but is effective when administered orally, which progesterone itself is not. Other substances showing progestational activity are androgens such as testosterone and synthetic methyltestosterone. In all cases the keto group at C-3 and the double bond between C-4 and C-5 seem to be essential to activity.

Progesterone is definitely known to undergo reduction in the body to *pregnanediol* (formula XXIV), which may be detected in the urine as a glucuronide in which the glucuronic acid residue is attached to the oxygen on C-3. Pregnanediol differs from progesterone in that both keto groups are reduced to the alcohol and that the double bond of ring A is hydrogenated. The maximal urinary output (4 to 5 mgm. daily) of pregnanediol in the human menstrual cycle is during the luteal phase, about a week before menstruation. During pregnancy it increases to a maximum of 60 to 100 mgm. daily in the eighth month. Pregnanediol is also found in the urine of males with embryonal malignant tumors of the testicle. The biosynthesis of progesterone in such a tumor has been verified (77).

A woman in the eighth month of pregnancy, fed cholesterol tagged with deuterium, was found to excrete in her urine pregnanediol with a sufficient

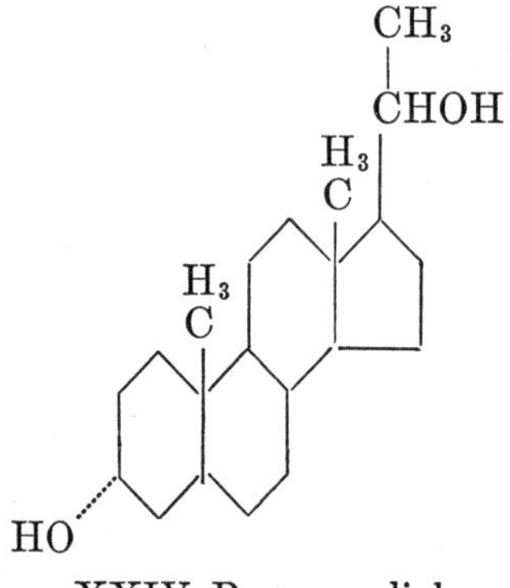

XXIV. Pregnanediol

deuterium content to indicate that one-half to two-thirds of it arose by the degradation of cholesterol (2). Presumably, the conversion of cholesterol to pregnanediol went by way of progesterone, and this is the most direct evidence yet available to connect cholesterol and the ovarian hormones. Earlier work had indicated that cholesterol, after incubation with liver slices, yielded products possessing estrogenic activity, but this line of attack has not been followed up. As in the case of the estrogens, an important site of progesterone catabolism is the liver.

The Testes

Androgens. The androgens are synthesized primarily in the interstitial tissue of the testes, and their manufacture is under the control of the pituitary gonadotropins. The androgens control the development of the masculine secondary sexual characteristics, such as hair distribution and change in voice, as do the estrogens in analogous fashion for the female. Furthermore, they maintain the normal functional condition of the accessory sex organs of the male, especially the prostate and the seminal vesicles. The epithelia of these organs become atrophied in castrated animals, and are restored to normal structure and secretory activity by administration of androgens. These responses, and also the growth of the comb in capons, are used in biological assay for androgenic substances. Together, the testicular and ovarian hormones are referred to as the *sex hormones* or, more rarely, the *sexogens.*

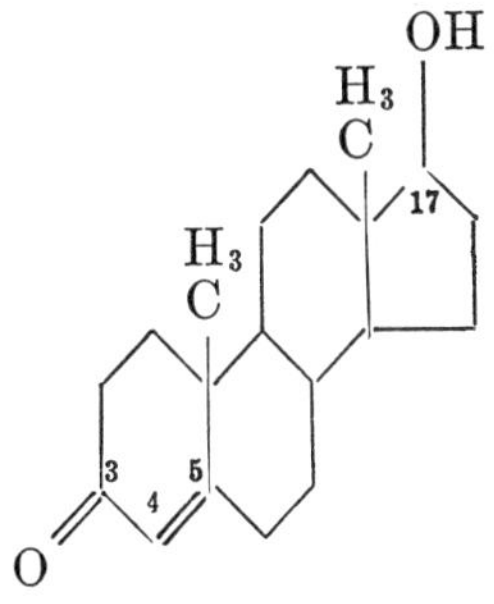

XXV. Testosterone

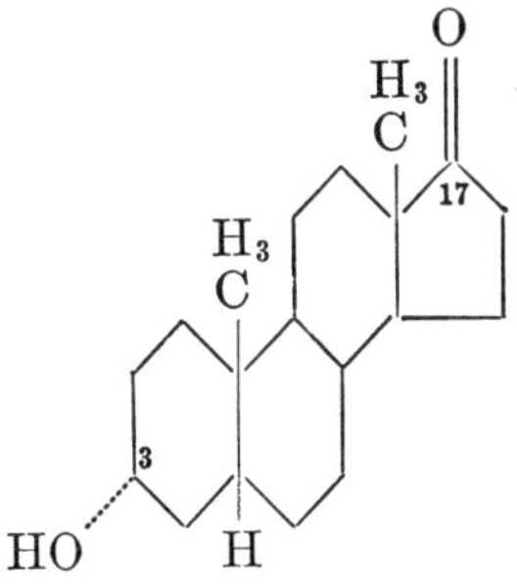

XXVI. Androsterone

Estrone and estriol, and also other estrogens with structures differing from animal estrogens have been isolated from plants. Androgenic materials have been obtained from plants, but their structure has not been determined.

Androgenic material can be demonstrated in extracts of ovarian tissue. Several androgens have been crystallized from adrenal cortex extracts, with chemical structures different from the androgens of testicular origin. The androgen characteristic of the testes is *testosterone* (formula XXV). Two androgens of considerably lower potency had been isolated from male human urine and characterized before testosterone was isolated from bull testes. These are *androsterone* (formula XXVI) and *dehydroepiandrosterone* (formula XXVII). Androsterone is only one-sixth as active as testosterone. It will be noted that testosterone seems to be estrone in reverse as far as the oxygen-containing groups on the CPP nucleus are concerned. While estrone is a 3-hydroxy, 17-keto compound, testosterone is a 3-keto, 17-hydroxy compound. An additional difference is, of course, that ring A of

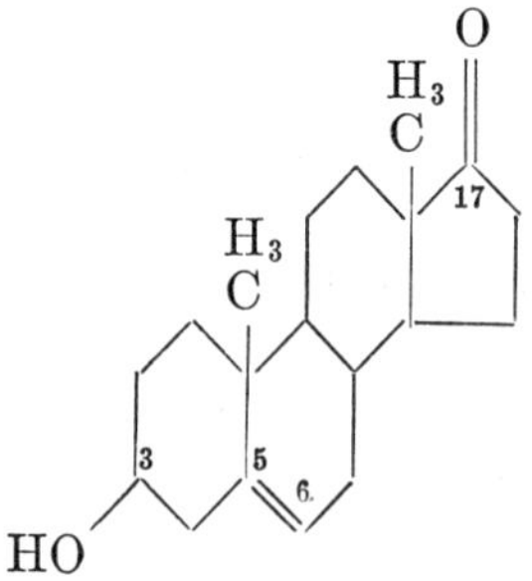

XXVII. Dehydroepiandrosterone

testosterone is not aromatic as is that of estrone, and the C-10 angular methyl is present. The ring system of testosterone is exactly that of progesterone and the only difference between these two compounds is the nature of the grouping on C-17, a hydroxyl in the former and a methylketo in the latter. Testosterone and *17-methyltestosterone* (formula XXVIII) display progestational activity. The androgens may alternatively be named as derivatives of the hydrocarbons *androstane* (formula XXIX).

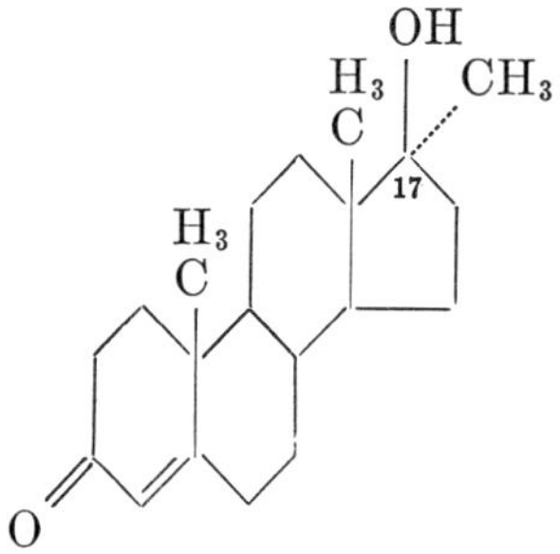

XXVIII. 17-Methyltestosterone

Androsterone, like estrone, is a 3-hydroxy, 17-keto compound, but it differs from estrone in being fully saturated and possessing the C-10 angular

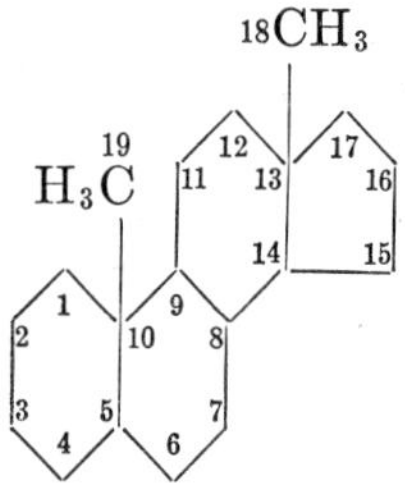

XXIX. Androstane

methyl. Dehydroepiandrosterone differs from androsterone in possessing one double bond between carbons 5 and 6, and having its C-3 hydroxyl group in the beta configuration rather than the alpha configuration of androsterone. A number of other androgens have been isolated from urine, testes, and adrenal cortex, displaying varying combinations of keto and hydroxy groups with saturated rings or with one double bond appearing in some one of the four rings (and in one case in two of the four rings), but none of these is a particularly active androgen.

In addition to the effects upon secondary sex characteristics and upon accessory sex organs of the male, many androgens possess a *protein anabolic effect*, which can be measured by body weight gain, by nitrogen retention, or in experimental animals by isolation and weighing of a single selected muscle, such as the *levator ani* in the rat. There are several derivatives of testosterone which possess effective protein anabolic activity, as measured by the *levator ani* test, with weak androgenic effect compared with androsterone and testosterone.

Perfusion of the isolated testis with acetate tagged with a radioactive carbon isotope yields testosterone which contains the radioactive carbon; this confirms the ability of the testis to synthesize testosterone (45). The administration of testosterone either orally or intramuscularly is followed by the excretion of androgens and other 17-ketosteroids in the urine. The urinary androgens are in a water-soluble, inactive form, having been conjugated with sulfuric or glucuronic acid. Dehydroepiandrosterone sulfate and androsterone sulfate have been isolated from the urine (42). The most common additional 17-ketosteroid in human urine is etiocholanol-3(alpha)-one-17, which differs from androsterone in that rings A and B are *cis* in the former and *trans* in the latter. Testosterone undergoes metabolic alteration in the prostate (35) and in the liver and other tissues (76).

The functions of spermatogenesis by the seminiferous tubules and of androgen production by the interstitial cells of Leydig are closely related, but may vary independently of each other. They are stimulated by different pituitary hormones, FSH for spermatogenesis, and ICSH for androgen production. Administration of large amounts of androgen (e.g., 50 mgm. testosterone daily) will inhibit spermatogenesis in man, probably by inhibition of formation of pituitary FSH, and may even produce histologically demonstrable degeneration of the seminiferous tubules. A similar effect, also attributed to suppression of pituitary secretion, occurs during periods of administration of chorionic gonadotropin (40). Following inhibition by testosterone or chorionic gonadotropin, a gradual recovery of spermatogenic tissue may occur, with eventual increase of spermatozoa counts above the pre-treatment level.

17-Ketosteroids. The 17-ketosteroids are, properly speaking, all compounds containing the CPP nucleus with an oxygen doubly bound to C-17.

In this sense, estrone, androsterone, and dehydroepiandrosterone, among the natural sexogens, may be considered 17-ketosteroids. The term is usually applied, however, only to the neutral 17-ketosteroids which remain in the urine extract after treatment with aqueous alkali to remove phenolic steroids such as estrone. The neutral 17-ketosteroids arise chiefly from substances originally secreted by the testes and the adrenal cortex. They are, however, of considerable interest to the physician because of the diagnostic applications of urinary 17-ketosteroid values. These values are determined by photometric methods. The most commonly used methods take advantage of the fact that 17-ketosteroids will react with *m*-dinitrobenzene in an alkaline alcoholic medium to produce strongly colored complexes (31). The total daily 17-ketosteroid excretion averages 17 mgm. in normal men and 12 mgm. in normal women.

All significant studies of the urinary excretion of androgens have been tabulated and summarized by Dorfman and Shipley (16). On account of variations in procedure, the averages from different laboratories vary, for normal men, from 19 to 99 International Units (I.U.) per day: one I.U. is equivalent in androgenic activity to that of 0.1 mg. of androsterone. The corresponding averages for women vary from 14 to 47 I.U. per day. The range of individual values is of course much wider, being 10 to 225 I.U. for men and 5 to 85 I.U. for women. Children under 5 years have about 1/20 the normal adult excretion of androgens and of 17-ketosteroids, and at age 10 to 15 about 1/3. At ages above 50 the androgen and 17-ketosteroid excretions decrease, reaching values about 1/4 normal at age 80.

Deviations from the normal urinary excretion of androgens and 17-ketosteroids occur in certain clinical conditions. In male castrates, both androgens and 17-ketosteroids show marked decreases, while in women, after ovariectomy, the androgen excretion decreases while 17-ketosteroid remains unchanged. In pituitary and adrenocortical insufficiency, androgens and 17-ketosteroid excretion decrease in both sexes. In the case of adrenocortical insufficiency (Addison's disease) the decrease is more marked in the female. Conversely, adrenal cortical hyperactivity as in adrenal cancer results in dramatic increases in 17-ketosteroid excretion. The highest titer of 17-ketosteroids obtained from the urine of a human being was found in that of a woman with metastatic adrenal cancer. She excreted 2.1 grams of 17-ketosteroids per day (12), a value some two hundred times the normal.

The Adrenal Cortex

Corticoids. Among the more than 30 steroids isolated from the adrenal cortex are certain typical estrogens, such as estrone, and androgens, such as *adrenosterone* (formula XXXIX) which is characteristic of the adrenal cortex and should not be confused with androsterone. Progesterone also appears among the adrenal steroids. The steroids which are more numerous and more representative of the essential functions of the adrenal cortex are the corticoids, which may be roughly divided according to their physiological effects into *mineralocorticoids* and *glycocorticoids*. This division is not strict, since some corticoids belong in both categories.

The mineralocorticoids promote the renal tubular absorption of Na^+, Cl^-, HCO_3^-, and water, and the excretion of K^+ and HPO_4^{--}. This is briefly designated as the *electrolyte effect*. The compound 11-deoxycorticosterone best exemplifies this action. This is the familiar synthetic steroid marketed as deoxycorticosterone acetate or DOCA, and used in the therapy of adrenocortical deficiency—Addison's disease. Deoxycorticosterone (DOC) is not a major component of the steroids obtained directly by fractionation of the adrenal cortex, although it is present there. The effect upon water and salts is mediated chiefly by the kidney, but other related activities have been demonstrated which appear to be entirely extra renal. Synthetic DOCA inhibits the output of Na^+ in sweat, and increases diffusion of Na^+ and Cl^- into the peritoneal cavity. The 11-deoxycorticoids in general are the typical mineralocorticoids and are also more effective than the 11-oxycorticoids (see below) in the maintenance of life in adrenalectomized animals and in patients with Addison's disease, and in the stimulation of recovery of fatigued muscle in adrenalectomized rats.

A still more potent mineralocorticoid, detected in the blood, as well as in the adrenals of man and experimental animals, is *aldosterone*, a steroid unique among those naturally occurring, because of the possession of an *angular aldehyde* in the place of the angular methyl usually attached to carbon 13. In solution, it exists in the form of a hemiacetal as shown in formula XXX (24). Its activity is qualitatively similar to that of deoxy-

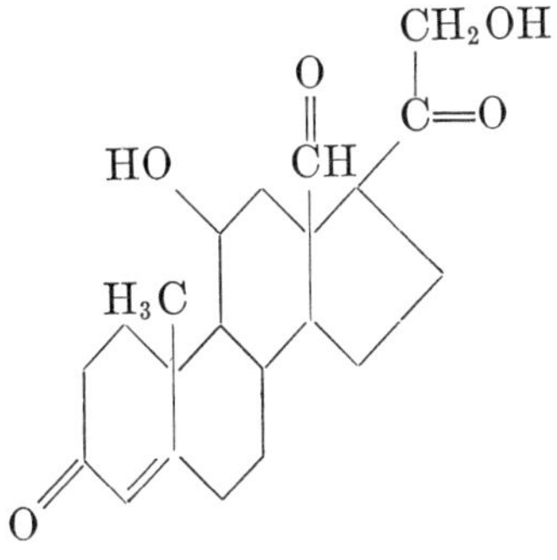

aldehyde form 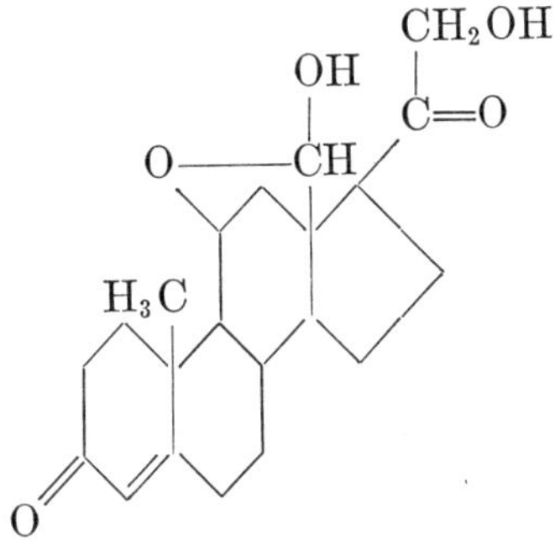 hemiacetal form

XXX. Aldosterone

corticosterone, equal to it quantitatively when assayed for growth-promoting activity, and at least 25 times as active in the sodium-retention assay. It is probably the chief hormone involved in the regulation of sodium, potassium, and chloride metabolism, perhaps of magnesium metabolism as well. Aldosterone and other mineralocorticoids are produced in the *zona glomerulosa* of the adrenal cortex (25). Certain adrenocortical tumors produce excessive amounts of aldosterone (or a similar hormone) and cause the disease known as *primary aldosteronism*, characterized by muscular weakness, hypertension, alkalosis, and elevated blood plasma sodium concentrations with depression of blood potassium.

Glycocorticoids typically bear an alcoholic or ketonic oxygen at C-11 and are therefore designated as 11-oxycorticoids. Aldosterone, a highly potent mineralocorticoid, also has glycocorticoid activity approximately equal to that of cortisol, a typical 11-oxycorticoid (70). Glycocorticoids cause deposition of glycogen in the liver of the adrenalectomized animal, prevent insulin convulsions in intact rats, and increase the urinary loss of glucose in partially depancreatized and in adrenalectomized-depancreatized rats, or in intact rats who are forcibly overnourished. This *metabolic effect* in the human may be summarized as the maintenance of glycogen stores and the blood sugar level, promotion of intestinal absorption of fat, and the acceleration of the formation of carbohydrate from protein.

Other effects of the 11-oxycorticoids include decrease of circulating eosinophils and lymphocytes, atrophy of lymphoid tissue, and inhibition of the growth and proliferation of fibroblasts. These effects are noted in patients under treatment with large doses of cortisone. The inhibition of fibroblasts is manifested by delayed healing of surgical or other wounds in mesodermal structures, while epithelial layers heal as usual. The period of delay is only about 2 weeks, even with continued administration of cortisone; normal healing then takes place. Also related to fibroblast inhibition are the failure to form granulation tissue, and the reduction of pain, swelling, walling-off, and pus formation in local infections. Inhibition of hyaluronidase activity by 11-oxycorticoids has been observed with evidence that the substrate rather than the enzyme is altered (60). The manifestations of allergy and sensitization are minimized. The biosynthesis of mucopolysaccharides is inhibited (33). Cortisone and cortisol (see below) are the most effective naturally-occurring glycocorticoids, but artificially halogenated derivatives, particularly *fluorohydrocortisone*, show greater activity.

The above effects are probably involved in the favorable results of ACTH or 11-oxycorticoid treatment of the "collagen diseases" and of certain types of arthritis. No full rationale of such therapy exists. In many types of infectious disease, administration of 11-oxycorticoids (cortisone or hydrocortisone) or of ACTH seriously disturbs the usual host-parasite relationship, the effects described above contributing to the disturbance, which usually favors the pathogenic micro-organism to the disadvantage of the patient.

The classification of corticoids into mineralocorticoids and glycocorticoids is not sharp. Most corticoids possess both properties in varying degree and many will also show estrogenic, androgenic, or progestational activity. The activity of corticoids may be tested in a number of ways, such as (a) by their influence on Na retention, (b) by their effect upon the length of survival of an adrenalectomized animal, (c) by their effect in preventing insulin convulsions in rats, and (d) by their influence on the rate of glycogen deposition in adrenalectomized rats. It has been found that the position of individual corticoids on the scale of activity depends upon the test used. Methods for assay of corticoids for characteristic electrolyte and metabolic effects have been reviewed by Dorfman (14).

Most of the definitely known corticoids are steroid molecules containing 21 carbon atoms—that is, the 17 carbons of the ring system, the 2 carbons of the two angular methyls at C-10 and C-13, and the 2 carbons of a 2-carbon side chain at C-17. The hydrocarbon skeleton, from which such steroids can be named, is *pregnane* (formula XXXI). It should be noted that of the sexogens, progesterone is a pregnane derivative, and at least one of the corticoids, deoxycorticosterone, is reported to have progestational activity comparable to that of progesterone in the adrenalectomized cat (34). Progesterone itself is also included in the list of corticoids.

The 21-carbon corticoids may be divided further according to the number of oxygens contained in the molecule. Of those possessing five oxygen atoms, the two most important are *11-dehydro-17α-hydroxycorticosterone* (also known as Compound E, or cortisone) (formula XXXII) and *17-hydroxycorticosterone* (Compound F) (hydrocortisone or cortisol) (formula XXXIII). All of these corticoids have their oxygens on carbons 3, 11, 17, 20, and 21. The oxygens on C-17 and C-21 are invariably present as hydroxyl groups, but any or all of the other three may be present in either

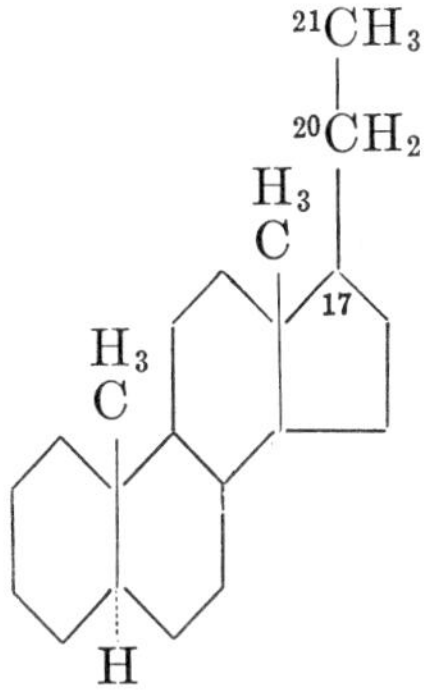

$^{21}CH_3$ — $^{20}CH_2$ — (C-17); H_3C (C-13); H_3C (C-10); H

XXXI. Pregnane

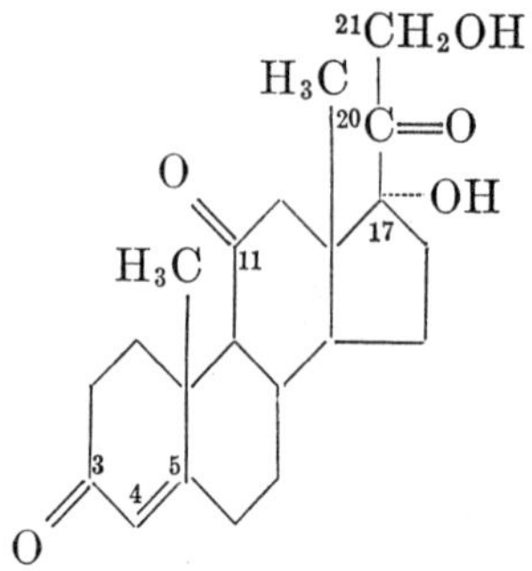

XXXII. 11-Dehydro-17-hydroxycorticosterone (cortisone)
(Compound E) Δ^4-Pregnene-17α,21-diol-3,11,20-trione

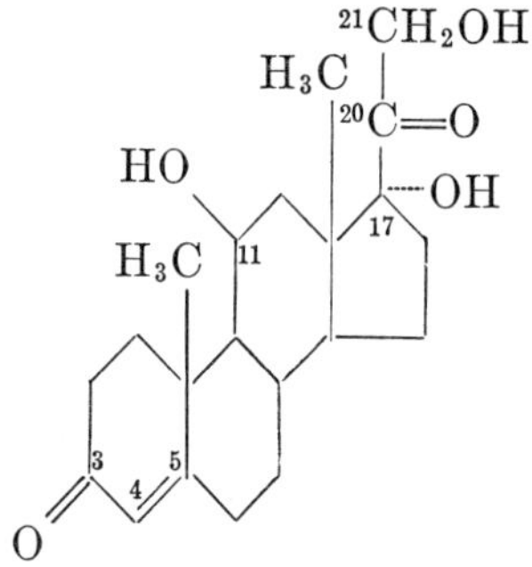

XXXIII. 17-Hydroxycorticosterone (cortisol or hydrocortisone)
(Compound F)

hydroxy or keto form. Where the corticoid is a 3-ketosteroid, as in the two mentioned, there is a double bond between carbons 4 and 5. Note that aldosterone (see above) also contains 5 oxygen atoms.

Certain other corticoids possess four oxygen atoms apiece. Those which possess oxygens at carbons 3, 17, 20, and 21 are known as the 11-deoxy group, since when compared with the corticoids containing five oxygens the oxygen in the 11 position is missing. In the 17-deoxy group the four oxygens are at carbons 3, 11, 20, and 21. Of these corticoids, three are of some importance. Two belong to the 17-deoxy group which is also known, after its most important member, as the corticosterone group. The two are *corticosterone* (formula XXXIV) itself, and *11-dehydrocorticosterone* (formula XXXV). The third belongs to the 11-deoxy group and is *11-deoxy-17-hydroxycorticosterone* (formula XXXVI).

In most of the corticoids which possess three oxygens apiece, the oxygen is situated at carbons 3, 17, and 20. In *11-deoxycorticosterone* (formula XXXVII), the most important of the group, the oxygens are on carbons 3, 20, and 21.

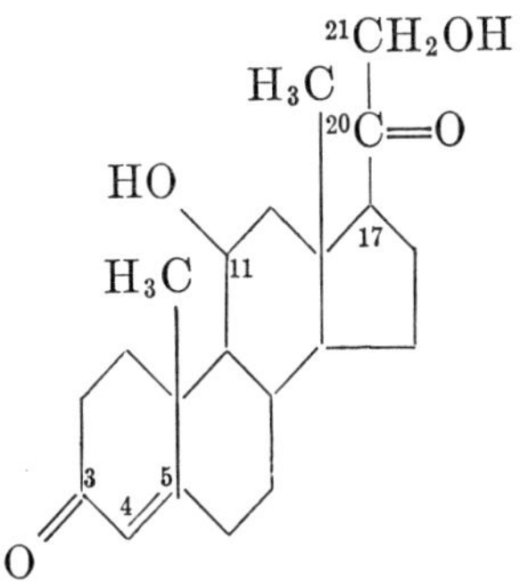

XXXIV. Corticosterone (Compound B)

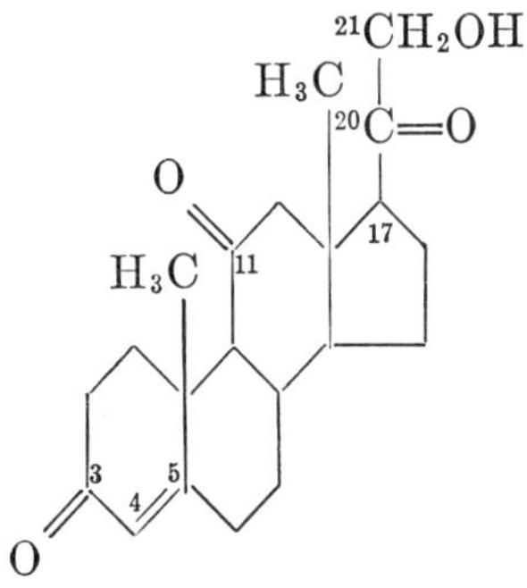

XXXV. 11-Dehydrocorticosterone (Compound A)

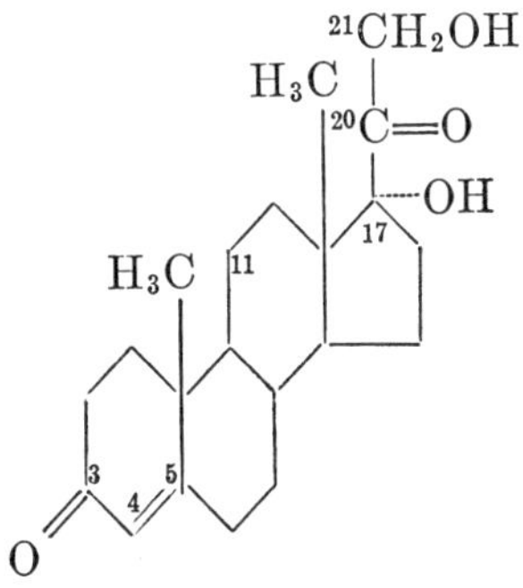

XXXVI. 11-Deoxy-17-hydroxycorticosterone (11-Deoxycortisol) (Substance S)

Finally of the 21-carbon corticoids which contain only two oxygens apiece, on carbons 3 and 20, one is progesterone.

Of the corticoids with less then 21 carbons in the molecule, at least three possess 19 carbons, having no 2-carbon side chain on carbon 17. At least two of these show androgenic properties, *androstene-3,17-dione* (formula XXXVIII) being as effective as androsterone in this respect, and *adrenosterone* (formula XXXIX) one-fifth as effective. Lastly, we have estrone, which because of the phenolic character of ring A and the consequent absence of one angular methyl has only 18 carbons on its molecule.

The measurement of the small amounts of corticoids circulating in the blood is arduous and still somewhat uncertain. Cortisol and corticosterone have been found in the largest amounts, averaging 80 micrograms per liter by one method and 25 micrograms per liter by another. Cortisone and 11-dehydrocorticosterone are present in about half the concentration and aldosterone in about $\frac{1}{35}$ the concentration of cortisol.

Metabolism. The 17-ketosteroids of the urine have their origin in corticoids as well as in androgens, and, in fact, arise predominantly from

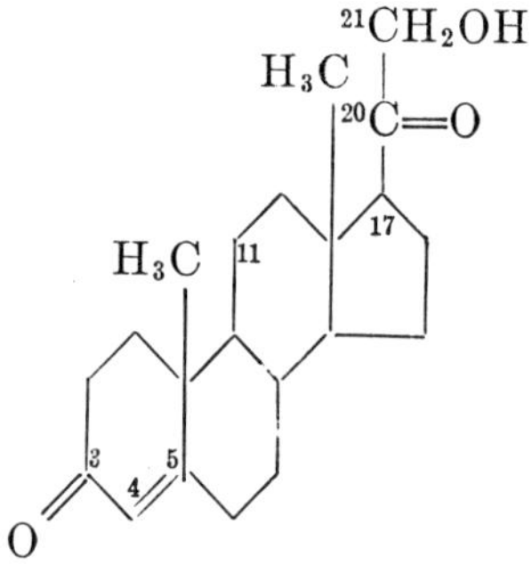

XXXVII. 11-Deoxycorticosterone (DOC)

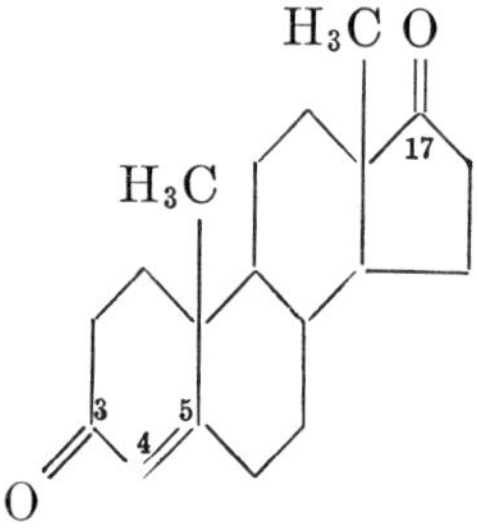

XXXVIII. Δ^4-Androstene-3,17-dione 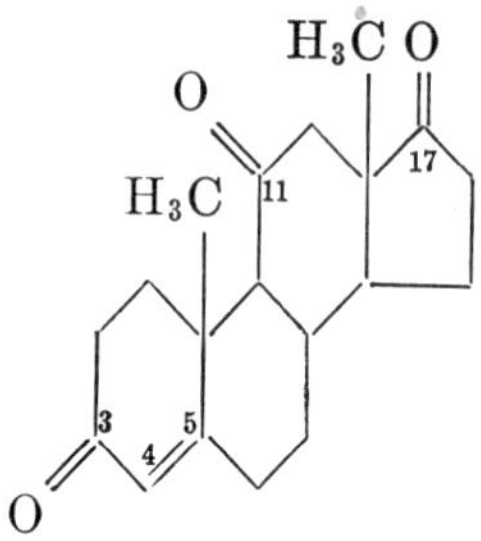 XXXIX. Adrenosterone

the former. This last is demonstrated by the fact that in Addison's disease, which involves hypofunction of the adrenal cortex, 17-ketosteroid excretion drops well over 50 per cent (72). A typical urinary 17-ketosteroid is *iso-androsterone*, which differs from androsterone only in the orientation of the hydroxyl on carbon-3.

Cortisone, cortisol, and cortical metabolites other than the 17-ketosteroids have also been found in the urine. Several chemical reagents, including antimony trichloride (53), have been adapted for the estimation of non-ketonic steroids. Urinary *formaldehydogenic corticoids* yield formaldehyde upon oxidation with periodic acid, and include C-21 steroids with alpha-ketol or alpha-glycol side chains.

The manner and order in which the corticoids are formed in the adrenal cortex have been intensively investigated (28). The parent substance is acetate, which is a common intermediate (in the form of acetylcoenzyme A) in the breakdown of carbohydrates, fats and proteins in the body. There is evidence for two pathways of corticoid biosynthesis.

One pathway proceeds by the formation of cholesterol, which is stored in the adrenal cortex. From cholesterol is formed Δ^5-pregnenolone; this reaction is definitely accelerated by ACTH. Enzyme systems have been identified in the adrenal cortex catalyzing the conversion of Δ^5-pregnenolone to

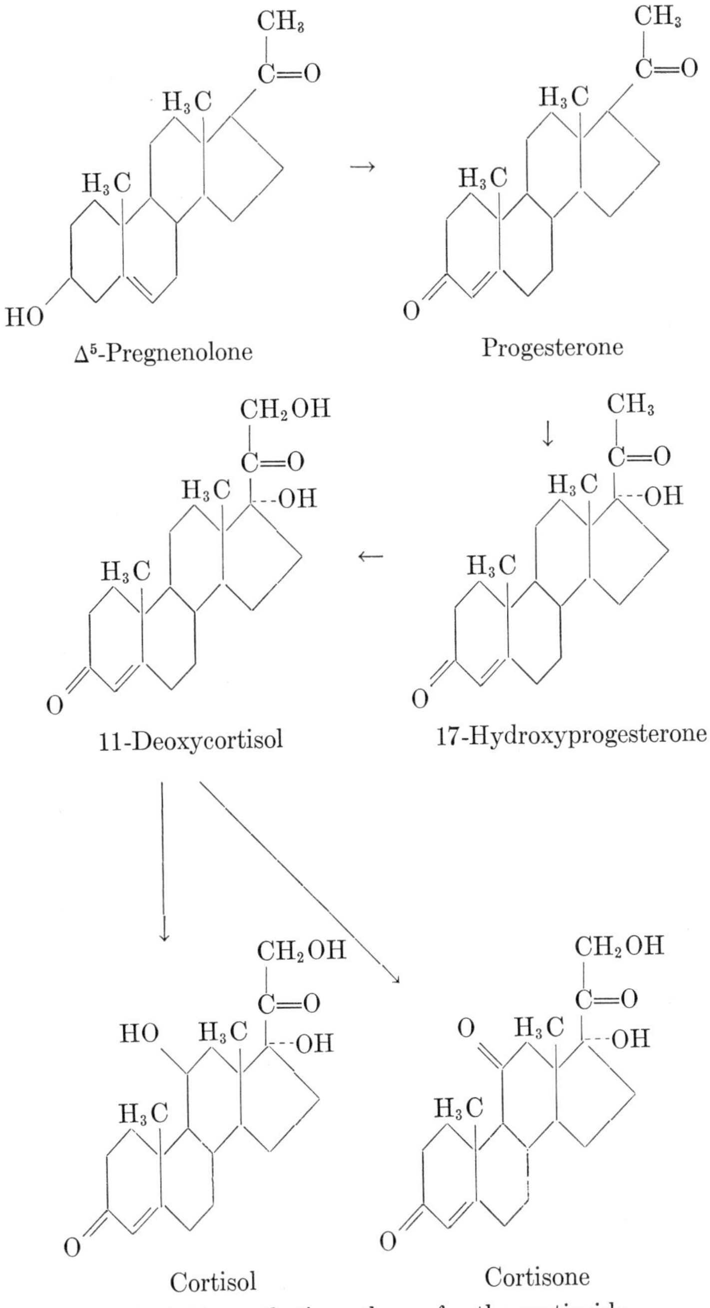

CH_3
C=O
H_3C
H_3C
HO

Δ⁵-Pregnenolone

→

CH_3
C=O
H_3C
H_3C
O

Progesterone

↓

CH_2OH
C=O
H_3C ---OH
H_3C
O

11-Deoxycortisol

←

CH_3
C=O
H_3C ---OH
H_3C
O

17-Hydroxyprogesterone

↓ ↘

CH_2OH
C=O
HO H_3C ---OH
H_3C
O

Cortisol

CH_2OH
C=O
O H_3C ---OH
H_3C
O

Cortisone

XL. A biosynthetic pathway for the corticoids

progesterone, then to 17-hydroxyprogesterone, then to 11-deoxycortisol, and thence to cortisol and cortisone (formula XL). Evidence also exists for the formation of corticosterone from progesterone with DOC as an intermediate.

The other pathway of corticoid biosynthesis does not lead to the formation of cholesterol, but rather to the direct synthesis of corticoids from acetate. When C^{14}-acetate is supplied to the adrenal, the C^{14} concentration in cortisol and cortisone is higher than that in adrenal cholesterol, which leads to the conclusion that cholesterol is not an obligatory intermediate. This direct pathway of corticoid formation is independent of ACTH stimulation.

NEUROHORMONES

The neurohormones constitute an additional and important class of hormones that are neither proteins, peptides or steroids. They are derivatives of amino acids. The neurohormones have been described on pages 159 ff.

HORMONE-ENZYME RELATIONSHIPS

It has already been stated that the high activity of minute quantities of hormones made it seem likely that these chemicals exerted their influence through the activation or inhibition of enzyme systems. Unfortunately, evidence as to specific hormone-enzyme relationships is as yet scant. Attempts at observing the effect of various hormones on specific enzyme systems *in vitro* have usually resulted in flat failure or in the collection of data of doubtful significance. A reason for this may well be that the role played by hormones is not so simple as that of mediating a single reaction, but is rather a complex one involving control of the interrelationships of many enzymatic reactions. If that were the case a hormone would be expected to exhibit its effect only within the intact cell, and this is indeed what seems to happen.

Dorfman (13), discussing the steroid hormones only, formulated the following possible actions in their relationship with enzymes: (a) as coenzymes or prosthetic groups of enzymes; (b) as accelerators or inhibitors of enzymes or enzyme systems; (c) direct or indirect effects upon accelerators or inhibitors; and (d) changing enzyme concentrations in tissues. The same possibilities may be considered for non-steroid hormones. No instance of the first possibility—that hormones act as actual components of enzyme systems—has yet been demonstrated. Examples of the other three possibilities are given in the following sections. Other examples may be found by consulting review articles (13, 39), but many of these examples are contradictory, and few have any clear relation to the known physiological actions of the hormones.

Glucokinase

The nearest approach to an *in vitro* demonstration of a connection between a hormone and a specific enzyme system is in the case of glucokinase, the enzyme which catalyzes the reaction of ATP and glucose to form ADP and glucose-6-phosphate (see Chapter 14). It was reported in 1945 (49), that this reaction could be inhibited by anterior pituitary extract, and that such inhibition could be counteracted by insulin, both *in vitro* and *in vivo*. The particular anterior pituitary hormone involved in the inhibition is not yet certainly known, but strong evidence points to the growth hormone. Later it was shown that adrenal cortical extracts could also inhibit the glucokinase reaction and that insulin could release this inhibition also. As yet, insulin has not been found to have an effect on any other enzyme involved in the glycose-glycogen cycle, and it should be stressed that even in the case of glucokinase the role of insulin is not one of stimulation but one simply of a release from inhibition. If insulin has a more positive role, it is either upon an enzyme system not yet studied or on intact cells only. Stadie (64), however, has characterized as unsatisfactory all experimental evidence purporting to show physiologically significant effects of hormones on enzymes, other than in intact cells. He presents evidence to indicate that insulin acts by first combining with some component of the cellular structure and suggests that some anterior pituitary substance (not ACTH or growth hormone) negates the effect of insulin by preferentially combining with the cellular component in question.

Phosphorylase

Adrenalin, in quantities too small to produce the characteristic rise in blood pressure, will bring about breakdown of glycogen in liver and muscle with concomitant rises in blood glucose and blood lactic acid. (Glucose is the end product of glycogen breakdown in liver and lactic acid the end product of glycogen breakdown in muscle.) This action is very rapid, the glucose and lactic acid of blood showing significant increases within three minutes. Together with these phenomena, there is a decrease of blood phosphate, while hexose phosphate accumulates in muscle. These various phenomena could all be explained if one were to suppose an increased activity of phosphorylase in the tissues as a result of adrenalin injections, since accelerated phosphorolysis of glycogen would lower the concentration of glycogen and phosphoric acid and would raise the concentration of glucose, hexose phosphate and lactic acid.

Two forms of phosphorylase have been isolated from muscle, phosphorylase *a* and phosphorylase *b*. Phosphorylase *a* is obtained from muscle of deeply anesthetized animals and is active without addition of adenylic acid. Phosphorylase *b* is obtained from fatigued muscle and is inactive unless adenylic acid is added. *In vivo* there is apparently an equilibrium between the two forms. Evidence has been presented (67) to show that adrenalin shifts this balance in favor of the active enzyme, phosphorylase *a*, thus accelerating glycogen breakdown and initiating the other chemical changes mentioned above. This action of adrenalin has been demonstrated in the absence of intact cells (18a).

Prostatic Acid Phosphatase

Phosphomonoesterases with an optimal pH below 7 are commonly called acid phosphatases. Such enzymes can be demonstrated in nearly all types of animal cells and in many plants and micro-organisms (73). The human prostate produces an acid phosphatase which can be identified by specific inhibitors (22) and differentiated from other acid phosphatases which reach the blood plasma from other cells or tissues. Under normal conditions the prostatic acid phosphatase makes up only a very small portion of the plasma acid phosphatase, but if there is cancer of the prostate, and particularly if cancer originating in the prostate spreads widely throughout other parts of the body, a notable increase in plasma acid phosphatase, identifiable as of prostatic origin, often occurs.

One of the most striking and best known examples of the effect of hormones upon the activity of an enzyme in the human body is the effect of androgens and of estrogens upon the production of acid phosphatase by the prostate. These effects are reflected in the plasma acid phosphatase activity, which is increased by androgens, and decreased following castration or the administration of estrogens. Such observations have been made for the most part in patients with cancer of the prostate, where castration, estrogens, or both are recognized forms of therapy. Hormone-induced variations in production of acid phosphatase by the prostate are associated with atrophy or hypertrophy of the secreting cells.

It can be seen from this section that the whole subject of hormone-enzyme relationships is in an obscure state. The few glimmers of light that are now beginning to appear, while interesting and hopeful, are far from sufficient to illuminate even a corner of the subject.

REFERENCES

1. Astwood, E. B. Mechanism of action of antithyroid compounds. Brookhaven Symposia in Biol. No. 7, 61–73, 1954.
2. Bloch, K. The biological conversion of cholesterol to pregnanediol. J. Biol. Chem., **157:** 661–666, 1945.
3. Brand, E., *et al.* On the structure of thyroglobulin. J. Biol. Chem., **128:** xi, 1939.
3a. Bromer, W. W., *et al.* The amino acid sequence of glucagon. J. Am. Chem. Soc., **78:** 3858–3860, 1956.
4. Cantarow, A., *et al.* Studies on inactivation of estradiol by the liver. Endocrinology, **33:** 309–316, 1943.

5. CANTAROW, A., AND TRUMPER, M. *Clinical Biochemistry*. 4th ed. Philadelphia, W. B. Saunders Co., 1949.
6. CHOW, B. F., *et al.* Effects of digestion by proteolytic enzymes on the gonadotropic and thyrotropic potency of anterior pituitary extracts. J. Endocrinol., **1:** 440–469, 1939.
7. CLAESSON, L., *et al.* Crystalline human chorionic gonadotropin and its biological action. Acta endocrinol., **1:** 1–18, 1948.
8. COHN, E. J., *et al.* Studies in the physical chemistry of insulin. II. Crystallization of radioactive zinc insulin containing two or more zinc atoms. J. Am. Chem. Soc., **63:** 17–22, 1941.
9. DEDUVE, C. Glucagon. The hyperglycaemic glycogenolytic factor of the pancreas. Lancet, **2:** 99–104, 1953.
10. DEROBERTIS, E. Proteolytic enzyme activity of colloid extracted from single follicles of the rat thyroid. Anat. Rec., **80:** 219–231, 1941.
11. DIXON, H. B. F. Adrenocorticotropic hormone and the control of adrenal secretion. Proc. Roy. Soc. Med., **48:** 903–907, 1955.
11a. DIXON, H. B. F. Melanophore-stimulating activity of corticotropin. Biochim. et Biophys. Acta, **19:** 392–394, 1956.
12. DORFMAN, R. I. Biochemistry of androgens; *in* PINCUS, G., AND THIMANN, K. V., eds. *The Hormones*. Vol. I, pp. 467–458. New York, Academic Press, 1948.
13. DORFMAN, R. I. Steroids and tissue oxidation. Vitamins and Hormones, **10:** 331–370, 1952.
14. DORFMAN, R. I. The bioassay of adrenocortical hormones. Recent Progress in Hormone Research, **8:** 87–112, 1953.
15. DORFMAN, R. I. Bioassay of steroid hormones. Physiol. Rev., **34:** 138–166, 1954.
16. DORFMAN, R. I., AND SHIPLEY, R. A. *Androgens*. New York, John Wiley & Sons, 1956.
17. DU VIGNEAUD, V., *et al.* The synthesis of an octapeptide amide with the hormonal activity of oxytocin. J. Am. Chem. Soc., **75:** 4879–4880, 1953.
18. EDINGER, T. The pituitary body in giant animals fossil and living: A survey and a suggestion. Quart. Rev. Biol., **17:** 31–45, 1942.
18a. ELLIS, S. The metabolic effect of epinephrine and related amines. Pharmacol. Rev. **8:** 485–562, 1956.
19. ENGEL, F. L. The endocrine control of metabolism. Bull. N. Y. Acad. Med., **29:** 175–201, 1953.
20. ENGEL, P., AND ROSENBERG, E. Estrogen-inactivating liver extracts. Endocrinology, **37:** 44–46, 1945.
21. FAWCETT, D. M., *et al.* Tyrosine iodinase. J. Biol. Chem., **209:** 249–256, 1954.
22. FISHMAN, W. H., AND LERNER, F. A method for estimating serum acid phosphatase of prostatic origin. J. Biol. Chem., **200:** 89–97, 1953.
23. FOÀ, P. P., *et al.* Anterior pituitary growth hormone and pancreatal secretion of glucagon. Proc. Soc. Exptl. Biol. Med., **83:** 758–761, 1953.
24. GAUNT, R., *et al.* Aldosterone,—a review. J. Clin. Endocrinol. and Metab., **15:** 621–646, 1955.
25. GIROUD, C. J. P., *et al.* Secretion of aldosterone by the *zona glomerulosa* of rat adrenal glands incubated *in vitro*. Proc. Soc. Exptl. Biol. Med., **92:** 154–168, 1956.
25a. GOTTSCHALK, A., AND ADA, G. L. Separation and determination of the component sugars of mucoproteins. Biochem. J., **62:** 681–696, 1956.
26. GREER, M. A., AND ASTWOOD, E. B. The antithyroid effect of certain foods in man as determined with radioactive iodine. Endocrinology, **43:** 105–119, 1948.

27. Gross, J., and Pitt-Rivers, R. Recent knowledge of the biochemistry of the thyroid gland. Vitamins and Hormones, **11:** 159–172, 1953.

27a. Harris, J. I., and Roos, P. Amino acid sequence of a melanophore-stimulating peptide. Nature, **178:** 90, 1956.

28. Hechter, O., and Pincus, G. Genesis of the adrenocortical secretion. Physiol. Rev., **34:** 459–496, 1954.

29. Hellerstein, S., *et al.* Effect of hypothalamic extract on leucocytes of infant rats. Am. J. Physiol., **171:** 106–113, 1952.

30. Hoch, F. L., and Lipmann, F. Uncoupling of respiration and phosphorylation by thyroid hormones. Proc. Natl. Acad. Sci. U. S., **40:** 909–921, 1954.

31. Holtorff, A. F., and Koch, F. C. The colorimetric estimation of 17-ketosteroids and their application to urine extracts. J. Biol. Chem., **135:** 377–392, 1940.

32. Howard, K. S., *et al.* Structure of β-corticotropin: final sequence studies. J. Am. Chem. Soc., **77:** 3419–3420, 1955.

33. Layton, L. L. Cortisone inhibition of mucopolysaccharide synthesis in the intact rat. Arch. Biochem. Biophys., **32:** 224–226, 1951.

34. Leathem, J. H., and Crafts, R. C. Progestational action of desoxycorticosterone acetate in spayed-adrenalectomized cats. Endocrinology, **27:** 283–286, 1940.

35. Lemon, H. M., *et al.* Metabolism of testosterone by neoplastic human prostate. J. Clin. Endocrinol. and Metab., **13:** 948–956, 1953.

36. Lerner, A. B., and Lee, T. H. Isolation of homogeneous melanocyte-stimulating hormone from hog pituitary gland. J. Am. Chem. Soc., **77:** 1066–1077, 1955.

37. Lerner, A. B., *et al.* The mechanism of endocrine control of melanin pigmentation. J. Clin. Endocrinol. and Metab., **14:** 1436–1490, 1954.

37a. Li, C. H., and Chung, D. Amino acid composition of hypophyseal growth hormone as determined by paper chromatography of the dinitrophenol amino acids. J. Biol. Chem., **218:** 33–40, 1956.

38. Li, C. H., *et al.* Isolation and properties of the anterior-hypophyseal growth hormone. J. Biol. Chem., **159:** 353–366, 1945.

38a. Li, C. H., *et al.* Action of chymotrypsin on hypophyseal growth hormone. J. Biol. Chem., **218:** 41–52, 1956.

39. Lieberman, S., and Teich, S. Recent trends in the biochemistry of the steroid hormones. Pharmacol. Reviews, **5:** 285–380, 1953.

40. Maddock, W. O., and Nelson, W. O. Effects of chorionic gonadotropin in adult men. J. Clin. Endocrinol. and Metab., **12:** 985–1015, 1952.

40a. McDonald, K. R., and Weise, V. K. Effect of arginine-vasopressin and lysine-vasopressin on plasma 17-hydroxycorticosteroid levels in man. Proc. Soc. Exper. Biol. Med., **92:** 481–483, 1956.

41. Miller, G. L., and Andersson, K. J. F. The molecular weight of insulin. J. Biol. Chem., **144:** 459–464, 1942. An ultracentrifuge study of reduced insulin. *Ibid.*, 465–473.

42. Munson, P. L., *et al.* Isolation of dehydroisoandrosterone sulfate from normal male urine. J. Biol. Chem., **152:** 67–77, 1944.

43. Pearlman, W. H. The chemistry and metabolism of the estrogens; *in* Pincus, G., and Thimann, K. V., eds. *The Hormones*. Vol. I, pp. 351–400. New York, Academic Press, 1948.

44. Pedersen-Bjergaard, G., and Pedersen-Bjergaard, K. Sex hormone analyses. III. Estrogenic and gonadotropic substances in the urine from a woman with normal menstrual cycles and normal pregnancies. Acta endocrinol., **1:** 263–281, 1948.

44a. PHILLIPS, A. H., AND LANGDON, R. G. Influence of thyroxine and other hormones on hepatic triphosphopyridine nucleotide (TPN)—cytochrome reductase activity. Biochim. et Biophys. Acta, **19:** 380–382, 1956.
45. PINCUS, G. Some basic hormone problems. J. Clin. Endocrinol. and Metab., **12:** 1187–1196, 1952.
46. PITT-RIVERS, R. Mode of action of antithyroid compounds. Physiol. Rev., **30:** 194–205, 1950.
47. PITT-RIVERS, R., *et al.* Conversion of thyroxine to 3,3′,5-triiodothyronine in vivo. J. Clin. Endocrinol. and Metab., **15:** 616–620, 1955.
48. POPENOE, E. A., *et al.* Oxytocic activity of purified vasopressin. Proc. Soc. Exp. Biol. Med., **81:** 506–508, 1952.
49. PRICE, W. H., *et al.* The effect of anterior pituitary extract and of insulin on the hexokinase reaction. J. Biol. Chem., **160:** 633–634, 1945.
50. PURVES, H. D., AND GRIESBACH, W. E. The site of thyrotrophin and gonadotrophin production in the rat pituitary studied by McManus-Hotchkiss staining for glycoprotein. Endocrinology, **49:** 244–264, 1951.
51. REINEKE, E. P., *et al.* The effect of progressive iodination followed by incubation at high temperature on the thyroidal activity of iodinated proteins. J. Biol. Chem., **147:** 115–119, 1943.
52. ROCHE, J., *et al.* Caractérisation des iodohistidines dans les protéines iodées (thyroglobuline et iodoglobuline). Biochem. et Biophys. Acta, **8:** 339–345, 1952.
53. ROSENCRANTZ, H. An antimony trichloride reagent suitable for the detection and estimation of non-ketonic steroids. Arch. Biochem. Biophys., **44:** 1–8, 1953.
54. RYAN, K. J., and ENGEL, L. L. The interconversion of estrone and estradiol by human tissue slices. Endocrinology, **52:** 287–291, 1953.
55. SAHYUN, M. Crystalline insulin of low zinc content. J. Biol. Chem., **138:** 487–490, 1941.
56. SALTER, W. T. The chemistry and physiology of the thyroid hormone; *in* PINCUS, G., AND THIMANN, K. V., eds. *The Hormones*. Vol. II, pp. 181–300. New York, Academic Press, 1950.
57. SALTER, W. T., *et al.* Goitrogenic agents and thyroidal iodine. Their pharmacodynamic interplay upon thyroid function. J. Pharmacol. & Exper. Therap., **85:** 310–323, 1945.
58. SCHARRER, E., AND SCHARRER, B. Hormones produced by neurosecretory cells. Recent Progr. Hormone Res., **10:** 183–240, 1954.
59. SEGALOFF, A. The effect of partial hepatectomy on the inactivation of alpha-estradiol. Endocrinology, **38:** 212–213, 1946.
60. SEIFTER, J., *et al.* Evidence for the direct effect of steroids on the ground substance. Ann. N. Y. Acad. Sc., **56:** 693–697, 1953.
61. SLUYTERMAN, L. A. Ae. Electrophoretic behavior in filter paper and molecular weight of insulin. Biochim. et Biophys. Acta, **17:** 169–176, 1955.
62. SNAPPER, I. Parathyroid hormone and mineral metabolism. Bull. N. Y. Acad. Med., **29:** 612–624, 1953.
63. SPROTT, W. E., AND MACLAGEN, N. F. Metabolism of thyroid hormones. The deiodination of thyroxine and triiodothyronine in vitro. Biochem. J., **59:** 288–294, 1955.
64. STADIE, W. C. The combination of insulin with tissue. Ann. N. Y. Acad. Sc., **54:** 671–684, 1951.
65. STAUB, A., *et al.* Purification and crystallization of the hyperglycemic-glycogenolytic factor (HGF). Science, **117:** 628, 1953.

66. STUNKARD, A. J., *et al.* The mechanism of satiety—effect of glucagon on gastric hunger contractions in man. Proc. Soc. Exptl. Biol. Med., **89:** 258–261, 1955.
67. SUTHERLAND, E. W. The effect of the hyperglycemic factor and epinephrine on enzyme systems of liver and muscle. Ann. N. Y. Acad. Sc., **54:** 693–706, 1951.
68. TABACHNICK, I. I. A., AND BONNYCASTLE, D. D. The effect of thyroxine on the coenzyme A content of some tissues. J. Biol. Chem., **207:** 757–760, 1954.
69. TAUROG, A., *et al.* The monoiodotyrosine content of the thyroid gland. J. Biol. Chem., **184:** 83–97, 1950.
70. THORN, G. W., *et al.* Highly potent adrenal cortical steroids: structure and biologic activity. Ann. Intern. Med., **43:** 979–1000, 1955.
71. ULRICH, F., *et al.* The effects of hypophyseal growth hormone on the metabolism of Ca^{45} in hypophysectomized rats. Endocrinology, **49:** 213–217, 1951.
72. VENNING, E. H., AND BROWN, J. S. L. Excretion of glycogenic corticoids and of 17-ketosteroids in various endocrine and other disorders. J. Clin. Endocrinol., **7:** 79–101, 1947.
73. WALKER, B. S., *et al.* Acid phosphatases. Am. J. Clin. Pathol., **24:** 807–837, 1954.
74. WAUGH, D. F. Regeneration of insulin from insulin fibrils by the action of alkali. J. Am. Chem. Soc., **70:** 1850–1857, 1948.
75. WHITE, A., *et al.* Prolactin. J. Biol. Chem., **143:** 447–464, 1942.
76. WOTIZ, H. H., *et al.* Metabolism of testosterone by human tissue slices. J. Biol. Chem., *in press.*
77. WOTIZ, H. H., *et al.* Steroid biosynthesis by surviving testicular tumor tissue. J. Biol. Chem., **216:** 677–687, 1955.

PART III

Growth

CHAPTER 9

Nucleoproteins and Growth

Growth in its most general sense is not characteristic of life alone. Crystals grow in evaporating solutions and deltas grow out to sea. These latter types of growth are, however, of the nature of physical accretions, where substances of a given type, such as KCl or fine silt, are added unchanged to an accumulation thereof. Living growth is characterized by the fact that substances *differing* from those composing the living organism are first changed to a suitable form and then assimilated. Living growth is also characterized, in multicellular organisms at least, by an increase in complexity with growth. Whereas a small crystal of potassium chloride can only grow to a large crystal of potassium chloride without changes in structure other than those involved in dimensional increases, the fertilized ovum grows, not to a large fertilized ovum, but to a complex and differentiated organism.

In considering the chemical basis of living growth it will be necessary to deal with *nucleoproteins*; these are compound proteins characterized by the possession of nucleic acids as prosthetic groups. *Nucleic acids* are complex substances among whose hydrolysis products are phosphoric acid (whence their acid reactions), pentoses, purines, and pyrimidines, the latter two being heterocyclic compounds the nature of which will be discussed. A detailed monograph of the chemistry of nucleic acids is that of Chargaff and Davidson (13).

NUCLEIC ACIDS

Nomenclature

There are two general types of nucleic acids. One is found, for the most part, in the nuclei of cells, the other mainly in the cell cytoplasm. One of the chemical differences between these two nucleic acids is that the former type contains a deoxypentose, the latter a pentose. For this reason the two are termed *deoxypentosenucleic acid* and *pentosenucleic acid*. Where the deoxypentose and the pentose have been isolated and characterized, they have turned out to be *D-2-deoxyribose* and *D-ribose*, respectively (formula I) (47) and, indeed, there is reason to believe that in pentosenucleic acids, particularly, the only sugar occurring is ribose. For that reason, the two nucleic acids are frequently referred to as *ribonucleic acid* and *deoxyribonucleic acid*. Deoxyribonucleic acid is most frequently abbreviated as DNA, less frequently as DRNA, or dorna. Ribonucleic acid is referred to as RNA.

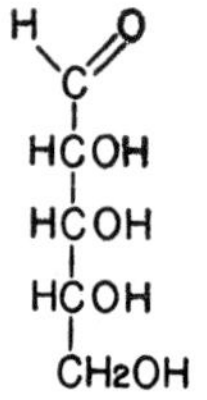

STRAIGHT-CHAIN FORMULA 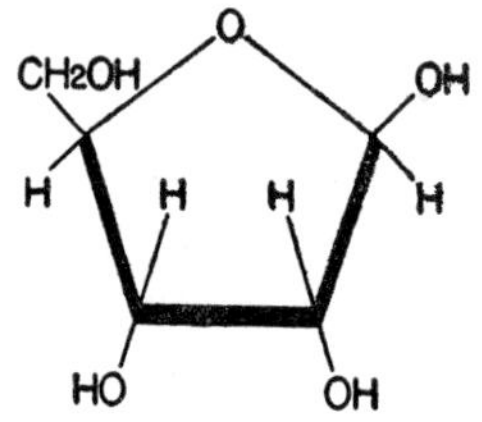　　　RING FORMULA

D-RIBOSE

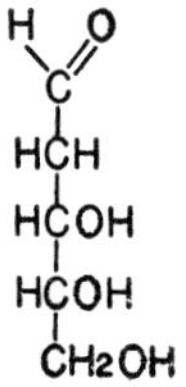

STRAIGHT-CHAIN FORMULA 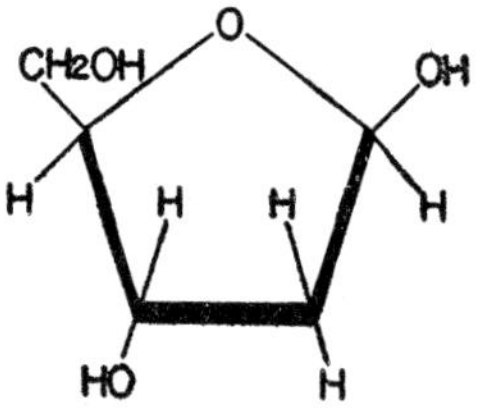　　　RING FORMULA

D-2-DEOXYRIBOSE

I

Earlier nomenclature of the nucleic acids was based on the sources from which they were derived and some acquaintance with the terms used for them is necessary if only because references to them in the literature might otherwise be confusing. DNA is most easily isolated from cells high in nuclear content—glandular sources such as spleen or thymus being excellent for the purpose. For this reason DNA has been frequently called *thymus nucleic acid* or *thymonucleic acid*. Similarly, RNA is most easily isolated from cells poor in nuclear material and rich in cytoplasm, such as yeast, and has therefore been called *yeast nucleic acid*. Again, because of their occurrence in the nucleus and cytoplasm, respectively, DNA and RNA are sometimes called *chromonucleic acid* and *plasmonucleic acid*, the "chromo" prefix of the first term being a reference to the chromosomes of the nucleus, where the DNA is localized.

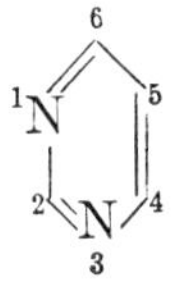

II. Pyrimidine

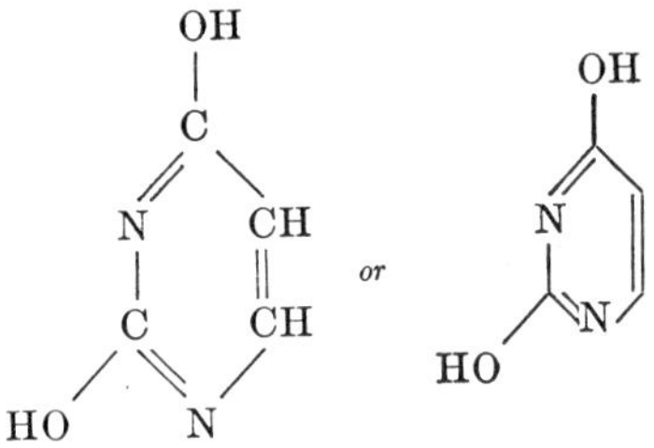

III. Uracil (enol form)

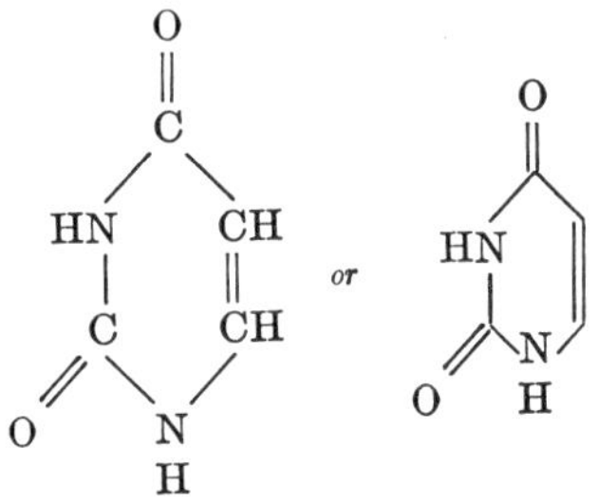

IV. Uracil (keto form)

Hydrolysis Products

Pyrimidines. Among the hydrolysis products of nucleic acids are three pyrimidines. A pyrimidine (formula II) is a heterocyclic compound characterized by the possession of a six-membered unsaturated ring containing two nitrogens separated by a carbon. The pyrimidines naturally occurring in nucleic acids are hydroxy and amino derivatives. The hydroxy derivatives are usually presented in their tautomeric keto forms. Thus, *uracil*, 2,6-dihydroxypyrimidine could be pictured as the *enol* form (formula III), in which the double bond pattern of pyrimidine is maintained, or the *keto* form (formula IV), involving a shift of two hydrogens and two double bonds. In solution, the two forms are undoubtedly in equilibrium, with the *keto* form predominant. To make more apparent the structural similarities of the various pyrimidines with one another, the *enol* forms will be used wherever possible in this book, without in any way implying the invalidity of the *keto* forms.

Two other pyrimidines occurring in nucleic acids in major quantities are *thymine* (formula V), which is 5-methyl uracil, and *cytosine* (formula VI), in which the 6-hydroxy group of uracil is replaced by an amino group. The distribution of these pyrimidine bases (so-called because of the weakly

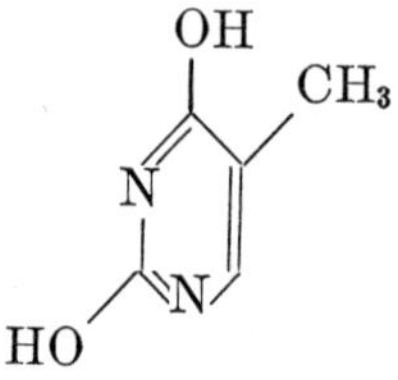

V. Thymine

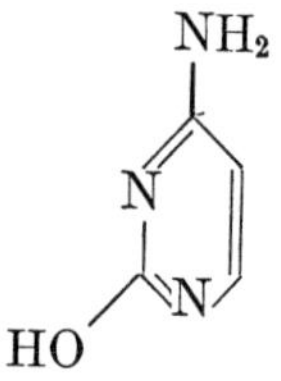

VI. Cytosine

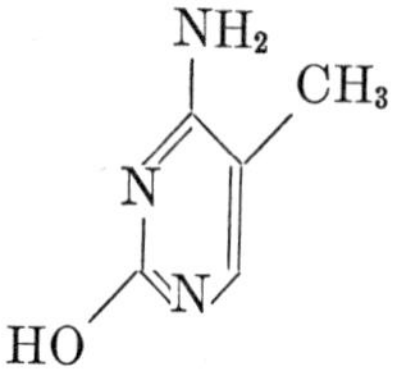

VII. 5-Methylcytosine

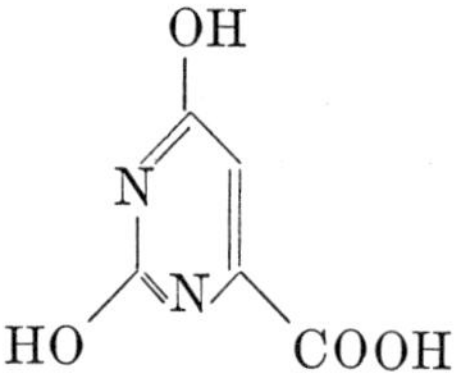

VIII. Orotic acid

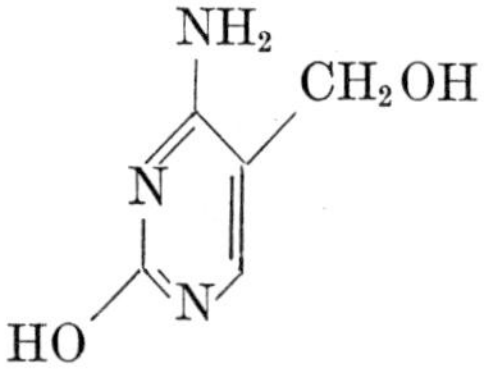

IX. 5-Hydroxymethylcytosine

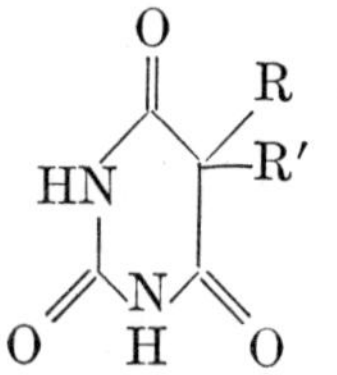

X. Barbiturates (*keto* form)

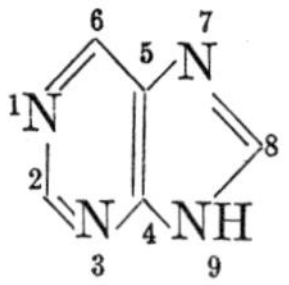

XI. Purine

basic properties of the ring nitrogens) is important. Cytosine occurs in both RNA and DNA, uracil in RNA only, and thymine in DNA only. Small quantities of *5-methylcytosine* (formula VII) have been located in some nucleic acids of animal origin (80), and *orotic acid* (4-carboxy uracil, formula VIII) has been found in the mold, neurospora. In certain strains of bacteriophage, *5-hydroxymethylcytosine* (formula IX) is found and cytosine is absent (34). Thiamine (vitamin B_1), however, contains a 2,5-dimethyl-6-amino pyrimidine as part of its molecule (see page 795), and somewhat more complex pyrimidine compounds are the various synthetic barbiturates so important as sedatives, analgesics, and anesthetics (formula X).

Purines. Two purines are also found among the hydrolysis products of nucleic acids. A purine (formula XI) is a heterocyclic compound characterized by the possession of two unsaturated rings, one six-membered and one five-membered, with two carbon atoms in common and two nitrogens in each ring.

The two purines found in nucleic acids are *adenine* (formula XII), 6-aminopurine, and *guanine* (formula XIII), 2-amino-6-hydroxypurine. Both adenine and guanine are found in both RNA and DNA. No other purines are found in any significant quantities in nucleic acids. Three purines are, however, significant in the metabolism of nucleic acids. These are *hypoxanthine* (formula XIV), *xanthine* (formula XV), and *uric acid* (formula XVI), which are, respectively, 6-hydroxypurine, 2,6-dihydroxypurine, and 2,6,8-trihydroxypurine. Uric acid is so called because of the weakly acidic properties of the hydroxyl groups, which are apparent if the formula is written in the enol form.

The alkaloids of tea, cocoa, and coffee are purine derivatives. Among these are caffeine (1,3,7-trimethylxanthine), theobromine (3,7-dimethyl-

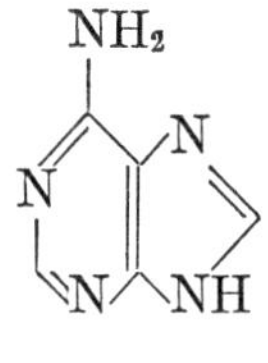

XII. Adenine

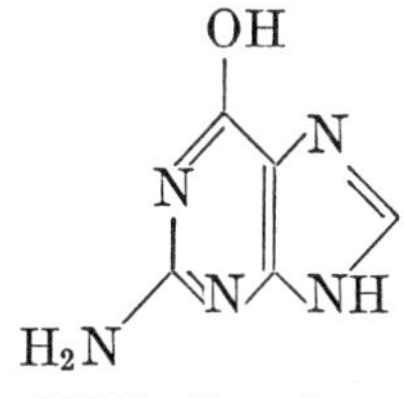

XIII. Guanine

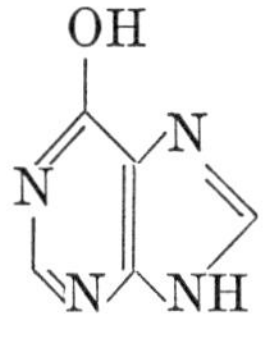

XIV. Hypoxanthine

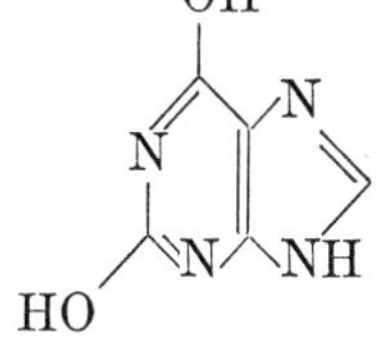

XV. Xanthine

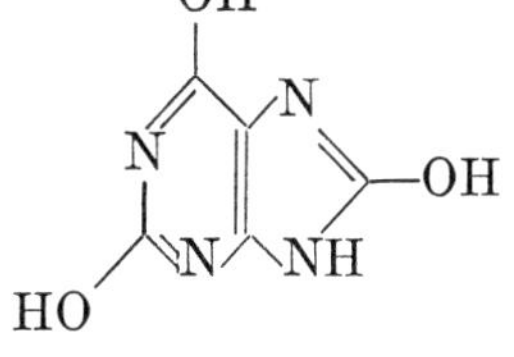

XVI. Uric acid

xanthine), and theophylline (1,3-dimethylxanthine). These are used as stimulants and diuretics.

Nucleosides. The combination of a nitrogenous base such as a purine or pyrimidine with a carbohydrate via an N-glycoside linkage is termed a nucleoside. In nucleic acids, purines and pyrimidines are so linked with ribose or deoxyribose, and by careful hydrolysis nucleosides can be separated.

Individual nucleosides derive their names from the purine and pyrimidine bases they contain, the purine nucleosides adding the suffix *osine,* the pyrimidine nucleosides the suffix *idine.* The four nucleosides occurring in RNA-hydrolyzates would thus be named *adenosine, guanosine, cytidine, uridine.* In DNA-hydrolyzates, *deoxyadenosine, deoxyguanosine, deoxycytidine* and *thymidine* are found. Of limited occurrence in DNA-hydrolysates, *5-methyldeoxycytidine* and *5-hydroxymethyldeoxycytidine* have also been identified. It should be especially noted that cytosine is the pyrimidine base itself, and not a nucleoside. Nucleosides can, in general, be classified as *ribosides* or *deoxyribosides.*

Two purine nucleosides, not occurring in nucleic acids, but nevertheless of interest in nucleic acid metabolism, are *inosine,* the riboside of hypoxanthine and *xanthosine,* the riboside of xanthine.

All natural ribonucleosides are beta-D-ribofuranosides, and all natural deoxyribonucleosides are beta-2-deoxy-D-ribofuranosides. The sugar is attached at nitrogen 3 in the pyrimidine nucleosides (formula XVII) and

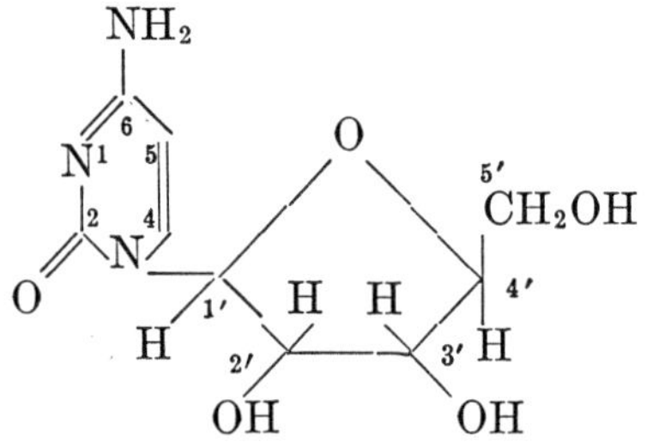

XVII. Cytidine

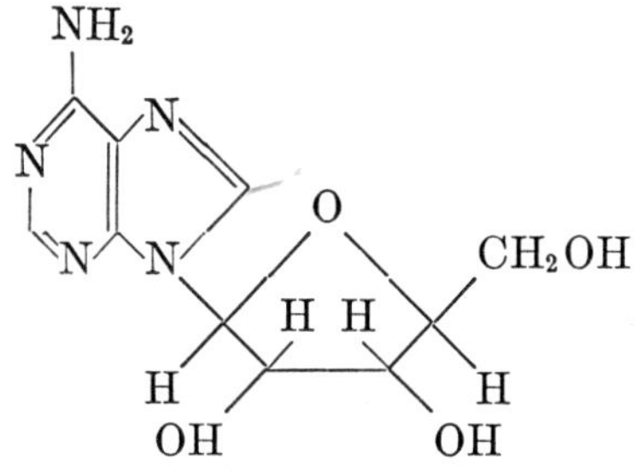

XVIII. Adenosine

at nitrogen 9 in the purine nucleosides (formula XVIII). Note that the numbering of the sugar portion of the molecule is indicated by the use of primed numbers.

Nucleotides. The phosphoric ester of a nucleoside is a nucleotide. The nucleosides of nucleic acids occur in combination with phosphoric acid and upon hydrolysis by alkali or by enzymes it is possible to obtain nucleotides as fragments. The major naturally-occurring nucleotides obtained from the hydrolysis of nucleic acids are *adenylic acid*, *guanylic acid*, *cytidylic acid*, *uridylic acid*, and *thymidylic acid*. The nature of the purine or pyrimidine base is in each case obvious from the name. Again, the names do not in themselves indicate the nature of the carbohydrate portion, so that it is more precise to sepak of adenine ribose phosphate (or adenine ribose nucleotide) and adenine deoxyribose phosphate (or adenine deoxyribose nucleotide).

The point of attachment of the phosphate group to the nucleoside is always on the hydroxyl of one of the carbohydrate carbons, but is not invariably on a particular one. Both adenosine-2′-phosphate and adenosine-3′-phosphate have been isolated from among the hydrolysis products of RNA, (6, 14) and similar isomers of yeast cytidylic acid are also thought to exist (46). Adenosine-5′-phosphate and its derivatives can be obtained from muscle. This distinction between the sources of adenosine-3′-phosphate and adenosine-5′-phosphate has caused them to be referred to, in the literature, as *yeast adenylic acid* (formula XIX) and *muscle adenylic acid* (formula XX), respectively.

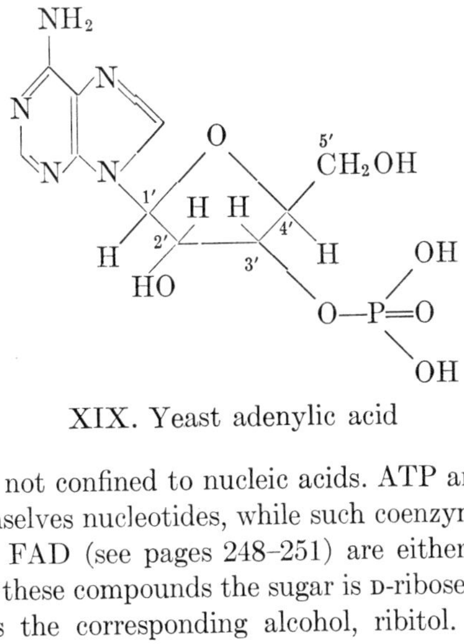

XIX. Yeast adenylic acid

Nucleotides are not confined to nucleic acids. ATP and ADP (see pages 225–226) are themselves nucleotides, while such coenzymes as DPN, TPN, FM, UDPG, and FAD (see pages 248–251) are either nucleotides or dinucleotides. In all these compounds the sugar is D-ribose, except in FM and FAD, where it is the corresponding alcohol, ribitol. Glucose occurs in

UDPG but not linked as part of a nucleoside. In DPN, TPN, and FAD, adenine occurs, linked to ribose precisely as in nucleic acids, while in UDPG, it is uracil that is so linked. That purines or pyrimidines are not

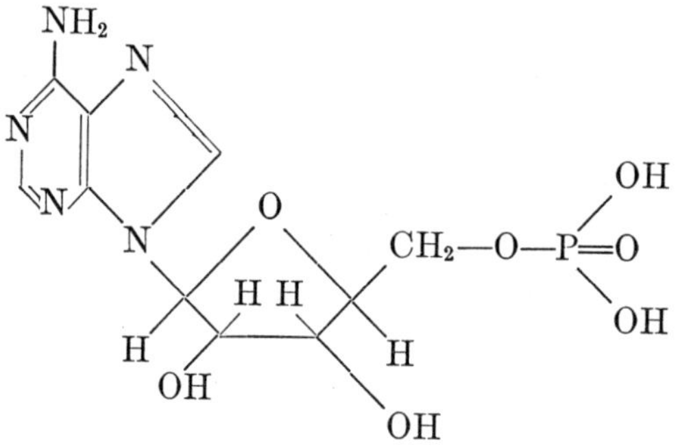

XX. Muscle adenylic acid

essential to the formation of a nucleoside link is shown in DPN and TPN where a pyridine derivative forms an N-glycoside bond, and in FM and FAD where an isoalloxazine derivative does so.

Oligonucleotides. Gentle hydrolysis of nucleic acids yields fragments still larger than the nucleotides just discussed. These consist of a variable number of nucleotides connected by phosphate ester linkages. These are referred to as *dinucleotides*, *trinucleotides*, *tetranucleotides*, and so on, depending on the number of nucleotides present in the fragment. As a group, these constitute the *oligonucleotides*. Where the number of nucleotides contained in the fragment is large or indefinite, it is called a *polynucleotide*. The nucleic acids themselves are in this sense polynucleotides, and may be classified as polydiesters of phosphoric acid. Phosphoric acid forms the internucleotide link, which is between the 3′ carbon of one sugar and the 5′ carbon of another (75) (formula XXI). In RNA, the presence of an unesterified ribose hydroxyl at the 2′ position raises the possibility of branching in the polynucleotide backbone (74).

Polynucleotide Structure

Nucleic acid molecular weights. Early studies on nucleic acids involved their isolation by alkaline extraction. Under such conditions, considerable depolymerization into relatively small oligonucleotides took place, so that the final product tended to have an average molecular weight roughly corresponding to that of a tetranucleotide, that is, about 1200. For that reason, it was once thought that the nucleic acids were actually tetranucleotides in their native state, each molecule containing one each of the four bases. Methods of extraction involving milder conditions, such as the use of neutral dilute saline solutions, have succeeded in providing

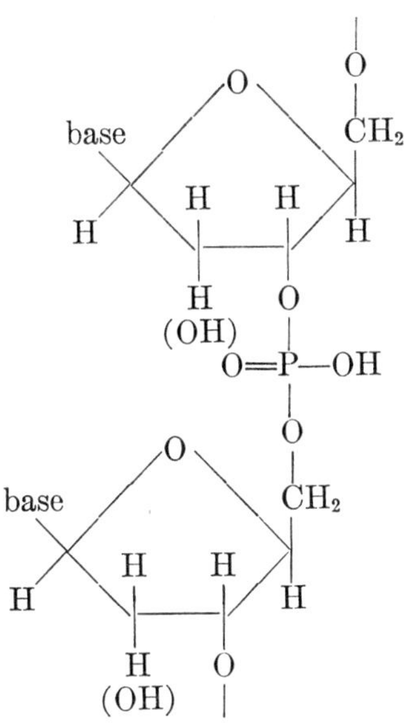

XXI. Internucleotide link in nucleic acids

preparations more nearly native. These were found to be astonishingly high in molecular weight by experiments involving diffusion and the ultracentrifuge, while studies in viscosity and streaming birefringence showed them to be markedly asymmetric as well. The latter property seems to be borne out by electron-microscope studies (3).

Of the two, DNA is the more complex molecule, molecular weights up to nearly 8,000,000 having been reported (57). The over-all length of such a molecule is 650 millimicrons. Axial ratios are from 400 to 500 (58). For RNA preparations, values over 500,000 (26) have been reported. In size, at least, nucleic acids are thus readily seen to be of the order of complexity of proteins.

From the pronounced asymmetry of the nucleic acid molecules, (the RNA preparation referred to just above possessed an axial ratio of 70) it is most often assumed that the molecules are an extended series of nucleotides rather than a more complex two- or three-dimensional network.

Base distribution. So far we have presented a somewhat simple picture, analogous in a way to that of an extended polypeptide chain consisting of four different amino acids. From what is becoming known of the diverse functions of the nucleic acids, more complexity would be expected and certainly heterogeneity of nucleic acids even within cells of the same tissue of the same animal has been observed (17, 55). This heterogeneity shows itself most clearly in the distribution of purine and pyrimidine bases within the nucleic acid molecule.

The results of enzymatic studies (using the specific nucleic acid hydrolyzing enzymes *ribonuclease* and *deoxyribonuclease*) seem to show that the bases are distributed neither evenly nor randomly. The nucleases catalyze the hydrolysis of the internucleotide link. They are specific for some links and not others. Analysis of the enzymatic digests of the DNA from wheat and sperm (65, 66) yield a number of di-, tri- and pentanucleotides, which, apparently, resist further hydrolysis. Fragments of the polynucleotide chain, sufficiently large to be retained by a dialyzing membrane, are left completely untouched after exhaustive enzymatic treatment. This "core" is largely purine nucleotide in nature, while the small, dialyzable fragments split off by enzymatic action are comparatively rich in the pyrimidine nucleotides (48, 63). Assuming the polynucleotide chain is unbranched, it would follow that the molecule contains regions composed predominantly of purine nucleotides and others of pyrimidine nucleotides (71, 77).

Results from various laboratories agree in indicating that the several bases are not present in nucleic acids in equimolar porportions. Two main groups of deoxypentosenucleic acids have been distinguished (12); one group from animal tissues and yeast in which adenine and thymine preponderate and a second from various bacteria in which guanine and cytosine are the major constituents. These are termed the AT group and the GC group. In both cases, the molar ratio of total purine to total pyrimidine is about one.

Very little is known about the order of occurrence of the different bases along the polynucleotide backbone, less than is known in the case of the position of amino acids along the polypeptide backbone. Markham and Smith (49) describe a nucleotide chain, which they suppose to exist as a cyclonucleotide within a RNA molecule. The structure of the chain is:

ACUCUCCAGAGCUCCAAGUUGUUCCGCCUAGCA

where A, C, U, and G stand for adenylic acid, cytidylic acid, uridylic acid and guanylic acid respectively. The order seems no more periodic than do the amino acid chains worked out for insulin (see page 77).

Fine structure. There is more to nucleic acid structure than simply a long nucleotide chain, just as there is more to protein structure than simply a long polypeptide chain. Nucleic acids may be denatured as proteins may under conditions where the molecular weight is left unaffected, and weak, secondary bonds have been broken (18, 72).

Watson and Crick (78) have proposed, on the basis of x-ray diffraction studies, that DNA is composed of two helical structures with a common axis. The two structures are held together by hydrogen bonds between a purine of one helix and a pyrimidine of another. Since the two structures are separated from one another by a constant distance that makes the hydrogen bonding between a purine and a pyrimidine possible, two purines would not have room to exist side by side, while two pyrimidines would be too far apart to make hydrogen bonding possible. Hydrogen bonding can take place most easily between adenine and thymine and between guanine and cytosine (formula XXII) (where the cyosine may be replaced in part or in whole by 5-methylcytosine or 5-hydroxymethylcytosine). Watson and Crick make the supposition then that an adenine on either chain is balanced by a thymine on the other (and vice versa), while a guanine on either chain is balanced by a cytosine (or similar pyrimidine) on the other (and vice versa).

Thymine-Adenine Cytosine-Guanine

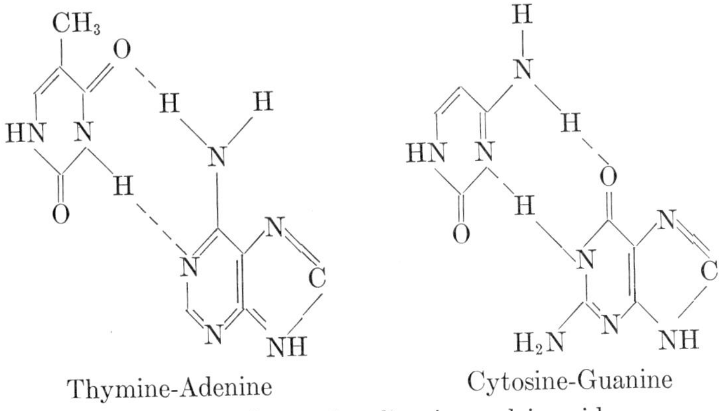

XXII. Hydrogen bonding in nucleic acids

If this is so, it would account for the fact that the purine/pyrimidine ratio in DNA preparations is always close to 1, and it would also account for the existence of AT and GC varieties of DNA. The base distribution is free to vary except for the provision that adenine and thymine remain equimolar and guanine and cytosine remain equimolar. If the adenine-thymine pair is preponderant, the AT variety results, otherwise the GC.

The breaking of hydrogen bonds between the two helices, with resultant slippage, tangling or other loss of specific configuration, would account for denaturation without change in molecular weight under mildly disruptive conditions.

Determination

Chemical methods. Both DNA and RNA are of universal occurrence in all nucleated cells whether plant, animal, or bacterial. Even viruses, which are subcellular organisms, contain at least one or the other of the nucleic acids, and often both.

Methods for detecting the gross quantity of nucleic acids in various tissues usually involve colorimetric determinations of either phosphate groups or of the sugar components of the nucleic acids. In the former case, the tissue must be first fractionated in order to remove acid-soluble phosphate and phospholipids which would otherwise interfere (64). In the latter case the two nucleic acids may be distinguished from one another by the use of appropriate reagents. Deoxy-sugars are in general more active than their fully oxygenated relatives so that DNA will react with some substances with which RNA will not. The most widely used chemical test for DNA consists of heating it in the presence of diphenylamine under appropriate conditions, yielding a blue color. No color is obtained with RNA. The specific detection of RNA is the harder problem. Orcinol and phloroglucinol show reactions with both DNA and RNA, but are more sensitive with respect to the latter.

Ultraviolet microspectrophotometry. For the determination of the intracellular distribution of nucleic acids, ordinary chemical methods are insufficient. Caspersson (11) has described a spectrophotometric technique which can record photographically the light absorption of various portions of tissue slices. Ultraviolet light is used since the unsaturated ring systems of the purines and pyrimidines absorb strongly in the range between 250 and 280 millimicrons, and a microscope attachment allows individual cells to be photographed. The only significant interference in this method is from the aromatic amino acids of proteins, since their ring systems also absorb in the indicated range. At equivalent concentrations their absorption is only about 10 per cent that of purines and pyrimidines, so that in most cases protein interference can be ignored.

Caspersson's results indicated the presence of nucleic acids both in the nucleus and in the cytoplasm, thus definitely establishing the fact that "nucleic acid" is somewhat of a misnomer and that they are not confined, as had been originally thought, to the cell nucleus. Furthermore, nucleic acids were not uniformly distributed throughout the cell. In the nucleus, they are associated with the chromosomes, and in the cytoplasm they tend to exist as aggregates. When such subcellular particles as mitochondria or microsomes are isolated by differential centrifugation of cytoplasm they are indeed found to be rich in nucleic acids.

Histochemical stains. The intracellular location of DNA and RNA can be separately determined by the use of specific histochemical stains. Of these the most famous is the *Feulgen stain*. The reagent used is the red dye, basic fuchsin, which has been decolorized through addition of sulfurous acid. If this is applied to tissue slices which have been exposed to the hydrolytic action of hydrochloric acid, the nuclear regions of each cell stain a brilliant red-purple, since the deoxyribose freed by acid hydrolysis of DNA converts the decolorized basic fuchsin to a colored dye. The less active ribose of RNA does not react in this manner.

Both nucleic acids are basophilic, and dyes such as methylene blue will stain intracellular regions which are rich in either nucleic acid. It is possible to locate RNA within the cell by staining serial slices before and after their exposure to some enzyme capable of catalyzing the depolymerization of RNA. The enzyme ribonuclease is specific in this respect for RNA and will not affect DNA. Once depolymerized, the RNA is easily washed out of the slice, and those regions which are stained before and not after enzymatic treatment are presumed to contain RNA (69).

On the basis of microspectrophotometry and histochemistry it has been fairly well established that DNA is located exclusively within the nucleus of the cell in close association with its chromatin, while RNA is found in the cytoplasm, and to a small extent in the chromosomes.

Occurrence

Through the use of the various methods of analysis mentioned above, estimates have been made of the nucleic acid content of individual cells. In a particular animal, the DNA content of cellular nuclei does not vary greatly from tissue to tissue. Isolated nuclei from beef liver, spleen and kidney, for instance, contained 6 to 7 $\times$ 10^{-6} micrograms per nucleus (44) and this seems to be about the figure for the majority of cells in all tissues studied. In each tissue, particularly in pancreas and liver, a small number of cells were found with DNA contents twice, four, and eight times normal (73). This is in agreement with the theory that in any given species the DNA content per set of chromosomes is constant and that the multiple values represent varying degrees of polyploidy (see page 428). In egg cells and sperm cells, with only half the normal complement of chromosomes, the DNA content is only half normal also.

Nucleic Acid Metabolism

Although the word *metabolism* has been used frequently in previous chapters, this is the first occasion where it has become necessary to deal systematically with a series of metabolic reactions. It will be well, therefore, to define our terms. Living growth, as has already been stated, is a process characterized by the utilization of substances differing more or less from the substance in the growing body. The body must be capable of taking the relatively complex components of its foodstuffs, converting them into smaller and simpler molecules, and then recondensing or refashioning these simpler molecules into complex substances that are characteristic of itself. Chemical reactions in the body which form large molecules from small ones and which, generally, store chemical energy in so doing, are *anabolic* in nature. The formation of glycogen from glucose and proteins from amino acids are examples of anabolism. Such reactions are also referred to as *biosyntheses*. In similar fashion, chemical reactions in the body which form small molecules from large ones and which, generally, release chemical energy in so doing, are *catabolic* in nature. The utilization by the body of glycogen or fat to provide energy is an example of catabolism, as are the various digestive processes. It is obvious that in the normal, mature body which is maintaining constant weight, the processes of anabolism and catabolism are on the average in balance. During periods of growth, anabolic processes predominate; while under conditions of starvation, the reverse is true. The sum of all reactions which are either anabolic or catabolic in nature is indicated by the term *metabolism*.

Anabolism. The biosynthesis of nucleic acids has been studied by feeding isotopically-labeled substances to rats and then determining the isotope content of their tissue nucleic acids. In general, it appears that purines and pyrimidines are most readily incorporated into tissue nucleic acids when they are fed in the form of nucleotides. When the nucleoside or the free base is fed, a large part of the labeled atoms, which are N^{15} in this case, appear in the form of allantoin, the final product of the degradation of purines in the rat. Closely related purines such as xanthine, hypoxanthine, and isoguanine (2-hydroxy-6-aminopurine) after ingestion likewise appear mainly in excreted allantoin. In general, it can be concluded that purines, as such, after absorption into the body take part only in catabolic processes, while the purines of nucleic acids are almost exclusively the product of biosynthesis from simpler substances.

To determine the nature of the simpler substances, various possible precursors labeled with C^{14} or N^{15} were fed to pigeons and their excreted uric acid was then broken down so that each atom in the purine ring could be separately studied for the extent of labeling. In this way it was found that the carbons in positions 2 and 8 of the purine ring were derived from formate ion, and the carbon in position 6 from carbonate ion (31). The C—C—N group of positions 4, 5, and 7 were derived from glycine. The nitrogen in position 1 is derived from aspartic acid, while those in positions 3 and 9 are derived from the amide nitrogen of glutamine (44a). The final ring closure involves the introduction of a formate carbon at position 2, with the aid of leucovorin (see page 811) and ATP (7, 24). Ring closure probably takes place after condensation of ribose-1-phosphate with 4-amino-5-imidazole carboxamide to form 5-imidazole carboxamide riboside (40) (formula XXIII). According to this scheme, the free purine does not occur at any stage and that would account for the fact that ingested purines are not incorporated into nucleic acid. There is no point at which they could enter the series of reactions.

The carbon 2 of the pyrimidine ring is derived from carbonate (31) while the methyl group of thymine is derived from formate (76). Purines are

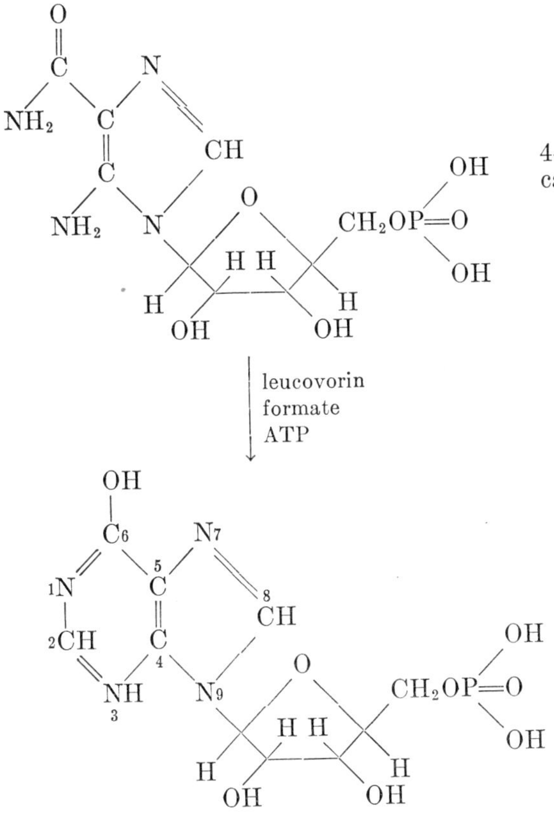

4-Amino-5-imidazole carboxamide ribotide

↓ leucovorin formate ATP

Inosinic acid

XXIII Purine ring formation

not pyrimidine precursors, a fact known from the experiments already described involving ingestion of labeled purines. Nor can ingested uracil, cytosine, or thymine themselves be incorporated into nucleic acids, although pyrimidine nucleosides and pyrimidine nucleotides can be so incorporated (60). Labeled orotic acid (uracil-4-carboxylic acid), after ingestion, has been found to increase the isotopic content in the pyrimidines of liver RNA of rats, but not in the purines. The acyclic precursor of pyrimidines is oxaloacetic acid, from which ureidosuccinic acid is eventually formed, a conversion in which the high energy compound, carbamyl phosphate (see page 607) is involved. Upon ring closure, ureidosuccinic acid becomes dihydroorotic acid, which is dehydrogenated to orotic acid (formula XXIV) (7). Nucleotide formation from orotic acid requires ATP and ribose-5-phosphate. These react to form AMP and 5-phosphoribosylpyrophosphate (PRPP), the reaction being catalyzed by a Mg^{++} activated enzyme found in mammalian liver (41). The PRPP, in turn, reacts with orotic acid to form orotidine-5′-phosphate (orotidylic acid) which, upon decarboxylation, becomes uridylic acid (formula XXV) (45).

Oxaloacetic acid —transamination→ Aspartic acid —$H_2N{-}C(=O){\sim}O\text{Ⓟ}$ (carbamyl phosphate)→ Ureidosuccinic acid —$-H_2O$→ Dihydroorotic acid —$-2H$→ Orotic acid

XXIV. Pyrimidine ring formation

The anabolism of the pentoses is dealt with in Chapter 14.

Catabolism. Ingested nucleic acids are broken down by a variety of digestive enzymes to smaller portions capable of being absorbed through the intestinal wall. The nucleic acids are freed from the nucleoproteins of which they are the prosthetic groups by the action of the acid of the stomach and the proteases of the gastric and pancreatic juices. The polynucleotide structure of the nucleic acid is then hydrolyzed to smaller units by various nucleases (polynucleotidases). A *nuclease* is an enzyme which catalyzes the depolymerization of polynucleotides by splitting the C—O—P linkage between two adjacent nucleotides. For this reason they are spoken of as depolymerizing enzymes or *depolymerases*. However, like all enzymes, they catalyze the reverse reaction as well, so that they are also *polymerases* (33). From the nature of the reaction they catalyze, it can be seen that they are hydrolyzing enzymes and more specifically, phosphodiesterases. Nucleases are known which attack RNA only, and others which attack only DNA. The former is most frequently known as *ribonuclease*, the latter as *deoxyribonuclease* or *dornase*. The pancreas and pancreatic juice are par-

ATP + → AMP +

5-Phosphoribosylpyrophosphate
(PRPP)

—Pyrophosphoric acid

+

Orotic acid

Orotidylic acid

$-CO_2$

Uridylic acid

XXV. Nucleotide biosynthesis

ticularly rich in both ribonuclease and deoxyribonuclease, and it is therefore probable that the first steps in nucleic acid depolymerization take place in the duodenum. Both nucleases have been prepared in crystalline form from beef pancreas (42, 43). Ribonuclease is the more stable of the two and has the rather low molecular weight, for a protein, of 14,000 (32).

Nucleic acids are not depolymerized all the way to the mononucleotide stage by pancreatic nuclease action. Gutman (27) presents a scheme in which the nucleic acids are depolymerized only to various oligonucleotides by the depolymerases of pancreatic juice. The oligonucleotides are then further hydrolyzed to the mononucleotide stage by the action of various specific nucleases in the intestinal mucosa. Mononucleotides can be dephosphorylated by the alkaline phosphatase of the small intestine to the various nucleosides. Both nucleotides and nucleosides can be absorbed in the intestine.

Of this absorbed material, nucleotides can be incorporated, to some extent, into nucleic acid. The more general state of affairs is, however, a continued catabolism both for ingested nucleotides and for degradation products of tissue nucleic acid. This takes place chiefly in the liver.

Nucleosides in general can be broken down to ribose (or deoxyribose) and purine (or pyrimidine) by the action of *nucleosidases*. Traces of these are present in the small intestine which are specific only for purine nucleosides. Purine nucleosidases are found in greater concentration in spleen, lungs, liver, and heart. Pyrimidine nucleosidases undoubtedly exist but little is as yet known about them. It has been shown that the action of nucleosidase is not a hydrolysis but a phosphorolysis which results in the production of base plus ribose-1-phosphate (39) (formula XXVI). For this reason Kalckar suggests that the enzyme be called *nucleoside phosphorylase*. The reaction is reversible and its direction is controlled by the concentration of free phosphate. Ribose-1-phosphate is reversibly converted to ribose-5-phosphate, and this in turn can reversibly form ribulose-5-phosphate, which by forming glyceraldehyde-3-phosphate can be catabolized by glycolysis (see Chapter 14).

The catabolic fate of pyrimidines and purines differs widely. The pyrimidine ring is reduced, then broken in the liver (9, 61). One suggested scheme of catabolism is shown in formula XXVIa (20a). The purine ring, on the other hand, remains intact. Guanine is deaminated by the action of *guanase* to xanthine. Xanthine can also be obtained by an alternative route in which adenosine and guanosine are first deaminated by the respective *nucleoside deaminases* to inosine and xanthosine. These latter are then split by purine nucleosidase to hypoxanthine and xanthine. Hypoxanthine is oxidized in the liver to xanthine by the enzyme *xanthine dehydrogenase*, which also catalyzes the further oxidation of xanthine to uric acid. In human beings

Cytidine + H_3PO_4 → Cytosine + Ribose-1-phosphate

XXVI

Thymine $\xrightarrow{+2H}$ Dihydrothymine $\xrightarrow{+H_2O}$ Beta-ureidoisobutyric acid $\xrightarrow{+H_2O,\ -CO_2,\ -NH_3}$ Beta-aminoisobutyric acid

XXVIa-Catabolism of the pyramidine ring

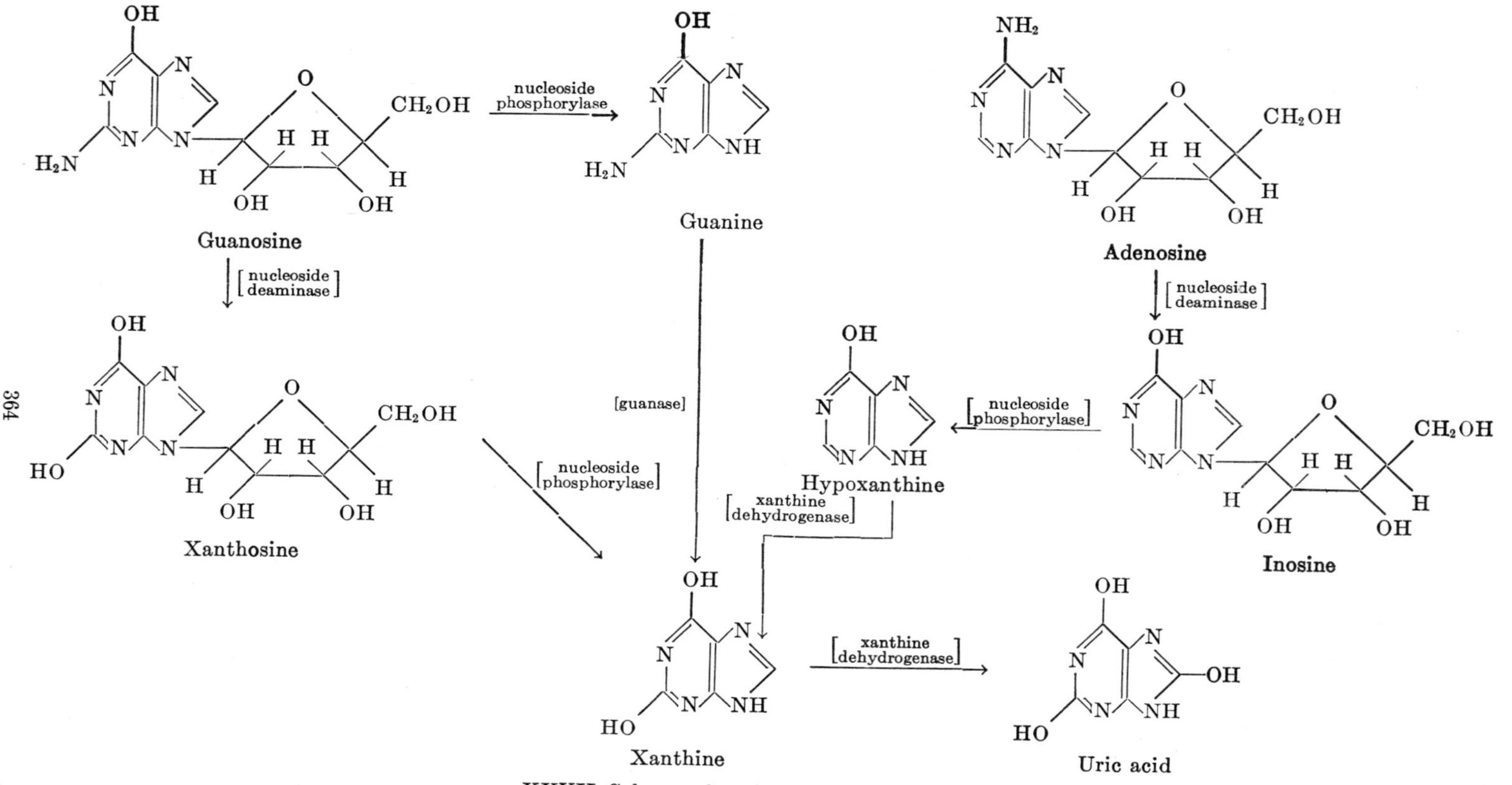

XXVII. Scheme of purine metabolism in man

uric acid is the final product of purine metabolism and is excreted in the urine as such (formula XXVII).

Gout. The occurrence of gout is the direct result of the fact that uric acid is the final product of purine metabolism in the human species. Uric acid is highly insoluble in water, and to be removed successfully in the urine it must be present in low concentrations. When, for any reason, uric acid (or perhaps closely related precursors) is produced in quantities greater than the kidney will excrete, there may be deposition of the substance in the joint cartilages. The joint usually first affected is that of the big toe. This deposition has very painful consequences. In one study six normal subjects excreted a mean of 390 mgm. of uric acid per day, while five young gouty patients excreted 567 mgm. of uric acid per day (21). When both normal and gouty subjects were maintained on a purine-free diet, the plasma uric acid concentrations were found to be 4.1 mgm. per 100 ml. for the normals and 7.7 mgm. per 100 ml. for the gouty cases.

The reasons for the increased activity of purine biosynthesis in the gouty subjects are not yet known, but gout would nevertheless be impossible with any naturally-occurring rate of uric acid production if man were equipped with the enzyme *uricase*. Uricase occurs in all species of mammals other than man and the anthropoid apes. It is deficient, but not absent, in one breed of dog, the Dalmatian coachhound. Uricase catalyzes the conversion of uric acid to allantoin (formula XXVIII). Allantoin, while

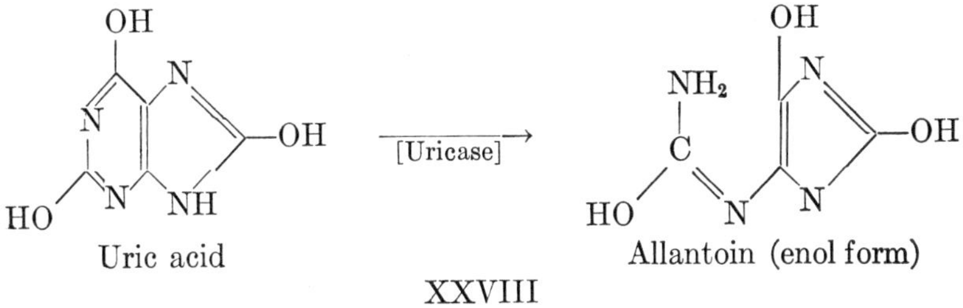

Uric acid —[Uricase]→ Allantoin (enol form)

XXVIII

not soluble in the ordinary laboratory sense of the word, is, despite that, some twenty times as soluble as uric acid. At 20°C., 100 ml. of water will dissolve 60 mgm. of allantoin, but only about 3 mgm. of uric acid. It should be pointed out that uric acid is not a complete dead end in man even though uricase is absent. It has been found that in normal men, more than one-fifth of uric acid injected intravenously is degraded to other nitrogenous substances (81). Fully 78 per cent, however, was excreted unchanged.

Gutman (27) distinguishes between two varieties of gout—chronic tophaceous gout and acute gouty arthritis. That the manifestations of the former are due to the deposition of uric acid or urate in tissues he thinks certain, but he finds reason to doubt that urate is the actual villain in the latter case. Thus an acute attack of gout can not be induced in either normal or gouty subjects by feeding or injecting uric acid. Furthermore, in gouty patients acute gout often involves joints that do not show extensive urate deposits under x-ray and, in fact, as urate accumulates the joint becomes less painful and acute symptoms are transferred to other joints as yet relatively free of urate deposits. Gutman concludes that acute gouty arthritis "is evoked by some precursor of uric acid, as yet unidentified", rather than by uric acid itself.

The situation is made more serious and less amenable to correction by such a simple expedient as putting the patient on a low-purine diet, since the body synthesizes uric acid from the simple molecules already discussed in connection with the biosynthesis of purines. Dietary lipid, protein, and carbohydrate can all be looked upon as uric acid precursors. A case of xanthinuria in which a xanthine stone was passed and high concentrations of the uric acid precursor, xanthine, with only a trace of uric acid, appearing in the urine has been reported (16).

NUCLEOPROTEINS

Nature of Protein Moiety

In spermatozoa. Because the spermatozoon is little more than a small bag of tightly packed nucleoprotein (in the form of chromosomes) equipped with a cytoplasmic tail for locomotive purposes, it has been much used as a source of DNA. For similar reasons, it has been possible to investigate the chemical nature of sperm deoxypentosenucleoprotein as a whole and that of fish sperm in particular because of the easy availability of large quantities thereof. It was early discovered that at least part of the protein of sperm deoxypentosenucleoprotein was highly basic and of exceedingly simple structure for a protein. Such proteins are, according to the classical classification, histones or even, in the case of the sperm of certain fish such as salmon or herring, protamines.

The protamines, in particular, are unusually simple proteins. One of them, salmine, the protamine of salmon sperm, has already been described on page 95 as consisting of only seventy-two amino acids, of which no less than 50 are arginine. The protamines from the sperm of other species of fish are of similar nature. Felix (20) finds them to be heterogeneous; the clupeine of herring sperm, for instance, being made up of at least six components. Protamine molecular weights range from 2,000 to 12,000, with an average molecular weight of 5,000. Felix suggests that the amino acid sequence is a repetitive one. The N-terminal sequence is proline-alanine and the C-terminal sequence arginine-arginine. Between these is a hexapeptide unit, arg-arg-arg-arg-M-M (where M stands for some neutral amino acid) repeated a number of times. The various protamines would differ in the number of times the unit was repeated and in the nature of the neutral amino acids in the individual units.

Most types of sperm cell, including those of man and other mammals (15), contain histone. No known type of sperm contains both histone and protamine. The proportion of arginine in histone, while high enough to make the molecule markedly basic, is considerably lower than in protamines. Only 25 per cent of the histone nitrogen is attributable to arginine, as compared with 90 per cent in the case of protamines. Histones also contain tyrosine and sulfur-containing amino acids, which protamines do not, and are large enough in molecular weight to be retained by a dialysis bag, which protamines are not.

The union between nucleic acid and the protamine or histone in sperm deoxypentosenucleoprotein is electrovalent in nature, the bond existing between the acidic negatively charged phosphoric acid groups of the nucleic acid and the basic positively charged guanidine groups of the protamine or histone arginine. As a result, the two substances can be separated relatively easily by varying the pH appropriately or by increasing the ionic strength of the solution.

Biochemists have been somewhat disturbed on occasion over the comparative simplicity of these protein constituents of spermatozoa. Stedman and Stedman (68) have reported the presence in sperm of a more complex and more representative protein, differing from histones chemically in its tryptophane content and slightly acidic nature.

In genes. The nucleoproteins of somatic cells have been less well characterized than those of spermatozoa, largely because of the greater complexity of cellular composition and the lower concentration of nucleoprotein. The deoxypentosenucleoprotein of the cell nucleus has aroused particular attention because of its association with chromosomes, and the distinct possibility that the genes themselves are large but single nucleoprotein molecules.

The deoxypentosenucleoproteins of cell nuclei are usually isolated as nucleohistones and for a long time it was thought that histones were the predominant protein of somatic cell nuclei as they were of fish spermatozoa. The possibility remained, however, that nucleohistones were artifacts. The large and relatively unstable native nucleoproteins might be disrupted into nucleic acid and protein moieties by the preparative techniques, however mild. Where protein is not then removed by precipitating agents, nucleic acid when isolated would, due to its own acidic nature, tend to remove from the protein mixture the basic portions—that is, the histones. The association of the nucleic acid and histone would thus be purely a result of chemical manipulation and, in fact, nucleohistones can be separated and recombined very easily by changes in acidity. For this reason some have discarded the term nucleohistone as being too strongly indicative of a naturally-occurring substance and have substituted the more artificial-sounding term, *histone nucleate.*

In 1943, Stedman and Stedman (67) reported the isolation of the tryptophane-containing acidic protein mentioned above. They gave it the name *chromosomin.* It was found to be a constituent of all cell nuclei tested including those of sperm. A similar protein called *chromosin* has been isolated from nuclei by Mirsky and Pollister (53).

In viruses. Further knowledge concerning nucleoproteins can be obtained from a study of viruses. There is no glib general definition for the term *virus.* It is applied to those agents of disease which (a) can pass through a standard filter capable of holding back bacteria, and which (b) fail to multiply in any medium other than living cells of a susceptible host. With the advance of knowledge the boundary between viruses as so defined and bacteria proper has become less distinct. Viruses thus vary in size and complexity of structure from the equivalent of large protein molecules almost to the equivalent of small bacteria. For an introduction to viruses the student is referred to Burnet (8).

Chemically, all known viruses contain at least one type of nucleoprotein. Many possess both types and various other constituents as well. The advantage of viruses as a means of studying nucleoproteins resides in the fact that in viruses smaller amounts of non-nucleoprotein material may be found as compared with amounts found in bacteria and other cells. Indeed, the smaller viruses, particularly those infecting plants, are virtually pure nucleoprotein which can actually be prepared in crystalline form without destroying the infectivity of the molecule. Turnip yellow mosaic virus contains as much as 37 per cent RNA and no DNA.

The protein content (not counting the nucleic acid) of the viruses thus far studied is always more than 50 per cent. In the case of the tobacco mosaic virus it is about 94 per cent (25). In comparison, the protein content of the chromosomes of the germinal cells of the mouse, and therefore, presumably, of gene nucleoprotein, is 80 per cent (52). In viruses, the protein is not the basic histone type, but is a comparatively acidic, globulin-like substance, much more closely resembling chromosomin. In some viruses, at least, the protein is localized in the outer regions or shell of the virus, while the nucleic acid occurs in a central core running the length of the virus particle (28, 62). The protein shell governs the immunological behavior of the virus while the nucleic acid, which alone penetrates the host cell, governs the type of disease produced (20b). The nucleic acid and protein of tobacco mosaic virus may be separated and infective particles may be reconstituted, to a minor extent, from a mixture of these two non-infective portions of the virus (20c). Virus nucleic acid is composed of polynucleotide chains arranged in helical form (see page 354) and the protein of the tobacco mosaic virus, for instance, shows evidence, on the basis of x-ray diffraction studies, of being composed of repeated subunits arranged helically about the nucleic acid as a core. In the tobacco mosaic virus, there are about 1200 such subunits per virus molecule, the molecular weight of the subunit being approximately 35,000.

Nucleoproteins and Autoreproduction

Living growth consists fundamentally of the synthesis of protein molecules. Since protein molecules are so complex, the production of such molecules in the specific form required for a particular function in a particular tissue of a particular organ would seem a formidable undertaking. The fact that specific protein molecules are produced innumerable times and almost always correctly is perhaps one of the most astonishing features of living tissue. Furthermore, if we postulate that proteins are synthesized by the action of specific enzymes, we must ask what synthesizes these enzymes which are themselves proteins, what synthesizes the enzyme-synthesizers, and so on for as long as we have breath to ask. To cut short this endless chain of questions we must postulate that some biochemical substance must have the property of *autoreproduction*—the ability to reproduce itself exactly when placed in the proper medium. *The simplest substances known to be autoreproducible are the genes and the viruses and both of these are largely or entirely nucleoprotein in nature.* To understand growth, therefore, it will be helpful to consider the role of nucleoproteins as the entities which dictate in some manner the exact nature of the specific proteins to be produced.

In nuclei. A vast amount of indirect evidence (see Chapter 11) in the field of genetics has led to the conclusion that chromosomes consist of chains of smaller structures which act as carriers of hereditary traits. These structures have been termed *genes*. Although genes have never been isolated in the chemical sense, it is likely that they are complex deoxypentose-nucleoprotein macromolecules. The gene is characterized by two important properties:

(1) It is autoreproducible. In the process of mitotic division it duplicates itself *exactly*, so that each daughter cell contains a set of identical genes (except in gametogenesis). In this way, specific biochemical characteristics are perpetuated through the entire structure of a living organism.

(2) It acts as a catalyst for the synthesis of a specific enzyme. Experiments with the mold *Neurospora crassa* have shown that strains varying in their requirements for various food factors can be produced after exposure to ultraviolet light (5). The new requirement for a particular food factor arises from the failure of the altered strain of mold to perform a certain chemical synthesis that had previously given it no trouble. This failure was undoubtedly due to the inability of the organism to synthesize an appropriate enzyme. A study of many such strains has resulted in the

conclusion that the loss of the ability to synthesize an enzyme was due to the change induced in a single gene as a result of the absorption of an energetic ultraviolet photon. Such experiments are considered to support the theory that each individual gene is responsible for the production of a single specific enzyme (35), and according to this "one gene-one enzyme" hypothesis, the number of different enzymes in a cell and the number of functioning genes are identical. Such a theory is not, however, universally accepted (37).

It should be obvious that if genes act as enzyme synthesizers, this in itself would enable them to control all species and individual differences. Specific proteins are synthesized under the catalytic influence of specific enzymes. The loss of a single enzyme, the one capable of oxidizing tyrosine to the catechol derivative, will result in an albino offspring.

Any break in the hereditary pattern from parents to offspring which can not be accounted for by the simple reshuffling of genes inherited from mother and father (see Chapter 11) is termed a mutation. The procedure mentioned, where under ultraviolet bombardment an individual neurospora capable of synthesizing, for instance, the amino acid lysine from inorganic constituents produces asexually a new individual which is incapable of such a synthesis, is in effect an artificially produced mutation. Most of the mutations dealt with in such experiments are fundamentally changes in specific gene molecules. Such changes can be produced artificially by energetic radiation (ultraviolet, x-rays, gamma rays, neutrons, alpha-particles) and also by a variety of chemicals such as colchicine, nitrogen mustards, and methyl cholanthrene.

The methods whereby mutations are produced in the laboratory are drastic ones which apparently render a gene unable to perform a function. Spontaneous *back-mutations* have been observed in Neurospora in which an altered gene regains its function or in which perhaps a second gene is so changed as to take over the role of the first. The causes of "spontaneous" mutations in organisms during the course of evolution remain a field for speculation. The effect of cosmic rays has been suggested as well as the radiation effects of radioactive isotopes of biochemically significant elements which either occur naturally, as K^{40}, or are continually produced in small quantities as a result of cosmic ray bombardment, as C^{14} (1). Experimental proof for any of these hypotheses is as yet unsatisfactory.

If nucleic acid is the primary genetic material, the problem arises as to how it directs the synthesis of enzymes and other proteins, as well as its own autoreproduction? Our information is still too limited to permit a definite answer to this question, but certain current speculations may foreshadow the outlines of the true explanation.

It has been suggested (79) that the two helical polynucleotide components of the nucleic acid (see page 354) somehow unwind and separate. Each then might catalyze the synthesis of its polynucleotide partner, each adenine group inducing a thymine group to take up a position opposite, and vice versa, while guanine and cytosine behaved in the same complementary fashion. Thus one molecule of nucleic acid would become two identical molecules.

However, no mechanism for the hypothetical "untwisting", which would require considerable energy, has been suggested. Furthermore, a mechanism which merely accounts for the self-duplication of RNA or DNA is not enough, for the problem of protein synthesis remains. Gamow (22) points out that the shape of the empty spaces between the nucleotides in the double helix of nucleic acid would be determined by the arrangments of the four surrounding nucleotides. Since DNA consists chiefly of four different nucleotides (see page 350), it is possible to show that 20 different arrangements of four adjacent nucleotides are possible and that there could exist a one-to-one correspondence between these 20 types of internucleotide spaces and the 20 amino acids found generally in cellular proteins. The number and order of the amino acids in a protein would thus be uniquely determined by the number and order of the nucleotides in the double helix of a particular DNA molecule. Conversely, the protein so formed could act as a template for the building up of a molecule of DNA identical with the original.

Gamow's suggestion would seem to imply an alternate production of nucleic acid and protein molecules, instead of the more direct duplication of nucleoprotein in a single step. Kacser (38) has proposed a mechanism which would account for this. This author suggests that the chromosome is a "solid phase" adsorption complex between protein and nucleic acid. The interface between these two complementary structures constitutes the functional part of the assembly. If the protein and nucleic acid portions become separated at one end of the assembly, amino acids might begin to deposit on the nucleic acid portion, to build up the corresponding protein, and nucleotides might deposit on the protein portion. This process might continue up the assembly, zipper-fashion, until the original portions were completely separated and two new identical portions of nucleoprotein resulted.

This mechanism would account for the self-duplication of nucleoproteins, but there is no evidence that a molecule of nucleic acid is produced each time a molecule of protein is synthesized. Haurowitz (29) has suggested a way in which molecules of protein could be duplicated. According to him, while a protein is held in its expanded state by a molecule of nucleic acid, each amino acid residue attracts another amino acid of the same sort, analogous to what happens in the growth of a crystal. These new amino acids are then linked together by enzymes of the cathepsin type and the new molecule is cast off, leaving the protein-nucleic acid complex free to cause the synthesis of another molecule. This picture has the advantage that it could account for the production of the modified gamma-globulins known as antibodies (chapter 22).

Haurowitz (30) has estimated the time required to synthesize a molecule of protein. He finds that in the case of diphtheria antitoxin (molecular weight, 160,000), each active synthesizing site requires not over two seconds to make a protein molecule.

Ultraviolet microspectrophotometry has made it possible to study the DNA distribution within a chromosome. Caspersson's group has found that portions of the chromosome show a fine structure consisting of bands rich in nucleic acids, with the regions between relatively free of nucleic acid. The nucleic acid composes about 30 per cent of the banded regions, the absolute quantity within a single band being estimated at between five and fifty times 10^{-11} mgm. The remainder of the band is protein, rich in diamino acids, which may therefore be histone in character. The protein in the interband spaces is poorer in diamino acids and richer in tyrosine and tryptophane and thus more nearly corresponds to chromosomin. The banded regions of the chromosomes Caspersson calls *euchromatin*. Interspersed between euchromatin sections of the chromosomes are regions which lack the banded structure and which are termed *heterochromatin*. These are rich in proteins containing a high proportion of diamino acids.

During prophase the nucleus loses protein to the cytoplasm while the nucleic acid content increases slightly, soon reaching a maximum. In early prophase the nucleic acid-protein ratio of the nucleus is approximately 1:20; in late prophase, 1:5; and at metaphase, 1:3. During telophase the reverse process takes place as the tight dense chromosomes of metaphase spread out and become distributed through a rounded nucleus once more, this being accompanied by a proliferation of protein. The picture that can be drawn is that of the nucleic acids of chromosomes directing the synthesis of protein material up to maximum growth, then initiating the process of mitosis, during which the synthesized protein is squeezed out of the nucleus and two cells are formed, the nucleic acids of each beginning the process of protein synthesis anew. The point in this process at which the nucleic acids reproduce is not, apparently, in metaphase, when the chromosome number is doubled, but in interphase at a point just before the onset of mitosis (19, 59).

It has been generally accepted that the nucleic acid of the nucleolus was largely or entirely RNA. Monty *et al.* (54a), working with isolated nucleoli from cat and rat liver cells report nucleolus nucleic acid to be mainly DNA and the overall composition of nucleoli to be much like that of the remainder of the nucleus. Since RNA is the characteristic nucleic acid of the cytoplasm, it has been suggested that RNA is formed by the DNA of the genes (22a). The question, however, of whether RNA is formed from DNA, or vice versa, or whether each is formed independently is still highly controversial. Mazia (51) suggests that the DNA of the nucleus is primarily the genetic control of the cell, while the RNA is the metabolic control. It should be pointed out that RNA also occurs in chromosomes to the extent of about 10 per cent of the total nucleic acid (54). Since pentosenucleoproteins can possess the property of autoreproduction, as is demonstrated in tobacco mosaic virus, in the chromosomes they may synthesize cytoplasmic RNA.

One finding contrary to the theory that DNA is the primary genetic material, and acts as a template for protein formation and self-duplication, is the observation that DNA has not been found in the pronuclei of eggs of certain echinoderms, coupled with the fact that these eggs can be induced to develop parthenogenetically with no contribution of DNA from sperm. These findings have led to the theory (50) that RNA is the primary genetic substance and that the function of DNA is to control, by competitive inhibition, the activity of RNA. In this connection, see also (3a).

In cytoplasm. The RNA of cytoplasm, like the DNA of the nucleus, occurs largely in the form of particulate structures. These structures include the *mitochondria*, which are comparatively large granules one to three microns in diameter, and the much smaller *microsomes*, which are 0.06 to 0.15 microns in diameter and which may be particles of mitochondria broken off in the process of isolation (23). The mitochondria, which have been much studied, are major components of the cell cytoplasm. In the liver cell, mitochondria account for 30 per cent of the total cellular nitrogen. Rat liver mitochondria (70) are 63 per cent protein plus pentosenucleoprotein and 29 per cent lipid. The lipid moiety is 79 per cent phospholipid (nearly half of which is lecithin) and 4.4 per cent cholesterol. Mitochondria contain the enzyme systems involved in aerobic oxidations and oxidative phosphorylations. The RNA may act as a "skeleton" along which the enzyme molecules linked by phospholipid molecules are arranged in some specific order.

Evidence for the synthetic function of RNA resides in the fact that many investigators have found that tissue in a state of active growth is particularly rich in RNA. Thus embryonic kidney tissue will absorb ultraviolet much more strongly than will adult kidney. Further, a tissue of which only part is growing will show a localized concentration of RNA in the growing portion. A plant rootlet, for instance, will absorb ultraviolet more strongly as the tip is approached and most strongly at the outer surface of the tip. The same applies to cells which produce secretions rich in protein. The cytoplasm of nerve cells is also rich in RNA and there is experimental evidence that in the process of neural stimulation protein is broken down and expended at accelerated rates and must be resynthesized (36).

The relationship between the directive properties of the DNA of the chromosomes and those of the RNA of the cytoplasmic granules is not certain. The more accepted view has the genes as the only primary autoreproducers in the cell. Since the RNA of the cytoplasmic granules was formed under genic influence, its influence on protein synthesis is secondary. Any change in the RNA structure of the cytoplasm would thus come under the heading of a non-hereditary change, similar to cutting the tail off a rat. The cell itself, when a molecule of RNA is inactivated, may be handicapped by its absence, but during mitosis when a flood of new RNA enters the cytoplasm from the nucleolus, the daughter cells find the lost unit restored just as the tailless rat remains tailless but produces tailed offspring.

Another point of view is that there are cytoplasmic hereditary factors more or less independent of the genes (10). According to this view the cytoplasmic granules not only reduplicate themselves, but once altered they may continue to autoreproduce the new form not only in the cell itself but in the daughter cells after mitosis. Such autoreproductive hereditary factors in the cytoplasm are called *plasmagenes*.

In bacteria and yeasts. While bacteria and yeasts do not have nuclei as highly organized as those found in the cells of animals and higher plants, nuclear material does occur. Bacteria can even be said to have genes, 250 being reported present in the colon bacillus (4). The greater complexity of the nucleus in higher cells may simply be the expression of the need for a more refined mechanism for cell division as the number of genes to be transmitted in exact form becomes greater.

Studies on bacteria have given rise to new conceptions as to the relative importance of the nucleic acid and protein moieties of the nucleoprotein molecule in the control of enzyme make-up. The earlier view was that the protein was the complex and specific portion of the molecule, while the contribution of the nucleic acid was comparatively minor, being restricted perhaps to furnishing a framework for the genes. The fact that spermatozoa, the prime carriers of at least those genic factors contributed by the male parent, seem poor in proteins (presumably in the interest of smaller mass and consequent greater mobility) by including a large proportion of histone, while carrying nucleic acid in its full complexity, was somewhat disturbing to this line of thought. So also were the developing ideas as to the multiplicity and specificity of nucleic acids. Most startling however was the discovery that some biochemical characteristics of bacteria could be directed by an extract of DNA alone.

It was long known that an extract prepared from a "smooth" variety of certain strains of Pneumococcus could induce the transformation of a "rough" variety of the same bacillus into a type-specific "smooth" strain, and that the new "smooth" strain bacillus would then breed true. In 1944, it was found that the active principle of the transforming extract was a deoxypentosenucleic acid with a molecular weight of the order of half a million (2). It was easily inactivated by the action of deoxyribonuclease. "Smooth" pneumococci differ from the "rough" strains by possessing a polysaccharide capsule; they have a tendency to change to the "rough" form spontaneously. Since only the DNA of "smooth" pneumococcus can perform the transformation, the experiment is evidence of the extreme specificity of nucleic acids.

Boivin (4) has performed similar experiments on colon bacilli and concludes that DNA is the primary instrument of heredity, postulating at least one specific acid, for instance, for each strain of that bacillus.

In virus. Viruses which have invaded a cell can impose their own synthesizing capacities upon it. Thus, to cite an extreme example, a bacterium infected by a bacteriophage consisting largely of deoxynucleoprotein will cease manufacturing its own characteristic RNA and DNA, and will proceed to manufacture only viral DNA (34) without any decrease in its metabolic activity as measured by its rate of oxygen utilization.

Bacteriophages attach to their host bacterial cells by a complementary arrangement of COO^- on the viral surface and NH_3^+ groups on the bacterial surface (56). Even slight changes in the pattern would make the difference between sensitive and resistant strains of bacteria and between virus strains of high and low infectivity. Only the nucleic acid core of the bacteriophage penetrates the host, the protein portion remaining behind. It is suggested (4a) that within the cell it occupies a definite site on the bacterial chromosome. If so, this points up the view of a virus as an exogenous gene. Viruses, in their proliferation within susceptible cells, seem frequently to produce new strains in a process analogous to mutation, as indicated by changes in their infectivity.

Zahler (82) claims that in the relatively large bacteriophage viruses twenty-five genes or their equivalents may be present. He estimates 200,000 phosphorus atoms to be present in a T2 bacteriophage, and allows 8,000 per gene. He considers the nucleic acid of a single gene to have a molecular weight of 3,000,000.

REFERENCES

1. Asimov, I. The radioactivity of the human body. J. Chem. Ed., **32:** 84–85, 1955.
2. Avery, O. T., *et al.* Studies on the chemical nature of the substance inducing transformation of pneumococcal types. Induction of transformation by a desoxyribonucleic acid fraction isolated from Pneumococcus type III. J. Exper. Med., **79:** 137–157, 1944.
3. Bayley, S. T. An electron microscope study of nucleic acid. Nature, **168:** 470–471, 1951.
3a. Ben-Ishai, R., and Volcani, B. E. Dependence of protein synthesis on ribo-

nucleic acid formation in a thymine-requiring mutant of Escherichia coli. Biochim. et Biophys. Acta, **21:** 265–270, 1956.

4. Boivin, A. Directed mutation in colon bacilli, by an inducing principle of desoxyribonucleic nature: Its meaning for the general biochemistry of heredity. Cold Spring Harbor Symp., Quant. Biol., **12:** 7–17, 1947.

4a. Boyd, J. S. K. Bacteriophage. Biol. Revs. Cambridge Phil. Soc., **31:** 71–107, 1956.

5. Bonner, D. Biochemical mutations in neurospora. Cold Spring Harbor Symp., Quant. Biol., **11:** 14–24, 1946.

6. Brown, D. M., *et al.* Investigations on the problem of the ribonucleoside-2′-phosphates. J. Cellular Comp. Physiol., **38:** Suppl. **1:** 11–15, 1951.

7. Buchanan, J. M. and Wilson, D. W. Biosynthesis of purines and pyrimidines, Fed. Proc., **12:** 646–650, 1953.

8. Burnet, F. M., *Virus as Organism.* Cambridge, Mass., Harvard University Press, 1945.

9. Caren, R., and Morton, M. E. Pyrimidine metabolism in normal man studied with N^{15}-labeled uracil. J. Clin. Endocrinol. and Metab., **13:** 1201–1229, 1954.

10. Caspari, E. Role of genes and cytoplasmic particles in differentiation. Ann. N. Y. Acad. Sci., **60:** 1026–1037, 1955.

11. Caspersson, T. O. *Cell Growth and Cell Function.* New York, W. W. Norton & Co., 1950.

12. Chargaff, E., *et al.* Bacterial desoxypentosenucleic acids of unusual composition. J. Am. Chem. Soc., **72:** 3825, 1950.

13. Chargaff, E., and Davidson, J. N. *The Nucleic Acids* (in 2 volumes). New York, Academic Press, 1955.

14. Cohn, W. E. Some results of the applications of ion-exchange chromatography to nucleic acid chemistry. J. Cellular Comp. Physiol. **38:** Suppl. **1:** 21–40, 1951.

15. Dallam, R. D., and Thomas, L. E. Chemical studies of mammalian sperm. Biochim. et Biophys. Acta, **11:** 79–89, 1953.

16. Dent, C. E., and Philpot, G. R. Xanthinuria. An inborn error or deviation of metabolism. Lancet, **266:** 182–185, 1954.

17. De Lamirande, G., *et al.* Heterogeneity of intracellular ribonucleic acid in rat-liver cells. J. Biol. Chem., **214:** 519–524, 1955.

18. Doty, P., and Rice, S. A. Denaturation of deoxypentose nucleic acid (DNA). Biochim. et Biophys. Acta, **16:** 446–448, 1955.

19. Fautrez, J., and Fautrez-Firlefyn, N. Deoxyribonucleic acid content of the cell nucleus and mitosis. Nature, **172:** 119–120, 1953.

20. Felix, K. Protamines, nucleoprotamines and nuclei. Am. Scientist, **43:** 431–449, 1955.

20a. Fink, K., *et al.* Metabolism of thymine (methyl-C^{14} or -2-C^{14}) by rat liver in vitro. J. Biol. Chem., **221:** 425–433, 1956.

20b. Fraenkel-Conrat, H. The role of the nucleic acid in the constitution of active tobacco mosaic virus. J. Am. Chem. Soc. **78:** 882–883, 1956.

20c. Fraenkel-Conrat, H., and Williams, R. C. Reconstitution of active tobacco mosaic virus (TMV) from its inactive protein and nucleic acid components. Proc. Natl. Acad. Sci. U. S., **41:** 690–698, 1955.

21. Friedman, M., and Byers, S. O. Increased renal excretion of urate in young patients with gout. Am. J. Med., **9:** 31–34, 1950.

22. Gamow, G. Possible relation between deoxyribonucleic acid and protein structure. Nature, **173:** 318, 1954.

22a. Goldstein, L., and Plaut, W. Direct evidence for nuclear synthesis of cytoplasmic ribonucleic acid. Proc. Natl. Acad. Sci. U. S., **41:** 874–880, 1955.
23. Green, D. E. The cyclophorase system, *in* Edsall, J. T., ed. *Enzymes and Enzyme Systems*, pp. 15–46. Cambridge, Mass., Harvard University Press, 1951.
24. Greenberg, G. R. Mechanisms involved in the biosynthesis of purines. Fed. Proc., **12:** 651–659, 1953.
25. Greenstein, J. P. Nucleoproteins. Advances Protein Chem., **1:** 209–287, 1944.
26. Grinnan, E. L., and Mosher, W. A. Highly polymerized ribonucleic acid: preparation from liver and depolymerization. J. Biol. Chem., **191:** 719–726, 1951.
27. Gutman, A. B. Some recent advances in the study of uric acid metabolism and gout. Bull. New York Acad. Med., **27:** 144–164, 1951.
28. Hart, R. G. Electron-microscopic evidence for the localization of ribonucleic acid in the particles of tobacco-mosaic virus. Proc. Natl. Acad. Sci. U. S., **41:** 261–264, 1955.
29. Haurowitz, F. *Chemistry and Biology of the Proteins.* New York, Academic Press, 1950.
30. Haurowitz, F. Proc. Third Int. Cong. Biochem., Brussels, 1955, page 104.
31. Heinrich, M. R., and Wilson, D. W. The biosynthesis of nucleic acid components studied with carbon 14. I. Purines and pyrimidines in the rat. J. Biol. Chem., **186:** 47–60, 1950.
32. Hirs, C. H. W., *et al.* The amino-acid composition of ribonuclease. J. Biol. Chem., **211:** 941–950, 1954.
33. Hoppel, L. A., *et al.* Nucleotide exchange reactions catalyzed by ribonuclease and spleen phosphodiesterase. II. Synthesis of polynucleotides. Biochem. J., **60:** 8–15, 1955.
34. Hershey. A. D., *et al.* Nucleic acid economy in bacteria infected with bacteriophage T2. I. Purine and pyrimidine composition. J. Gen. Physiol., **36:** 777–789, 1953.
35. Horowitz, N. H., and Leupold, U. The one-gene one-enzyme hypothesis. Cold Spring Harbor Symposia Quant. Biol., **16:** 65–74, 1951.
36. Hydén, H. Protein and nucleotide metabolism in the nerve cell under different functional conditions. Symp., Soc. Exper. Biol., **1:** 152–162, 1947.
37. Judah, J. D., and Spector, W. G. Reaction of enzymes in injury. Brit. Med. Bull., **10:** 42–46, 1954.
38. Kacser, H. Molecular organization of genetic material. Science, **124:** 151–154, 1956.
39. Kalckar, H. M. The biological synthesis of purine compounds. Symp., Soc. Exper. Biol., **1:** 38–55, 1947.
40. Korn, E. D., and Buchanan, J. M. Biosynthesis of the purines. VI. Purification of liver nucleoside phosphorylase and demonstration of nucleoside synthesis from 4-amino-5-imidazole-carboxamide, adenine, and 2,6-diaminopurine. J. Biol. Chem., **217:** 183–191, 1955.
41. Kornberg, A., *et al.* Enzymic synthesis of purine nucleotides. J. Biol. Chem., **215:** 417–427, 1955.
42. Kunitz, M. Crystalline ribonuclease. J. Gen. Physiol., **24:** 15–32, 1940.
43. Kunitz, M. Isolation of crystalline desoxyribonuclease from beef pancreas. Science, **108:** 19–20, 1948.
44. Leuchtenberger, *et al.* Comparison of the content of desoxyribosenucleic acid (DNA) in isolated animal nuclei by cytochemical and chemical methods. Proc. Natl. Acad. Sci., U.S. **37:** 33–38, 1951.

44a. LEVENBERG, B., *et al.* Biosynthesis of the purines. X. Further studies in vitro on the metabolic origin of nitrogen atoms 1 and 3 of the purine ring. J. Biol. Chem., **220:** 379–390, 1956.

45. LIEBERMAN, I., *et al.* Enzymic synthesis of nucleoside diphosphates and triphosphates. J. Biol. Chem., **215:** 429–440, 1955.

46. LORING, H. S., AND LUTHY, N. G. The isolation in crystalline form and characterization of two isomeric cytidylic acids derived from yeast nucleic acid. J. Am. Chem. Soc., **73:** 4215–4218, 1951.

47. MACDONALD, D. L., AND KNIGHT, C. A. The identity of the purine-bound pentose of some strains of tobacco mosaic virus. J. Biol. Chem., **202:** 45–50, 1953.

48. MAGASANIK, B., AND CHARGAFF, E. Structure of ribonucleic acids. Biochim. et Biophys. Acta, **7:** 396–412, 1951.

49. MARKHAM, R., AND SMITH, J. D. Structure of ribonucleic acids. III. The end groups, the general structure, and the nature of the core. Biochem. J., **52:** 565–571, 1952.

50. MARSHAK, A., AND MARSHAK, C. Biological role of deoxyribonucleic acid. Nature, **174:** 919–920, 1954.

51. MAZIA, D. Physiology of the cell nucleus. Modern Trends Physiol. and Biochem., pp. 77–122, 1952.

52. MELLORS, R. C., *et al.* Quantitative cytology and cytopathology. III. Measurement of the organic mass of sets of chromosomes in germinal cells of the mouse. Cancer, **7:** 873–883, 1954.

53. MIRSKY, A. E., AND POLLISTER, A. W. Chromosin, a desoxyribose nucleoprotein complex of the cell nucleus. J. Gen. Physiol., **30:** 117–147, 1946.

54. MIRSKY, A. E., AND RIS, H. The chemical composition of isolated chromosomes. J. Gen. Physiol., **31:** 7–18, 1948.

54a. MONTY, K. J., *et al.* Isolation and properties of liver cell nucleoli. J. Biophys. Biochem. Cytol., **2:** 127–145, 1956.

55. OLMSTED, P. S., AND VILLEE, C. A. Nucleic acid composition of human liver-cell fractions. J. Biol. Chem., **212:** 179–186, 1955.

56. PUCK, T. T., AND TOLMACH, L. J. The mechanism of virus attachment to host cells. IV. Physicochemical studies on virus and cell surface groups. Arch. Biochem. Biophys., **51:** 229–245, 1954.

57. REICHMANN, M. E., *et al.* The molecular weight and shape of deoxypentose nucleic acid. J. Am. Chem. Soc., **74:** 3203–3204, 1952.

58. REICHMANN, M. E., *et al.* A further examination of the molecular weight and size of deoxypentose nucleic acid. J. Am. Chem. Soc., **76:** 3047–3053, 1954.

59. ROELS, H. Mitosis and deoxyribosenucleic acid content of the nucleus. Nature, **173:** 1039–1040, 1954.

60. ROLL, P. M., *et al.* The metabolism of yeast nucleic acid in the rat. J. Biol. Chem., **180:** 333–340, 1949.

61. RUTMAN, R. J., *et al.* The catabolism of uracil in vivo and in vitro. J. Biol. Chem., **210:** 321–329, 1954.

62. SCHRAMM, G., *et al.* Infectious nucleoprotein from tobacco mosaic virus. Nature, **175:** 549–550, 1955.

63. SCHMIDT, G., *et al.* The action of ribonuclease. J. Biol. Chem. **192:** 715–726, 1951.

64. SCHNEIDER, W. C. Phosphorus compounds in animal tissues. I. Extraction and estimation of desoxypentose nucleic acid and of pentose nucleic acid. J. Biol. Chem., **161:** 293–303, 1945.

65. SMITH, J. D., AND MARKHAM, R. The enzymic breakdown of desoxyribonucleic acids. Biochim. et Biophys. Acta **8:** 350–351, 1952.

66. SMITH, J. D., AND MARKHAM, R. Polynucleotides from deoxyribonucleic acids. Nature, **170:** 120–121, 1952.
67. STEDMAN, E., AND STEDMAN, E. Chromosomin, a protein constituent of chromosomes. Nature, **152:** 267–269, 1943.
68. STEDMAN, E., AND STEDMAN, E. The chemical nature and functions of the components of cell nuclei. Cold Spring Harbor Symp., Quant. Biol., **12:** 224–236, 1947.
69. STOWELL, R. E. Nucleic acids in human tumors. Cancer Res., **6:** 426–435, 1946.
70. SWANSON, M. A., AND ARTOM, C. Lipide composition of the large granules (mitochrondria) from rat liver. J. Biol. Chem., **187:** 281–287, 1950.
71. TAMM, C., *et al.* Distribution density of nucleotides within a deoxyribonucleic acid chain. J. Biol. Chem., **203:** 673–688, 1953.
72. THOMAS, R. Denaturation of deoxyribonucleic acid. Biochim. et Biophys. Acta, **14:** 231–240, 1954.
73. THOMSON, R. Y., AND FRAZER, S. C. The deoxyribonucleic acid content of individual rat-cell nuclei. Exptl. Cell Research, **6:** 367–383, 1954.
74. TODD, A. R. The nucleotides. Some recent chemical research and its biologic implications. Harvey Lectures, **47:** 1–20, 1953.
75. TODD, A. R. Chemical structure of the nucleic acids. Proc. Natl. Acad. Sci. U. S., **40:** 748–755, 1954.
76. TOTTER, J. R., *et al.* Incorporation of isotopic formate into the nucleotides of ribo- and desoxyribonucleic acids. J. Am. Chem. Soc., **73:** 1521–1522, 1951.
77. VOLKIN, E., AND COHN, W. E. The structure of ribonucleic acids. II. The products of ribonuclease action. J. Biol. Chem., **205:** 767–782, 1953.
78. WATSON, J. D., AND CRICK, F. H. C. The structure of DNA. Cold Spring Harbor Symp. Quant. Biol., **18:** 123–131, 1953.
79. WATSON, J. D., AND CRICK, F. H. C. Genetical implications of the structure of deoxyribosenucleic acid. Nature, **171:** 964–967, 1953.
80. WYATT, G. R. Occurrence of 5-methyl-cytosine in nucleic acids. Nature, **166:** 237–238, 1950.
81. WYNGAARDEN, J. B., AND STETTEN D., JR., Uricolysis in normal man. J. Biol. Chem., **203:** 9–21, 1953.
81a. YATES, R. A., AND PARDEE, A. B. Pyrimidine biosynthesis in Escherichia coli. J. Biol. Chem., **221:** 743–756, 1956.
82. ZAHLER, S. A. Nucleotide content of bacteriophage genetic units. Science, **111:** 210, 1950.

CHAPTER 10

Cancer

Cancer is commonly spoken of as a disease of growth. This is true in rather an inverted sense. The cancer cell is one which, although proficient at the art of dividing, has in some manner lost the mechanism of growth cessation. It is around the ability to stop growing rather than the ability to grow that the disease centers. From the theoretical standpoint this should be emphasized.

Clinically, the growth of malignant areas is their most outstanding characteristic and the most easily noticed. Such growth is not in itself, however, truly characteristic. Malignant growth does *not* proceed at a rate greater than that of normal growth. Embryonic tissue grows as quickly. Even in adult life, when visible growth seems to have ceased, many tissues retain the ability to regenerate rapidly. Injury to the skin, for instance, will usually stimulate surrounding areas to rapid mitosis. Such regenerative growth may take place at a rate equal to that of the most malignant tumors. The big difference lies in this, that when the regenerating skin has accomplished its purpose, rapid growth ceases and mitosis proceeds only at the rate necessary to replace epidermis. A skin cancer, however, continues growth indefinitely.

Cancer is also a disease of organization. In all multicellular organisms cell specialization exists. The more complicated the organism, the greater the degree of specialization. Various cells become more and more adapted to the proper carrying out of relatively few functions and sacrifice their independence in so doing. The multicellular organism becomes a society of cells, few of which could exist independently for more than a few moments. That this is the price paid for what we can not help but look at as "progress" in the evolutionary scheme of things is fairly obvious. An analogy can be made between this state of affairs and various human societies. We can compare a Stone Age tribal society and that of a modern city such as New York, and note that the greater comfort of the latter implies a greater interdependence of man upon man. How many New Yorkers could long survive in a state of nature, no longer protected by the artificial safeguards built up by the co-operation of a specialized society?

Unfortunately, the fact of organization among cells exposes new vulnerabilities. Cells are potentially immortal as is shown in the case of unicellular organisms and in the germ plasm of multicellular organisms. Despite this potentiality among their individual members, cell societies in many cases age and die through breakdown of their organization. The analogy to a human society will hold here, too. A blizzard can paralyze city life for days, threaten mass starvation if needed food is delayed by impassable roads, and do infinite damage although its direct effect upon the health of the individual man and woman within the city may be nil.

Cancer generally involves a breakdown in the organization of the cell society. Cells which have become cancerous tend to lose by degrees their specialized function, and to multiply without regard to their own welfare or to that of the whole organism. A consequence, perhaps, of this reaction against specialization, this retreat to independence, is the property of transmissibility shown by many cancerous growths. That is, a tumor or a portion thereof can be removed surgically and implanted in another organism; and providing it can elicit from the new host a supporting structure of connective tissue and an adequate supply of blood vessels, it will continue to grow and multiply. Normal tissues when so transplanted grow little if at all, so that the phenomenon emphasizes the almost complete autonomy of cancer in relation to the organism. It is as though it were a portion of the flesh turned parasite.

To summarize all this, cancers may be defined as growths of cells which are largely independent of the organism which supplies their nutrition—cells possessing a frequently atypical structure and having no definite growth limits. The "independence" of cancers is not complete (13). Thus, cancers often retain certain of the functions of the normal tissue from which they sprang and their growth and metabolism can be affected by administration of hormones. Nevertheless, it is also true that as a cancer develops, the general tendency is toward loss of organization and increased independence.

All tissues of all multicellular organisms are subject to cancer, although some are much more prone to the disease than others, and certain phenomena among unicellular organisms as well bear some resemblance to cancer.

Cancer is not the most general term applicable to these disorders of growth. *Tumor* (from the Latin *tumere*, to swell) may be applied to any swelling; *neoplasm* (from the Greek *neo* and *plasma*, meaning new formation) refers to non-physiological growth or multiplication of cells. *Benign tumors* are growths which, although they may continue to grow slowly over long periods of time, remain restricted in area, do not spread to other parts of the body, and have little or no tendency to recur after surgical removal. The dangers of benign tumors are limited to the effects of pressure and obstruction by the tumors. A *malignant tumor*, to which the term cancer should be restricted, is one whose growth is unrestricted and which has a tendency to infiltrate neighboring tissues, and as a result of transmission of cancer cells via the blood or lymphatic system even to invade far removed portions of the body. Portions of a tumor which have taken root in distant parts of the body are *metastases*, and a tumor resulting from such a process is a *metastatic tumor*. The two main classes of malignant tumors are the carcinomas and the sarcomas. *Carcinoma* is a tumor arising from epithelial tissue, whether pavement or glandular, while *sarcoma* is one arising from connective and muscle tissues. Either name may be modified by a prefix designed to indicate the particular tissue from which the tumor derives.

One highly important factor, perhaps obscured by the broad and general nature of these terms, must be emphasized before any further consideration of the subject takes place. Cancer is not a single disease. Cancers differ from one another in properties of all sorts, not only in accordance with the various tissues from which they originate but with the species or organism in which they occur. To illustrate this, let us use a biochemical example. In the rat, hepatoma cells are characterized by very low catalase activities as compared with cells of normal rat liver. In the mouse, hepatoma cells also show decreases in catalase concentration but this is not as marked as in the rat. Furthermore, the catalase deficiency in hepatomas varies with the particular strain of the rat.

The importance of this generality in connection with *human* cancer is obvious. In no other human disease has so much work been done on animals and so little on human beings. It is easy, often too easy, to assume that results gained from a study of mouse cancer will be applicable to human cancer. An indication that this might not invariably be so is the fact that although polycyclic hydrocarbons are the classical carcinogens for rats, mice, and rabbits, attempts to induce cancer by their use on monkeys have so far failed (36). With all this in mind, the student is cautioned that except where specifically stated otherwise the material in this chapter has been derived from animal experimentation.

CHEMICAL CONSTITUTION OF CANCER CELLS

There is no one chemical constituent or group of constituents which may be found in all normal cells and which does not occur in all cancer cells, or vice versa. Attempts to show that concentrations of certain cell constituents in cancer cells are generally higher or lower than in normal cells have been only partially successful, since to every general rule formulated, there always turn out to be exceptions.

The morphology of the cancer cell, as well as its chemical make-up, varies with the stage and location of the disease. There is a progressive change, involving a loss of organization of the cell. Cancer cells are more irregular in size, shape, and staining reactions than are the normal tissue cells from which they stem, they show various abnormalities in the structure of their organelles and in the process of mitosis. So marked are the abnormalities that, despite attempts at chemical diagnostic methods, the surest method of diagnosis of cancer remains the biopsy, the microscope, and the trained expert eye of the pathologist.

To turn to more chemical considerations, cancer cells usually have a higher water content, the percentage of water generally increasing with the rate of growth. This, immediately, is an example of a variation which need not be in any way significant with regard to cancer. In general, normal cells which are actively metabolizing tend to have higher water contents than those whose role in the body is more sedentary and sluggish. The water content of cancer cells may therefore be simply an attribute of *active* cells rather than of *cancer* cells. A similar statement may be made about the fact that cancer cells tend to be comparatively rich in RNA (17), a property they have in common with normally growing cells.

There is probably no element, organic grouping, or molecule which could conceivably be found in tissues which has not been analyzed for over and over again in cancers. Unfortunately the particular variations from the normal, found in such analyses, are not consistent nor are the reasons for the variations known. An individual variation is sometimes correlated with a particular property of cancerous cells. For example, a deficiency of calcium frequently is found in cancers of various types. Experiments have shown that intercellular adhesiveness is dependent upon the presence of calcium (11); and calcium deficiency may therefore help to account for the lesser adhesiveness of cancer cells and the tendency of individual cells to break away from the cancerous mass and set up independently as metastases in other parts of the body. Cancer cells show no consistent difference from the normal cells of their tissue of origin either in the purine and pyrimidine content of their nucleic acids (8) or the amino acid composition of their proteins.

Enzymology of Cancer

The enzyme pattern. Greenstein (21) has suggested an approach which would take into consideration not single constituents but the over-all make-up of the cell with particular emphasis on those arbiters of chemical function, the enzymes. The specialization of various cells is reflected in the enzymes they contain, so that, to use Greenstein's examples, bone tissue and intestinal mucosa can be distinguished from other tissues by the fact that they are rich in alkaline phosphatase and poor in catalase and arginase. The two can be distinguished from one another by the fact that intestinal mucosa is also rich in esterase while bone tissue is not. Similarly, cardiac muscle and skeletal muscle can be distinguished by the markedly higher content of cytochrome oxidase in the former. In general, one can assume without much difficulty that each tissue has its own distinct enzyme pattern by which it can be differentiated from all other tissues. The question then arises as to whether there is any significance in the change of this pattern, rather than in any of its components, in cancer cells.

The technical difficulties in the way of such investigations are manifold. It is usually impossible to obtain a sample of a single tissue. The analysis of epithelial tissue is easily rendered meaningless by the presence in the sample of an indeterminate amount of connective tissue which would itself possess an entirely different enzyme pattern. Muscle tissue analyses are thrown off by the presence of blood. This list can be extended indefinitely. The case with regard to cancers is considerably worse. Cancers, when of a size admitting of chemical analyses, frequently contain sizable necrotic areas. Because of the undiscriminating growth of a cancer without regard to vascularization, inner portions may literally die of lack of food and oxygen, even while the outer rim is still heedlessly expanding. The inclusion of such metabolically inert material in the analysis of a cancerous mass would destroy the significance of the pattern. Worse still, since the growth of a cancer into adjacent normal tissue (a property known as invasiveness) is often ragged and irregular, it is frequently difficult to dissect out a cancer without including varying amounts of normal tissue, which in the case of metastases may have an enzyme pattern radically different from the cancer's normal tissue of origin. A great deal of work in the analysis of cancerous and normal tissue has been undertaken without sufficient consideration of such possible interfering factors; and for this reason careful histological controls should be included in order that the analyses of any tissue contain as part of the data such information as the percentages of the whole which are cancerous, normal tissue of origin, connective tissue, necrotic areas, and so on. This can be done only by the microscopic study of representative slices of the tissue to be analyzed.

Certain general conclusions have been drawn from enzyme studies. The loss of cellular specialization in cancer extends to specialized enzyme pattern. Cancer cells arising from tissue which is unusually high in a given enzyme tend to show lower concentrations. Similarly, cancers become richer in enzymes in which the normal tissue of origin is unusually poor. Furthermore, this tendency would seem to be progressive and is more marked in the more advanced cancers and particularly so in cancers that have been transplanted from host to host a number of times (22). Thus, whereas cancers in their early stages resemble their tissue of origin rather closely, differences increase with the progression of the disease so that cancers of varying origin tend to converge and approach one another in enzyme pattern. It is as though, chemically as well as morphologically, the cell retrogresses to a primitive and unspecialized ancestral pattern.

Specific enzyme systems in cancer. The most striking change among the individual enzyme systems of cancer cells is the decline in the content of the various enzymes involved in aerobic oxidation, such as cytochrome oxidase, succinic acid dehydrogenase and D-amino acid oxidase. This is not to say that cancerous tissue is lower in these enzymes than is normal tissue in general. We can only say that the over-all content is substantially lower than that of the normal tissue of origin in virtually every case although certain enzymes of the tricarboxylic acid cycle (see Chapter 14) may not decline (52). The heme-containing enzymes are an extreme case of this. The catalase activity of liver not only declines markedly in hepatomas but also decreases frequently under the influence of distant cancer in the organism which has not established itself in the liver (39).

Alterations in other enzyme systems are neither as uniform nor as marked in cancers taken in general. The activity of enzymes which are involved in the metabolism of nucleic acids or of proteins continues high in most cancers. Examples of such enzymes are the nucleases, arginase, certain peptidases, and xanthine dehydrogenase. Beta-glucuronidase is high in tumors, as it is in embryonic and proliferating tissue in general (34). Hyaluronidase is not present in cancers in any unusual concentration. This is significant since it has been suggested several times that hyaluronidase or some closely related enzyme might be partly responsible for the invasiveness of cancers (10).

Enzyme pattern of the host. A cancer represents, in a sense, an alien element within the body. Although derived from the host's own cells, its autonomy of growth and its failure to fit itself to the needs of the body lead to conditions of sufficient abnormality to elicit specific responses from the host even at early stages of the disease. Such systemic responses become significant where the tumor approaches 5 per cent of the body weight, but the nature of the responses is not sufficiently invariant to be of diagnostic value. Reference has already been made to the fact that there is a decline in liver catalase under the influence of distant cancers. In the mouse a wide variety of non-liver cancers have been shown to accomplish this, and the effect was found to become more marked as the disease progressed. Furthermore, the effect was reversible since removal of the cancer by surgery restored the catalase function of liver to normal.

The interest in such responses to the presence of cancer by the host is great, since by observation of such responses one could conceivably detect a cancer before it had grown big enough for direct observation and all too frequently too big or too metastatic for treatment. Unfortunately, periodic observations of liver biopsies for their catalase content is impractical as a routine clinical test for the presence of cancer.

The tissue most easily observed is, of course, blood; and it is for this reason that host responses to cancer have been most sought for there. Blood catalase activity frequently declines in the presence of cancer, but the change is so small as to be of little or no diagnostic value. Anemia, which often accompanies cancer, has too many other possible causes to be diagnostic. Differences in the properties of blood of cancerous and normal patients are frequently reported in the literature. As examples, the plasma of cancerous patients was reported to have a lower ability to decolorize methylene blue than did that of normal patients; serum albumin of cancerous patients usually coagulated more readily than did that of normal patients; and plasma of cancerous patients inhibited trypsin activity more than did that of normal patients.

All blood tests for cancer so far reported are far removed from perfection. Comparatively large percentages of both false positives and false negatives occur, and repeated tests of their efficiency by workers in the field have proven increasingly disappointing. One study of five such tests (15) ended in the conclusion that none of them was suitable even as screening tests (see also (26)).

Electrophoretic analysis of plasma for the purpose of cancer detection has been disappointing. Except in multiple myeloma, no distinctive cancer patterns of plasma proteins have been found. In multiple myeloma the globulin concentration is typically increased. This disease also offers an example where a urinary constituent reveals the presence of a particular cancer. In that disease the urine may contain a group of proteins of an average molecular weight of 37,000, with the unusual property of precipitating from solution at 45 to 58°C., and of redissolving at 100°C. These are called *Bence-Jones proteins*.

A useful diagnostic tool for the detection of a single type of cancer involves the measurement of the acid phosphatase activity of blood serum (24). In cases of metastatic carcinoma of the prostate this is often markedly increased.

Metabolism of Cancer Cells

It is tempting to assume that there is something superior about the metabolism of cancer cells, some greater versatility or new source of energy that gives them an unfair advantage over normal cells. This is based on the undoubted facts that (a) cancer cells can invade and overgrow normal tissue; (b) cancer cells can grow at alien sites in the organism or even in alien organisms; and (c) cancer cells can multiply under conditions of food and oxygen supply that would be most unfavorable for normal cell growth. These facts are not, however, inconsistent with the thought that cancer cells prosper not through greater efficiency but through a capacity to adapt themselves to a "lower standard of living". From this point of view, the cancer cell, through its loss of the high degree of specialization that characterizes normal cells, no longer requires so specialized an environment and can therefore, despite a lowering of general efficiency, compete favorably with its normal brother. Unfortunately, investigations to date have not revealed any one item in the over-all pattern of metabolism that is startlingly absent or unexpectedly present in the tumor cell and which may be seized upon to explain the unrestrained growth associated with cancer (20).

The cytochrome system. The oxygen consumption of normal tissue slices is markedly increased when excess metabolite, such as succinic acid, is added to the system. Apparently, reserve supplies of cytochrome *c* and cytochrome oxidase are available to handle oxidation of the additional material. Cancerous tissue slices show considerably less increase of oxygen consumption upon addition of metabolite. It is almost as though, with characteristic improvidence, cancer cells had no reserve supply of the cytochrome system and were content with whatever amount they could get by on. A less anthropomorphic view is that cancer involves an impairment of the metabolic pathways involved in the biosynthesis of heme. One of the important building-blocks of heme is glycine (see page 616), and there is some evidence that glycine metabolism, notably glycine-serine interconversion (see page 619) is slowed in tumors (20). A more general impairment of metabolism may be reflected in the fact that mitochondria (which contain the cytochrome system as well as other enzymes involved in aerobic respiration) and cytoplasmic particulates in general are low in tumor cells as compared with normal cells (33).

In order to differentiate among various components of the cytochrome system, the oxygen consumption of tissue slices was studied after cytochrome *c* as well as metabolite had been added to the system (23). It was here found that the respiration of cancerous tissue approached that of normal tissue. From this it is concluded that malignant tissue is relatively more deficient in cytochrome *c* than in cytochrome oxidase and that it is the concentration of the former that is the limiting factor under the conditions of the experiment.

Glycolysis. The deficiency of the cytochrome system, which limits the power of the cancer cell to conduct aerobic oxidations, is accompanied by a greater tendency for the cell to rely upon glycolytic mechanisms for necessary energy. Glycolytic reactions are a comparatively primitive method of energy production and much less efficient than aerobic oxidation (see

Chapter 14). All tissues can glycolyze; cancers differ from most, but not all, normal tissues in that they will glycolyze even in the presence of adequate supplies of oxygen. While most of the experiments upon which these conclusions are based were conducted *in vitro*, utilizing such techniques as Warburg manometry (see Chapter 6), confirmatory *in vivo* evidence exists. The pH of growing tumors *in situ* was measured (31), and was found to be distinctly lower than that of normal tissue, about 7.0 as compared with 7.4. The pH of the tumor was found to drop still lower, to as little as 6.3, after administration of glucose either subcutaneously or intraperitoneally, without any signs of systemic acidosis; this indicates the drop in pH to be confined to the tumor. This is strong evidence in favor of the *in vivo* glycolysis of tumors, since glycolysis involves the production of lactic acid as an end product (through the hydrogenation of pyruvic acid in the reducing medium of tissue) which is a relatively strong organic acid and easily able to lower tissue pH by the amounts experimentally measured.

Although this tendency in favor of glycolysis is so far the characteristic that most clearly differentiates the majority of cancers from the majority of normal tissues, it is still not the final specific difference so long sought by cancer investigators. The degree of predominance of glycolysis does not consistently differentiate cancerous from normal tissue.

CARCINOGENESIS

The study of cancer was profoundly stimulated when, in 1915, Yamagiwa and Ichikawa first announced the artificial induction of cancer in an experimental animal. This was done by applying coal tar to rabbits' ears over long intervals until skin cancers appeared at the sites of application. This meant that investigators no longer had to confine their animal studies to the relatively few spontaneous cancers they might find, but could produce the cancers in quantity and in accordance with need. Furthermore, they could study cancer development in all its stages from the pre-cancerous through the primary tumor to secondary transplants. Again, from a study of the chemical nature of the cancer-producing agents, or *carcinogens*, there was always the possibility that the metabolic changes involved in carcinogenesis might be understood.

Unfortunately, complications developed. Various carcinogens show species and tissue differences in their actions. Therefore, the application of the results of animal experiments to human cancer is unsafe, though not entirely useless. Mider (34a) points out that only six clearly identifiable agents have as yet been shown to cause cancer in man, as opposed to laboratory animals. He lists these as 1) x-rays, 2) ultraviolet radiation, 3) arsenic or potassium arsenite, 4) beta-naphthylamine, 5) benzidine and 6) 4-amino-diphenyl. A given carcinogen differs in its effect upon a given animal depending on the method of application, the solvent used, the interval between applications, the age, diet, and physical condition of the animal, and so on. It remains difficult therefore to interpret the results obtained in any general manner. Lastly, so many carcinogens have been described (approximately one thousand) of such varying chemical nature that generalizations as to their action is all but impossible.

Carcinogens

The term carcinogen may in its broadest sense be applied to any factor, chemical or otherwise, capable of inducing cancer. Most of the agents used in the laboratory for the artificial induction of cancer are non-physiological in the sense that the animal would not ordinarily be exposed to them in the natural course of its life. Such agents fall into two chief classes—chemicals, and short-wave radiation.

Polycyclic hydrocarbons. The relationship between coal tar and carcinogenesis was first noted in the fact that skin cancer became an occupational disease among workers in the growing coal tar industry of the 19th century, and, as has been stated, the first artificial cancers were induced by coal tar. The isolation from coal tar of pure chemical carcinogens revealed the active substances to be polycyclic aromatic hydrocarbons. To this group belong a wide variety of active carcinogens which remain even today the most used agents for experimental carcinogenesis.

Two such hydrocarbons isolated from coal tar and found to be strongly

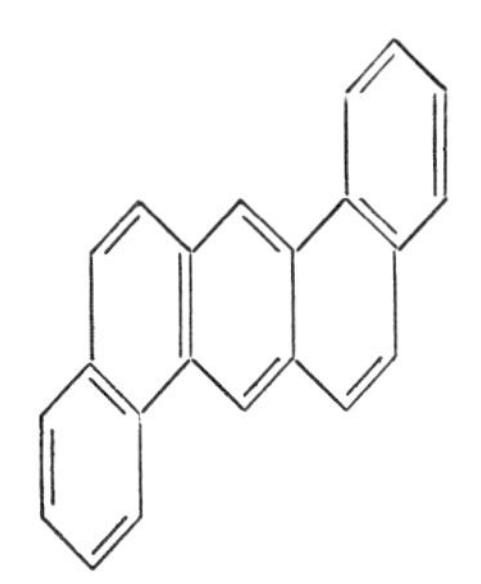

I. 1,2,5,6-Dibenzanthracene

II. 3,4-Benzpyrene

carcinogenic are *1,2,5,6-dibenzanthracene* (formula I) and *3,4-benzpyrene* (formula II). Other polycyclic ring structures, such as 3,4-benzphenanthrene and chrysene, which themselves were weakly carcinogenic, if at all, yield active carcinogens upon alkylation. Thus, *2-methyl-3,4-benzphenanthrene* (formula III) and *5,6-dimethylchrysene* (formula IV) are active carcinogens. Compounds similar to these have been located in the smoke of burning cigarettes in investigations attempting to determine possible relationships between smoking and the increased incidence of lung cancer in recent decades (12).

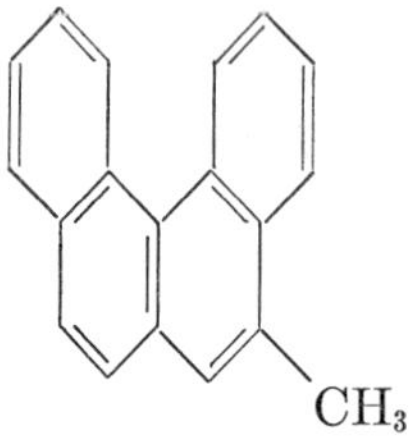

III. 2-Methyl-3,4-benzphenanthrene

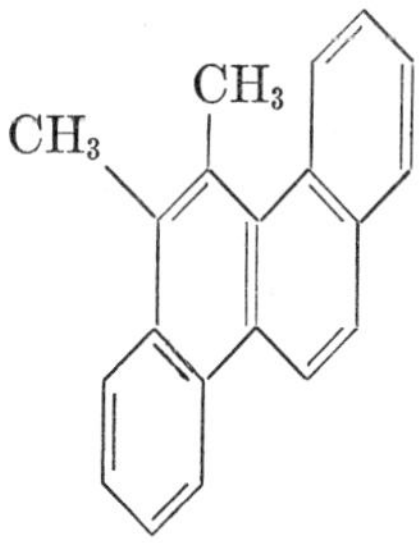

IV. 5,6-Dimethylchrysene

One of the most active, and certainly the most interesting, of this type of carcinogen is *20-methylcholanthrene* (formula V), often called merely methylcholanthrene. It should be seen at once that this hydrocarbon can be viewed as a steroid derivative in which rings A, B, and C have been aromaticized, and in which the C-17 side chain has been condensed with ring C to form still a fourth benzene ring. The name of the compound is derived from the steroid ring-numbering convention (see page 129) which

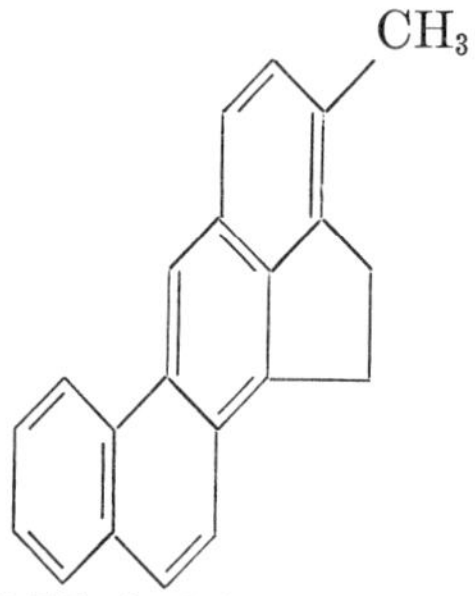

V. 20-Methylcholanthrene

places the methyl substituent found on carbon 20. Methylcholanthrene can be prepared in the laboratory from deoxycholic acid, one of the bile acids.

Heterocyclic analogs of the polycyclic hydrocarbons may also be carcinogenic. The nitrogenous compounds 3,4,5,6-dibenzcarbazole (formula VI) and 1,2,5,6-dibenzacridine (formula VII) are examples. These carcinogens are water-insoluble and are either applied to the skin as solutions in benzene or other fat solvents, inducing carcinomas, or are injected subcutaneously as solutions in more physiological solvents, such as lard or sesame oil, inducing sarcomas. It must not be thought that cancer follows the touch of these compounds unfailingly or quickly. Applications must be continued, usually over a period of months, for the effect to take place.

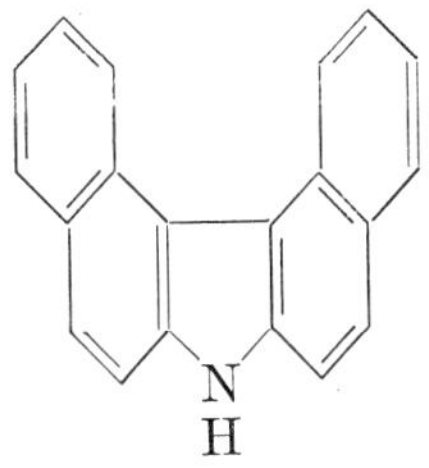

VI. 3,4,5,6-Dibenzcarbazole

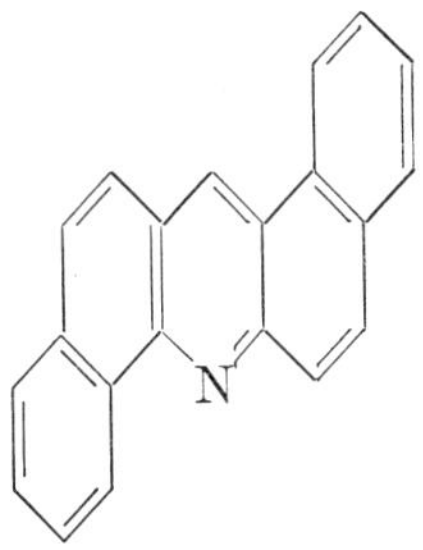

VII. 1,2,5,6-Dibenzacridine

Attempts have been made to correlate the carcinogenic properties of compounds with their chemical structure (40). Suspicion has fallen most readily upon the "K-region", that is, the reactive double bond between carbons 9 and 10 of the phenanthrene ring (formula VIII) as being essential

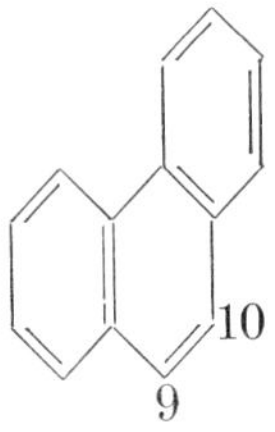

VIII. The K-region

to carcinogenesis. However, no such theories have yet proven entirely satisfactory.

The carcinogenic activity of a given compound varies with the size and frequency of the dose, the manner of application (whether painted or injected), the solvent used, the presence of additional substances not in themselves carcinogenic, and the nature of the animal's diet. Thus, 10-methyl-1,2-benzanthracene is highly active when administered subcutaneously, but only weakly active when applied to the skin, while the reverse is true of 2-methyl-3,4-benzanthracene. Although 1,2-benzanthracene is not carcinogenic when applied externally or injected subcutaneously, it will cause hepatomas in some rats if included in their diet. Again, methylcholanthrene in acetone is a more effective carcinogen than when dissolved in benzene, while in solution in anhydrous lanolin carcinogenic power is almost entirely lost. The application of croton oil to areas of skin being painted with 3,4-benzpyrene hastens the development of cancer, while the simultaneous use of brombenzene or unsaturated dibasic acids with that carcinogen retards the effect. A high-fat diet often facilitates artificial induction of cancers, while a diet low in the sulfur-containing amino acids may retard the development of some kinds of induced cancers.

Studies of carcinogens have revealed a form of variation of response more intriguing and perhaps more important, theoretically, than any of those listed so far. When carefully inbred strains of mice were used as subjects for applications of carcinogens, it was found that different strains varied in the frequency and rapidity with which individual members developed cancers of certain types. These strains usually showed a higher incidence of spontaneous cancers of those types as well. The use of animal strains of known cancer-forming characteristics is thus highly important if significant results are expected in laboratory investigations.

Micro-organisms exposed to dilute solutions of a carcinogen such as methylcholanthrene exhibit an accelerated mitotic cycle (47). On the other hand, yeasts exposed to camphor, which is not ordinarily considered a

$$C_6H_5{-}N{=}N{-}C_6H_4{-}N(CH_3)_2$$

IX. p-Dimethylaminoazobenzene

carcinogen, show inhibited mitoses, but continue growth to produce larger-than-normal cells which display morphological changes characteristic of cancer cells (49). Whether such cancer-like phenomena among micro-organisms are truly analogous to cancer in the multicellular organism remains, of course, a matter for speculation; such experiments seem to show cancer as a truly basic cell disease.

Other chemicals. A chemical carcinogen, second only to methylcholanthrene in experimental usefulness, is *p-dimethylaminoazobenzene* (formula IX), a representative of a group of carcinogens, completely different from the polycyclic hydrocarbons. p-Dimethylaminoazobenzene induced hepatomas in rats (but not in mice) when included in the diet. The nature of the remainder of the diet affected tumor incidence, carrots and rice stimulating the carcinogenic effect and liver depressing it. This carcinogen is also frequently known as *butter yellow.*

A related carcinogen of great versatility is *o-aminoazotoluene*, which not only induces hepatomas on feeding, but also results in various forms of cancer when applied externally or injected subcutaneously. It is effective on mice as well as on rats.

A peculiar carcinogen, chemically in a class by itself, is *N-acetyl-2-aminofluorene* (formula X). It is active only when administered orally, leading to the conclusion that not itself, but a product of its metabolism, is the true carcinogen. Furthermore, on ingestion it gives rise to a large variety of tumors, exceeding in versatility even o-aminoazotoluene in this respect.

Carcinogenic properties are not restricted to organic compounds. Tumors, such as bone sarcomas, can be induced by beryllium salts and beryllium-zinc salts (1). Other metals, such as cobalt, nickel, and chromium, have also been found to give rise to tumors under appropriate conditions (25, 29).

Radiation. As in the case of chemical carcinogens, the ill effects of short-wave radiation was first discovered by tragic experience among the pioneer workers with x-rays and radioactive materials. Ultraviolet radiation is the least energetic radiation definitely known to be carcinogenic. Exposure to ultraviolet was first shown to cause skin cancer in mice in 1928 (16). Rats and mice are both susceptible to ultraviolet, albino varieties forming cancers more readily. The induced cancers thus formed are in the dermis (i.e., sarcomas), whereas human cancers attributed to the effect of ultraviolet are in the epidermis (carcinomas). The difference is thought to be due to the variation in the resistance of the skin to penetration by the radiation. The wave length of maximum carcinogenicity of ultraviolet light is from 260 to 340 millimicrons.

Cancer caused by radiation more penetrating than that of ultraviolet

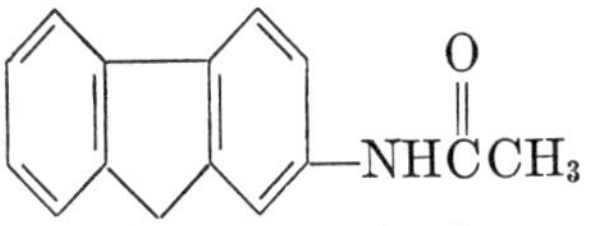

O
‖
—NHCCH$_3$

X. N-Acetyl-2-aminofluorene

is more apt to affect the blood and bones. Leukemia and osteogenic sarcoma are frequently the result of continued exposure to x-rays or gamma-rays. Radium and strontium-90 are similar chemically to calcium, and are deposited in the bones if accidentally ingested. Even small quantities so deposited may induce osteogenic sarcoma as a result of the localized radiation.

It is obvious that the importance of carcinogenic factors such as these is increasing rapidly in the modern world as a result of prospects for both atomic warfare and atomic-powered industry.

Diet and occupation. No specific diet or food factor or combination of food factors have been found to be carcinogenic. There is some reason to think, however, that excessive caloric intake may be correlated with a somewhat increased cancer incidence (7). Strong (48) points out that man and domestic animals are more cancer-prone than are wild animals and suggests that one possible explanation for it is that man and domestic animals tend to eat more than do wild animals.

The effect of specialized occupations upon human beings is extremely important in cancer production (28) and was certainly the first cancer "cause" noted by investigators. In the 18th century, it was noticed that chimney sweeps were more prone than the general population to cancer of the scrotum. Workers in industries using coal tar more frequently developed cancer of the skin. Similarly, cancers of the bladder are unusually common among workers in aniline dye industries. In each case, the long continued effect of a carcinogen in the soot, coal tar, or aniline dye is to be suspected. Much of the same may be said of people who are exposed, in the course of their work, to the effects of x-radiation and gamma-radiation. Among them there is a definitely higher incidence of leukemia and osteogenic sarcomas. This is of particular interest to the physician for personal reasons. The incidence of leukemia among radiologists is ten times greater than among medical practitioners in other fields, while among physicians as a group it is 1.7 times greater than in the general population (27). A short-wave radiation to which all humans are more or less subject is, of course, the ultraviolet of sunlight.

Viruses

None of the carcinogenic agents hitherto discussed can be considered the primary cause of cancer. They are too varying in nature and, probably, too indirect in their action. It is more reasonable to suppose that their common action is to induce some change in some cell component or components and that this changed component is the true fundamental carcinogen. Suspicion as to the nature of the cell component which undergoes this change falls most readily upon the nucleoproteins. Thus, Boyland (4) points out the effect of various categories of carcinogens upon DNA. X-rays and the nitrogen mustards depolymerize DNA; beryllium, zinc and other metals precipitate it; urethane inhibits its synthesis; various aliphatic carcinogens cross-link with it and various aromatic carcinogens form DNA-complexes. In every case, DNA is altered and is no longer in its native state.

Nucleoproteins are a natural suspect since they are capable of autoreproduction and are thought to control the chemical characteristics of the individual cell (see Chapter 9) which means that a cancer cell, with its changed chemical nature and its capacity to maintain its cancerous nature during reproduction, must possess a nucleoprotein or nucleoproteins differing from its normal ancestors. The fact that strains of mice can be bred which are more susceptible to cancer, or less so, than the general population indicates that there are genic differences such that the nucleoproteins of one strain may be more readily changed to the cancerous variety by the action of particular carcinogens than those of the other. The fact that ultraviolet radiation is carcinogenic may be significant in view of the fact that the purine and pyrimidine rings of the nucleic acids are about the strongest absorbers of ultraviolet in the body, while the effect of the more energetic radiations such as x-rays may well be secondary in nature, working through the formation of short-lived free radicals which, in turn, affect the nucleoproteins. There is also reason to think that the chemical carcinogens affect nucleoproteins.

With all this in mind, it is reasonable to suppose that it might be possible to extract from cancerous material suspensions of changed nucleoprotein which on injection into healthy tissue would be capable of inducing the cancerous change there as well. Such suspensions have indeed been found and because of their resemblance to viruses both in size and in chemical nature as nucleoproteins have been termed *tumor viruses*.

The first such cell-free cancer-inducing agent of biological origin was discovered in 1911 by Peyton Rous. He found that extracts of spontaneous chicken sarcomas, when injected intramuscularly into other chickens, gave rise to sarcoma in the new hosts at the site of injection in about ten days. The properties of the extract indicate the active agent to be a pentose-nucleoprotein in nature. A pentosenucleoprotein was isolated from chicken tumor I by differential centrifugation (9). When partially purified, this was shown to be capable of inducing tumor in another chicken in doses as low as 4×10^{-13} grams dry weight. Lipid is associated with the nucleoprotein, and its removal renders the complex non-carcinogenic, although the lipid itself is not a carcinogen. The tumor virus is rather species specific, thus indicating that the nucleoprotein complex will operate only in cells of approximately the chemical nature to which it is accustomed. In general, only the tissue from which it has been extracted will be affected in a new host. Chicken-tumor virus acts only with difficulty in such allied fowls as ducks, turkeys, and guinea hens (14). It must be injected in large amounts and within a short period after the fowl hatches. Similar tumor viruses have been found for spontaneous papillomas of the wild cottontail rabbit, except that here a deoxypentosenucleoprotein is present (2).

It was observed (3) that the offspring of a mouse belonging to a strain in which there was a high incidence of spontaneous mammary tumor did *not* tend to develop the tumor if the young was not allowed to suckle its mother but was nourished on the milk of a foster mother belonging to a strain with low incidence of mammary cancer. The reverse is also true. A newborn mouse which would ordinarily be expected, for hereditary reasons, to be free of mammary cancer will, with high probability, develop this cancer if fed on the milk of a cancer-prone mother. This indicates the presence of a *milk factor*, or mammary tumor inciter, in the mouse. Virus-like particles have indeed been located in the milk from mice of cancer-prone strains (18) which were absent in milk from cancer-resistant strains. These particles were found to contain an as yet uncharacterized nucleic acid.

If cancer were due to virus infection, antibodies should appear in the blood of cancer patients. The finding of such antibodies would, however, not prove the virus origin of cancer, for if cancer cells merely represent mutated normal cells (see below), they might contain new antigens against which the host could form antibodies. Be that as it may, antitumor antibodies have apparently been demonstrated.

Graham and Graham (19) tested the sera of 48 patients for antibodies to their own tumors. Twelve patients were found to have titers of 1:16 to 1:128. It is of interest that these were patients who seemed to be resisting the progress of the neoplasm better than the others. The majority of the patients who failed to show circulating antibodies had far advanced lesions.

Further evidence in support of the virus hypothesis comes from the observations of Nungester and Fisher (35) that injection of the antigens found in the pellet produced by high-speed centrifugation of 6C3HED lymphosarcoma of mice produced a protective antiserum.

Theories of Carcinogenesis

There is a growing tendency to accept the supposition that carcinogenesis is the result of the presence within the cell of an abnormal nucleoprotein. The quarrel now is whether the abnormal nucleoprotein is exogenous or endogenous—that is, whether it is a body foreign in all its stages to the normal cell, or a body produced from a normal constituent of that cell. To those who hold the former view, cancer is a virus disease which is fundamentally similar to other virus diseases. To account for the fact that cancer strikes in so erratic a fashion and is, as far as we know, not contagious, it is sometimes postulated that the cancer virus infects all cells and only becomes virulent in relatively rare cases where the cell metabolic processes are disturbed by the various carcinogenic agents. Those who believe in the endogenous origin of the abnormal nucleoprotein consider the cancer cell to be the result of a mutation. In view of the chemical similarity between genes and viruses, it might be suggested that the difference between the virus and cell mutation theories is partly, at least, semantic.

Cancer and mutations. A mutation might be looked on as the result of an inexact self-duplication of a gene. In random changes, most mutated genes would be lethals or at least would cause the loss of some characteristic previously present. This has been demonstrated over and over again in such work as that on Neurospora (see page 369). If the gene controlling an enzyme synthesis were changed by mutation in such a direction as to cause growth without limit, cancer might result.

Spontaneous carcinogenesis could then be the result of a change in nucleoprotein structure on the random basis that one in so many gene autoreproductions is inexact and one in so many imperfections is cancerous. Naturally, on such a basis the incidence of cancer would increase with age, as it does, since the older an organism is the more autoreproductions have taken place and the greater the chance for the cancerous mutation. The incidence of cancer would be higher where more mitoses take place in a given time, as in epithelial tissue.

The incidence of cancer would also be increased if autoreproduction of nucleoproteins were interfered with so as to augment the chance of inexact self-duplication. Damage to the gene could interfere with its capacity to duplicate its old self, and damage to nucleic acids is known to result from the impact of high-energy radiation (43). If all this were so, one would expect that chemical agents that affected genes and increased mutation rates (mutagens) would also be carcinogens. In actual fact, though many mutagens are not carcinogens and many carcinogens are not mutagens, a sizable number of compounds are indeed both (5, 42). These include mustard gas, nitrogen mustards, urethane, certain carbamates, and some of the polycyclic hydrocarbons, such as 20-methylcholanthrene. Quantitative correlation of the two types of properties has not been successfully carried out; that is, a more intense mutagen is not necessarily a more efficient carcinogen or vice versa. In this connection, however, it should be remembered that carcinogenic properties are measured usually on vertebrates, and mutagenic properties on micro-organisms and insects.

If a mutation is the result of a change in a nucleoprotein, it can be looked upon as the result of a mild denaturation effect. Most carcinogens are protein denaturants (34) and denaturation theories of carcinogenesis have been advanced.

Evidence exists (6) that certain mutations in some bacterial cultures take place at rates independent of the rate of cell division and constant with time, which is hard to reconcile with the inexact duplication hypothesis of mutation.

Cytoplasmic factors. The cell changes discussed in the previous section involve genes which, being located in the nucleus, may therefore be considered organelles which are relatively well protected against environmental influences. The mitochondria, organelles of the cytoplasm which contain nucleic acid, are more exposed and presumably more vulnerable. Furthermore, the most general change in the enzyme patterns of cancerous cells involves the decrease in concentration of the enzymes of aerobic oxidation which are located within the mitochondria (and chicken sarcoma virus particles contain RNA, characteristic of cytoplasm, rather than DNA, characteristic of nuclei). It is not beyond possibility therefore that the cancerous change may involve not the DNA of the gene but the RNA of the mitochondrion and that a decreased effectiveness in the biosynthesis of the cytochromes, the flavo-enzymes, and the pyridino-enzymes may be one of the immediate results of the change.

This would imply, however, that such cytoplasmic particulates would be capable of maintaining their new cancerous identity as they themselves autoreproduced, and as the cell multiplied, despite the lack of change in the chromosomal gene make-up. Whether such autonomy on the part of cytoplasmic particulates is possible is a matter of much controversy (44). Those who believe that the cytoplasm can play a role in determining hereditary traits give those cytoplasmic factors the name of *plasmagenes*.

One significant line of investigation in this respect involves certain strains of *Paramecium aurelia*, whose presence in a culture was found to be poisonous to other strains (38). These "killer" strains accomplish this effect by the production and liberation of paramecin, an antibiotic which proves to be a deoxypentosenucleoprotein (50). Paramecin is produced by cytoplasmic particulates ("kappa"-particles) which apparently multiply at rates independent of that of the cell itself. If the cell is in a medium under which conditions of rapid growth are possible, cell growth and multiplication outstrip kappa-particle duplication, so that each generation has less paramecin-producing material and is consequently less virulent as a killer. Eventually, a cell is produced without any kappa-particles and a new "sensitive" strain results despite the fact that the genetic constitution is precisely the same as it was in the original killer strain. If at any time short of the complete loss of kappa-particles the micro-organism is placed in a medium where growth proceeds only slowly, kappa-particles duplicate at a rate faster than cell division so that each cell generation becomes more virulent. This represents the one definitely known case in the animal kingdom for cytoplasmic heredity independent of nuclear mechanisms.

Enzymatic theories. Regardless of the nature of the changed nucleoprotein which presumably initiates cancer—whether it is exogenous and a virus, or endogenous and either a gene or a plasmagene—it should make its effect known in the form of a change in the enzyme pattern of the cell. As has been stated, the only such change consistent in cancer cells is the deficiency of the enzymes involved in aerobic respiration. Such a deficiency would affect the tricarboxylic acid cycle (see Chapter 14). The compounds involved in the cycle represent links in the catabolism of carbohydrate and also in the anabolism of much of the protein molecule. Ordinarily, there is a balance. Cancer, however, may be the result of a disturbed balance between these catabolic and anabolic processes (37). With the oxidizing enzymes deficient, aerobic oxidation of carbohydrate slows, the cell turns to other methods of obtaining energy, such as glycolysis, and the compounds of the tricarboxylic acid cycle, now being oxidized at a reduced rate, are more available for protein biosynthesis. It is this last which would account for the unrestrained growth characteristic of cancerous tissue.

CARCINOTHERAPY

The definitive methods of fighting cancer—extirpating it through surgery or destroying it by radiation—are of course beyond the scope of this book. The search for biophysical or biochemical alternatives to surgery is unflagging, because many cancers are unfortunately, through their position or their metastatic state, completely inoperable.

Nutrition

Since cancer incidence shows a positive correlation with a state of good nourishment and since a cancer in its continuous growth would be expected to require a steady supply of metabolites, attempts have been made to control tumor growth by decreasing over-all food intake, or by sharply restricting one or more food factors. Unfortunately, no method of dietotherapy has succeeded in doing more than delaying, not abolishing, the appearance of cancers in experimental animals, and none has succeeded in more than slowing, not stopping, the growth of cancerous masses.

Even these partial advantages are gained through diets so inadequate as to interfere seriously with the normal functioning of the body. Mice fed upon a calorie-restricted or cystine-deficient diet show decreases in incidence of mammary cancer as compared to well nourished controls, but they also show irregularity in estrus, in some cases being completely anestrous, while mammary tissue failed to grow in the virgin and atrophied in the breeding female (51). Once a cancer is established, its growth can be slowed by drastic restriction of calories or of such food factors as riboflavin or pantothenic acid. In each case, however, body weight falls and the weight of the cancer grows larger in proportion. In general, cancers, being under no obligation to control or regulate their own growth for the good of the body as a whole, will continue to grow however restricted the host's diet. If no other source of energy or food factor exists, there are always normal tissues to be stripped so that in the end the host suffers more than the cancer. With respect to moderately restricted diets over prolonged periods as a general means of cancer prophylaxis, Greenstein says—

> "When the vagaries of human nature, the possible carcinogenic hazards to which the individual is exposed and the demands of modern human civilization are all taken into consideration, it would be impracticable, not to say absurd, to suggest the prolonged abjuration of the few pleasures which life grants in the hope of avoiding cancer at some unknown and distant future" (21).

Radioactive Isotopes

Since the development of the cyclotron, and particularly of the atomic pile, and the consequent mass production of artificial radioactive isotopes, clinical uses of these isotopes have increased tremendously. Radioactive isotopes, when introduced into the body, act as sources of penetrating radiation which in proper dosage may harm and even kill cells. This is ordinarily a most undesirable procedure. However, where the isotopes can be so distributed in the body that the effect of the radiation is largely confined to cancer cells, the result may be beneficial.

The most successful such therapy involves the use of radioactive iodine in some cases of cancer of the thyroid. Iodine, after administration, is quickly and almost exclusively concentrated in the thyroid where the cancer cells then come under concentrated attack (along with normal cells, of course), while the rest of the body is relatively unaffected (45). Where the cancer has advanced to the point where thyroid tissue has lost the specialized ability to accumulate iodine, as is usually the case in thyroid metastases, the treatment becomes ineffective.

Unfortunately, no other element is concentrated by the body to the extent of iodine, so that there is no prospect as yet that isotope therapy will replace the classical x-irradiation. Less marked concentrations are found in the case of radioactive strontium and radioactive phosphorus. The former localizes in the bones and may be helpful in the case of osteosarcomas. The latter is found in greater than average concentration in bone marrow, liver, spleen, and kidney. Since these tissues usually show the greatest infiltration of leukemic cells, radioactive phosphorus therapy is used in leukemia. Moreover, specialized techniques have been worked out in which injection of radioactive isotopes in non-absorbable form near a cancer is employed. For a summary of such clinical devices in connection with cancer see Kamen (32).

Hormones

The effect of certain hormones on growth processes in the body leads to the hope that their use may serve to control the growth of cancer. The nearest approach to success has been achieved with the sex hormones. In cancer of the prostate any treatment which results in decreasing the effectiveness of male sex hormones tends to atrophy the prostate and cause the cancer to regress at least temporarily. Such treatment includes administration of estrogens and castration (30). The treatment is not always successful. For one thing, a later recurrence may take place due to compensatory production of male sex hormones by such steroid manufacturing glands as the adrenal cortex.

Chemotherapy

The chemotherapeutic attack upon cancer has had limited success. Rhoads (41) suggests that the most hopeful prospects for chemotherapeutic success lie among those chemicals which, in one way or another, inhibit nucleic acid anabolism. Assuming that the nucleic acids of cancer cells differ from those of normal cells then it is conceivable that certain compounds might interfere with the growth of cancer cells while leaving normal cells untouched. Thus, 2,6-diaminopurine and 8-azaguanine (formula XI),

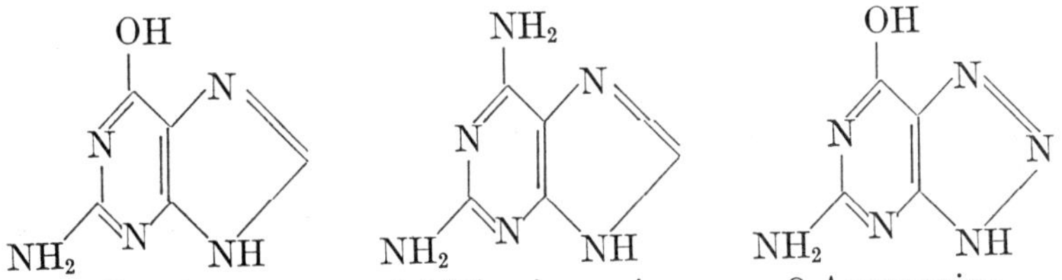

Guanine 2,6-Diaminopurine 8-Azaguanine

XI. Guanine and its antagonist

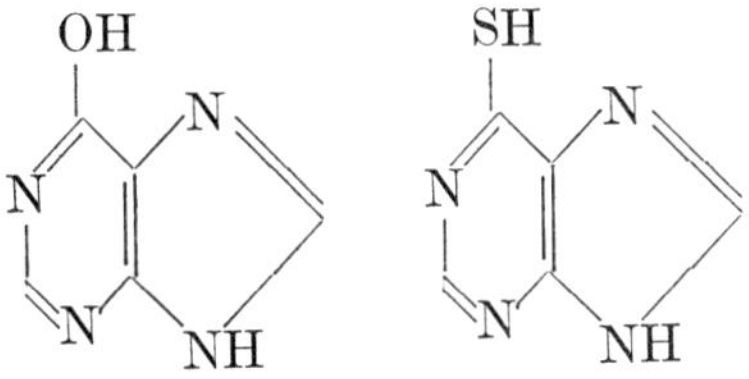

Hypoxanthine 6-Mercaptopurine

XII.

both of which may be viewed as guanine antagonists, have retarded the growth of certain cancers in mice. Folic acid antagonists, which would inhibit the activity of folic acid in purine ring formation, have also had their hopeful moments.

The compound 6-mercaptopurine (formula XII) has been found useful, to some degree, in the treatment of leukemia. It presumably acts as a hypoxanthine antagonist, blocking the synthesis of inosinic acid (hypoxanthine ribotide) from 4-amino-5-imidazolecarboxyamide ribotide (see page 359) (46).

REFERENCES

1. Barnes, *et al.* Beryllium bone sarcomata in rabbits. Brit. J. Cancer, **4:** 212–222, 1950.
2. Beard, J. W., *et al.* The nature of a virus associated with carcinoma in rabbits. Surg. Gynec. Obst., **74:** 509, 1942.
3. Bittner, J. J. Some possible effects of nursing on the mammary gland tumor incidence in mice. Science, **84:** 162, 1936.
4. Boyland, E. Different types of carcinogens and their possible modes of action: a review. Cancer Res., **12:** 77–84, 1952.
5. Boyland, E. Mutagens. Pharmacol. Revs., **6:** 345–364, 1954.
6. Braun, W. *Bacterial Genetics*. Philadelphia, Penna., W. B. Saunders Co., 1953, page 67.
7. Bullough, W. S. Mitotic activity and carcinogenesis. Brit. J. Cancer, **4:** 329–336, 1950.
8. Chargaff, E., and Lipshitz, R. Composition of mammalian desoxyribonucleic acids. J. Am. Chem. Soc. **75:** 3658–3661, 1953.

9. Claude, A. Particulate components of normal and tumor cells. Science, **91:** 77–78, 1940.
10. Coman, D. R. Mechanism of the invasiveness of cancer. Science, **105:** 347–348, 1947.
11. Coman, D. R. Cellular adhesiveness in relation to the invasiveness of cancer: electron microscopy of liver perfused with a chelating agent. Cancer Res., **14:** 519–521, 1954.
12. Commins, B. T., *et al.* Polycyclic hydrocarbons in cigaret smoke. Brit. J. Cancer, **8:** 296–302, 1954.
13. Dunphy, J. E. Changing concepts in the surgery of cancer. N. Eng. J. Med., **249:** 17–25, 1953.
14. Duran-Reynals, F. The infection of turkeys and guinea fowls by the Rous sarcoma virus and the accompanying variations of the virus. Cancer Res., **3:** 569–577, 1943.
15. Eriksen, N., *et al.* Studies of various tests for malignant neoplastic diseases. I. The reduction of methylene blue by plasma. J. Nat. Cancer Inst., **11:** 705–728, 1951. II. The Gruskin intradermal test. *Ibid*, 729–732. III. The Hoff-Schwartz intradermal test. *Ibid*, 733–738. IV. The effect of zinc ion upon the serum alkaline phosphatase activity. *Ibid*, 739–756. V. The heat "coagulation" of plasma. *Ibid*, 757–772.
16. Findlay, G. M. Ultraviolet light and skin cancer. Lancet, **2:** 1070–1073, 1928.
17. Goldberg, L., *et al.* The nucleic acid content of mouse ascites tumor cells. Exptl. Cell Res., **1:** 543–570, 1950.
18. Graff, S., *et al.* Isolation of mouse mammary carcinoma virus. Cancer, **2:** 755–762, 1949.
19. Graham, J. B., and Graham, R. M. Antibodies elicited by cancer in patients. Cancer, **8:** 409–416, 1955.
20. Greenberg, D. M. Isotopic tracer studies on the biochemistry of cancer. Cancer Res. **15:** 421–436, 1955.
21. Greenstein, J. P. *Biochemistry of Cancer. 2nd ed.* New York, N. Y., Academic Press, 1954.
22. Greenstein, J. P., and Leuthardt, F. M. Enzymatic activity in primary and transplanted rat hepatomas. J. Nat. Cancer Inst., **6:** 211–217, 1946.
23. Greenstein, J. P., *et al.* Chemical studies on human cancer. I. Cytochrome oxidase, cytochrome c, and copper in normal and neoplastic tissues. J. Nat. Cancer Inst., **5:** 55–76, 1944.
24. Gutman, A. B. Serum "acid" phosphatase in patients with carcinoma of the prostate gland. J. A. M. A., **120:** 1112–1116, 1942.
25. Heath, J. C. Cobalt as carcinogen. Nature, **173:** 822–823, 1954.
26. Henry, R. J., *et al.* Inaccuracy of four chemical procedures as diagnostic tests for cancer. J. Am. Med. Assoc., **147:** 37–39, 1951.
27. Henshaw, P. S., *et al.* Incidence of leukemia in physicians. J. Nat. Cancer Inst., **4:** 339–346, 1944.
28. Hueper, W. C. Environmental cancers: a review. Cancer Res., **12:** 691–697, 1952.
29. Hueper, W. C. Experimental studies in metal cancerigenesis. IV. Cancer produced by parenterally introduced metallic nickel. J. Nat. Cancer Inst., **16:** 55–67, 1955; VII. Tissue reactions to parenterally introduced powdered metallic chromium and chromite ore. *Ibid.*, 447–462.
30. Huggins, C. Prostatic cancer treated by orchiectomy: The five year results. J. A. M. A., **131:** 576–581, 1946.

31. Kahler, H., and Robertson, W. v. B. Hydrogen-ion concentration of normal liver and hepatic tumors. J. Nat. Cancer Inst., **3:** 495–501, 1943.
32. Kamen, M. D. *Radioactive Tracers in Biology.* 2nd edit. New York, N. Y., Academic Press, 1951.
33. Laird, A. K. Cell fractionation of normal and malignant tissues. Exptl. Cell Res., **6:** 30–44, 1954.
34. Lemon, H. M., *et al.* Biochemistry of human cancer. N. Eng. J. Med., **251:** 937–944, 975–980, 1011–1017, 1954.
34a. Mider, G. B. Some developments in cancer research. J. Chronic Diseases, **4:** 296–320, 1956.
35. Nungester, W. J., and Fisher, H. The inactivation *in vivo* of mouse lymphosarcoma 6C3HED by antibodies produced in a foreign host species. Cancer Res. **14:** 284–288, 1954.
36. Pfeiffer, C. A., and Allen, E. Attempts to produce cancer in Rhesus monkeys with carcinogenic hydrocarbons and estrogens. Cancer Res., **8:** 97–109, 1948.
37. Potter, V. R., and Heidelberger, C. Alternate metabolic pathways. Physiol. Rev., **30:** 487–572, 1950.
38. Preer, J. R. A study of some properties of the cytoplasmic factor "Kappa" in Paramecium aurelia, variety 2. Genetics, **33:** 349–404, 1948.
39. Price, V. E . and Greenfield, R. E. Liver catalase. II. Catalase fractions from normal and tumor-bearing rats. J. Biol. Chem., **209:** 363–376, 1954.
40. Pullman, A., and Pullman, B. Electronic structure and carcinogenic activity. Advances in Cancer Res., **3:** 117–169, 1955.
41. Rhoads, C. P. Rational cancer chemotherapy. Science, **119:** 77–80, 1954.
42. Scherr, G. H., *et al.* The mutagenicity of some carcinogenic compounds for Escherichia coli. Genetics, **39:** 141–149, 1954.
43. Scholes, G., *et al.* Action of x-rays on nucleic acids. Nature, **164:** 709–710, 1949.
44. Schultz, J. The question of plasmagenes. Science, **111:** 403–407, 1950.
45. Seidlin, S. M., *et al.* Spontaneous and experimentally induced uptake of radioactive iodine in metastases from thyroid carcinoma: A preliminary report. J. Clin. Endocrinol., **8:** 423–432, 1948.
46. Skipper, H. E. Mechanism of action of 6-mercaptopurine. Ann. N. Y. Acad. Sci., **60:** 315–321, 1954.
47. Spencer, R. R., and Melroy, M. B. Effect of carcinogens on small free-living organisms. I. Eberthella typhi. J. Nat. Cancer Inst., **1:** 129–134, 1940.
48. Strong, L. C. A new theory of mutation and the origin of cancer. Yale J. Biol. & Med., **21:** 293–299, 1949.
49. Thomas, P. T. Experimental imitation of tumour conditions. Nature, **156:** 738–740, 1945.
50. van Wagtendonk, W. J. The action of enzymes on paramecin. J. Biol. Chem., **173:** 691–704, 1948.
51. Visscher, M. B., *et al.* The influence of caloric restriction upon the incidence of spontaneous mammary carcinoma in mice. Surgery, **11:** 48–55, 1942.
52. Wenner, C. E., *et al.* Metabolism of neoplastic tissue. II. A survey of enzymes of the citric acid cycle in transplanted tumors. Cancer Res., **12:** 44–49, 1952.

CHAPTER 11

Reproduction and Heredity

All living organisms may produce offspring which in many ways duplicate the chemical make-up of their parents; in this process chemical phenomena are involved. The chemistry of reproduction is far from being completely understood. Certain of the facts of heredity which can not yet be fully explained chemically seem to be potentially capable of a chemical interpretation, and this is briefly discussed here. We also present the bare outline of formal genetics. Furthermore, hereditary diseases are numerous, and the wise physician will want to know something about the way in which a child derives them from its immediate parents or remote ancestors. We are going to assume that the physiology and cytology of reproduction are already known to the student, and only such portions of the cytologic process will be presented as seem essential to our argument.

BIOCHEMISTRY OF REPRODUCTION

New human beings are begun when a single spermatozoon from a man reaches and fertilizes an ovum from a woman. Fertilization takes place in the upper part of the fallopian tube. Approximately three hours are required for the sperm to travel from the external os of the cervix to the site of fertilization. The spermatozoa originate from the spermatogenic cells of the epithelial lining of the seminiferous tubules of the testicles, and undergo maturation during their passage through the tortuous 20 feet of tubule which makes up either of the epididymes. In the acid environment of the epididymis the spermatozoa possess only minimal motility.

Seminal Fluid

The seminal fluid or semen ejaculate consists of a suspension of spermatozoa in the *seminal plasma*, which is a mixture of the secretions of the epididymes, vasa deferentia, seminal vesicles, prostate, bulbo-urethral (Cowper's) glands, and urethral glands (glands of Littré). Although there are 60 million or more spermatozoa per ml. of ejaculate from the normal fertile man, the actual volume of spermatozoa plus epididymal secretion is less than 5 per cent of the total volume. The volume of the ejaculate is usually between 2 and 6 ml., with the major portion contributed by the

TABLE XI-1

Average composition of prostatic fluid (after Huggins)

	mEq./kgm. H_2O	grams/100 ml.
Na^+	156	0.36
K^+	30	0.12
Ca^{++}	30	0.06
Citric acid	156	1.0
Cl^-	38	0.14
Inorganic phosphate	1	0.003
Bicarbonate	8	0.07
Protein		2.5
H_2O	55.5	93–98

See also Hudson and Butler (22).

secretions of the prostate and seminal vesicles. The pH of fresh ejaculate is approximately that of blood plasma, 7.4, but may increase to pH 8 with loss of CO_2. The pH of prostatic fluid is said by some to be highly alkaline but has been measured as about 6.4 (23). The chemical composition of prostatic fluid is given in table XI-1. Prostatic fluid does not coagulate on boiling. It contains 255 to 1727 King and Armstrong units of acid phosphata e per ml (24). Acid phosphatase can be found in voided male urine in much larger amounts than in the urine obtained by ureteral catheterization. The difference is the contribution of the resting secretion of the prostate. In health, very little acid phosphatase appears in the blood. In cancer of the prostate, particularly if these are metastases, the acid phosphatase activity of the plasma is greatly increased. This fact is of major importance in the diagnosis and clinical management of prostatic cancer. Fructose is present with a concentration range of 0.20 to 0.80 per cent, the mean value being in the neighborhood of 0.30 per cent (34). This is utilized as a source of energy by the spermatozoa. Seminal fructose levels of less than 0.135 per cent may indicate a hypogonadal state (27). Sperm and prostate are richer in zinc than is any other body tissue or fluid (36).

Traces of choline are detectable in fresh semen. The concentration of choline increases with lapse of time after ejaculation, up to over 2 per cent. *Glycerophosphorylcholine* is the precursor, formed in the seminal vesicles, and hydrolyzed by phosphatases of the prostatic secretion (32).

Human semen coagulates promptly after ejaculation. The clot liquefies in about 15 minutes. The protein which clots is a product of the seminal vesicles. The proteolytic enzyme which dissolves the clot has been identified in human prostatic secretion. Motility of spermatozoa is limited during the clotted stage and is restored with liquefaction.

In seminal plasma, or in certain artificial media such as buffered egg yolk, spermatozoa will retain their fertilizing power for considerable time if stored at temperatures just above freezing. The secretions of the vagina are acid (pH 2.8 to 5.0), and spermatozoa do not survive well in an acid medium. It is therefore believed that in a majority of cases insemination does not occur unless the spermatozoa are ejaculated directly against the external os of the cervix. Men and women vary in fertility, however, and pregnancy has been known to result when this condition was not fulfilled. The high buffer capacity of semen presumably helps compensate for vaginal acidity. The acidity of the human vaginal secretion is the result of the presence of lactic acid. This acid is produced by lactobacilli, the activity of which appears to be favored by the presence of estrogens.

Meaker and Glaser found the pH of the cervical secretion to be from 8.0 to 9.0, being above 8.5 in 80 out of the 100 cases examined by them. The alkalinity of this secretion was not notably influenced by age, parity, menstrual cycle, endocervicitis, or viscosity of the endocervical mucus. Others however have reported a cyclic production of a cervical mucus at about midcycle, which renders the cervix penetrable by spermatozoa for 4 to 10 days. If this is so, it may be the basis of the so-called "safe period" during other parts of the cycle, but it has been observed that impregnation *can* occur at any time. What is still in doubt is the relative frequency of conception during different parts of the cycle. The hormones involved in the production of spermatozoa and in the menstrual cycle are described in Chapter 8.

The human egg, like other cells, consists primarily of a denser mass composed in great part of deoxypentose nucleoproteins, lipoproteins, phospholipids, and small amounts of pentosenucleoproteins; this mass is the nucleus and is surrounded by a less dense mass of protoplasm, the cytoplasm. The nucleus of the cell contains a certain number of bodies which are called chromosomes, which can be observed when the nucleus is in the process of division. The numbers of chromosomes vary in mature cells from three pairs in certain plants and some species of Drosophila, to more than one hundred in certain moths, crayfish, and some plants. In man there are 23 pairs of chromosomes in the body cells. In the formation of the gametes the number of chromosomes is reduced to half.

Fertilization

Hyaluronidase. The human ovum is surrounded by a mass of ovarian follicular cells, the *cumulus oophorus*. The compact innermost portion of the cumulus is the *corona radiata*, the cells of which are presumed to have a nutritive function for the ovum. In the rabbit, several ova are bound together in a clot of cumulus cells and mucus. The mucus clot can be dissolved *in vitro* by the enzyme hyaluronidase, or by spermatozoa. There is hyaluronidase in the sperm of man, rabbit, and most mammalian species.

Hyaluronidase will not, however, bring about the detachment of the cells making up the corona, which are removed by a substance contributed by the fallopian tube. The presence of cumulus cell masses has been considered a barrier to the entry of sperm into ova, but instances of sperm entry into rabbit ova have been observed with the cumulus mass still attached (11). Clinical reports have been conflicting on the improvement of human fertility by the use of bovine hyaluronidase as an adjuvant to human sperm, although it seems reasonable to suppose the hyaluronidase of the spermatozoa to be an important part of their mechanism for the penetration of the ovum. There is no hyaluronidase in the ejaculate in cases where no spermatozoa are produced (16). Seminal fluid also contains significant quantities of proteases and peptidases (33).

Parthenogenesis. The development of an unfertilized ovum into a viable organism is a well recognized occurrence in certain non-mammalian species. Mammalian ova have repeatedly been activated by high or low temperatures, by chemical reagents, or by foreign sperm. Such activation may continue *in vitro* through several stages of division. Activated rabbit ova have been placed surgically into the fallopian tubes of pseudopregnant rabbits, and rabbit ova have been activated by low temperature while in the tube. A small proportion of such experiments have resulted in the production of living young at term (11).

Fertilizin. It was demonstrated in 1913 (31) that waters which had been in contact with eggs of the sea urchin or certain other species were able to cause the agglutination of the homologous spermatozoa, and Lilly coined the term fertilizin for the active agent in this egg-water. Since then fertilizin has been obtained from eggs of various species of invertebrates, and has also been reported to occur in eggs of vertebrates. The fertilizins seem to originate from the gelatinous coat of the egg which slowly goes into solution as the egg stands in water. There is some evidence that this agglutinating activity of fertilizin on the spermatozoa has some connection with the fertilization process, perhaps serving to cause the spermatozoon to stick to the gelatinous surface of the egg, from this point it can begin its process of boring into the egg to reach the nucleus. Fertilizin has been obtained in electrophoretically homogeneous form from certain eggs and some of its chemical properties have been studied. It is highly acidic, contains about 25 per cent of sulfate or a sulfuric ester such as is found in chondroitin sulfate or mucoitin sulfate. The preparation gives both protein and polysaccharide reactions and could be considered as belonging to the group of mucoproteins. However, according to Tyler (54), it is not possible to separate it into a protein and a polysaccharide constituent. He believes this is because the various sugars and amino acids are interlinked, forming a compound which is neither a protein nor a polysaccharide. He considers fertilizin to belong to the group of compounds which include the human blood group substances. Fertilizin was found to have a minimum molecular weight of 82,000.

Amniotic Fluid

The amniotic fluid is considered to be composed of extracellular fluid of both fetal and maternal origin, to which is added hypotonic fetal urine. It is in equilibrium with the maternal and fetal blood plasma, and contains the usual inorganic ions found in extracellular fluid in approximately the same concentrations (see Chapter 17). Highly variable protein levels have been reported, ranging from 0.2 to 1.5 per cent. The glucose is lower than that of blood plasma, usually about half the blood value. The uric acid concentration is variable, but usually higher than that of the maternal blood. The volume of amniotic fluid at term is usually between 0.5 and 1.3 liters.

CELL DIVISION

Chromosomes

When a cell divides, two descendant cells are produced, each of which derives its nucleus and its cytoplasm from the nucleus and cytoplasm of the mother cell. The division of the cytoplasm is relatively simple. A furrow appears on the surface of the cell around its whole circumference. This furrow gradually cuts deeper and deeper into the cell until it separates the cytoplasm into two halves.

While the cytoplasm is dividing, the nucleus is also dividing, involving a rather more elaborate series of events. The nucleus in the resting cell seems to be a relatively undifferentiated vesicle, but during division well defined structures become visible. They may have varied shapes; they may be round, or short rods, or long rods, or they may be V-shaped or J-shaped. These structures pick up certain stains more intensely than the rest of the cell, in some cases almost exclusively, and they are therefore called chromosomes, from Greek roots meaning "color" and "body."

After chromosomes have become distinctly visible, the nuclear membrane disappears and the nucleus as a separate entity is gone. The chromosomes are associated with a structure called the spindle which develops inside of the cell. By the time the spindle is fully formed, the chromosomes all appear double, the two identical parts lying side by side but separated by a clear space except for being joined in one short region. Finally the pairs of chromosomes separate, one member going into one half of the dividing cell and one member into the other. Thus the divided cells each have a complete set of chromosomes, which results from this division. During this stage of division the chromosomes have a large amount of the highly stainable deoxypentosenucleoprotein, and we have reason to believe that this nucleoprotein is of the highest importance in the process of heredity. In some way the chromosomes of the original cells have managed to duplicate themselves by building a copy out of the surrounding protoplasm of the cell.

Chromosome Components

Chromosomes, as they are visible during cell division, have been shown to be coiled like spiral springs. The relatively slender threads of which these are formed are, in many cases, too narrow to be resolved by ordinary microscopic methods. When these threads can be seen, however, they appear to be covered by a succession of fine beads called *chromomeres*. Chromomeres are of different sizes at different points, and their arrangement is characteristic for each chromosome. The pattern enables the chromosome to be identified. For instance, one chromosome might have two successive medium-sized chromomeres at one end, then a larger one, then a series of three very small ones. This is shown in the diagram of a specific human chromosome (figure XI-1).

From the diagram one also sees that the distances between the chromomeres may vary as well as their size and staining capacities.

Fig. XI-1. Chromomere pattern of a strand of a typical chromosome

A better idea of the linear arrangement of the chromosomal elements is found by study of the giant chromosomes from the salivary glands of the larvae of certain species of flies. Here the chromosome, instead of appearing as a slender thread, looks like a rather wide cylinder marked by large numbers of cross-bands or discs. It has been suggested that these giant chromosomes are the result of the lining up of large numbers of chromosome threads in parallel.

This linear arrangement and characteristic pattern of the cross-bands in the giant chromosome or of the chromomeres in the threadlike single chromosomes are of great importance in connection with the linear arrangement which must be postulated to explain the known facts of genetics.

GENETICS

It has been found that inheritance is particulate in nature, that is, characteristics from mother and father do not blend in the germ plasm (although in some cases they appear to blend in the offspring), but remain separate and may segregate again in the offspring of the next generation. This simple fact is the clue to an understanding of modern genetical theory. Our knowledge of inheritance in an exact sense all goes back to the experiments of the Austrian monk, Gregor Mendel, who first worked out the essential laws by experiments with the garden pea. Mendel's work made no impression on his contemporaries and the publication of it in 1866 in the Proceedings of the Natural History Society of Brünn attracted no attention despite the fact that copies of this journal reached various parts of the world including the United States. Nevertheless the work was so carefully and brilliantly conceived that when it was rediscovered at the beginning of this century by Correns, De Vries, and von Tschermak (53), it was found to explain the mechanism of inheritance—not only in the garden pea but in all plants and animals, including man, which have subsequently been studied.[1]

Mendel's Laws

Particulate nature of inheritance. *Dominance.* Among hybrid or crossbred offspring each individual exhibits just one of each pair of contrasted ancestral characters, to the total (or almost total) exclusion of the other. Intermediate forms do not appear.

[1] A translation into English of Mendel's paper has been made by the Royal Horticultural Society of London, and is available at the Harvard University Press under the title, "Experiments in Plant Hybridization," by Gregor Mendel. This translation is reproduced as an appendix in *Principles of Heredity* by E. W. Sinnott, L. C. Dunn, and Th. Dobzhansky (McGraw-Hill, 1950).

Mendel called the character that prevailed *dominant*, and the character that was suppressed (or apparently suppressed) *recessive*. Thus the first important result was the discovery that, *dealing with pure lines*, crosses between a plant with the dominant character and a plant with the recessive character yielded offspring which, as regards the character in question, all resembled the dominant parent. If we designate the appearance of the parents as D (dominant), and R (recessive), Mendel's first result may therefore be expressed thus: The cross, Dominant by Recessive, produces Dominant-appearing offspring. Or in symbols—

$$D \times R = D$$

Later work has shown that *complete dominance* is much less common than was originally thought. In fact, some effects of each gene of the pair are generally discernible in the hybrids (8).

In the next generation the crossbred plants which had been produced by crossing D and R and which were all apparently like D, were allowed to fertilize themselves; and it was then found that their offspring exhibited both of the two original forms, showing on the average three D's to one R. We may take as an example the tall x dwarf cross: 1064 second-generation plants were produced; 787 were tall (D), and 277 were dwarfs (R).

When any of these "recovered" dwarfs (i.e., recessive descendants of a group of plants all "Dominant" in appearance) were allowed to fertilize themselves, they gave rise to dwarfs, R, only, a process which could be continued for any number of generations. In other words, the recessive character bred true. The hereditary unit tending to produce it (we call such units *genes*) had not been altered by its association in the hybrid with the dominant (D) gene for tallness.

On the other hand, when the dominant-appearing descendants were allowed to fertilize themselves, one-third of them produced pure dominants which in subsequent generations, when allowed to fertilize themselves, gave rise to dominant descendants only; two-thirds of them, however, were an impure group which produced once again the characteristic mixture of dominants and recessives in the proportion of 3:1. These results are shown in figure XI-2. The result of the initial hybridization is a first generation (F_1) which resembles the dominant parent. They may be represented by the symbol (D), since they look like the dominant, even though they carry the possibility of producing offspring characterized by the recessive character—that is to say, the recessive character has remained latent in the inheritance.

The genetic formulas of the various individuals are evidently as follows, where D represents the dominant gene and R the recessive:

Type as identified by inspection	*Genetic type as identified by breeding experiments*
D	DD or DR
R	RR

From this it may be seen that the pure dominant individual, which possesses only D genes, can produce only D offspring when it fertilizes itself and R can produce only R. The dominant-appearing individuals (D), however, being genetically mixed, can produce D, (D) and R offspring in the ratio of 1:2:1.

The student is warned that these ratios are statistical in nature, and apply strictly to large numbers only. If we obtain only four offspring from the mating of two heterozygotes (DR × DR), various outcomes are possible. The laws of heredity determine merely the probability of a given type of individual. In this mating the probability of producing a DR individual is $\frac{1}{2}$, and of a DD or RR individual, $\frac{1}{4}$ for each. Consequently, the probability of four RR offspring is $(\frac{1}{4})^4$, or $\frac{1}{256}$. This is a relatively small probability, yet this event (four recessive offspring in succession from two hybrid parents) would occur on the average in about one-half of one per

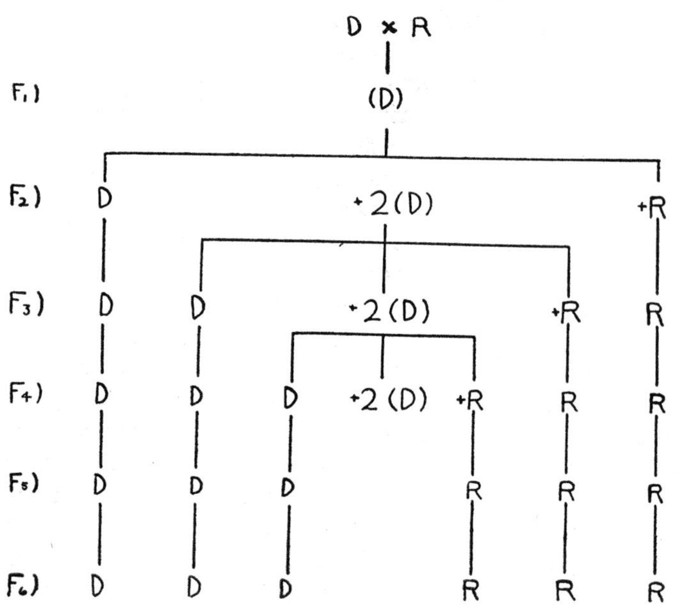

FIG. XI-2. Diagram Illustrating "Mendelian splitting" or segregation which occurs when hybrids mate. The homozygous dominants and recessives breed true, but the heterozygotes (which, in the present case, resemble the dominant parent) continue to produce, in addition to heterozygotes, homozygous recessives and homozygous dominants.

cent of the matings. The probability of the result we think of as proving the laws of Mendelism, three dominant-appearing individuals and one recessive, would be $4(3/4)^3(1/4) = 27/64$. It is thus 108 times as likely as the run of four recessives in a row, but nevertheless far from certain (35). The student will perceive that there are also various other possible outcomes of this mating, each with its own probability. It is only when large numbers of offspring are observed that the laws of Mendel are found to be obeyed with great accuracy.

It will be seen from the above that if we have a married couple, both members of which are presumed to carry the recessive gene for amaurotic family idiocy, we can not tell them that their first three children will be normal and the fourth an idiot, and that they can therefore safely have three children, as has amazingly enough been actually suggested! Instead, we can merely state that *if* both parents do in fact carry this gene, the chance that any one of their offspring will be an idiot is 1/4, or one in four.

Independent assortment of unit characters. Mendel's second law states that each pair of hereditary factors showing this dominant-recessive relationship behaves quite independently of every other pair. This is called the law of independent assortment. This second law is not without exceptions, for reasons which will presently appear. However, it does hold for many factors, and unless the genes were connected together in some way in the organism, it would be expected to hold for all factors influencing inheritance. Reproductive cells (gametes) having all possible combinations of genes would be expected. The fact that the law does not always operate suggests that there may be some physical connection between certain genes, or in other words, the genes may be transmitted in groups during the process of reproduction.

To take an actual instance of independent assortment: Mendel found that the character pair yellow-green (of which yellow is dominant) and the pair round-wrinkled (of which round is dominant) assorted independently. Either gene pair, considered alone, in the F_2 generation produced dominants and recessives in the expected 3:1 ratio. Mendel crossed a round yellow race with a wrinkled green one, and, as expected, got nothing but round yellow seeds in the first (F_1) generation. If such plants were then crossed with plants which were genetically pure (*homozygous*) for both of the recessive genes in question (i.e, wrinkled green peas), there were four possible types of offspring, namely, round-yellow, round-green, wrinkled-yellow, and wrinkled green; the genes, assorting independently, produced the four possible types of offspring in equal numbers.

Mendel represented the dominant gene for round as A, and the recessive gene for wrinkled as a, the dominant for yellow as B, and the recessive for

TABLE XI-2

Results of crossing AaBb with aabb (Mendel)

FEMALE GAMETES	MALE GAMETES			
	AB	*Ab*	*aB*	*ab*
ab	*AaBb*	*Aabb*	*aaBb*	*aabb*

green as *b*. This would give us, for his hybrid (heterozygous) pea, the formula *AaBb*. The pure recessives would be *aabb*. They could produce only one kind of gamete, namely, *ab*. The heterozygotes could produce four kinds of gamete, in equal numbers, *AB*, *Ab*, *aB*, and *ab*. Combining these with the gametes of the wrinkled green peas, we get four possible combinations, and therefore four possible types of offspring, in equal numbers, namely, *AaBb*, *Aabb*, *aaBb*, and *aabb*, as shown in table XI-2. This was confirmed by experiment.

If the reader has followed this reasoning, he should have no difficulty in predicting the outcome of a cross of two plants heterozygous for both genes (or, what amounts to the same thing, the results of allowing a doubly heterozygous plant to fertilize itself). Each plant produces the four possible gametes, *AB*, *Ab*, *aB*, and *ab*, in equal numbers. It is a matter of chance which ovum is fertilized by which pollen grain, so we get all the possible combinations, as shown in table XI-3.

Since round is dominant over wrinkled, and yellow is dominant over green, not all these sixteen different genetic types will be distinguishable in the offspring. We find, in fact, that just four types can be distinguished, and they are classified as shown in table XI-4.

It will be noted that not only are there just four distinguishable types of offspring, but since each type of gamete is produced as frequently as any other, the four types of offspring will occur in the ratio shown, which is 9:3:3:1. The actual ratio observed by Mendel (36) was 315 round and

TABLE XI-3

Combinations resulting from the cross of individuals heterozygous for two characters

FEMALE GAMETES	MALE GAMETES			
	AB	*Ab*	*aB*	*ab*
AB	*AABB*	*AABb*	*AaBB*	*AaBb*
Ab	*AABb*	*AAbb*	*AaBb*	*Aabb*
aB	*AaBB*	*AaBb*	*aaBB*	*aaBb*
ab	*AaBb*	*Aabb*	*aaBb*	*aabb*

TABLE XI-4

Phenotypically different offspring from the mating shown in table XI-3

AB (ROUND, YELLOW)	*Ab* (ROUND, GREEN)	*aB* (WRINKLED, YELLOW)	*ab* (WRINKLED, GREEN)
AABB	*AAbb*	*aaBB*	*aabb*
AABb	*Aabb*	*aaBb*	
AaBB	*Aabb*	*aaBb*	
AaBb			
AABb			
AaBb			
AaBB			
AaBb			
AaBb			

yellow, 101 wrinkled and yellow, 108 round and green, and 32 wrinkled and green.

The use of capital letters for the dominant gene and small letters for the recessive was continued by Mendel's successors; but when it was found that sometimes there were more than two genes which could occur at a chromosome locus, the system ceased to be so satisfactory. Consequently many genes are designated by various combinations of letters and numbers.

No Blending Inheritance

Inheritance is particulate in nature, as we have seen in reviewing Mendel's work. That is, the germ plasm is passed from parent to offspring in the form of discrete particles, and not as a portion of a more or less uniform mixture of the germ plasm of mother and father. These units of heredity, or of germ plasm, are known as genes. It is customary to consider a characteristic, such as blue eyes, diabetes, or hemophilia, as due to the action of a single gene pair, one member coming from the father and one from the mother. In some cases the father contributes no gene at that locus, so the action of the mother's gene, although it is recessive, is unhampered. It is probable that nearly every gene in the inherited constitution of the individual has *some* effect on nearly every character, although in some cases the effect must be slight.

Genes seem to control all the vital processes of an organism, although some examples of "cytoplasmic" inheritance have been found to extend over a few generations in certain lower forms. It has not been proved, and probably can never be proved, that all the characteristics of a given organism are determined by genes. As has been pointed out (47), there is a reason why we shall never completely prove the genic nature of some of the really major characteristics of any form of life. Since these major characteristics, such as having lungs, a heart, blood, and so forth, are absolutely vital to the life of the organism, a change in the genes which control them is almost certain to lead to a condition which will interfere with normal development, or in other words will produce what geneticists call a *lethal.* A lethal gene is a gene which kills the organism which inherits it. It usually requires a double dose, since lethals are generally recessive. It can be seen that dominant lethals could not persist beyond one generation. Changes or mutations altering for the worse the absolutely essential features of an organism will be eliminated almost immediately by the action of natural selection, and each species will be homozygous for the normal (non-lethal) aspect of genes of such major importance.

The importance to biology of the concept of particulate inheritance can hardly be overestimated, but until the discoveries of Mendel and those who confirmed his work became known, it was assumed by practically all laymen, and by biologists too, even by Darwin, that inheritance was of a blending character. For example, it was supposed that a parent with a black skin and a parent with a white skin would always produce children with brown skins, and that these brown children, mated with similar brown children, would of course have brown offspring, for it was thought that the two characters, or types of "blood", were forever mixed, just as ink and milk poured together into the same container can never again be separated.

When considerable numbers of genes are involved, as in skin color,[2] heredity does at first sight appear to be of a blending character. Genetic analysis of such characters is difficult. The genius of Mendel led him to select for study varieties which differed from each other in only a few characters, genetically as well as apparently, so that he readily recovered the parental types from crosses among the offspring. All inherited characters which have been fully analyzed have always been found to depend on particulate genes which retain their individuality even in hybrids. This applies even to those apparent cases of blending inheritance which have been adequately investigated.

We may discuss briefly an example which, without careful investigation, might seem to be an example of the blending type of inheritance. Among chickens there are a number of genetically pure color types, each of which breeds true. But crossing black Andalusian with splashed white fowl produces a slaty grey intermediate; the so-called Blue Andalusians (2). It seems almost self-evident to the uninstructed observer that the genetic material of the two parents has been blended in the offspring to produce an intermediate and novel type, but this is not true. Breeding experiments

[2] Davenport proposed a theory involving only two pairs of genes for skin color in man; more are probably involved, however.

reveal that "blending" is not the correct explanation. Crosses of Blue Andalusians among themselves produce both black and splashed-white offspring as well as the Blue Andalusian, and it is absolutely impossible to obtain a stock of Blue Andalusians which will breed true. The two types of hereditary material have not blended, but during a period of coexistence in a certain individual they combine their effects to produce a new type of appearance. The genetic materials, however, have remained unchanged, and when they again emerge by themselves in the offspring, they produce the same effects they produced in the original parents. A cross of Blue Andalusians gives the expected 1:2:1 ratio.

Since the Blue Andalusians do not look exactly like either the black or the splashed-white parent, we miss in this case the complete dominance which Mendel observed in his peas. There are many other examples. Thus a cross between a "Chinese" primula which has wavy crenated petals and a "Star" primula with simply notched petals gives progeny intermediate between the two parents; and yet, as the next generation shows, the case is one of Mendelian inheritance—that is, the two characteristics have stayed distinct in the germ plasm, and both parental types, as well as the mixed variety, appear among the offspring.

In many cases the hybrid, while exhibiting on the whole the character of the dominant, may show also some influence of the recessive character, but not enough to warrant our speaking of the result as a "blend". Thus, when white (dominant) Leghorn poultry are crossed with brown (recessive) Leghorn, most of the offspring have some "ticks" of color. When these are inbred they produce one-quarter brown and three-quarters pure white or white with a few "ticks".

When dealing with characters which can only be present or absent, it is often not possible to measure degrees of dominance, although in situations which at first seem to belong to this category (as in the case of the blood groups of man) some information as to degrees of dominance has been obtained. If we are dealing with quantitative characters (that is, characters which may be counted, measured, or in some way expressed in numerical terms) it is possible to express the degree of dominance more exactly. The eye of the fruit fly, Drosophila, is made up of a number of separate elements or facets. The number of these elements in any individual eye can be counted and the effect of various genes on them can be determined. Figure XI-3 shows the effect of a mutant gene which is called "Bar", when it is present in the homozygous and heterozygous condition, and also shows the results of facet counts on the homozygous "wild" type without the mutant gene. In this figure, the vertical axis represents the percentage frequencies of each of the three types, and the horizontal axis, the facet

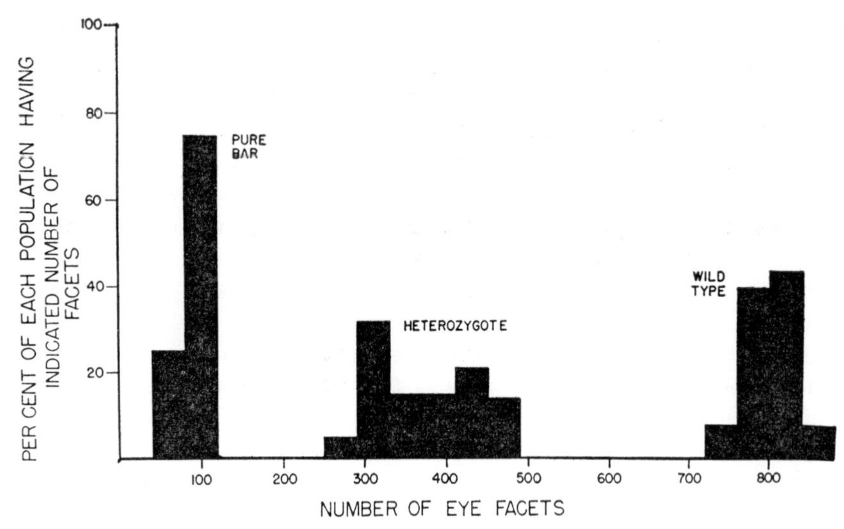

FIG. XI-3. Quantitative effects of a gene

numbers. The three *genotypes*[3] are sharply distinguished although there is a certain normal variability within each type. Dominance is clearly incomplete in this case, since the heterozygote is decidedly different from either of the homozygotes. It has been suggested (55) that early in development a facet-forming substance is produced in the young fly and that the Bar gene starts up a new chemical reaction which breaks down this substance. In a later stage, the actual process of facet formation occurs, but the number of facets formed is determined by the amount of facet substance which still remains.

Blending inheritance seems superficially so logical that it was accepted almost without question even by such pioneers in biometry as Galton and Boas, although the latter abandoned it in his later papers. But it is totally wrong, and as Fisher (14) has pointed out, almost the whole of the revolutionary effects of Mendelism can be seen to result from a knowledge of the particulate character of the hereditary elements. Although Darwin accepted the fusion or blending theory of inheritance, just as other men of his time did, and although he almost certainly never heard of Mendel's work, it is interesting to note that he did at certain times feel the need for a non-blending theory. In a letter to Huxley which was probably dated 1857, these sentences occurred (14):

"Approaching the subject from the side which attracts me most, that is, inheritance, I have lately inclined to speculate, very crudely and indistinctly, that propa-

[3] The *genotype* of an individual is a summary of his genetic constitution; the *phenotype* is a statement of his outward appearance, or what you can determine by direct tests.

gation by true fertilization will turn out to be a sort of mixture, and not true fusion, of two distinct individuals, or rather of innumerable individuals, as each parent has its parents and ancestors. I can understand on no other view the way in which crossed forms go back to so large an extent to ancestral forms. But all of this, of course, is infinitely crude."

Fisher points out that this idea was never developed by Darwin, probably because of the rush of work preceding and following the publication of *The Origin of Species.*

Chetverikov (12) and Fisher (14) have also pointed out that on the assumption of blending inheritance the heritable variability of the species is halved each generation It was realized even by Darwin that domesticated species displayed far too much variability for this to be true, and modern studies have shown that wild organisms exhibit about the same degree of variability as the domesticated. The hypothesis of blending inheritance, which at first sight seemed so reasonable, must now be completely rejected.

Chromosomes as Grouped Genes

Mendel supposed that his hereditary factors were present in pairs in a mature organism, but single in the reproductive cells or gametes. Soon after Mendel's work was rediscovered, it was noticed that there was a close parallel between this hypothetical situation and the actual position of the chromosomes.

It had been observed that the pairs of chromosomes present in the cells of the mature organism separated themselves during the formation of gametes. The chromosomes can not be identical with the hypothetical units of heredity of Mendel because there are not nearly enough of them in any organism to account for the great variety of the known genetic factors. Furthermore, it is now known that the individual genes of any organism are transmitted in groups during the reproductive process, and it is generally believed that the number of these groups is equal to the number of chromosome pairs. Each group of genes within a particular chromosome then assorts independently of other groups (Mendel's second law). Members of any given group usually stay together.

In the formation of reproductive cells (gametes), each pair of chromosomes splits up, and thus every gamete possesses one member of each pair. In the process of gametogenesis the chromosomes are not simply separated into two numerically equal groups. At least one member of each chromosome pair must be present in every adult cell for that cell to be able to develop and function normally. Therefore every gamete will contain one chromosome from each pair.

From the chromosomes of the gametes, combined after fertilization, appear new combinations of the various chromosomes. Since genetic experiments have shown it to be purely a matter of chance which chromosome of each pair enters into any gamete, we are justified in stating that the chromosomes are shuffled and redealt in single sets. The members of the different pairs recombine at random and all possible combinations can have equal probability. Knowing the number of pairs of chromosomes in any given organism, we may compute the number of possible different combinations. For two pairs we have 2^2 which means four possible combinations. For three pairs we have 2^3 equaling 8 combinations, and for four pairs 2^4 equaling 16 combinations. For n pairs 2^n gives the number of combinations.

In man there are 46 chromosomes per adult cell, or 23 pairs, so the number of combinations is therefore 2^{23}. This gives the staggering total of 8,388,608 possible combinations; of which only two are exactly the same as either original parental combination. Therefore, the chances that a human being will repeat either of the parental combinations exactly is only one in 4,194,304. (This calculation takes no account of crossing-over (page 425)).

Genes

Dominants and alleles. From the foregoing, it should be clear that each organism inherits two genes affecting each of its various characteristics, one from the mother and one from the father, and that in respect to the more vital characteristics of the organism, in all probability these two genes are substantially identical. Genes which affect minor external or less important characteristics may sometimes be different in the two parents. For example, one parent may have had a particular shade of brown eyes and the other blue eyes, and the eye color in each case was determined by the presence of a gene, or combination of genes, capable of occupying a certain place or locus in the serial arrangement of genes on one particular chromosome.

When the two genes occupying corresponding loci in a chromosome pair are different, a number of different results are possible. One possibility is that the two genes may both exert their respective effects without much mutual interference. This is observed, for example, in the M and N blood types where either gene, if present, will produce its characteristic response in the red blood cells, whether or not the other gene of the pair is present. In some cases we may find that the effect of the two genes together is more or less intermediate between the effect of either gene occurring in double dose, as in the Blue Andalusian fowl. In still other cases, we may find that the presence of one gene almost completely obscures the effect of the other and some particular feature of the organism is determined almost entirely by the one gene. This is the phenomenon observed by Mendel which he called dominance. Other genes, in different loci, may sometimes alter the expression of a certain gene. Such genes are called "modifiers". Substantially complete dominance is so common that Sewall Wright (59) has made the following rough generalizations:

1. Dominance of one allele of a pair is the rule. Genes capable of occupying the same locus in members of a chromosome pair, both acting together in the process of development, are referred to as *allelomorphs* or *alleles*.
2. The recessive allele is usually less advantageous to the species than is the dominant.
3. Organisms possessing only the recessive genes are generally found to lack something which the dominant gene would have caused to be produced. At least this is true in most of the cases where we have some knowledge of what the primary chemical effects are.
4. The recessive genes are usually less abundant in natural wild populations than their dominant allelomorphs.
5. In cases in which new genes are known to have occurred suddenly—that is, by mutation—they are usually recessive.

All of these principles are subject to much qualification, depending on the particular case being studied.

Since we speak from time to time about the changes in genes (which we refer to as mutations), it will be well to emphasize here and now the very great constancy of genes, as a rule and under ordinary circumstances. The recessive gene suffers no modification even if it is carried unexpressed for many successive generations in heterozygous individuals which show only the dominant phenotype. Raymond Pearl (42) observed the transmission of a single gene in Drosophila through three hundred generations. He arranged his experiment so that in each generation the gene was always associated with a degenerated allelomorph of itself. If we translate into terms of human generations, such an experiment would require a period of time beginning long before the Bronze Age, say about 7000 B.C., and extending up to the present time. Pearl observed no changes in this particular Drosophila gene in any phase of the experiment.

Multiple alleles. In Drosophila a series of genes has been observed, affecting eye color, which may produce colors ranging from the normal red of the Drosophila eye to coral, which is a very little lighter, to eosin, cherry, apricot, buff, tinged, and through shades of ivory to white.

Where only two genes are available from any crossing, only three genetically different types of offspring are possible: an individual which is homozygous for the first gene, an individual homozygous for the second gene, and the individual which is heterozygous, that is, which has one gene of each kind. When a number of genes form the series of allelomorphs, the possible types are greater (51), and the number of different genotypes with n allelomorphs is equal to $n \frac{(n + 1)}{2}$. Thus the eight allelic Rh genes now known in man (page 437) makes up 36 different genotypes.

Among the allelomorphic gene series which are known, most members of the series affect the same characteristic of the organism. But this does not necessarily have to be so. In Drosophila, for instance, there is a gene which produces a disarrangement of the normal order of the rows of facets in the compound eye. Another gene which can occupy the same locus in the chromosome seems to produce no effect on the eye but causes little scalloped incisions in the tips of the wings. The fly which is heterozygous for these two genes is completely normal phenotypically and has neither the facet disarrangement nor the notches on the wings.

The knowledge that there may be a number of genes all having some effect on the same characteristic enables us to interpret the so-called continuous variability often observed in characters. Thus, although it is comparatively easy to classify men into four different blood groups which do not overlap, it is not possible to classify mankind into sharply different groups in regard to height (aside from perhaps the Pygmies), and we observe all sorts of gradations between tall men and short men, although stature is partly determined by heredity. To a certain extent skin color also furnishes an example of this type of continuous variability. When cases of supposed continuous variability are thoroughly investigated genetically, however, they are found to be the effects of a considerable number of genes, acting generally two at a time, or in some cases acting several at a time, but modified by environment.

Number and size of genes. Considering the minuteness of the head of the spermatozoon, which we know must contain at least one out of every pair of genes which the adult organism is going to receive (ignoring cases where the Y chromosome has certain genes missing), we can see that a certain upper size limit can be set merely from considerations of space. In the chromosomes of Drosophila more than four hundred loci for recessive mutations have been recognized and others are constantly being discovered. It will, of course, be appreciated that the total number of genes possessed by this organism must be very much in excess of those which will come under laboratory observation as a result of mutation. The number of genes has been estimated as between a minimum of somewhat over two thousand and a maximum of something over fourteen thousand with a probable number of about five thousand. The size of the gene in Drosophila must be between 10^{-8} and 10^{-5} cubic microns. Genes seem to be rodlike in shape, with a length of approximately 125 millimicrons and a diameter of 5 to 20 millimicrons. These dimensions are not very different from those of the fibrous protein molecules. It may well be, therefore, that the gene is a nucleoprotein macromolecule (see Chapter 9) and it has been estimated, in fact, (43) that a gene contains at least 1000 to 8000 nucleotide pairs. It will be noted that it is similar in size to some virus particles. It may or may not be a coincidence that both are capable of reproducing themselves inside the living cell, and nowhere else.

In some organisms the genes are larger than in Drosophila, and in other organisms, as probably in man, they are considerably more numerous, say four or five times as many as in Drosophila (49).

The chromosomes of which we have already spoken are serial arrangements of a number of these genes. They may be looked upon, according to Huxley (25), as "super-molecules" built up out of genes and capable of breakage at points between the genes. Some non-genetic material is probably also present in the chromosomes.

Mechanism of Gene Action

Certain genes having vital and important effects have been observed; for instance, in man a dominant mutation has occurred which results in the congenital absence of hands and feet (38). Few readers who have seen the pathetic picture of a family so afflicted would be willing to state that this is a gene having only a minor effect (fig. XI-4). Another illustration of an important effect which a single gene can have is furnished by Fisher's discovery (15) that the gene which, when heterozygous, produces the orna-

Fig. XI-4. Hereditary absence of hands and feet, presumably due to the action of a single gene (after Mohr).

mental feature called crest in the common domestic fowl, when homozygous produces cerebral hernia, a condition which is very often fatal to the fowl.

Exactly how each gene produces its effects is not known. There is no general rule. One hypothesis for the mechanism of the formation of the blood group antigen A, for instance, would be that the gene for A was simply one molecule, or several molecules, of the group A substance and that in some way it could cause itself to be duplicated in the cells of the body during development. There is no adequate way of testing such a hypothesis, since the hypothetical amount of A constituting a gene is far below the limits of detectability, but it is at any rate extremely unlikely that every single chemical substance present in the mature organism is determined by a special gene. There are simply too many substances present at the end. The role of genes is more likely to be found in some effect on chemical reactions and rates of reaction in the developing organism.

An illustration from a lower form of life will assist us to understand how the determining factor in the formation of the chemical substance is not necessarily that substance itself. Pneumococci can be separated into more than forty groups, based on the presence of different polysaccharides in the capsule which surrounds them. The substance which determines the production of a given carbohydrate in the capsule is itself a high molecular weight deoxypentosenucleic acid which apparently contains none of the carbohydrate at all. The carbohydrate alone is unable to stimulate the pneumococcus to produce more of the carbohydrate, but the nucleic acid, added to organisms which would normally develop without capsules, causes the production of a capsule containing the particular polysaccharide it sponsors. How it does this we are unable to say, and of course we do not know whether the action of genes in the higher forms is similar to this process. The general question of the physiology of the gene and the mechanism of gene action has been discussed by Wright (59), Beadle (3) and Horowitz (21) in reviews to which the reader may turn for further information.

Boivin (5) states: "In bacteria—and, in all likelihood, in higher organisms as well—each gene has as its specific constituent not a protein but a particular desoxyribonucleic acid which, at least under certain conditions (directed mutations of bacteria), is capable of functioning *alone* as the carrier of hereditary character; therefore, in the last analysis, each gene can be traced back to a macromolecule of a special desoxyribonucleic acid."

Ways in which a molecule of nucleic acid could autoreproduce itself, or direct the synthesis of a molecule of protein, have been discussed on page 371.

In a great many cases a particular environment is indispensable for the normal manifestation of the gene effect. Here we will simply emphasize once more that we must sharply distinguish in our minds between the genotype, which is determined purely by the genetic constitution, and the phenotype, which is determined partly by the genetic constitution and partly by the environment.

Linkage. Since the genes are arranged in serial fashion in the chromosomes, and since the chromosomes assort independently in inheritance, it is natural to predict that all the genes in a single chromosome will be found associated with each other in the offspring possessing that chromosome. This "sticking together" of the genes in inheritance, due to their being in the same chromosome, is called linkage, and we must now examine in somewhat more detail what linkage is and what it is not.

Linkage is the association together of two or more genes *in the process of heredity*. This does not imply that the several characteristics determined by these genes will necessarily be observed to be associated with each other in each individual of the population. For example, if the gene for blood group A and the gene for blue eyes were known to be located in the same chromosome, we should not necessarily expect therefore to find a higher percentage of the blue-eyed group A people than of blue-eyed group B people in the population. *Any such association in the population of chemical or morphological traits does not prove genetic linkage between the genes responsible, and linkage is not customarily detected by any such simple means* (45, 50).

The reason two traits caused by genetically-linked genes are not necessarily associated in the population is that parts of chromosomes may be interchanged during cell division and the formation of gametes, and thus a gene originally present on one chromosome of a pair may come to be on the opposite chromosome of the pair. This step is called "crossing-over" and it occurs frequently enough to destroy, within relatively few generations, any original association of the characteristics in the populaton. To return to our hypothetical illustration, crossing-over will sooner or later produce chromosomes which will have the blood group gene A associated with genes for darker eye color and chromosomes for blood groups other than A will be formed which will have on them the gene for blue eyes. The result will eventually be a mixture of physical types in the same proportions which would be found if the genes responsible were on different chromosomes, so that a mere inspection of the population will not enable us to tell whether or not these two gene series are genetically linked.

In practice, genetic linkage is detected by observing that the two genes are almost always associated in inheritance. At first sight such observation may seem impossible, since crossing-over has usually resulted in the production of all the different combinations of the two gene series. However, an inspection of the accompanying illustration (table XI-5) will show that,

TABLE XI-5

Backcross demonstrating linkage in maize

S La/su la ♂ x su la/su la ♀

EGGS	SPERMS (POLLEN)			
	S La	su la	(S la)	(su La)
su la	S La/su la 45.5%	su la/su la 45.5%	S la/su la 4.5%	su La/su la 4.5%

The symbol S stands for a dominant gene producing starchy kernels, su for a recessive producing sugary kernels; la stands for a recessive gene producing "lazy" sprawling-type of growth of the stalks, La for its dominant normal allele. The types of sperm in parentheses are produced by crossing-over. In the absence of crossing-over, linkage would be complete and only two of the four possible types of offspring would be produced. The inequality in frequencies of the various types is a measure of the degree of linkage.

under ordinary circumstances, linkage of the two genes will be demonstrated by the fact that from any given combination of parents only certain of the theoretically possible types of offspring will be produced, whereas if the genes were on different chromosome pairs and thus assorted independently, all the possible types would be produced with the theoretical frequencies.

A type of genetic linkage which is particularly easy to detect is sex linkage. This occurs when a gene is carried in one of the so-called sex chromosomes. It will do no harm if we oversimplify the real situation a bit and state that in man, as in a number of other organisms, sex is determined by a particular one of the twenty-four pairs of chromosomes. The female is produced from a fertilized ovum containing two such chromosomes, both of the same kind (called the X chromosomes); and the male is produced when the two members of the chromosome pair are not alike, one being the so-called Y chromosome. From data compiled on organisms that have been rather thoroughly studied, such as the fruit fly, it has been concluded that the Y chromosome, the presence of which determines maleness (or we might perhaps better say that it takes two X chromosomes to determine femaleness), contains very few genes of any sort. In fact it seems to be practically inert, genetically. This means that genes in the X chromosome, even when they are recessive, will not have their expression repressed in the male, since there exist no corresponding genes in the Y chromosome which is paired with the X chromosome, and thus there is nothing to repress their effect. As an example of the effect of such a gene, we may mention the deficiency in clotting power of the blood which is called hemophilia. Another example is the inheritance of "ordinary" red-green color blindness.

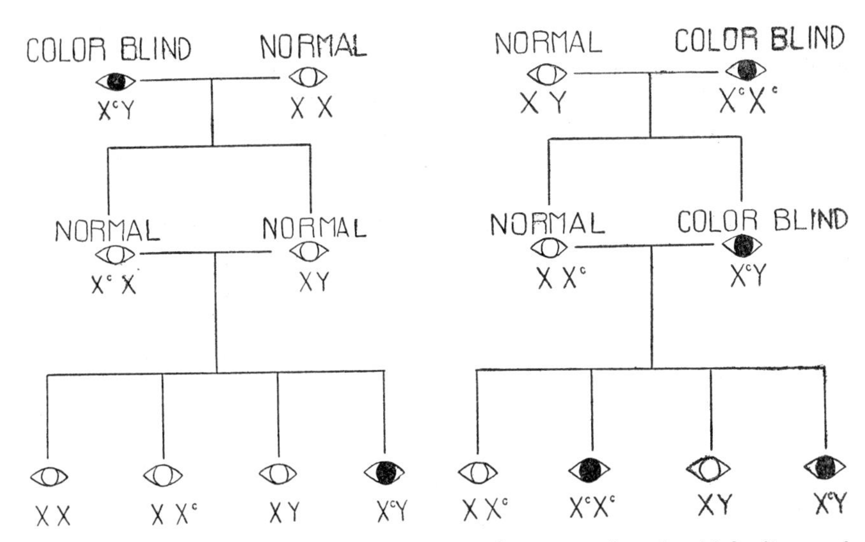

Fig. XI-5. Diagrammatic representation of the supposed mode of inheritance of ordinary (red-green) color blindness. The symbol X indicates a normal X chromosome; X^c indicates an X chromosome which carries the gene for color blindness.

It seems clear that the gene for color blindness (of the sort to which we have referred) is in the X chromosome and that it is a recessive. A woman who has only one such gene therefore will still have normal color vision, because of the dominant action of the normal gene in the other X chromosome, but a man having one such gene will be color-blind because of the absence of the dominant gene in the Y chromosome, which would otherwise suppress the action of the color-blind gene. A mating between a color-blind male and a homozygous normal woman would produce normal males and normal females. It will be seen from figure XI-5, however, that the female offspring will carry in one of their X chromosomes color-blind genes. Consequently, on the average, half of their sons will be color-blind. Color blindness is fairly common in men, running at least as high as 5 per cent in some populations, but it is quite rare in women. An inspection of the diagram will show that for a woman to be color-blind, it is necessary for her father to be color-blind and for her mother either to carry the color-blind gene in a single dose, or to be herself color-blind, which is of course much less likely.

At first sight it might be thought that the existence of sex linkage disproves the earlier statement that linkage is not detected by association of two traits in the population, for there is unquestionably an association between maleness and color blindness in the population. The point is that such association between maleness and color blindness is not necessarily due to sex linkage, but may be due to the effect of sex on the expression of an autosomal gene (i.e., one not upon the X and Y chromosomes). If color blindness, for example, were caused by an autosomal gene which was dominant in males but recessive in females, much the same picture of association would be produced. And human traits have been studied, as for example baldness (20) and taste reaction to phenylthiocarbamide (10, 13), in which the expression of the gene seems to be better in one sex than in the other. In the case of taste reaction to phenylthiocarbamide the gene seems to be expressed better in the female, but in the case of baldness the gene seems better expressed in the male.

Sex-linked inheritance can be distinguished from autosomal sex-limited or sex-influenced inheritance only by the study of descendants of affected males (46). To determine whether an apparently sex-linked character could reach full expression in females, homozygous individuals from matings between female carriers and affected males would have to be examined.

More than twenty examples are known of genes which are apparently carried in the X chromosome of man. It is clear that any two of these genes will be themselves genetically linked since they are carried on the same chromosome, and the study of human pedigree supports this idea.

Expressivity and penetrance. There are two terms applying to the manifestation of genes which we at times have occasion to use. The first is called *expressivity* and is a measure of the amount or kind of effect shown in an individual possessing the gene. The second is called *penetrance* and is frequently measured as the percentage of individuals who, when they possess the gene, show any effect from it. Naturally, most of the genes which have been chosen for experimental work with the lower forms show a very high or complete penetrance. The frequency with which the gene produces its effect depends both on the environment and on the genotype, the effectiveness in the heterozygous condition often being less. A good example of environmental influence is found in the gene "giant" in Drosophila. Here lack of sufficient food for the larvae results in fewer giants. The expressivity of this gene is nearly uniform, all giants being about the same size. Such genes would seldom be used in experimental work with lower forms. In man, however, we can not pick and choose our material to the same extent, but must take it as it comes, and we shall have to deal with some genes which do not show complete penetrance. In the case of deleterious genes, of course, natural selection would tend to postpone more and more the age of onset of the symptoms, possibly by the selection of suitable modifying genes.

Mutation. Doubling or further multiplying the ordinary (diploid) number of chromosomes in the cell nuclei of a species produces *polyploids* and is one type of mutation, particularly common in plants. In this book we shall restrict the term mutation to changes in the composition of individual genes. The first mutation to be recorded in an animal appeared in 1791 in a male lamb belonging to the flock of Seth Wright, a Massachusetts farmer. As a result of this mutation, the lamb had very short bowed legs, and a special breed was developed from it by deliberate selection because it was an advantage to farmers to have short-legged sheep which were not able to jump the stone walls which surround the New England sheep pastures. This early breed eventually became extinct, but the same mutation later appeared a second time, this time in Norway, and the short-legged breed was reconstructed. Since that time numerous other mutations have been observed in animals, and we have some reason to believe that we know the exact origin of one or more of the mutations since found in man. A mutation of a normal to a hemophilia-producing gene is thought by some (18) to have occurred in the person of Queen Victoria, who transmitted it to many of her descendants, including members of the old Russian and Spanish royal families.

The self-propagation of genes (see page 371) is one of the most remarkable things about them. Muller (40) has commented that not only is this self-propagation in itself remarkable, but the study of mutations reveals the still more remarkable fact that after mutation, when the chemical structure of the gene has changed, the gene still has the property of propagating itself, *its new self*. Although a given gene may be changed in various ways, there is generally a strong tendency for any given gene to undergo changes of some particular kind (41), so that it usually mutates in some one direction rather than in another. Muller suggests that an animal is generally in such good equilibrium with its environment (as the result of countless generations of natural selection) that any change is likely to be a change for the worse. How extensive the change within an individual gene has to be before we recognize it as a mutation is still unknown. Muller thinks, however, that eventually it should be possible to decide whether the gene is composed of several molecules (or unit particles), one of which may change at a time. At present we do not know the exact nature of this alteration. We do know that it can be affected by outside agencies such as x-rays (39) and by the administration of toxic substances such as colchicine, the nitrogen mustards, and various other compounds (21).

There seem to be certain normal rates of mutation for each gene. These may be so low that mutations are not observed often enough to enable us to state what the normal rate of mutation is, although more extensive observations would nearly always enable us to ascertain this. There have not been many direct reports of new gene mutations (19, 38) in man, but there is evidence (19) from reliable pedigrees that certain human gene mutations are occurring with some frequency.

Haldane (19) estimated that the normal allele of the hemophilic gene mutates to the hemophilic gene about once in thirty thousand individuals per generation. Without such mutations, since hemophilia is a very serious disadvantage and since the action of natural selection is to eliminate such a gene, the trait would certainly have disappeared long ago. Gunther and Penrose (17) estimated that the normal allele of the epiloia gene (epiloia is a rare disease in which mental defect and epilepsy are associated with tumor formation in the brain, the skin, and certain viscera) mutates to the gene for epiloia at a somewhat lower rate.

BLOOD GROUPS

The A, B, O Blood Groups

The four classical blood groups, O, A, B, AB, were discovered in 1900 to 1902 by Dr. Karl Landsteiner and pupils. The red blood corpuscles of certain individuals are acted upon by substances present in the plasma of certain other persons in such a way to form clusters and clumps. These clumps are at first so small that they can be seen only under the microscope, but when the reaction is strong they grow to a size easily discernible by the naked eye. The chemical substances in the red corpuscles which permit their being agglutinated in this way are the agglutinogens. Corresponding agglutinins anti-A and anti-B are found in plasma.

The division of all persons into four blood groups depends upon the fact that the two different blood corpuscle agglutinogens, A and B, can be present singly or together, or can be absent. If we designate the absence of both by O, we have four possibilities: O, A, B, and AB. The relation of the serum agglutinins anti-A and anti-B to the characteristic blood corpuscle substances of the individual is given by the Landsteiner rule: *There is always found that agglutinin or agglutinins which could co-exist physiologically with the blood corpuscle characteristic which is present.* Thus, for example, anti-A is found in the presence of O and B, but not of A. These relations are illustrated by table XI-6, in the last column of which the blood group

TABLE XI-6

The human blood groups

BLOOD GROUP	AGGLUTINOGEN IN CORPUSCLES	AGGLUTININ IN SERUM
O	O	anti-A and anti-B
A	A	anti-B
B	B	anti-A
AB	AB	—

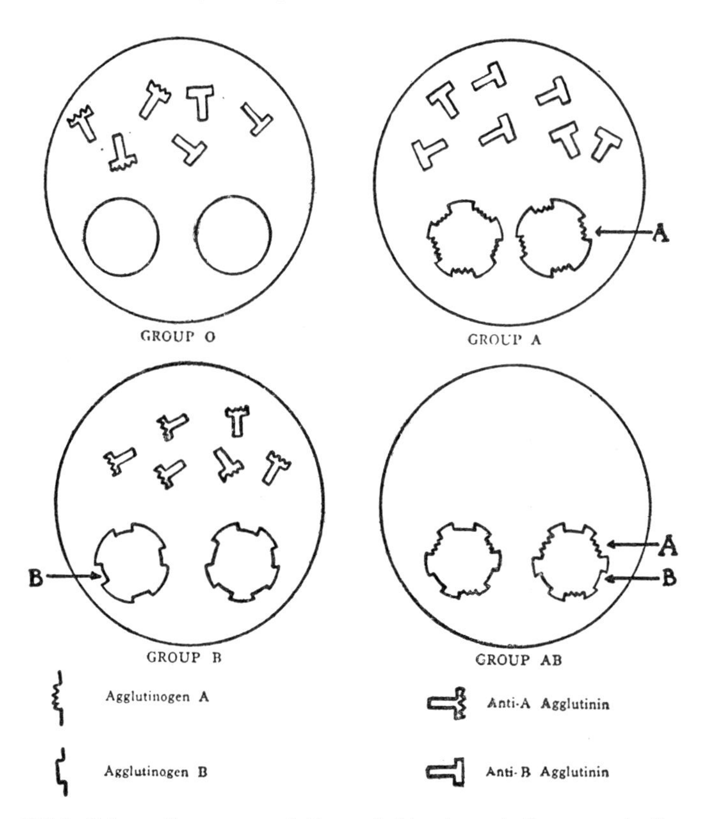

FIG. XI-6. Schematic representation of blood agglutinogens (antigens) and agglutinins.

serum characteristics are presented (see fig. XI-6). In determining blood groups, the corpuscle suspension of the person to be tested is allowed to react, preferably in the test tube (48, 57), with sera which have the property of agglutinating cells containing blood group factors A and B, respectively. Typical agglutination is shown in figure XI-7. The group is determined by the simple scheme shown in table XI-7. Blood grouping tests can also be carried out on glass slides instead of in test tubes. This method is often used for clinical tests in hospitals. An idea of the appearance of the reactions to the naked eye and under the microscope is given in figure XI-8.

Inheritance. The mechanism of inheritance was shown by Bernstein (4) to depend upon a series of three allelic genes which are designated as A, B, and O. Since each person must possess some combination of two of these genes, six different combinations—OO, AA, AO, BB, BO, and AB—are possible. Since, however, the factor O seems to be recessive to both A and B (or at least its presence in the genetic make-up of the individual

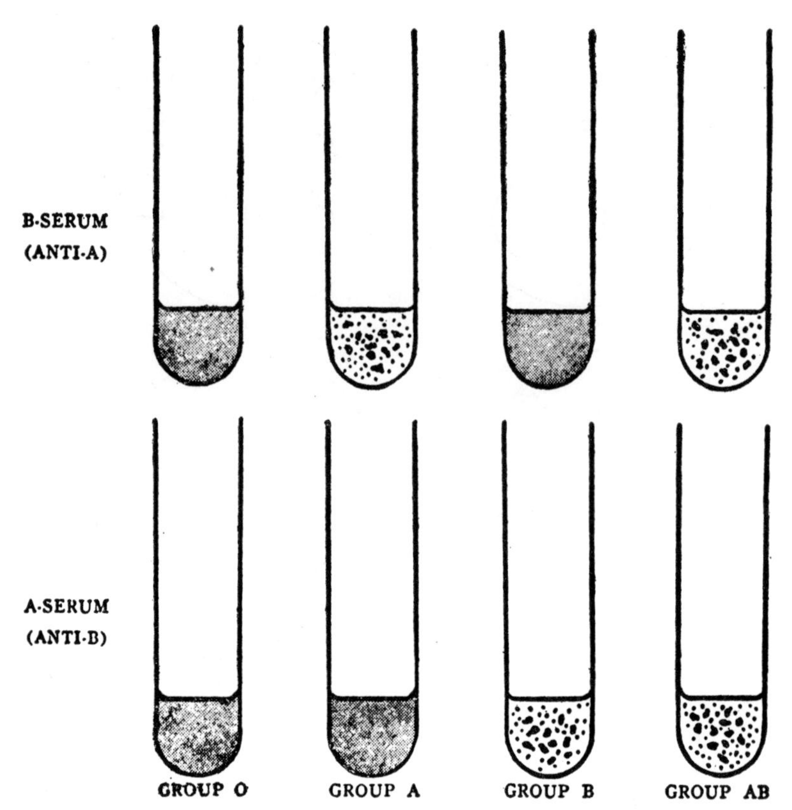

FIG. XI-7. Schematic representation of blood grouping as carried out in test tubes

TABLE XI-7

Determination of groups with two test sera, anti-A and anti-B

	KNOWN SERUM ANTI-A	KNOWN SERUM ANTI-B	GROUP
Agglutination of the unknown blood corpuscles	−	−	O
	+	−	A
	−	+	B
	+	+	AB

does not interfere with the full expression of the A or B characteristic), the heterozygotes AO and BO are indistinguishable from the homozygotes AA or BB, respectively, and the six gene combinations produce only four distinguishable groups, as shown in table XI-8.

Group AB is homogeneous in the sense that all individuals in the group are alike in having one A gene and one B gene. An individual of group

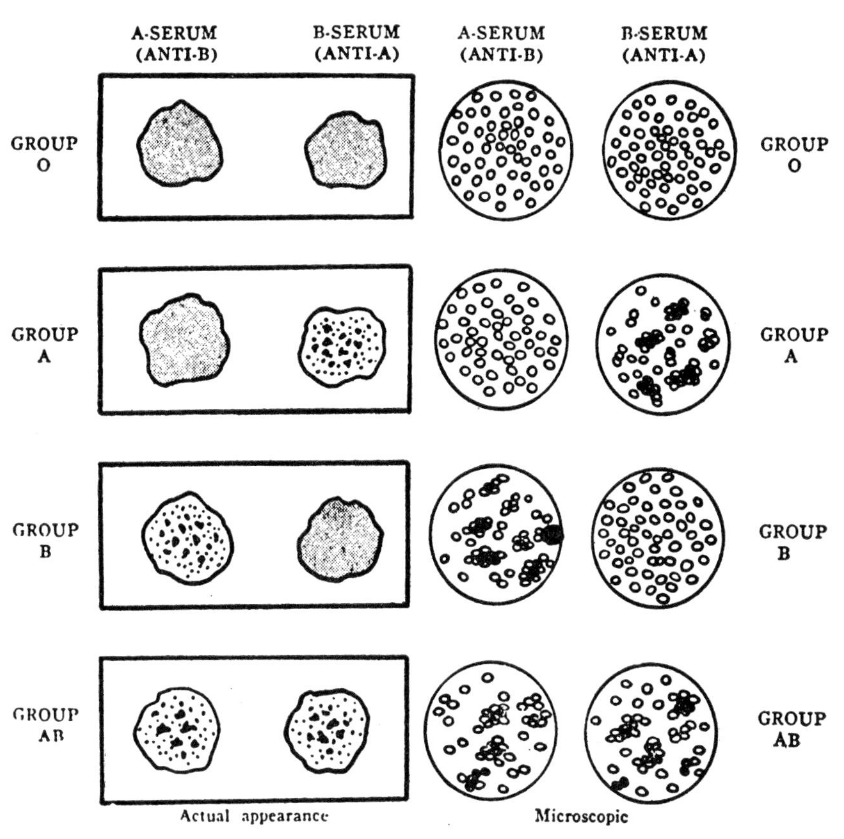

FIG. XI-8. Schematic representation of blood grouping on slides

TABLE XI-8

Genotypes and phenotypes of blood group genes

GENOTYPE	BLOOD GROUP (PHENOTYPE)
OO	O
{AA AO	A
{BB BO	B
AB	AB

TABLE XI-9

Inheritance of blood groups, as found in actual studies (48)

GROUPS OF PARENTS	NUMBER OF FAMILIES	NUMBER OF CHILDREN IN GROUP				
		O	A	B	AB	Total
O x O	1405	3355	(14)	(9)	0	3378
O x A	2647	2486	3389	(10)	(2)	5887
O x B	1365	1315	(7)	1690	(1)	3013
O x AB	504	(6)	607	612	(3)	1228
A x A	1270	516	2354	0	(1)	2871
A x B	1299	561	908	734	782	2985
A x AB	419	0	478	238	279	995
B x B	536	188	(1)	975	0	1164
B x AB	304	(2)	171	383	225	781
AB x AB	57	0	26	36	63	125
Totals	9806	8429	7955	4687	1356	22,427

Numbers in parentheses are apparent exceptions to the accepted theory of blood group inheritance. They are doubtless due either to faulty technic in making the tests, or to illegitimacy.

AB can therefore never become the parent of a child of group O. The blood group of an individual never changes. Table XI-9 shows how nearly the experimental results conform to this, the Bernstein theory.

From the absence of normal children who have a blood group incompatible with that of their mother (and the same argument applies to the M and N types discussed on page 435), we may infer that the blood group genes do not mutate at any rapid rate, for otherwise children incompatible with their mothers would have been observed. Blood groups thus may be more "conservative" characters than albinism and hemophilia, both of which seem to be recruited steadily by mutations from the normal gene.

The genes A and B cause the production in the red blood cells of the chemical substances agglutinogens A and B, respectively. The chemical nature of these substances has been investigated by Kabat (26) and by Morgan (1). As isolated, they are mucopolysaccharides.

Racial Classification

Tens of thousands of individuals have been "blood grouped" since 1900. The very early tests did not seem to reveal very much difference in the frequencies of the four groups in various nationalities, but during the World War I two Polish doctors named Hirszfeld, working on the Macedonian battlefront, made the discovery that the percentages of the four groups varied considerably in different races. As characters for use in anthropological classification, the blood groups offer several advantages (9): (a) they are inherited in a known way according to Mendelian principles; (b) they are not altered by illness, medical treatment, of differences in climate or food; (c) their frequency in a population is a very stable characteristic; (d) they probably arose very early in the course of man's evolution; (e) there is a considerable correlation between geography and the distribution of the blood groups; and (f) the blood groups are sharply distinguishable "all-or-none" characters which do not grade into each other.

Blood group gene B is absent in certain populations, and some of the more recently discovered blood group factors seem virtually confined to one racial group, but in general, the differences found between different human populations are merely differences in the *frequencies* of the various genes. These differences have been produced by the action of mutation, selection, and *genetic drift* (random variations in gene frequencies in very small populations). The action of these evolutionary agencies has been continuously opposed, however, by the race mixture which has constantly gone on.

By making use of blood groups and other genetically determined traits, it has proved possible (9) to divide the human race into thirteen groups: (a) Early Europeans, (b) Lapps, (c) Northwest Europeans, (d) East and Central Europeans, (e) Mediterraneans, (f) Africans, (g) Asians, (h) Indo-Dravidians, (i) American Indians, (j) Indonesians, (k) Melanesians, (l) Polynesians, (m) Australian Aborigines.

Other Blood Groups

M, N blood types. The A, B, and O series of allelomorphs does not by any means exhaust the list of genes which influence the blood agglutinogens. Another series discovered in 1927 is known as M and N (28). These letters represent inherited agglutinogens detectable in human blood by the use of agglutinins formed by rabbits injected with M-positive (or N-positive) human red cells. Each human being has either two M genes, two N genes, or one of each; three types are thus determined as shown in table XI-10.

Unlike the A, B groups, the M and N blood types, as they are called, have but little importance in the performance of blood transfusions, since

TABLE XI-10

The M, N blood types

GENOTYPE	PHENOTYPE
MM	M
NN	N
MN	MN

agglutinins capable of reacting specifically with these agglutinogens are rarely found in normal human blood.

Subdivisions of blood groups A and AB. The blood factor A is not always exactly the same in different individual human bloods. There are two main varieties of A which are designated as A_1 and A_2 (48, 57). Of these, the latter gives weaker reactions with the average anti-A reagent. The variations are expressed also in the group AB, so that we find two subgroups, A_1B and A_2B, where the difference in the A antigen is very much the same as it was in group A, save that in certain instances the A reaction in individuals of subgroup A_2B may be much weaker than in any individual of group A_1, or even A_2. It has been proposed (52) that the subgroups of A are inherited in the following way (see table XI-13): Instead of the gene series A, B, and O, suppose we have a series of four allelomorphs, A_1, A_2, B, and O, with A_1 being dominant over A_2, and A_1, A_2, and B all being dominant over O. This theory on the whole seems to fit most of the observations satisfactorily, although it does not account for the rather wide variability in sensitivity to anti-A agglutinins of cells of individuals belonging to the subgroups A_2 and A_2B (7).

A serum which subdivides both antigens M and N into two varieties has been found (44), so that instead of two genes at the M, N locus, there would seem to be four, tentatively designated as M, MS, N, and NS.

The Rh series of genes. In 1940 Landsteiner and Wiener (29) discovered that the serum of a rabbit which had been injected with the blood of a Rhesus monkey would agglutinate certain human bloods, and not agglutinate others, irrespective of the A, B, O group, or M, N type of these bloods. The new factor was designated Rh, from the first two letters of Rhesus. Human bloods tested with this serum could be differentiated into two types, Rh positive, and Rh negative. Agglutinins which reacted with the Rh antigen had been previously observed (6), for on looking back it can be seen that this is the most probable explanation of certain blood incompatibilities encountered by some of the earlier workers, and several such cases (6) were retested and shown to be due to Rh incompatibility.

The Rh factor might not have excited much attention had it not been for the demonstration by Levine and others (30) that Rh incompatibility between mother and child was very often the cause of a rare but fairly serious disease of infants known as erythroblastosis fetalis, or hemolytic disease of the newborn. Wiener and others (58) also showed that incompatibilities in which the recipient had received repeated transfusions were often due to the Rh factor.

The way in which Rh incompatibility operates to produce erythroblastosis is briefly this: If the mother is Rh negative and the fetus Rh positive (having inherited the Rh factor from the father), Rh-containing cells or Rh substance may get through the placenta from the circulation of the fetus into that of the mother. The mother's agglutinin-forming mechanisms may respond (in about one out of 20 such incompatible pregnancies and one out of 200 or 400 total pregnancies) by the production of anti-Rh agglutinins. Agglutinins diffuse readily through the placenta, thus some of the anti-Rh formed by the mother diffuses into the fetal circulation. There it combines with the fetal red cells, causing hemolysis and damaging the liver and other organs.

At first it was believed that Rh was inherited as a simple Mendelian pair, rh and Rh, the latter gene causing the Rh-positive condition, and being dominant over rh. Three different genotypes would be possible, rhrh (Rh negative), and Rhrh and RhRh (both of the latter Rh positive). On this theory, the inheritance would be very simple, the various matings and their possible outcomes being:

RhRh x RhRh → 100 per cent RhRh
RhRh x Rhrh → 50 per cent RhRh, and 50 per cent Rhrh
Rhrh x Rhrh → 25 per cent RhRh, 50 per cent Rhrh, and 25 per cent rhrh
RhRh x rhrh → 100 per cent Rhrh
Rhrh x rhrh → 50 per cent Rhrh, and 50 per cent rhrh
rhrh x rhrh → 100 per cent rhrh

However, it was soon found that the Rh factor was antigenically and genetically complex. Wiener now supposes that at least eight allelomorphic genes are involved, as follows:

$$R^1, R^2, R^o, r', r'', r^y, R^z, r$$

The first seven are dominant over rh, but not over each other. The genes R^1 and R^2 are "double acting" genes, the former producing the antigens rh′ and Rh^o, the latter rh″ and Rh^o.

The discovery of an agglutinin which reacted regularly with Rh negative bloods led to the postulate that Rh-negative bloods contained an Hr factor (Hr being the letters Rh in reverse). But this agglutinin also reacts with some of the Rh-positive bloods. These observations led the British workers to propose an entirely different system of Rh nomenclature (44) (see table XI-12).

A table of the Rh-Hr genes and their reactions with the various antisera according to Wiener's nomenclature is shown in table XI-11.

According to Wiener's nomenclature the antisera designated by the British as anti-C, anti-D, and anti-E (table XI-12) are designated anti-rh′, anti-Rh_0, and anti-rh″ respectively, and the sera called by the British anti-c, anti-d, and anti-e are designated as anti-hr′, anti-Hr_o, and anti-hr″. The only blood which is correctly called completely Rh negative would

TABLE XI-11

Rh genes and their reactions with anti-Rh and anti-Hr antisera, according to Wiener's nomenclature

GENE	REACTION WITH SERUM					
	anti-rh′	anti-Rh$_o$	anti-rh″	anti-hr′	anti-Hr$_o$	anti-hr″
r	−	−	−	+	+	+
r′	+	−	−	−	+	+
r″	−	−	+	+	+	−
r^y	+	−	+	−	+	−
R^o	−	+	−	+	−	+
R^1	+	+	−	−	−	+
R^2	−	+	+	+	−	−
R^z	+	+	+	−	−	−

react with none of the three anti-Rh sera, anti-C, anti-D, or anti-E, and, as can be seen from tables XI-11 and XI-12, would react with all three of the anti-hr sera. If we consider all 6 of the known sera, then, there is no blood which does not react with at least three of them.

Application of Blood Groups in Cases of Disputed Parentage

Knowing the rules of inheritance of the blood groups, it is possible to establish in some cases that a man falsely accused of being the father of an illegitimate child is in reality innocent. It is obvious that we can not expect to prove by blood groups that he is the father, since it is always possible that some other man having the same blood group is the father.

TABLE XI-12

Possible chromosome types and reactions with anti-rhesus sera, according to the British nomenclature

CHROMOSOME	REACTION WITH SERUM*					
	anti-C	anti-D	anti-E	anti-c	anti-d	anti-e
cde	−	−	−	+	+	+
Cde	+	−	−	−	+	+
cdE	−	−	+	+	+	−
CdE	+	−	+	−	+	−
cDe	−	+	−	+	−	+
CDe	+	+	−	−	−	+
cDE	−	+	+	+	−	−
CDE	+	+	+	−	−	−

* The symbol + indicates a positive reaction (agglutination); − indicates a negative reaction.

TABLE XI-13

Supposed mechanism of inheritance of subgroups

MATING	CHILDREN POSSIBLE
A_2 x O	O, A_2
A_2 x A_2	O, A_2
A_2 x B	O, A_2, B, A_2B
A_2B x O	A_2, B
A_2B x A_2	A_2, B, A_2B
A_2B x B	A_2, B, A_2B
A_2B x A_2B	A_2, B, A_2B
A_1B x O	A_1, B
A_1B x A_1	A_1, B, A_2B, A_1B
A_1B x A_2	A_1, B, A_2B
A_1B x A_2B	A_1, B, A_2B, A_1B
A_1B x B	A_1, B, A_1B
A_1B x A_1B	A_1, B, A_1B
A_1 x O	O, A_1, A_2
A_1 x A_1	O, A_1, A_2
A_1 x A_2	O, A_1, A_2
A_1 x B	O, A_1, A_2, B, A_1B, A_2B
A_1 x A_2B	A_1, A_2, B, A_1B, A_2B

The rules of heredity of blood groups may be applied to problems of disputed paternity. First of all we may give four examples of possible types of mating and the types of children which would result from them. All that is necessary is to know that each parent contributes one gene of a pair to the offspring and that the factors A and B may be considered as dominant over O (table XI-14).

By similar tables one may work out all of the possible matings and their consequences, and a compilation of these results enables us to draw up table XI-15. It will be noted that some types of children can not be born to mothers of certain types. For instance, a mother of blood group O can not have an AB child. Consequently, we may abstract from table XI-15 the information which enables us to establish the non-paternity of an accused man, and this is shown in table XI-16.

In carrying out tests in such cases it is important that competent workers using known sera of known potency and specificity should do the work.

The M and N types are inherited as outlined above and may be applied in exactly the same way to the exclusion of paternity in the case of an innocent man. This is illustrated in table XI-17. Testing for the M and N factors is slightly more difficult than for the A and B groups, and not all workers are able to prepare adequate testing reagents, so these tests should certainly never be entrusted to any except experts.

TABLE XI-14

Diagrammatic examples of the hereditary transmission of the factors that determine certain blood groups

I MATING AA x AA	II AA x AO	III AO x BO	IV OO x AB
A A	A O	B O	A B
A AA AA	A AA AO	A AB AO	O AO BO
A AA AA	A AA AO	O BO OO	O AO BO
Progeny: 100% group A	100% A	25% O 25% A 25% B 25% AB	50% A 50% B

On the top of each diagram are designated the factors possessed (and transmissible) by one parent; on the left, those of the other; which parent does not matter, because this transmissibility has been shown to be independent of sex. The letters within the horizontal lines show the possible progeny resulting.

The fundamental law is that a factor can not be present in the blood of a child unless it was present in the blood of at least one of its parents (4). Diagrams of all the 21 possible types of mating have been tabulated (48, 57).

TABLE XI-15

Blood groups of offspring possible or impossible from any mating combination

MATING COMBINATION	ALLEGED FATHER	KNOWN MOTHER	CHILDREN POSSIBLE FROM THEIR MATING	CHILDREN NOT POSSIBLE FROM THEIR MATING. DECISIVE FOR NON-PATERNITY	IMPOSSIBLE FROM THIS MOTHER IN ANY MATING
1	O	O	O	A, B, (AB)	AB
2	O	A	O, A	B, AB	
3	O	B	O, B	A, AB	
4	O	AB	A, B	AB, (O)	O
5	A	O	O, A	B, (AB)	AB
6	A	A	O, A	B, AB	
7	A	B	O, A, B, AB		
8	A	AB	A, B, AB	(O)	O
9	B	O	O, B	A, (AB)	AB
10	B	A	O, A, B, AB		
11	B	B	O, B	A, AB	
12	B	AB	B, A, AB	(O)	O
13	AB	O	A, B	O, (AB)	AB
14	AB	A	A, B, AB	O	
15	AB	B	A, B, AB	O	
16	AB	AB	A, B, AB	(O)	O

The letters designate the blood groups of the respective individuals (48). Those in parentheses (column 5) could not be children of the corresponding mothers (column 3) in any mating. Since no such child could exist to raise a problem of proof, these instances are omitted from table XI-16 which summarizes the net indications of non-paternity deducible from these blood groupings.

TABLE XI-16

Combinations allowing the man to establish non-paternity, omitting instances of impossible mother-child combinations (condensed from table 28)

PUTATIVE FATHER	KNOWN MOTHER	KNOWN CHILD
O	O	A, B
O	A	B, AB
O	B	A, AB
O	AB	AB
A	O	B
A	A	B, AB
B	O	A
B	B	A, AB
AB	O	O
AB	A	O
AB	B	O

TABLE XI-17

Exclusion of paternity on the basis of characteristics M and N

CHILD	MOTHER	FATHER NOT POSSIBLE
M	M or MN	N
N	N or MN	M
MN	M	M
MN	N	N

From the sketch of the inheritance of the Rh types it is obvious that disputed parentage may be tested by determination of these types also. However, the number of possible matings is very large and a table corresponding to those given would occupy far too much space. Furthermore, relatively few laboratories in the country are equipped to do these delicate tests, therefore it seems best to be content with a statement that the tests and their interpretation should be made only by experts.

Several states have passed laws making it possible for the court to order the participants of a suit involving paternity to submit to blood grouping tests, and directing that these tests may be received as evidence if they exclude paternity. They, of course, can not be construed as proving paternity. Other states which do not have such laws usually admit the evidence when it is offered.

REFERENCES

1. Aminoff, D., Morgan, W. T. J., and Watkins, W. M. The isolation and properties of the human blood-group A substance. Biochem. J., **46:** 426–439, 1950.

2. Bateson, W., and Punnett, R. C. On the interrelations of genetic factors. Proc. Roy. Soc. London, s. B., **84:** 3-8, 1911.
3. Beadle, G. W. Biochemical genetics. Chem. Rev., **37:** 15-96, 1945.
4. Bernstein, F. Zusammenfassende Betrachtungen über die erblichen Blutstructuren des Menschen. Z. indukt. Abstammung-u. Vererbungslehre, **37:** 237-270, 1925.
5. Boivin, A. Directed mutation in colon bacilli, by an inducing principle of desoxyribonucleic nature: its meaning for the general biochemistry of heredity. Cold Spring Harbor Symp. Quant. Biol. **12:** 7-17 (1947).
6. Boyd, W. C. Rh blood factors: an orientation review. Arch. Path., **40:** 114-127 1945.
7. Boyd, W. C. Assay of blood grouping sera: Variations in reactivity of cells of different individuals belonging to groups A and B. Ann. New York Acad. Sci., **46:** 927-937, 1946.
8. Boyd, W. C. *Genetics and the Races of Man.* Boston, Little, Brown & Co., 1950.
9. Boyd, W. C. Anthropologie und Blutgruppen. Klin. Wochschr. **34:** 993-999, 1956.
10. Boyd, W. C., and Boyd, L. G. Sexual and racial variations in ability to taste phenyl thiocarbamide, with some data on the inheritance. Ann. Eugenics, **8:** 46-51, 1937.
11. Chang, M. C., and Pincus, G. Physiology of fertilization in mammals. Physiol. Rev., **31:** 1-26, 1951.
12. Chetverikov, S. S. On certain features of the evolutionary process from the viewpoint of modern genetics. J. Exper. Biol. (Russian), **2:** 3-54, 1926.
13. Falconer, D. S. Sensory thresholds for solutions of phenyl thiocarbamide. Results of tests on a large sample made by R. A. Fisher. Ann. Eugenics, **13:** 211-222, 1947.
14. Fisher, R. A. *Genetics and the Theory of Natural Selection.* Oxford, Eng., Clarendon Press, 1930.
15. Fisher, R. A. Crest and hernia in fowls due to a single gene without dominance. Science, **80:** 288-289, 1934.
16. Greenberg, B. E., and Gargill, S. L. The relation of hyaluronidase in the seminal fluid to fertility. Human Fertil., **11:** 1, 1946.
17. Gunther, M., and Penrose, L. S. The genetics of epiloia. J. Genetics, **31:** 413-430, 1935.
18. Haldane, J. B. S. *New Paths in Genetics.* New York and London, Harper & Brothers, 1942.
19. Haldane, J. B. S. The mutation rate of the gene for hemophilia and its segregation ratios in males and females. Ann. Eugenics, **13:** 262-271, 1947.
20. Harris, H. The inheritance of premature baldness in men. Ann. Eugenics, **13:** 172-181, 1946.
21. Horowitz, N. H. Biochemical genetics of Neurospora. Adv. in Genetics **3:** 33-71, 1950.
22. Hudson, P. B., and Butler, W. W. S. A study of the enzyme acid phosphatase and its possible role in intermediary carbohydrate metabolism of the prostate gland and its secretions in dog and man. J. Urol., **63:** 323-333, 1950.
23. Huggins, C. The prostatic secretion (review). Harvey Lect., **42:** 148-193, 1946-1947.
24. Huggins, C. The physiology of the prostate gland. Physiol. Rev., **25:** 281, 1948.
25. Huxley, J. L. *Evolution, the Modern Synthesis.* New York and London, Harper & Brothers, 1942.

26. Kabat, E. A., *et al.* Immunochemical studies on blood groups. IV. J. Exper. Med., **85:** 685–699, 1947.
27. Landau, R. L., and Loughead, R. Seminal fructose concentration as an index of androgenic activity in man. J. Clin. Endocrin., **11:** 1411–1424, 1951.
28. Landsteiner, K., and Levine, P. On individual differences in human blood. J. Exper. Med., **47:** 757–775, 1928.
29. Landsteiner, K., and Wiener, A. S. An agglutinable factor in human blood recognized by immune sera for rhesus blood. Proc. Soc. Exper. Biol. & Med. **43:** 223–224, 1940.
30. Levine, P., and Katzin, E. M. Isoimmunization in pregnancy and the varieties of isoagglutinins. Proc. Soc. Exper. Biol. & Med., **45:** 343–346, 1940.
31. Lillie, F. R. *Problems of Fertilization.* Chicago, Ill., University of Chicago Press, 1919.
32. Lundquist, F. Glycerophosphorylcholine as a precursor of free choline in mammalian semen. Nature, **172:** 587–588, 1953.
33. Lundquist, F., *et al.* Purification and properties of some enzymes in human seminal plasma. Biochem. J., **59:** 69–79, 1955.
34. MacLeod, J. Biochemistry of the male genital tract. Ann. N. Y. Acad. Sci., **54:** 796-85, 1952.
35. Mather, K. *Statistical Analysis in Biology.* New York, Interscience Publishers, 1947.
36. Mawson, C. A., and Fischer, M. I. Zinc and carbonic anhydrase in human semen. Biochem. J., **55:** 696–700, 1953.
37. Mendel, G. J. *Experiments in Plant Hybridization* (Trans. by Royal Horticultural Society, London). Boston, Harvard University Press, 1948.
38. Mohr, O. L. *Heredity and Disease.* New York, W. W. Norton & Co., 1934.
39. Muller, H. J. Artificial transmutation of the gene. Science, **66:** 84–87, 1927.
40. Muller, H. J. The gene as the basis of life. Proc. Internat. Congr. Plant Sc., **1:** 897–921, 1929.
41. Muller, H. J. Types of visible variations induced by x-rays in Drosophila. J. Genetics, **22:** 299–334, 1930.
42. Pearl, R. Biology and social trends. J. Wash. Acad. Sc., **25:** No. 6: 253–296, 1935.
43. Pontecorvo, G., and Roper, J. A. Resolving power of genetic analysis. Nature, **178:** 83–84, 1956.
44. Race, R. R., and Sanger, R. *Blood Groups in Man.* Oxford, Eng., Blackwell 1950.
45. Rife, D. C. A common misconception concerning gamma behavior of linked factors. Human Biol., **11:** 546–548, 1939.
46. Rundales, R. W., and Falls, H. F. Hereditary (? sex-linked) anemia. J. Am. M. Sci., **211:** 641–658, 1946.
47. Russell, E. S. *The Interpretation of Development and Heredity.* Oxford, Eng., Clarendon Press, 1930.
48. Schiff, F., and Boyd, W. C. *Blood Grouping Technic.* New York, Interscience Publishers, 1942.
49. Spuhler, J. N. On the number of genes in man. Science, **108:** 279–280, 1948.
50. Strandskov, H. H. Human genetics and anthropology. Science, **100:** 570–571, 1944.
51. Taylor, G. L., and Race, R. R. Human blood groups. Brit. M. Bull., **2:** 160–164, 1944.

52. Thomsen, O., *et al.* Über die möglichkeit der Existenz zweier neuer Blutgruppen; auch ein Beitrag zur Beleuchtung sogenannter Untergruppen. Acta path. et microbiol. scandinav., **7:** 157–190, 1930.
53. von Tschermak, E. The rediscovery of Gregor Mendel's work. J. Hered., **42:** 163–171, 1951.
54. Tyler, A. Properties of fertilizin and related substances of eggs and sperm of marine animals. Am. Naturalist, **83:** 195–219, 1949.
55. Waddington, C. H. *An Introduction to Modern Genetics.* New York, Macmillan Co., 1939.
56. Walsh, R. J., and Montgomery, C. M. A new human iso-agglutinin subdividing the MN blood groups. Nature, **160:** 504, 1947.
57. Wiener, A. S. *Blood Groups and Transfusion.* 3rd edit. Springfield, Ill., Charles C Thomas, 1943.
58. Wiener, A. S., and Peters, H. R. Hemolytic reactions following transfusions of blood of homologous group, with three cases in which same agglutinogen was responsible. Ann. Int. Med., **13:** 2306–2322, 1940.
59. Wright, S. The physiology of the gene. Physiol. Rev., **21:** 487–527, 1941.

PART IV

Metabolism

CHAPTER 12

Food and Diet

The various foods of man, excepting water and purely mineral items, are the anabolized products of other living organisms. The various animal products in man's diet, whether it be whale blubber, ants, honey, milk, lobsters, or pork chops, are derived from organisms that, in turn, must produce their tissues from raw materials in the form of the green plant or from still other animals that feed on the green plant. No matter how long the chain of animals that eat and are eaten, the primary source of food can be traced, eventually, to the green plant.

Whence does the green plant obtain its fat, protein, and carbohydrate? Upon what does it feed? The answer is that its own raw materials, or food, consist of carbon dioxide from the air, and the water and water-soluble mineral content of the soil. From these relatively simple materials it constructs the complex substances discussed in the first part of this text. The synthesis of protein from carbon dioxide, water, nitrates, and sulfates, to take the most startling example, is an anabolic process which requires the input of considerable energy. This is certain, since the oxidation of proteins back to carbon dioxide, water, and inorganic material liberates considerable energy. Where does the plant obtain this energy? The answer is that plant anabolism is made possible through the absorption and utilization of the energy of sunlight. The process which utilizes and stores this energy is known as *photosynthesis*.

PHOTOSYNTHESIS

Chlorophyll

Photosynthesis in green plants is dependent upon the presence of *chlorophyll*, a green pigment. There is significance, therefore, in the adjective "green" when we speak of plants in connection with photosynthesis, since plants which do not have chlorophyll, for instance, the various mushrooms and yeasts, can not photosynthesize. Accompanying chlorophyll are various carotenoids, the yellow or red color of which is masked by the chlorophyll-green. The chlorophyll content of most leaves is about one per cent of the dry weight, while the carotenoids are present in concentrations as little as one-eighth to as much as one-half that of chlorophyll.

Chlorophyll in its native state within the *chloroplast* (that plant cell organelle which contains the pigment) is probably in close association with protein, lipid, or both. The catalyst of the photosynthetic process may thus be viewed as a compound lipoprotein of which chlorophyll and various carotenoids are the prosthetic groups, or as an enzyme of which these are the coenzymes.

Nor is it an accident that the prosthetic groups are colored. Since plant anabolism uses light as the source of energy, some mechanism must exist for the absorption of light. The structure of chlorophyll is such that it absorbs light in the violet and red, the non-absorbed portion of the spectrum being reflected for the most part and appearing green to our eyes. The function of the carotenoids may well be to absorb some of the light rejected by the chlorophyll. The pigments would thus behave, primarily, as "energy traps", while the lipoprotein moiety would be the photosynthetic enzyme proper. It is also suggested (8a) that the carotenoids may act as anti-oxidants protecting the cell against the destructive effects of peroxide groups formed through side-reactions during the course of photosynthesis.

Despite all efforts the enzyme system, as an undenatured whole, has not yet been isolated from the chloroplast. Chlorophyll by itself, although comparatively easy to isolate in quantity, is powerless to effect the photosynthetic transformation, so that photosynthesis has not yet been duplicated in the test tube, and visions of a future in which man's ultimate food supply can depend on his chemical industries alone are as yet remote.

The chemical structure of chlorophyll (formula I) is remarkably similar to that of the heme of hemoglobin (see Chapter 5). It is built upon the porphyrin nucleus but differs from heme in the following respects.

(1) The carboxyl group of pyrrol nucleus IV is esterified with the long chain alcohol *phytol*. Phytol ($C_{20}H_{39}OH$) is built up of isoprene units and is, therefore, related to the carotenoids, a fact which may possibly be significant in view of the close association of carotenoids and chlorophyll within the leaf. The porphyrin and phytol portions of the molecule are respectively polar and non-polar, so that the former can associate with the protein and the latter with the lipid in the photosynthesizing lipoprotein. Pyrrol nucleus IV also contains two more hydrogen atoms than does the analogous ring in heme.

(2) The propionic acid group of pyrrol nucleus III is oxidized and condensed with a —CH= group to form a five-carbon ketone ring. It is further esterified with methyl alcohol.

(3) The two-carbon substituent on pyrrol nucleus II is an ethyl group rather than a vinyl group as in heme.

(4) The metal associated with the four nitrogens of the pyrrols is magnesium, and not iron as in the case of the hemes.

Despite these differences, the similarities between the two substances are far more striking and it is interesting to note the economical manner in

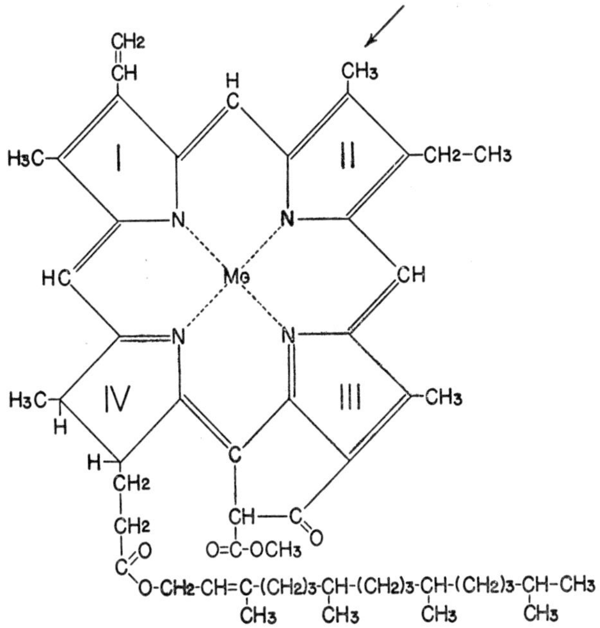

I. Chlorophyll *a*

Note: Replacement of the $-CH_3$ group (indicated by the arrow) with a $-C{\overset{O}{\diagup}}\!\!\!\!\underset{H}{\diagdown}$ group will result in a chlorophyll *b* molecule.

which evolutionary forces adapt a single molecular framework to such vastly different yet equally important chemical roles in plants and animals.

Chlorophyll is not a chemical individual. The one-carbon substituent on pyrrol nucleus II may be a methyl group, as in heme, in which case we have *chlorophyll a*, or it may be an aldehyde group, in which case we have *chlorophyll b*. Chlorophyll *a* is slightly more soluble in fat solvents than chlorophyll *b*. Their solubility pattern differs from that of the simple lipids in that they are fairly soluble in 95 per cent alcohol, and only difficultly soluble in petroleum ether. The chlorophylls *a* and *b* vary also in their absorption spectra. Chlorophyll *a* absorbs almost twice as strongly in the red, for instance. The ratio of the *a* and *b* varieties in most higher plants is about three to one. In algae there are species variations and, in addition, other closely related varieties of chlorophyll.

The Photosynthetic Reaction

As a result of photosynthesis within the chloroplast of the plant cell, carbon dioxide and water are converted into carbohydrate and molecular oxygen:

$$n\,CO_2 + nH_2O \xrightarrow[\text{chlorophyll}]{\text{light}} C_nH_{2n}O_n + n\,O_2 \quad \text{Equation XII-1}$$

This is an energy-consuming reaction. To form carbohydrate from carbon dioxide and water, the plant cell makes use of radiant energy of the visual spectra, chiefly of the longer wave lengths.

Note that a molecule of oxygen is produced for each molecule of carbon dioxide used up and that photosynthesis thus serves not only to prepare the food supply for all life, but, in addition, renews the oxygen supply of the atmosphere as quickly as it is consumed by catabolic processes. It is thought, in fact, that the earth's primeval atmosphere (before the advent of life) contained no free oxygen and was rich in carbon dioxide, and that it was mainly the photosynthetic activity of plant life as it developed that established the composition of the present atmosphere. It has been estimated that photosynthetic processes in the earth's present load of plant life is sufficient to renew all the oxygen in the air in a little over two thousand years and to decompose all the water in the oceans in about two million years. It is also interesting to note that the contribution of marine algae to the utilization of atmospheric carbon dioxide is eight times that of land vegetation (30).

The over-all photosynthetic reaction shown in equation XII-1 can be broken up into a light-catalyzed reaction, and a "dark reaction" which does not require light quanta to continue:

$$n\,CO_2 + 2n\,H_2O \xrightarrow[\text{chlorophyll}]{\text{light}} C_nH_{2n}(OH)_{2n} + n\,O_2 \quad \text{Equation XII-2}$$

$$C_nH_{2n}(OH)_{2n} \xrightarrow{\text{“dark reaction”}} C_nH_{2n}O_n + n\,H_2O \quad \text{Equation XII-3}$$

Experiments with isotopic oxygen have shown that the liberated oxygen is derived from the water molecule, not from the carbon dioxide. Thus, the reaction shown in equation XII-2 can be in turn divided into two portions:

a) The oxidation of water to oxygen, and

b) The reduction of carbon dioxide to carbohydrate.

Oxidation of water. It is the oxidation of water that is the light-catalyzed portion of the reaction. In the process, hydrogen atoms are transferred from water to 6,8-thioctic acid (lipoic acid), chlorophyll catalyzing the reaction in the presence of light (20):

$$2H_2O + 2\ \underset{\text{thioctic acid}}{\text{(S—S ring)}(CH_2)_4COOH} + \text{light energy} \xrightarrow{\text{chlorophyll}} 2\ \underset{\text{reduced thioctic acid}}{\text{(SH SH)}(CH_2)_4COOH} + O_2$$

Equation XII-4

The hydrogen atoms in the reduced thioctic acid so produced may now follow one of two pathways (3). They may recombine with oxygen via an enzyme chain including the various hydrogen-carrying coenzymes described in Chapter 7, with the production of three molecules of ATP for every two hydrogen atoms so reacting. The second pathway followed by the hydrogen atoms involves the reduction of carbon dioxide. For every two hydrogen atoms used in carbon dioxide reduction, an atom of oxygen is evolved. Manganous ion may be involved in this reaction (18a).

Reduction of carbon dioxide. This is also an energy-consuming reaction. The energy needed here is derived not from light quanta, however, but from the energy obtained by hydrolyzing the ATP produced as described immediately above. Three molecules of ATP are required for the reduction of one molecule of CO_2 (31).

Efforts have been numerous and strenuous to determine the nature of the first reaction involved in the reduction of carbon dioxide and the nature of the organic product formed. From the fact that one molecule of oxygen is formed for every molecule of carbon dioxide consumed, and that the empirical formula for the organic residue must therefore be CH_2O, it was early concluded that photosynthesis must result in the production of carbohydrate, and that lipids and proteins must be produced from carbohydrates by non-photosynthetic mechanisms. The nature of the first carbohydrate produced was by far the more difficult problem and before the days of radioactive tracers guesses ranged from such simple products as formaldehyde to such complex ones as starch. Monosaccharides, disaccharides, and polysaccharides all exist in the leaf and such is the speed of photosynthesis that it was difficult to show that any of them were formed before any of the others.

Calvin (8), using radioactive carbon 14 in the raw material, marine algae as the photosynthesizing cells, and paper chromatography as the means of separation of products, attempted to identify the primary products by allowing photosynthesis to proceed for very short periods after supplying the radioactive carbon dioxide before killing and analyzing the plants. It is interesting to note that in a minute and a half 90 per cent of the carbon dioxide had been reduced and no less than fifteen different radioactive organic products could be isolated. In five seconds of photosynthesis five compounds could be detected—malic acid, aspartic acid, pyruvic acid, 3-phosphoglyceric acid, and 2-phosphoglyceric acid. Sixty-five per cent of the radioactive carbon occurred in the last two compounds. The phosphoglyceric acid is formed from CO_2 by means of a carbon dioxide acceptor which is a phosphorylated five-carbon keto-sugar, *ribulose-1,5-diphosphate*. Carbon dioxide adds on to it to form an unstable 6-carbon intermediate, which then breaks up to form two molecules of 3-phosphoglyceric acid (formula II) (42).

$$\begin{array}{c} CH_2O\textcircled{P} \\ | \\ C{=}O \\ | \\ HCOH \\ | \\ HCOH \\ | \\ CH_2O\textcircled{P} \end{array} + CO_2 \rightarrow \left[\begin{array}{c} CH_2O\textcircled{P} \\ | \\ HOOC{-}C{-}OH \\ | \\ C{=}O \\ | \\ HCOH \\ | \\ CH_2O\textcircled{P} \end{array}\right] \xrightarrow{+\ H_2O} 2\ \begin{array}{c} CH_2O\textcircled{P} \\ | \\ HCOH \\ | \\ COOH \end{array}$$

Ribulose-1,5-diphosphate 3-Phosphoglyceric acid

II. Carbon dioxide fixation in photosynthesis

Two fates can then befall the 3-phosphoglyceric acid. A portion accepts the hydrogen evolved in the photochemical breakdown of water (see above) and is reduced to a triosephosphate, consuming in the process three molecules of ATP for each molecule of 3-phosphoglyceric acid reduced. The triose phosphate condenses to hexose phosphate from which in turn can be formed polysaccharides, fats and the precursors of amino acids.

Another portion of the 3-phosphoglyceric acid can by several different routes be reconverted to ribulose-1,5-diphosphate (5). This involves a type of reaction, catalyzed by the enzyme *transketolase*, in which a ketol group ($HOCH_2CO$—) is transferred from one phosphorylated sugar to another. An example of such a reaction which is thought to take place as part of the ribulose-1,5-diphosphate cycle is given in formula III.

It has become customary to speak of the various sugars engaged in such reactions, particularly where the exact configuration may be uncertain, according to the number of carbon atoms concerned. Thus, triose phosphate is a C_3 compound, *sedoheptulose-7-phosphate* a C_7 compound and *ribulose-*

5-phosphate a C_5 compound. The reaction shown in formula III can therefore be represented as a C_3 plus C_7 equals C_5 plus C_5 reaction. The addition

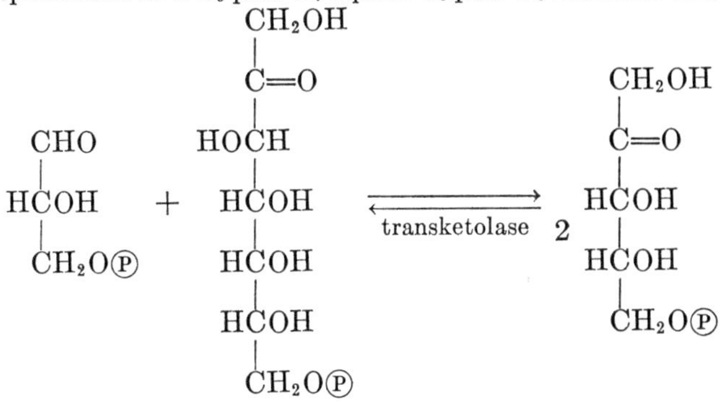

$$
\begin{array}{ccccc}
 & CH_2OH & & & \\
 & | & & & CH_2OH \\
 & C{=}O & & & | \\
CHO & | & & & C{=}O \\
| & HOCH & & & | \\
HCOH \quad + & HCOH & \underset{\text{transketolase}}{\rightleftharpoons} & 2 & HCOH \\
| & | & & & | \\
CH_2O\text{Ⓟ} & HCOH & & & HCOH \\
 & | & & & | \\
 & HCOH & & & CH_2O\text{Ⓟ} \\
 & | & & & \\
 & CH_2O\text{Ⓟ} & & & \\
\text{triose phosphate} & \text{sedoheptulose-7-phosphate} & & & \text{ribulose-5-phosphate}
\end{array}
$$

III. Transketolation

of carbon dioxide to ribulose-1,5-diphosphate (see formula II) is a C_1 plus C_5 equals C_3 plus C_3 reaction.

In summary, then, the triose phosphate (C_3) formed from phosphoglyceric acid can add to itself to form a glucose or fructose phosphate (C_3 plus C_3 equals C_6). The C_6 can react with the C_3 to form a C_4 and a C_5. The C_4 can react with the C_3 to form a C_7. The C_7 can react with a C_3 to form two C_5's.

If we sum up all the reactions mentioned in the previous paragraph, we have 5 C_5 being converted to 3 C_1. The ribulose-5-phosphate formed is phosphorylated, in the presence of *phosphoribulokinase* and ATP to ribulose-1,5-diphosphate (16a), which is ready to take up another carbon dioxide, split to phosphoglyceric acid and begin the cycle all over again.

With all this in mind, we can prepare an outline scheme of photosynthesis as in formula IV (see also (9)).

The efficiency of photosynthesis, taking into consideration the quantity of light energy actually absorbed by the green leaf, is remarkably high according to the investigations by Warburg and Burk (40). These investigators found that 4 quanta of red light were absorbed for each molecule of carbon dioxide reduced. Since the reduction of 1 mol of carbon dioxide to carbohydrate consumes 112 Kcal. and since 4 einsteins (1 einstein equals 1 mol quantum equals 6.026×10^{23} quanta) of red light contain 160 Kcal., this is equivalent to saying that 65 per cent of the light energy is stored as chemical energy. Under optimal conditions, the process might thus attain nearly 100 per cent efficiency (7). However, other workers maintain that some 9 quanta of red light are necessary per mol of CO_2 reduced, which would make the estimated efficiency 30 per cent.

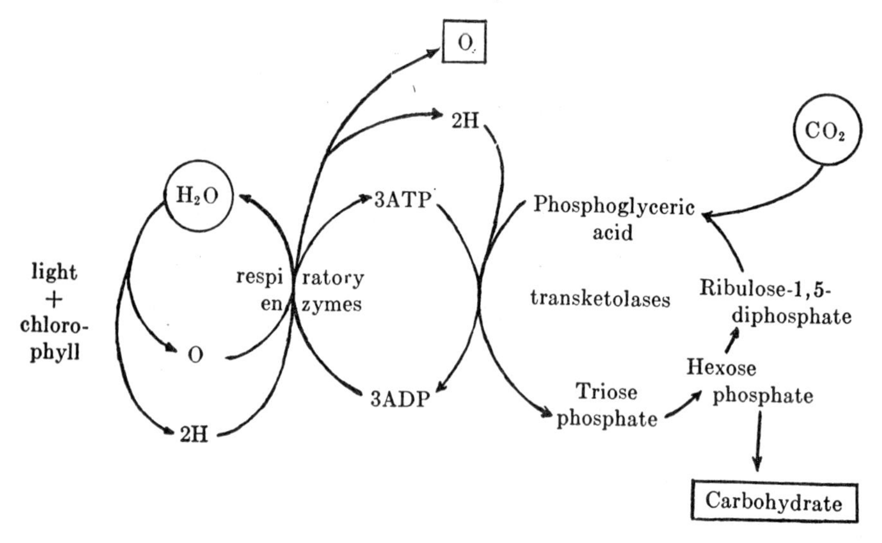

Note: Items enclosed in a circle are used up in the process; items enclosed in a rectangle are produced.

IV. Scheme of photosynthesis

It remains true, however, that such high efficiencies result from careful control of laboratory conditions—i.e., the use of red light, which is well absorbed by the chloroplast, in quantities just sufficient to induce optimum reaction. If the efficiency is calculated taking into account absorption of sunlight which usually irradiates the leaf to excess and contains wave lengths between the red and violet which are but little utilized in the plant, the efficiency of the process is much lower. Rabinowitsch estimates that the efficiency in direct sunlight is of the order of 3 per cent where absorbed light only is considered, or 2 per cent when the total incident light (only two-thirds of which is absorbed) is taken into consideration.

FOOD FACTORS

Animal species, while capable of extensive biosyntheses, are by no means as versatile as the green plant. In other words the food factors required by animals are more numerous and generally more complex than those needed by plants. *Food factors* or *dietary essentials* may be defined as the simplest chemical substances—i.e., those of the lowest molecular weight—which will adequately nourish an organism. Those food factors which are required by the body in comparatively small quantities, such as the vitamins and some, but not all, minerals may be differentiated from the rest as *micronutrients.*

Carbohydrates and Lipids

Glucose. In any consideration of food factors, carbohydrates and lipids may be considered together since their metabolism follows, in part, a common pathway. They occur in foods in forms usually too complex for direct use, as starch or disaccharides on the one hand or as glycerides on the other. In the case of carbohydrates, digestive processes break down ingested starch and disaccharides to the monosaccharides, glucose, fructose, and galactose, which are all readily absorbed from the intestine and are interconvertible within the body. Glucose and fructose are used in parenteral or intravenous feeding. Either can supply all the body's needs for carbohydrate. In a practical sense, therefore, these monosaccharides may be considered the body's carbohydrate food factors; and since the sugar in blood is characteristically glucose, it can be chosen as the carbohydrate food factor.

We may go further than this. Glucose can be used in the body as the raw material for the biosynthesis of both glycerol and most of the fatty acids, so that glucose is the lipid food factor as well. This can be demonstrated on a large scale by the fact that ingestion of large quantities of carbohydrates will result in the formation of large quantities of fat. In the case of livestock this is a useful and desirable process; in the case of human beings, it is somewhat less so. The interconversion of carbohydrates and fat is made possible by the fact that in the catabolism of these substances a "two-carbon fragment" is reached through a series of reversible reactions, with either type of substance as a starting material (although the body can not gain carbohydrate at the expense of fatty acid—see Chapters 14 and 15). On the intracellular level, then, the two-carbon fragment would be the carbohydrate-lipid food factor.

Fatty acid food factors. Not all the fatty acids can be synthesized in the body from carbohydrates or from other fatty acids. Apparently some species of animals, at least, find it difficult to form multiple double bonds in the fatty acids. Rats develop dermatitis when fed an otherwise-adequate diet which is deficient in the naturally-occurring fatty acids more unsaturated than oleic acid. Addition to this diet of linoleic acid removes the symptoms. Hansen and Burr (11) summarized the current state of knowledge of these unsaturated fatty acid deficiencies in humans. The results on human beings were in no way conclusive due to the small number of subjects and the general difficulty of conducting dietary experiments on such a poor laboratory animal as *Homo sapiens*. However, deficiency seemed to interfere with the maintenance of skin integrity and resulted in such manifestations as prickly heat and eczema.

Hansen and Burr list linoleic acid, linoleyl alcohol, linolenic acid, arachidonic acid (formula V), docosahexenoic acid, and hexahydroxystearic acid as relieving the dermatitis in small laboratory animals. Of these linoleic acid, linolenic acid, and arachidonic acid are normal components of the diet. Together they are referred to often as *essential fatty acids.* The word "essential" is not to be given undue physiological significance. The fact

$$CH_3(CH_2)_4CH\ ({=}CH\ CH_2CH{=})_3CH\ (CH_2)_3COOH$$

V. Arachidonic acid

that oleic acid is not an "essential" fatty acid does *not* mean that the body can do without it as part of its structure, but means merely that it is not needed as such in the diet because the body can synthesize it from other fatty acids or even from glucose.

Proteins

"Complete" and "incomplete" proteins. Whereas starches and lipids from various species of plants and animals differ only to a minor degree in their nutritional value, proteins are unique in that certain of them are incapable of supporting growth when they are the only protein constituent of a diet otherwise adequate. Thus, rats which were fed upon a diet containing casein as the only source of protein thrived and grew at a normal rate. When gliadin, the prolamine of wheat, was substituted for casein they grew only slightly; and when zein, the prolamine of maize, was the protein source, the animals actually declined in weight.

The reason for the nutritional deficiency of some proteins is absence of, or near absence of, certain amino acids in their chemical make-up. Gliadin possesses a lysine content of one per cent as compared with 6 per cent in casein, and moreover only a little over one per cent of tryptophane as compared with twice that quantity in casein. Zein lacks lysine and tryptophane entirely. From this it would appear likely that the rat can not synthesize lysine or tryptophane from other food substances but must find these ready made in the diet. That not all amino acids are of similar importance in the diet is indicated by the fact that casein is quite deficient in glycine, serine, and cystine, possessing 0.5 per cent or less of each, yet remains a protein of excellent nutritional properties. When tryptophane and lysine are added to zein, the "enriched" protein allows growth.

In general, complete proteins, i.e., those which are capable of supporting growth when they are the sole contributors to the protein of the diet, are from animal sources. Examples are casein and beta-lactoglobulin from milk, ovalbumin and ovovitellin from egg, and the various proteins of meat, fish, and poultry. In addition, glycinin of soybean, excelsin of Brazil nut, and glutenin of wheat qualify. Incomplete proteins, those incapable of supporting growth, are the prolamines of wheat, barley, and rye. Zein and gelatin are examples of particularly incomplete proteins. Naturally, the classification of a protein as complete or incomplete depends upon its quantity in the diet. It must be present in sufficient amounts so that even the necessary amino acid in which it is poorest is adequate. Thus, edestin, to be "complete" when fed to rats, must be fed in quantities 25 per cent greater than the amount which would be adequate for casein. If the quantity of edestin is reduced to that of casein it becomes partially incomplete.

Amino acid food factors ("essential amino acids"). The use of mixtures of purified amino acids to replace protein in experimental diets has borne out the concept of amino acids as food factors. Ten of the amino acids are indispensable in the diet of the young growing rat. These are: (1) tryptophane, (2) lysine, (3) methionine, (4) threonine, (5) phenylalanine, (6) leucine, (7) isoleucine, (8) valine, (9) histidine, and (10) arginine.

Given adequate quantities of each of these ten, the growing rat could biosynthesize all the other amino acids needed for protein production. It can not be overemphasized that "non-essential" amino acids such as glycine, alanine, and glutamic acid, despite their nutritional label, are integral and necessary parts of proteins—as vital to life as any of the others. Their "non-essentiality" consists only of this: that they can be synthesized by the body from other customary components of the diet. The "essential" amino acids can not so be synthesized and must be found as such in the diet. It is preferable to think of the ten amino acids listed as the amino acid food factors for the growing rat and to avoid the confusing word "essential" altogether. The adult rat, where growth has ceased, no longer requires arginine in the diet. The rat apparently can synthesize arginine at a rate fast enough to balance the normal rate of destruction of that amino acid, but not fast enough to form additional arginine-containing protein as would be required in a growing organism.

As a by-product of these researches into amino acid nutrition, threonine was discovered (25). When mixtures of all amino acids known at that time (in 1935) failed to permit growth in rats, the existence of another amino acid food factor was suspected, and investigations followed which revealed the existence of threonine. It was early realized that the list of amino acid food factors varied with the species of animal investigated. Thus, the chick, in addition to the food factors required by the young rat, must be supplied dietary glycine as well. Human needs in this direction can not therefore be determined by animal experimentation alone. Investigations into the nature of man's amino acid food factors were begun as early as 1941 (16). More recently, Rose (33) announced a complete list of such food factors for man. He fed healthy graduate students on a diet consisting of maize starch, sucrose, butterfat, maize oil, inorganic salts, vitamins, and purified amino acid mixtures. The adequacy of the diet to supply the essential needs for protein synthesis was judged by its ability to maintain nitrogen equilibrium. When the loss of nitrogen through excretion was greater than that ingested (negative nitrogen balance), it meant that the particular combination of amino acids in the artificial diet could not entirely compensate for all the amino acids being lost in normal protein catabolism. If the situation could not be corrected by simple increase in the quantity of the amino acid mixture being ingested, it meant that one or more amino acid food factors were missing.

By such methods it was found that the list of amino acid food factors for adult, non-growing humans is: (1) valine, (2) leucine, (3) isoleucine, (4) threonine, (5) methionine, (6) lysine, (7) phenylalanine, and (8) tryptophane. Note that compared with the list of amino acid food factors for the growing rat, both arginine and histidine are missing.

It should be kept in mind, however, that experiments such as these are of necessity relatively short-term affairs. There is no certainty that the apparent absence of, say, histidine from the list is not simply due to the fact that the body is possessed of an efficient mechanism for conserving histidine when necessary and that prolonged deprivation might not eventually result in undesirable effects. The most we can say is that the food factors listed include those amino acids the absence of which in the diet results in undesirable effects almost immediately. In addition to negative nitrogen balances, absence of any of the food factors resulted in failure of appetite, increasing fatigue and nervous irritability, even when the subject was not aware that his diet was deficient. Another point to remember is that these experiments were conducted on full grown men. The dietary requirements of infants in the first half year of life include 60 mgm./kgm./day of threonine and 90 mgm./kgm./day of phenylalanine (29). These are about eight times the quantity (on a per kg. basis) of that required by adults. Albanese (1) states that certain dietary proteins require to be supplemented, in the diet of infants, with sources of cystine and tyrosine.

Some essential amino acids can be replaced in the diet by their alpha-hydroxy or alpha-keto analogs. Thus, imidazole lactic acid (formula VI), or imidazole pyruvic acid (formula VII) can replace histidine in the diet of the rat. Similarly, indole lactic acid (formula VIII) or indole pyruvic acid (formula IX) can replace tryptophane. The same is true for the analogs of leucine, isoleucine, valine and methionine. In each case, the complete amino acid can be formed by transamination (see Chapter 16). Since lysine and threonine do not take part in reversible transamination reactions, however, these amino acids can not be replaced in the diet by hydroxy or keto analogs.

Given ample supplies of phenylalanine, the mammalian body is capable of adding the para-hydroxy group to form tyrosine at a rate sufficient to fulfill its needs indefinitely even in the entire absence of dietary tyrosine. The reverse is not true, so that of the two, phenylalanine is the food factor

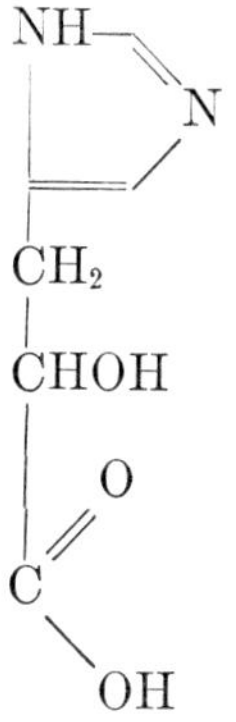

NH—
N
CH_2
CHOH
O
C
OH

VI. Imidazole lactic acid

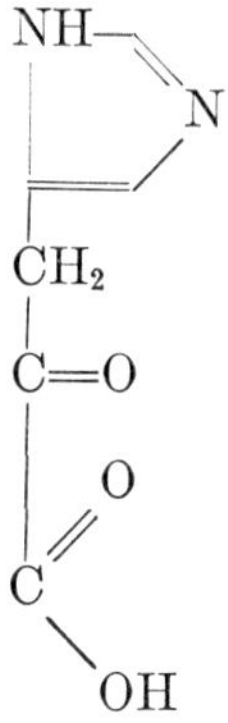

NH—
N
CH_2
C=O
O
C
OH

VII. Imidazole pyruvic acid

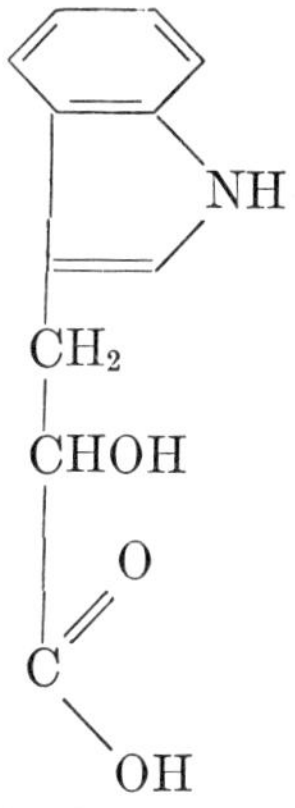

NH
CH_2
CHOH
O
C
OH

VIII. Indole lactic acid

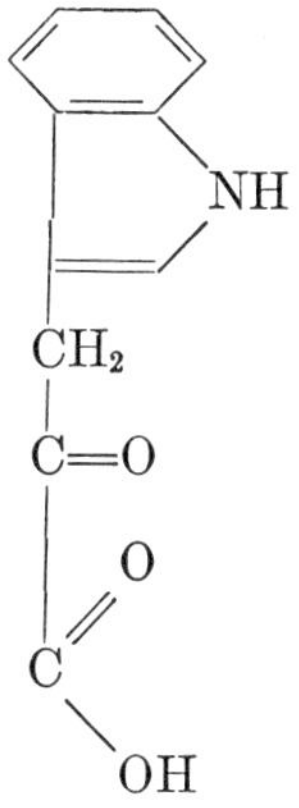

NH
CH_2
C=O
O
C
OH

IX. Indole pyruvic acid

and tyrosine is not. Nor can the body synthesize tyrosine from any amino acid other than phenylalanine. If, then, to a diet containing only the amino acid food factors as protein source tyrosine is added, then the requirement of phenylalanine is decreased, since now only enough phenylalanine is needed for its own sake; none additional is necessary for the production of tyrosine. In fact, the dietary requirement of phenylalanine in the presence of ample tyrosine is only 25 to 30 per cent of the requirement otherwise (34). This effect of added tyrosine is known as *sparing action*. Cystine, which is not a food factor, has a similar sparing action on methionine, which is a food factor and which the body can use as a precursor for cystine. The dietary requirement for methionine in the presence of ample cystine is only 20 per cent of the requirement otherwise (34).

Rose has presented values for the minimum daily human requirement of each amino acid food factor where all other amino acids are amply supplied in the diet (see table XII-1). Rose states, however, that the recommended daily intake in the formulation of diets should be twice the minimum daily requirement. Rose also investigated the possible utilization of D-amino acids by the body. He found that D-methionine was as effective as the normal L-form in the maintenance of nitrogen equilibrium, and that D-phenylalanine could be partially utilized. The D-forms of the other amino acid food factors, however, were entirely unutilizable by man.

Later work tends to show that the figures given in table XII-1 err, indeed, on the side of generosity. In a series of studies on young women, Leverton *et al.* find the minimum daily requirement to be half to two-thirds of Rose's values (19a).

It is possible to have too much of an essential amino acid. Supplementing the diet of the young rat with leucine resulted in a slowing of growth unless isoleucine was also added to the diet. These two very similar amino acids are, apparently, in competitive antagonism with one another (13). In general, not only is each amino acid required in ample quantity, but all are required in proper proportion to one another (2). A deficiency of one interferes with the utilization of all. Other amino acids are also involved (9a).

Other Organic Food Factors

The organic groupings other than amino acids associated with proteins in the form of prosthetic groups or coenzymes are in many cases synthesized from simpler components of the diet. Thus, the nucleic acids are synthesized ultimately from glucose, glycine, and phosphate; mucopolysaccharides have glucose as their most important precursor and porphyrins are composed largely of glycine fragments. These substances just mentioned are needed

TABLE XII-1

Nutritional requirements of essential amino acids

AMINO ACID	MINIMUM DAILY REQUIREMENT
	gms.
L-Tryptophane	0.25
L-Phenylalanine	1.10
L-Lysine	0.80
L-Threonine	0.50
L-Valine	0.80
L-Methionine	1.10
L-Leucine	1.10
L-Isoleucine	0.70

by the body in relatively large quantities, and it would almost seem as though the organism dared not risk obtaining sufficient supplies of the ready-made material, or comparatively large fragments of it in the diet.

In the case of many of the coenzymes, however, only traces are needed and here the body tends to rely on the diet for certain essential portions. Thus in the case of DPN (see page 248), which is made up of a residue of adenine, two of ribose, two of phosphate, and one of nicotinamide, the body reserves the ability to synthesize ribose and adenine from carbohydrate and amino acid food factors, since both of these structures are needed in quantity in nucleic acids. Phosphorylated compounds are very widespread in the body and here likewise the organism always retains the ability to add phosphate to any organic substance capable of esterification. The nicotinamide, however, is a derivative of pyridine, a ring system which does not occur in the body except in this and two other coenzymes, none of which is needed in more than trace quantities. The cells of the human body have lost the ability to synthesize a structure needed in such small quantities and rely on dietary supplies. The simplest compound from which the cells can form the needed nicotinamide is nicotinic acid, which is therefore the food factor. Niacin and niacinamide are common synonyms for nicotinic acid (formula X) and nicotinamide (formula XI).

X. Nicotinic acid (niacin)

XI. Nicotinamide (niacinamide)

An organic substance needed in traces in the diet for the proper functioning of the body is most frequently termed a *vitamin*. The name dates back to the turn of this century when the physiological effects of certain dietary deficiencies were noted without the biochemists of the day being able to determine the chemical nature of the food factors. Early investigations led to the belief that one of the mysterious missing factors was an amine (and, surely enough, the one then being studied, thiamine, was an amine), so that the name of vitamine (i.e., life amine) was given them. Later it was found that not all "vitamines" are amines but the name had become too firmly embedded in the literature for any change to be made other than the minor one of dropping the final "e" and thus mutilating the suffix out of a complete resemblance to the amine it often was not. Because of the mystery that surrounded these trace substances at the beginning, both medical men and the lay public took to regarding vitamins with something of awe and wonder. Actually, the vitamins are no more mysterious than, for instance, the amino acid lysine, and are food factors in the same sense that lysine is—the only difference being that the vitamins are needed in much smaller quantities.

A vitamin is not always a distinct chemical individual. Just as indole lactic acid can substitute for tryptophane, since it supplies the needed indole ring, niacin can substitute for niacinamide. Once the body is supplied with the needed carboxylated pyridine ring, it can manage to form the amide from it. The carboxyl group must exist on the ring, however. The body can not use pyridine itself as a niacinamide precursor. When several closely related chemical substances can each function as a particular vitamin, the individuals are known as *vitamers* (the suffix being inspired by the more common term, isomer). It is for this reason that vitamin assays, whether chemical or microbiological, must avoid being too specific, since the vitamin content of a particular foodstuff may be in the form of a mixture of vitamers and a test which is positive to one vitamer and negative to another will not give an accurate measure of its vitamin content.

The system whereby vitamins are distinguished by letters of the alphabet also belongs to the early days of investigation when the chemical nature of these substances was unknown. Two factors, the absence of which gave rise to two separate syndromes, could only be differentiated by being termed vitamin A and vitamin B. This system of nomenclature is quite unsatisfactory. Factors have been given letter designations and then identified as something known and the letter rendered unusable for future designations. This happened to "Vitamin F", which was named, then abandoned, and is now thought to be identical with the unsaturated fatty acid food factor. Originally, the letters were awarded in alphabetical order, a system which extended consecutively up to and including "Vitamin H", but in other cases new vitamins were named according to the initial letter of the effect they brought about, as vitamin K which is the "Koagulations-vitamin" or vitamin P which is thought to be involved with capillary permeability. In addition, almost every lettered vitamin was eventually found to be multiple, not only in the sense of vitamerism, but in that some contained groups of factors possessing greatly different chemical natures and physiological functions. "Vitamin B" was particularly malignant in this respect. Since early separations involved extraction by different solvents under various conditions, what was called "Vitamin B" was actually a mixture of many water-soluble, nitrogen-containing food factors. Various fractions were isolated from "Vitamin B", which were labeled "Vitamin B_1", "Vitamin B_2", and so on. The latest vitamin of major importance to be isolated is the cobalt-containing "Vitamin B_{12}", which is also known as cyanocobalamine. Different workers, isolating the same factor, simultaneously bestowed different names on it. What some workers called "Vitamin

B_2", for instance, was identical with what others called "Vitamin G". At the present time most of the trace organic food factors have been chemically identified and have received names. It is well to use the names rather than the letter designation wherever possible.

The individual vitamins are listed and discussed systematically in Chapter 21.

Mineral Food Factors

There must be at least one food factor containing each of the chemical elements necessary to life. Carbon, hydrogen, oxygen, nitrogen, and sulfur are supplied in large quantities by the carbohydrate, lipid, protein, and water of the diet. The remaining elements necessary to life are usually thought of nutritionally as *minerals.* They are distinguished from the elements we have been concerned with thus far in that they are available to the organism when ingested in their ionic forms.

Phosphorus represents an intermediate element. Although it can be utilized by the body when ingested as phosphate ion and is considered therefore one of the mineral elements of food, it is chiefly present in the diet as nucleic acid, phospholipid, phosphoprotein, and phosphorylated sugars and sugar derivatives. Other elements needed in comparatively large quantities are calcium, which with phosphate comprises the major portion of the teeth and bones; sodium, which is the major cation of plasma and extracellular fluid; chlorine, which is the major anion of plasma and interstitial fluid; potassium, which is the major cation of intracellular fluid; magnesium, which is present in both intracellular and extracellular fluid; and iron, which is a vital constituent of hemoglobin and the other heme catalysts.

In addition to these, a number of minerals are part of various enzymes, coenzymes or hormones and are therefore essential to life in trace quantities. Thus, iodine is an essential part of thyroxine, zinc of the enzyme carbonic anhydrase, copper of butyryl CoA dehydrogenase, molybdenum of xanthine dehydrogenase and cobalt of cyanocobalamine. Fluorine, and manganese are also essential in traces. These trace minerals, together with the vitamins, are included by the term, micronutrients. In the case of all these mineral elements, from phosphorus on down, the food factors involved are ions themselves. Mineral metabolism is taken up in Chapter 17.

MALNUTRITION

Malnutrition results when the tissues receive inadequate amounts of any of the food factors. The commonest reason for such a failure of food factor supply is their deficiency in the diet. In the world as a whole, the most common by far of the varieties of malnutrition is *undernutrition,* a condition in which the food supply is insufficient to meet the energy needs of the body, so that a state of chronic starvation results.

The most important single cause of dietary inadequacy is, of course, economic. In general, high protein foods are more expensive than carbohydrate foods. Families which are forced to devise means for maintaining a food supply at the least possible expense almost unavoidably choose a diet which, even if adequate in its caloric content, is deficient in essential amino acids, the B vitamins, calcium, and often in other factors as well.

Where money is no object, there are still the deleterious effects of artificial overselection of food items as a matter of taste, habit, carelessness, or faddism. Radical "reducing diets", particularly where not undertaken with medical supervision, are often dangerously deficient. Dangerous food habits which are much more widespread than the individual peculiarities just referred to and which are undesirable aspects of American diet as a whole are overconsumption of highly refined foods and, particularly, the immoderate use of sugar. The overrefinement of the grains used in bread-making results in loss of those portions which contain the major part of the B vitamin content. This, fortunately, is being offset in recent years by the greater public awareness of the problem and by the use of vitamin concentrates in "enriched" flour.

Malnutrition may also be caused by conditions other than direct dietary deficiencies. A diet, ordinarily adequate, may become deficient if bodily requirements are increased, either for physiological reasons as in pregnancy and during lactation or for pathological reasons as under conditions of hyperthyroidism. Where the individual is normally engaged in strenuous physical labor, requirements of all nutrients, including the B vitamins, are sharply increased.

It is not to be forgotten, moreover, that ingestion does not automatically mean that the food constituents are available for use by the body. There is the hurdle of absorption yet to overcome. Chronic gastrointestinal disease, by reducing the opportunity for absorption, may bring about deficiency diseases even where the diet itself is adequate. At least one pathological condition, that of pernicious anemia, is brought about by the failure of the body to absorb cyanocobalamine, even though adequate quantities are always available in the intestine where it is formed through bacterial action (see Chapter 21).

Destruction of vitamins in otherwise adequate diets during the cooking process, either by heating for too long a period or by excessive soaking in water, is a factor that must be taken into consideration. Ascorbic acid is particularly liable to destruction by heat; the B vitamins and the various minerals are liable to extraction in boiling. Less commonly, destruction of vitamins may proceed after ingestion. Thus ascorbic acid is reasonably stable only in quite acid solutions where the pH is 5 or less, and is oxidized at a rapidly increasing rate as its surroundings become more alkaline. In achlorhydric patients where the pH of the gastric juice is much higher than normal the amount of ascorbic acid available is seriously reduced.

The deleterious effect of prolonged cooking upon certain vitamins is not to be taken as an indication of the undesirability of cooking as a general rule. Cooking is nutritionally valuable for several reasons. Subjectively, it tends to improve the flavor and increase the palatability of foods. Objectively, it starts the process of hydrolysis of both proteins and starch, so that cooked food is generally more easily digestible than raw food, and it also serves to kill bacteria and parasites present in the food item. As an example of the last very important item we may cite the well known danger of trichinosis following the ingestion of insufficiently cooked pork.

In some cases, particularly where the vitamins are concerned, the diet may not actually be as deficient as a simple analysis of the food stuffs may indicate. Certain food factors, notably vitamin K, cyanocobalamine, folic acid, and biotin, are produced by intestinal bacteria in amounts sufficient for human needs so that, strictly speaking, although these substances are true food factors and can not be synthesized within the human body, they need not be present in the human diet. In this respect, at least, the relationship of the intestinal bacteria to their human host is symbiotic rather than parasitic. Long courses of oral antibiotic therapy, by diminishing the production of vitamins of bacterial origin, may lead to deficiency unless there is supplementation.

On the other hand, intestinal flora may contribute to dietary deficiency. In animals, notably swine, small quantities of antibiotics added to the feed (5 to 50 grams per ton) result in an increased growth rate in the young animal. Aureomycin and terramycin seem most effective, bringing about a 15 per cent increase in growth rate. The mechanism is most likely one of inhibiting the growth of intestinal flora resulting in better disease control or possibly in diminishing the fraction of ingested nutrients consumed by the bacteria and increasing the fraction available to the host animal (6).

Again, it is unrealistic to consider each food factor by itself. The metabolic interrelationships of carbohydrate, lipid, and protein are such that the need for any one of them is affected by the quantity of the other two in the diet. A good diet would not be merely one in which there was at least an adequate amount of each, but one in which the quantities of all three were well balanced. Among the micronutrients, such interrelationships may also exist. The tocopherols, by their activity in poising the oxidation-reduction potential, increase the stability of other vitamins which would otherwise be subject to destruction by oxidation (15). Such an effect, where two substances together are more effective than the sum of each separately, is called *synergism.*

Carbohydrate and Lipid

Some 50 to 60 per cent of the caloric intake of the average American is in the form of carbohydrate. The percentage tends to be higher among the less prosperous groups. A diet which is higher than average in carbohydrate need not necessarily be one which is deficient in other respects, although it is more apt to be so than one which is high in protein, particularly where an excessive amount of the carbohydrate is in the form of highly refined grains and sugars.

The function of both carbohydrate and fat in the body is to provide energy. Carbohydrate, from carbohydrate or protein in the diet, is constantly being utilized by the various tissues of the body and is the ultimate power house which keeps the body machinery working. Carbohydrate is stored in the animal body as glycogen, mainly in the liver and muscles, but the total amount so stored is small in terms of calories, being less than would be required to sustain normal activity a single day. Soskin and Levine (37) set the total body carbohydrate of a hypothetical normal man with a body weight of 70 kgm., a muscle weight of 35 kgm., a liver weight of 1.8 kgm., and a blood plus extracellular fluid volume of 21 liters, to be 370 grams. Muscle contains 0.7 per cent glycogen, giving a total of 245 grams in all the muscles. Liver contains 6 per cent glycogen, giving a total of 108 grams. The remaining 17 grams is in blood and extracellular fluid in the form of glucose. This mass of carbohydrate has a caloric equivalent of 1517 Kcal., allowing 4.1 Kcal. per gram, and is sufficient to maintain a human being engaged in a sedentary occupation for 13 hours. Caloric reserves in excess of this are in the form of simple lipid which can be stored in almost indefinite quantities. This so-called depot fat takes up a smaller volume per calorie than does carbohydrate and is thus the more compact form for storing calories in anticipation of a long term need.

Since carbohydrate can readily be converted into fat in the body, the amount of lipid in an adequate diet need not be high. In fact, it need be present only in quantities sufficient to insure an adequate supply of fat-soluble vitamins and, possibly, of such unsaturated fatty acids as linoleic and arachidonic acids.

The nutritionally useful carbohydrates include the various starches, glycogen, and their hydrolysis products, the dextrins, maltose, and glucose, also fructose and galactose, together with their glucose-containing disaccharide derivatives, sucrose and lactose. Before absorption, the polysaccharides and disaccharides are hydrolyzed to monosaccharides which can be stored in the form of glycogen. An important glucose-containing polysaccharide which is *not* directly available for human use is cellulose. It is ingested in very appreciable quantities along with most plant food, particularly those of the leafy vegetables, but is excreted in the feces unchanged except for minor bacterial decomposition. Its only function within the alimentary canal is the rather dubious one of supplying "roughage"—that is, stimulating by its bulk and by the roughness of its texture the natural peristalsis of the intestine. Indirectly, however, cellulose is of great importance to human nutrition, since herbivorous animals possess alimentary canals of much greater capacity comparatively than those of omnivorous animals such as man, so that food retention is longer and intestinal bacteria have ample opportunity to hydrolyze cellulose to glucose. Cattle are thus able to subsist on grass and hay and to convert these cheap items into expensive meat and dairy products. Presumably, if the human possessed an alimentary canal of adequate capacity, he could utilize cellulose directly; but the current method of indirect utilization of cellulose is likely to remain widely popular.

Overweight. The excessive intake of carbohydrate may be harmful in two ways. By satisfying the appetite, it may reduce consumption of foods containing proteins and micronutrients, thus leading to deficiency states in those respects. Secondly, by increasing the over-all caloric intake it may induce a more than optimal weight (overweight). Simple overweight may be considered the result of eating more food than is required, with metabolic factors relatively uninvolved. This has been distinguished from *obesity* (10) a more serious disease involving overweight, in which any of a number of metabolic factors may be involved and which, unlike simple overweight, is often irreversible. The extrinsic cause of obesity is still excessive caloric intake but here disorders of the fat mobilization or deposition would tend to aggravate the result. Strains of mice with inherited tendencies toward obesity have been studied. In one of these strains it seems that pancreatic hormones are involved. The obese mice secrete a hyperglycemic-glycogenolytic anti-insulin hormone. The inhibition of insulin by this hormone has the effect of a partial pancreatectomy. Compensating overactivity of the insulin-secreting cells of the pancreas may conceivably lead to failure and account for the correlation long noted between overweight and diabetes.

The effects of a hyperglycemic hormone on food intake may be considerable. Mayer (24) has proposed a *glucostatic mechanism* whereby food intake is regulated. Put simply, he believes the sensation of hunger to be the result of a lowered glucose utilization. This could be due to a decrease in blood glucose as a result of a fasting interval or to hormonal interference with glucose utilization as in insulin-lack (diabetes) or glucagon-excess (hereditary obesity). In the latter case, hunger may accompany a high blood-glucose level which must then be forced up still higher at the cost of an excessively large meal before the sensation of satiety is achieved. See, however, (5a) for arguments against the glucostatic mechanism.

To reduce overweight it is necessary to restrict caloric intake to a point where it is below caloric expenditure. This is an unavoidable consequence of simple arithmetic. Weight reduction may be achieved either by reducing intake (dieting), or by increasing expenditure (exercise). Of the two, the latter is by far the less satisfactory. A sudden addiction to violent exercise, particularly in the case of a middle-aged or sedentary person, may involve considerable and undesirable strain on the circulatory system. Furthermore, increased exercise is accompanied by increased appetite with the all too frequent result that the net effect on weight loss is zero, or negative. That this is not too surprising is demonstrated by the fact that the calories expended in climbing twenty flights of stairs, each ten feet high, may be restored by the consumption of one slice of bread, and the calories expended by walking a mile, by the consumption of an ounce of cream. Refraining from eating that slice of bread or that ounce of cream is obviously the easier method of accomplishing the result.

In reducing caloric intake, the greatest danger is that the individual, having spent happy years adding inches to his waistline, is suddenly determined to remove the accumulation in a matter of days. While a fierce and unrelenting regimen of lettuce and fat-free crackers will not expose an overweight person to the immediate danger of starvation, there is the very good chance that deficiencies in proteins and micronutrients may develop far more quickly than the excess weight is reduced.

The exact dietary rules best qualified to effect weight reduction are not yet beyond dispute. Soskin and Levine (37) recommend one gram of protein per day per kilogram of "ideal" body weight and see no harm in protein intakes in excess of this. It goes without saying that protein in the diet must be such that at least minimal quantities of the various essential amino acids be supplied (see page 457). The use of vitamin concentrates and mineral supplements is necessary in drastic reducing diets.

The relative proportions of lipids and carbohydrate intake in reducing diets is of importance. A drastic reduction of "starchy" foods may be self-defeating, if lipids are not simultaneously restricted. A diet relatively high in lipid and low in carbohydrate and protein will induce ketosis, so that however restricted the caloric intake, the carbohydrate and protein content should be high enough to prevent this. Soskin and Levine suggest that the carbohydrate content of the reducing diet be two to three times (in calories) that of the lipid content. Short-range fluctuations of weight during dieting may result from changes in the water content of the body and in the first stages of dieting fat loss may often be obscured or exaggerated. Once the desired weight has been attained, the patient should avoid reverting to his pre-diet food habits, or in time the job will be all to do over again.

Protein

The minimum amount of protein needed to maintain nitrogen equilibrium (and hence the minimum amount necessary for an adequate diet) is still a matter of controversy. Early investigations, before the significance of the amino acid food factors was known, yielded results that were naturally conflicting since various proteins would yield various minima, depending upon the weight of protein which would contain an adequate amount of the amino acid food factor in which it was poorest. In the case of proteins such as gelatin or zein, the weight required would approach the infinite. Taking the amino acid food factors into account, Newburgh maintains that 60 grams of protein per day will suffice to maintain nitrogen balance (28). As for the amino acid food factors themselves, Rose has calculated the minimum intake required for nitrogen balance in the adult male, *where there is an adequate dietary supply of the other amino acids*, as 6.35 grams (see table XII-1).

Protein requirement is increased in various pathological conditions. There is serious protein loss in cases of hemorrhage, since one liter of blood contains 148 grams of hemoglobin and 60 grams of plasma protein, or a total of 208 grams of protein. The normal rate of tissue protein breakdown is accelerated during infection, and after various injuries and after operations (41). Proteins may be lost in exudates, following burns or wounds, or via the urine in albuminuria. In the former case, losses may amount to 50 grams per day or even more, and in the latter, to 25 grams. Protein deficiencies may also result from metabolic disturbances which reduce the body's ability to synthesize protein from amino acids.

The most obvious result of chronic protein malnutrition is loss in weight. This is to be distinguished from loss in weight due to uncomplicated caloric deficiency in that the loss is represented not by decline in depot fat but by actual wasting of functional tissue. Protein deficiency is usually accompanied by disturbances of the water balance in part due to the lowered osmotic pressure following loss of serum albumin. Water tends to be retained while tissues may waste up to 25 per cent of their original weight, the consequent disproportion of water content resulting in *nutritional edema*. The bloated abdomens of famine victims are characteristic. Restoration of a diet adequate in protein, by leading to an initial loss of excess water, may actually result in a decrease in weight at first.

TABLE XII-2

Food and Nutrition Board, National Research Council recommended daily dietary allowances, revised 1953

	AGE	WEIGHT	HEIGHT	CALORIES	PROTEIN	CALCIUM	IRON	VITAMIN A	THIAMINE	RIBO-FLAVIN	NIACIN	ASCORBIC ACID	VITAMIN D
	Years	*kg. (lb.)*	*cm. (in.)*		*gm.*	*gm.*	*mg.*	*I.U.*	*mg.*	*mg.*	*mg.*	*mg.*	*I.U.*
Men	25	65 (143)	170 (67)	3200	65	0.8	12	5000	1.6	1.6	16	75	
	45	65 (143)	170 (67)	2900	65	0.8	12	5000	1.5	1.6	15	75	
	65	65 (143)	170 (67)	2600	65	0.8	12	5000	1.3	1.6	13	75	
Women	25	55 (121)	157 (62)	2300	55	0.8	12	5000	1.2	1.4	12	70	
	45	55 (121)	157 (62)	2100	55	0.8	12	5000	1.1	1.4	11	70	
	65	55 (121)	157 (62)	1800	55	0.8	12	5000	1.0	1.4	10	70	
	Pregnant (3rd trimester)			Add 400	80	1.5	15	6000	1.5	2.0	15	100	400
	Lactating (850 ml. daily)			Add 1000	100	2.0	15	8000	1.5	2.5	15	150	400
Infants	0–1/12												
	1/12–3/12	6 (13)	60 (24)	kg.x120	kg.x3.5	0.6	6	1500	0.3	0.4	3	30	400
	4/12–9/12	9 (20)	70 (28)	kg.x110	kg.x3.5	0.8	6	1500	0.4	0.7	4	30	400
	10/12–1	10 (22)	75 (30)	kg.x100	kg.x3.5	1.0	6	1500	0.5	0.9	5	30	400
Children	1–3	12 (27)	87 (34)	1200	40	1.0	7	2000	0.6	1.0	6	35	400
	4–6	18 (40)	109 (43)	1600	50	1.0	8	2500	0.8	1.2	8	50	400
	7–9	27 (59)	129 (51)	2000	60	1.0	10	3500	1.0	1.5	10	60	400
Boys	10–12	35 (78)	144 (57)	2500	70	1.2	12	4500	1.3	1.8	13	75	400
	13–15	49 (108)	163 (64)	3200	85	1.4	15	5000	1.6	2.1	16	90	400
	16–20	63 (139)	175 (69)	3800	100	1.4	15	5000	1.9	2.5	19	100	400
Girls	10–12	36 (79)	144 (57)	2300	70	1.2	12	4500	1.2	1.8	12	75	400
	13–15	49 (108)	160 (63)	2500	80	1.3	15	5000	1.3	2.0	13	80	400
	16–20	54 (120)	162 (64)	2400	75	1.3	15	5000	1.2	1.9	12	80	400

Cream is that portion of milk which rises on standing. Since fat is lighter than water, that portion is a milk-fat concentrate. The fat content of cream varies according to the method of separation and is usually less than 35 per cent. The vitamin A of milk is concentrated along with the fat. The fat-poor portion of milk remaining after cream has been removed is *skim milk*, which has considerable food value despite the unfavorable connotations attached to the name. It contains, in large measure, the proteins and minerals of milk and is a valuable addition to the diet of those attempting to reduce.

By churning, the fat globules of cream, fresh or fermented, can be made to coalesce, forming *butter*, and leaving behind the aqueous portions of cream in the form of *buttermilk*. Butter is a still more concentrated form of milk fat than is cream, the fat content being over 80 per cent and is a correspondingly richer source of vitamin A. Milk is also often used in making oleomargarine, a high-fat product closely resembling butter in taste and texture, and differing in that the source of its fat is not primarily milk fat but other less expensive fats. Oleomargarine does not naturally have the vitamin content of butter. Fat-soluble vitamins are customarily added, however, to table oleomargarine.

Infant feeding. Human milk is the best food for the human infant, and mothers should be advised by the physician to nurse their children at the breast. When mothers are unwilling or unable to nurse, their young infants have to be fed on a substitute, which is commonly made by diluting cow's milk to bring the protein concentration down and adding sugar to bring the carbohydrate content up.

Bottle-fed babies store greater quantities of nitrogen and calcium than do breast-fed babies. The nitrogen retention, particularly, results in a muscle mass approximately 25 per cent greater in the bottle-fed babies. However, tetany, a manifestation of hypocalcemia, is much more frequent in the bottle-fed infant, and rickets, in which calcium phosphate deposition is interfered with, is apt to be more severe (4).

This is one indication that no matter how we attempt to make the proportions of the various food constituents in human and cow's milk identical, certain finer differences remain. Casein makes up $1/3$ of the protein of human milk and a little over $5/6$ of the protein of cow's milk (17). While the two proteins are both complete, casein forms a curd when subjected to the action of the gastric juices. The result is that cow's milk is digested more slowly than human milk. This is reflected in the fact that the feces of a bottle-fed baby are generally more abundant and of a harder consistency than are those of a breast-fed baby.

Colostrum is secreted by the mammary gland shortly before parturition and for a few days thereafter. It is comparatively poor in fat—about two-thirds that of ordinary milk—but has twice the protein content of milk. The additional protein is a globulin which appears identical with the gamma-globulin of plasma. Protective antibodies, representing the immunities developed by the mother, are contained in this globulin fraction.

The two food factors in which both cow's milk and human milk are deficient are iron and often vitamin D. The deficiency of iron is not serious for the baby since he is born with an iron reserve sufficient for about six months, by which time there will be a dietary supplement of such iron sources as egg yolk and cereal preparations. Iron-deficiency anemia is most commonly encountered in infants in the second year of life (36). Under optimal conditions, including the exposure of the human mother or the dairy cow to sunlight, vitamin D may be present in human or dairy milk in amounts adequate to prevent rickets in infants and children. In those parts of the world where adequate exposure to sunlight is impractical, supplementary feeding of infants with vitamin D is standard practice.

Bottle-fed babies are apt to show an undesirably low level of blood ascorbic acid within ten days after birth. Supplementary ascorbic acid is therefore desirable at a very early age. In addition, the supplementation of the infant's diet by foods other than milk is desirable.

Eggs

Where milk is the specialized and complete food of the young mammal, the egg of the bird represents the specialized and complete food for the developing bird embryo. The egg yolk contains most of the egg's nutritional value. The most important protein of egg yolk is ovovitellin, which, like the casein of milk, is a phosphoprotein and a good source of all the amino acid food factors. Of these factors, it has only one-third the valine content of casein and is about 25 per cent poorer in lysine.

The lipid content of egg yolk is high, comprising nearly one-third of the whole. Most of the lipid is in the form of phospholipids (mainly lecithin) and cholesterol. With respect to both protein and lipid, egg yolk approximates meat in chemical composition. As far as minerals and vitamins are concerned, egg yolk is superior to meat and approaches, and in some respects surpasses, milk in value. Although calcium is only half as concentrated as in milk, eggs are twice as rich in phosphorus because of their phospholipid content. One great advantage eggs have over milk lies in their content of iron and copper, for both of which egg yolk is one of the richest natural sources, being nearly three times as rich in iron as lean beef and about twice as rich in copper. Egg yolk is an excellent source for vitamin A and for the B vitamins. It contains significant quantities of vitamin D, but is a poor source of ascorbic acid.

Egg white is a 10 per cent colloidal solution of egg albumin, also known as ovalbumin, with small amounts of mucoprotein accompanying it. Because of the comparative lack of other materials in egg white, egg albumin has been one of the proteins most investigated in the laboratory and it is significant that the German word for protein is *Eiweiss*. Egg albumin is a nutritionally complete protein. Raw egg white contains the protein *avidin*, which is capable of forming a stable complex with biotin, the complex possessing none of biotin's food factor properties. The feeding of large amounts of raw egg white or of avidin to laboratory animals can produce a biotin deficiency. The biotin-inhibiting properties of egg white are destroyed on cooking.

Meat

The chemical composition and nutritive value of meats varies only slightly among the more common sources in the American diet, and what is said here will hold for beef, veal, mutton, lamb, pork, and poultry.

Fresh lean meat, i.e., muscle tissue, is about one-fifth protein, mostly myosin. Myosin is a complete protein which is nutritionally on a par with the proteins of milk and eggs. The lipid content of meat varies with the cut and with the animal source, but this is not a vital factor except in the case of those on reducing diets. This variable lipid content is of course the simple lipid of the depot fat. The carbohydrate content of lean meat is low and confined largely to the muscle glycogen.

Lean meat is poor in calcium, containing only one-tenth the calcium concentration of milk, but is about as rich a source of phosphorus as is egg yolk. Lean meat, like eggs and unlike milk, is a good source for iron and copper and is low in vitamin C. In addition it is poor in vitamins A and D. Nor are these last named fat-soluble vitamins found to any appreciable extent in the depot fat of animals. Lean meat, however, is a good source of the B vitamins, particularly thiamine and riboflavin. Lean pork, for instance, is one of the richest natural sources of thiamine.

Special mention should be made of the glandular organs of the animals usually used for meat—the liver in particular. The liver is the chemical factory of the body and is therefore particularly rich in enzymes, and consequently in coenzymes. Since the B vitamins occur in the body as parts of coenzymes, liver is rich in these food factors, particularly in riboflavin. Fresh liver is the food richest in riboflavin. The fat-soluble vitamins seem to be stored in the liver. This is particularly true in the case of the vitamins A and D. Storage can take place to a fantastic degree where the diet is rich in these vitamins. Liver also possesses small but significant amounts of ascorbic acid. The use of liver extracts in combating pernicious anemia is due to the cyanocobalamine in them.

Fish and seafood in general closely resemble ordinary meats as far as the quantity and quality of the proteins are concerned. The livers of certain fish such as cod and halibut, while not usually eaten directly, are the richest known natural sources of the vitamins A and D. Fish liver oils are used in infant dietaries for the prevention of rickets, but are not preferable to vitamin-enriched milk, since the latter can be fed to the infant from the start whereas fish oils are not usually prescribed until the second month of life. Seafood differs from ordinary meats in being higher in mineral content. The most significant item here is its high iodine content. Seafood generally has ten times the concentration of iodine of land food. Since iodine is necessary in the biosynthesis of thyroid hormone, the populations of regions where the iodine content of the soils and consequently of the land plants may be negligible often suffer from endemic goiter. Such problems have been solved by small quantities of iodine in city reservoirs or by the use of iodized salt; these problems do not appear at all where seafood is a significant portion of the diet. Salt water fish, as would be expected, are richer in minerals in general and iodine in particular than are fresh water fish.

Although meat of various sorts is an excellent food highly prized by most people, there are large populations in Asia and Africa for whom meat, by and large, is an unattainable luxury. A vegetarian diet, however, whether imposed by necessity or adopted through choice, may be adequate for all nutritional needs (12).

Grains

Most grains or cereals are seeds produced by members of the grass family. If any foods may be considered the basic substratum of the human diet, it is these; and there is certainly no quarrel with the proverbial definition of bread as the "staff of life". Grains owe their importance to several factors. They are relatively cheap and available, low in water content, and may be stored over periods of time without spoiling, a fact not true of most of the animal foods so far discussed. In the United States, cereal products account for more than one-fourth of the total caloric intake, and of the cereals wheat products constitute 75 per cent. In the world as a whole, rice is even more important than wheat since the vast populations of the Orient subsist to a large extent upon that grain. Other important grains are maize, oats, barley, rye, and millet.

The wheat kernel consists of three parts: (1) an outer coat of *bran*, which comprises 12 to 15 per cent of the whole; (2) the *germ*, which is the actual embryo plant and forms 2 to 3 per cent of the whole; and (3) the *endosperm*, which forms the remaining major portion of the kernel and represents the initial food supply of the young plant. In the preparation of white flour, the bran and germ are discarded in the milling process, so that the flour is essentially finely ground endosperm. Such flour is poor in protein as compared with meat, both quantitatively and qualitatively. In amount, protein comprises 10 per cent of wheat flour as compared with 20 per cent in meat or eggs. In quality, the protein of wheat is less well endowed with amino acid food factors. The glutenin of wheat is complete, but gliadin is partially incomplete. About 75 per cent of wheat flour is carbohydrate in the form of starch. The lipid content is negligible.

Wheat germ is ordinarily discarded in the milling process, largely because its relatively high percentage of oil results in a tendency to become rancid and thus interferes with the keeping qualities of flour. This is unfortunate since it is richer in protein, minerals, and vitamins than is the endosperm. Because of the fact that it forms such a small portion of the whole kernel, its contribution to the protein content is negligible, but in the case of the micronutrients it is most important. It is particularly rich in the B complex. Thus, whole wheat flour which contains wheat germ is over four times as rich in thiamine as refined flour despite the fact that in terms of mass the germ contributes only some 3 per cent of the flour. Ascorbic acid is absent in wheat as are vitamins A and D.

Wheat, in common with most plant foods, is rich in minerals with the exception of sodium. Plants, in general, are potassium-rich and sodium-poor as compared with animals so that man and herbivorous animals are forced to supplement their diet with sodium chloride. Mineral content is most concentrated in the bran, which is discarded in the usual milling process. Some two-thirds of the carbohydrates of bran are indigestible pentosans, so that bran is much used as "roughage" in the home treatment of constipation. While undoubtedly often efficacious, the rough fibrous nature of bran makes it unfit for use in large quantities without medical supervision. The irritating effect of bran may seriously aggravate disorders of the alimentary canal.

Bread, while composed mostly of wheat flour—at least in the United States, possesses additives which affect its nutritive value, usually for the better. Milk is frequently used in the preparation of bread, the nutritional value of this practice needing no comment. The use of yeast in preparing bread contributes to its vitamin content, as does the practice of manufacturing whole wheat bread in which germ and bran are included. In recent years, enrichment with various vitamins has become common.

Rice contains less protein (8 per cent as compared with 10 per cent) and more carbohydrate than wheat. Rice is therefore somewhat less suited as a basic diet than is wheat. The bran layers contain most of the thiamine content of rice; but again whole rice does not keep as well as does polished rice so that the economic pressures in favor of the latter are sometimes overwhelming. Of historical interest is the fact that the first avitaminosis definitely characterized and cured was beriberi, found to occur among East Indians subsisting on a diet composed largely of white (polished) rice, and cured by the addition of brown (unpolished) rice.

Maize differs from wheat in containing an appreciable (4 per cent) quantity of lipid material. From this corn oil is derived. Maize is an incomplete food in itself since its proteins, while equal quantitatively to those of wheat, are far less complete. It is also much poorer than wheat in the B vitamins.

Fruits and Vegetables

Fruits, with a few tropical exceptions, contain no starch when fully ripened and most of the carbohydrate is in the form of utilizable sugars. The mineral content is generally lower than in vegetables. Fruits are characterized first by being among the richest natural sources of free monosaccharides (usually glucose or fructose), and of sucrose, which, together with their content of organic acids (usually malic acid or citric acid) and related compounds that contribute to that elusive element we call flavor, make them very pleasant eating. Secondly, certain fruits represent our most valuable sources of ascorbic acid. The juices of orange, lemon, grapefruit, pineapple, and tomato are recognized as excellent sources of that food factor. Orange juice is for this reason commonly added to the infant diet at a very early age.

Tomato juice is a rich source of provitamin A (see page 135), while pineapple and orange juices are fair sources. Fruits are poor in the B vitamins in general and in minerals. Certain fruits, such as olives and nuts, are remarkable and useful for their high lipid content.

Vegetables, particularly the leafy vegetables, are second only to fruits as sources of ascorbic acid and are superior to fruits as sources of the B vitamins, provitamin A, and minerals. The importance of leafy vegetables, in general, can however be overemphasized. With a few exceptions they are very deficient in protein, and the caloric content of leafy vegetables is trifling. Spinach, for instance, although rich in iron, is no richer than many other foods and considerably poorer than is meat. It may contribute to loss of calcium from the body, despite its high calcium content, simply because of its still higher content of oxalate ion which forms an insoluble and unutilizable salt with calcium. The oxalate content of various leafy vegetables is a factor to be considered in the cases of those people with a tendency to form oxalate kidney stones (see page 769).

Starchy or mealy vegetables such as potatoes, beans, or peas differ in that they are richer in carbohydrate than most vegetables. Potatoes are a poor but not entirely insignificant source of ascorbic acid. In fact, in parts of Europe where potatoes form a major portion of the diet and where citrus fruits are virtually unknown this vegetable alone stands between man and scurvy. In potatoes as in other ascorbic acid-containing vegetables the vitamin concentration is higher in the outer portions, so that too enthusiastic peeling is nutritionally undesirable. Beans and peas are unusually high in protein, lima beans reaching a value of 7.5 per cent and soybeans nearly twice that concentration.

REFERENCES

1. Albanese, A. A. *Protein and Amino Acid Requirements of Mammals.* New York, Academic Press, 1950.
2. Almquist, H. J. Proportional requirements of amino acids. Arch. Biochem. Biophys., **48:** 482–483, 1954.
3. Arnon, D. I. The chloroplast as a complete photosynthetic unit. Science, **122:** 9–16, 1955.
4. Bakwin, H. Infant feeding. J. Clin. Nutrition, **1:** 349–354, 1953.
5. Benson, A. A. Photosynthesis: first reactions. J. Chem. Educ., **31:** 484–487, 1954.
5a. Bernstein, L. M., and Grossman, M. L. An experimental test of the glucostatic theory of regulation of food intake. J. Clin. Invest., **35:** 627–633, 1956.
6. Braude, *et al.* The value of antibiotics in the nutrition of swine: a review. Antibiotics & Chemotherapy, **3:** 271–291, 1953.
7. Burk, D. Photosynthesis: a thermodynamic perfection of nature. Fed. Proc., **12:** 611–625, 1953.
8. Calvin, M. The path of carbon in photosynthesis, VI. J. Chem. Ed., **26:** 639–657, 1949.
8a. Calvin, M. Function of carotenoids in photosynthesis. Nature, **176:** 1215, 1955.
9. Calvin, M. The photosynthetic carbon cycle. J. Chem. Soc., **1956:** 1895–1915, 1956.
9a. Elvehjem, C. A. Amino acid balance in nutrition. J. Am. Dietet. Assoc., **32:** 305–308, 1956.
10. Fertman, M. B. Newer concepts of experimental obesity. Arch. Internal Med., **95:** 794–805, 1955.
11. Hansen, A. E., and Burr, G. O. Essential fatty acids and human nutrition. J. A. M. A., **132:** 855–859, 1946.
12. Hardinge, M. G., *et al.* Nutritional studies of vegetarians. J. Clin. Nutrition, **2:** 73–82, 1954.
13. Harper, A. E., *et al.* Leucine-isoleucine antagonism in the rat. Arch. Biochem. Biophys., **51:** 523–524, 1954.
14. Hayes, M. A. Water-soluble vitamin requirements in surgical convalescence. Ann. Surg., **140:** 661–667, 1954.
15. Hickman, K. C. D., and Harris, P. L. Tocopherol interrelationships. Advances Enzymol., **6:** 469–524, 1946.
16. Holt, L. E., Jr., *et al.* Nitrogen balance in experimental tryptophane deficiency in man. Proc. Soc. Exper. Biol. & Med., **48:** 726–728, 1941.
16a. Hurwitz, J. *et al.* Spinach phosphoribulokinase. J. Biol. Chem., **218:** 769–783, 1956.
17. Jeans, P. C. Feeding of healthy infants and children; *in A. M. A. Handbook of Nutrition*, 2nd edit., pp. 275–298. New York, N. Y., Blakiston Co., 1951.
18. Jenness, R., *et al.* Nomenclature of the proteins of bovine milk. J. Dairy Science, **39:** 536–541, 1956.
18a. Kessler, E. The role of manganese in the oxygen-evolving system of photosynthesis. Arch. Biochem. Biophys., **59:** 527–529, 1955.
19. Levenson, S. M., *et al.* The healing of soft tissue wounds; the effects of nutrition, anemia, and age. Surgery, **28:** 905–935, 1950.
19a. Leverton, R. M. *et al.* The quantitative amino acid requirements of young women. I. Threonine; II. Valine; III. Tryptophan; IV. Phenylalanine, with and without tyrosine; V. Leucine. J. Nutrition, **58:** 59–81, 83–93, 219–229, 341–353, 355–365, 1956.
20. Levitt, L. S. The role of magnesium in photosynthesis. Science, **120:** 33–35, 1954.

21. Luckey, T. D., *et al.* The physical and chemical characterization of rat milk. J. Nutrition, **54:** 345–359, 1954.
22. Macy, *et al. The Composition of Milks.* Bulletin of the National Research Council, No. 119, January, 1950.
23. Mann, G. V., and Stare, F. J. Nutritional needs in illness and disease; *in A. M. A. Handbook of Nutrition,* 2nd edit., pp. 351–382. New York, Blakiston Co., 1951.
24. Mayer, J. Glucostatic mechanism of regulation of food intake. New Eng. J. Med., **249:** 13–16, 1953.
25. McCoy, R. H., *et al.* Feeding experiments with mixtures of highly purified amino acids. VIII. Isolation and identification of a new essential amino acid. J. Biol. Chem., **112:** 283–302, 1935.
26. Morton, R. K. The lipoprotein particles in cow's milk. Biochem. J., **57:** 231–237, 1954.
27. National Academy of Sciences—National Research Council, Washington, D. C., Publication 302, *Recommended Dietary Allowances,* Revised 1953.
28. Newburgh, L. H. Obesity; *in* Jolliffe, N., *et al.,* eds. *Clinical Nutrition,* pp. 689–743. New York, Paul B. Hoeber, Inc., 1950.
29. Pratt, E. L., *et al.* The threonine requirement of the normal infant. J. Nutrition, **56:** 231–251, 1955; Snyderman, S. E., *et al.* The phenylalanine requirement of the normal infant. *Ibid.,* 253–263, 1955.
30. Rabinowitsch, E. I. *Photosynthesis.* Vol. I. New York, Interscience Publishers, 1945.
31. Racker, E. Synthesis of carbohydrate from carbon dioxide and hydrogen in a cell-free system. Nature, **175:** 249–251, 1955.
32. Rhoads, J. E. Supranormal dietary requirements of acutely ill patients. J. Am. Dietet. Assoc., **29:** 897–903, 1954.
33. Rose, W. C. Amino acid requirements of man. Fed. Proc., **8:** 546–552, 1949.
34. Rose, W. C., and Wixom, R. L. The amino acid requirements of man. XIII. The sparing effect of cystine on the methionine requirement. J. Biol. Chem., **216:** 763–773, 1955; XIV. The sparing effect of tyrosine on the phenylalanine requirement. *Ibid.,* **217:** 95–101, 1955.
35. Shank, R. E. Revisions of the recommended dietary allowances. J. Am. Dietet. Assoc., **30:** 105–110, 1954.
36. Smith, N. J., *et al.* Iron storage in the first five years of life. Pediatrics, **16:** 166–173, 1955.
37. Soskin, S., and Levine, R. Carbohydrate malnutrition; *in* Jolliffe, N., *et al.,* eds. *Clinical Nutrition,* pp. 208–235. New York, Paul B. Hoeber, Inc., 1950.
38. Soupart, P., *et al.* Amino-acid composition of human milk. J. Biol. Chem., **206:** 699–704, 1954.
39. Stieglitz, E. J. Nutrition problems of geriatric medicine; *in A. M. A. Handbook of Nutrition,* 2nd edit., pp. 327–350. New York, Blakiston Co., 1951.
40. Warburg, O., and Burk, D. The maximum efficiency of photosynthesis. Arch. Biochem., **25:** 410–443, 1950.
41. Wilkinson, A. W., *et al.* Nitrogen metabolism after surgical operations. Lancet, **1:** 533–537, 1950.
42. Wilson, A. T., and Calvin, M. The photosynthetic cycle. Carbon dioxide-dependent transients. J. Am. Chem. Soc., **77:** 5948–5957, 1955.

CHAPTER 13

Digestion

It requires but little anatomical knowledge to realize that the body is built around a hollow tube, the digestive tract. Considered in its entirety its functions are extremely simple. It stores foodstuffs for a limited time, prepares foodstuffs for absorption, accomplishes their absorption, and rejects the unabsorbed remainder as feces. Along with these primary functions, it acts to a very limited degree as an excretory channel and as a location for the formation of certain vitamins by bacterial action.

Most foodstuffs are not absorbable in nutritionally significant quantities in the forms in which they are usually ingested. The massive molecules of proteins must be hydrolyzed to amino acids, or at least to small and simple peptides, before they can enter the body in significant amounts. Among the carbohydrates only the simple sugars or monosaccharides are absorbable. Some difference of opinion exists among physiologists as to the degree to which unsplit fat can be absorbed. It is, however, a fact that a very effective enzymatic mechanism exists with apparently the sole function of hydrolyzing fat into its components, fatty acids and glycerol. Clinical experience has taught us that if this mechanism of fat hydrolysis is inactivated, the absorption of fats from the digestive tract is drastically diminished. Digestion is a convenient term which includes all of the hydrolytic cleavages which occur in the gastrointestinal tract. Digestion converts crude protein, carbohydrate, and fat into simpler components which can be effectively absorbed.

Our discussion of digestion and absorption will be limited to those mechanisms which act to a sufficient degree to contribute significantly to the nutrition of the body. From the point of view of nourishment, for example, native animal and vegetable proteins are unabsorbable until they are completely or almost completely hydrolyzed. On the other hand, minute amounts of animal or vegetable protein can certainly be absorbed through the mucous membranes of the mouth, of the rectum, and probably of most other portions of the alimentary canal. Such absorption in minute amounts is amply demonstrated by the clinical fact that in a person who is hypersensitive to a given protein, the ingestion of that protein or its introduction in an enema may, with striking promptness, induce the symptoms by which his allergy is manifested. These symptoms may be vasomotor rhinitis, bronchial spasm, urticaria, angioneurotic edema, or even vasomotor syncope. Such absorption also occurs in non-allergic individuals. This can be shown by the procedure known as local passive transfer. One spot on the skin of the normal subject is sensitized by the endermal injection of a small amount of the allergic patient's serum. That single spot on the normal subject can be caused to itch and redden by giving him the allergenic protein by mouth or enema. However, the amounts of specific protein required to bring about these manifestations are remarkably small. We are concerned here only with those mechanisms involved in the transfer of large amounts of foodstuff from the digestive tract to the body proper.

Our discussion will deal almost exclusively with enzymes and their activators. We shall, as might be expected, emphasize the chemical aspects of digestion. There are, of course, also mechanical aspects the consideration of which belongs to the domain of physiology.

THE SALIVARY GLANDS AND THEIR SECRETION

The salivary glands differ from each other somewhat in the type of secretion produced. The parotid glands produce a thin saliva rich in starch-splitting enzyme and containing relatively little mucin. The sublingual and submandibular glands produce a thick secretion, highly mucinous and with little enzymatic action. Little information is available concerning the secretion of the buccal glands on account of the difficulty of isolating their secretion. We shall concentrate our attention chiefly upon the parotid gland since it has been studied most intensively. It is a compound tubular type of gland where secretory cells cluster around minute branches of collecting ducts. When we examine stained sections with the microscope we find fine basal granulations in the cells. These have been identified as ribonucleic acid. We recognize that the most significant product of these cells is a protein with enzymatic, in this case, starch-splitting activity. We have here therefore, an example of the association of cytoplasmic ribonucleic acid with protein synthesis.

One should recall that the innervation of the salivary glands is both sympathetic and parasympathetic. In the case of the parotid gland, the parasympathetic stimuli reach the gland from the auriculotemporal branch of the trigeminal nerve with some fibers from the glossopharyngeal. Use of electrical stimulation or of a parasympathomimetic drug like pilocarpine increases the secretory rate and the volume. A moderate vasodilatation can be observed. Sympathetic innervation is from the upper cervical chain passing through the superior ganglion. This has no stimulatory action upon the parotid; from the other salivary glands sympathetic stimulation yields a scanty saliva of increased viscosity. In the parotid, vasoconstriction and contraction of the myoepithelial or basket cells can be observed; the basket cells lie immediately peripheral to the secreting acinar cells. These two actions lead to an actual decrease in the size of the gland.

Rate and Amount of Secretion

Although the methods used may be criticized to some extent as being unphysiological, Lashley (39) observed basal secretory rates in normal adults of from 0.5 to 8.0 ml. per hour per single gland. He used a simple metallic collection disc held over the aperture of Stenson's duct by moderate suction. The disc contained a compartment for collection and a compartment for storage which latter was attached to a measuring device. Salivary secretion whether basal or reflexly stimulated has been shown to require the oxidation of glucose (2). A comparison of the composition of blood entering and leaving the parotid gland in dogs showed that from 0.8 to 2.9 mgm. glucose were utilized per gram of gland per hour. This was under basal or unstimulated conditions. The mean value was 2.1 mgm. This was compared with an independent measure of the oxygen consumption where the mean value was 1.2 ml. per gland per hour which is the equivalent of 1.8 mgm. glucose. With stimulation an increase of 1.5 mgm. glucose per gland per hour per ml. saliva secreted was observed.

It is impracticable to attempt an accurate measurement of the total 24-hour salivary output. A reasonable estimate is about a liter and a half per day. Since the salivary glands themselves weigh only 65 grams, this indicates a rather remarkable activity upon their part. Mixed human saliva is of variable composition depending upon the relative contributions of the different sets of salivary glands. It consists chiefly of water with up to 0.4 per cent protein (of which about 12 per cent is amylase), 0.2 per cent salts, and 0.1 per cent organic matter other than protein. The specific gravity ranges between 1.002 and 1.008. Its viscosity is from 18 to 35 times that of water; and the pH, as determined in our laboratory on 45 healthy young adults, was 7.58 with a standard deviation of 0.36 pH units. In this series the pH was measured promptly but without special precautions against loss of carbon dioxide. Schmidt-Nielsen (60) reported that with special precautions against loss of carbon dioxide the pH of fresh resting parotid saliva varied between 5.45 and 6.06 with an average of 5.81. For mandibular saliva the figures were 6.02 to 7.14, average 6.39. Various stimuli gave higher pH values. Loss of carbon dioxide resulted in a rapid rise in pH to a limiting value of about 7.9.

Functional Components of Saliva

The secretion of the parotid and other salivary glands contains in somewhat diminished amounts all the diffusible substances present in the blood.

In addition to these components, which need not be discussed in detail, we find characteristically in the parotid secretion the enzyme, *salivary amylase*, or according to the older terminology, *ptyalin*. Among the salivary glands the parotid is the chief if not the exclusive producer of this enzyme. Amylase is present in the saliva of newborn human infants and at all subsequent ages. It has been crystallized (47). Its Q_{10} value in the physiological range is from 2.0 to 2.3; its temperature optimum, 40°C.; its pH optimum, 6.6. It is inactive below pH 4. It loses 15 per cent of its activity on dialysis against dilute ammonia; the activity is restored by 0.01 M NaCl. The action of the enzyme upon starch is a hydrolytic one in which successive units of maltose are split off, forming dextrins.

Amylase is a characteristic secretion of the parotid glands. A characteristic component of the secretion of the other sets of salivary glands is *mucin*. This mucoprotein is most significant for its physical property of increasing the viscosity of the saliva and thereby increasing its lubricating effect. Mucin gives the customary qualitative reactions of proteins, is not heat coagulable, can be precipitated with acid, and requires full saturation with ammonium sulfate for salting out. The prosthetic group of mucin is mucoitin sulfuric acid or mucoitic acid which can be hydrolyzed to sulfuric, acetic, and glucuronic acids plus a hexosamine. In the intact molecule it can be demonstrated that the acetic acid is combined with the hexosamine as acetylglucosamine. Mucoitic acid is very similar to the chondroitic acid of cartilage which, however, contains acetylgalactosamine.

Salivary buffer capacity. Dreizen *et al.* (19) found that the buffer capacity of the saliva was higher in caries-resistant as compared with caries-susceptible individuals. Their highest buffer capacities were found in malnourished patients relatively free from dental caries.

There can be no doubt that the physical and chemical properties of the saliva and the nature of its bacterial flora have some bearing upon the problem of dental decay. Caries does not occur in unerupted teeth; no matter how hereditary, nutritional, or endocrine factors may alter the susceptibility of the teeth to decay, teeth do not decay until they are exposed to the saliva. The initiating factor may, therefore, be chemical or bacteriological, or a combination of the two. Miller long ago proposed that the carious state was the result of a chemico-parasitic process; the action of acids produced by the fermentation of carbohydrates in the mouth.

Nord (54), summarizing the researches in this field since the time of Miller, states that caries is produced by the activities of micro-organisms which reach the amelo-dentinal junction through a defect in the enamel and then proceed to grow along the line. The production of acid or hydrolytic enzymes or both by micro-organisms may result in a shrinkage of the dentine forming a space which will be filled by dental lymph. This offers an excellent environment for microbial growth. The enamel is attacked from the inside which is less resistant.

Salivary calculus. This term is applied to two separate and distinct forms of deposits originating from the saliva. More commonly it designates the coating (tartar) of calcium and magnesium phosphates, with some admixture of carbonates and organic debris, which forms upon the teeth—natural or artificial—of many individuals. The salivary origin of this deposit is indicated by its presence in greatest amount on the surfaces nearest the outlets of the salivary glands. Alternatively, a salivary calculus may be a stone in a salivary gland or in its duct. The presence of such stones, singly or in greater numbers, is known as *sialolithiasis*. Stones occur most frequently in the submandibular gland or duct. Their composition is similar to that of the salivary deposit upon dental surfaces.

GASTRIC DIGESTION

The actual observation that gastric juice can dissolve meat was made and recorded by Spallanzani in 1783. Several famous patients, whose direct fistulous openings from the stomach to the outside world attracted the scientific interest of their physicians, have contributed painfully but effectively to our knowledge of gastric physiology and chemistry. Let us call the roll. Alexis St. Martin, whose opportunity to serve science came as the result of a gunshot wound, was studied by Beaumont at the then isolated frontier post of Mackinac and was reported in 1833 (4). Beaumont's chief contributions, aided by his not always co-operative patient, dealt chiefly with the inorganic composition of gastric juice. It was from a specimen of gastric juice from Alexis St. Martin that the definitive identification of HCl as the acid component was made. Much later Fred V. was studied by Carlson whose first report on this famous patient appeared in 1912 (10). The observations made by Carlson on Fred have been the backbone of our knowledge of gastric motility. Moving ahead to 1943, we find Tom, whose physicians, Wolf and Wolff (69), filled in with great detail the relationships outlined by Cannon (9). Finally we have the unnamed woman aged twenty-four who was studied by Crider and Walker at St. Louis, who demonstrated gastric responses to emotional stimuli opposite to those of Tom and her other male predecessors. Anger and related emotions in the male subjects were with great consistency accompanied by unusual secretory activity of a gastric mucosa which became hyperemic with the onset of the emotional state. Motility was increased. In the anonymous St. Louis woman similar emotional situations led to pallor of the mucosa with decreased gastric motility and secretion (14). The authors properly point out that if the observed differences are characteristic, such differences in gastric response to emotion may explain the much greater incidence of peptic ulcer in men as compared with women.

Gastric Acidity

An outstanding feature of the secretion of the stomach is its acidity. Let us discuss this before considering the enzymatic functions; we will show later that the activity of one of the significant enzymes is dependent upon the maintenance of acidity.

Volume of gastric secretions. A continuous spontaneous or basal secretion can be demonstrated in the majority of human stomachs. Bengt Ihre (33) found the rate to vary from 15 to 117 ml. per hour in 17 normal men, and from 14 to 42 ml. per hour in 6 normal women. Subjects with high basal secretion were usually found to have high values for secretion under various stimuli which will be discussed later. The basal secretion is not constant. There are regularly recurring outbursts of secretory activity occurring during fasting about every 2 or 3 hours and lasting 10 to 30 minutes, and accompanying phases of accelerated motor activity, first observed by Carlson and called "hunger contractions". Wolf and Wolff found them not to be associated with hunger sensations more than 50 per cent of the time.

Neural stimulation of gastric secretion. Although it is a familiar observation that the suggestion of food increases salivary secretion and intensifies the sensation of hunger, it requires unusual experimental conditions to demonstrate the concomitant increase in rate of gastric secretion. Passing a stomach tube in a normal subject is usually a distasteful process, and such a subject is resistant to suggestions by description or by odor which would, without the tube, be appetizing. Under hypnosis (33) it has been possible to demonstrate two- to threefold increase in rate of secretion as a result of "positive suggestion," with accompanying increase of acidity comparable to that following a test meal. Both hypnosis and normal sleep bring about a moderate increase in the basal secretion. Sham feeding of experimental animals with gastric fistulae or pouches yields large outputs of gastric juice, but such experiments give little information about physiological secretions in the human. Numerous reports on human subjects particularly suited for this type of experiment by the presence of esophageal stricture and gastric fistula show that "psychic" secretion exists. Sham feeding of one such subject demonstrated increases in volume, acidity, and pepsin output in proportion to the acceptability of the sham-fed meal to the patient (34).

Neural stimuli reach the stomach by way of the vagus nerve. Vagal stimulation, direct or reflex, increases volume, acidity, pepsin, and mucin of the gastric secretion. Complete vagotomy abolishes or decreases the acid secretion in about 75 per cent of human cases in which the operation has been performed for the relief of peptic ulcer (67). Levels of blood sugar below normal cause vagal stimulation of central origin, with resulting increased volume and acidity of the gastric secretion. Measurement and titration of gastric juice secreted after injection of 20 units of insulin, which usually decreases the blood sugar to 0.050 per cent or less, tests the completeness of the operation. Absence of secretory response to insulin indicates completeness of the operation. Vagotomy, or alternatively the use of anticholinergic drugs, may be used therapeutically to depress excessive acid gastric secretion. *Mechanical stimulation* of the gastric mucosa other than by distension does not induce acid secretion. The output of mucus is increased.

Chemical stimulation of gastric secretion. Meat extracts, broths, and products of protein hydrolysis such as proteoses and peptones cause prompt and copious acid secretion when they are placed in the stomach. This stimulation may be partly local, by direct action upon the secreting cells. But in animals with experimentally produced denervated gastric pouches (Heidenhain pouches), the introduction of such substances into the main stomach causes secretion in the pouch. Since the only connection between main stomach and pouch is by way of the circulating blood, the stimulating substances must either themselves travel in the blood, or they must act to liberate a hormone from the mucosa of stomach or intestine. The stimulating substances do exert a mild secretagogue effect when injected intravenously, but this effect is not quantitatively comparable to the effect of their direct contact with gastric mucosa.

Edkins (21) made extracts of minced pyloric mucosa with peptone, glucose, or HCl. Injection of these extracts had strong stimulatory action upon acid secretion, compared with negative results with plain water extracts of pyloric mucosa. Presumably an active substance was formed by the interaction of the pyloric mucosa with peptone, glucose, or HCl. This active substance was designated *gastrin.* It was long confused with histamine (formula I), which is present in crude tissue extracts, and which will by itself produce a remarkable increase in the volume and acidity of gastric secretions. Histamine and gastrin are, however, separate substances with similar action. Extracts have been obtained from canine pyloric mucosa with gastrin activity. The effective dose of such extracts contains less than

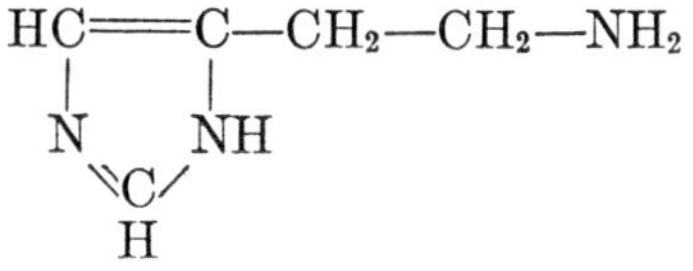

```
HC===C—CH2—CH2—NH2
 |    |
 N    NH
  \\C/
   H
```

I. Histamine

one-sixth the threshold dose of histamine as measured by the cat blood pressure method (26). The effect of gastrin is solely upon gastric secretion. The mechanisms by which neural stimulation and hormonal stimulation of gastric acid secretion operate are distinct and independent. Vagal stimulation does not act by liberation of gastrin. This is demonstrated by the failure of vagally denervated gastric pouches in dogs to secrete acid when the main stomach is secreting acid under vagal stimulation (35). The structure and chemical composition of gastrin has not yet been established. Gastrin is usually designated as a hormone, although not all physiologists are agreed that it possesses all the requisites (28).

Caffeine prolongs the secretion of acid gastric juices, and usually increases the acid output. In doses of 500 mgm. caffeine has been used as a "test meal" in the clinical study of gastric secretion (50). *Alcohol* in dilute solution increases the volume and acidity of gastric juice whether given by mouth, by stomach tube, by rectal instillation, or intravenously. The direct contact of concentrated alcoholic preparations with the gastric mucosa tends more to cause mucus production. A suitable alcohol test-meal for the stimulation of gastric secretion for analytical study consists of 50 ml. of 7 per cent alcohol, or 200 ml. of 5 per cent alcohol; both of these are commonly used in clinical study. Injection of 0.25 mgm. of *histamine* produces in the average subject a prompt and maximal output of acid gastric juice. The maximum is usually reached within half an hour. If the hypothesis is accepted that histamine is a part of the normal mechanism for the stimulation of gastric secretion, it would be expected that antihistaminic drugs would depress acidity. Actually no definite or significant effect of these drugs upon gastric acidity has been observed. Parasympathomimetic drugs stimulate the output of both acid and enzymes. *Atropine* blocks postganglionic cholinergic effects and has an action opposite to the parasympathomimetic drugs. *Cholinesterase inhibitors* such as prostigmine permit increased duration of acetylcholine activity, with increase in gastric motility and secretion.

Physiological inhibitors of gastric secretion. Although the intravenous injection of glucose has only a minimal effect upon gastric secretory function, the direct instillation of hypertonic *glucose* solutions into the duodenum definitely decreases the volume and acidity of the gastric juice (49). Intravenous administration of *amino acid* mixtures has been observed to cause decreased gastric motility and secretion (14). The presence of *fat* in the small intestine is inhibitory to gastric secretion and motility. The depressant effect of fat or sugar has been shown by Ivy and his co-workers to be the result of the release of an inhibiting substance or chalone to which they have given the name *enterogastrone*. It has not been fully purified or identified, but can be separated from secretin (see page 499) and from cholecystokinin (see page 503). Another substance with a depressant action upon gastric secretion and motility has been isolated from urine and called *urogastrone*. It acts in a manner very similar to enterogastrone but is not identical with it (25).

Emotional disturbances may inhibit or excite gastric secretion. Wolf and Wolff (69) report that in their male subject diminution of motility, circulation, and acid production was associated with mental reactions of avoidance of or withdrawal from an emotionally loaded situation. Stimulation of gastric function occurred when the emotional response was one of fighting back. More frequent occurrence of the avoidance reaction was observed by Crider and Walker (14) in their female patient.

Formation of hydrochloric acid. Different varieties of cells compose the mucous glands in the stomach. Of these cell-types two have definitely established functions: the body chief cells secrete pepsinogen; the parietal cells secrete hydrochloric acid. Pure parietal secretion is a slightly hypertonic solution of pure hydrochloric acid at a concentration of about 0.17 N and a corresponding pH of 0.87. It is impossible to isolate the pure secretion of normally functioning parietal cells. The figures just quoted are the result of indirect evidence, which will be outlined briefly.

The subcutaneous or intramuscular injection of histamine has been found to stimulate the acid secretion of the parietal cells with minimal action on the other cells composing the gastric glands. Linde *et al.* (41) tied off the stomachs of anesthetized cats and placed therein measured amounts of glycine solution which was later removed completely. The volume increment was measured, also the comparative titration values against standard alkali, and from these figures was calculated the HCl concentration of the parietal secretion. A value of 0.17 N was the minimum figure. Higher figures were presumed to result from errors introduced by absorption of the glycine solution.

This minimal figure was in agreement with the work of Hollander (32). Instead of tying off the stomach in his experimental dogs, he constructed Pavlov pouches. A Pavlov pouch is a portion of the secretory mucosa of the stomach surgically rearranged to communicate directly with the outside of the body but retaining its original innervation from the vagus nerve. Distinction should be made between this and the Heidenhain pouch mentioned previously, which is surgically entirely separated from the main portion of the stomach and deprived of all innervation, vagus or otherwise.

When histamine was administered to dogs with Pavlov pouches, an inverse linear relationship was observed between the concentration of HCl and neutral chloride, with the neutral chloride becoming zero at 0.17 N HCl, which appears to confirm this value of the concentration of HCl in pure parietal secretion.

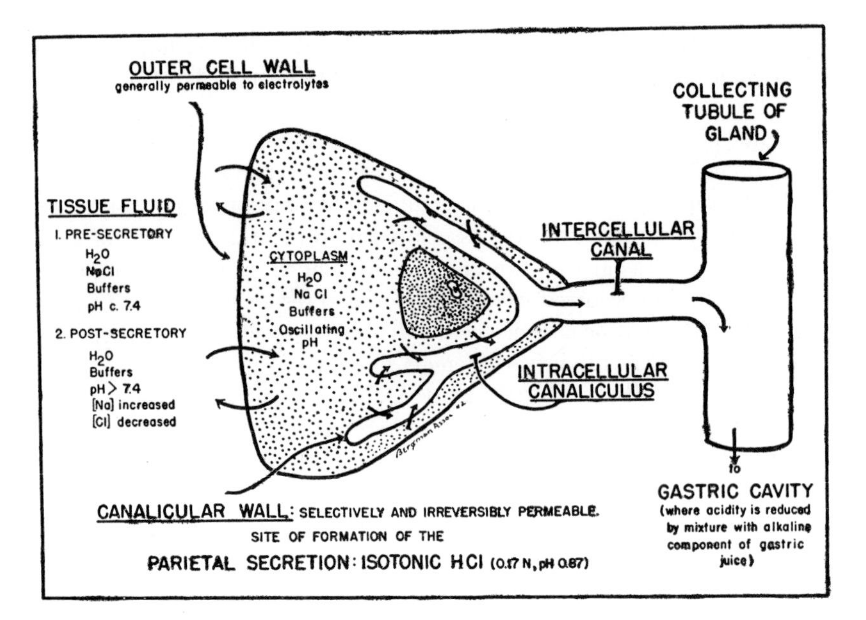

FIG. XIII-1. Schematic representation of the process of HCl formation in the parietal cell of the stomach. (Courtesy of Dr. Hollander.)

The chemical reactions at the wall of the intracellular canaliculus:

$$NaCl + H_2O \rightarrow HCl + NaOH$$
$$NaOH + NaH_2PO_4 \rightarrow H_2O + Na_2HPO_4$$

(and similarly for other buffer salts in the cytoplasm and lymph).

The energy factors:

Chemical work—in the cytoplasm and lymph
Electrical work—at the canalicular wall
Osmotic work—at the canalicular wall (zero)
Mechanical work—from the intracellular canaliculus to the open end of the gland tubule.

Since there is strong evidence that pure parietal secretion contains water, hydrochloric acid, and nothing else, it follows, according to Hollander, that somewhere in the parietal cell there must be a membrane which is permeable to water, hydrogen ion, halide ions, and to nothing else. Figure XIII-1 is an idealized picture of a single parietal cell. The intracellular canaliculi would seem to be a reasonable location for the logically demanded membrane of such strictly limited permeability. If we accept this idea that the canalicular membrane is permeable only to water, hydrogen ions, and halide ions, then any metabolic process which results in the extrusion of water by the cell would result in water secretion through the canaliculus containing those ions, hydrogen and halide, which can pass that membrane. Escape of these ions would leave an excess of sodium ions and hydroxide ions, which together with all other solutes present would be extruded through the less selective portion of the cell in equilibrium with extracellular fluid. In summation then, Hollander's theory is that of a membrane hydrolysis of sodium chloride. The slightly hypertonic concentration of hydrochloric acid probably represents a slight hypertonicity of the intracellular fluid as compared with extracellular fluid. Hollander postulates no osmotic work performed in the process just described. Work must be done, however, by the cell in excreting water. It had been previously demonstrated by others that there is increased glucose consumption and oxygen utilization during the process of active secretion of gastric hydrochloric acid.

Some further light has been shed on the process going on within the parietal cell by the work of Patterson and Stetten (55). They devised an ingenious apparatus for the continuous measurement of the pH difference across the stomach wall. They worked with isolated rat stomach, and found that continuous oxygenation was necessary to maintain acid production. Acid formation was inhibited by cyanide, fluoride, arsenite, and iodoacetate. Tetramethyl *p*-phenylene diamine, which is an inhibitor of DPN and TPN-linked enzyme systems, had an inhibiting effect at low concentrations. Others have shown that carbonic anhydrase and the niacin-containing DPN and TPN are abundantly present in parietal cells. Animals with niacin deficiency show decreased acid secretion. There is increased bicarbonate in the venous drainage of the stomach during periods of active acid secretion. A difference in electrical potential can be demonstrated across the gastric wall in which the mucosal side is negative to the serosal side in an external circuit. These observations indicate that there is an orientation of the parietal cells which Patterson and Stetten consider as resulting from a stratification of enzyme systems inside the cell. Figure XIII-2 indicates their concept of this stratification. By this mechanism a local high concentration of hydrogen ions may be developed at an area close to the exit from the cell on the gastric side. Hydrogen ions are produced by the oxidation of carbohydrate intermediates utilizing a pyridine nucleotide coenzyme. The carbohydrate intermediate is represented by RH_2 in the diagram. Note that two hydrogens are liberated from the carbohydrate intermediate, one of them appearing as hydrogen ion, the other being passed on to the flavoprotein system which serves as a reductant of the neighboring cytochromes. An electron passed on from cytochrome to cytochrome oxidase can catalyze the formation of hydroxyl ion from molecular oxygen. If the system were not stratified, the hydroxyl ion would be immediately neutralized by the hydrogen ion formed in the first step. Actually, the hydroxyl ions react with CO_2 in the presence of carbonic anhydrase. This forms bicarbonate ion which diffuses into the blood stream as sodium bicarbonate. An "alkaline tide" can be observed in the urine after meals while hydrochloric acid is being secreted by the stomach. An absence of alkaline tide indicates failure to secrete hydrochloric acid. Patterson and Stetten have been unable to add histological proof to their concept of the stratification of these enzymes. With this single exception, all other points in this theory correspond to experimentally observed facts. In summary the main points of their theory are:

1. That the hydrogen ions arise from the reduction of pyridine nucleotide.
2. That in the parietal cell the enzymes of the oxidation-reduction systems are arranged in successive strata.

The importance of the enzyme carbonic anhydrase in the secretion of gastric acid is indicated by the inhibition of basal or of histamine-stimulated acid secretion in the intact human stomach by the specific carbonic anhydrase inhibitor, 2-acetylamino-1,3,4,-thiadiazole-5-sulfonamide (Diamox) given intravenously (36).

Many other theories have been proposed to explain the formation of HCl by the parietal cells. For a detailed critical presentation of the evidence bearing on these theories, see (30).

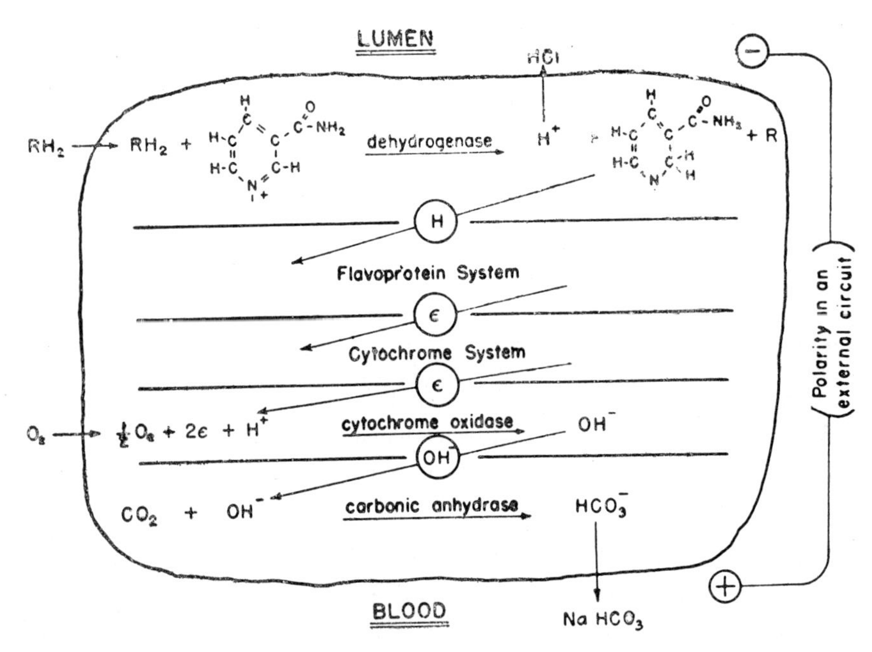

Fig. XIII-2. Stratification of enzymes in the parietal cell, according to Patterson and Stetten. From Transactions of the First Conference on Metabolic Relations, Josiah Macy, Jr. Foundation, February, 1949.

Gastric Mucin

In its general composition, the mucin of the gastric juice is similar to the mucin of saliva (see page 486). The secretion of mucin is increased by vagal stimulation and by local stimulation of the gastric mucosa, either mechanically or by irritant chemicals. The buffer action of gastric mucin is responsible for only a small part of the difference in acidity between pure parietal secretion and mixed gastric secretion. The difference depends rather upon inorganic buffers which are found in the native secretion and which can be removed by dialysis.

The gastric juice contains both dissolved and undissolved mucus, the latter being visible as strings or shreds. The dissolved mucus may be precipitated from gastric juice with acetone. Both the dissolved and the undissolved mucus consist of protein with polysaccharide prosthetic groups.

Clinical Gastric Analysis

Specimens of gastric contents obtained by means of a stomach tube always contain a mixture of the secretions of the different glands of the gastric mucosa. The concentration of HCl in such specimens never approaches the value of 0.17 N or 170 mEq. per liter which is characteristic of pure parietal cell secretion. Furthermore, contamination from saliva, food residues, and regurgitated duodenal contents almost invariably has occurred. The pH may vary between 0.9 and 2.5 excluding those cases where hydrochloric acid is absent. An occasional person, apparently normal otherwise, will show absence or minimal secretion of HCl. This situation occurs more frequently in older age groups. *Titratable acidity* of gastric contents is usually expressed in "units" or "degrees of acidity", defined as the number of ml. of 0.1 N NaOH required to neutralize 100 ml. of gastric contents. By chance, these old-fashioned "units" are identical numerically with milliequivalents of acid per liter. The latter more descriptive term should be used. *Free acidity* is the result of titration with an indicator such as Töpfer's reagent (dimethylaminoazobenzene, red at pH 2.9 to yellow at pH 4), which has a turning point at a low pH value. Such a titration measures only strong acids, and in gastric contents measures only the HCl which has not been neutralized or bound. *Total acidity* is the result obtained with an indicator such as phenolphthalein (pK = 9.7), which includes not only free HCl but also acid salts, organic acids, and acid bound to protein.

Fasting gastric juice usually contains less than 25 mEq. of free acid and less than 35 mEq. of total acid per liter. The volume of contents usually obtainable from a normal fasting stomach is less than 50 ml. Volumes of residual gastric contents greater than 120 ml. and with free acidity over 50 mEq. per liter raise strong suspicion of peptic ulcer.

The volume and acidity of the gastric contents increase following a test meal of low volume and low buffer capacity. One hour after a simple test meal such as arrowroot cookies with 40 ml. of water the gastric content of normal adults averages about 100 ml. in volume, with 40 mEq. of free acid and 55 mEq. of total acid per liter. Variations among individuals are very great. Similar values of free acidity can be obtained with a test meal of dilute alcohol (see page 490). Stimulation with histamine injections gives free acidities which are usually about 10 mEq. higher, but which occasionally are very much higher, up to a limit of about 120 mEq. per liter. When alcohol or histamine is the stimulus, there is usually no notable difference between free and total acidity. If collections are made at short intervals after the test meal, a curve of acidity against time can be plotted, which is of some diagnostic value (7).

Achlorhydria is the inability of the gastric mucosal glands to produce any hydrochloric acid whatever. Simple titration is not adequate to demonstrate achlorhydria. Roth and Bockus (58) have emphasized that an adequate dosage of histamine of tested potency must be used, with concomitant production of objective histamine effects, such as headache, flush, or slightly lowered blood pressure. Salivary contamination of the gastric collections must be rigidly excluded. Samples should be taken every 15 minutes for 2 hours, and should be checked for decrease of pH as well as for titratable acid. Stimulation with caffeine followed by histamine is suggested. A patient should not be declared achlorhydric until several attempts to secure an acid response under the most favorable conditions have failed. The demonstration of achlorhydria under these conditions excludes the diagnosis of peptic ulcer.

An adequate dosage of histamine for eliciting acid gastric response, if this is possible, contains 0.1 mgm. of histamine base per 10 kilograms of body weight. Histamine base is not injected as such, but as a salt; 0.166 mgm. of histamine dihydrochloride or 0.275 mgm. of histamine diphosphate contains 0.1 mgm. of histamine base. Achlorhydria is a typical finding in pernicious anemia, hence the rigid demonstration or exclusion of achlorhydria becomes an important diagnostic aid. Simple hypochlorhydria or subnormal acidity has no special diagnostic import. It is true that acid production is impaired or destroyed in cancer of the stomach, but such impairment is not an early sign and is of minor diagnostic value in comparison with x-ray and gastroscopic studies.

Hyperchlorhydria and *hypersecretion* are characteristic statistically of patients with the tendency to develop peptic ulcer. Excessive vagal stimulation, possibly a result of emotional stresses, appears to explain these excesses of gastric activity. Gastric analysis is of less value than x-ray in the diagnosis of peptic ulcer, since the acidity ranges of normal subjects and of ulcer patients overlap widely.

Gastric Enzymes

Pepsin, the characteristic gastric proteinase, is the product of the action of hydrogen ion upon pepsinogen, which is secreted by the body chief cells of the gastric mucosal glands. Pepsin has a molecular weight of 35,000 and its molecule is made up of at least two polypeptide chains. Crystalline pepsin is not a single molecular species. It has been shown to contain at least two proteolytic components with different solubilities. Pepsinogen is converted autocatalytically into pepsin at pH 4.6 or in more acid solutions. Pepsin catalyzes proteolysis most effectively at strongly acid pH values with some variation according to the protein used as substrate and to the state of that protein. The more denatured a protein, the higher the pH at which digestion takes place. The pH optimum for the peptic hydrolysis of native protein is 0.8 to 1, while at pH values higher than 2, the hydrolysis is slow. Denatured proteins, however, are readily hydrolyzed by pepsin at a pH of 3 to 4 which is the usual pH range of the chyle (12). Pepsin shows slight enzyme activity at pH 4, none at pH 5. At pH 6 it is irreversibly inactivated. The initial step in gastric digestion of native proteins would thus be an acid-catalyzed denaturation. Pepsin catalyzes the hydrolysis of the already denatured proteins to fragments of molecular weight less than 2000, designated as *proteoses* if they can be salted out by full saturation with ammonium sulfate, and as *peptones* if they can not.

Studies with synthetic substrates show that pepsin preferentially catalyzes the hydrolytic cleavage of peptide linkages involving the amino group of tyrosine or phenylalanine. Protamines, which are deficient in tyrosine and phenylalanine, are not hydrolyzed by peptic action.

There is a substance in the urine with properties similar to pepsin. This *uropepsin* is pepsinogen which has entered the circulating blood from actively secreting gastric gland cells, and has been excreted by the kidney. The uropepsin activity of the urine is very slight compared to the pepsin activity of the gastric juice, but the two increase or decrease together in disease. Patients with peptic ulcer often show increased uropepsin activity. Uropepsin is increased following administration of ACTH (64).

In the course of the peptic proteolysis of *casein*, the chief protein of milk, insoluble *calcium paracaseinate* is precipitated if adequate Ca^{++} is present. This is the case if milk is ingested, since milk is rich in Ca^{++}. This hydrolysis of casein to paracasein and peptone has often been attributed to a special enzyme rennin. Such an enzyme has not been demonstrated in human gastric juice. The clotting of milk which occurs upon its entry into the infant or adult stomach may be adequately explained by a combination of (a) the hydrolysis of casein in the presence of pepsin, with production of soluble paracasein and the precipitation of insoluble calcium paracaseinate, and (b) the precipitation of isoelectric casein by the free HCl of the gastric juice. *Rennet* is a milk-clotting preparation obtained by extraction of the fourth stomach (*abomasum*) of nursing calves. This preparation has been used since prehistoric times in the clotting of milk for the first step in the making of most varieties of cheese. The milk-clotting enzyme *rennin* is present in rennet, has been crystallized, and is distinct from pepsin.

Gastric lipase has little physiological significance. It is more consistently demonstrable in the gastric secretions of infants and children than of adults. It can be differentiated from regurgitated pancreatic lipase (which is often present in gastric contents) by a difference in pH optima. The optimal pH for gastric lipase varies from pH 5.5 to pH 7.9 (61), with a shift of the optimum 1.5 to 2.0 units to the acid side in the presence of calcium salts.

DIGESTIVE PROCESSES IN THE INTESTINE

Located in the duodenum and in the upper part of the jejunum, the duodenal glands (Brunner's glands) secrete a mucinous fluid which has a titratable alkalinity of about 0.03 N. The entry of acid gastric effluent into the intestine stimulates duodenal secretion, apparently by a liberation of a hormone designated as *duocrinin* (28). In this manner a part of the acidity is neutralized. Further neutralization is accomplished by the pancreatic juice, which is about 0.10 N in titratable alkalinity, and to a less extent by the bile, which in the human is about 0.01 N in titratable alkalinity, and by the secretion from the intestinal glands (crypts of Lieberkühn) which is of variable alkalinity. In spite of the considerable volume of alkaline secretions which enters the intestine, the intestinal contents do not become more than slightly basic in reaction. The average pH in the duodenum is 5.7 in fasting subjects, and slightly more acid, averaging 4.7, after meals. In the lower portion of the ileum the fasting pH is about 7.2, which is comparable to the pH of the feces on a normal mixed diet.

The external or digestive secretion of the pancreas is absent or at best intermittent in the fasting state. The total daily secretion (31) averages more than 1100 ml./day, a value which is decreased to 200 ml./day during fasting. Vagal stimulation has little effect upon the volume output, but increases the concentration of enzymes in the secretion. The stimulation of the copious production of alkaline pancreatic juice is supplied by the hormone *secretin*, which is liberated by the contact of acid with the mucosa of the duodenum. Secretin can be obtained for experimental purposes by acid extraction of intestinal, preferably duodenal tissue. It is a polypeptide of basic reaction, and has been crystallized. It is used in diagnostic studies as a pancreatic stimulant, producing as suggested above a large volume of juice, of high alkalinity but low enzyme concentration. Secretin also increases the volume of bile secreted by the liver. A second hormone, *pancreozymin*, can also be demonstrated in extracts of duodenal mucosa. The action of pancreozymin is similar to the result of vagal stimulation, causing increased output of pancreatic digestive enzymes with little alteration in volume or pH of the fluid. The action of pancreozymin is not one of vagus stimulation, since the action of the hormone is not abolished by atropine or other drugs which block cholinergic responses.

Although the secreting cells of the pancreatic acini resemble in a general way the cells of the salivary glands, the secretions are distinctly different. The pancreatic secretion contains no mucin, has a higher content of solids—about 1.3 per cent—is less viscous and more alkaline, pH 7.0 to 8.7. The alkalinity is the result of the relatively high concentration of bicarbonate, 65 to 102 mEq./liter. The total osmolar concentration of the pancreatic juice after intravenous injection of secretin is approximately equal to that of plasma. Bicarbonate and chloride concentrations show a reciprocal relationship, the sum of the two being nearly constant at 155 to 157 mEq./liter (18). Sodium and potassium are somewhat more variable, a value of 140 mEq./liter being representative for sodium and 7 mEq./liter for potassium. Although there appears to be only one type of externally secreting cells, the pancreatic secretion contains several enzymes or enzyme precursors, which will be listed and described.

Trypsin and *chymotrypsin*, the two chief proteases of pancreatic juice, are secreted in the form of the inactive zymogens, which are *trypsinogen* and *chymotrypsinogen*, respectively. Trypsinogen is autocatalytically converted to trypsin. It is also converted by *enterokinase*, an enzyme of the intestinal secretion. In either case a portion of the N-terminal end of the polypeptide chain of trypsinogen is hydrolyzed away. (Compare the conversion of fibrinogen to fibrin, see page 171.) The peptide removed from the trypsinogen molecule is termed *trypsin inhibitor*. Its structure is simple, consisting of an N-terminal valine and a C-terminal lysine with 5 or 6 residues of aspartic acid in between (17).

Chymotrypsinogen is converted to the active chymotrypsin by trypsin, but here the activation process is somewhat less simple. Chymotrypsinogen is essentially a cyclopeptide (5). The cyclopeptide is broken by hydrolyzing the arginyl-isoleucine link in a sequence so far determined to be leucyl-seryl-arginyl-isoleucyl-valine. This results in the arginine residue becoming a C-terminal amino acid of the now open-chain peptide and the isoleucine the N-terminal amino acid. Then the dipeptide, seryl-arginine, is hydrolyzed away from the C-terminal end, leaving behind active chymotrypsin.

Trypsin and chymotrypsin are proteinases or endopeptidases with optimal pH values between 8 and 9. They, as well as pepsin, are specific only for certain well defined types of peptide links. The peptide linkages selectively hydrolyzed in the presence of trypsin are those involving arginine or lysine; in the presence of chymotrypsin, those involving phenylalanine, tyrosine, tryptophane, or methionine. The nature of adjoining or near-by side chains determines whether a given peptide link is hydrolyzed or not (see page 243). Certain proteins which are resistant to pepsin, such as protamines and histones, are hydrolyzed in the presence of trypsin. Trypsin acts slowly upon undenatured collagen, albumin, globulin or hemoglobin, but acts rapidly upon the products of peptic digestion. The action of chymotrypsin is limited to products of peptic and tryptic digestion, except that milk-clotting comparable to that with pepsin occurs with chymotrypsin.

The end products of physiological tryptic and chymotryptic hydrolysis are peptides of lower molecular weight than the proteoses and peptones which are produced by peptic digestion. There is some liberation of free amino acids, particularly those with aromatic groups. The *carboxypeptidase* of the pancreatic juice catalyzes the hydrolysis of end peptide linkages with free carboxyl groups. Such hydrolysis liberates free amino acids. To summarize digestion of proteins by the enzymes of pancreatic juice, native proteins are slowly hydrolyzed and the products of peptic digestion are rapidly hydrolyzed to peptides of molecular weight less than 1000, with some liberation of free amino acids. Simultaneously with the hydrolysis of of protein, nucleic acids liberated by protein hydrolysis are hydrolyzed to oligonucleotides with the aid of *polynucleotidase*.

Immunological procedures applied to cytochemistry (46) have shown that chymotrypsinogen and procarboxypeptidase are located only in the apex of the acinar cell of the pancreas and in the pancreatic juice within the acinar lumen and duct systems. The nucleases occur in the cytoplasm of acinar as well as of other cells.

Pancreatic amylase or *amylopsin* has an optimal pH for activity between 6.3 and 7.2, requires Cl^- for its normal activity, and yields maltose by the hydrolysis of starch or dextrins. Human pancreatic amylase has been crystallized from aqueous solutions at pH 6.4 to 6.6, yielding a product with activity identical with that of human salivary amylase (48). The amylase activity of the pancreatic juice of very young infants is so low as to be negligible, hence it is not advisable to feed starchy foods at early ages. The amylase values increase slowly, and do not reach adult levels until the age of about two years.

Pancreatic lipase, also called *steapsin*, is a preformed enzyme with optimal pH between 7 and 8. It promotes the hydrolysis of fats to fatty acids and glycerol, attacking the fatty acids in the alpha-position preferentially, so that the steps involved are: triglyceride, 1,2-diglyceride, 2-monoglyceride, glycerol (46a). Although it is a true enzyme and not a zymogen, its activity is enhanced by bile salts and by Ca^{++}. This effect of bile salts is partly brought about by the emulsification of fats, which will be discussed in a later section. Bile salts also change the pH optimum of pancreatic lipase to values between 6 and 7, more congruent with the pH of the intestinal contents (8). Since pancreatic lipase is already active, its escape from the pancreatic ducts as a result of injury to the pancreas is followed by hydrolysis of fat in the tissues with which it makes contact. Such hydrolysis of tissue fats is spoken of as *fat necrosis.* The proteolytic enzymes of the pancreas can also cause damage, since they are self-activating, and their activation is catalyzed by contact with tissues. In addition to the ability to hydrolyze fats, pancreatic juice has enzymatic action on the hydrolysis or synthesis of cholesterol esters and of lecithin. The cholesterol esterase, like the lipase of the pancreas, is more effective in the presence of bile salts (65).

Increases in the lipase and amylase activity of blood serum occur in certain diseases of the pancreas, but normal values have been reported following total pancreatectomy (52). The trypsin-like proteolytic enzyme of the blood, plasmin (see page 175), is not generally considered to be of pancreatic origin. A simple test for fecal trypsin serves as a screening test for pancreatic deficiency (63). Amylase is normally present in the urine and may be increased in diseases of either parotid gland or pancreas.

In hereditary fibrocystic disease of the pancreas, trypsin, amylase, and lipase are absent from the pancreatic secretion. Since other proteinases are available, the absence of trypsin does not prevent the utilization of proteins. Monosaccharides and disaccharides are satisfactorily utilized. Most troublesome is the failure to hydrolyze fats, which interferes with their subsequent absorption and with the absorption of fat-soluble vitamins. This disease occurs in one in 600 live births in Caucasoids, is seldom observed in Negroes, and has never been noted in Mongoloids.

The digestive enzymes of the intestinal juice. The intestinal glands (crypts of Lieberkühn) produce and secrete the activator of trypsinogen, *enterokinase.* These glands also possess numerous other enzymatic activities, but there remains considerable doubt as to whether under physiological conditions the digestive enzymes are secreted into the lumen of the intestine as a part of the intestinal juice or whether the substrates enter the intestinal cells and are there subjected to enzymic action. By-passing the academic discussion concerning this point, it is safe to say that the enzymes of the intestinal cells are capable of accomplishing the hydrolysis of the nutritional disaccharides, sucrose, maltose, and lactose, and are probably capable of accomplishing the hydrolysis of starch. The lipolytic activity of the intestinal cells or their secretion is minimal, as indicated by the failure of fat hydrolysis following exclusion of the pancreatic secretion from the intestine. The intestinal cells contain or contribute *aminopeptidases*, *tripeptidases*, and *dipeptidases*, which in conjunction with the carboxypeptidase of the pancreatic juice can complete the hydrolysis of the products of peptic and tryptic digestion, yielding amino acids as the final product. The term *erepsin*, formerly applied to those peptidases when their activity was supposed to be that of a single enzyme, is no longer good usage. Intracellular proteinases and peptidases are also sometimes referred to as *cathepsins*. There is specificity among the dipeptidases, and tripeptidases are without significant action on dipeptides or tetrapeptides. *Arginase*, which catalyzes the hydrolysis of arginine into ornithine and urea, is typically a liver enzyme but has been identified in the intestinal cells. *Polynucleotidase* is present in intestinal cells as well as in the pancreatic secretion, catalyzing the hydrolysis of nucleic acid or fragments of nucleic acid into mononucleotides. The phosphate radical is hydrolyzed away from mononucleotides by *nucleophosphatase*, leaving purine or pyrimidine nucleosides. *Purine nucleosidase* has been identified in intestinal mucosa. The intestinal juice or *succus entericus* is a mucinous secretion, the amount of mucin depending upon local mechanical or chemical stimulation rather than upon neural or hormonal stimuli. *Enterocrinin* is liberated from intestinal mucosa upon contact with acid. Crystallized as a flavianate (29), it is a potent stimulator of the flow of intestinal juice, but does not correspondingly increase the output of digestive enzymes.

Intestinal lysozyme. Lysozyme depolymerizes certain amino polysaccharides which are obtained from definite species of bacteria, for example, *Micrococcus lysodeikticus* and *Sarcina lutea.* Not only does this enzyme destroy by hydrolysis these specific mucopolysaccharides *in vitro*, but it also attacks them as part of the parent organisms, thereby lysing them. It can, therefore, be classified as an antibiotic (see Chapter 22). In the human body we find lysozyme in the lachrymal secretion, in the mucus secretion of the respiratory passages, of the stomach, and of both large and small intestine. It is also present in human milk and in abnormal accumulations of body fluids such as the transudates which result from circulatory disturbances and the exudates which occur as the result of infections. Lysozyme can be obtained in crystalline form using the white of egg as a starting material, and lysozymes of plant and bacterial origin are also known. No substrate for lysozyme has been found in the tissues of the human body although considerable investigation has been directed along these lines. The lysozyme of tears has been shown to have a local damaging effect upon human colonic mucosa exposed through a colostomy opening (27). Increased amounts of lysozyme have been reported in the stools of patients with ulcerative disease of the intestine, in the gastric juice and in the mucous membrane of pylorus and duodenum of patients with peptic ulcer. Cultures of intestinal organisms do not produce lysozyme in amounts sufficient to explain intestinal lysozyme as a bacterial product. The emotional state of the patient or subject seems to have some bearing upon the amount of lysozyme eliminated in the stools (22). Lysozyme does not hydrolyze the mucus of gastric juice nor does it disturb the production of mucus by surface epithelial cells (68).

Intestinal gases consist of swallowed air, which may be altered in its composition by exchange with blood gases, and which receives additions, chiefly of H_2, CO_2, and CH_4, from bacterial action in the large intestine. Kirk (37) found an average output of colonic gases of 1.48 ml. per minute from twenty normal subjects, using collection periods of 3 to 10 hours. The average composition of the collected flatus was 9.0 per cent CO_2, 3.9 per cent O_2, 7.2 per cent CH_4, 20.9 per cent H_2, and 59.0 per cent N_2. The concentration of H_2S ranged from zero to 0.0017 per cent. Combustible gases, H_2 and CH_4, may be present in explosive concentrations.

BILE

No digestive enzymes are secreted by the liver into the bile. The bile contains certain substances which in other ways contribute to the digestive process, as well as some substances which are purely waste products. Since the bile notably influences the composition of the intestinal contents, it seems proper at this point to discuss its components and their origins. The average daily human output of bile is estimated at 500 ml. per 24 hours. This output has never been measured in the human under strictly physiological conditions. The human gall bladder contains approximately 50 ml. of bile, which is there subjected to a process of concentration by the removal of water. The normal gall bladder adds nothing to the bile except mucin. During the time that bile remains in the gall bladder the concentrations of most of its components increase from four to tenfold. Contraction of the gall bladder is stimulated by a hormone *cholecystokinin*, which is liberated from upper intestinal mucosa by contact with fats, fatty acids, peptone, or dilute HCl.

The *bile acids* are steroids which in addition to one or more hydroxyl groups on the CPP nucleus possess a side chain on carbon 17 which ends with a carboxyl group. *Cholic acid* (formula II), the most abundant bile acid, has hydroxyl groups on carbons 3, 7, and 12, of which the hydroxyl group on carbon 3 is alpha in configuration, unlike those of the sterols.

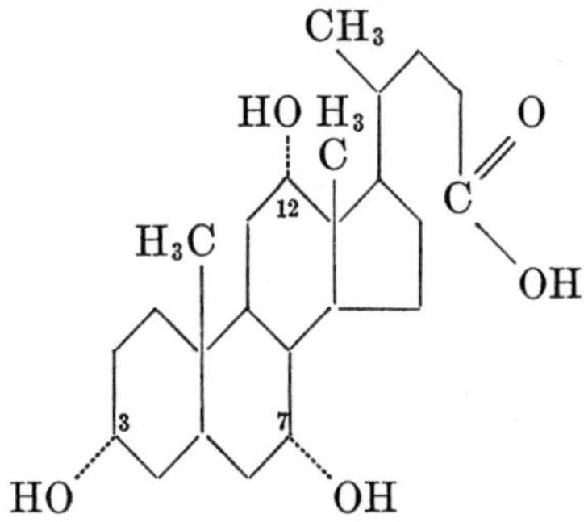

II. Cholic acid

Two other bile acids, found in man, are *deoxycholic acid* (formula III) and *chenodeoxycholic acid* (formula IV), the former possessing hydroxyls on carbons 3 and 12, the latter on carbons 3 and 7. (9a)

Cholic acid in its free state has been reported as a minor bile constituent under pathological conditions, but usually it is found almost entirely in conjugation with glycine and to a lesser extent in man with the amino sulfonic acid, *taurine* (formula V). These conjugates are termed *glycocholic*

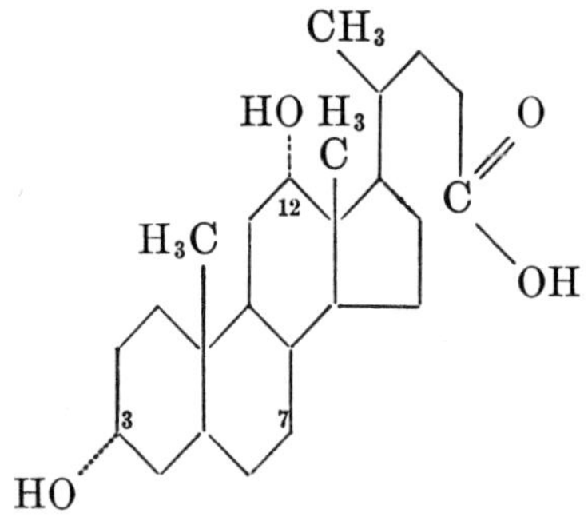

III. Deoxycholic acid

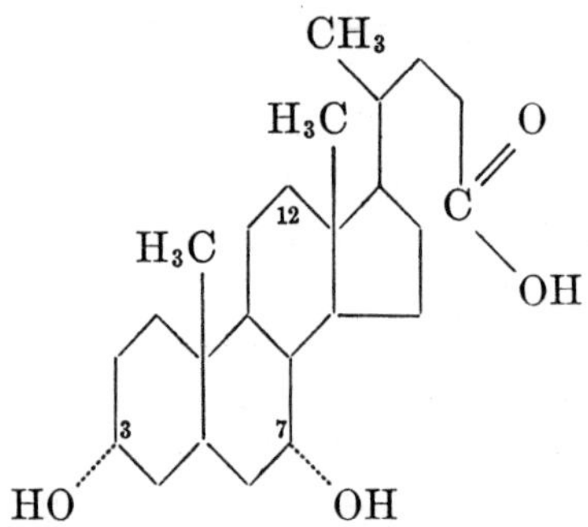

IV. Chenodeoxycholic acid

acid and *taurocholic acid*, respectively. Conjugation takes place by the reaction, first, of cholic acid with coenzyme A (see page 804) in the presence of ATP. The resulting high-energy *cholylcoenzyme A* reacts with taurine to form taurocholic acid. In the bile, these acids exist chiefly as ionized salts, the *bile salts* (formulas VI and VII).

Cholic acid is formed in the liver from cholesterol. The conjugation with glycine or with taurine also takes place in the liver. Taurine is formed from cysteine or other S-containing precursors. Cholate can be measured in the blood (62), and has been found to range between 0.2 and 3.0 mgm. per

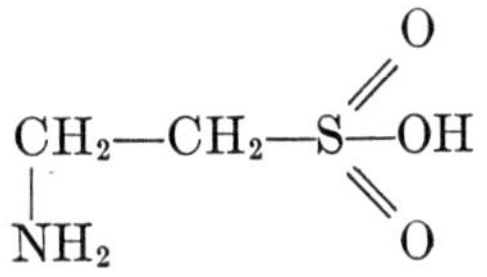

V. Taurine

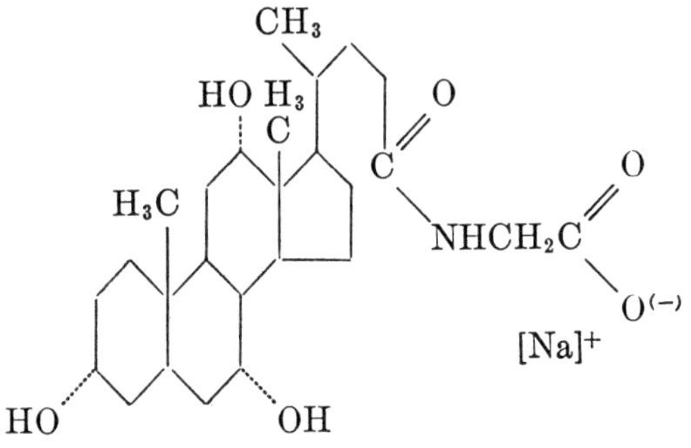

VI. Sodium glycocholate

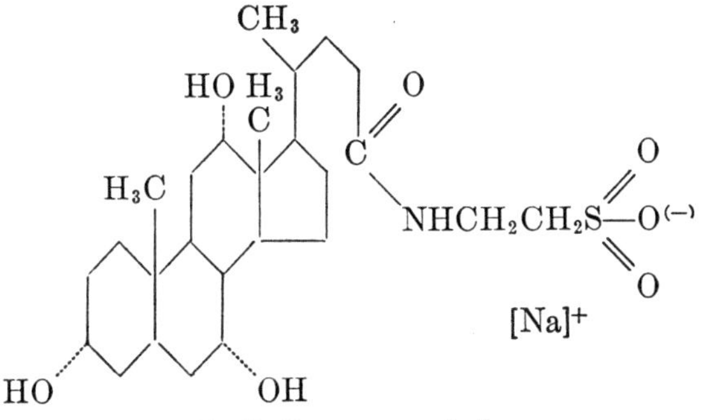

VII. Sodium taurocholate

100 ml., without great variation in individuals. Values are increased in hepatitis and particularly in obstructive jaundice.

The bile acids and bile salts are concerned in the digestion and absorption

of lipids. The fats and sterols, notoriously insoluble in aqueous solutions, are dispersible in aqueous solutions of bile acids or bile salts. This is partly a matter of the lowering of the interfacial tension permitting stability of fine emulsions, and partly the formation of soluble complexes. Thus the bile acids and their salts aid in the emulsification of lipids prior to digestive hydrolysis. Fine emulsions of lipids expose a greater surface to the action of the hydrolyzing enzymes, and therefore digestion of such emulsions proceeds faster than the digestion of coarse suspensions. After hydrolysis, soluble complexes of bile acids with fatty acids form, facilitating the absorption of fatty acids from the intestine. Exclusion of bile from the intestine greatly decreases both the hydrolysis and absorption of fats. About 90 per cent of the total bile acid output is returned to the body in the form of complexes with absorbed fatty acids, constituting the *enterohepatic circulation* of bile acids. The presence of bile salts is necessary for the effective absorption of carotene (see page 135) and the fat-soluble vitamins.

Bile salts are *surface active* since they possess polar and non-polar groups. In taurocholic acid, for instance, the CPP portion of the molecule, being largely hydrocarbon, is oil-soluble and water-insoluble. The taurine portion is highly polar and consequently oil-insoluble and water-soluble. If a small quantity of bile salt is added to water, the individual molecules will tend to orient themselves on the surface in such a way that a monomolecular film is formed in which the taurine (or glycine) portion is immersed in the water, while the CPP nucleus remains in the air. Thus the surface tension of the system approaches that of CPP/air, which is much lower than the original value for water/air. For this reason, extension of the area of water surface requires a considerably smaller input of energy in the presence of bile salts than in their absence. Thus, if water containing bile salt is shaken, a relatively stable foam is produced. Other substances, notably the *soaps* (alkali salts of fatty acids), likewise possess surface activity and such compounds are referred to collectively as *detergents*.

Actually, surface activity is but a special case of a more general phenomenon. If bile salts are added to a system containing two phases, an aqueous and an oily one, they tend to collect at the phase boundary. Again the taurine or glycine portion of the molecule is oriented so as to dip into the water. The CPP portion, on the other hand, extends into the oil. Actually, the characteristic property of such substances is not merely surface activity, but rather "phase boundary activity." Here again, the interphase energy is lowered and extensions of its surface now require comparatively little energy. Upon shaking such a system, an emulsion tends to form. In such an emulsion a multitude of tiny oil droplets is suspended in the aqueous medium, each droplet being surrounded by a layer of properly oriented bile salt. The possibility of a system composed of water droplets in an oily medium is not excluded. The choice between the two possibilities rests on the relative quantity of each phase and on the nature of the surface-active agent. It is by such emulsifying tendencies that ordinary soap, for instance, renders greasy or oily particles more "soluble" in water, and thus exhibits cleansing properties. Similarly, bile salts aid the digestion and absorption of fats and fat-soluble vitamins by their emulsification.

Lipids of the bile may occur in concentrations well over one per cent in healthy gall bladder bile. The concentration in liver bile is variable and much less. Phospholipids, neutral fat, and free fatty acids have been reported in highly variable amounts. *Cholesterol* is not always the most abundant lipid in normal bile, but its pathological interest has caused much study to be centered upon it. It is present in normal liver bile in concentrations somewhat below those observed in blood. Liver bile contains from 0.02 to 0.15 per cent cholesterol. In normal gall bladder bile the concentration may reach 0.6 per cent. In patients with cholesterosis of the gall bladder (deposits of cholesterol in the mucous membrane) or with cholesterol gallstones, the concentration of cholesterol in the fluid portion of gall bladder bile may be 2 per cent or more. Under these circumstances, crystals of cholesterol may be detectable by microscopic examination of gall bladder bile obtainable from the intact patient by intubation of the duodenum followed by stimulation of gall bladder contraction by oleic acid or magnesium sulfate. *Gallstones* have been observed only in man and in domesticated animals. In the human patient, gallstones are usually composed in whole or in part of cholesterol. The disturbances leading to cholesterol stone formation are multiplex and beyond the scope of this text. They include the possibilities of overproduction of cholesterol, diminished solubility of cholesterol as a result of decreased acidity, bile acid content, or fatty acid content of the bile, and local alterations in the gall bladder as a result of infection. A less common type of gallstone in man, but more frequent in cattle and hogs, is composed chiefly of calcium salts and bile pigment.

Calcium is present in liver bile in a concentration equal to or slightly less than that in blood plasma, 5 milliequivalents per liter. Its presence in the intestine enhances the activity of pancreatic lipase and possibly of other digestive enzymes. It is probable that some Ca^{++} is absorbed from the gall bladder, since normally only a sixfold concentration occurs there.

Mucin is contributed chiefly by the gall bladder mucosa. Liver bile contains practically no mucin, but bile from the gall bladder contains up to 4 per cent. When the cystic duct is obstructed the mucinous secretion accumulates in the gall bladder, constituting the "white bile" of surgical terminology. If the sphincter of Oddi is rendered non-functional in experimental animals, the gall bladder does not fill with bile and the mucinous secretion collects and may be analyzed (66). This secretion contains no cholesterol.

Bile pigments are waste products derived from the porphyrin ring, chiefly of hemoglobin. Concentrations of bile pigment are observed in liver bile up to 0.07 per cent and in gall bladder bile up to one per cent. Higher values have been reported post mortem (22). The bile pigments take no known part in digestion or absorption of foodstuffs. They have considerable pathological significance, however, and warrant detailed consideration. Their metabolic origin is from protoporphyrin, particularly the protoporphyrin IX of hemoglobin (see page 191). Bile pigment formation is the channel for disposal of the porphyrin portion of the hemoglobin of worn-out red cells.

In hemoglobin and other heme derivatives we are dealing with a metal complex of protoporphyrin IX. The metal is iron, and the metalporphyrin complex is coupled to the protein globin. When sterile blood is allowed to stand, bile pigment and iron appear in increasing amounts (3). Conversion of blood pigment to bile pigment takes place in sterile blood at a rate too slow to account for the daily output. Since blockade of the reticulo-endothelial systems leads to diminished production of bile pigments in isolated spleen and liver, and since it is known from histological studies that cells of this system are phagocytic for red cells, the chief site of bile pigment formation is considered to be within reticulo-endothelial cells, including the Kupffer cells of the liver.

Biliverdin (formula VIII), which is protoporphyrin IX with the alpha methene bridge oxidatively ruptured, is the bile pigment first formed from

```
        M  V    M  P    P  M    M  V
HO—      N    C    N    C    N    C    N    —OH
              H    H    H         H
```

VIII. Biliverdin

the heme of hemoglobin. The ring breakage occurs while the porphyrin is still combined with iron and globin, and the bile pigment remains so combined, with the ring possibly still closed by a hydrogen bond. Such compounds of bile pigment with iron and protein are designated by Lemberg as verdohemochromes. This particular compound, where the protein is globin, is *verdohemoglobin*. The iron of verdoheme compounds is labile and is next split off, leaving biliverdin bound to the protein globin, forming *biliverdinglobin*. Biliverdinglobin acts as a hydrogen acceptor to lactic acid dehydrogenase and other dehydrogenase systems acting upon carbohydrate intermediates. By accepting two hydrogens biliverdinglobin becomes *bilirubinglobin*. Simple biliverdin added to blood is not reduced.

The measurement of bilirubin (formula IX) in the blood plasma or serum is usually carried out by some modification of the Hijmans van den

M V M P P M M V

HO—N—C—N—C—N—C—N—OH

H H H_2 H H

IX. Bilirubin

Bergh reaction, which consists of adding a mixture of sulfanilic acid and sodium nitrite to a solution of bilirubin. The bilirubin is thereby coupled with the diazotized sulfanilic acid, forming a much more deeply colored substance, *azobilirubin*. This reaction is weak or negative when applied directly to normal human serum—the *direct* Hijmans van den Bergh reaction. It is definitely positive when applied to normal human serum previously treated with alcohol—the *indirect* Hijmans van den Bergh reaction. Both direct-reacting and indirect-reacting bilirubin have been crystallized (51) and show different properties. Indirect-reacting bilirubin becomes direct-reacting by standing in solution at pH 10.5, and the reverse change occurs at acid pH values. The difference between the two types of bilirubin is not related to combination with protein, but rather to the formation of a complex of bilirubin with a metal (Mg, Ca, Cu, Fe, or Mn). Such complexes form in bile, where high pH and high metal ion concentration favor the reaction. The metal is bound to the pyrrole nitrogens, distinguishing the complexes from simple salts of bilirubin. Direct-reacting bilirubin is the metal complex, indirect-reacting bilirubin the metal-free bile pigment (51) Direct bilirubin changes to indirect bilirubin with versene (EDTA). This is further evidence that direct bilirubin in serum is a bilirubin-metal complex (11). There is also evidence that bilirubin conjugated with glucuronic acid is direct-reading (59).

Direct-reacting bilirubin increases in blood serum in cases of obstructive jaundice or regurgitation jaundice, when bile pigment which has once passed through the liver cells is returned to the blood. The most common cause of such jaundice in a pure form is obstruction of the common bile duct, either by a gallstone or by pressure of a pancreatic tumor. The Hijmans van den Bergh reaction can be applied quantitatively (44), showing 0.2 to 0.8 mgm. of total bilirubin per 100 ml. of normal human serum, all or nearly all of which is indirect-reacting. In pure obstructive jaundice or regurgitation jaundice the increased pigment in the serum is direct-reacting bilirubin; in pure hemolytic jaundice the increase is in indirect-reacting bilirubin.

Urobilins, which are fecal and urinary pigments, and *urobilinogens* which are colorless derivatives of bile pigments, result from further reduction of bilirubin in the intestine, chiefly by the action of bacteria. The vinyl side chains of bilirubin are reduced to ethyl, and one or more of the remaining methene bridges become methylene bridges. The terminology of individual compounds in these groups is complex (40). For practical purposes it seems adequate to designate them simply as the colored urobilins and the colorless urobilinogens. The inclusive fecal output of these substances is from 50 to 280 mgm. per day. A few milligrams of these substances are normally excreted in the urine, as a result of absorption of the substances from the intestine and their circulation in the blood. The bulk of the absorbed urobilins and urobilinogens is not excreted in the urine, but is re-excreted in the bile. Absence of urobilinogen or urobilin in the urine is therefore, in a jaundiced patient, evidence that the jaundice is obstructive in its origin, in other words that bile pigment is not entering the intestine. This test may be misleading if there is obstructive jaundice with *E. coli* infection above the obstruction in the bile ducts. In such a case reduction of bilirubin will occur at the site of the infection and urobilins and urobilinogens will appear in the urine even though there is complete obstruction. Also in very severe jaundice bilirubin may enter the intestine by diffusion in amounts adequate to form these reduction products. The usual clinical test made on urine specimens is for urobilinogen, by adding a solution of *p*-dimethylaminobenzaldehyde in strong HCl.

Bilirubin itself is not a normal urinary component. It appears in the urine in jaundice, particularly in regurgitation jaundice. It is commonly tested for by simple inspection, noting an abundant yellow or brown foam as opposed to the scanty white foam of normal urine.

The mechanisms just described for the formation of bile pigment from the heme of hemoglobin explain the production of almost 90 per cent of the normal output of bile pigment. The origin of the remainder is not altogether clear. Some may come from the breakdown of other heme compounds such as myoglobin, catalase, and the cytochromes. When isotopically tagged glycine is administered to a human subject, the incorporation of glycine into the heme of hemoglobin can be demonstrated, and also the formation of isotopically tagged bile pigment after the lapse of the normal life span of the red cells. A portion of the isotope appears in bile pigment, however, in the early days of the experiment, indicating formation of bile pigment other than from the heme of degraded erythrocytes. This early output of isotope in bile pigment is of greater proportion in pernicious anemia, and still greater in congenital porphyria. In a case of congenital porphyria (42) 31 per cent of the bile pigment excreted was demonstrated by the isotope technique to originate from sources other than the destruction of mature circulating red cells. This suggests that the body utilizes glycine to form porphyrin derivatives which are not utilized in the production of the normal heme compounds, and that these are rather promptly converted to bile pigment and excreted. The fact that isotopically tagged protoporphyrin, injected intravenously into the dog, yields tagged bile pigment (43) makes this suggestion highly probable.

ABSORPTION

Although the stomach is not anatomically adapted to the function of absorption, moderate amounts of many simple substances enter the bloodstream by way of the gastric mucosa. Most foodstuffs are absorbed from the small intestine, which is well adapted to this function by its great length and by its specialized epithelial lining. Because of the villi the absorbing area is much larger than would be expected from the length and size of the small intestine. Weight is maintained and absorption not significantly interfered with if the proximal 50 to 70 per cent of the small intestine of a dog is removed. Loss of the distal 50 per cent interferes with fat absorption and causes loss of weight (38). Absorbed foodstuffs may leave the small intestine by one of two paths. The first path is by way of the blood capillaries in the walls of the intestine and especially those in the villi, thence to the mesenteric veins and the portal vein. Any absorbed material which takes this path passes through the liver before it enters the general circulation. This is important because the liver is able to adapt the concentration and the chemical structure of many of the absorbed materials for utilization in other tissues. The second path of absorption is by way of the lymph vessels of the intestine, the large lacteals, and the thoracic duct. The materials which take this path reach the blood more slowly but more directly because the thoracic duct empties into the venous system near the heart.

Absorption of Inorganic Substances

Not enough water is removed by the small intestines to cause any notable decrease in the fluidity of the intestinal contents. Thus the material which passes through the ileocecal valve has about the same consistency as the chyme, which passes through the pylorus. This is not because no water is absorbed in the small intestine, but rather because copious secretions pour into the intestine and contain in them sufficient water to compensate for that which has been absorbed.

The digestive secretions also contain considerable amounts of inorganic ions (Na^+, K^+, Ca^{++}, Cl^-, $HPO_4^=$) which are added to those of the ingested foodstuffs, and which are in large measure reabsorbed. Balance experiments, and experiments with isolated intestinal loops, indicate that water and most inorganic ions are able to move freely in both directions across the intestinal barrier. The rate of absorption of Ca^{++} is measurably slower than that of most other inorganic ions. The reasons for this difference in rate are obscure (53). When the diet is adequate in Ca^{++}, its rate of absorption depends more upon the vitamins D (see page 829) than upon any other known factor. The rate of absorption of iron is determined by a special mechanism involving ferritin (see page 695).

Absorption of Protein

The amino acids formed by digestive hydrolysis of protein foods are diffusible and sufficiently water-soluble to allow their prompt entry into both blood and lymph vessels of the intestinal villi. In the case of at least some L-amino acids, such as alanine, phenylalanine, and histidine, active absorption takes place (i.e., absorption against a concentration gradient (24)). The mechanism by which this is achieved requires energy, naturally, and active absorption ceases if the respiration of the intestinal cells is interfered with. Glutamic acid and aspartic acid are not actively absorbed but glutamine and asparagine are. Glutamine is absorbed unchanged, while asparagine is hydrolyzed to aspartic acid and ammonia in the process. The greater portions of the amino acids are carried by the blood vessels, which fact may simply reflect the more rapid rate of blood flow as compared with lymph flow. The removal of amino acids from the intestinal contents is so effective that the limiting factor appears to be the rate of amino acid liberation by protein hydrolysis in the intestine.

Proteoses and peptones are also water-soluble and more or less diffusible. They are taken up by intestinal cells, within which further hydrolysis occurs. Analytically detectable increases of polypeptide concentration occur in the portal blood of experimental animals following protein meals. The changes in amino acid concentration of systemic blood following protein absorption (see Chapter 16) are more uniform and predictable, leading to the conclusion that the absorption of protein is chiefly in the form of amino acids.

There is nevertheless evidence that molecules larger than amino acids may be absorbed. For instance, strepogenin (70), which is a polypeptide, passes into the blood. Small amounts of intact protein are absorbed.

Absorption of Carbohydrates

We have seen that carbohydrates are digested to monosaccharides. Sugars other than monosaccharides are not absorbed from the intestine unless taken in relatively high concentration, and if so absorbed, they are promptly eliminated by the kidneys. For instance, if sucrose is eaten in large amounts, it may be detected in the blood and in the urine. Normally, however, the sugars which are absorbed are monosaccharides, and the pathway of absorption is through the portal circulation to the liver.

These simple sugars are absorbed at different rates. Experiments with intact rats (15) and with isolated guinea pig intestine (16) agree that glucose and galactose are absorbed most rapidly, and at approximately the same rate. Fructose is absorbed at about half the rate of glucose or galactose, with conversion of a large part of the fructose to glucose (in the guinea pig experiments) during the passage of the sugar across the intestinal mucosa. Mannose, sorbose, and pentoses are absorbed at much slower rates, consistent with simple diffusion.

The much greater speed with which glucose and galactose are taken up suggests that there must be some special mechanism for their absorption. A plausible theory which has not yet been supported by completely adequate evidence is that the sugars are phosphorylated in the intestinal mucosal cells. If glucose and galactose are so phosphorylated, this would explain their rapid rate of absorption because they would be converted, promptly after their diffusion into the cells, to less diffusible substances, and diffusion into the cells from the higher concentration in the intestine would continue unrestricted. The idea that glucose and galactose may be phosphorylated is supported by the observation that monoiodoacetic acid and phlorizin, which prevent phosphorylation, also delay absorption. The hypothesis of phosphorylation of sugars as a part of their mechanism of intestinal absorption is controverted by the observation (13) that 1-deoxyglucose and 6-deoxyglucose are concentrated in hamster intestinal serosa. There are no known enzyme systems which would catalyze the phosphorylation of these deoxysugars. Absorption of glucose or galactose is not quantitatively associated with transfer of phosphate, so it must be assumed that phosphorylation and dephosphorylation take place within the intestinal cells, and that the cellular phosphate is conserved and re-cycled.

The Absorption of Lipids

Free fatty acids may form a small part of the diet, but for the most part they are combined as neutral fat or more complex lipids. The salts of fatty acids, which are called soaps, are probably rarely present as they do not exist as such except at a pH higher than 9, and this degree of alkalinity does not occur in the gastrointestinal tract. The mechanism of absorption of lipids has been hotly debated and there is still more than one point of view. There is no question but that free fatty acids are liberated by the action of digestive lipases. This hydrolysis has been considered to be complete, or nearly so, with no difference between animal and vegetable fats. Absorption of the fatty acids occurs exclusively in the small intestine, aided by the bile acids. The fatty acids are, with few exceptions, water-insoluble, but can combine with bile acids to form water-soluble complexes which can enter the surface of the absorbing cells of the intestine. Presumably the water-soluble complex is broken up within the cell membrane. The bile acid portion is in part extruded back onto the intestinal surface of the cell and in part returned to the liver by way of the portal circulation. The fatty acids are esterified to fats and phospholipids within the intestinal cell and removed chiefly (70 per cent or more) by way of the intestinal lymphatics and the thoracic duct. The intestinal lymphatics are the main channel for the absorption of 12 carbon and longer fatty acids. Those with shorter chains, when traced by C^{14} tagging, appear only in limited amounts in the lymphatics, and for the most part are transported by the blood from the site of absorption (6). *Glycerylphosphorylcholine* can be identified in the intestinal cells and can be presumed to combine there with some of the fatty acids and pass them along in the form of lecithin, the remainder being resynthesized to triglycerides. Further formation of phospholipid occurs in the liver (see page 596).

Frazer (23) has dissented from the concept of fat absorption just stated. He postulates partial hydrolysis of fats (30 per cent or less) and the absorption of a complex of residual fatty acid monoglycerides and diglycerides with bile acids by way of the lymphatics, and direct transport of fatty acids by way of the portal vein to the liver.

Investigation of the contents of the intestinal lymphatics of animals following feeding of glycerides differently labeled in the glycerol and fatty acid portions (51) showed that 25 to 45 per cent of the fat was completely hydrolyzed, and was resynthesized, the labeled fatty acids appearing in the lymphatics esterified with unlabeled glycerol. The labeled glycerol, set free in this hydrolysis, was not used in the resynthesis, but was independently metabolized. The remaining 55 to 75 per cent of the original labeled fat was hydrolyzed to monoglycerides. The phospholipids of the lymphatic contents were found to be formed approximately half from absorbed monoglycerides and half from absorbed fatty acids and unlabeled glycerol.

Phospholipids, lecithin in particular, may be hydrolyzed by pancreatic enzymes before absorption. It has been demonstrated that phospholipids also may be resynthesized in the intestinal walls. Cholesterol appears to be absorbed in two forms, in a soluble complex with bile salts, and as cholesteryl esters formed with higher fatty acids. These esters are soluble and are regularly found in the blood serum and have also been found in chyle. Plant sterols, even when very similar structurally to cholesterol, are poorly absorbed. Relatively little cholesterol as compared with fat is absorbed by way of the lymphatics. The absorption of cholesterol is not a critical process in human nutrition, since the body can synthesize its own cholesterol in adequate supply (see Chapter 15).

Sprue is a disease in which absorption of foodstuffs from the small intestine is notably diminished, with resulting nutritional failure, glossitis, and macrocytic anemia. The absorption of fatty acids, glycerol, and sugars is particularly affected. The stools are bulky and contain excess fat (*steatorrhea*). Diminished phosphorylation may be demonstrated, but the cause of the disturbance has not been clearly defined, and may be multiple (20) The defect in fat absorption in *celiac disease*, a sprue-like condition affecting infants is apparently related to the presence of wheat gluten in the diet (1).

"DETOXICATION"

Within the alimentary canal, and particularly in the large intestine, food is subject to bacterial action. This results in the production of certain food factors such as vitamin K and several of the B vitamins. Bacterial action also results in the production of compounds that are of no value to the body, and are not absorbed in quantity under normal conditions. Indeed, these products are frequently unpleasant in odor and sometimes even toxic. An example of such products are indole and skatole (formula X), degrada-

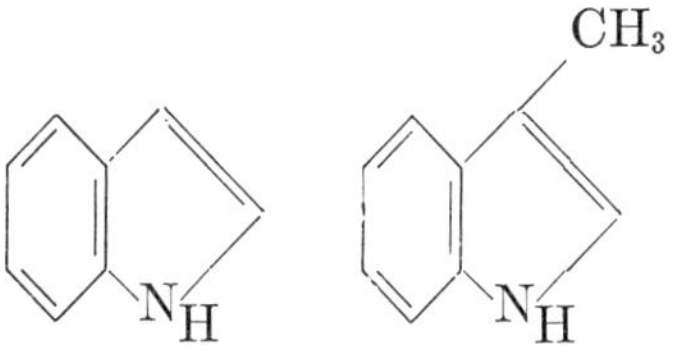

CH_3

NH NH

Indole Skatole

X. Putrefactive products of triptophane

tion products of tryptophane through bacterial action, which are responsible in large part for the characteristic odor of feces.

Small quantities of indole and skatole are absorbed and excreted by way of the kidneys. The same is true of other undesirable putrefactive products that find their way into the blood stream. The same is also true of many foreign compounds or drugs, not ordinarily part of the body's internal environment, which are introduced by accidental ingestion, or deliberately for their therapeutic value.

These foreign substances, before excretion, frequently undergo chemical change. Indole and skatole, for instance, are conjugated with sulfuric acid and excreted in the urine in the form of sulfate esters, the so-called *ethereal sulfates*. The changes undergone in this or other ways, often, but not always, result in compounds that are less toxic than the original or easier to excrete or both. For this reason the process has come to be known as "detoxication".

Actually, the term is misleading and should be used with caution. The process by which a particular toxic substance is changed in the body prior to excretion is largely random. If it fits an enzyme system, the toxic substance will undergo whatever chemical change that enzyme system is equipped to catalyze. If this change reduces toxicity, well and good. For instance, brombenzene, if administered to a dog, is combined with cysteine (formula XI) to form the relatively nontoxic para-bromophenylmercapturic acid.

$$C_6H_5Br + HSCH_2CH(NH_2)COOH + CH_3COOH \xrightarrow{-2H} Br\text{-}C_6H_4\text{-}SCH_2CH(NHCOCH_3)COOH + H_2O$$

Brombenzene, Cysteine, Acetic acid → Para-bromophenyl-mercapturic acid

XI. "Detoxication" by conjugation with cysteine

The reactions included in "detoxication" are oxidation, reduction, hydrolysis and conjugation. Conjugation is by far the most important and the above cases are examples. (The formation of the mercapturic acid also involved an oxidation.)

Cysteine and glycine are the two amino acids important in detoxication. Glycine is well known for the manner in which it will conjugate with ingested benzoic acid to form hippuric acid (formula XII). Hippuric acid is formed in the liver, the kidneys, and possibly elsewhere. When the liver is damaged, as by disease or chemical poisoning, the output of hippuric acid in the urine after a test-feeding of benzoic acid is greatly diminished. Lowered hippuric acid excretion may also, of course, be the result of renal damage.

More important than conjugation with either of these amino acids is conjugation with glucuronic acid. Glucuronic acid does not occur free in the body and, consequently, conjugation does not take place directly. Glucuronic acid, however, does occur as part of a uridine coenzyme (see page 251) through the action of which it can be transferred to various ac-

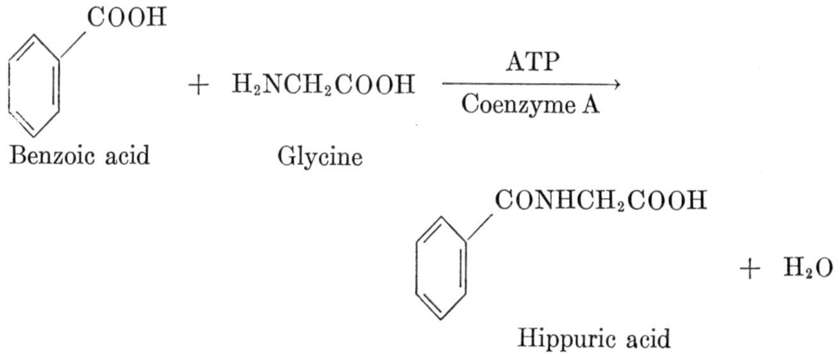

COOH

$+ H_2NCH_2COOH \xrightarrow[\text{Coenzyme A}]{\text{ATP}}$

Benzoic acid Glycine

$CONHCH_2COOH$

$+ H_2O$

Hippuric acid

XII. "Detoxication" by conjugation with glycine

ceptor compounds. This may be the mechanism which, physiologically, is used in the biosynthesis of mucopolysaccharides. If the enzyme system, as is quite possible, is not particularly specific with respect to the nature of the acceptor molecule, a number of foreign compounds can substitute for the physiological acceptor molecules. *Glucuronides* will in this way be formed. Phenol is thus conjugated in part to phenol glucuronide and excreted in this form (formula XIII). The bonding of glucuronic acid to one

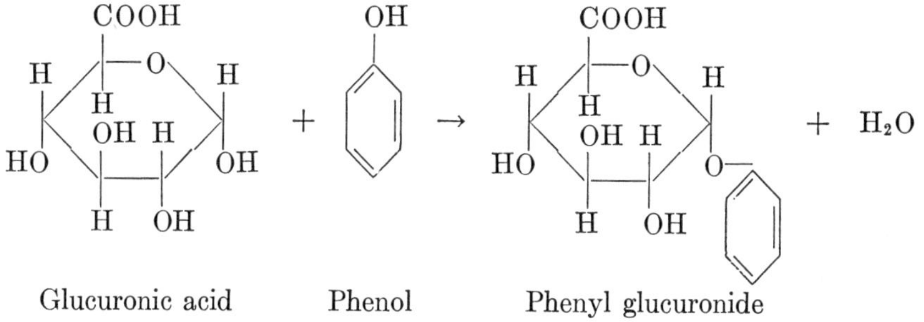

COOH OH COOH

H —O H H —O H

H H

OH H + → OH H + H_2O

HO OH HO O—

H OH H OH

Glucuronic acid Phenol Phenyl glucuronide

XIII. "Detoxication" by conjugation with glucuronic acid

of the active groups of the toxic molecule makes the latter less reactive with other substances and therefore tends to reduce the toxicity. The polar groups of the glucuronic acid contribute to the water-solubility of the conjugate and facilitate excretion.

Calcium carbonate and calcium phosphate, insoluble in water at neutral or slightly alkaline hydrogen-ion concentrations, are solubilized in the presence of glucuronides (45) at comparable pH values. Solutions of complex uronosides such as L-menthol-D-glucuronide and 8-hydroxyquinoline-D-glucuronide have greater solvent effect than free glucuronic acid or its salts. The deliberate production of glucuronides by administration of salicylic acid derivatives to patients has been used to produce a urine in which calcium phosphate will remain dissolved where, in the absence of glucuronides, it would precipitate, thus inhibiting the formation of calcium phosphate calculi (56).

REFERENCES

1. ANDERSON, C. M., *et al.* Coeliac disease. Gastro-intestinal studies and the effect of dietary wheat flour. Lancet, **262:** 836–842, 1952.
2. ANREP, G. V., AND CANNAN, R. K. The metabolism of the salivary glands. II. The blood sugar metabolism of the submaxillary gland. J. Physiol., **56:** 248–258, 1922.
3. BARKAN, G., AND WALKER, B. S. The red blood cell as a source of the iron and bilirubin of the blood plasma. J. Biol. Chem., **131:** 447–454, 1939.
4. BEAUMONT, W. *Experiments and Observations on the Gastric Juice and the Physiology of Digestion.* Plattsburg, New York, F. P. Allen, 1833.
5. BETTLEHEIM, F. R., AND NEURATH, H. The rapid activation of chymotrypsinogen. J. Biol. Chem., **212:** 241–253, 1955.
6. BLOOM, B., *et al.* Intestinal lymph as pathway for transport of absorbed fatty acids of different chain lengths. Am. J. Physiol., **166:** 451–455, 1951.
7. BOCKUS, H. L. *Gastroenterology.* Vol. I. *The Esophagus and Stomach,* pp. 198–230. Philadelphia and London, W. B. Saunders Co., 1943.
8. BORGSTRÖM, B. Effect of taurocholic acid on the pH-activity curve of rat pancreatic lipase. Biochim. et Biophys. Acta, **13:** 149–150, 1954.
9. CANNON, W. B. The influence of emotional states on the alimentary canal. Am. J. Med. Sci., **137:** 480–487, 1909.

9a. CAREY, J. B. JR. Chenodeoxycholic acid in human blood serum. Science, **123:** 892, 1956.

10. CARLSON, A. J. Contributions to the physiology of the stomach. I. The character of the movements of the empty stomach in man. Am. J. Physiol., **31:** 151–168, 1912.
11. CHILDS, B. The nature of direct bilirubin. Bull. Johns Hopkins Hosp., **97:** 333–342, 1955.
12. CHRISTENSEN, L. K. The pH optimum of peptic hydrolysis. Arch. Biochem. Biophys., **57:** 163–173, 1955.
13. CRANE, R. K., AND KRANE, S. M. On the mechanism of the intestinal absorption of sugars. Biochim. et Biophys. Acta, **20:** 568–569, 1956.
14. CRIDER, R. J., AND WALKER, S. M. Effect of intravenously administered amino acids on the stomach of a woman with a gastric fistula. Arch. Surg., **57:** 10–17, 1948.
15. CORI, C. F. The fate of sugar in the animal body. I. The rate of absorption of hexoses and pentoses from the intestinal tract. J. Biol. Chem., **66:** 691–715, 1925.
16. DARLINGTON, W. A., AND QUASTEL, J. H. Absorption of sugars from isolated surviving intestine. Arch. Biochem. Biophys., **43:** 194–207, 1953.
17. DAVIE, E. W., AND NEURATH, H. Identification of the peptide split from trypsinogen during auto-catalytic activation. Biochim. et Biophys. Acta, **11:** 442, 1953.
18. DREILING, D. A., *et al.* The secretion of electrolytes by the human pancreas. Gastroenterology, **30:** 382–390, 1956.
19. DREIZEN, S., *et al.* Buffer capacity of saliva as measure of dental caries activity. J. Dent. Res., **25:** 213–222, 1946.
20. DURANT, T. M., AND ZIBOLD, L. A. Sprue: A consideration of etiology, differential diagnosis, and management. M. Clin. North America, **33:** 1671–1680, 1949.

21. Edkins, J. S. The chemical mechanism of gastric secretion. J. Physiol., **34:** 133–144, 1906.
21a. Elliott, W. H. Enzymic synthesis of taurocholic acid: a qualitative study. Biochem J., **62:** 433–436, 1956.
22. Elton, N. W. Bilirubin concentrations in the human gallbladder. Am. J. Clin. Path., **6:** 81–90, 1936
23. Frazer, A. C. Mechanism of intestinal absorption of fat. Nature, **175:** 491, 1955.
24. Fridhandler, L., and Quasterl, J. H. Absorption of amino acids from isolated surviving intestine. Arch. Biochem. Biophys., **56:** 424–440, 1955.
25. Friedman, M. H. F. Urinary gastric secretory depressants (urogastrone). Vitamins and Hormones, **9:** 313–353, 1951.
26. Friedman, M. H. F., and King, E. N. Presence of a specific gastric hormone (gastrin) in the dog's pyloric mucosa. Fed. Proc., **6:** 107, 1947.
27. Grace, W. J., *et al.* Studies of the human colon. I. Variations in concentration of lysozyme with life situation and emotional state. Am. J. Med. Sci., **217:** 241–251, 1949.
28. Grossman, M. I. Gastrointestinal hormones. Physiol. Rev., **30:** 33–90, 1950.
29. Heggeness, F. W., and Nassett, E. S. Purification of enterocrinin. Am. J. Physiol., **167:** 159–165, 1951.
30. Heinz, E., and Öbrink, K. J. Acid formation and acidity control in the stomach. Physiol. Rev., **34:** 643–673, 1954.
31. Hildes, J. A., *et al.* The water and electrolyte excretion of the human pancreas. Gastroenterology, **21:** 64–70, 1952.
32. Hollander, F. The composition and mechanism of formation of gastric acid secretion. Science, **110:** 57–63, 1949.
33. Ihre, B. Human gastric secretion. Acta med. scandinav., Suppl. 95, 1938.
34. Janowitz, H. D. A quantitative study of the gastric secretory response to sham feeding in a human subject. Gastroenterology, **16:** 104–116, 1950.
35. Janowitz, H. D., and Hollander, F. Critical evidence that vagal stimulation does not release gastrin. Proc. Soc. Exper. Biol. and Med., **76:** 49–52, 1951.
36. Janowitz, H. D., and Hollander, F. Gastric carbonic anhydrase revived. Gastroenterology, **30:** 536–537, 1956.
37. Kirk, E. The quantity and composition of human colonic flatus. Gastroenterology, **12:** 782–794, 1949.
38. Kremen, A. J., *et al.* An experimental evaluation of the nutritional importance of proximal and distal small intestine. Ann. Surg., **140:** 439–448, 1954.
39. Lashley, K. S. Reflex secretion of the human parotid gland. J. Exper. Psychol., **1:** 461–493, 1916.
40. Lemberg, R., and Legge, J. W. *Hematin Compounds and Bile Pigments.* New York, Interscience Publishers, 1949.
41. Linde, S., *et al.* Experiments on the primary acidity of the gastric juice. Acta physiol. scandinav., **14:** 220–232, 1947.
42. London, I. M., *et al.* Porphyrin formation and hemoglobin metabolism in congenital porphyria. J. Biol. Chem., **184:** 365–371, 1950.
43. London, I. M., *et al.* Conversion of protoporphyrin to bile pigment. Fed. Proc., **10:** 217, 1951.
44. Malloy, H. T., and Evelyn, K. A. The determination of bilirubin with the photoelectric colorimeter. J. Biol. Chem., **119:** 481–490, 1937.
45. Mandl, I., *et al.* Solubilization of insoluble matter in nature. II. Part played by salts of organic and inorganic acids occurring in nature. Biochim. et Biophys. Acta, **10:** 540–569, 1953.

46. MARSHALL, J. M., JR. Distributions of chymotrypsinogen, procarboxypeptidase, deoxyribonuclease, and ribonuclease in bovine pancreas. Exptl. Cell Res., **6:** 240–242, 1954.
46a. MATTSON, F. H., AND BECK, L. W. The specificity of pancreatic lipase for the primary hydroxyl groups of glycerides. J. Biol. Chem., **219:** 735–740, 1956.
47. MEYER, K. H., *et al.* Amylolytic enzymes. X. Isolation and crystallization of alpha-amylase from human saliva. Helvet. chim. acta, **31:** 2158–2164, 1948.
48. MEYER, K. H., *et al.* Purification and crystallization of human pancreatic amylase. Arch. Biochem., **18:** 203–205, 1948.
49. MUIR, A. Carbohydrate metabolism and gastric secretory activity. Quart. J. Med., **18:** 235–261, 1949.
50. MUSICK, V. H., *et al.* A simplified caffeine gastric test meal for the diagnosis of peptic ulcer. J. A. M. A., **141:** 839–841, 1949.
51. NAJJAR, V. A., AND CHILDS, B. The crystallization and properties of serum bilirubin. J. Biol. Chem., **204:** 359–366, 1953.
52. NARDI, G. L. Metabolic studies following total pancreatectomy for retroperitoneal leiomyosarcoma. New Eng. J. Med., **247:** 548–550, 1952.
53. NICOLAYSEN, R., *et al.* Physiology of calcium metabolism. Physiol. Rev., **33:** 424–444, 1953.
54. NORD, CH. F. L. The cause of dental caries. Brit. Dent. J., **81:** 309–316, 1946.
55. PATTERSON, W. B., AND STETTEN, DEW., JR. A study of gastric HCl formation. Science, **109:** 256–258, 1949.
56. PRIEN, E. L., AND WALKER, B. S. Salicylamide and acetylsalicylic acid in recurrent urolithiasis. J. A. M. A., **160:** 355–360, 1956.
57. REISER, R., *et al.* The intestinal absorption of triglycerides. J. Biol. Chem., **194:** 131–138, 1952.
58. ROTH, J. L. A., AND BOCKUS, H. L. Criteria for histamine achlorhydria. Gastroenterology, **15:** 374–377, 1950.
59. SCHMID, R. Direct-reacting bilirubin, bilirubin glucuronide, in serum, bile, and urine. Science, **124:** 76–77, 1956.
60. SCHMIDT-NIELSEN, B. The pH of parotid and mandibular saliva. Acta physiol. scandinav., **11:** 104–110, 1946.
61. SCHØNHEYDER, F., AND VOLQVARTZ, K. Gastric lipase in man. Acta physiol. scandinav., **11:** 349–360, 1946.
62. SHERLOCK, S. P. V., AND WALSHE, V. Blood cholates in normal subjects and in liver disease. Clin. Sci., **6:** 223–234, 1948.
63. SHIVELY, J. A., AND MARKEY, R. L. A simple test for trypsin in stool. Am. J. Clin. Path. Tech. Sect., **22:** 1220–1222, 1952.
64. SMYTH, G. A. Activation of peptic ulcer during pituitary adrenocorticotropic hormone therapy. J. A. M. A., **145:** 474–477, 1951.
65. SWELL, L., *et al.* Role of bile salts in activity of cholesterol esterase. Proc. Soc. Exptl. Biol. Med., **84:** 417–420, 1953.
66. WALKER, B. S., AND WHITAKER, L. R. An atraumatic method for the study of gallbladder secretion. Rev. Gastroenterol., **2:** 129–132, 1935.
67. WALTERS, W., AND BELDING, H. H. III. Physiological effects of vagotomy. J. A. M. A., **145:** 607–613, 1951.
68. WANG, K. J., *et al.* Action of lysozyme on gastrointestinal mucosa. Arch. Path., **49:** 298–306, 1950.
69. WOLF, S., AND WOLFF, H. G. *Human Gastric Function.* 2nd edit. New York, Oxford University Press, 1947.
70. WOOLLEY, D. W. Streptogenin activity of peptides of glutamic acid. Fed. Proc., **6:** 424, 1947.

CHAPTER 14

Carbohydrate Metabolism and Diabetes

Metabolism is a word with broad coverage. It includes everything which happens to foodstuffs in the body, from the time and place of entrance to the time and place of exit. In this broad sense, metabolism includes digestion, absorption, intermediary metabolism, and excretion. The word is often used in a more restricted sense, covering only intermediary metabolism —those chemical changes which occur after absorption of a substance into the body and until the substance or its waste products are irrevocably committed to an excretory channel. By tradition, and to aid understanding of this complex process, it is customary to subdivide metabolism according to the type of raw material involved. Following this useful custom, we shall deal in this chapter with carbohydrate metabolism and in later chapters with lipid, protein, water, mineral, and oxygen metabolism. These subdivisions are helpful but not altogether rational. We shall see that the metabolic pathways of different substances cross and recross, and that many significant interconversions are possible.

Diabetes mellitus is the most important of the metabolic diseases. It involves the fundamental processes of carbohydrate, fat, and protein metabolism, with devastating effect upon human efficiency, longevity, and fertility. It has rowelled investigators into intense and sustained activity, and at the same time guided them through the maze of the metabolic pattern by the very disturbances it produced. In no other disease has the chemical approach been so rewarded in increase of understanding, in improvement of the patient's situation, and in the uncovering of new problems.

The word *diabetes* derives from the Greek, meaning "a passing through". In classical Greek, diabetes means a siphon. In English, diabetes means the chronic excretion of an excessive volume of urine. Modified by an adjective meaning tasteless, *diabetes insipidus* is a disease characterized by the excretion of large volumes of very dilute urine. This uncommon disease involves the antidiuretic hormone of the posterior lobe of the pituitary, and will be discussed in Chapter 17.

In ordinary medical terminology, when diabetes is mentioned, the adjective *mellitus*, meaning "honeyed", is appended or, if no other modifier is used, is understood. *Diabetes mellitus* is a disease characterized by the excretion of glucose in the urine and an increase in the concentration of glucose in the blood. The designation of this disease as a diabetes is proper, since there is usually an increase in the volume of urine. Diabetes mellitus has been recognized as a serious medical problem for more than two thousand years. In the Sushruta Samhita, which has been ascribed to the 6th century B.C. (80), diabetes is recognizably described, and the medical student of that time was taught that

> "Secretions or discharges . . . should be tested with the organ of taste. The sweet . . . taste of the discharges should be inferred from the fact of their being or not being swarmed with hosts of ants or flies . . ." (7).

Dobson in 1775 separated a sugar from diabetic urine by evaporation. Chevreul did not isolate glucose from grapes until 1815. The two sugars were found identical.

Diabetes is no rare disease. There are more than a million diabetics in the United States, where diabetes stands ninth among the causes of death and is responsible for one death out of 40. Although there are significant differences in the onset and course of diabetes depending upon age, race, and nutritional status, it appears in all population groups, and is significant in all specialties and subspecialties of medical practice.

The objective signs of diabetes are glycosuria and hyperglycemia, usually accompanied by polyuria. *Glycosuria* is the presence of glucose in the urine, demonstrable by any one of several simple chemical tests, of which Benedict's (see below) is most commonly used. *Hyperglycemia* is elevation of the concentration of glucose in the blood above its normal level of 0.1 per cent to levels which in untreated diabetes usually lie between 0.2 and 0.5 per cent. In less than 5 per cent of cases of diabetes does the blood sugar ever go over 1.0 per cent. One case with a blood sugar slightly over 2.0 per cent has been reported (40). *Polyuria*, or an increase of urinary volume above the normal upper limit of 2 liters in 24 hours, is a direct result of the excretion of glucose in the urine; the glucose increases the effective osmotic concentration of the urine and opposes reabsorption of water in the renal tubules. The relationship between glucose output and urine volume is not mathematically exact. Patients may have glycosuria with normal urine volume.

Less objective signs of diabetes are *polydipsia* and *polyphagia*—increased intake of fluids and food. The untreated diabetic is thirsty in response to the water depletion resulting from polyuria and is in a metabolic state showing many of the characteristics of starvation. This metabolic state will be explained in this chapter, where the normal metabolism of carbohydrate will be considered in comparison with its waste in the diabetic. In no other disease are simple chemical tests so useful to the clinician, both for purposes of diagnosis and as a means of checking the progress of treatment, as in diabetes.

CLINICAL ANALYSIS FOR CARBOHYDRATES

All monosaccharides reduce Cu^{++} to Cu^{+} in solution, over a wide range of pH. *Benedict's qualitative solution* is widely used for the detection of sugar in urine. It is prepared by dissolving 173 grams of sodium citrate, and 100 grams of anhydrous sodium carbonate in 800 ml. of hot water. To this is added slowly 17.3 grams of crystalline copper sulfate dissolved in 100 ml. of water. The volume is finally made up to 1 liter at 20°C. with water. Five ml. of Benedict's solution is placed in a test tube and pre-heated to boiling. Five-tenths ml. of urine is added with mixing. The mixture is kept boiling for one minute or held in a boiling water bath for five minutes, then allowed to cool slowly to room temperature. The presence of 0.05 per cent of hexose or pentose in the urine will be indicated by a yellow precipitate of Cu_2O. Before the precipitate settles, it will appear as a greenish turbidity. Higher concentrations of sugar are indicated by a more copious precipitate appearing more quickly and redder in color. This alkaline copper solution is reduced by all hexoses and pentoses under the conditions specified, and also by *maltose*, which does not occur in urine, and by *lactose*, which may be present in the urine of lactating women. L-Xylulose (which appears in the urine of patients with the rare metabolic disease, *essential pentosuria*) (22), fructose, and high concentrations of glucose will reduce Benedict's solution even at room temperature, if allowed to stand overnight.

Benedict's solution may be prepared in a modification suitable for quantitative estimations of the sugars which reduce Benedict's qualitative solution. The quantitative modification contains per liter:

	grams
Copper sulfate crystalline	18.0
Sodium carbonate anhydrous	100.0
Sodium citrate	200.0
Potassium thiocyanate	125.0

Use the same steps in preparation as for the qualitative solution, adding the copper sulfate last. This solution is not suitable for qualitative testing. When it is heated with a reducing sugar, the Cu^{+} formed is precipitated as white CuSCN. To titrate the sugar in urine, exactly 25 ml. of *Benedict's quantitative solution* is pipetted into a porcelain dish of about 100 ml. capacity. Solid sodium carbonate is added in an amount sufficient to saturate the solution at the boiling temperature. The solution is kept boiling, and the urine is added slowly from a burette, at a rate just enough to replace the loss of volume of the boiling solution. Stirring is necessary to prevent caking of the reactants in the dish. The end point is the disappearance of blue or green color from the system. In titration of simple sugar solutions, the end point is distinctly seen in a nearly colorless mixture. When urines are titrated, brown pigments usually obscure the end point and make the determination less exact. The calculation of the concentration of reducing sugar is based upon the fact that 25 ml. of Benedict's quantitative solution is completely reduced by 50 mgm. of glucose, 52 mgm. of fructose, 54 mgm. of galactose, 67 mgm. of lactose, or 74 mgm. of maltose.

There are many other tests for sugar which involve reduction of Cu^{++}. In organic chemistry, *Fehling's solutions* are used for detection of aldehydes and ketones. Benedict's and other reagents involving Cu^{++} reduction are modifications of Fehling's test for more specific purposes. The Benedict reagent is less sensitive than Fehling's to certain normal urinary components such as uric acid and creatinine, hence is more specific for sugars in the urine. Reduction of Bi^{++} to metallic bismuth is the basis of another set of sugar reagents (e.g., Nylander's solution) little used in America, but often mentioned in European publications. Ag^{+} is reduced similarly to metallic silver. The deposition of a silver mirror on a glass surface by the reduction of ammoniacal silver salts with glucose is an amusing demonstration, but less reliable than Cu^{++} reduction as a clinical test.

Reduction of Cu^{++} in acid solutions is much more rapid with monosaccharides than with the reducing disaccharides. If time and temperature are controlled, acid Cu^{++} solutions such as *Barfoed's* can be used to differentiate lactose and maltose from the monosaccharides.

The principle of Cu^{++} reduction can also be applied to the measurement of the blood sugar, which is chiefly glucose. The classical American method of blood sugar estimation is that of Folin and Wu (26). It involves the preparation of a clear protein-free, fat-free blood filtrate by the addition to diluted blood of equivalent amounts of sodium tungstate and sulfuric acid. Blood proteins are precipitated by the liberated tungstic acid; blood lipids are adsorbed to the protein precipitate, which is separated by filtration. A measured volume of blood filtrate is mixed with an alkaline cupric tartrate solution, heated for a definite time, and then reacted with a solution of phosphomolybdic acid. In acid solution, reduction of phosphomolybdic acid to a blue complex molybdate by Cu_2O occurs. The amount of Cu_2O depends upon the amount of reducing sugar originally present in the blood filtrate. Colorimetric comparison is made against a standard glucose solution simultaneously treated exactly like the blood filtrate. This is one of the simplest of blood sugar methods. It has the disadvantage of being distinctly non-specific. The values obtained are higher than the true sugar concentrations of the blood, since the alkaline cupric tartrate is reduced by other substances in the blood than reducing sugars. By this method, the normal range of fasting blood sugar is 0.08 to 0.12 per cent. The true glucose value, as measured by more specific but less simple methods, is 0.06 to 0.08 per cent.

Reduction of substances other than copper may be applied to quantitative blood sugar measurements. Ferricyanide may be reduced by glucose to ferrocyanide and the concentration measured colorimetrically as prussian blue. This reaction is the basis of a micro-method (25) whereby sugar can be measured in 0.05-ml. samples of blood, which can be obtained by skin puncture. For estimations of urinary sugar, reduction of organic compounds is commonly used, for example the reduction of picric acid (5) and of dinitrosalicylic acid (87). Test papers for urinary glucose are commercially available, impregnated with the enzymes, *peroxidase* and *glucose dehydrogenase*, plus an oxidizable substrate such as *o*-tolidine. Glucose dehydrogenase catalyzes the conversion of glucose to gluconic acid and H_2O_2, which latter, in turn, oxidizes the *o*-tolidine to a blue product. This final reaction is catalyzed by the peroxidase. Note that such test papers respond only to glucose and not to other sugars possibly present in the urine.

Most sugars undergo partial decomposition at temperatures below their melting points. This leads to inexactness in melting-point determinations. However, the identification of sugars by melting point can be done accurately by first preparing a derivative of the sugar with phenylhydrazine or a substituted phenylhydrazine. The derivatives, which may be phenylosazones or phenylhydrazones, have sharp melting points and may be readily identified after recrystallization. All monosaccharides and most disaccharides will yield identifiable osazones. Glucose and fructose both give the same osazone, glucosazone. Sucrose does not form a specific osazone but slowly yields glucosazone as a result of hydrolysis. It is a common practice to attempt identification by examining crystalline derivatives under the microscope. This is reliable in skilled hands and under favorable conditions, but is full of pitfalls for the occasional analyst. The proper choice of reagents for the identification of a particular sugar is discussed in specialized monographs (69).

A number of color reactions have been devised for sugars and for sugar derivatives which depend upon the liberation of furfural or furfural derivatives when carbohydrates are decomposed with strong acid. All carbohydrates give the *Molisch reaction*: a few drops of 5 per cent alcoholic solution of alpha-naphthol or thymol are added to the solution in question, which is then stratified over concentrated sulfuric acid; in the presence of carbohydrate a violet color develops at the plane of contact. *Seliwanow's reagent* is 0.05 per cent resorcinol in 12 per cent aqueous HCl solution. It is specific for ketohexoses under the conditions specified. Fructose being the only important ketohexose in metabolism, it is used as a qualitative test for that sugar. To one volume of solution in question is added 5 volumes of the reagent and the mixture is heated in boiling water. The appearance within a few minutes of a red color, a condensation product of resorcinol with hydroxymethylfurfural, is a positive result. Oligosaccharides and polysaccharides which yield fructose on hydrolysis, (e.g., sucrose and inulin) will also show a positive Seliwanow reaction. The *orcinol reaction* for pentoses is carried out in a strongly acid solution, and depends upon furfural formation. The simple procedure described in many laboratory manuals is unreliable. Drury (20) has listed the proper precautions to be used in pentose studies with orcinol.

The dicarboxylic acids produced by oxidation of simple sugars with nitric acid are all soluble except the mucic acid produced from *galactose.* This test is used to identify galactose free or in combination as in lactose.

More specific than most color or reduction tests are the *microbiological* procedures for identification of sugars. Most strains of ordinary bakers' or brewers' yeast, *Saccharomyces cerevisiae,* will bring about the fermentation of glucose, fructose, maltose, and sucrose to ethyl alcohol and CO_2 but yield negative results with galactose. It is possible, however, to adapt a strain of yeast so that it will ferment galactose. In all microbiological identifications, it is highly important to run controls with known sugars, since sudden changes in the metabolic capabilities of a pure strain may occur, to say nothing of the possibility of contamination of the culture with extraneous organisms. Many other organisms may be used in sugar identification (69). Conversely, the ability to metabolize certain carbohydrates and not others is a common taxonomic criterion used by microbiologists.

The rotation of the plane of polarization of plane-polarized light by sugar solutions offers another measurement which can be utilized for analytical purposes.

THE ANABOLISM OF CARBOHYDRATE

Three of the monosaccharides are of dietary significance. These are glucose, fructose, and galactose. Only the first two of the three occur as such in ordinary diet and there only to a limited extent. The usefulness of other dietary carbohydrates is determined by the degree to which they are converted to one or more of the useful monosaccharides in the digestive processes. Starch, sucrose, and lactose (which are digested to glucose, to glucose and fructose, and to glucose and galactose, respectively) represent by far the major portion of nutrient carbohydrate in the diet. The monosaccharide, mannose, could be of significance in human metabolism. It is absorbable and can be converted into glycogen. However, it is present in foodstuffs only in the carbohydrate portion of serum and egg proteins and is quantitatively insignificant in human nutrition. The sugar alcohol, sorbitol, is slowly absorbed from the intestine and transported to the liver, where it is converted to fructose by the action of sorbitol dehydrogenase. Sorbitol is present in some fruits and vegetables.

Absorption transfers the monosaccharide end products of digestion from the lumen of the intestine to the blood of the intestinal capillaries. The venous blood of the intestine passes by way of the portal vein to the liver. The simple sugars in the liver face three immediate alternatives—escape through the liver into the hepatic vein and the general circulation, storage in the liver as glycogen, and catabolism in the liver. The last will be considered in a later section along with catabolism of carbohydrate in the body as a whole.

Escape through the liver is demonstrated by the increase in circulating blood sugar following ingestion of a test dose of sugar, or following an ordinary meal. The circulating blood sugar is chiefly glucose. Its concentration is approximately 0.1 per cent, more exactly 0.06 to 0.08 per cent in the fasting subject. Most mammalian species show a similar blood sugar level. Glucose and other simple sugars are distributed equally in the water of plasma and the water of blood cells. Since the cells contain less water than plasma, the concentration of sugar in plasma is greater than in cells or in whole blood. Following the ingestion of a test dose of more than 100 grams of glucose, the average person will show an elevation of blood sugar above the "renal threshold" of 0.18 per cent, and glucose will be excreted in the urine. The increase of blood sugar following an ordinary meal is seldom more than 0.04 per cent, which does not induce glycosuria. There is, however, a normal output of reducing sugars in the urine, less than 0.9 gram per 24 hours, not enough to cause reduction of Benedict's solution.

There is a measurable difference in the sugar level of arterial and venous blood. This is called the A-V difference and depends upon (a) the current rate of utilization of glucose by the organ supplied, and (b) the rate of blood flow through that organ. Glucose which reaches muscular or other cells must travel from the blood by way of the extracellular fluid. It is difficult and unrewarding to attempt the measurement in the human subject of the glucose content of tissue fluids. Sufficient experimental evidence exists to enable us to make the inference that the glucose content of tissue fluids is very close to that of blood, and that fluctuations in the glucose level of one will be closely followed by similar fluctuations in the other.

Glycogenesis

The liver can store glycogen up to 6 per cent of its weight, or a total of a little over 100 grams. Liver glycogen can be formed from all three of the

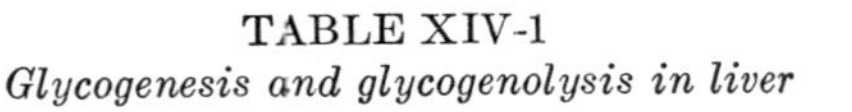

TABLE XIV-1
Glycogenesis and glycogenolysis in liver

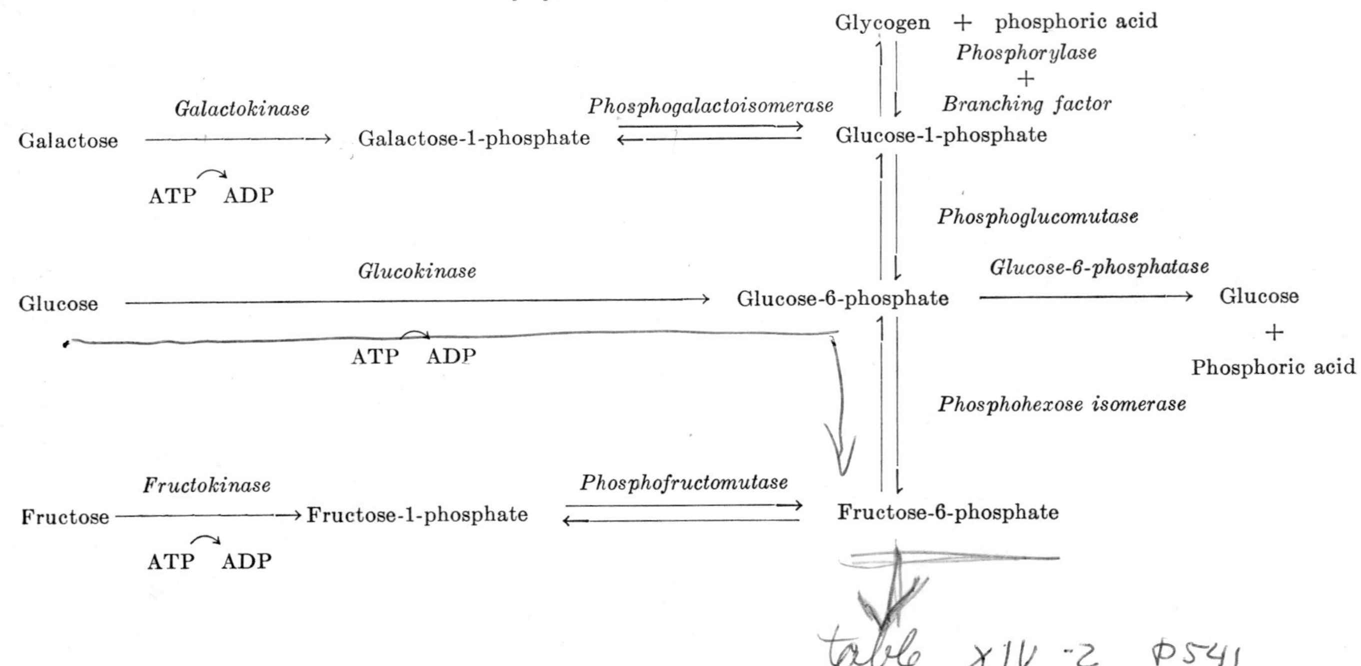

simple nutrient sugars—glucose, fructose, and galactose; the process is called glycogenesis (table XIV-1). The first step in glycogen formation from glucose is the transfer of a phosphate group from adenosine triphosphate (ATP) to the sugar, forming glucose-6-phosphate (formula I). The ATP is thereby converted to adenosine diphosphate (ADP). Fructose is also phosphorylated in the liver, with the formation of fructose-1-phosphate (formula II) (16). Enzymes known as *glucokinase* and *fructokinase*, or collectively as *hexokinase*, catalyze these reactions, which take place

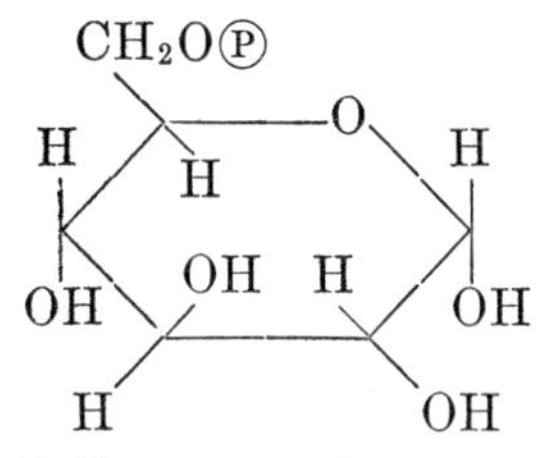

I. Glucose-6-phosphate

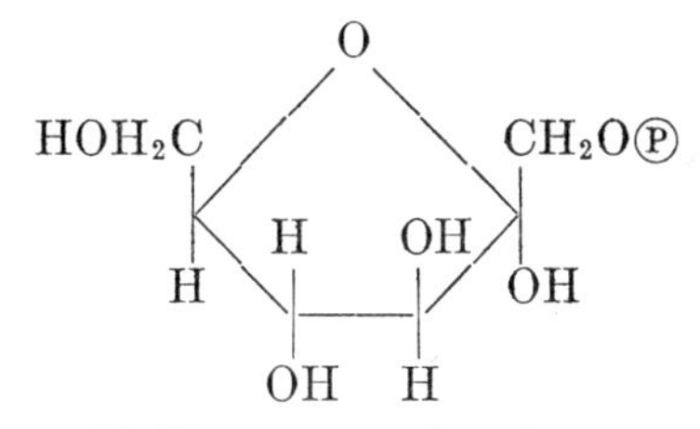

II. Fructose-1-phosphate

within the cells of the liver. In the fermentation of sugars by yeast, a single hexokinase is effective for glucose, fructose, and mannose. Simple sugars diffuse freely in and out of cells, phosphate esters of sugars do not. Thus the hexokinase reaction permits the capture of glucose by cells. The hexokinase reaction is not reversed, since it involves the transfer of phosphate from a high-energy to a low-energy bond.

Fructose-1-phosphate is in equilibrium with fructose-6-phosphate, the conversion being catalyzed by an enzyme designated as *phosphofructomutase*. Fructose-6-phosphate (formula III) is reversibly convertible to glucose-6-phosphate, catalyzed by *phosphohexose isomerase*. Glucose-6-phosphate and glucose-1-phosphate (formula IV) are interconvertible under the influence of *phosphoglucomutase*.

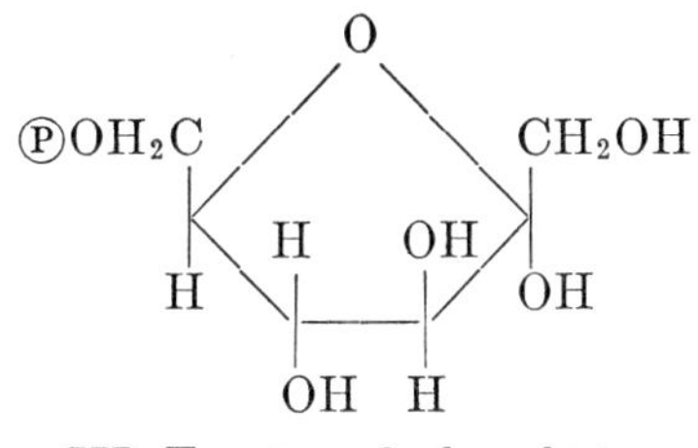

III. Fructose-6-phosphate

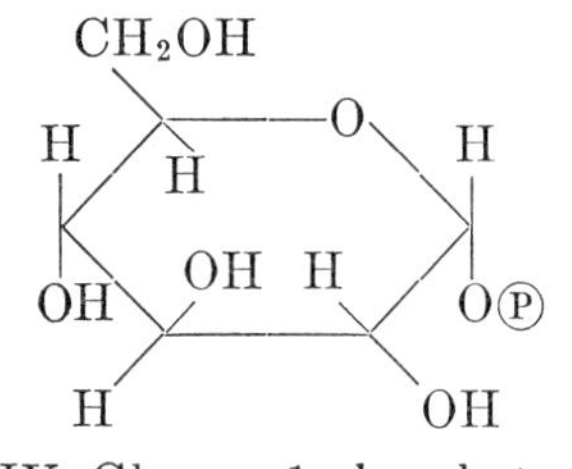

IV. Glucose-1-phosphate

Glucose-1-phosphate is converted to glycogen and inorganic phosphate reversibly, catalyzed by *phosphorylase* and *branching factor* (15). Phosphorylase (perhaps more accurately termed transglycosidase, see page 238) is specific for the formation of the 1-4 amylose type of linkage. The introduction of the 1-6 amylopectin type of linkage at the branching points of glycogen requires a second enzyme. The second enzyme, the branching factor, converts 1-4 to 1-6 linkages, thus acting as an intramolecular transglycosidase. Phosphate is not required.

The equilibrium under physiological conditions is at about 77 per cent glycogen to 23 per cent glucose-1-phosphate. Such an equilibrium indicates that the change in free energy is not sufficiently great to require the intervention of a high-energy phosphate bond and ATP is not utilized in the conversion of glycogen to glucose-1-phosphate. The easy interconversion of glycogen and glucose-1-phosphate is quite different from the hydrolysis of glycogen or starch by amylase. No detectable polysaccharide formation occurs when glucose or maltose is exposed to the action of amylase. Glycogen is at the same stage of utilization, energically speaking, as the hexose phosphates with which it is in equilibrium, and one step ahead of glucose which must be esterified by expenditure of a high-energy phosphate bond before it can be metabolized in the usual manner by glycolysis. As might also be judged from the even equilibrium, glycogen is in a state of dynamic interchange. This can be demonstrated by tracer studies which indicate the liver glycogen has a half life of about a day.

Note that in the utilization of galactose, there is no formation of galactose-6-phosphate; this ester has never been found in biological systems. Galactose-1-phosphate (formula V) has been isolated from the livers of galactose-fed animals; this ester is converted to glucose-1-phosphate in liver by an enzyme system, (*phosphogalactoisomerase*) which consists of at

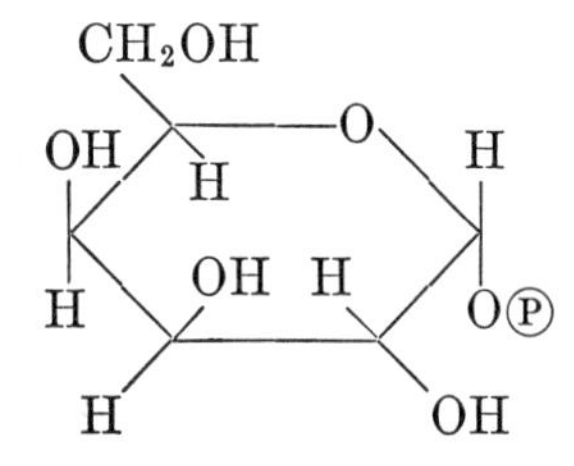

V. Galactose-1-phosphate

least two enzymes. The first stage of the reaction, catalyzed by *uridyl transferase*, which has uridinediphosphoglucose (see page 251) as a coenzyme, produces glucose-1-phosphate:

$$\text{Galactose-1-Ⓟ} + \text{UDPG} \rightleftarrows \text{Glucose-1-Ⓟ} + \text{UDPGal}$$

The second stage regenerates UDPG from UDPGal (uridinediphosphogalactose) by what appears to be a Walden inversion. The enzyme catalyzing this change is *galactowaldenase*. Since DPN is required, the reaction is more probably an oxidation-reduction rather than a simple Walden inversion (57). In this case, the carbon 4 shifts reversibly from its L-configuration in galactose to its D-configuration in glucose.

The rate of glycogenesis is one of the significant factors in determining the outcome of *sugar tolerance tests*. Such tests consist of the administration to a fasting subject of a known amount of a pure sugar, followed by repeated measurements of that sugar in the blood or the urine, or both. The *glucose tolerance test* is of particular significance in the diagnosis of diabetes, where glycogenesis is inhibited by mechanisms to be described later. If 50 to 100 grams of glucose is given by mouth to a normal subject with fasting blood sugar of about 0.10 per cent, the venous blood sugar will rise within the next hour to about 0.15 per cent and will have started its return to normal before the end of the hour without having induced glycosuria. By the end of the third hour the blood sugar will have returned to normal, probably to a little below the original fasting level as a result of the stimulus of the ingested sugar to glycogenesis. The same dose of glucose given to an untreated diabetic, whose fasting blood sugar would be above the normal value already, induces a rise of blood sugar increasing for about two hours, and not returning to the original level by the end of the third hour. Normal subjects who have been on a low carbohydrate diet may show a diabetic type of tolerance curve. The same is frequently true of obese people without diabetes. In the latter case, weight reduction usually results in a return of the glucose tolerance test pattern to normal (65).

The *galactose tolerance test* involves the administration of a test dose of 40 grams of galactose. Following such a dose, the normal person eliminates less than three grams of galactose in the urine during the 5-hour period following the test dose. In liver disease involving defective glycogenesis, more galactose is excreted. The shape and timing of the blood galactose curve in the normal is comparable to that of the glucose tolerance test, increasing in height and duration when larger doses of galactose are given. With the 40-gram dose, blood galactose levels may go as high as 0.08 per cent in the normal. With an 80-gram dose, blood galactose may go slightly above 0.10 per cent. The galactose tolerance test is in common clinical use as a measure of the glycogen function of the liver.

Experimental hepatectomy in animals is followed by a sudden fall in blood sugar, since the stored liver glycogen is removed and hepatic gluconeogenesis (see page 534) is stopped. It is somewhat surprising how little effect liver disease in man has in altering blood sugar, except in terminal stages. The reserve capacity of the liver is so great that, except under stress, little change in carbohydrate metabolism is noted. Blood sugar changes in response to fasting, to insulin, and to sugar administration are likely to be exaggerated and protracted in liver disease. Lowest blood sugar levels occur before breakfast, following the longest fasting period. The anabolism of galactose is specifically a liver function—hence the use of galactose tolerance tests.

Glycogenolysis

Glycogen of liver cells can break down again to glucose, but not to fructose or galactose. The transformation of glycogen to glucose occurs only in liver cells, not in cells of other tissues or organs. This breakdown is called glycogenolysis, and takes place in three steps. First, glycogen reacts with phosphate ion to yield glucose-1-phosphate. This is a reversal of the last step in glycogenesis. As in that step, the enzyme is phosphorylase. Phosphorylase splits off only the glucose units in the outer straight chain portions of the molecule. Action stops when the 1-6 branch points are reached. What is left is a *limit dextrin*. A second enzyme, *amylo-1,6-glycosidase* ("debranching factor") catalyzes the hydrolysis (phosphate not being involved) of these 1-6 links. Phosphorylase can next proceed with its work until a new limit dextrin is formed. Note that branching factor and debranching factor do not catalyze reactions that are the reverse of each other. Branching factor converts a 1-4 link to a 1-6 link. Debranching factor hydrolyzes a 1-6 link, but does not form a 1-4 link.

Glucose-1-phosphate once formed, is converted to glucose-6-phosphate. This is another reversal, catalyzed by phosphoglucomutase. Finally, a phosphatase specific for glucose-6-phosphate and found only in liver cells converts glucose-6-phosphate irreversibly to glucose. Unlike glycogen and the glucose phosphates, glucose can leave the liver cell and contribute to the blood sugar. *In vitro*, potassium ions have been shown to promote glycogenesis, and sodium ions to promote glycogenolysis (33) in rabbit liver slices. The storage of glycogen in the human binds 0.36 meq. potassium for each gram of glycogen. This potassium is released with glycogenolysis.

The ability of the liver to store ingested carbohydrate as glycogen and later to release it as glucose was first noted a century ago by the French physiologist Claude Bernard. The question for which he experimentally sought the answer was, "What is the source of the sugar in diabetic urine?" In those years there was no suggestion that the pancreas was at fault. Claude Bernard studied normal animals, trying to determine which organs contributed glucose to the blood. This he found that the liver could do. In animals not absorbing sugars from the intestine, there was a higher level of glucose in the blood of the hepatic vein than in that of either the portal vein or the hepatic artery. It was a disappointment when he found but little glucose in the livers of his animals. Further chemical study led to the discovery of glycogen and its storage function in the liver. This was the first of many discoveries in normal biochemistry inspired by a primary interest in diabetes.

Glycogen in Muscle

The mechanism for the production of glycogen from glucose in muscle is identical with that in the liver. The first two steps in its breakdown are also the same, forming first glucose-1-phosphate and then glucose-6-phosphate; the necessary enzymes, phosphorylase and phosphoglucomutase, are at hand in muscle. No conversion of glucose-6-phosphate to glucose can occur, however, since the enzyme glucose-6-phosphatase does not occur in muscle. Glucose-6-phosphate can not itself leave the muscle cell by diffusion; its only possible alternatives are glycolysis (see page 539) or resynthesis to glycogen by reversal of the steps mentioned above.

The formation of glycogen from fructose is more direct in muscle than in liver. Muscle fructokinase forms fructose-6-phosphate directly, omitting the intermediate formation of fructose-1-phosphate (see table XIV-1).

Effects of Hormones Upon Glycogen Storage

The relative amounts of glucose-6-phosphate converted in the liver to glucose and to glycogen are determined in part by the activity of liver glucose-6-phosphatase. *Insulin* decreases liver glucose-6-phosphatase activity, favoring glycogen storage (1). *Cortisol* has an opposite effect. *Adrenalin* promotes the breakdown of glycogen to glucose-6-phosphate. In the liver, and with adequate liver glycogen, this results in increased liberation of glucose to the blood. Muscle glycogen is converted to glucose-6-phosphate which, for lack of the specific phosphatase, can not be converted to glucose. The glucose-6-phosphate undergoes anaerobic glycolysis (see page 539), forming lactic acid which can escape from the cell. Transported by the blood to the liver, the lactic acid is converted to glycogen. Thus in the well fed animal adrenalin decreases liver glycogen, and in the fasting animal may increase it.

On the basis of experimental evidence summarized by Cori (14), he has proposed an integrated concept of the actions of three hormones on hexokinase: the *anterior pituitary* produces a specific inhibitor of liver and muscle hexokinase; the *adrenal corticoids* intensify and protract the pituitary inhibition of hexokinase; *insulin* releases hexokinase from inhibition, favoring thereby the formation of glucose-6-phosphate from glucose, and thus indirectly favoring the formation of glycogen. Subsequent studies have for the most part been consistent with this triple concept (4), although there has been some evidence which is directly contradictory (83). This mechanism for the endocrine control of hexokinase, or more specifically of glucokinase, will be introduced repeatedly into the remainder of this discussion of carbohydrate metabolism as a good working hypothesis and not as a set of facts fully proven.

Glycogen Storage Disease

This is a very rare metabolic error, inherited as a Mendelian recessive, in which the liver is greatly enlarged and its cells are gorged with glycogen which may reach concentrations as high as 10 to 15 per cent of the wet weight of the organ. (In other forms of glycogen storage disease, the glycogen accumulation occurs predominantly in cardiac and skeletal muscle.) The glycogen content of the granular leukocytes is increased, which is a valuable point in diagnosis. The blood sugar is low and there is likely to be ketonuria (see page 594) after several hours fasting. Both are obvious concomitants of the inability to make use of liver glycogen.

Homogenates of liver taken at autopsy from patients with glycogen storage disease will not hydrolyze glycogen under conditions which permit its hydrolysis by homogenates of normal liver. On the other hand, liver glycogen isolated from patients with the disease is rapidly hydrolyzed by homogenates of normal liver. This would indicate that the biochemical flaw in glycogen storage disease lies in liver enzyme deficiency rather than in abnormal glycogen structure. The specific deficiency may be that of glucose-6-phosphatase. However, liver glycogen in some cases may be of abnormal structure (63).

Gluconeogenesis (Glyconeogenesis)

Other substances than hexoses and glycogen can form glucose in the body, and glucose so formed behaves no differently from glucose derived directly from the carbohydrates of the diet. The formation of glucose from precursors which are not hexoses or hexosans is called gluconeogenesis, and is almost exclusively a function of the liver. (The alternative term, glyconeogenesis, is justified by the fact that glucose formation implies glycogen formation as well.) Gluconeogenesis has, however, been observed in the kidneys of hepatectomized animals. Glucose can be formed from lactic acid, pyruvic acid, and all the intermediates between glucose or glycogen and lactic acid in the glycolytic mechanism (see table XIV-2). Glucose can also be formed from the acids of the citric acid cycle (see page 548), from mannitol and sorbitol, from glycerol (with triose phosphate as an intermediate), and from the amino acids listed as glycogenic on page 615. Gluconeogenesis from these amino acids can occur only after the amino group has been removed by oxidative deamination or by transamination (see Chapter 16). Gluconeogenesis from *fatty acids* does not occur, meaning that glycogen or glucose can not be increased in the whole animal at the expense of fatty acids. Tagged carbon atoms of administered fatty acids may appear in glucose or glycogen, but there is no net gain in glucose or glycogen since an equal number of carbon atoms is oxidized to CO_2 in the process. This problem of the interconversion of fatty acid and carbohydrate will be taken up in more detail in the next chapter. The common intermediate in gluconeogenesis is glucose-6-phosphate. It will be recalled that this ester can not cross cell boundaries, and must be used or transformed in the cell where it is formed.

The glycocorticoids of the adrenal have been shown to promote gluconeogenesis from amino acids. Recalling the previously mentioned adrenocortical augmentation of the inhibitory effect of the pituitary upon hexokinase—the first enzyme involved in glucose utilization—we can see that the combined effect of the glycocorticoids is to retard utilization and speed formation of glucose, in short to maintain the level of blood glucose. Gluconeogenesis is also stimulated by thyroxin, and by lack of insulin. In the normal person, gluconeogenesis prevents hypoglycemia during prolonged fasting.

Production of Physiological Sugars

Fructose. This is the characteristic sugar of the seminal plasma. Its origin from blood glucose has been demonstrated by Mann and Parsons (55) who found in diabetic rabbits levels of fructose in semen proportional to the glucose of the blood. Normal human semen contains highly variable amounts of fructose, up to 0.8 per cent or 40 mgm. per ejaculate (47). The values are higher in human diabetics. Seminal fructose is secreted by the seminal vesicles, which are dependent upon androgen stimulation for their maintenance. Fructose disappears from the seminal fluid following castration or hypophysectomy, and reappears with administration of testosterone or gonadotrophin, respectively.

Lactose. This has already been mentioned exclusively as the sugar of milks. Its precursors are glucose and glycogen. Since either of these substances will serve as a substrate for lactose production, it is probable that the mechanism is one of phosphorolytic glycogenolysis. No parallelism between blood sugar and milk sugar comparable to that described for seminal fructose has ever been reported. The insulin requirement of diabetic women is often decreased during lactation, while in normal lactating women blood sugar tends to be low.

Pentoses. Mammals can not phosphorylate dietary pentoses and therefore can not metabolize them. In fact, there is some evidence that, in rats at least, xylose, when incorporated in the diet for prolonged periods, is toxic (10). Ribose-5-phosphate must therefore be biosynthesized from an already phosphorylated hexose, glucose-6-phosphate (18). Glucose-6-phosphate is first oxidized to 6-phosphogluconolactone by the enzyme glucose-

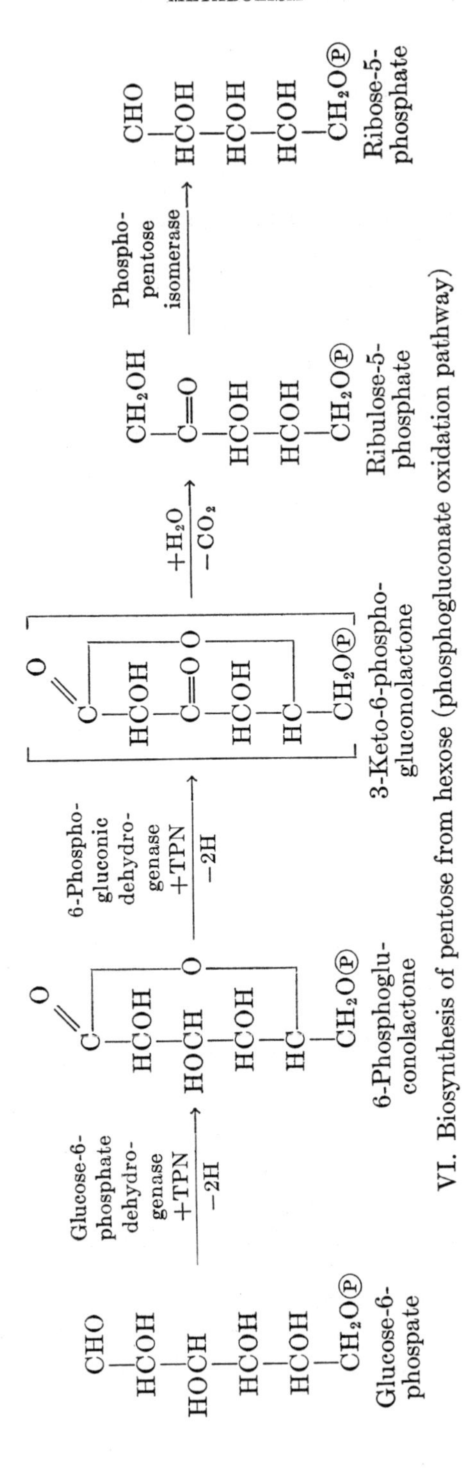

VI. Biosynthesis of pentose from hexose (phosphogluconate oxidation pathway)

6-phosphate dehydrogenase. This enzyme has been found in liver, erythrocytes, kidney, bone-marrow, spleen, heart and brain. It is accompanied by 6-phosphogluconic dehydrogenase which oxidatively decarboxylates 6-phosphogluconolactone to ribulose-5-phosphate, possibly by way of a 3-ketohexose intermediate (35). This is the step in which the hexose derivative is degraded to the 5-carbon stage. The dehydrogenases involved in these two reactions both utilize TPN as coenzyme. In the third and final step, ribulose-5-phosphate is converted to ribose-5-phosphate by phosphopentose isomerase. This scheme of pentose anabolism is summarized in formula VI, where the sugars are presented in straight-chain fashion so that the interrelationships may be more obvious. Note that the introduction of the keto-group at carbon 3 of the glucose chain destroys the L-configuration of the hydroxyl group in that position and permits the introduction of a D-hydroxyl in the final step. There is, however, some evidence that *xylulose-5-phosphate* (formula VII) is the first pentose formed and that this is converted to ribulose phosphate, a step which would involve a second inversion at carbon 4 of the pentose molecule (54).

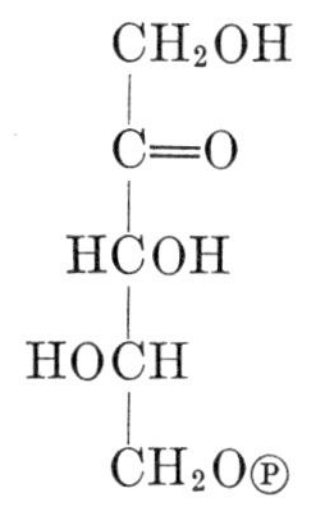

VII. Xylulose-5-phosphate

Deoxyribose is, apparently, not formed by an analogous mechanism but is the product of a condensation of a triose with a two-carbon compound (53) a mechanism which may also represent an alternate route for the formation of ribose (6).

Uronic acids. These are sugar acids which form part of the structure of a number of substances important to the body. *Glucuronic acid* is the uronic acid of animal tissues, and is a component of hyaluronic acid, condroitin and mucoitin sulfuric acids, and heparin. Free glucuronic acid has never been clearly demonstrated in any tissue or body fluid. Glucuronic acid is formed by the enzymatic DPN-linked oxidation of uridinediphosphoglucose (UDPG) to uridinediphosphoglucuronic acid (UDPGA). The glucuronic acid is then passed on to an alcohol or phenol group on an acceptor molecule, leaving free uridinediphosphate (UDP). Separate enzyme systems for the oxidation of UDPG, and for the transfer of glucuronic acid from UDPGA, have been demonstrated in liver (86a).

An enzyme, *beta-glucuronidase*, catalyzes the hydrolysis of numerous glucuronides, including those of substances of extraneous origin such as menthol and borneol and those of physiological origin such as estriol and pregnanediol (23). High activity of this enzyme is observed in liver, spleen, lung, endocrine tissues, and leukocytes. Some degree of activity is observed in all fresh tissues. Uterine beta-glucuronidase activity is subnormal in ovariectomized mice, but is increased by injections of natural or synthetic estrogens. The activity is greatly diminished in the human uterus after the menopause. The transfer of glucuronic acid from a conjugate to an acceptor can also be catalyzed by beta-glucuronidase (24).

Amino sugars. Glucosamine and its N-acetyl derivative are components of the group-specific agglutinogens of the red blood cells. Acetylglucosamine also occurs in mucoitin sulfuric acid, hyaluronic acid, and heparin. Galactosamine is characteristic of chondroitin sulfuric acid. The L-enantiomorph of glucosamine is a structural component of the antibiotic, streptomycin. D-glucosamine is phosphorylated by ATP in beef brain extracts by the same enzymic mechanism which promotes the phosphorylation of glucose and fructose (31).

Uridinediphosphate is involved in the production of amino sugars, or in the formation of polysaccharide chains involving amino sugars (27) or possibly in both processes. Both UDP-N-acetylglucosamine and UDP-N-acetylgalactosamine have been isolated from bovine liver (70). The source of the amino group is glutamine (13), and the acetyl group is introduced from acetyl CoA.

Formation of Fat from Sugars

The capacity of the body to store glycogen is sharply limited. The liver storage is an amount equivalent to about 500 Kcal., and the maximal storage in the muscles about 2500 Kcal. The total is scarcely adequate to meet the energic needs of one day of strenuous activity. The only significant mechanism whereby carbohydrate can be stored indefinitely is by conversion to fat. This transformation is demonstrated grossly by the fact that hogs can be fattened for market on a low-fat high-carbohydrate diet. It has been demonstrated more elegantly by measuring the rate of conversion of tagged glucose to fatty acids. In order that fatty acids may be formed from sugars, the latter must be first degraded to two-carbon fragments, and some foodstuff must be simultaneously utilized to provide energy. This latter requisite is implicit in the fact that fats yield approximately 9 Kcal. per gram on complete combustion, whereas sugars yield only about 4 Kcal. per gram. The extra energy stored in fat must derive from the only possible source—the chemical energy of foodstuffs (see section on biosynthesis of lipids, Chapter 15). The formation of two-carbon fragments from sugars will be taken up in the next section, which deals with catabolism of carbohydrates. The conversion of two-carbon fragments to fatty acids is accelerated by insulin (8). It has long been recognized clinically that insulin in small doses was helpful in promoting the deposition of fat.

THE CATABOLISM OF CARBOHYDRATE

Glucose reaches the blood, and thence passes freely to the extracellular fluids and into the cells, from three sources: (a) direct absorption of glucose from the alimentary tract; (b) glycogenolysis; and, (c) gluconeogenesis. The quantity of glucose contributed to blood sugar by the breakdown of glycogen is normally the smallest, compared with the other two sources. It becomes more significant in situations of acutely increased demand. The utilization of glucose by the cells for the production of energy will occupy us in this section.

Glycolysis

There is a series of reactions, involving phosphoric esters of glucose, fructose, galactose, and their split products, which makes a limited amount of energy available from these sugars without utilization of molecular oxygen. The intermediates, end products, and specific enzymes for this series of reactions have been identified in many different tissues in many different organisms. The whole process has repeatedly been demonstrated in organisms ranging from bacteria and yeasts, through protozoa to all phyla of animals. The series was first worked out in yeast and in frog and mammalian muscle. This sequence of reactions is known as *glycolysis*, sometimes for emphasis as anaerobic glycolysis, and sometimes as the *Embden-Meyerhof pathway*. Skeletal muscle and brain initiate glucose utilization exclusively through glycolysis. Other tissues glycolyze in addition to using other pathways such as phosphogluconic oxidation (see page 535).

The sequence of glycolytic reactions as far as the formation of fructose-6-phosphate has already been described under glycogenesis. These steps have been summarized in table XIV-1. The sequence can start in any tissue with glycogen, glucose, or fructose. Direct glycolysis of galactose is probably limited to liver, as is glycogenesis from galactose.

The next step, which is the beginning of glycolysis proper, requires a second molecule of ATP, which gives up a terminal phosphate group, becoming ADP. (The first molecule of ATP, it will be recalled, was expended in the initial phosphorylation of absorbed monosaccharide, see page 529.) The phosphate attaches irreversibly to fructose-6-phosphate, forming fructose-1,6-diphosphate (formula VIII), commonly called hexose diphosphate or HDP. In this particular reaction, it has been shown that ATP is not the only energy-donor that can be utilized. Uridinetriphosphate (UTP)

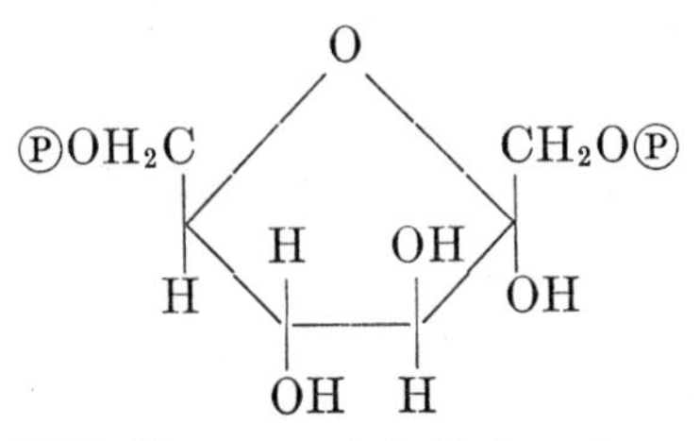

VIII. Fructose-1,6-diphosphate

and inosinetriphosphate (ITP) do about as well (50). The enzyme which catalyzes this phosphorylation is different from the hexokinases, and is named *phosphohexokinase*, or better *phosphofructokinase*. Fructose-1,6-diphosphate next cleaves into two triose phosphates: glyceraldehyde-3-phosphate (formula IX) and dihydroxyacetone phosphate (formula X). This cleavage is reversible and is specifically catalyzed by an enzyme known both as *aldolase* and as *zymohexase*. The two isomeric triose phosphates are interconvertible, and form an equilibrium mixture. Their interconversion is catalyzed by *phosphotriose isomerase*. Dihydroxyacetone phosphate has no further function in the glycolytic process. It can, however, be enzymatically reduced to alpha-glycerophosphate (Karrer ester), which is utilized in phospholipid synthesis, and which can form glycerol upon hydrolysis catalyzed by a phosphatase.

CHO
|
HCOH
|
CH$_2$OⓅ

IX. Glyceraldehyde-3-phosphate

CH$_2$OH
|
C=O
|
CH$_2$OⓅ

X. Dihydroxyacetone phosphate

The metabolic cycle is now ready for the generation of a high-energy phosphate bond. Note that so far two such high-energy bonds (table XIV-2) have been used up by conversion of ATP to ADP, and that the bonds on the hexose and triose phosphates so far have all been low energy.

Glyceraldehyde-3-phosphate reacts with inorganic phosphate to form 1,3-diphosphoglyceric acid. The enzyme involved has as its active group the —SH of a glutathione (GSH) prosthetic group (73). DPN is bound to the SH and the reaction that takes place is shown in formula XI. Note that the high-energy acylmercaptan bond is formed before the high-energy

TABLE XIV-2

Summary of glycolysis

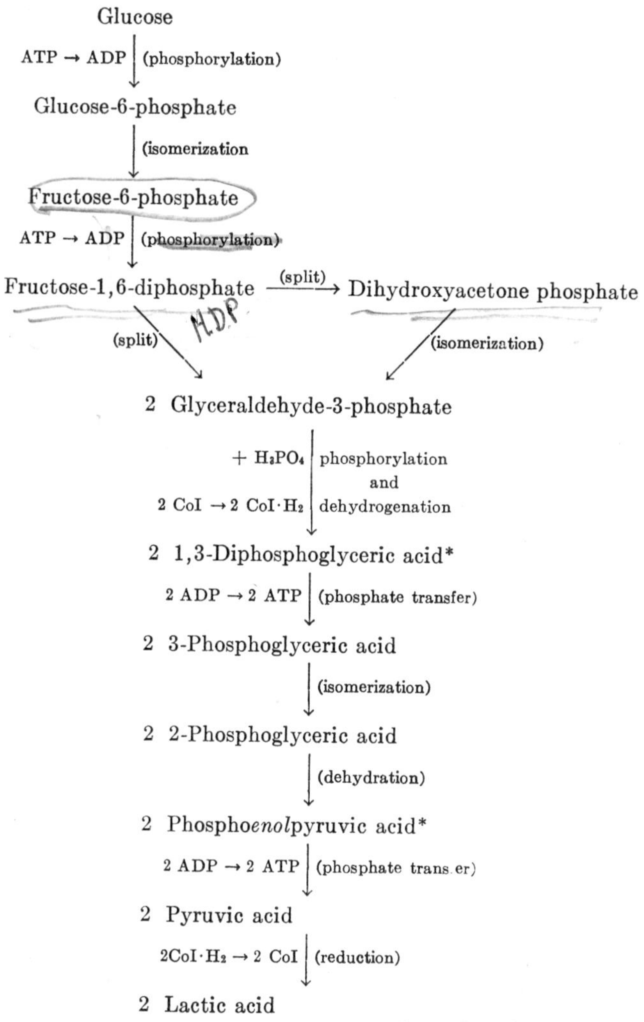

Glucose

ATP → ADP ↓ (phosphorylation)

Glucose-6-phosphate

↓ (isomerization

Fructose-6-phosphate

ATP → ADP ↓ (phosphorylation)

Fructose-1,6-diphosphate —(split)→ Dihydroxyacetone phosphate

(split) ↘ ↙ (isomerization)

2 Glyceraldehyde-3-phosphate

+ H_3PO_4 ↓ phosphorylation and dehydrogenation

2 CoI → 2 CoI·H_2

2 1,3-Diphosphoglyceric acid*

2 ADP → 2 ATP ↓ (phosphate transfer)

2 3-Phosphoglyceric acid

↓ (isomerization)

2 2-Phosphoglyceric acid

↓ (dehydration)

2 Phospho*enol*pyruvic acid*

2 ADP → 2 ATP ↓ (phosphate transfer)

2 Pyruvic acid

2CoI·H_2 → 2 CoI ↓ (reduction)

2 Lactic acid

* Compounds containing high-energy phosphate bonds.

phosphate bond. The energy for the formation of these bonds is derived from the dehydrogenation of the glyceraldehyde-3-phosphate.

The high energy phosphate of 1,3-diphosphoglyceric acid is transferred to ADP, converting it to ATP. The transfer is catalyzed by *phosphoglyceric*

$$\begin{array}{l} CHO \\ | \\ HCOH \\ | \\ CH_2O(P) \end{array} + DPN{-}S{\cdot}G\text{-enzyme} \rightarrow \begin{array}{l} \quad O \\ \;\,// \\ C{\sim}S{\cdot}G \text{ enzyme} \\ | \\ HCOH \\ | \\ CH_2O(P) \end{array} + DPN{\cdot}H_2$$

Glyceraldehyde -3-phosphate; high-energy acylmercaptan intermediate

$$\downarrow + H_3PO_4$$

$$\begin{array}{l} \quad O \\ \;\,// \\ C{\sim}O(P) \\ | \\ HCOH \\ | \\ CH_2O(P) \end{array} + SH{\cdot}G\text{-enzyme}$$

1,3-Diphospho-glyceric acid

XI. Formation of high-energy phosphate bond

phosphokinase, also called *phosphopherase*. Since each hexose molecule has used two molecules of ATP to reach this state, and each triose molecule (two from each hexose) has generated one molecule of ATP from ADP, the books are now even in regard to high-energy phosphate bonds.

The 3-phosphoglyceric acid (formula XII), which remains after the delivery of the high-energy phosphate, rearranges, catalyzed by *phosphoglyceromutase*, into 2-phosphoglyceric acid (formula XIII). This substance then loses water reversibly, in the presence of the enzyme *enolase*, to form phospho*enol*pyruvic acid (formula XIV). Here again a high-energy phosphate bond is generated; the phosphate group starts as an ester of a secondary alcohol and finishes as a high-energy enol ester. The phosphate can now be passed on with the aid of *pyruvic phosphokinase* to ADP, converting it to ATP; the residue is pyruvic acid. The ledger of high-energy phosphate bonds now shows a credit balance. Each triose molecule has generated one

$$\begin{array}{l} COOH \\ | \\ HC{-}OH \\ | \\ CH_2O(P) \end{array} \qquad \begin{array}{l} COOH \\ | \\ HC{-}O(P) \\ | \\ CH_2OH \end{array} \qquad \begin{array}{l} COOH \\ | \\ C{-}O{\sim}(P) \\ \| \\ CH_2 \end{array}$$

XII. 3-Phospho-glyceric acid; XIII. 2-Phosphogly-ceric acid; XIV. Phospho-*enol*pyruvic acid

extra high-energy bond; for each original hexose this makes two such bonds gained. Each hexose molecule has used two high-energy phosphate bonds from ATP, and has returned four such bonds to ADP.

All animal cells and tissues, as well as many plant cells and most of the familiar micro-organisms, use pyruvic acid both in synthetic processes and as a source of energy. In alkaline solutions, even of such low alkalinity as prevails in animal tissues, pyruvic acid exists as a keto-enol equilibrium mixture (formula XV). The existence of the two forms increases the re-

$$\begin{array}{c} CH_3 \\ | \\ C{=}O \\ | \\ COOH \end{array} \quad \rightleftharpoons \quad \begin{array}{c} CH_2 \\ \| \\ C{-}OH \\ | \\ COOH \end{array}$$

XV. Pyruvic acid

activity of pyruvic acid. It is able to take part in numerous condensations, dismutations, and oxido-reductive reactions, in the majority of which thiamine pyrophosphate or cocarboxylase is a necessary coenzyme.

In the fermentation of hexoses by yeast, two additional steps occur in the glycolytic process. Pyruvic acid loses CO_2 to form acetaldehyde, and the acetaldehyde is reduced to ethyl alcohol. These steps are absent or insignificant in animal metabolism.

In animal tissues reduction of pyruvic acid to *lactic acid* occurs under conditions of low oxygen concentration, since under such conditions DPN is available chiefly in the reduced form. The reaction is catalyzed reversibly by *lactic acid dehydrogenase,* with oxidation of DPN. Lactic acid is the end product of glycolysis, when the process is carried on under anaerobic conditions. Compare the empirical formula of lactic acid ($C_3H_6O_3$) with that of glucose ($C_6H_{12}O_6$) and you will see that the over-all glycolytic reaction is the splitting of glucose into two equal parts. Oxygen is not introduced, nor is carbon dioxide produced. Lactic acid is a metabolic blind alley in the animal organism. It can only be reconverted to pyruvic acid or excreted. In the brain, increased concentration of lactic acid and decreased concentration of high-energy phosphate has been observed in animals killed during hypoxic states. Similar observations have been made in animals in states of shock (see Chapter 17), in convulsions, and following injury. A purely emotional disturbance, such as fright (75), can bring about an increase in lactic acid even in rats treated with curare which prevents muscular production of excess lactic acid. Brain lactic acid concentrations are decreased in rats during sleep.

Lactic acid and pyruvic acid coexist in the blood. Since both are diffusible substances, they presumably coexist in the cells in proportions similar to those in blood. The average levels in the blood for a resting and fasting human subject are 8.2 mgm. lactic acid and 0.78 mgm. pyruvic acid per 100 ml. of blood (37). Increases of both substances have been consistently demonstrated after mild exercise and after glucose ingestion.

It should be borne in mind that at body pH, carboxylic acids exist as the negatively charged ions. Though this fact is ignored here, as being a complication that contributes little to the exposition, it is nevertheless common (and more precise) to read of pyruvate and lactate in the literature rather than of pyruvic acid and lactic acid. The same is true of carboxylic acids occurring as intermediates in other portions of the general metabolic scheme, and malate, fumarate, or succinate may be read in place of malic, fumaric and succinic acids.

Inhibitors of glycolysis. The cytochrome oxidase system, which operates in aerobic oxidations is inhibited by salts of hydrazoic acid, such as *sodium azide* (NaN_3). Azides also have the property of dissociating glycolysis from the transfer of phosphate-bond energy. Glycolysis continues, but those cell activities which depend upon it for energy cease. The uncoupling occurs while 1,3-diphosphoglyceric acid is still bound to phosphoglyceric phosphokinase (82). The chief inhibiting effect of *fluorides* upon glycolysis is exerted by action upon enolase. This is indicated by the accumulating of 2-phosphoglyceric acid in fluorided preparations. *Iodoacetic acid* inhibits triosephosphate dehydrogenase.

Alternate Metabolic Pathways

In outlining the scheme of glycolysis we must avoid giving the impression that the utilization of any given substance by the body rigidly follows one and only one path. The routes of intermediate metabolism may vary from tissue to tissue and in the same tissue under differing conditions (71).

A part of the glycolytic pathway can be by-passed by means of a degradation of glucose to the three-carbon stage through aerobic mechanisms. The first portion of this alternate pathway has already been described (see page 535) where the reactions are listed beginning with glucose-6-phosphate and ending with ribose-5-phosphate. In proceeding from glucose-6-phosphate to ribose-5-phosphate, note that there has been a net loss of two hydrogen atoms, and that what was originally the carbon 1 of the glucose molecule has been lost as CO_2.

The ribose-5-phosphate thus produced can be utilized for nucleic acid anabolism or it can be catabolized further. The ribose of purine nucleotides can be utilized by red cells (19) at 4°C., the temperature used most commonly for storage of blood for transfusion. The addition of inosine or adenosine to stored blood lengthens the time of satisfactory storage. Red cell ghosts can also utilize the ribose of purine nucleotides (51).

The first step of the catabolism of ribose-5-phosphate involves transketolation (see page 452) with ribulose-5-phosphate (its precursor) to produce *sedoheptulose-7-phosphate* and glyceraldehyde-3-phosphate (formula XVI). Glyceraldehyde-3-phosphate is a member of the anaerobic glycolytic pathway, and it may enter that pathway or it may react with sedoheptulose-7-phosphate again, transferring a 3-carbon fragment this time to form fructose-6-phosphate and erythrose-4-phosphate (formula XVII). This transfer of a $CH_2OH—CO—CHOH—$ is referred to as a *transaldolation* and is catalyzed by a *transaldolase* (36).

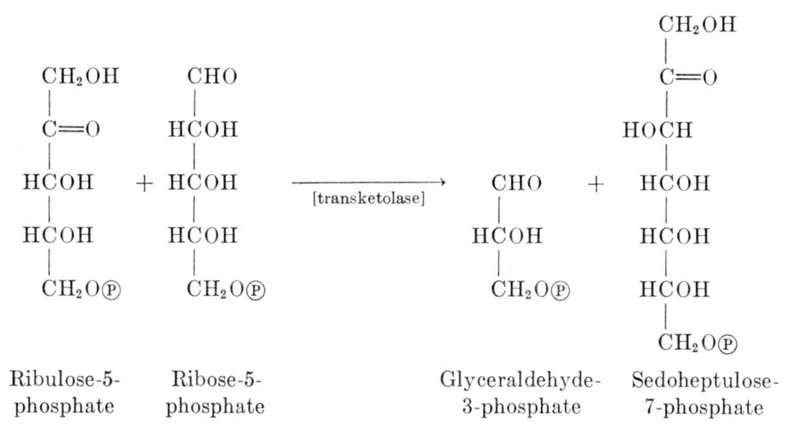

XVI. Phosphogluconate oxidation pathway—first transketolation

Sedoheptulose-7-phosphate		Glyceraldehyde-3-phosphate	→ [transaldolase]	Erythrose-4-phosphate		Fructose-6-phosphate
CH_2OH						
$C{=}O$						CH_2OH
$HOCH$						$C{=}O$
$HCOH$	+	CHO	→	CHO	+	$HOCH$
$HCOH$		$HCOH$		$HCOH$		$HCOH$
$HCOH$		$CH_2O(P)$		$HCOH$		$HCOH$
$CH_2O(P)$				$CH_2O(P)$		$CH_2O(P)$

XVII. Phosphogluconate oxidation pathway—transaldolation

The fructose-6-phosphate is a member of the anaerobic glycolytic pathway and can be converted to glucose-6-phosphate and may then be oxidized to ribulose-5-phosphate once more, losing a carbon as carbon dioxide.

It can be shown that this lost carbon is the carbon 2 of the original glucose molecule.

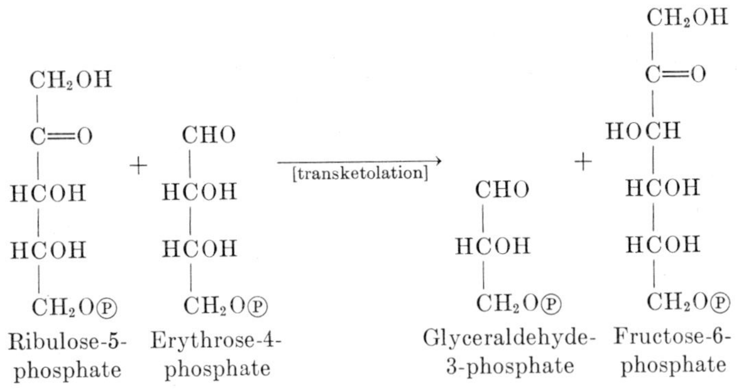

XVIII. Phosphogluconate oxidation pathway—second transketolation

Meanwhile the erythrose-4-phosphate can react with a molecule of ribulose-5-phosphate to form, by way of another transketolation, glyceraldehyde-3-phosphate and fructose-6-phosphate (formula XVIII). Again, fructose-6-phosphate can travel by way of glucose-6-phosphate to ribulose-5-phosphate, losing as carbon dioxide what had once been carbon 3 of the original glucose molecule.

What has now been described has been called by a number of names, of which the most descriptive is the *phosphogluconate oxidation pathway*. Alternate names are Dickens (or Warburg-Lipmann-Dickens) shunt, or direct oxidative pathway. The reactions involved are summarized in the diagram shown in formula XIX.

The phosphogluconate oxidation pathway differs from the main glycolytic pathway in that half the glucose molecule is aerobically oxidized. Aerobic oxidation, as we shall see, produces considerably more energy than does anaerobic glycolysis. The shunt is useful for that reason where the tissue oxygen supply is adequate but, of course, inoperative where the oxygen supply is low. Tissues in which the oxygen supply is under a continuous heavy drain, such as skeletal muscle (86) or brain (9) make use of glycolysis only. Glandular tissues, such as liver (41), where the oxygen supply is not subject to overload, make considerable use of the phosphogluconate oxidation pathway. In perfused rat liver, 56 per cent of the glucose oxidized goes by the phosphogluconate oxidation pathway (62a). For a detailed discussion of the significance of alternate pathways of carbohydrate catabolism, see (92a).

Another example involves fructose-1-phosphate. Ordinarily, this enters the glycolytic pathway by conversion to fructose-1,6-diphosphate by the enzyme phosphofructomutase (see table XIV-1). However, in the liver,

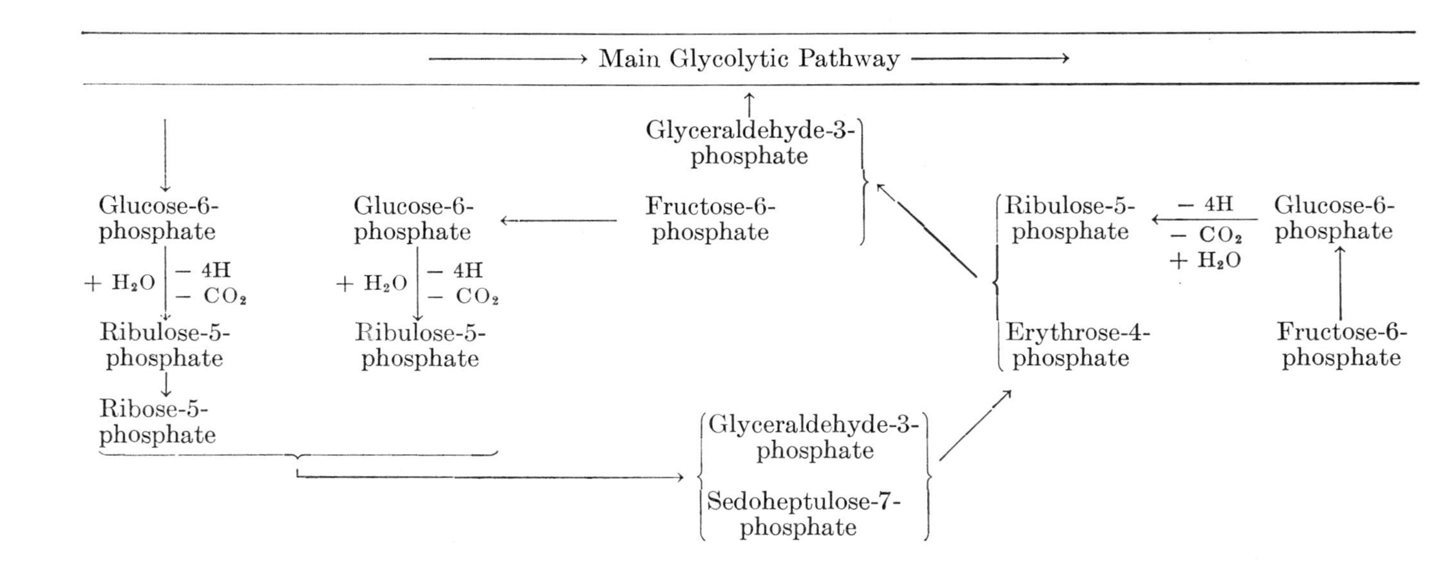

Over-all equation:

$$\text{Glucose-6-phosphate} + 3\,O_2 \rightarrow \text{Glyceraldehyde-3-phosphate} + 3\,CO_2 + 3\,H_2O$$

XIX. Phosphogluconate oxidation pathway—summary

fructose-1-phosphate can be split by aldolase to glyceraldehyde and dihydroxyacetone phosphate. The latter compound is already on the mainstream of glycolysis. Glyceraldehyde can be placed upon that mainstream by being phosphorylated (with the expenditure of an ATP high-energy phosphate and the catalytic influence of *triosekinase*) to glyceraldehyde-3-phosphate (34). In some micro-organisms, glucose is oxidized without previous phosphorylation. *Notatin*, or glucose dehydrogenase converts glucose to delta-gluconolactone (see page 109) which forms gluconic acid. A similar oxidation occurs in mammalian liver, but its importance in metabolism is doubtful (48).

Tricarboxylic Acid Cycle

This is also referred to as the citric acid cycle or the Krebs cycle. It is a series of reactions in which the carbon atoms of pyruvic acid are oxidized one at a time to CO_2, with stepwise liberation of energy which is transferred to mechanical or chemical effectors by the generation of high-energy phosphate bonds. Not only is the pyruvic acid derived by glycolysis of hexose sugars so oxidized with release of available energy, but so also are metabolic intermediates of lipids and proteins oxidized in the tricarboxylic acid cycle and energy similarly made available. Such inclusiveness makes this oxidative mechanism singularly important; added to this we have the fact that the *cyclophorase* system of enzymes, which catalyzes the cycle, can be demonstrated in muscle, liver, brain, and kidney. While complete positive evidence is lacking for tissues other than those named, there is quite general agreement that the tricarboxylic acid cycle is the chief mechanism for the oxidative utilization of foodstuffs as sources of energy. In micro-organisms, many modifications of the tricarboxylic acid cycle have been observed. The cycle will be presented here in summary form; a detailed discussion of the history of the concept, and the evidence upon which it has been established, has been given by Krebs in his Harvey Lecture (44).

The cycle begins and ends with oxaloacetic acid (formula XX). This acid can be formed independently of the cycle by the reaction of carbon dioxide with phospho*enol*pyruvic acid (formula XXI). This is essentially the addi-

$$\begin{array}{l} COOH \\ | \\ C{=}O \\ | \\ CH_2 \\ | \\ COOH \end{array}$$

XX. Oxaloacetic acid

tion of CO_2 to pyruvic acid. Oxaloacetic acid can also be formed by the oxidative deamination of aspartic acid (see Chapter 16).

$$CH_2{=}\overset{\overset{\text{O}Ⓟ}{|}}{C}{-}COOH + CO_2 + IDP \rightarrow \underset{\underset{COOH}{|}}{CH_2}\overset{\overset{O}{\|}}{C}COOH + ITP$$

Phosphoenolpyruvic acid — Oxaloacetic acid

XXI. Formation of oxaloacetic acid

Pyruvic acid and oxaloacetic acid are condensed to form citric acid through intermediates involving *lipothiamide pyrophosphate* (LTPP) and *coenzyme A* (CoA). For the formula of LTPP, see page 796, and for that of CoA, page 804. The active group of LTPP is a cyclic disulfide link, which may be represented here as $\underset{S}{\overset{S}{|}}\!\!>\!\text{LTPP}$, while CoA has as its active group a sulfhydryl, so that it may be represented as HS—CoA.

The entry of pyruvic acid into the tricarboxylic acid cycle is accomplished in three major steps, a decarboxylation and two transacetylations (66). The decarboxylation (formula XXII) results in the formation of a high-energy acyl-mercaptan bond.

$$CH_3\overset{\overset{O}{\|}}{C}COOH + \underset{S}{\overset{S}{|}}\!\!>\!\text{LTPP} \xrightarrow{\text{decarboxylation}} CH_3\overset{\overset{O}{\|}}{C}{\sim}S{-}\text{LTPP} + CO_2$$
$$\qquad\qquad\qquad\qquad\qquad\qquad\qquad\qquad\qquad\quad / \\ \qquad\qquad\qquad\qquad\qquad\qquad\qquad\qquad\quad HS$$

Pyruvic acid — Acetyllipothiamide pyrophosphate

XXII. Entry of pyruvic acid into the tricarboxylic acid cycle—step 1

With the removal of CO_2, what is left of the pyruvic acid is an acetyl group. This is transferred to CoA (formula XXIII) to form *acetylcoenzyme A*, which also contains the high-energy acyl-mercaptan bond. The energy for the formation of this bond comes from the over-all loss of two hydrogen atoms in the reactions presented in formulas XXII and XXIII, with the net result that LTPP is reduced to $LTPP \cdot H_2$. The reduced LTPP is dehydrogenated to the oxidized form once more by passing its hydrogen atoms to DPN, which in turn passes them on through the usual respiratory chain by way of the cytochromes to molecular oxygen.

It is acetylcoenzyme A (sometimes referred to as *active acetate* or *the two-carbon fragment*) that transfers the acetyl group to oxaloacetic acid to form citric acid and initiate the tricarboxylic acid cycle (formula XXIV).

$$CH_3C(=O)\sim S\text{—}LTPP(\text{—}HS) + HS\text{—}CoA \xrightarrow{\text{transacetylation}} CH_3C(=O)\sim SCoA + (HS)_2LTPP$$

Acetyllipothiamide pyrophosphate — Acetylcoenzyme A — Reduced lipothiamide pyrophosphate

XXIII. Entry of pyruvic acid into the tricarboxylic acid cycle—step 2

$$\underset{\text{Oxaloacetic acid}}{COOH\text{—}C(=O)\text{—}CH_2\text{—}COOH} + \underset{\text{Acetyl-coenzyme A}}{CH_3\text{—}C(=O)\sim S\text{—}CoA} + H_2O \rightarrow \underset{\text{Citric acid}}{COOH\text{—}HOC(CH_2COOH)\text{—}CH_2\text{—}COOH} + \underset{\text{Coenzyme A}}{HS\text{—}CoA}$$

XXIV. Entry of pyruvic acid into the tricarboxylic acid cycle—step 3

Wherever citric acid is found in biological systems, there also are two other tricarboxylic acids, *cis-aconitic acid* and *isocitric acid*, which form an equilibrium mixture (formula XXV). The enzyme, *aconitase*, catalyzes the reversible dehydrations. Chemical equilibrium is never attained since citric acid is always being formed while isocitric acid is always being removed by dehydrogenation as will be described below. Although the concentration of each acid does not vary greatly under ordinary conditions. this is not chemical equilibrium but is what is known as a *steady state*.

$$\underset{\text{Citric acid}}{COOH\text{—}C(OH)(CH_2COOH)\text{—}CH_2\text{—}COOH} \underset{+H_2O}{\overset{-H_2O}{\rightleftharpoons}} \underset{\textit{Cis}\text{aconitic acid}}{COOH\text{—}C(=CHCOOH)\text{—}CH_2\text{—}COOH} \underset{-H_2O}{\overset{+H_2O}{\rightleftharpoons}} \underset{\text{Isocitric acid}}{COOH\text{—}HC(CH(OH)COOH)\text{—}CH_2\text{—}COOH}$$

XXV. Citric acid equilibrium mixture

Isocitric acid is reversibly oxidized to oxalosuccinic acid (formula XXVI), catalyzed by *isocitric dehydrogenase*, with reduction of the oxidized form of TPN.

Oxalosuccinic acid is reversibly decarboxylated in the presence of Mn^{++}

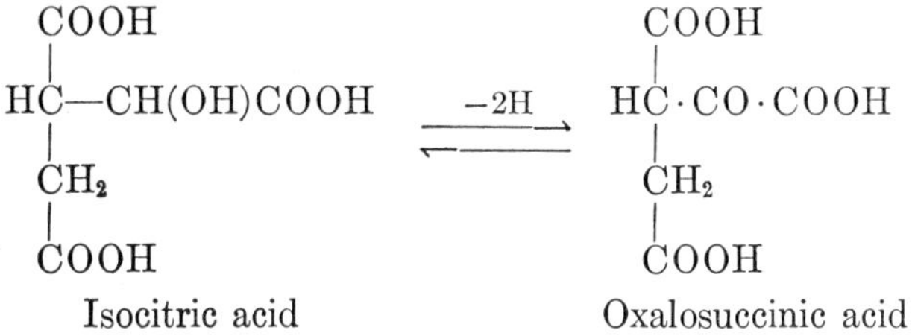

$$\begin{array}{c} COOH \\ | \\ HC—CH(OH)COOH \\ | \\ CH_2 \\ | \\ COOH \\ \text{Isocitric acid} \end{array} \xrightleftharpoons{-2H} \begin{array}{c} COOH \\ | \\ HC \cdot CO \cdot COOH \\ | \\ CH_2 \\ | \\ COOH \\ \text{Oxalosuccinic acid} \end{array}$$

XXVI

and oxalosuccinic carboxylase, which has been shown to be identical with isocitric acid dehydrogenase (62), to form CO_2 and *alpha-ketoglutaric acid* (formula XXVII).

Alpha-ketoglutaric acid undergoes an oxidative decarboxylation by a mechanism similar to that described earlier in connection with pyruvic acid (66). Here, too, coenzyme A is involved. The first step in the reaction is the decarboxylation of alpha-ketoglutaric acid and the condensation of the resulting product, *succinic acid*, with CoA (formula XXVIII), a process catalyzed by *alpha-ketoglutaric acid dehydrogenase*. Notice that in the process, the high-energy acyl-mercaptan bond of *succinylcoenzyme A* is formed. The energy for the formation of that bond is derived from the

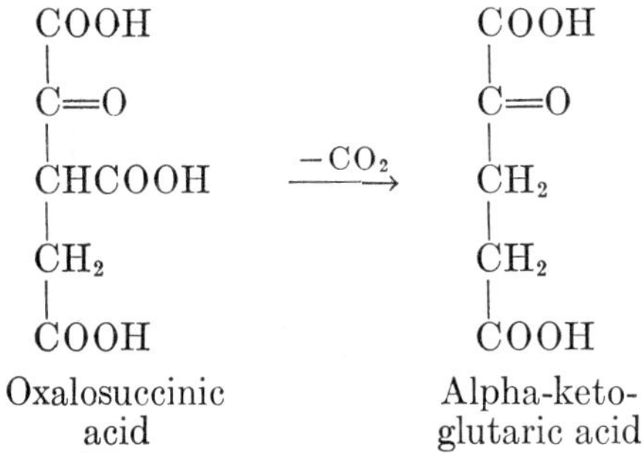

$$\begin{array}{c} COOH \\ | \\ C=O \\ | \\ CHCOOH \\ | \\ CH_2 \\ | \\ COOH \\ \text{Oxalosuccinic} \\ \text{acid} \end{array} \xrightarrow{-CO_2} \begin{array}{c} COOH \\ | \\ C=O \\ | \\ CH_2 \\ | \\ CH_2 \\ | \\ COOH \\ \text{Alpha-keto-} \\ \text{glutaric acid} \end{array}$$

XXVII

$$\begin{array}{c} COOH \\ | \\ CH_2 \\ | \\ CH_2 \\ | \\ C=O \\ | \\ COOH \\ \text{Alpha-ketoglutaric} \\ \text{acid} \end{array} + HS—CoA \xrightarrow{-2H} \begin{array}{c} COOH \\ | \\ CH_2 \\ | \\ CH_2 \\ | \\ C=O \\ \wr \\ SCoA \\ \text{Succinyl-} \\ \text{coenzyme A} \end{array} + CO_2$$

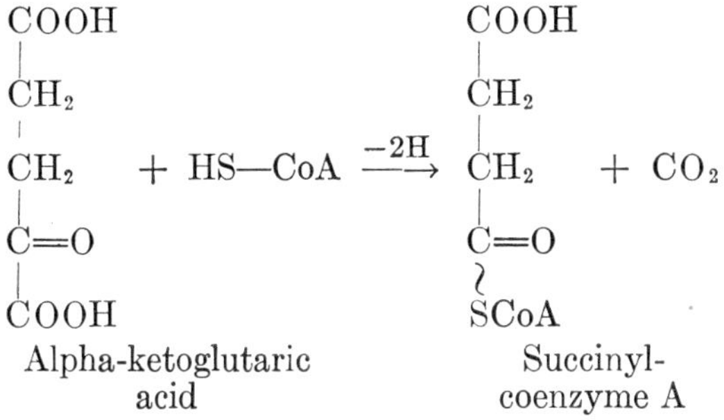

XXVIII. Decarboxylation of alpha-ketoglutaric acid—step 1

simultaneous dehydrogenation that proceeds, the two hydrogen atoms liberated being used to reduce a molecule of DPN and entering, in that way, the respiratory chain.

The succinylcoenzyme A, under the influence of "*P enzyme*", is then hydrolyzed to form succinic acid and CoA, the energy of the acyl-mercaptan bond being utilized to form a high-energy phosphate bond (formula XXIX). As is usual in the case of reactions involving phosphate transfer, Mg^{++} is required as activator. This is the only reaction in the tricarboxylic acid cycle which, like two of the reactions in the glycolytic chain, results in the direct formation of a high-energy phosphate bond. There is evidence (79) that ATP is not formed directly in this reaction but that GTP is first formed from GDP and that then the reaction GTP + ADP → ATP + GDP takes place.

$$\begin{matrix} COOH \\ | \\ CH_2 \\ | \\ CH_2 \\ | \\ C{=}O \\ \wr \\ SCoA \end{matrix} + ADP + H_3PO_4 \rightarrow \begin{matrix} COOH \\ | \\ CH_2 \\ | \\ CH_2 \\ | \\ COOH \end{matrix} + ATP + CoA$$

Succinyl-coenzyme A → Succinic acid

XXIX. Decarboxylation of alpha-ketoglutaric acid—step 2

Succinic acid is now reconverted to oxaloacetic acid by the *dicarboxylic acid cycle* (formula XXX). In this cycle, succinic acid is first dehydrogenated under the influence of *succinic acid dehydrogenase* to fumaric acid, which is hydrated in the presence of *fumarase* to malic acid, which is then dehydrogenated through the catalytic activity of *malic acid dehydrogenase* to oxaloacetic acid. Of the various dehydrogenases involved in the tricarboxylic acid cycle, succinic acid dehydrogenase is unique in not being a pyridinoenzyme. Neither DPN nor TPN is involved in the dehydrogena-

$$\begin{matrix} COOH \\ | \\ CH_2 \\ | \\ CH_2 \\ | \\ COOH \end{matrix} \xrightarrow{-2H} \begin{matrix} COOH \\ | \\ CH \\ \| \\ CH \\ | \\ COOH \end{matrix} \xrightarrow{+H_2O} \begin{matrix} COOH \\ | \\ CHOH \\ | \\ CH_2 \\ | \\ COOH \end{matrix} \xrightarrow{-2H} \begin{matrix} COOH \\ | \\ C{=}O \\ | \\ CH_2 \\ | \\ COOH \end{matrix}$$

Succinic acid → Fumaric acid → Malic acid → Oxaloacetic acid

XXX. Dicarboxylic acid cycle

tion of succinic acid. Succinic acid dehydrogenase is a metallo-flavo-enzyme, the metal involved being iron (81). The molecular weight of this enzyme is about 200,000, and four atoms of iron and one flavin dinucleotide are contained in each molecule.

The tricarboxylic acid cycle is summarized in formula XXXI, and the over-all equation is given in formula XXXII. The liberated hydrogens indicated in formula XXXII find their way eventually into combination with molecular oxygen by way of the respiratory enzyme chain. The five

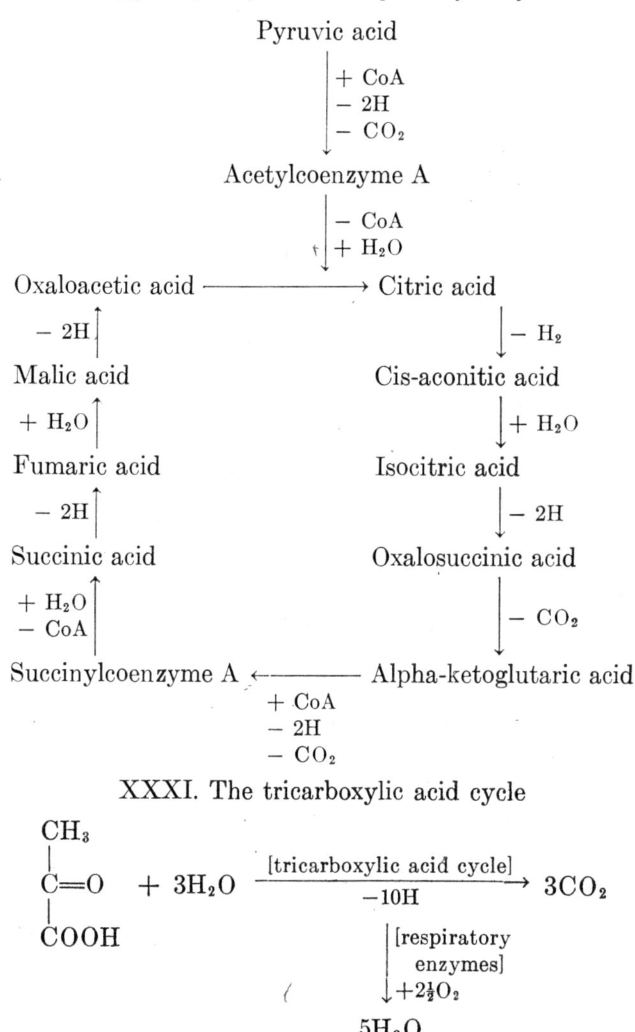

XXXI. The tricarboxylic acid cycle

XXXII. Over-all equation of the tricarboxylic acid cycle

water molecules thus formed more than make up for the three invested and the net result is that each molecule of pyruvic acid is oxidized to three molecules of carbon dioxide and two of water.

All the steps involved in both glycolysis and the tricarboxylic acid cycle are either directly or indirectly reversible. Those which are not directly reversible are naturally those involving an ATP hydrolysis. There the reverse reaction does *not* involve re-formation of ATP but is a phosphatase-catalyzed hydrolysis with the production of inorganic phosphate. Thus, glucose-6-phosphate is hydrolyzed to glucose by glucose-6-phosphatase and fructose-1,6-diphosphate is hydrolyzed to fructose-6-phosphate by fructose-1,6-diphosphatase (16).

The similarity in the metabolic route followed by pyruvic acid and alpha-ketoglutaric acid, described in the text above and indicated in formula XXXI is such that competitive inhibition between the two steps has been observed and the relative concentrations of pyruvic acid and alpha-ketoglutaric acid may be one factor controlling the rate at which the tricarboxylic acid cycle proceeds (60a).

The *cyclophorase system* is the group of enzymes and coenzymes necessary and sufficient to catalyze the reactions composing the tricarboxylic acid cycle and is associated with mitochondria. When mitochondrial preparations are broken up mechanically, numerous enzymes are liberated which catalyze individual reactions of the tricarboxylic acid cycle, but the orderly sequence of the cycle as a whole no longer is maintained. In the normal operation of the cycle no accumulation of intermediates occurs, and each enzyme works effectively even at minute concentrations of substrate. The entering materials appear to be taken up by the appropriate enzymes and then passed on in order to other enzymes which are probably located in definite positions on the surface of the mitochondrion. Unlike the situation when enzymes act in solution, interaction between enzyme and substrate does not appear to depend upon random collision.

Energetics of carbohydrate catabolism. The total decrease in free energy in passing from glucose to carbon dioxide and water is about 690 Kcal. per mol glucose (45). Of this, the glycolytic conversion of glucose to two lactic acid molecules is responsible for 50 Kcal. per mol glucose or about 7 per cent of the total. The conversion of lactic acid to carbon dioxide and water releases the remaining 320 Kcal. per mol of lactic acid (or 640 Kcal. per mol of original glucose).

In order for this free energy to be used by the body, some of it at least must be converted into high-energy phosphate bonds. We have thus far indicated only three places where high-energy phosphate bonds are formed during the conversion of glucose to CO_2 and H_2O. Two of these are in the glycolytic chain: in the formation of 1,3-diphosphoglyceric acid and in the formation of phospho*enol*pyruvic acid. One is in the tricarboxylic acid cycle in the hydrolysis of succinylcoenzyme A. Since a single glucose molecule gives rise to two of each of the compounds named, six high-energy phosphate bonds are formed. Allowing for the two high-energy phosphate bonds invested in the glucose molecule at the beginning of the glycolytic chain, we account for four high-energy phosphate bonds per molecule of glucose. Many more such bonds must be formed, however, if the body's energy needs are to be supplied.

If the free energy losses involved in the individual steps of carbohydrate catabolism are measured (46), it is seen that reactions involving gains and losses of carbon dioxide or water result in no great free energy changes. Thus the hydration of *cis*-aconitic acid to isocitric acid involves a free energy decrease of only 0.45 Kcal. per mol while that for the decarboxylation of oxalosuccinic acid to alpha-ketoglutaric acid is only 8.6 Kcal. per mol.

The case is quite different in reactions involving a dehydrogenation. There free energy losses of from 35 to 70 Kcal. per mol are involved. There are six places in the glycolytic chain and the tricarboxylic acid cycle where dehydrogenations take place. These are:

1. The dehydrogenation of glyceraldehyde-3-phosphate to 1,3-diphosphoglyceric acid.
2. The dehydrogenation of pyruvic acid and coenzyme A to acetylcoenzyme A.
3. The dehydrogenation of isocitric acid to oxalosuccinic acid.
4. The dehydrogenation of alpha-ketoglutaric acid and coenzyme A to succinylcoenzyme A.
5. The dehydrogenation of succinic acid to fumaric acid.
6. The dehydrogenation of malic acid to oxaloacetic acid.

In each of these reactions, enough free energy is lost to enable more than one high-energy phosphate to be formed by a process termed *oxidative phosphorylation*. In dehydrogenations catalyzed by pyridinoenzymes, three high-energy phosphate bonds are formed for every pair of hydrogen atoms removed. Since the hydrogen atoms are eventually combined with molecular oxygen by way of the respiratory enzymes: pyridinoenzymes → metalloflavo-enzymes → cytochromes → oxygen, it is tempting to suppose that one high-energy phosphate bond is formed at each stage, and indeed, there is some evidence tending to place formation of one of the bonds in the reactions by which hydrogen atoms are transferred from the cytochromes to oxygen. Another point in favor of this is that the dehydrogenation of succinic acid to fumaric acid, which is the only one of the six dehydrogenations that does not involve a pyridinoenzyme but which begins at the metallo-flavo-enzyme stage, is also the only dehydrogenation to produce two high-energy phosphate bonds rather than three.

In passing from glyceraldehyde-3-phosphate to carbon dioxide and water, there are thus 17 high-energy phosphate bonds formed by oxidative phosphorylation (two at dehydrogenation step number 5 in the list above and three at each of the others). Since each glucose molecule gives rise to two glyceraldehyde-3-phosphate molecules, there are 34 such bonds formed in going from glucose to CO_2 and H_2O. Add to these the four bonds mentioned earlier as having been formed directly and the total number is 38.

If we consider the free energy of formation of a high-energy phosphate bond to be 4.7 Kcal. per mol (see page 228), then the formation of 38 of them requires 178.6 Kcal. This represents just a little over one-quarter of the total free energy loss in the oxidation of glucose.

In the series of reactions involved in anaerobic glycolysis, there is only one dehydrogenation and that is the first on the list on page 555. There is no net gain in high-energy phosphate bonds because of that step, however since DPN is re-oxidized and the dehydrogenation cancelled in the reduction of pyruvic acid to lactic acid. The net gain in high-energy phosphate bonds in anaerobic glycolysis, that is, in the conversion of glucose to lactic acid, is, therefore, only two per glucose molecule. Of the 50 Kcal. free energy lost in the process, only 9.4 is stored in high-energy phosphate bonds, an efficiency of less than 20 per cent.

In passing from lactic acid to carbon dioxide and water, there is first the dehydrogenation of lactic acid to pyruvic acid by lactic acid dehydrogenase, a pyridinoenzyme. This involves the production of three high-energy phosphate bonds. After that, the oxidation of pyruvic acid proceeds by way of dehydrogenations 2 to 6 in the list on page 555. The number of high-energy phosphate bonds produced in the oxidation of glucose is thus the same whether lactic acid is formed in the intermediate stages or not.

It has been stated that the tricarboxylic acid cycle is reversible, and this is true. However, it is obvious that such a reversal is an energy-consuming process. Allowing for the 25 per cent efficiency with which the cycle converts the free energy of glucose oxidation into high-energy phosphate, it takes at least four cycles moving in the catabolic direction to supply the energy for one cycle moving in the anabolic direction. Whatever the direction of the cycle in localized parts of the body at some particular time or other, it can be seen then that, on the whole, the net motion of the tricarboxylic acid cycle must be in the catabolic direction.

Fixation of carbon dioxide. When CO_2 or bicarbonate containing an isotopic carbon atom is administered to experimental animals, some of the carbon isotope finds its way into organic compounds of the animal's body.

Such incorporation of CO_2 takes place by four known reactions:

1. CO_2 plus phosphoenolpyruvic acid, yielding oxaloacetic acid;
2. CO_2 plus alpha-ketoglutaric acid, yielding oxalosuccinic acid;
3. CO_2 plus ornithine, yielding citrulline (see page 606).
4. CO_2 plus acetone, yielding acetoacetic acid.

Undoubtedly other physiological reactions utilize CO_2 synthetically. For example, CO_2 carbon appears in purines and pyrimidines in amounts not fully explicable by the reactions cited. The appearance of isotopic carbon from CO_2 in glycogen and glucose is explained as the result of gluconeogenesis from oxalosuccinic or oxaloacetic acids. Isotopic carbon from these substances can appear in fats, since fats may be synthesized in the body from carbohydrate.

Muscular Contraction

The catabolism of carbohydrate is the immediate source of the energy needed in muscular contraction. The first chemical reaction known to take place in muscle stimulated by a nerve impulse is the hydrolysis of ATP to ADP. This hydrolysis is catalyzed by the enzyme *adenosine triphosphatase*, or ATP-ase. The energy thus released is used to polymerize G-actin to F-actin through a postulated high-energy intermediate, G-actin $\sim$ Ⓟ. One mol of ATP is consumed in the polymerization of one mol of G-actin (57,000 grams). Assuming a gram of muscle to contain 25 to 30 mg. of actin, muscular contraction involves the hydrolysis of 4 to 5×10^{-7} mols ATP per gram (59, 60). ATP-ase is not absolutely specific but will catalyze the hydrolysis of other polyphosphates, such as GTP and UTP (41a).

The ATP-ase of muscle is closely associated with actomyosin, and, in particular with the H-meromyosin fragment (see page 150). Lipoprotein granules have been isolated from skeletal muscle (68) with ATP-ase activity. This activity, however, is considered to be distinct from actomyosin ATP-ase.

Only *after* at least some of the ATP present in muscle is hydrolyzed to ADP, does phosphocreatine take part in the process. It reacts with ADP, forming ATP once again by phosphate transfer, being itself converted to creatine. This reaction is reversible, since the high-energy bonds in the two compounds are about equivalent in energy content. With ATP disappearing in the process of muscular contraction, however, the reaction is forced in the direction indicated. The relationship of the two reactions may be shown by means of a method of representation introduced by Baldwin (2) and sometimes referred to as "Baldwin cycles" (formula XXXIII). It will be noted that a molecule of ATP is alternately hydrolyzed and re-formed, so that a great deal of energy may be transferred through a relatively small amount of ATP which acts as an "energy shuttle".

Why the necessity for this two-step hydrolysis process? In the first place, phosphocreatine acts as a high-energy phosphate bond reserve. One hundred grams of rat muscle, for instance, contains 20 milligrams of high-energy

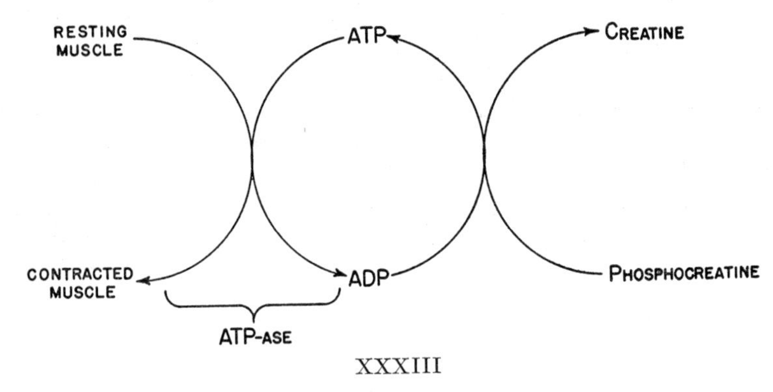

XXXIII

phosphate derivable from the conversion of ATP to ADP and 75 milligrams, nearly four times as much, of high-energy phosphate derivable from the conversion of phosphocreatine to creatine. It might be supposed that since the phosphate bonds of ATP and phosphocreatine are virtually equivalent, the entire 95 milligrams of phosphate might as well be all one or all the other—and the process thus simplified. Actually, there is an important difference between the two compounds. The ATP-ase in muscle catalyzes the hydrolysis of ATP only; it has no effect upon phosphocreatine. Thus, if all the high-energy phosphate were in the form of phosphocreatine then, utilizing actual muscle biochemical mechanisms, none of its energy could be made available for contraction. On the other hand, if the high-energy phosphate were all in the form of ATP, the effect of the ATP-ase would be to explode that energy too rapidly. The initial impulse which sets off the ATP is limited by the fact that only a small quantity of ATP is present. Action, thereafter, is continued in a more controlled fashion by allowing the phosphocreatine-ADP reaction to act as a bottleneck.

The total supply of ATP and phosphocreatine together is insufficient to support muscular contraction for more than a very limited time. However, ATP can be resynthesized through the glycolytic breakdown of muscle glycogen. If a muscle is poisoned with iodoacetate (which inhibits glycolysis), ATP resynthesis is prevented and on repeated stimulation of the muscle, its ATP will be entirely converted to ADP. ADP, as a last gasp, may be reconverted *in part* to ATP through the action of myokinase, an enzyme found in muscle which builds up a molecule of ADP to ATP at the expense of hydrolyzing a second molecule to adenylic acid (AMP) (formula XXXIV). This is possible since ADP still has one high-energy phosphate bond which can be utilized for the transfer. However, this is only

ADP + ADP $\xrightarrow{\text{[myokinase]}}$ ATP + Adenylic acid (AMP)

XXXIV

the most temporary of solutions since the AMP formed is deaminated to inosinic acid and can then no longer re-enter the phosphate cycle (formula XXXV). At each step of resynthesis of ATP by this method, therefore, half the remaining available phosphorus supply is lost and very quickly the muscle goes into final rigor. This is precisely what happens in *rigor mortis*. The extent of hardening after death parallels the decline in ATP content of the muscle.

In a muscle in normal contraction, glycogen is broken down and ATP is resynthesized through the formation of high-energy phosphate bonds during glycolysis. Phosphoenolpyruvic acid formed during glycolysis can donate its phosphate group to ADP, resynthesizing ATP and itself being converted to pyruvic acid. Pyruvic acid can be eventually oxidized to carbon dioxide and water if there is adequate molecular oxygen. Where, as in a muscle during strenuous exercise, the oxygen supply is insufficient to oxidize the pyruvic acid as quickly as it is formed from glycogen breakdown, pyruvic acid is reduced to lactic acid (formula XXXVI) and the latter accumulates.

NH_2 · N · N · N · N · O · H H · H · OH · OH · H · $—CH_2O—P(=O)(OH)_2$ $+ \; H_2O$

Adenylic acid

↓

OH · N · N · N · N · O · H H · H · OH · OH · H · $—CH_2O—P(=O)(OH)_2$ $+ \; NH_3$

Inosinic acid

XXXV

$$\underset{\text{Pyruvic acid}}{CH_3C(=O)—C(=O)OH} \;\underset{-\,2H}{\overset{+\,2H}{\rightleftharpoons}}\; \underset{\text{Lactic acid}}{CH_3CH(OH)C(=O)OH}$$

XXXVI

What, then, is the function of the phosphocreatine since the phosphoenolpyruvic acid formed from glycogen breakdown can substitute for it? Actually, glycolysis is a comparatively slow starting reaction and the need for muscular activity may be very urgent indeed. The phosphocreatine acts as an energy reserve which is ready at the moment and which can bridge the gap between the time of stimulus and the time when glycolysis is proceeding with sufficient rapidity. When muscular activity is over, there is a short period of *anaerobic recovery* during which glycogen continues to break down and lactic acid to accumulate—but the high-energy phosphate bonds are expended, not in resynthesizing ATP which for the moment is no longer being broken down, but in resynthesizing the phosphocreatine from creatine and thus restoring the original high-energy bond situation.

While the energy available within the muscle from glycogen breakdown is considerably more than the sum of that available from muscle ATP and phosphocreatine, it too is insufficient in extreme cases. A strenuously exercised muscle is eventually depleted of glycogen, crammed with lactic acid, and weary to exhaustion. It has been living on its capital, and the capital is gone. In order for the muscle to recover, the lactic acid must be reconverted to glycogen. This takes energy, and the energy is gained by the complete oxidation, aerobically, of lactic acid to carbon dioxide and water. Until this step, the various chemical changes in contracting muscle have not required molecular oxygen, glycolysis being an *anaerobic* process. The muscle after prolonged stimulation has thus incurred what is termed an *oxygen debt*, and this is paid off during the period of *aerobic recovery.*

The oxidation of lactic acid is carried on by first dehydrogenating it to pyruvic acid after which it enters the tricarboxylic acid cycle.

It will be seen that ultimately the energy is derived from the aerobic oxidation of glycogen to carbon dioxide and water (formula XXXVII). In muscles performing moderate activity, aerobic oxidation keeps up with the energy demands.

The exact way in which the energy of ATP hydrolysis is converted into muscular contraction is a matter of hot debate with about as many theories advanced as there are investigators. The possible function of ATP in the conversion of G-actin to F-actin has already been mentioned. Another possibility (61) is that the actomyosin fiber is maintained in an extended state by the mutually repulsive forces of positively-charged divalent ions adsorbed at intervals along the fiber. Under the influence of the nerve impulse, the ATP molecules (present as negative ions at the pH of their environment) are also adsorbed and neutralize these charges. The actomyosin molecule, unhampered by repulsive forces, then contracts since a long flexible fiber is more stable in a coiled state than in an extended one.

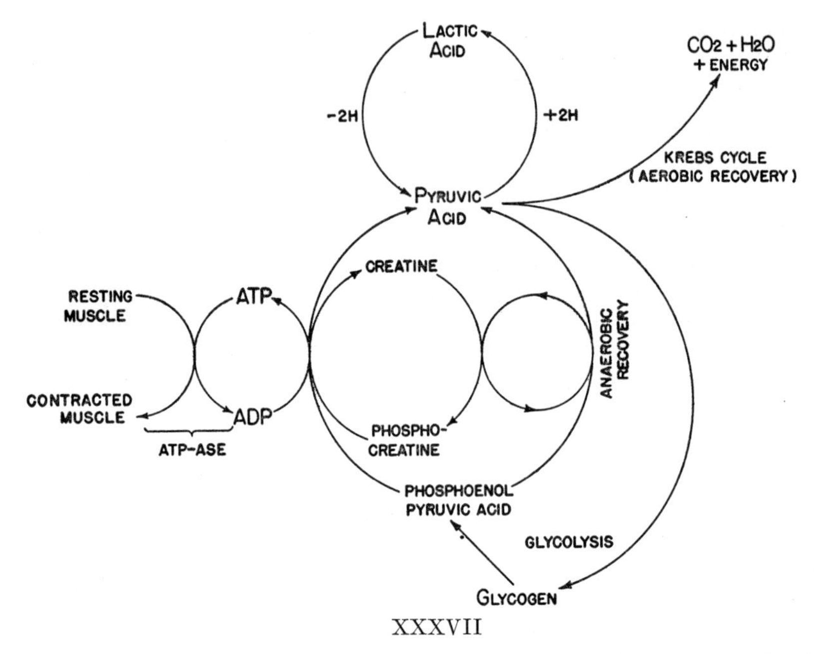

XXXVII

The ATP-ase in hydrolyzing the ATP forms ADP which is not adsorbed by the actomyosin and is therefore released. The positive net charge is restored, the fiber extends once more and the cycle is ready to begin again.

For detailed discussion of various theories of muscular contraction, see (61a, 64, 90).

DIABETES

Experimental diabetes in dogs was first produced by removal of the pancreas. This was accomplished by Mering and Minkowski in 1889. The organ fundamentally at fault in human diabetes mellitus is also the pancreas, specifically the beta cells of the islets of Langerhans (91). The sole known hormone produced by the beta cells is *insulin*. Diabetes in the human can always be attributed either to an absolute deficiency of insulin or to a relative deficiency. Relative deficiency may result from excessive activity of hormones antagonistic to insulin, arising from the adrenal cortex, anterior pituitary, or thyroid; less commonly insulin may be inactivated by insulinase or by specific antibodies.

The most striking manifestations of insulin deficiency are disturbances of carbohydrate utilization; glycosuria and hyperglycemia. The metabolism of fat is also highly abnormal, with failure to convert sugar to fat, and failure to oxidize fatty acids completely. Diabetic acidosis, or *ketosis*, results from the accumulation of acetoacetic acid, and its derivatives beta-hydroxybutyric acid and acetone; acetoacetic acid is a normal intermediate in the oxidative utilization of fatty acids. The lipids of the blood, including fats, phospholipids, and cholesterol, are increased. Protein metabolism, particularly protein synthesis, is disturbed, with excessive gluconeogenesis from protein the chief abnormality. Although we formally classify diabetes as a disease of carbohydrate metabolism, this distinction is historical and somewhat artificial. It is best considered as a disease of metabolism in general.

Insulin

Insulin is a protein hormone, with a molecular weight of about 36,000 (17) at pH 7; in acid solutions it dissociates into fragments of molecular weight 12,000, which are still physiologically active. The usual explanation is that the active unit of insulin is the "monomer" with a molecular weight of 12,000. Association takes place (favored by an increase in pH and in ionic strength and a decrease in temperature) through electrovalent bonding to the "trimer". Aggregates as high as the "pentamer" (molecular weight, 60,000) may be present (85). The 12,000 molecular weight subunit consists of two pairs of polypeptide chains (see page 77). On the basis of studies of dinitrophenyl-substituted insulin (30), it has been suggested that the true active subunit of insulin has a molecular weight of only 6,500, and consists of but one of each type of polypeptide chain known to be present. In alkaline solutions insulin quickly loses activity. Proteolysis destroys activity; chymotrypsin, for example, catalyzes the cleavage of insulin into peptides of molecular weight about 800 each, plus an inactive protein residue with 80 per cent of the original cystine. On account of the destructive action of some of the proteolytic digestive enzymes, insulin effects are minimal and unreliable when the drug is taken by mouth. The subcutaneous route is ordinarily used, the intravenous occasionally.

Banting and Best produced in 1921 the first successful insulin by ligation of the pancreatic ducts of dogs, causing atrophy of the externally secreting cells which produce proteolytic and other digestive enzymes and leaving only the islands of Langerhans. This precluded the destructive action of proteolytic enzymes which had previously prevented the extraction of insulin in water solution. Large scale production avoids this maneuver; whole pancreases from animals slaughtered for meat are extracted with strongly acidified alcohol. The extract is salted out with ammonium sulfate and the insulin precipitated at its isoelectric point (about pH 5).

Insulin is assayed in units; three units will depress the blood sugar of a fasting 2 kgm. rabbit to 0.045 per cent. Pure crystalline zinc insulinate contains 22 units per mgm. The dosage of insulin for each patient is a matter of individual study for his physician. The dosage is established, with the patient on a constant diet, by blood sugar measurements before and after insulin administration. Since there is no direct chemical or mathematical relationship between units of insulin administered and grams of carbohydrate metabolized, attempts at determining and utilizing such a relationship are fallacious and may be dangerously misleading. Overdosage of insulin produces hypoglycemia, with accompanying weakness, tremor, sweating, and often unconsciousness. Relief of hypoglycemia is by administration of sugars, preferably glucose, by mouth if the patient is conscious, intravenously if the patient is unconscious. A 20-gram dose of glucose by either route is usually adequate.

Insulin in human blood has been measured by bioassay on alloxan-diabetic (see page 573) hypophysectomized adrenalectomized rats (11). Fasting human blood contains about 0.1 unit of insulin per liter, increasing to as much as 0.34 unit per liter after glucose ingestion. The normal human pancreas contains about 2,000 units per kilogram, the average diabetic pancreas about 20 per cent of this figure. Application of the same bioassay technique to the blood plasma of human diabetics divides such patients into two distinct groups. One group has no demonstrable free plasma insulin, the other has a normal amount.

The functions of insulin. After administration of insulin, blood glucose is lowered, glycogenesis both in liver and muscle is increased and the glycolysis of glucose is accelerated. Little insulin is required to achieve these effects, a concentration of 5×10^{-6} units per ml. tissue having been shown to result in observable increase in glucose utilization (92). The manner in which insulin exerts its effect has been hotly debated. A widely-accepted explanation is that insulin either stimulates, or releases from inhibition by other hormones, the enzyme glucokinase. This would promote the phosphorylation of glucose to glucose-6-phosphate, decreasing blood sugar levels and stimulating both glycogenesis and glycolysis.

Certain other actions of insulin seem less closely related to the release of glucokinase from inhibition. Insulin has been shown to restore to normal the synthesis of phosphocreatine and the oxidation of pyruvic acid through the tricarboxylic acid cycle when these functions are depressed, as they are in experimental animals made diabetic by the injection of alloxan (28). In normal cats, insulin increases the incorporation of radioactive P^{32} into ATP and into phosphocreatine (78). Insulin promotes the formation of fatty acids from two-carbon fragments, and favors the building of proteins from amino-acids (43).

Among the other steps in carbohydrate metabolism for which insulin stimulation has been claimed in attempts to explain observed effects for which hexokinase stimulation is insufficient cause, is that involving the condensation of pyruvic acid and oxaloacetic acid (89). This, by bringing in the tricarboxylic acid cycle would automatically involve virtually the entire field of intermediate metabolism; fat and protein as well as carbohydrate metabolism, and diabetes mellitus is indeed a disorder of metabolism in general rather than of carbohydrate metabolism alone.

Insulin forms a chemical combination with muscle (rat diaphragm) and simultaneously increases the rate of glucose uptake and glycogen synthesis of the muscle. If the rat is previously rendered diabetic by alloxan (see page 573) or if the rat is previously injected with either crude anterior pituitary extract or purified pituitary growth hormone, the ability of the muscle to combine with insulin is diminished (84). Even *in vitro* crude pituitary extracts diminish the insulin-combining power of muscle specimens, but purified growth hormone does not. A possible explanation for this sort of insulin action is that in combining with tissue, it increases the permeability of cell membranes to glucose by accelerating an enzyme-catalyzed transport mechanism (77). In the absence of insulin or where its effects are counteracted by pituitary hormones, blood glucose would accumulate while cell utilization of glucose, either in glycogenesis or glycolysis would decrease.

Considerable experiment and discussion has been lavished upon the question whether diabetes is the result of *underutilization* of glucose in the tissues, or of *overproduction* of glucose by excessive gluconeogenesis and hepatic glycogenolysis. It is now apparent that both mechanisms are involved. Insulin administered to the diabetic both increases tissue utilization and checks hepatic output of glucose. The more fundamental effect appears to be that upon tissue utilization. Whether by the release of glucokinase from inhibition or by some other mechanism, insulin promotes the uptake of glucose by tissue cells. The uptake of fructose does not appear to require insulin. The measurement of cellular uptake of sugars has been accomplished by the measurement of the simultaneous decrease in plasma inorganic phosphate, which is required in the phosphorylation processes accompanying cellular uptake. A dog rendered diabetic by total pancreatectomy shows no change in tolerance for fructose as compared with a normal dog. The expected decrease in plasma inorganic phosphate occurs following fructose injection. Following glucose injection into such a pancreatectomized dog, no fall in plasma inorganic phosphate is observed unless insulin is supplied (49), or unless glucose is administered in greatly increased concentration and over longer periods of time, thus forcing its entry into tissue cells.

Insulin and other hormones. A hypophysectomized dog can be killed promptly by the same dose of insulin which would be well tolerated by a normal dog. Diabetes produced in dogs by removal of the pancreas is mitigated following hypophysectomy. Such pancreatectomized and hypophysectomized dogs (called "Houssay dogs" after the pioneer experimenter in this field) develop severe diabetes after injection of anterior pituitary extracts. In rats removal of the pituitary produces a 50 per cent increase in muscle hexokinase activity (4), with a further increase of similar magnitude upon administration of insulin.

Hyperglycemia and glycosuria can be induced in adult intact cats and dogs by the daily administration for 3 to 8 days of Young's (93) anterior pituitary diabetogenic factor. The initial response to daily injections of the diabetogenic pituitary extract is an increase in body weight with retention of nitrogen. Glycosuria appears after several days. The increase in weight is not the result of water retention but is associated with synthesis of muscle protein. This is in contrast to the failure of protein synthesis usually observed in human diabetes. Young has not been able consistently to evoke diabetic symptoms in puppies and kittens by the use of his extracts which are effective in adult dogs and cats. The pituitary diabetogenic factor has been identified with the growth hormone since preparations of the latter in states of high purity were shown to induce diabetic responses in dogs and in a human subject (42). However, preparations of pure growth hormone have been described (72) which do not induce diabetes, so that the existence of a distinct diabetogenic pituitary hormone may be indicated.

The anterior pituitary is concerned with adaptation to starvation; it retards sugar utilization and increases gluconeogenesis. Fasting intact rats economize their carbohydrate reserves, showing scarcely any change in the total glycogen of the body between 24 and 48 hours of fasting. Hypophysectomized fasting rats lose carbohydrate at a rate eight times more rapid than that of the normal fasting rat (14). Insulin is concerned with adaptation to food intake. It promotes storage of foodstuffs as glycogen, as fat, and as protein. It comes into action with increase of blood sugar.

The glycocorticoids of the adrenal augment the diabetogenic effects of the anterior pituitary. Administration of large doses of ACTH produces temporary diabetes in the human. This adrenal type of diabetes is characterized by a considerable degree of insulin resistance. Adrenocortical deficiency shows a decrease in glycocorticoid output, along with that of other corticoids. Blood sugar is low, and there is a depression of gluconeogenesis.

Patients with hyperthyroidism often show a diabetic glucose tolerance curve, which is partly explained by increased rate of absorption of glucose from the intestine. There is, however, an actual decrease in the rate of glycogen formation from intravenously injected glucose (91). This effect of thyroid overactivity has not been adequately explained in terms of known mechanisms of glucose utilization.

Insulinase is an enzyme system identified in liver, kidney, and muscle.

Both endogenous and injected insulin are inactivated by insulinase (4). Insulin injected into the portal vein of dogs is less effective than insulin injected into the femoral vein. Liver insulinase is decreased during fasting and restored by feeding. Certain sulfanilamide and aryl sulfonyl urea derivatives which are used in the treatment of diabetes are presumed to act by inhibiting insulinase. (58a)

Antibodies which neutralize the hormonal action of insulin have been demonstrated in a small proportion of rabbits after repeated injection, and arise spontaneously in a small proportion of insulin-resistant human diabetics following treatment (52). Allergy to injected insulin as manifested by itching at the site of injection or by urticaria is more frequently related to species-specific animal protein than to organ-specific insulin. Allergic manifestations can sometimes be obviated by the use of more highly purified insulin, or insulin prepared from a different species. The antibody involved in allergic responses is not identical with the antibody which neutralizes hormone action. Patients with insulin allergy typically show the normal metabolic effects of insulin.

A *pancreatic hyperglycemic hormone* (glucagon) is secreted by the alpha cells of the islets of Langerhans. It is a protein, and functions by promoting hepatic glycogenolysis. This contrary-acting substance is present in small amount in most available preparations of insulin.

The Diabetic Patient

The diabetic is likely to have ancestors or blood relatives who are diabetics. The mode of occurrence of diabetes in families suggests that diabetes may be inherited as a single recessive gene (88) which is not sex-linked, although other genes (modifiers) may influence the severity of the disease and the time of onset.

At the time when the disease first becomes manifest, the patient is likely to be overweight. With progress of the disease, weight is usually lost as a result of the loss of potential calories by glycosuria. Weakness occurs proportional to the weight loss. Diabetics, particularly if inadequately treated, are susceptible to infections. Staphylococcal infections of the skin are far more threatening to the diabetic than to the normal person. Tuberculous infection is likely to become active in the diabetic.

Carotenemia. Diabetics often show a pigmentation of the skin, often a yellowish color of serum much greater than normal. This has been shown to result from the accumulation of several lipochrome pigments, particularly carotene. Oral administration of carotene to diabetics results in a higher blood level of carotene, maintained for a greater length of time than in non-diabetics. The assumption has been made that the liver in diabetes is unable to convert carotene into vitamin A as rapidly as normal, hence the accumulation of carotene in blood and tissues.

Arteriosclerosis, which involves lipid infiltration followed by calcification of the arterial walls, is perhaps a disease or perhaps a part of the normal process of ageing. In diabetics it appears earlier and progresses faster than in the general population. In many diabetics it is a major complication, and may lead to death from vascular occlusion. Arteriosclerosis begins as an abnormality of lipid metabolism, and as such will be considered in the next chapter. There is usually a high blood lipid content in severe diabetes. Cholesterol is elevated in the blood of untreated diabetics; the elevation often persists after treatment. Since high blood cholesterol levels are statistically correlated with the development of arteriosclerosis, the use of low fat diets is being more and more recommended.

Ketosis is the most urgent complication of diabetes, and may lead quickly to coma and death. It is a serious failure of fat oxidation, and causes a severe acidosis. It will be discussed in these two aspects in the appropriate chapters. In brief, ketosis is the failure to oxidize fats completely. An intermediate metabolic product, acetoacetic acid, together with its reduction product, beta-hydroxybutyric acid, and its decarboxylation product, acetone, accumulate in the body and appear in the urine. The resulting coma is in part the result of increased blood and tissue acidity and in part a direct result of the action of high concentrations of these "ketone bodies" on the nervous system. Dehydration and loss of ions from body fluids, both the result of vomiting and polyuria, add to the dangers of diabetic ketosis. The most effective prevention and treatment of diabetic ketosis is by the use of insulin. Ketosis also occurs as a result of starvation, in which case food is required and insulin is contraindicated. Note for the present that both situations in which ketosis may develop involve a subnormal rate of glucose utilization.

Low fertility is characteristic of untreated diabetic women. The chief contributing factor is fetal or neonatal death. Diabetic mothers show low levels of estrogens and of pregnanediol excretion, and high levels of chorionic gonadotrophin. The fetus is frequently oversized and edematous.

Principles of diabetic treatment. Our present state of knowledge offers no means for the permanent cure of diabetes. The physician treating a diabetic patient attempts to maintain a reasonably normal blood sugar level, avoiding hypoglycemia and excessive hyperglycemia (over 0.15 per cent). He also tries to keep the patient's body weight normal. He plans to protect the patient from ketosis, which is the cause of diabetic coma, and if possible from arteriosclerosis, which is the direct or indirect cause, since the discovery of insulin, of most diabetic deaths.

Education of the patient concerning the nature of the disease, the means available for its control, and particularly the hopeful outlook for the properly controlled patient, is perhaps the most important feature of successful treatment. The disease demands full-time attention, and few patients can afford the full-time services of a medical attendant. Fortunately, the procedures which are the daily necessities of the diabetic are not too complicated to be learned by the child who can comprehend his school work or the adult who is capable of earning a living. In medical centers, diabetic education is carried on in organized classes. Under other circumstances, the instruction is individualized and given by the physician or a competent assistant.

Diet is the most important daily variable to be brought under control. Restriction of food intake has been an important part of the treatment of diabetes for as long as the metabolic nature of the disease has been recognized. Such dietary restriction is usually voluntary on the part of the patient, encouraged and advised by his physician. At times it has been enforced by circumstances entirely out of the control of either. During the periods of food rationing necessitated in Britain and in Germany by World Wars I and II, striking decreases in death rates from diabetes occurred. The British figures also show smaller decreases in diabetic mortality corresponding to the two periods of economic depression between these wars (58). In general, the diabetes mortality is higher in countries with a high standard of living, (21) where the general opportunity for high caloric intake is greatest.

The cardinal principle of the voluntary dietary treatment of diabetes is to avoid overfeeding, but the diabetic diet must be adequate in all the dietary essentials. It is impossible to write down an arbitrary diet suitable for all diabetics. It is important, particularly if insulin is being used, that the diet be constant from day to day as far as total calories from carbohydrate and carbohydrate-forming foods are concerned. It is customary to have food portions weighed, at least until the patient is accustomed to the diet and can estimate weights with the help of common household measures. The scrupulous weighing of the diet to the nearest gram may become exasperating during the protracted lifetime of the properly treated diabetic. The degree of latitude which can or should be allowed to an individual patient in this respect depends so much upon the personalities of the patient, his family, and his physician, that no definite rules can be laid down.

The complete elimination of *carbohydrates* from the diabetic diet is unwise. Reduction of carbohydrate alone favors ketosis—a serious and too often fatal overloading of the mechanisms of fat oxidation. Present-day diabetic diets usually contain 150 to 300 grams of carbohydrate per day. *Protein* should not be restricted below the normal requirement of a healthy person, approximately 1 gram per kilogram of body weight per day. Increase to 1.25 grams per kilogram is probably advantageous, but adds to the expense of the diet. The greater protein needs of growing children should be met as in health. *Fat* allowance bears some relation to the patient's nutritional state; 50 grams per day is an almost standard figure. It can be decreased for the obese diabetic or increased if it is otherwise difficult to maintain normal weight. The vitamin and mineral requirements, particularly in children, may demand the use of vitamin concentrates and of milk. For further information on diabetic diets, see (12).

Exercise is a means of lowering blood sugar by increasing its utilization. The effect is considerable in active children, so that unusual or protracted exercise may cause a child who is taking insulin to develop hypoglycemia. Just as a diabetic's diet and insulin should be measured daily, so should his exercise be as constant as possible from day to day. Additional exercise should be compensated by extra carbohydrate or possibly less insulin. Patients require less insulin upon discharge from the hospital and resumption of activity. Study of blood levels of lactic and pyruvic acids in diabetic patients indicates that these substances are formed more slowly after glucose ingestion than in the normal subject. The response of the diabetic patient to exercise is, however, normal in regard to the production of lactic and pyruvic acids. Insulin appears to be necessary for glycolysis in the resting subject, but not for glycolysis during exercise (37). Physiological experimentation thus confirms clinical experience that exercise can, to a degree at least, substitute for insulin.

Insulin may not be required by mild diabetics, and in fact is usually not prescribed if the patient can tolerate a diet adequate in calories and containing 150 grams of carbohydrate without consistent glycosuria. There is considerable variance among specialists in diabetes in their manner of use of insulin, and in their attitude towards strict control of diet and exercise. A procedure acceptable to many authorities, and widely used, is to start an adult patient on an adequate measured diet, as described, with a small dose of protamine zinc insulin (about 12 units) injected before breakfast. *Protamine zinc insulin* is insulin modified with 1.25 mgm. of protamine derived from fish sperm and 0.2 mgm. of zinc for each 100 units of insulin. Protamine zinc insulin differs from regular insulin, which has an action lasting twelve hours, with peak activity at between three and four hours, by having its action spread out in time by the slow rate of dissociation of the compound of insulin with protamine. Protamine zinc insulin has a low peak of activity at twenty-four hours or more, so that the minimum blood sugar level will be just before breakfast, following the longest fasting period of the twenty-four hours. Since glycosuria is dependent upon blood sugar level, the urine at that time should contain the least sugar. The dose of protamine zinc insulin is increased, usually 4 units a day, until the early morning urine is free from sugar. In most cases the blood sugar, taken before breakfast, will then be normal or not greatly elevated. The urine and blood sugars are then measured one hour after the noon meal. If the urine is sugar-free and the blood sugar not over 0.150 per cent, the insulin dosage is adequate. If it is not, more insulin is needed, but not as protamine zinc insulin. Further increase of the single dose of protamine zinc insulin would lead to hypoglycemia in the early hours of the morning. Instead, regular insulin is given ahead of the protamine insulin, and thirty minutes before breakfast. Again, the dose is small at the start, usually 8 units, and is increased daily by 4 units until the blood sugar is normal four hours after the injection. If urine is now sugar-free and blood sugar is not over 0.150 per cent one hour after the noon meal, the dosage is adequate. If not, a second dose of regular insulin is demanded, thirty minutes before the noon meal, which is increased as before, checking blood and urine sugar one hour after the evening meal. In some cases, three injections of regular insulin may be needed. *Globin insulin* and *NPH insulin* are intermediate in duration of activity between regular and protamine zinc insulin. Patients who require two or more doses of regular insulin in addition to the single dose of protamine zinc insulin can often be carried on a single dose of globin or NPH insulin. Regular insulin can be added to NPH insulin without serious loss of the promptness of action of the regular insulin, permitting a single injection daily even for those patients who require additional regular insulin. *Lente insulin* is a suspension of minute particles of zinc insulin with a sufficiently high zinc content to make the insulin almost insoluble at blood pH. The time of action of lente insulin closely approximates that of NPH insulin so that the two can be used interchangeably.

NON-DIABETIC MELITURIAS

The demonstration of reducing sugars in the urine by Benedict's or a similar test does not of itself indicate diabetes. *Glucose* may occur in the urine of a non-diabetic as a result of (a) heavy ingestion of sugars, (b) emotional stress, (c) hyperthyroidism, (d) hyperactivity of anterior pituitary or adrenal cortex (but this type of glycosuria may have many points in common with diabetes), (e) intracranial damage, particularly of the hypothalamic region, from trauma, vascular accident, infection, or tumor, and (f) as a concomitant of severe infections, intoxications, and chronic diseases; see Joslin (10) for elaboration of this last rather indefinite class. *Lactose* occurs in the urine post-partum and during lactation. *Pentose* (L-*xylulose*) or *fructose* may be found in the urine in rare cases of metabolic anomaly and in small amounts in normal urine.

Renal glycosuria is distinct from these categories and from diabetes. Glucose is excreted in the urine at normal levels of blood sugar as a result of failure of the renal tubule cells to reabsorb glucose from the glomerular filtrate. Such a situation is often designated as a "lowered glucose threshold" in the kidney. For the time, we can state that the normal renal threshold for glucose is about 0.16 per cent. Above this blood sugar level glycosuria occurs, below it there is no glycosuria. This is a crude and oversimplified statement, but is usually true.

Renal glycosuria is of fairly frequent occurrence accompanying diseases such as nephrosis where there is degeneration of renal tubule cells. In many instances, however, it occurs as an isolated phenomenon, with no symptoms and no impairment of nutrition or general health. The glycosuria may be constant or periodic. A moderate degree of renal glycosuria is frequently observed during pregnancy, disappearing after delivery, and returning with further pregnancies. Renal glycosurics may develop diabetes, but the existence of renal glycosuria does not increase the probability of developing diabetes. The differentiating observation between diabetes and renal glycosuria is the blood sugar—high in diabetes, normal in renal glycosuria. In borderline cases, the glucose tolerance test is usually decisive. Patients with uncomplicated renal glycosuria require no active treatment. Restriction of diet is not helpful, and may result in undesirable undernutrition. It is standard practice to check blood sugars at regular intervals throughout such a patient's lifetime.

The *de Toni-Fanconi syndrome* is a more serious failure of renal tubulal absorption in which there is not only failure to reabsorb sugar, resulting in glycosuria, but also failure to reabsorb amino acids, Ca^{++}, HPO_4^{--}, and water, resulting in growth failure, rickets, and polyuria. It is characteristically a disease of children, rickets and failure of growth usually appearing at about the second year of life.

Galactose diabetes, also called *galactosemia*, is a rare but serious familial inability to metabolize galactose. It causes severe malnutrition, which can be avoided by substituting other foods for milk and the most noticeable gross symptom is the development of cataracts (76). The metabolic fault lies in the lack of the enzyme, uridyl transferase (38) (see page 530), which catalyzes the conversion of galactose-1-phosphate to glucose-1-phosphate. The accumulation of galactose-1-phosphate and galactose may result in heightened competition with glucose for enzyme systems held in common. The cytotoxic effect of galactose would thus be due to the induced underutilization of glucose with damage, notably, to the brain (irreversible mental retardation) and to the lenticular epithelium (cataracts). The disease is quite comparable to diabetes in that there is hypergalactemia combined with galactosuria, and fatty infiltration of the liver. In considering the diagnosis of galactose diabetes, it should be kept in mind that normal infants may have significant amounts of galactose, up to 40 per cent of the total blood sugar, in the blood shortly after a milk feeding (32).

ALLOXAN DIABETES

Certain substances injected intravenously will cause destruction of the beta cells of the islets of Langerhans. These substances include alloxan (formula XXXVIII) and alloxantin (formula XXXIX).

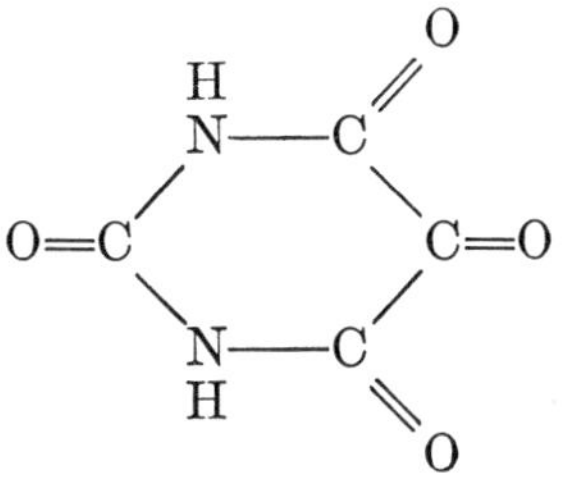

```
              O
       H     //
       N———C
      /     \
  O=C        C=O
      \     /
       N———C
       H     \\
              O
```

XXXVIII. Alloxan

Methyl alloxan and dimethyl alloxantin are similarly effective. Alloxan has a similar effect when administered by other routes. The characteristic response of the experimental animal to injection of alloxan is (a) hyperglycemia lasting one hour or less, then (b) hypoglycemia lasting several hours, and finally, (c) permanent hyperglycemia. The temporary hypoglycemia is pancreatic in origin, probably the result of liberation of preformed insulin by the degenerating beta cells.

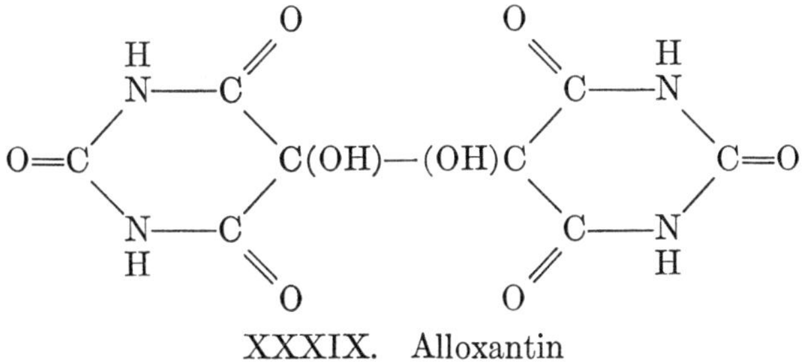

```
              O                 O
       H     //                  \\      H
       N———C                      C———N
      /     \                    /     \
  O=C        C(OH)—(OH)C                C=O
      \     /                    \     /
       N———C                      C———N
       H     \\                  //      H
              O                 O
```

XXXIX. Alloxantin

So far, alloxan has been useful as an additional means of producing experimental diabetes. Some of the facts learned by its use may in time help elucidate the cause of ordinary human diabetes. For example, glutathione is protective against destruction by alloxan of the beta cells, as well as against the production of diabetes by ACTH. Blood glutathione decreases along with successful induction of alloxan diabetes. Animals with depleted blood glutathione are more susceptible to alloxan. Dehydroascorbic acid (formula XL) may be considered as having a structure not unlike alloxan. One injection of 1.1 grams per kgm. body weight will produce hyperglycemia in rats. Three daily injections will produce permanent diabetes. Sulfhydryl compounds such as cysteine and glutathione, if given intravenously two

```
         HO  H        O
         |   |       //
HOCH2—C—C——C
         |   |       \
         H   |        C=O
             |       /
             O——C
                   \\
                    O
```

XL. Dehydroascorbic acid

minutes before the injection of dehydroascorbic acid will prevent its diabetogenic action. If given ten minutes afterwards, the sulfhydryl compounds have no protective effect. For this reason, Patterson and Lazarow (67) conclude that dehydroascorbic acid irreversibly blocks the sulfhydryl groups of some important enzyme. It is quite possible that sulfhydryl groups are particularly important in the synthesis of insulin with its unusually high cystine content (29).

2,3-*Diketo*-L-*gulonic acid* (formula XLI) which is the hydrated form of dehydroascorbic acid is found in the blood of diabetics but not of nondiabetics (56).

```
         OH H      O
         |  |     //
HOCH2—C—C—C
         |  |     \
         H  OH     C=O
                  /
                 C
                / \\
              HO    O
```

XLI. 2,3 Diketo-L-gulonic acid

SPONTANEOUS HYPOGLYCEMIA

Low levels of blood sugar comparable to those observed after insulin overdosage sometimes occur when insulin is not given. The symptoms are similar to those of insulin overdosage. In mild cases the patient is weak, hungry, and uneasy, often showing blurring of vision, perspiration, tremor, and palpitation. More severe cases may stagger drunkenly or collapse. Such hypoglycemic states may result from numerous causes; the commonest are (a) reactive hyperinsulinism, (b) tumor or hyperplasia of the islets of Langerhans, and (c) diseases of the liver.

Reactive hyperinsulinism is an exaggerated physiological response to increased blood sugar levels. It is likely to occur in conscientious persons who are hyperreactive to everyday frustrations. It has also been observed in patients who have been submitted to gastroenterostomy or total or subtotal gastric resection. The latter group represents the result of too sudden an entrance of carbohydrate into the absorptive portion of the small intestine. Both groups tend to have hypoglycemic attacks two to four hours after meals, and do well on a high protein, low carbohydrate diet.

Islet-cell tumors or hyperplasia lead to relatively constant insulin output, most notably indicated by low fasting blood sugar levels, in contrast to the postprandial hypoglycemia of the reactive type. Patients with tumor or hyperplasia are usually treated surgically.

Liver disease, if extensive, can lead to a hypoglycemic state comparable to that observed in experimental animals after hepatectomy. The obvious cause is inability of the severely damaged liver to carry on the functions of glycogenesis, glycogenolysis, and particularly gluconeogenesis. In this situation treatment is exactly opposite to that of reactive hypoglycemia. Since hypoglycemia occurs after relatively long periods of fasting, meals should be frequent and should be high in carbohydrate. The protein content of the diet should be governed by the nature and stage of the hepatic disease. In the majority of diseases involving destruction of liver cells, a high protein allowance is advantageous.

Hypoglycemia may also be the result of decreased secretion of pituitary or adrenal hormones, or of glucagon. In glycogen storage disease, hypoglycemia may result from lack of the action of the specific glucose-6-phosphatase of the liver.

METABOLISM OF ALCOHOL

Ethyl alcohol is a foodstuff, a drug, or a poison according to the amount used, and the manner of its use. There is a certain advantage in the intravenous administration of alcohol to surgical patients during the first postoperative day (74), or longer if indicated. Alcohol is given usually in 5 per cent solution in water or isotonic salt solution containing 5 per cent glucose. At this concentration, and at a rate of 15 ml. alcohol (300 ml. of 5 per cent solution) per hour, a moderate analgesic and sedative drug effect of alcohol is achieved, together with the supply of caloric needs both by alcohol and by glucose.

When alcohol tagged with C^{14} was administered to rats in dosages of 1 gram per kgm. of body weight (3), 75 per cent of the C^{14} was recovered as CO_2 within five hours, and 90 per cent in ten hours. Rats habituated to drinking 10 per cent alcohol instead of water for five months showed no increase in the rate of alcohol oxidation. Tissue slices of rat liver and kidney converted alcohol to CO_2 rapidly, heart and diaphragm oxidized alcohol slightly, and brain not at all.

Liver contains an alcohol dehydrogenase, which catalyzes the conversion of ethyl alcohol to acetaldehyde, with DPN as the hydrogen acceptor. Acetaldehyde is converted to acetic acid, catalyzed by an aldehyde dehydrogenase. Acetic acid is oxidized by way of the citric acid cycle. Alcohol, upon complete oxidation, yields 7 Kcal. per gram. There is no evidence that the energy liberated in the conversion of alcohol to acetic acid is available for performance of work via phosphate-bond formation. The oxidation of acetic acid from alcohol in the citric acid cycle is strictly comparable to the combustion of acetic acid from other sources, is phosphate-linked, and is a source of energy.

The early stages of oxidation of alcohol in the human body proceed at a limited rate which is reasonably constant in each individual. In different individuals the rate varies, so that the blood alcohol concentration falls by 0.01 to 0.02 per cent per hour. Two ounces of 100-proof whiskey taken by a person of average size will build up a maximal concentration of alcohol in the blood of about 0.05 per cent. Concentrations below this level produce no evidences of intoxication except in a few unusually susceptible subjects. Six ounces or more of 100-proof whiskey can raise the blood alcohol concentration above 0.15 per cent, at which level all subjects show clear evidence of impairment of neuromuscular function. Blood levels of 0.15 per cent or above have been accepted in courts of law as *prima facie* evidence of alcoholic intoxication in cases involving violation of motor vehicle laws. At blood alcohol levels between 0.05 per cent and 0.15 per cent there can usually be demonstrated by laboratory methods some degree of delay in reactions involving judgment or discrimination, and some subjects are definitely and obviously intoxicated. Alcohol concentrations can be measured directly on blood samples by a distillation method or more conveniently by estimation from the alcohol content of expired air (39). Blood alcohol concentrations of the order of 0.5 per cent are lethal.

REFERENCES

1. Ashmore, J., *et al.* Studies on carbohydrate metabolism in rat liver slices. VI. Hormonal factors influencing glucose-6-phosphatase. J. Biol. Chem., **218:** 77–88, 1956.
2. Baldwin, E. *Dynamic Aspects of Biochemistry.* 2nd edit. Cambridge, England, Cambridge University Press, 1952.
3. Bartlett, G. R., and Barnet, H. N. Some observations on alcohol metabolism with radioactive ethyl alcohol. Quart. J. Studies on Alcohol, **10:** 381–397, 1949.
4. Beaser, S. B. Diabetes Mellitus. New Eng. J. Med., **243:** 81–85, 1950.
5. Benedict, S. R., and Osterberg, E. A method for the determination of sugar in normal urine. J. Biol. Chem., **48:** 51–57, 1921.

6. Bernstein, I. A. Synthesis of ribose by the chick. J. Biol. Chem., **205:** 317–329, 1953.
7. Bhishagratna, K. L. *An English Translation of the Sushruta Samhita,* Vol. I, p. 75. Calcutta, India, Wilkins Press, 1907.
8. Bloch, K., and Kramer, W. Effect of pyruvate and insulin on fatty acid synthesis *in vitro*. J. Biol. Chem., **173:** 811–812, 1948.
9. Bloom, B. Catabolism of glucose by mammalian tissues. Proc. Soc. Exptl. Biol. Med., **88:** 317–318, 1955.
10. Booth, A. N., *et al.* Effects of prolonged ingestion of xylose on rats. J. Nutrition, **49:** 347–355, 1953.
11. Bornstein, J. Normal insulin concentration in man. Australian J. Exper. Biol. & Med. Sc., **28:** 93–97, 1950.
12. Caso, E. K. Calculation of diabetic diets. J. Am. Dietet. Assoc., **26:** 575–583, 1950.
13. Castellani, A., *et al.* Enzymatic formation of hexosamine in epiphyseal cartilage homogenate. Nature, **178:** 313, 1956.
14. Cori, C. F. Enzymatic reactions in carbohydrate metabolism. Harvey Lect., Series XLI, 253–272, 1946.
15. Cori, G. T., and Cori, C. F. Crystalline muscle phosphorylase. IV. The formation of glycogen. J. Biol. Chem., **151:** 57–63, 1943.
16. Cori, G. T., *et al.* The metabolism of fructose in liver. Isolation of fructose-1-phosphate and inorganic pyrophosphate. Biochim. et Biophys. Acta, **7:** 304–317, 1951.
17. Creeth, J. M. Sedimentation and diffusion studies on insulin: the maximum molecular weight. Biochem. J., **53:** 41–47, 1953.
18. Dickens, F. Alternative routes of carbohydrate oxidation. Brit. Med. Bull., **9:** 105–109, 1953.
19. Donohue, D. M., *et al.* Preparation and transfusion of blood. J. A. M. A., **161:** 784–788, 1956.
20. Drury, H. F. Identification and estimation of pentoses in the presence of glucose. Arch. Biochem., **19:** 455–466, 1948.
21. Dublin, L. I., and Marks, H. H. Mortality from diabetes throughout the world. Diabetes, **1:** 205–217, 1952.
22. Enklewitz, M., and Lasker, M. The origin of L-xyloketose (urine pentose). J. Biol. Chem., **110:** 443–456, 1935.
23. Fishman, W. H. Beta-glucuronidase. Advances Enzymol., **16:** 361–409, 1955.
24. Fishman, W. H., and Green, S. Glucosiduronic acid synthesis by β-glucuronidase in a transfer reaction. J. Am. Chem. Soc., **78:** 880–882, 1956.
25. Folin, O., and Malmros, H. An improved form of Folin's micro method for blood sugar determinations. J. Biol. Chem., **83:** 115–120, 1929.
26. Folin, O., and Wu, H. A system of blood analysis. Supplement I. A simplified and improved method for determination of sugar. J. Biol. Chem., **41:** 367–374, 1920.
27. Glaser, L., and Brown, D. H. The enzymic synthesis in vitro of hyaluronic acid chains. Proc. Natl. Acad. Sci. U. S., **41:** 253–260, 1955.
28. Goranson, E. S., and Erulkar, S. D. The effect of insulin on the aerobic phosphorylation of creatine in tissues from alloxan-diabetic rats. Arch. Biochem., **24:** 40–48, 1949.
29. Guzman Barron, E. S. The importance of sulfhydryl groups in biology and medicine. Texas Repts. Biol. Med., **11:** 653–670, 1953.

30. Harfenist, E. J., and Craig, L. C. The molecular weight of insulin. J. Am. Chem. Soc., **74:** 3087–3089, 1953.
31. Harpur, R. P., and Quastel, J. H. Relations between acetylcholine synthesis and metabolism of carbohydrates and D-glucosamine in the central nervous system. Nature, **164:** 779–782, 1949.
32. Hartman, A. F. Pathologic physiology in some disturbances of carbohydrate metabolism. J. Pediatrics, **47:** 537–570, 1955.
33. Hastings, A. B., *et al.* Incorporation of isotopic carbon dioxide in rabbit liver glycogen *in vitro*. J. Biol. Chem., **177:** 717–726, 1949.
34. Hers, H. G., and Kusaka, T. Metabolism of fructose-1-phosphate in liver. Biochim. et Biophys. Acta, **11:** 427–437, 1953.
35. Horecker, B. L., *et al.* The enzymic conversion of 6-phosphogluconate to ribulose-5-phosphate and ribose-5-phosphate. J. Biol. Chem., **193:** 383–396, 1951.
36. Horecker, B. L., and Smyrniotis, P. Z. Purification and properties of yeast transaldolase. J. Biol. Chem., **212:** 811–825, 1955.
37. Horwitt, M. K., *et al.* Lactic and pyruvic acids in the blood after glucose and exercise in diabetes mellitus. Am. J. Physiol., **156:** 92–99, 1949.
38. Isselbacher, K. J., *et al.* Congenital galactosemia, a single enzymic block in galactose metabolism. Science, **123:** 635–636, 1956.
39. Jetter, W. W. A critical survey of various chemical methods for determining the alcohol content of body fluids and tissues with their physiological and medicolegal significance. Quart. J. Studies on Alcohol, **2:** 512–543, 1941.
40. Joslin, E. P., *et al.* *The Treatment of Diabetes Mellitus.* 8th edit. Philadelphia, Penna., Lea and Febiger, 1946.
41. Katz, J., *et al.* The occurrence and mechanism of the hexose monophosphate shunt in rat-liver slices. J. Biol. Chem., **214:** 853–868, 1955.
41a. Kielley, W. W. *et al.* The hydrolysis of purine and pyrimidine nucleoside triphosphates by myosin. J. Biol. Chem., **219:** 95–101, 1956.
42. Kinsell, L. W., *et al.* Accentuation of human diabetes by "pituitary growth hormone". Proc. Soc. Exptl. Biol. Med., **83:** 683–686, 1953.
43. Krahl, M. E. Incorporation of C^{14}-amino acids into glutathione and protein fractions of normal and diabetic rat tissues. J. Biol. Chem., **200:** 99–109, 1953.
44. Krebs, H. A. The tricarboxylic acid cycle. Harvey Lect., 1948–49.
45. Krebs, H. A. Some aspects of the energy transformations in living matter. Brit. Med. Bull., **9:** 97–104, 1953.
46. Krebs, H. A. The tricarboxylic acid cycle; *in Chemical Pathways of Metabolism*, Vol. I., p. 109–172. New York, Academic Press, 1954.
47. Landau, R. L., and Loughead, R. Seminal fructose concentration as an index of androgenic activity in man. J. Clin. Endocrinol., **11:** 1411–1424, 1951.
48. Leloir, L. F. Enzymic isomerization and related processes. Advances Enzymol., **14:** 193–218, 1953.
49. Levine, R., *et al.* Nature of the action of insulin on the level of serum inorganic phosphate. Am. J. Physiol., **159:** 107–110, 1949.
50. Ling, K.-H., and Lardy, H. A. Uridine- and inosinetriphosphates as phosphate donors for phosphohexokinase. J. Am. Chem. Soc., **76:** 2842–2843, 1954.
51. Lionetti, F., *et al.* The effect of adenosine upon the esterification of phosphate by erythrocyte ghosts. J. Biol. Chem., **220:** 467–476, 1956.
52. Lowell, F. C., and Franklin, W. Induced insulin resistance in the rabbit. J. Clin. Invest., **28:** 199–206, 1949.
53. McGeown, M. G., and Malpress, F. H. Synthesis of deoxyribose in animal tissues. Nature, **170:** 575–576, 1952.

54. McGeown, M. G., and Malpress, F. H. Synthesis of ribose in animal tissues. Nature, **173:** 212–213, 1954.
55. Mann, T., and Parsons, U. Studies on the metabolism of semen. Biochem. J., **46:** 440–450, 1950.
56. Marcovich, A. W., and Marcovich, J. F. Diketogulonic acid and diabetes mellitus. J. Lab. Clin. Med., **42:** 681–684, 1953.
57. Maxwell, E. S. Diphosphopyridine nucleotide, a cofactor for galacto-waldenase. J. Am. Chem. Soc., **78:** 1074, 1056.
58. Meakins, J. C. *The Practice of Medicine.* 5th edit. St. Louis, Mosby, 1950.
58a. Mirsky, I. A. The role of insulinase and insulinase-inhibitors. Metabolism, Clin. and Exper., **5:** 138–143, 1956.
59. Mommaerts, W. F. H. M. The molecular transformation of actin. III. The participation of nucleotides. J. Biol. Chem., **198:** 469–475, 1952.
60. Mommaerts, W. F. H. M. Stoichiometric and dynamic implications of the participation of actin and adenosine triphosphate in the contractile process. Biochim. et Biophys. Acta, **7:** 477–478, 1951.
60a. Montgomery, C. M., and Webb, J. L. Metabolic studies on heart mitochondria. I. The operation of the normal tricarboxylic acid cycle in the oxidation of pyruvate. J. Biol. Chem., **221:** 347–357, 1956.
61. Moles, M. F., and Botts, J. A model for the elementary processes in muscle action. Arch. Biochem. Biophys., **37:** 283–300, 1952.
61a. Morales, M. F., *et al.* Elementary processes in muscle action: an examination of current concepts. Physiol. Revs., **35:** 475–505, 1955.
62. Moyle, J., and Dixon, M. Identity of triphosphopyridine nucleotide-linked isocitric dehydrogenase and oxalosuccinic carboxylase. Biochim. et Biophys. Acta, **16:** 434–435, 1955.
62a. Murphy, J. R., and Muntz, J. A. The metabolism of glucose in the perfused rat liver. J. Biol. Chem., **224:** 987–997, 1957.
63. Najjar, V. A. The metabolism of carbohydrates, fats and bile pigments by the liver and the alterations in hepatic disease; a review of recent advances. Pediatrics, **15:** 444–466, 1955.
64. Needham, D. M. Adenosine triphosphate and the structural proteins in relation to muscle contraction. Advances Enzymol., **13:** 151–198, 1952.
65. Newburgh, L. H., and Conn, J. W. A new interpretation of hyperglycemia in obese middle-aged persons. J. A. M. A., **112:** 7–11, 1939.
66. Ochoa, S. Enzymic mechanisms in the citric acid cycle. Advances Enzymol., **15:** 183–270, 1954.
67. Patterson, J. W., and Lazarow, A. Sulfhydryl protection against dehydroascorbic acid diabetes. J. Biol. Chem., **186:** 141–144, 1950.
68. Perry, S. V. The adenosine triphosphatase activity of lipoprotein granules isolated from skeletal muscle. Biochim. et Biophys. Acta, **8:** 499–509, 1952.
69. Pigman, W. W., and Goepp, R. M., Jr. *Chemistry of the Carbohydrates.* New York, Academic Press, 1948.
70. Pontis, H. G. Uridine diphosphate acetylgalactosamine in liver. J. Biol. Chem., **216:** 195–202, 1955.
71. Potter, V. R., and Heidelberger, C. Alternative metabolic pathways. Physiol. Rev., **30:** 487–512, 1950.
72. Raben, M. S., and Westermeyer, V. W. Differentiation of growth hormone from the pituitary factor which produces diabetes. Proc. Soc. Exptl. Biol. Med., **80:** 83–86, 1952.

73. RACKER, E., AND KRIMSKY, I. Mechanism of action of glyceraldehyde-3-phosphate dehydrogenase. Nature, **169:** 1043–1044, 1952.
74. RICE, C. O., *et al.* Parenteral nutrition in the surgical patient as provided from glucose, amino acids, and alcohol. The role played by alcohol. Ann. Surg., **131:** 289–306, 1950.
75. RICHTER, D., AND DAWSON, R. M. C. Brain metabolism in emotional excitement and in sleep. Am. J. Physiol., **154:** 73–79, 1948.
76. RITTER, J. A., AND CANNON, E. J. Galactosemia with cataracts. New Eng. J. Med., **252:** 747–752, 1955.
77. ROSS, E. J. Insulin and the permeability of cell membranes to glucose. Nature, **171:** 125, 1953.
78. SACKS, J. The effect of insulin on phosphorus turnover in muscle. Am. J. Physiol., **143:** 157–162, 1945.
79. SANADI, D. R., *et al.* Guanosine triphosphate, primary product of phosphorylation coupled to the breakdown of succinyl coenzyme A. Biochim. et Biophys. Acta, **14:** 434–436, 1954.
80. SARTON, G. *Introduction to the History of Science*, Vol. I, p. 76. Baltimore, Md., The Williams & Wilkins Co., 1927.
81. SINGER, T. P., *et al.* Observations on the flavine moiety of succinic dehydrogenase. Arch. Biochem. Biophys., **60:** 255–257, 1956.
82. SPIEGELMAN, S., *et al.* Phosphate metabolism and the dissociation of anaerobic glycolysis from synthesis in the presence of sodium azide. Arch. Biochem., **18:** 409–436, 1948.
83. STADIE, W. C., *et al.* The effect of insulin and adrenal cortical extract on the hexokinase reaction in extracts of muscle from depancreatized cats. J. Biol. Chem., **184:** 617-626, 1950.
84. STADIE, W. C., *et al.* Hormonal influences on the chemical combination of insulin with rat muscle (diaphragm). Am. J. Med. Sc., **218:** 275–280, 1949.
85. STEINER, R. F. Reversible association processes of globular proteins. I. Insulin. Arch. Biochem. Biophys., **39:** 333–354, 1952.
86. STETTEN, D., JR., AND TOPPER, Y. J. The metabolism of carbohydrates. Am. J. Med., **19:** 96–110, 1955.
86a. STROMINGER, J. L., *et al.* Enzymatic formation of uridine diphosphoglucuronic acid. J. Biol. Chem., **224:** 79–90, 1957.
87. SUMNER, J. B. A specific reagent for the determination of sugar in urine. J. Biol. Chem., **65:** 393–395, 1925.
88. THOMPSON, M. W., AND WATSON, E. M. The inheritance of diabetes mellitus. Diabetes, **1:** 268–275, 1952.
89. VILLEE, C. A., *et al.* Metabolism of C^{14}-labeled glucose and pyruvate by rat diaphragm muscle *in vitro*. J. Biol. Chem., **195:** 287–297, 1952.
90. WEBER, H. H., AND PORTZEHL, H. Muscle contraction and fibrous muscle proteins. Advances Protein Chem., **7:** 161–252, 1952.
91. WILDER, R. M. Reflections on the causation of diabetes mellitus. J. A. M. A., **144:** 1234–1239, 1950.
92. WILLEBRANDS, A. F., *et al.* Quantitative aspects of the action of insulin on the glucose and potassium metabolism of the isolated rat diaphragm. Science, **112:** 227–228, 1950.
92a. WOOD, H. G. Significance of alternate pathways in the metabolism of glucose. Physiol. Revs., **35:** 841–859, 1955.
93. YOUNG, F. G. The relationship of the anterior pituitary gland to diabetes mellitus. Acta med. scandinav., **135:** 275–288, 1949.

CHAPTER 15

Lipid Metabolism and Ketosis

The fats and related lipids distinguish themselves among the foodstuffs by their high caloric value. Each gram of fat yields 9.3 Kcal., which is more than twice the energy yield (4.1 Kcal.) of a gram of starch or protein. The amount of fat included in human dietaries is highly variable, but usually remains within the limits of 50 and 150 grams per day if circumstances permit free choice. Small amounts of certain unsaturated fatty acids, arachidonic, linoleic, or linolenic, appear to be necessary in human nutrition (see page 455). Since these essential fatty acids occur in many edible oils as well as in egg yolk and meats, an adequate intake is assured from the usual diet.

Emulsions of fat, homogenized to particles less than one micron in diameter and stabilized with phosphatides, may be sterilized in the autoclave and given intravenously to patients unable to take adequate nourishment by mouth. Such lipid emulsions have low osmotic effects; therefore high concentrations (1000 to 1600 Kcal. per liter) of lipid can be used (46). Highly caloric (1200 Kcal. per liter) solutions for intravenous feeding can also be prepared using a combination of glucose and alcohol. Even on fat-free diets, approximately 2 grams of lipid material are eliminated daily in the feces (29). The origin of such endogenous fecal fat is chiefly from intestinal bacteria.

BLOOD LIPIDS

According to Thannhauser (45), normal human blood plasma may contain from zero to 200 mgm. of neutral fat per 100 ml., from 150 to 250 mgm. of phospholipids, chiefly phosphoglycerides, and from 150 to 260 mgm. of cholesterol, 70 to 75 per cent of which is in the form of fatty acid esters and the remaining 40 to 70 mgm. is free or unesterified. In healthy men there is a mean annual increase in total blood serum cholesterol of 2.3 mgm. per 100 ml. between the ages of 17 and 45 (24). The mean value at age 22 is 178.7 mgm. per 100 ml. All figures cited above refer to plasma or serum collected before breakfast. Values in excess of these may be observed following lipid-rich meals (alimentary hyperlipemia), or when fasting is prolonged beyond the usual overnight period, or after protracted exercise. The figures given here refer to the population of the United States of America. Notable differences are found in other populations (25).

The different fatty acids which are combined (in the forms given above) in the blood plasma form a mixture which is approximately one-third saturated fatty acids and two-thirds unsaturated. Among the unsaturated fatty acids, oleic acid is the most abundant, making up about half the unsaturated fraction. Linoleic acid makes up about one-third of this fraction, and arachidonic acid about one-tenth. Linolenic acid, along with pentaenoic and hexaenoic acids, is present in measurable amounts (10). The distribution in blood cells is quite different, with no linolenic acid, and in comparison to plasma lipids about twice the proportion of arachidonic acid and about one-quarter of the linoleic acid.

The lipids of the plasma, listed above, do not occur free, but rather are components of a series of *lipoproteins* which make up 3 to 5 per cent of plasma protein and are large ellipsoidal molecules of small axial ratio with molecular weights in some cases as high as 3,000,000 (2a). The lipoproteins can be divided into alpha-lipoproteins and beta-lipoproteins, the latter being the larger molecule and the richer in lipid (77 per cent of the whole). Of the lipid portion in beta-lipoproteins, cholesterol, both free and esterified, makes up somewhat more than half; phosphoglycerides, the remainder. The alpha-lipoproteins are richer in phosphoglycerides and poorer in cholesterol (36). The beta/alpha lipoprotein ratio is higher in diabetics than in normal individuals (1).

Another convenient method of classifying the plasma lipoproteins is based upon their rate of flotation (17) under defined experimental conditions in the ultracentrifuge. Lipoproteins are of lower density than serum or the aqueous salt solution used in ultracentrifugal measurements, hence they move against the centrifugal field. Flotation, or negative sedimentation, can be measured in the same units, Svedbergs, as sedimentation (1 Svedberg = 10^{-13} cm./sec./dyne/gram.) The Svedberg of flotation, or negative Svedberg, is symbolized as S_f. A single lipoprotein, homogeneous in its flotation rate, could be characterized by this rate, as for example an S_f 5 lipoprotein. The mixture of lipoproteins in blood serum has been divided into classes, each of which includes the lipoproteins with S_f values between two arbitrary limits, for example the S_f 12–20 class including all lipoproteins with flotation rates between S_f 12 and S_f 20. Since flotation rate is slowed down by increased lipoprotein concentration, classes corrected for concentration effects are designated as "standard", for example the standard S_f 12–20 class. Four chief classes are described on the basis of their flotation rates:

S_f 3-8—Normal components of all sera. Relatively constant for each individual, vary among individuals. *Do not fluctuate with meals or disease.*
S_f 10-20—Stable on given diet, not affected by individual meals. Contains 30 per cent cholesterol.
(S_f 12-20—Statistically related to obesity and the occurrence of atherosclerosis (18).)
(S_f 20–30—Sometimes found with the S_f 10-20 group.)
S_f 30-70—Major fraction during alimentary lipemia. Greatly modified by meals. Some types contain cholesterol.
(S_f 35-100—Statistically related to obesity (18).)
S_f over 75—Represent the chylomicrons and other large aggregates. Increased after fat meals, and are part of alimentary lipemia.

The concentration of each of these four classes increases with age. There is no sex difference in blood levels below the age of 20. Between the age of 20 and 50, the concentration in men is significantly higher than that in women. Beyond 50, the average concentration in men, but not in women, decreases, in part because of the removal of men with particularly high values through vascular disease (16).

Hyperlipemia or abnormal increase of neutral fat in blood plasma is a characteristic pathological change in severe diabetes, in nephrosis, in certain types of anemia, and following the administration of alcohol and other anesthetic drugs. There is also a familial derangement of fat metabolism designated as essential hyperlipemia. Milkiness of the plasma is noted when hyperlipemia is excessive, but Thannhauser (45) has observed increases in neutral fat up to 150 per cent of normal which were detectable only by chemical analysis. Lipemic plasma has been used as an intravenous nutrient mixture.

Injections of heparin will clear the milky plasma of alimentary hyperlipemia, as will the injection of plasma from any animal injected with heparin. Plasma lipoproteins of high S_f are converted to lipoproteins of lower S_f, and free fatty acids are liberated into the plasma. These effects are attributed to a "clearing factor" or lipoprotein lipase (not identical with plasma esterases or with pancreatic lipase), which is formed or activated when heparin interacts with plasma and tissue fluid or extract. The release of fatty acids is inhibited by L-thyroxine and to a less degree by L-triiodothyronine (41).

Blood cholesterol increases with hyperlipemia and with hypoproteinemia. There is a very high blood cholesterol in untreated diabetes, decreasing with insulin therapy, but not often receding to the normal level. The blood cholesterol is typically elevated in hypothyroidism.

BODY LIPIDS

Lipids in the body originate from ingested lipids, and by the transformation of carbohydrate and protein into lipid material. Every cell of the body contains lipids, which are a part of the cellular structure (see Chapter 3). Such *tissue lipid* remains intact regardless of diminished caloric intake. Phospholipids are important components of the cellular lipid. In addition, there is *storage lipid*, composed chiefly of neutral fats, and making up the adipose tissue. Certain locations in the body function as depots for fat; about half of the stored fat is found in the subcutaneous layer, the remainder is divided into perirenal, mesenteric, omental, and subperitoneal deposits with a minimal amount in the fascial planes between the muscles. The proportion of body weight contributed by fat can be calculated from body specific gravity. Studies (4) on healthy men of comparable height and weight show that young men (average age 22) have 13 per cent fat, while middle-aged men (average age 50) have increased the fat percentage to 20. Cholesterol averages about 0.3 per cent of body weight, varying from 0.12 per cent in red cells to 4.5 per cent in the adrenal.

The fatty acids of human subcutaneous fat are as follows according to Calandra and Cattaneo (21) with the percentage of the total fatty acids given for each: oleic, 38.7; linoleic, 24.8; palmitic, 20.8; eicosadienoic, 3.3; palmitoleic, 3.2; gadoleic, 2.8; stearic, 2.2; erucic, 1.8; myristic, 1.5; arachidonic and myristoleic, 0.4 each; arachidic, 0.1. In addition to triglycerides of the above fatty acids there is about one per cent of cholesterol. Each animal species has a characteristic composition of storage fat, which is constant if the animal is fed chiefly upon carbohydrate and protein, with a minimum of dietary fat. If any animal is fed upon a high fat diet, his depot fat will show a variation in composition in the direction of the composition of the fat which has been fed. Fatty acids which are not normally found in the storage fat of the animal will occur there under these circumstances. Stored fat is not metabolically inert, but is in constant exchange with blood lipids. Hence abnormalities of the composition of storage fat will gradually disappear upon cessation of excessive fat feeding. Similar alterations in fatty acid composition of milk fats can be observed in lactating animals fed upon high fat diets. All cells, including those of adipose tissue (37), contain lipases and esterases which facilitate the hydrolysis and resynthesis of lipids.

Atherosclerosis is an accumulation and deposition of lipids, chiefly cholesterol in combination with fatty acids and protein, in the intimal layer of the walls of arteries and arterioles. Such deposits are common during and after middle age, so common that there is a difference of opinion among pathologists whether simple atherosclerosis is a disease or a part of growing old. In its more advanced forms, it is certainly a disease, and in the United

States and countries with similar diet and economics, the most important of diseases. By roughening the inner wall of blood vessels, atherosclerosis increases the probability of formation of intravascular clots which may obstruct the flow of blood to important organs, such as the myocardium. Advanced atherosclerosis makes vessel walls more rigid, increasing the peripheral resistance to blood-flow and thereby increasing arterial blood pressure. Atherosclerotic changes may progress insidiously until brought to attention by cerebral hemorrhage, or by cardiac or renal failure.

Atherosclerosis is generally considered to be encouraged by a diet high in lipids, and is certainly known to be accelerated by chronic hyperlipemia, as in diabetes (35). Measurement of S_f 10-20 or S_f 12-20 lipoproteins, of total cholesterol, and of the ratio of cholesterol to lipid phosphorus gives increased values in the sera of rabbits rendered atherosclerotic by cholesterol feeding. There is a tendency to high values in patients with coronary disease or other manifestations of atherosclerosis. No one of these measurements is, however, a clear-cut discriminator between such patients and healthy persons of similar age. The wide range of values found limits the diagnostic usefulness of these analyses. No correlation exists between the cholesterol content of the diet and the level of cholesterol in the blood plasma. A diet low in cholesterol will bring about a temporary decrease in plasma cholesterol, which returns to its previous level within less than a year (14). The gradual return of the plasma cholesterol to its usual value must be attributed to an increased rate of cholesterol synthesis within the body. *Lipidoses* are disturbances of lipid metabolism within tissue cells. Several such metabolic diseases exist, and are discussed in detail by Thannhauser (45).

OXIDATION OF FATS

Glycerol can be phosphorylated and enter the glycolytic series of reactions, and therefore can form glucose and glycogen. Fatty acids, in order to be oxidized in the Krebs cycle (see page 548), must first be converted to acetylcoenzyme A which can react with oxaloacetic acid to form citric acid. Further oxidation then proceeds by the same tricarboxylic acid cycle described for carbohydrate. Isotopically tagged carbon atoms of administered fatty acids appear in the acids which compose the tricarboxylic acid cycle. The differences in oxidative mechanisms of fats as compared with carbohydrates occur, therefore, in the earlier steps. It is important to recall, in considering these early processes, that the nutritionally significant fatty acids all have an even number of carbon atoms.

It has been known for over half a century that the carbon atoms are removed from fatty acids in multiples of two, but it was not until 1951 that the details of the process were worked out. It is now known that, *in vivo*, the oxidation of a fatty acid to the two-carbon stage involves coenzyme A (see page 804) in every step of the process (31).

The first step in the oxidation of fatty acids is the condensation of such an acid with coenzyme A to form the high-energy acyl-mercaptan link (formula I). ATP is required for this reaction to take place, being converted to adenylic acid (AMP) and pyrophosphate (PP) in the process. At least three enzymes, designated as *thiokinases*, catalyzing this reaction, have been shown to exist, each specific for fatty acids of given chain lengths. One of these has as its substrates acetic acid and propionic acid; the second, fatty acids with from 4 to 11 carbons in the chain; and the third, fatty acids with from 12 to 22 carbons in the chain.

$$\underset{\text{Fatty acid}}{RCH_2CH_2CH_2COOH} + HSCoA + ATP \xrightarrow{\text{[thiokinase]}} \underset{\text{Acylcoenzyme A}}{RCH_2CH_2CH_2CO \sim SCoA} + AMP + PP$$

I. Condensation of fatty acids with coenzyme A

Each acylcoenzyme A formed then enters the *fatty acid oxidation cycle*, which, in four steps, shortens the length of the fatty acid carbon chain by two, producing acetylcoenzyme A as the other product. In the first of these steps, the acylcoenzyme A is dehydrogenated under the influence of an *acyl dehydrogenase* (formula II). This group of enzymes has as coenzyme, FAD, which is reduced in the process. At least three such enzymes have been identified (8). One is the brilliant green copper-containing *butyryl dehydrogenase*, which has maximal activity upon butyryl coenzyme A and no activity when the carbon chain is longer than eight carbons. Two yellow flavoenzymes are known which will dehydrogenate fatty acids with longer carbon chains. They are known as Y_1 and Y_2. Y_1 contains iron. Each of these three transfers the hydrogens to still a fourth flavoenzyme (which, like Y_1, contains iron) which then passes the hydrogens on, probably by way of the cytochromes, to molecular oxygen. The ratio of copper to FAD

$$\underset{\text{Acylcoenzyme A}}{RCH_2CH_2CH_2CO \sim SCoA} \xrightarrow[\text{acyl dehydrogenase}]{FAD} \underset{\text{Alpha, beta-unsaturated acylcoenzyme A}}{RCH_2CH{=}CHCO \sim SCoA}$$

II. Fatty acid oxidation cycle—step 1

in butyryl dehydrogenase is 2 (33), that of iron to FAD in the iron-containing enzymes, 1/6 or less. All these flavoenzymes may form a complex (19).

The second step of the fatty acid cycle involves the hydration of the double bond of the unsaturated compound formed as a result of the first step. The enzyme involved is *crotonyl hydrase* or *alpha,beta-unsaturated acyl hydrase*. The product is the beta-hydroxyacylcoenzyme A (formula III). Note that the first two steps of the cycle result in the oxidation of the beta-carbon of the fatty acid. It is for this reason that fatty acid oxidation is said to proceed through a process of *beta-oxidation*.

$$RCH_2CH{=}CHCO \sim SCoA \xrightarrow[\text{crotonyl hydrase}]{H_2O} RCH_2\underset{\displaystyle OH}{\underset{|}{C}}HCH_2CO \sim SCoA$$

Alpha,beta-unsaturated acylcoenzyme A → Beta-hydroxyacylcoenzyme A

III. Fatty acid oxidation cycle—step 2

The third step is another dehydrogenation to the keto-derivative catalyzed by an enzyme which is termed either *beta-hydroxyacyl dehydrogenase* or *beta-keto reductase* (formula IV). The coenzyme of beta-keto reductase is DPN. The beta-hydroxyacylcoenzyme A can exist in both a D and an L form, since at this stage, the beta-carbon has become asymmetric. The enzyme, beta-keto reductase, is, apparently, specific for the D form.

$$RCH_2\underset{\displaystyle OH}{\underset{|}{C}}HCH_2CO \sim SCoA \xrightarrow[\text{beta-hydroxyacyl dehydrogenase}]{DPN} RCH_2COCH_2CO \sim SCoA$$

Beta-hydroxyacylcoenzyme A → Beta-ketoacylcoenzyme A

IV. Fatty acid oxidation cycle—step 3

Finally, in the fourth and last step of the cycle, the beta-ketoacylcoenzyme A reacts with another molecule of coenzyme A to undergo a *thioclastic cleavage* (that is, one which is analogous to a hydrolytic cleavage except that the atoms of a substituted H_2S are added rather than those of water), producing acetylcoenzyme A and an acylcoenzyme A just like that which originally entered the fatty acid oxidation cycle except that its carbon chain is two carbon atoms shorter (formula V). The enzyme that catalyzes this reaction is *beta-ketothiolase*.

$$RCH_2COCH_2CO \sim SCoA \xrightarrow[\text{beta-ketothiolase}]{\text{HSCoA}}$$
Beta-ketoacylcoenzyme A

$$RCH_2CO \sim SCoA + CH_3CO \sim SCoA$$
Acylcoenzyme A — Acetylcoenzyme A

V. Fatty acid oxidation cycle—step 4

The fatty acid cycle is summarized in formula VI, and its connection with the tricarboxylic acid cycle shown. Note the central position occupied by coenzyme A. Remember, too, that with each turn of the fatty acid oxidation cycle, two carbon atoms are cut off the fatty acid carbon chain. In the case of stearic acid, for instance, with its 18 carbons, eight turns of the cycle are required for complete conversion to acetylcoenzyme A. The eighth turn of the cycle would begin with butyrylcoenzyme A ($CH_3CH_2CH_2CO \sim SCoA$), proceed to the unsaturated derivative crotonylcoenzyme A ($CH_3CH{=}CHCO \sim SCoA$) (whence the name, crotonyl hydrase, for the enzyme which catalyzes the hydration of such compounds), then to betahydroxybutyrylcoenzyme A ($CH_3CH(OH)CH_2CO \sim SCoA$) and then to acetoacetylcoenzyme A ($CH_3COCH_2CO \sim CoA$). Acetoacetylcoenzyme A is converted by the addition of coenzyme A under the influence of betaketothiolase to two molecules of acetylcoenzyme A (formula VII) and the oxidation of stearic acid to acetylcoenzyme A is complete.

The molecules of acetylcoenzyme A released by each repetition of the

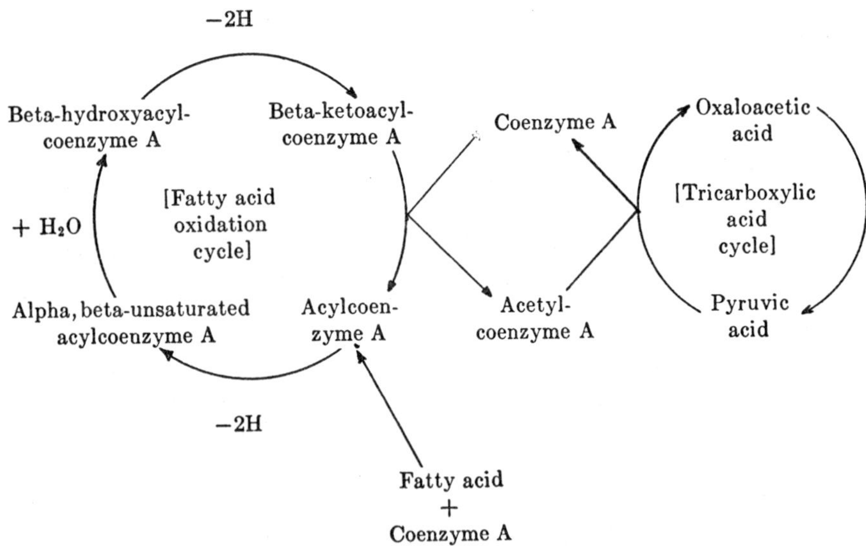

VI. Fatty acid oxidation cycle—summary

series may either (a) enter the tricarboxylic acid cycle as shown in formula VI, or (b) react with another molecule of acetylcoenzyme A to form acetoacetylcoenzyme A, releasing coenzyme A, a reaction which is the reverse of that shown in formula VII.

$$CH_3COCH_2CO \sim SCoA \xrightarrow[\text{beta-ketothiolase}]{\text{HSCoA}} 2CH_3CO \sim SCoA$$

Acetoacetylcoenzyme A — Acetylcoenzyme A

VII. Thioclastic cleavage of acetoacetylcoenzyme A

Approximate turnover numbers (see page 260) have been calculated with butyric acid as substrate (32), assuming molecular weights of 100,000 for all enzymes and equimolecular concentrations, for the enzymes of the five stages as follows: 200 for activation, 500 for the first dehydrogenation, 20,000 for hydration, 10,000 for the second dehydrogenation, and 700 for cleavage. Thus the beta-ketoacyl CoA derivatives, being capable of formation much faster than they are metabolized, would be present under conditions of a steady state at a higher concentration than any of the other forms, and would account for the largest proportion of the CoA molecules involved at any time in the process of fatty acid oxidation.

In the liver and in other tissues, there is a specific deacylase which releases CoA from acetoacetyl CoA. Free acetoacetic acid is therefore formed. Acetoacetic acid is not rapidly or effectively converted to acetoacetylcoenzyme A by liver cells, but being water-soluble can enter the circulation and

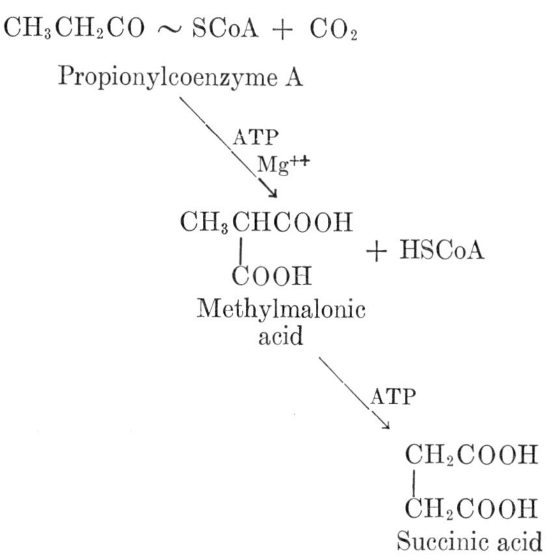

VIII. Fate of odd-carbon fatty acids

be utilized by extra-hepatic tissues, particularly muscle and kidney. Activation of acetoacetic acid prior to oxidation in extra-hepatic tissues, but not in liver, occurs with the aid of an enzyme which transfers CoA from succinyl CoA to acetoacetic acid. After such activation, acetoacetic acid is oxidized by way of the citric acid cycle, which in turn regenerates succinyl CoA by the reaction of alpha-ketoglutaric acid with CoA and DPN, forming simultaneously CO_2 and $DPN \cdot H_2$.

Fatty acids with an odd number of carbon atoms are oxidized by the same mechanisms as described for even-carbon fatty acids, except that propionylcoenzyme A is finally left. This enters the tricarboxylic acid cycle by a pathway not involving the formation of acetoacetic acid. There is evidence that propionylcoenzyme A is converted to methylmalonic acid (formula VIII) in the presence of ATP and Mg^{++} and that this isomerizes to succinic acid in the presence of ATP and coenzyme A (12). Succinic acid is, of course, an intermediate of the tricarboxylic acid cycle. It will be recalled that odd-carbon fatty acids form no significant part of the dietary intake.

Preparations of mitochondria, obtained from rat liver by methods described by Lehninger (9), are highly effective catalysts for the aerobic oxidation of fatty acids as well as for the reactions of the Krebs tricarboxylic acid cycle. The activity of the mitochrondrial enzyme system *in vitro* requires the presence of ATP and of Mg^{++}, and furthermore requires "sparking" or "priming" by the presence of some intermediate in the Krebs cycle, such as fumaric acid or alpha-ketoglutaric acid. Such an intermediate need be present only in catalytic amount to start fatty acid oxidation, which then maintains itself and regenerates the necessary intermediates.

In rat liver slices, ammonium ions increase the formation of acetoacetic acid. This is explained by the decrease in alpha-ketoglutaric acid resulting from its conversion by ammonium ions into glutamic acid (described in the next chapter). This diversion of alpha-ketoglutaric acid limits the formation of oxaloacetic acid, which is necessary for the entry of acetylcoenzyme A into the tricarboxylic acid cycle. With decreased oxidation, acetylcoenzyme A accumulates and condenses to acetoacetic acid.

Both saturation and desaturation of fatty acids can be demonstrated in the liver, desaturation being usually predominant. One double bond can be introduced into a saturated fatty acid in the liver with no apparent difficulty. Introduction of further double bonds seems to be impossible. The introduction of a double bond into saturated fatty acids is not a function of liver only, but has also been demonstrated in adipose tissue (49). So far, emphasis has been placed upon the liver as the chief organ involved in the early stages of fatty acid oxidation. Experiments with eviscerated rats (15) indicate that free fatty acids are effectively and completely oxidized by tissues other than liver. Brain, skeletal muscle, and other tissues of the rat will oxidize palmitic acid (48).

Another type of fatty acid oxidation, in which the terminal methyl group is oxidized to carboxyl, has been demonstrated in intact men and dogs (47). Such *omega-oxidation* takes place most notably with saturated fatty acids not more than 12 nor less than 8 carbon atoms in length. The feeding of such a fatty acid leads to the appearance in the urine of dicarboxylic acids with either the same number of carbon atoms as the original fatty acid, or less than that number by multiples of two. It therefore is evident that dicarboxylic acids within the 8 to 12 carbon atom range can undergo beta-oxidation with splitting off of two-carbon fragments. This concept has been confirmed by the direct feeding of dicarboxylic acids to dogs and recovering them in the urine along with other dicarboxylic acids shorter by 2 or 4 carbon atoms. The output of dicarboxylic acids in the urine, on a constant fatty acid intake, is increased by increasing the carbohydrate of the diet. This is in contrast with the urinary output of ketone acids, which will be shown later to be diminished or abolished by adequate carbohydrate intake. There is no evidence that omega-oxidation is a significant pathway in the oxidation of the customary dietary fatty acids.

It is possible that there exists an alternate fatty acid oxidation cycle in which *pantetheine* takes the place of coenzyme A. (Pantetheine is a "half-molecule" of coenzyme A, see page 804, lacking the nucleotide portion but retaining the pantothenic acid moicty and the SH group.) This alternate cycle is not linked to the tricarboxylic acid cycle and may be primarily concerned with the formation of the alpha-beta unsaturated link and the consequent synthesis of polyunsaturated compounds such as the isoprenoids (44).

Conversion of fatty acids to carbohydrate does not occur in animals as a significant physiological process. The evidence in regard to gluconeogenesis from fatty acids can be summarized in two main points: (1) tagged carbon atoms administered in fatty acids do appear in glucose and glycogen; but (2) no net gain in glucose or glycogen follows the feeding of fatty acids to starving or diabetic animals. To explain these apparently contradictory facts in terms of known oxidative mechanisms, we must presume that there is no significant direct conversion of acetyl CoA to pyruvic acid. Numerous other anabolic pathways are open to the two-carbon fragment or acetyl group of acetyl CoA, but none of these leads to carbohydrate formation. When a two-carbon fragment enters its only known catabolic reaction, initiating the Krebs cycle by condensation with oxaloacetic acid, carbohydrate-forming intermediates are produced but two carbons are oxidized. The two carbons which are oxidized are not those of the two-carbon fragment; therefore tagged carbons of two-carbon fragments derived from fatty acids may appear later in carbohydrates derived from Krebs cycle intermediates, but since two other carbons have been oxidized there can result no net gain of carbohydrate.

Energetics of fatty acid oxidation. Each turn of the fatty acid cycle includes two dehydrogenations, one involving a flavoenzyme and one a pyridinoenzyme. Assuming that two high-energy phosphate bonds are formed in the first case and three in the second, we end with five molecules of ATP formed in each turn of the cycle.

Stearic acid requires 8 turns of the cycle to be converted to 9 molecules of acetylcoenzyme A. Each of these 9 molecules of acetylcoenzyme A is converted to CO_2 and H_2O by a turn of the tricarboxylic acid cycle, and each produces 12 molecules of ATP in this way. The situation is summarized in formula IX.

$$\left.\begin{array}{c} C_{17}H_{35}COOH + 8O_2 + 9CoA\text{-}SH + 40ADP + 40H_3PO_4 \\ \Big\downarrow \text{[Fatty acid oxidation cycle]} \\ 9CoA\text{-}S \sim COCH_3 + 57H_2O + 40ATP \end{array}\right\} \text{step 1}$$

$$\left.\begin{array}{c} 9CoA\text{-}S \sim COCH_3 + 18O_2 + 108ADP + 108H_3PO_4 \\ \Big\downarrow \text{[Tricarboxylic acid cycle]} \\ 9CoA\text{-}SH + 117H_2O + 18CO_2 + 108ATP \end{array}\right\} \text{step 2}$$

$$\left.\begin{array}{c} C_{17}H_{35}COOH + 26O_2 + 148ADP + 148H_3PO_4 \\ \Big\downarrow \text{[Total oxidation} \\ 18CO_2 + 174H_2O + 148ATP \end{array}\right\} \text{sum of steps 1 and 2}$$

IX. ATP formation in fatty acid oxidation

KETOGENESIS AND KETOSIS

Acetoacetic acid can be reduced in liver to beta-hydroxybutyric acid. These two are known clinically as the *ketone acids*. Acetoacetic acid will slowly decompose spontaneously to acetone and CO_2 (formula X). The

$$\begin{array}{ccccc} CH_3 & & CH_3 & & CH_3 \\ | & & | & & | \\ HCOH & & C{=}O & & C{=}O \quad \text{Acetone} \\ | & \leftrightharpoons & | & \leftrightharpoons & | \\ CH_2 & & CH_2 & & CH_3 \\ | & & | & & + \\ COOH & & COOH & & CO_2 \end{array}$$

β-Hydroxybutyric acid — Acetoacetic acid

X

ketone acids plus acetone make up the group of substances known clinically as the *ketone bodies*, which are normal intermediates in fatty acid metabolism. Accumulation of the ketone bodies in tissues and body fluids in abnormal amounts is known as *ketosis* and most commonly results from starvation or from diabetes mellitus.

The normal concentration of ketone bodies in the blood is about 0.5 mgm./100 ml. The normal daily urinary output of ketone bodies is less than 500 mgm. During fasting, the blood ketones increase for the first few days, while the urinary excretion may reach 10 grams a day and remain between 5 and 10 grams for the duration of the fast. If fat alone is now ingested, blood levels and urinary outputs of ketone bodies will increase, but not sufficiently to account for any great proportion of the ingested fat. Adequate carbohydrate intake promptly restores blood and urinary ketones to normal values. In severe diabetes, the output of ketone bodies may rise to several hundred grams per day.

Clinical experience has taught us that fatty acids, plus certain amino acids which are metabolized in a manner comparable to fatty acids (see Chapter 16), are ketogenic, which means that they increase the production of ketone bodies. Carbohydrate, glycerol, and the majority of amino acids are *antiketogenic*, which means that the metabolism of these substances diminishes the tendency to excessive ketone body production. Even in normal subjects, it is possible to induce ketosis by feeding a diet in which the ketogenic substances are present in great excess over the antiketogenic substances. Such a *ketogenic diet* is of value in the treatment of epilepsy, particularly in children, for reasons which are not fully understood. Other therapeutic applications of the ketogenic diet, such as in urinary tract infections, have been superseded by more effective measures. If it should be desired to prescribe a ketogenic diet, or conversely to prescribe a diet which would minimize the formation of ketone bodies, the Schaffer K/A ratio still remains a valuable check on the ketogenic potentialities of a diet, even though the theory upon which it was based is now known to be oversimplified. In this formula K refers to ketogenic substances, and A to antiketogenic substances (formula XI). If the value of this ratio for any diet exceeds 2, that diet is ketogenic and a person subsisting on that diet will probably show *ketonemia* (more than one mgm. of ketone bodies, expressed as acetone, per 100 ml. of blood) and *ketonuria* (the elimination of more than 20 mgm. of ketone bodies, expressed as acetone, in the 24-hour urine). In fasting men, blood concentrations of ketone bodies have been shown to increase progressively during periods of starvation lasting up to 5 days (22). With longer periods of starvation, ketosis has been observed to diminish.

The presence or absence of ketosis depends upon the relative rates of the reactions shown in an abbreviated form in formula XII. Note that the

$$\frac{K}{A} = \frac{(2.4 \times \text{grams protein}) + (3.43 \times \text{grams fat})}{(3.2 \times \text{grams protein}) + (0.57 \times \text{grams fat}) + (5.56 \times \text{grams carbohydrate})}$$

XI

conversion of acetylcoenzyme A to citric acid, the normal enegy-yielding pathway via the tricarboxylic acid cycle, requires oxaloacetic acid. Although this acid is regenerated by the tricarboxylic acid cycle, there are some losses, as by transamination to aspartic acid. If glycolysis is proceeding at an adequate rate, oxaloacetic acid may be formed by addition of CO_2 to the glycolytic intermediate, phospho*enol*pyruvic acid. Thus, deficiency of carbohydrate intake, by slowing the glycolytic formation of phospho*enol*-pyruvic acid, slows the oxidative utilization of acetylcoenzyme A, favoring the excess formation of acetoacetic acid. Excess intake of fats promotes the formation of acetylcoenzyme A without providing phospho*enol*pyruvic acid for oxaloacetic acid formation, thus also favoring the formation of acetoacetic acid. In diabetes, there is probable interference with both the glycolytic and oxidative pathways. The ingestion of ammonium salts is ketogenic by increasing the rate of loss of oxaloacetic acid through transamination to aspartic acid.

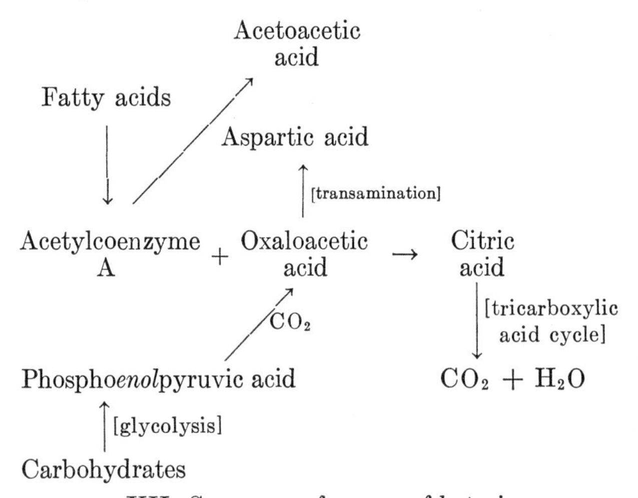

XII. Summary of causes of ketosis

Ketonuria. A state of clinical ketosis is most commonly confirmed by the finding of one or more of the ketone bodies in the urine. The common nitroprusside tests for "acetone" in the urine are really tests for acetoacetic acid. To be sure, they will indicate the presence of acetone, but they are at least five times more sensitive to acetoacetic acid. When urine specimens are allowed to stand, acetoacetic acid continues to break down into acetone and CO_2, which renders the nitroprusside tests less responsive to the same original amount of ketone bodies. The most satisfactory of the numerous nitroprusside procedures is that of Rothera (13): 5 ml. of fresh urine is saturated with solid ammonium sulfate and mixed with 10 drops of freshly prepared 2 per cent sodium nitroprusside solution; this is then mixed with 10 drops of concentrated ammonia water and allowed to stand 15 minutes. The presence of acetoacetic acid, or of larger concentrations of acetone, is indicated by the characteristic blue-purple color.

Diabetic ketosis is of common occurrence in diabetics who are untreated or inadequately treated. The Rothera test is sensitive to low concentrations of acetoacetic acid in the urine, and may be positive in cases where the patient shows no symptoms referable to ketosis. A positive test for acetoacetic acid in the urine should not, however, be ignored. Symptoms often develop rapidly and proceed to coma. Pre-coma symptoms and signs include anorexia, nausea, vomiting, abdominal pain, increased pulmonary ventilation, and drowsiness. These symptoms arise in part from acidosis (see Chapter 18) and in part from toxic effects of the ketone bodies (11). There is often an acetone odor on the breath. Treatment in this stage consists of increased insulin dosage or a re-evaluation of the patient's diet, usually in the direction of increased carbohydrate—or both. In cases of actual coma an emergency exists and treatment is best carried on in a hospital.

BIOSYNTHESIS OF LIPIDS

We can safely assume that the knowledge of the fattening of animals on diets relatively low in fat was put to use by our early pastoral ancestors. In 1860 Lawes and Gilbert (27) set up the first balance sheet of ingested protein, fat, and carbohydrate as compared with the formation of these same substances in pigs, and demonstrated conclusively that triglycerides could be formed from carbohydrate. This work has been amply confirmed by many types of investigation, including the use of isotopic labeling, so that there is no question that neutral fat can be formed from carbohydrate. The same conclusion has been drawn from experiments on micro-organisms and on higher plants. The increase of fat with simultaneous decrease in carbohydrate which occurs in the ripening of seeds has been observed in many species. The yeast *Torulopsis lipofera*, grown in a glucose medium, forms fat at a rate of 4 to 11 per cent of its dry weight in 5 hours (27), far exceeding the rate of one per cent of body weight per day observed in abundantly nourished swine.

Since the formation of fat from carbohydrate involves reduction, it inexorably follows that some substance must be simultaneously oxidized. Also, since fat formation involves energy storage, the stored energy must be supplied by an energy-yielding process. Both of these requirements are met by the oxidation of carbohydrate. The biosynthesis of fat from carbohydrate has never been observed to proceed under anaerobic conditions. Some anaerobic organisms can synthesize fat from alcohol or other noncarbohydrate precursors. Most micro-organisms, including yeasts, require oxygen for fat synthesis.

In the animal body the synthesis of fat is not limited to any particular organ. Evidence of lipid formation has been observed in liver, skin, intestine, and brain (27). The lipid synthesis of the brain is most rapid at the time of myelinization, which utilizes fatty acids locally synthesized, rather than those supplied in the diet. Adipose tissue has, in common with other tissues, the ability to form glycogen from glucose and to convert glucose to lipids.

The main pathway of *fatty acid synthesis* is by the condensation of two-carbon fragments, in the form of acetyl CoA, derived either from fatty acid breakdown or from pyruvic acid. In liver slices, the conversion of added tagged acetic acid to fatty acids is demonstrable. Such fatty acid synthesis is increased in the presence of added glucose or pyruvic acid, with a still greater increase if insulin is also present. Lengthening of fatty acid chains occurs by the successive addition of two carbon atoms at the carboxyl end. This is, in essence, simply the reversal of the fatty acid oxidation cycle, with the same enzymes catalyzing the individual steps in reverse. The reversal of the two oxidative steps requires *in vitro* the presence of appropriate reducing agents (19) and *in vivo* the activity of a simultaneous oxidative process.

In animals given D_2O in amounts sufficient to maintain its concentration in body fluids at about 2 per cent, deuterium in the saturated fatty acid of the animals eventually reaches a concentration which indicates that half the hydrogen of biosynthesized saturated fatty acids comes from body water (40). Unsaturated fatty acids contain less deuterium under these experimental conditions, showing that saturated fatty acids are biosynthesized and subsequently converted to unsaturated fatty acids, presumably by the action of fatty acid dehydrogenase. It has already been noted (page 455) that multiple-unsaturated fatty acids such as linoleic and linolenic acids are not biosynthesized in animals.

Synthesis of Compound Lipids

The formation of *glycerol* directly by the reduction of glyceraldehyde is catalyzed by a dehydrogenase, utilizing DPN, and identified in rat and pig liver (51). Unphosphorylated glyceraldehyde is formed by the cleavage of fructose-1-phosphate by 1-phosphofructaldolase. Glycerol can be phosphorylated to alpha-glycerophosphate by *glycerokinase*, an enzyme which has been located in rat liver. ATP, and to a lesser extent UTP, can be utilized as the phosphate donor (5).

An enzyme which catalyzes the linkage of stearyl CoA to alpha-glycerophosphate has been identified in liver extracts (28). This is one step in phospholipid synthesis. The introduction of choline into phospholipids may involve its phosphorylation, catalyzed by *choline phosphokinase*, with conversion of ATP to ADP. The enzyme has been purified from yeast (50), but similar activity has been demonstrated in preparations from mammalian liver, brain, intestine, and kidney. The incorporation of phosphorylcholine into lecithin may proceed by way of *cytidine diphosphate choline* as intermediate. The reaction in rat liver mitochondria is specifically activated by CTP, ATP being ineffective in this respect (23). In guinea-pig liver mitochondria, however, ATP is required (39).

Sphingol (see page 127) is synthesized by the condensation of a 16-carbon acylcoenzyme A with serine, followed by a decarboxylation and reduction (42) (formula XIII).

$$\underset{\text{Acylcoenzyme A}}{RCO{\sim}SCoA} + \underset{\text{Serine}}{\begin{array}{c} COOH \\ | \\ CHCH_2OH \\ | \\ NH_2 \end{array}} \rightarrow RCO{-}\begin{array}{c} COOH \\ | \\ CCH_2OH \\ | \\ NH_2 \end{array} \xrightarrow{-CO_2} RCO{-}\begin{array}{c} CHCH_2OH \\ | \\ NH_2 \end{array} \xrightarrow{+2H} \underset{\text{Sphingol}}{\begin{array}{c} OH \\ | \\ RCH \end{array}{-}\begin{array}{c} CHCH_2OH \\ | \\ NH_2 \end{array}}$$

XIII. Biosynthesis of sphingol

Synthesis of Sterols

Both plant and animal cells synthesize the cyclopentenophenanthrene ring which characterizes the sterols and steroids. Man is not capable of absorbing or utilizing the sterols which are formed by plants. Vitamin D_2, which is irradiated ergosterol, is absorbable and utilizable, but it should be noted that the sterol ring is broken in the process of its formation from ergosterol, a plant sterol, by irradiation. The parent plant sterol, ergosterol, is not absorbed. Cholesterol is absorbed and metabolized. The chief pathway of cholesterol metabolism is conversion to bile acids, and the chief excetory product is glycocholic acid (see page 505).

Cholesterol formation from two-carbon fragments (30) has been studied extensively in liver slices, although other tissues can also carry out the synthesis. A water-soluble enzyme system capable of forming cholesterol has been obtained from rat liver. Tagged acetic acid is incorporated into cholesterol and each carbon atom in the cholesterol molecule can be traced to either the methyl or the carboxyl carbon of acetate. Coenzyme A is involved in the synthesizing system, as is shown by parallelism between CoA concentration and rate of cholesterol formation in liver slices from pantothenic acid-deficient rats (26). The rate of cholesterol biosynthesis is increased in alloxan-diabetic rats (20). Squalene (see page 137), which may possibly be on one of the pathways of cholesterol synthesis, has also been shown to be formed from acetate (3). As explained on page 334, steroid hormones may be biosynthesized from cholesterol, or directly from two-carbon fragments.

Abnormal Fat in Liver Cells

Ordinary yeast supplied with ethyl alcohol will convert it to fat, presumably by way of acetaldehyde and acetic acid. Yeast supplied with glucose, which of course is converted by the yeast into ethyl alcohol, grows more vigorously but utilizes less of the sugar in fat production. In many types of cells, fat formation is associated with senescence or with diminishing food supply. Damage to human liver cells is often indicated to the examining pathologist by an accumulation of fatty acids and neutral fat within the cells. These cells have as a normal function the production of the phospholipids which circulate in the blood and constitute the most available form of lipid for tissue utilization. This function of hepatic cells is depressed, and abnormal intracellular fat deposits are produced, by certain poisons, including organic halides, carbon tetrachloride, and salts of heavy metals. Failure of normal carbohydrate metabolism, as in diabetes, may lead to similar accumulations of abnormal fat in liver cells. Infections, not only of the liver, but also of remote parts of the body, may do similar harm. Liver damage of this type, often proceeding to connective tissue replacement of hepatic cells (cirrhosis), has long been recognized as associated with alcoholism. Many ardent consumers of alcohol, however, show no evidence of liver cell abnormality. Current clinical and biochemical thought considers these cases of "alcoholic fatty livers" as examples of a deficiency disease brought about by the limited food intake so often observed among alcoholics. High carbohydrate diets have been found helpful in limiting the deposition of fat in liver cells in such cases, but high protein diets are even more effective. The actual deficiency appears to be in *choline* and its dietary precursors, which include the amino acid methionine. The function of methionine appears to be that of a donor of methyl groups, permitting the conversion of ethanolamine to choline (see page 125). Ethanolamine itself may be formed by the reduction of glycine or by the decarboxylation of serine. Thus, choline can be formed within the body, provided an adequate supply of methyl donors is available from dietary sources. *Formic acid* or the formate group, when tagged in the form of deuterio-

C^{14}-formate, can be demonstrated to form methyl groups of rat tissue choline without losing its hydrogen (38). *Betaine* (formula XIV), an effective methyl donor, appears to be a product of the oxidation of choline by the choline oxidase enzyme system of the liver, of which a specific portion is choline dehydrogenase. The choline oxidase system is necessary for transmethylation from choline. The carbon atom 2 of the imidazole ring of histidine also contributes to the carbon of labile methyl groups (43).

Choline is designated as a *lipotropic substance*, which means that it is necessary for the normal production of phospholipids, which constitute the major portion of the indispensable lipids of cells. In the absence of choline, neutral fat can not be converted into useful phospholipids. The rate of phospholipid synthesis in normal subjects has been measured (7) and found to be highly variable among different subjects, but relatively constant in the same subject over a period of months. Administration of choline or methionine to the normal subject does not increase the rate of phospholipid formation, but such an increase can often be observed following the administration of large doses of choline or methionine to a patient with chronic hepatitis or cirrhosis. It appears that much of the benefit derived from the high protein diet in such cases was the result of the methionine so supplied.

$$^{(-)}OOCCH_2\overset{(+)}{N}(CH_3)_3$$

XIV. Betaine

Fatty livers can also be produced experimentally in pancreatectomized animals treated with insulin, and they occur spontaneously in human and particularly in juvenile diabetics. Human fatty livers have been reported with over 30 grams fat per 100 grams wet weight (2). The normal mean value is 5.7 grams. Methionine, choline, and betaine can be used successfully in combating this type of fatty liver, as well as the type caused by choline deficiency. Inositol is lipotropic, but only in a limited manner and in connection with diets low in fat (1a). Certain pancreatic extracts (6) are also effective lipotropic agents, but their mechanism is not understood. They do not contain sufficient amounts of known lipotropic agents to explain their activity.

A most extreme example of fatty degeneration of liver cells is seen in advanced cases of *kwashiorkor*, a disease of children caused by multiple dietary deficiencies, of which deficiency of protein seems to be the most significant. In cases which have come to autopsy liver cells were found to consist of a single mass of fat which had pushed the nucleus aside and left only a thin shell of surrounding cytoplasm. In cases which recover, some degree of connective tissue replacement of liver cells (cirrhosis) occurs and persists. Kwashiorkor was identified and named in the Gold Coast of

Africa, but has appeared in many localities where starchy foods, particularly maize, manioc, or cassava, are the dietary staples (34), and there is lack of the proteins of high biological value supplied by such foods as milk, meat, and eggs.

REFERENCES

1. Baker, R. W. R., *et al.* A study by paper electrophoresis of the serum lipoproteins in diabetic and non-diabetic subjects. Quart. J. Med. (new series), **24:** 295–305, 1955.

1a. Best, C. H., *et al.* Statistical evaluation of lipotropic action of inositol. Biochem. J., **48:** 448–452, 1951.

2. Billing, B. H., *et al.* The value of needle biopsy in the chemical estimation of liver lipids in man. J. Clin. Invest., **32:** 214–225, 1953.

2a. Bjorklun, R., and Katz, S. The molecular weights and dimensions of some human serum lipoproteins. J. Am. Chem. Soc., **78:** 2122–2126, 1956.

3. Bloch, K., *et al.* Biological synthesis of cholesterol. Harvey Lectures, **48:** 68–88, 1954.

4. Brozek, J., and Keys, A. Age changes in body composition during maturity. J. Gerontol., **6:** no. 3, suppl., 67–68, 1951.

5. Bublitz, C., and Kennedy, E. P. Synthesis of phosphatides in isolated mitochondria. III. The enzymic phosphorylation of glycerol. J. Biol. Chem., **211:** 951–961, 1954.

6. Canepa, J. F., *et al.* A study of lipotropic factors derived from the pancreas. Am. J. Physiol., **156:** 387–395, 1949.

7. Cornatzer, W. E., and Cayer, D. The effect of lipotropic factors on phospholipide turnover in the plasma of normal persons as indicated by radioactive phosphorus. J. Clin. Invest., **29:** 534–541, 1950.

8. Crane, F. L. Flavoproteins involved in the first oxidative step of the fatty acid cycle. Biochim et Biophys. Acta, **17:** 292–294, 1955.

9. Edsall, J. T. *Enzymes and Enzyme Systems*. Cambridge, Mass., Harvard University Press, 1951.

10. Evans, J. D., *et al.* Polyunsaturated fatty acids in normal human blood. J. Biol. Chem., **218:** 255–259, 1956.

11. Fisher, P. The role of the ketone bodies in the etiology of diabetic coma. Am. J. M. Sci., **221:** 384–397, 1951.

12. Flavin, M. Metabolism of propionic acid in animal tissues. Nature, **176:** 823–826, 1955.

13. Friedemann, T. E., *et al.* An assessment of the value of the nitroprusside reaction for the determination of ketone bodies in urine. Quart. Bull. Northwestern Univ. M. Sch., **20:** 301–310, 1946.

14. Gertler, M. M., *et al.* Diet, serum cholesterol, and coronary heart disease. Circulation, **2:** 696–704, 1950.

15. Geyer, R. P., *et al.* Extrahepatic lipid oxidation in the rat. Fed. Proc., **10:** 188–189, 1951.

16. Glazier, F. W., *et al.* Human serum lipoprotein concentrations. J. Gerontol., **9:** 395–403, 1954.

17. Gofman, J. W., *et al.* Blood lipids and human atherosclerosis. Circulation, **2:** 161–178, 1950.

18. Gofman, J. W., and Jones, H. B. Obesity, fat metabolism, and cardiovascular disease. Circulation, **5:** 514–517, 1952.

19. GREEN, D. E. Oxidation and synthesis of fatty acids in soluble enzyme systems of animal tissues. Clinical Chem., **1:** 53–67, 1955.
20. HOTTA, S., AND CHAIKOFF, I. L. Cholesterol synthesis from acetate in the diabetic liver. J. Biol. Chem., **198:** 895–899, 1952.
21. HOUSSAY, B. A., *et al. Human Physiology.* New York, McGraw-Hill, 1950.
22. KARTIN, B. L., *et al.* Blood ketones and serum lipides in starvation and water deprivation. J. Clin. Invest., **23:** 824–835, 1944.
23. KENNEDY, E. P., AND WEISS, S. B. Cytidine diphosphate choline: A new intermediate in lecithin biosynthesis. J. Am. Chem. Soc., **77:** 250–251, 1955.
24. KEYS, A., *et al.* The concentration of cholesterol in the blood serum of normal men and its relation to age. J. Clin. Invest., **29:** 1347–1353, 1950.
25. KEYS, A., *et al.* Serum cholesterol and other characteristics of clinically healthy men in Naples. Arch. Internal Med., **93:** 328–336, 1954.
26. KLEIN, H. P., AND LIPMANN, F. The relation of coenzyme A to lipide synthesis. II. Experiments with rat liver. J. Biol. Chem., **203:** 101–108, 1953.
27. KLEINZELLER, A. Biosynthesis of lipids. Advances Enzymol., **8:** 299–341, 1948.
28. KORNBERG, A., AND PRICER, W. E., JR. Enzymatic synthesis of phosphorus-containing lipides. J. Am. Chem. Soc., **74:** 1617, 1952.
29. LEWIS, G. T., AND PARTIN, H. C. Fecal fat on an essentially fat-free diet. J. Lab & Clin. Med., **44:** 91–93, 1954.
30. LIEBERMAN, S., AND TEICH, S. Recent trends in the biochemistry of the steroid hormones. Pharmacol. Rev., **5:** 285–280, 1953.
31. LYNEN, F. Participation of coenzyme A in the oxidation of fat. Nature, **174:** 962–965, 1954.
32. MAHLER, H. R. Role of coenzyme A in fatty acid metabolism. Fed. Proc. **12:** 694–702, 1953.
33. MAHLER, H. R. The fatty-acid oxidizing system of animal tissues. IV. The prosthetic group of butyryl coenzyme A dehydrogenase. J. Biol. Chem., **206:** 13–26, 1954.
34. MEIKLEJOHN, A. P., AND PASSMORE, R. Nutrition and nutritional disease. Ann. Rev. Med., **2:** 129–154, 1951.
35. MORRISON, L. M. Arteriosclerosis. J. A. M. A., **145:** 1232–1236, 1951.
36. PAGE, I. H. Atherosclerosis, an introduction. Circulation, **10:** 1–27, 1954.
37. RENOLD, A. E., AND MARBLE, A. Lipolytic activity of adipose tissue in man and rat. J. Biol. Chem., **185:** 367–375, 1950.
38. RESSLER, C., *et al.* Studies *in vivo* on labile methyl synthesis with deuterio-C^{14}-formate. J. Biol. Chem., **197:** 1–5, 1952.
39. RODBELL, M., AND HANAHAN, D. J. Lecithin synthesis in liver. J. Biol. Chem., **214:** 607–618, 1955.
40. SCHOENHEIMER, R. *The Dynamic State of Body Constituents.* Cambridge, Mass., Harvard University Press, 1942.
41. SHORE, B. Effect of thyroxine and related compounds on heparin-activated fatty acid liberating enzyme. Proc. Soc. Exptl. Biol. Med., **90:** 398–400, 1955.
42. SPRINSON, D. B., AND COULON, A. The precursors of sphingosine in brain tissue. J. Biol. Chem., **207:** 585–592, 1954.
43. SPRINSON, D. B., AND RITTENBERG, D. The metabolic reactions of carbon atom 2 of L-histidine. J. Biol. Chem., **198:** 655, 1952.
44. STERN, J. R. An alternate fatty acid cycle involving thioesters of pantetheine. J. Am. Chem. Soc., **77:** 5194–5195, 1955.
45. THANNHAUSER, S. J. Lipidoses: Diseases of the cellular lipid metabolism. *Oxford*

Medicine, Vol. 4, Pt. 2, Chapter VII-A. New York, Oxford University Press, 1949.

46. Van Itallie, T. B., *et al.* Clinical use of fat injected intravenously. Arch. Int. Med., **89:** 353–357, 1952.
47. Verkade, P. E., *et al.* Researches on fat metabolism. XII. The influence of carbohydrate on diaciduria. Biophys. et Biochem. Acta, **2:** 38–56, 1948.
48. Volk, M. E., *et al.* Oxidation of endogenous fatty acids of rat tissues *in vitro*. J. Biol. Chem., **195:** 493–501, 1952.
49. Wertheimer, E., and Shapiro, B. The physiology of adipose tissue. Physiol. Rev., **28:** 451–464, 1948.
50. Wittenberg, J., and Kornberg, A. Choline phosphokinase. J. Biol. Chem., **202:** 431–444, 1953.
51. Wolf, H. P., and Leuthardt, F. Über die Glycerindehydrase der Leber. Helv. chim. acta, **36:** 1463–1467, 1953.

CHAPTER 16

Protein Metabolism and Starvation

The subject matter of protein metabolism is more varied and more complicated than that of either fat or carbohydrate metabolism. The reason is not far to seek. The digestive end products of carbohydrates (the monosaccharides) and of lipids (fatty acids, mainly) are each members of a closely-knit chemical group. In each case, there is therefore a single main line of metabolism. Alternate pathways exist but they are recognizably merely variations on a theme.

The digestive end products of proteins are some twenty amino acids, the individual chemical nature of whose side-chains varies from hydrocarbon (both aliphatic and aromatic) through alcohols and phenols, to thiols, disulfides, thioethers and from amines and guanidines to several heterocyclic groupings (see Chapter 1). Each of these amino acids has been found to have its own metabolic pathway; each presents problems not entirely duplicated by any of the others; and several of them have specific roles in the body.

However, in one respect at least, all the amino acids (with the exception of proline and hydroxyproline) are alike. They all have an amino group and a carboxyl group attached to one of the carbons. The most general aspect of protein metabolism, therefore, concerns these amino and carboxyl groups and it is that which we will consider first.

GENERAL CATABOLISM OF THE AMINO ACIDS

The amino acids which are liberated by digestive hydrolysis of proteins are taken up by the portal vein, and are distributed in the general circulation. Following a protein meal, the amino-nitrogen of the blood increases by 2 to 6 mgm. over its fasting level of 4 to 6 mgm. per 100 ml., and remains at an elevated level for about 6 hours.

The utilization of amino acids by the body may be *structural,* in that they may be incorporated into the various proteins of the body. This aspect of protein metabolism will be discussed later in the chapter (see page 650).

The utilization of amino acids may also be *energic*, in that they are converted into nitrogen-free intermediates, which can be catabolized by the same mechanisms which make available the energy of fat and carbohydrate.

Deamination

In the formation of nitrogen-free intermediates, one of the necessary steps is obviously that of removing the alpha-amino group. This is accomplished in two stages in most of the amino acids. The first is a dehydrogenation, catalyzed by the *amino acid dehydrogenases*, of the amino acid to an imino acid; the second is a non-enzymatic hydrolysis of the imino acid to a keto acid (formula I).[1]

$$\underset{\text{Amino acid}}{\mathrm{R\overset{\overset{NH_2}{|}}{C}HCOOH}} \xrightarrow[\text{[amino acid dehydrogenase]}]{-2\mathrm{H}} \underset{\text{Imino acid}}{\mathrm{R\overset{\overset{NH}{\|}}{C}COOH}} \xrightarrow{+\mathrm{H_2O}} \underset{\text{Keto acid}}{\mathrm{R\overset{\overset{O}{\|}}{C}COOH}} + \mathrm{NH_3}$$

I. Deamination

There are two amino acid dehydrogenases with group specificity present in the body. One is D-amino acid dehydrogenase and the other L-amino acid dehydrogenase which, as the names imply, are restricted in their catalytic effects to D-amino acids and L-amino acids respectively. Although the D-amino acids are "unnatural" and occur in the body to a very minor extent, the D-amino acid dehydrogenase is much the better known of the two. This is partly due to the fact that the D-amino acid dehydrogenase is easily extracted from tissue preparations whereas the L-amino acid dehydrogenase remains more firmly bound to the cell particulates and is therefore less easily studied.

Both amino acid dehydrogenases are flavoenzymes, the coenzyme in the case of D-amino acid dehydrogenase being FAD, that of L-amino acid dehydrogenase being FM. In either case, the flavin accepts the hydrogen removed from the amino acid and passes it on *via* the cytochrome system to molecular oxygen (see page 278). Deamination is thus an aerobic process.

D-Amino acid dehydrogenase is the less specific of the two. It will deaminate all the D-amino acids but D-glutamic acid. L-Amino acid dehydrogenase will oxidize all the L-amino acids but L-serine, L-threonine, L-aspartic acid, L-glutamic acid, L-lysine and L-arginine. Neither enzyme will deaminate the optically inactive amino acid, glycine.

To deaminate some amino acids, specific enzymes are required. Thus, glycine is deaminated by *glycine dehydrogenase*, which deaminates glycine

[1] For simplicity, non-ionized formulas of amino acids are used throughout this chapter.

to glyoxylic acid (formula II). The coenzyme of glycine dehydrogenase is also FAD. (The enzymes discussed so far in this section are frequently

$$\underset{\text{Glycine}}{H_2NCH_2COOH} \xrightarrow[\text{[Glycine dehydrogenase]}]{-2H} \underset{\text{Iminoacetic acid}}{HN{=}CHCOOH} \xrightarrow{+H_2O} \underset{\text{Glyoxylic acid}}{OCHCOOH}$$

II. Deamination of glycine

termed "amino acid oxidases" and "glycine oxidase" but "dehydrogenase" fits better with the classification described on page 275.)

In similar fashion, L-glutamic acid may be deaminated by L-glutamic acid dehydrogenase, an enzyme for which DPN or TPN is the coenzyme. In this case, the product is alpha-ketoglutaric acid (formula III).

$$\underset{\text{Glutamic acid}}{\begin{array}{c} COOH \\ | \\ CH_2 \\ | \\ CH_2 \\ | \\ CHNH_2 \\ | \\ COOH \end{array}} \xrightarrow[\text{[L-Glutamic acid dehydrogenase]}]{-2H} \underset{\text{Alpha-iminoglutaric acid}}{\begin{array}{c} COOH \\ | \\ CH_2 \\ | \\ CH_2 \\ | \\ C{=}NH \\ | \\ COOH \end{array}} \xrightarrow[-NH_3]{+H_2O} \underset{\text{Alpha-ketoglutaric acid}}{\begin{array}{c} COOH \\ | \\ CH_2 \\ | \\ CH_2 \\ | \\ C{=}O \\ | \\ COOH \end{array}}$$

III. Deamination of glutamic acid

The alpha-amino group in other cases, notably that of serine, threonine, and cysteine, is removed by means other than oxidative deamination. This will be discussed in the section devoted to the special catabolism of the amino acids (see page 623).

It should be noted that deamination alone suffices in the case of some amino acids to convert them into intermediates on the main-stream of carbohydrate catabolism. Alanine is converted to pyruvic acid and glutamic acid to alpha-ketoglutaric acid. Each of these products may directly enter the tricarboxylic acid cycle (see page 548), serving either as a source of energy through further catabolism or as a source of glycogen through the reverse steps of carbohydrate anabolism.

Urea formation. The ammonia which is split off from amino acids in the process of deamination is ultimately converted to urea and eliminated in the urine in that form. In theory, we can derive urea by the addition of two molecules of ammonia to one of carbonic acid, with the loss of two molecules of water. This reaction will not proceed uncatalyzed *in vitro* and is accomplished in the body in a roundabout manner (formula IV). Let us start in the eastern portion of the diagram. The amino acid, arginine,

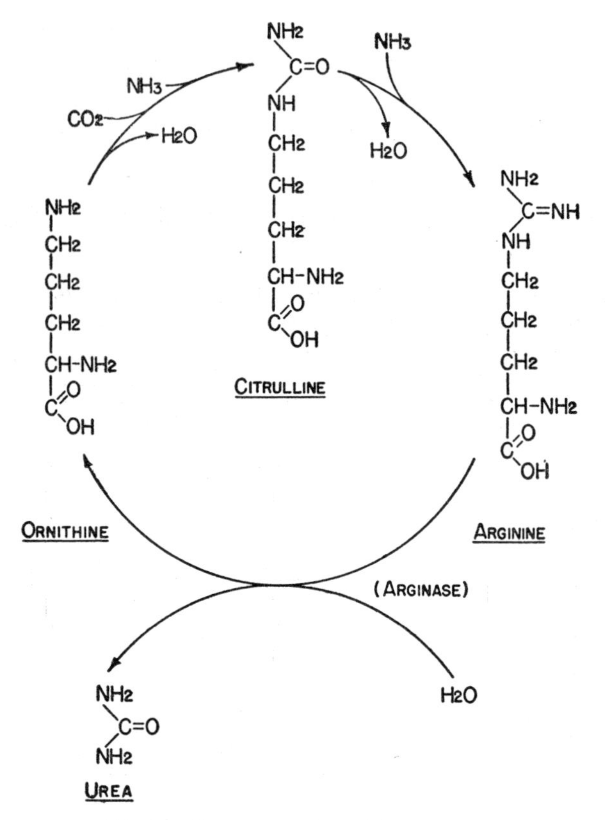

IV. The Krebs and Henseleit urea cycle

is the specific substrate for an enzyme, *arginase*, which is present in mammalian livers, but not in those of birds, snakes or lizards. This is significant because mammals excrete most of their waste nitrogen in the form of urea whereas the other animals mentioned excrete theirs chiefly as uric acid. Catalyzed by arginase, arginine is hydrolyzed to *ornithine* and *urea*. The ornithine so formed adds CO_2 and NH_3 to form *citrulline*, which in turn adds NH_3 to form arginine. Note that in these two steps the components of urea are added and that in each of the steps a molecule of water is given off. Arginine is now ready, in the presence of arginase, to liberate urea and be reconverted to ornithine to start the cycle again.

The manner in which ammonia is added to ornithine first and then to citrulline is not direct. Glutamic acid and aspartic acid are both involved. Glutamic acid, apparently, reacts first with carbon dioxide and then with ammonia to form carbamylglutamic acid (formula V). Note that the first reaction involves the formation of a carbamino group. This is a well established type of reaction and we shall see that it plays a role in carbon dioxide transport (see page 727). Alternatively, carbamylglutamic acid can arise from glutamine and carbonic acid (4). The carbamylglutamic acid then transfers its carbamyl group to the delta-amino group of ornithine, forming citrulline and regenerating glutamic acid. The transfer of the amide group requires the expenditure of a high-energy phosphate group from ATP and involves the formation of an intermediate phosphate ester *carbamyl phosphate* (37) (formulas VI and VII).

$$
\begin{array}{l}
\mathrm{COOH} \\ | \\ \mathrm{CH_2} \\ | \\ \mathrm{CH_2} \\ | \\ \mathrm{CHNH_2} \\ | \\ \mathrm{COOH}
\end{array}
\xrightarrow{+CO_2}
\begin{array}{l}
\mathrm{COOH} \\ | \\ \mathrm{CH_2} \\ | \\ \mathrm{CH_2} \\ | \\ \mathrm{CHNHCOOH} \\ | \\ \mathrm{COOH}
\end{array}
\xrightarrow[-H_2O]{+NH_3}
\begin{array}{l}
\mathrm{COOH} \\ | \\ \mathrm{CH_2} \\ | \\ \mathrm{CH_2} \\ | \\ \mathrm{CHNHCONH_2} \\ | \\ \mathrm{COOH}
\end{array}
$$

Glutamic acid — N-Carboxy-glutamic acid — Carbamyl-glutamic acid

V. Glutamic acid and the urea cycle (first stage)

$$
\begin{array}{l}
\mathrm{COOH} \\ | \\ \mathrm{CH_2} \\ | \\ \mathrm{CH_2} \\ | \\ \mathrm{CHNHCONH_2} \\ | \\ \mathrm{COOH}
\end{array}
+ \mathrm{ATP} \rightarrow
\begin{array}{l}
\mathrm{COOH} \\ | \\ \mathrm{CH_2} \\ | \\ \mathrm{CH_2} \\ | \\ \mathrm{CHNH_2} \\ | \\ \mathrm{COOH}
\end{array}
+ \text{Ⓟ}\!-\!\mathrm{O}\!\sim\!\overset{\overset{\displaystyle O}{\|}}{\mathrm{C}}\!-\!\mathrm{NH_2} + \mathrm{ADP}
$$

Carbamylglutamic acid — Glutamic acid — Carbamyl phosphate

VI. Glutamic acid and the urea cycle (second stage)

$$
\text{Ⓟ}\!-\!\mathrm{O}\!\sim\!\overset{\overset{\displaystyle O}{\|}}{\mathrm{C}}\!-\!\mathrm{NH_2} +
\begin{array}{l}
\mathrm{NH_2} \\ | \\ \mathrm{CH_2} \\ | \\ \mathrm{CH_2} \\ | \\ \mathrm{CH_2} \\ | \\ \mathrm{CHNH_2} \\ | \\ \mathrm{COOH}
\end{array}
\rightarrow
\begin{array}{l}
\mathrm{NHCONH_2} \\ | \\ \mathrm{CH_2} \\ | \\ \mathrm{CH_2} \\ | \\ \mathrm{CH_2} \\ | \\ \mathrm{CHNH_2} \\ | \\ \mathrm{COOH}
\end{array}
+ \mathrm{H_3PO_4}
$$

Carbamyl phosphate — Ornithine — Citrulline

VII. Ornithine to citrulline in the urea cycle

Citrulline gains the second molecule of ammonia by condensing with aspartic acid to form argininosuccinic acid (again with the aid of ATP and the possible formation of *phosphocitrulline* as an intermediate (50) which then splits to form arginine and fumaric acid (formula VIII) (51)

From what has been said, it can be seen that the over-all reaction of the urea cycle is:

$$CO_2 + 2\ NH_3 + 2\ ATP \rightarrow NH_2CONH_2 + 2\ ADP + 2\ H_3PO_4$$

The urea cycle is thus an energy-consuming process that requires the expenditure of two mols of high-energy phosphate per mol of urea produced.

All the steps in the urea cycle are probably physiologically reversible. Some certainly are, since urea tagged with radioactive carbon may be injected into animals and more than 20 per cent of the carbon will be eliminated as CO_2 in the expired air, the remainder appearing as urea in the urine.

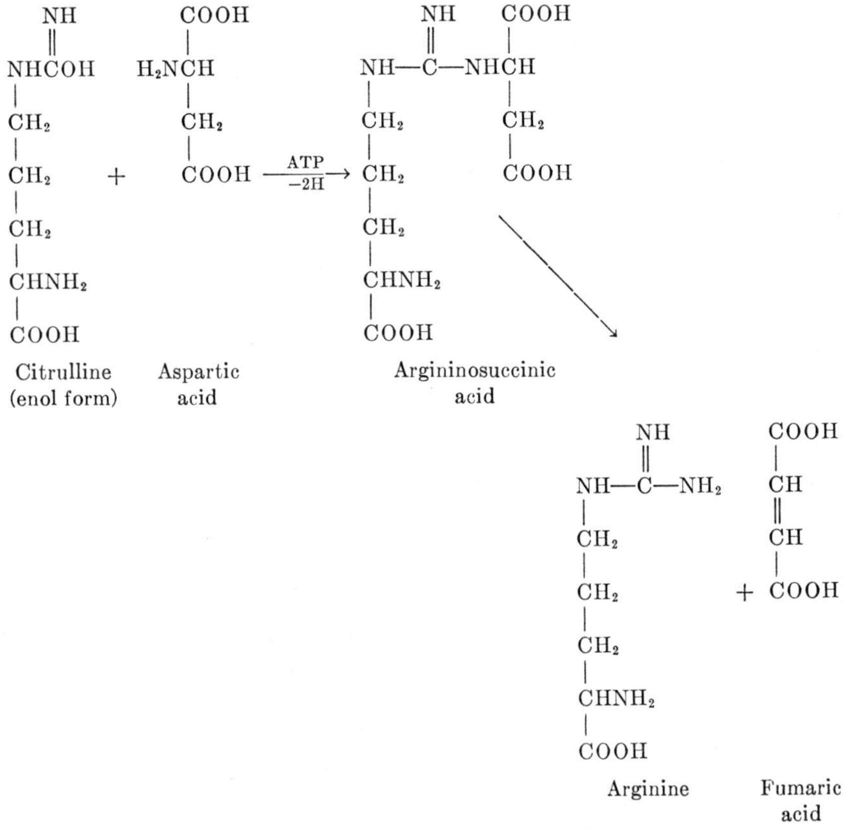

```
   NH          COOH                  NH     COOH
   ||           |                    ||      |
 NHCOH       H2NCH              NH—C—NHCH
   |            |                 |          |
  CH2          CH2               CH2        CH2
   |            |                 |          |
  CH2    +     COOH  —ATP/−2H→   CH2        COOH
   |                              |
  CH2                            CH2            \
   |                              |              \
 CHNH2                          CHNH2             \
   |                              |                ↘
  COOH                           COOH

Citrulline    Aspartic         Argininosuccinic
(enol form)     acid                 acid

                                        NH             COOH
                                        ||              |
                                   NH—C—NH2            CH
                                    |                   ||
                                   CH2                 CH
                                    |                   |
                                   CH2             +  COOH
                                    |
                                   CH2
                                    |
                                  CHNH2
                                    |
                                   COOH

                                  Arginine          Fumaric
                                                      acid
```

VIII. Aspartic acid and the urea cycle

Urea formation proceeds in the liver. Experiments on hepatectomized dogs showed that in that species at least, no significant amount of urea is formed elsewhere. The following results were observed after hepatectomy: (a) fall in blood sugar; (b) fall in blood urea, with disappearance of urea from the urine within 36 hours; (c) no conversion of injected amino acids or ammonium compounds into urea; and (d) increase of blood ammonia. These results were foreshadowed in the early 1890's by the work of Pavlov, Nencki, and others who established in experimental animals an *Eck fistula*, which connected the portal vein to the inferior vena cava, thereby short-circuiting the portal circulation through the liver. Such animals showed diminished urea output in the urine.

Ammonia storage. If growing rats are fed only the essential amino acids, and those only in the amounts necessary to give maximum growth when the other non-essential amino acids are abundantly supplied, growth is very slow. The lack of nitrogen is the limiting factor in this situation. In place of the non-essential amino acids, urea or ammonium salts, such as diammonium citrate, may serve as a source of nitrogen (56). Such utilization of ammonium salts must take place promptly since ammonia is distinctly toxic. The ammonia concentration in normal blood is 0.8 micrograms per milliliter (10a). Uptake of ammonia by the brain occurs when blood ammonia concentration exceeds 1 microgram per milliliter and coma develops with higher concentration (6).

It is useful for the body to have a mechanism whereby ammonia is not merely excreted as urea, but may also be stored for future use in a non-toxic but chemically available form. It is so stored by the enzymatic formation of the amide of glutamic acid, *glutamine*. The formation of glutamine from glutamic acid requires ATP (25). The deamination of glutamine is catalyzed by the widely distributed enzyme, *glutaminase* (formula IX). The analogous formation of asparagine (formula X), the amide of aspartic

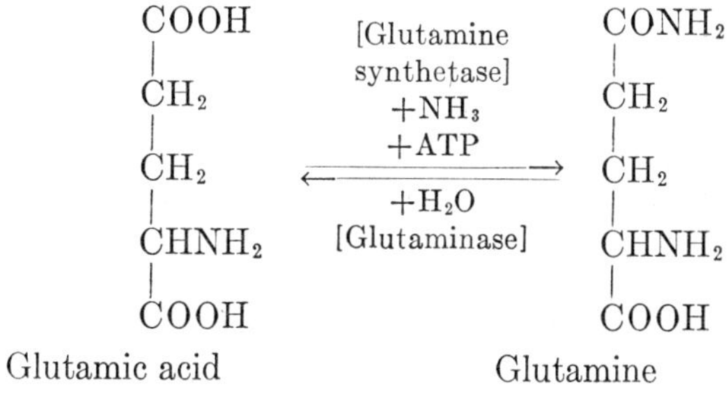

IX. Ammonia storage

acid, is an auxiliary method of ammonia storage. The amino acid analysis of most proteins shows the presence of appreciable quantities of amide

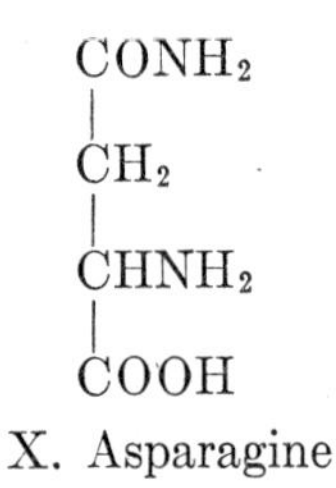

X. Asparagine

nitrogen (76) which are presumably derived from the asparagine and glutamine of the molecule.

Transamination

Under the influence of enzymes called *transaminases*, in which pyridoxal phosphate (see page 806) is the coenzyme, the alpha-amino group of an L-amino acid can be transferred to a keto acid acceptor. Transamination is probably a more important type of reaction in protein metabolism than is deamination.

The best-known keto acid acceptor is alpha-ketoglutaric acid, and the classic transamination reaction is that of a mixture of alanine and alpha-ketoglutaric acid, which, in the presence of chopped liver or muscle tissue, form pyruvic acid and glutamic acid, probably through the intermediate formation of a Schiff base with the pyridoxal phosphate coenzyme (formula XI). The process is reversible.

By means of transamination, amino acids which do not possess a specific amino acid dehydrogenase, can get rid of their amino group.Thus, aspartic acid, which is not oxidatively deaminated by either L-amino acid dehydrogenase or by any specific dehydrogenase, will transaminate with alpha-ketoglutaric acid to form oxaloacetic acid and glutamic acid (formula XII).

The enzyme which catalyzes the reaction shown in formula XII is *glutamic-oxaloacetic transaminase* (G-OT) which is widely distributed in tissues and can be detected in the blood plasma. When there is local death of tissue (as in obstruction of a portion of the coronary circulation, or damage to the liver by virus infection, poisoning with carbon tetrachloride, or metastases of cancer to the liver), large quantities of this transaminase leave the damaged cells and increase the transaminase activity of the blood plasma. Measurement of glutamic-oxaloacetic transaminase activity of blood serum is therefore of value in the differential diagnosis of myocardial infarction and of diseases involving destruction of liver cells (81).

Myocardial tissue is low in *glutamic-pyruvic transaminase* (G-PT), while liver has G-PT activity comparable to liver G-OT. Elevation of G-PT in blood plasma is therefore more specific for liver disease than elevation of G-OT (82).

An important amino group donor is glutamine. Transamination reactions

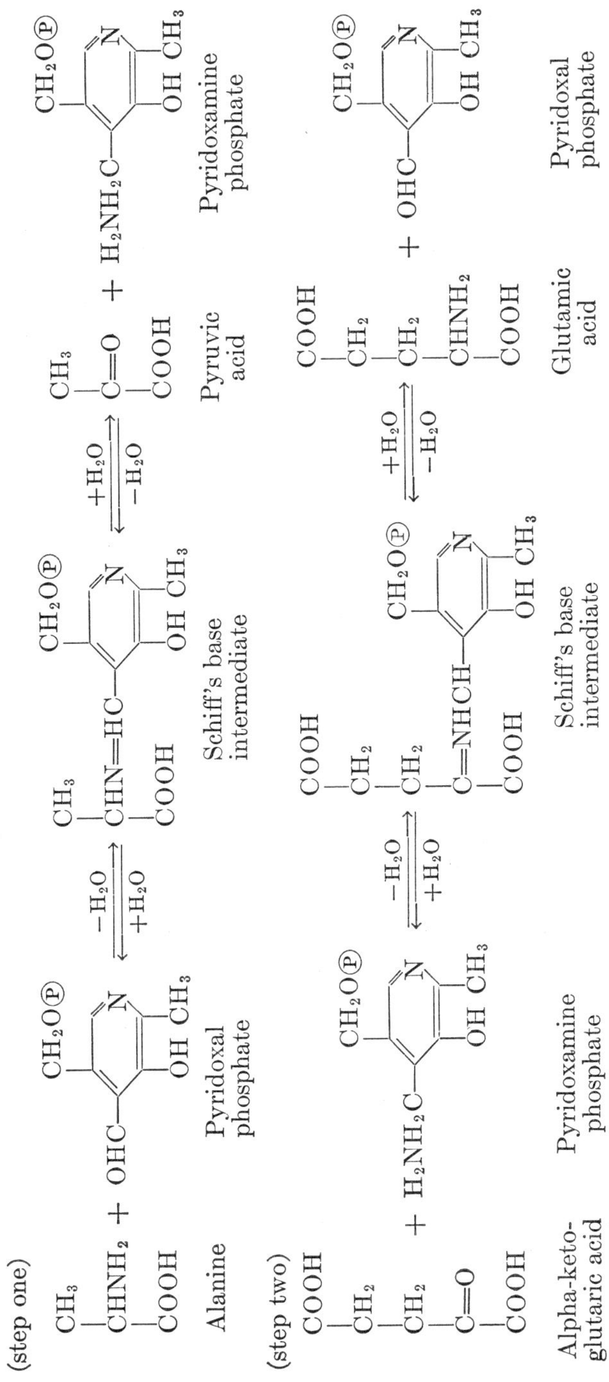

(step one)

CH_3-CHNH_2-COOH + $OHC-C_5HN(CH_2O\text{Ⓟ})(OH)(CH_3)$ $\underset{+H_2O}{\overset{-H_2O}{\rightleftharpoons}}$ $CH_3-CHN{=}HC-C_5HN(CH_2O\text{Ⓟ})(OH)(CH_3)$ with $COOH$ $\underset{-H_2O}{\overset{+H_2O}{\rightleftharpoons}}$ $CH_3-C{=}O-COOH$ + $H_2NH_2C-C_5HN(CH_2O\text{Ⓟ})(OH)(CH_3)$

Alanine + Pyridoxal phosphate ⇌ Schiff's base intermediate ⇌ Pyruvic acid + Pyridoxamine phosphate

(step two)

$COOH-CH_2-CH_2-C{=}O-COOH$ + $H_2NH_2C-C_5HN(CH_2O\text{Ⓟ})(OH)(CH_3)$ $\underset{+H_2O}{\overset{-H_2O}{\rightleftharpoons}}$ $COOH-CH_2-CH_2-C{=}NHCH-C_5HN(CH_2O\text{Ⓟ})(OH)(CH_3)$ with $COOH$ $\underset{-H_2O}{\overset{+H_2O}{\rightleftharpoons}}$ $COOH-CH_2-CH_2-CHNH_2-COOH$ + $OHC-C_5HN(CH_2O\text{Ⓟ})(OH)(CH_3)$

Alpha-keto-glutaric acid + Pyridoxamine phosphate ⇌ Schiff's base intermediate ⇌ Glutamic acid + Pyridoxal phosphate

XI. Transamination

$$\begin{array}{l} COOH \\ | \\ CH_2 \\ | \\ CHNH_2 \\ | \\ COOH \end{array} + \begin{array}{l} COOH \\ | \\ CH_2 \\ | \\ CH_2 \\ | \\ C{=}O \\ | \\ COOH \end{array} \xrightleftharpoons[\text{(transaminase)}]{} \begin{array}{l} COOH \\ | \\ CH_2 \\ | \\ C{=}O \\ | \\ COOH \end{array} + \begin{array}{l} COOH \\ | \\ CH_2 \\ | \\ CH_2 \\ | \\ CHNH_2 \\ | \\ COOH \end{array}$$

XII. Aspartic acid transamination (over-all reaction)

involving glutamine are catalyzed by *glutamine transaminase* which is specific for glutamine but which will utilize at least thirty different ketoacids as amine-group acceptors (19). Transaminations involving glutamine differ from those described above in not being directly reversible due to the deamidation of the alpha-ketoglutaramic acid formed (formula XIII). Glutamine thus serves as a means of amino acid synthesis, and its usefulness in this respect may account for its relatively high concentration in tissue and body fluid. Asparagine behaves similarly.

$$\begin{array}{l} CONH_2 \\ | \\ CH_2 \\ | \\ CH_2 \\ | \\ C{=}O \\ | \\ COOH \end{array}$$

XIII. Alpha ketoglutaramic acid

Most amino acids pass through transamination reactions during the major pathways of their metabolism. Presumably, those keto acid analogs of the essential amino acids which can accept amino groups can replace the amino acids themselves in the diet, provided the body has ample nitrogen reserve. Thus, phenylpyruvic acid can replace phenylalanine in the diet, since it can be transaminated to phenylalanine. Lysine and threonine, however, must be supplied in the diet as such. Their keto acid analogs do not accept amino groups by transamination.

Decarboxylation

Decarboxylation is another general catabolic reaction of the amino acids. The *decarboxylases* that have been most studied are those that occur in bacteria. Thus, cadaverine is formed by the bacterial decarboxylation of lysine (formula XIV). There are specific decarboxylases for each of a num-

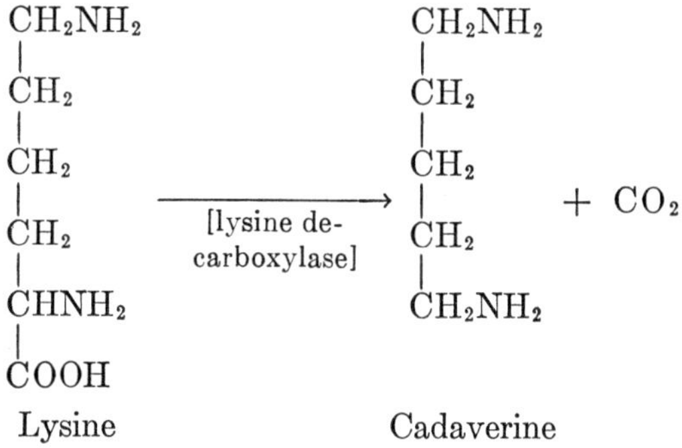

$$\begin{array}{c} CH_2NH_2 \\ | \\ CH_2 \\ | \\ CH_2 \\ | \\ CH_2 \\ | \\ CHNH_2 \\ | \\ COOH \end{array} \xrightarrow[\text{carboxylase]}]{\text{[lysine de-}} \begin{array}{c} CH_2NH_2 \\ | \\ CH_2 \\ | \\ CH_2 \\ | \\ CH_2 \\ | \\ CH_2NH_2 \end{array} + CO_2$$

Lysine Cadaverine

XIV. Decarboxylation of lysine

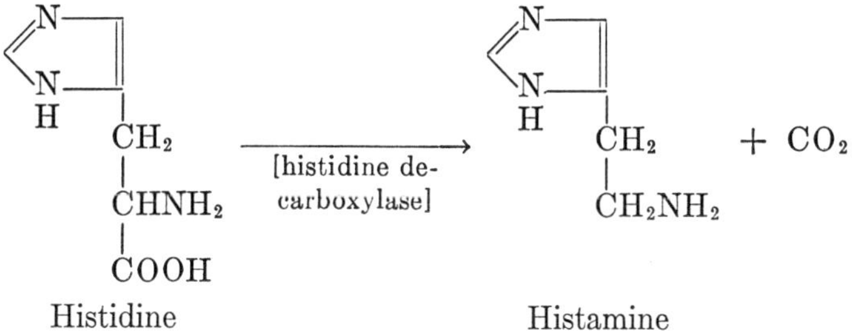

N—
N—
H
$$CH_2 \xrightarrow[\text{carboxylase]}]{\text{[histidine de-}} CH_2 + CO_2$$
$$CHNH_2 \qquad CH_2NH_2$$
$$COOH$$

Histidine Histamine

XV. Decarboxylation of histidine

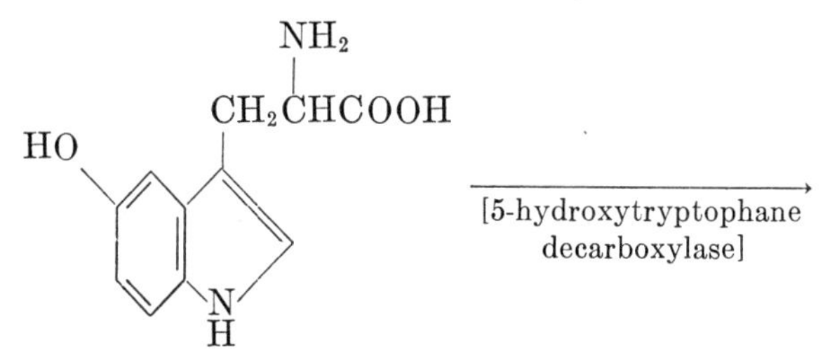

[5-hydroxytryptophane
decarboxylase]

5-Hydroxytryptophane

NH_2
CH_2CH_2
HO
N
H
$+ CO_2$

Serotonin

XVI. Decarboxylation of 5-hydroxytryptophane

ber of amino acids, and for each of these pyridoxal phosphate is the coenzyme.

The details concerning decarboxylase activity in mammalian tissue are less well known. Decarboxylations take place in the catabolism of several of the amino acids. For instance, the physiologically-active compound, histamine (see page 157), is derived from histidine by the action of histidine decarboxylase (formula XV) (61) and a 5-hydroxytryptophane decarboxylase, which serves to catalyze the formation of serotonin (see page 163), has been located in hog and guinea-pig kidney (14) (formula XVI). Similarly,

$$CH_2COOH-CHNH_2-COOH \xrightarrow[\text{decarboxylase]}]{\text{[aspartic acid}} CH_2COOH-CH_2NH_2 + CO_2$$

Aspartic acid → Beta-alanine

$$CH_2COOH-CH_2-CHNH_2-COOH \xrightarrow[\text{decarboxylase]}]{\text{[glutamic acid}} CH_2COOH-CH_2-CH_2NH_2 + CO_2$$

Glutamic acid → Gamma-aminobutyric acid

XVII. Decarboxylation of dicarboxylic amino acids

$$R-CHNH_2-COOH + OHC-C_5N(CH_2O\text{Ⓟ})(OH)(CH_3) \xrightarrow{-H_2O} R-CH(COOH)-N{=}HC-C_5N(CH_2O\text{Ⓟ})(OH)(CH_3)$$

Amino acid + Pyridoxal phosphate → Schiff's base

$$\downarrow -CO_2$$

$$R-CH_2NH_2 + OHC-C_5N(CH_2O\text{Ⓟ})(OH)(CH_3) \xleftarrow{+H_2O} R-CH_2N{=}HC-C_5N(CH_2O\text{Ⓟ})(OH)(CH_3)$$

Amine + Pyridoxal phosphate ← Decarboxylated Schiff's base

XVIII. Pyridoxal phosphate and decarboxylation

aspartic acid and glutamic acid can be decarboxylated to beta-alanine and gamma-aminobutyric acid, respectively (formula XVII). The optically inactive beta-alanine occurs in the food factor, pantothenic acid (see page 804) and also occurs in carnosine and anserine (see page 154). Gamma-aminobutyric acid is found in the central nervous system, particularly in the gray matter. The glutamic acid decarboxylase of the nervous system also occurs chiefly in the gray matter. Enzyme systems exist in brain and liver which catalyze the transamination of both gamma-aminobutyric acid and beta-alanine with alpha-ketoglutaric acid (54).

The action of the pyridoxal phosphate coenzyme in the decarboxylation reaction is similar to its reaction in transamination in that an intermediate Schiff's base is probably formed. It is the Schiff's base that is decarboxylated. The decarboxylated base is then hydrolyzed to form the free amine (formula XVIII).

CATABOLISM OF THE INDIVIDUAL AMINO ACIDS

Once the alpha-amino group of an amino acid has been removed either by transamination or deamination, the remainder of the molecule may follow one of two general metabolic paths. The first path, that followed by the majority of the amino acids, ends in the conversion of the molecule to one of the compounds of the tricarboxylic acid cycle. The removal of the amino group from alanine, aspartic acid and glutamic acid accomplishes this in this one step. In the case of serine, threonine, cysteine, histidine, methionine, valine, arginine, and proline, the result is attained by somewhat more circuitous routes. Since carbon atoms from all these amino acids arrive at the tricarboxylic acid cycle by some pathway not involving acetoacetic acid, and by way of that can reach glucose and glycogen, these amino acids are referred to as *glucogenic.*

When the deaminated residue of an amino acid has been metabolically converted to a ketone body (see page 593), it cannot contribute to glycogen storage. Acetoacetic acid is formed from leucine, isoleucine, phenylalanine, and tyrosine, and these amino acids are, therefore, *ketogenic.*

The decision which establishes a particular amino acid as glucogenic or ketogenic has usually been based upon experiments involving the feeding of the amino acid to an animal rendered diabetic by pancreatectomy or glycosuric by the use of the glucoside, phlorizin. In such an animal the feeding of a glucogenic amino acid leads to an increased output of urinary glucose, the feeding of a ketogenic amino acid leads to increased output of ketone bodies. The ratio of glucose (dextrose) to nitrogen, (the D/N ratio), in the urine of a phlorizinized animal is 3.65 when the animal is fed a diet consisting entirely of protein. Since protein contains approximately 16 grams of nitrogen per 100 grams protein, the urinary glucose supplied by

100 grams of protein would be 3.65 × 16 or approximately 58 grams of glucose. This is the basis of the customarily applied dietetic rule that 58 per cent of protein is convertible to glucose in the body.

In the case of two of the major amino acids, tryptophane and lysine, it is still uncertain as to whether they are glucogenic or ketogenic. Glycine is a special case.

In considering the catabolism of the individual amino acids there is no set rule as to the order in which they ought to be considered. Glycine, as the simplest amino acid, may be placed first. The others we shall take up in groups based upon the chemical nature of the side chains involved.

Glycine

Glycine is glucogenic on the basis of experiments with phlorizinized dogs. However, glycogenesis or urea formation directly from glycine take place only to a limited extent (3). Glycine, more than any other amino acid, becomes incorporated into more complex molecules, through the catabolism of which the carbon and nitrogen atoms originating in glycine eventually find themselves in either glycogen or urea.

This metabolic versatility of glycine is in contrast to its rather unremarkable role as a protein constituent. Proteins whose primary function is structural or connective are rich in glycine. Collagen (see page 141) is 25 per cent glycine. The more complex proteins of metabolic significance, however, are relatively low in glycine.

Glycine conjugates with certain aromatic acids forming products which are excreted in the urine. The classic example is the formation of hippuric acid from benzoic acid (see page 517). This type of reaction represents the formation of a link similar to the peptide bond occurring in proteins. The reaction is energy-consuming and must therefore be coupled with ATP hydrolysis. The presence of coenzyme A is required and *benzoylcoenzyme A* is an intermediate.

Another example of the formation of a glycylamide linkage is the conjugation of glycine with bile acids to form bile salts such as glycocholic acid (see page 505). Glycine also occurs as one of the amino acids in the tripeptide, glutathione (see page 62). The glycine backbone (C—C—N) is incorporated intact in the purine ring (see page 358).

Porphyrin biosynthesis. Glycine is the source of the four nitrogen atoms of the porphyrin ring (see page 188) and of eight of the carbon atoms as well. In porphyrin biosynthesis, glycine condenses with a succinyl derivative intermediate in the citric acid cycle, possibly succinylcoenzyme A (see page 552) sometimes referred to as "active succinate", to form a beta-keto acid which decarboxylates to *delta-aminolevulinic acid* (53) (formula XIX). Note that the carbon dioxide so lost was originally part of the glycine molecule so that of the two carbon atoms of glycine only the methylene carbon appears in the porphyrin ring.

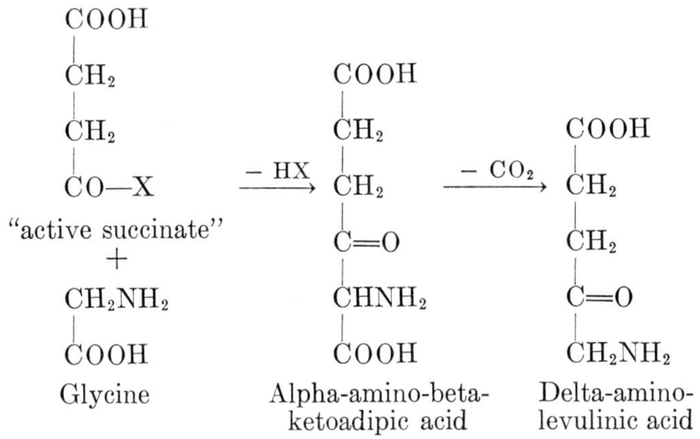

$$\begin{array}{c} COOH \\ | \\ CH_2 \\ | \\ CH_2 \\ | \\ CO{-}X \\ \text{"active succinate"} \\ + \\ CH_2NH_2 \\ | \\ COOH \\ \text{Glycine} \end{array} \xrightarrow{-\,HX} \begin{array}{c} COOH \\ | \\ CH_2 \\ | \\ CH_2 \\ | \\ C{=}O \\ | \\ CHNH_2 \\ | \\ COOH \\ \text{Alpha-amino-beta-} \\ \text{ketoadipic acid} \end{array} \xrightarrow{-\,CO_2} \begin{array}{c} COOH \\ | \\ CH_2 \\ | \\ CH_2 \\ | \\ C{=}O \\ | \\ CH_2NH_2 \\ \text{Delta-amino-} \\ \text{levulinic acid} \end{array}$$

XIX. Formation of delta-aminolevulinic acid

Two molecules of delta-aminolevulinic acid condense to form the pyrrole derivative, *porphobilinogen* (formula XX), in which the carbon atoms derived from glycine are indicated by asterisks. The manner in which porphobilinogen is converted to heme is still under dispute. If four molecules of porphobilinogen were simply to condense into a porphyrin ring, a uroporphyrin would result. Decarboxylation of the acetic acid side chains would give a coproporphyrin, and subsequent dehydrogenation and decarboxylation of two of the propionic acid side chains would yield a protoporphyrin. There is as yet no firm evidence as to the stage in porphyrin ring formation at which these decarboxylations and dehydrations ocur.

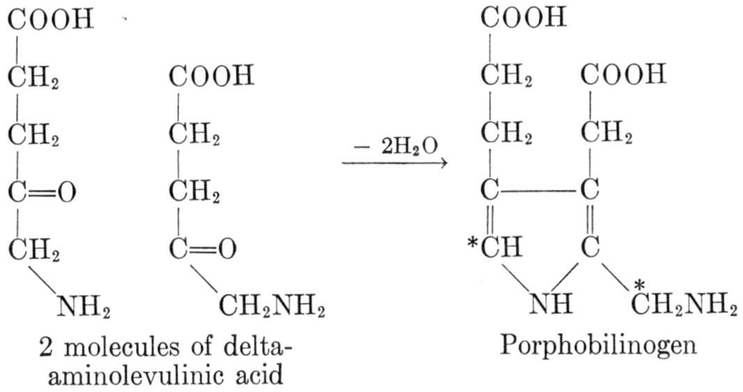

$$\begin{array}{cc} COOH & \\ | & COOH \\ CH_2 & | \\ | & CH_2 \\ CH_2 & | \\ | & CH_2 \\ C{=}O & | \\ | & C{=}O \\ CH_2 & \diagdown \\ \diagdown & CH_2NH_2 \\ NH_2 & \end{array} \xrightarrow{-\,2H_2O} \begin{array}{ccc} COOH & & \\ | & & COOH \\ CH_2 & & | \\ | & & CH_2 \\ CH_2 & & | \\ | & & \\ C & {-\!-\!-} & C \\ \| & & \| \\ {}^{*}CH & & C \\ & \diagdown \; \diagup & \diagdown \\ & NH & {}^{*}CH_2NH_2 \end{array}$$

2 molecules of delta-aminolevulinic acid — Porphobilinogen

XX. Formation of porphobilinogen

Any theory of porphyrin biosynthesis must also account for the fact that the isomeric form produced is preponderantly of the III series (see page 189). If four porphobilinogen rings were to condense by the same mechanism at all points of union, a porphyrin belonging to the radially symmetrical I series would result. In actual fact, however, one porphobilinogen ends up in reverse position, and in the final porphyrin ring the carbon atoms derived from glycine are distributed asymmetrically (formula XXI). Theories to account for this have been advanced (63), but none has been definitely established.

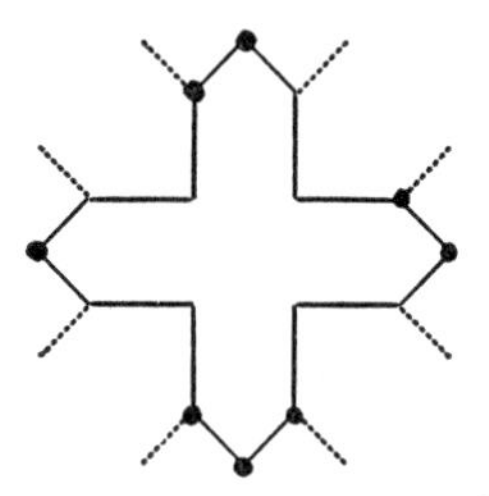

XXI. Distribution of glycine carbons in porphyrin ring

Glycine derivatives. Two methylated derivatives of glycine found in mammalian tissue are *sarcosine* and *glycylbetaine* (formula XXII). (Betaines as a class are fully N-methylated amino acids which are much more basic than the parent compound. Glycylbetaine is often referred to simply as betaine.) The N-methyl groups of sarcosine and the betaines are metabolically derived from compounds, such as methionine, which contain labile methyl groups (see page 629).

Sarcosine	Glycylbetaine
CH_2COOH	$CH_2COO^{(-)}$
|	|
$NHCH_3$	$^{(+)}N(CH_3)_3$

XXII. Methylated derivatives of glycine

Another glycine derivative of importance in metabolism is the amidinated glycine compound, *glycocyamine* or guanidinoacetic acid. Glycocyamine is formed in the kidney by the interaction of glycine and arginine (formula XXIII). Since the reaction involves a transfer of the amidine group, the process is called *transamidination* and the enzyme involved, glycine transamidinase.

Glycine degradation. There is an increase in urea excretion after glycine feeding or injection. An enzyme, glycine dehydrogenase (see page 604), deaminates glycine to glyoxylic acid. However, experiments with perfused organs and tissue slices indicate that the loss of the glycine amine group, either by deamination or transamination is minor. That this is so is also indicated by the manner in which the C—N link of glycine is maintained intact in its incorporation into the purine and porphyrin ring systems.

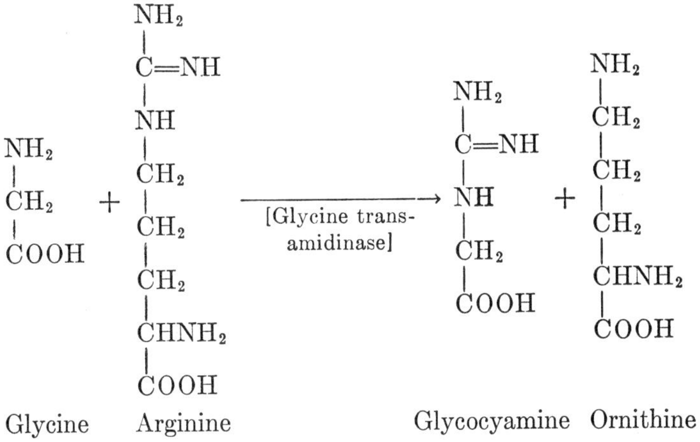

$$\underset{\text{Glycine}}{\begin{array}{l}NH_2\\|\\CH_2\\|\\COOH\end{array}} + \underset{\text{Arginine}}{\begin{array}{l}NH_2\\|\\C{=}NH\\|\\NH\\|\\CH_2\\|\\CH_2\\|\\CH_2\\|\\CHNH_2\\|\\COOH\end{array}} \xrightarrow[\text{amidinase]}]{\text{[Glycine trans-}} \underset{\text{Glycocyamine}}{\begin{array}{l}NH_2\\|\\C{=}NH\\|\\NH\\|\\CH_2\\|\\COOH\end{array}} + \underset{\text{Ornithine}}{\begin{array}{l}NH_2\\|\\CH_2\\|\\CH_2\\|\\CH_2\\|\\CHNH_2\\|\\COOH\end{array}}$$

XXIII. Transamidination

Glycine can serve as a source of "one-carbon fragments" in the form of formic acid. This has been shown by isotope experiments in which tagged carbons in the alpha position of the glycine molecule but not the carboxyl carbon were detected in isolated formate. The formic acid may be formed by the oxidative decarboxylation of glyoxylic acid (formula XXIV). The transfer of formic acid groups arising from glycine (and other sources) involves the intermediate formation of a folic acid complex (see page 812).

As formic acid, glycine can participate in a number of reactions, notably

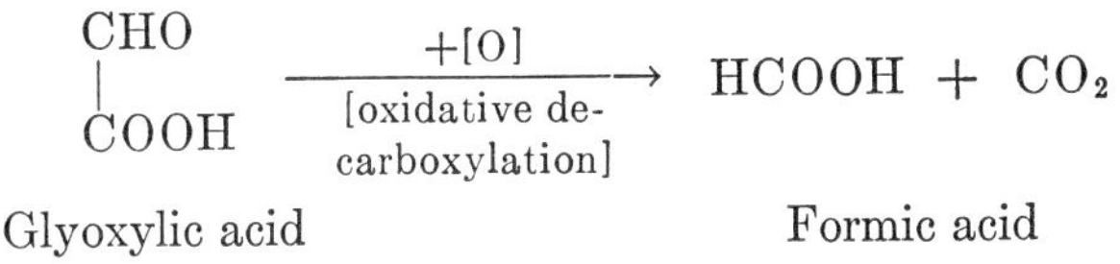

$$\underset{\text{Glyoxylic acid}}{\begin{array}{l}CHO\\|\\COOH\end{array}} \xrightarrow[\text{[oxidative de-carboxylation]}]{+[O]} \underset{\text{Formic acid}}{HCOOH} + CO_2$$

XXIV

$$\underset{\text{Formic acid}}{HCOOH} + \underset{\text{Glycine}}{\begin{array}{l}CH_2COOH\\|\\NH_2\end{array}} \underset{+H_2O}{\overset{-H_2O}{\rightleftharpoons}} \underset{\text{Formylglycine}}{\begin{array}{l}CHO\\|\\CHCOOH\\|\\NH_2\end{array}} \underset{-2H}{\overset{+2H}{\rightleftharpoons}} \underset{\text{Serine}}{\begin{array}{l}CH_2OH\\|\\CHCOOH\\|\\NH_2\end{array}}$$

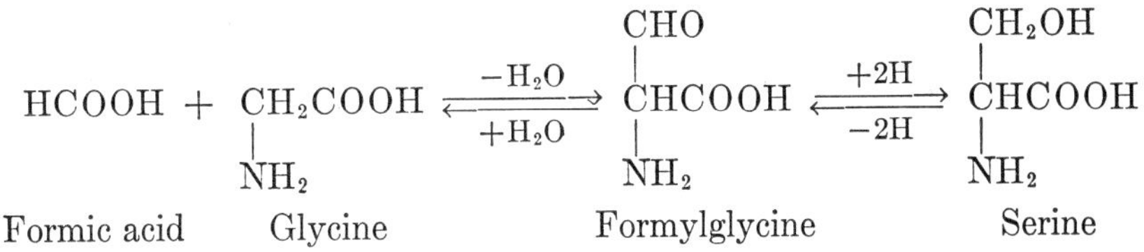

XXV. Glycine-serine interconversion

in the formation of serine (formula XXV) (64). Since serine is glucogenic, this represents one of the routes whereby glycine can form glucose. The formation of serine in the body does not, however, require the prior formation of glycine. Pyridoxal phosphate is required as coenzyme for the glycine-serine interconversion (1a).

Synthesis of glycine. In man (and in all animals studied, with the lone exception of the chick) glycine is not a dietarily essential amino acid, which indicates that it can be synthesized in the body. When benzoic acid is fed, 0.5 to 0.7 grams of glycine in the form of hippuric acid can be excreted per hour. This quantity is too great to be explained entirely by utilization of preformed protein-linked glycine. In normal human subjects, intravenous injection of 1.77 grams of sodium benzoate results in an output of 0.760–0.282 grams of benzoic acid (in the form of hippuric acid) in the urine during the first hour. At the end of this hour, the glycine of the blood plasma has fallen by 5.1–6.7 per cent. Patients with rheumatoid arthritis or other collagen diseases put out comparable amounts of hippuric acid during the first hour, but show a mean decrease in plasma glycine more than double that of normal subjects (43).

An important metabolic source of glycine is serine (see formula XXV). Although the direct conversion of non-nitrogenous compounds such as acetic acid and pyruvic acid to glycine by the addition of ammonia is a minor route, the process takes place chiefly through the prior formation of serine.

Amino Acids with Aliphatic Hydrocarbon Side chains

Alanine. Of all the amino acids, alanine follows the simplest metabolic scheme. By either transamination (see formula XI) or deamination it is converted to pyruvic acid and, as such, enters the tricarboxylic acid cycle. In this way it can be used for energic breakdown or glycogenesis. Ingestion of alanine by a diabetic animal produces a quantitative increase in glucose output, although only a small portion of the carbon of tagged alanine actually appears in glycogen. Here, as with other glucogenic amino acids except glycine, it is assumed that the deaminated derivatives, in this case pyruvic acid, are oxidized, thereby sparing glucose or glucogenic metabolites. Since the transamination of alanine is reversible, the amino acid is readily synthesized from carbohydrate *via* pyruvic acid, and is therefore non-essential in the diet.

Valine, leucine, and isoleucine. These three amino acids differ from all the others in that the molecules contain a branched carbon chain. Their first steps in catabolism are transaminations with alpha-ketoglutaric acid to form keto acids which are then oxidatively decarboxylated to branched acids with one carbon less than the original amino acids (formula XXVI). From there on, the paths diverge.

Valine, having reached the isobutyric stage, is further degraded to propionic acid according to the scheme shown in formula XXVII (5). The conversion of isobutyric acid to methylmalonic acid semialdehyde proceeds first by condensation with coenzyme A to form *isobutyrylcoenzyme A*. The remaining changes proceed through a series of steps similar to those in the fatty acid oxidation cycle (see page 586) (55a). Propionic acid is convertible to pyruvic acid (see page 590) or to succinic acid (formula XXVIII) and so the glucogenic character of valine is explained. Only three of the five carbon atoms of valine appear in glycogen, two being lost as CO_2. The richest dietary source of valine is the milk proteins.

The isovaleric acid produced from leucine is split into a 3-carbon and a 2-carbon fragment (formula XXIX) (15). The 2-carbon fragment follows the metabolic pathway of the similar fragments derived from fatty acid

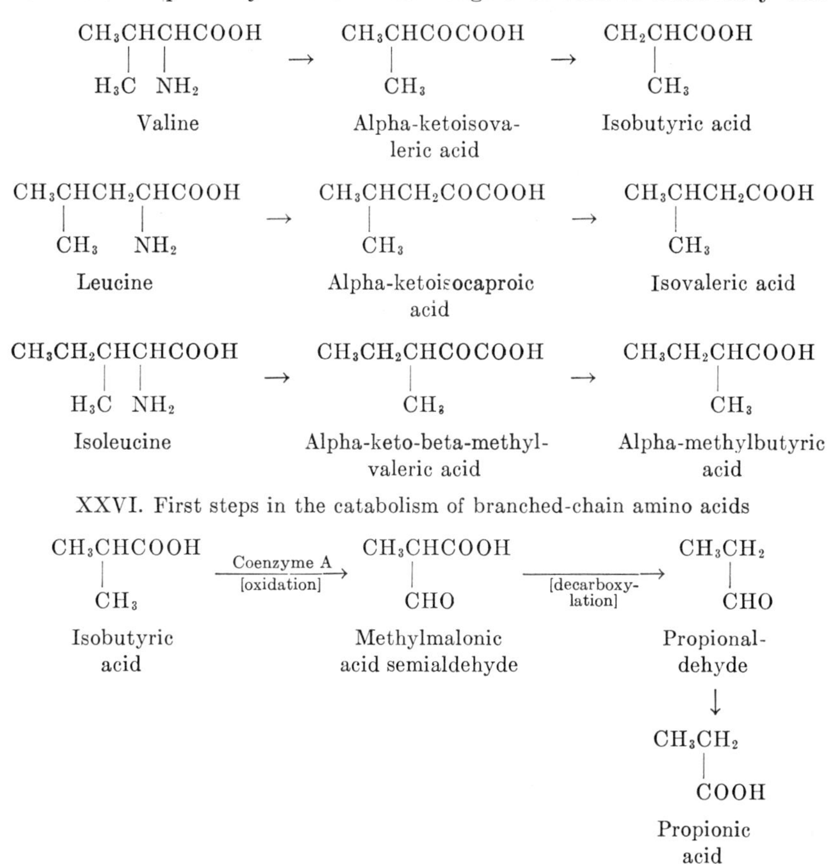

XXVI. First steps in the catabolism of branched-chain amino acids

XXVII. Later stages in catabolism of valine

catabolism (see page 589) and may therefore form acetoacetic acid. The 3-carbon fragment is converted to acetoacetic acid by oxidative carboxylation. This explains the ketogenic character of leucine.

The alpha-methylbutyric acid produced from isoleucine also splits into a 3-carbon and a 2-carbon fragment (formula XXX). This is accomplished

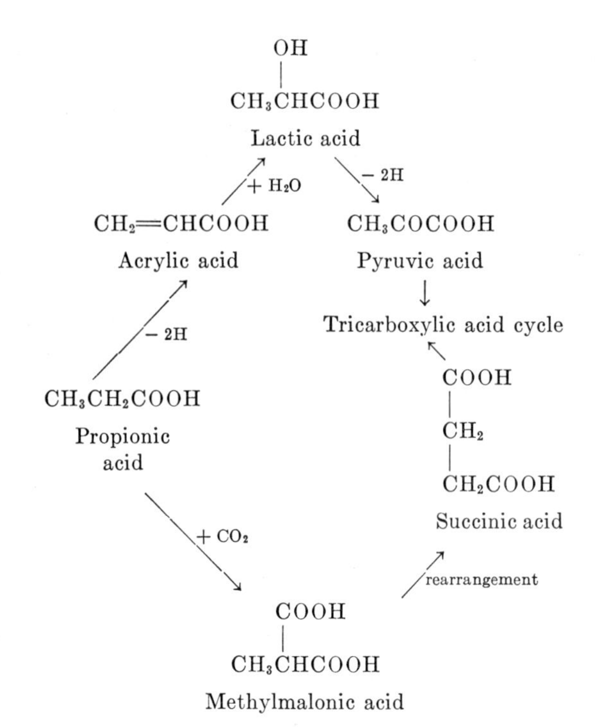

XXVIII. Entry of propionic acid into tricarboxylic acid cycle

$$CH_3CH(CH_3)\,|\,CH_2COOH \rightarrow \begin{cases} -CH_2COOH & \text{(ketogenic)} \\ CH_3CH(CH_3)- & \text{(ketogenic)} \end{cases}$$

XXIX. Further catabolism of leucine

$$CH_3CH_2\,|\,CH(CH_3)COOH \rightarrow \begin{cases} -CH(CH_3)COOH & \text{(glucogenic)} \\ CH_3CH_2- & \text{(ketogenic)} \end{cases}$$

XXX. Further catabolism of isoleucine

by beta-oxidation (see page 587) on the longer carbon chain, and involves the formation of alpha-methylbutyrylcoenzyme A. Steps similar to those in the oxidation of straight chain fatty acids lead to the production of acetylcoenzyme A (ketogenic) and butyrylcoenzyme A (glucogenic). During this series of changes, the branched-chain structure of the CoA thiol esters persists down to the stage just ahead of the final split into acetylcoenzyme A and propionylcoenzyme A. This substance is alpha-methylacetoacetylcoenzyme A (formula XXXI). It has been suggested (55) that the reverse process, condensation of acetylcoenzyme A with propionylcoenzyme A to form alpha-methylacetoacetylcoenzyme A, provides an explanation for the formation of branched chain compounds in the body.

$$CH_3CH_2\underset{\displaystyle CH_3}{\underset{|}{C}}HCO{\sim}SCoA$$

Alpha-methylbutyrylcoenzyme A

$$CH_3CO\underset{\displaystyle CH_3}{\underset{|}{C}}HCO{\sim}SCoA$$

Alpha-methylacetoacetylcoenzyme A

XXXI. Intermediates in isoleucine catabolism

Both leucine and isoleucine are dietarily essential amino acids. Leucine can act as an antimetabolite for isoleucine (29) as though both compete, at least on occasion, for the same enzyme. An excess of one may therefore increase the dietary requirement for the other.

Amino Acids with Alcohol Side Chains

Serine. We have already discussed the interconversion of serine and glycine (see page 619) and because of this, serine shares a great many of the metabolic pathways characteristic of glycine. Serine can give rise to the formic acid one-carbon fragment, as can glycine, and without prior conversion to glycine. The non-nitrogenous precursor of serine has been shown to be one of the 3-carbon intermediates in glycolysis, a triose or glyceric acid, but not pyruvic acid (2).

The decarboxylation of serine yields ethanolamine (formula XXXII). Serine, and through it, glycine, are therefore of prime importance in phospholipid biosynthesis. In the phospholipids, the hydroxyl groups of serine, ethanolamine and choline are all phosphorylated, and it is noteworthy that in the phosphoproteins, the phosphate groups appear to occur as esters of the hydroxyl group of the serine side chains.

$$\underset{\displaystyle OH}{\underset{|}{C}}H_2\underset{\displaystyle NH_2}{\underset{|}{C}}HCOOH \xrightarrow[\text{[decarboxylation]}]{} \underset{\displaystyle OH}{\underset{|}{C}}H_2\underset{\displaystyle NH_2}{\underset{|}{C}}H_2 + CO_2$$

Serine → Ethanolamine

XXXII. Formation of ethanolamine

Serine is *anaerobically* deaminated by the enzyme, *serine deaminase*, to pyruvic acid (formula XXXIII). Serine is therefore glucogenic, and since the reactions are reversible, it is not a dietarily essential amino acid. Anaerobic deamination is characteristic of the amino acids with alcohol side-chains, serine and threonine, and those only. Oxidative deamination, however, will not occur in serine and threonine which are not substrates for L-amino acid dehydrogenase.

Serine is involved in the formation of cysteine from methionine (see page 632).

Threonine. Like serine, threonine may be anaerobically deaminated.

$$\begin{array}{c}CH_2CHCOOH\\ \mid\quad\mid\\ OH\ NH_2\\ \text{Serine}\end{array} \xrightarrow{-H_2O} \begin{array}{c}CH_2{=}CCOOH\\ \mid\\ NH_2\\ \text{Alpha-aminoacrylic acid}\end{array}$$

$$\downarrow$$

$$\begin{array}{c}CH_3CCOOH\\ \|\\ NH\\ \text{Alpha-iminopropionic acid}\end{array} \xrightarrow{+H_2O} \begin{array}{c}CH_3CCOOH\\ \|\\ O\\ \text{Pyruvic acid}\end{array}$$

XXXIII. Anaerobic deamination of serine

The enzyme is *threonine deaminase*, and is probably identical with serine deaminase. The anaerobic deamination of threonine follows the steps described for serine and the over-all change is that shown in formula XXXIV. The alpha-ketobutyric acid can be oxidatively decarboxylated to propionic acid so that threonine is glucogenic. Transamination yields alpha-aminobutyric acid, a demonstrable product when threonine is incubated in rat liver homogenate (42).

$$\begin{array}{c}CH_3CHCHCOOH\\ \mid\quad\mid\\ HO\ \ NH_2\\ \text{Threonine}\end{array} \rightarrow \begin{array}{c}CH_3CH_2CCOOH\\ \|\\ O\\ \text{Alpha-ketobutyric acid}\end{array} + NH_3$$

XXXIV. Anaerobic deamination of threonine

Threonine is a dietarily essential amino acid. In the rat, deficiency of threonine depresses the rate of synthesis of phospholipids and nucleoproteins in the liver and causes fatty livers. Threonine can therefore be designated as a lipotropic substance (66). Rats fed on diets deficient in threonine showed loss of weight, atrophy of the testes and alterations in the chromophilic cells of the anterior pituitary (62). There is some evidence in favor of threonine being a precursor of glycine, splitting into that amino acid and acetic acid. This reaction cannot be one that occurs to any great extent, or else it is not reversible, as otherwise threonine would not be dietarily essential any more than serine is.

Amino Acids with Carboxyl Side Chains

Aspartic acid. Some of the metabolic reactions of aspartic acid have already been discussed. Thus, by transamination it forms oxaloacetic acid, which can enter the tricarboxylic acid cycle. Aspartic acid is therefore glucogenic and is not dietarily essential. Transamination is the only known means whereby the amino group is removed from aspartic acid.

Aspartic acid can be decarboxylated to beta-alanine (see page 614) and it participates in ammonia storage by its amidation to asparagine (see page 609). Its role in the urea cycle has been described (see page 608).

Beta-alanine is rapidly metabolized, probably by deamination to formylacetic acid, which is decarboxylated to acetaldehyde, which forms acetate by oxidation (formula XXXV) (49).

$$NH_2CH_2CH_2COOH$$

Beta-alanine

↓ oxidative deamination

$$CHOCH_2COOH$$

Formylacetic acid

↓ oxidative decarboxylation

$$HOOCCH_3$$

Acetic acid

XXXV. Catabolism of beta-alanine

Glutamic acid. The metabolism of glutamic acid is similar to that of aspartic acid. By deamination or transamination, it will reversibly form alpha-ketoglutaric acid which is a component of the tricarboxylic acid cycle, so that glutamic acid is glucogenic and not dietarily essential. It is decarboxylated to gamma-aminobutyric acid (see page 614) and amidated to glutamine (see page 609). Its role in the urea cycle has been described (see page 607).

Glutamic acid is particularly important in transamination since its keto-acid derivative, alpha-ketoglutaric acid, is the most common ammonia acceptor involved in the process. Glutamic acid is one of the amino acids occurring most frequently in proteins and also is one of the most frequently found among the free amino acids of the body. One-third of the free amino acids of human blood plasma is either glutamic acid or glutamine.

Glutamic acid and glutamine are present in high concentrations in brain as compared to other tissues. In the presence of glucose, brain slices will take up glutamic acid even when there is already more glutamic acid within the cell than outside. Potassium ion is taken up at the same time. Aspartic acid will not replace glutamic acid in this respect, nor will glutamine or any other amino acid. This function of glutamic acid in aiding potassium ion transport is important in view of the role of potassium in nerve conduction (see page 156). The theory has been advanced (78) that a major function of glutamic acid in the brain is the removal of ammonia, which is particularly toxic to neural tissue, by the formation of glutamine. Research on the possible connection of dietary glutamic acid and mental function is summarized in (77).

Glutamic acid is one of the three amino acids in glutathione. Isotopic studies have yielded evidence of the interrelationship of glutamic acid and the amino acids, proline and arginine (60), (see page 641), and with histidine (see page 644).

Amino Acids with Sulfur-Containing Side Chains

There are three sulfur-containing amino acids: cysteine, cystine, and methionine. Of these, cysteine and cystine form a redox pair (formula XXXVI) and may therefore be considered together.

Cysteine and cystine. Cysteine is one of the three amino acids in glutathione, and it is the one which is responsible for its characteristic property of forming a redox system exactly analogous to the cysteine/cystine system. The oxidized form of glutathione is shown in formula XXXVII and contains two molecules of glycine, two of glutamic acid and one of cystine. The glutathione redox system may be conveniently symbolized as 2GSH/GSSG. Cysteine and glutathione confer a certain degree of resistance to ionizing radiation on account of the ease with which the thiol group (—SH) is oxi-

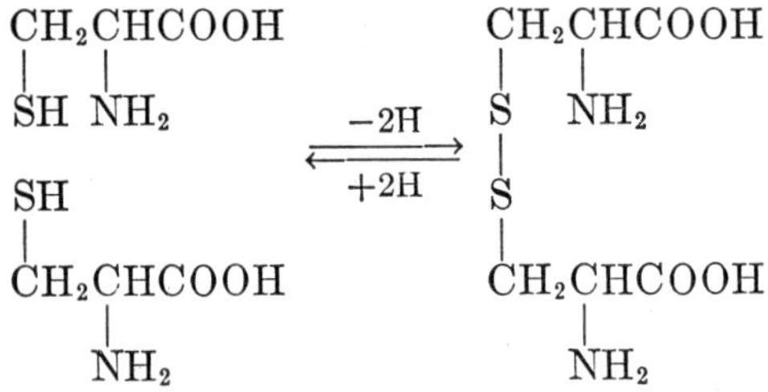

$$\begin{array}{l} CH_2CHCOOH \\ |\quad\ \ | \\ SH\ \ NH_2 \\ \\ SH \\ | \\ CH_2CHCOOH \\ \quad\ \ | \\ \quad\ NH_2 \end{array} \quad \underset{+2H}{\overset{-2H}{\rightleftharpoons}} \quad \begin{array}{l} CH_2CHCOOH \\ |\quad\ \ | \\ S\quad NH_2 \\ | \\ S \\ | \\ CH_2CHCOOH \\ \quad\ \ | \\ \quad\ NH_2 \end{array}$$

2 cysteines — Cystine

XXXVI. The cysteine/cystine redox system

dized. Cysteine-protected rats who have received 800 r (see page 889) total body x-irradiation have more neutrophils in the blood and more developing myeloid cells in the bone marrow, as compared with saline-injected controls who received the same irradiation (57).

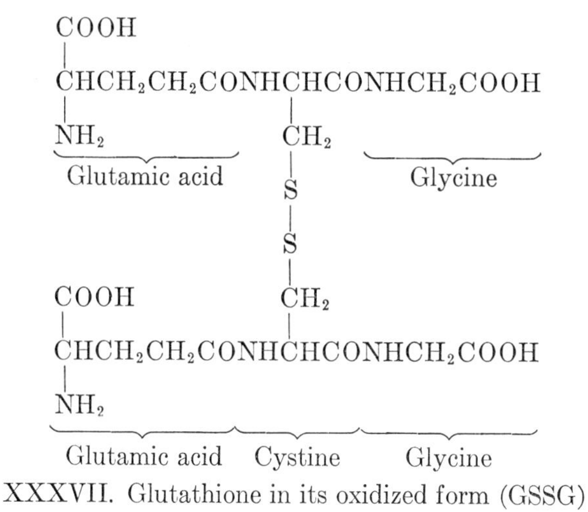

XXXVII. Glutathione in its oxidized form (GSSG)

The special role of cystine in protein structure, i.e., the function of disulfide bridges in connecting parallel peptide chains, has been discussed on page 88 and its application to a protein such as keratin on page 147. The breaking of the disulfide bridge is one of the possible causes of protein denaturation, and the appearance of the resulting thiol group is one of the classical consequences of denaturation. One protein, other than the keratins, which contains an unusually high percentage of cystine is insulin.

The thiol group is important in the action of some enzymes. Glutathione is the coenzyme of glyceraldehyde-3-phosphate dehydrogenase, where it functions as a precursor of the high-energy acyl-mercaptan bond (see page 229).

In the hereditary condition, *cystinuria*, cystine, lysine, arginine and ornithine (69) are excreted in the urine to an extent 50 to 100 times normal. Interest in the disease is centered on the excreted cystine, as is evidenced by the name, because cystine is the least soluble of the naturally-occurring amino acids and its presence in the urine of cystinurics occasionally gives rise to bladder stones. (Cystine was the first natural amino acid discovered, as a consequence of being found in such a stone.)

The excretion of cystine in cystinurics continues during fasting and is not increased by the feeding of cystine, the sulfur of which appears in the urine in the form of sulfate. Cystinuria is increased by the feeding of cysteine or of methionine, but less so if the subject is on a high-protein diet. Methionine tagged with radioactive S appears in the urine of cystinurics as cystine.

Dent and co-workers (21) believe cystinuria to be due to the decreased reabsorption by the renal tubules of cystine, lysine, arginine and ornithine and doubt the existence of any causative error in amino acid metabolism. They suggest that the reason ingested cysteine will cause a rise in urinary cystine while ingested cystine will not, is that cysteine, being soluble, is absorbed through the intestines quickly and in quantities too great for the cystinurics to handle, whereas cystine, being quite insoluble, is absorbed slowly.

A more serious anomaly of cystine metabolism is cystine storage disease, in which there is renal damage and bony abnormalities similar to rickets (35).

The sulfur of the sulfur-containing amino acids eventually appears in the urine in the form of inorganic sulfate. In the process, cysteine is oxidized to *cysteine sulfinic acid* which is further oxidized to *cysteic acid* by an enzyme system requiring coenzyme III (67). Cysteine sulfinic acid can be de-

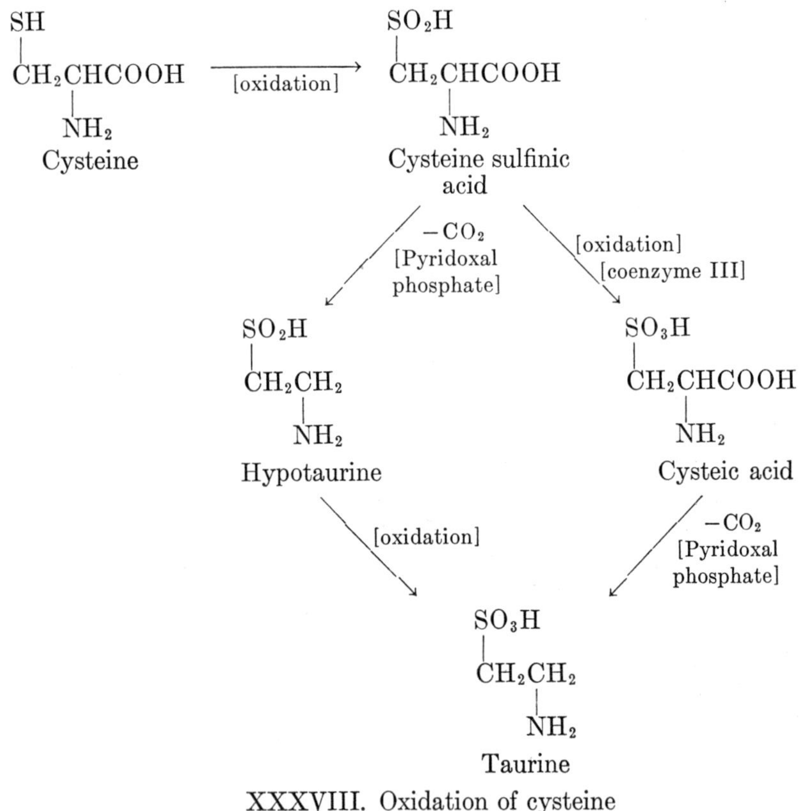

XXXVIII. Oxidation of cysteine

carboxylated (33) and then oxidized to *taurine* (formula XXXVIII) which is found conjugated in the bile salts (see page 505). The cysteic acid, formed from cysteine as shown in formula XXXVIII can be hydrolyzed to alanine and inorganic sulfate (formula XXXIX).

$$\underset{\text{Cysteic acid}}{\begin{matrix} SO_3H \\ | \\ CH_2CHCOOH \\ | \\ NH_2 \end{matrix}} \xrightarrow{+ H_2O} \underset{\text{Alanine}}{\begin{matrix} CH_3CHCOOH \\ | \\ NH_2 \end{matrix}} + SO_4^{=} + 2H^+$$

XXXIX. Hydrolysis of cysteic acid

Through the action of *desulfurases* (also called *desulfhydrases*), cysteine can be converted to pyruvic acid with the removal of both H_2S and NH_3 (formula XL). This would explain the fact that both cysteine and cystine are glucogenic. Limited amounts of sulfide, such as are produced in desulfhydration, are oxidized to sulfate by mechanisms of which the details are lacking, and appear as such in the urine. Sulfate is also produced from cysteine sulfinic acid, which is transaminated to beta-sulfinyl pyruvic acid, which is in turn split to pyruvic acid and sulfite. The sulfite is then oxidized to sulfate. The last two steps are catalyzed by manganous ion (formula XXXIXa) (68a).

$$\underset{\text{Cysteine sulfinic acid}}{\begin{matrix} SO_2H \\ | \\ CH_2CHCOOH \\ | \\ NH_2 \end{matrix}} \xrightarrow{\text{transamination}} \underset{\text{Beta-sulfinyl-pyruvic acid}}{\begin{matrix} SO_2H \\ | \\ CH_2COCOOH \end{matrix}} \xrightarrow[{[O]}]{Mn^{++}}$$

$$\underset{\text{Pyruvic acid}}{CH_3COCOOH} + SO_3^{=} \xrightarrow[{[O]}]{Mn^{++}} SO_4^{=}$$

XXXIXa. Formation of sulfate from cysteine sulfinic acid

The coenzyme associated with the desulfurases is pyridoxal phosphate which is thus the cofactor for enzymes catalyzing at least three widely different chemical reactions: transaminations (see page 610), decarboxylations (see page 612) and desulfhydration.

Methionine. Methionine is distinguished by a sulfur-bound methyl group which is a labile methyl capable of being transferred to other compounds. The process is called *transmethylation* and takes place under the influence of enzymes known as *transmethylases.*

An example of a transmethylation reaction is that whereby methionine

$$\underset{\text{Cysteine}}{\underset{\text{SH}\ \ \text{NH}_2}{\text{CH}_2\text{CHCOOH}}} \xrightarrow{-\text{H}_2\text{S}} \underset{\text{Alpha-amino-acrylic acid}}{\underset{\text{NH}_2}{\text{CH}_2{=}\text{CCOOH}}}$$

$$\downarrow$$

$$\underset{\text{Alpha-iminopropionic acid}}{\underset{\text{NH}}{\text{CH}_3\text{CCOOH}}} \xrightarrow[-\text{NH}_3]{+\text{H}_2\text{O}} \underset{\text{Pyruvic acid}}{\underset{\text{O}}{\text{CH}_3\text{CCOOH}}}$$

XL. Desulfhydration of cysteine

reacts with glycocyamine to form creatine (N-methylglycocyamine), itself being demethylated to *homocysteine*, (formula XLI). Methionine is similarly responsible for the methylation of a number of substances in the body: glycine to sarcosine and to glycylbetaine, ethanolamine to choline, carnosine to anserine (see page 154), nicotinamide to N^1-methylnicotinamide (see page 803), *nor*adrenalin to adrenalin (see page 161). Methionine can also act as a source of 1-carbon groups necessary for final closure of the purine ring (see page 358) and for the methyl side chain of thymine (32).

$$\underset{\text{Methionine}}{CH_3{-}S{-}CH_2{-}CH_2{-}CHNH_2{-}COOH} + \underset{\text{Glycocyamine}}{NH_2{-}C({=}NH){-}NH{-}CH_2{-}COOH} \xrightarrow[\text{[transmethylase]}]{\text{ATP, folic acid}} \underset{\text{Homocysteine}}{SH{-}CH_2{-}CH_2{-}CHNH_2{-}COOH} + \underset{\text{Creatine}}{NH_2{-}C({=}NH){-}NCH_3{-}CH_2{-}COOH}$$

XLI. Transmethylation

The actual methyl-transferring compound is not methionine itself but a complex of methionine with adenosine, S-Adenosylmethionine (formula XLII), and the formation of this compound requires ATP (11). The labile methyl in the complex is bound to the positively charged tri-substituted sulfur (sulfonium ion). After transfer the compound is hydrolyzed to adenosine and homocysteine.

Apparently, the only methyl groups which can be transferred by enzymatic mechanisms are those bound to tertiary sulfonium ions or to quater-

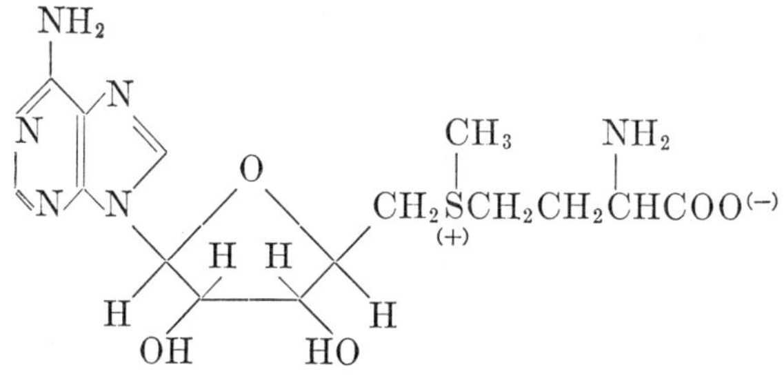

NH_2

N N O CH_3 NH_2

N N $CH_2\overset{(+)}{S}CH_2CH_2CHCOO^{(-)}$

H H

H H

OH HO

XLII. S-Adenosylmethionine

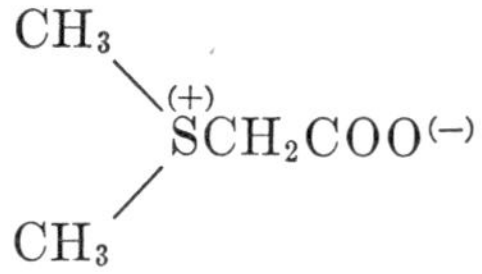

$$\begin{matrix} CH_3 \\ & \searrow \overset{(+)}{S}CH_2COO^{(-)} \\ & \nearrow \\ CH_3 \end{matrix}$$

XLIII. Dimethylthetin

nary ammonium ions. Of the compounds we have mentioned as being produced by transmethylation from methionine, two, choline, and glycylbetaine, are quaternary ammonium compounds (see formulas on page 125 and 618). Choline or glycylbetaine can therefore remethylate homocysteine to methionine. Choline does not transfer a methyl group directly, but is first converted to betaine (48).

The animal body can utilize sources of active methyl groups other than methionine, which can be replaced in the diet by homocysteine plus choline or betaine. Synthetic sulfonium compounds, such as dimethylthetin (formula XLIII) can also act as methyl donors.

The ethyl analog of choline, triethylcholine, or the ethyl analog of methionine, ethionine, will inhibit growth in rats, probably by forming ethyl analogs (70) of essential metabolites which normally carry methyl groups. In mice, the presence of triethylcholine will diminish the formation of acetylcholine (38). These are examples of competitive inhibition (see page 257).

We have already seen that the carbon skeleton of cysteine and cystine can be derived from carbohydrate. The sulfur atom can be obtained from homocysteine (demethylated methionine) by the reaction of the latter with serine to form the intermediate, *cystathionine*, which hydrolyzes to cysteine and homoserine (formula XLIV). This reaction is activated by pyridoxal phosphate (7). It is for this reason that cysteine or cystine in the diet (although dietarily non-essential) exert a sparing action on methionine, since the latter then need not be expended on the biosynthesis of the other sulfur-containing amino acids.

Alpha-ketobutyric acid, (i.e., desulfhydrated homocysteine), has been detected after the incubation of cystathionine with rat liver slices (13). The

$$\underset{\text{Homocysteine}}{\begin{array}{l} CH_2SH \\ | \\ CH_2 \\ | \\ CHNH_2 \\ | \\ COOH \end{array}} + \underset{\text{Serine}}{\begin{array}{l} HOCH_2 \\ | \\ CHNH_2 \\ | \\ COOH \end{array}} \xrightarrow{-H_2O} \underset{\text{Cystathionine}}{\begin{array}{ll} CH_2-S-CH_2 & \\ | \qquad\quad | & \\ CH_2 \qquad CHNH_2 & \\ | \qquad\quad | & \\ CHNH_2 \quad COOH & \\ | & \\ COOH & \end{array}} \xrightarrow{+H_2O} \underset{\text{Homoserine}}{\begin{array}{l} CH_2OH \\ | \\ CH_2 \\ | \\ CHNH_2 \\ | \\ COOH \end{array}} + \underset{\text{Cysteine}}{\begin{array}{l} HSCH_2 \\ | \\ CHNH_2 \\ | \\ COOH \end{array}}$$

XLIV. Methionine as precursor of cysteine

same product is formed by the oxidative deamination of homoserine (47a). Oxidative decarboxylation would produce propionic acid which would be in accord with methionine's observed glucogenic character (formula XLV).

$$\underset{\text{Homocysteine}}{\begin{array}{l} SH \\ | \\ CH_2 \\ | \\ CH_2 \\ | \\ CHNH_2 \\ | \\ COOH \end{array}} \xrightarrow[\text{+}H_2O \text{ [desulf-hydration]}]{-H_2S \; -NH_3} \underset{\text{Alpha-keto butyric acid}}{\begin{array}{l} CH_3 \\ | \\ CH_2 \\ | \\ C{=}O \\ | \\ COOH \end{array}} \xrightarrow[\text{[oxidative decarboxylation]}]{} \underset{\text{Propionic acid}}{\begin{array}{l} CH_3 \\ | \\ CH_2 \\ | \\ COOH \end{array}}$$

XLV. Possible catabolism of homocysteine

Labile methyl groups can be oxidized to formic acid. Conversely, labile methyl groups can be formed from formic acid. Since formic acid can arise from the catabolism of amino acids such as glycine and serine, it might seem that there would therefore be no need for preformed labile methyl groups in the diet. There is evidence that this is indeed so and that, in rats at least (23) homocysteine will meet the dietary requirements even in the absence of methyl donors such as choline, provided that vitamin B_{12} and folic acid are present in adequate quantities.

Amino Acids with Aromatic Side Chains

Phenylalanine and tyrosine. These may be considered together. Phenylalanine is dietarily essential. Tyrosine is formed from phenylalanine so that it is not dietarily essential, but its presence in the diet spares phenylalanine.

The metabolism of phenylalanine proceeds only by way of conversion to tyrosine (44). The enzyme system which catalyzes this conversion is *phenylalaninase*. For activity, phenylalaninase requires both DPN and oxygen. Since pyridinoenzymes are generally anaerobic dehydrogenases, the oxygen requirement seems to indicate the presence of more than one enzyme (47b). The fact that phenylalaninase is inhibited by cyanide and azide points to the presence of a metal activator such as iron or copper. It would seem then that oxygen is introduced into phenylalanine, and tyrosine formed by an enzyme system somewhat similar to that of the respiratory chain (see page 280).

In patients with *phenylpyruvic oligophrenia* (also known as *phenylketonuria*), an inborn error of metabolism which is invariably accompanied by mental deficiency, phenylalanine, together with the related substances, phenylpyruvic acid, phenyllactic acid and phenylacetic acid, are excreted in the urine. The biochemical lesion here is apparently the inability of the patient to form tyrosine from the phenylalanine, allowing the latter amino acid and its abnormal derivatives to accumulate. The concentration of phenylalanine in the plasma may be increased as much as 40 times above the normal value of about one milligram per 100 ml. Phenylpyruvic acid, absent from normal urines, is excreted by these patients in amounts up to 2 grams per day. Correction of these biochemical abnormalities, and improvement in mental status, have been repeatedly observed in phenylketonuric patients maintained on diets low in phenylalanine (34).

The catabolism of tyrosine proceeds through its transamination to parahydroxyphenylpyruvic acid, which is in turn oxidized to 2,5-dihydroxyphenylpyruvic acid by the enzyme *hydroxyphenylpyruvase*, which requires the presence of ascorbic acid. In premature infants, which are generally ascorbic-acid-deficient, or in children or adults with dietary ascorbic acid deficiency, a condition known as *hydroxyphenyluria* exists in which tyrosine and its deaminated derivatives are found in the urine. There is evidence, however, to indicate that ascorbic acid does not activate the reaction directly, but serves only to activate some cofactor previously inactivated by the process used to isolate the enzyme systems (40a).

2,5-Dihydroxyphenylpyruvic acid is oxidatively decarboxylated to homogentisic acid, in the presence of *dihydroxyphenylpyruvic decarboxylase*. In normal human beings, the benzene ring of homogentisic acid is oxidatively cleaved with the aid of the enzyme *homogentisase* or *homogentisic acid oxidase*, which may be an iron-containing enzyme (17). In people with the inborn metabolic error known as *alkaptonuria*, this enzyme or some cofactor is apparently missing and homogentisic acid is excreted unchanged in the urine. Alkaline urine containing homogentisic acid darkens on exposure to air. In association with alkaptonuria, and sometimes independently, may be observed *ochronosis*, which is the pigmentation of cartilage and sclera with homogentisic acid oxidation products.

The *maleylacetoacetic acid*, formed in normal humans by the oxidative cleavage of homogentisic acid, is converted to its *trans* isomer, *fumarylacetoacetic acid* (39), by *maleylacetoacetate isomerase*, an enzyme which requires glutathione (GSH) as cofactor (24a). Fumarylacetoacetic acid is then hydrolyzed to acetoacetic acid and fumaric acid (52) by *fumarylacetoacetate hydrolase*. Thus, since fumaric acid is a member of the tricarboxylic acid cycle, phenylalanine and tyrosine, though usually considered ketogenic, contribute four of their nine carbons to the main-stream of carbohydrate metabolism and glycogenesis, four to the formation of ketone bodies, while the ninth and last is converted to CO_2.

The catabolic scheme presented here for phenylalanine and tyrosine is summarized in formula XLVI.

Tyrosine is the precursor of a number of important physiological substances in the body. Diiodotyrosine, triiodothyronine and thyroxine, all important in thyroid function (see page 310) are formed by iodination of tyrosine-containing proteins *in vivo* and *in vitro*. Another hormone for which tyrosine is the precursor is adrenalin (see page 161). Its manner of formation is as yet uncertain, but a possible mechanism is presented in formula XLVII.

The common black animal pigment, melanin, is formed by the oxidation and condensation of tyrosine, initiated by the enzyme *tyrosinase*. The mechanism of melanin formation as presented by Lerner (44) is shown in formula XLVIII. In the inherited metabolic error, *albinism*, tyrosinase is lacking and melanin is completely absent from the skin, hair, eyes, and other usual situations. Melanin itself is, apparently, a poly-indole quinone, and a portion of its molecule is shown in formula XLIX.

Tryptophane. Tryptophane is dietarily essential. On the basis of animal experiments, tryptophane has not been conclusively shown to be either glucogenic or ketogenic, and its status in that respect is in doubt. From the urine of rats, rabbits and dogs, compounds such as kynurenine, kynurenic acid and xanthurenic acid have been isolated after the administration of tryptophane. Kynurenine, but not hydroxykynurenine, has been detected in the urine of normal humans who have consumed an excess of tryptophane. Under febrile conditions, the situation is reversed; hydroxykynu-

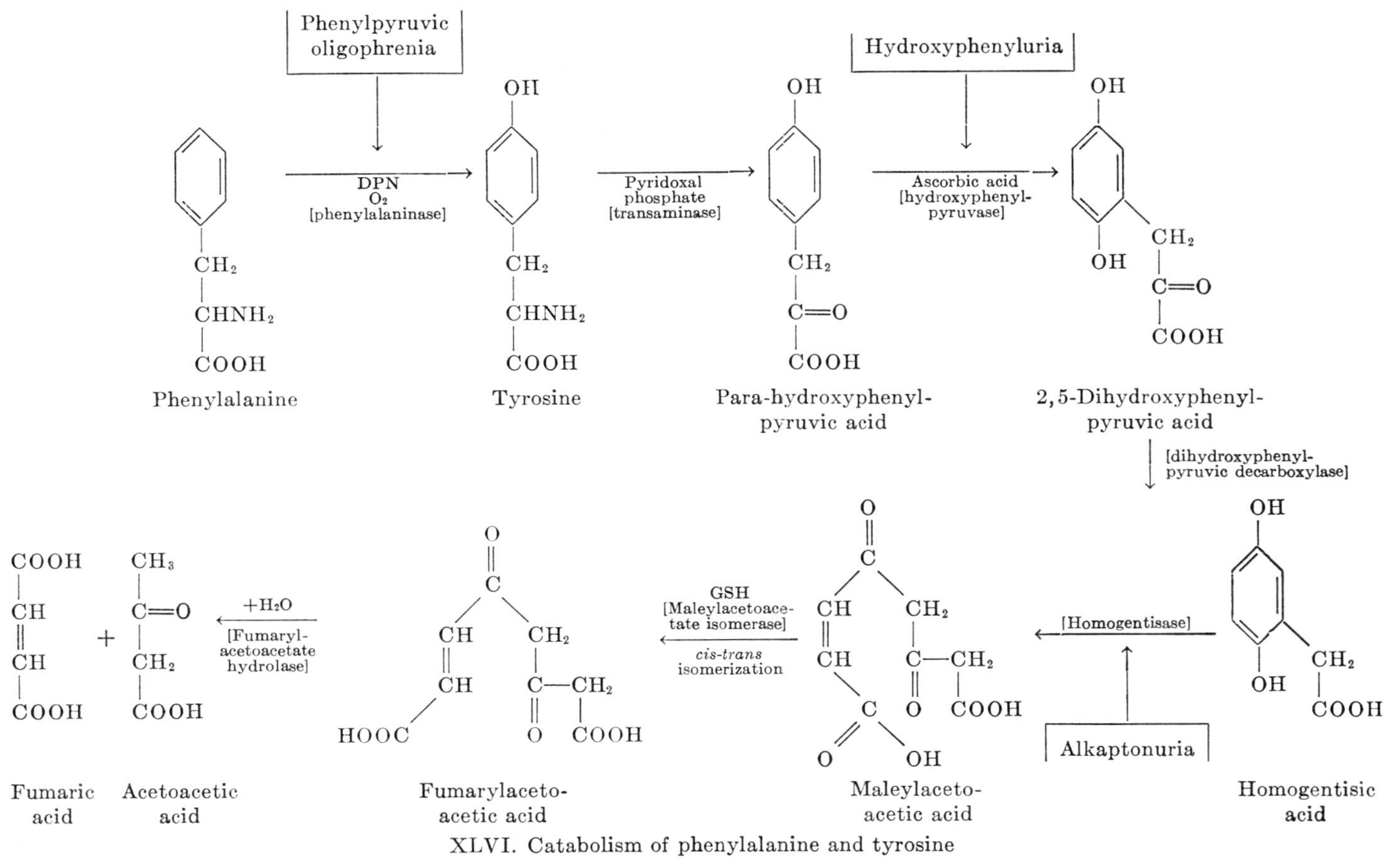
Phenylpyruvic oligophrenia
Hydroxyphenyluria
DPN
O2
[phenylalaninase]
Pyridoxal phosphate [transaminase]
Ascorbic acid [hydroxyphenyl-pyruvase]
Phenylalanine
Tyrosine
Para-hydroxyphenyl-pyruvic acid
2,5-Dihydroxyphenyl-pyruvic acid
[dihydroxyphenyl-pyruvic decarboxylase]
[Homogentisase]
Alkaptonuria
Homogentisic acid
GSH
[Maleylacetoacetate isomerase]
cis-trans isomerization
Maleylaceto-acetic acid
Fumarylaceto-acetic acid
+H2O
[Fumaryl-acetoacetate hydrolase]
Fumaric acid
Acetoacetic acid

XLVI. Catabolism of phenylalanine and tyrosine

Tyrosine ($HO-C_6H_4-CH_2-CHNH_2-COOH$) —[tyrosinase]→ Dihydroxyphenylalanine (Dopa) ($(HO)_2C_6H_3-CH_2-CHNH_2-COOH$) —[decarboxylation]→ Hydroxytyramine ($(HO)_2C_6H_3-CH_2-CH_2NH_2$) —[oxidation]→ Noradrenalin ($(HO)_2C_6H_3-CHOH-CH_2NH_2$) —[transmethylation]→ Adrenalin ($(HO)_2C_6H_3-CHOH-CH_2NHCH_3$)

Tyrosine —[decarboxylation]→ Tyramine ($HO-C_6H_4-CH_2-CH_2NH_2$) —[oxidation]→ Hydroxytyramine

XLVII. Formation of adrenalin

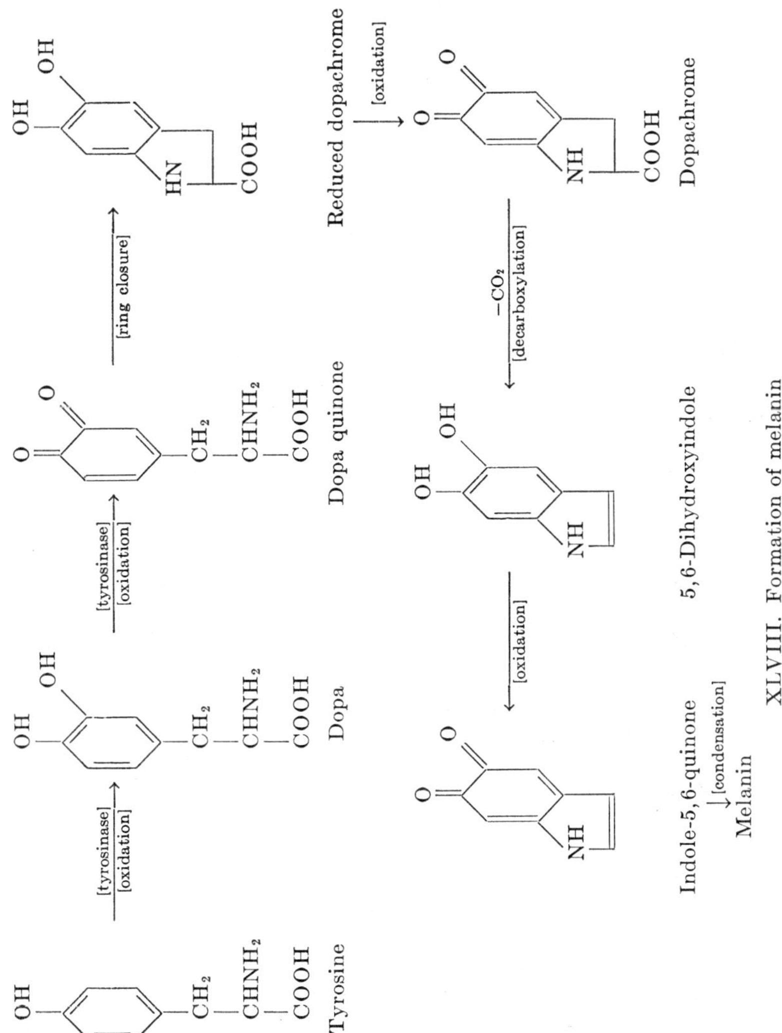

XLVIII. Formation of melanin

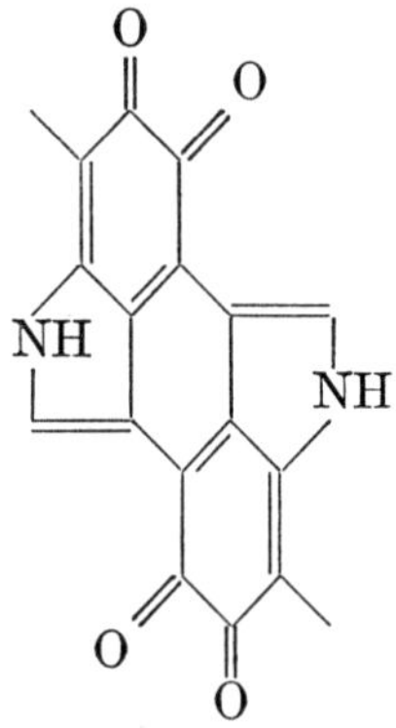

XLIX. Repeating unit of the melanin molecule

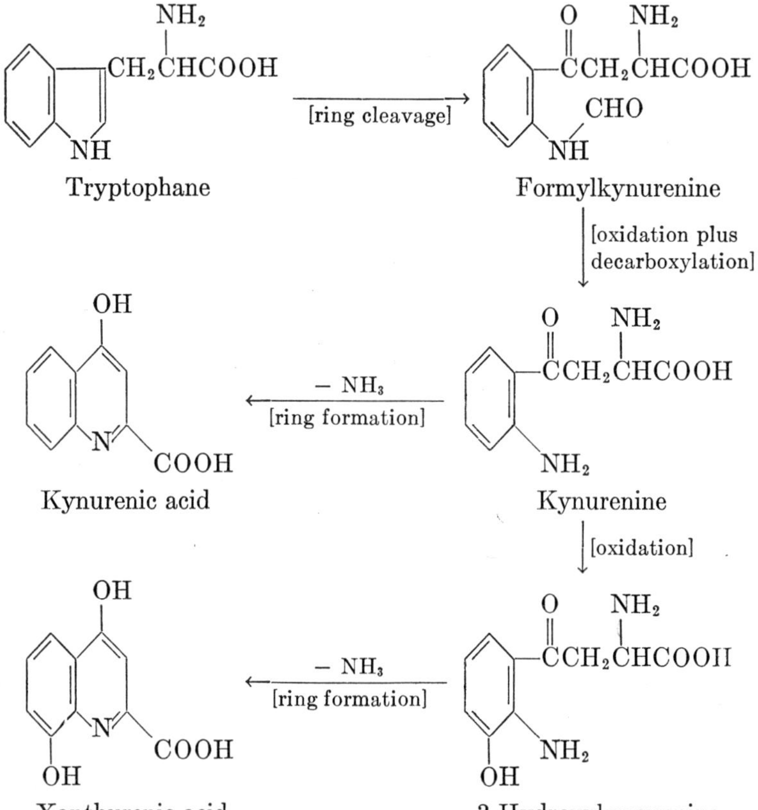

Tryptophane —[ring cleavage]→ Formylkynurenine

Formylkynurenine —[oxidation plus decarboxylation]→ Kynurenine

Kynurenine —($-NH_3$) [ring formation]→ Kynurenic acid

Kynurenine —[oxidation]→ 3-Hydroxykynurenine

3-Hydroxykynurenine —($-NH_3$) [ring formation]→ Xanthurenic acid

L. Catabolism of tryptophane

renine is excreted and little or no kynurenine (18). A possible route for the formation of these tryptophane degradation products is shown in formula L.

The manner in which these are further catabolized is not known, except for the observation that kynurenine is a precursor of nicotinic acid, one of the B vitamins (see page 801). This was shown in experiments on Neurospora and rats (31) but the route of the conversion has not yet been worked out in detail. One of the intermediate steps is apparently the cleavage of kynurenine (or hydroxykynurenine) to anthranilic acid (or hydroxyanthranilic acid) and alanine (formula LI) by the enzyme, kynureninase, present in mammalian liver. If this last reaction represented the main-stream of tryptophane catabolism, it would indicate that at least three of its carbon atoms were available for glucogenesis.

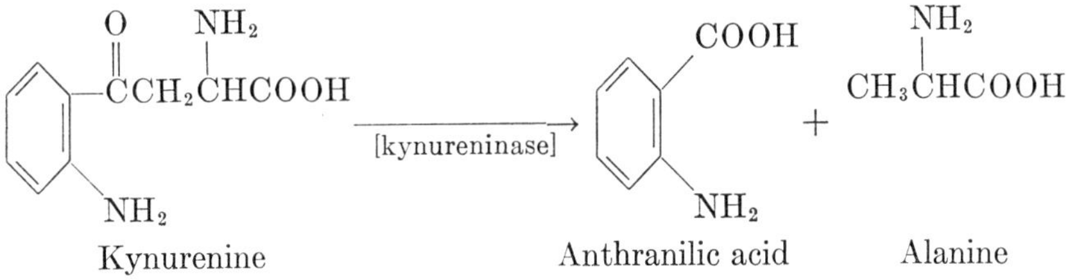

O NH$_2$
—CCH$_2$CHCOOH
NH$_2$
Kynurenine

[kynureninase] →

COOH
NH$_2$
Anthranilic acid

+

NH$_2$
CH$_3$CHCOOH
Alanine

LI. Cleavage of kynurenine

COOH
NH$_2$
OH
3-Hydroxy-anthranilic acid

O_2 →

COOH
CHO
HOOC NH$_2$

↓ $-CO_2$

COOH
CHO
NH$_2$

↙ $-H_2O$

COOH
N
Nicotinic acid

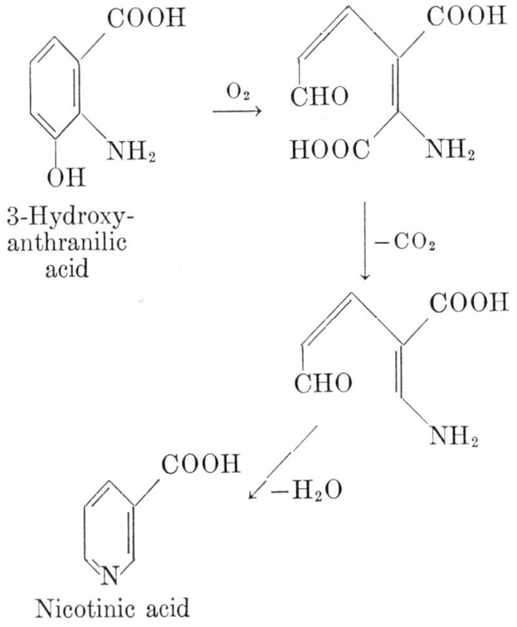

LII. Formation of nicotinic acid

The synthesis of nicotinic acid from tryptophane proceeds in mammals and probably in man as well to an extent sufficient to meet a significant part (though not all) of the physiological requirement. From experiments with Neurospora, it seems likely that the precursors of nicotinic acid are the anthranilic acids formed by the action of kynureninase on the kynurenines (68). The anthranilic acid ring is cleaved oxidatively and subsequent condensation forms a pyridine ring (formula LII). Dietary deficiency of tryptophane in rats leads to a specific decrease in certain enzyme activities in the liver (80). The oxygen uptake of homogenates from the livers of the deficient rats was decreased, as were the specific activities of xanthine dehydrogenase and succinic acid dehydrogenase. The pellagragenic qualities of maize may be due as much to the poverty of its proteins in tryptophane as to its lack of adequate quantities of nicotinic acid.

Tryptophane is oxidized in the liver to 5-hydroxytryptophane, which is decarboxylated by a specific decarboxylase to serotonin (see page 163). The catabolism of serotonin may proceed by way of oxidative deamination to 5-hydroxyindole acetic acid (formua LIIa) (76a).

CH_2COOH
HO
NH

LIIa. 5-Hydroxyindole acetic acid

Carcinoid tumors (argentaffinoma) of the small intestine are characterized by their high content of serotonin, which may be released into the circulation, producing changes in the vascular bed which may cause damage to skin and heart. Patients with carcinoid tumors excrete significantly increased amounts of 5-hydroxyindole acetic acid (68b).

Amino Acids with Basic Side Chains

Arginine. Arginine possesses a guanidino group on its side chain. The role of this guanidino group in urea formation has been described on page 606 and in the formation of glycocyamine and creatine on page 619. Histones and protamines are particularly rich in arginine (see page 366). The amino acids, ornithine and citrulline, which appear as intermediates in the urea cycle are not among the amino acids that make up the structure of the proteins of the human body. Phosphoarginine appears in invertebrate muscle, where it fulfills the function of phosphocreatine in vertebrate muscle.

Arginine has been shown, by isotope experiments, to be a precursor of proline and glutamic acid. The suggested scheme of directly or indirectly reversible transformations is shown in formula LIII. Since glutamic acid is glucogenic and is not dietarily essential, it follows that the same holds true for both arginine and proline.

Lysine. Like tryptophane, lysine is not found to be either glucogenic or

```
NH2
|
C=NH
|
NH               NH2
|                |
CH2              CH2                      CHO                 COOH
|  ------------> |   ----------------->   |   ------------>   |
CH2 [deaminidation] CH2  [oxidative       CH2   [oxidation]    CH2
|                |    deamination]        |                   |
CH2              CH2                      CH2                 CH2
|                |                        |                   |
CHNH2            CHNH2                    CHNH2               CHNH2
|                |                        |                   |
COOH             COOH                     COOH                COOH
Arginine         Ornithine           Glutamic acid         Glutamic
                                     semialdehyde            acid
                                          |
                                          | [ring formation]
                                          v
                                     (ring)=N  ------------>  (ring) NH
                                              [reduc-
                                               tion]
                                          |                   |
                                        COOH                 COOH
                                 Pyrroline carboxylic        Proline
                                         acid
```

LIII. Interrelationship of arginine, glutamic acid, and proline

ketogenic in animal experiments. Of all the amino acids, it is the least involved in interrelationships with other amino acids. Once deaminated, it can not accept an amino group by transamination in the body, hence can not be replaced in the diet by its corresponding keto acid. Apparently, it is oxidized to alpha-aminoadipic acid by a roundabout route that involves oxidative deamination of the alpha-amino group, and then its restoration at the expense of the epsilon-amino group by way of ring formation and ring cleavage (59) (formula LIV). After that is done, oxidative deamination produces alpha-ketoadipic acid and oxidative decarboxylation produces glutaric acid (formula LV). The further catabolism of glutaric acid can take place in the rat by way of alpha-ketoglutaric acid into the tricarboxylic acid cycle (58).

$$CH_2NH_2-CH_2-CH_2-CH_2-CHNH_2-COOH \text{ (Lysine)} \xrightarrow[\text{deamination}]{\text{oxidative}} CH_2NH_2-CH_2-CH_2-CH_2-C{=}O-COOH \xrightarrow{-H_2O} \text{(cyclic imine, COOH)}$$

$$\text{(cyclic imine, COOH)} \xrightarrow{+2H} \text{Pipecolic acid (NH, COOH)} \xrightarrow{-2H} \text{(unsaturated ring, NH, COOH)}$$

$$\xrightarrow{+H_2O \text{ and rearrangement}} CHO-CH_2-CH_2-CH_2-CHNH_2-COOH \xrightarrow{\text{oxidation}} COOH-CH_2-CH_2-CH_2-CHNH_2-COOH \text{ (Alpha-aminoadipic acid)}$$

LIV. Catabolism of lysine—early stages

Lysine is dietarily essential. Infants on a lysine-poor diet maintain normal plasma nitrogen levels, but with an increase in the arginine content of the plasma proteins, suggesting that arginine substitutes for lysine (1). Lysine deficiency, like that of threonine, leads to fatty livers in rats. 5-Hydroxylysine, which occurs in collagen, is formed directly from lysine (65).

Under the influence of bacterial enzymes, lysine is decarboxylated and yields the diamine, cadaverine, which like the similar putrescine, derived from ornithine, is toxic but can be oxidatively detoxicated by diamine dehydrogenase.

Histidine. Histidine occurs in combination with beta-alanine in the form of carnosine, which is present in the muscles of most vertebrates, and also in anserine, which is a methyl carnosine (see page 154). Another histi-

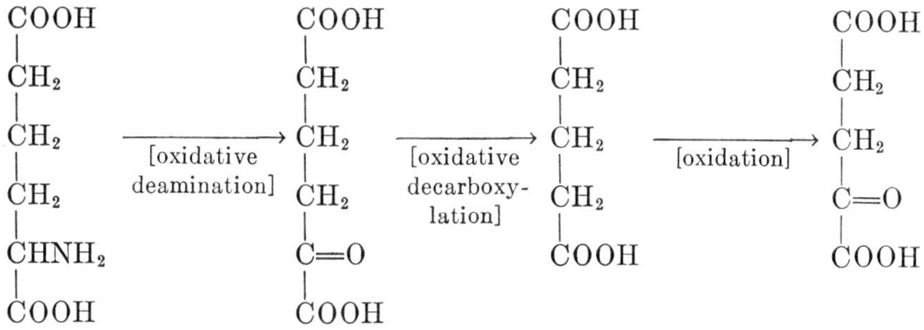

$$\begin{array}{c} COOH \\ | \\ CH_2 \\ | \\ CH_2 \\ | \\ CH_2 \\ | \\ CHNH_2 \\ | \\ COOH \end{array} \xrightarrow[\text{deamination]}]{\text{[oxidative}} \begin{array}{c} COOH \\ | \\ CH_2 \\ | \\ CH_2 \\ | \\ CH_2 \\ | \\ C{=}O \\ | \\ COOH \end{array} \xrightarrow[\text{decarboxylation]}]{\text{[oxidative}} \begin{array}{c} COOH \\ | \\ CH_2 \\ | \\ CH_2 \\ | \\ CH_2 \\ | \\ COOH \end{array} \xrightarrow{\text{[oxidation]}} \begin{array}{c} COOH \\ | \\ CH_2 \\ | \\ CH_2 \\ | \\ C{=}O \\ | \\ COOH \end{array}$$

Alpha-amino-adipic acid | Alpha-ketoadipic acid | Glutaric acid | Alpha-ketoglutaric acid

LV. Catabolism of lysine—later stages

dine compound of possible physiological significance is ergothioneine which is widespread in animal tissue, particularly liver. This is a thiol derivative of histidylbetaine (formula LVI).

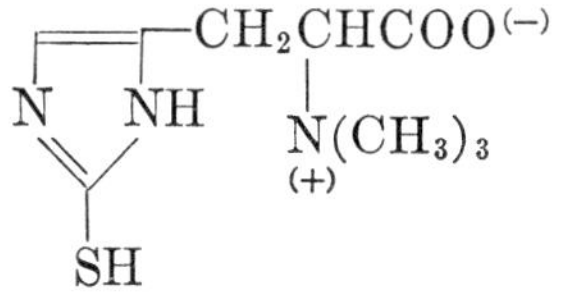

$$\text{(imidazole ring: N, NH; 2-SH)}-CH_2CH(N^{(+)}(CH_3)_3)COO^{(-)}$$

LVI. Ergothioneine (thiolhistidylbetaine)

The liver enzyme, *histidase*, deaminates histidine to *urocanic acid*, which is eventually catabolized to glutamic acid with loss of ammonia and formic acid, through stages in which the enzyme, *urocanase*, is involved (formula LVII.) This formation of glutamic acid accounts for the observed glucogenic character of histidine.

An additional route of histidine catabolism has been described (61a). It involves the methylation of the ring nitrogen furthest from the 3-carbon side-chain.

The enzyme, histidine decarboxylase, which is present in traces in the liver and kidney, is capable of converting histidine to the very toxic amine, histamine. Histamine seems to be actively concerned with the manifestations of various types of allergic reactions (9).

Imino Acids

The two imino acids in proteins are proline and hydroxyproline, the latter occurring in significant quantities only in collagen. The relationship of proline, arginine and glutamic acid has already been described (see page 641). Hydroxyproline is formed from proline in the body by an as yet unelucidated mechanism. Dietary hydroxyproline, isotope experiments

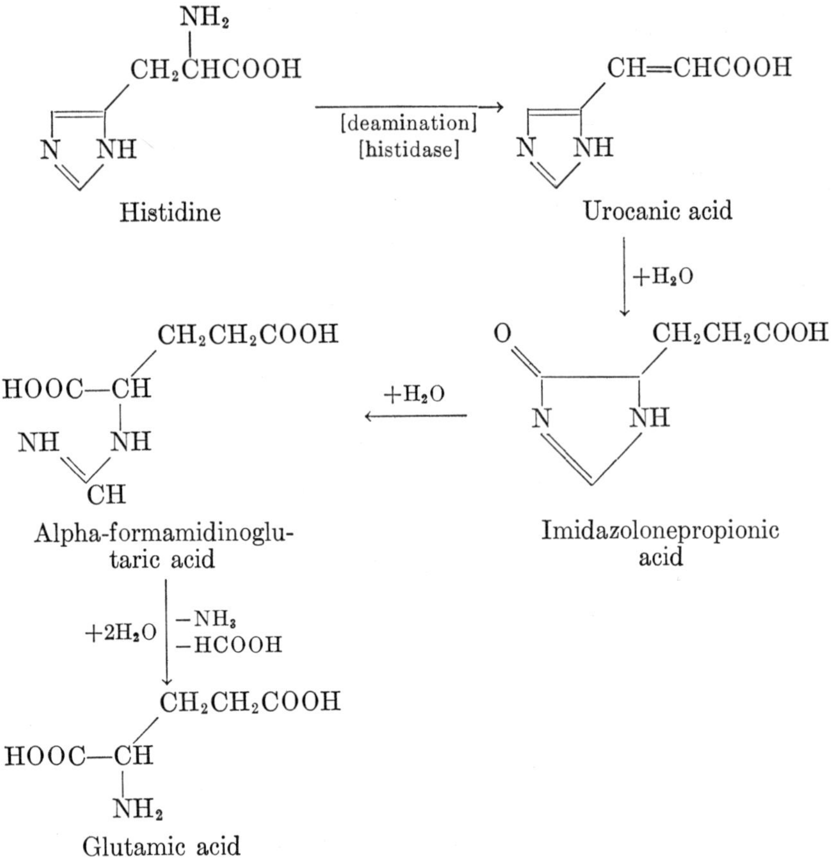

NH_2

$CH_2CHCOOH$

N NH

Histidine

→ [deamination] [histidase]

$CH{=}CHCOOH$

N NH

Urocanic acid

↓ $+H_2O$

O CH_2CH_2COOH

N NH

Imidazolonepropionic acid

← $+H_2O$

CH_2CH_2COOH

HOOC—CH

NH NH

CH

Alpha-formamidinoglutaric acid

↓ $+2H_2O$ $-NH_3$ $-HCOOH$

CH_2CH_2COOH

HOOC—CH

NH_2

Glutamic acid

LVII. Conversion of histidine to glutamic acid

show, is not incorporated into body protein (71), nor is it converted into proline.

Because of the relationship with glutamic acid, proline is glucogenic and not dietarily essential. Hydroxyproline, being formed from proline is likewise not dietarily essential. Whether hydroxyproline is glucogenic or ketogenic is yet uncertain. Evidence for both glucose formation and ketone body formation exists (41). A possible route for the catabolism of proline and hydroxyproline involves the enzyme, *proline oxidase* (28), which can oxidatively cleave the proline and hydroxyproline ring (formula LVIII) to produce 5-carbon keto acids. The further catabolism of the aminovaleric acids has not yet been completely worked out, but the conversion of the alpha-keto-delta aminovaleric acid arising from proline to the semi-aldehyde of glutamic acid (perhaps by an internal transamination) seems probable. Proline would thus be connected with the glucogenic glutamic acid

Proline → [Proline oxidase] → Alpha-keto-delta-aminovaleric acid ($CH_2—CH_2—NH_2$; CH_2; C=O; COOH)

Hydroxyproline → [Proline oxidase] → Alpha-keto-gamma-hydroxy-delta-amino-valeric acid (OH; CH—CH_2—NH_2; CH_2; C=O; COOH)

LVIII. Ring cleavage of the imino acids

by a second metabolic path. The presence of the hydroxyl on the gamma-carbon of the aminovaleric acid derivative arising from hydroxyproline is perhaps significant. A deamination and an oxidative decarboxylation, which are both common processes in amino acid catabolism, would result in the formation of either beta-hydroxybutyric acid or acetoacetic acid, both of which are ketone bodies. The aminovaleric acid might, however, also end in glutamic acid by reasonable conversions. The existence of both paths might explain the conflicting evidence as to the glucogenicity or keto-genicity of hydroxyproline.

Summary of Amino Acid Catabolism

It may be helpful, at this point, to gather together in concise form the information presented in the last section, and present the main route of catabolism for the carbon backbone of each amino acid.

1. *Glycine* is primarily concerned in anabolic reactions. Its only important catabolic reaction involves oxidative deamination and oxidative decarboxylation to formic acid.

2. *Alanine* is transaminated or oxidatively deaminated to pyruvic acid and enters the tricarboxylic acid cycle.

3. *Valine* is transaminated and twice oxidatively decarboxylated to form propionic acid, which, on oxidation to pyruvic acid, enters the tricarboxylic acid cycle.

4. *Leucine* is transaminated and oxidatively decarboxylated to form a branched 5-carbon acid. This is split to a 3-carbon fragment and a 2-carbon fragment, both of which give rise to acetoacetic acid.

5. *Isoleucine* is transaminated and oxidatively decarboxylated to form a branched 5-carbon acid. This is split to a 3-carbon fragment and a 2-carbon fragment. The 3-carbon fragment is converted to pyruvic acid and enters the tricarboxylic acid cycle. The 2-carbon fragment is converted to acetoacetic acid.

6. *Serine* is anaerobically deaminated to pyruvic acid which enters the tricarboxylic acid cycle.

7. *Threonine* is anaerobically deaminated to alpha-ketobutyric acid which is oxidatively decarboxylated to propionic acid which, on conversion to pyruvic acid, enters the tricarboxylic acid cycle.

8. *Aspartic acid* is transaminated to oxaloacetic acid which enters the tricarboxylic acid cycle.

9. *Glutamic acid* is transaminated or oxidatively deaminated to alpha-ketoglutaric acid which enters the tricarboxylic acid cycle.

10. *Cysteine* is desulfhydrated and hydrolytically deaminated to pyruvic acid which enters the tricarboxylic acid cycle.

11. *Cystine* is reduced to cysteine and shares its fate.

12. *Methionine* is demethylated to homocysteine, the lost labile methyl being eventually oxidized to formic acid. The homocysteine is desulfhydrated and hydrolytically deaminated to alpha-ketobutyric acid which is oxidatively decarboxylated to propionic acid which, on conversion to pyruvic acid, enters the tricarboxylic acid cycle.

13. *Phenylalanine* is converted to tyrosine and shares its fate.

14. *Tyrosine* undergoes transamination, oxidation, rearrangement and oxidative decarboxylation, being converted to homogentisic acid. The benzene ring of homogentisic acid is oxidatively cleaved to maleylacetoacetic acid which is converted to fumarylacetoacetic acid, then hydrolytically cleaved to acetoacetic acid and fumaric acid. The fumaric acid enters the tricarboxylic acid cycle.

15. *Tryptophane* undergoes ring cleavage followed by oxidation and decarboxylation to kynurenine. Certain substances are known to arise from kynurenine, but its ultimate catabolic fate is uncertain.

16. *Arginine* is hydrolyzed to urea and ornithine. The ornithine can be converted to glutamic acid by oxidative deamination followed by oxidation. Thereafter it shares the fate of glutamic acid.

17. *Lysine* is twice oxidatively deaminated and once oxidatively decarboxylated to form glutaric acid which may enter the tricarboxylic acid cycle after conversion to alpha-ketoglutaric acid.

18. *Histidine* is deaminated to urocanic acid, which on hydrolytic ring

cleavage, followed by hydrolysis, will form glutamic acid and formic acid. Thereafter, it shares the fate of glutamic acid.

19. *Proline*, by oxidative ring cleavage, is converted to glutamic acid. Thereafter it shares the fate of glutamic acid.

20. *Hydroxyproline* may be converted by oxidative ring cleavage to an aminovaleric acid derivative which in turn may possibly give rise to either glutamic acid (and through it, glucose) or to ketone bodies.

The preceding listing includes main catabolic routes only and does not mention the numerous secondary reactions such as the conversion of serine to ethanolamine or to glycine and formic acid, the decarboxylation of aspartic acid to beta-alanine, formation of adrenalin and thyroxine from tyrosine and many more.

The amino acids can be divided into two groups: (A) those whose carbon skeletons can not be synthesized in the body from carbohydrate precursors and (B) those whose carbon skeletons can so be synthesized.

Group A includes, naturally, all the dietarily essential amino acids plus tyrosine. Tyrosine, though not dietarily essential can be formed in the body only from the dietarily essential phenylalanine and in no other way. Group B includes all other amino acids.

We can tabulate the ultimate fate of the carbon atoms of the amino acids, assuming each amino acid to follow its most direct and important catabolic route. In doing this, we shall omit glycine, whose reactions are primarily anabolic, and hydroxyproline, concerning whose catabolism there are still conflicting notions.

In considering table XVI-1, it seems noteworthy that the distinction between groups A and B is not primarily on the basis of whether the amino acids are glucogenic or ketogenic. While all the amino acids in group B are glucogenic, not all in group A are ketogenic. In other words, an amino acid may be converted to glucose without in any way implying the ability of the body to synthesize that amino acid from glucose. What is startling is that in the case of every amino acid in group A, and in the case of *no* amino acid in group B, an oxidative decarboxylation takes place. To be sure, decarboxylations take place in alternate or subsidiary routes in such group B amino acids as aspartic acid, glutamic acid, histidine and, notably, glycine, but in each case it does *not* represent the only route of catabolism for that amino acid. In the case of the group A amino acids no catabolic routes are known which by-pass the oxidative decarboxylation.

It is tempting then to conjecture that is the oxidative decarboxylation step which, in amino acid catabolism, is irreversible, and it is the presence of that step in all catabolic routes of a given amino acid that renders that amino acid dietarily essential or, in the case of tyrosine, derivable only from a dietarily essential amino acid.

The reader should remember that this analysis refers only to the carbon atoms and does not involve the question of the origin of the sulfur atoms of cysteine and cystine, for instance, which, of course, must be supplied as cysteine, cystine, or methionine in the diet.

Nitrogenous End Products of Amino Acid Catabolism

Of the nitrogenous urinary components the history of which can be traced back to the amino acids, the most abundant is urea. Urea formation has already been discussed on page 605.

The most important remaining end product of amino acid catabolism is *creatinine* which arises from *creatine phosphate* by ring-formation with loss of phosphoric acid (12) (formula LIX). The formation of creatinine from creatine phosphate is apparently irreversible in the body.

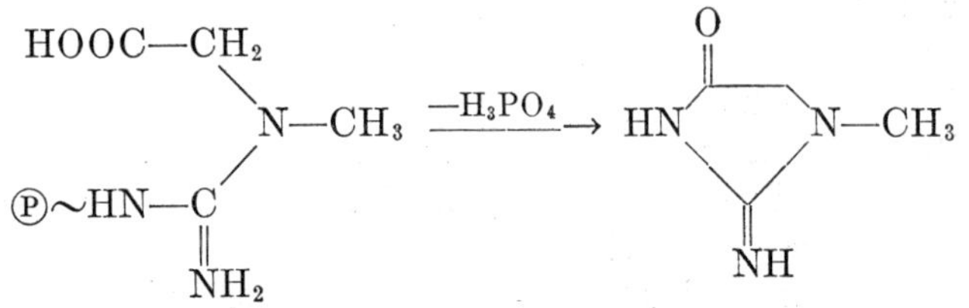

$$\text{HOOC—CH}_2\text{—N(CH}_3\text{)—C(=NH}_2\text{)—HN}\sim\text{Ⓟ} \xrightarrow{-H_3PO_4} \text{HN—C(=O)—CH}_2\text{—N(CH}_3\text{)—C(=NH)— (ring)}$$

Creatine phosphate — Creatinine

LIX. Formation of creatinine

Creatine is formed from glycine first by transamidination with arginine (see page 619), followed by transmethylation with methionine (see page 630). Its phosphorylated derivative, phosphocreatine, occurs in muscle tissue where it is of importance in contraction because of the high energy phosphate bond it contains (see page 153). The concentration of creatine is high in liver, kidney, and testicle, as well as in cardiac and skeletal muscle, but its function in non-muscular tissue has not been clearly established. Presumably, it acts as a carrier of high-energy phosphate bonds there also.

Creatine is present in normal male blood serum in amounts between 0.17 and 0.58 mgm. per 100 ml., in female blood serum between 0.35 and 0.93 mgm. (74). Small quantities of creatine in urine (60 to 150 mgm./day) have been recorded for normal males, while the value in females is on the average over twice as high (79). In hyperthyroidism, in states involving undernutrition, and very typically in diseases of the muscles, abnormal amounts of creatine appear in the urine. The demethylated creatine precursor, glycocyamine (see page 619), in amounts of about 20 mgm. per day, is a normal urinary constituent (8). Its excretion in myasthenia gravis is not increased (75) while that of creatine is.

Creatine given to a human subject by feeding or injection can not be

TABLE XVI-I

*Fate of carbon atoms of the amino acids**

Amino acid	NO. CARBONS ENDING AS:						
	Total carbons	Glucose precursors	Ketone bodies	Formic acid	Urea	Carbon dioxide	Unknown
A. Amino acids whose carbon skeletons can not be synthesized from carbohydrate in the body							
†Valine	5	3				2	
†Leucine	6		5			1	
†Isoleucine	6	3	2			1	
†Threonine	4	3				1	
†Methionine	5	3		1		1	
†Phenylalanine	9	4	4			1	
Tyrosine	9	4	4			1	
†Tryptophane	11	3				1	7
†Lysine	6	5				1	
B. Amino acids whose carbon skeletons can be synthesized from carbohydrate in the body							
Alanine	3	3					
Serine	3	3					
Aspartic acid	4	4					
Glutamic acid	5	5					
Cysteine	3	3					
Cystine	6	6					
Arginine	6	5			1		
Histidine	6	5		1			
Proline	5	5					

* Glycine and hydroxyproline are not included. See text.
† Dietarily essential amino acids.

entirely traced. In the male, there is a slight transitory rise in the serum creatine level, with little or no creatinuria. In the female the serum level rises higher and stays elevated for a longer time, and creatinuria occurs. In both sexes some storage of creatine in muscle and viscera can be demonstrated, but not enough to account for the amount given. In myopathic patients, on the other hand, injected creatine is almost quantitatively excreted in the urine. The feeding of creatine precursors, such as glycine, arginine, or glycocyamine, will cause little change in creatine output by normal subjects, but will cause increases in the myopathic patient. There is no evidence of any defect in transamidination or transmethylation in myasthenics (75).

The N-methyl groups of both creatine and creatinine are not significantly oxidized to carbon dioxide but are excreted unchanged (47). In this respect, they differ from the N-methyl groups of sarcosine, choline and glycylbetaine, and from the S-methyl group of methionine.

Creatinine is present in normal blood serum in concentrations from 0.9 to 1.65 mgm. per 100 ml. (74). Increases in blood creatinine occur only in serious renal insufficiency. The urinary excretion of creatinine is reasonably constant in the individual from day to day, and among individuals varies from 1 to 2 grams, with a very rough proportionality to body weight. The urinary output is independent of protein intake or creatine intake. Injected creatinine is excreted in the urine to the extent of 75 to 96 per cent of the amount injected. The rest disappears, possibly by diffusion into the intestinal tract where it is utilized by bacteria.

PROTEIN ANABOLISM

The synthesis of protein from amino acids might conceivably occur by simple reversal of proteolysis, catalyzed by proteinase, the direction being determined by mass action. Experimental attempts at *in vitro* synthesis of proteins by reverse enzyme action have resulted only in the production of *plasteins*, which are insoluble polypeptides. The molecular weights of the protein-like substances produced by the action of chymotrypsin on peptic digests of various proteins has been estimated as high as 250,000 to 500,000 (73).

In addition to the direct condensation of amino acids, the process of *transpeptidation* is of undoubted importance in protein synthesis. Transpeptidation can be viewed as different from hydrolysis in the nature of the acceptor molecule. Thus, in the hydrolysis of glycylglycine, water is the acceptor molecule, two molecules of glycine being formed. In the transpeptidation of glycylglycine, an amino acid such as alanine may be the acceptor molecule, forming glycine and glycylalanine (formula LX).

Apparently, proteinases and peptidases are specific for the substrate rather than for the acceptor molecule so that they may catalyze either hydrolysis or transpeptidation (36, 41a) depending in part upon pH. At pH values in the neighborhood of 5, the hydrolytic process is predominant. At pH values between 7 and 8, transpeptidation takes place preferentially. It should be noted that the latter pH range is physiological.

Transpeptidation can be used not only to alter the composition of a peptide chain, but also to lengthen it (27). For example, if glycylglycine were first ammoniated to glycylglycylamide, it could then be transpeptidated to glycylglycylalanine with ammonia reformed (formula LXI). The process

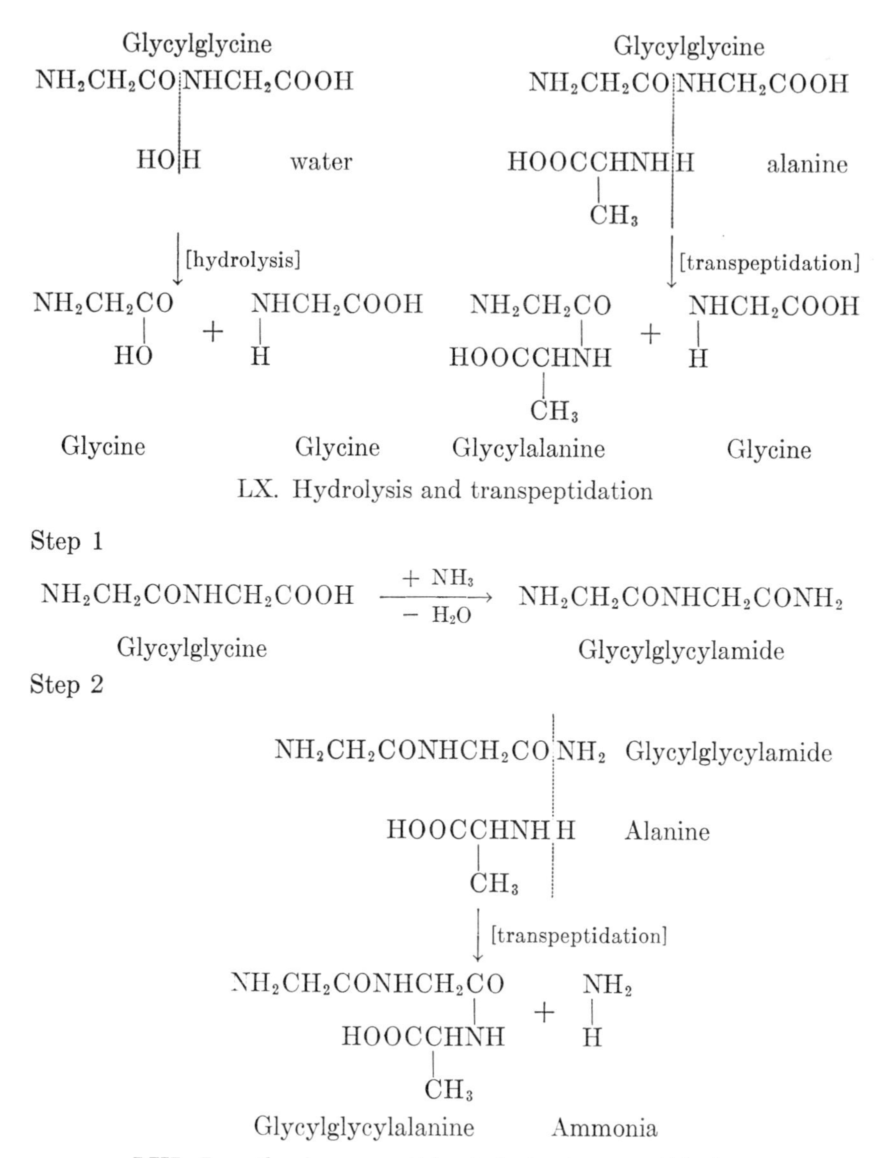

LX. Hydrolysis and transpeptidation

LXI. Lengthening a peptide chain by transpeptidation

can be repeated indefinitely, the nature of the product depending on the specificity of the enzymes involved (26).

The hydrolysis of peptide bonds involves a decrease in free energy of 0.5 to 4.0 Kcal. per mol of peptide bond (depending on the nature of the amino acid residues involved) and the formation of these bonds naturally involves the input of a similar quantity of free energy. This means that peptide synthesis must be coupled with the hydrolysis of ATP. Lipmann (45) has succeeded in synthesizing amides and substituted amides of carboxylic acids in the presence of pigeon liver extracts where ATP was part of the system, but not in its absence. True peptides were not formed, but glutamine was formed from glutamic acid and ammonia and hippuric acid from benzoic acid and glycine. Glutamine and hippuric acid are enough like peptides to allow the supposition that peptide formation proceeds analogously. Lipmann concluded from his observation of an equimolecular relationship between the amide bond formed and the free phosphate liberated that one high energy phosphate is utilized for each peptide bond formed.

It is possible that the role of ATP in peptide bond synthesis is to form as intermediate the high-energy aminoacyladenylate (20) (formula LXII).

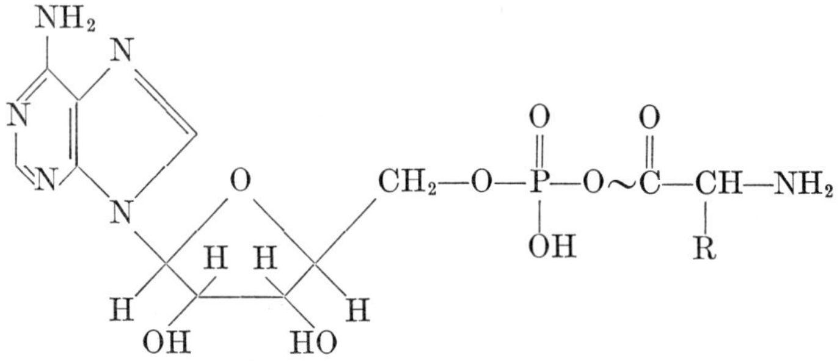

$$\text{Adenine–ribose–}CH_2\text{—O—}\overset{O}{\overset{\|}{P}}(OH)\text{—O}\sim\overset{O}{\overset{\|}{C}}\text{—}CH(R)\text{—}NH_2$$

LXII. Aminoacyladenylate

Nucleic acids, and DNA in particular, are thought, as a result of many lines of evidence (see Chapter 9), to be the chemical mediators of the cell and to control the biosynthesis, for instance, of specific cell enzymes. The complexity of nucleic acid structure is quite probably as great as that of protein structure and the various specific DNA molecules of the cell, which autoreproduce from cell generation to cell generation, may well represent templates upon which protein molecules are fashioned. The discussion in Chapter 9 did not consider the problem of energy input involved in converting amino acids to proteins. Dounce (22) suggests that nucleic acids are first phosphorylated at each nucleotide, at the expense of ATP, forming a series of high-energy pyrophosphate links. Such a high-energy template could then not only dictate the specific structure of the protein or nucleic acid whose synthesis it catalyzes but supply the energy requirement as well.

STARVATION

Starvation, which is endemic through the major portion of this planet and has been so through all human history, is primarily an aspect of protein insufficiency in the diet. Leaving inorganic ions and micronutrients, such as polyunsaturated fatty acids and vitamins, to one side, it is quite possible (though expensive) for a human being to subsist entirely on a diet of protein. Carbohydrates and lipids, on the other hand, although valuable as energy sources, cannot replace protein. At most, they can spare protein by reducing the degree to which amino acids need be degraded for energy and increasing the degree to which they can be retained for incorporation into structural and metabolic protein. Tissue growth and tissue maintenance depend upon protein synthesis.

Minimal Requirement of Protein

The body has a certain ability to adapt itself to deficiencies in dietary protein. In young people, for instance, growth rate is reduced. Fetal growth takes place at the expense of maternal tissue where dietary protein is restricted.

During prolonged fasting the excreted nitrogen can fall below 3 grams per day, corresponding to less than 20 grams of protein. Twenty grams of protein per day would thus seem to be the irreducible minimum but it has never been shown that as little as that is sufficient for a maintenance diet.

The Food and Nutrition Board of the National Research Council recommends 1 gm. of protein per day per kilogram body weight for adults, and greater amounts in younger people, recommending up to 4 gm. per day per kilogram body weight for infants. In the latter half of pregnancy, the recommendation rises to 1.6 gm. per day per kilogram body weight and during lactation the corresponding figure is 2 grams. Elderly people are more liable to protein deficiency than younger people and 2 grams per day per kilogram body weight is recommended (40). In all these cases, it is presumed that the protein contains adequate proportions of all the dietarily essential amino acids, and it is usual to recommend that half to three-quarters of the ingested protein be of animal origin. The daily requirements of the various dietarily essential amino acids have been listed in Chapter 12.

Protein Deficiency

The prolonged use of a diet inadequate in protein leads inevitably to growth failure in children, and to loss of body protein in adults. Actual loss of weight may not occur if the diet is calorically adequate or excessive, and fat may be stored as adipose tissue even in protein-deficient states. Nutritional edema (see page 669) may mask losses of true body weight. Chemical studies of severely underproteinized persons will show low levels of hemoglobin and of plasma protein, although these findings may be minimized by decreased plasma volume.

Protein deficiency may occur not only from inadequate protein in the diet, or use of proteins lacking essential amino acids, but also from pathological states affecting the digestion, absorption, or utilization of proteins. Examples of such causes of protein deficiency, occurring in patients to whom adequate diets are economically available, would be obstructions or functional derangements of the gastrointestinal tract, and the metabolic alterations produced by infections, liver damage, or hyperthyroidism.

Intravenous feeding with proteins is a directly practicable procedure if the protein used is a plasma protein, preferably albumin, from the same species as the recipient. Dogs can be kept in nitrogen balance by the intravenous injection of adequate amounts of dog plasma as the sole source of nitrogen. The biological half-life of intravenously injected globulin (from the same species and tagged with I^{131}) is 4.6 ± 0.8 days for rabbits, 13.1 ± 2.8 days for human adults, and 20.3 ± 4.2 days for children under 8 years of age. Foreign proteins behave similarly until antibodies are produced (see page 845), at which time their rate of disappearance is accelerated (16). Human plasma albumin, in amounts of 37.5 grams daily by vein, has maintained nitrogen balance in human subjects for as long as 16 days (24), with no other nitrogenous food. Larger amounts were necessary when plasma albumin was fed. These results imply that amino acids stored as plasma protein are nutritionally available. It is not yet clear whether or not complete hydrolysis of plasma protein precedes such nutritional utilization. High cost makes intravenous alimentation with human albumin undesirable for general use. Protein hydrolysates, which contain a mixture of amino acids, are less effective weight for weight than plasma albumin in maintaining nitrogen balance, but are definitely more economical.

Hypoproteinosis of childhood. Present-day American parents have been well indoctrinated with the necessity for supplying vitamins in the diets of their children, and for supplying minerals, particularly calcium. Lynch and Snively (46) report that many children seen in office practice have been fed chiefly on milk and carbohydrate food, to the neglect of meat and eggs. Milk contains protein, but in relatively dilute solution, so that more than a quart a day is required for adequate protein intake in older children.

Such underproteinized children tend to lack appetite, to fail in normal growth, and to be subject to frequent gastrointestinal and other ailments. Exclusion or strict limitation of milk, with insistence upon more concentrated forms of protein at every meal, has been a successful therapeutic approach.

Effects of Prolonged Fasting

Terence MacSwiney, the mayor of Cork, fasted 74 days in protest during his incarceration, and died of starvation. This is the longest period of complete fasting of which we have accurate knowledge as to the exact duration.

TABLE XVI-2

Sunderman's observations on a man who had fasted 45 days

DATE	DATE IN RELATION TO DAY OF FAST	BODY WEIGHT	SERUM COMPONENTS															URINE EXCRETION	
			Chloride	CO_2	Inorganic phosphorus	Total base	Calcium	Magnesium	Total protein	Albumin	Globulin	Total cholesterol	Esterified cholesterol	Urea nitrogen	Uric acid	Creatinine	Sugar (serum)	Gonadotrophins	17-ketosteroids
		lb.	*mEq./l.*	*vols. %*	*mgm./100 ml.*	*mEq./l.*	*mgm./100 ml.*	*mEq./l.*	*gm./100 ml.*	*gm./100 ml.*	*gm./100 ml.*	*mgm./100 ml.*	*mgm./100 ml.*	*mgm./100 ml.*	*mgm./100 ml.*	*mgm./100 ml.*	*mgm./100 ml.*	*mouse u./24 hrs.*	*mgm./24 hr.*
4/15	41st day of fast	101		—	—	—	—	—	—	—	—	—	—	—	—	—	—	none	—
4/19	45th day of fast	97	74.0	100	3.4	141	12.2	2.7	6.2	5.0	1.2	198	184	24	9.0	1.4	63		
4/24	5th day after fast	113	107.8	63	—	146	10.7	1.6	5.4	2.8	2.6	—	—	14	1.1	1.0	—		
6/1	43rd day after fast	134	98.7	61	—	148	9.8	1.9	6.7	4.0	2.7	190	110	12	4.1	1.0	83	7.2	4.7
Normal range of values . . .			99–104	55–60	3–4	143–148	9–11	1.9–2.0	6–7	3.3–4.3	2.2–2.8	170–190	90–114	9–17	3–4	1.0–1.2	80–110	4–16	8–15

Sunderman (72) has reported studies on a subject who had fasted 45 days. The loss in weight during the fast was 40.5 pounds, starting at 137.5 and ending at 97. No chemical studies were made before or during the fast, which was undertaken for religious reasons. Biochemical data at the close of the fast were compared with similar measurements made 43 days after customary food intake had been resumed. Some of the results are shown in table XVI-2. Note that at the end of the fast Cl^- had been lost from the blood plasma and replaced by HCO_3^-, and that the blood cholesterol was almost entirely in the ester form. For another study of starvation in an individual case, see (10).

General Metabolic Phenomena of Starvation

Half of the world is chronically underfed. This half of the world's population subsists on diets in which cereals or potatoes make up 80 to 100 per cent. Such a diet contains an inadequate quantity of proteins, and the proteins there contained are of relatively poor nutritive quality. In this large population group, there is found both *subnutrition*, which is the lack of an adequate quantity of food, and *malnutrition*, which involves deficiency of food factors.

During starvation, the body, not receiving enough calories for maintenance of body structure and normal activity, draws upon its stores of fats and carbohydrates. Survival under conditions of complete starvation depends, to a major degree, upon the amount of adipose tissue available. Glycogen stores are usually exhausted within 24 hours. As first the carbohydrates and then the fats are depleted, the body's proteins are progressively drawn upon and wasting of tissue occurs. On a cellular level, changes in composition, at least in the early stages, are confined to the cytoplasm (30). In early stages of starvation, decreases in the protein content of spleen, liver, and genitalia can be observed. Skeletal muscle is drawn upon at later stages, while the heart, adrenals, and central nervous system show the least protein loss of all tissues. Amino acids derived from less significant tissues are apparently used for the maintenance of these essential structures. If starvation proceeds to the stage of loss of muscle proteins, creatine appears in increased amount in the urine.

The basal metabolic rate in states of prolonged severe starvation decreases below the value predicted on the basis of age, weight, height, and similar criteria. Such depressions of basal metabolic rate may fall as low as 70 per cent of normal and still be reversible by adequate food intake, particularly of protein foods.

Ketosis can be brought about by periods of fasting as short as 24 hours, and is certain to appear within the first few days of acute starvation. Such ketosis is mild compared to that of severe diabetes, since the mechanisms for the oxidative utilization of foodstuffs are not impaired or overloaded, as they are in the diabetic. After the exhaustion of stored glycogen, the body proteins may in part be converted into glycogen. The ketone acids are excreted partly as ammonium salts and partly paired with alkali metal cations. Sodium loss is most noticeable early in the ketosis of starvation. Later there is potassium loss paralleling the atrophy of muscles and other tissues.

As starvation progresses, other nitrogenous components of the body decrease along with the tissue proteins. The blood plasma, for instance, which normally shows a total protein concentration consisting of about 4.7 per cent albumin and 2.5 per cent globulin (A/G ratio of about 1.9), shows values decreased to 2 to 3 per cent albumin and a globulin of around 2.5, yielding an A/G ratio of about 1.0 or less. It will be noted that the albumin decreases more than does the globulin, possibly reflecting a greater significance to the organism of the globulin fraction. A frequent and almost invariable consequence of severe protein deficiency is edema. This is often so pronounced that the victims, instead of being thin as expected, have distended abdomens. The development of such edema of starvation is in part the result of the lowered colloid osmotic pressure of the plasma, which in turn results from the decrease in plasma albumin. The non-protein nitrogenous constituents of the blood are not generally affected, but may show moderate increase in prolonged starvation. This may, in part, be attributed to diminished renal function as a result of protein loss from the kidneys. The amino acid nitrogen of the blood remains surprisingly constant. Creatinine of blood or urine shows no consistent change, although creatinuria may occur as has already been mentioned.

The total nitrogen of the urine, of course, is much decreased since the source of the nitrogenous components is from protein metabolism. Urinary urea decreases along with the total excreted nitrogen, and in starvation is proportionally decreased even more, falling from a normal 90 per cent of the total to about 70 per cent. The ammonia in the urine is increased, consistent with the increase of excretion of ketone acids.

As the protein supplies of the body fall, and with them the plasma protein level, it might be expected that the resistance of the body to disease would decrease, since antibodies (see Chapter 22) have been identified with the gamma globulins. Observations made during World War II on man did not bear out this contention. The body appears to continue to be able to produce antibodies even when it is prevented by inadequate protein supplies in the diet from making normal amounts of other proteins.

REFERENCES

1. Albanese, A. A. Effect of a lysine-poor diet on the composition of human plasma proteins. J. Biol. Chem., **200:** 787–792, 1953.

1a. Alexander, N., and Greenberg, D. M. The purification and properties of the serine-forming enzyme system. J. Biol. Chem., **220:** 775–785, 1956.

2. Arntein, H. R. V., and Keglević, D. A comparison of alanine and glucose as precursors of serine and glycine. Biochem. J., **62:** 199–205, 1956.

3. Bach, S. J. *The Metabolism of Protein Constituents in the Mammalian Body*. Oxford, England, Clarendon Press, 1952.

4. Bach, S. J., and Smith, M. Role of glutamine in urea synthesis. Nature, **176:** 1126–1127, 1955.

5. Berg, C. P. Physiology of the D-amino acids. Physiol. Rev., **33:** 145–189, 1953.

6. Bessman, S. P., and Bessman, A. N. The cerebral and peripheral uptake of ammonia in liver disease with an hypothesis for the mechanism of hepatic coma. J. Clin. Invest., **34:** 622–628, 1955.

7. Binkley, F., *et al.* Pyridoxine and the transfer of sulfur. J. Biol. Chem., **194:** 109–113, 1952.

8. Borsook, H., *et al.* The formation of glycocyamine in man and its urinary excretion. J. Biol. Chem., **138:** 405–410, 1941.

9. Boyd, W. C. *Fundamentals of Immunology.* 3rd edit. New York, Interscience, 1956.

10. Brožek, J. Starvation and nutritional rehabilitation. A quantitative case study. J. Am. Dietet. Assoc., **28:** 917–926, 1952.

10a. Calkins, W. G. Blood ammonia in normal persons. J. Lab. Clin. Med., **47:** 343–348, 1956.

11. Cantoni, G. L. S-Adenosylmethionine: a new intermediate formed enzymically from L-methionine and adenosine triphosphate. J. Biol. Chem., **204:** 403–416, 1953.

12. Caputto, R. The biological transformation: creatine into creatinine. Arch. Biochem. Biophys., **52:** 280–281, 1954.

13. Carroll, W. R., *et al.* α-Ketobutyric acid as a product of the enzymatic cleavage of cystathionine. J. Biol. Chem., **180:** 375–382, 1949.

14. Clark, C. T. 5-Hydroxytryptophan decarboxylase: preparation and properties. J. Biol. Chem., **210:** 139–148, 1954.

15. Coon, M. J. The metabolic fate of the isopropyl group of leucine. J. Biol. Chem., **187:** 71–82, 1950.

16. Coons, A. H. Labelled antigens and antibodies. Ann. Rev. Microbiol., **8:** 333–352, 1954.

17. Crandall, D. I. Homogentisic acid oxidase. II. Properties of the crude enzyme in rat liver. J. Biol. Chem., **212:** 565–582, 1955.

18. Dalgliesh, C. E., and Tekman, S. Excretion of kynurenine and 3-hydroxykynurenine by man. Biochem. J., **56:** 458–463, 1954.

19. Davis, B. D. Some aspects of amino acid biosynthesis in microörganisms. Fed. Proc., **14:** 691–695, 1955.

20. Demoss, J. A., *et al.* The enzymatic activation of amino acids via their acyl-adenylate derivatives. Proc. Natl. Acad. Sci. U. S., **42:** 325–332, 1956.

21. Dent, C. E., *et al.* The pathogenesis of cystinuria. I. Chromatographic and microbiologic studies of the metabolism of sulfur-containing amino acids. J. Clin. Invest., **33:** 1210–1215, 1954.

22. Dounce, A. L. Duplicating mechanism for peptide chain and nucleic acid synthesis. Enzymologia, **15:** 251–258, 1952.

23. Du Vigneaud, V., *et al.* The biological synthesis of "labile methyl groups." Science, **112:** 267–271, 1950.
24. Eckhart, R. D., *et al.* Comparative studies on the nutritive value of orally and intravenously administered human serum albumin in man. J. Clin. Invest., **27:** 119–134, 1948.
24a. Edwards, S. W., and Knox, W. E. Homogentisate metabolism: the isomerization of maleylacetoacetate by an enzyme which requires glutathione. J. Biol. Chem., **220:** 79–91, 1956.
25. Elliott, W. H. Studies on the enzymic synthesis of glutamine. Biochem. J., **49:** 106–112, 1951.
26. Fox, S. W., *et al.* Enzymic synthesis of peptide bonds. VI. The influence of residue type on papain-catalyzed reactions of some benzoylamino acids with some amino acid anilides. J. Am. Chem. Soc. **75:** 5539–5543, 1953.
27. Fruton, J. S., *et al.* Elongation of peptide chains in enzyme-catalyzed transamidation reactions. J. Biol. Chem., **190:** 39–53, 1951.
28. Fruton, J. S., and Simmonds, S. *General Biochemistry.* New York, John Wiley and Sons, 1953.
29. Harper, A. E., *et al.* L-Leucine, an isoleucine antagonist in the rat. Arch. Biochem. Biophys., **57:** 1–12, 1955.
30. Harrison, M. F. Effect of starvation on the composition of the liver cell. Biochem. J., **55:** 204–211, 1953.
31. Haskins, F. A., and Mitchell, H. K. Evidence for a tryptophane cycle in Neurospora. Proc. Natl. Acad. Sci., **35:** 500–506, 1949.
32. Herrman, R. L., *et al.* The synthesis of purines and thymine from methionine in the rat. J. Am. Chem. Soc., **77:** 1902–1904, 1955.
33. Hope, D. B. Pyridoxal phosphate as the coenzyme of the mammalian decarboxylase for L-cysteine sulfinic acid and L-cysteic acids. Biochem. J., **59:** 497–500, 1955.
34. Horner, F. A., and Streamer, C. W. Effect of a phenylalanine-restricted diet on patients with phenylketonuria. J. A. M. A., **161:** 1628–1630, 1956.
35. Jackson, H. F., and Clark, B. E. Cystinosis. A. M. A. Am. J. Dis. Children, **85:** 531–544, 1953.
36. Johnston, R. B., *et al.* Catalysis of transpeptidation reactions by chymotrypsin. J. Biol. Chem., **187:** 205–211, 1950.
37. Jones, M. E., *et al.* Carbamyl phosphate, the carbamyl donor in enzymatic citrulline synthesis. J. Am. Chem. Soc., **77:** 819–820, 1955.
38. Keston, A. S., and Wortis, S. B. The antagonistic action of choline and its triethyl analogue. Proc. Soc. Exper. Biol. & Med., **61:** 439–440, 1946.
39. Knox, W. E., and Edwards, S. W. The properties of maleylacetoacetate, the initial product of homogentisate oxidation in liver. J. Biol. Chem., **216:** 489–498, 1955.
40. Kountz, W. B., *et al.* Nitrogen balance studies under prolonged high nitrogen intake levels in elderly individuals. Geriatrics, **3:** 171–183, 1948.
40a. La Du, B. N. Jr., and Zannoni, V. G. The tyrosine-oxidation system of liver. III. Further studies on the oxidation of *p*-hydroxyphenylpyruvic acid. J. Biol. Chem., **219:** 273–281, 1956.
41. Lang, K., and Schmid, G. Proline oxidase. Biochem. Ztschr., **322:** 1–8, 1951.
41a. Levin, Y., *et al.* Hydrolysis and transpeptidation of lysine peptides by trypsin. Biochem. J., **63:** 308–316, 1956.
42. Lien, O. G., Jr., and Greenberg, D. M. Identification of alpha-aminobutyric acid enzymically formed from threonine. J. Biol. Chem., **200:** 367–371, 1953.

43. LEMON, H. M., *et al.* Abnormal glycine metabolism in rheumatoid arthritis. J. Clin. Invest., **31:** 993–999, 1952.
44. LERNER, A. B. Metabolism of phenylalanine and tyrosine. Advances Enzymol., **14:** 73–128, 1953.
45. LIPMANN, F. Mechanism of peptide bond formation. Fed. Proc., **8:** 597–602, 1949.
46. LYNCH, H. D., AND SNIVELY, W. D., JR. Hypoproteinosis of childhood. J. A. M. A., **147:** 115–119, 1951.
47. MACKENZIE, C. G., AND DU VIGNEAUD, V. Biochemical stability of the methyl group of creatine and creatinine. J. Biol. Chem., **185:** 185–189, 1950.
47a. MATSUO, Y., *et al.* Metabolic pathways of homoserine in the mammal. J. Biol. Chem., **221:** 679–687, 1956.
47b. MITOMA, C. Partially purified phenylalanine hydroxylase. Arch. Biochem. Biophys. **60:** 476–484, 1956.
48. MUNTZ, J. A. The inability of choline to transfer a methyl group directly to homocysteine for methionine formation. J. Biol. Chem., **182:** 489–499, 1950.
49. PIHL, A., AND FRITZSON, P. The catabolism of C^{14}-labeled β-alanine in the intact rat. J. Biol. Chem., **218:** 345–351, 1955.
50. RATNER, S. Urea synthesis and metabolism of arginine and citrulline. Advances Enzymol., **15:** 319–387, 1954.
51. RATNER, S., AND PETRACK, B. The mechanism of arginine synthesis from citrulline in kidney. J. Biol. Chem., **200:** 175–185, 1953.
52. RAVDIN, R. G., AND CRANDALL, D. I. The enzymatic conversion of homogentisic acid to 4-fumarylacetoacetic acid. J. Biol. Chem., **189:** 137–149, 1951.
53. RIMINGTON, C. Porphyrins. Endeavour, **14:** 126–135, 1955.
54. ROBERTS, E., AND BREGOFF, H. M. Transamination of γ-aminobutyric acid and β-alanine in brain and liver. J. Biol. Chem., **201:** 393–398, 1953.
55. ROBINSON, W. G., *et al.* Tiglyl coenzyme A and α-methylacetoacetyl coenzyme A; intermediates in the enzymatic degradation of isoleucine. J. Biol. Chem., **218:** 391–400, 1956.
55a. ROBINSON, W. G., *et al.* Coenzyme A thiol esters of isobutyric, methacrylic, and β-hydroxyisobutyric acids as intermediates in the enzymatic degradation of valine. J. Biol. Chem., **224:** 1–11, 1957.
56. ROSE, W. C., *et al.* The utilization of the nitrogen of ammonium salts, urea and certain other compounds in the synthesis of non-essential amino acids *in vivo*. J. Biol. Chem., **181:** 307–316, 1949.
57. ROSENTHAL, R. L., *et al.* Hematologic changes in rats protected by cysteine against total body x-irradiation. Am. J. Physiol., **166:** 15–19, 1951.
58. ROTHSTEIN, M., AND MILLER, L. L. The metabolism of L-lysine-6-C^{14}. J. Biol. Chem., **206:** 243–253, 1954.
59. ROTHSTEIN, M., AND MILLER, L. L. The conversion of lysine to pipecolic acid in the rat. J. Biol. Chem., **211:** 851–858, 1954.
60. SALLACH, H. J., *et al.* The *in vivo* conversion of glutamic acid into proline and arginine. J. Am. Chem. Soc., **73:** 4500, 1951.
61. SCHAYER, R. W. Biogenesis of histamine. J. Biol. Chem., **199:** 245–250, 1952.
61a. SCHAYER, R. W., *et al.* Ring nitrogen methylation; a major route of histamine metabolism. J. Biol. Chem., **221:** 307–313, 1956.
62. SCOTT, E. B., AND SCHWARTZ, C. Histopathology of amino acid deficiencies. II. Threonine. Proc. Soc. Exper. Biol. Med., **84:** 271–276, 1953.
63. SHEMIN, D., *et al.* The succinate-glycine cycle. I. J. Biol. Chem., **215:** 613–626, 1955.

64. SIEKEVITZ, P., AND GREENBERG, D. M. The biological formation of formate from methyl compounds in liver slices. J. Biol. Chem., **186:** 275–286, 1950.
65. SINEX, F. M., AND VAN SLYKE, D. D. The source and state of the hydroxylysine of collagen. J. Biol. Chem., **216:** 245–250, 1955.
66. SINGAL, S. A., *et al.* The effect of threonine deficiency on the synthesis of some phosphorus fractions in the rat. J. Biol. Chem., **200:** 875–882, 1953.
67. SINGER, T. P., AND KEARNEY, E. B. A new pyridine nucleotide coenzyme for biological oxidations. Biochim. et Biophys. Acta, **8:** 700–701, 1952.
68. SINGER, T. P., AND KEARNEY, E. B. Pyridinenucleotide coenzymes. Advances Enzymol., **15:** 79–139, 1954.
68a. SINGER, T. P., AND KEARNEY, E. B. Intermediate metabolism of L-cysteine sulfinic acid in animal tissues. Arch. Biochem. Biophys., **61:** 397–409, 1956.
68b. SJOERDSMA, A., *et al.* Simple test for diagnosis of metastatic carcinoid (argentaffinoma). J. A. M. A., **159:** 397, 1955.
69. STERN, W. H. Excretion of amino acids in cystinuria. Proc. Soc. Exper. Biol. & Med., **78:** 705–708, 1951.
70. STEKOL, J. A., AND WEISS, K. On de-ethylation of ethionine in the rat. J. Biol. Chem., **185:** 577–583, 1950.
71. STETTEN, M. R. Some aspects of the metabolism of hydroxyproline, studied with the aid of isotopic nitrogen. J. Biol. Chem., **181:** 31–37, 1949.
72. SUNDERMAN, F. W. Studies in serum electrolytes. XIV. Changes in blood and body fluids in prolonged fasting. Am. J. Clin. Path., **17:** 169–180, 1947.
73. TAUBER, H. Synthesis of protein-like substances by chymotrypsin. J. Am. Chem. Soc., **73:** 1288–1290, 1951.
74. TIERNEY, N. A., AND PETERS, J. P. The mode of excretion of creatine and creatine metabolism in thryoid disease. J. Clin. Invest., **22:** 595–602, 1943.
75. TORDA, C., AND WOLFF, H. G. Glycocyamine elimination in patients with myas thenia gravis. J. Lab. & Clin. Med., **31:** 1174–1178, 1946.
76. TRISTRAM, G. R. The amino acid composition of proteins; *in* NEURATH, H., and BAILEY, K., eds. *The Proteins*. Vol. I, Part A. New York, Academic Press, 1953.
76a. UDENFRIEND, S., *et al.* Biogenesis and metabolism of 5-hydroxyindole compounds. J. Biol. Chem., **219:** 335–344, 1956.
77. WAELSCH, H. Glutamic acid and cerebral function. Advances Prot. Chem., **6:** 301–342, 1951.
78. WEIL-MALHERBE, H. Significance of glutamic acid for the metabolism of nervous tissue. Physiol. Rev., **30:** 549–568, 1950.
79. WILDER, V. M., AND MORGULIS, S. Creatinuria in normal males. Arch. Biochem. Biophys., **42:** 69–71, 1953.
80. WILLIAMS, J. N., JR., AND ELVEHJEM, C. A. The effects of tryptophan deficiency upon enzyme activity in the rat. J. Biol. Chem., **183:** 539–544, 1950.
80a. WRIGHT, B. E. The role of polyglutamyl pteridine coenzymes in serine metabolism. II. A comparison of various pteridine derivatives. J. Biol. Chem., **219:** 873–883, 1956.
81. WRÓBLEWSKI, F. The significance of alterations in serum glutamic oxaloacetic transaminase in experimental and clinical states. Trans. N. Y. Acad. Sci., Ser. II., **18:** 444–450, 1956.
82. WRÓBLEWSKI, F., AND LADUE, J. S. Serum glutamic pyruvic transaminase in cardiac and hepatic disease. Proc. Soc. Exptl. Biol. Med., **91:** 569–571, 1956.

CHAPTER 17

Electrolytes and Water: Edema and Shock

In addition to the carbon, hydrogen, nitrogen, and sulfur which, together with oxygen, make up the metabolic pathways discussed in the last few chapters, certain other elements are essential in the diet. Absence or deficiency of these elements can check growth and cause specific losses of function. These elements are *sodium*, *potassium*, *chlorine*, *calcium*, *magnesium*, *phosphorus*, *iron*, *copper*, *cobalt*, *iodine*, *fluorine*, *zinc*, *molybdenum*, and *manganese*. These all can be utilized by the body if taken in the form of salts, hence can be expressed as ions. To this list of metabolically significant ions should be added *ammonium*, *carbonate*, and *sulfate*, which are derived by metabolic action upon foodstuffs. The *bromide* ion is present consistently in human plasma, but as far as we know it has no significance. Minute and variable amounts of many other elements are present in the body without known function.

Units of Concentration

Concentrations of these ions or of compounds involving them have been commonly stated in milligrams per 100 ml. More convenient and useful units are the milliequivalent, the millimol, and the milliosmol. The most useful of these is the *milliequivalent*. An equivalent, meaning an equivalent weight in grams, of an electrolyte or of an ion is the molecular weight of the compound, or the sum of the atomic weights of the atoms composing the ion, expressed in grams and divided by the electrovalence of the ion or by the number of electrons transferred in the formation of the electrovalent bond of the compound. For example, an equivalent of NaCl is 58.454 grams, one molecular weight, since only one electron is involved in its electrovalent bond. Similarly an equivalent of Na^+, K^+, H^+, Cl^-, or Br^- is one atomic weight in grams, and the equivalent of NH_4^+ is the sum of the atomic weights in grams of one nitrogen and four hydrogens. In the case of Na_2SO_4 the molecular weight in grams must be divided by two to give the equivalent and the same division must be made in the case of bivalent ions like SO_4^{--} and Ca^{++}. H_3PO_4 or $PO_4^{\equiv}$ must similarly have their weights divided by three to give the conventional equivalent. We will see later that a compromise is made in expressing the equivalent concentration of phosphate in the blood plasma since this acid is never completely neutralized under physiological conditions. The advantages of using this particular unit are: (a) that one equivalent of any acid will exactly combine with one equivalent of any base; and (b) in calculation of total acidities or basicities of mixtures, concentrations expressed in equivalents may be added together.

The milliequivalent, being the thousandth part of an equivalent, is of a size better adapted for the expression of concentrations in blood and body fluids of electrolytes and their ions. To convert the old-time units (mgm. per 100 ml.) into milliequivalents (mEq.) per liter, multiply by:

$$\frac{10 \cdot \text{number of electrons}}{\text{sum of atomic weights}}$$

It is now standard clinical practice to express concentrations of electrolytes and ions in mEq. per liter, and dosages of intravenous electrolytes in milliequivalents.

The *mol* is defined as the molecular weight in grams; the *millimol* as the thousandth part of a mol. These units are frequently used in expressing amounts and concentrations of non-electrolytes, particularly in physicochemical measurements.

The *osmol* is the mol divided by the number of ions formed in the dissociation of the electrolyte in question, and the milliosmol is its thousandth part. For non-electrolytes the osmol is equal to the mol. The osmolar or milliosmolar concentration of a solution of known composition has a fixed and calculable value; this value does not usually correspond exactly with the experimentally measured osmotic pressure. The osmotic pressure developed experimentally depends not only upon the total osmolar concentration but also upon the degree of semipermeability of the experimental membrane to the solutes involved.

The concept of *ionic strength* was defined in Chapter 1. It should perhaps be definitely stated that ionic strength is not identical with total ionic concentration expressed in any of the above units.

Body Fluid Compartments

Total water can be estimated in small animals by desiccation at low temperatures. In larger animals and man it is best measured by the intravenous injection of a known amount of a substance which is water-soluble and to which all cell membranes are permeable. The ideal test substances are the oxides of deuterium or tritium, but analytical procedures for these substances require elaborate instrumentation. Of non-isotopic test substances, antipyrine (72) has been most successfully used. With either antipyrine or deuterium oxide, about two hours must be allowed for equilibration. Then the concentration of the test substance is determined in the water of a sample of the subject's blood plasma.

$$\text{Liters total body water} = \frac{\text{mgm. test substance injected} - \text{mgm. excreted}}{\text{mgm. test substance per liter plasma water}}$$

For a review of methods of measuring body water and body fat, see (71a).

Steele (74) has measured the total body water of 51 male hospital patients and personnel, some old or obese, and found it to vary from 40 to 68 per cent (with a mean of 53 per cent). Similar measurements on 31 women gave results of from 30 to 53 per cent (mean 45 per cent). The variability is undoubtedly the result of the varying amount of adipose tissue, which is low in water content. The total water of the body constitutes about 70 per cent of the weight of the fat-free tissue, or "lean body mass".

The adult containing an average amount of adipose tissue will be found to contain about 35 liters (50 per cent of 70 kgm.) of total free body water. If a substance to which cell membranes are not permeable, such as radioactive sulfate (85), is used as the test substance, a similar calculation will give the volume of *extracellular fluid*. In our same average adult this volume will be about eleven liters. Subtracting this from the total free body water we get 24 liters as the approximate volume of the *intracellular fluid*. Again repeating the experiment, and using as a test substance an organic dye which does not escape quickly from the blood-vascular system, we can estimate the *blood plasma volume*. In our same average lean subject this will be about three liters. Subtracting this from the volume of extracellular fluid, we obtain about eight liters, the volume of *interstitial fluid*—fluid which lies around the tissue cells, but outside their walls and also outside the blood vessels. To summarize we find in the average (70 kgm.) adult:

Intracellular fluid		24 liters
Extracellular fluid		
Plasma	3 liters	
Interstitial fluid	8 liters	
Total extracellular fluid		11 liters
Total free body water		35 liters

Normal Water Balance

Water is freely diffusible throughout the body and moves from one compartment to another as impelled by hydrostatic and osmotic forces. The intracellular compartment (excluding blood cells and endothelial cells) can exchange only with interstitial fluid, which in turn can exchange with plasma. Water enters or leaves the body directly only to or from the plasma compartment. Water is produced, however, within cells by the oxidation of the hydrogen of metabolites; this is called metabolic water, or water of oxidation (60). Combining this with the interchanges of water between plasma and the outside world, a balance-sheet of water turnover can be set up. The approximate figures for our average 70-kgm. man would be:

Intake per 24 hours	liters	Output per 24 hours	liters
Fluids	1.2	Lungs	0.5
Water in food	1.0	Skin	0.5
Metabolic water	0.3	Urine	1.4
		Feces	0.1
Total	2.5		2.5

Although the figures given represent a purely hypothetical average subject, and even though the individual items are obviously subject to wide variation, certain significant facts are illustrated. 1) The normal person is in a state of water equilibrium, with neither a positive balance (water gain) nor a negative balance (water loss) over the 24 hours. 2) The state of water balance can not be accurately established by the crude comparison of fluids taken in and urine excreted.

In addition to the equilibrium of body water as determined by intake and output, there is an osmotic equilibrium which concerns the several compartments. Water can move freely from any compartment to one adjacent, hence the compartments tend to have the same total osmotic pressure, which has the rather astonishing value of 7.9 atmospheres or 6000 mm. of mercury. This enormous force is never available, since many solutes also pass freely through compartmental partitions. One substance which does not pass freely is protein. The intracellular fluid is high in protein content, and the cellular boundary is impermeable under physiological conditions to protein. The outlying interstitial fluid is low in protein. This difference in concentration of a non-diffusible substance conditions an osmotic flow of water and diffusible solutes from interstitial fluid to the interior of the cell, which is checked by the limit of the capacity of the cell. A distensive force is thereby applied to the cell, which is opposed by the compressive force of the stretched elastic cell boundary and the surrounding elastic connective tissue. In addition to water brought in by osmotic force, metabolic water is also produced within cells by the oxidation of the hydrogen of metabolites. By successive slight alterations of the forces making up this equilibrium, an ebb and flow of cell water can be postulated, bearing with it soluble nutrients and waste products. Assuming purely static conditions, water and other substances to which the cell wall is permeable enter and leave the cell by diffusion.

Exchanges between blood plasma and interstitial fluid are less restrained. The capillaries are not uniformly impermeable to protein; depending upon anatomical location, the capillary wall varies from an almost complete barrier to proteins (as in the ciliary body of the eye) to a highly permeable meshwork (as in the hepatic sinusoids). The protein concentration of interstitial fluid is therefore variable in different organs but always less than that of blood plasma. This difference in *colloid osmotic pressure* conditions an osmotic flow of water and diffusible solutes from interstitial fluid to plasma. This flow is augmented by tissue pressure and opposed by capillary blood pressure. On account of frequent small variations in these pressures true equilibrium is probably never established in living tissues. It seems legitimate to state that despite minor fluctuations the relative volumes of plasma and interstitial fluid remain within physiological limits as a result of a balance of these four pressures—colloid osmotic pressure of plasma and tissue pressure opposing capillary pressure and colloid osmotic pressure of the interstitial fluid.

Since blood pressure is higher as arterial blood enters the capillary, ejection of fluid occurs through the capillary wall. Pressure falls along the capillary, allowing a re-entry of fluid as a result of osmotic force. It is important to consider this exchange of fluid as a rapid and dynamic turnover rather than simply as a balance of pressures. The rapidity of exchange between plasma and extracellular fluid is indicated by the measurement of the diffusion of short-life radioactive Na^{24} across the vascular wall (7). The results in human subjects indicated that 32 per cent of the total plasma Na^+ leaves the capillary blood every minute.

Dehydration and Thirst

When water intake is stopped, water output is diminished. The urinary volume will fall to about 500 ml. per 24 hours, but the output through skin and lungs will not be significantly altered. These unescapable water losses add up to about 1500 ml. per day. Since the channels of water loss lead directly from the plasma, this compartment will be the first to lose water; this water is replaced promptly from the interstitial fluid. Since water moves freely among the compartments, as soon as there is a significant increase in concentration of extracellular solutes, water will move from the intracellular fluid to the extracellular fluid as a result of the difference in concentration. Hence in water deprivation there will be loss of water from all compartments, osmotic pressures will remain equal in all compartments; but since the intracellular compartment contains the most water, it will lose the most.

Thirst is a sensation resulting from loss of cellular water or the presence of an excess in the body of a solute in hypertonic concentration (1). It is the chief symptom for the first two days of water deprivation in an otherwise healthy person, and increases in intensity with continuing water lack. By the third day dehydration of muscular and neural cells has reached a degree where such cells show functional changes, manifested in the patient by weakness and confusion. Disability increases with progressive cellular dehydration. Death occurs when 15 per cent of the body weight has been lost, usually within ten days of complete water deprivation.

Dehydration is a constant threat to the patient with *diabetes insipidus*. Reabsorption of water in the renal tubules fails in this disease on account of deficiency of the postpituitary antidiuretic substance. Verney (82) has shown in dogs that liberation of the antidiuretic substance from the neurohypophysis follows the stimulation of osmoreceptors in the central nervous system by intracarotid injections of hypertonic solutions of NaCl. This water-conserving mechanism normally goes into action whenever the salt concentration is increased. In diabetes insipidus this mechanism fails and water is maximally excreted regardless of the concentration of the blood plasma. The daily urine volume exceeds 6 liters; the specific gravity is 1.006 or less. The patient becomes rapidly dehydrated, suffers intense thirst, and gains relief by drinking very large volumes of water. Diabetes insipidus is evidence of damage by injury or disease to a functional unit which includes the posterior lobe of the pituitary, the paraventricular and supraoptic nuclei in the hypothalamus, and the supraopticohypophyseal tract. The disease may result from damage to the osmoreceptors, the postpituitary itself, or the interconnecting tract.

Salt intoxication is comparable to water depletion. Excretion of salt by the human kidneys requires a liter of water for each 300 mEq. of sodium chloride excreted. Drinking salt solutions of greater concentration than this excretory limit results in loss of the water obligated in the excretion of the excess salt. The desperate expedient of augmenting scanty supply of drinking water with urine is therefore of no physiological advantage whatever. The use of sea water for drinking in similar desperate situations is not only useless but actually harmful since 1) sea water is more concentrated than the physiological excretion limit, so that extra water is removed from the body for its excretion, and 2) its use is likely to produce vomiting and diarrhea with extra water losses. The drinking of alcoholic liquors under circumstances of water deficiency is also harmful, since alcohol produces water loss by inhibition of the pituitary antidiuretic mechanism (44). The symptoms of salt ingestion in excess of water intake are identical with those of water depletion, and arise by the same mechanisms from loss of cellular fluid.

Excessive sweating involves some loss of salt but a greater loss of water, since sweat is a hypotonic solution with respect to plasma. The rate of sweating may reach 3 liters/hour under extreme conditions and total daily sweat losses of 10 to 12 liters have been reported for men working in the desert (66). Sweating without compensatory water intake produces a situation comparable to water deprivation, and will accelerate the progress of the syndrome of water deprivation. More commonly, losses by sweating are made up by increasing the water intake. In this situation, salt (NaCl) deficiency can develop. A similar cause of salt deficiency is loss of *gastro-intestinal fluids* when replacement is made by water without salt. The salt concentration of the fluids lost by diarrhea, vomiting, or surgical drainage is comparable to that of extracellular fluid. Deficiency of the adrenal cortical hormones (Addison's disease) is characterized by excessive excretion of Na^+; in ketonuria from diabetes or starvation there is similar depletion since the ketone acids are excreted partly as sodium salts, and the loss may be augmented by vomiting; polyuria from other causes, such as chronic nephritis, may involve salt loss.

Patients with *salt depletion* from any of the causes mentioned show a consistent grouping of symptoms. In all instances, there is diminished volume of extracellular fluid both in the plasma and interstitial compartments. There is no warning sensation comparable to thirst. Urine volume is not consistently diminished as in water deprivation; the Na^+ and Cl^- output in the urine is greatly decreased. Early symptoms are weakness and faintness passing into stupor; appetite is lost early while vomiting complicates the more advanced stages and adds to the salt loss. If this vicious cycle is not checked, the patient goes into an apathetic confused state with decreased blood pressure and increased pulse rate. Death results from the diminished volume and increased viscosity of the blood.

Water intoxication (heat cramps or miner's cramps) occurs when much water is taken by a person with salt depletion or without compensatory salt intake. The usual story is one of rapidly drinking large volumes of water after severe or protracted sweating. It has been reported in hospital patients who have been given dilute (isotonic) glucose solution intravenously faster than their kidneys could handle the water. The cells acquire excess water, while extracellular fluid volume is diminished (45). Prevention and treatment consist of replacing the deficient salt.

Edema

Edema is the retention of both water and electrolytes—an increase in the volume of interstitial fluid. In amount it may involve as much as 100 pounds of water and a pound of sodium chloride (57). It may be local or general; it may permeate connective tissue or accumulate in cavities. Numerous physical and chemical mechanisms are known to have a part in its formation. These mechanisms may act singly or in any combination.

Increased venous and capillary blood pressure may result from cardiac insufficiency or venous obstruction. Excessive fluid is filtered through the capillary wall. Edema of cardiac origin is characterized by its mobility—it responds to gravity and sinks to the lower parts of the body, shifting slowly after changes of position. It may also gather in the pleural and peritoneal cavities. The localization of the edema of venous obstruction is determined by the site of the obstruction, and is less influenced by gravity. *Ascites*, accumulation of fluid in the peritoneal cavity, in cirrhosis of the liver is in part the result of constriction of the portal circulatory bed by scar tissue which has replaced normal liver structure.

Decreased plasma protein diminishes the colloid osmotic pressure of the plasma and permits greater net outflow to the interstitial fluid. This mechanism is also operative in cirrhosis of the liver. On account of its greater concentration in plasma and lower molecular weight, albumin has more osmotic effect than globulin. In cirrhosis and other diseases involving destruction of liver cells, plasma albumin decreases faster than globulin. Typically a hypoproteinemia develops, with increased proportion of globulin to albumin—in clinical jargon, a decreased or reversed A/G ratio. This mechanism is probably responsible for ascites in early stages of cirrhosis; portal obstruction adds to the difficulty later.

The edema of *malnutrition* is also the result of failure to produce plasma albumin; in this instance the lack is of protein food. The plasma volume is diminished, which gives to the blood analyst a falsely optimistic view of the nutritional state by partly compensating for the greatly depressed levels of albumin, hemoglobin, and red cells.

The *nephrotic syndrome* is characterized by low levels of plasma protein and the excretion of large amounts of albumin in the urine. The syndrome may exist by itself, or as a stage in the development of several varieties of renal disease. Nephrotic edema is widespread through the body, as indicated by measurements of extracellular fluid volume. The rate of protein loss in the urine is usually adequate to explain the hypoproteinemia and the resulting edema, but this is not always the case and there are a number of unresolved questions about this disease.

Decreased elimination of water and electrolytes occurs in acute nephritis as a result of decreased glomerular filtration and in cardiac insufficiency as a result of decreased renal blood supply. The ions involved are those of extracellular fluid, particularly sodium.

Increased permeability to protein of the capillary wall effectively decreases the colloid osmotic pressure of the plasma even in the absence of hypoproteinemia. This is probably the mechanism of the formation of sharply localized areas of edema, for example, the ordinary mosquito bite or the sudden *angioneurotic edema* which is a common manifestation of clinical allergies. It may also take part in the formation of the edema of nephritis; in this disease there are visible signs of damage to blood vessels.

Lymphatic obstruction causes localized edema; the fluid is usually high in protein content, since the water and ions can be more readily reabsorbed into the blood capillaries. The edema is localized since that portion which moves by force of gravity or of increased tissue pressure away from the obstructed area is promptly absorbed.

It is plain that the multiplex disturbances which produce edema can not be corrected by any single therapeutic measure. In a particular patient, several of the mechanisms may be working together. A malnourished patient, for example, may be edematous both from hypoproteinemia and from cardiac insufficiency of thiamine deficiency. Aside from the correction of dietary deficiencies, effective therapeutic measures include 1) improvement of cardiac efficiency by adequate nutrition, or by cardiac drugs such as digitalis; 2) decrease of demand on the heart muscle by restricting activity; 3) restoration of plasma protein by injection of human plasma, or better, of plasma albumin concentrate (even in nephrotic edema where albumin is being wasted, this may decrease gastrointestinal edema and improve the appetite for and digestion of protein foods); 4) restriction of salt intake or use of ion-exchange resins by mouth to bind sodium ions and thereby limit their absorption from the gastrointestinal tract; 5) the use of diuretic drugs to increase urinary output of water and salts; 6) chemotherapy and other effective measures against infection; 7) surgical removal or by-passing of venous or lymphatic obstruction; and 8) surgical removal of collections of edema fluid (86).

Shock

Shock is a state in which all physiological functions are depressed and disorganized as a result of failure of adequate arterial blood supply from the heart. Shock can arise from many causes, most of which are physical rather than chemical in their nature. The changes produced by shock are both physical and chemical. Any form of damage to the body which can result in a decrease of the general arterial blood supply can cause shock. Blalock has classified shock according to the physiological system responsible for the primary disturbance: (a) hematogenic shock results from loss of circulating blood and the operative mechanism is the reduction of blood volume; (b) neurogenic shock results from physical damage to the nervous system or from the action of drugs upon the nervous system as in spinal anesthesia with the reduction of blood pressure the primary operative mechanism; (c) vasogenic shock follows vasodilatation mediated otherwise than through neural mechanisms with lowering of blood pressure again the effective mechanism as in histamine shock or anaphylactic shock; (d) cardiogenic shock where again lowered blood pressure is the effective mechanism but the cause is failure of the pumping action of the heart. Combinations of these mechanisms may occur; vasogenic shock may complicate any other type of shock by the release from liver and muscle under hypoxia of a vasodepressor material (VDM) which has been identified as ferritin. Whether the blood pressure fails or the blood volume, the series of events which follows is similar. The failure of normal arterial blood supply leads to stagnant hypoxia. There is an immediate depression of all physiological functions. Most notably, consciousness is lost and body temperature becomes subnormal. With continued hypoxia and metabolic depression, if the patient survives, there is a predominance of hydrolytic and anaerobic chemical processes and a depression of the anabolic, synthetic, and oxidative mechanisms. Blood sugar rises at first from the breakdown of glycogen, the rate of glycogenolysis being augmented by the liberation of adrenalin. Anaerobic glycolysis continues, but since the oxidative functions are depressed, there is accumulation of lactate and pyruvate. The high-energy adenosine polyphosphates diminish in all tissues, particularly the liver and kidney. Inosinic acid appears in the plasma. Renal function is depressed as indicated by increased levels of non-protein nitrogen in the circulating blood, while the accumulation of acid metabolites tends to depress the pH of blood and body fluids. Disintegration of tissue cells is evidenced by liberation of potassium, a typically intracellular ion, into the extracellular fluids and the blood plasma and by an increased content of the polypeptides of blood. Hepatic hypoxia depresses the deamination of amino acids. During this stage of tissue breakdown the body temperature tends to rise from its initial depressed level and usually to go above the normal. There is no evidence of extensive loss of cellular water in shock. Increase of plasma volume at the expense of interstitial fluid is a physiological response in compensatory opposition to the mechanisms which originally induced the shock state. Hemodilution is therefore considered a favorable sign in shock patients. It should be emphasized that even where there is no external bleeding, traumatic shock results in a primary loss of blood volume. In burns, plasma is lost by exudation through the burned area; in contused or crushed regions of the body, fluid is constantly escaping from the capillaries and accumulating in the injured region. Prediction of the water balance subsequent to the appearance of shock is unreliable. Extra fluid losses may occur through sweating, increased respiratory rate, and occasionally by vomiting. On the other hand, the depression of renal function usually keeps urinary outputs at a low value.

The cardinal points in the treatment of a patient in shock are a prompt decision implicating the physiological disturbances contributory to the shock state and the quickest possible restoration of adequate arterial blood supply. The longer a patient remains in a state of shock, the greater is the probability that the physiological and biochemical changes may become irreversible. It is, of course, imperative to check or minimize hemorrhage as promptly as possible. The use of the head-down, shock, or Trendelenburg position usually brings about an elevation of blood pressure which is desirable. Restoration of blood volume should be accomplished as promptly as possible, preferably with whole blood, second best with blood derivatives such as human blood plasma, human plasma albumin, human red cell suspension, or human ascitic fluid. The use of simple glucose or saline solutions, if blood or blood derivatives are not available, may restore consciousness and reverse the direction of the unfavorable physiological changes. Since these simple solutions do not contribute to the colloid osmotic pressure of the plasma, the benefits derived from their use are likely to be transitory. *Plasma expanders*, formerly misleadingly designated as "blood substitutes", are colloidal preparations which, by their osmotic effect, attract interstitial fluid into the blood vascular system (34). Although plasma expanders are not substitutes for blood or plasma, their use may carry a patient over a critical period until blood or plasma can be given. The colloidal material should be of such a particle size that it will not be rapidly excreted by the kidneys, but still small enough to exert adequate osmotic pressure. The chemical nature of the ideal plasma expander should permit slow metabolism rather than indefinite storage in blood and tissues. Certain preparations of *gelatin* appear to meet these specifications most closely. Other plasma expanders in tentative use are *dextran* (a glucose polysaccharide produced by *Leuconostoc mesenteroides*), and the synthetic *polyvinylpyrrolidone* or PVP.

Analysis of Compartmental Fluids

It is a simple matter to obtain a sample of blood plasma. It is next to impossible to secure samples of normal human interstitial fluid or of the intracellular fluid of tissues; analysis of these fluids has been by indirection. (a) Direct analysis of such cells as those of blood, pus, or sperm, which can be isolated from their surrounding fluids, indicates a fundamental difference in mineral composition from that of blood plasma. Na^+ and Cl^- are the chief ions of blood plasma; these ions are absent or in low concentration in the cells—in their place are found K^+ and HPO_4^{--}. This difference between cell and surrounding fluid, although subject to great quantitative variations, is consistent in all forms of life and in the lifeless earth. "Potassium is of the soil and not the sea; it is of the cell and not the sap" (22). (b) By making the assumptions that Cl^- is limited to extracellular fluid and that the concentration of Cl^- is the same in plasma and in interstitial fluid, then

$$\frac{\text{Total mEq. } Cl^- \text{ in tissue sample}}{\text{mEq. } Cl^- \text{ per liter plasma}} = \text{Liters extracellular fluid in sample}$$

(In practice, since we know that both assumptions are inexact, corrections are applied to both numerator and denominator of the above fraction in accordance with our knowledge. In muscle we know, for example, that there appears to be 3 mEq. of Cl^- per liter of intracellular water.) Knowing the volume of extracellular fluid in the sample, and still assuming uniform distribution of ions throughout the extracellular fluid, we can work out the partition of any ion, such as Mg^{++}, as follows

mEq. extracellular Mg^{++} in sample =

(liters extracellular fluid in sample) $\times$ (mEq. Mg^{++} per liter plasma)

and

mEq. intracellular Mg^{++} in sample =

(total mEq. Mg^{++} in sample) $-$ (mEq. extracellular Mg^{++} in sample)

(c) The direct analysis of ultrafiltrates of plasma or of normal or pathological fluids which are derived from the blood plasma by filtration through capillary walls (cerebrospinal fluid, aqueous humor, glomerular fluid) or which are accumulations of interstitial fluid (ascitic fluid, lymph, blister fluid) yields varying results, since each of these fluids has a different physiological history after it leaves the blood vascular system. There is a certain consistency about their composition which reflects their common origin. Average results of analyses by the above methods of specimens of human origin are shown in table XVII-1, together with figures on the ions of digestive juices. The data were compiled from a number of sources. The conventions used in calculating mEq. are those indicated by Gamble (27): for HPO_4^{--} the two electrons are calculated as 1.8 since 20 per cent of the ion is in the form of $H_2PO_4^-$ at pH 7.4; the Van Slyke factor, 2.43 is used to convert per cent protein to mEq. per liter; the factor for converting volumes per cent CO_2 to mEq. HCO_3^- per liter is 0.45. Conversion of units per liter serum to units per liter serum water is by use of the factor 1.07 for serum of normal specific gravity. For full details on the water and electrolyte composition of the body fluids, and of individual tissues as well, see (55).

The small differences in concentrations of electrolytes in plasma-derived fluids as compared with each other and as compared with plasma can be explained in part by differences in protein content. While it is doubtful

TABLE XVII-1

Approximate concentrations of osmotically significant ions in human body fluids; milliequivalents per liter of water

	Na^+	K^+	Ca^{++}	Mg^{++}	Protein	Organic acids	Cl^-	HCO_3^-	$HPO_4^=$	$SO_4^=$
Blood plasma	144	4.5	5.0	1.8	18	6	105	28	2.0	0.7
Cerebrospinal fluid	143	2.9	2.5	2.5			125	21		
Aqueous humor	143	4.7	2.8	1.0			108	32		
Vitreous humor	127	5.6	3.8	1.6			115			
Gastric juice	20	8					146			
Bile	142	8					109	38		
Pancreatic juice	140	7					40	116		
Jejunal juice	140	6					111	30		
Thoracic duct lymph	144	4.8	5.6	2.2			114			
Sweat	83	5	5	5			86			
Edema fluid	135	4.6								
Pleural fluid	137	4.3								
Synovial fluid	146	4.8	3.6	1.8			114	22		
Intracellular fluids										
Skeletal muscle	7	155			74		3		60	
Cardiac muscle	15	142			75		7		67	
Liver	5	145			89		12		98	
Brain	27	184			46		3		102	

that a true Gibbs-Donnan equilibrium (see Chapter 1 and Chapter 6) is ever achieved in the body, the conditions for approaching such an equilibrium exist at the capillary boundary, where plasma of high protein content is separated from interstitial fluid of low protein content by a membrane of limited permeability to protein. The inequality of concentrations of the non-diffusible protein anion demands an unequal distribution of diffusible ions; assuming complete non-diffusibility of protein, the calculated distribution ratio

$$r_{sf} = \frac{\text{serum anion concentration}}{\text{fluid anion concentration}} = \frac{\text{fluid cation concentration}}{\text{serum cation concentration}}$$

is 0.96, which is approximated closely by Na^+, K^+, Cl^-, and HCO_3^-. The theory predicts that the diffusible cations will be less concentrated and the diffusible anions more concentrated in the derived fluid than in the original plasma. With K^+, Na^+, HCO_3^-, and Cl^- this is the case in edema fluid, in accumulations of pleural and pericardial fluid, and in plasma ultrafiltrates.

Because of their clinical usefulness, the figures in table XVII-1 for blood plasma should be permanently and exactly memorized and the others remembered as larger or smaller in comparison. No attempt has been made to maneuver the average figures in the table to make total anions equal total cations; in an actual fluid this must be so. The apparent ion deficit in gastric juice is made up by H^+, in pancreatic juice by OH^-; deficits in the neutral fluids may be blamed upon inadequacies of analytical and statistical methods.

SODIUM

Our daily Na^+ intake is approximately five grams or 220 mEq., and in a healthy person output matches intake. This daily turnover of Na^+, which depends almost entirely upon the Na^+ content of the diet, is subject to great variation as a result of individual tastes and habits. Restriction of Na^+ intake has been carried to the extreme of 150 mgm. per day in the rice diet (43). Under this regime the urinary Na^+ output falls to about 10 mgm. per day. The concentration of Na^+ in the blood serum does not significantly change but the blood plasma volume diminishes. With increased Na^+ intake the rough parallelism of output with intake is maintained up to about three times the average figures stated. The excretion of such large amounts of Na^+ requires an increase in water excretion above the normal which is approximately proportional to the excess Na^+ excreted. Water taken with salt is, however, excreted more slowly than water taken alone. The experimental use of salt intakes of the order of 480 mEq. per day caused a temporary gain of weight by water retention even though the urinary output increased to about 3 liters per 24 hours (7). The maximum concentration at which the human kidney can excrete sodium chloride seems to be about 300 mEq. per liter. Ingestion of sea water or of any saline solution above this critical concentration will require that water be removed from the body in the excretory process.

For the relationship between Na^+ and bioelectric phenomena, see page 156.

Analysis in Blood Serum

The mean value for sodium in the serum of normal individuals has been determined in our laboratory as 145 mEq. per liter with a standard deviation of 2.2 mEq. per liter. The method used (14) consisted of precipitation as uranyl zinc sodium acetate and color development with sulfosalicylic acid and sodium acetate. This is but one of numerous modifications of a standard gravimetric method for sodium analysis. Most clinical analyses for Na^+ were done by some modification of the uranyl zinc sodium acetate method until the introduction of the flame photometer. This is an instrument specially designed to measure the energy of the atomic line spectra of characteristic wave lengths emitted by sodium and potassium when excited by high temperatures. When properly standardized this instrument yields analytical results concordant with those obtained by the more tedious chemical examination and with a considerable saving of time. In a series of 107 normal specimens, a mean value of 144 mEq. of Na^+ per liter was obtained with a standard deviation of 3.6 mEq. per liter (56). The Na^+ content of cerebrospinal fluid, aqueous humor, or other extracellular fluids available for analysis approximates 96 per cent of the plasma value, as predicted by the Gibbs-Donnan theory. It has already been pointed out that the intracellular fluids are low and variable in Na^+ content.

From the established Na^+ content of the extracellular fluid and from its estimated total volume, about 11 liters, we can easily calculate that the total Na^+ in the extracellular fluid of the body is a little less than two equivalents. Actual analysis of the ash of the human body indicates the presence of about one equivalent more. This, neglecting the small intracellular fraction of Na^+, is located in the skeleton. The mineral reinforcement of bone contains about one mol of Na^+ for 30 mols of Ca^{++}. In summary, the total body Na^+, which for our average man is about 60 grams or a little less than three equivalents, is divided approximately one third in the bones and two thirds in the extracellular fluids; a small and variable amount of Na^+ is found within the cells.

Clinical Significance of Sodium Analyses

The previous discussion has emphasized the constancy of the Na^+ content of blood plasma and extracellular fluid even under conditions of physiological stress. Variation in either direction beyond the normal limit of Na^+ concentration is indicative of serious disturbance of either the osmotic balance or the acid-base balance. Shifts of water generally occur in such a way as to maintain the Na^+ concentration of the extracellular fluid. The actual concentration of Na^+ will not change until the imbalance is excessive. Hence the finding of a depressed level of plasma Na^+ (*hyponatremia* (49)) is indicative of a severe degree of salt depletion or water excess. Conversely, it is only in extreme states of water deficiency or salt excess that an elevated Na^+ concentration (*hypernatremia* (69)) will be observed in the blood serum. When such findings are noted an emergency exists and prompt correction is indicated. For example, if a person has been losing electrolytes by excessive diarrhea or vomiting to such a degree that the serum Na^+ is depressed, it would be futile and dangerous to attempt to restore the volume of extracellular fluid with 5 per cent glucose solution. From the point of view of osmotic balances this would be the equivalent of the injection of pure water. Diuresis would occur with the loss of still more electrolyte. The indication here is the restoration of fluid volume with isotonic sodium chloride solution. Other clinical situations which can contribute to salt depletion include excessive sweating (which can occur on the operating table), surgical drainage of gastrointestinal fluids, excessive dietary salt restriction, excessive use of diuretic drugs, and adrenocortical deficiency. In Addison's disease, as a result of a deficient supply of mineralocorticoids, the concentration of Na^+ and also of Cl^- in the blood serum is distinctly below normal. The plasma volume is greatly decreased, in terminal cases as much as 45 per cent. The concentration and total excretion of the sodium and chloride ions are greatly increased in the urine and also in the sweat.

The opposite situation to Addison's disease is hypercorticoadrenalism, in which situation increased concentrations of Na^+ in the blood serum have sometimes been observed. Infants in whom water depletion exists (as the result of inadequate water intake during sickness or as the result of diarrhea) frequently have concentrations of Na^+ greater than 150 mEq. per liter in their blood plasma. Fluid mixtures of low sodium content (15 to 40 mEq. per liter) are recommended (22a) in the repair of such cases of hypernatremic dehydration. Note that diarrhea can lead either to hyponatremia or hypernatremia which makes accurate determination of the Na^+ content of blood plasma highly important as a guide in treatment.

POTASSIUM

The predominance of Na^+ among the cations of extracellular fluids yields to K^+ within the boundaries of the cells. Potassium ion within the cells accounts for 99 per cent of body K^+. The observed fact that cells accumulate K^+ in preference to Na^+ has been explained by the active extrusion of Na^+ (see page 156). In human red cells, K^+ accumulation is most effective at pH 7.4 and in glucose concentrations of 20 to 200 mgm. per 100 ml. Procedures which check cell metabolism, e.g., heavy x-irradiation or poisoning with fluoride, abolish the ability to accumulate K^+ beyond the small amount explained by Gibbs-Donnan equilibrium. Concentrations of K^+ vary in different tissues; rates of uptake, measured by the use of radioactive K^{42}, also vary; uptake is more rapid in growing cells; with disintegration of cells K^+ is released; in general, the concentration in milliequivalents of K^+ within cells is comparable to that of Na^+ in the extracellular fluid (table XVII-1). The approximate proportion to protein in muscle is 3 mEq. K^+ per gram of nitrogen. The total potassium in the body of the average man is 3.8 equivalents or 150 grams (61); the daily turnover averages about 80 mEq., ranging between 2.5 and 4 grams; more than four-fifths of the daily excretion is in the urine, the remainder in the stools and sweat. Each normal evacuation of the bowel carries out 10 to 20 mEq. of K^+ and 2 to 5 mEq. of Na^+. With abnormal increased water content of the stools, the sodium loss increases and may be as much as 100 mEq. per liter of watery dejecta. The potassium loss does not increase as much or as consistently, but is significant in diarrhea and in surgical cases where intestinal fluids are being drained away.

Measurement of the exchangeable K^+ can be made by injection of a known amount of $K^{42}Cl$, and determining the proportion of K^{42} in the urinary K^+ after allowing 24 hours for equilibration. On account of the short half-life of K^{42}, corrections for radioactive decay must be made (see page 893). The mean value for exchangeable K^+ is 46.3 ± 4.31 mEq. per kilogram of body weight in men, and 31.5 ± 2.90 mEq. per kilogram in women (2). The amount of exchangeable K^+ is related to the lean body mass and particularly to skeletal muscle. Exchangeable K^+ is less than total K^+ by 7 to 25 per cent with a mean value of 15.2 per cent. The total K^+ of the body is determined by use of the total body gamma-monitor, assuming that the whole response is due to K^{40} (67). Increase in total body potassium accompanies growth and the repair of deficiency states.

The maintenance of a higher concentration of K^+ within cells, as compared with plasma or extracellular fluid, requires energy which derives from the metabolic activities of the cells. One theory of the mechanism of K^+ uptake against a concentration gradient, the concept of the sodium pump, has been explained on page 156.

The blood plasma contains 4.52 mEq. of K^+ per liter with a standard deviation of 0.45 mEq. (56) as determined by the flame photometer. These values are in good agreement with others previously obtained by tedious and tricky chemical micromethods. Lower values have been observed in aqueous humor and cerebrospinal fluid, in rough agreement with Gibbs-Donnan predictions. Blood cells average 95 mEq. per liter; whole blood kept in storage at low temperatures loses K^+ from cells to plasma, while hemolysis liberates large quantities. It is essential for potassium analyses that serum be used which is unhemolyzed and which was promptly separated from the cells. When blood is stored at 37°C., K^+ enters the cell from the plasma until glycolysis is complete, then reverses the direction of its movement. Addition of glucose prolongs the retention of K^+ by red cells (13). Increased acidity of the blood increases the concentration of K^+ in plasma and decreased acidity decreases the concentration of K^+ independently of the concentration of K^+ within the cells. Each change of 0.1 in plasma pH is associated with an inverse change of 0.6 mEq. per liter in plasma K^+ (7a).

Potassium Deficiency and Excess

It is not possible to diagnose a state of K^+ deficiency with certainty on a basis of blood plasma levels, since they are sometimes depressed when cellular K^+ is depleted and sometimes not. If however extra K^+ is given to a normal person he will excrete it almost quantitatively, retaining only about 0.1 mEq. per kilogram of body weight (75); patients who have lost large quantities of gastrointestinal fluid, and who have been maintained chiefly by parenteral fluids with minimal K^+ intake, will retain much larger amounts. This is taken to indicate K^+ deficit. K^+ depletion occurs with restricted intakes since renal excretion of K^+ never falls to levels as low as those of Na^+ with minimal intakes. The minimal excretion observed under complete K^+ deprivation was 6 mEq. per day—compare this with the less than 0.5 mEq. of Na^+ excreted on the rice diet. Protracted self-medication with laxatives has produced K^+ depletion in otherwise healthy people (70). In one such case, urinary K^+ fell to 4 mEq. per day, with serum K^+ concentrations of 2.3 mEq. per liter or less. The measurement of exchangeable K^+ is of value in determining states of K^+ deficit, as in inadequately regulated diabetes. In K^+-depleted rat muscle, it has been observed that from 8 to 40 per cent of the missing positive charge is made up for by lysine in the form of lysinium ion (16).

It is difficult to assess the total results of K^+ deficiency in human patients, where the deficiency is usually a complication and not a primary disease. It occurs most frequently in patients who have been parenterally nourished. Since K^+ is present in all cells, animal or vegetable, and in milk, no diet intended to be eaten in the normal manner is likely to be deficient in this ion unless purposely made so out of purified materials. Rats maintained on such a diet are apathetic, dwarfed, and diarrheic; the heart muscle degenerates and voluntary muscles are weak or paralyzed; the final stage is paralysis of the intestinal musculature. The effects of K^+ deficiency in rats are accentuated and made more dangerous, including the development of severe myocardial necroses, if Na^+ was simultaneously in good supply (8).

Familial periodic paralysis is an inborn error of K^+ metabolism inherited as a Mendelian dominant in which at unpredictable intervals the patient, often during sleep, develops flaccid paralysis of the muscles of the trunk and extremities, lasting up to a few days. Plasma K^+ concentrations during attacks fall to about half their normal value (*hypokalemia*). There is no unusual excretion of K^+; what appears to occur is a movement of K^+ from extracellular to intracellular fluid. Attacks often follow the ingestion of unusual amounts of sugars or the experimental injection of insulin. This is best explained by the fact that both sugar ingestion and insulin administration increase the deposition of glycogen and each gram of glycogen deposited binds approximately ⅓ mEq. K^+. Attacks can usually be terminated by total fasting or, more promptly, by the oral administration of 5 to 10 grams of KCl. The best regime for prevention of attacks is a diet high in protein and low in carbohydrate and fat. On such a diet, the dietary

K^+ intake can be less than 1 gram/day without attacks. Without dietary control, prevention of attacks requires the nightly use of a 5 gram dose of KCl.

Potassium intoxication can occur when the concentration of K^+ in the plasma is elevated above 7 mEq. per liter (42). This may occur in Addison's disease, in severe renal inadequacy, and as a result of overdosage of potassium salts. The chief effect is upon the function of the heart muscle and is best demonstrated in electrocardiograms. The characteristic sequence of changes from the normal is: 1) increased height and steepness of the T wave; 2) increased width of the QRS complex; 3) widening and later loss of the P wave; 4) gross intraventricular conduction defects; 5) cardiac arrest with irregular waves of low voltage. Useful therapeutic measures include the attempt to restore electrolyte balance by intravenous injection of hypertonic (3 per cent) NaCl solution, or of hypertonic (25 per cent) glucose solution containing one unit regular insulin per 2 grams glucose, and more fundamentally the removal of the cause of inadequate renal function. Numerous mechanical devices for increasing extrarenal excretion (see page 786) have been used, and may be lifesaving if the renal damage is not irreversible.

It is difficult to provoke K^+ intoxication in the human subject whose renal function is normal. Where renal function is questionable, it is wise to avoid the use of potassium compounds as drugs, or the use of foods with high K^+ content. The manifestations of K^+ intoxication appear to be the result of increased K^+ concentration in plasma (*hyperkalemia*) and presumably in extracellular fluid, and not to increased K^+ within the cells of the heart. In the experimental animal one can increase cellular K^+ by large injections of K^+ salts. If water is available, the intracellular levels are quickly restored to normal. Chemically significant elevations of total body K^+ have not been observed even in patients with hyperkalemia. When a gain in whole body K^+ has been observed, it could be attributed either to growth or to the correction of a state of potassium deficiency.

CHLORIDE

The 70-kgm. adult contains about 3 equivalents or 110 grams of Cl^-, no significant portion of which is in the cells, or in any form of organic combination. The red discs of the blood contain 77 to 91 mEq. per liter. The Cl^- of venous blood plasma is 99 to 105 mEq. per liter with a mean of 102 mEq.; values for arterial plasma are consistently 3 mEq. higher—this is the chloride shift, a significant mechanism in CO_2 transport which will be explained in the next chapter. The Cl^- content of whole blood depends upon the proportions of cells and plasma; plasma or serum analyses for Cl^- may be clinically useful; whole blood Cl^- is not.

The daily turnover of Cl^- is even more variable than that of Na^+. The chief determinant is the NaCl of the diet, which varies from negligible amounts in the rice diet up to 15 grams daily for a heavy salt eater; expressed as Cl^- this would be 9 grams or 330 mEq. In the absence of sweating, normal excretion of Cl^- is almost entirely in the urine; the daily fecal output is only about 3 mEq.

The chloride of extracellular fluids (table XVII-1) approximates the concentration predicted by the Gibbs-Donnan equilibrium. In the cerebrospinal fluid, the Cl^- content is 119 to 128 mEq. per liter which is higher than the predicted value. With pathological increase of protein content, as in tuberculous meningitis, the spinal fluid Cl^- decreases.

Bromide

In the blood of dogs or men on an ordinary diet there is almost always a detectable amount of bromide; the concentration is 0.1 mEq. per liter or less. This ion serves no known function in the body although there is evidence that it is required for growth in chicks (6a). Its presence becomes medically significant only when the concentration is considerably increased. When a bromide is taken in excessive doses or over a prolonged period of time for its sedative effect, its concentration in the blood plasma may reach levels of from 10 to 20 mEq. per liter. At these levels impairment of the patient's mental functions, particularly memory and concentration, is likely to appear, often accompanied by a spotty, purplish discoloration of the skin with elevations very similar to the papules of ordinary acne. With even more enthusiastic use of the drug, still higher concentrations of circulating bromide may be achieved, as an accompaniment of which the patient shows definite psychotic changes including disorientation, lethargy, and confusion. There is an unpredictable effect upon reflexes usually shown by disturbances of gait and co-ordination. Recovery from such bromide intoxication is usually spontaneous within a few weeks if the drug is no longer taken. The distribution of bromide in the body is identical with that of chloride. It will be recalled that bromide is commonly used to measure the volume of the extracellular fluid space. Excretion of bromide and recovery from bromide intoxication can be accelerated by the administration of sodium or ammonium chloride (11).

THE BONE MINERALS

The ions Ca^{++}, Mg^{++}, F^-, and HPO_4^{--} circulate in low concentrations in extracellular fluid, in equilibrium with the solid mineral reinforcement of the bone, or *bone salt*, which is deposited in early life by precipitation of insoluble compounds involving these ions. Throughout life there is continuous interchange of these ions between bone salt and the interstitial fluid, and between interstitial fluid and plasma. Bone salt is being constantly dissolved and reprecipitated. The mechanisms controlling gain and loss of bone salt have been explained by Albright and Reifenstein (3); the normal excretory function of the kidney keeps body fluids slightly undersaturated in respect to Ca^{++} and HPO_4^{--}, which are the chief ions of bone salt; on account of this undersaturation Ca^{++} and HPO_4^{--} are constantly being dissolved from bone salt; if nutritional and hormonal status is normal, bone salt is deposited by crystallization from interstitial fluid in response to local mechanical stress; the mechanical stress will commonly occur where bone salt has been dissolved. This combination of mechanisms keeps the skeleton "sufficiently strong but not needlessly bulky". The hardest part of this theory to explain is the deposition of bone salt as a response to mechanical stress, but such deposition is a commonplace of clinical observation. If a broken bone is set at an abnormal angle, bone salt is deposited to meet the abnormal stress. Similar deposition of bone salt in response to stress is observed in diseases, such as rickets, which cause bone deformity (see fig. XVII-1).

Composition of Bone Salt

X-ray diffraction tells us that the bone salt is crystalline. Varied interpretation of the diffraction patterns has led to opposing concepts of the structure of bone salt. Fankuchen (18) explains the discrepancy as a result of (a) difficulties in getting sharp patterns without pre-treatment which may alter structure, and (b) "the fact that most of the possible structures give similar diffraction patterns". The idea that bone salt is comparable in its structure to the apatite series of minerals has been the outgrowth of the work of many investigators.

The crystal lattice of bone salt is considered by Hendricks and Hill (38) to be that of *hydroxyapatite*, the formula of which can be simply approximated as $Ca_5\ (PO_4)_3\ OH$. The composition of dental enamel agrees well with that of hydroxyapatite. The over-all composition of the mineral portion of bone is less basic, and more comparable to a neutral calcium phosphate. Hendricks and Hill suggest that in bone, excess $HPO_4^{=}$ forms compounds at the hydroxyapatite crystal surface, as do also Na^+, Mg^{++}, HCO_3^-, and citrate. As the hydroxyapatite crystal grows, such ions which are present but cannot be incorporated into the crystal lattice of hydroxyapatite eventually cover the crystal surface and stop crystal growth. These surface constituents may make up 10 per cent of the mineral matter of bone (37).

Formation of Bone Salt

Since the calcium and phosphate ions of extracellular fluid can be shown by use of radioactive tracers to be in constant exchange with similar ions

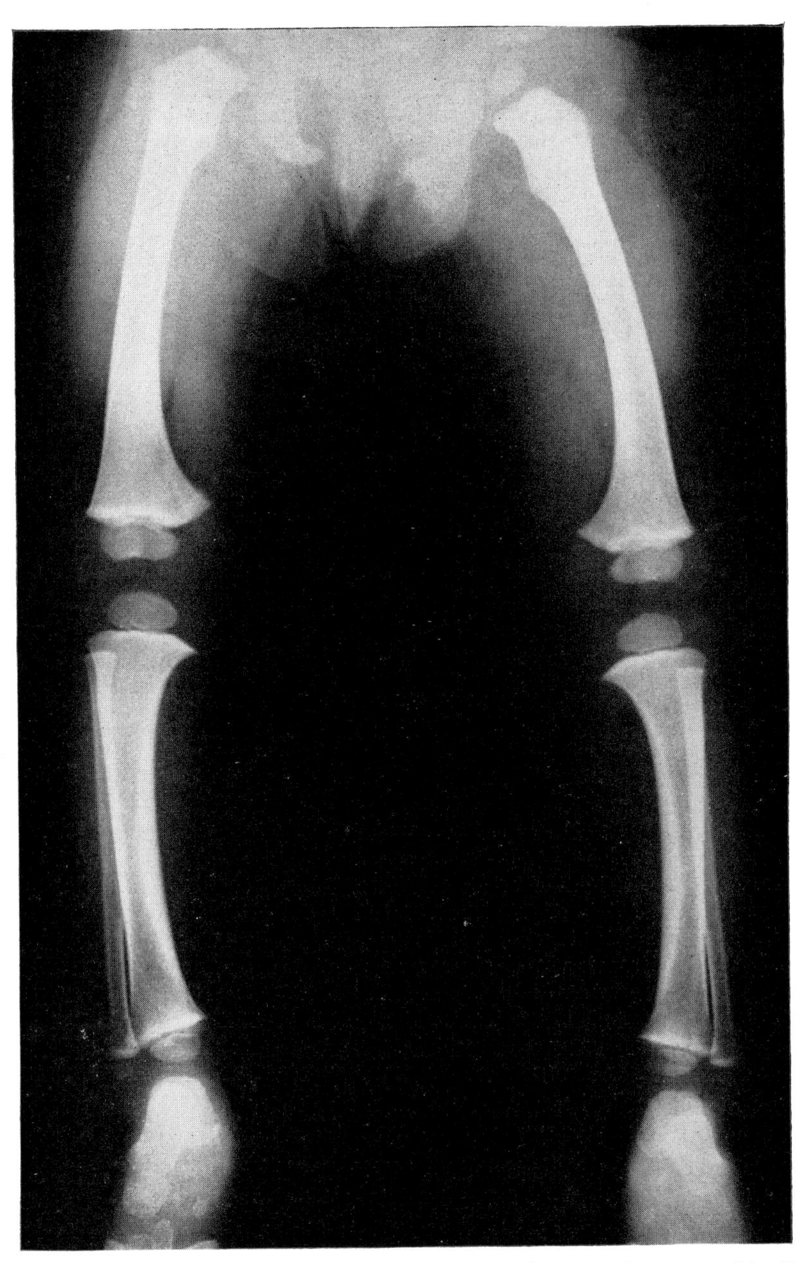

FIG. XVII-1. Deposition of bone to meet abnormal stresses in a case of healing rickets (see page 832) in a child 1½ years old. (Courtesy of Dr. George Levene.)

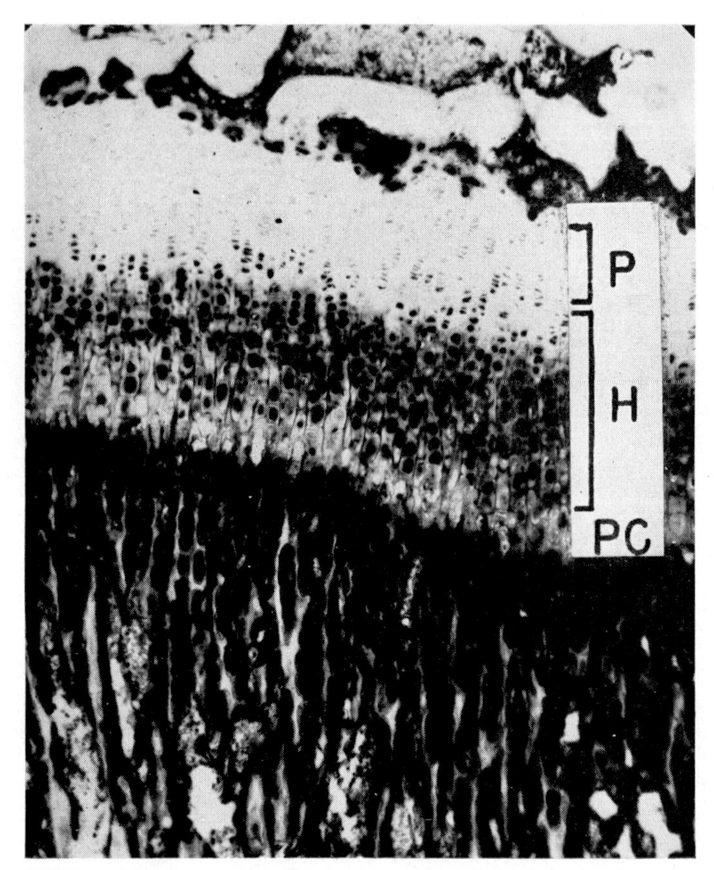

FIG. XVII-2. Distribution of alkaline phosphatase in longitudinal section of epiphyseal plate of tibia of 28-day rat. Dark areas indicate alkaline phosphatase. "In the zone of proliferation (P) there is no enzyme in the matrix and little in the cells. In the hypertrophic zone (H) there is an abundance of enzyme in the cells and it also appears in the interstitial substance. In the zone of provisional calcification (PC) there is no enzyme in the degenerating cells but much in the interstitial substance" (30). Photograph by courtesy of Dr. Clary J. Fischer.

composing bone salt, it follows that bone salt is being constantly deposited and constantly dissolved. This local process in the bone must depend, as regards direction and rate, upon local concentrations of calcium and phosphate ions. There are two known mechanisms which can alter local concentrations of phosphate ion in the interstitial fluid of the bone.

Alkaline phosphatase can be demonstrated by the Gomori stain in the cells of the endosteum and the inner layer of the periosteum of young rats (30). The osteoblasts and blood vessels of the Haversian canals are rich in phosphatase; osteoclasts and osteocytes have varying and limited phosphatase content. The distribution of phosphatase in the ossification centers at the epiphyseal plates is of particular interest; this is shown in figure XVII-2. Phosphatase catalyzes the hydrolysis of non-ionizing organic esters phosphoric acid, yielding free phosphate ion. Plasma contains, and the of interstitial fluid of bone can therefore be presumed to contain, such esters. Phosphatase is presumed to promote calcification by accelerating a local increase in phosphate ion concentration, derived from phosphate esters. Children have more alkaline phosphatase in the blood plasma than adults. At any age, disease which involves increased activity of osteoblasts in laying down organic bone matrix will result in increased plasma alkaline phosphatase. This is true, as in rickets, even if the organic matrix is not being adequately calcified. Alkaline phosphatase is measured in plasma by allowing a known volume of plasma to react at pH 8.6 with a salt of an organic ester of phosphoric acid, such as sodium glycerophosphate. The liberated phosphate ion is measured colorimetrically and the phosphatase activity expressed in arbitrary units (6). The plasma of normal adults will show 3 to 5 Bodansky units of alkaline phosphatase activity per 100 ml.; the plasma of growing children, up to 10 Bodansky units. Increased alkaline phosphatase activity is observed not only in bone diseases with increased osteoblastic activity but also in hepatic diseases and in biliary tract obstruction. Do not confuse alkaline phosphatase with acid phosphatase (page 405).

Glycogenolysis may be the source of additional phosphoric ester. Catalyzed by *phosphorylase*, glycogen reacts with phosphate ion to yield glucose-1-phosphate. The fetal cartilages, which later are to calcify into bone, contain large stores of glycogen. The glycogen disappears at the time of deposition of bone salt. Inhibitors of glycogenolysis will inhibit calcification in cartilage. In cartilage, at least, glycogenolysis appears to be an essential preparation for calcification (31). It has not been determined whether this conversion of phosphate ion by phosphorylase into organic ester form, to be reconverted later into phosphate ion by phosphatase, has the effect of increasing local phosphate ion concentration, or puts phosphate in ester form where it is needed, or simply supplies energy for the general process of calcification. It does not appear to function in the formation of dental enamel (5), which is of similar but not identical composition as compared to bone salt.

The enzymatic activities just described are only a part of the mechanism of bone salt deposition. Changes take place in the acid mucopolysaccharides of calcifying cartilage. Uncalcified cartilage shows the property of metachromasia (purple coloring of toluidine blue) which is not shown by the organic matter of even freshly calcified bone. If the matrix of freshly calcified bone is treated with 1 per cent nitric acid for 5 minutes, it regains its metachromasia. The presumption has been made that Ca^{++} becomes attached to the acid groups of mucopolysaccharides during calcification so that chemical reaction with toluidine blue is blocked. It is possible therefore that acid mucopolysaccharides have a part in the binding of Ca^{++} and its eventual deposition as bone salt.

Calcium

In quantity, calcium surpasses all other minerals in the body. Most of the calcium is in the skeleton; from 796 to 1518 grams of calcium have been found in the skeleta of adult European or American males (63). The soft tissues and body fluids contain in total about 4 grams. This relatively picayune amount is of high physiological consequence. The concentration in blood plasma is maintained at a remarkably steady level by the interaction of several regulatory mechanisms and is in equilibrium with the stored calcium in bone salt.

Physiologists and pharmacologists look upon Ca^{++} as an antagonist to Na^{+}, Mg^{++}, and particularly to K^{+}. Calcium salts will temporarily promote the excretion of Na^{+} and have been used in the treatment of edema. The restorative effect of calcium salts on animals anesthetized by injections of magnesium salts has been demonstrated to generations of students of pharmacology. The damping of neuromuscular irritability by Ca^{++} has at least a double mechanism. Ca^{++} inhibits the dephosphorylation of ATP. This applies to muscle directly, and also to nerve where the nerve impulse mobilizes K^{+}, increasing the synthesis of acetylcholine in the presence of ATP. Ca^{++} decreases such acetylcholine synthesis (21).

Calcium in Blood Plasma

The calcium of blood is a plasma and not a cell component. The plasma concentration is 5 mEq. per liter and varies normally only 0.2 mEq. from this figure. Analysis is usually done on serum to avoid interference by anticoagulants. The usual method (10) is direct precipitation as calcium oxalate, separation and washing in the centrifuge, solution of the precipitate in warm dilute sulfuric acid, and titration with permanganate. This procedure measures total calcium.

Only about 40 per cent of the total plasma calcium is in the form of the ion; calcium ion concentration has been measured directly by its effect upon the frog heart. The remainder, except for a negligible diffusible non-ionized fraction, is combined with protein in a form which does not diffuse or ionize. The proportion of ionized and of protein-bound calcium depends upon the concentrations of protein and of total calcium. A nomogram has been published which facilitates this calculation (59). It appears to be the ionized calcium which is significant in the mechanism of blood coagulation (see page 172), as well as in neuromuscular irritability.

Increased total and ionized plasma calcium is found in hyperparathyroidism (page 691) and following overdosage of vitamin D. An increase in total plasma calcium without alteration of the calcium ion concentration is characteristic of hyperproteinemia; a decrease of total calcium may occur with diminished plasma protein. Decreases involving both total and ionized plasma calcium occur in hypoparathyroidism, infantile tetany, in hyperphosphatemia from impaired renal excretion, and in some unusual cases of osteomalacia and of rickets, where normal plasma calcium is the usual finding. Ingestion or injection of calcium salts produces only transitory changes in plasma calcium.

Calcium and Citric Acid

Human blood contains from 1.5 to 4.0 mgm. citric acid per 100 ml.; the normal human adult excretes 0.2 to 1 gram of citric acid daily in the urine. There is a considerable skeletal store of citric acid. In general, physiological and pathological calcium deposits contain citrate which forms a soluble complex with Ca^{++} (35). Citrate ion will dissolve many Ca^{++} salts which are otherwise insoluble in neutral solutions. Administration of a carbonic anhydrase inhibitor to rats diminishes the urinary output of citrate nearly to zero. If urinary calcium output is high in rats so treated, calcium salts are precipitated in the kidney (33).

Injection of solutions containing Na^{+} and citrate ion at pH 7.4 will increase the serum calcium level and the urinary calcium output in dogs. Injected citrate is rapidly removed from the circulation in normal dogs or in dogs with ligated ureters; in nephrectomized dogs the elevation of plasma citrate persists 2 or 3 days or until death (26). This suggests that the kidney in dogs is the chief organ for removal of citrate and that the removal is predominantly metabolic and occurs only to a minor degree by direct excretion of citrate.

When sodium citrate is given by mouth, the effect is quite the opposite. The concentration of calcium in the blood and the urinary calcium output both decrease. This may be the result of the complexing of calcium ion in the intestine, which favors the absorption of phosphate and therefore the deposition of calcium and phosphate ions from blood into bone salt.

Calcium Deficiency

It is difficult to find in the clinical literature descriptions of well authenticated cases of unmixed calcium deficiency. As in most nutritional deficiencies, lack of calcium in the diet is likely to accompany other dietary shortages and the resultant disease reflects the multiple depletions. Consult table XII-2 in Chapter 12 for recommended amounts of calcium in the diet. Adults remain in balance, or in positive balance, as far as calcium intake and output are concerned, in long-term studies on a daily intake of 900 mgm. (63). Many adults adapt well to lower intakes, remaining in a balance brought about by lower fecal outputs.

Osteomalacia is a failure or delay of deposition of bone salt. It can occur as the result of simple calcium deficiency, but such cases are rare in American experience (89). It can also occur from phosphate deficiency or vitamin D deficiency. *Rickets* is osteomalacia complicated by its occurrence during early childhood, the stage of most rapid calcification of bone. When osteomalacia occurs as a clear result of calcium deficiency there is still likely to be some contributing abnormality of absorption or utilization. Fecal losses of calcium are significant in sprue (see page 514) and other diseases where fecal fat is excessive. Pregnancy and lactation have long been recognized as causing a loss of calcium; osteomalacia may be the sequel of a series of pregnancies. Since the fetus at birth contains only about 20 grams of calcium, the calcium requirement added by pregnancy is not excessive. The drain of lactation is greater; the infant must retain from 60 to 85 grams of additional calcium during the first nine months of extra-uterine life. A quart of human milk contains about 400 mgm. of calcium and many nursing mothers average a quart or more daily. It is customary to prescribe extra calcium, usually as dairy milk, for the nursing mother to minimize or overcome negative calcium balance.

The bones in osteomalacia become more pliable and deform under weight bearing. There may be a reduction of all components of bone but the loss of bone salt is greater than that of organic matrix. Osteoblastic activity is increased which is reflected in an increased alkaline phosphatase of the plasma. Osteomalacia is to be distinguished from *osteoporosis*, which is failure of osteoblastic activity in the formation of the organic matrix.

Tetany may occur in osteomalacia or other clinical situations where the ionized Ca^{++} of the plasma is decreased. It may be induced by vitamin D treatment of severe rickets. It is characteristic of hypoparathyroidism which usually results from removal of or injury to the parathyroids during surgical procedures. When tetany is the result of simple Ca^{++} deficiency, therapy with calcium salts is usually adequate. Such therapy is often inadequate in parathyroprival tetany, and treatment with parathyroid hormone or with dihydrotachysterol becomes necessary.

Excretion of Calcium

Nearly a gram of Ca^{++} may be poured into the gastrointestinal tract as a component of the 8 to 10 liters of digestive fluids secreted daily. Much of this may be reabsorbed, but some is always lost. Urinary excretion of Ca^{++} occurs at plasma levels above a threshold of 3.5 mEq. per liter (3). Hence, urinary and fecal loss of Ca^{++} continue even though the intake may be below a maintenance level. The normal state is one of calcium balance; total losses can be expected to equal total intakes in healthy and well nourished subjects. The normal urinary excretion is between 0.1 and 0.3 grams per 24 hours; the remaining output is fecal. For details of an adequate analytical method for the study of Ca^{++} excretion, see Fiske and Logan (24). The Sulkowitch reagent, composed as follows

Oxalic acid	2.5 grams
Ammonium oxalate	2.5 grams
Glacial acetic acid	5.0 ml.
Distilled water to make	150.0 ml.

is useful in office screening of patients for abnormalities of urinary calcium excretion, 2 ml. of the reagent being added dropwise to 5 ml. of urine. The concentration of Ca^{++} in the urine can be judged by the rapidity and amount of precipitation of calcium oxalate.

Phosphorus

Phosphorus in the body exists only in the form of the phosphate radical which occurs in both organic and inorganic combination. Like calcium, phosphate is found in greatest amount and highest concentration in the skeleton. There is about 500 grams of skeletal P in the average adult male. The remainder of the body contains about 75 grams P which is divided among a large group of compounds including inorganic phosphates, nucleoproteins, nucleotides, phospholipids, and phosphoric esters. The average adult intake of P on an adequate diet is estimated at about 2 grams.

The National Research Council has not included a figure for P in its recommended allowances. Sherman (71) has estimated the "rock-bottom" adult requirement at 0.88 gram per day. Nutritionists rather generally increase this figure by 50 per cent in stating a desirable intake. Pregnancy and lactation demand increases proportional to those of calcium. The allowance for children is not significantly less—never below one gram—on account of the combined demands of skeletal and muscular growth. Bottle-fed infants are figured at 100 mgm. P per kgm. of body weight.

Phosphorus Compounds in the Blood

The blood plasma of the adult contains free phosphate ion commonly called "inorganic phosphate" in a mean concentration of 2 mEq. per liter with a standard deviation of 0.2 mEq. At the blood pH of 7.4, 80 per cent of this is in the form of HPO_4^{--} and 20 per cent is $H_2PO_4^-$. Expressed in the older units, the mean value is 3.5 mgm. per 100 ml.

The concentration in the plasma of children is significantly higher. Newborn infants have almost twice the adult value for plasma inorganic phosphate, and from ages 3 to 11, normal levels of children exceed the adult level by about 50 per cent (73). A value which would be normal for a healthy adult would be significantly low in a child. The practice of vitamin D supplementation in infant feeding has made it unnecessary to continue previously established differentiations in plasma inorganic phosphate levels between breast-fed and bottle-fed babies, and differences according to latitude and seasons. In active rickets the inorganic phosphate usually falls to half the normal infant level or less—a value of 0.6 mgm. per 100 ml. has been observed—and the alkaline phosphatase activity increases up to four times the normal value. The phosphatase increase is often the earliest chemical change in rickets and the last to return to normal under treatment. Human rickets is usually the result of vitamin D deficiency (see page 832), often complicated by other deficiencies. Experimentally in animals, rickets can be produced by the single deficiency of calcium or phosphate or vitamin D. Experimental rickets produced by calcium deficiency is highly atypical and more comparable to osteomalacia. If phosphate in the experimental diet is deficient, rather typical rickets appears even though the vitamin D intake is adequate.

Lower values of plasma inorganic phosphate are characteristically observed in uncomplicated primary hyperparathyroidism, and in diseases involving failure of phosphate absorption. Increases of plasma inorganic phosphate occur in hypoparathyroidism and in renal disease or urinary obstruction.

The inorganic phosphate of the blood must not be considered as inert or as purely a waste product. It is derived by hydrolysis of phosphate groups from foodstuffs or metabolites. Phosphate ions from plasma may diffuse freely into interstitial fluid, and thence may enter cells, becoming available for the numerous phosphorylations involved in growth and metabolism. Such significant structural substances as nucleoproteins and phospholipids are synthesized in the cell with phosphate ion as a necessary raw material. Tracer studies have shown rapid incorporation of phosphate ion into phospholipids. Phosphate ion is significantly concerned in acid-base regulation which will be dealt with in the next chapter.

The inorganic phosphate is of such clinical importance that the other forms of phosphate occurring in the blood are often forgotten. Their significance is physiological rather than clinical. We find in plasma about 0.5 mgm. of ester P and about 8 mgm. of phospholipid P per 100 ml. Red cells contain about 50 mgm. of ester P and about 15 mgm. of phospholipid P per 100 ml. The colorimetric method of Fiske and Subbarow (25) is suitable for the measurement of all fractions of phosphate in blood and urine.

Excretion of Phosphate

Like calcium, phosphate is excreted both in urine and feces. The distribution between the two excretory routes is highly irregular in the normal person, with the greater part usually in the feces. The urinary excretion is chiefly as inorganic phosphate with only about one per cent in the form of esters or other organic compounds (83).

Parathyroid Activity and the Bone Minerals

The first effect noted when parathyroid hormone is given to a parathyroid-deficient patient is an increase in the output of urinary inorganic phosphate. There is associated increase of urinary volume with increased chloride output. The water loss is greater than the salt loss; the patient becomes thirsty. This sequence starts within an hour after an adequate dose of parathyroid hormone.

Within eight hours a fall in blood plasma inorganic phosphate occurs, which is quite abrupt. At about this same time a slower increase in blood plasma Ca^{++} begins and continues for several hours. The final change, increase in urinary Ca^{++}, is delayed until after the plasma Ca^{++} has noticeably risen.

These same end results are found in the *hyperparathyroid* patient—excessive excretion of both Ca^{++} and phosphate in the urine, diminished inorganic phosphate and increased Ca^{++} in the blood plasma. The continued loss of Ca^{++} and phosphate constitutes a drain upon the skeletal stores. Eventually decalcification reaches a point where areas of bone salt loss may be seen by x-ray; the areas may be local and sharply defined, or may be diffuse and visible only as a general loss of density. Hyperparathyroid patients are particularly subject to phosphatic calculi in the urinary tract. Hyperparathyroidism is the result of tumor (adenoma or carcinoma) or hypertrophy involving parathyroid tissue. The disease is not rare, but is sufficiently uncommon to arouse considerable interest when a case is discovered on a hospital service. The diagnosis is not as simple as this brief outline implies. The treatment is usually by surgery. For serious study of this disease an excellent clinical summary and bibliography is available (3).

There is no disagreement among experts concerning the facts of parathyroid action as outlined. It has not been definitely established whether the major function in the human is stimulation of renal excretion—particularly of phosphate ion—or a stimulation of bone-salt solution. The clinical observations seem to favor the renal mechanism. Experiments with animals indicate that bone salt is dissolved even when renal action has been eliminated by nephrectomy. The authors of the monograph cited

(3) are outspokenly in favor of the renal mechanism, but give fair consideration to the experimental evidence which prevents its acceptance as complete explanation of parathyroid effects. We are at present in a position where we must accept the idea of both a renal action and a bone-dissolving action of parathyroid hormone, without insisting that either is the cause of the other. It is certain that the two actions vary in their importance in different species; the renal action is most significant in the human and were it not for experiments on dogs it might be accepted as a complete explanation. Correlated with the bone-dissolving effect, increase in the mucoprotein of blood and urine has been observed after injection into rats of parathyroid hormone (17).

There is also an effect of parathyroid hormone on the level of citric acid in blood serum (50). Serum citrate rises after injection of parathyroid extract, and is decreased in thyroparathyroidectomized rats.

Parathyroid deficiency in the human is usually the result of actual or functional loss of the glands during or after surgery. Rarely the glands degenerate without apparent cause. The signs are the reverse of those described as induced by the hormone. Blood phosphate is increased, the Ca^{++} depressed so that tetany is the most threatening symptom; urine volumes are low, with diminished excretion of phosphate and other electrolytes.

Comparison of parathyroid excess with parathyroid deficiency is the most striking example of the reciprocal relationship between Ca^{++} and phosphate concentrations in plasma. Presumably on account of a solubility-product relationship with the hydroxyapatite in the bone, increase of one of these ions leads to a decrease of the other. This relationship in plasma cannot be stated mathematically with precision, probably because the changes occur through the medium of the interstitial fluid of bone, samples of which can not be obtained for analysis.

Magnesium

It is proper to consider magnesium among the bone minerals since more of the magnesium of the body is concentrated in the skeleton than in the remaining soft tissue. The total magnesium content of the body is about 21 grams or seven fourths of an equivalent. The total skeleton contains about 11 grams of magnesium; the muscles contain about 6 grams, with the remainder divided among the other tissues and body fluids. Since bone ash contains less than one per cent of magnesium as compared to over 38 per cent of calcium, its relative quantitative importance is obviously minor. Its function as an enzyme activator is of much greater physiological significance. Certain peptidases and phosphatases require Mg^{++} for maximal activity as do virtually all reactions involving ATP.

Inhibition or acceleration of a number of enzymes involved in neuromuscular activity has been observed to be related to Mg^{++} concentrations. The most significant effect of increased Mg^{++} is to decrease the amount of acetylcholine liberated by the motor nerve impulse. More exactly, this effect depends upon the Ca^{++}/Mg^{++} ratio (62). In addition to the effect upon muscle, there is a direct depressant action upon the central nervous system. Although it is not common practice, surgical anesthesia can be induced in man by intravenous injection of magnesium salts. Although epsom salt has a venerable therapeutic history, its use is not entirely without danger of magnesium poisoning, particularly in young children. The danger also exists when epsom salt is given in high concentration as an enema (19). The effects upon nerve and muscle functions manifest themselves in magnesium intoxication as coma and flaccid paralysis. When Mg^{++} concentrations of blood plasma are decreased, tetany, muscular tremors, and possibly convulsions occur.

The concentration of total Mg^{++} in human blood serum is 1.8 mEq. per liter with a standard deviation of 0.12 mEq. (39, 84). Analytical methods are described in the papers cited. Measurements of total serum magnesium are not particularly rewarding in clinical study except where magnesium poisoning may be expected. In one case of coma and flaccid paralysis following an epsom salt enema the serum total Mg^{++} three hours after the enema was 17 mEq. per liter (19). In more commonplace clinical situations fluctuations in serum Mg^{++} are relatively small. The serum concentration may rise in nephritic patients up to about double the normal value (84). Between 17 and 31 per cent of the normal serum Mg^{++} is bound and not ultrafiltrable. The percentage of bound Mg^{++} is increased in hyperthyroidism (15). This bound portion of magnesium disappears in myxedemic patients.

The problem of specific isolated magnesium deficiency does not arise in human nutrition. Any diet composed of natural foodstuffs and meeting the minimum of other requirements is bound to be adequate in magnesium. Intakes as low as 0.22 grams per day yield a positive Mg^{++} balance in the normal. The excretion of magnesium, like that of calcium, is divided between urinary and fecal channels. The average 24-hour urinary excretion of magnesium lies between 8 and 9 mEq. The variation in this figure is very considerable. For a detailed quantitative study of magnesium retention and excretion under diverse conditions, the reports of Tibbetts and Aub (76) should be consulted.

Fluoride

The property which sets fluoride in a different class from the other ions of the body is its extreme toxicity. Fatal poisoning has resulted from the ingestion of as little as 0.2 grams of sodium fluosilicate (29); violent painful spastic muscular contractions alternate with periods of flaccid palsy. Death is from paralysis of the respiratory muscles. Degenerative changes are found *post mortem* in the cells of kidney, liver, and gastrointestinal mucosa. Similar poisoning has been reported with somewhat larger doses of simple fluorides. Fluoride in toxic concentrations specifically inhibits anaerobic glycolysis (Chapter 14). This effect alone is sufficient to explain its acute toxicity, but does not account for the curious results of chronic ingestion of sublethal doses.

If the drinking water of a community contains 1.5 mgm. of fluoride per liter, 10 per cent or more of the children using the water during the period of tooth development will show a mild degree of hypoplasia of dental structure called *mottled enamel.* With 6 mgm. fluoride per liter of drinking water the incidence approaches 100 per cent and intensity of mottling is maximal. With 1.0 mgm. of fluoride per liter of drinking water, mottled enamel does not occur but the number of cavities in the permanent teeth of the school children (age 12 to 14) of the community is approximately one third of the number of cavities in similar groups from low-fluoride (less than 0.2 mgm. per liter) communities (58). Increase of fluoride to 2.0 mgm. per liter of drinking water produces mottled enamel but without further significant decrease in caries. The increased resistance to decay is probably the result of actual alteration of enamel structure by incorporation of fluoride. If the intake of fluoride is long maintained at a level of 0.2 mgm. per kgm. body weight per day, *osteosclerosis* or abnormal calcification of periosteum, endosteum, and ligaments is likely to occur. This has been noted chiefly among workers exposed to the dust of fluoride minerals.

There is in normal blood somewhat less than 0.1 mgm. fluoride per 100 ml., and a normal urinary content of about 1 mgm. per liter (29). In mottled-enamel areas the urinary fluoride averages 3 mgm. per liter. Fluoride is normally concentrated in bones and teeth; the amount found is quite variable, being usually between 0.02 and 0.05 per cent in dry defatted bone and below 0.03 per cent in teeth. Most studies indicate a lower fluoride content of carious teeth compared to sound teeth, but the differences have not always been statistically significant. Bone will take up fluoride *in vitro* from very dilute solutions. Conversion of bone salt to fluor-apatite has been demonstrated in fossil bones.

Since the only demonstrable effect of fluoride deficiency is increased susceptibility to dental decay, the designation of fluoride as an essential dietary component is doubtful. A number of communities have raised the fluoride content of public water supplies to one mgm. per liter, which is optimal in caries prevention but below the level of mottled enamel production.

THE METABOLIC MINERALS

Certain ions are indispensable to the metabolic machinery in that they form compounds necessary to carry out definite operations. Usually they are part of individual enzymes or hormones or vitamins; some function in a manner not yet fully understood.

Iron

The most significant iron compound of living organisms is *heme*, which in most forms of life is a component of important oxidizing enzymes such as the cytochromes, catalase, and peroxidase. In vertebrates, in earthworms, and in the root nodules formed by nitrogen-fixing bacteria is also found hemoglobin. Its function in bacterial nitrogen-fixation is not clear. In those animals which possess it, it is a carrier of oxygen and indirectly of carbon dioxide. Its structure has been explained in Chapter 5; its respiratory function will be described in the chapter immediately to follow.

To revert to the familiar 70-kgm. average adult male, we find that each liter of his blood contains about 500 mgm. of iron in the form of hemoglobin. His blood volume will be a little over 5 liters, and allowing for immature red cells in the bone marrow and non-circulating red cells in the spleen, he will have about 3 grams of iron in the form of blood hemoglobin. To gain an idea of the distribution in the human body of iron in other forms, we are forced to draw an analogy from the distribution in the dog (32). If the proportions are similar in man, there will be about 0.4 grams of iron in the form of muscle hemoglobin, and about 0.8 grams as functional tissue iron involving such compounds as cytochromes and catalase. About 1.1 grams of iron will be in storage, the greater part of it as *ferritin* in liver, spleen, and bone marrow. Ferritin is a protein which can contain up to 23 per cent ferric iron. Its recognized function is that of a storage form of ferric iron, in equilibrium with the Fe^{+++} of the blood plasma.

Hemosiderin, a colloidal ferric hydroxide-phosphate of variable iron content up to 35 per cent, is present in small amounts in the bone marrow of normal individuals with adequate iron stores. Excessive tissue deposits of iron in the form of hemosiderin are the result of abnormal absorption of iron or may follow blood transfusions.

About one per cent per day of the red cell hemoglobin iron is released by degradation of hemoglobin to bile pigment. This iron is not excreted, but is retained for hemoglobin formation. In health, red cell formation keeps pace with red cell destruction. During the period of functional life of the red cell there is no exchange of hemoglobin iron with plasma iron (23).

The Absorption of Iron

The anemic dog is a highly satisfactory test animal for the study of iron absorption. If iron compounds are injected intravenously into such a dog, the iron will be utilized practically quantitatively for the formation of hemoglobin. If iron compounds are given by mouth to such a dog, the amount of hemoglobin formed is a direct measure of the iron actually absorbed. If radioactive iron is used as a tracer, it can be detected in the circulating red cells 4 hours after it is fed. Complete conversion of absorbed radioactive iron into hemoglobin occurs within a week. Normal non-anemic dogs do not absorb iron and convert it to hemoglobin with any such effectiveness. Since iron excretion has been shown to be negligible, it appears that the amount of iron in the body is regulated through the mechanism of absorption.

The only form of iron which is absorbed from the gastrointestinal tract in nutritionally significant quantities is the Fe^{++} ion. Ferrous and ferric hydroxides and chelated iron of foodstuffs are converted to ions and dissolved in the gastric juice. In the acid medium of the stomach, ferric ions are reduced to ferrous with the simultaneous oxidation of ascorbic acid and other reducing agents present in foodstuffs. Under ordinary conditions, less than 10 per cent of the iron of ingested foodstuff is absorbed.

Iron which is combined in heme or heme derivatives is not available for absorption and is not easily made available in the digestive tract. No specific mechanism has been discovered for the digestive liberation of iron from heme. An unpredictable and irregular amount of iron is so liberated by the action of micro-organisms. Since bacterial action in the gastrointestinal tract is absent or minimal in the stomach and upper intestine, it becomes clear that heme compounds are not reliable sources of nutritional iron. This has been borne out by clinical experience—treatment of iron-deficiency anemia with foods rich in heme-combined iron is highly unsatisfactory compared with the dramatic improvement following the use of simple ferrous salts by mouth. Since iron-deficiency anemia implies complete exhaustion of iron reserves, it has been suggested that oral iron therapy be continued for a time after the hemoglobin concentration is normal (40).

The rate of absorption of ferrous iron from the gastrointestinal tract depends upon the need for iron in hemoglobin synthesis. The entire regulatory mechanism is not clear, but in part it is mediated through ferritin in the cells lining the gastrointestinal tract. When iron absorption is active, the ferritin concentration of intestinal mucosal cells increases. At high cellular ferritin concentrations absorption of iron is blocked. The block is

TABLE XVII-2

Pathways, equilibria, and valence changes in iron metabolism

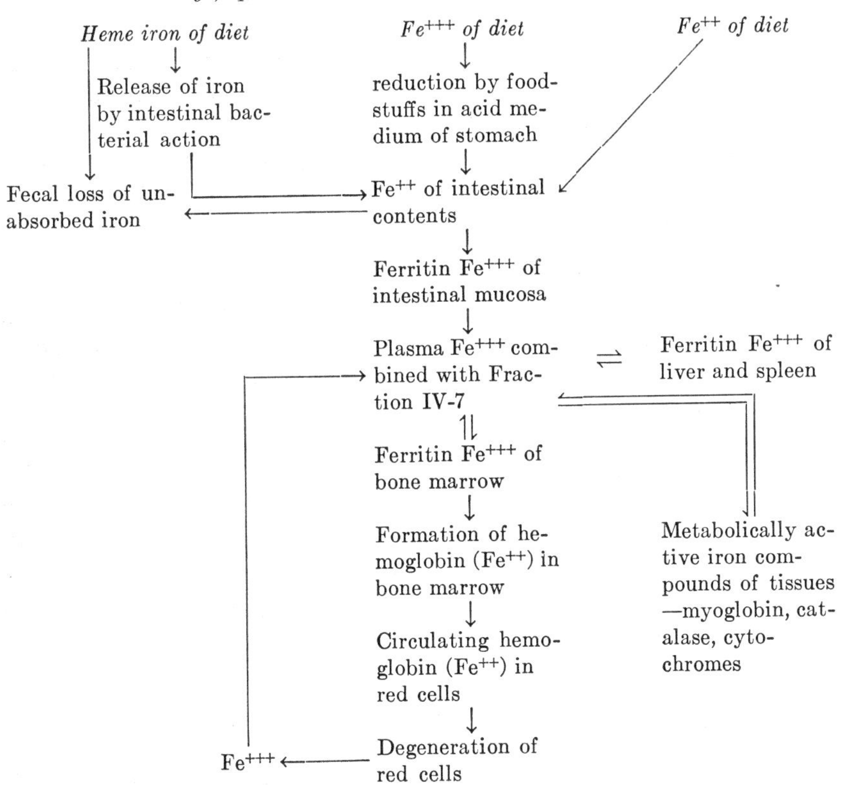

slowly removed as the iron of stored ferritin leaves the cell and enters the plasma (28). Deficiency in the B-vitamin *pyridoxine* leads to increased iron absorption in experimental animals and an increase in total body iron.

Blood Plasma Iron

Iron is transported in the plasma in the form of Fe^{+++} attached to siderophilin, the iron-binding beta$_1$-globulin. Each molecule of siderophilin can bind two atoms of Fe^{+++}, but the total iron-binding capacity (TIBC) of the plasma is less than half utilized in normal individuals. Only in abnormal instances of excessive iron absorption (*hemochromatosis*) or following transfusion, does the plasma iron approach full saturation of the siderophilin, approximately 3 mgm. Fe^{+++} per liter of plasma. The mean normal blood plasma iron is 1.05 mgm. per liter of plasma, with a standard deviation of 0.33 mgm. (9).

Plasma iron can have its origin from ferritin of the intestine or any of the storage organs, or it may be liberated directly by the breakdown of blood pigment into bile pigment. The maximum value for plasma iron is likely to occur at about 8 A.M. and the minimum at about 6 P.M., reversed in night workers. These diurnal variations coincide with similar variations in plasma bilirubin (47), suggesting that hemoglobin is broken down more rapidly during sleep than during waking hours. Iron is removed from plasma by all metabolically active tissues, particularly by the bone marrow for hemoglobin formation and by the liver and spleen for storage as ferritin.

Excretion of Iron

The body is remarkably conservative in regard to iron. A minute amount is excreted in the bile, subject to possible reabsorption in the intestine. Normal cell-free urine contains no iron (52). There is a regular and continuous desquamation of cells which, of course, contain iron. Such desquamation occurs from the surface of the skin and from the urinary and intestinal tract. The combined excretion of iron by all of these channels has been estimated as about 0.5 to 1.5 mg./day. This rate of iron loss is so insignificant that nutritional iron deficiency in the adult male is extremely uncommon. In women the menstrual loss is about 300 mgm. of iron a year. A loss of blood equivalent to 400 mgm. or more of iron may be expected with each childbirth. Hemorrhage from accident or disease may, of course, impose an iron loss upon any person at any time.

Iron Deficiency

In spite of the very small rate of loss of iron from the body under normal circumstances, iron-deficiency anemia is one of the commonest deficiency diseases in clinical practice. The causes of iron deficiency are seldom purely dietary. Some significant iron loss by bleeding usually appears in the patient's history. If it does not appear, it should be assiduously sought out. Long-continued regular losses of small amounts of blood, as from bleeding hemorrhoids, can produce an iron deficiency just as effectively as a major hemorrhage. Adolescent girls are likely to be deficient in iron on account of the combined effect of the menarche and the high iron requirement for growth at that age. In general, women are more frequently iron-deficient than men on account of the menstrual loss, losses of blood at childbirth, and the contribution during pregnancy of about 350 mgm. of iron to the fetus.

The characteristic change in the blood in iron-deficiency anemia is a decrease in the hemoglobin content of the red blood discs. There is also a decrease in their size and a less notable decrease in their number. The bone marrow shows histological evidence of overactivity, with abundant normoblasts. The characteristic physical sign is pallor, the characteristic symp-

TABLE XVII-3

Recommended daily dietary allowances of iron

	Iron (mgm.)
Man (70 kgm.)	12
Woman (56 kgm.)	12
Pregnancy (latter half)	15
Lactation	15
Children up to 12 years	
Under 1	6
1–3	7
4–6	8
7–9	10
10–12	12
Children over 12 years	15

tom, fatigue. The treatment is by oral administration of ferrous salts in small but frequent doses—small to avoid gastrointestinal irritation and frequent to maintain absorption. The toxicity of injected iron salts sets a limitation on the use of the intravenous route for iron therapy, but some iron preparations are suitable for such use. In severe cases transfusions of whole blood or of red cells are given. Prevention is by minimizing blood losses and by adequate intake of iron-containing foods, supplemented by ferrous salts during pregnancy and in other situations where iron losses are predictable.

Iron Requirements

The allowances stipulated by the National Research Council (see table XVII-3) are generous in regard to iron. Since there is always some doubt about the absorbability of iron in foodstuffs, it is probably advantageous to allow a wide margin of safety. Hahn (32) estimates from uptake experiments with radioactive iron that, even for the period of most rapid growth, 5 mgm. per day of absorbable iron is adequate. The actual requirements for growth, menstruation, and pregnancy have been calculated (36) and fall well under Hahn's estimate.

The newborn infant has a store of about 60 mgm. of iron in the liver, provided the mother had an adequate diet. During the first six months of his life, the infant is likely to live on an iron-deficient diet since the iron content of milk is nutritionally negligible. At six months the iron of the liver of an exclusively milk-fed infant has decreased to about 15 mgm., 45 mgm. have gone into hemoglobin formation, yielding 13.2 grams of hemoglobin. More significantly, the normal child is born with a very high hemoglobin content of the blood, 22 grams per 100 ml., which decreases by half during the first two months. This decrease is partly by dilution and partly by hemoglobin destruction; the iron from hemoglobin destruction is available for the building of new hemoglobin.

At about six months the iron needs of the infant become critical. If there has been no significant iron intake, the reserves stored before birth have been fully utilized. It is therefore important that iron-rich foods should be included in the diet as early as they can be tolerated.

Copper

Copper, cobalt, zinc, manganese and molybdenum are the five *essential trace metals.* These are each present in the body to a total extent of less than one gram (in a 70 kg. man, there is estimated to be about 960 mg. zinc, 150 mg. copper, 18 mg. molybdenum, 12 mg. cobalt, and 11 mg. manganese). They are necessary to the organism as cofactors for various essential enzymes. See a review by Schroeder (69a) for their relationship to human health and for the possible effects of "abnormal ions", such as Cd^{++} and Al^{+++} which enter the body in small quantities and may exert effects through competitive inhibition of the essential trace metals.

In rats and hogs, adequate tissue concentrations of copper are necessary for optimal absorption of iron, transport of iron, and utilization of iron in hemoglobin synthesis. In sheep, copper has been shown to be necessary for the folding of the keratin molecule which produces the normal wool structure, and also necessary in preventing a serious disease in lambs, characterized by demyelinization in the cerebrum and the motor tracts of the spinal cord. In weanling pups, diets deficient in copper produce bone disorders, associated with anemia and graying of hair, though serum levels of calcium, phosphorus and vitamin D are maintained at normal values (4). The copper-deficiency anemia in dogs is characterized by a reduction in red cell number but, in contrast to iron-deficiency anemia, the hemoglobin content of the individual erythrocyte is normal (81).

No state of copper deficiency has been demonstrated in man, if low values of tissue copper are taken as the criterion. *Tyrosinase,* which functions in the formation of melanin, is a copper-containing enzyme, as are also *butyryl dehydrogenase* (see page 586), and *ceruloplasmin,* a plasma oxidase (41).

The adult human body contains between 100 and 150 mgm. copper; liver, kidney, and brain have the highest concentrations. Blood plasma contains about one mgm. of copper per liter carried in association with an alpha-globulin. This value increases gradually during pregnancy, is approximately doubled near term, and returns to normal within two months after delivery (20). Administration of estrogens to human subjects increases the ceruloplasmin of the plasma to values higher than have been observed under any other circumstances (68). The fetus stores copper, particularly in the liver.

Human red cells contain about one mgm. of copper per liter. The red blood cell copper concentration remains constant, but plasma copper increases in concentration (*hypercupremia*) in a number of disease states with no evidence for the reason therefor (87). Patients with Wilson's disease (hepatolenticular degeneration) have an excessive total body copper, and excrete excessive copper in the urine. Ceruloplasmin in the plasma is diminished. The fecal excretion of copper is lower than normal, probably indicating increased absorption of copper from the gastrointestinal tract. Balance experiments indicate that the copper requirement of the average adult is about 2 mgm. a day, which is fulfilled by almost any conceivable diet.

Cobalt

Cobalt, as a component of vitamin B_{12} (see page 816), is a necessary micronutrient in the human dietary. The body utilizes cobalt, as far as we know, only in the form of vitamin B_{12} or closely related compounds. Intestinal synthesis of this vitamin by bacterial action is, however, a significant source, and presumably cobalt taken in other forms is in part incorporated into vitamin B_{12}. Cobalt, other than as vitamin B_{12}, has so far found little rational use in the therapy of human diseases, although certain anemias in experimental and domestic animals have been attributed to cobalt deficiency and have been alleviated by cobalt. When given subcutaneously in large doses, cobalt salts (but not vitamin B_{12}) will correct the anemia which develops in hypophysectomized rats (12). The human requirement has not been definitely established, but is certainly very small, less than a microgram a day.

Zinc

Zinc is present in all human organs in amounts varying roughly from 20 to 180 micrograms per gram of wet tissue. Whole human blood contains 8.8 ± 2.0, and plasma contains 3.0 ± 1.6 micrograms per ml. (77). The white cells of the blood contain about 25 times as much zinc (per cell) as the red cells. A zinc-containing protein of white cells has been purified and found to contain 0.3 per cent zinc (78). The function of the zinc of the white cells is not known, but that in the red cells is accounted for by *carbonic anhydrase* (see page 725), an enzyme which contains zinc. No figure for the daily human requirement of zinc has been determined. Average human diets contain 10 to 15 mgm. per day, one mgm. or less is excreted in the urine and the remainder in the feces. Growth failure, hair loss, and skin eruptions have been noted in rats brought up on zinc-deficient rations. Some suggestions of zinc deficiency in the human have arisen in connection with beriberi and other diseases involving chronic deficiency of several known food factors, but no specific zinc-deficiency disease has been described in man.

Carbonic anhydrase is one of the few compounds in which zinc is known to be necessary for its physiological function. Yeast alcohol dehydrogenase has been reported to contain 4 atoms of zinc per molecule (79), and pancreatic carboxypeptidase contains 1 atom of zinc per molecule (80).

Manganese

Several enzymes, including glutamotransferase (46) and some of the peptidases, are activated by manganous ion. In the latter case, other divalent ions may substitute but in the former the Mn^{++} requirement is absolute. Mn^{++} probably plays a key role in oxidative phosphorylation, acting as a link between enzyme systems with pyridine or flavin coenzymes and enzyme systems with ATP as coenzyme (51).

Molybdenum

Molybdenum is an essential micronutrient for all living organisms. The need for molybdenum is particularly notable in *Azotobacter* and other nitrogen-fixing micro-organisms. The equally important nitrate reductase of green plants and of some fungi also contains molybdenum. Abnormally large amounts of molybdenum in forage cause poisoning in cattle, preventable by increase of copper intake.

Molybdenum is one of the metals that can form part of the prosthetic group of a metallo-flavo-enzyme (see page 281). The best known example is that of xanthine dehydrogenase, and enzyme which contains both molybdenum and FAD (53). Iron has also been reported as a constituent, the ratio of iron, FAD, and molybdenum being 8:2:1 (65). Aldehyde dehydrogenase has also been found to be a molybdo-flavo-enzyme (54).

Iodine

In 1895, Baumann found iodine in thyroid tissue in organic combination and surmised that the physiological potency of thyroid was related to the iodine content. Oswald, in 1901, isolated an iodine-containing protein, thyroglobulin, from thyroid. In 1916, Kendall obtained thyroxine, an iodine-containing amino acid component of thyroglobulin. In 1926, Harington devised methods for improving the yield and proposed the accepted structural formula. In 1953, Gross and Pitt-Rivers reported the more potent hormone, 3,5,3′-triiodothyronine (see page 310).

About 5 per cent of the iodine of thyroid is present as iodide ion (48). Of the total iodine of the hydrolysate of rat thyroid approximately 15 per cent is in the form of monoiodotyrosine, 30 per cent diiodotyrosine, and 20 per cent or more thyroxine. The proportion of 3,5,3′-triiodothyronine to thyroxine is about one molecule to twenty. Iodine is 10,000 times more concentrated in thyroid than in any other tissue or organ. Aside from this remarkably high concentration in the thyroid gland, iodine behaves much like any other halogen as far as distribution throughout extracellular fluid is concerned. Following large doses it appears in the gastric juice, replacing chloride. It is excreted in the urine. Since the time of Baumann we have known that the average concentration of iodine in the thyroid gland is 40 mgm. per 100 grams of fresh tissue; this makes the total iodine content of the thyroid about 10 mgm. The thyroid takes up iodine from the blood and from perfusion fluid. This uptake reaches a maximum in 10 minutes or less following an increase in the iodine content of the blood or perfusion fluid. For uptake to occur, the iodine must be in the form of iodide ion; iodine in the form of thyroxine, diiodotyrosine, or iodate is not specifically concentrated by the thyroid. Iodine in these forms is slowly converted to iodide in the blood, after which it is absorbed by the thyroid (48). The thyroid becomes saturated with iodine when the concentration in the gland is increased above the normal by 10 to 20 mgm. per 100 grams of fresh tissue; this applies to rats, guinea pigs, rabbits, and dogs. Animals maintained on an iodine-deficient diet will fix iodine in the thyroid rapidly and maximally.

The use of radioactive iodine has cleared up many problems concerning iodine metabolism in the thyroid gland. The average daily human absorption of iodide is 0.05 to 0.1 mgm. The use of a radioactive tracer shows that in such dosage 50 per cent of the iodide is in the thyroid gland within 48 hours. Iodide entering the thyroid quickly attaches to tyrosine forming diiodotyrosine groups. The further formation of thyroxine is slower, and is mediated by an enzyme system involving peroxidase and possibly xanthine oxidase. When radioactive iodine-containing amino acids are detected in the thyroid, they are found in combination as thyroglobulin, with about 60 per cent of the iodine in the form of diiodotyrosine. The iodine leaving the thyroid and entering the blood is 10 per cent in diiodotyrosine, the remainder being thyroxine and 3,5,3′-triiodothyronine. The colloid of the thyroid follicles contains a proteolytic enzyme which is present in greater amount when the thyroid secretion is experimentally stimulated by thyrotrophic hormone. The release of iodine-containing amino acids into the blood is considered to be the result of their liberation from thyroglobulin by proteolysis.

The iodothyronines and iodotyrosines are broken down in the blood or tissues, and the liberated iodide returned to the thyroid to be worked over. Thyroxine rapidly disappears after injection, although it is not destroyed by incubation with blood *in vitro*. The liver is active in the destruction and removal of thyroxine. Thyroxine has a greater effect in hepatectomized than in normal animals. Injected diiodotyrosine in small doses is rapidly destroyed in the body, releasing its iodine as iodide.

The iodine of the blood consists of two fractions: (a) inorganic iodide, which is unaffected by thyroid function but increases with administration of iodine compounds; and (b) protein-bound iodine, which directly reflects thyroid activity. The inorganic iodide normally varies between 0.5 and 4 micrograms per 100 ml. human plasma. The protein-bound iodine is normally between 3 and 8 micrograms per 100 ml. plasma; it includes the iodine of circulating thyroxine, and is decreased in hypothyroidism and increased in hyperthyroidism. Urinary excretion of a minute dose of radio-iodide is 37 per cent within 18 hours, the fecal excretion in the same time, 17 per cent. Thyroglobulin is physiologically active upon oral administration, as are iodinated proteins. Since proteins are hydrolyzed in the digestive tract, it is apparent that the hydrolytic products are physiologically active.

Ordinary albumin, casein, and other proteins may be iodinated in such a way that the products are effective in a manner comparable to thyroglobulin. Crystalline thyroxine has been isolated from iodinated proteins of known biological activity (64). The amount of thyroxine isolated was proportional to the biological activity of the proteins.

The enzymatic conversion of iodide ion by the thyroid gland to iodine-containing amino acids is inhibited in the rat when the plasma level of iodide ion goes above 20 to 35 micrograms per 100 ml. (88). Although the organic binding of iodine ceases, iodide ion is still taken up and stored by the thyroid. Such a mechanism in the human would explain the favorable therapeutic effects of iodine and iodide in hyperthyroidism. Drugs other than iodine which depress thyroid action include thiocyanates, which inhibit the entry of iodine into the thyroid, and particularly thiouracil and its derivatives, which inhibit the enzyme system which converts iodide into thyroxine, hence increase iodine storage. The thyrotrophic hormone of the anterior pituitary increases the ability of the thyroid to retain iodide, but hypophysectomy does not abolish this ability (48). It is concluded that the thyrotrophic hormone increases the size of thyroid cells but does not modify their iodine-fixing function. The pituitary thyrotrophic hormone does appear to be necessary, however, for the final conversion into thyroxine. There is some evidence for the formation of diiodotyrosine and thyroxine outside the thyroid gland. Radioactive diiodotyrosine and thyroxine have been demonstrated in thyroidectomized rats following the administration of radio-iodine.

The iodine content of soil and foodstuffs is highly variable with geographic distribution. Only in foodstuffs of marine origin can iodine be said to be uniformly plentiful. The actual requirement of iodine by the human body is small compared to that of other minerals of comparable significance. An intake of 0.1 mgm. daily is adequate. Lack of iodine is reflected in disturbances of thyroid structure and function. Most commonly this is a simple increase in the size of the thyroid causing no disturbances other than cosmetic and possibly mechanical. Less frequently there is degeneration of the thyroid with loss of function. There is no unanimity of opinion concerning the question of iodine deficiency as a cause of hyperthyroidism nor of iodine deficiency as a contributory cause of malignancy of the thyroid.

Simple goiter is enlargement of the thyroid gland without marked alteration of its physiological function. Its best recognized cause is deficiency of iodine in the diet (89). It is endemic in certain regions where the soil and native foodstuffs are poor in iodine. A further cause may be dietary. Certain foodstuffs, for example turnips, contain "anti-thyroid factors" which may presumably interfere with the metabolism of iodine. This theory has some verification by animal experimentation and may explain why the entire population of goiter belts is not afflicted. There are no particular biochemical changes in the body in simple or endemic goiter. The signs of hypo- or hyperthyroidism are absent. Hypothyroidism may develop in cases of simple goiter, or following prolonged medication with iodides, which in high concentration block the synthesis of thyroid hormones. For a discussion of the medical management of cases of simple goiter see Youmans (89).

REFERENCES

1. Adolph, E. F., *et al.* Multiple factors in thirst. Am. J. Physiol., **178:** 538–562, 1954.
2. Aikawa, J. K., *et al.* Isotopic studies of potassium metabolism in diabetes. J. Clin. Invest., **32:** 15–21, 1953.
3. Albright, F., and Reifenstein, E. C. *The Parathyroid Glands and Metabolic Bone Disease.* Baltimore, The Williams & Wilkins Co., 1948.
4. Baxter, J. H., and Van Wyk, J. J. A bone disorder associated with copper deficiency. I. Gross morphological, roentgenological and chemical observations. Bull. Johns Hopkins Hosp., **93:** 1–23, 1953.
5. Bevelander, G., and Johnson, P. L. The histochemical localization of glycogen in the developing tooth. J. Cell. & Comp. Physiol., **28:** 129–135, 1946.
6. Bodansky, A. Phosphatase studies; determination of serum phosphatase. Factors influencing accuracy of determination. J. Biol. Chem., **101:** 93–104, 1933.

6a. Bosshardt, D. K. *et al.* Effect of bromine on chick growth. Proc. Soc. Exper. Biol. Med., **92:** 219–221, 1956.

7. Burch, G., *et al.* Rates of sodium turnover in normal subjects and in patients with congestive heart failure. J. Lab. & Clin. Med., **32:** 1169–1191, 1947.

7a. Burnell, J. M. *et al.* The effect in humans of extracellular pH change on the relationship between serum potassium concentration and intracellular potassium. J. Clin. Invest., **35:** 935–939, 1956.

8. CANNON, P. R., *et al.* Sodium as a toxic ion in potassium deficiency. Metabolism, **2:** 297–312, 1953.
9. CARTWRIGHT, G. E., *et al.* Studies on free erythrocyte protoporphyrin, plasma iron and plasma copper in normal and anemic subjects. Blood, **3:** 501–525, 1948.
10. CLARK, E. P., AND COLLIP, J. B. A study of the Tisdall method for the determination of blood serum calcium with a suggested modification. J. Biol. Chem., **63:** 461–464, 1925.
11. CORNBLEET, T. Bromide intoxication treated with ammonium chloride. J. A. M. A., **146:** 1116–1119, 1951.
12. CRAFTS, R. C. The effects of cobalt, liver extract, and vitamin B_{12} on the anemia induced by hypophysectomy in adult female rats. Blood, **7:** 863–873, 1952.
13. DANOWSKI, T. S. The transfer of potassium across the human blood cell membrane. J. Biol. Chem., **139:** 693–705, 1941.
14. DARNELL, M. C., AND WALKER, B. S. Determination of sodium in biological fluids. Indust. & Engin. Chem., Analyt. Ed., **12:** 242–244, 1940.
15. DINE, R. F., AND LAVIETES, P. H. Serum magnesium in thyroid disease. J. Clin. Invest., **21:** 781–786, 1942.
16. ECKEL, R. E., *et al.* Lysine as a muscle cation in potassium deficiency. Arch. Biochem. Biophys., **52:** 293–294, 1954.
17. ENGEL, M. D., AND CATCHPOLE, H. R. Excretion of urinary mucoprotein following parathyroid extract in rats. Proc. Soc. Exper. Biol. Med., **84:** 336–338, 1953.
18. FANKUCHEN, I. X-ray studies on compounds of biochemical interest. Ann. Rev. Biochem., **14:** 207–224, 1945.
19. FAWCETT, F. W., AND GENS, J. P. Magnesium poisoning following an enema of epsom salt solution. J. A. M. A., **123:** 1028–1029, 1943.
20. FAY, J., *et al.* Studies on free erythrocyte protoprophyrin, serum iron, serum iron-binding capacity and plasma copper during normal pregnancy. J. Clin. Invest., **28:** 437–491, 1949.
21. FELDBERG, W. Present views on the mode of action of acetylcholine in the central nervous system. Physiol. Rev., **25:** 596–642, 1945.
22. FENN, W. O. The role of potassium in physiological processes. Physiol. Rev., **20:** 377–415, 1940.
22a. FINBERG, L. AND HARRISON, H. E. Hypernatremia in infants. Pediatrics, **16:** 1–14, 1955.
23. FINCH, C. A., *et al.* Iron metabolism: erythrocyte iron turnover. J. Lab. & Clin. Med., **34:** 1480–1490, 1949.
24. FISKE, C. H., AND LOGAN, M. A. Determination of calcium by alkalimetric titration. J. Biol. Chem., **93:** 211–226, 1931.
25. FISKE, C. H., AND SUBBAROW, Y. The colorimetric determination of phosphorus. J. Biol. Chem., **66:** 375–400, 1925.
26. FREEMAN, F., AND CHANG, T. S. Role of the kidney and of citric acid in production of a transient hypercalcemia following nephrectomy. Am. J. Physiol., **160:** 335–340, 1950.
27. GAMBLE, J. L. *Chemical Anatomy, Physiology, and Pathology of Extracellular Fluid.* 5th edit. Cambridge, Mass., Harvard University Press, 1950.
28. GRANICK, S. Structure and physiological functions of ferritin. Physiol. Rev., **31:** 489–511, 1951.
29. GREENWOOD, D. A. Fluoride intoxication. Physiol. Rev., **20:** 582–616, 1940.
30. GREEP, R. O., *et al.* Alkaline phosphatase in odontogenesis and osteogenesis and its histochemical demonstration after demineralization. J. Am. Dent. A., **36:** 427–442, 1948.

31. GUTMAN, A. B., AND YU, T. F. Further studies of the reaction between glycogenolysis and calcification in cartilage. Metabolic Interrelations, Trans. 1st Conf., pp. 11–26. New York, Josiah Macy, Jr. Foundation, 1949.
32. HAHN, P. F. The use of radioactive isotopes in the study of iron and hemoglobin metabolism and the physiology of the erythrocyte. Advances M. & Biol. Physics, **1:** 288–319, 1948.
33. HARRISON, H. E., AND HARRISON, A. C. Inhibition of urine citrate excretion and the production of renal calcinosis in the rat by acetazoleamide (Diamox®) administration. J. Clin. Invest., **34:** 1662–1670, 1955.
34. HARTMANN, F. W., AND BEHRMANN, V. G. The present status of plasma expanders. J. A. M. A., **152:** 1116–1120, 1953.
35. HASTINGS, A. B., *et al.* The ionization of calcium, magnesium, and strontium citrates. J. Biol. Chem., **107:** 351–370, 1934.
36. HEATH, C. W. Iron nutrition. J. A. M. A., **120:** 366–370, 1942.
37. HENDRICKS, S. B. The nature of bone salt. Ann. N. Y. Acad. Sci., **60:** 660, 1955.
38. HENDRICKS, S. B., AND HILL, W. L. The nature of bone and phosphate rock. Proc. Natl. Acad. Soc., **36:** 731–737, 1950.
39. HOFFMAN, W. S. A colorimetric method for the determination of serum magnesium based on the hydroxyquinoline precipitation. J. Biol. Chem., **118:** 37–45, 1937.
40. HOLLY, R. G. The value of iron therapy in pregnancy. Lancet, **74:** 211–214, 244, 1954.
41. HOLMBERG, C. G., AND LAURELL, C. B. Oxidase reactions in human plasma caused by coeruloplasmin. Scand. J. Clin. & Lab. Invest., **3:** 103–107, 1951.
42. KEITH, N. M., AND BURCHELL, H. B. Clinical intoxication with potassium: its occurrence in severe renal insufficiency. Am. J. M. Sci., **217:** 1–12, 1949.
43. KEMPNER, W. Treatment of heart and kidney disease and of hypertensive and arteriosclerotic vascular disease with the rice diet. Ann. Int. Med., **31:** 821–856, 1949.
44. KLEEMAN, C. R. Alcohol diuresis. II. The evaluation of ethyl alcohol as an inhibitor of the neurohypophysis. J. Clin. Invest., **34:** 448–455, 1955.
45. LADELL, W. S. S. The changes in water and chloride distribution during heavy sweating. J. Physiol., **108:** 440–450, 1949.
46. LAJTHA, A., *et al.* Manganese-dependent glutamotransferase. J. Biol. Chem., **205:** 553–564, 1953.
47. LAURELL, C. B. Diurnal variation of serum-iron concentration. Scand. J. Clin. & Lab. Invest., **5:** 118–121, 1953.
48. LEBLOND, C. P. Iodine metabolism. Advances M. & Biol. Physics, **1:** 353–386, 1948.
49. LETTER, L., *et al.* The low-sodium syndrome—its origins and varieties. Bull. N. York. Acad. Med., **29:** 833–845, 1953.
50. L'HEUREUX, M. V., AND ROTH, G. J. Influence of parathyroid extract on citric acid of the serum. Proc. Soc. Exper. Biol. Med., **84:** 7–9, 1953.
51. LINDBERG, O., AND ERNSTER, L. Manganese, a cofactor of oxidative phosphorylation. Nature, **173:** 1038–1039, 1954.
52. LINTZEL, W. Zur Frage des Eisenstoffwechsels; über das Harneisen. Ztschr. f. Biol., **87:** 157–166, 1928.
53. MACKLER, B., *et al.* Metalloflavoproteins. I. Xanthine oxidase, a molybdoflavoprotein. J. Biol. Chem., **210:** 149–164, 1954.
54. MAHLER, H. R., *et al.* Metalloflavoproteins. III. Aldehyde oxidase, a molybdoflavoprotein. J. Biol. Chem., **210:** 465–480, 1954.
55. MANERY, J. F. Water and electrolyte metabolism. Physiol. Rev., **34:** 334–417, 1954.

56. Marinis, T. P., *et al.* Sodium and potassium determinations in health and disease. J. Lab. & Clin. Med., **32:** 1208–1216, 1947.
57. Marriott, H. L. Water and salt depletion. Brit. M. J., **1:** 245–250, 285–290, 328–332, 1947.
58. McClure, F. J. Fluorine and other trace elements in nutrition. J. A. M. A., **139:** 711–716, 1949.
59. McLean, F. C., and Hastings, A. B. Clinical estimation and significance of calcium-ion concentration in the blood. Am. J. M. Sci., **189:** 601–613, 1935.
60. Morrison, S. D. A method for the calculation of metabolic water. J. Physiol., **122:** 399–402, 1953.
61. Mudge, G. H. Potassium imbalance. Bull. N. Y. Acad. Med., **29:** 846–864, 1953.
62. Nicolau, J. del C., and Engbaek, L. The nature of neuromuscular block produced by magnesium. J. Physiol., **124:** 370–384, 1954.
63. Nicolaysen, R., *et al.* Physiology of calcium metabolism. Physiol. Rev., **33:** 424–444, 1953.
64. Pitt-Rivers, R. The chemical assay of biologically active iodinated proteins: isolation of thyroxine. Biochim. et Biophys. Acta, **3:** 675–678, 1949.
65. Richert, D. A., and Westerfield, W. W. The relation of iron to xanthine oxidase. J. Biol. Chem., **209:** 179–189, 1954.
66. Robinson, S., and Robinson, A. H. Chemical composition of sweat. Physiol. Rev., **34:** 202–220, 1954.
67. Rundo, J., and Sagild, U. Total and exchangeable potassium in humans. Nature, **175:** 774, 1955.
68. Russ, E. M., and Raymunt, J. Influence of estrogens on total serum copper and coeruloplasmin. Proc. Soc. Exper. Biol. Med., **92:** 465–466, 1956.
69. Schoolman, H. M., *et al.* Clinical syndromes associated with hypernatremia. Arch. Int. Med., **95:** 15–23, 1955.
69a. Schroeder, H. A. Trace metals and chronic diseases. Advances in Internal Med., **8:** 259–303, 1956.
70. Schwartz, W. B., and Relman, A. S. Metabolic and renal studies in chronic potassium depletion resulting from overuse of laxatives. J. Clin. Invest., **32:** 258–271, 1953.
71. Sherman, H. C. *Calcium and Phosphorus in Foods and Nutrition.* New York, Columbia University Press, 1947.
71a. Siri, W. E. Gross composition of the body. Advances in Biol. and Med. Phys., **4:** 239–280, 1956.
72. Soberman, R., *et al.* The use of antipyrine in the measurement of total body water in man. J. Biol. Chem., **179:** 31–42, 1949.
73. Stearns, G., and Warweg, E. The partition of phosphorus in whole blood and serum; the serum calcium and plasma phosphatase from birth to maturity. J. Biol. Chem., **102:** 749–765, 1933.
74. Steele, J. M. Body water in man and its subdivisions. Bull. N. Y. Acad. Med., **27:** 679–688, 1951.
75. Tarail, R., and Elkinton, J. R. Potassium deficiency and the role of the kidney in its production. J. Clin. Invest., **28:** 99–113, 1949.
76. Tibbetts, D. M., and Aub, J. C. Magnesium metabolism in health and disease. J. Clin. Invest., **16:** 491–515, 1937.
77. Vallee, B. L., *et al.* The relationship between carbonic anhydrase activity and zinc content of erythrocytes in normal, in anemic, and other pathological conditions. Blood, **4:** 467–478, 1949.

78. VALLEE, B. L., *et al.* Metalloproteins. Soluble zinc-containing protein extracted from human leucocytes. Arch. Biochem. Biophys., **48:** 347–360, 1954.
79. VALLEE, B. L., AND HOCH, F. L. Zinc, a component of yeast alcohol dehydrogenase Proc. Natl. Acad. Sci. U. S., **41:** 327–338, 1955.
80. VALLEE, B. L., AND NEURATH, H. Carboxypeptidase, a zinc metalloenzyme. J. Biol. Chem., **217:** 253–261, 1955.
81. VAN WYK, J. J., *et al.* The anemia of copper deficiency in dogs compared with that produced by iron deficiency. Bull. Johns Hopkins Hosp., **93:** 41–46, 1953.
82. VERNEY, E. B. Absorption and excretion of water; antidiuretic hormone. Lancet, **2:** 739, 781, 1946.
83. WALKER, B. S., AND WALKER, E. W. The organic fraction of urinary phosphorus. J. Lab. & Clin. Med., **18:** 164–166, 1932.
84. WALKER, B. S., AND WALKER, E. W. Normal magnesium metabolism and its significant disturbances. J. Lab. & Clin. Med., **21:** 713–721, 1936.
85. WALSER, M., *et al.* Body fluids in hypertension and mild heart failure. J. A. M. A., **160:** 858–864, 1956.
86. WILD, J. B. Mechanical drainage of massive edema. J. A. M. A., **159:** 26–27, 1955.
87. WINTROBE, M. M., *et al.* The function and metabolism of copper. J. Nutrition, **50:** 395–419, 1953.
88. WOLFF, J., AND CHAIKOFF, I. L. Plasma inorganic iodide as a homostatic regulator of thyroid function. J. Biol. Chem., **174:** 555–564, 1948.
89. YOUMANS, J. B. Mineral deficiencies. J. A. M. A., **143:** 1252–1259, 1950.

CHAPTER 18

Respiration and Acidosis

Each individual cell respires. Aerobic oxidations, which supply the major portion of cellular energy, require molecular oxygen as the ultimate hydrogen acceptor. Carbon dioxide is produced by the decarboxylation of organic acid metabolites, as in the citric acid cycle. Internal, or cellular, respiration is the gas exchange of individual cells, their uptake of oxygen and output of carbon dioxide. Blood and extracellular fluids transfer oxygen and carbon dioxide between cells and pulmonary alveolar capillaries. This can be called intermediary respiration. In the alveoli of the lung, gas exchange occurs between blood and alveolar air. This, together with pulmonary ventilation —the physical movement of air in and out of the lungs, can be designated as external respiration.

GASEOUS ENVIRONMENT

The Gas Laws

At a constant temperature the product of the pressure and the volume of a confined sample of a perfect gas is constant. If the temperature is varied, the pressure-volume product varies in proportion to the absolute temperature. These relationships can be summarized

$$PV = nRT$$

where P is pressure, V is volume, n is number of mols of gas, T is the absolute temperature, and R is the "gas constant", which must be expressed in units conforming to the units used in measuring pressure and volume. If pressure is given in atmospheres and volume in liters, the value of R is .082 liter-atmospheres per degree per mol.

This "perfect gas equation" is derived from the kinetic theory of gases and assumes that the individual gas molecules occupy no significant portion of the total volume, and that there is no attraction nor repulsion between individual gas molecules. These assumptions do not hold for any actual gas, therefore this is an approximate, and not an exact, physical law. The deviations from the perfect gas law are greatest when (a) the gas is near its liquefying point, or (b) the gas is under high pressure. A more exact but more troublesome form of the perfect gas equation involves corrections for molecular volume and for intermolecular forces—the van der Waals corrections. The physiologically significant gases closely approximate the behavior of a perfect gas at physiological temperatures, hence van der Waals corrections are seldom necessary in physiological studies. For the application and derivation of the van der Waals corrections and the derivation of the perfect gas equation, the interested student may consult texts of physical chemistry.

The perfect gas equation applies also to a mixture of gases, so that at constant volume and temperature the fraction of the total pressure exerted by one component is equal to its *mol fraction*. By mol fraction is meant the mols of the component in question divided by the total mols of all components of the gas mixture. If the pressure is constant, the fraction of the total volume occupied by one component is equal to its mol fraction. To summarize, assuming a mixture of gases under constant volume, pressure, and temperature, the mol fraction of any single gas equals its pressure fraction and equals its volume fraction. *Partial pressure* is the actual pressure exerted by a single component, and equals total pressure multiplied by mol fraction. Note that the percentage by volume of a single component equals 100 multiplied by mol fraction. These relationships can be exemplified by a consideration of outdoor air and of alveolar air.

Outdoor Air

If the water vapor is removed from a sample of outdoor air, and the sample is then analyzed, its composition will be found to be as shown in the first two columns of table XVIII-1. These values are constant.

If the dry outdoor air is at the same pressure as that of the atmosphere, which has a mean pressure at sea level of 760 mm. Hg, the partial pressure exerted at mean atmospheric pressure by each component will be found by multiplying its mol fraction by 760 mm. Hg. These values make up the third column of table XVIII-1.

The partial pressures of O_2 and CO_2 determine the direction of movement of these respiratory gases in the lungs, body fluids, and tissues. Each gas

TABLE XVIII-1

Composition of outdoor air

	PER CENT BY VOLUME	MOL FRACTION	PARTIAL PRESSURE
			mm. Hg
Oxygen	20.94	0.2094	159.1
Carbon dioxide	0.04	0.0004	0.3
Nitrogen and rare gases	79.02	0.7902	600.6

moves with the gradient of partial pressure from regions of high partial pressure to regions of lower partial pressure.

The actual partial pressure of oxygen in outdoor air may occasionally be higher than the value given in the table. This would occur in dry weather with high barometric pressure. More frequently, the actual partial pressure of oxygen is somewhat less than the quoted figure, since outdoor air usually contains water vapor which contributes to the total volume and pressure. Physiologically, we are more concerned with pressure than with volume, since pressure is the motive power of gas exchange in lungs and tissues. We therefore calculate the correction for water vapor in terms of pressure. The vapor pressure of water has been experimentally determined over the entire range of temperatures compatible with life. If air is saturated with water vapor, the barometric pressure is corrected by subtracting the known vapor pressure of water (often called "aqueous vapor tension") at the observed temperature from the observed barometric pressure. The actual partial pressure of oxygen or any other atmospheric gas is then calculated by multiplying its mol fraction by the corrected barometric pressure.

If air is neither dry nor saturated with water vapor (this is the usual situation with outdoor air), the vapor pressure of water at the observed temperature must first be multiplied by the fraction of saturation with water vapor. This fraction is the *relative humidity*, which is usually stated as a percentage. It can be measured by the use of a sling psychrometer or similar instruments measuring the differences between the temperatures indicated by a dry-bulb and a wet-bulb thermometer.

The corrections just described are necessary to establish the exact value of the partial pressure or mol fraction or partial volume of oxygen or carbon dioxide in outdoor air. In the majority of quantitative respiration experiments it is not necessary to establish this value. The volume of dry air inspired during a given time can ordinarily be calculated from the nitrogen of the expired air collected during that time. This calculation depends upon the fact that the dissolved nitrogen of the blood and tissue fluids is in saturation equilibrium with the nitrogen of the air, and there is neither gain nor loss of nitrogen by the body during respiration under normal partial pressures of nitrogen.

During respiration under conditions of greatly increased air pressure, as in the caissons used in tunneling under bodies of water, the tissues and body fluids dissolve more nitrogen than at ordinary atmospheric pressures, following Henry's Law—at constant temperature the solubility of a gas in a liquid is proportional to the partial pressure of the gas. Since molecular nitrogen is metabolically inert, such an increase in dissolved nitrogen produces no disturbance. The danger of high air pressures lies in too rapid a return to normal pressure. Bubbles of gas, chiefly nitrogen, form in the tissues, causing local or referred pain. Nitrogen is more soluble in fatty than in non-fatty tissue, which accounts for the particular tendency for such bubbles to form in fascial planes and in the fat of the popliteal region. *Aeroembolism*, or the formation of bubbles within the blood stream, is less likely to occur, since the arterial blood is in equilibrium with alveolar and atmospheric air during decompression. Aeroembolism is potentially fatal, and may occur with extremely rapid decompression, or as a result of local spasm of blood vessels or their compression by extravascular bubbles. The attacks of localized pain resulting from decompression are known to caisson workers as "the bends". Trouble is avoided by lowering the pressure in stages, each stage decreasing the excess pressure by one half, until the excess pressure is not more than one atmosphere. Decompression from one atmosphere can be accomplished safely in one stage.

At any considerable height above sea level, the mean barometric pressure is distinctly less. The sea-level pressure of 760 mm. Hg is reduced to 623 mm. Hg at a height of 1 mile and to 518 mm. Hg at a height of 2 miles. The great majority of humanity lives within a mile of sea level. Groups who live at unusual heights acclimate themselves to the low oxygen partial pressure by physiological adaptation such as an increase in the blood erythrocyte count and in the diameter and tortuosity of the smaller blood vessels, and by biochemical adaptations such as a reduction of the alkaline reserve (see page 732) to maintain blood within the normal pH range and an increase in the readiness with which hemoglobin will release oxygen to the tissues (24).

The possibilities and degree of such adaptation, however, are limited and the advent of air flight at high levels has introduced new problems to medicine. In flying above a height of 10,000 feet, supplementary oxygen is usually supplied. At 40,000 feet, even 100 per cent oxygen, at the pressure normal to those heights, is inadequate. Pressurized equipment is then called for.

In normal persons, a sudden decrease in oxygen partial pressure either as the result of a too-rapid ascent or as the consequence of sudden leakage of pressurized equipment, will result in a loss of consciousness as a result of cerebral hypoxia. A person exposed to altitudes of 26,000 feet remains conscious for about 220 seconds, and this period decreases about 20 seconds with each successive thousand feet. In addition, a condition similar to bends may result during rapid ascents in aircraft from ground level to altitudes above 30,000 feet and during relatively slow ascents above this altitude. Another uncomfortable concomitant of travel at high altitudes and one which is a direct result of the gas laws, is the distention of stomach and intestines due to the expansion of trapped gases within them.

Alveolar Air

The air in the alveoli of the lungs differs in composition from the atmospheric air. In the first place, it is nearly or fully saturated with water vapor at body temperature. In calculations of partial pressures it is usually assumed to be fully saturated. Actual measurements indicate a lag in reaching body temperature, full saturation, or both. The discrepancy during normal respiration is not more than 2 mm. Hg of water vapor pressure. During forced breathing the lag may be greater.

At 37°C. the partial pressure of water vapor at saturation is 47 mm. Hg. The total pressure of air in the bronchial tree, including the alveoli, is identical with the barometric pressure outside during all phases of normal pulmonary ventilation. Therefore the partial pressure of any single component of the alveolar air equals the mol fraction of that component multiplied by (barometric pressure minus 47) mm. Hg.

Since the entire volume of air in the lungs is not exchanged with each inspiration and expiration, the air which enters the alveoli is a mixture of freshly inspired air and air which has already exchanged gases with the blood. Direct analysis of alveolar air is impossible in the intact human subject. No great error is introduced by the assumption that the last portion of air expelled during a forced expiration is of a composition close to that of alveolar air. Analyses of such "end expiratory air" in normal men at rest and at sea level show an average oxygen partial pressure of 97.4 mm. Hg (6). Analysis of the arterial blood of the same subjects gave an average value of 97.1 mm. Hg. An indirect method has been devised (17) for calculating the partial pressures of alveolar CO_2 and O_2 from the partial pressures of CO_2 in arterial blood and of CO_2 and O_2 in the expired air. The results obtained by this method are designated as the "effective" partial pressures and represent a physiological mean pressure. Using this method a greater partial pressure gradient can be demonstrated for oxygen between alveolar air and arterial blood. This difference averages 9 mm. Hg at rest and 16.5 mm. Hg after achieving a steady state during the performance of work. Lowered oxygen content of inspired air to a simulated altitude of 16,500 feet produced no significant alteration of the gradients. Average values of the partial pressures of the respiratory gases in alveolar air and arterial blood in the resting human subject at sea level may be seen in table XVIII-2. A variation of ± 7 mm. Hg for oxygen and of ± 5 mm. Hg for carbon dioxide may occur. The partial pressures of carbon dioxide in alveolar air and arterial blood are shown as identical in the table. Actually the blood shows about 0.5 mm. Hg more carbon dioxide than the alveolar air. The slight differences in partial pressure are not so much the result of failure of equilibration in the alveoli as from the fact that all the blood returning from the lungs has not passed through alveolar capillaries.

TABLE XVIII-2

The respiratory gases in air and in blood

(Figures other than for outdoor air are averages)

	OXYGEN	CARBON DIOXIDE
Mol fraction in dry outdoor air	0.2094	0.0004
Partial pressure in dry outdoor air (mm. Hg)	159.1	0.3
Mol fraction in alveolar air, assuming full saturation with water vapor at 37°C	0.15	0.056
Partial pressure (mm. Hg) in alveolar air	106	40
Volume contained in 100 ml. arterial blood (in ml. corrected to standard conditions)	19.6	48.2
Partial pressure (mm. Hg) in arterial blood	97	40
Volume contained in 100 ml. venous blood (in ml. corrected to standard conditions) (at rest)	14	54.8
Partial pressure (mm. Hg) in venous blood at rest	40	46
Partial pressure (mm. Hg) in tissue fluids at rest	30 (maximum)	50 (maximum)

Increased pulmonary ventilation, independently of the rate of carbon dioxide production, increases the rate of removal of carbon dioxide from the entire bronchial tree, thus reducing the concentration and therefore the partial pressure of carbon dioxide in the alveolar air. By voluntary forced breathing the alveolar partial pressure of carbon dioxide can be reduced to about 15 mm. Hg. By the same maneuver the alveolar partial pressure of oxygen is increased to around 150 mm. Hg.

OXYGEN

Because the partial pressure of oxygen in the alveoli is higher than that in the venous blood that enters the alveolar capillaries, a *diffusion gradient* exists and oxygen diffuses across the alveolar membranes into the blood. This gain of oxygen converts venous into arterial blood which possesses an oxygen pressure equal or nearly equal to that in the alveoli.

Physical Solution in the Blood

Oxygen, on entering the blood, is dissolved in the blood plasma. A small portion remains dissolved in the plasma and is thus transported in physical solution. A larger portion leaves the plasma and unites with hemoglobin, the chromoprotein of the red blood discs. Note that all the transported oxygen must first be dissolved in plasma. Referring again to Henry's Law, the amount of oxygen which dissolves in a given volume of plasma is proportional to the partial pressure of oxygen in the gas phase with which the plasma is in equilibrium. For reasons noted in the preceding section, the dissolved oxygen in the blood may not quite reach equilibrium with the oxygen of the alveolar air.

The *absorption coefficient* of oxygen in blood plasma at 37°C. is 0.027. The absorption coefficient is the volume (measured at the standard conditions of 0°C. and one atmosphere) of a specified gas which will be dissolved by one ml. of a specified liquid at a pressure of one atmosphere. Therefore the volume of oxygen dissolved in arterial blood plasma, calculated from its partial pressure is

$$0.027 \times \frac{97}{760} \times 100 = 0.34 \text{ ml. (at standard conditions) per 100 ml. plasma}$$

But the total volume of oxygen which can be extracted in a Torricellian vacuum from arterial blood is 19.6 ± 1.2 ml. per 100 ml. blood (15). This indicates that most of the oxygen in arterial blood is transported in a state other than simple solution.

Chemical Combination in the Blood

Hemoglobin, a compound protein with four ferroheme molecules as prosthetic groups (see Chapter 5) can combine loosely with four molecules of oxygen, one at each heme. The resultant oxygenated compound is *oxyhemoglobin.* The reaction may be written as follows:

$$Hb + 4\ O_2 \rightleftarrows Hb(O_2)_4$$

In the equation, the usual chemical conventions have been followed so that Hb represents one molecule of hemoglobin and $Hb(O_2)_4$ represents one molecule of oxyhemoglobin. In many physiological and medical writings Hb represents a quarter of a hemoglobin molecule and oxyhemoglobin is written HbO_2.

Assuming that oxygen follows the perfect gas equation, one mol of oxygen would occupy about 22,400 ml. at standard conditions. Since one mol of hemoglobin (67,200 grams) combines with 4 mols of O_2 (89,600 ml. at standard conditions), one gram of hemoglobin combines with 1.33 ml. of oxygen (measured at standard conditions). The usual content of hemoglobin in the red discs of 100 ml. of human blood is about 15 grams. The oxygen capacity of the hemoglobin of normal human blood is therefore approximately 15 (1.33) or 20 ml. of oxygen measured under standard conditions per 100 ml. of blood; this expression is often abbreviated to 20 volumes per cent. The hemoglobin of arterial blood is not fully saturated with oxygen under normal conditions of respiration. As stated earlier, the oxygen content of arterial blood is about 19 volumes per cent, including both the small amount

(0.3 volumes per cent) dissolved in plasma and the larger amount combined with hemoglobin. The partial pressure of oxygen in arterial blood (97 mm. Hg) is related only to the dissolved oxygen. Oxygen combined with hemoglobin exerts no pressure. Normal newborn infants are hypoxemic at birth. The per cent saturation of the blood 2 minutes after birth is only 38 as compared with a value of 95 for adults. Within 5 minutes, however, it has risen to 67; within 17 minutes to 89; and in an hour and a half to 94 (7).

The reversible combination of oxygen with hemoglobin requires that the hemoglobin molecule be intact. The function of O_2 transport can not be maintained by heme or globin separately, nor by heme combined with proteins other than globin, nor by heme combined with denatured globin. The iron of hemoglobin remains in the ferrous (Fe^{++}) state throughout the respiratory cycle. The union of oxygen with hemoglobin to form oxyhemoglobin is thus not an oxidation. Although this union is by linkage with the iron of the heme groups, no electron transfer occurs, and the reaction is designated as an oxygenation and never as an oxidation. Unfortunately, unoxygenated hemoglobin is inconsistently called reduced hemoglobin.

The amount of oxygen which combines with a given amount of hemoglobin depends, when other conditions are held constant, upon the partial pressure of oxygen. The relationship between partial pressure of oxygen and degree of oxygenation of hemoglobin is conveniently represented in *saturation curves*. Figure XVIII-1 shows the saturation or the percentage of the total hemoglobin of a solution of hemoglobin actually converted to oxyhemoglobin plotted against partial pressures of oxygen in gas mixtures equilibrated with the hemoglobin solution. Note that with increasing partial pressure of oxygen the percentage of saturation increases. In obtaining such a curve experimentally, all variables other than the two appearing in the plot are held constant. Temperature and pH particularly affect the saturation curves of hemoglobin.

If a similar experiment is carried out with blood rather than with a solution of hemoglobin, the shape of the curve becomes more sigmoid, see figure XVIII-2, and the location of the curve depends upon the partial pressure of CO_2. Increasing the CO_2 of the gas mixture at a given partial pressure of oxygen favors the dissociation of oxyhemoglobin into unoxygenated hemoglobin and oxygen. This peculiarity of the behavior of oxyhemoglobin makes the unloading of oxygen more complete in the tissue capillaries, where the partial pressures of CO_2 are high. In the lungs, release of CO_2 to the alveolar air takes place, lowering the partial pressure of CO_2 and thereby increasing the ability of the hemoglobin to combine with oxygen at its prevailing partial pressure. The oxygen dissociation curve during the first two months of life lies somewhat to the left of the adult normal (i.e., in the direction of decreased dissociation). This may be due to the presence

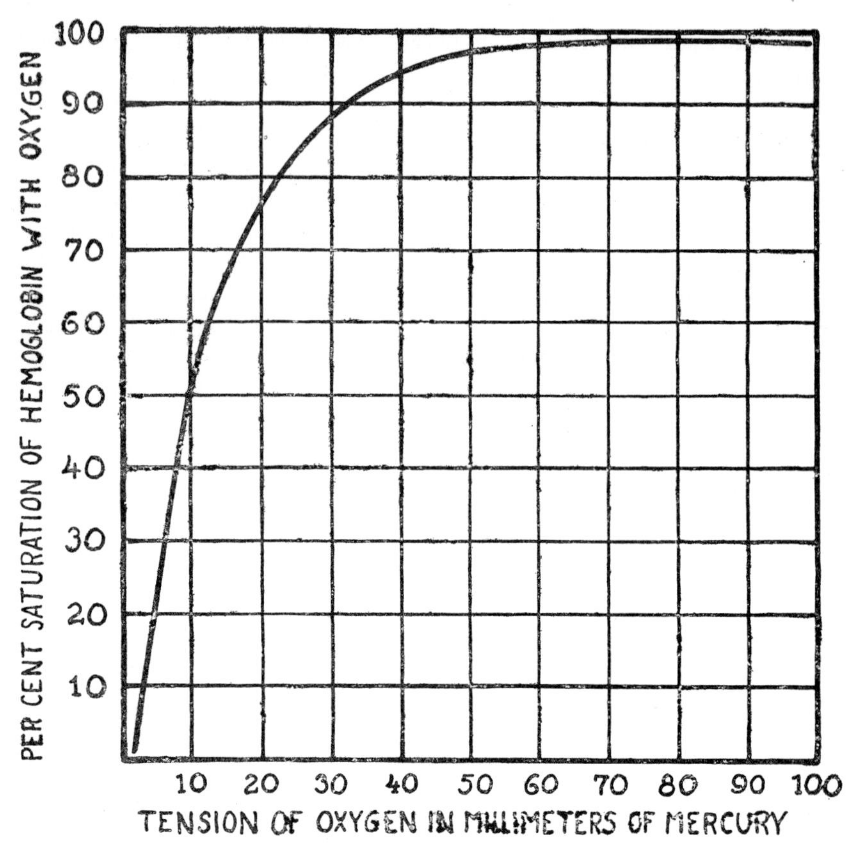

FIG. XVIII-1 Oxygen saturation curve of hemoglobin solution

of fetal hemoglobin (18). The oxygen dissociation curve for hemoglobin S (see page 197) lies to the right of the normal curve. Blood from patients with sickle-cell trait has a normal oxygen dissociation. It is suggested that these differences in affinity are not the result of the difference in structure among hemoglobins A, F, and S, but of the difference of dialyzable factors present in fetal, adult, and sicklemic blood (1, 4).

A number of physiological adaptations are indicated by the shape of the dissociation curve of oxyhemoglobin. It is clear that under physiological conditions 100 per cent saturation of hemoglobin with oxygen is never attained. This could be accomplished at partial pressures of oxygen comparable to those prevailing in outdoor air, but such concentrations of oxygen do not occur in the pulmonary alveoli. At the partial pressures of oxygen in alveolar air, the saturation of hemoglobin is about 95 per cent. At partial pressures of this magnitude, 80 mm. Hg or over, the curve has

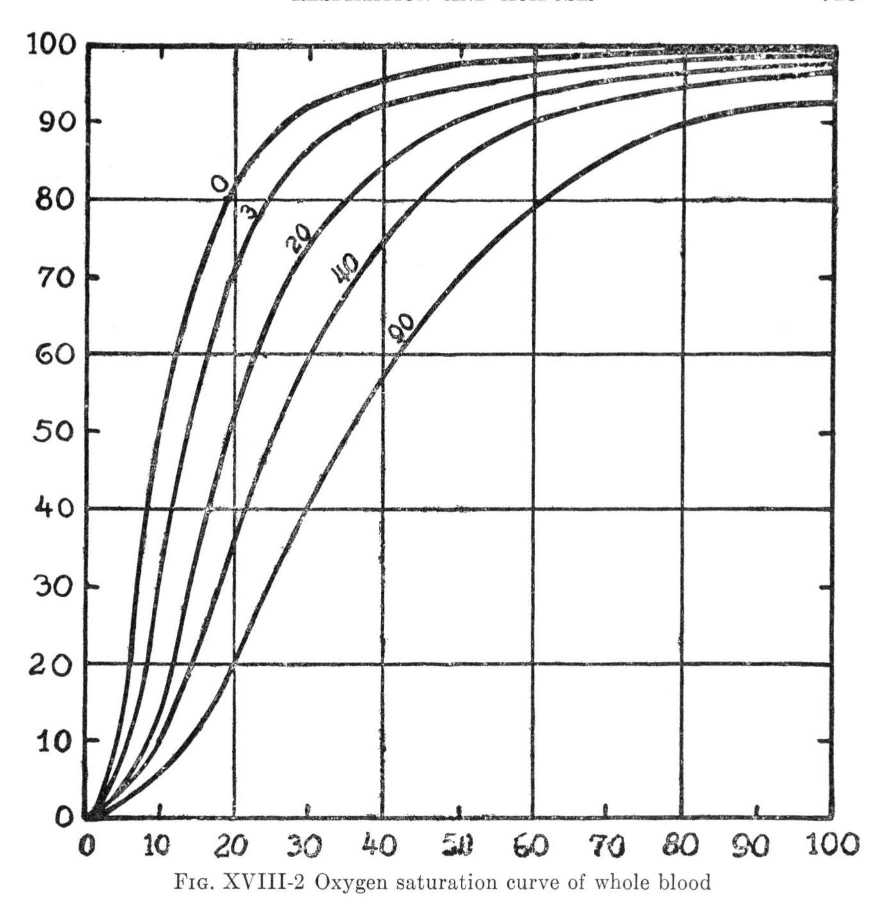

FIG. XVIII-2 Oxygen saturation curve of whole blood

distinctly flattened. Variations in alveolar oxygen above this figure have little effect upon the oxygen content of arterial blood. The steep slope of the dissociation curve between partial pressures of 20 and 60 mm. Hg corresponds to the difference in oxygen partial pressure between arterial blood and tissues, and permits effective unloading of oxygen in regions where oxygen is deficient. Just as full saturation is not attained physiologically, complete deoxygenation of the blood does not occur. The venous blood of a resting subject is still 60 to 70 per cent saturated with oxygen, and with activity this figure may fall to 25 per cent. The sigmoid or "S" shape of the whole blood curve is obviously favorable to oxygenation at partial pressures above 60 mm. Hg, and to deoxygenation at partial pressures below 50 mm. Hg. The tendency to deoxygenation is less in the curve of the pure hemoglobin solution, and is less at temperatures below human body temperature. A further advantage of being a warm-blooded animal lies in the increased rate of oxygenation and dissociation with temperature. At body temperature adjustment of oxygen saturation to altered partial pressure of oxygen occurs in about 0.01 second.

The volume of oxygen taken up by the circulating blood in the lungs is normally larger than the volume of carbon dioxide given off. The ratio is

$$\frac{\text{volumes } CO_2 \text{ output}}{\text{volumes } O_2 \text{ intake}}$$

the *respiratory quotient* (RQ), and has a value of unity only in the theoretical case where only carbohydrate is being metabolized. Examination of the formula for any utilizable carbohydrate will show that it already possesses enough oxygen to take care of the oxidation of its hydrogen to water. Additional oxygen is necessary only for the oxidation of the carbon of the molecule to CO_2. With all other foodstuffs, some oxygen is required for the oxidation of hydrogen in addition to that required for the oxidation of carbon. Hence for all foodstuffs other than carbohydrate the respiratory quotient is less than unity. In the fasting subject, the respiratory quotient is 0.82 ± 0.05; on an ordinary diet the value is about 0.85, decreasing to a limit of 0.7, which value indicates the exclusive metabolism of fat.

Although the gaseous nitrogen of the atmosphere is not metabolized by the body, there is a difference in the concentration (usually expressed as a difference in the partial pressure) of nitrogen in expired as compared with inspired air. One might be led to believe from the crude figures that the body produced nitrogen gas by the metabolism of nitrogenous foodstuffs. This is definitely not so. The apparent increase in nitrogen concentration is the result of the greater amount of oxygen uptake compared with carbon dioxide output. Only in the hypothetical case where the respiratory quotient is unity, indicating utilization of a purely carbohydrate diet, would the nitrogen content and partial pressure be the same in expired as in inspired air.

Although nitrogen takes no part in metabolism, a considerable amount is transported in the dissolved state in the blood plasma. This amount is proportional to the partial pressure of nitrogen in the alveolar air, and amounts to about 1.5 liters (under standard conditions) in the total body fluids and tissues of an average person.

Muscle hemoglobin or *myoglobin* is not identical with blood hemoglobin and has a very different dissociation curve. It is still 60 per cent saturated at oxygen partial pressures of 5 mm. Hg. It loses its oxygen rapidly at partial pressures below that value. Its physiological function is that of a reserve supply of oxygen in those rare situations when the blood reaching the muscle is almost completely deoxygenated.

Hypoxia

Deficiency of oxygen supply to tissues, locally or generally, is designated as hypoxia, or if extreme, as anoxia. Either may be the result of (a) deficient uptake of oxygen by the blood in the lungs, or (b) deficient delivery of oxygenated blood to the tissues. The second category falls entirely into the domain of circulatory physiology and pathology, and will not be considered in this book. Concerning the first category, chemical changes in the blood may be causative or contributing factors in the production of hypoxic states.

Failure of the blood to become adequately oxygenated in the lungs may be the result of a purely environmental abnormality, such as the low partial pressure of oxygen in the atmosphere at high altitudes. The commonest clinical causes of failure of oxygenation of blood in the lungs are mechanical, involving disease of the respiratory tract or paralysis of the muscles of respiration. *Cyanosis*, a blue color of the skin, is a clinical sign which is often the first indication of hypoxia, of either pulmonary or circulatory origin. It appears when the hemoglobin of the blood in the skin capillaries contains approximately 5 grams per 100 ml. in the unoxygenated form. Normally the content of unoxygenated hemoglobin in the capillary blood is about 2.5 grams per 100 ml. The critical level of unoxygenated hemoglobin required to produce cyanosis is variable on account of differences in thickness, pigmentation, and vascularity of the skin. Of greatest chemical interest are those causes of hypoxia which involve alteration of the hemoglobin molecule. A number of noxious substances combine with or otherwise modify hemoglobin, depriving it of its property of reversible union with oxygen. Some of these modifications of hemoglobin are more effective in producing cyanosis than is unoxygenated hemoglobin itself.

Methemoglobin, or ferrihemoglobin, is formed when the iron of the heme radical of hemoglobin is oxidized to the ferric (Fe^{+++}) state. The conversion of hemoglobin to methemoglobin is reversible and does not in itself result in any damage to the red cell. Methemoglobin, however, can not carry out the function of oxygen transport. Small amounts of methemoglobin, less than 0.2 grams per 100 ml. of blood, are sometimes detected in the blood of normal subjects. The presence of larger amounts (methemoglobinemia) may be the result of the intake of nitrates, nitrites, or of any one of a rather considerable list of organic compounds (5). A very limited number of cases of congenital methemoglobinemia have been reported. The presence of 1.5 grams of methemoglobin per 100 ml. of blood is sufficient to produce cyanosis. Methemoglobin may be detected qualitatively by a dark band in its absorption spectrum at 630 millimicrons. This band can be caused to disappear by the addition of a little dilute cyanide solution. Quantitative estimation of methemoglobin can be done simply by the use of the photoelectric colorimeter (12).

Hemoglobin which has been removed from the red cells and maintained in water solution exposed to the air undergoes a slow conversion to methemoglobin. Hemoglobin which has been liberated from red cells into the plasma is partly converted to methemoglobin. In plasma equilibrium is reached at about 50 per cent conversion. In the red cell, however, under normal conditions hemoglobin remains almost entirely in its normal condition with the iron in the ferrous state. There is evidence (23) that the reduction of methemoglobin to normal hemoglobin is mediated in the red cell by an enzyme system which involves a flavin enzyme, *methemoglobin reductase*, and apparently both DPN and TPN. Reduction of methemoglobin is accompanied by oxidation of glucose or lactate. Alternative substrates, shown to be effective in dog erythrocytes, are fructose, mannose, galactose, fumarate, and malate. A comparatively low content of the flavin enzyme has been observed in the red cells of patients with congenital methemoglobinemia.

A very considerable degree of cyanosis produced by methemoglobinemia can be tolerated without the production of incapacitating symptoms. The fatal concentration in man is not known. Finch (13) has reported up to 50 per cent conversion to methemoglobin without the production of severe symptoms. He recommends the slow intravenous injection of one mgm. of methylene blue per kgm. body weight in one per cent solution as the most satisfactory method for the treatment of severe methemoglobinemia. Its action is catalytic to the normal enzymatic mechanism of the erythrocytes. Congenital methemoglobinemia requires daily treatment with either methylene blue by mouth in dosages up to 300 mgm. daily or ascorbic acid in dosages up to 500 mgm.

Methemoglobin combines with the cyanide ion producing *cyanmethemoglobin*. Dosages of cyanides otherwise lethal may be combated by the intravenous injection of sodium nitrite in dosages of 0.5 grams. This results in the formation of methemoglobin, which will combine with some of the cyanide and prevent its more dangerous combination with cytochrome oxidase. The injection of nitrites should be followed by the injection of sodium thiosulfate in 25-gram dosage, which will combine with the slowly liberated cyanide ion to form thiocyanate which is not dangerously toxic and which is readily excreted. If sodium nitrite in a preparation suitable for intravenous administration is not available, inhalation of amyl nitrite is reported to be a satisfactory substitute. It should be recalled that this drug has a strong depressor action and may in itself cause unconsciousness.

Sulfhemoglobin is even more effective than methemoglobin in the production of cyanosis. The presence of 0.5 grams of sulfhemoglobin per 100 ml. of blood will produce a detectable bluish color in the patient's skin. Sulfhemoglobin is formed by the reaction of hydrogen sulfide with oxyhemoglobin. The chemical change in the hemoglobin molecule which is thus brought about is not clearly understood. Sulfhemoglobin can be identified by its absorption band at 618 millimicrons which does not disappear by the addition of cyanide. The use of cyanide differentiates this band from that of methemoglobin which is located very close to it. The formation of sulfhemoglobin, unlike that of methemoglobin, appears to be irreversible. Clinically, sulfhemoglobin is observed usually only in small and relatively insignificant amounts. It is most characteristic of so-called *enterogenous cyanosis*, which results from formation of abnormal amounts of hydrogen sulfide by bacterial action in the intestine. Enterogenous cyanosis can be produced easily in rabbits by the feeding of powdered sulfur (2), with demonstrable production of sulfhemoglobin.

Carbon monoxide hemoglobin resembles oxyhemoglobin very closely in its absorption spectrum, and in the type of union between the gas and the hemoglobin molecule. Like the reaction of hemoglobin with oxygen, the reaction of hemoglobin with carbon monoxide is reversible, but the equilibrium is much more in favor of the formation of carbon monoxide hemoglobin. At very low concentrations of carbon monoxide in the inspired air, carbon monoxide hemoglobin tends to increase in the blood. The rate of increase depends upon the concentration of carbon monoxide in the air and upon the rate of pulmonary ventilation. Approximate equations have been established experimentally (14) which indicate the expected saturation of the blood after an exposure of t minutes: with the subject at rest, the per cent saturation of the blood with carbon monoxide will be $3t$ (per cent CO in inspired air); for light activity (pulse 80, ventilation 9.5 liters per minute), the coefficient becomes 5; for light work (pulse 110, ventilation 18 liters per minute), the coefficient is 8; and for heavy work (pulse 135, ventilation 30 liters per minute), the coefficient is 11. These generalizations apply when the concentration of CO in the inspired air is 0.02 per cent or more. At 0.01 per cent CO in the inspired air, the blood reaches a limit of saturation at about 7 per cent. By saturation with CO is meant the per cent of the total hemoglobin combined as carbon monoxide hemoglobin with proportional loss of oxygen capacity of the blood. The effect upon oxygen transport is actually more than proportional, since the presence of carbon monoxide hemoglobin alters the dissociation curve of the remaining oxyhemoglobin, shifting the curve to the left and making it less sigmoid. This makes the remaining oxyhemoglobin less effective in unloading oxygen in the tissues. A 50 per cent loss of hemoglobin by chronic hemorrhage may not in itself be incapacitating, but 50 per cent saturation of hemoglobin with CO is close to the point of unconsciousness from hypoxia. The presence of methemoglobin has an effect upon the dissociation curve of the remaining oxyhemoglobin qualitatively similar but quantitatively less than that of carbon monoxide hemoglobin. If both substances are present, the effects are additive (8).

Oxygen Therapy

The inhalation of pure oxygen does not, of itself, result in any important increase of the oxygen transported by the blood in the normal individual, despite the fact that the partial pressure of oxygen is increased to five times that in the atmosphere. This is not surprising if it is remembered that under normal conditions, we are working in the flat portion of the hemoglobin saturation curve (see figure XVIII-2) where, try as we might, an increase from 95 per cent saturation to 100 per cent saturation is all that can be achieved. (There is also a five-fold increase in the very minor contribution made to oxygen transport by solution of the gas in plasma.)

Where hypoxia exists, however, because of retarded circulation involving a larger than normal loss of oxygen during the interval between successive blood passages through the alveolar capillaries, or because of inflammation or edema of the lung involving a heightened barrier to gas transport across the alveolar membrane, the use of pure oxygen results in much more dramatic increases in the oxygenation of the blood since there we are working in the steep portion of the hemoglobin saturation curve.

Oxygen Toxicity

During the later years of the 19th century, Paul Bert and Lorrain Smith presented evidence that oxygen at partial pressures considerably above the normal had a toxic effect. In Paul Bert's experiments a convulsive state was produced in his experimental birds. Lorrain Smith demonstrated pulmonary irritation when experimental animals breathed oxygen at moderately high partial pressures over prolonged periods. This work was largely neglected until about 1910 when human experimentation re-emphasized the fact that oxygen can be toxic. The detailed effects of high partial pressures of oxygen are given in a review by Donald (10). In brief, exposure to oxygen at partial pressure of 450 mm. Hg will produce toxic effects. The time required is shorter the higher the partial pressure of oxygen and is extremely variable among different human subjects. The toxic effects appear to arise as the result of two separate mechanisms: (a) depression of the mechanisms of carbon dioxide transport in the blood resulting in increased blood carbonic acid with production of a respiratory acidosis; and (b) the oxidative destruction of certain important metabolic enzymes such as D-amino acid dehydrogenase, xanthine dehydrogenase, and particularly, pyruvate dehydrogenase.

The brain appears to be the most sensitive of all the tissues of the body to oxygen toxicity, with the result that common symptoms of acute oxygen toxicity are of the nature of neurological manifestations. Twitching, particularly of the lips, is an early sign which with continued exposure becomes more generalized until a convulsive state is reached. From experiments *in vitro* we learn that next to the brain the spinal cord is most sensitive, followed in order by liver, testis, kidney, lung, and muscle. As a result of our knowledge of possibilities and dangers of the toxicity of oxygen at high pressures it is possible to control the partial pressures of oxygen during diving operations in such a manner as to minimize such dangers.

CARBON DIOXIDE

Properties

Carbon dioxide does not approximate a perfect gas as closely as do oxygen and nitrogen. One mol of a perfect gas at 0°C. and 1 atmosphere would occupy 22.414 liters. Real gases usually occupy a smaller volume than this per liter because of the small, but significant, cohesive forces between the individual molecules of the gas. One mol of nitrogen occupies 22.402 liters, one mol of oxygen 22.394 liters and one mol of carbon dioxide 22.264 liters. These values play a part in the interconversion of various units used in blood-gas measurements.

Two such systems of units are the ml. of a given gas (measured under standard conditions) per 100 ml. of plasma or blood (sometimes called volumes per cent) and millimols of a given gas per liter of plasma or blood. One millimol of a perfect gas is equivalent to 22.41 ml. of the gas under standard conditions. One millimol per liter is therefore equal to 2.241 ml. of gas per 100 ml. To convert ml. per 100 ml. (the usual experimentally observed quantity) into millimols per liter (the quantity of greater theoretical significance), we must divide the former by 2.241 in the case of the perfect gas. For oxygen and carbon dioxide, the respective conversion figures are 2.239 and 2.226.

Carbon dioxide is much more soluble in water than is oxygen. At 0°C. and 1 atmosphere pressure, 1 liter of water will dissolve 48.9 ml. of oxygen and 1713 ml. of carbon dioxide. The role of physical solution in carbon dioxide transport is naturally much greater than in oxygen transport.

Again unlike oxygen, carbon dioxide reacts chemically with the water in which it dissolves. It forms *carbonic acid*, H_2CO_3. The velocity of this reaction—the hydration of carbon dioxide—when uncatalyzed is so slow that only an insignificant amount of carbonic acid is formed from CO_2 in the extracellular fluid. Within certain cells, including the red cells of the blood, carbon dioxide is rapidly hydrated by the action of a specific zinc-containing enzyme, *carbonic anhydrase*. This enzyme is present in red cells but not in plasma or interstitial fluid. Within red cells, therefore, rapid hydration of CO_2 to H_2CO_3 or dehydration of H_2CO_3 to CO_2 can take place. The direction of the reaction is determined by the partial pressure of CO_2.

Carbonic acid can ionize to form *bicarbonate ion*:

$$H_2CO_3 \rightleftarrows H^+ + HCO_3^-$$

The theoretical pK (see page 12) of this system is 3.6. However, under physiological conditions, carbonic acid is always in equilibrium with its anhydride, carbon dioxide, so that the usual situation would be more accurately portrayed by the following equation:

$$H_2CO_3 \begin{array}{l} \rightleftarrows CO_2 + H_2O \\ \rightleftarrows H^+ + HCO_3^- \end{array}$$

Since the carbonic acid/carbon dioxide equilibrium is heavily in favor of the latter, only a small fraction of the total original carbonic acid is available for ionization, so that the actually observed pK under physiological conditions is 6.11.

At the pH of the blood, 7.4, bicarbonate ion has a concentration twenty times that of undissociated carbonic acid. Bicarbonate ion, in addition to being the conjugate base of carbonic acid, is an acid in its own right since it can ionize to form carbonate ion, $CO_3^=$. The pK of this ionization is, however, 9.76, so that the concentration of carbonate ion in blood is at all times insignificant.

Carbon dioxide can react with amino groups to form *carbamino compounds*:

$$RNH_2 + CO_2 \rightleftarrows \underset{\text{carbamino compound}}{RNHCOOH}$$

This reaction occurs very rapidly and no special enzyme is required. Carbon dioxide will combine in this fashion with simple amino compounds and with the negative ions of amino acids, but not with ammonium ions or the isoelectric ions of amino acids. Note also that the carbamino reaction involves CO_2 directly. Carbonic acid or bicarbonate ion will not form carbamino compounds.

Transport

Carbon dioxide originates in the tissues by the decarboxylation of organic acid metabolites, as in the tricarboxylic acid cycle (see Chapter 14).

The CO_2 diffuses freely in solution from its intracellular site of formation into interstitial fluid and blood plasma. In these liquids, one of three things can happen to it.

1. It may remain in the plasma as dissolved CO_2. In the absence of carbonic anhydrase, the rate of hydration is slow. The dissolved carbon dioxide content of the blood (about 2 to 2.5 ml. per 100 ml. blood) is the only portion of the total carbon dioxide content that contributes to the partial pressure of that gas. As in the case of oxygen, the over-all movement of CO_2 is determined by differences of partial pressure, which is highest in the tissues where CO_2 is produced, and decreases as it passes through tissue fluids and blood and reaches the alveolar air. For values of these partial pressures, see table XVIII-2

2. Carbon dioxide may combine with the plasma proteins to form carbamino compounds. The plasma proteins, however, do not possess many free amino groups capable of reacting with CO_2, so that the contribution this process makes to CO_2 transport is a small one, amounting to not more than 1.1 ml. per 100 ml. of plasma. Carbamino compounds of plasma proteins are not affected by the state of oxygenation of the blood, so that their concentration is similar in arterial and venous blood.

3. The largest part of the carbon dioxide does not remain in the plasma at all, but diffuses into the erythrocytes, where, again, three things can happen to it.

a. Some CO_2 remains in solution in the erythrocyte water content.

b. A considerable portion of the CO_2 combines with hemoglobin to form carbamino compounds. Approximately one-fifth of the total CO_2 of the blood is carried as carbaminohemoglobin. Hemoglobin is more important in this respect than are the plasma proteins for two reasons. First, it is present in greater concentration than any other blood protein and second, it possesses a larger content of lysine, with its free amino group, than any other blood protein. Oxyhemoglobin does not form carbamino groups with the same facility that hemoglobin does. A given quantity of oxyhemoglobin can carry, roughly, only one-third as much carbon dioxide in the form of carbamino groups as can hemoglobin. This means that in venous blood where the hemoglobin/oxyhemoglobin ratio is comparatively high, carbon dioxide is more readily transported by this mechanism than in arterial blood, where the ratio is comparatively low.

c. The greatest part of the CO_2 entering the erythrocytes is hydrated under the influence of carbonic anhydrase to form carbonic acid, which in turn ionizes to form bicarbonate ion. The bicarbonate ion rediffuses back into the plasma. Erythrocyte bicarbonate and plasma bicarbonate remain at equilibrium. Since the cation with which bicarbonate is chiefly balanced within the red cell, K^+, cannot quickly leave the red cell, other anions must enter the red cell in order to maintain electroneutrality. This anion deficit is chiefly made up by Cl^-. This *chloride shift* is reversed in the lungs.

Oxyhemoglobin is a stronger acid than is hemoglobin. This means that at a given pH, it will dissociate to a greater extent. The additional negatively charged groups on the oxyhemoglobin molecule formed by this additional dissociation will replace an equivalent amount of HCO_3^-, while the latter combines with the liberated H^+ to form carbonic acid. In the presence of carbonic anhydrase, carbonic acid breaks down to CO_2 and water. This means that in the alveolar capillaries, as hemoglobin is oxygenated to oxyhemoglobin, CO_2 is liberated and is diffused into the alveolar air. A change in the opposite direction occurs in the tissues.

Since the resulting hydrogen ion excess in arterial blood is consumed by reacting with bicarbonate ion and since the hydrogen ion deficit in venous blood is replaced by ionization of carbonic acid, the net result is that there is no significant change in the pH of the blood. This sequence of events is therefore known as the *isohydric change* of hemoglobin.

When a sample of normal human arterial blood is acidified and placed in a Torricellian vacuum, it will give up all of its free and combined CO_2. This amounts to 48.2 ml. per 100 ml. of blood with a standard deviation of 1.4 ml., measured under standard conditions of temperature and pressure (15). Venous blood taken from a resting subject has a total CO_2 content of 54.8 ml. per 100 ml. of blood with a standard deviation of 1.6 ml. The arterio-venous difference of approximately 6.6 ml. of CO_2 represents the amount eliminated from 100 ml. of blood during its passage through the lungs, and the amount picked up by the same amount of blood during its circuit through the tissues. When the subject exercises, the arterio-venous difference is greatly increased.

It has been calculated (9) that hemoglobin is responsible for the carriage of 83 per cent of the total CO_2 transported by blood, either directly in the form of carbamino groups, or indirectly by means of the isohydric change during the hemoglobin/oxyhemoglobin conversion. In the latter case, CO_2 ends up in the plasma, so that volume for volume, plasma and erythrocytes contain approximately equal quantities of total CO_2.

Regulation of Respiration

Chemoreceptors located in the carotid and aortic glomi are primarily sensitive to declines in oxygen partial pressure. Nerve impulses from the chemoreceptors reflexly regulate the activity of the respiratory center in the medulla. Changes in pH and carbon dioxide partial pressure exert their effects directly on the respiratory center.

Increases in arterial oxygen partial pressure above the normal 97 mm.

Hg have virtually no effect. Decreases in arterial oxygen partial pressure have a stimulating effect upon respiration, increasing the rate and depth of ventilation of the lungs.

The respiratory rate is much more markedly affected by small deviations from the normal in arterial pH and CO_2 partial pressure. Increased concentration of carbon dioxide or lowered pH (the two are usually associated) results in an increased respiratory rate. This, by flushing the lungs with atmospheric air more vigorously, reduces the partial pressure of CO_2 in the alveoli and hence in the blood as well, the latter process serving to raise the blood pH. Decreased concentration of CO_2 or raised pH depresses respiration, allowing carbon dioxide to accumulate and the pH to fall.

The sensitivity of respiration rate to CO_2 partial pressure is such that inspired air containing as little as 1 per cent CO_2 increases the rate perceptibly. Air containing 3 to 4 per cent CO_2 suffices to double the respiration rate, and a content of 9 per cent CO_2 will increase the rate tenfold. Up to 5 per cent CO_2 in the air is readily tolerated. Between 5 and 10 per cent there is increasing discomfort. Above 10 per cent, respiratory stimulation has passed its peak and the rate of breathing begins to decline. At 30 per cent, CO_2 is narcotic and at 40 per cent, fatal in relatively short time.

The use of oxygen mixed with 5 per cent CO_2 in the treatment of asphyxiated persons is usually unwarranted. Since breathing has slowed or has even been suspended, arterial and alveolar CO_2 partial pressures in the patient are already far above normal.

Decreases in alveolar CO_2 partial pressure have as radical an effect on respiration rate as do increases, but in the opposite direction. A fall in alveolar concentration from the normal 40 mm. Hg to 30 mm. Hg reduces the respiration rate to one-fourth normal. It is possible, even easy, to reduce the alveolar CO_2 partial pressure to such low values by voluntary forced breathing, that respiration, if allowed to return to involuntary control, will cease altogether for periods as long as two minutes.

THE REGULATION OF pH

Between the approximate limits of pH 6.8 and pH 7.8 it is possible for human cells to remain alive. The much narrower limits of pH 7.32 and pH 7.46 mark the extremes of variation observed in the blood of the healthy resting human subject (15). The newborn infant is at the lower limit of this range, 7.33, but in 24 hours, the pH of its blood has risen to 7.43 (16) as CO_2 tension decreases and oxygen saturation of the blood increases. This constancy of pH is the result of the co-ordinated action of three mechanisms: (a) the buffers of the cells and the extracellular fluids; (b) the elimination by the lungs of CO_2 ; (c) the elimination in the urine of acids or bases, including the formation by the kidney of ammonia from glutamine and from amino acids in response to the stimulus of increased acidity. Mechanism (c), which involves kidney action, is discussed in Chapter 20. It should be pointed out here that renal mechanisms are much slower in their response than is the mechanism of excretion of excess CO_2 by the lungs. Renal formation of ammonia is effective against excess acidity only. The others are involved in the defense against both acidity and alkalinity. Mechanism (b) was just discussed in the section on the regulation of respiration. It is mechanism (a), then, that chiefly concerns us here.

Buffers in Blood

The most important blood buffer is the bicarbonate ion/carbonic acid system (HCO_3^-/H_2CO_3) which has a pK of 6.1 in the presence of CO_2. Knowing that, and the fact that the normal pH of blood is 7.4, it is possible to use the Henderson-Hasselbalch equation (see page 12) to calculate the ratio of the concentrations of HCO_3^- and H_2CO_3 in the blood. Thus:

$$\text{pH} = \text{pK} + \log \frac{(\text{conjugate base})}{(\text{acid})}$$

$$7.4 = 6.1 + \log \frac{(HCO_3^-)}{(H_2CO_3)}$$

$$\log \frac{(HCO_3^-)}{(H_2CO_3)} = 1.3$$

$$\frac{(HCO_3^-)}{(H_2CO_3)} = 20$$

This means that of the hydrated carbon dioxide in the blood, 95 per cent is in the form of bicarbonate ion and 5 per cent in the form of carbonic acid. The HCO_3^-/H_2CO_3 system is nowhere near its maximum efficiency as a buffer at this pH, but this disadvantage is offset by the responsiveness of the system to the gain or loss, by pulmonary mechanisms, of the dissolved carbon dioxide with which it is in equilibrium.

The tribasic acid, H_3PO_4 can ionize in three stages with pK values of 2.0, 6.8, and 12.4. Only the middle pK, that for the $HPO_4^{=}/H_2PO_4^-$ system is close enough to the pH of the blood for significant quantities of both the acid and the conjugate base to exist. Using the Henderson-Hasselbalch equation, we can easily show that of the phosphate ions present in blood, 80 per cent are $HPO_4^{=}$ and 20 per cent $H_2PO_4^-$.

The remaining buffers of the blood include the various blood proteins. Protein buffer systems are more complex than the inorganic buffers mentioned above. Instead of two or three dissociable hydrogens per molecule, there may be dozens. The ionizable groups in a protein molecule include the ϵ-NH_4^+ group of lysine, the guanidinium group of arginine, the imidazolium group of histidine and the beta or gamma carboxyl groups of aspartic acid and glutamic acid respectively. Of these, the ammonium and guanidinium groups with pK values of over 9 and the carboxyl groups with pK values of under 4 are ineffective as buffers at the pH of blood. The imidazolium group of histidine with a pK of 7.6 works very well, however.

Of the human hemoglobin molecule, 8.1 per cent by weight has been shown to be histidine (27). This represents 35 imidazolium groups per hemoglobin molecule. Thus a third function may be listed for this versatile blood protein. In addition to its activities in oxygen transport and carbon dioxide transport, it also serves as the chief protein buffer of the blood.

The pK value of any grouping within a molecule is much affected by the nature of the neighboring atoms and groups due to phenomena such as resonance. Some of the histidine groups of hemoglobin are bound, for instance, to the iron atoms in the molecule. When hemoglobin is oxygenated, the iron atoms form new bonds with oxygen molecules. The electronic configuration of these new bonds influences neighboring bonds in such a way that the imidazolium pK is lowered. In other words, oxyhemoglobin is a stronger acid than hemoglobin, a fact important, as we have explained, in carbon dioxide transport.

The clinical literature concerning regulation of pH in the human body uses the terms "acid" and "base" according to notions that belong to a comparatively early stage of acid-base nomenclature and that are quite out of accord with modern concepts. Thus, chloride ion is usually viewed as "acidic" because it is dimly associated with hydrochloric acid. Similarly, sodium and potassium ions are considered "basic" since they are vaguely reminiscent of the so-called bases, sodium and potassium hydroxide. Actually, according to the Brønsted-Lowry viewpoint (see page 10), sodium and potassium ions are neither acidic nor basic, while chloride ion, far from being an acid, is actually an exceedingly weak base (although at the pH of blood, it is best regarded as neither acid nor base.)

In modern terminology, there are two important inorganic acids in the blood, H_2CO_3 and $H_2PO_4^-$ and two bases, HCO_3^- and $HPO_4^{=}$. In addition, there are hemoglobin, oxyhemoglobin and plasma proteins, which can act as polybasic acids ionizing to yield polyacidic bases.

The buffers of blood, although the best known and most easily studied, possess less than a fifth of the buffer capacity of the body as a whole. Schwartz, Jensen and Relman (21) report that H^+ administered to human subjects was retained largely in tissue and bone where it was mainly exchanged for K^+.

A tendency toward increased alkalinity takes place primarily during the metabolism of carboxylic acids such as acetic, malic, citric and succinic acids, which occur abundantly in foodstuffs in the ionized form. When these negative ions are converted to carbon dioxide and water, their contribution to electroneutrality is compensated for by the formation of bicarbonate ion. Carbonic acid is weaker than the carboxylic acids it replaces (compare its pK of 6.11 with that for acetic acid, 4.7, malic acid, 3.4 and 5.0, citric acid, 3.1, 4.7 and 5.4, and succinic acid, 4.2 and 5.6). The result is a tendency toward an increase in pH.

The metabolic conversion of foodstuffs into compounds of greater acidity occurs in various ways. The daily output of CO_2 is the equivalent, if it were all hydrated to H_2CO_3 (which it is not) of 20 to 40 liters of normal acid. Sulfuric acid and phosphoric acid are produced by the oxidation of the S of the diet, chiefly proteins, and the P of the diet, phospholipids, phosphoproteins and nucleic acids.

Because the bicarbonate ion, HCO_3^-, is the most important of the bases in blood, the term, alkaline reserve, has been used to designate blood bicarbonate. Blood bicarbonate can be estimated by determining the CO_2 capacity of the plasma.

This measurement gives the quantity of CO_2 present in the plasma when the plasma is saturated with CO_2 at the oxygen and CO_2 partial pressures of normal alveolar air. While all the CO_2 of the plasma is not present as bicarbonate ion, the fraction so combined is the largest and most variable fraction. The other fractions, dissolved CO_2 and free H_2CO_3, are smaller and relatively constant. Hence the total CO_2 capacity varies with the amount of HCO_3^- present. The CO_2 capacity is measured by first equilibrating plasma with either actual alveolar air from the lungs of the analyst or a mixture of gases with the composition of alveolar air. A measured sample of the equilibrated plasma is then introduced into a mercury-filled tube where it is mixed with acid and subjected to a Torricellian vacuum. The extracted CO_2 is either subjected to atmospheric pressure and the volume measured, or else in a slightly different instrument is brought to a definite volume and its pressure measured. The CO_2 capacity may be calculated either as volumes per cent of CO_2 (see page 725) or as milliequivalents HCO_3^- per liter plasma. The latter terminology is more common in current use. The normal average value is 60 volumes per cent, of which 95 per cent is actually from bicarbonate ion, the remainder being dissolved CO_2 or free H_2CO_3. Expressed in milliequivalents of HCO_3^- per liter (or identically in millimols of CO_2 per liter), the corresponding value is 27. The CO_2 *combining power* of plasma is the CO_2 capacity from which has been subtracted a correction of 1.2 mEq. per liter. This correction represents the concentration of undissociated H_2CO_3 in plasma at a partial pressure of

CO_2 of 40 mm. Hg. The corrected value represents the bicarbonate content of the blood. Total CO_2 *content* of the true plasma of blood as it is drawn from the subject is normally 24 to 33 mEq. per liter, with an average value of 28.2 mEq. (22). True plasma is plasma separated from the cells of the blood under conditions of strict exclusion of air, so that no loss of CO_2 can take place. The CO_2 content of true plasma is the single most useful measurement in the evaluation of clinical states of acid-base imbalance. Actually, no single analytical determination will give a fully reliable assessment of such clinical states (22).

Acid-Base Imbalance

In uncompensated acidosis and alkalosis, the normal HCO_3^-/H_2CO_3 ratio in the body fluids is altered. The normal value, as we have already shown, is about 20. If this ratio is decreased, whether by decreased concentration of bicarbonate ion or increased concentration of carbonic acid, the pH is lowered and a condition of *acidosis* results. If the ratio is increased, either by increased concentration of bicarbonate ion or decreased concentration of carbonic acid, the pH is raised and a condition of *alkalosis* results. It is obvious from this definition that two broad types of both acidosis and alkalosis can exist, depending upon whether carbonic acid or bicarbonate ion is the substance affected. In *compensated acidosis* and *compensated alkalosis*, the tendency toward a pH shift is counterbalanced by renal or respiratory mechanisms. Thus, in chronic acidotic conditions there may be increased renal reabsorption of HCO_3^- (25). In cases of such compensation, the absolute quantities of both HCO_3^- and H_2CO_3 may be considerably altered but the ratio (and consequently the pH) is normal.

The carbonic acid concentration of the blood can be directly affected by disorders of respiration. Failure of pulmonary elimination of CO_2 results in increased concentrations of CO_2 and H_2CO_3 in tissues and body fluids. The acidosis resulting from this heightened concentration of carbonic acid is *respiratory acidosis*. In respiratory acidosis, the CO_2 content is always increased above normal but the blood bicarbonate may remain normal or may be moderately increased. It is possible in respiratory acidosis for the CO_2 content of blood plasma to be greater than its CO_2 capacity. This apparent impossibility is explained by the fact that CO_2 capacity is arbitrarily measured at a partial pressure of CO_2 equal to that in normal alveolar air. With failure to eliminate CO_2, the plasma is subject to higher partial pressures of that gas.

Respiratory acidosis results from and accompanies serious disease of the respiratory organs, such as chronic pulmonary fibrosis and emphysema, in which the danger from hypoxia far outweighs the danger of acidosis. Patients with respiratory acidosis do not always respond well to therapy with high concentrations of oxygen (3). Long continued increases of acidity or of CO_2 partial pressure in the arterial blood will sometimes depress the respiratory center, so that it no longer responds to these changes, which are stimulatory to pulmonary ventilation under usual circumstances. In this situation, pulmonary ventilation is maintained chiefly by the stimulus of hypoxia. If oxygen is given to a patient in this state, pulmonary ventilation diminishes and the partial pressure of CO_2 in the blood increases further, aggravating the pre-existing respiratory acidosis. A fall in pH to a level of about 7 results in a state of coma, followed by functional failure of the heart and respiratory centers. In either acidosis or alkalosis of respiratory origin, prompt partial compensation occurs by exchange of H^+ for Na^+ between cellular and extracellular fluid (11). Under these circumstances, exchange of K^+ is negligible.

Alkalosis resulting from decreased concentration of carbonic acid in the body fluids due to overventilation, which may occur in febrile or psychotic states, or during inhalation anesthesia, is called *respiratory alkalosis*. The most serious result of alkalosis of any type is muscular hyperirritability, which at a plasma pH of about 7.6 becomes a state of tetany.

Acid-base imbalance resulting from deviations in bicarbonate concentration are termed *metabolic acidosis* and *metabolic alkalosis* as the concentration is decreased or increased respectively. The reason for the adjective "metabolic" will require some explanation.

Approximate concentrations of cations in blood plasma in mEq./l. are Na^+, 144; K^+, 5; Ca^{++}, 5; and Mg^{++}, 2. The total cation concentration is thus 156 mEq./l. The anion concentrations, using the same units are HCO_3^-, 27; Cl^-, 104; $HPO_4^=$, 2; $SO_4^=$, 1; organic acid anions, 6; and protein anions, 16. The total anion concentration is thus also 156 mEq./l. as is to be expected from the necessity of maintaining electroneutraility. (The student is warned against pairing off particular cations and anions in the plasma against one another, as in speaking of "sodium chloride" in the blood. The ions exist as ions and not as molecules.)

Of the various plasma ions, the concentration of protein is controlled by the balance of protein anabolism and catabolism, possibly complicated by losses of protein, as in hemorrhage or albuminuria. Of the remaining ions, all are controlled by renal mechanisms, while bicarbonate ion is also controlled by respiratory mechanisms. As we have already stated, renal mechanisms are much slower than respiratory ones. This means that any deviation from the normal on the part of any ion other than bicarbonate ion tending to upset electroneutrality is compensated for by the one ion, bicarbonate ion, whose concentration can be altered relatively quickly. Decreases of any or all of the cations result in a compensatory decrease of bicarbonate ion, which in turn reduces the bicarbonate/carbonic acid ratio below 20 with a resulting acidosis, while increases in any or all of the cations result conversely in alkalosis. Similarly, an increase in any anion other than bicarbonate is compensated for by a decrease in bicarbonate ion (acidosis) and a decrease in any anion other than bicarbonate results in alkalosis. Since the concentrations of the various ions may be altered as a consequence of various metabolic reactions in the body, the terms metabolic acidosis and metabolic alkalosis are applied.

Metabolic acidosis is a more urgent clinical problem than is respiratory acidosis. Loss of cations may result from excessive excretions or direct loss of body fluid. The CO_2 content of the plasma is invariably below normal. There is either an absolute or relative decrease of the alkaline reserve and a decrease in the CO_2 capacity of the plasma, since the lost cations imply a lowered concentration of bicarbonate ions and hence a lowered ability of the plasma to transport carbon dioxide. Causes of metabolic acidosis other than loss of fluids include ketosis, failure of excretion of acid waste products by diseased kidneys or failure of renal circulation, and failure of gluconeogenesis from lactic acid in terminal stages of liver disease and in asphyxial states, resulting in accumulation of lactic acid. Note that acetoacetic acid and beta-hydroxybutyric acid, which accumulate in ketosis, both have a pK of 3.6, while lactic acid has a pK of 3.8. The presence of any of these would shift the H_2CO_3/HCO_3^- equilibrium toward the undissociated acid and would thus have an acidotic influence.

The oral administration of such drugs as ammonium, calcium, or magnesium chlorides results in *therapeutic acidosis.* The conversion of ammonium ion to urea by the liver produces hydrogen ion as one of the end products:

$$2NH_4^+ + CO_2 \rightarrow (NH_2)_2CO + H_2O + 2H^+$$

Calcium ions and, to a lesser extent, magnesium ions are absorbed more slowly than are chloride ions in the intestine. Since electroneutrality must be maintained, the deficiency of Cl^- caused by its more rapid diffusion into the body is made up for by diffusion into the intestine of HCO_3^-. The consequent decrease of HCO_3^- in body fluids results in acidosis. The general therapeutic approach to situations involving metabolic acidosis includes removal of the cause, and replacement of cations and water.

Metabolic alkalosis may result from loss of gastric HCl as a result of severe vomiting. Here the alkalotic effect is due to the loss of H^+ and not to the loss of Cl^-. In achlorhydric individuals, the loss of Cl^- in vomiting does not bring about an alkalosis, since it is not accompanied by H^+ loss. Metabolic alkalosis may be therapeutic, as from the direct administration of bicarbonate ion, usually in the form of sodium bicarbonate. Potassium deficiency can cause alkalosis in the blood because of K^+/H^+ exchanges. Thus in chronic diarrhea, K^+ may diffuse from the cell to the blood and from the blood to the intestinal tract where it will be lost in the feces. H^+ replaces the lost K^+ in the cell by diffusing into it from the blood. The result is that the cells become acidotic while the blood becomes alkalotic. In metabolic alkalosis, respiratory compensation is minimal and usually not significant (20).

Historically, acidosis was originally defined as "a condition in which the concentration of bicarbonate in the blood is reduced below the normal level" (26). It has been proposed (19) that we return to this definition, since bicarbonate is a valid measure of the *alkaline reserve*, being the base available in the blood for neutralization of acids stronger than carbonic. If this proposal is widely accepted, one can no longer use the terms acidosis and alkalosis to describe changes in the ratio of carbonic acid to bicarbonate, as we have done above. The terms, *acidemia* and *alkalemia*, would then be used to describe pH changes. Note that with this terminology, primary respiratory changes in acid-base balance would necessarily be designated as respiratory acidemia and alkalemia, since there is no significant immediate effect of respiratory disturbances on blood bicarbonate. Compensatory changes would bring about an alkalosis in response to a respiratory acidemia and an acidosis in response to a respiratory alkalemia.

REFERENCES

1. Allen, D. W., *et al.* The oxygen equilibrium of fetal and adult human hemoglobin. J. Biol. Chem., **203:** 81–87, 1953.
2. Barkan, G., and Walker, B. S. Sulfhemoglobin formation and labile iron *in vitro* and *in vivo*. Science, **96:** 66–68, 1942.
3. Beale, H. D., *et al.* Delirium and coma precipitated by oxygen in bronchial asthma complicated by respiratory acidosis. New Eng. J. Med., **244:** 710–714, 1951.
4. Becklake, M. R., *et al.* Oxygen dissociation curves in sickle cell anemia and in subjects with the sickle cell trait. J. Clin. Invest., **34:** 751–755, 1955.
5. Bodansky, O. Methemoglobinemia and methemoglobin-producing compounds. Pharmacol. Rev., **3:** 144–196, 1951.
6. Comroe, J. H., Jr., and Dripps, R. D., Jr. The oxygen tension of arterial blood and alveolar air in normal human subjects. Am. J. Physiol., **142:** 700–707, 1944.
7. Crehan, E. L., *et al.* A study of the oxygen saturation of arterial blood of normal newborn infants by means of a modified photoelectric oximeter: preliminary report. Proc. Staff Meetings Mayo Clinic, **25:** 392–397, 1950.
8. Darling, R. C., and Roughton, F. J. W. The effect of methemoglobin on the equilibrium between oxygen and hemoglobin. Am. J. Physiol., **137:** 56–68, 1942.
9. Davenport, H. W. *The ABC of Acid-Base Chemistry*. 2nd ed. Chicago, University of Chicago Press, 1949.
10. Donald, K. W. Oxygen poisoning in man. Brit. M. J., **1947, I:** 667–672, 712–717, 1947.
11. Elkinton, J. R., *et al.* Effects in man of acute experimental respiratory alkalosis and acidosis on ionic transfers in the total body fluids. J. Clin. Invest., **34:** 1671–1690, 1955.

12. Evelyn, K. A., and Malloy, H. T. Microdetermination of oxyhemoglobin, methemoglobin, and sulfhemoglobin in single sample of blood. J. Biol. Chem., **126:** 655–662, 1938.
13. Finch, C. A. Methemoglobinemia and sulfhemoglobinemia. New Eng. J. Med., **239:** 470–478, 1948.
14. Forbes, W. H., *et al.* The rate of carbon monoxide uptake by normal men. Am. J. Physiol., **143:** 594–608, 1945.
15. Gibbs, E. L., *et al.* Arterial and cerebral venous blood: arterial-venous differences in man. J. Biol. Chem., **144:** 325–332, 1942.
16. Graham, B. D., and Wilson, J. L. Chemical control of respiration in newborn infants. Am. J. Diseases Children, **87:** 287–297, 1954.
17. Lilienthal, J. L., Jr., *et al.* An experimental analysis in man of the oxygen pressure gradient from alveolar air to arterial blood during rest and exercise at sea level and at altitude. Am. J. Physiol., **147:** 199–216, 1946.
18. Morse, M., *et al.* The position of the oxygen dissociation curve of the blood in normal children and adults. J. Clin. Invest., **29:** 1091–1097, 1950.
19. Nash, T. P., Jr. Acid-base terminology. Scientific Monthly, **82:** 255–257, 1956.
20. Roberts, K. E., *et al.* Evaluation of respiratory compensation in metabolic alkalosis. J. Clin. Invest., **35:** 261–266, 1956.
21. Schwartz, W. B., *et al.* The disposition of acid administered to sodium-depleted subjects: The renal response and the role of the whole body buffers. J. Clin. Invest., **33:** 587–597, 1954.
22. Singer, R. B., and Hastings, A. B. An improved clinical method for the estimation of disturbances in the acid-base balance of human blood. Medicine, **27:** 223–242, 1948.
23. Spicer, S. S., *et al.* Studies in vitro on methemoglobin reduction in dog erythrocytes. J. Biol. Chem., **177:** 217–230, 1949.
24. Stickney, J. C., and Van Liere, E. J. Acclimatization to low-oxygen tension, Physiol. Rev., **33:** 13–34, 1953.
25. Sullivan, J., *et al.* Renal response to chronic respiratory acidosis. J. Clin. Invest, **34:** 268–276, 1955.
26. Van Slyke, D. D., and Cullen, G. E. Studies of acidosis. I. The bicarbonate concentration of the blood plasma; its significance and its determination as a measure of acidosis. J. Biol. Chem., **30:** 289–346, 1917.
27. Vickery, H. B. The histidine content of the hemoglobin of man and of the horse and sheep, determined with the aid of 3,4-dichlorobenzenesulfonic acid. J. Biol. Chem., **144:** 719–730, 1942.

CHAPTER 19

Heat and Work

It has long been known that the production of heat by friction shows that energy in the form of work can be converted into heat. The experiments of Joule and later workers showed that no matter how the work was applied, if heat loss was guarded against and unavoidable losses corrected for, exactly the same amount of heat always appeared in correspondence to the expenditure of a given amount of work. To give concrete examples, it was found that the work involved in lifting a one-pound weight 778 feet was the equivalent of the heat required to increase the temperature of one pound of water one degree Fahrenheit. Or, in metric units, 427 kgm. meters was the equivalent of one Kcal. of heat. By showing that work and heat were interconvertible in a fixed ratio, these measurements served to prove the validity of the law of conservation of energy.

Previous to this the invention of the steam engine had already showed that it was possible to bring about the reverse transformation of heat into work. But systematic study of the problem revealed a very important difference. The transformation of work into heat can be arranged so as to give a 100 per cent yield, but the reverse is not true. Steam engines never convert more than a fraction of the heat of their fuel into work, and the second law of thermodynamics (see page 213) shows they never can. Any engine, given an amount of heat Q, and taking in its steam at a temperature T_2, and discharging it at a lower temperature T_1, would never give an amount of work W greater than

$$W = \frac{Q(T_2 - T_1)}{T_2}$$

where the temperatures T_2 and T_1 are given in degrees absolute and Q and W are both expressed in identical units—say calories.

It can easily be seen that for actual heat engines W can never be more than a fraction of Q (since T_1 is never zero), and in fact no saturated steam engine operating with a boiler pressure of, for example, 163 pounds per square inch can ever convert more than 18.5 per cent of the heat into work. Even this is computed on the basis of the heat actually in the steam coming into the engine. This is of course by no means all of the heat potentially available from the fuel. Some heat is lost by radiation, and even more goes up the flue with the smoke. The efficiency of the boiler in transferring the heat of combustion to the engine may not be over 70 per cent. Thus the over-all maximum efficiency of the installation would be only 70 per cent of 18.5, or about 13 per cent. This is for a typical stationary steam plant. Steam locomotives probably do not utilize over 8 per cent of the energy available potentially in the fuel they burn. Steam engines with a very high temperature of operation (T_2) may show an efficiency as high as 24 per cent, and one form of the internal combustion engine, the diesel, may sometimes give efficiencies as high as 40 per cent.

It was natural for the early physiologists and biochemists, seeing the triumphs of the new science of thermodynamics in the field of engineering, to ask themselves—How efficient is the human body as a converter into work of the heat available from the combustion of the foodstuffs? For it was apparent that all animals, including man, must eat in order to live and work, and must eat more when the work is heavy. An obvious way of answering this question was to determine the calorific value of the foodstuffs burned in the body during a given period of work and to compare this with the work produced, expressed in heat units. The answer obtained was that the human body was more efficient mechanically than a steam locomotive, but certainly not 100 per cent efficient. This was not too surprising, for it was known that the efficiency of the locomotive is low, and thermodynamics, as we have seen, shows that 100 per cent conversion of heat into work is impossible by any machine.

However, although it was perfectly natural for the early workers in our field to ask themselves the above question and to attempt to solve it in the manner indicated, it was nevertheless entirely the wrong question. The body has no power whatever of converting heat into work. A heat engine has, for it makes use of some of the heat produced by burning its fuel to warm and thus expand a confined gas. The expansion of this confined gas does work, and because of the conversion of some of the energy contained in it into work, the gas ends up at a lower temperature (T_1) than at the start of the work cycle. The two temperatures of the gas are the T_2 and T_1 of the above equation. But the body operates in an entirely different way. It is not a heat engine, and the laws of heat engines do not apply to it. The more general laws of thermodynamics, originally based on a study of heat engines, do apply however, and we can predict now that even the body is never going to be able to convert 100 per cent of the energy potentially available in its foodstuffs into work. It is easy to see why the body can not be a heat engine. For one thing, it keeps all its parts at approximately the same temperature, and thus the temperature gradient

T_2 minus T_1, which is essential for the operation of the heat engine, is lacking.

How does the body produce work? In the previous chapters it became clear that the body makes use of the energy of its foodstuffs by oxidizing them at constant temperature by a series of enzyme reactions, and that the energy of oxidation of the foodstuffs is transferred to high-energy phosphate bonds which carry it to the effector organs where it is turned into work. ATP is obviously one of the important compounds in this process. So the body uses the energy of foodstuffs, not by converting it into heat and then converting part of this heat into work as does a heat engine, but in some way by converting part of the chemical energy directly into work. But how?

In spite of the large amount of work which has been done on this interesting question, the exact way in which the muscle accomplishes the energy transformation is not completely understood. As an example of current thinking on the subject, we may present a simple and plausible model proposed by Morales and Botts (12, 13). These authors suggest that in the extended state the molecular filaments of actomyosin are held in a configuration which is a compromise between the contracting tendency of random molecular movement and the extending tendency of repulsion due to a net charge. This net charge is supposed to be positive, due to the adsorption by the actomyosin of bivalent cations. The anion of ATP adsorbs on the myosin, and discharges it, thus reducing or abolishing the repulsion; the filaments then contract, gaining "configurational entropy" and acting in this respect like contracting rubber. Enzymatic dephosphorylation now converts ATP to ADP, which is not so strongly adsorbed by myosin. This process removes the negative charge which was contributed by the ATP, and the repulsion of the re-exposed positive charges on the myosin re-extends the muscle to its original state.

Hill (8) has objected to the suggestion that the immediate source of the energy of muscle contraction is a gain in entropy, on the grounds that such a gain in an adiabatic process would mean an absorption of heat, whereas actually a liberation of heat is observed. It is true that a gain in entropy in such a process means an absorption of heat but Morales and Botts reply that since there must obviously be also some mechanism for resetting and refiring the muscle, it might well be that this process liberates more heat than that absorbed due to the entropy change.

A conception of the muscle as a "chemical entropy engine" which would not involve changes in temperature has also been proposed (17).

If the body functions as some sort of chemical engine, the very interesting question arises, how efficient is it? How much of the theoretically possible energy of the foodstuffs does it convert into work? The maximum possible work to be obtained from a chemical reaction depends upon the quantity called in thermodynamics the free energy of the reaction. Some electrical cells can convert a very high percentage of the free energy of the chemical reaction that runs them into electrical energy. Therefore, since the body is essentially an oxidation-reduction cell producing work (instead of electrical energy, for the most part) directly, we might expect rather high mechanical efficiencies from it. For instance, we could compare the body to an electric battery coupled with an electric motor, although the analogy is probably not perfect, since the body seems to convert the chemical energy of its foodstuffs directly into work (although producing a small amount of electricity also), whereas the electric cell produces primarily electric energy which is converted by the motor into work. In an ordinary dry cell metallic zinc is consumed, and thus acts as the "fuel" of the cell, very much as the foodstuffs of the body act as fuels. In both cases the temperature remains nearly constant during the operation of the system, and in both cases the mechanical efficiency can theoretically be quite high. The essential reaction in a dry cell is

$$Zn - 2e = Zn^{++}$$

The metallic zinc is oxidized to its ions, and electrons are transferred as a result. Various complexities which need not concern us here make it difficult to compute the efficiency of this cell exactly, but the high efficiency of electric cells is indicated by the study of storage batteries such as those used in automobiles, where one can compare directly the amount of electrical energy which went into charging the battery with the amount recovered on discharging it. The chemical reactions vary with the type of cell used, but the efficiency of the ordinary lead storage battery, when discharged slowly, has been found to be as high as 95 per cent. Obviously a high proportion of the free energy of the chemical reaction involved in the discharge of the cell is converted into work. Undesirable secondary reactions keep the efficiency from being higher, and also limit the useful life of the cell, under ordinary conditions, to about two years.

Electric motors may be constructed which convert 80 per cent or more of the electrical energy fed into them into mechanical work. Thus such a motor, coupled to a good storage battery, might yield a mechanical efficiency of 0.80×95, or 76 per cent. One might possibly expect the human body to show similar high efficiencies.

Various measurements have been made of the efficiency of the human body while it was performing work, and the results indicate that the body is mechanically more efficient than a steam locomotive (a result which surprised the early biochemists but does not now surprise us), but that it is not as efficient as a battery-motor combination, or even the best internal combustion engines (diesels). A few words about the methods of computing the efficiency of the human body, and some explanation of the results, seem to be in order.

For any machine we have the relation

$$\text{Input} = \text{Output} + \text{Work lost by friction} + \text{Other losses}$$

The efficiency is the ratio output:input, usually expressed as a percentage.

We can measure the work output of the human body by the weight lifted through a given height, or in other ways. We can measure the input into the body if we know the calorific value of the various foodstuffs when oxidized in the body. A maximum value is obtained by burning the foods directly with oxygen in the bomb calorimeter, and measuring the heat evolved. The following results are obtained for the three classes of foodstuffs:

Class of foodstuff	*Kcal./gram dry weight*
Carbohydrate	4.10
Fat	9.45
Protein	5.65

Since the first two classes of foodstuffs are oxidized as completely in the body as in the bomb calorimeter, we can obtain the calorific values for human utilization merely by correcting for the imperfect absorption of the foods. It has been estimated that on the average 98 per cent of the carbohydrate is absorbed, 95 per cent of the fat, and 92 per cent of the protein. To obtain the calorific value of the proteins in the body we must allow also for the fact that they are not completely oxidized. In a classic experiment whose results are still quoted, Rubner (10) found that about 75 per cent of the potential calorific value of proteins was realized in the animal body (see page 744). Making these corrections, and rounding off the results to whole numbers, since the precise figures may vary somewhat with the particular food within any one of the classes, we obtain the following very important and easily memorized values for the calorific value of the three classes of food within the animal body:

Class of foodstuff	*Kcal./gram dry weight*
Carbohydrate	4
Fat	9
Protein	4

It is possible to build a calorimeter big enough to accommodate a human being, and such a machine makes it possible to measure the heat evolved per given time interval by a patient. In a closed system such as this calorimeter any work done by the patient does not affect the outside world as work, for it is converted by friction into heat within the calorimeter. This applies to the indispensable vital work such as breathing, heart action, kidney action, and involuntary muscular motions. If we know the amounts and proportions of the three classes of foodstuffs being utilized by the patient, we may compute how much heat he should be producing, and compare this with the observations.

In order to find out how much of the three different classes of food the patient is oxidizing, we must measure three things. To estimate the amount of protein metabolized during the interval, we must know how much nitrogen is excreted in the urine. All of the urinary nitrogen was once protein, and by assuming the usual conversion factor of 6.25, we may compute the amount of protein metabolized in the time the patient spends in the calorimeter and from this, the heat produced from protein metabolism (see page 744). We must also know the amount of oxygen consumed by the patient; and third, we must know the amount of carbon dioxide produced. The reasons for wanting these last two items of information will be immediately explained.

Respiratory Quotients

Different classes of food require varying amounts of oxygen to produce a given amount of carbon dioxide. From a glance at the formula of a typical carbohydrate (page 103) and a typical fat (page 120), it is evident that carbohydrates produce more CO_2 per mol of oxygen used, since they already contain enough oxygen to oxidize the hydrogen to water and only need oxygen for the carbon. Fats need oxygen for both the carbon and hydrogen (except for the comparatively small amount of oxygen in the ester linkage, they have no molecular oxygen).

The *respiratory quotient* (RQ) of a substance is the ratio of the volume of carbon dioxide produced in the oxidation of a substance to the volume of oxygen used. When we can write the chemical reaction involved, we can compute the RQ. For instance, in the case of glucose, we have the equation

$$C_6H_{12}O_6 + 6O_2 \rightarrow 6CO_2 + 6H_2O$$

Recalling that Avogadro's law states that all gases at standard conditions of pressure and temperature occupy the same volume per mol, we easily see that the ratio

$$\frac{\text{Liters } CO_2 \text{ produced}}{\text{Liters } O_2 \text{ used}} = 6/6 = 1.00$$

For any given fat the theoretical RQ can also be computed. For instance, if we take tristearin we have

$$C_{57}H_{110}O_6 + 81.5O_2 \rightarrow 57CO_2 + 55H_2O$$

TABLE XIX-1
Metabolism of protein

	C	H	N	O
Composition of 100 grams of dried muscle..	52.38	7.27	16.65	22.68
Urine contains 42.45 grams................	9.41	2.66	16.28	14.10
Feces contain 2.94 grams..................	1.47	0.21	0.37	0.89
Total excreted in urine and feces..........	10.88	2.87	16.65	14.99
Remainder................................	41.50	4.40	—	7.69
Allowing for intramolecular water.........	41.50	3.43	—	—
Oxygen needed for remaining H............				27.22
Oxygen needed for remaining C............				110.58
Total oxygen needed.......................				137.80

So the RQ equals 57/81.5 or 0.699. Corresponding calculations give an RQ of 0.703 for tripalmitin and 0.713 for triolein. Mixed body fats give an RQ of about 0.707. It is generally assumed that a fair average value for the RQ of fats is 0.71.

The computation of the RQ of proteins for their metabolism in the human body is rendered more complicated by the fact that they are not completely oxidized. It seems best to consult an actual experiment. Rubner, one of the founders of the science of animal calorimetry, fed an animal dried muscle which had been analyzed chemically. One hundred grams of the dry material contained 5.5 grams of ash. In an experiment calculated by Loewy (10), an animal eliminated, of 100 grams of the organic part of protein, 42.5 grams in the urine and 2.94 grams in the feces. The detailed computations are shown in table XIX-1. The CO_2 produced is calculated by adding the carbon remaining (41.50 grams) and the oxygen required to oxidize it (110.58 grams). We obtain 152.08 grams of CO_2. Now the weight of one liter of oxygen at standard conditions of temperature and pressure is 1.4290 grams, and the weight of one liter of carbon dioxide is 1.9767 (7). Using these values, we find

$$RQ = \frac{152.08/1.9767}{137.80/1.4290} = 0.7978$$

Lusk, using an older value for the density of carbon dioxide, obtained from the above experimental data a value of 0.8016 for the RQ of protein. This is the value which appears in nearly all the books on the subject. To estimate the calorific value of dried muscle in the animal body Rubner first burned some in the bomb calorimeter, thus obtaining the total heat value. By similarly burning the urine and feces resulting from the ingestion of 100 grams of the dried muscle and subtracting, Rubner was able to calculate that although the total calorific value of the 100 grams of dried muscle was 534.5 Kcal., 129.77 Kcal. was lost in the urine and feces, leaving a fuel value available to the body of 404.73 Kcal. Slight deductions for the heat present in the protein in its dissolved state but lost on drying, and for the heat of solution involved in dissolving urea and other urinary constituents had to be applied. This left 400.06 Kcal. as the maximum energy obtainable from 100 grams of the solids of dried meat. It is from this that we obtain the 4 Kcal. per gram given earlier (page 742) as the calorific value of protein in the body. It can be seen that the heat available in the body from protein is about 75 per cent of that available in the bomb calorimeter. About 25 per cent of the heat is lost because of the needs of the body to excrete the nitrogen from proteins in suitable forms.

The most efficient utilization of proteins would demand the excretion of nitrogen in the form of ammonia, and this some fishes do. Mammals, however, not having a whole ocean into which to excrete, have to choose the less toxic urea, and birds, being hatched out of eggs, must conserve water by excreting the more insoluble uric acid. The over-all loss of energy potentially available from our food resulting from the excretion of urea instead of ammonia is smaller than might be imagined, less than 3 per cent (16).

From this experiment we may calculate that each gram of nitrogen appearing in the urine means 5.923 liters of oxygen consumed and corresponds to 4.723 liters of carbon dioxide produced. Earlier calculations, based on the assumption that the atomic weight of hydrogen is 1.00000, led to the values 5.94 and 4.76, and these figures often appear in the literature. One liter of oxygen, when used to oxidize protein, produces 4.463 Kcal. of heat (10). An earlier and slightly higher value appears in some American books (15). We can compute the heat which is produced by the metabolism of protein by multiplying the number of grams of nitrogen secreted in the urine by $5.923 \times 4.463 = 26.433$. Also, since 100 grams of the dried muscle in Loewy's experiment caused the production of 16.28 grams of nitrogen in the urine, we assume a conversion factor of N to p of $1/0.1628 = 6.142$.

If we record the oxygen consumed, carbon dioxide produced, and urinary nitrogen, all for the same period of time, for a patient, we can calculate the amount of heat produced, from our knowledge of the respiratory quotients and calorific values of protein, carbohydrate, and fat. From the urinary nitrogen we then calculate the metabolic heat due to protein, and the oxygen consumed and carbon dioxide produced. Deducting these from the total values observed, we have the oxygen consumed and carbon dioxide produced by the fat and carbohydrate being oxidized. Since the respiratory quotients of fat and carbohydrate are different, each observed respiratory quotient between that of pure fat and that of pure carbohydrate corresponds to the utilization of some particular mixture of fat and carbohydrate. For each such mixture the utilization of one liter of oxygen produces a definite amount of heat. This means that a knowledge of the non-protein RQ enables us to compute the non-protein calories which the patient is p oducing. This, added to the protein-produced calories, gives the total heat production.

As we have seen, the RQ of pure carbohydrate is 1.00, and it was formerly assumed that the RQ of fat was 0.707. In oxidizing 1.232 grams of carbohydrate, one liter of oxygen is consumed and 5.047 Kcal. of heat produced. It was formerly assumed that one liter of oxygen oxidized 0.502 grams of fat with the production of 4.686 Kcal. of heat. Mixtures of carbohydrate and fat in various proportions would give RQ values intermediate between 0.707 and 1.00 and calorific values between 5.047 and 4.686.

More recent analyses of (human) fat lead to somewhat different values for the respiratory quotient and calorific value of fat. Cathcart and Cuthbertson (5a) analyzed fat from the panniculus adiposus abdominalis from seven subjects and found that 0.522 gram, utilizing one liter of oxygen, gave 4.749 Kcal. of heat. A composite result from analyses of fat from skeletal muscle and liver, which Cathcart and Cuthbertson believe to be more representative of the fat being oxidized by the fasting patient, indicated that 0.516 gram of fat utilized one liter of oxygen with the production of 4.735 Kcal. of heat and had an RQ of 0.718. Therefore, according to modern ideas, mixtures of carbohydrate and fat in various proportions would give respiratory quotients ranging from 0.718 to 1.00 and calorific values ranging from 4.735 to 5.047.

Abramson (1) showed a simple way of dealing with the problem. If we make use of the above values of the oxygen consumption, carbon dioxide production and calorific value per liter of oxygen of the three classes of foodstuffs, accepting the modern values for fat, we can write three equations

$$f/0.516 + c/1.232 + 5.923\,N = O_2 \qquad [1]$$

$$0.718f/0.516 + c/1.232 + 0.7978(5.923\,N) = CO_2 \qquad [2]$$

$$H = 4.735f/0.516 + 5.047c/1.232 + 4.463(5.923\,N) \qquad [3]$$

where f represents grams of fat, c carbohydrate, N urinary nitrogen, O_2 and CO_2 oxygen and carbon dioxide in liters, and H stands for heat in Kcal. Since N, O_2, and CO_2 are known from observation, we can solve for c and f in equations [1] and [2]. We obtain

$$f = 1.830\,(O_2 - CO_2) - 2.191\,N \qquad [4]$$

$$c = 4.369\,CO_2 - 3.137\,O_2 - 2.065\,N \qquad [5]$$

We already have (where p represents grams of protein)

$$p = 6.142\ N \qquad [6]$$

Substituting in equation [3], we have

$$H_f = 16.791\ O_2 - 16.791\ CO_2 - 20.109\ N \qquad [7]$$

$$H_c = -12.850\ O_2 + 17.897\ CO_2 - 8.459\ N \qquad [8]$$

$$H_p = 26.434\ N \qquad [9]$$

where H_f, H_c, and H_p stand for the portions of the metabolic heat being produced by fat, carbohydrate, and protein, respectively. Adding these, or substituting the values of c and f from equations [4] and [5] into [3], we obtain

$$H = 3.940\ O_2 + 1.105\ CO_2 - 2.134\ N \qquad [10]$$

From this equation the total heat being produced can be found, and from equations [7], [8], and [9] the actual amounts being contributed by fat, carbohydrate, and protein can be calculated.

As an example we may take some results obtained by Abramson, who found that in one hour the subject L. S. excreted 0.385 grams of nitrogen in the urine, used 14.02 liters of oxygen, and produced 11.38 liters of carbon dioxide. The RQ in this case is 0.812. From equation [10] we find

$$H = 55.25 + 12.59 - 0.82 = 67.02 \text{ Kcal.}$$

produced by the patient per hour.

From equations [7], [8], and [9] we find this total heat is made up of $H_f = 36.59$, $H_c = 20.25$, and $H_p = 10.18$ Kcal. We see that this particular subject was producing 54.59 per cent of his heat by the oxidation of fat, 30.27 per cent by the oxidation of carbohydrate, and 15.19 per cent by the oxidation of protein.

From equations [4], [5], and [6] we can obtain, if we desire, the weights in grams of the respective foodstuffs oxidized in one hour, which turn out to be

$$f = 3.99 \text{ grams}$$

$$c = 4.94 \text{ grams}$$

$$p = 2.36 \text{ grams}$$

Abramson showed that these equations could all be solved by nomograms, and figure XIX-1 shows a nomogram which represents a combination and

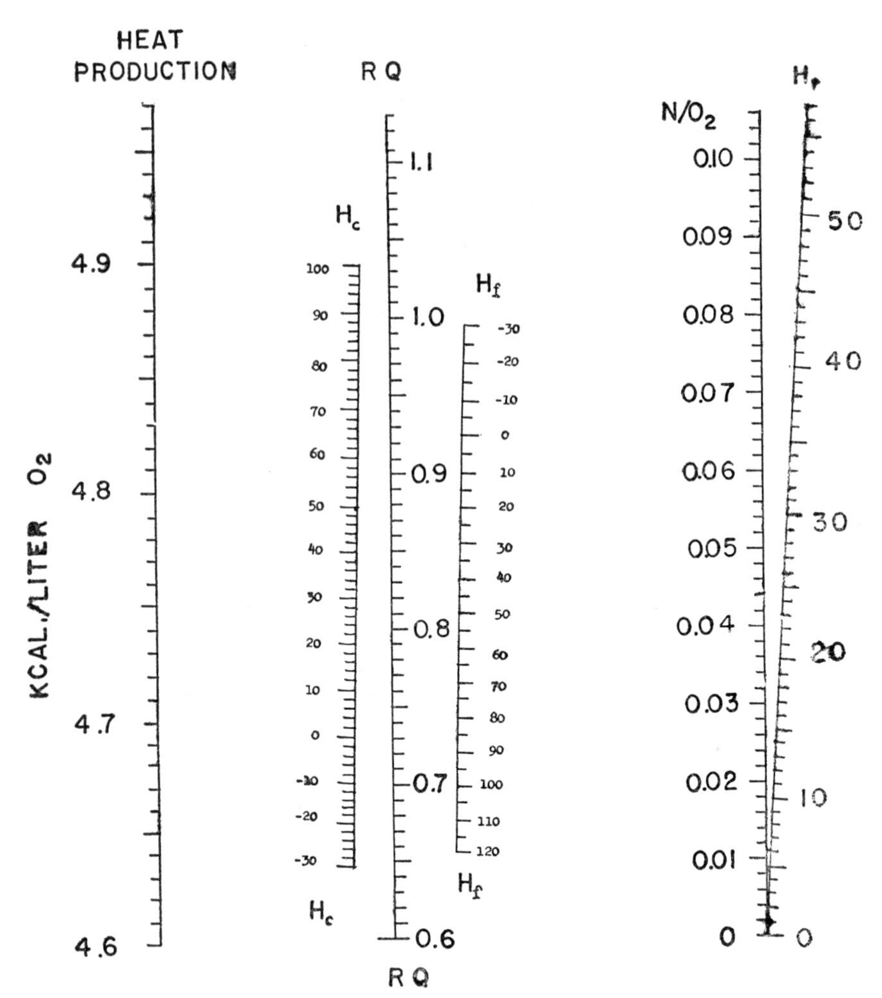

FIG. XIX-1. Nomogram for calculating the heat produced in kilocalories per liter of oxygen, and percentage of this heat produced from protein, fat, and carbohydrate, when ratio (N/O_2) of urinary nitrogen (in grams) to the oxygen consumed (in liters), and RQ are known. Connect observed values of N/O_2 and RQ with ruler or stretched thread. Intersection of ruler or thread with other scales gives the heat production per liter of oxygen (left-hand scale), percentage of the heat coming from carbohydrate (H_c), from fat (H_f), and from protein (H_p).

simplification of the two constructed by him. This covers the part of the range likely to be of any clinical interest.

Now, if the heat actually produced by a subject, measured in a human calorimeter, is compared with that calculated in the above manner from the nitrogen and carbon dioxide produced and the oxygen consumed, it is found that for the subject at rest the results agree very closely. Furthermore, Atwater and Benedict (10) showed that if the subject does work, the heat actually produced is less than that calculated, by exactly the amount of the work done. This was regarded, in the words of Lusk, as "a splendid confirmation of the law of the conservation of energy." In other words, the body obeys the laws of chemistry and physics, and functions as a machine. This discovery seems almost self-evident today, but at the time when the influence of the vitalists, who believed that mere scientific principles could never explain the workings of a living organism, was still very powerful, it was regarded as a great triumph for the chemists.

The important thing about these facts for us today is their application to the diagnosis of metabolic disease. Before we pass on to this subject, however, we may pause to answer one question which was brought up earlier, namely, how efficient is the human body in converting the energy of foodstuffs into work?

Mechanical Efficiency of the Human Body

Experiments can be carried out in which a man does work inside a calorimeter. In one accurately controlled experiment the patient did work by pumping the pedals of a stationary bicycle, and the work done as well as the heat produced were recorded. During the performance of the mechanical equivalent of some 550 Kcal. of work 4550 Kcal. of heat were produced. Both the heat and the work must have come from the oxidation of foodstuffs, so the total input to the human machine in this experiment was about 5100 Kcal., of which 10.8 per cent was manifested as work. This is a little better than the efficiency of a locomotive, but not as good as a good diesel, and not at all to be compared with the mechanical efficiency of a battery-motor combination. If the body does indeed function as a chemical cell, what is the explanation of such a low efficiency?

In the first place the body must keep itself alive, which is a more complicated requirement than any ordinary electric cell has to meet. The breathing, pumping of the heart, perspiration, and possibly other vital functions had to go on during the work period, not merely as usual, but at a greatly accelerated rate. This must have used up a good deal of the energy the body produced in the form of internal work, and caused this part of the work to be converted into heat. Thus, this work does not appear in our tabulation of the work performed. Another factor contributing to lowered efficiency is the internal viscosity of the muscles themselves, which must be overcome during work. It is reasonable to suppose that these "frictional" losses are proportionally greater when the body is producing work at a high rate, as in riding a bicycle or running a race.

In the second place, Szent-Györgyi (19) has argued from observations on the reversible contraction of muscle at various temperatures that the maximum efficiency of a muscle unit (autone) is 50 per cent. From this we are led to believe that the efficiency of the human body will never be found to be greater than 50 per cent.

Zuntz, followed by later workers (10), attempted to measure the efficiency with which the muscles of the human body can convert the energy of foodstuffs into work by comparing the heat output of the body while working without any external load with the total energy output (heat plus work) when working under load. Even this probably does not make nearly adequate allowance for the added burden on the heart and lungs during work, and the energy consumed in overcoming internal friction, so that we are getting a minimum estimate of efficiency from such measurements. Nevertheless, one subject, studied while walking up a five-degree slope, achieved a mechanical efficiency of over 40 per cent for a short time, which puts him in a class with the best internal combustion engines we can construct. We have seen above that the maximum efficiency of muscle is thought to be only a little greater than this. In regard to mechanical efficiency, then, the human body and a diesel engine are about on a par.

Cost of Work from Various Sources

There is another aspect to comparisons of the efficiency of the human body with that of other chemical cells and with heat engines, and that is the economic one. How much does it cost to produce one Kcal. of work by the various methods? When we examine this we find that the human body, because of the cost of its fuels, can not compete even with the steam locomotive, although it does produce work more cheaply than the ordinary dry cell combined with an electric motor.

Books on thermodynamics do not take any account of the cost of producing work by various methods, since they take it for granted that the engineer in charge of a plant has sense enough to get his heat by burning the cheapest fuel locally available and knows that electric cells such as the dry cell are far too expensive to use for the large scale production of work. But it is instructive for our purposes to make some comparisons of the cost of work from different systems, and this is done in table XIX-2. In making these calculations, it has been assumed that sheet zinc costs 50 cents per pound (the cost of other components of the dry cell being ignored), cane sugar (a good fuel for the human body) something less than 10 cents per pound, diesel oil 13.4 cents per gallon, and coal $25 per ton. The mechanical efficiency of the diesel and of the human body have been estimated at 40 per cent, the steam locomotive at 8 per cent, the dry cell-motor combination at 80 per cent (which is probably too high).

In addition the table contains figures for the production of work from

TABLE XIX-2

Costs (in cents) of 1000 Kcal. of energy in the form of work, as obtained from different sources

MATERIAL (FUEL)	METHOD	COST OF 1000 KCAL.
		cents
Zinc metal	Oxidation in dry cell, plus electric motor	130.0
Sugar	Enzymatic oxidation in the human body	13.0
Coal	Steam locomotive	4.06
Fuel oil	Oxidation in diesel engine	0.86
(Coal)	(Electrochemical oxidation in hypothetical $C + O_2$ cell)	0.45

coal by a process which possesses theoretical interest, but has never been realized practically. This involves the direct use of carbon and oxygen in a cell making use of the reaction

$$C + O_2 \rightarrow CO_2$$

Such a cell, if it could be constructed, would be the most efficient method of getting work from coal (20). However, although many clever inventors, such as Edison, have worked on this problem, it still seems far short of solution, and the results are included here merely for use in comparisons.

No figures are given for the work obtainable from radioactive materials, as this source of energy does not yet compete commercially with coal and oil. Opinions differ as to the probable extent of the peacetime use of nuclear and thermonuclear energy in the future (9, 18). Opinions also differ as to the extent of the coal and oil reserves which will be exploitable economically.

An estimate of the total world reserves of fuel energy is given by Baumeister (3) from which table XIX-3 is taken in modified form. Fission fuel reserves represent 22 times the energy stored in fossil fuels and some con-

TABLE XIX-3

Estimated world reserves of fuel energy

FUEL	ENERGY
	kilocalories
Fossil fuels	
Petroleum	2×10^{18}
Natural gas	0.25×10^{18}
Coal	18×10^{18}
Total fossil fuels	20×10^{18}
Fission fuels	440×10^{18}

sider even this estimate too low. Furthermore, there is the possibility that direct utilization of solar energy will become a reality in the foreseeable future. Barring atomic war, then, man's expanding industrial civilization seems in no immediate danger of running out of energy. Even if all our coal and oil were gone, and nuclear power were not plentiful enough, there would still be the possibility of growing starchy or woody plants in the tropics, fermenting the carbohydrate to alcohol and burning the alcohol in efficient internal combustion engines.

Indirect calorimetry. We have already mentioned that the rate at which the patient produces heat, since it measures his metabolic rate, is of clinical importance. It would not be convenient, however, for each physician to have in his office a big calorimeter elaborate enough to contain a human being and to measure his heat production. As pointed out above, however, the heat production can be predicted quite satisfactorily if we can measure the nitrogen and carbon dioxide produced and the oxygen consumed, for a given period of time. These measurements involve much simpler apparatus, and are not beyond the realm of clinical possibility.

However, it will pay us to examine more closely the calculation by which we obtained the total heat from the nitrogen, carbon dioxide, and oxygen values. The results for subject L. S. were,

$$H = 55.25 + 12.59 - 0.82 \text{ Kcal.} \qquad [11]$$

The first and second terms in this equation were calculated from the oxygen and carbon dioxide, respectively; the last term came from the value of the urinary nitrogen.

A striking fact is at once evident. The last term is much smaller than the other two. In this case, it amounts to 1.15 per cent of the total heat produced. The reason for this is that the oxygen and carbon dioxide figures have taken almost complete account of the protein part of the metabolism, and only a small correction is required for the nitrogen. It may easily be shown that the third term will never be very large. In the most extreme case possible, which would be that of a patient producing all of his heat by metabolizing protein, the value of this third term would be $2.134/26.434 = 8.1$ per cent. Since patients in the basal state are rarely producing more than 20 per cent of their total calories from protein, the error as a rule will amount to less than $0.2\ (8.1) = 1.6$ per cent, if we neglect the term calculated from the urinary nitrogen. For clinical purposes this is a small error, and the earlier workers generally neglected the nitrogen term in their calculations, thus obtaining results equivalent to calculating the heat from the following equation

$$H = 3.941\ O_2 + 1.106\ CO_2 \qquad [12]$$

It is nevertheless possible to avoid most of the error caused by neglecting this term, by taking account of the fact that patients in the basal state are generally producing a fairly uniform portion of their heat from protein. Weir (21) suggests that this fraction of the total heat which is due to protein is 12.5 per cent; Benedict (4), in the most careful studies ever carried out on this subject, found in fourteen experiments that his subjects during the first day of their fast were producing from protein various fractions of their total heat which averaged to 12.65 per cent, with a standard deviation of 2.99. A more recent report (11) makes the value 19 per cent. On the second day they were producing somewhat more of their heat from protein. If, for the sake of round numbers, we assume that the average patient is producing 12.5 per cent of his total heat from protein, we can correct for the omission of the third term of equation [12] by diminishing both coefficients by one per cent, obtaining

$$H = 3.900\ O_2 + 1.094\ CO_2 \qquad [13]$$

The error resulting from the use of this equation is very small.

Taking account of the fact that $CO_2/O_2 = RQ$, we find

$$H = (3.900 + 1.094\ RQ)\ O_2 \qquad [14]$$

which enables us to calculate the heat produced per liter of O_2 if we know the RQ.

For calculations of patients in the basal state, this can be simplified even further, for it has been found that the RQ of all such normal patients is 0.82, plus or minus 5 per cent either way. This gives us the simple relation

$$H = 4.797\ O_2 \qquad [15]$$

which enables us to calculate the heat produced merely from the oxygen consumed.

It will be noticed that the part of the coefficient in equation [15] which is affected by variations in RQ amounts only to about one-fifth of the total. This means that even if the patient's RQ is not exactly 0.82, we shall make a relatively small error by assuming it is. To be exact, if the true RQ is 0.77, the correct coefficient in [15] would be 4.738; if the correct RQ is 0.87, the correct coefficient would be 4.847. The variation is only a little more than one per cent either way.

These simplifications enable us to calculate the heat output of the resting post-absorptive patient merely from a measurement of the rate at which he is consuming oxygen. The apparatus for doing this is much simpler and easier to use than the elaborate and expensive calorimeters we have been discussing. Basically, all that is needed is a means of measuring the oxygen consumed by the patient as he breathes.

It is customary to make no correction for the urinary nitrogen in making such calculations for clinical purposes, and to assume (although this is known to be only a half truth) that the patient is metabolizing that mixture of fat and carbohydrate which has an RQ of 0.82. The heat produced is then computed from the oxygen consumed and the calorific value of the mixture of fat and carbohydrate which has an RQ of 0.82. Using the calorific values of fat and carbohydrate current at the time, the early workers obtained a value of 4.825 Kcal. per liter of oxygen. This value was used in calculating the heat produced by the patient. The basal metabolism machines commercially available in this country are calibrated for this figure. Omitting the correction required to take account of the fact that protein is also being burned (5) this gives an estimate of heat production which is consistently too high, but from equation [5] we see that the error is only 0.58 per cent, which for clinical purposes is negligible.

Basal Metabolic Rates

The rate at which the human subject is utilizing oxygen without the influence of food or muscular work is the basis for computing the basal metabolic rate (BMR). It is assumed that the BMR measures the energy required for the fundamental processes which must go on in the body to maintain life. In order to measure BMR one must use a subject, resting but awake, who has had no food for 14 to 18 hours and has indulged in no muscular activity for approximately half an hour.

This basal metabolism is not quite equivalent to the minimal metabolism which is the mere maintenance of life, as is shown by the fact that during sleep the rate of oxygen consumption measured is somewhat lower than during the so-called basal state.

Measurement of the basal metabolic rate has clinical importance and may be briefly described here. The patient, having taken no food since the preceding evening, comes to the laboratory before breakfast with a minimum expenditure of work, lies in complete muscular and mental repose for about 30 minutes, and then breathes pure oxygen from a closed system through a facial mask or some similar device.

The apparatus used consists of a device which contains oxygen, and records changes in its volume. The patient is connected to the apparatus in such a way that the air returned by the patient passes through soda-lime, which removes the CO_2 produced. From the rate of O_2 consumption, the rate of heat production is computed as described above.

There is a close correlation between the insensible loss of weight from the body (by evaporation of water and loss of CO_2) and the BMR (14), so that a measurement of this insensible weight loss by use of a good balance large enough to support a human being enables the BMR to be calculated quite

TABLE XIX-4

BMR (Kcal./m²/hr.)

AGE	MALE	FEMALE
1	53	53
5	50	50
10	45	43
15	42	37
20	39	35
40	36	34
60	34	32
80	32	31

readily. BMR determinations on many normal persons have enabled us to establish accurate standards for the BMR (15).

In the early days of animal calorimetry Rubner proposed that the rate of heat liberation was proportional to the surface area of the body, and elaborate measurements of the surface area were made and nomograms and other aids in calculating the surface area of the body were devised. However, the relationship is only a rough one. The basal metabolism is more nearly proportional to the three fourths power of the body weight.

The normal BMR is now calculated from measurements on the patient, usually weight and sitting height, allowing for age and sex. The normal BMR to be expected varies somewhat with the sex of the patient, with age, condition of nutrition, and so on. A detailed table of BMR values obtained for both sexes at different ages by various workers, in Kcal. per square meter body surface per hour is given in (6). A simplified version is given in table XIX-4. Note that the values decrease steadily with age and that after the age of 5 are consistently higher in men than in women.

Various internal glands—especially the thyroid—affect the BMR. If the normal functioning of the thyroid is decreased, it may lower the BMR by as much as 25 per cent and occasionally even more. Correspondingly, an overactive thyroid may increase the BMR as much as 40 per cent above normal. Other endocrine glands, such as the pituitary, may also cause disturbances of the BMR. It is also increased by fever, dyspnea, leukemia, and polycythemia. The BMR is decreased in starvation, hypothyroidism, and Addison's disease.

The close relationship between thyroid activity and BMR has resulted in increased interest in the analysis of the iodine content of blood. The iodine is the measure of the circulating thyroid hormone which consists of one or more of the iodothyronines (see page 310) bound to one or more of the serum proteins. The measurement of this protein-bound iodine (PBI) can be related to the BMR. The techniques involved are simpler and less tedious for the patient than are the older BMR measurements based on oxygen uptake and bid fair to replace them. (5b, 19a).

There is a certain variation shown in the BMR in persons in normal health. It is generally considered that this normal variation amounts to ± 10 per cent of the mean value as computed from the usual standards. Unless deviations exceed this, they are not considered indicative proof of an abnormal condition. Several investigators, however, have concluded that the accepted standards are too high, and that the rates in most normal subjects fall between $+5$ and -20 per cent (9a).

Specific Dynamic Action

The ingestion of food raises the metabolic rate. This is true of all foods, but there is a more pronounced effect from protein than from fat or carbohydrate. This tendency to increase the rate of oxidation is called the specific dynamic action (SDA). It is not due to the increased muscular activity of the digestive system, for it is not produced by feeding non-digestible substances or substances which yield no calories as cellulose and meat extracts, although these do stimulate the activity of the gastrointestinal musculature.

The Effect of Muscular Work

The operation of the muscles of the body is responsible for the liberation of a considerable amount of the body's heat. The human body at rest may be compared to an idling motor. Heat is being produced, whether it is needed or not, at about the rate of 80 Kcal. per hour (2). Various forms of activity increase this, depending on the degree of exertion. For instance, a patient sitting and reading produces about 105 Kcal. per hour. For other forms of exertion, the following results have been reported: typewriting, 140; swimming, 400 to 500; running, 570; sawing wood, 480. The body is capable of producing heat at a more rapid rate than this (over 1000 Kcal. per hour) when the exercise is heavy and performed at such a rate that most of the energy appears as heat, as in rapid stair climbing or a 100-yard dash, but effort at such rates can not be sustained.

The Daily Requirement of Calories

Computations, based upon the basal metabolism of normal subjects and estimates of their normal requirements in the way of extra energy, have led to a common estimate of about 2500 Kcal. per day as the average fuel requirement of a sedentary man. With increased activity, very much more is of course required. A bicycle rider, riding a six-day race, needs about 10,000 Kcal. per day. It has already been noted that most of the calories of the diet appear as heat, and only a fraction as work.

REFERENCES

1. Abramson, E. Computation of results from experiments with indirect calorimetry. Acta Physiol. Scand., **6:** 1–19, 1945.
2. Adolph, E. F. *Physiology of Man in the Desert.* New York, Interscience Publishers, 1947.
3. Baumeister, T. Atomic power; its prospects and realities. Trans. N. Y. Acad. Sci., **18:** 718–731, 1956.
4. Benedict, F. G. *The Influence of Inanition on Metabolism.* Washington, The Carnegie Institute of Washington, 1907.
5. Boyd, W. C. Error in calibration of commercial BMR machines. J. Applied Physiol., **6:** 711–715, 1954.
5a. Cathcart, E. P., and Cuthbertson, D. P. The composition and distribution of the fatty substances of the human subject. J. Physiol. **72:** 349–360, 1931.
5b. Dailey, M. E. and Skahen, J. R. A statistical appraisal of the serum protein-bound iodine as a test of thyroid function. New Eng. J. Med., **254:** 907–909, 1956.
6. Du Bois, E. F. Energy metabolism. Ann. Rev. Physiol., **16:** 125–134, 1954.
7. Glasstone, S. *Text-book of Physical Chemistry.* 2nd. edit. New York, Van Nostrand Co., 1946.
8. Hill, A. V. A discussion on muscular contraction and relaxation: their physical and chemical basis. Proc. Roy. Soc., *B*, **137:** 40–50, 1950.
9. Hoyle, F. *The Nature of the Universe.* Oxford, England, Blackwell, 1951.
9a. Leiner, M. C. and Abramowitz, S. Basal metabolic rate. J.A.M.A. **162:** 136, 1956.
10. Lusk, G. *The Science of Nutrition.* Philadelphia and London, W. B. Saunders Co., 1928.
11. McCance, R. A., and Strangeways, W. M. B. Protein catabolism and oxygen consumption during starvation in infants, young adults and old men. Brit. J. Nutrition, **8:** 21–32, 1954.
12. Morales, M., and Botts, J. A model for the elementary process in muscle action. Arch. Biochem. Biophys., **37:** 283–300, 1952.
13. Morales, M., and Botts, J. Outline of a theory of muscle action, and some of its experimental basis. Faraday Discussion, 1953, No. 13.
14. Newburgh, L. H., and Johnston, M. W. The insensible loss of water. Physiol. Rev., **22:** 1–18, 1942.
15. Peters, J. P., and Van Slyke, D. D. *Quantitative Clinical Chemistry.* 1st edit., 1932; 2nd. edit., 1946. Baltimore, The Williams & Wilkins Co.
16. Pilgrim, R. L. C. Waste of carbon and of energy in nitrogen excretion. Nature. **173:** 491, 1954.
17. Pryor, M. G. M. The molecular mechanism of contraction. Proc. Roy. Soc., *B*, **137:** 71–73, 1950.
18. Putnam, P. C. *Energy in the Future.* New York, Van Nostrand, 1953.
19. Szent-Györgyi A.; *in* Barron, E. S. G., Ed. *Trends in Physiology and Biochemistry.* New York, Academic Press, 1952.
19a. Thompson, H. L. *et al.* Method for protein-bound iodine: the kinetics and the use of controls in the ashing technique. J. Lab. Clin. Med. **47:** 149–163, 1956.
20. Thompson, M. D. K. *Theoretical and Applied Electrochemistry.* New York, Macmillan Co., 1925.
21. Weir, J. B. de v. New methods for calculating metabolic rate with special reference to protein metabolism. J. Physiol., **109:** 1–9, 1949.

CHAPTER 20

Excretion and Some of Its Disturbances

Carbon dioxide is produced continuously by the human body at a rate varying with its mass and activity. The removal of CO_2 by the lungs has already been described. Certain other substances are produced by the body with equal consistency. We have seen that urea, uric acid, and creatinine are end products of normal and indispensable metabolic processes. Similarly, the formation of SO_4^{--} and HPO_4^{--} from S and P of foodstuffs is as inevitable as the production of CO_2. Water is also a metabolic product, but not in amounts adequate to meet the obligations of the excretory mechanisms. The elimination of CO_2 by the lungs, of solutes by the kidney, and of metabolic heat by both these organs and by the skin, all require the output of water, in excess of the amount produced by metabolic processes. The kidney is the chief organ for the elimination from blood and body fluids of excess solids, with the notable exception of the degradation products of hemoglobin, which leave the body as bile pigment derivatives, chiefly by way of the intestinal tract. In renal failure, the intestine and even the skin perform excretory functions which are, in those organs, insignificant during health.

The kidney is, however, no simple sluiceway through which a stream of water indiscriminately bears the effluvia of metabolism. The kidney, more than any other organ, regulates the total osmotic pressure of all body fluids, and participates in the regulation of the concentration of each and every soluble and diffusible component of the extracellular fluid—the internal environment in which our component cells live and operate. Let us select a single example of this precise regulatory function. The kidneys maintain the concentration of cations in plasma at an excess of 25 to 27 mEq. per liter above the sum of the concentration of non-volatile anions. This excess of cations, by the law of electroneutrality, retains in the plasma an equivalent concentration of HCO_3^- (the only available volatile anion), which is thereby maintained at 25 to 27 mEq. per liter. Meanwhile the physiological control of respiration is keeping the plasma H_2CO_3 at a concentration between 1.25 and 1.35 mEq. per liter. Filling in these values, and the physiological pK of H_2CO_3, we can set up the Henderson-Hasselbalch equation for the bicarbonate system in plasma

$$pH = 6.1 + \log \frac{26}{1.30} = 7.4$$

Thus the actual maintenance of the pH of the plasma and interstitial fluid at its constant value of 7.4 depends upon the opposed actions of lungs and kidneys (31). The mechanisms of acid-base regulation by the kidney will be dealt with in a later section. This example was selected to illustrate the *regulatory* function of the kidney, which is a refinement of the excretory function. In brief, the kidney responds to high concentrations of many substances in the plasma by excreting them at a more rapid rate. By so doing, the kidney not only regulates concentrations of individual ions and molecules, but also stabilizes the pH and osmotic pressure of blood plasma and indirectly of other body fluids.

There are three steps in the formation of urine by an individual nephron: (a) *filtration* in the glomerulus; (b) *absorption* of water and certain solutes in the tubules; and, (c) *excretion* of water and certain solutes by the tubule cells. The amount and composition of the urine formed by these steps is dependent upon physical forces, such as pressure and velocity of blood, as well as upon the chemical composition of the blood. Glomerular filtrate contains the diffusible components of the blood plasma in concentrations similar to those in the water of plasma. Work is done by the kidney against osmotic pressure in the conversion of glomerular filtrate into the more concentrated urine. The difference in osmotic pressure between blood and urine, as estimated from freezing-point determinations, averages about 25 atmospheres. For each liter of urine this means a minimal energy requirement of 25 liter-atmospheres or 0.6 Kcal., the equivalent of the total combustion of 0.15 grams of glucose. The actual metabolic activity of the kidney is many times greater than this minimal figure. The less the glomerular filtrate undergoes concentration as it passes through the renal tubules, the less the energy required. The process of concentration is not uniform for the different urinary solutes. Different substances are concentrated to different degrees. The SO_4^{--} ion is ordinarily concentrated about 90 times, the Cl^- ion only twice, the Na^+ ion has approximately the same concentration in urine as in blood plasma, while glucose is normally completely reabsorbed in the tubules and does not appear in any appreciable amount in the urine.

Physicians, now and throughout the recorded history of medicine, consider study of the urine an integral and important part of the physical examination of the patient. To be sure, much of the attention given to urinoscopy in ancient and medieval medicine depended upon the fact that urine was the only body fluid easily and painlessly available, and some of the archaic diagnostic and prognostic manipulations of the specimen were heavily tainted with sympathetic magic. We still find it diagnostically useful to examine the urine, although we now can intubate the vena cava and cannulize the cisterna magna.

The purposeful collection of urinary specimens over a measured time is a matter of some importance. The 24-hour period is usually most convenient in all types of long term metabolic studies, as for example in the diagnosis of hyperparathyroidism (see page 691). Shorter metabolism periods suffice for most of the renal function tests to be described later. Reducing sugars are more likely to be detected from one to two hours after a meal, chorionic gonadotrophins in the concentrated early morning urine, albumin when physical activity is resumed after a night's sleep; other instances of timing will appear as we go along. When quantitative methods are to be applied, timing must be exact.

Another important and often neglected factor in the outcome of urine examinations is proper preservation of specimens. One hour's standing in an overheated utility room or laboratory can turn a glycosuric urine specimen from strongly Benedict positive to Benedict negative. The microorganisms which catalyze this fermentation of sugar are ubiquitous, and often inhabit the urinary bladders of diabetics. With bacterial multiplication, urines may develop heat-coagulable proteins; cells, casts, and other microscopically visible components will decompose; and as a result of changes in pH crystals may deposit. Refrigeration is usually adequate to control undesirable chemical changes. Toluene or chloroform, one ml. per liter, will temporarily restrain the multiplication of yeasts and bacteria. Formaldehyde is an almost perfect preservative for cells and casts, but in high concentrations interferes with some chemical tests.

PHYSICAL CHARACTERISTICS OF THE URINE

Considerable information of diagnostic value can be obtained by direct observation of a urine specimen, aided only by a few simple instruments. The normal amber *color* of urine is the result of the presence of several normal pigments, which will be discussed in a later section. The depth of the amber color is a rough index of the extent to which the urine has been concentrated, which in turn is dependent to some extent on the patient's state of hydration. Pathological or unusual coloring of the urine may result from drugs, bile pigments, blood pigments, homogentisic acid, melanin, or oxidation products of phenols. The presence of porphyrins in abnormal amount is not always indicated by abnormal color of the urine, but yields a red fluorescence under filtered ultraviolet light. Normal urine is clear and transparent, except for a slight and slowly gathering sediment of epithelial cells and mucoprotein from the urinary tract (this sediment is the "nubecula" of the old-time urinoscopists, who made quite a fuss about it). However, the urine of a healthy person may contain a precipitate of calcium and magnesium phosphates if the specimen happens to be alkaline. This may result from an alkaline-ash diet, as most vegetarian diets are. The urine of horses and cattle is usually turbid with such phosphates. During the time of active gastric secretion, the urine becomes less acid and may become basic, or alkaline urine may be simply the result of enthusiastic use of sodium bicarbonate for its supposed therapeutic effects. Any urine specimen, on standing in a warm place, may support the growth of microbes which hydrolyze urea to ammonium carbonate, and may thereby turn alkaline and deposit phosphates. Phosphatic sediments will clear promptly with acidification, and are usually without direct pathological significance. All sediments other than those already mentioned are abnormal (not necessarily pathological) and should be investigated. Certain crystals, such as calcium oxalate and uric acid, are often seen in normal urine under the microscope, but do not occur normally in amounts sufficient to cause visible turbidity unless the urine is refrigerated. Abnormal turbidity occurs from pus, blood, excessive mucus, and less commonly from insoluble compounds. The odor of urine yields little useful information, unless it be notably putrid, which in a fresh specimen would indicate active bacterial contamination within the urinary tract, or a vesicointestinal fistula. The normal odor of fresh urine is contributed chiefly by aromatic acids such as phenylacetic and by traces of mercaptans contributed by vegetable foodstuffs such as onions and asparagus. On standing, urines acquire an ammoniacal odor from microbial conversion of urea.

The daily *volume* of urine of adults living in the United States is usually between 0.8 and 1.6 liters. Less than 0.5 liters is normally accumulated in the bladder during the night's sleep. The output is of course subject to great variation from differences in fluid intake and in sweating. The increased urine output in cold weather is proverbial. By purposeful increase of fluid intake, a normal person can raise his output to 8 liters or more in a day. Further variations arise from differences in diet: proteins form urea, which obligates additional water for its excretion; the caffeine of a cup of coffee has a measurable diuretic effect. Emotional states affect urinary volume in ways which are not uniform among different individuals. The urine volume is characteristically increased in certain diseases including diabetes mellitus, diabetes insipidus, most chronic nephritides, and during convalescence from acute infections. The volume is characteristically decreased during fevers, following loss of fluids (see Chapter 17), in acute nephritis, and in the majority of circulatory diseases.

The normal 24-hour urinary output of 0.8 to 1.6 liters can be approximated by considering the output as one ml. per minute. This rate is subject to continuous and wide variation. No definite limits set off normal rates of output from *diuresis*, or increased rate on one hand, or from *antidiuresis*, or less commonly *oliguresis*, meaning decreased rate on the other. These terms are applied to transitory changes in rate of urine excretion, as measured by volume output. If the changes are of long duration, an excretion of more than 2 liters per 24 hours is called *polyuria* and one of less than 0.6 liters, *oliguria*. *Anuria* refers to periods where no urine is formed at all.

Increased water content of the body stimulates increased urine volume output, acting through the posterior pituitary mechanism discussed in Chapter 17. The presence of osmoreceptor cells in the central nervous system is postulated. These cells are stimulated by lowered osmotic pressure (increased water content) of the blood of the internal carotid circulation, and respond by inhibiting the secretion of the antidiuretic hormone by the *pars nervosa* of the hypophysis. The action of the antidiuretic hormone is to promote absorption of water by the renal tubular cells. When this action is inhibited completely, as a result of high fluid intake or in diabetes insipidus, the water content of the urine is comparable to that of glomerular filtrate, and the specific gravity of the urine is 1.010 or less.

The *specific gravity* of the urine is usually measured with a simple hydrometer. This is accurate enough for all purposes except the most refined investigative work, provided the hydrometer has been properly calibrated, and is periodically checked against solutions of known specific gravity. The temperature correction, 0.001 added for each 4°C. above the calibration temperature of the hydrometer, should always be included. In the interpretation of specific gravities in clinical diagnosis, it is advisable to make another correction for large amounts of proteins, if such be present, by subtracting 0.003 for each one per cent of protein found. Most normal urine specimens have specific gravities between 1.008 and 1.030 and most 24-hour collections between 1.016 and 1.022, but variations beyond these limits occur. It can not be too strongly emphasized that variation of specific gravity throughout the 24 hours is characteristic of normal regulatory function of the kidney, and that fixation of specific gravity means deterioration of such functional ability. For this reason, most single specific gravity measurements are not contributory to diagnosis, while repeated measurements are informative. More of this matter will come up later. The specific gravity of the urines of young infants tends to be low by adult standards, and does not increase in severe dehydration to the extent observed in the adult (23). The ability to produce a concentrated urine also decreases in elderly people even in the absence of renal disease. Various "coefficients" have been proposed by which the total solids of the urine may be presumably calculated from the specific gravity. These calculations are in error whenever the urine varies from a fixed ratio of its chief components, urea and NaCl. Since such variations are inherent in normal renal function, the arbitrary conversion of known specific gravity into doubtful total solids adds no information and may introduce confusion.

ACIDITY AND ALKALINITY

The limits of pH observed in human urine are 4.5 and 8, with the mode at about pH 6. The chief sources of the usual acidity of human urine are proteins and phospholipids of the diet, which yield sulfuric and phosphoric acids as end products of their metabolism. Increase of these anions in the plasma brings about a compensatory decrease in HCO_3^- (see page 733), in other words a tendency towards metabolic acidosis. Excretion of alkali and alkaline earth cations associated with urinary SO_4^{--} and HPO_4^- would proceed at a rate faster than their supply in the body, and would cause depletion of cations were it not for two important conservative mechanisms functioning in the kidney: (a) the excretion of an acid urine, using H^+ as a balancing cation; and (b) the formation and excretion of NH_4^+ as a balancing cation.

The normal daily output of acid in the urine, as measured by Folin's (11) method of titration, is from 15 to 40 mEq. To this should be added the acid neutralized by ammonia formed in the kidney and excreted as ammonium salts; this normally amounts to 40 to 60 mEq. more. In acidotic states, such as diabetic ketosis, very much higher values for the sum of these two acid fractions are observed, up to 700 mEq. in very severe acidosis. This increase in acidity in diabetic ketosis is largely accounted for by the ketone acids. In normal urine, organic acids make up less than 4 per cent of the titratable acidity.

Herbivorous animals and human vegetarians ingest salts of organic acids—malates, citrates, acetates, tartrates—in much greater amounts than in the normal mixed diet of the human. The acid radicals of such organic salts are oxidized eventually to CO_2 and eliminated via the lungs. To maintain electroneutrality these must be replaced by HCO_3^-. If so, the urine will be alkaline. This explains the increase in alkalinity of human urine following ingestion of acid fruits or fruit juices.

The mechanism whereby the kidney excretes urine which is definitely more acid or more alkaline than the blood plasma is not fully understood although the problem has been subjected to much analysis. The urine exaggerates pH changes of the plasma. If the plasma pH is above 7.4, the pH of the urine will be still higher; similarly, with plasma pH below 7.4, the pH of the urine will be still lower. Much fluctuation in urinary pH occurs without measurable change in plasma pH. Pitts (31) postulates ionic exchange in the renal tubule cell, H^+ leaving the cell for the lumen of the tubule in exchange for Na^+ or other metallic cations. The H^+ is considered to arise within the cell by a process involving the oxidation of a metabolite with production of CO_2, from which H_2CO_3 is formed, under the catalysis of carbonic anhydrase. The H_2CO_3 is the source of the H^+ exchanged for Na^+, and $NaHCO_3$ leaves the cell and enters the venous blood. Note the similarity of this mechanism with that proposed for the formation of gastric HCl (see page 491).

Approximately 60 per cent of the ammonia formed by the renal tubule cells arises from the glutamine of the plasma, by the action of *glutaminase* which converts glutamine to ammonia and glutamic acid. Tracer studies have shown that the rest of the ammonia is derived from other amino acids, with only a negligible amount from urea. Acidity of the urine is a stimulus for ammonia production, acting locally on the tubule cell. In extreme acidosis, 75 per cent of the H^+ excreted may be combined with ammonia as NH_4^+.

COMPOSITION OF NORMAL URINE

The single solute usually present in greatest amount in human urine is *urea*, the major end product of protein metabolism in mammals. The manner of its formation from the amino groups of amino acids and from CO_2 by way of the arginine-ornithine-citrulline cycle of Krebs and Henseleit has been described in Chapter 16. The 24-hour urine of the adult will contain from 10 to 35 grams of urea, the amount being proportional to the protein intake. Urea does not disappear from the urine on a protein-free diet, since a minimal amount is produced under such circumstances by the catabolism of body proteins. Analysis of the urine for urea has its chief value in tests of renal function, such as the urea clearance test, to be described in a later section. Most analytical methods for urea involve the enzyme *urease*, obtained from vegetable sources such as jack bean meal. This enzyme catalyzes the conversion of urea, by addition of one molecule of water, to ammonia and CO_2. The ammonia so formed can be measured by titration or by Nesslerization and colorimetry (11). A correction must be made for the ammonia already present in urine, or the ammonia may alternatively be removed by previous treatment with permutit. Urea can also be measured by direct specific colorimetric methods which do not involve enzymes (1).

Urea and the other nitrogen-containing solutes of the urine can be measured as a group by applying the Kjeldahl method for total nitrogen. The *total nitrogen* of a 24-hour specimen may run as low as 3 grams or higher than 20 grams of nitrogen, depending upon the protein intake. Urines collected from subjects on the usual American diet run about 11 grams of nitrogen. On a low-protein diet, about 60 per cent of the total nitrogen is contributed by urea, on a high protein diet about 90 per cent. The reason for this variation is the relative constancy of the other nitrogen-containing solutes.

One such nitrogenous component is *ammonia*, which is present in acid urines entirely as *ammonium salts*. Under ordinary conditions ammonia contributes 0.2 to 0.8 grams of nitrogen to the total nitrogen of the 24-hour urine. The ammonia content of urine is clinically significant in the evaluation of states of acidosis, and in making corrections of urea analyses when the urease method is used. The aeration and permutit methods described by Folin (11) have not been superseded. From 0.3 to 1.0 grams of *uric acid* is present in the usual 24-hour urine, contributing one third of its weight to the total nitrogen. With unusually high purine intake in the diet, the uric acid may be as much as 2 grams. Uric acid is the major end product of purine metabolism in man and in a very limited number of other mammalian species (see Chapter 9). On account of the low solubility of uric acid and urates, they often separate as precipitates from normal urines upon cooling. Such precipitates are colored with urinary pigments—uric acid itself is colorless. Uric acid will be considered in a later section as one of the components of stones of the urinary tract. Urinary uric acid is also of significance in the diagnosis and follow-up of cases of gout (see page 365). Analysis for uric acid is by colorimetric methods (11). A total of less than 100 mgm. of allantoin and of purines other than uric acid is excreted per 24 hours. The *creatinine* of urine, 1 to 2 grams per 24 hours, contributes less than one gram to the total nitrogen. The metabolic origin of creatinine has already been considered (see page 648). Analysis for creatinine is quite simple (11) but has little clinical significance. Creatinine is the most constant of the urinary constituents in its rate of output. *Creatine* is a variable urinary constituent and makes up roughly 6 per cent of the total creatinine.

The excretion of amino acids in the urine averages about 1 gram per day, of which glycine, histidine, and 3-methylhistidine (formula I) make up 70

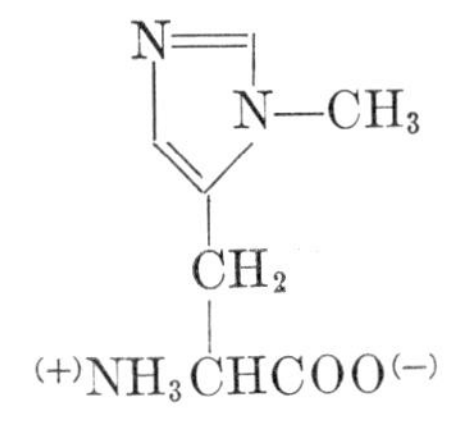

N=|
| N—CH_3
CH
|
CH_2
|
$^{(+)}NH_3CHCOO^{(-)}$

I. 3-Methylhistidine

per cent (39). 3-Methylhistidine is a hydrolysis product of anserine (see page 154) but aside from that is not known to occur in the body.

Total amino acids can be measured most simply by a colorimetric method (11), the individual amino acids identified most directly by paper chromatography. Amino acids in the urine may be free or conjugated. Some 2 grams per day of conjugated amino acids are excreted; glycine, glutamic acid, and aspartic acid being the amino acids most frequently involved (39). The most interesting of the conjugated amino acids is *hippuric acid* (formula II), which is the conjugation product of glycine with benzoic acid. Its formation in liver and kidney and its excretion in the urine varies from 0.1 to 1.0 grams per day, depending upon the amount of benzoic acid and

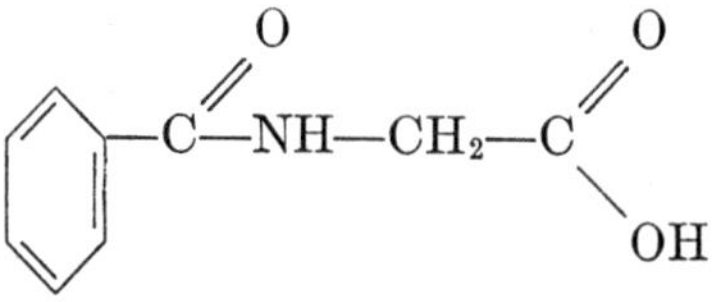

$$C_6H_5{-}C({=}O){-}NH{-}CH_2{-}C({=}O){-}OH$$

II. Hippuric acid

benzoic acid precursors (aromatic compounds with side chains containing an odd number of carbon atoms) in the diet. Certain fruits, such as cranberries, are rich in benzoic acid, which is also used as a preservative for food products with a legal limitation of 0.1 per cent. Hippuric acid is of historical interest as the first compound shown to be synthesized in the body. This was demonstrated by Wöhler in a liver perfusion experiment. The hippuric acid of the urine can be measured by direct precipitation, or by precipitation as benzoic acid after hydrolysis, followed by weighing or titration. The clinical value of such analyses is chiefly in connection with tests of liver function where measured amounts of benzoic acid are administered and their percentage recovery as hippuric acid in the urine is taken as an indication of the functional capacity of the liver. This assumption is not strictly valid, since hippuric acid synthesis can also take place in kidney.

Increased amino acid excretion is characteristic of liver disease of sufficient severity to impair deamination and urea formation and also of the Fanconi syndrome (see page 572). Certain errors of metabolism are reflected by the presence of abnormally high quantities of some amino acids and not others, such as cystine and the basic amino acids in cystinuria (see page 627). Amino acid excretion is increased in post-operative patients, in rachitics and in pregnant women, threonine being most markedly increased in the last case (26). For a review of the clinical aspects of aminoaciduria see (6).

Some 20 to 100 micrograms of primary amines other than amino acids are excreted per day in the normal urine. At least 10 such compounds have been detected by chromatography (8).

Urinary *indican* (formula III), an alkali salt of indoxyl sulfuric acid, ap-

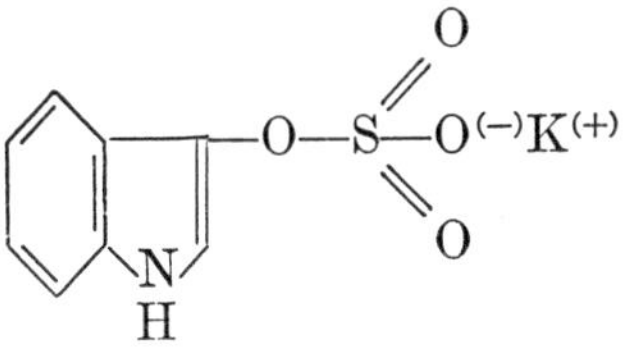

III. Urinary indican (potassium salt of indoxylsulfuric acid)

pears in the urine in amounts up to 20 mgm. daily. From tryptophane indol is formed by the action of intestinal bacteria. The oxidation to indoxyl and the conjugation with sulfuric acid probably takes place in the liver. In those days when "intestinal autointoxication" was thought to be a major cause of human misery, the fluctuations in the indican content of urine were assiduously studied. Several simple color tests for indican are described in texts of clinical pathology, but are seldom applied in current practice. *Indol acetic acid*, a component of one of the urinary pigments, is similar to indican in metabolic origin. Several other urinary pigments, of known or unknown composition, contain nitrogen, as do the small amounts of B vitamins and their derivatives excreted in the urine.

Finally, we can close our listing of nitrogen-containing urinary components with mention of the urinary proteins, which are normally less than 40 mgm. per day and include certain *enzymes*. The urine shows weak amylolytic, lipolytic, and proteolytic activity, the last chiefly as *uropepsin*, which is identical with pepsinogen. Uropepsin is of gastric origin and its assay may prove useful as an indication of the peptic activity of the stomach at the time. The urinary lipase is probably of pancreatic origin, the amylase of pancreatic and salivary origin. Increase of urinary amylase may be of differential diagnostic value in cases of mumps and of pancreatitis. *Urokinase* specifically converts plasminogen to plasmin.

The urinary excretion of *acid phosphatase* in girls and women of all ages averages about 50 King-Armstrong units per 23 hours. Young boys have a similar rate of acid phosphatase excretion, but at puberty a sharp increase occurs. Maximal excretion occurs during the fourth decade, reaching a maximum of about 350 units per day. In general, adult men have approximately 5 times the urinary acid phosphatase output of adult women. The excess of acid phosphatase excretion in men originates from prostatic secretion, as indicated by a number of indirect evidential points. Prostatic fluid, as obtained by massage, increases in acid phosphatase concentration during puberty. Psychosexual stimuli increase the acid phosphatase output in the urine of men, but not of women. Urine obtained from adult male subjects by catheterization contains much less acid phosphatase than urine voided normally by the same subjects (41).

Most of the nitrogen-containing urinary components mentioned in the preceding paragraph are derived from protein. There are a few organic acids found in small amounts in the urine which are also protein derivatives, but contain no nitrogen. They include phenyl acetic, phenyl propionic, *p*-hydroxyphenyl lactic, and *p*-hydroxyphenylpyruvic acids. All of these are quite normal, and quite without clinical significance other than that they are excreted in increased amounts in scurvy. *Homogentisic acid* (dihydroxyphenyl acetic acid) if present indicates a hereditary abnormality of tyrosine metabolism, *alkaptonuria*. Homogentisic acid (formula IV) has the properties of blackening by autoxidation in alkaline solutions, and of reducing Benedict's and other alkaline copper reagents for sugars. The urine of alkaptonurics, if made alkaline, turns black on standing exposed

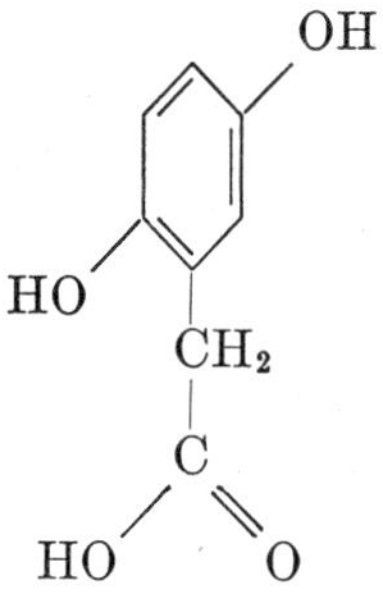

OH

HO

CH_2

C

HO O

IV. Homogentisic acid

to air, darkening from the top downward. The formation of homogentisic acid from tyrosine is demonstrated by variations in output in alkaptonurics when tyrosine is fed or withheld. The feeding of *p*-hydroxyphenyl propionic acid or of its alpha-hydroxy derivative does not increase the output of homogentisic acid, whereas the alpha-keto derivative will do so, indicating that the last mentioned substance is an intermediate in the formation of homogentisic acid. The normal oxidation of tyrosine ultimately involves breaking of the ring (see page 635). Small amounts of tyrosine normally escape ring breakage and are excreted as *p*-hydroxyphenyl acetic and *p*-hydroxyphenyl propionic acids. In alkaptonuria, the ring of tyrosine is not disrupted.

Ochronosis, which is gray to brown or black pigmentation of sclera, cartilage, or occasionally of the skin of the face, may occur with alkaptonuria, or independently, as a result of a similar failure of tyrosine metabolism. The presence of *phenylpyruvic acid* in the urine is associated with imbecility or idiocy in the metabolic disease, phenylketonuria or phenylpyruvic oligophrenia, see p. 633.

In addition to aromatic organic acids, *phenols*, present in amounts of the order of 300 mgm. per 24 hours in urine, are derived from aromatic amino acids. Such formation of phenols from the aromatic rings of phenylalanine, tyrosine, and tryptophane is chiefly by bacterial deamination and decarboxylation in the intestine, and to a lesser degree by similar reactions in the body tissues. Certain phenols, indoxyl and skatoxyl, contain nitrogen, while others such as phenol and *p*-cresol do not. The phenols are usually found in the urine conjugated with sulfuric acid, as in the case of indican.

Oxalic acid and calcium oxalate. Crystals of calcium oxalate dihydrate, identified under the microscope by their characteristic "envelope" appearance, are of very common and consistent occurrence in the urine sediments of healthy people. Since calcium oxalate is extremely insoluble, such precipitation is to be expected from a solution containing both Ca^{++} and the oxalate ion. Urine contains Ca^{++} in appreciable concentrations. The amount of oxalate is much less, the range in normal urine being from 14 to 56 mgm. per 24 hours (42). The source of the urinary oxalate is partly from oxalic acid and oxalates of the diet, and partly from oxalate formed by metabolic processes within the body (17). Ingestion of spinach, which contains 0.8 to 0.9 per cent oxalic acid, is followed by increased oxalate output in the urine within 6 hours, except in achlorhydric patients, who showed such an increase only if hydrochloric acid was given with the spinach. Some excretion of oxalate occurs, however, in subjects maintained on a diet low in oxalate, and in fasting subjects. Calcium oxalate in the urine has an importance for students of medicine far out of proportion to the small amounts excreted. Being solid, it can form stones (*calculi*) in the urinary tract, and even without calculus formation can be the cause of renal irritation with consequent bleeding, pain, or both. The harmful effects of excessive urinary oxalate (*oxaluria*) briefly suggested here are explained more fully in a review by Jeghers and Murphy (18) together with a discussion of possible metabolic origins of oxalic acid in the body. It appears that carbohydrate, rather than protein or fat, in the diet promotes the formation of oxalate. The foodstuffs which directly contribute are certain vegetables: beet tops, spinach, New Zealand spinach, sorrel, Swiss chard, and rhubarb stalks—all of these containing 0.5 per cent or more of oxalate expressed as oxalic acid. Much of the oxalic acid is already combined in the plant as insoluble calcium oxalate, hence these figures can be corrected downward as far as absorbable oxalate is concerned. The amount absorbed will also vary inversely with the amount of calcium present in the other foods eaten at the same time. The feeding of these high-oxalate vegetables to unwilling infants and young children as a source of calcium is probably wasted effort; conversely, such vegetables, in reasonable amounts, have been demonstrated to have no adverse effect on the calcium metabolism of healthy children or adults. High oxalate vegetables should be excluded, as a reasonable precautionary measure, from the diets of those who have had renal calculi or renal colic demonstrably the result of oxaluria. Rhubarb stalks make a tasty pie filling, harmless in moderate amounts to healthy people, causing only a transient increase in the urinary content of calcium oxalate crystals. *Rhubarb leaves contain toxic amounts of oxalic acid,* of the order of 1.2 per cent, and are too dangerous to be eaten, even in moderate amounts. Fatal cases have been few, and have involved the eating of large amounts, in one case (18) half a peck of the cooked leaves. Nonfatal cases have been characterized by renal colic, blood in the urine, convulsions, and sometimes gastrointestinal symptoms. Others have eaten rhubarb leaves (amount not stated) with no subsequent difficulties. Lest there be misunderstanding, it should be stated that only those vegetables specifically mentioned are of high oxalate content. Beet roots contain only a little over 0.1 per cent of oxalate, and green vegetables such as celery, turnip leaves, endive, kale, escarole, asparagus, broccoli, cabbage, Brussel sprouts, and cauliflower contain still less, as do other root vegetables, seeds, and fruits. Tea leaves are reported to contain about 0.2 per cent and powdered cocoa about 0.4 per cent oxalate, calculated as oxalic acid. In the guinea pig, about 60 per cent of urinary oxalate appears to originate from ascorbic acid (25).

Other Organic Constituents

Lipids occur only in minute concentrations in urine. Normal urine contains chemically detectable amounts of several of the steroid hormones and of metabolic products of steroids. The epithelial cells and leukocytes of normal urine contain lipid material, as do all cells. The fat droplets occasionally seen in urine sediments are usually contaminants from dirty receptacles or from lubricants used in catheterization. True *chyluria,* the presence of fat emulsified in the urine, occurs only when there is a fistula between the urinary bladder and the abdominal lymphatics. This may occur as a result of malignancy or from infestation of the lymphatics with the parasitic worm, *Filaria.* Except after heavy intake of carbohydrates, sugars are not present in normal urine in amounts sufficient to reduce Benedict's solution. The very small amounts of sugars present under fully normal circumstances can be measured by a modification of Folin's blood sugar method (11). *Glucuronic acid* in the form of conjugates with many drugs or with steroid hormones and their metabolism products occurs in the urine in a small and highly variable amount. *Citric acid,* a product of carbohydrate metabolism, is excreted in the urine in amounts of 0.2 to 1.2 grams per day. *Ascorbic acid* (vitamin C) is excreted in amounts which depend upon the intake and the subject's state of saturation (see page 822). The fat-soluble vitamins are not urinary components. Sugars and organic acids of urine are rapidly altered or destroyed by bacterial action if specimens are allowed to stand.

Inorganic Constituents

Among the *anions* of the urine, Cl^- leads in quantity with a daily excretion in adults usually between 5 and 10 grams (140 to 280 mEq.), varying almost exactly with the intake of Cl^- in the diet. Diminished excretion as compared with intake occurs when Cl^- is lost through channels other than the urine, as by sweating, vomiting, or diarrhea, or when fluid is being abnormally retained in the body as in the formation of edema or of large inflammatory exudates. The chloride of the urine is usually determined by titration of excess silver nitrate with thiocyanate after precipitation of the Cl^- as AgCl (11). Although the output of chloride in the urine gives no direct diagnostic information, such measurements are often helpful in the assessment of the patient's state of fluid balance. *Phosphate* ions leave the body both in the urine and by way of the feces. The urinary output, expressed as P, is usually about one gram. The output is so variable in its total amount and in its distribution between urinary and fecal channels that it is ordinarily impossible to decide whether phosphate output is normal, diminished, or increased without placing the subject on a known diet and conducting a balance experiment. The removal of the parathyroids leads to definite retention of phosphate, with diminished urinary output (see page 691), and conversely in hyperparathyroidism there is increased urinary excretion of phosphate as well as of calcium. Positive phosphate balance occurs during growth and during pregnancy. Measurement of urinary phosphate is usually carried out colorimetrically by the use of a molybdic acid reagent, which forms phosphomolybdic acid in acid solution. Phosphomolybdate can be reduced to a blue molybdenum molybdate complex by several reducing agents which do not reduce molybdic acid itself (11). *Sulfur* is excreted in the urine in three groups of compounds—inorganic SO_4^{--}, conjugated sulfate (chiefly with phenolic compounds), and neutral sulfur. The last group mentioned includes compounds where the sulfur is present in forms other than the sulfate radical; examples of such compounds are cystine and taurine. On a diet normal in protein content inorganic SO_4^{--} makes up about 90 per cent of the total sulfur excretion and the other two groups about 5 per cent each. On a low protein intake the percentage of the total sulfur excreted as inorganic SO_4^{--} may fall to about 60 per cent, with neutral sulfur remaining about the same in absolute amount, and therefore constituting a greater portion of the percentage. Since output of phenols is less on a low protein diet, the absolute amount of conjugated sulfate would also be less, but with the percentage possibly increased. Inorganic sulfate can be measured gravimetrically by precipitation with $BaCl_2$. If the urine is previously submitted to acid hydrolysis, conjugated sulfates are released as SO_4^{--} and precipitation gives the sum of inorganic and previously conjugated SO_4^{--}. If the urine is evaporated to dryness and ignited with a suitable oxidizing agent, then redissolved and precipitated with $BaCl_2$, the result measures the total sulfur, the sum of all three fractions. On the basis of these three analyses, the fractions of urinary sulfur can be determined by subtraction. The total sulfur of the urine ranges from 0.6 to 2 grams per day, expressed as S, and varying with the protein content of the diet. In place of the tedious gravimetric methods, a precipitation with benzidine may be carried out on a semi-micro scale (24) followed by titration. Sulfate analyses in the urine have little clinical application. *Bicarbonate* ion is present in urine, in concentrations increasing with the pH. *Fluoride* is present in traces, varying with the intake (see page 694).

Of the urinary cations, Na^+ is quantitatively predominant, 3 to 5 grams (130 to 220 mEq.) being excreted daily. In Addison's disease (adrenocortical failure) the Na^+ output is excessive from failure of the renal tubules to reabsorb this ion. Adrenal mineralocorticoids or synthetic deoxycorticosterone will correct this situation and in excess will bring about a similar loss of K^+. The output of K^+ is 1 to 3 grams (26 to 77 mEq.) daily. The urinary output of K^+ by the normal kidney continues even under fasting conditions, whereas Na^+ in the urine drops to very low levels in the absence of Na^+ intake. Of the approximately one gram of Ca^{++} eliminated daily, from 10 to 40 per cent is usually eliminated in the urine, the remainder in the feces. The daily urinary output of Mg^{++} ranges from 32 to 307 mgm. with a mean of 103 mgm. Some Mg^{++} is eliminated in the feces. There is no iron in the fluid part of normal urine (20), but minute amounts are present in the cells of the urinary sediment. Lead is not a physiological substance, but the environment of civilized life offers so many opportunities for ingestion of lead that it can usually be detected in the urine. Amounts as high as 0.15 mgm. per 24 hours have been observed in the urines of subjects who had no occupational exposure to lead and no symptoms of lead poisoning.

URINARY PIGMENTS

The pigments of normal urine have not been well or completely characterized. *Uroerythrin* was proposed over a century ago as a name for the red pigment which colors deposits of uric acid and urates. This pigment is red only in acid solution, being yellow-green in alkali. It has absorption bands at 546–520 and 506–481 millimicrons. *Urochrome*, which is usually designated as the chief normal pigment of urine, is not a single substance. Descriptions of this pigment correspond to a mixture of pigments, containing among others uroerythrin and urobilin. *Urorosein* was the name given to a red pigment obtained when certain pathological urines were strongly acidified, and which has been tentatively identified with *indirubin* (formula V), which is a normal urinary component, excreted in increased amount in cases of leukemia (12).

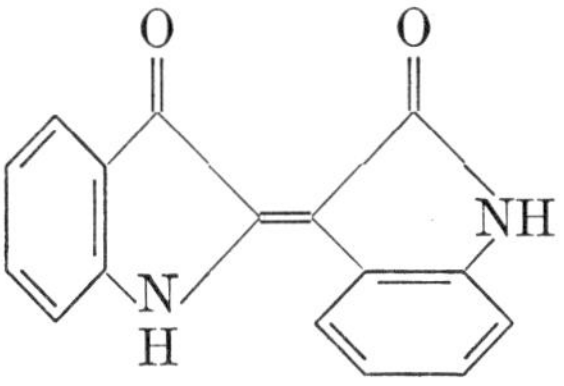

V. Indirubin

Rosein is the term proposed by Meiklejohn for a red pigment, not necessarily a single chemical entity, developing in acid or acidified urine, and extractable by amyl alcohol but not by chloroform or toluene, and decolorized by light and by alkali. Oxidation of indole-3-acetic acid (formula VI) produces a rosein-like pigment with a weak absorption band at 500, plus a stronger band at 530 millimicrons, comparable to the absorption bands of uroerythrin and urorosein. Indolacetic acid has been identified in urine,

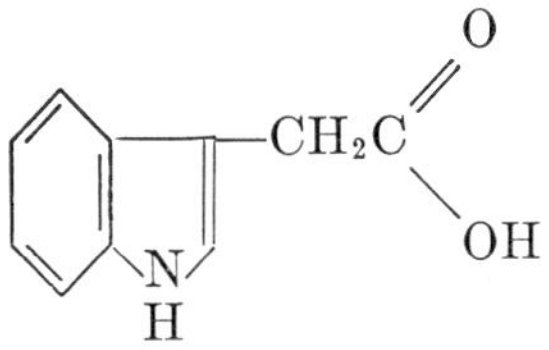

VI. Indole-3-acetic acid

chiefly for the reason that it is a plant growth hormone or auxin. Rosein and auxin output in the urine are increased simultaneously by liver disease and by nutritional deficiencies. A *roseinogen* has been extracted by Meiklejohn from the urine of a patient with cirrhosis. The reactions of the roseinogen are those of indole-3-acetic acid except that the roseinogen is more soluble. It has been suggested that the increased solubility is achieved by conjugation with glycine. Roseinogen has not been crystallized.

Urobilin and *urobilinogen* are bile pigment derivatives and occur normally. Urobilin has a brown color; urobilinogen is colorless. Urobilinogen responds to the familiar Ehrlich aldehyde reagent which is paradimethylaminobenzaldehyde in hydrochloric acid. Both urobilin and urobilinogen have fluorescent zinc salts which can be used as a further means of identification. Chief clinical interest in these substances lies in their absence in obstructive jaundice, in which situation no bile pigment derivatives can be absorbed from the small intestine since no bile pigment is there. Absence of these two substances from the urine, then, is an indication of obstructive jaundice. In their place we find unchanged bile pigment, particularly bilirubin. In clinical laboratories this is usually identified by the familiar shaking-and-foam test which is actually more dependable than most proposed chemical tests for bile pigments. Of the chemical tests, most satisfactory is Harrison's which uses paper strips saturated with barium chloride and dried. The paper specified is S & S, No. 470. These papers are dipped into the urine for ten seconds and a drop of Fouchet's reagent at the surface line gives a green color if bilirubin is present. Fouchet's reagent is 25 per cent trichloracetic acid containing 0.9 per cent ferric chloride. A quantitative modification of this test, utilizing spectrophotometric measurement of the color, has been described (40).

Melanin is a rare abnormal pigment sometimes found in the urine of patients with the highly malignant melanotic type of tumors. Melanin in itself is black. It is often excreted in the form of a colorless precursor which darkens on exposure to the air. Since the harmless homogenistic acid behaves in the same way, differentiation here is important. Direct identification of homogentisic acid can be easily done by utilizing the fact that it darkens ordinary photographic paper in strong alkaline solution in full daylight. Techniques for the concentration and identification of melanin have been summarized by Rothman (35). The pigment isolated from melanotic urine has properties similar to the pigment obtained from human red hair.

Porphyrins. Normal urine contains a small amount, 0.1 mgm. per 24 hours or less, of coproporphyrin, and a very much smaller amount of uroporphyrin (29). Excess porphyrin may be the result of a primary disease, porphyria, or may occur as a symptom of numerous disorders, including chronic alcoholism, liver or hemolytic disease, poisoning by quinine, cinchophen, various hypnotics, sulfonamide drugs, and various inorganic poisons, particularly lead.

De Langen and ten Berg (9) found that increased excretion of coproporphyrin in the urine was an early sign of lead poisoning. In such cases porphyrin is not excreted in amounts sufficient to produce a visible change in the color of the urine. As a semi-quantitative screening test, a few drops of glacial acetic acid and 2 ml. of ether are added to 20 ml. of the suspected urine. The mixture is shaken and examined under ultraviolet light, in which the ether layer fluoresces. With urines of normal porphyrin content, the fluorescence is light blue to green, but gradually deepening red fluorescence appears as the porphyrin content increases. In case of an equivocal result a quantitative method can be employed, which is also described in the paper cited. A similar porphyrin excretion test for lead exposure has been utilized in American industrial medical services (7).

Increased urinary excretion of coproporphyrin is also characteristic of regurgitation jaundice (see page 509) resulting either from biliary-tract obstruction or from liver damage. There have also been a few reported cases of individuals who excrete large amounts (1 to 6 mgm. per day) of coproporphyrin without demonstrable cause other than a hypothetical metabolic error. These patients show no consistent symptoms associated with their coproporphyrinuria. In all these instances of excess coproporphyrin excretion, the porphyrin is the type III isomer.

Patients with *congenital porphyria* excrete a different porphyrin, uroporphyrin, Type I, along with coproporphyrin I, often show photosensitivity of the skin, and red fluorescence of teeth and bone. This is a rare disease and most of the diagnosed cases have been written up. Much of our knowledge of this disease comes from the careful study of a single patient, Petry. Petry died (aged 32) on January 21, 1925, and was autopsied the same day by Borst. His organs were analyzed for porphyrins and other pigments by Königsdörffer. Their combined report (4) gives full details of the histological and histochemical (fluorescence-microscopical) findings, with a chapter based upon their findings discussing congenital porphyria as a "*Konstitutionsanomalie*". Deficient formation of hemoglobin, myoglobin, and cytochromes could not be demonstrated, nor could deficient mechanisms for destruction of blood pigments be shown, nor could they prove disturbance in intermediary bile pigment metabolism, nor show that the abnormal porphyrins were by-products of bile pigment metabolism. They conceived the fundamental error in congenital porphyria to be the persistance of an early phylogenetic and ontogenetic developmental stage, in which the independent synthesis of porphyrin by erythroblasts occurs. Erythroblasts are produced in excess, and are also destroyed at an excessive rate, with liberation of the porphyrin. Uptake of dietary porphyrins, or of porphyrins formed by intestinal bacterial synthesis or bacterial degradation of chlorophyll, was considered but could not be demonstrated.

Acute porphyria, with abdominal, neurological, and psychiatric symptoms, may appear at any age. This disease is not definitely congenital although there appears to be an inherited predisposition evidenced by familial incidence. Both causation and therapy in this disease are still obscure. Uroporphyrin and coproporphyrin, mainly Type I, are excreted in varying and sometimes very large amounts (30 mgm. per day), accompanied by *porphobilinogen*, (see page 617). Of all the porphyrias, this is the most serious, sometimes causing death at the first attack, in other instances leaving the patient in a paralyzed state. Procedures for the identification of the several urinary porphyrins are summarized by Ham (15).

ABNORMAL URINARY CONSTITUENTS

The presence of *sugars* above the limit of the normal concentrations and of the ketone bodies (see page 593) in amounts greater than 20 mgm. per day in the urine is abnormal. The implications of these substances as abnormal urinary components have already been discussed. Normal urine contains very little protein, the mean excretion being 39 mgm. per 24 hours (34). This minute amount of normal urinary protein includes serum globulin and serum albumin, plus mucoproteins. Larger amounts of protein are abnormal. *Coagulable protein* (usually called *albumin*, and actually mostly albumin with some admixture of serum globulin) can be detected by the familiar heat-and-acetic-acid test in any amount greater than normal. Technical note: the urine should be acidic for the preliminary heating, but not excessively acidified, since soluble acid metaprotein may be formed. Sulfosalicylic acid, usually applied in urine examination as a ring test, is a popular but quite expensive protein precipitant. Nitric acid is traditional for this purpose and is notably cheaper, but has certain well recognized disadvantages—the staining of the analyst's skin and possible damage to clothing and woodwork, the precipitation of urea nitrate or uric acid or both from concentrated urines (dilute and try again), the precipitation of *resin acids* (components of numerous pharmaceutical preparations, cough syrups in particular), and the formation by oxidation of a dark pigment band which obscures the albumin ring. Robert's reagent (one volume of concentrated nitric acid plus 5 volumes saturated aqueous magnesium sulfate solution) eliminates the last mentioned difficulty and is equally sensitive to albumin as nitric acid. If large numbers of urines are to be tested for albumin, a very satisfactory timesaving device is to test all specimens by the ring test, using Robert's reagent, then recheck the positives with the heat-and-acetic-acid test to eliminate the false positives.

Albuminuria. The presence of albumin, i.e., heat-coagulable protein, in the urine is always abnormal, but is not diagnostic of any particular disease, and indeed may occur in a healthy person under conditions of physical stress. Evidence gained from animal studies shows that the glomerular filtrate contains small amounts of plasma proteins, which indicate that the glomerulus is not a perfect ultrafilter. The renal tubule cells normally reabsorb this protein, making normal urine practically protein-free. In fact, it has been estimated that reabsorption of protein takes place at the rate of 5 mgm./hour, which is equivalent to a daily filtration and reabsorption of one-third the plasma content of protein (36). Albuminuria may therefore arise from increased protein leakage through damaged glomeruli, or from failure of protein reabsorption by non-functional tubule cells, or both. There is no evidence that tubule cells extrude protein into the urine. The highest outputs of urinary albumin are seen in those diseases where tubules are known to be affected.

Albuminuria may occur in the following situations:

(1) From stress. This is usually physical stress, such as a marathon run, or long immersion in cold water. Some possibility exists that albuminuria may result from emotional stress.

(2) From postural defect. Some healthy people, usually adolescents or young adults, have albuminuria when up and about and no albuminuria when resting in a horizontal position. This appears to be the result of a change in kidney circulation, and has in some cases been associated with lumbar lordosis, or with a movable kidney. Such albuminuria is often associated with the late growth period, and may disappear completely with the attainment of full growth. In other instances, such intermittent albuminuria may be associated with renal disease.

(3) From pre-renal disease. Circulatory diseases, and occasionally fevers, may cause local passive congestion or other circulatory disturbance in the kidneys, leading to albuminuria.

(4) From renal disease. This is of course the most common cause of albuminuria, and the disease may be practically any to which the kidney is subject, including congenital malformations, nephritis, nephrosis, tuberculosis, tumors, infarcts, and poisonings. The heaviest albuminurias occur in nephrosis and in amyloid disease, with as much as 5 per cent coagulable protein in the urine. With most other lesions less than one per cent albumin is found.

(5) From post-renal sources. Admixture of blood, pus, sperm, or other albuminous material from the lower genito-urinary tract gives rise to albuminuria.

Mucoprotein is not heat coagulable, and is not strictly an abnormal substance in the urine. It is secreted by the cells of the urinary tract, and is normally present in small amounts. In any irritation or infection of the urinary tract it is greatly increased in amount and is often visible grossly as shreds or microscopically as threads.

Proteoses of the urine are a rather doubtful and clinically unimportant category of proteins. They are not heat coagulable, but will precipitate with nitric acid, or by saturation with ammonium sulfate after removing heat coagulated albumin by filtration. Proteoses have been reported in the urine in almost every type of disease, hence they have no diagnostic significance. Some contribution of proteose to urine may occur from prostatic secretions.

Bence-Jones protein. On November 1, 1845, Henry Bence-Jones, physician to St. George's Hospital, examined urine specimens from a 47-year old grocer "who had been out of health for thirteen months". The urine, if heated and left to cool, became solid. "This solid redissolved by heat, and again formed on cooling." On January 2 of the next year the patient died. "The following day I saw that the bony structure of the ribs was cut with the greatest ease, and that the bodies of the vertebrae were capable of being sliced off with the knife." This patient was a case of what we now call *multiple myeloma* (2), and the curiously behaving solid we now call the Bence-Jones protein. In a urine slightly acidified with acetic acid, the Bence-Jones protein shows its presence by precipitation at temperatures between 45 and 60°C., and redissolves at 100°C. If albumin is also present it may be filtered off at 100°C., and the urine upon cooling will again throw down the Bence-Jones protein at about 60°C. The Bence-Jones protein is found in the urine of a little over half of the pathologically proven cases of multiple myeloma, which is a neoplastic disease characteristically involving the red marrow bones—sternum, ribs, and vertebrae. The tumor cells resemble plasma cells. The x-ray appearance is of multiple rounded defects. The disease is usually rapidly fatal. The Bence-Jones protein may be eliminated in the urine in very large quantities, up to 70 grams in a day. It is considered by most authorities to be a tumor-cell modification of serum globulin. Immunological evidence leads to the conclusion that it is related to human proteins, but this conclusion has been challenged by Dent and Rose (10), who make the rather daring suggestion that multiple myeloma is a virus disease, and the Bence-Jones protein a constituent of the virus. The Bence-Jones protein is not specific for multiple myeloma. It has been reported in many other bone diseases, also in myxedema, nephritis, and chronic leukemia (there is, however, a leukemic type of multiple myeloma). These occurrences without association with multiple myeloma are rare, and some are doubtful.

Blood. In speaking of blood in the urine we distinguish *hematuria* as the presence of red blood discs in the urine as the result of hemorrhage into the urinary tract. The observation of hematuria is made or confirmed with the microscope. As opposed to hematuria, *hemoglobinuria* is the presence of free, dissolved hemoglobin in the urine. This is evidence of hemolysis occurring in the blood-vascular system. There are numerous causes of such hemolysis, including transfusions of incompatible blood, severe infections with hemolytic micro-organisms such as the hemolytic streptococcus, malaria, typhoid, and yellow fever. Hemolysis may occur with severe burns or chilling, congenital or acquired hemolytic jaundice, and numerous types of poisoning, e.g., fava beans. Some few individuals develop hemoglobinuria after exercise (13). The color of urine is not a reliable guide to the presence of hemoglobin. Ordinarily dependence is placed on one or another of the modifications of the peroxidase reaction. The urine normally contains no peroxidase. True enzymatic peroxidase is present in white cells. The hemoglobin of red cells has a peroxidase action. In the presence of hydrogen peroxide, hemoglobin, acting like a peroxidase, catalyzes the oxidation of a number of organic substances. The one commonly used in testing for blood is benzidine, less commonly guaiac or *o*-tolidine, all of which yield blue oxidation products. Since leukocytes contain peroxidase, the presence of large amounts of pus in the urine may give a false positive reaction for hemoglobin. This reaction of leukocyte peroxidase will be absent if the urine is boiled, whereas the pseudoperoxidase action of hemoglobin is not affected. The easiest differentiation is by microscopic identification of the leukocytes. Compounds closely related to hemoglobin may also occur, such as methemoglobin, rarely from a congenital metabolic fault, more commonly from oxidizing agents such as nitrites or chlorates used as drugs, and also myoglobin, the red respiratory pigment of muscle which is found in the urine after crush injuries. These pigments give a peroxidase reaction similar to that of hemoglobin. They can be distinguished best by the use of the spectroscope.

UROLITHIASIS

The formation of stones in the urinary tract occurs by the precipitation, usually in crystalline form, of urinary solutes, together with the aggregation of the precipitated material into a compact mass bound together by an organic matrix (5). The substances found in urinary concretions are relatively few in number, and most of them have been recognized as stone formers for many years. Table XX-1 shows the substances found in 1000 calculi examined by Prien (32). Certain of these warrant more detailed description.

Calcium oxalate has already been discussed as a normal urinary component. The crystals observed in many normal urine specimens are calcium oxalate *dihydrate*, which does occur in calculi. The *monohydrate* is a more common stone component, as shown by the table. Calcium oxalate calculi are characteristically found in patients with normal acid urines, and are not associated with any other disease or abnormality than oxaluria.

Calcium oxalate monohydrate stones are usually of "hempseed", "mulberry", or "jackstone" shape, and with a smooth brown surface. Stones containing the dihydrate (pure or mixed) are more likely to be rough, often with projecting crystalline spicules, which add to the trauma produced by the stone in traversing the narrower portions of the urinary tract. Such stones are often stained with blood pigment. Calcium oxalate can be crudely and simply identified as a stone component by the fact that after ignition

TABLE XX-1

Substances found in 1000 urinary calculi

Substance	Number	Total		Percent
Pure calcium oxalate				
Calcium oxalate monohydrate	137			
Calcium oxalate dihydrate	4			
Calcium oxalate (mixed)	186	327	=	32.7%
Calcium oxalate plus apatite				
Calcium oxalate monohydrate plus apatite	72			
Calcium oxalate dihydrate plus apatite	47			
Calcium oxalate (mixed plus apatite)	224	343	=	34.3%
Apatite (pure)	34	34	=	3.4%
$MgNH_4PO_4 \cdot 6H_2O$ (pure)	3	3	=	.3%
$MgNH_4PO_4 \cdot 6H_2O$ plus apatite	155	155	=	15.5%
$MgNH_4PO_4 \cdot 6H_2O$ plus apatite plus calcium oxalate (mixed)	32	32	=	3.2%
Calcium hydrogen phosphate dihydrate				
Pure	2			
Mixed	17	19	=	1.9%
Uric acid				
Pure	47			
Mixed	11	58	=	5.8%
Cystine				
Pure	22			
Mixed	7	29	=	2.9%
		1000	=	100.0%
Sodium acid urate				
Microscopic amount in one mixed uric acid calculus				

(to dull red heat until all visible carbon is burned away) of a small bit of the powdered stone, it will effervesce with a drop of dilute HCl, and will not do so prior to ignition. The heating converts the oxalate to carbonate. This observation will not of course differentiate the two hydrates. For exact crystallographic methods of identification of calcium oxalate and of other stone components, the paper of Prien and Frondel (33) should be consulted.

Apatite stones, with composition similar to that of bone salt (see Chapter 17), together with the other *phosphatic* calculi, are associated in a general way with alkaline, infected urines, although this association is by no means uniform. The alkaline earth phosphates as a group are less soluble as pH increases. Infection of the urinary tract, particularly with organisms which convert urea into ammonium carbonate, often produces an alkaline urine and causes the precipitation of apatite and other phosphatic deposits. Excess excretion of phosphate, as in hyperparathyroidism (see page 691), may also lead to the production of such calculi, even in acid urine and in the absence of infection. A consistently alkaline-ash diet, or the frequent use of sodium bicarbonate, may result in an alkaline urine and formation of phosphatic calculi. Such calculi have been found in cattle, also stones containing calcium carbonate, which is not found in human uroliths. Human phosphatic calculi may form rapidly and grow to such size as to fill the renal pelvis and calyces ("staghorn" calculi).

Uric acid will be noted as one of the less common stone formers. The stones are usually small and colored with urine pigment. There is no significant association of the formation of uric acid calculi with diet or with any other disease, with the exception of gout (see page 365) and here the association is not striking. The kidneys of newborn babies often contain heavy deposits of urates, the so-called "uric acid infarcts", which soon disappear. Their presence is not considered pathological. *Xanthine* calculi did not appear in Prien's series, but have been occasionally reported (30).

Cystine calculi are smooth, relatively soft, may reach considerable size, particularly in the bladder, and are present only in cases of *cystinuria.* (see page 627). Since cystine is the least soluble of the naturally occurring amino acids, precipitation frequently occurs, crystals of cystine can usually be found in the urine of cystinurics, and almost inevitably a calculus will form. Such calculi are usually large and single, and are commonly found in the bladder. Crude identification is easy by the "burnt feathers" odor on ignition.

Renal concretions of indigo or cholesterol are pathological curiosities which remain inadequately explained. Some of the radiolucent (i.e., x-ray transparent) calculi formerly classified as "urostealiths" have been shown to consist of the uncalcified or possibly decalcified organic matrix (5).

THE MEASUREMENT OF RENAL FUNCTION

A very large number of kidney function tests have been used by clinicians and physiologists. Most of these tests fall into one of three classes:

(1) Those which are based upon the excretion of normal urinary components;

(2) Those which are based upon the excretion of foreign substances; and

(3) Those which are based upon the study of a single component, either normal or foreign, simultaneously in the blood and in the urine. All of these test the excretory function of the kidney. The regulatory function of the kidney is demonstrated by the maintenance of normal levels of significant substances, e.g., urea, uric acid, or inorganic phosphate, in the blood, and by the maintenance of a normal acid base balance. Analysis for blood components is not in itself usually designated as a renal function test, but is a valuable diagnostic aid.

Concentration and dilution tests are the chief representatives of the first category of tests, where normal urinary constituents are measured. Usually, instead of selecting a single component, measurement is made of the specific gravity, which is a rough index of all components. One procedure commonly used combines a concentration and a dilution test: the patient remains on his usual diet and collects 2-hour specimens punctually during the waking day, and collects the "night urine" as a single 12-hour sample. At least a pint of fluid is taken with each of the three meals, but no solid food between meals, and no food or fluids during the 12-hour "night" period. This procedure contrasts a day with ample fluid with a night of no fluid. In the patient with adequate powers of urinary concentration and dilution, the variation in specific gravity between the highest and the lowest in the several specimens should be at least 0.009; the night urine should have a specific gravity of at least 1.018 and a volume not over 500 ml. For a description of pure concentration and pure dilution tests consult Ham (15). Hayman and his collaborators (16) plotted the number of glomeruli counted post-mortem against the maximal specific gravity of the urine measured before death. The value for the maximal specific gravity fell consistently with the number of glomeruli until the latter figure reached 750,000 per kidney. At lower values of the glomerulus count the specific gravity remained fixed at 1.010. The normal kidneys contain about a million glomeruli each.

Phenolsulfonephthalein (PSP or phenol red) offers many advantages as a foreign substance useful in renal function testing. The substance keeps indefinitely in solution, can be sterilized by boiling, causes no pain or local reaction upon injection, and is excreted mainly through the kidneys. Although rate of urine volume output does not greatly affect the rate of excretion of PSP, except with severe oliguria, it is best to provide adequate urine volumes during the test by giving the patient 600 ml. of water. The usual dose is 6 mgm. of PSP dissolved in one ml. saline. If given by subcutaneous or intramuscular injection a latent period or "appearance time" of 3 to 10 minutes elapses before the dyestuff appears in the urine. Its presence can be detected by alkalizing the urine with NaOH, which turns the dye from its acid yellow color to its alkaline red color. Quantitative estimation of the amount excreted can be made by comparison with known color standards. Normally from 40 to 60 per cent of the injected dose is eliminated in the first 70 minutes after subcutaneous or intramuscular injection and from 60 to 85 per cent in the total test period of 130 minutes. Total eliminations less than 25 per cent for the test period indicate serious los of excretory ability. After intravenous injection there is no latent period, and from 28 to 51 per cent of the dose is eliminated within 15 minutes, in patients with normal renal function. In cases where unilateral renal disease is suspected, the outputs of the two kidneys can be studied separately by the use of ureteral catheters. The PSP test has certain limitations. In obese individuals, in cases of hypothyroidism, and in situations where circulation is slow, the PSP output is diminished, suggesting renal damage which may not exist. The PSP test is not sensitive to early or minimal changes in renal disease, where fixation of specific gravity and clearance tests (see next paragraph) may already be informative. During acute diseases, such as influenza, the PSP output may be temporarily decreased, with a return to normal with recovery.

Clearance tests are the most widely used methods involving simultaneous measurement of blood concentration and urine output of a given substance. Such tests measure the excretory function of the kidney, usually disregarding the regulatory function. The simplifying assumption is made that the action of the kidney is to "clear" the blood of a certain substance. Clearance is defined as the *calculated volume* of plasma cleared of a given substance in one minute. Actually plasma is not usually cleared of any substance by passage through the kidney. Clearance is spoken of most accurately as a virtual volume, or more bluntly as a fictitious volume.

The general formulation of clearance involves no mathematics other than simple arithmetic. Clearance, in ml. per minute, is

$$\frac{UV}{P}$$

where U and P represent respectively the concentrations of the given substance in urine and in plasma, measured in the same units, and V is the volume of urine per minute. *Glucose* normally has a clearance of zero, since it is completely reabsorbed by the renal tubule cells, giving U a value of zero.

Urea clearance was the original application of the clearance principle, and is the test most commonly used clinically. Urea is only partially returned to the plasma by the renal tubule cells. The assumption that simple diffusion of urea occurs from glomerular filtrate into tubule cell fits most data better than the assumption of active reabsorption of urea by the cell, although there is not complete agreement on this point (3). At any rate, a portion of the urea is excreted, hence urea clearance can be calculated. With copious water excretion the amount of urea excreted in unit time is independent of the volume output. Urea clearance under these conditions is called *maximum clearance* and is calculated by the simple formula already stated. The normal maximum clearance of urea is 60 to 100 ml., with a mean of 75. At rates of water excretion less than 2 ml. per minute, the urea output has been observed to increase in direct proportion to the square root of the urine volume. In this event, the square root of V is used in place of V in the formula, giving the *standard clearance*, defined as the "volume of blood, which one cc. of urine excreted in 1 minute suffices to clear of urea" (27). The value of standard clearance ranges from 41 to 65 ml. normally, with a mean at 54. Unlike the maximum clearance, the standard clearance has no direct physiological significance, but is an empirical expedient for clinical purposes only. Note that both clearances can not be calculated from the same data. In calculating urea and other clearances it is advisable in persons of unusual body configuration to multiply V by a correction for surface area

$$\frac{1.73}{\text{actual surface area of patient}}$$

The value of 1.73 is the average adult surface area in square meters. The surface area of the patient can be obtained by the use of prepared weight-height-area tables. Clearance, as well as blood volume and kidney weight, has been found to parallel surface area more closely than height, weight, or any other accessible physical measurement.

Inulin clearance is considered to be the result of glomerular filtration with no reabsorption, therefore the maximum clearance of inulin represents the *filtration rate*, which in men has a mean value of 127 ml. of glomerular filtrate produced per minute, in women, 117 ml. These figures represent the upper limit of clearances by glomerular filtration alone. If the clearance of any substance is greater than the simultaneously measured inulin clearance, it is concluded that that substance is excreted by the tubules. *Thiosulfate* has an identical clearance with inulin. Neither inulin nor thiosulfate will leave extracellular fluid to enter cells.

Creatinine clearance, at normal blood levels, is identical in normal dogs with that of inulin. After ingestion of creatinine, the clearance may increase to a maximum of 175, indicating that tubular excretion is taking place in addition to glomerular filtration. In the normal human subject the clearance of endogenous creatinine is usually 10 to 20 per cent higher than the simultaneous clearance of inulin. In damaged human kidneys there is no constant relationship between the clearances of inulin and of endogenous creatinine.

Diodrast clearance, at relatively low diodrast (3,5-diiodo-4-pyridone-N-acetic acid) concentrations in the blood, is the result of glomerular filtration plus maximal tubular excretion, and may reach a value of 740 ml., which represents the physiologically functional blood flow through the kidney per minute. The 740 ml. of plasma represents 1200 ml. of whole blood, about one fourth of the cardiac output per minute. Of all the substances so far mentioned, diodrast alone is actually "cleared" from the plasma by the kidney. Another such substance is *p-aminohippuric acid* (PAH) which has a clearance identical with that of diodrast.

The maximal rate of tubular reabsorption of glucose, or glucose Tm, is observed at concentrations of glucose in the arterial plasma above a saturation value which in men is approximately 0.3 per cent and in women, 0.26 per cent. Note that this value is considerably higher than the glucose threshold, above which glycosuria occurs. The threshold is quite variable in different individuals and somewhat variable in the same individual. In 80 per cent of normal subjects it falls within the range of 0.14 to 0.19 per cent. The glucose threshold is usually measured in venous blood, but this does not account for the great difference between the threshold and the saturation value for glucose absorption. According to Smith (38), certain nephrons with high glomerular activity become saturated at lower levels of plasma glucose, therefore permitting escape of glucose into the urine. Glucose Tm is calculated by multiplying the saturation value by the filtration rate or inulin clearance. It was 375 $\pm$ 79.7 mgm. glucose per minute in 24 normal men studied by Smith and his colleagues, and 303 $\pm$ 55.3 mgm. in 11 normal women. In patients with renal disease, glucose Tm is proportional to the number of intact nephrons, since both glomeruli and tubules are involved.

The maximal diodrast or PAH excretion (diodrast Tm or PAH Tm) is a measure of tubular function alone, since the part played by glomerular filtration is constant and negligibly small. The diodrast Tm or PAH Tm minus the glucose Tm measures the number of aglomerular tubules. Since inulin excretion is purely glomerular, inulin clearance minus glucose Tm measures the number of glomeruli with functionless tubules. For further consideration of renal clearances and the physiological information which they yield, consult the monograph by Smith (38).

The generation of high-energy phosphate bonds, which normally accompanies the aerobic oxidations catalyzed by the cyclophorase system (see page 548), is inhibited by 2,4-dinitrophenol. This same drug greatly reduces the renal tubular excretion of *p*-aminohippuric acid, of diodrast, and of phenol red, with no significant change in renal hemodynamics (28).

These results in dogs suggest that phosphate-bond energy is made use of in tubular excretion.

The concept of clearance has been applied to excretion by channels other than the kidney (21). If the small intestine is perfused with 2 per cent sodium sulfate at a rate between 23 and 30 ml. per minute, the concentration of urea in the outflow is about equal to the urea concentration of the patient's plasma. The urea clearance is therefore about equal to the ml. of perfusing fluid per minute.

The *artificial kidney* is a device by which blood is submitted to dialysis outside the body and then returned to the circulation. The solution against which the blood is dialyzed contains glucose and the significant ions of the blood plasma. Such an artificial kidney can remove as much as 10 grams of urea nitrogen per hour from the patient's blood.

RENIN AND EXPERIMENTAL HYPERTENSION

If the blood supply to one kidney of an experimental animal is decreased, an increase in blood pressure occurs which returns to the preoperative level within 6 weeks. Such decrease of blood supply was first accomplished by Goldblatt, using a clamp which constricted but did not close the renal artery. A similar result can be achieved by enclosure of the kidney within a tight covering, or by inducing injury with resulting scar tissue contraction.

If an animal is made hypertensive by the above procedure, and the unoperated normal kidney is removed, or if both kidneys are rendered ischaemic, the hypertension is permanent. In dogs with bilateral ischaemic kidneys, such hypertension may persist as long as 6 years, with no loss of renal excretory function as measured by the usual functional tests.

Houssay and his co-workers demonstrated, by such procedures as transplanting an ischaemic kidney to the neck, obtaining venous blood directly from such a kidney, and injecting the blood into another animal, that the hypertension in animals with ischaemic kidneys resulted from the presence of a pressor substance in the blood. The pressor substance is the product of the reaction of a renal enzyme, *renin*, upon its specific substrate, *hypertensinogen*, which is one of the plasma globulins. The product of this enzymatic action is a polypeptide designated as *hypertensin* or *angiotonin*. One polypeptide fraction, *hypertensin I*, has been purified and found to have an isoelectric point of pH 7.7 and to consist of aspartic acid, proline, valine, isoleucine, leucine, tyrosine, phenylalanine and arginine in equimolecular proportions plus a double supply of histidine (37).

The production and storage of renin seems to be limited to the renal cortex. A reduction in pulse pressure, with the substitution of a steady for a pulsate blood flow, seems to be the major physiological factor in producing or liberating renin. The actual pressor substance, the polypeptide, undergoes destruction by further enzymatic activity. The concentration of pressor substance in the blood depends upon the balance between the two enzymatic processes by which it is formed and destroyed. In order to detect hypertensin in blood, it is necessary to check the action of the destructive enzymes by immediately cooling the blood to 0°C. (14). The enzymes which destroy angiotonin most effectively are those obtained from kidney extract, which contains many unidentified enzymes. Destruction is accelerated by the addition of oxidized cytochrome, both *in vitro* and *in vivo*, as evidenced by reduction of pressure of normal rats. Diminished response to repeated doses, or *tachyphylaxis*, of renin is explained as the result of using up the available hypertensinogen. The origin of hypertensinogen has been located in the liver, since the substance can not be demonstrated after hepatectomy, or when the liver is severely damaged by carbon tetrachloride poisoning.

In rats pretreated with deoxycorticosterone, and allowed abundant salt intake, injection of renin, or repeated injection of angiotonin produces proteinuria and necrotic lesions in the walls of arterioles and small arteries (22).

Hypertensin has been demonstrated by bioassay in patients with normal blood pressure. A small, but statistically significant, increase in blood hypertensin has been observed in patients with clinically benign forms of hypertension (19), and much greater increases in hypertensive patients in the malignant phase.

The degree to which renin participates in the regulation of human blood pressure and in the causation of human hypertension has not been established. In a limited number of human cases, unilateral renal disease has been accompanied by hypertension which was relieved by the removal of the diseased kidney, when the contralateral kidney was normal.

REFERENCES

1. Archibald, R. M. Colorimetric determination of urea. J. Biol. Chem., **157:** 507–518, 1945.
2. Bayrd, E. D., and Heck, F. J. Multiple myeloma. A review of eighty-three proved cases. J. A. M. A., **133:** 147–157, 1947.
3. Bjering, T. Renal excretion of urea. Acta med. scandinav., **136:** Suppl. **234:** 33–40, 1949.
4. Borst, M., and Koenigsdoerffer, H. *Untersuchungen über Porphyrie mit besonderer Berücksichtigung der Porphria Congenita.* Leipzig, S. Hirzel, 1929.
5. Boyce, W. H., and Garvey, F. K. The amount and nature of the organic matrix in urinary calculi. A review. J. Urol., **76:** 213–227, 1956.
6. Brick, I. B. The clinical significance of aminoaciduria. New Eng. J. Med., **247:** 635–644, 1952.
7. Brooks, A. L. An appraisal of a urinary porphyrin test in detection of lead absorption. Indust. Med. & Surg., **20:** 390–392, 1951.

8. Davies, D. F., *et al.* Primary amines. II. Their natural occurrence in urine of normotensive and hypertensive subjects. J. Lab. & Clin. Med., **43:** 620–632, 1954.
9. de Langen, C. D., and ten Berg, J. A. G. Porphyrin in the urine as a first symptom of lead poisoning. Acta med. scandinav., **130:** 37–44, 1948.
10. Dent, C. E., and Rose, G. A. The Bence-Jones protein of multiple myelomatosis: its methionine content and its possible significance in relation to the etiology of the disease. Biochem. J., **44:** 610–618, 1949.
11. Folin, O. *Laboratory Manual of Biological Chemistry.* 5th Ed. New York, Appleton-Century, 1934.
12. Friedmann, E., *et al.* Contribution to the chemistry of leucaemic urine. Biochim. Biophys. Acta, **5:** 45–52, 1950.
13. Gilligan, D. R., and Blumgart, H. L. March hemoglobinuria. Medicine, **20:** 341–395, 1941.
14. Gollan, F., *et al.* Hypertensin in systemic blood of animals with experimental renal hypertension. J. Exper. Med., **88:** 389–400, 1948.
15. Ham, T. H. *A Syllabus of Laboratory Examinations in Clinical Diagnosis.* Cambridge, Mass., Harvard University Press, 1950.
16. Hayman, J. M., Jr., *et al.* Renal function and the number of glomeruli in the human kidney. Arch. Int. Med., **64:** 69–83, 1939.
17. Herkel, W., and Koch, K. Untersuchungen zur Oxalsäure. Ausscheidung insbesondere bei Nierensteinkranken. Deutsches Arch. klin. Med., **178:** 511–537, 1936.
18. Jeghers, H., and Murphy, R. Practical aspects of oxalate metabolism. New Eng. J. Med., **233:** 208–215, 238–246, 1945.
19. Kahn, J. R., *et al.* The assay of hypertensin from the arterial blood of normotensive and hypertensive human beings. J. Exper. Med., **95:** 523–529, 1952.
20. Lintzel, W. Zur Frage des Eisenstoffwechsels. IV. Ueber das Harneisen. Ztschr. Biol., **87:** 157–166, 1928.
21. Maluf, N. S. R. Urea-clearance by perfusion of the intact small intestine in man. J. Urol., **60:** 307–315, 1948.
22. Masson,G.M.C.,*etal.* Angiotonin induction of vascular lesions in desoxycorticosterone-treated rats. Proc. Soc. Exper. Biol. Med., **84:** 284–287, 1953.
23. McCance, R. A. Renal function in early life. Physiol. Rev., **28:** 331–348, 1948.
24. McKittrick, D. S., and Schmidt, C. L. A. Determination of sulfate by the benzidine method. Arch. Biochem., **6:** 411–417, 1945.
25. Meiklejohn, A. P. The physiology and biochemistry of ascorbic acid. Vitamins and Hormones, **11:** 61–96, 1953.
26. Miller, S., *et al.* Urinary excretions of ten amino acids by women during the reproductive cycle. J. Biol. Chem., **209:** 795–801, 1954.
27. Möller, E., *et al.* Studies of urea excretion. II. The relationship between urine volume and the rate of urea excretion by normal adults. J. Clin. Invest., **6:** 427–465, 1928.
28. Mudge, G. H., and Taggart, J. V. Effect of 2,4-dinitrophenol on renal transport mechanisms in the dog. Am. J. Physiol., **161:** 173–180, 1950.
29. Nicholas, R. E. H., and Rimington, C. Qualitative analysis of the porphyrins by partition chromatography. Scandinav. J. Clin. & Lab. Invest., **1:** 2–11, 1949.
30. Pearlman, C. K. Xanthine urinary calculus. J. Urol., **64:** 799–800, 1950.
31. Pitts, R. F. Acid-base regulation by the kidneys. Am. J. Med., **9:** 356–372, 1950.

32. PRIEN, E. L. Studies in urolithiasis. II. Relationships between pathogenesis, structure, and composition of calculi. J. Urol., **61:** 821–836, 1949.
33. PRIEN, E. L., AND FRONDEL, C. Studies in urolithiasis. I. The composition of urinary calculi. J. Urol., **57:** 949–991, 1947.
34. RIGAS, D. A., AND HELLER, C. G. The amount and nature of urinary proteins in normal human subjects. J. Clin. Invest., **30:** 853–861, 1951.
35. ROTHMAN, S. Studies on melanuria. J. Lab. & Clin. Med., **27:** 687–692, 1942.
36. SELLER, A. L., *et al.* Filtration and reabsorption of protein by the kidney. J. Exper. Med., **100:** 1–9, 1954.
37. SKEGGS, L. T., JR., *et al.* Amino-acid composition and electrophoretic properties of hypertensin I. J. Exper. Med., **102:** 435–440, 1955.
38. SMITH, H. W. *The Kidney Structure and Function in Health and Disease.* New York, Oxford University Press, 1951.
39. STEIN, W. H., AND CAREY, G. C. A chromatographic investigation of the amino-acid constituents of normal urine. J. Biol. Chem., **201:** 45–58, 1953.
40. THOMA, G. E., AND KITZBERGER, D. M. Spectrophotometric method for determination of urinary bilirubin. J. Lab. & Clin. Med., **22:** 1189–1192, 1948.
41. WALKER, B. S., *et al.* Acid phosphatases. Am. J. Clin. Path., **24:** 807–837, 1954.
42. WIDMARK, E. M. P. Oxalic acid in the urine. Acta med. scandinav., Suppl., **26:** 340, 1928.

PART V

Pathology

CHAPTER 21

Vitamins and Vitamin Deficiency Diseases

The term "vitamine" was originally given by Casimir Funk to the impure substance which he concentrated from rice polishings and which prevented the paralysis which Eijkman had observed in fowl nourished principally on polished rice. The more modern term *vitamin* is applied to a chemically heterogeneous group of organic compounds which must be supplied in the diet or by intestinal bacterial synthesis to maintain health. Continued deficiency of a vitamin leads to severe illness. Substances which may be vitamins for one species may not be vitamins for some other species. This may be because the second species does not require this substance or it may be that the second species is able to synthesize it by its own metabolic processes. Certain vitamins may carry on their essential metabolic function as components of more complex molecules. An example would be thiamine, which forms a part of the essential coenzyme, thiamine pyrophosphate. In such instances, which are numerous, the simplest substance capable of supplying the physiological need is designated as the vitamin. In the case just mentioned, thiamine is the vitamin and not thiamine pyrophosphate, although the presence of either in the diet in adequate amounts would meet the physiological requirements. A disease resulting from the deficiency of a vitamin is called an *avitaminosis*. The avitaminosis resulting from the lack of vitamin C would be called avitaminosis C, and so for the other vitamins. Actually, since a diet deficient in one vitamin, unless expressly so designed experimentally, is likely to be deficient in other vitamins as well, most human cases of vitamin deficiency disease turn out to be multiple avitaminoses, and therefore can not be so simply, clearly, and exactly designated.

CONDITIONED VITAMIN DEFICIENCIES

Even when all vitamins and other food factors are available to a given person in amounts usually considered adequate, he or she may develop a nutritional deficiency.

Increased requirement for a vitamin or a group of vitamins may be the result of growth, pregnancy, lactation, or physical exertion, or may be secondary to diseases such as fever or hyperthyroidism which accelerate metabolic processes. The need for certain vitamins increases with increased carbohydrate intake, hence patients who are being nourished chiefly by intravenous injections of glucose are commonly given thiamine and other water-soluble vitamins in proportion to the amount of glucose given.

Limited food intake when adequate food is available may be voluntary or the result of disease. Lack of appetite is characteristic of many physical and mental ills, and restriction of diets is a very common form of medical treatment. The physician who prescribes diets must keep in mind the vitamins and other food factors, and must make up with vitamin preparations the deficiencies in the foodstuffs allowed. Many people, not always the elderly, the poor, or the notably peculiar, subject themselves to strange limitations of diet which may lead to deficiency disease. A properly taken medical history includes an account of the patient's actual diet.

Absorption or utilization of vitamins may be impaired in gastrointestinal, hepatic, or endocrine disease. The use of mineral oil, cathartics, or adsorbing agents in therapy, or the surgical removal of portions of the gastrointestinal tract may limit the ability to absorb vitamins as well as other foodstuffs.

Excretion of vitamins may be accelerated as a result of excessive fluid output through skin or urinary tract. *Destruction* of vitamins may occur excessively in the alimentary tract or in the body in abnormalities of the digestive secretions (e.g., achlorhydria) or as the result of poisoning with heavy metals or certain synthetic organic compounds.

THE VITAMIN B COMPLEX

On account of the fundamental physiological importance of certain members of this group, we here abandon the traditional alphabetical order and give the group of B vitamins first consideration. The term vitamin B in itself no longer has significance. It was once applied to the crude material obtained from rice polishings, which prevented the development of beriberi in man and of similar diseases in experimental fowl and other animals. This substance, now known in its pure form as thiamine or vitamin B_1, is but one of a rather large group of substances which make up the vitamin B complex. They are water-soluble, organic compounds which are necessary components of all living cells. Their function, either alone or as structural components of more complex molecules, is in catalytic systems where they function as coenzymes. The substances which make up the B complex exist and have similar functions in single celled organisms and in all the cells of all species of plants and animals. In certain species, however, they are produced adequately or even abundantly within the organism. For such species they are, of course, not vitamins. The concept of the B complex is anthropocentric, and is limited to those substances falling within this general category which are not synthesized by man or by common laboratory animals and therefore are dietary requirements for these species. If one of these substances is synthesizable by a given organism, it can be termed, along with the true vitamins for that organism, as an *essential metabolite*.

Thiamine and Lipoic Acid

Thiamine, known also as vitamin B_1 and in the European literature as aneurine, is the substance which was present in the early crude mixture obtained from rice bran which was effective in the prevention of polyneuritis in animals and the corresponding human disease, beriberi. Thiamine itself is 2,5-dimethyl-6-aminopyrimidine united with 4-methyl-5-hydroxyethylthiazol. The formula shown in formula I is that of the chloride-

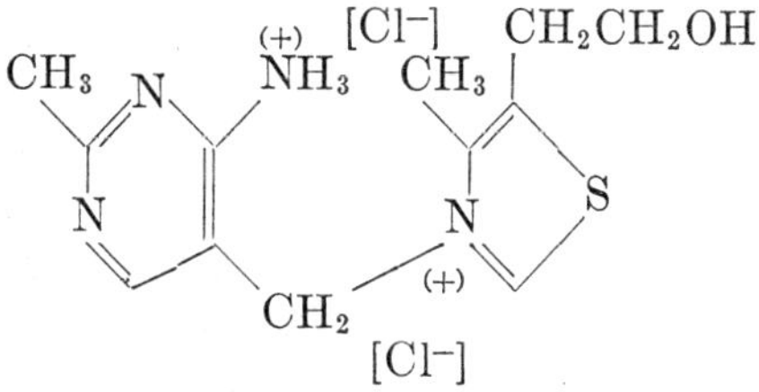

(+) NH_3 [Cl⁻] CH_3 CH_2CH_2OH; CH_3 N N; N (+) S; CH_2 [Cl⁻]

I. Thiamine chloride hydrochloride

hydrochloride, the form in which it is commonly marketed. Thiamine functions as a part of the catalytic system of living cells in the form of its pyrophosphate ester, *cocarboxylase* (formula II). This coenzyme is involved in the decarboxylation of pyruvic acid and other alpha-keto acids.

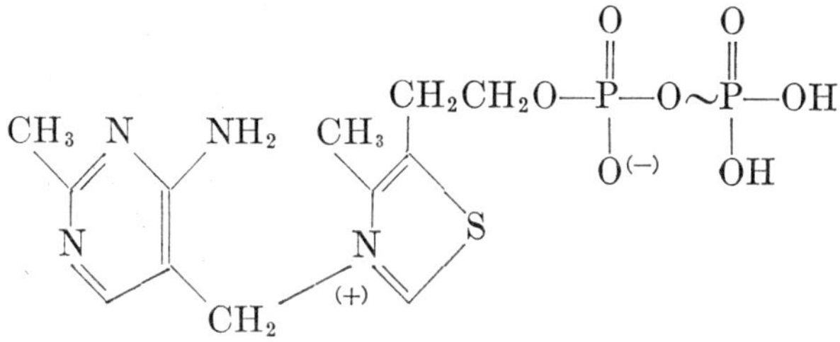

CH_3 N NH_2 CH_3 $CH_2CH_2O{-}P(=O)(O^{(-)}){-}O{\sim}P(=O)(OH){-}OH$; N N (+) S; CH_2

II. Thiamine pyrophosphate (TPP) or cocarboxylase

The simple non-oxidative decarboxylation of pyruvic acid to form acetaldehyde requires only thiamine pyrophosphate as the coenzyme. In the more complex situation in animal and bacterial metabolism, where the decarboxylation is oxidative and yields an acetyl group, a different coenzyme is required. This coenzyme contains thiamine pyrophosphate with lipoic acid joined by a peptide link to the 6-amino group of the pyrimidine portion of thiamine, and is known as *lipothiamide pyrophosphate* (LTPP) (formula III).

—$CH_2CH_2CH_2CH_2$
S—S
C=O
CH_2CH_2O—P—O~P—OH
CH_3 N NH CH_3
O (−) OH
N N S
(+)
CH_2

III. Lipothiamide pyrophosphate (LTPP)

Lipoic acid is a generic term for a group of dithio acids which form such conjugates. Synonyms for lipoic acid are *pyruvate oxidation factor*, *acetate replacement factor* (since it permits growth of *L. casei* in acetate-free media), and *protogen* (a growth factor for protozoa). One substance which definitely possesses lipoic acid activity is *6-thioctic acid* or cyclic 6,8-dithiooctanoic acid, which is shown attached to thiamine pyrophosphate is formula III.

When pyruvic (or alternatively alpha-ketoglutaric) acid is decarboxylated oxidatively, the actual product is acetyl (or succinyl) LTPP, the acyl group being linked to the lipoic acid portion of LTPP through a high-energy acyl-mercaptan bond. The acyl groups are then transferred to CoA, forming acetyl (or succinyl) CoA. In this series of reactions, the disulfide ring is alternately broken and re-formed, the redox reaction being catalyzed by DPN, as shown on page 549. Lipoic acid is also involved as electron acceptor in the chlorophyll-catalyzed photolysis of water (see page 451) while TPP has been reported to be a necessary coenzyme for various transketolases (43).

Lipoic acid has been obtained from liver. It is fat-soluble, but the conjugated form, LTPP, is water-soluble. Lipoic acid is not yet officially classified as a vitamin. No disease resulting from lipoic acid deficiency has been confirmed, nor is it certain that lipoic acid must be supplied in the animal diet. Lipoic acid occurs in green plants, in yeast, and in many bacteria as well as in liver.

In thiamine deficiency, chemical analysis shows increased amounts of both pyruvic and lactic acid in the blood and in the urine as well as diminished amounts of thiamine. Analysis for thiamine is more specific and more informative diagnostically. The average concentration of thiamine in human blood is 3.4 micrograms per 100 ml., with a standard deviation of 1.1 micrograms (30). Thiamine is converted into cocarboxylase within cells by the action of a phosphokinase, utilizing ATP as the source of phosphate. Thiamine can enter and leave cells freely, and appears in the urine provided the intake is adequate. Cocarboxylase, once formed, remains and functions within the cell, and is not a normal urinary component. In well-nourished subjects the output of thiamine in the urine averages 230 micrograms per day. Values below 150 micrograms are considered sub-optimal, and less than 100 are indicative of a deficient intake (8). The minimal daily adult intake for prevention of beriberi is about 0.4 milligrams. The daily intake advised by the National Research Council is 1.0 to 1.6 milligrams for adults.

Thiamine deficiency has been a common cause of heart disease in the United States (56). The cardiac disturbances can reasonably be attributed to the inability of the heart muscle to make adequate oxidative utilization of the normal metabolites, pyruvic and lactic acids. In early or mild deficiencies the heart rate is increased. Later, enlargement of the heart develops with decreased diastolic blood pressure, and eventual congestive failure. Electrocardiographic changes in beriberi are not specific. Slowing of the heart rate may occur in advanced cases but this is more commonly observed in experimental animals.

The neuritis of thiamine deficiency starts with degeneration of the myelin sheaths, which may be followed by fragmentation of axons. The motor function of nerves may be depressed in varying degrees from barely perceptible muscular weakness to complete paralysis. Disturbances of function of the sensory nerves are variable and often bizarre in their distribution. There may be anesthesia, hyperesthesia, paresthesia, or even severe cramp-like pain. Both motor and sensory neural disturbances usually appear first in the lower extremities and in mild or moderate cases are often limited to that area. Advanced cases often show generalized edema—the so-called "wet beriberi".

Alcoholism has been a frequent causative factor of thiamine deficiency in the United States. Those patients who find themselves impelled into protracted periods of heavy drinking usually take but little food during these periods and that food is usually of poor nutritional value. From this situation develops the typical "alcoholic polyneuritis" which is really an acute form of beriberi and if not too advanced, responds well to thiamine therapy. Even in advanced cases some useful degree of remyelinization and regeneration often occurs with adequate treatment. The American alcoholic population has demonstrated a notable decrease in beriberi and in other deficiency diseases referable to the B complex since the vitamin enrichment of white bread has become customary (14).

A much greater incidence of beriberi unrelated to alcoholism is recorded in those oriental populations who are forced to rely mainly upon polished rice in their diet. Most of the oriental cases are chronic and relatively mild. The severity is increased in localities and at times of economic hardship. Raw fish and raw clams contain a thiaminase which may destroy the thiamine of foodstuffs during the early stages of the digestive process. This mechanism is of little significance in human nutrition but is economically important as the cause of the "Chastek" paralysis of ranch foxes fed on raw fish.

Of foods commonly used in human diets, yeast is the most abundant source of thiamine as well as of most of the other vitamins of the B group. Large amounts of yeast were used in the making of old fashioned home-made bread, providing adequate amounts of the B vitamins. Although modern baking methods do not use such large proportions of yeast, the deficiency is made up by the purposeful enrichment of bread with added thiamine along with other vitamins and valuable nutritive substances. Grains in their native state are excellent sources of thiamine with the highest concentration occurring in the germ. Of meats commonly used, pork is the best source, with about five times the concentration of thiamine as compared to beef or mutton. These latter meats, along with fish and fresh fruits, vegetables, and nuts, are significant dietary contributors of thiamine. Measurement of the thiamine content of foodstuffs can be accomplished biologically by the estimation of the amount required to prevent polyneuritis in experimental animals maintained on a deficient diet or by measurement of the growth of micro-organisms which require thiamine as a growth factor. A satisfactory chemical method of thiamine determination (9) involves the oxidation of thiamine to thiochrome in alkaline potassium ferricyanide solution. The concentration of thiochrome is measured by its fluorescence in ultraviolet light. A satisfactory colorimetric method depends upon the reaction of thiamine in alkaline medium with diazotized *p*-amino acetophenone (37). Thiamine in the dry form and in acid solution is stable to heat. It is less stable as the alkalinity of the solution is increased. These properties are responsible for varying degrees of destruction of thiamine in foodstuffs during cooking.

The toxicity of thiamine is low enough to render the administration of ordinary prophylactic or therapeutic doses a perfectly safe procedure. Intravenous dosages of thiamine hydrochloride in concentrations of 100 mgm. per ml. and in total dosage of 126 mgm. per kgm. of body weight have caused death in rabbits by paralysis of the respiratory center (18), and at least one death apparently the result of intravenous overdosage with the same substance has occurred in man.

Riboflavin

Riboflavin (formula IV) is 6,7-dimethyl-9(*d*-ribityl)-isoalloxazine. It is no longer commonly called by its original designations, vitamin B_2 or G. In the form of FAD (flavin adenine dinucleotide) it is the prosthetic group of the flavin enzymes which are found in all animal cells. Examples of such flavin enzymes are xanthine dehydrogenase and D-amino acid dehy-

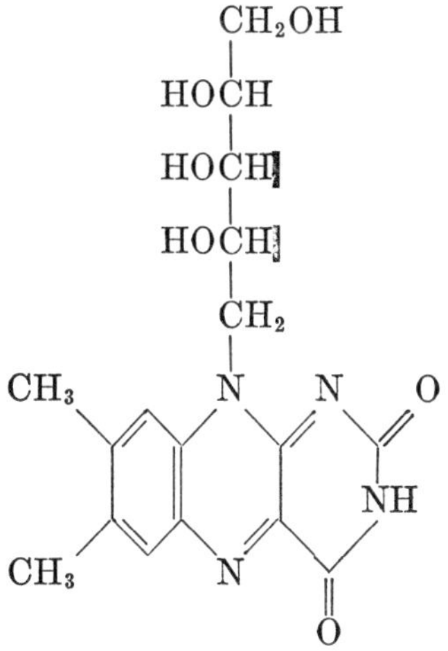

CH_2OH
HOCH
HOCH
HOCH
CH_2
CH_3 N N O
NH
CH_3 N
O

IV. Riboflavin

drogenase. Other types of flavin enzymes occur in which the prosthetic group is FM (flavin mononucleotide or riboflavin phosphate), for example the L-amino acid oxidase of rat kidney.

The flavin coenzymes are formed in animal tissues from ingested and absorbed riboflavin, but the enzymes catalyzing their formation have not yet been characterized. A very specific *flavokinase*, catalyzing the formation of FM and utilizing ATP or ADP, has been demonstrated in yeast, and a separate enzyme for conversion of FM to FAD. The flavin coenzymes are firmly bound to the protein apoenzymes, forming *flavoproteins*, which function as hydrogen carriers (see page 276).

Rats placed on a diet deficient in riboflavin but otherwise adequate fail to grow. They shed much or all of their fur, and develop areas of symmetrical dermatitis. A very characteristic sign is the ingrowth of blood vessels into the normally avascular cornea of the eye. In human ariboflavinosis, as in the rat, the cornea becomes vascularized and may eventually become opaque. The tongue takes on a magenta color and is often fissured. Painful lesions develop at the corners of the mouth accompanied by hyperactivity of the sebaceous glands in the skin of the face. Riboflavin deficiency is a common human affliction, but is usually associated with other deficiencies. It is common in portions of Africa, the Orient, and in the West Indies. It is less frequently met with in the United States. The underlying biochemical lesion in riboflavin deficiency is, naturally, a failure of those oxidations catalyzed by flavoprotein enzymes. Thus, failure of xanthine oxidase activity can be demonstrated in the livers of riboflavin-deficient rats. Also an over-all deficiency of riboflavin in the whole body can be demonstrated in deficient animals as compared to normally nourished controls. In human cases of riboflavin deficiency it is possible to demonstrate diminished urinary output of the substance. In the urines of normal individuals, Cleland (8) found daily excretions of 130 to 800 micrograms of riboflavin with a mean value of 340. This compares well with the results of a similar series in which riboflavin was determined by a microbiological method (15), where the range was 65 to 980 micrograms with an average of 345. The chemical method referred to, like most methods for riboflavin, depends upon the fluorescence of this substance under ultraviolet light. Microbiological assay is carried out by measurement of the growth rate of organisms such as *Lactobacillus casei* and *Leukonostoc mesenteroides* which require an external source of riboflavin.

The National Research Council's recommended adult intake of riboflavin is from 1.4 to 1.6 milligrams per day. The experimentally determined minimal requirement for an adult male on a 2200 Kcal. daily food intake is between 1.1 and 1.6 mgm. daily (24). As with most of the B vitamins, yeast is the richest source, supplying from 2 to 8 milligrams per 100 grams of dry material. Among other foods, milk is the best everyday source. Significant contributions of riboflavin are made to the diet by green leafy vegetables, eggs, meats, and by most fruits and vegetables. Wheat germ is a rich source, which fact makes whole wheat products very satisfactory contributors of this vitamin to the diet. Riboflavin is quite stable to heat, but is destroyed by exposure to light. In addition to the dietary sources of riboflavin, some contribution is made by the production of this vitamin as a result of the action of intestinal bacteria. As far as human nutrition is concerned, the contribution is probably small. In cattle, sheep, and goats, where bacterial action in the rumen takes place early in the digestive process and before any significant amount of absorption has occurred, bacterial production of this vitamin takes on a much greater significance. In cattle the output of riboflavin in the day's milk has been found to be as much as ten times the riboflavin of the day's diet (57). It should be pointed out in connection with all the B vitamins that while we consider them as deriving chiefly from plant sources, the plants may in turn derive them from bacteria and fungi resident in the soil and possibly symbiotically associated with the rootlets of the plants.

Measurement of very low levels of riboflavin can not be accomplished by the relatively crude methods used for urine or foodstuffs. By the use of a very sensitive fluorometer, the level in human blood serum of free riboflavin plus riboflavin phosphate has been found to be 0.8 micrograms per 100 ml., with a standard deviation of 0.08 micrograms. In addition to this, FAD was present in an amount corresponding to three times that figure. The riboflavin content of white cells and platelets in the same subjects was 252 micrograms per 100 ml., with a standard deviation of 11 micrograms. The corresponding figures for red cells were 22.4 micrograms per 100 ml., with a standard deviation of 1.3 micrograms (5).

Nicotinic Acid

Following the rule of selecting the simplest effective structure as the vitamin, *nicotinic acid* (formula V) is named rather than its amide, *nicotinamide* (formula VI). Both of these are equally effective as vitamins.

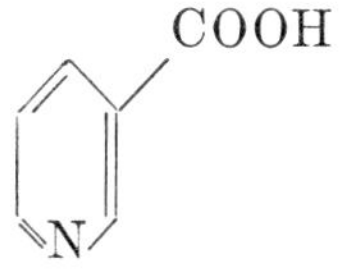

COOH CONH$_2$

N N

V. Nicotinic acid 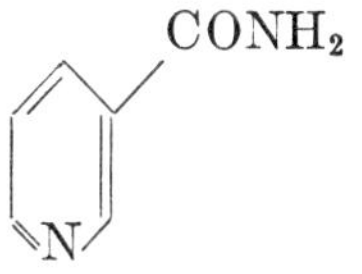 VI. Nicotinamide

Niacin and *niacinamide* are synonyms for these two compounds. Nicotinamide is a structural component of three important coenzymes (coenzyme I or DPN, coenzyme II or TPN, and coenzyme III, see page 248). Most of the nicotinic acid in tissues is present in the form of the nicotinamide coenzymes. In general, the concentrations of the nicotinamide coenzymes in tissues are diminished in states of nicotinic acid deficiency, and to a degree corresponding to the severity of the deficiency.

The National Research Council's recommendation for adult intake of nicotinic acid is 10 to 16 mgm. per day. Comparison of this figure with the results of numerous attempts to establish the actual human requirement indicates that the recommendation is probably above the minimum. Intakes of 30 mgm. or more of nicotinic acid produce a definite pharmacological effect somewhat comparable to that of histamine. There is flushing, particularly of the face and hands, sometimes with itching and burning. Aside from this apparently harmless effect, which is not observed with nicotinamide, these vitamins show no toxic effects until the dose is increased in experimental animals to amounts greater than one gram per kgm. of body weight.

The human manifestation of nicotinic acid deficiency is *pellagra*, which is the most prevalent of the serious avitaminoses in the United States.

As usually seen, pellagra is not the result of pure nicotinic acid deficiency, but carries with it evidences of deficiency of the other B vitamins. The predisposing cause of pellagra is almost invariably alcoholism or poverty, or both. There is a seasonal variation in the incidence of pellagra in temperate climates which is out of phase with the seasonal changes in the supply of B vitamins. This has been explained (47) on the basis of seasonal differences in metabolic activity. Before the demonstration by Goldberger that pellagra was a disease of nutritional deficiency curable by feeding of yeast, the incidence of pellagra was high in prisons and asylums and particularly in regions where corn (*Zea mays*) meal was the dietary staple, unmitigated by fresh meats or vegetables. The symptoms of pellagra have long been mnemonically described as "dermatitis, diarrhea, and dementia". The dermatitis tends to be symmetrical and to present sharply marked-off areas of red, thickened skin, which later turn brown and scaly. It is likely to be restricted to areas which are exposed to light or to friction. The skin areas surrounding the external genitalia are particularly prone to pellagrous dermatitis, and may be affected without involvement of other areas. The tongue is red and swollen, later cracked and peeling. The diarrhea is both an early and a persistent manifestation and is one of the chief contributing causes of death in pellagrins. Achlorhydria and a macrocytic hyperchromic anemia often develop. Persistent nausea is a common symptom. The dementia of pellagra may range from the mildest of psychoneuroses to the most severe manic or stuporous states. The association of these psychotic manifestations with pellagra is verified by their prompt favorable response to therapy with nicotinic acid, usually within a week. Dosages of nicotinic acid in the therapy of pellagra may run as high as one gram per day, and are more effective if accompanied by the other components of the B complex. It is very difficult to develop nicotinic acid deficiency in rats. Monkeys develop experimental pellagra, pigs and dogs have frequently developed pellagra, or the canine equivalent "blacktongue", spontaneously as well as experimentally.

As it is for all the other B vitamins, yeast is a rich source of nicotinic acid. Among other foods, liver leads the list in nicotinic acid content. Good sources are meats, fish, eggs, whole wheat, unpolished rice, and peanuts. Nicotinic acid is not destroyed by ordinary cooking or by canning. There is considerable evidence that many animals, including man, can meet a part of the nicotinic acid requirement by synthesizing it with tryptophane as the starting substance. Pellagra-producing diets in man have been characterized by low content of tryptophane as well as of nicotinic acid. Sixty mg. of tryptophane is nutritionally the approximate equivalent of one mgm. of nicotinic acid (23).

Both nicotinic acid and nicotinamide are excreted in human urine along with N^1-methylnicotinamide and its 6-pyridone derivative (formula VII). The total excretion varies with the intake, and has been reported as high as 45 milligrams and as low as 1.7 mgm. per day in presumably well nourished human subjects. The mean value is 12.8 mgm. per day (8). There is very little nicotinic acid demonstrable in blood plasma. There is somewhat over one mgm. per 100 ml. of red blood cells, chiefly in the form of coenzymes. The coenzymes I, II, and III do not appear in the urine.

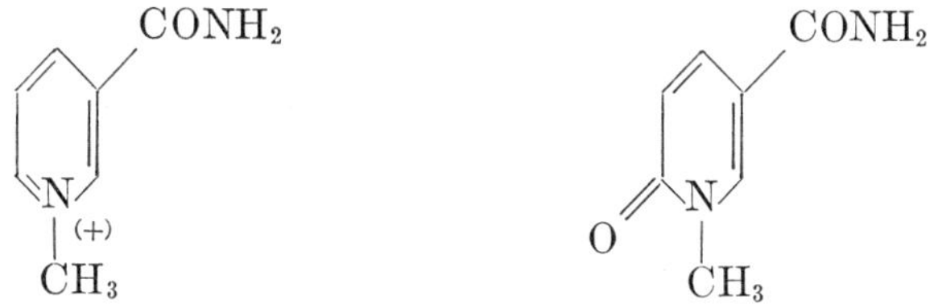

$CONH_2$ / N$^{(+)}$ / CH_3	$CONH_2$ / O / N / CH_3
N^1-methylnicotinamide	N^1-methylnicotinamide-6-pyridone

VII. Metabolic products of nicotinamide

The microbiological assay utilizing a strain of *Lactobacillus arabinosus* has been so far the most satisfactory of analytical procedures. It is equally responsive, mol for mol, to nicotinic acid, nicotinamide, and the coenzymes. A number of colorimetric methods have been proposed which lack the specificity of the microbiological assay. N^1-methylnicotinamide in the urine can be measured relatively simply by a fluorometric method (9). Separation of nicotinic acid from nicotinamide is most easily accomplished by paper chromatography.

Pantothenic Acid

The structure of this substance as shown in formula VIII can be described as a dipeptide of beta-alanine and a substituted butyric acid. The production of a state of deficiency of pantothenic acid in human subjects requires the combination of a deficient diet with the use of a pantothenic acid antagonist such as omega-methylpantothenic acid (3). Such a deficiency state is characterized by adrenocortical deficiency and by peripheral neuropathies. The deficient diets which lead to the development of beriberi, ariboflavinosis, and pellagra are also usually deficient in pantothenic acid. Pantothenic acid has been found to be a useful adjunct in the treatment of these deficiencies and of the various manifestations of avitaminosis associated with alcoholism.

A very characteristic syndrome develops in rats maintained on a diet deficient in pantothenic acid. There is parenchymatous damage to kidney and heart and particularly to the adrenals. The external signs include dermatitis, inflammation around the mouth and nose, and in dark-colored rats, graying of the hair. The formation of antibodies (see page 845) is inhibited (41). Unlike normal rats, rats deficient in pantothenic acid do not develop fatty livers on a diet containing one per cent cholesterol (17).

Biosynthesis of pantothenic acid takes place in the leaves of green plants and in many micro-organisms. *Pantonine* (formula VIII), an alpha-amino acid corresponding to the alpha-hydroxy acid (*pantoic acid*) conjugated with beta-alanine in pantothenic acid, has been identified in *E. coli*, and is

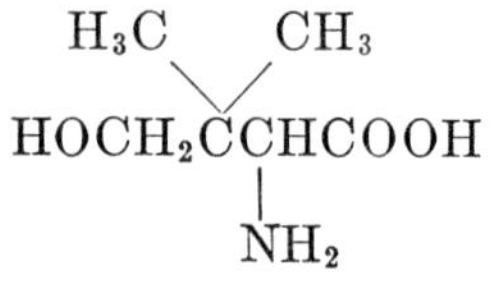

$$\begin{array}{c} H_3C \quad CH_3 \\ HOCH_2C\,CHCOOH \\ NH_2 \end{array}$$

Pantonine

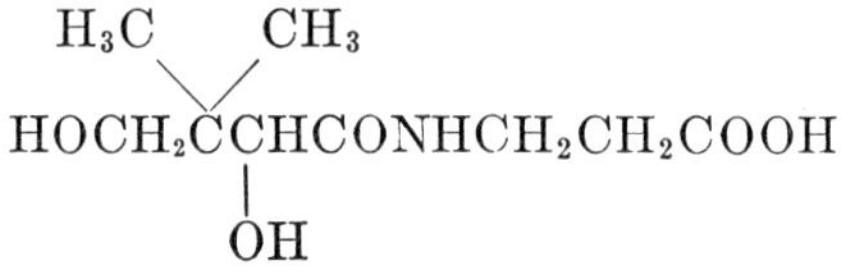

$$\begin{array}{c} H_3C \quad CH_3 \\ HOCH_2C\,CHCONHCH_2CH_2COOH \\ OH \end{array}$$

Pantothenic acid

$$\begin{array}{c} H_3C \quad CH_3 \\ HOCH_2C\,CHCONHCH_2CH_2CONHCH_2CH_2SH \\ OH \end{array}$$

Pantetheine

$$\begin{array}{l} \quad H_3C \quad CH_3 \\ \quad CH_2C\,CHCONHCH_2CH_2CONHCH_2CH_2SH \\ O \qquad OH \\ HO{-}P{=}O \qquad OH \\ O \qquad P{=}O \\ HO{-}P{=}O \quad O \quad OH \quad OH \\ O \quad H\,H \quad N \quad N \\ H \quad O \quad H \quad N \\ \qquad N \\ \qquad NH_2 \end{array}$$

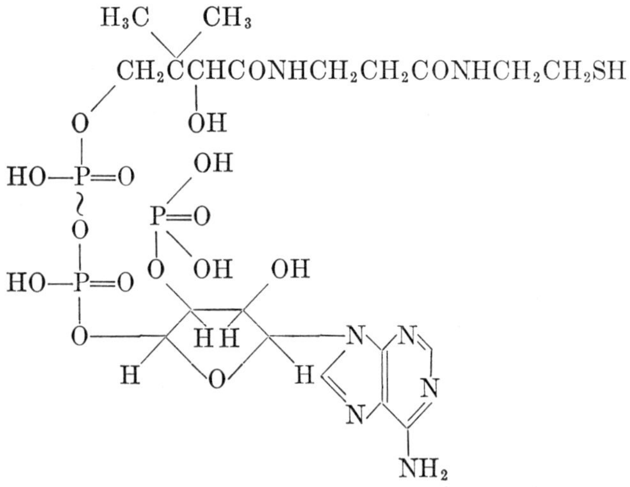

Coenzyme A (CoA)

VIII. Coenzyme A and its precursors

considered to be an intermediate in the production of pantothenic acid by that organism. An enzyme system has been extracted from *E. coli* which will condense pantoic acid with beta-alanine (35). ATP is required for this condensation, as in other instances of peptide bond synthesis, but CoA is not required, as it is in hippuric acid synthesis (see page 616). The entire requirement in cattle seems to be met by bacterial production within the rumen, and up to 60 per cent of the requirement of rats appears to be supplied by microbial action in the cecum. It is not known to what extent the human requirement is met from bacterial sources in the intestine, and actually there is considerable doubt as to the exact human requirement. Williams (57) recommends a daily intake of 9 to 12 milligrams, based on a total caloric intake of 2500 Kcal. No toxic effects have been observed from large doses of pantothenic acid administered to man or to experimental animals. The daily urinary output of pantothenic acid is in the neighborhood of 3 mgm. Human blood contains about 30 micrograms per 100 ml.

As with other B vitamins, yeast and liver are the most concentrated natural sources of pantothenic acid. The supply in other foodstuffs is much the same as with the other B vitamins, except that wheat germ is not a satisfactory source. Pantothenic acid withstands moist heat in neutral solutions, but is rapidly hydrolyzed at higher pH values.

Not all micro-organisms produce pantothenic acid, some require it, and some require more complex substances containing pantothenic acid. *Lactobacillus arabinosus* responds only to free pantothenic acid, and is used in bioassay methods for this reason. *Lactobacillus bulgaricus* has a specific requirement for a fragment of the CoA molecule, the *Lactobacillus bulgaricus* factor (LBF) or *pantetheine* (see formula VIII). This substance is not stable in air, but oxidizes to form a disulfide (comparable to the oxidation of 2 molecules of cysteine to one of cystine) known as pantethine. This disulfide and also mixed disulfides of pantetheine with other thiols show LBF activity. CoA, a derivative of pantetheine, can also form mixed disulfides.

Coenzyme A (or CoA, see formula VIII) is the metabolically active form of pantothenic acid. It accounts for all the cellular pantothenic acid in animals, and most of the pantothenic acid in plant and microbial cells. It consists of pantetheine linked by its primary alcohol group through pyrophosphate to a diphosphoadenosine. The —SH group on the pantetheine portion is the business end of CoA, where high-energy bonds form with acetyl and other acyl radicals.

In the animal body pantothenate is condensed with cysteine in the presence of ATP to form pantothenylcystine, which is converted to pantetheine (21). Pantetheine is converted to phosphopantetheine, catalyzed by *pantetheine kinase*, utilizing ATP. The phosphopantetheine condenses with another molecule of ATP, eliminating pyrophosphate, and forming de-phospho-CoA, which is converted to CoA by a third molecule of ATP The synthesis of CoA probably takes place in all organs, since CoA is required by all cells and no CoA has been detected circulating in the plasma.

The functions of CoA in the transfer of 2-carbon fragments have been described in previous chapters. In brief summary, we recall that the initiation of the Krebs cycle by the union of a 2-carbon fragment with oxaloacetic acid requires at least one activating enzyme utilizing ATP and one condensing enzyme utilizing CoA. The 2-carbon fragment is in the form of acetyl CoA when it condenses with oxaloacetic acid, and CoA is liberated at the time of the condensation. CoA is also involved in the oxidation and the synthesis of fatty acids, and acetylations, such as the acetylation of sulfonamide drugs and the formation of acetyl choline. Succinyl CoA liberates succinic acid in the Krebs cycle, and also functions in porphyrin synthesis by condensation with glycine. The significance of CoA in steroid synthesis, and in the formation of certain peptide bonds, has been demonstrated, although the details are not yet fully worked out.

Colorimetric analysis for CoA depends upon its function as a coenzyme in the acetylation, with decreased color, of 4-aminoazobenzene (19). Spectrophotometric assay can be done quickly by measurement of the rate of reduction of DPN by alpha-ketoglutaric acid catalyzed by a CoA-dependent enzyme system (55).

Vitamins B_6

From the point of view of human nutrition *pyridoxal*, *pyridoxine*, and *pyridoxamine* are qualitatively and quantitatively equivalent. These substances together constitute a group often known as vitamin B_6 (see formula IX). Pyridoxine is the form of vitamin B_6 most abundant in cereal sources.

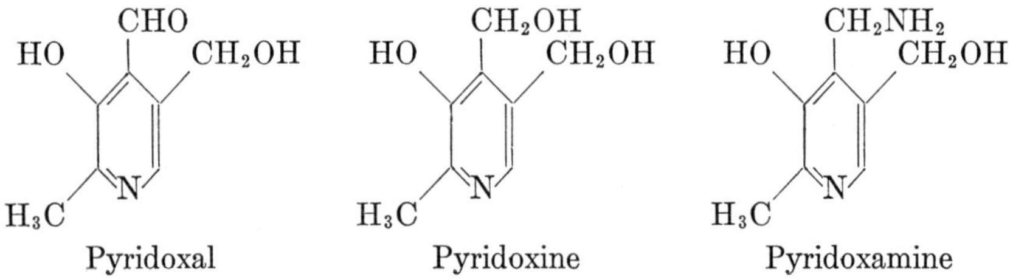

CHO; HO; CH_2OH; N; H_3C	CH_2OH; HO; CH_2OH; N; H_3C	CH_2NH_2; HO; CH_2OH; N; H_3C
Pyridoxal	Pyridoxine	Pyridoxamine

IX. The B_6 vitamins

Pyridoxal and pyridoxamine phosphates are the metabolically active forms in which vitamin B_6 occurs in animal cells and tissues. Pyridoxal is phosphorylated to pyridoxal phosphate at the expense of ATP in the presence of a pyridoxal kinase (27). Pyridoxamine phosphate and pyridoxal phosphate are interconverted when either acts as coenzyme for transaminases (see page 610). Pyridoxal phosphate is also a coenzyme for enzymes catalyzing the transfer of sulfur (see page 629) and is similarly involved in the decarboxylation of certain amino acids. Significant decarboxylations in human metabolism are exemplified by the formation of gamma-aminobutyric acid from glutamic acid, and of 3,4-dihydroxyphenylethylamine (a precursor of adrenalin) from dihydroxy-phenylalanine (dopa). The activity of lactic acid dehydrogenase is depressed in rats deficient in vitamin B_6 (52). Vitamin B_6 also seems to be necessary for the utilization of essential fatty acids (4).

The biochemical lesions which would be expected to occur in vitamin B_6 deficiency can be demonstrated in experimental animals. Transamination can be shown to be deficient and abnormalities in the metabolism of tryptophane, particularly its conversion to nicotinic acid, can be demonstrated. Lack of the B_6 group seems quite definitely involved in the symptoms observed in pellagra and beriberi; persistent neurological disturbances have responded well to therapy with pyridoxine, as have neurological symptoms associated with diseases not primarily of nutritional origin. There is some controversy as to whether the lesions in the corners of the mouth, already described as characteristic of riboflavin deficiency, may not be the result of deficiency of pyridoxine and its relatives. Pyridoxine deficiency in the young rat is characterized by *acrodynia*, which is a dermatitis obviously painful to the animal, developing chiefly on the paws, tail, ears, nose, and mouth. Rats and other experimental animals may develop epileptiform attacks as a result of restriction of the B_6 group. Human infants occasionally develop convulsive disorders which are apparently the result of vitamin B_6 deficiency (38). Pyridoxine, however, has not proven to be a reliable drug in the treatment of epilepsy in man.

A number of colorimetric methods have been proposed for the determination of the B_6 group. In general, these methods have proved non-specific in distinguishing one member of the group from another, and all of these substances from other color-reactive substances in foods and tissues. Methods based upon the prevention of acrodynia in young rats on a vitamin B_6-deficient diet have been widely used. Microbiological assay is carried out using yeast, *Neurospora*, or *Streptococcus faecalis* as the test organism.

Human subjects depleted of vitamins B_6 show abnormalities of amino acid metabolism, particularly of tryptophane (54). Between 2 and 3 mgm. of pyridoxine in the daily diet will prevent such disturbances. No toxic effects of overdosage have been reported in man. Toxic reactions have been produced in animals only with enormous overdosages of orders of magnitude of 3 grams per kgm. of body weight, or more. Although all three forms of vitamin B_6 may be demonstrated in the urine, these substances are excreted chiefly in the form of their metabolic product, 4-pyridoxic acid. The daily excretion of pyridoxine in all forms is less than one milligram.

Effective dietary sources of the vitamin B_6 group are in general similar to those of the other B vitamins. Whole grain cereals, fresh meats, fresh vegetables and milk are the major sources in the human diet.

Biotin

Biotin is a rather curious member of the B complex because it is produced so abundantly by the synthetic action of intestinal flora that pure nutritional deficiency of biotin is not only seldom observed but difficult to produce. Like the other members of the B group, it has been found to be a constituent of all cells where its presence has been investigated. It is also of interest that two vitamins, originally thought different, were later shown to be identical with a single substance, biotin.

In 1901, Wildiers observed that a substance found in beer wort and in growing yeast cultures was required to bring about the growth of yeast in an otherwise purely synthetic medium. Note that this substance is produced by growing yeast, although it is also a necessity for yeast growth. Wildiers called this substance *bios*. Considerable effort was expended in the isolation and analysis of bios; it was eventually found to consist of several components, one of which was isolated in 1935 and named biotin.

The other experimental pathway leading to the discovery of biotin began with the observation by Bateman in 1916 of "egg-white injury". Raw egg white was found to contain a protein designated as *avidin* which combined irreversibly with a previously unrecognized vitamin which was, for a time, designated as vitamin H. Deficiency of this vitamin could be produced effectively only by egg-white injury, which was characterized in rats by dermatitis, loss of hair, baldness around the eyes producing the so-called "spectacle eyes", and eventually emaciation and death. About ten years later Boas found that a diet high in certain foods such as liver, kidneys, egg yolk, yeast, or milk would protect animals against egg-white injury. The identification of so-called vitamin H with biotin was accomplished in 1940; and in 1942 the formula for biotin was tentatively proposed. Later a slightly modified formula was confirmed by synthesis (formula X). Also in that same year, 1942, biotin deficiency was reported in 4 human subjects who had been taking about 30 per cent of their total calories in the form of egg white over a period of 7 to 8 weeks (51). These

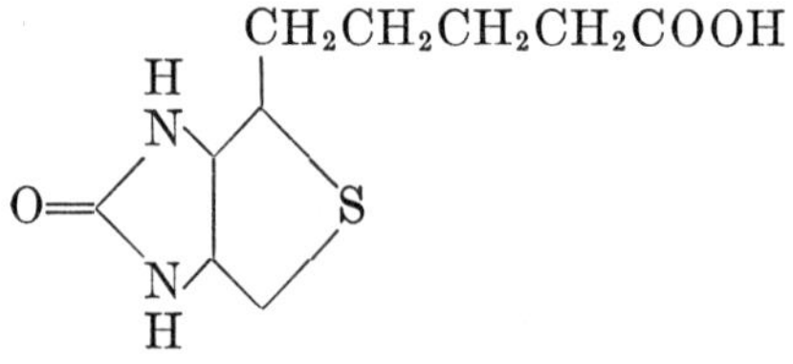

$CH_2CH_2CH_2CH_2COOH$

H
N
O= S
N
H

X. Biotin

men and women showed a striking ashy pallor, a dry skin with scaling, extreme lassitude and somnolence muscular pains and failure of appetite. Relief from these symptoms took place within a few days after injections of biotin, 150 to 300 micrograms per day.

Assay of biotin (57) is accomplished most satisfactorily by methods involving either the relief of egg-white injury in experimental animals or the growth of yeast or of lactic acid bacteria. Chemical methods are unsatisfactory because of the remarkably low concentrations in which biotin occurs in most foodstuffs, cells, and body fluids. The biotin required in mammalian nutrition has its most important origin in bacterial action. The production of biotin deficiency requires either the administration of raw egg white or the inhibition of bacterial action in the intestine of the experimental animal or subject by the administration of an antibacterial substance such as one of the sulfonamide drugs. Men, cattle, and rats all excrete more biotin than is contained in their diet. It appears probable that all micro-organisms require biotin, and that most of them are capable of synthesizing it. When such organisms are grown in a biotin-free medium, their rate of growth is limited by the rate of biotin synthesis.

Biocytin (epsilon-N-biotinyl-L-lysine, see formula XI) is the predominant

$CH_2CH_2CH_2CH_2CONHCH_2CH_2CH_2CH_2CHCOOH$

NH_2

H
N
O=
S
N
H

XI. Biocytin

form of biotin in many natural products, such as autolysed yeast. Biocytin is utilized by some micro-organisms and not by others. When biocytin is given to human subjects, it is quickly converted to biotin. An enzyme, *biocytinase*, has been identified in human blood (58).

Numerous functions have been assigned to biotin-containing coenzymes in the metabolism of yeast and other micro-organisms. In man and other animals, biotin appears to be associated with the early stages of the tricarboxylic acid cycle and the fixation of carbon dioxide. Two definite reactions in which it appears chiefly to be involved are the reversible decarboxylations of oxaloacetic acid and oxalosuccinic acid, yielding CO_2 and, respectively, pyruvic acid and alpha-ketoglutaric acid. Biotin may act not as a coenzyme for, but rather in the formation of, enzymes which decarboxylate 4-carbon acids, or produce them by carboxylation. In liver preparations from biotin-deficient rats, the ability to convert glutamic acid, NH_3, and CO_2 to carbamyl glutamate (see page 607) is diminished.

On account of the difficulty of completely eliminating intestinal synthesis of biotin in human experimentation, it has not been possible exactly to evaluate the human requirement. Estimates have varied from as little as 2 micrograms to as much as 300 micrograms a day. The average dietary intake appears to be somewhat less than 50 micrograms per day.

The urine contains considerable quantities (0.02 to 0.03 micrograms per milliliter) of biotin and biotin derivatives as measured by microbiological assay using *Neurospora crassa* as the test organism. All biotin thus measured is still capable of combining with avidin and must therefore contain the cyclic urea ring which has been shown necessary for such union. Biotin and a number of its derivatives yield positive responses in the *Neurospora crassa* assay (59). Output, on account of intestinal synthesis, is from two to six times the intake. As might be expected, biotin is widely distributed in nearly all types of naturally occurring foodstuffs. Concentrations in foodstuffs are usually low, the richest sources being yeast, egg yolk, liver, and kidney.

An enzyme system, tentatively designated as *biotin oxidase*, has been demonstrated in guinea pig kidney cortex, which removes 2-carbon fragments from the side chain of biotin under aerobic conditions (2). This enzyme system is not identical with the one catalyzing the oxidation of lower fatty acids, although the biotin oxidase system is inhibited noncompetitively by fatty acids.

Folic Acid

This substance (see formula XII), known also as pteroylglutamic acid (PGA), is the simplest compound which will have a curative or preventive effect upon a specific deficiency disease in mammals or birds, which has as its chief characteristic a macrocytic anemia (reduction in hemoglobin and red cells per unit volume of blood, but increased size and hemoglobin content of individual red cells).

Birds will develop such an anemia on a diet lacking folic acid or its more complex derivatives which will be described later. Growth and feathering are also poor, and the oviducts of deficient pullets do not respond to stilbestrol.

Mammals in general do not develop the deficiency disease on a deficient diet, since folic acid is produced adequately by bacterial synthesis in the mammalian intestine. The disease can be produced in mammals by a deficient diet plus suppression of intestinal bacterial growth (as with the antibacterial agent succinylsulfathiazole), or by a diet deficient in both folic acid and pantothenic acid. In monkeys, however, macrocytic anemia develops in long-term experiments involving only deficiency of folic acid in the diet. Such experiments led to the discovery of "vitamin M" in liver and

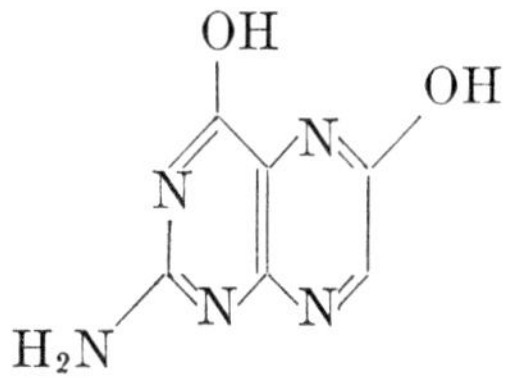

Xanthopterin

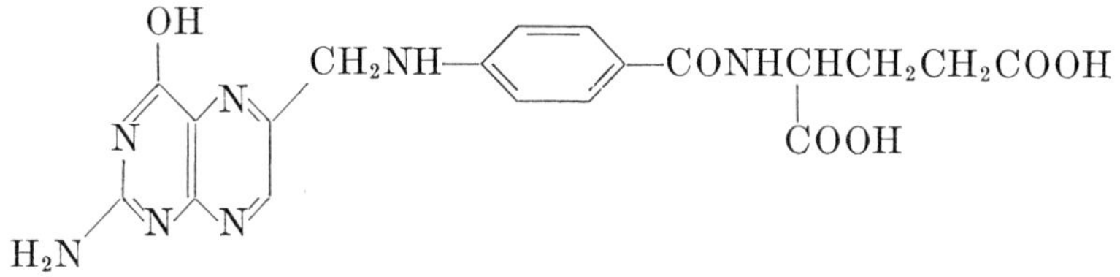

Folic acid (pteroylglutamic acid)

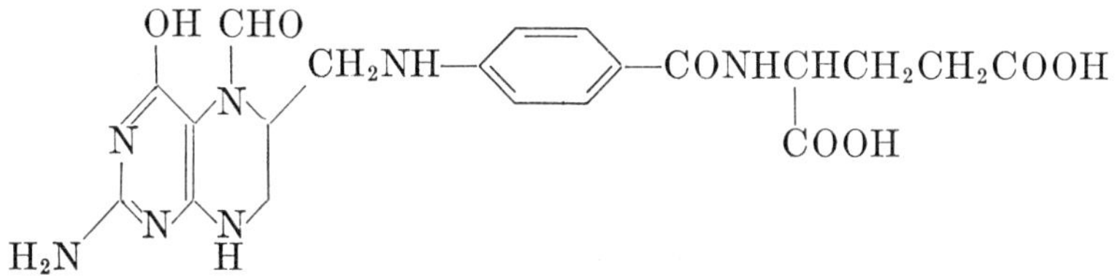

Leucovorin (CF)

XII. Folic acid and related compounds

yeast, its differentiation from other B vitamins, and its eventual identification with folic acid.

Man is most likely to develop folic acid deficiency as a result of gastrointestinal disease. Folic acid has been found effective in the treatment of those human macrocytic anemias which develop in pregnancy, in states of malnutrition, in tropical and non-tropical sprue. Folic acid is not as valuable as liver extract or vitamin B_{12} in the treatment of pernicious anemia, a type of macrocytic anemia known only in the human species. The anemia of pernicious anemia is corrected under adequate folic acid therapy, but certain associated degenerative changes in the nervous system are not checked, as they are by liver extract or vitamin B_{12} .

Xanthopterin (see formula XII) is simpler in structure than PGA. It does not qualify as the vitamin, since it has no effect in chicks with folic acid deficiency. It does, however, cure a similar anemia in Chinook salmon more effectively than does PGA. It is somewhat effective, but less so than PGA, in folic acid-deficient monkeys. Other forms of folic acid are more complex (the term "folic acid" is often used in a generic sense to include all substances with similar metabolic activity to PGA).

Storage forms of folic acid include PGA (also called vitamin B_c or *Lactobacillus casei* factor), and pteroyl di- and polyglutamates. The "fermentation *L. casei* factor", isolated from a bacillus which sometimes contaminates molds used in the production of riboflavin, is pteroyl triglutamic acid. The various folic acid conjugates and pteroyl polyglutamic acids have been reviewed in detail by Sargent (46). Their physiological activity is usually increased by hydrolysis to PGA, which is accomplished by an enzyme (vitamin B_c conjugase) present in many tissues.

Leucovorin (formula XII), *folinic acid,* and the *citrovorum factor* (CF), are all synonyms for the derivative of folic acid which is actually functional in enzyme systems, although it is not known to be the actual coenzyme, which may have a still more complex structure. In CF the glutamic acid is the natural L-isomer, and the formyl group is not necessary for vitamin activity, since it is added and removed during metabolic changes. The structure of CF is that of 5-formyl-5,6,7,8-tetrahydro PGA. In the intact animal liver, and to a less degree in bone marrow and possibly in other tissues, PGA is converted to CF. The conversion takes place in liver homogenates if oxygen is excluded (40). The yield is increased by ascorbic acid, which is presumed to act in the reduction of PGA to the 5,6,7,8-tetrahydro form (29).

Leucovorin (CF) is, as stated above, not necessarily the entire coenzyme in which folic acid functions. The structure of this coenzyme is not known. The coenzyme has been designated as CoF, referring to its chief function, the transfer of formyl groups, or one-carbon fragments. There is some indication that *N^{10}-formyltetrahydrofolic acid* (formula XIII) which differs from CF in the position of the CHO group may have general transformylation properties (28). This substance occurs in the urine of normal adults after oral administration of folic acid plus ascorbic acid.

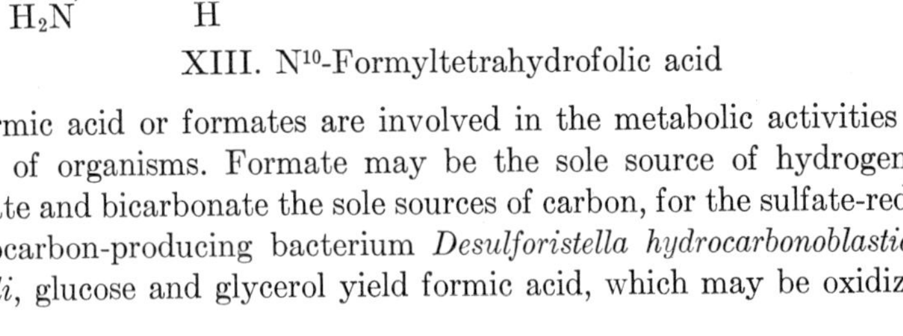

COOH
OH
CH_2—N—⟨benzene⟩CONHCHCH$_2$CH$_2$
N N
CHO COOH
N N
H_2N H

XIII. N^{10}-Formyltetrahydrofolic acid

Formic acid or formates are involved in the metabolic activities of all types of organisms. Formate may be the sole source of hydrogen, and formate and bicarbonate the sole sources of carbon, for the sulfate-reducing hydrocarbon-producing bacterium *Desulforistella hydrocarbonoblastica.* In *E. coli,* glucose and glycerol yield formic acid, which may be oxidized by formic dehydrogenase, or converted to carbon dioxide and hydrogen gas by formic hydrogenlyase. Tubercle bacilli increase their oxygen consumption in the presence of formate, as well as of lactate or acetate. The ergot of rye yields formic and other organic acids as minor metabolic products. Lymphocytes immobilized in a cellophane bag implanted in the abdominal cavity of an experimental animal will incorporate C^{14}-tagged formate.

Sources of formate in the animal body include: (a) serine, which yields formic acid and glycine (see page 623); (b) glycine, which is converted to glyoxylic acid (see page 619), which oxidatively decarboxylates to formic acid and CO_2; (c) labile methyl groups (formula XIV); and (d) histidine (see page 644). Labile methyl groups include the N-methyls of choline and

$$H_3C—N\begin{matrix} \diagup H \\ \diagdown R \end{matrix}$$

$$\downarrow (-2H)$$

$$H_2C{=}N—R$$

$$\downarrow (+H_2O)$$

$$HCHO + H_2N—R$$

XIV. N-methyl to formaldehyde

of betaines, and the S-methyl of methionine. Conversely, formate can be converted to the methyl groups of choline, methionine, and creatine, as shown by work with isotopic tracers. The methyl group of methanol is not a direct source of labile methyl groups, but is available after oxidation to formate. When rats are made deficient in folic acid, the incorporation into tissue choline of C^{14} given as methanol is diminished (53).

Certain particular transfers of one-carbon fragments, catalyzed by an enzyme system containing the hypothetical CoF, have been verified by the use of bacterial cells grown in total absence of *p*-aminobenzoic acid, folic acid, and similar growth factors, but supplied with the end products ordinarily derived through the action of such enzyme systems. The apoenzymes concerned along with CoF in these reactions have not yet been isolated or purified.

The formation of certain *amino acids* requires the transfer of a one-carbon fragment. Thus the addition of formate to glycine to form *serine* requires the hypothetical CoF. Similar one-carbon transfers occur in the formation of *histidine*, *leucine*, and *lysine*, and in the addition of a one-carbon fragment to homocysteine to form *methionine*.

Purines are formed by a ring closure for which a one-carbon fragment is required. The particular compound in which ring closure occurs varies with different species. In several bacterial species, inhibition of growth with sulfanilamide (an antagonist to *p*-aminobenzoic acid, preventing its incorporation into PGA) results in the accumulation of 4-aminoimidazole-5-carboxamide (formula XV), which on ring closure with a one-carbon frag-

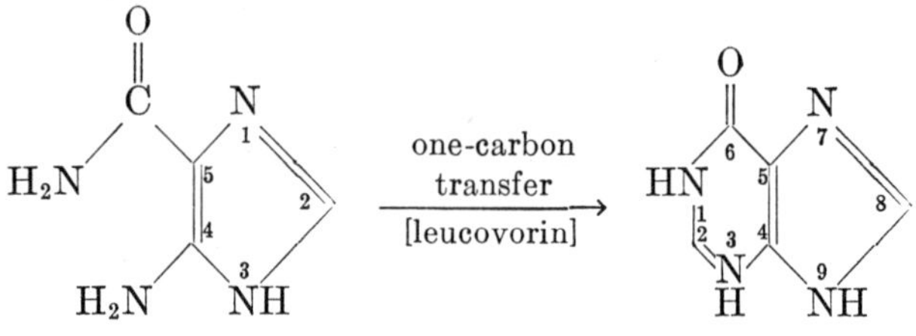

4-Aminoimidazole-5-carboxamide —one-carbon transfer [leucovorin]→ Hypoxanthine

XV. Purine ring formation

ment would have formed hypoxanthine (see page 359). This would involve incorporation of tagged formate into position 2 of hypoxanthine by the folic acid enzyme system. In rats, incorporation of tagged formate occurs at positions 2 and 8 of the adenine and guanine of liver nucleic acids, and this incorporation is decreased in folic acid deficiency (11). 4-Aminoimidazole-5-carboxamide is not an intermediate in purine formation in animals. The unclosed purine ring is combined as a riboside or ribotide before ring closure occurs.

The hypothetical CoF is also involved in the formation of the pyrimidine ring in *thymidine*, and of the benzimidazole ring in the *cobalamines* (see page 815), where one-carbon transfers are involved. Massive doses of thymine will substitute for catalytic quantities of folic acid as growth factors for certain species of lactobacilli, and in the prevention of macrocytic anemia in animals.

Folic acid deficiency can be induced in animals or in man by administration of *folic acid antagonists* such as 4-amino folic acid (aminopterin) or 4-amino-N^{10}-methyl folic acid (amethopterin). Such deficiency is characterized by leukopenia, aplastic anemia, and ulceration of mucous membranes. It is postulated that the folic acid antagonists compete with folic acid to enter a mechanism which normally forms a metabolic product such as CF, which exceeds PGA in biological effect by a factor of 10 to 1000 according to the testing procedure used. Folic acid antagonists are used in the therapy of some types of neoplastic disease. The symptoms induced by the use of the antagonists can not be reversed with folic acid, but respond to citrovorum factor. The citrovorum factor is effective in the same types of anemia which respond to folic acid. Folic acid is absorbed with ease following oral administration, and is excreted in the urine, partly unchanged and partly in the form of leucovorin.

Folic acid is another vitamin which we obtain in large measure by its production in the intestine by the action of bacteria. Daily human oral intakes range from about 40 micrograms to somewhat more than double that figure; somewhat less than 5 micrograms is put out daily in the urine, but 200 to 300 micrograms are present in the feces.

The human dietary requirement of folic acid and related compounds has not been definitely established on account of the contribution of intestinal bacterial synthesis. Folic acid appears to be quite non-toxic in any reasonable dosage for any of the laboratory animals. Fresh, green leafy vegetables and liver are the best food sources.

XVI. Cyanocobalamine

Cobalamines

Crystalline vitamin B_{12}, or *cyanocobalamine*, can be obtained from liver and, in larger yield, from *Streptomyces griseus* as dark red needles with about 17 per cent water of crystallization. The solid form is stable, as are solutions at neutral or slightly acid pH. Its structure has now been worked out (formula XVI). The empirical formula is $C_{63}H_{90}O_{14}N_{14}PCo$ and the molecular weight is 1357. A heavily substituted porphyrin-like ring forms the main portion of the molecule. It is not a true porphyrin since one methene bridge is missing. In addition, the metallic atom to which the ring is chelated is not iron as in heme or magnesium as in chlorophyll, but cobalt. The cobalt is attached by one of its valences to a nucleotide which contains a benzimidazole ring in the place of the more usual purine. Part of the evidence on which this structure rests consists of the computing of a three-dimensional electron density projection by an electronic computer, illustrating the impact of modern "mechanical brains" on biochemistry (22).

Cyanocobalamine is a neutral molecule and is represented in a schematic way in formula XVII. The cobalt of cyanocobalamine does not exchange with radioactive Co^{60} in solutions. When crystalline vitamin B_{12} is subjected to irradiation and aeration, the CN^- is replaced by OH^-, forming *hydroxocobalamine* or vitamin B_{12a}. Large doses of hydroxocobalamine will counteract the effect of otherwise lethal doses of potassium cyanide in mice, the injected B_{12a} being converted to B_{12} and excreted in the urine (39). Nitrous acid converts B_{12} or B_{12a} into B_{12c} or nitritocobalamine (see formula XVII).

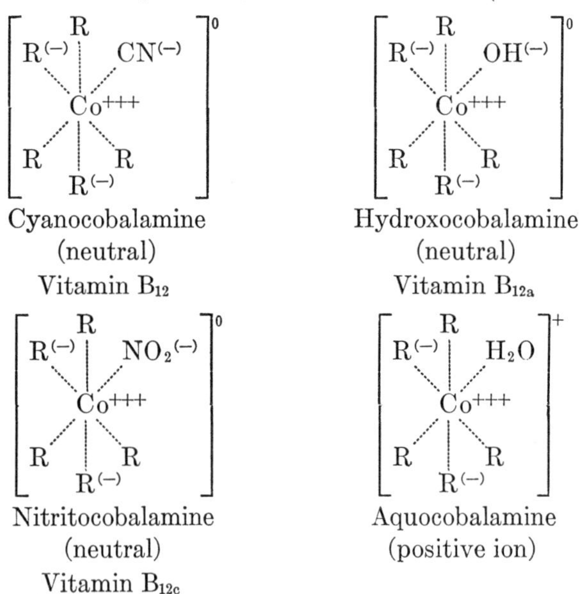

XVII. The cobalamines

The positive *aquocobalamine* ion is the form in which cobalamines occur as salts of strong acids (see formula XVII). *Ammonia cobalichrome* has NH_3 in place of the H_2O of aquocobalamine. Other cobalichromes have amino acids, peptides, or proteins in place of the H_2O. *Pseudovitamins* B_{12} occur in the feces of many animals, including man. These compounds contain adenylic acid in place of the 5,6-dimethylbenzimidazole ribose phosphate. They are supposed to be intermediates in the bacterial synthesis of B_{12}. They have no vitamin activity for animals, but show activity for many micro-organisms.

The primary source of vitamin B_{12} is the metabolic activity of bacteria and actinomycetes. The chief commercial sources are *Streptomyces griseus*, *Bacillus megatherium*, and the livers of ruminants. Beef liver contains 40 to 60 micrograms per hundred grams (6). Organisms isolated from the bovine rumen produce several different cobalamines. In the synthesis of B_{12} by *E. coli*, *p*-aminobenzoic acid is a necessary precursor or cofactor, for which neither folic acid nor leucovorin will substitute.

The chief medical interest of vitamin B_{12} lies in the fact that it is more effective than any other substance in the treatment of *pernicious anemia*, which can be considered as the deficiency disease resulting from failure in the human to absorb vitamin B_{12}. The failure is the result of the lack of a thermolabile *intrinsic factor* in the gastric juice. This intrinsic factor was demonstrated in normal human gastric juice by Castle (7), and has been chemically characterized as a mucoprotein with a molecular weight of 15,000 (32). The crucial test for its presence involves the incubation of meat with gastric juice. Meat incubated with normal gastric juice and fed to patients with pernicious anemia produces a remission. Meat incubated with the gastric juice of pernicious anemia patients has no effect. The intrinsic factor of normal gastric juice was postulated by Castle as reacting with an *extrinsic factor* present in meat and other foodstuffs, to produce an *erythrocyte-maturing factor*, the absence of which induced pernicious anemia.

We now know that vitamin B_{12} is both the extrinsic factor and the erythrocyte-maturing factor, and that the function of the intrinsic factor is to promote absorption of the vitamin. The mechanism of this action of intrinsic factor is still obscure. Vitamin B_{12} is stored in the liver, and accounts for the proved efficacy of liver and certain liver extracts in the treatment of pernicious anemia. Note that pernicious anemia is not the result of dietary deficiency of vitamin B_{12}, but rather the result of absorptive failure, which in turn appears to reflect a physiological disturbance of gastric secretion. The pernicious anemia patient is consistently achlorhydric. The anatomical background for this secretory defect is an atrophy of normal glandular structure in the stomach, which is not corrected by any known therapy. One case of an anemia closely resembling pernicious anemia and plainly the result of dietary B_{12} deficiency has been reported (42). This patient had normal gastric HCl and intrinsic factor.

Considerable amounts of vitamin B_{12} may be present in the feces of patients with pernicious anemia. Such vitamin B_{12}, for the most part formed by bacterial synthesis in the intestine, has escaped absorption and therefore has no antianemic value. The actual daily requirement of the human body is probably less than one microgram per day. Intramuscular injection of a dose calculated on the basis of one microgram per day, but given at intervals of 3 to 4 weeks, is adequate to maintain normal red cell production in typical pernicious anemia patients, and to prevent progress of the associated neurological lesions. Doses of over 50 micrograms at a time are probably wasteful on account of rapid excretion in the urine. No toxic effects follow the injection of doses of 300 micrograms. Oral administration of vitamin B_{12} in preparations containing intrinsic factor of animal origin is effective, but requires somewhat larger dosage. Still larger oral dosage without intrinsic factor is adequate for most patients (44) and is more economical.

Assay methods for vitamin B_{12}, aside from its action in human pernicious anemia, include growth measurements on rats, microbiological procedures, and colorimetric or fluorimetric determination of the 5,6-dimethylbenzimidazole component. A very sensitive bioassay method which will measure the B_{12} in body fluids, utilizes the growth of the protozoon *Euglena gracilis*. Distilled water from storage tanks or wash-bottles contains up to 2 micrograms of B_{12} per liter, formed by bacteria or actinomycetes (45).

Vitamin B_{12} has been found to be one of the important components of the *animal-protein factor* necessary for the growth of chicks. Liver, yeast, milk, meat, fish, and eggs are valuable dietary sources of vitamin B_{12}. Biosynthesis of this vitamin by bacterial action is probably its ultimate source. The meat of cattle, sheep, and horses is richer in vitamin B_{12} than that of swine. Bacterial action in the rumen of sheep and cattle, and in the cecum of horses, is more prolonged than in the less capacious digestive tracts of the hog or the human.

The maturation of erythrocytes is not the only function, or the fundamental function of vitamin B_{12}. Like the rest of the B vitamins, B_{12} is probably a universal cell component, and is certainly necessary for the growth of many species of micro-organisms. The essential function of vitamin B_{12} is, according to present evidence, in the biosynthesis and transfer of certain methyl groups, particularly the methyl group of thymine. Both vitamin B_{12} and folic acid are necessary for normal red cell formation.

B Vitamins of Doubtful Status

Carnitine (formula XVIII) has long been known to be a constituent of muscle, and to contribute up to 3 per cent of the solid matter of meat

$$
\begin{array}{l}
CH_3 \\
\quad\diagdown^{(+)} \\
CH_3—NCH_2CHCH_2COO^{(-)} \\
\quad\diagup \qquad\quad | \\
CH_3 \qquad\quad OH
\end{array}
$$

XVIII. Carnitine (vitamin B_T)

extract. The function of carnitine in mammalian tissue is not known beyond the fact that it can act as a methylating agent like other betaines (see page 631). Only certain flour moths and mealworms in their larval state require carnitine in their diet. There is no indication that it is a vitamin for man or any mammalian species. The name vitamin B_T was applied to this growth factor for *Tenebrio molitor*, the common mealworm, before the chemical structure was determined. The mealworm requires carnitine only in catalytic amounts, but carnitine is present in liver and milk, as well as in muscle extract, in appreciable quantities. Carnitine is synthesized in the developing chick embryo, and in all animals studied exclusive of those few insects which require vitamin B_T in their larval stage.

Choline and *inositol* are sometimes listed among the B vitamins, although their chief physiological importance is as lipotropic substances (see page 599). Both of these substances are consistently found in all types of cells, but usually in concentrations much higher than that of the B vitamins, and not making up a portion of any known coenzyme. Choline, as a constituent of phosphoglycerides, and capable of being formed in large amounts in normal metabolic processes, seems to be clearly out of the vitamin class, although it is used therapeutically in the treatment of a deficiency disease, alcoholic fatty liver. Inositol is a component of Wildiers' bios, along with biotin, and has been shown to be required by numerous microbial species. It is certainly an essential structural component in man and other animals, forming a part of the phosphoinositides of liver, nervous system, and other tissues. It is synthesized in the intestine and possibly in the body, and no specific deficiency disease has been attributed to lack of inositol in the human diet. It has been reported essential, however, to the growth of human cells in tissue culture (11a).

Para-aminobenzoic acid has also been listed among the B vitamins, but has no known function in animals other than as a component of folic acid, and can not substitute for folic acid in all the functions of the latter substance.

VITAMIN C

Ascorbic acid, a sugar acid which is the only substance properly designated as vitamin C, has some points in common with the B vitamins. Like them, it is water-soluble and is present consistently in the cells of mature plants and animals, but the concentrations of ascorbic acid in seeds of plants and embryos of animals are often found to be zero by our most sensitive methods. Furthermore, the distribution of ascorbic acid among micro-organisms, including protozoa, is sparse and erratic, indicating no uniformly significant function.

The formula of ascorbic acid is shown here (formula XIX) as it occurs

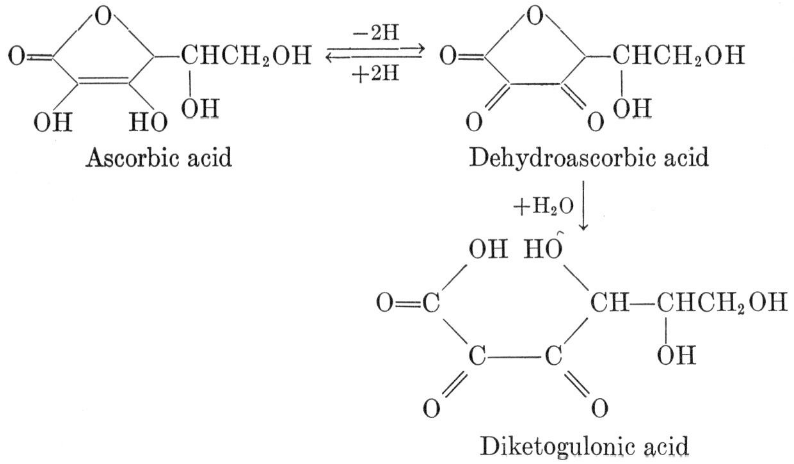

O= (ring O) —$CHCH_2OH$ (OH), OH HO — $\underset{+2H}{\overset{-2H}{\rightleftharpoons}}$ — O= (ring O) —$CHCH_2OH$ (OH), O O

Ascorbic acid — Dehydroascorbic acid

$+H_2O$ ↓

OH HO

O=C CH—$CHCH_2OH$

C—C OH

O O

Diketogulonic acid

XIX. Ascorbic acid and related compounds

in natural sources, in equilibrium with its oxidized form, *dehydroascorbic acid*. The reduced form predominates. Dehydroascorbic acid is also in part irreversibly hydrolyzed to diketogulonic acid. A larger part is completely metabolized to CO_2 and H_2O, but some appears as urinary oxalate.

The rich dietary sources of ascorbic acid, available for human use, include the citrus fruits, berries, green vegetables, apples, and with less richness most other fruits and vegetables. Meats in general contain suboptimal amounts of vitamin C, although it is present in all animal tissues. Ascorbic acid can be crystallized as odorless, colorless plates, M. P. 192°C. It is very soluble in water, insoluble in the fat solvents. The crystals are stable, solutions are not. Oxidation occurs on exposure of neutral or alkaline solutions or foodstuffs to air.

The oxidative inactivation of ascorbic acid is accelerated by heat and by the presence of catalytic traces of copper. Acidic foodstuffs, including most fruits, hold their vitamin C content well during normal periods of storage. Some non-acidic vegetables, such as potatoes, cabbage, and turnips, are customarily stored actually in the living state and keep their ascorbic acid content at a significant level. Ascorbic acid is commercially synthesized from glucose, and is available in pure solid form. Ascorbic acid, in equilibrium with its first oxidation product, dehydroascorbic acid, is the only natural substance which will protect against *scurvy*. A few synthetic homologues and stereoisomers possess some protective power.

Ascorbic acid functions as a coenzyme in the oxidation of *p*-hydroxyphenylpyruvic acid, the product of deamination of tyrosine (49). Ascorbic acid also inhibits the catalysis of hyaluronic acid decomposition by hyaluronidase, and inhibits xanthine dehydrogenase *in vitro* (13).

Scurvy or scorbutus has been known for centuries, and its association with famines and with long sea voyages recognized. Lind published in 1753 the report of the first clinical investigation ever deliberately planned, in which it was proved that citrus fruits cure scurvy. At present in the United States scurvy occurs most commonly in children 6 months to 2 years of age, who have been bottle-fed without supplementation with vitamin C. Scurvy is also a threat to those persons, usually elderly, who live alone and prepare only the simplest possible meals for themselves.

Scurvy in experimental animals and in premature infants is characterized by incomplete oxidative metabolism of tyrosine, indicated by the presence in the urine of tyrosine itself and intermediates such as *p*-hydroxyphenylpyruvic acid and homogentisic acid. This defect is remedied by small doses of ascorbic acid, but is also remedied by administration of folic acid, which does not protect against the other manifestations of scurvy. It will be recalled that ascorbic acid is necessary for conversion of folic acid to citrovorum factor. A more damaging metabolic defect in scurvy is the failure of production of the intercellular materials of connective and supporting tissue, including both collagen and mucopolysaccharides. In clinical or experimental scurvy, capillary hemorrhage is a characteristic sign, brought about by the lack of intercellular binding substances. In growing bones or healing fractures, formation of organic matrix is suppressed with resulting deformities and failure of calcification, although calcium metabolism is not primarily affected. Repair of wounds is delayed through failure of formation of scar tissue. Bleeding of the gums is an early and classical symptom. Bleeding under the skin or conjunctiva leads to visible signs. Internal hemorrhages may become quite extensive before causing such symptoms as bloody stools or painful joints.

An experimenter (10) subsisted for six months on a diet containing no vitamin C. The ascorbic acid of his plasma reached zero after 41 days and remained there. The first sign observed referable to scurvy was the development of perifollicular hyperkeratotic papules on the skin of the buttocks and calves. The papules contained ingrown hairs, which were frequently fragmented. This sign had previously been considered a specific manifestation of vitamin A deficiency, but the experimenter had been taking a minimum of 30,000 I.U. of vitamin A daily, and his blood level of vitamin A was found normal. The papules appeared on the 135th day of the diet at which time the ascorbic acid content of the white blood cells was zero. After 161 days, perifollicular hemorrhages were observed in the skin. During the entire experiment no bleeding of the gums occurred, which can perhaps be explained by good oral hygiene. There was no anemia, and no increase of capillary fragility as measured by standard clinical tests. Small hemorrhages did appear in the skin of the lower extremities after a long period of standing. A surgical wound was made in the right mid-back of the experimenter after 3 months on the deficient diet. Biopsy ten days later demonstrated normal healing. After 182 days on the diet a similar surgical wound was made in the left mid-back. Biopsy ten days later demonstrated no healing except in the skin. Microscopic examination of sections showed newly formed fibroblasts but no intercellular substance between them. Ingrowth of capillaries was lacking. After the biopsy, the diet was maintained but one gram of ascorbic acid was given intravenously each day. Ten days later another biopsy demonstrated healing, with intercellular substance being formed and capillaries growing in. Subjectively the experimenter felt tired and weak after the third month of the diet, and increasingly so until the termination of the deficiency. All signs and symptoms cleared rapidly after daily vitamin C injections were given.

Others have repeated these observations on other volunteer human subjects and have for the most part confirmed the findings outlined above. Hyperkeratosis of hair follicles is usually, but not always, the first external sign, followed by perifollicular bleeding. If acne is present, it is likely to be made worse. Not all subjects have escaped without swollen and bleeding gums. Delayed healing of wounds has been repeatedly confirmed. Many subjects do not report fatigue or weakness.

All plants and most animals can synthesize their own vitamin C. It is a true vitamin only for primates and guinea pigs. The guinea pig requires about 10 mgm. of ascorbic acid per day to remain in good health and man requires the same amount as a bare minimum. This minimum should be trebled to provide a reasonable factor of safety. The National Research Council recommendation of 75 mgm. is excessive, but in no sense harmful. No toxic effects of ascorbic acid have ever been recorded either for guinea pigs or men, other than a mild diuresis following enormous dosages. The ascorbic acid of human blood plasma is 1.0 to 1.4 mgm. per 100 ml. when the subject is "saturated" with the vitamin (36), which means that any excess over this level is excreted in the urine. Some ascorbic acid is excreted, however, at lower plasma levels. A plasma level of 0.6 mgm. or more can be considered evidence of adequate vitamin C supply. In deprivation experiments, the ascorbic acid of the white blood cells, normally over 15 mgm. per 100 ml., approaches or equals zero before signs of scurvy appear. "Saturation" requires a dosage of about 100 mgm. daily. In the feeding of young infants, it is customary to start at the age of two weeks with 30 ml. of orange juice daily, and to raise the intake to 60 ml. daily at the age of three months. Thirty ml. of fresh orange juice contain approximately 18 mgm. of ascorbic acid.

The body under stress requires more ascorbic acid. It is characteristic of infectious disease that the plasma ascorbic acid is at a low level and that greater intakes are required to produce saturation (12). Similar findings have been reported in patients with burns and with fractures. This observation has been extended in experimental animals by the demonstration of depletion of the ascorbic acid of the adrenal under stress, provided the adenohypophysis is present and intact. The conclusion that ascorbic acid is required and used in the synthesis of cortical steroids is not, however, consistent with the continued production of steroids by scorbutic guinea-pigs, whose adrenals are depleted of ascorbic acid (48).

Analysis for ascorbic acid is frequently done by titration with 2,6-dichlorophenolindophenol, which acts both as oxidizing agent and redox indicator, being blue in the oxidized form and colorless when reduced. This titration must be carried out in solutions more acid than pH 5.5, since ascorbic acid is autoxidizable in more basic solutions. This titration is of course not specific for ascorbic acid, but measures the totality of substances present and capable of reducing the blue dye. Such reducing substances are not significantly present in urine or blood plasma. A more specific colorimetric analytical method involves the formation of a colored hydrazone of dehydroascorbic acid with 2,4-dinitrophenylhydrazine (9).

FLAVONES AND COUMARINS

The designation vitamin P is sometimes given to a rather large group of substances of plant origin which act to increase the strength of capillaries. The discovery of this group of compounds grew from the observation that purified preparations of vitamin C were ineffective in the control of bleeding in vascular purpura, a disease in which capillary fragility is increased, but that control was established by the use of crude and unpurified vitamin C preparations. The first material which was reported to exert vitamin P activity was a yellow preparation called *citrin*, obtained from fruits. Citrin turned out to be a mixture of several flavone derivatives.

Flavone is usually considered to be the parent substance of the P vitamins. Some of the active substances are substituted flavones, as for example rutin (formula XX). Other substances with vitamin P activity have only a part of the flavone structure and are more properly designated as coumarins. An example of this group is esculin (formula XXI) which is about five hundred times as active as citrin. All the P vitamins are of plant origin. Rutin is obtained from buckwheat, and esculin from chestnuts. Not

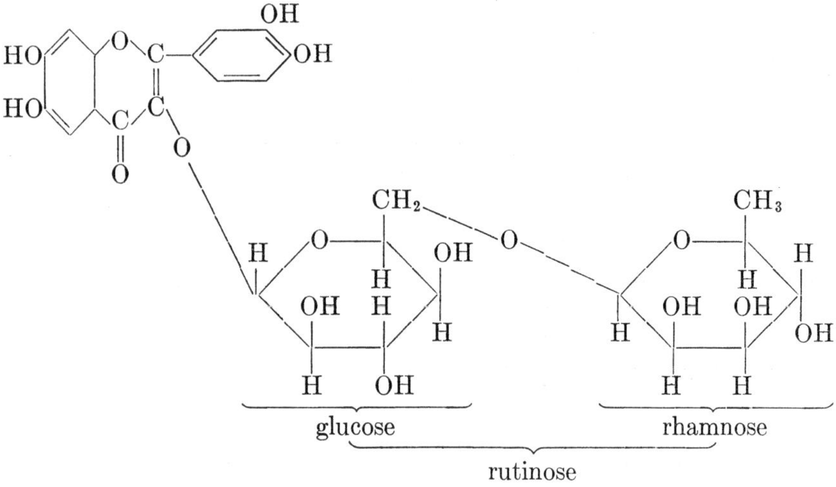

glucose rhamnose

rutinose

XX. Rutin

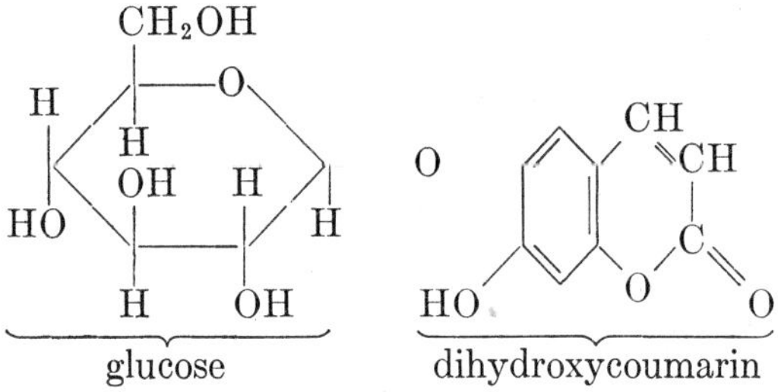

glucose dihydroxycoumarin

XXI. Esculin

all flavone or coumarin derivatives show vitamin P activity. Those which do show such activity are designated as *bioflavonoids.*

No unified or generally acceptable theory of the action of vitamin P has yet been proposed. Several possible theories of this action are discussed by Levitan (33) along with a consideration of the application of the vitamins P in the treatment of a number of human diseases characterized by capillary fragility. Two independent groups of clinical investigators (15a, 51a) find no evidence that bioflavonoids, or ascorbic acid, or both in combination, have any therapeutic value in the treatment of the common cold.

VITAMIN A

All vitamins so far discussed are classified as water-soluble, although riboflavin is actually rather sparingly soluble in aqueous solutions. With vitamin A we begin the study of the fat-soluble vitamins, which occur in the lipid portions of the foods in which they are found and which can be extracted with organic solvents. At least five substances possess what we are about to describe as vitamin A activity, but the substance known as vitamin A_1 is primarily significant in all mammals and in salt water fish. This substance we will designate as vitamin A or *axerophthol*, recognizing that it may be accompanied both in natural and synthetic sources by small amounts of vitamers and of related compounds with lower or absent biological activity.

Vitamin A is a terpene alcohol with the structure shown in formula XXII. The configuration about the double bonds is *trans* in each case. Isomers with one or more double bonds in the *cis* configurations have been prepared, but these have less physiological potency than the all-*trans* isomer (1). As an alcohol, vitamin A can exist free or in the form of fatty acid esters. The cyclic portion is a beta-ionone ring. If the terminal CH_2OH group is changed to CH and a double bond is left hanging free, we now have half a molecule of the substance beta-carotene. Ingested beta-carotene is converted into vitamin A in the intestinal wall, presumably by an oxidative cleavage at the central double bond, forming first an aldehyde group and then the

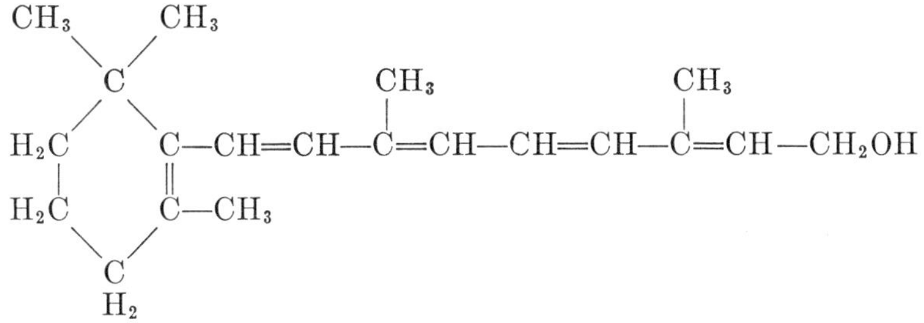

```
CH3     CH3
   \   /
     C                CH3                 CH3
   /   \               |                   |
H2C     C—CH=CH—C=CH—CH=CH—C=CH—CH2OH
 |      ||
H2C     C—CH3
   \   /
     C
     H2
```

XXII. Vitamin A_1 (Axerophthol)

alcohol by reduction. Thus, beta-carotene (formula XXIII) can properly be designated as a provitamin A, since two molecules of vitamin A can be obtained from one molecule of the parent carotene. Alpha-carotene, gamma-carotene, and the cryptoxanthines found in egg yolk and yellow corn are also provitamins A, but can yield only one molecule of vitamin A per molecule of original substance. In general vitamin A occurs in animal foods and the carotenes in plant foods. Butterfat, however, contains both.

Vitamin A has no significant solubility in water, is soluble in fats and fat solvents, and is stable to heat in the absence of air or oxygen. The vitamin A of cod liver oil is destroyed by heating the oil to the boiling point of water for 12 hours while air is being passed through the oil. This treatment does not destroy vitamin D. The provitamins or carotenoids have similar properties as far as solubility and stability are concerned. The ordinary processes of cooking and canning do not cause serious loss. The preservation of foods by freezing does not impair their vitamin A content, but dehydration of foods tends to cause appreciable loss of vitamin A.

```
CH3      CH3
   \    /
     C                    CH3                     CH3
   /   \                  |                       |
H2C     C—CH=CH—C=CH—CH=CH—C=CH—CH
 |      ||                                          ||
H2C     C—CH3                                       ||
   \   /                                            ||
     C                                              ||
     H2                                             ||
                                                    ||
CH3      CH3                                        ||
   \    /                                           ||
     C                    CH3                     CH3   ||
   /   \                  |                       |     ||
H2C     C—CH=CH—C=CH—CH=CH—C=CH—CH
 |      ||
H2C     C—CH3
   \   /
     C
     H2
```

XXIII. β-Carotene

Assay of vitamin A may be biological, involving measurements of the growth of rats, or by physical or chemical methods. Spectrophotometry is satisfactory for fairly concentrated preparations, and is based upon absorption maxima at 610 to 620 millimicrons for A_1 and 692 and 696 millimicrons for A_2. Colorimetric analysis is very sensitive, and can be applied to blood plasma, utilizing the reaction of vitamin A with activated glycerol dichlorohydrin (50).

Sources. As has been suggested already, vitamin A itself appears in foods of animal origin. The oils extracted from the livers of certain fish, particularly the cod, the halibut, and the shark, are the most concentrated sources of vitamin A. Egg yolks, milk, and butter are sources which are more common in the American diet. Oleomargarines sold for household use are customarily fortified with vitamin A. Dietary sources of the carotenes are the leafy vegetables such as lettuce and cabbage, particularly the outer and greener leaves. Green stalks such as asparagus and celery, and green seeds such as peas and green beans have high provitamin content; yellow vegetables and yellow fruits are uniformly valuable and rich sources. In general, the grains are not very satisfactory in supplying these provitamins but yellow maize is an important exception.

Pure beta-carotene has been accepted as the basis of the international unit of vitamin A. The unit is defined as the biological activity equal to that of 0.6 micrograms of beta-carotene. The international standard preparation of vitamin A contains 0.3 milligrams of beta-carotene dissolved in one gram of an inert vegetable oil. Each 2 milligrams of the international standard preparation will therefore contain one international unit of vitamin A activity.

Absorption and storage. Vitamin A is absorbed by the same mechanisms as the lipids of foodstuffs. The presence of bile in the intestine is necessary for such absorption. Vitamin A is stored in considerable amount in the liver, concentrations between 100 and 900 international units of vitamin A per gram having been found in 80 per cent of the livers of a series of 71 healthy individuals whose death was sudden and accidental (26). Absorption of the carotenes from the intestine is less efficient than that of the vitamin A alcohols. Not all of the carotenes are converted to vitamin A. Some of the color of skin and of blood plasma is from normal content of carotene. Vitamin A deficiency may occur not only from lack of the vitamin or its precursors in the diet, but also from failure of absorption from the intestine, as by excessive ingestion of mineral oil, or from failure of conversion of carotenes or storage of vitamin A or both by a diseased liver. The mean concentration of vitamin A in the blood plasma of adequately nourished men and women is close to 120 international units per 100 ml.; values for total carotenoids are higher and more variable.

Physiological functions. Vitamin A is concerned with growth, with the maintenance of epithelial tissues, and with vision. The first evidence for its existence was presented in 1913 by two independent teams of workers: certain natural fats and oils would stimulate growth in rats while other fats and oils with similar triglyceride composition would not. The necessity of vitamin A for growth has been repeatedly confirmed. The growth of the skeleton—not calcification, which is related to the vitamins D, but the actual growth of the organic matrix—is first affected by the lack of vitamins A in a young animal. Growth of soft tissues is less promptly checked, so that brain and cord may grow more rapidly than their bony cases and thus suffer damage by constriction. In animals deficient in vitamins A at any age, epithelial cells tend to atrophy and to be replaced from the basal cells by a stratified horny epithelium, high in keratin content. Lachrymal glands and conjunctiva are particularly sensitive to deficiency of the vitamins A, so that in experimentally deficient animals and in severe cases of human deficiency a dry-eyed state, *xerophthalmia*, is observed. The thick and dry conjunctiva is susceptible to infection and to mechanical damage. In a more advanced stage of deficiency, the cornea wrinkles and shrinks (*keratomalacia*) and may ulcerate and perforate. Other epithelial tissues which are affected in severe deficiency include the skin, salivary glands, respiratory tract, and genito-urinary tract.

The photoreceptor cells of the retinal rods, which function in dim light, contain visual purple, or rhodopsin. Visual purple is a compound of vitamin

A with a protein. The compound breaks down under illumination to visual yellow, which is composed of protein plus a yellow pigment, retinene. Visual yellow can either reform visual purple or be converted to vitamin A and a protein by a rather complicated process involving cozymase, nicotinic acid, and vitamin E. A large part of the liberated vitamin A is used again, but the process is not entirely reversible, and a supply of vitamin A is required for replacement. The visual purple normally regenerates in a few minutes of darkness or dim light. This is the basis of the phenomenon of *dark adaptation*, which is delayed or absent in subjects deficient in the vitamins A.

Avitaminosis A. The commonest manifestation of early or moderate vitamin A deficiency in man is *night blindness.* This is a rather extreme term for what is usually poor adaptation to dim light. Often there is no complaint on the part of the patient or else certain difficulty may be noted in doing close work with inadequate illumination. More commonly there is difficulty in vision when the patient suddenly leaves a brightly illuminated area for one which is relatively dim. Special photometric instruments have been devised for the measurement of dark adaptation and are useful in the diagnosis of moderate degrees of vitamin A deficiency. In using such instruments it should be recalled that conditions other than lack of vitamin A can lead to poor dark adaptation. Xerophthalmia is a much less common condition than night blindness in the human although easily produced in experimental animals. It is characterized by decreased production of tears and a dry conjunctivitis. In advanced cases there is no formation of tears whatever, and the conjunctiva becomes hard, brittle, and scaly. At this stage bacterial infection and mechanical damage are likely to occur, possibly resulting in complete functional loss of vision and extensive anatomical damage to the eye. In serious deficiencies similar changes can be detected in the epithelia of the respiratory tract, and of the genito-urinary tract. Even the skin may undergo excessive drying and cornification. The chemical abnormalities underlying these effects of vitamin A deficiency have not been clarified except in the instance of loss of dark adaptation (see page 165).

Hypervitaminosis A. Rather considerable excess of vitamin A has been taken over a protracted period without causing obvious ill effects. Evidence is accumulating, however, that damage to liver, skin, and skeleton may result from overdosage with vitamin A. The subject of hypervitaminosis A has been reviewed by Knudson and Rothman (31). On the basis of x-ray observation of hyperostoses, it seems probable that there are more instances of vitamin A excess in the United States than of vitamin A deficiency.

Human requirement. A very serious effort has been made to determine the vitamin A requirements of the normal human subject (26). The subjects for this experiment consisted of 20 male conscientious objectors and 3 woman volunteers. They were maintained on a diet from which were excluded such foods as dairy products, liver, and other organ meats, all fats, all fish, and all green and yellow vegetables and fruits. This diet was very low in vitamin A potency by biological assay with rats. Chemical analysis for carotenes demonstrated a maximum of 70 international units. The normal content of carotenes in the plasma of these subjects fell rather rapidly and approached zero levels during the first few weeks on the experimental diet. The vitamin A of the plasma, however, fell much more slowly. In one subject there was no decline in plasma vitamin A for 22 months. Taking the subjects as a group, a mean value of 88 international units per 100 ml. of plasma during the first two months of the experiment fell to 74 international units in 9 months, and in a reduced number of subjects was still 61 international units after 14 months. The response of these subjects to tests of dark adaptation demonstrated a greater relationship to the season of the year than to vitamin A depletion. In only three of the subjects could failure of dark adaptation be demonstrated as significantly different from that to be expected in relation to the season of the year. These three subjects had 40 or less international units of vitamin A in the blood plasma. In these three cases a daily dose of 1500 international units gradually improved dark adaptation and restored normal levels of plasma vitamin A. The daily requirement for vitamin A, obtained as a result of this entire experiment, is set at 3000 international units of vitamin A per day. No symptoms of vitamin A deficiency appeared in these subjects other than the 3 cases mentioned of unsatisfactory dark adaptation. Some of the subjects continued the experiment for two years. The conclusion was drawn that the human body mobilizes and expends the vitamin A reserves of the liver with notable economy during periods of vitamin A depletion. There was no evidence of bacterial synthesis of carotenes to any effective degree, although a number of micro-organisms, including *Staphylococcus aureus*, can perform this synthesis. The need for vitamin A is increased in diseases of the liver and of the gastrointestinal tract, during pregnancy and lactation, and in fevers.

VITAMINS D

There are several substances which have a vitamin D action. This action can be briefly summarized as the ability to protect the growing child or young animal against rickets. Two of these substances occur in effective concentrations in a limited number of foods, but their inactive precursors or provitamins are more widely distributed. The conversion of provitamins D to the effective vitamins takes place in nature by the action of solar vltraviolet radiation. Irradiation of *ergosterol*, a sterol of plant origin, yields uitamin D_2. Ergosterol was named from its first known source, which was ergot, a fungus growing on rye. It is also a product of yeasts, molds, and other fungi. Irradiation of ergosterol with either ultraviolet or cathode rays yields two intermediate products, lumisterol and tachysterol, prior

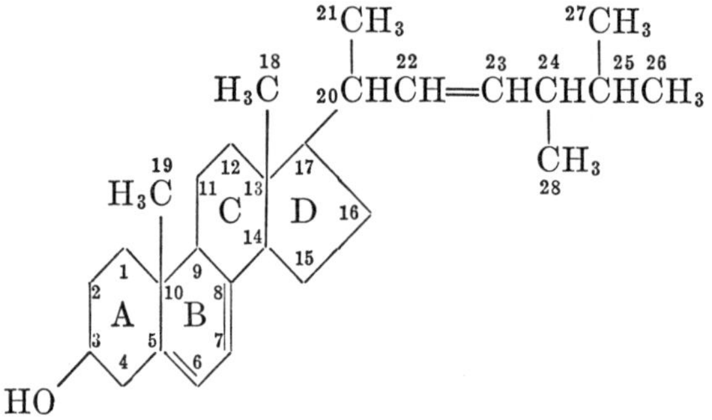

XXIV. Ergosterol (provitamin D_2)

to the formation of *calciferol* (vitamin D_2). The term vitamin D_1, now obsolete, referred to a molecular compound of calciferol and lumisterol. The formulas for ergosterol (formula XXIV) and for calciferol (formula XXV) are given, omitting those of the intermediate products and those of the further products of irradiation, isopyrocalciferol and pyrocalciferol, toxisterol, and suprasterols I and II. Irradiation of *7-dehydrocholesterol* (formula XXVI), an animal sterol, yields vitamin D_3 (formula XXVII), which is the form of vitamin D characteristic of cod liver oil. Other fish oils may contain vitamin D_2 in considerable amounts along with vitamin D_3. The chemical name of vitamin D_3, dimethyldihydrocalciferol, is seldom

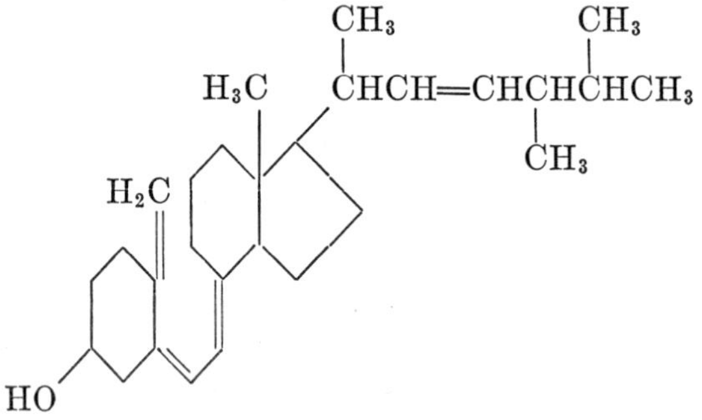

XXV. Calciferol (vitamin D_2, ergocalciferol)

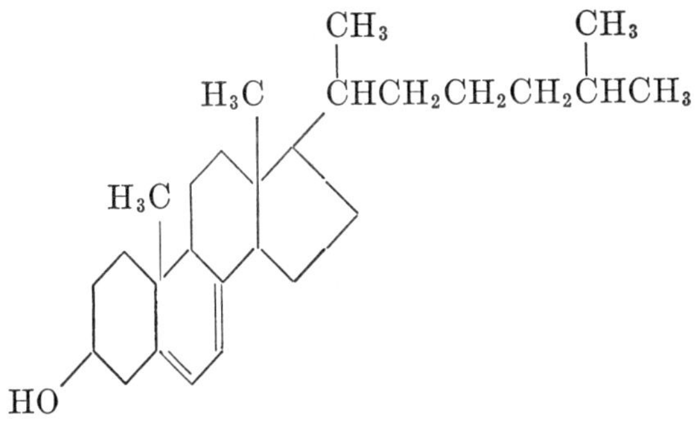

CH_3 CH_3

H_3C $CHCH_2CH_2CH_2CHCH_3$

H_3C

HO

XXVI. 7-Dehydrocholesterol (provitamin D_3)

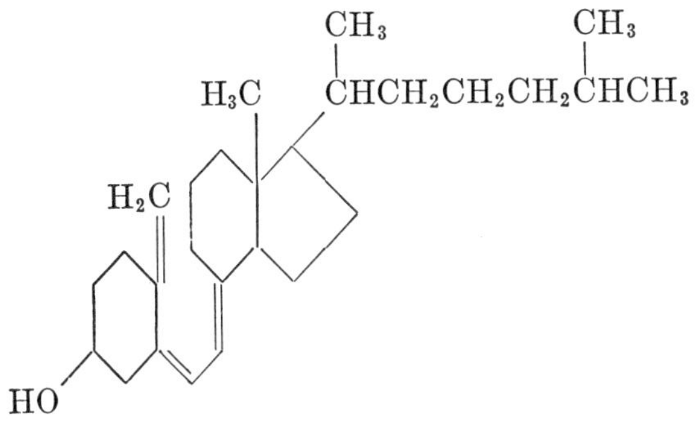

CH_3 CH_3

H_3C $CHCH_2CH_2CH_2CHCH_3$

H_2C

HO

XXVII. Vitamin D_3 (cholecalciferol)

used. Both vitamins D_2 and D_3 are available in crystalline form and in solution for therapeutic and prophylactic uses.

The higher plants contain insignificant amounts of vitamins D or none at all. Yeasts and molds contain abundant provitamin D_2 which becomes activated in sunlight, which probably explains the vitamin content observed in some plant products such as cocoa shells and hay. Most animal tissues likewise contain little vitamin D, but there are certain striking exceptions to this statement: the livers and other viscera of many species of fish, as well as the fat of these fish, the fats of fish-eating animals, the eggs of all birds, and the milks of all milk-yielding animals all contain vitamin D, chiefly in the form of D_3. Fish oils vary widely in vitamin D content; the liver oil of the bluefin tuna contains 40,000 international units per gram; of halibut and mackerel, about 1000; of cod, about 100; of sturgeon, none. The international unit is the vitamin D activity of 0.025 micrograms of ergocalciferol, an amount which serves also as the U. S. P. and A. O. A. C. unit. The vitamin D content of eggs varies with the diet of the bird. Milks, human and bovine, usually contain less than 40 units to the quart. Nearly all brands of evaporated milk are fortified with vitamin D,

400 units to the reconstituted quart. Many fresh milks are similarly fortified, 400 units to the quart. The provitamins D are much more abundantly distributed than the vitamins. Land animals in general form their own vitamins D from provitamins, which are most concentrated in the skin where they may be converted by sunlight. The manner of synthesis of the provitamins by organisms is not fully known. The known facts about the synthesis of steroids in general are summarized in Chapter 15. Certain simple organisms, as for example *Aspergillus niger*, can accomplish the total synthesis of ergosterol with sodium acetate as the sole source of carbon. This is in line with the evidence for the synthesis of cholesterol in the animal from acetyl groups.

Properties. The vitamins D are soluble in fats and fat solvents, insoluble in water. They have a characteristic absorption spectrum in the ultraviolet, which is used in methods of vitamin D analysis of fish oils and other rich sources. There are also certain reactions with antimony trichloride and with aluminum chloride in the presence of pyrogallol which have been made the basis of quantitative analyses. The vitamins D are stable at all temperatures commonly used in the preparation of foods, even in the presence of air. Rancidity developing in fats or oils impairs the vitamin D content.

The bioassay of vitamin D is described in detail in the United States Pharmacopoeia. In brief summary, young rats depleted of vitamin D are fed various amounts of the test material while control rats on the same diet are fed standard cod liver oil. There are several tests applied to the animals to demonstrate cure of rickets. One of these, the "line test", utilizes the proximal end of the tibia or the distal end of the radius or ulna. The bone is sectioned longitudinally and treated with 2 per cent silver nitrate for one minute, which converts calcium phosphate to silver phosphate. After washing and exposure to light, calcified areas will show a black stain. The indication of healing rickets is a line of new calcification through the junction of epiphysis and diaphysis. If chicks are used for assay, the ash content of the tibiae is measured. Assays on different animal species are usually not in agreement. By the rat test, vitamins D_2 and D_3 have the same potency, 40,000,000 international units per gram. In chicks, 35 units (as measured by rat assay) of D_2 must be given to produce the effect of one unit of D_3. In the prevention or cure of human rickets, no difference has been shown between D_2 and D_3.

Function. It was indicated at the outset that vitamin D activity is concerned with the prevention and cure of rickets. Rickets is osteomalacia (see page 688) occurring in growing bone, and can result from deficiency of calcium, phosphate, or most commonly from deficiency of vitamin D. The deposition of bone salt requires obviously the presence of calcium and of phosphate in the diet, but with abundant amounts of these minerals present and deficiency of vitamin D, calcification will still not take place in a normal manner. The orderly advance of capillaries into the cartilage, and the uniform degeneration of cartilage cells just before the capillaries reach them, are replaced by a confused and irregular advance of calcification, with islands of persisting cartilage cells. Organic matrix of bone is laid down but it is incompletely calcified, and can be distorted by weight-bearing or by the pull of muscles.

Administration of adequate vitamin D, even if the supply of calcium and of phosphate is minimal, will normalize this situation in the area where normal calcification would have been taking place if the deficiency had not occurred. The abnormal region of irregular bone formation and delayed calcification is slowly repaired later. The line of healing can be shown by the x-ray on account of its greater content of bone salt than the adjacent regions. Vitamin D is involved in calcification through its influence upon at least two portions of the process: the absorption of calcium and secondarily of phosphate from the intestine, and the actual process of ossification as localized in the bone. The chemical mechanism of its action in either of these locations is not definitely known. The healing of rickets begins after administration of vitamin D, even though the diet contains no calcium or phosphate. An activating effect of phosphorylated vitamin D_2 upon alkaline phosphatases of bone, kidney, and intestine has been demonstrated (16). Vitamins D have been shown to increase the citrate (see page 687) concentration in blood and in tissues other than liver, and the urinary output of citrate. These effects are most noticeable in animals with experimentally-induced rickets.

The deformities produced while rickets is present may persist and be made permanent as calcification becomes normal. The bones are stiffened in their deformed position as the abnormal area belatedly calcifies. The location and nature of the deformities depend not only upon the severity of the deficiency but upon the age of the child at the time of the disease. *Craniotabes* is characteristic of rickets at an early age (3 to 8 months) and consists of thinning and softening of the bones of the skull, or of limited areas of those bones. The cranial sutures are abnormally wide and the affected bone areas indent easily with pressure, returning to their original shape by their inherent elasticity. The permanent deformities likely to result from rickets active in the cranium are the "Olympian brow" and the "square head." These occur from pressure upon the sides and back of the head exerted by the weight of the head itself as the child lies supine or on either side. Deformities which occur at later ages are those of the thorax, of the spine, the pelvis, and the lower extremities.

The deformities briefly noted above are preceded by definite biochemical changes, of which the earliest is the increase in plasma alkaline phosphatase (see page 684) above the normal infantile limits of 3 to 12 Bodansky units. This change is also the most persistent of the biochemical abnormalities, and lasts well into the stage of healing, denoting increased osteoblastic activity. Plasma *inorganic phosphate* is characteristically decreased in human rickets; in this respect note that infants and children have normally higher values for inorganic phosphate than the adult. No normal baby will have an inorganic phosphate level below 5 mgm. per 100 ml. plasma and this level may run as high as 7 mgm.; the normal range for adults is 3 to 4.5 mgm., and the values for older children are intermediate. Values of inorganic phosphate in infant plasma below 4 mgm. per 100 ml. are practically diagnostic for rickets. In severe cases they may fall below half this value. The calcium of the blood serum frequently remains normal in rickets and when there is a depressed value for calcium, tetany may be a complication (see page 688). These values refer only to human rickets in its usual form—rickets from vitamin D deficiency. They do not apply to rickets experimentally produced in animals, nor to those rare cases of human rickets resulting from calcium or phosphate deficiency in the presence of adequate vitamin D, nor to human rickets sequential to renal or other organic disease or to extreme general malnutrition. By definition rickets can not occur in the adult, since rickets is a disturbance of bone salt deposition in growing bone. Rickets can occur in the unborn or newborn child, but the most susceptible age is the middle of the first year of life. Other diseases can produce similar skeletal deformities, hence the value of biochemical diagnostic aids and of x-ray studies.

Requirements. Rickets does not exist where culture and climate permit mothers and infants to benefit by reasonable exposure to unfiltered sunlight. The infant who gets no exposure to sunlight requires 400 units a day of vitamin D_2 or D_3. With smaller amounts, less of the calcium of the diet is retained and skeletal growth is slower; with larger amounts up to around 1800 units a day there is no further increase in growth or percentage of calcium retained. If amounts larger than 1800 units a day are given over long periods of time there is an adverse effect upon appetite and general growth. These facts have been established by study of large groups of infants under well controlled conditions. The requirements of older children are less definitely demonstrable by controlled experimental study, but it is well known that rickets can occur as long as growth continues. In such studies as have been done, it appears that throughout the growth period the same intake of 400 units per day permits optimal utilization of calcium. It should be kept in mind that the requirements of calcium and phosphate are greater in older children and in adolescents, and that vitamin D will not compensate for mineral deficiencies. Once full growth has been achieved, no need for vitamin D beyond the minimal amounts gained from the normal diet and from normal exposure to sunlight has been demonstrated. It is considered advisable for those who are on restricted diets and who lead an indoor life to take a small supplementary ration of vitamin D, but such advice is common-sense precaution rather than scientifically established necessity. Causes other than vitamin D deficiency appear to explain adequately the calcium losses from the bones of elderly people, and simple vitamin D therapy is not commonly successful in such cases. The one situation in adult life where vitamin D becomes essential is during the latter half of pregnancy and the following period of lactation. It is probable that 400 units per day is adequate here, but many medical authorities have advised the use of 800 units. Even if the mother is receiving such a supplement, the breast-fed baby should be given the usual supplementary vitamin D, since the output of vitamin D in the milk is variable and not fully reliable.

The treatment of actual rickets involves more than the supply of adequate intakes of calcium, phosphate, and vitamin D, although these are of course essential. Rickets will heal if the normal requirements of these substances are provided, but healing will be more rapid if larger doses of vitamin D, from 1000 to 4000 units daily, are used. Still larger doses are used under special circumstances, for discussion of which the textbooks and literature of pediatrics should be consulted. With 1000 units a day, plasma phosphate will reach normal in about 10 days, and evidence of incipient healing can be demonstrated by x-ray in about 20 days.

Toxicity of vitamin D must be kept in mind whenever large doses of vitamin D are considered for infants or children, or whenever administration of supplementary vitamin D is considered for the adult. It has already been noted that doses of over 1800 units a day may, if long continued, produce loss of appetite and slowing of growth in infants. Much larger doses have been administered to adults daily for supposed therapeutic effects in various diseases, particularly of the skin and the joints. In some such patients calcification of bursae has occurred, also deposition of calcium salts in kidneys, heart, bronchi, and blood vessels, with serious disturbances of function. There have been a few fatalities, both in infants and adults, attributed to overdosage of the vitamins D; the immediate cause of death was failure of kidney function.

The immediate effect of an overdose of a D vitamin is an increase in blood plasma calcium concentration, apparently the result of increased absorption of Ca^{++} from the intestine. Abnormal deposition of calcium salts occurs as the blood calcium concentration decreases. Normocytic, normochromic anemia is characteristic of chronic overdosage with D vitamins (49a).

VITAMINS K

The naturally-occurring vitamins K_1 and K_2 are derivatives of 1,4-naphthoquinone (formula XXVIII) which itself possesses moderate antihemorrhagic activity, as does phthiocol (formula XXIX) from the tubercle

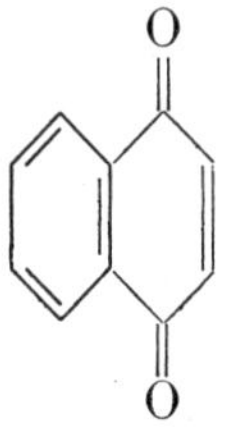

XXVIII. 1,4-Naphthoquinone

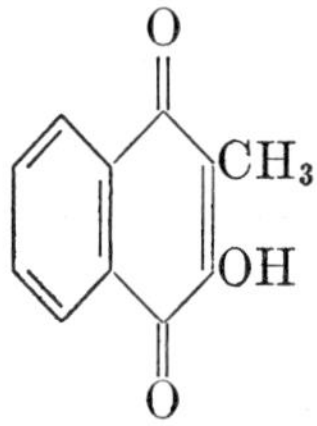

CH_3, OH

XXIX. Phthiocol

bacillus. In the natural vitamins K, the hydroxyl group of phthiocol is substituted by a long side chain which is phytyl in vitamin K_1 (formula XXX), and difarnesyl in vitamin K_2 (formula XXXI). Vitamin K_1 occurs chiefly in the green leaves of plants, K_2 in micro-organisms, especially bacteria. The latter is the form produced by bacterial action in the intestines of men and animals. In the above substances, conversion of the quinone to the corresponding hydroquinone brings about no loss of activity. Any substitution in the benzene (not the quinone) ring causes complete loss of activity. The 2-methyl group is necessary for a high level of activity. The substituent on the 3-position can be phytyl or difarnesyl, as in the natural vitamins, or a simpler aliphatic chain, or a hydroxyl group, as in phthiocol, with little alteration in vitamin activity. If the 3-position is unsubstituted, we have

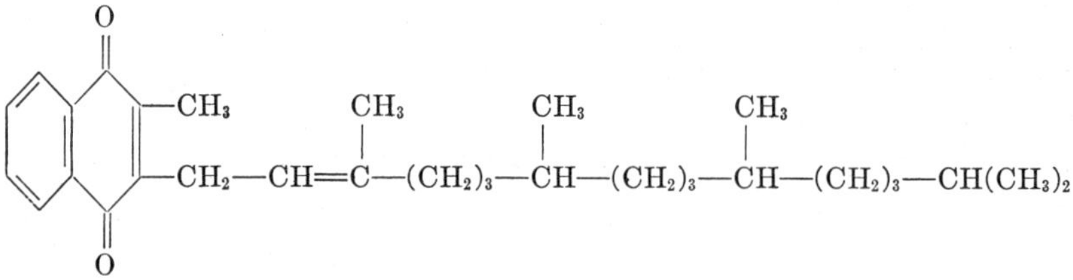

$$-CH_3$$
$$-CH_2-CH{=}\underset{|}{\overset{CH_3}{C}}-(CH_2)_3-\overset{CH_3}{CH}-(CH_2)_3-\overset{CH_3}{CH}-(CH_2)_3-CH(CH_3)_2$$

XXX. Vitamin K_1 (Phytonadione)

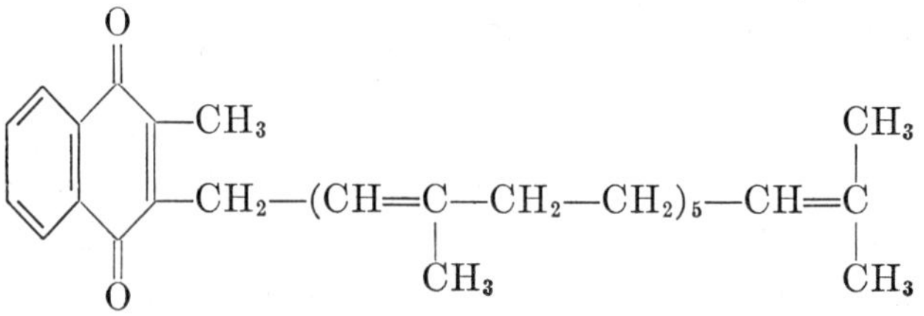

$$-CH_3$$
$$-CH_2-(CH{=}\underset{CH_3}{C}-CH_2-CH_2)_5-CH{=}\overset{CH_3}{\underset{CH_3}{C}}$$

XXXI. Vitamin K_2

a synthetic vitamin K, 2-methyl-1,4-naphthoquinone, or menadione (formula XXXII), which is 2 to 4 times more active than natural vitamin K_1.

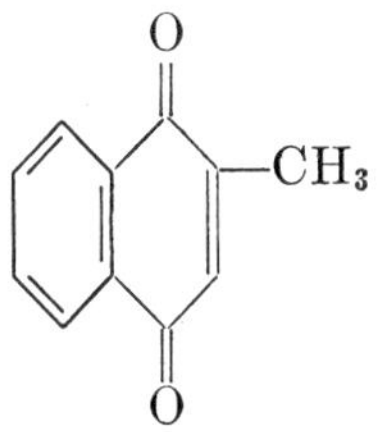

O

$—CH_3$

O

XXXII. Menadione

One microgram of synthetic vitamin K constitutes one A. O. A. C. unit; no international nor U. S. P. unit has been adopted.

One or another of the vitamins K will function to catalyze the formation of prothrombin in the liver. Deficiency of the vitamins K leads to deficient circulating prothrombin, and consequent prolonged clotting time. This defect can be remedied by administration of a vitamin K, provided the liver is not too diseased to produce prothrombin. The K vitamins have no coagulant action *in vitro*. Since natural vitamin K_1 is present in most normal diets and since vitamin K_2 is produced by the action of intestinal bacteria, deficiency of K vitamins is uncommon in otherwise healthy people. Deficiency may result from failure to absorb the vitamin, as in disturbances of lipid absorption such as in sprue or protracted dysentery, or in the absence of bile from the intestine as in obstructive jaundice. Protracted use of antibacterial drugs, such as sulfonamides, may dangerously cut down the bacterial population of the intestine and thus cause production of vitamin K_2 to fail. Much less commonly the deficiency may result from lack of adequate food, or bluntly, starvation.

Each newborn infant starts extra-uterine life with no intestinal bacteria. This situation is spontaneously remedied, but several days may be required for vitamin K_2 production to get well under way in the intestine. *Hemorrhagic disease of the newborn* results from vitamin K deficiency, and appears characteristically on the second or third day. If death does not occur from loss of blood or intracranial hemorrhage, recovery occurs within a few days as vitamin K_2 is produced by the increasing bacterial flora of the intestine. This disease can be prevented by small doses of vitamin K—for example, one mgm. of the synthetic product—given to the mother weekly during the final few weeks of pregnancy and daily when labor is expected to start. The newborn infant may be given 0.5 mgm. at birth and one mgm. daily in divided doses for the critical 3 days. Water-soluble products, such as synthetic vitamin K combined with $NaHSO_3$ may be given orally to infants.

Whatever the cause, deficiency of the vitamins K is manifested by low prothrombin levels in the blood, prolonged clotting time, and hemorrhage. The hemorrhage does not occur without previous injury, but the injury may be so slight as to be trivial were it not for the delay in blood clotting. Prothrombin may be estimated in the blood by measurement of the clotting time in the presence of excess thromboplastin, by techniques described in detail in manuals of clinical pathology.

Vitamin K deficiency caused by exclusion of bile from the intestine can be treated by oral administration of bile salts with additional vitamin K. Other disturbances of vitamin K absorption call for parenteral administration of vitamin K, which usually shows a favorable response within 6 hours. The daily requirement has not been established, but a few milligrams is adequate for the correction of ordinary deficiencies. Antagonists of vitamin K, such as *dicoumarol* (formula XXXIII), inhibit production

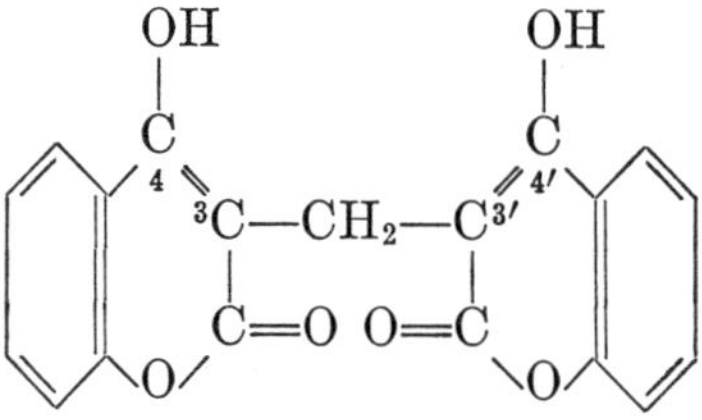

XXXIII. Dicoumarol [3,3′-Methylene-bis(4-hydroxycoumarin)]

of prothrombin and large doses of vitamin K of the order of 40 mgm. are required to overcome such inhibition. Vitamins K_1 and K_2 are non-toxic. Menadione has caused vomiting, albuminuria, and porphyrinuria in doses of 180 mgm. or more. Such large dosage serves no useful therapeutic purpose.

TOCOPHEROLS

A physiological action observed in experimental animals but not in man and designated as that of "vitamin E" is characteristic of three known compounds, all derivatives of chromane. The best known and most potent is alpha-tocopherol (formula XXXIV). The international unit of vitamin E

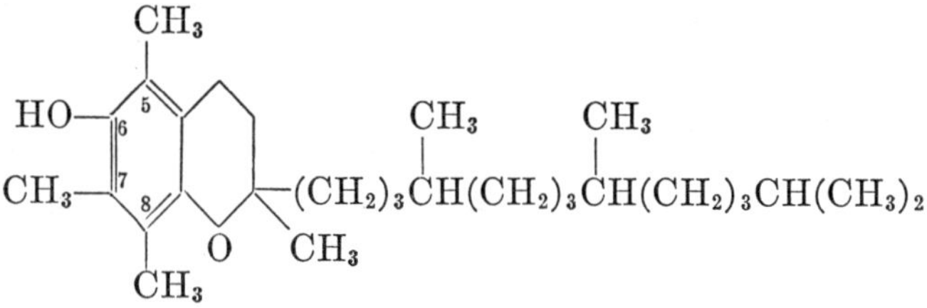

XXXIV. alpha-Tocopherol

activity is that of 1 mgm. of synthetic alpha-tocopherol acetate. Beta-tocopherol lacks the methyl group on carbon 7, gamma-tocopherol lacks the methyl group on carbon 5 and is isomeric with beta-tocopherol. Alpha-tocopherol has twice the potency of beta- or gamma-tocopherol. These three tocopherols are all soluble in fats and fat solvents, very sparingly soluble in water, and stable to the heat involved in cooking. They are converted by oxidizing agents to inactive quinones, but are not oxidized under ordinary conditions of food preparation. They are also destroyed, as far as vitamin effect is concerned, by ultraviolet radiation.

The tocopherols occur chiefly in plants, especially in certain vegetable oils, including the oils of wheat germ, cottonseed, and rice germ. There is no vitamin E in olive oil, very little in peanut oil. Other sources include lettuce and alfalfa, where the vitamin is present in the lipids of the leaves. Small amounts are present in foods of animal origin, particularly beef liver, but there is no evidence of synthesis of vitamins E in the animal. Fats and oils containing tocopherols develop rancidity by oxidation more slowly than those which contain no vitamin E or other *antioxidant.* Vitamin E, by its own preferential oxidation, inhibits the formation of peroxides of unsaturated fatty acids (25). Other antoxidants include certain phenols, naphthols, and quinones. Tocopherol, synthetically produced from alkylated hydroquinone and phytyl halide, is often added to fats and foodstuffs to delay spoilage by oxidative rancidity. It will be recalled that vitamin A and the carotenes are susceptible to oxidative destruction. They can be stabilized *in vitro* by tocopherols, and such stabilization may be a function of vitamin E in the gastrointestinal tract and in tissues. Several chemical tests have been devised for the determination of vitamin E. Spectrophotometric analysis is highly satisfactory in pure or nearly pure solutions, utilizing an absorption maximum in the ultraviolet at 294 millimicrons. Biological assay may be based upon the prevention of testicular atrophy in the rat (20).

Bulls, cocks, and male rats and mice depleted of vitamin E by a deficient diet show a decrease in motility of spermatozoa, an increase in the proportion of abnormally formed spermatozoa, and eventually degeneration of the germinal epithelium. The last effect is not reversible by vitamin E administration, while the earlier effects may be so normalized. In female rats made deficient in vitamin E there occurs progressively as the deficiency continues: (a) prolongation of gestation; (b) stillbirths; and (c) death and resorption of the unborn young. These effects cease with administration of vitamin E. With further prolongation of deficiency irreversible uterine degeneration occurs. Hens deficient in vitamin E produce eggs in which the embryo chicks die before reaching the hatching stage. In animals of both sexes, degeneration of nerves and striated muscles has been observed in severe deficiencies, associated with creatinuria. Affected tissues first show increased oxygen consumption, but later degenerate and are replaced by scar tissue. In such deficient animals, transamination is diminished up to 50 per cent. The muscle lesions of vitamin E deficiency in rats are prevented by alpha-tocopherylhydroquinone (which lacks antisterility effect) as well as by alpha-tocopherol (34).

Although sterility and abortion in cattle and hogs have been in some instances overcome by the use of vitamin E, its application in the treatment of human infertility has led to no clear-cut favorable results. It is unlikely that such deficiency plays any significant part as a cause of sterility in men or women. Neither has vitamin E been notably effective in the treatment of muscular dystrophies in the human. The best therapeutic successes have been in *primary fibrositis*, a disease characterized by fibrillar nodules near joints, and by creatinuria. The vitamins E are nontoxic. No human requirement can be stated, since no signs or symptoms definitely related to tocopherol deficiency have been observed in man. Human blood plasma contains one mgm. or less of vitamins E per 100 ml., and the plasma level in man does not correlate with the dietary intake.

REFERENCES

1. Ames, S. R., *et al.* Biochemical studies on vitamin A. XIV. Biopotencies of geometric isomers of vitamin A acetate in the rat. XV. Biopotencies of geometric isomers of vitamin A aldehyde in the rat. J. Am. Chem. Soc., **77:** 4134–4136; 4136–4138, 1955.
2. Baxter, R. M., and Quastel, J. H. The enzymic breakdown of D-biotin *in vitro*. J. Biol. Chem., **201:** 751–764, 1953.
3. Bean, W. B., and Hodges, R. E. Pantothenic acid deficiency induced in human subjects. Proc. Soc. Exper. Biol. Med., **86:** 693–698, 1954.
4. Beaton, J. R., and Goodwin, M. E. Studies on the effect of vitamin B_6 deprivation on carbohydrate metabolism in the rat. Can. J. Biochem. Physiol., **32:** 684–688, 1954.
5. Burch, H. B., *et al.* Fluorometric measurements of riboflavin and its natural derivatives in small quantities of blood serum and cells. J. Biol. Chem., **175:** 457–470, 1948.
6. Campbell, J. A., and McLaughlan, J. M. Vitamin B_{12} and the growth of children. A review. Can. Med. Assoc. J., **72:** 259–263, 1955.
7. Castle, W. B. Observations on etiologic relationship of achylia gastrica to pernicious anemia. I. Effect of administration to patients with pernicious anemia of contents of normal human stomach recovered after ingestion of beef muscle. Am. J. Med. Sci., **178:** 748–764, 1929.
8. Cleland, J. B. Biochemical aids in the diagnosis of deficiency of the vitamin B complex. M. J. Australia, **1:** 468–476, 1951.
9. Consolazio, C. F., *et al.* *Metabolic Methods*. St. Louis, C. V. Mosby Company, 1951.
10. Crandon, J. H., *et al.* Experimental human scurvy. New Eng. J. Med., **223:** 353–367, 1940.

11. Drysdale, G. R., *et al.* The relationship of folic acid to formate metabolism in the rat; formate incorporation into purines. J. Biol. Chem., **193:** 533–538, 1951.
11a. Eagle, H. *et al.* *Myo*-inositol as an essential growth factor for normal and malignant human cells in tissue culture. Science, **123:** 845–847, 1956.
12. Faulkner, J. M., and Taylor, F. H. L. Vitamin C and infection. Ann. Int. Med., **10:** 1867–1873, 1937.
13. Fiegelson, P. The inhibition of xanthine oxidase *in vitro* by trace amounts of L-ascorbic acid. J. Biol. Chem., **197:** 843–850, 1952.
14. Figueroa, W. F., *et al.* Lack of avitaminosis among chronic alcoholics. Its relation to fortification of cereal products and the general nutriture of the population. J. Clin. Invest., **29:** 812, 1950.
15. Fitzpatrick, J., and Tompsett, S. L. The daily excretion of riboflavin, biotin, pantothenic acid, and nicotinic acid derivatives by normals. J. Clin. Path., **3:** 69–71, 1950.
15a. Franz, W. L. *et al.* Blood ascorbic acid level in bioflavonoid and ascorbic acid therapy of common cold. J.A.M.A. **16:** 1224–1226, 1956.
16. Goldsmith, G. A., *et al.* Recent advances in nutrition and metabolism. A. M. A. Arch. Int. Med., **90:** 513–561, 1952.
17. Guehring, R. R., *et al.* Cholesterol metabolism in pantothenic acid deficiency. J. Biol. Chem., **197:** 485–493, 1952.
18. Haley, T. J., and Flesher, A. M. A toxicity study of thiamine hydrochloride. Science, **104:** 567–568, 1946.
19. Handschumacher, R. E., *et al.* An improved enzymatic assay for coenzyme A. J. Biol. Chem., **189:** 335–342, 1951.
20. Herraiz, M. L., and Radice, J. C. The biological assay of vitamin E by the "male rat" test. Ann. N. Y. Acad. Sci., **52:** 88–93, 1949.
21. Hoagland, M. B., and Novelli, G. D. Biosynthesis of coenzyme A from phosphopantetheine and of pantetheine from pantothenate. J. Biol. Chem., **207:** 767–773, 1954.
22. Hodgkin, D. C., *et al.* Structure of vitamin B_{12} . Nature, **178:** 64–66, 1956.
23. Horwitt, M. K. Niacin-tryptophan relationships in the development of pellagra. Am. J. Clin. Nutrition, **3:** 244–245, 1955.
24. Horwitt, M. K., *et al.* Correlation of urinary excretion of riboflavin with dietary intake and symptoms of ariboflavinosis. J. Nutrition, **41:** 247–264, 1950.
25. Hove, E. L. Antivitamin E stress factors as related to lipide peroxides. Am. J. Clin. Nutrition, **3:** 328–336, 1955.
26. Hume, E., and Krebs, H. A. Vitamin A requirements of human adults. Medical Research Council Special Report Series, No. 264. London, His Majesty's Stat. Off., 1949.
27. Hurwitz, J. The enzymic phosphorylation of pyridoxal. J. Biol. Chem., **205:** 935–947, 1953.
28. Jaenicke, L. Occurrence of N^{10}-formyltetrahydrofolic acid and its general involvement in transformylation. Biochim. et Biophys. Acta, **17:** 588–589, 1955.
29. Jukes, T. H. Folic acid and vitamin B_{12} in the physiology of vertebrates. Fed. Proc., **12:** 633–638, 1953.
30. Kirk, E., and Chieffi, M. Vitamin studies in middle-aged and old individuals. III. Thiamine and pyruvic acid concentrations. J. Nutrition, **38:** 353–360, 1949.
31. Knudson, A. G., Jr., and Rothman, P. E. Hypervitaminosis A. A. M. A. Am. J. Diseases Children, **85:** 316–334, 1953.
32. Latner, A. L. *et al.* Preparation of highly potent intrinsic factor mucoprotein. Biochem. J., **63:** 501–507, 1956.

33. Levitan, B. A. The biochemistry and clinical application of vitamin P. New Eng. J. Med., **241:** 780–789, 1949.
34. Mackenzie, J. B., and Mackenzie, C. G. Vitamin E activity of alpha-tocopherylhydroquinone and muscular dystrophy. Proc. Soc. Exper. Biol. Med., **84:** 388–392, 1953.
35. Maas, W. K., and Novelli, G. D. Synthesis of pantothenic acid by depyrophosphorylation of adenosine triphosphate. Arch. Biochem. Biophys., **43:** 236–238, 1953.
36. Meiklejohn, A. P., and Passmore, R. Nutrition and nutritional disease. Ann. Rev. Med., **2:** 129–154, 1951.
37. Melnick, D., and Field, H., Jr. Chemical determination of vitamin B_1. II. Method for estimation of the thiamine content of biological materials with the diazotized *p*-aminoacetophenone reagent. III. Quantitative enzymic conversion of cocarboxylase (thiamine pyrophosphate) to the free vitamin. J. Biol. Chem., **127:** 515–530, 531–540, 1939.
38. Molony, C. J., and Parmelee, A. H. Convulsions in young infants as a result of pyridoxine (vitamin B_6) deficiency. J. A. M. A., **54:** 405–406, 1954.
39. Mushett, C. W., *et al.* Antidotal efficacy of vitamin B_{12a} (hydroxo-cobalamin) in experimental cyanide poisoning. Proc. Soc. Exper. Biol. Med., **81:** 234–237, 1952.
40. Nichol, C. A. Metabolic alteration of pteroylglutamic acid. Proc. Soc. Exper. Biol. Med., **83:** 167–170, 1953.
41. Novelli, G. D. Metabolic functions of pantothenic acid. Physiol. Rev., **33:** 525–543, 1953.
42. Pollycove, M., *et al.* Pernicious anemia due to dietary deficiency of vitamin B_{12}. New Eng. J. Med., **255:** 164–169, 1956.
43. Racker, E., *et al.* Thiamine pyrophosphate, a coenzyme of transketolase. J. Am. Chem. Soc., **75:** 1010–1011, 1953.
44. Reisner, E. H., Jr., *et al.* Oral treatment of pernicious anemia with vitamin B_{12} without intrinsic factor. New Eng. J. Med., **253:** 502–506, 1955.
45. Robbins, W. J., *et al.* Euglena and vitamin B_{12}. Ann. N. Y. Acad. Sci., **56:** 818–830, 1953.
46. Sargent, F., II. "Folic acid": pteroylglutamic acid and related substances. New Eng. J. Med., **237:** 667–672, 703–707, 1947.
47. Sargent, F., II, and Sargent, V. W. Season, nutrition, and pellagra. New Eng. J. Med., **242:** 447–453, 507–514, 1950.
48. Sayers, G. The adrenal cortex and homeostasis. Physiol. Rev., **30:** 241–320, 1950.
49. Sealock, R. R., *et al.* The role of ascorbic acid in the oxidation of L-tyrosine by guinea pig liver extracts. J. Biol. Chem., **196:** 761–767, 1952.
49a. Scharfman, W. B. and Propp, S. Anemia associated with vitamin D intoxication. New Eng. J. Med., **255:** 1207–1212, 1956.
50. Sobel, A. E., and Snow, S. D. The estimation of serum vitamin A with activated glycerol dichlorohydrin. J. Biol. Chem., **171:** 617–632, 1947.
51. Sydenstricker, V. P., *et al.* Observations on "egg white injury" in man and its cure with a biotin concentrate. J. A. M. A., **118:** 1199–1200, 1942.
51a. Tebrock, H. E. *et al.* Usefulness of bioflavonoids and ascorbic acid in treatment of common cold. J.A.M.A. **162:** 1227–1233, 1956.
52. Tulpule, P. G., and Williams, J. N., Jr. The role of essential fatty acids in liver metabolism. J. Biol. Chem., **217:** 229–234, 1955.
53. Verly, W. G., *et al.* Effect of folic acid and leucovorin on synthesis of the labile methyl group from methanol in the rat. J. Biol. Chem., **196:** 19–23, 1952.

54. VILTER, R. W., *et al.* The effect of vitamin B_6 deficiency induced by desoxypyridoxine in human beings. J. Lab. & Clin. Med., **42:** 335–357, 1953.
55. VON KORFF, R. W. A rapid spectrophotometric assay for coenzyme A. J. Biol. Chem., **200:** 401–405, 1953.
56. WEISS, S., AND WILKINS, R. W. The nature of the cardiovascular disturbances in nutritional deficiency states (beriberi). Ann. Int. Med., **11:** 104–148, 1937.
57. WILLIAMS, R. J., *et al.* *The Biochemistry of the B Vitamins.* New York, Reinhold, 1950.
58. WRIGHT, L. D., *et al.* Biocytinase, an enzyme concerned with hydrolytic cleavage of biocytin. Proc. Soc. Exper. Biol. Med., **86:** 335–337, 1954.
59. WRIGHT, L. D., *et al.* Biotin derivatives in human urine. Proc. Soc. Exper. Biol. Med., **91:** 248–252, 1956.

CHAPTER 22

Infection

At some early stage of evolution, possibly before plants and animals had become clearly differentiated from each other, one or more species must have made the discovery that a very convenient source of food consists of the tissues of other organisms. In such tissues may be found materials, already pre-formed, suitable as energy sources and for building one's own tissues.

The whole of the animal kingdom depends upon eating plant life of some form or other directly or indirectly, and thus obtains the protein, fat, and carbohydrate which it needs. Only the green plants and the autotrophic bacteria are able to synthesize these substances from inorganic materials and carbon dioxide alone.

Among some of the species which had discovered the importance and desirability of eating other organisms as a source of food, the expedient of *parasitism* developed. After all, it is not economical to destroy the source of your food material completely, although some micro-organisms, perversely enough, do this to their hosts. We believe, however, they are in the minority. Instead, it is easier to live in or on the host, making use of the food supply which he provides directly or indirectly. This is parasitism, of which infection is an example.

However, the organisms which are thus forced to act as a host to parasites do not themselves remain idle in the evolutionary sense. Organisms in general have provided themselves with some kind of membrane or sheath, partly to protect themselves against invasion by other organisms from the outside. Other mechanisms, to be described in the next section, have also been developed. We are not without our defenses, therefore, against invasion by foreign organisms, and it is the chemical nature of these defenses which concerns us here. The invaders are generally classified either as animal parasites, or as bacteria or fungi (which are generally considered forms of plant life), or as viruses, which are smaller than either and apparently simpler, in some extreme cases consisting entirely of molecules of nucleoproteins.

RESPONSE TO INVASION

The human body has a number of mechanisms which oppose infection. The most important of these is phagocytosis, which means that leukocytes are mobilized and literally eat up the invading organisms. This, however, is rather a concern of pathology or immunology, and in spite of its great importance we shall not consider it further here. From the biochemical aspect we are interested in the chemical changes which take place in the body when it attempts to resist infection. The most important and most readily recognized of these changes is the production of antibodies.

Antibodies may be defined as the substances which appear in the blood plasma or tissues of an animal within a few days after the parenteral introduction, either naturally, as in disease, or artificially, into that animal of certain types of foreign substances. Artificially introduced foreign substances may be pure proteins, tissues or body fluids, bacterial cells, blood cells or cellular extracts or fractions. The particular property which distinguishes an antibody is that it reacts specifically with the material which was introduced. Any material which will evoke the production of antibodies is called an *antigen.* Antigens may be pure protein or polysaccharide in nature or they may be complex biological substances containing proteins or polysaccharides. An animal who has received injections of an antigen and who has subsequently formed specific antibodies is said, by convention, to be *immunized* to that particular antigen, and the serum containing specific antibodies is called an *immune serum.*

Antibodies and Their Production

Antibodies are plasma proteins belonging to the class of gamma globulins. Indeed, it has never been clearly established that there are any gamma globulins which are not antibodies. On the other hand, it has been impossible to prove that all the gamma globulins are antibodies. Some might conceivably be "normal" gamma globulins of unknown function; but since the body has to resist a large number of different types of infection, it therefore has constantly on hand a number of different antibodies in varying amounts. If we could measure and add up all the antibodies present, they might account for the whole of the gamma globulins. Since we can not in fact do this, the question remains rather academic.

In any case the properties of antibodies are not greatly different from those of gamma globulins in general. In man they have a molecular weight of about 160,000 and it has been calculated, for example, that the human antipneumococcus antibody (see page 50) has a short axis, 3.7 millimicrons in length, and a longer axis of 33.8 millimicrons (29).

Antibodies are presumably synthesized in the same way as other body proteins (see page 650), the antigen somehow acting to produce the differences from "normal" gamma globulin which are responsible for the specificity (5, 20). The idea that preexisting gamma globulins are converted to antibodies is contradicted by studies on isotopically labelled amino acids (18).

A serum euglobulin, called *properdin*, with a molecular weight some eight times that of the gamma globulins, has been isolated (32). It occurs in serum to the extent of not more than 0.03 per cent and in conjunction with complement (see page 852) and Mg^{++} appears to take part in a number of defense phenomena such as the destruction of bacteria (mainly gram-negative) and the neutralization of viruses.

Specificity. An outstanding characteristic of the antibodies is their specificity; that is, they tend to combine primarily with the invading organism or substances derived from it. However, there are limitations to this specificity even though it is so sharp that it allows the distinction of different species of blood or even different types of blood within the same species (see page 199). Antibodies, although specifically directed toward one invading organism or one of its components, are also capable as a rule of reacting with chemically related compounds. Thus, the antibodies produced by injecting rabbits with the egg albumin from the hen will also react with the egg albumin from the duck. Such cross-reactions are also observed with different species of pathogenic micro-organisms.

The general course of events in an infection is something like this. First, bacteria or other foreign organisms enter through the skin or a mucous surface and probably next pass to a regional lymph node. There they proceed to reproduce. Phagocytosis removes them from the blood as they are liberated into it, unless the disease is very serious, in which case bacteremia, meaning presence of bacteria in the blood, results. The phagocytic cells can carry the foreign invaders to various parts of the body, including the reticulo-endothelial system. Antibodies are formed in the reticulo-endothelial system. It is not at present possible to say exactly which cells form antibodies. Both lymphocytes and plasma cells have been characterized experimentally as antibody sources (19a).

Antibodies do not at once appear in the circulation following exposure of an animal to an antigen, and the animal does not become at once immune. Instead, there is a latent period, usually of the order of six days or somewhat longer. When antibodies do begin to appear in the blood, they combine with the antigens which caused their appearance and apparently make them more attractive to the phagocytes, so that the foreign invaders present in the blood are more rapidly removed, and may be removed from localized infections also. Also, antibodies may have the property of neutralizing toxic substances produced by the invaders, as for example the toxin of the diphtheria bacillus. The two principal functions of antibodies in infection seem to be neutralization of toxicity and sensitization of the invaders to phagocytosis.

Evidently antibodies must be constructed in a specific way in order to become antibodies which can combine firmly with special antigens. We do not know the exact nature of this modification, but it is supposed that it takes the form of a correspondence of positive and negative charges, and possibly hydrogen bonding sites, which enable the antibody to present a localized patch which matches one or more characteristic patches (complexes of positive and negative charges and sites of potential hydrogen-bond formation) on the antigen.

The exact number of these specialized sites on a molecule of antibody has not been determined. Some immunologists have written as if they considered that there were large numbers of such sites. It is more likely that the number of combining sites of antibody is not greater than two or three. It should be obvious that a difference of only two or three localized combining sites on the surface of a protein molecule of the size of gamma globulin would not make a very great difference in its chemical behavior. This is especially true if, as has been proposed by Pauling (31), the difference in the antibody molecule is not due to a difference in the order in which the amino acids follow each other in the polypeptide chain, but simply a difference in the way in which the polypeptide chain is folded. In this case the amino acid analysis of antibody would be expected to be the same as that of total gamma globulin, and up to date no differences have been found.

Increase of gamma globulin in infection. Since antibodies are gamma globulins, one would expect to see a marked increase of these proteins during infections, and this can usually be observed. The gamma globulin increase is not always exactly equal to the amount of specific antibody production, although it usually runs parallel to it. Boyd (6) found that the antibody in rabbits amounted to 40 to 70 per cent of the increase in globulin. Bjørneboe (3) was able, however, to produce antibodies in rabbits to such a degree that this accounted for all of the globulins present in their sera.

Agammaglobulinemia. In the blood of some male children, little or no gamma globulin can be detected, and this condition seems to be congenital, possibly sex-linked in its inheritance. Such children have a history of frequent severe bacterial infections and before the days of antibiotics probably did not survive childhood (17).

There is a similar condition in adults which affects both males and females and which seems to be acquired. Patients with agammaglobulinemia are benefited by injections of human gamma globulin.

Combination Between Antibody and Antigen

Antibodies formed in response to infection usually combine specifically with the invading organism or some portion of it. It is believed by immunochemists that they do this by purely chemical combination, the mechanism of which is not particularly different from that of the combination of other chemical compounds. The combination is, as a rule, a very firm one and it is difficult to split antibody and antigen apart after they have combined. The free energy change ($-\Delta F^\circ$) seems to be of the order of 5 to 10 Kcal. per mol (5). It is possible that the attraction between positively and negatively charged groups, such as the $—NH_3^+$ of lysine and the $—COO^-$ of aspartic acid and glutamic acid, plays a role in some antibody-antigen unions, but modern opinion inclines to the view that hydrogen bonding and van der Waals forces, especially the latter, are the main contributors to the force holding antibody and antigen together. The specificity of antibodies may depend upon their ability to fit portions of the surface of the antigen quite exactly, which would allow the short-range van der Waals forces to operate strongly. Some authors have even suggested that the antibody combining group consists essentially of a cavity which fits precisely over the projecting combining group of the antigen (21a, 31a).

Second stage of antibody-antigen combination. After the antibody and antigen have combined, a second stage, which is generally slower, may be observed. This may consist of the precipitation of certain soluble antigens, lysis in case the antigen is an invading micro-organism or a foreign red cell, or neutralization if the antigen is a toxin or a virus. The phenomenon of agglutination which consists of the aggregation into clumps of micro-organisms or foreign red cells is one of the easiest to observe and has already been described under blood groups (see Chapter 5).

Theories differ as to the mechanism by which this second stage is effected. The older theory, supported by Bordet (4) many years ago was that the second stage was a non-specific colloidal phenomenon due to the fact that the charges on the antigenic particles had been discharged by their combination with antibodies. A more recent theory proposed independently by Marrack (28) and by Heidelberger and Kendall (21) proposes that the second stage is merely a continuation of the first and is equally specific. Experiments designed to test this have not always been conclusive, but in some cases specificity of the second stage has certainly been demonstrated. Table XXII-1 shows this (1). It will be noted that the two distinguishable antigenic particles in this experiment (partially hemolyzed red cells

TABLE XXII-1

Specificity of hemagglutination (1)

ANTIGEN	ANTIBODY	RESULT
Ag + Ac	anti-A	Mixed clumps of ghost and intact cells
Ag + Bc	anti-B	Homogeneous clumps of normal cells
Ag + Bc	anti-A + anti-B	Homogeneous clumps of normal cells and homogeneous clumps of ghost cells

Ag = "ghost" (i.e., partially hemolyzed for identification) red blood cells of blood group A.

Bc = intact red cells of group B.

("ghosts") of blood group A and intact red cells of blood group B) did not form mixed clumps of agglutinated ghosts and intact cells when each was combined with its own specific antibody, as they should have if the forces bringing cells together in specific agglutination were solely secondary forces of a physical chemical nature which merely make the cells sticky. Instead the ghosts united only with other ghosts, and the intact cells only with other cells. When both ghosts and intact cells had a common antigenic receptor for the antibody, however, (top row) ghosts and intact cells did unite and the result was mixed clumps.

Proportions in which antibody and antigen combine. Antibody and antigen may combine in multiple proportions, which is not surprising if at least one of them has more than one combining group, and antigen at least does have, judging by the results of all analyses made to date. Antibody may have two or more combining groups, although the exact number is still not established. This means that the combination of antibody and antigen, although it really follows the stoichiometric rules of ordinary chemistry, may result in a very large number of compounds of different composition. This is shown in table XXII-2.

If insufficient antibody is added to an antigen, no visible results will follow, and it is customary to say that the reaction has been inhibited by excess of antigen. Excess of antibody does not always have the same effect,

TABLE XXII-2

Molecular compounds of thyroglobulin and its antibody observed (21)

EXTREME ANTIBODY EXCESS	ANTIBODY EXCESS END OF EQUIVALENCE ZONE	ANTIGEN EXCESS END OF EQUIVALENCE ZONE	ZONE OF PARTIAL INHIBITION	SOLUBLE COMPOUNDS IN INHIBITION ZONE
$A_{40}T$	$A_{14}T$	$A_{10}T$	A_2T	AT

A symbolizes antibody; T symbolizes thyroglobulin.

but with certain types of antibody it does. This is especially true with horse antibody. Rabbit antibody, which seems to resemble human antibody somewhat more closely, does not usually inhibit the reaction; instead, no matter how much antibody is added, a compound is formed which contains more and more antibody, although free antibody may be found in the supernatant fluid after the antibody-antigen compound has separated out.

FUNCTIONS OF ANTIBODIES IN DEFENSE

Antibodies aid the body in its defense against infection in a number of ways.

Neutralization

When antibody combines with a bacterial toxin in the right proportions, it neutralizes the toxin and renders it harmless, as can be demonstrated by mixing the two reagents *in vitro* and injecting the mixture into susceptible animals. Antibodies to certain enzymes will neutralize the enzymic activity. It seems not impossible that the antibody does not have to possess specific combining groups directed towards the toxic groups of the toxin, but detoxifies by merely covering up the toxic groups. This is suggested by the observations that certain toxins can be rendered non-toxic by limited chemical treatment but can still combine as well as ever with antitoxin, and by the observation that enzymes such as urease which act on substrates which are small molecules are not inactivated by their antibodies.

Neutralization depends partly upon the ratio in which the antigen and antibody are combined. In some cases, as with diphtheria toxin, the ratio giving neutralization corresponds well to the ratio which gives the most rapid precipitation or flocculation (8) when the reagents are mixed *in vitro*; in other cases the two ratios have been found to be different. A study of the mixtures which give the most rapid flocculation *in vitro* has been used to determine antibody and antigen concentrations, as in the Ramon titration of toxin and antitoxin, and the Dean and Webb optimal proportions method.

Sensitization (Opsonization)

The great clearing mechanism of the blood for foreign particles is phagocytosis, discovered by Metchnikoff. When micro-organisms reach the blood stream, the death or survival of the patient depends ultimately on whether or not the phagocytes succeed in freeing the blood stream of the invaders. This process is facilitated by antibodies, which by combining with the surface of the micro-organisms render them more attractive to the leukocytes. In addition to antibodies, thermolabile substances similar to complement (see below) are also involved in the mechanism of this sensitization of the micro-organisms to phagocytosis, which has been called *opsonization.*

Agglutination

Combination of antibody with bacterial and other cells often causes them to stick together in clumps. This is called agglutination (see page 431). It is likely that the agglutination is a by-product of the sensitization of the surfaces of the foreign cells, but it may aid the phagocytes in catching and devouring them by bringing them together in clumps and immobilizing them.

Bacteriolysis

Some micro-organisms visibly disintegrate and dissolve after antibody has combined with them. Others show no gross morphological changes, but are found to have been killed. These two results of antibody action are commonly grouped together under the heading bacteriolysis. It is obvious that this action of antibody is an important part of the defense mechanism of the body.

Some bacteria, such as gram-positive cocci, are not killed by combination with antibody. The reason for this is not understood.

Precipitation

Toxins and other soluble products of micro-organisms may be precipitated by the addition of antibody in the correct amounts. Particles of such precipitates, when formed *in vivo*, are soon removed from circulation by phagocytosis. The defensive role of the phenomenon is evident.

The precipitation (or flocculation) reaction is one of the most useful for the *in vitro* study of the reactions of antibodies and antigens, and in the hands of Heidelberger and his school (22) has greatly advanced our knowledge of the behavior of antibodies and antigens.

Flocculation Tests for Syphilis

If the blood of patients infected with syphilis contains antibodies to the spirochete, it might be possible to demonstrate them by a precipitation test, using a suitable extract of the infecting organism which might contain proteins, carbohydrates, and possibly lipids. However, it does not seem that the infecting organism of syphilis has yet been successfully cultivated in the laboratory, or at least no organism so cultivated has been shown to be the causative agent of the disease.

Nevertheless, in the early days of immunology it was discovered that *in vitro* tests for syphilis could be made. The first extracts were made from human syphilitic organs and were thought to contain an antigen characteristic of the spirochete. These were tested against with the blood of patients, and when a positive reaction occurred, it was considered that this was a sign that the patient had antisyphilis antibodies, and thus by inference, was infected with *Treponema pallidum.*

Later it was shown that extracts of normal organs would likewise react in this fashion, and this first led to doubts as to the validity of the test as a specific test for syphilis. However, it was shown clinically that except in the case of a few diseases such as malaria and vaccinia, a positive reaction was generally associated with a syphilitic infection. Thus the test, although based initially on a false assumption, has proved to be of practical value. The direct precipitation, or as it is usually called, flocculation test, which we have outlined is now used a good deal in the tests for syphilis, but historically it was preceded by another test which depends upon the use of complement, which we must now discuss.

Complement

If serum containing antibody is heated to 56°C. for half an hour, its ability to produce bactericidal and hemolytic reactions is lost, although it may still precipitate antigens and neutralize toxins. The lytic antibody has not been lost, however, and this can be demonstrated by actual measurement (say by micro-Kjeldahl determinations) of the antibody which combines with known amounts of antigen. It can also be shown that the addition of normal serum will restore the bactericidal or hemolytic power of the immune serum. Evidently some thermolabile material was destroyed in the heating process; this substance is called complement. Such substances are found in the normal serum of most species; the serum of the guinea pig is a particularly rich source.

At first it was natural to suppose that complement was a single substance, but fairly early observations indicated that this was not true. If electrolytes were removed by dialysis from complement-containing serum so that the euglobulins were precipitated, it was found that neither the supernatant nor the precipitate had the power of restoring the lytic power to a previously heated lytic serum. If, however, the precipitate and supernate were recombined and added, they did have this power. It was therefore obvious that complement could be separated into at least two components. Other experiments have shown that there are other components of the action of complement which may be removed by the absorbing power of yeast cells or by treatment of the serum with ammonia. Some of these more recently discovered fractions of complement are much more resistant to the action of heat than the fraction which was originally identified. It is customary to designate the four fractions of complement which are known at present as C'_1, C'_2, C'_3, and C'_4. The fraction now designated as C'_1 was originally called midpiece and C'_2 was at one time called endpiece. C'_1, C'_2, and C'_3 have been obtained in pure form as well defined proteins, although the C'_2 and C'_4 activity may reside in one and the same molecule of protein, which seems to be a mucoglobulin. C'_1 is a globulin with an isoelectric point of 5.2 to 5.4. C'_3 has not been obtained in pure form but seems to be possibly a phospholipid or a phosphoprotein. It will be convenient to lump all of these components together under the traditional name of complement in the discussion that follows.

Complement fixation. It was observed a long time ago that when antibody and antigen combine they use up complement in the process, and if the amount of complement in the mixture is suitably adjusted, the antibody-antigen reaction will quantitatively remove the complement activity. This is called *complement fixation.* Actually the first test for syphilis devised depended upon complement fixation rather than the direct demonstration of the reaction of supposed syphilitic antibodies with supposed syphilitic antigens. This test, known as the *Wassermann test,* is carried out in the following manner. The patient's serum together with a definite amount of guinea pig complement is placed in a test tube with a lipid extract, usually from normal beef heart (Bordet showed that extracts of normal organs worked as well as extracts of syphilitic organs). Other lipid substances such as cholesterol are sometimes added to increase the sensitivity of the test. If the patient has syphilis, the antibodies in his blood will react with this "antigen" and in the process complement will be used up. This reaction is, under ordinary circumstances, invisible. Next there is added to the test tube a mixture of sheep cells with a rabbit antibody which has the power of hemolyzing sheep cells but which has been heated to deprive it of its complement activity so that alone it will not cause lysis.

If, after these sheep cells combined with anti-sheep antibodies (which are spoken of as sensitized sheep cells) are added, they are observed to be hemolyzed, it follows that complement was still present in the mixture of the patient's serum and the lipid "antigen". This indicates, in turn, that there had been no reaction between the patient's serum and the lipid antigen, and the patient probably does not have syphilis. If, on the other hand, lysis of the added cells does not occur, it indicates that complement has been used up in the preliminary reaction between the patient's serum and the lipid antigen. Most such reactions are diagnostic of syphilis. It will be seen that the test is an indirect one. We add an indicator system—the sensitized sheep cells—to see whether or not one component of our system has been removed by a reaction which we otherwise would not be able to see.

The direct precipitation reaction for syphilis already described was developed *after* the Wassermann test by workers who accidentally made the observation that under suitable conditions, and with the addition of suitable adjuvants, it was possible to see directly the reaction between the antibody, or reagin as it is called, in syphilitic serum and the lipid antigen.

For a long time the complement fixation test was preferred and is still used to some extent. Many workers, however, have shifted to the more direct precipitation or flocculation test, and in many cases when a Wassermann reaction is referred to, what is actually meant is some modification of the original Sachs-Georgi precipitation test (13) (24) (26).

Complement fixation has also been utilized in the diagnosis of other diseases and in the detection of small amounts of blood. It has the advantage that it makes visible in a dramatic fashion—by the lysis or non-lysis of red cells—a phenomenon which otherwise is not very noticeable and which involves fairly small amounts of material.

DETRIMENTAL ANTIBODY RESPONSE

The production of antibodies may be considered part of the phenomenon of homeostasis, or, in Cannon's phrase, "the Wisdom of the Body" (11). But the homeostatic mechanisms of the body can go wrong and end up by being unhomeostatic, a phenomenon which Richards (34) has called *hyperexis*, or "the Stupidity of the Body". The body is a marvellous machine, but like any machine, it is capable of functioning at an unwelcome time or in an undesirable manner, as when a machine for sewing on buttons suddenly sews a button on your finger.

As an example of this general phenomenon, let us consider what happens when a foreign protein such as egg albumin is injected into the body. This is a contingency which nature, in the form of the forces of evolution, did not allow for, and it is not surprising that the body responds to the foreign antigen by the production of a perfectly useless antibody which will precipitate egg albumin *in vitro*. This antibody is useless because egg albumin is not toxic, and the body needs no defense against it.

Another example is furnished when the blood of one species is transfused into the circulation of another. Antibodies to the foreign plasma proteins and red cell constituents are produced by the recipient. Again this does no good, for the bloods of other species, unless they are very remote in the taxonomic scale, are not toxic. Even the introduction of blood from another individual of the same species may stimulate the production of antibodies directed towards the foreign red cells.

Not only does the production of antibodies in such situations do no good, it can lead to serious harm, as when an individual is given repeated transfusions of apparently compatible blood but eventually, as the result of his production of antibodies to red cell antigens present in the donor but not the recipient, responds to a new transfusion with a severe or fatal reaction. Another example of the misdirection of the antibody-forming mechanism is furnished by the stimulation of a pregnant woman by antigens from the fetus, leading to the production of antibodies which diffuse back into the fetal circulation and produce the disease *erythroblastosis fetalis* (pages 203, 436).

Evidence is accumulating that under certain circumstances some individuals, who presumably are particularly prone to produce antibodies, may actually form antibodies to certain antigens of their own tissues. This is thought to be a factor in the etiology of certain obscure diseases such as acquired hemolytic anemia.

Allergy

One of the commonest manifestations of the undesirable or overtime functioning of the antibody-forming mechanism in man is provided by the phenomena of hypersensitivity as in allergy. Allergy is a state of exaggerated response to common substances in the environment. Such responses lead to asthma, hay fever, and hives.

A typical example of the allergic diseases is the familiar seasonal paroxysmal attacks of coryza known as hay fever. Here the antibodies have been produced in response to antigens in the air. These are often mold spores or the pollens of plants. The patient finds that even slight exposure to the inciting agent results in sneezing, running of the nose, watering of the eyes, itching, and other symptoms. This is the result of the union of the antibodies he has developed with antigens in the material he inhales.

In some animals, and occasionally in man, fatal consequences follow the sudden union of antibodies with sensitizing antigens. We have already mentioned transfusion reactions. A more typical phenomenon in animals is anaphylaxis.

Anaphylactic shock may occur when certain animals (e.g., the guinea pig) receive a second injection of antigen 10 to 12 days after "active sensitization" with the same antigen, or 1 to 2 days after "passive sensitization" with the specific antibody. The second, or "shocking", dose must be considerably larger than the minute amounts needed for active sensitization, and is more effective if given suddenly by the intravenous route. Such anaphylaxis is a manifestation of the union of antigen with antibody *in vivo*. Sublethal shocking doses are followed by a long period—days or weeks—in which the animal is refractory to further shocks. Uterine or intestinal strips of sensitized animals, suspended in Ringer's solution, contract when antigen is added to the bath. Manifestations of anaphylaxis vary in different animal species. The guinea pig is subject to asphyxia, resulting from bronchial constriction. Dogs typically vomit and become paralyzed, their arterial pressure falls and the heart beat is accelerated. These symptoms appear to result chiefly from vasodilatation. Rabbits dying of anaphylaxis show constriction of the pulmonary arterioles, dilatation of the right heart, with acute heart failure. The *Arthus phenomenon* is local death of tissue at the site of injection of antigen into an immunized animal.

THERAPEUTIC USE OF ANTIBODIES

The most dangerous feature of certain infections, such as diphtheria and tetanus, is the toxin produced by the invading organism. Early attempts were made to supplement the body's natural defenses by the injection of antitoxic antibodies produced in animals, usually horses. This was attended by remarkably successful cures of cases of diphtheria which would otherwise have been fatal, and served to establish serum therapy firmly. Antitoxin to the toxin of tetanus was also soon introduced, and found to be of great value in preventing the development of tetanus following injuries of the sort which introduce dirt containing tetanus spores into the tissues.

Administration of antibodies other than antitoxins is also of great importance. An example of this is the use of human anti-measles antibody, usually obtained in the form of purified and concentrated gamma globulins from human plasma, in the prophylaxis of measles.

The therapeutic use of antibodies has been greatly decreased in recent years by two main causes. One of these is the use of chemotherapeutic agents such as antibiotics (see page 865); the other is the increasing use of active immunization.

IMMUNIZATION

The administration of foreign antibodies in the form of serum from another species is not without its drawbacks. For instance, the foreign proteins often produce allergic manifestations in the patient. Except when emergency exists, it is better if the patient is stimulated to produce his own antibodies. This not only has the advantage that the antibodies are not foreign proteins capable of leading to undesirable reactions in the patient, but also the patient equips himself with a certain amount of circulating protective antibody, and, which is much more important, with the means of producing more. Consequently, instead of waiting until a child gets diphtheria and then giving him horse antitoxin, it is much better to cause him to start producing anti-diphtheria antibodies on his own. The process of stimulating the production of antibodies to a disease is called active immunization.

A patient can be immunized by exposing him to the antigens of the infectious organism in one of a number of ways. One is to give him the disease, or a modified form of it. This is one of the most effective methods, but naturally not without risk. The two immunizations which are probably our most effective, those against smallpox and yellow fever, are examples of the successful use of a modified disease-producing agent.

Other ways of immunization include the injection of gradually increasing doses of a toxin produced by the micro-organism, the injection of a neutral mixture of such a toxin with a foreign antibody, the injection of detoxified toxin, and the injection of whole micro-organisms killed by heat or other agency. Absolute immunity is not always produced by these techniques, but they have been found to be reasonably successful in diseases such as diphtheria, typhoid, whooping cough, and tetanus.

The detoxication of toxins leads to modified proteins called toxoids, which are useful because they are still antigenic although they have lost their toxicity. This suggests again that the chemical groupings responsible for antigenicity are different from those conferring toxicity The method of producing toxoids which is most used involves incubating them with small quantities of formaldehyde. In addition to the reaction with the free amino groups (page 18), other more complicated reactions take place (15).

TOXINS

Some micro-organisms produce actively poisonous substances. Examples of this are the toxic protein produced by the diphtheria bacillus and that produced by *Clostridium botulinum*. It is customary to divide bacterial toxins into two classes, exotoxins and endotoxins. The exotoxins, typified by diphtheria toxin, are given off by the body of the bacterium during its growth and will remain in the culture medium if the organism is removed. The endotoxins appear to be intracellular constituents of the bacterial cell which are not freed during its lifetime. Some of them are probably identical with the protein-carbohydrate-lipid complexes which have been called Boivin antigens.

Some toxins are enzymes and their toxicity is due to their enzyme action. The mode of action of others is obscure. It has been suggested that diphtheria toxin acts by competitively inhibiting the synthesis of cytochrome b.

No reason is known why organisms produce toxins, because in many cases this seems to gain them no advantage, and in fact, if it brings about the death of the host this results sooner or later in the death of the invading micro-organisms or their immediate descendants.

The tubercle bacillus produces a soluble substance, *tuberculin*, which is not generally classed as a toxin because it seems to have no toxic effects on animals not previously exposed. However, it does have toxic effect on animals which have or have had tuberculosis and also on some of their cells when they are tested *in vitro*. Tuberculin, as isolated by growing the tubercle bacillus, contains three main components, a protein, a polysaccharide, and a nucleic acid.

Certain pathogenic bacteria contain a spreading factor which has been identified with the enzyme hyaluronidase (see page 144) which causes depolymerization of the hyaluronic acid which forms the ground substance of the skin and other tissues.

Although not toxins, other components of micro-organisms may have harmful effects during infection. The polysaccharides of pneumococcus are non-toxic, but since they combine with the special antibody which might otherwise cause the removal of the micro-organisms from the blood and thus protect against this disease, they make it easier for the pneumococci to gain a foothold and carry on their invasion of the body.

The structures of these polysaccharides have been studied; that of the Type 3 pneumococcus is the best understood. It is made up of aldobionic acid units which are composed of one molecule of glucuronic acid and one of glucose united by a glucoside link involving the aldehyde group of the glucuronic acid and carbon 4 of the glucose. This gives the compound which is known as cellobiuronic acid. Some of the other pneumococcus polysaccharides also contain cellobiuronic acid but may contain glucose or other carbohydrates.

VIRUSES

The viruses have certain characteristics in common which justify our treating them as a group. One of the most important is that they are capable of reproducing only inside of the cell of the susceptible host. Other characteristics are the frequent presence in the infected cells of *inclusion bodies*, the appearance of inflammation, mainly as a secondary phenomenon, and the proliferation or degeneration or both of the cells which are affected. At least two viruses are known which cause proliferation to the extent of producing tumors. There are Shope's papilloma and Rous's sarcoma (see Chapter 10).

Originally viruses were defined by their ability to go through pores in filters which would hold back bacteria, and were therefore called filtrable viruses, and also by the fact that unlike bacteria they could not be seen under the microscope. However, it is now known that they vary very much in size and that the vaccinia virus, for example, can be seen with microscopes using visible light. Even the smaller virus particles can be seen with the aid of the electron microscope.

Chemical Nature of Viruses

A study of the chemical nature of viruses has revealed that they vary a great deal in chemical composition. One which has been studied extensively, although it is of no importance in human disease, is the tobacco mosaic virus which causes a characteristic disease of the tobacco plant. This has been isolated in crystalline form. The smallest particles which were actively infective have not been found to be smaller than about 15 x 280 millimicrons. The tobacco mosaic virus appears to be composed entirely of protein and nucleic acid.

It is not known exactly how the virus gets inside the cells of the host, which it must do in order to produce an infection; but it is known that many viruses have the power of attaching themselves strongly to the surface of various types of cells, including erythrocytes. It is possible that this is the first stage in a process which later leads to a penetration of the cell and a proliferation of the virus inside of the cell. The presence of nucleoprotein in viruses, and the fact that some viruses seem to consist almost entirely of nucleoprotein, has suggested that there may be some analogy between viruses and genes. It will be noticed that both have the power of self-duplication inside of the cell and not outside of it.

Immunity to Viruses

Another characteristic associated with virus diseases is that a single attack of the disease confers an immunity which often lasts for the lifetime of the individual. However, there are a few notable exceptions to this. The influenza virus confers a very short immunity which lasts only one to three years according to Burnet (10), and the common cold is notorious for conferring an extremely brief immunity, if indeed any at all. However, these are exceptions which merely serve to point up the general rule, and all readers will be familiar with the immunity following virus diseases such as measles, smallpox, and chicken pox.

In the case of immunity to viruses, other factors in addition to antibodies may be operating. Although the evidence on this point is still controversial, one factor may be an increased resistance of the tissues themselves to the entry of the virus. Also, there is some evidence that at least in some virus diseases, some living virus remains present in the body for a long time, possibly many years, after recovery. The presence of this virus may stimulate the continuous production of sufficient antibody to keep the virus inactive and localized and to prevent reinfection.

ANIMAL PARASITES

Animal parasites, or their eggs, may be ingested directly with food or may be transmitted by an intermediate host (the mosquito, for example, transmits the malarial parasite). In still other cases intricate life cycles have been evolved, as in *Schistosoma mansoni*, whereby a free swimming form of the parasite is produced from the intermediate host and finds its way into water in which the human victim is exposed to it. There it reaches the host, penetrates the skin by the action of hyaluronidase, and by active motion eventually reaches whatever part of the body it finds most suitable for its development.

Animal parasites contain antigenic proteins and polysaccharides. Some parasites possess anti-enzymes or other protective devices to keep them from being destroyed by the enzymes of the host. As a general rule an attack by animal parasites does not confer immunity as effective as that conferred by bacterial or viral infections. Superinfection or reinfection is often possible, as for example in malaria.

Immunity to Animal Parasites

Circulating antibodies have often been detected in cases of parasitic infestation, although it has been more difficult to demonstrate them in the case of parasites living in the alimentary canal—possibly because less intact antigen from these parasites ever reaches the circulation to stimulate the production of antibodies.

During World War II attempts were made to work out a method of vaccinating against malaria, and some success was achieved in monkeys and birds but the mixtures used were too toxic to be used in human beings. Consequently, at the present time our main defense against malaria is the use of chemotherapeutic agents.

In the case of some parasites such as *Trichinella spiralis* it has been definitely demonstrated that infected animals have a considerable degree of resistance to reinfection. In other diseases such as schistosomiasis, the resistance to repeated infections seems to be very slight, and there is evidence that persons completely cured of the infection can be readily reinfected.

CHEMOTHERAPY AND METABOLITE ANTAGONISM

The clinical use of antibodies was discussed above. Here we may attempt to give a brief outline of chemotherapy, by which we mean the use of chemical substances other than antibodies to kill or inhibit specific pathogenic microorganisms or parasites within the body of the host, as contrasted to the use of tonics, diet, and so forth, merely to strengthen the body in general.

Chemotherapeutic agents are conventionally divided into two classes: (a) synthetic therapeutic agents; and (b) antibiotics. The distinction between these is not absolutely sharp, but all, or nearly all, of the agents in class (a) can be synthesized in the laboratory, although one of them, quinine, was not synthesized until 1951 and synthetic quinine is not available commercially. The agents of class (b) are the products of living organisms (usually, but not always, micro-organisms), but differ profoundly from antibodies in that they do not seem to be produced in specific response to an invading organism, and are much simpler compounds than antibodies, which are always protein in nature. Some of the antibiotics have been synthesized.

The chemotherapeutic agents which are now in use are mostly the result of deliberate search for substances which would kill or restrain invading organisms without doing serious damage to the host. Systematic searches of this sort began after it had been definitely established late in the nineteenth century that many diseases were the result of microbial invasion. The use of quinine in malaria and of mercury in syphilis was known before this search began. The first successful product of the search for chemotherapeutic agents was salvarsan, or 606, introduced by Ehrlich in 1909. Following this triumph of the application of chemistry to the therapy of infection, the search for chemotherapeutic agents was greatly intensified. However, except for a few substances which were mostly effective against tropical and trypanosomal diseases, disappointingly little additional progress was made until the 1930's, when the *sulfonamides* became established as effective antibacterial agents against streptococcal, gonococcal, meningococcal, and several other types of bacterial infection. The first of these drugs, now only of historical interest, was an organic dyestuff. Tréfouël and his collaborators soon demonstrated that the active group in this substance was the simple compound sulfanilamide (formula I).

$$H_2N—C_6H_4—SO_2NH_2$$

I. Sulfanilamide

Clinical use of sulfanilamide and its derivatives having substituents on the amide nitrogen led to the observation that their antibacterial action was inhibited by cell or tissue extracts or their hydrolysates, as for example by pus. Woods (38) studied yeast hydrolysates which inhibited the action of sulfanilamide, and obtained strong presumptive evidence that the inhibitor was *p*-aminobenzoic acid (formula II), which had not previously been

$$H_2N—C_6H_4—COOH$$

II. *p*-Aminobenzoic acid

$$NH_2CH_2—C_6H_4—SO_2NH_2$$

III. Homosulfanilamide

known to play any role in the biochemistry of living organisms. A surprising feature of the observations was the very great effectiveness of the inhibitor. One mol of *p*-aminobenzoic acid was found to inhibit the antibacterial effect of 5,000 or more mols of sulfanilamide. This suggested that *p*-aminobenzoic acid played some vital role in the life processes of the bacteria, and Woods proposed as a working hypothesis to explain the action of the sulfonamide drugs that (a) *p*-aminobenzoic acid or a closely related substance is essential for the growth of the sulfonamide-sensitive bacteria, and (b) sulfonamides, by virtue of their structural similarity to *p*-aminobenzoic acid combine with the catalytically active sites of the enzyme system which normally utilizes *p*-aminobenzoic acid in the bacterial cells, and thus competes with *p*-aminobenzoic acid for positions on the enzyme molecules.

This working hypothesis has not turned out to be a full explanation of all the antibacterial activities of the sulfonamide drugs, as is sharply indicated by the fact that *homosulfanilamide* (formula III) is an effective antibacterial drug, but is not inhibited by *p*-aminobenzoic acid. Wood's work, however, has stimulated many investigations into the general subject of metabolite antagonism, by which we mean the prevention of the normal utilization of a substance by the presence of an antagonist, which is usually a substance showing close similarity of structure.

Lampen and Jones related the need for *p*-aminobenzoic acid to its utilization in the formation of folic acid (25). Sulfonamides competitively prevent the incorporation of *p*-aminobenzoic acid into the folic acid molecule. Most organisms, possibly all organisms, require folic acid. Some require it ready made; these organisms are not sensitive to sulfonamides, since they do not build folic acid. The growth of such organisms can be antagonized by certain structural analogues of folic acid, such as aminopterin. The organisms which are characteristically sensitive to the sulfonamides are those which utilize *p*-aminobenzoic acid in the synthesis of the requisite folic acid. The inhibition of growth produced in such organisms by sulfonamides can be released either *competitively* by *p*-aminobenzoic acid, or *non-competitively* by folic acid.

The sulfonamides are not identical with the natural substrate of the enzyme and do not combine as firmly with it as does *p*-aminobenzoic acid itself. The dissociation constant of the sulfonamide-enzyme complex is higher. We may formulate this mathematically. In any competitive inhibition

$$E + S \rightleftharpoons ES \rightarrow E + P$$

$$E + I \rightleftharpoons EI$$

where E is the enzyme involved, S is its normal substrate, I is the competitive inhibitor, usually a structural analogue, and P is the product. By the mass action law, the dissociation constant of the enzyme-substrate complex is K_S

$$K_S = \frac{[E][S]}{[ES]}$$

and the dissociation constant of the enzyme-inhibitor complex is K_I

$$K_I = \frac{[E][I]}{[EI]}$$

Dividing K_I by K_S, we can get an inhibition constant, K

$$K = \frac{[I][ES]}{[S][EI]}$$

At 50 per cent inhibition

$$[ES] = [EI]$$

and

$$K = \frac{[I]}{[S]} \quad [1]$$

and

$$[I] = K[S]$$

Thus the effectiveness of a competitive inhibitor is determined by the ratio of the two dissociation constants, and the amount required to inhibit to any stated degree depends upon this constant and the amount of normal substrate already present. It is the ratio of substrate and inhibitor concentrations which is important, not their absolute amounts.

In non-competitive inhibition it is assumed that the inhibitor combines with catalytically inactive portions of the enzyme and somehow renders it inactive. We then have

$$ES + I \rightleftharpoons ESI \text{ (inactive)}$$

We can write the dissociation constant of this substrate-enzyme-inhibitor complex as

$$\frac{[ES][I]}{[ESI]} = K_I$$

At 50 per cent inhibition

$$[ES] = [ESI]$$

and

$$[I] = K_I \quad [2]$$

Comparing this with equation [1], we see that in non-competitive inhibition the effectiveness of the inhibitor depends upon the amount present and the dissociation constant of the substrate-enzyme-inhibitor complex. It is independent of substrate concentration.

If enough folic acid is used to supply the full growth requirement, there will be no inhibition whatever of *Streptococcus faecalis* by sulfonamide. There are, however, other organisms which are sensitive to sulfonamides, but their sensitivity is not overcome by folic acid. Hence, the mechanism described by Lampen and Jones does not explain all actions of sulfonamides. Most sulfonamide-sensitive pathogens fall into this unsatisfactory group.

While keeping within the bounds of his original hypothesis, Woods (39) has offered several possible explanations for the failure of folic acid to release all sulfonamide-sensitive organisms from sulfonamide inhibition: (a) some organisms may not be able to assimilate preformed folic acid; (b) the final substance normally synthesized may be different from the synthetic folic acid used experimentally—for example, certain organisms might produce the citrovorum factor (see page 812) without folic acid being even an intermediate; (c) *p*-aminobenzoic acid may have essential functions in some organisms not involving the synthesis of folic acid. This last point is borne out by the fact that certain nucleic acid derivatives, e.g., thymine, adenine, guanine, and xanthine, and a number of amino acids including methionine, may with some organisms replace *p*-aminobenzoic acid either as an essential metabolite or in releasing the organism from sulfonamide inhibition. It seems reasonable to consider these substances as products of the utilization of *p*-aminobenzoic acid by the bacterial cells, and that their formation is inhibited by sulfonamides.

Competitive inhibition of the growth of micro-organisms need not involve components of enzyme systems. Competitive antagonists to intermediates of carbohydrate metabolism, purine metabolism and to amino acids, have been prepared and demonstrated. For example, two effective inhibitors of the growth of *Escherichia coli* are thienylalanine (formula IV) and furylalanine (formula V). These substances are obviously related structurally

$$\text{C}_4\text{H}_3\text{S}\text{—CH}_2\text{—CH(NH}_2\text{)—COOH}$$

IV. Thienylalanine

$$\text{C}_4\text{H}_3\text{O}\text{—CH}_2\text{—CH(NH}_2\text{)—COOH}$$

V. Furylalanine

to phenylalanine, and inhibition of bacterial growth by either of these substances is released by phenylalanine. A constant ratio of inhibitor to phenylalanine is required to obtain 50 per cent inhibition, indicating that the inhibition is competitive (12). Cyclopentaneglycine is similar structurally to isoleucine (formula VI) and inhibits the growth of *E. coli* by interfering with utilization of isoleucine (19). Similarly, 6-mercaptopurine acts as a competitive antagonist for hypoxanthine and can retard leukocyte formation in leukemia (see page 401).

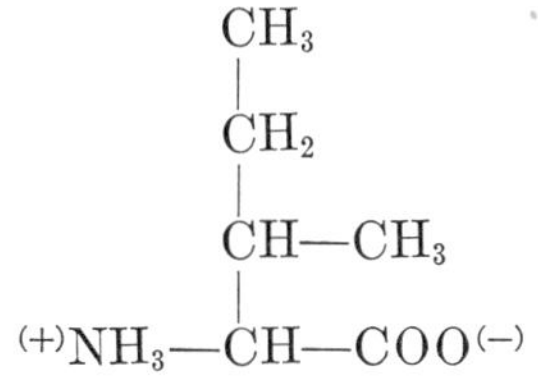

$$CH_3CH_2CH(CH_3)-CH(NH_3^{(+)})-COO^{(-)}$$

Isoleucine

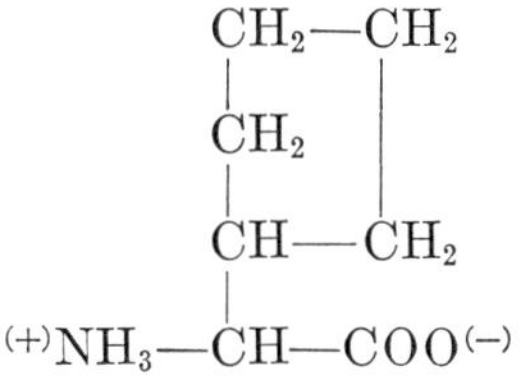

$$\begin{array}{l} CH_2-CH_2 \\ | \qquad\quad | \\ CH_2 \quad\ | \\ | \qquad\quad | \\ CH-CH_2 \\ | \\ {}^{(+)}NH_3-CH-COO^{(-)} \end{array}$$

Cyclopentaneglycine

VI. Amino acid antagonism

Similarly, several structural analogues have been synthesized which are inhibitory for individual members of the vitamin B complex (formula IX). Isonicotinic acid hydrazide (formula VII) is inhibitory to amine oxidase,

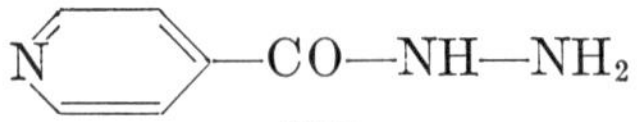

$$C_5H_4N-CO-NH-NH_2$$

VII

and 1-isonicotinyl-2-isopropylhydrazine (formula VIII) is about 10 times

$$C_5H_4N-CO-NH-NH-CH(CH_3)_2$$

VIII

as effective (30). These two compounds inhibit the diamine oxidase of mycobacteria, and have been employed in the chemotherapy of tuberculosis. An alternate concept of the action of isonicotinic acid hydrazide is that it chelates the iron of the cytochromes of the tubercle bacillus (11a).

ANTIBIOTICS

It was noted by early workers in bacteriology that certain organisms, such as the pyocyaneus bacillus (*Ps. aeruginosa*), would successfully invade and overgrow certain other microbial species *in vitro.* It could even be demonstrated that pyocyaneus would overgrow anthrax in the living body of an experimental animal. Such pitting of one living organism against another did not develop into a useful form of therapy, nor did the pyocyaneus bacillus ever become a source of useful antibiotics. Such antibacterial substances as pyocyaneus produces, and they are numerous, are too toxic to the animal body for effective use.

The earliest recorded use of antibiotics had nothing to do with the treatment of human disease, and took place without benefit of bacteriological science. Early in the 16th century, English brewers learned from their colleagues in the Low Countries that *hops,* when added to wort, resulted not only in a better flavor but also in a more dependable fermentation and

Food Factors	*Inhibitory Analogues*
Nicotinamide	Pyridine-3-sulfonamide
Thiamine	Pyrithiamine
Biotin	Desthiobiotin

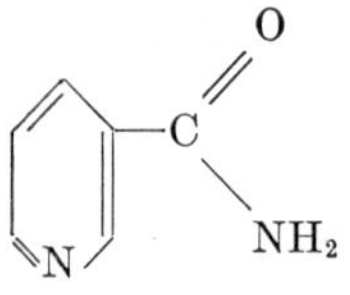

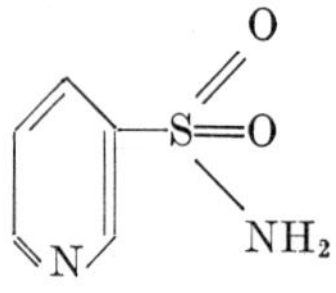

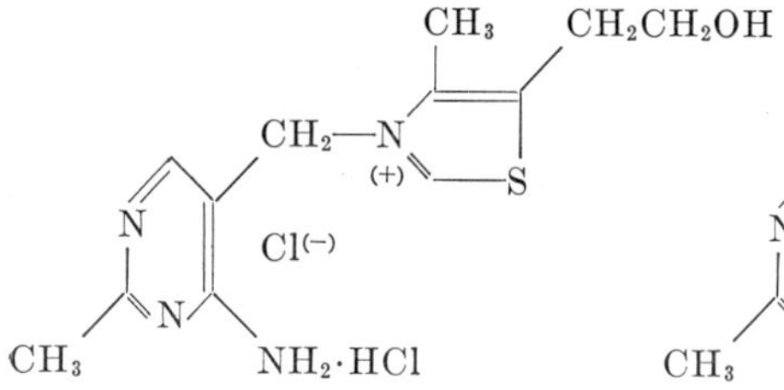

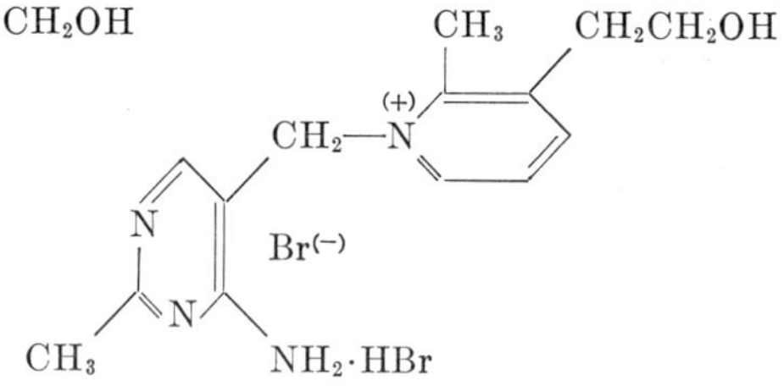

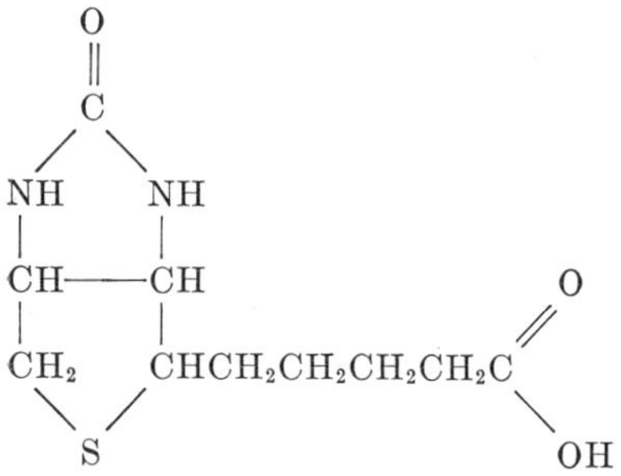

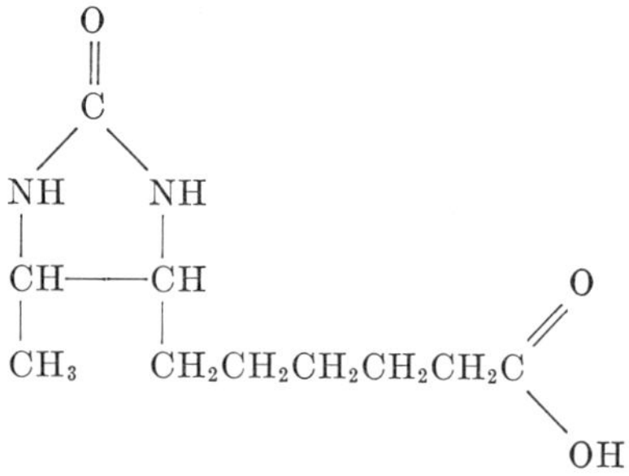

IX

better keeping qualities. "The continuance of the drink is alwaie determined after the quantitie of hops, so that being well hopped it lasteth longer," wrote Harrison in his late 16th century *Description of England.* The resin acids of hops are quite specifically inhibitory to the growth of several species of gram-positive bacteria which, in the absence of hops, tend to multiply in competition with yeast in the fermentation stage of brewing (2). The mechanism of the inhibition has not been explained.

Penicillin was the first truly successful medical antibiotic, and is still widely used. It was discovered in 1929 by Fleming, who observed the death of staphylococcal colonies near a contaminating colony of *Penicillium notatum,* and who also observed the antibacterial action of culture filtrates of the same mold.

There are actually several penicillins, organic acids for which we may write a general formula (formula X) in which R stands for a benzyl group

```
O                     S     CH3
 \\                  / \   /
   C—NH—CH—CH       C
  /        |   |      | \
R          |   |      |   CH3  O
           |   |      |        //
           C———N——————CH—C
          //              \
         O                  OH
```

X. Penicillin—general formula

in the penicillin most commonly used (Penicillin G). Other penicillins obtainable from cultures of *P. notatum*, or of *P. chrysogenum*, may have in place of the benzyl group *p*-hydroxybenzyl, Δ^2-pentenyl, *n*-amyl, or *n*-heptyl.

Penicillins inhibit, and in higher concentrations kill, growing cells of susceptible strains. Action on non-growing or resting cells is insignificant. Cells of *S. aureus* take up penicillin and bind it firmly. Free nucleotide begins to increase in growing cells immediately after the binding of penicillin. In tests made using radioactive penicillin, it was found not to be removable from the cells either by washing or by equilibration with non-radioactive penicillin (27). Penicillin does not penetrate into yeast cells, which are not affected by its presence.

Gram-positive bacteria as a group are more susceptible to penicillins than gram-negative bacteria. Staphylococci, which are gram-positive, can not synthesize glutamic acid but require it from an external source. Penicillins block the uptake of glutamic acid by susceptible staphylococcal strains. As these same strains became penicillin-resistant during cultivation in increasing concentrations of penicillin, they lost much of their ability to assimilate glutamic acid but were able to synthesize it and other necessary amino acids, in this respect becoming more like gram-negative bacteria (16).

The chemical antibacterial actions of the penicillins are accompanied by morphological changes in the affected bacteria. These changes are in some ways similar to those caused by ribonuclease and by cobra venom. Pratt and Dufrenoy (33) summarized and discussed the chemical and morphological evidence and suggest that the action of penicillin is due to excessive dehydrogenation, especially of the sulfhydryl groups. Since —SH groups seem to be essential to the activity of many enzymes, such a general theory might explain the multiple metabolic disturbances induced in susceptible cells by penicillin. The theory does not satisfactorily explain, however, the lack of effect of penicillin upon many organisms, such as protozoa, and upon isolated enzyme systems known to involve sulfhydryl enzymes.

The development of resistant bacterial mutants in penicillin-containing media should not be confused with the survival of "persisters", which are thought to survive because they happen to be in a resting state at the time of their exposure to penicillin. They are not resistant mutants since they yield progeny as sensitive as the original culture.

The ability of certain microbial strains to produce *penicillinase* constitutes a mechanism of natural resistance. Most naturally-resistant strains of staphylococci produce penicillinase, but most strains which have acquired resistance *in vitro* do not.

The direct toxic action of penicillin is negligible aside from the irritating effect when applied directly to central nervous system. Sensitivity reactions may be severe or fatal. Administration of penicillin is usually in the form of a calcium salt by mouth, or a sodium salt parenterally. No attempts will be made in this book to present the clinical and bacteriological indications or the dosage of penicillin or other antibacterial agents. For such information the reader should consult the current literature in the field of internal medicine.

Streptomycin (formula XI) is an organic base, produced by *Streptomyces griseus* and several more or less related organisms. Although more toxic

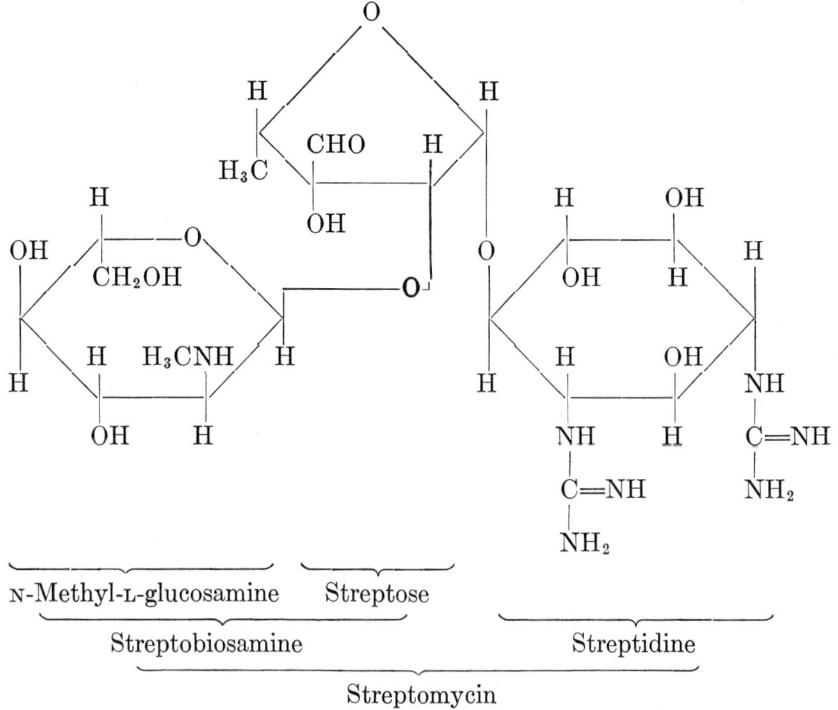

XI

than penicillin, it is useful against many organisms, including the tubercle bacillus, which are unaffected by penicillin. The discovery of streptomycin was the result of a deliberate survey of many micro-organisms for a product effective against gram-negative pathogenic bacteria, which grow freely in the presence of penicillin. This study was carried on by Selman Waksman and his associates.

In *E. coli*, streptomycin specifically inhibits an oxaloacetate-pyruvate condensation, which leads to the oxidation of these substances, but not by the usual path of citric acid formation (37). Variants of *E. coli* which are either resistant to streptomycin or dependent upon streptomycin do not have the ability to bring about appreciable oxidation of oxaloacetic and pyruvic acids. An organism which can grow in the presence of streptomycin or which requires streptomycin for its growth has developed other reactions, the nature of which is not known, which allow the cell to dispense with this particular condensation. Strains resistant or susceptible to streptomycin will form citric acid, but this does not appear to be the major metabolic pathway in *E. coli*. With *Shigella sonnei* utilization of carbohydrates and consumption of oxygen are also both inhibited along with inhibition of multiplication by streptomycin.

Neomycin is a water-soluble, thermostable, basic substance obtained from filtrates of *Streptomyces fradiae*. The main interest in the substance is its activity against streptomycin-resistant microbes, including the resistant strains of *M. tuberculosis*.

Chloramphenicol (formula XII) is now produced by a synthetic process

```
              NO2
               |
          (benzene ring)
               |
        HO—C—H          O   H
               |            //  /
        H—C—NH—C—C—Cl
               |                  \
            CH2OH               Cl
```

XII. Chloramphenicol

starting with *p*-nitrobromacetophenone, but was originally identified as a naturally-occurring antibiotic elaborated by a species of streptomyces. It is the first nitrobenzene compound to be found in living cells.

In vitro chloramphenicol acts as a non-competitive inhibitor of phenylalanine (40). Woolley cites this as another instance of the general rule that non-competitive antimetabolites are more effective as antibiotics than are competitive antimetabolites, since their activity is not abolished by a mere increase in the concentration of the normal specific metabolite. The effect of chloramphenicol is to inhibit the net synthesis of bacterial protein (42). As a result, probably, proteins necessary for the multiplication of the bacterial cell fail to be formed (41). A different and apparently unrelated action of chloramphenicol is the inhibition, at therapeutic concentrations, of bacterial esterases (35). At somewhat lower concentrations esterase activity is accelerated. Neither this nor any of the other specific inhibitory actions of chloramphenicol has been found to be competitive.

The broad-spectrum antibiotics, so called because of their effectiveness against a large number of pathogenic microbial species, include chloramphenicol, *tetracycline* (formula XIII), and naturally-occurring tetracycline derivatives (formula XIV), along with *erythromycin* and a number of others still under investigation. Erythromycin has a distinctive structure quite unlike other known antibiotics. Its molecule is made up of a 21-carbon lactone ring called *erythronolide*, attached to which are various substituents including a methylpentose and an amino methylpentose.

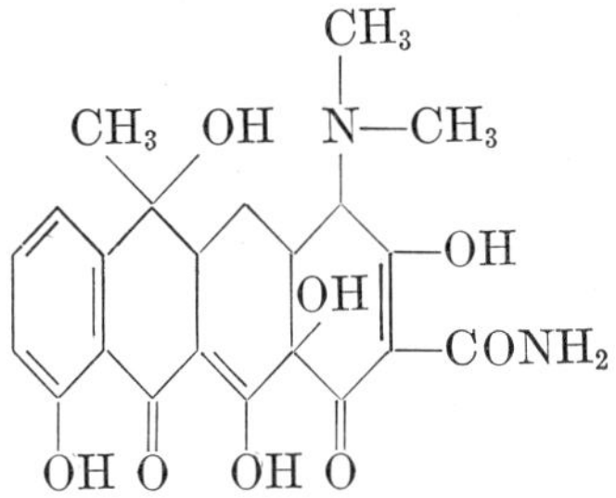

XIII. Tetracycline (Achromycin)

A group of antibiotics which is puzzling is the *polypeptide* antibiotics, which lack organic groupings recognizable as toxic or antagonistic to known metabolic processes. It is true that some polypeptide antibiotics contain

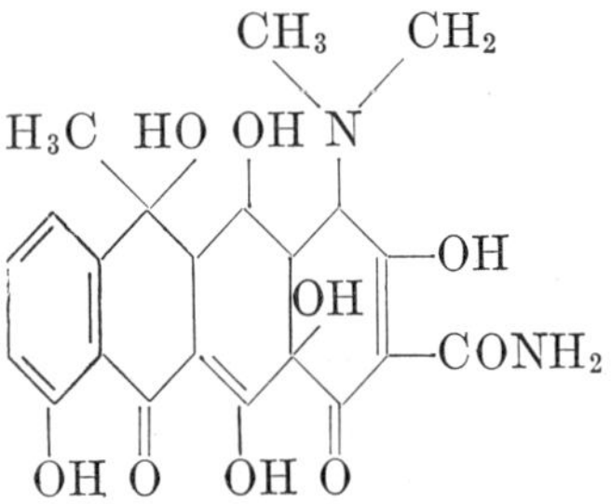

Oxytetracycline (Terramycin)

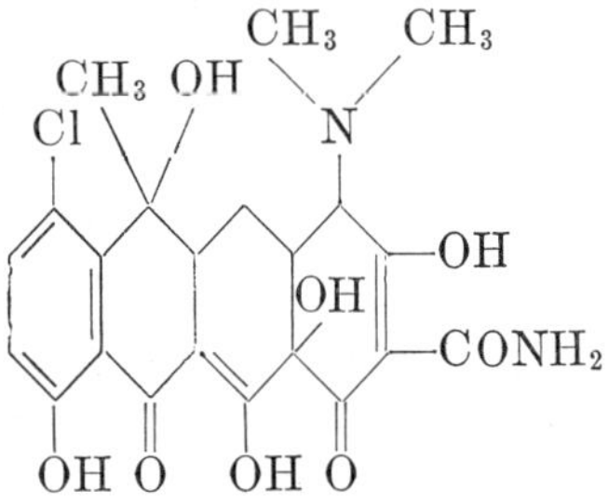

Chlorotetracycline (Aureomycin)

XIV

amino acids of the "unnatural" D series, but these amino acids themselves are not effective antibiotics and it is doubtful if their presence explains the action of the polypeptides. Among these compounds are *tyrocidine*, *gramicidin*, *bacitracin* and *polymixin*

In general, antibiotics produce their effects through the inhibition of some enzyme system vital to life, or, at least, growth (36). Thus, penicillin inhibits some early stage of RNA synthesis; streptomycin interferes with the combination of oxaloacetic acid and pyruvic acid, thus stalling the tricarboxylic acid cycle; chloramphenicol inhibits esterases; and the tetracyclines inhibit oxidative phosphorylation.

Man's struggle with infectious micro-organisms, viewed from a somewhat more detached point of view than that of a patient, is but one of the acts in the long drama of evolution. We saw at the beginning of this chapter how pathogenic organisms had adapted themselves to invasion of the living body and to the use of its resources for their own ends, and how the body countered with the mobilization of phagocytes and the production of specific antibodies. The introduction of chemotherapeutic agents and antibiotics merely marked the opening of another scene in the struggle. For a time it seemed that man's ingenuity threatened to drive pathogenic bacteria off the stage altogether. But our little enemies do not lack a sort of resourcefulness themselves. Intelligence is but one of the weapons, and a late one, in the struggle for existence. Micro-organisms possess instead the resources of plasticity and tremendous fecundity. They almost at once started to respond to the menace of the sulfonamides by developing resistance to these drugs. It has been stated that if no treatment other than sulfonamides had been developed for gonorrhea, chemotherapy of this disease would already be a thing of the past, so many resistant strains have developed.

Penicillin-resistant organisms are being isolated with increasing frequency (14), and the tubercle bacillus has shown a rapid development of resistance to streptomycin.

In addition to the development of resistance, there is the increasing frequency of mycotic infections in patients treated with antibiotics. Organisms such as *Monilia candida*, which are more closely related to the micro-organisms from which the antibiotics are obtained, are not inhibited by them, and in patients freed from other bacterial infections, fungi find fertile soil on which to multiply, being no longer held in check by the competition of more dominant micro-organisms (23). A greater menace is the development of the potentially highly virulent resistant strains of coagulase-positive staphylococci and hemolytic streptococci. It may be only temporarily that we have the upper hand in the age-old battle. It is certainly too soon to throw away any of our weapons, especially the subtle and flexible mechanisms of natural and acquired resistance of the human body. There is no prospect of driving the enemy from the field, or even of seriously diminishing his numbers. It is rather that we must prepare ourselves for a permanent state of all-out war, with no holds barred on either side.

REFERENCES

1. Abramson, H. A., *et al.* The specificity of the second stage of bacterial agglutination. J. Bact., **50:** 15–22, 1945.
2. Bishop, L. R. The resins of hops as antibiotics. Symp. Soc. Exper. Biol. III. *Selective Toxicity and Antibiotics*, pp. 101–104. New York, Academic Press, 1949.
3. Bjørneboe, M. The specific protein in rabbit antipneumococcal serum and its relation to the increase of serum protein during immunization. J. Immunol., **37:** 201–206, 1939.
4. Bordet, J. *Traité de l'Immunité.* Paris, Masson & Cie, 1920.
5. Boyd, W. C. *Fundamentals of Immunology.* 3rd edit. New York, Interscience, 1956.
6. Boyd, W. C., and Bernard, H. Quantitative changes in antibodies and globulin fractions of rabbits injected with several antigens. J. Immunol., **33:** 111–122, 1937.
7. Boyd, W. C., *et al.* The heat of an antibody-antigen reaction. J. Biol. Chem., **139:** 787–794, 1941.
8. Boyd, W. C., and Purnell, M. A. The essential difference between the two optimum proportions ratios. J. Exper. Med., **80:** 289–298, 1944.
9. Boyd, W. C., and Reguera, R. M. Hemagglutinating substances in various plants. J. Immunol., **62:** 333–339, 1949.
10. Burnet, F. M. *Virus as Organism.* Cambridge, Mass., Harvard University Press.
11. Cannon, W. B. *The Wisdom of the Body.* Philadelphia, W. W. Norton Co., 1939.

11a. Coleman, C. M. A proposed mode of action for isoniazid in the tubercle bacillus and other biological systems. Am. Rev. Tuberc., **69:** 1062–1063, 1954.

12. Dittmer, K. The structural bases of some amino acid antagonists and their microbiological properties. Ann. New York Acad. Sci., **52:** 1274–1301, 1950.
13. Eagle, H. *The Laboratory Diagnosis of Syphilis.* St. Louis, C. V. Mosby, 1937.
14. Finland, M. The present status of antibiotics in bacterial infections. Bull. N. Y. Acad. Med., **27:** 199, 1951.
15. French, D., and Edsall, J. T. Reactions of formaldehyde with amino acids and proteins. Advances. Prot. Chem., **2:** 277–335, 1945.
16. Gale, E. F., and Rodwell, A. M. Amino acid metabolism of penicillin-resistant staphylococci. J. Bact., **55:** 161–167, 1948.
17. Gitlin, D. Low resistance to infection: relation to abnormalities in γ-globulin. Bull. N. Y. Acad. Med., **31:** 359–365, 1955.
18. Green, H., and Anker, H. S. Synthesis of antibody protein. Biochim. et Biophys. Acta, **13:** 365–373, 1954.
19. Harding, W. M., and Shive, W. Cyclopentaneglycine, an inhibitory analog of isoleucine. J. Biol. Chem., **206:** 401–410, 1954.

19a. Harris, T. N. and Harris, S. The genesis of antibodies. Am. J. Med., **20:** 114–132, 1956.

20. Haurowitz, F. *Chemistry and Biology of Proteins.* New York, Academic Press, 1950.
21. Heidelberger, M. Quantitative absolute methods in the study of antigen-antibody reactions. Bact. Rev., **3:** 49–95, 1939.

21a. HOOKER, S. B. AND BOYD, W. C. A test of the alternation ("lattice") hypothesis with divalent and trivalent haptens. J. Immunol., **42:** 419–433, 1941.
22. KABAT, E., AND MAYER, M. M. *Experimental Immunochemistry*. Springfield, Ill., Charles C Thomas, 1948.
23. KLIGMAN, A. M. Are fungus infections increasing as a result of antibiotic therapy? J. A. M. A., **149:** 979–983, 1952.
24. KAHN, R. L. *The Kahn Test*. Baltimore, The Williams & Wilkins Co., 1928.
25. LAMPEN, J. O., AND JONES, M. J. The antagonism of sulfonamide inhibition of certain lactobacilli and enterococci by pteroylglutamic acid and related compounds. J. Biol. Chem., **166:** 435–448, 1946.
26. LEVINSON, S. A. Flocculation reactions in syphilis. With special reference to the Meinicke and Sachs-Georgi reactions. Am. J. Syphilis, **5:** 414–438, 1921.
27. MAASS, E. A., AND JOHNSON, M. J.: Penicillin uptake by bacterial cells. J. Bact., **57:** 415–422, 1949.
28. MARRACK, J. R. The chemistry of antigens and antibodies. Med. Research Council, Brit. Special Rept. Series. 1938.
29. NEURATH, H. The apparent shape of protein molecules. J. Am. Chem. Soc., **61:** 1841–1844, 1939.
30. OWEN, C. A., JR., *et al.* Enzymology of tubercle bacillus and other mycobacteria. I. Influence of streptomycin and other basic substances on the diamine oxidase of various bacteria. J. Bact., **62:** 53–62, 1951.
31. PAULING, L. A theory of the structure and process of formation of antibodies. J. Am. Chem. Soc., **62:** 2643–2657, 1940.
31a. PAULING, L. AND PRESSMAN, D. The serological properties of simple substances. IX. Hapten inhibition of precipitation of antiserums homologous to the *o*-, *m*-, and *p*-azophenylarsonic acid groups. J. Am. Chem. Soc., **67:** 1003–1012, 1945.
32. PILLEMER, L. The properdin system. Trans. N. Y. Acad. Sci., **17:** 526–530, 1955.
33. PRATT, R., AND DUFRENOY, J. *Antibiotics*. 2nd ed. Philadelphia, Lippincott, 1953.
34. RICHARDS, D. W. Homeostasis *versus* hyperexis: or Saint George and the dragon. Sc. Monthly, **77:** 289–294, 1953.
35. SMITH, G. N., *et al.* Inhibition of bacterial esterases by chloramphenicol (chlormycetin). J. Bact., **58:** 803–809, 1949.
36. UMBREIT, W. W. Mechanisms of antibacterial action. Pharmacol. Rev., **5:** 275–284, 1953.
37. UMBREIT, W. W., *et al.* The action of streptomycin. V. The formation of citrate. J. Bact., **61:** 595–604, 1951.
38. WOODS, D. D. The relation of *p*-aminobenzoic acid to the mechanism of the action of sulphanilamide. Brit. J. Exper. Path., **21:** 74–90, 1940.
39. WOODS, D. D. Biochemical significance of the competition between *p*-aminobenzoic acid and the sulphonamides. Ann. N. Y. Acad. Sci., **52:** 1199–1211, 1950.
40. WOOLLEY, D. W. A study of non-competitive antagonism with chlormycetin and related analogues of phenylalanine. J. Biol. Chem., **185:** 293–305, 1950.
41. WOOLLEY, D. W. Chloramphenicol. Trans. N. Y. Acad. Sci., Ser. II, **15:** 17–18, 1952.
42. WYSS, O., *et al.* Symposium on the mode of action of antibiotics. Bact. Rev., **17:** 17–49, 1953.

APPENDIX

Isotopes

Perhaps no technique of investigation has been so unexpected at the time of its first development, so powerful a tool after development, and so revolutionary in its immediate applications as that of isotopic tracer research in biochemistry. Its particular efflorescence in these initial years of the atomic age would make any textbook on human metabolism incomplete that did not contain fairly extensive reference to this unparalleled device for probing the metabolic pathways of the human body. The understanding of tracer methodology must begin with a consideration of the atomic nucleus.

THE ATOMIC NUCLEUS

Nuclear Structure

The atom is composed of two parts, a compact central *nucleus* carrying unit positive charges of varying number, and a number of "planetary" *electrons*, each carrying a unit negative charge. In the neutral atom the number of electrons is equivalent to the number of positive charges in the nucleus so that the atom as a whole is uncharged. Ordinary chemical reactions involve the outermost electron shells of an atom (see Chapter 7), and the chemical properties of individual elements depend upon the number and configuration of these outer shells. Ions are atoms which have lost or gained one or more electrons, acquiring positive or negative charges, respectively.

The nucleus is a very small portion of the atom in terms of volume; the diameter of an atom is of the order of 10^{-8} centimeters, whereas the nucleus is about 10^{-13} cm. in diameter. The volume of the nucleus is thus only 10^{-15} that of the entire atom. Nevertheless, almost all the mass of the atom is concentrated in that tiny nucleus, since the electrons—the wave packets of which fill virtually all the atom—possess very little mass indeed. It is this fact that lies behind the frequently heard statement that an atom is mostly empty space. The simplest atom is hydrogen, one form of which possesses the lightest known nucleus and only a single planetary electron. It has 99.94 per cent of its mass in the nucleus. Mendelevium, whose nucleus is more than 250 times as massive as that of hydrogen and which has 101 electrons, has 99.98 per cent of its mass in the nucleus.

The atomic nucleus is, with one exception, a composite structure; it is composed of still smaller entities of two kinds, *protons* and *neutrons*. In terms of mass these are about equal. The proton, however, has a unit positive charge—that is, a charge capable of neutralizing exactly that of one electron; while the neutron is uncharged. With the exception of the simplest known nucleus, that of ordinary hydrogen, which consists of a single proton and is the only known non-composite nucleus, all atomic nuclei contain both protons and neutrons.

For a given element the number of protons in the nucleus is fixed and is equal to the *atomic number* of that element. The atomic number of an element is defined as the number of unit positive charges in the nuclei of its atoms, which depends of course upon the number of protons within the nucleus. The atomic number of an element may also be defined as the number of planetary electrons present in the neutral atom.

For any given element the number of neutrons in the nucleus may vary within narrow limits. A variation in the number of the neutrons does not affect the chemical properties of the element, since neutrons are uncharged and therefore do not by addition to or subtraction from the nucleus alter the positive charge on the nucleus.

The atomic weight of a particular nucleus (or *nuclide*) depends upon the sum of the number of protons and neutrons (together called *nucleons*). The mass of all atomic and subatomic particles is based upon the arbitrary assignment to the most common isotope of oxygen of a mass number of exactly 16, a value which is treated as a pure number. On this basis the neutron and the proton have mass numbers of very nearly 1. More exactly, the proton has a mass number of 1.00813, while that of the neutron is 1.00896, and other atomic particles likewise differ by a fraction of a per cent from integral mass numbers. These small differences, while rendering possible the atomic bomb, are of no importance in biochemistry and mass numbers will be for the large part treated as though they were exact whole numbers. The electron has a mass number of about 0.00054, so that in most cases no serious error is introduced if its mass number is considered as simply zero.

The mass number of a nucleus is usually symbolized as A, while the atomic number is referred to as Z. A little thought will show that the number of protons in a given nucleus is equal to Z, while that of the neutrons, to $A - Z$.

Isotopes

It has already been said that the neutrons in the nucleus of a given atom may vary in number. Although this variation does not affect the chemical nature of the atoms in any general way, it does introduce important differences in their physical properties. Isotopes are atomic species, the nuclei of which are composed of a fixed number of both protons and neutrons, while in an element, as has been said, only the number of protons within the nuclei are fixed. Most elements are found to exist in nature as mixtures of different isotopes.

This may perhaps be understood more clearly if a specific case is considered—that of hydrogen. Most neutral atoms of hydrogen consist of a single proton as the nucleus and a single electron balancing its charge. The mass number of such a hydrogen atom is 1 if we consider that of the proton to be 1 and that of the electron to be 0. In any sample of hydrogen which may be obtained by ordinary chemical means from any source 99.98 per cent of the atoms have the configuration just described. (Certain procedures are known to the physical chemist which alter this percentage, but none of these comes under the classification of "ordinary chemical means".) The remaining 0.02 per cent of the hydrogen atoms possesses nuclei consisting of two particles, one proton and one neutron. Since the nuclear charge is still 1, such an atom would still have but one electron in its neutral state and would still be, chemically, hydrogen. Its mass number, however, would be 2, double that of the more common variety of hydrogen. A sample of hydrogen consisting only of atoms of the heavier variety would be twice as dense under similar conditions of temperature and pressure as would ordinary hydrogen. Such "heavy hydrogen" has been prepared and its physical properties, such as melting point, boiling point, specific heat, and so on, have been found to be quite different from those of ordinary hydrogen. Thus, the boiling points in degrees absolute for these hydrogen isotopes are 20.38 for the lighter one and 23.6 for the heavier.

The difference in properties between these isotopes is unusually great considering that they are isotopes of a single element. This is because isotopes never vary in mass number by less than 1. In the case of hydrogen, the mass number is only 1 to start with, and even the smallest possible difference in isotope mass number represents a 100 per cent increase in mass. In the case of oxygen, most of whose atoms have a mass number of 16, an isotope with a mass number of 17 would be less than 7 per cent more massive. As atoms become still more complex the percentage differences between isotopes become continually smaller.

Such are the differences between the two isotopes of hydrogen that they are usually given different names and symbols. The lighter isotope is known by the common name *hydrogen*, or more rarely *protium*, while the heavier isotope is called *deuterium*, and is often symbolized as D. There is still a third isotope of hydrogen, very rare and including only one atom in ten million or so, which has a mass number of 3, since its nucleus is composed of one proton and two neutrons. This "super-heavy" hydrogen is called *tritium*, and is sometimes symbolized as T. The isotopes of no other chemical element are assigned separate names or symbols.

A list of naturally occurring isotopes of elements of biochemical interest with their nucleon contents and frequency of occurrence is contained in table APP-1, the data of which were obtained from Sullivan's chart (16).

Isotopic differences persist when atoms form part of molecules. Water, for instance, may contain protium, deuterium, or tritium atoms, so that the following compounds may be formed: H_2O, HDO, HTO, D_2O, DTO, T_2O. Since three stable isotopes of oxygen exist, with mass numbers of 16, 17, and 18, it can be seen that three series of such kinds of water exist making a total of 18 in all. The water most commonly found in nature is that consisting of two protiums and an oxygen-16, with a molecular mass number of 18. (The molecular weight of a compound would more properly refer to the average molecular mass numbers of the various isotopic combinations existing within a compound. In substances of biochemical interest the molecular weight is very close to the molecular mass number of the most common isotopic combination.) Although the most massive water molecule that exists is one containing two tritiums and an oxygen-18, with a molecular mass number of 24, it is D_2O (sometimes called deuterium oxide), with a mass number of 20, which is commonly referred to as *heavy water*. The difference in mass numbers between ordinary water and heavy water is a little over 11 per cent, which is great enough to result in very easily measured changes in the physical properties of water. Thus, the boiling point, melting point, and maximum specific gravity of ordinary water are 100°C., 0°C., and 1.0000, while the corresponding figures for heavy water are 101.4°C., 3.82°C., and 1.1071.

A systematic symbolism for isotopes involves the use of a subscript before the symbol of the element to indicate the atomic number and a superscript after the symbol to indicate the mass number. A generalized such symbol would $_ZX^A$. As specific examples we might put forward the case of protium, deuterium, and tritium, which in this notation would be $_1H^1$, $_1H^2$, and $_1H^3$. Similarly, the two carbon isotopes would be $_6C^{12}$ and $_6C^{13}$, while the three oxygen isotopes would be $_8O^{16}$, $_8O^{17}$, and $_8O^{18}$. Since each element has an invariant atomic number, the initial subscript is not necessary and there is no ambiguity in referring simply to H^2 or C^{13}. Or, for that matter, the chemical symbol may be left out and only the two numbers given. In the biochemical literature the first of the two simplified notations is most frequently found.

Nuclear Stability

The atomic nucleus has within it the potentialities of instability. It contains only one kind of charged particle, the proton, and by the ordinary laws of electrostatic attraction one would expect that these protons would

TABLE APP-1

Naturally-occurring isotopes of elements of biochemical interest

ELEMENT	NUMBER PROTONS (Z)	NUMBER NEUTRONS (A − Z)	MASS NO. (A)	PER CENT OCCURRENCE WITHIN ELEMENT
Hydrogen	1	0	1	99.98
Hydrogen (deuterium)	1	1	2	0.02
Hydrogen (tritium)	1	2	3	trace
Carbon	6	6	12	98.9
Carbon	6	7	13	1.1
Nitrogen	7	7	14	99.62
Nitrogen	7	8	15	0.38
Oxygen	8	8	16	99.76
Oxygen	8	9	17	0.04
Oxygen	8	10	18	0.20
Fluorine	9	10	19	100.00
Sodium	11	12	23	100.00
Magnesium	12	12	24	78.6
Magnesium	12	13	25	10.1
Magnesium	12	14	26	11.2
Phosphorus	15	16	31	100.00
Sulfur	16	16	32	95.06
Sulfur	16	17	33	0.74
Sulfur	16	18	34	4.18
Chlorine	17	18	35	75.4
Chlorine	17	20	37	24.6
Potassium	19	20	39	93.3
Potassium	19	21	40	0.01
Potassium	19	22	41	6.7
Calcium	20	20	40	96.96
Calcium	20	22	42	0.64
Calcium	20	23	43	0.15
Calcium	20	24	44	2.06
Calcium	20	26	46	0.003
Calcium	20	28	48	0.19
Manganese	25	30	55	100
Iron	26	28	54	5.84
Iron	26	30	56	91.61
Iron	26	31	57	2.17
Iron	26	32	58	0.31
Cobalt	27	32	59	100.00
Copper	29	34	63	69.1
Copper	29	36	65	30.9
Zinc	30	34	64	48.89
Zinc	30	36	66	27.81
Zinc	30	37	67	4.11
Zinc	30	38	68	18.56
Zinc	30	40	70	0.62
Iodine	53	74	127	100.00

repel one another and that a structure composed only of positive charges would not exist. Actually, no atomic nucleus consists of more than one proton without the addition of neutrons as well. Although neutrons are uncharged, the laws of electrostatic attraction and repulsion between charges at distances comparable to those within an atomic nucleus (which are 10^5 times less than those which occur anywhere else in nature) are quite different from those met with elsewhere. Apparently at such short distances, various arrangements of protons and neutrons can become stable aggregates. The nature of the attractive forces between nucleons is a major field of investigation among nuclear physicists today and is beyond the scope of this book. We need only know that for a given number of protons in a nucleus only a certain number of neutrons will, when present, yield a stable aggregate.

In the case of certain nuclei, usually among those containing an odd number of protons, the number of neutrons required for stability is fixed and can not vary. Examples of such elements are: sodium, whose 11 protons must be joined with 12 neutrons, no more and no less, for stability; phosphorus, whose 15 protons must be joined with 16 neutrons; and iodine, whose 53 protons must be joined with 74 neutrons. Other nuclei, particularly those with an even number of protons, are more liberal in their neutronic requirements. Tin, with 50 protons, possesses stable nuclei of ten different mass numbers, containing 62, 64, 65, 66, 67, 68, 69, 70, 72, and 74 neutrons, respectively. This is another way of saying that tin has ten stable isotopes (though the one with 74 neutrons may be slightly unstable).

Not all stable proton-neutron combinations are equally stable, if we judge by the frequency with which such combinations occur in nature. Thus, as we see from table 47, 99.76 per cent of all oxygen molecules contain nuclei with eight protons and eight neutrons. While those nuclei containing eight protons and nine or ten neutrons are also stable in that they will exist unchanged indefinitely unless subjected to subatomic bombardment, they must have been less readily formed—i.e., presented less economical configurations from an energic viewpoint in the early formative times of the universe. The majority of the elements, even when they possess a number of isotopes, are predominantly in the form of one particular, and presumably, most stable isotope. This is particularly true among the simpler elements with chlorine as an exception since two isotopes of that element exist in the ratio 3:1, and boron, whose two isotopes are distributed in the ratio of 4:1.

The ratio of distribution of the various isotopes within an element, particularly for lighter elements, such as hydrogen, boron, carbon and oxygen varies somewhat with the source. Thus, the O^{16}/O^{18} ratio in nature varies about 4 per cent, while the B^{11}/B^{10} ratio varies some 3 per cent (8). Of particular interest to the biochemist are the variations found in living tissue. Thus, the C^{12}/C^{13} ratio in plants differs slightly with the environment in which the plants grow. Rain-forest plants have a ratio of about 91, while marine plants show ratios of 89–90 (17). The ratio in the general environment, as given by Sullivan, is 88.

Among the first twenty elements of the atomic table the most stable proton-neutron combinations are frequently those where the number of each is equal. Thus, the most common isotope of helium has two protons and two neutrons, that of carbon six of each, that of nitrogen seven of each, of oxygen eight of each, of magnesium twelve of each, of sulfur sixteen of each, and of calcium twenty of each. Exceptions exist, particularly among the nuclei containing odd numbers of protons, where the number of neutrons is one greater (as in sodium with 11 protons and 12 neutrons, or phosphorus with 15 and 16), or in the case of hydrogen, one less.

In general the most stable nuclei tend to be those containing an even number of protons and an even number of neutrons. Examples of these are He^4, C^{12}, O^{16}, Mg^{24}, Si^{28}, and S^{32}. Conversely, those nuclei with odd numbers of both protons and neutrons are least stable, and among elements of biochemical interest the only important case of such a nucleus is that of N^{14} with seven of each. Deuterium with one of each, and K^{40} with nineteen protons and twenty-one neutrons are present in their elements only to one or two hundredths of a per cent.

As the complexity of the nucleus increases and more and more protons are packed into its narrow limits, the proportion of neutrons must be continually increased to maintain stability. In table APP-1 we see that the most common isotope of iron has 26 protons and 30 neutrons (a neutron-proton ratio of 1.15), while in iodine the only stable combination is 53 protons and 74 neutrons (a neutron-proton ratio of 1.40). The most complex stable nucleus known is that of bismuth which contains 83 protons and 126 neutrons, a neutron-proton ratio of 1.52. Apparently, when the number of protons in a nucleus is higher than 83, no quantity of neutrons will suffice for stability, and stable nuclei more complex than bismuth do not exist.

Radioactivity

Nuclei more complex than that of bismuth, although not stable, do exist. The neutron-proton composition of those nuclei spontaneously changes, becoming eventually a stable form. This change or "decay" is the phenomenon called radioactivity. Radioactive nuclei decay in one or more of three ways. They may emit an alpha-particle, consisting of two protons and two neutrons, lowering the proton content and by decreasing their more numerous neutrons only an equal amount, increasing the neutron-proton ratio. The alpha-particle is, actually, the nucleus of a helium atom and may be represented chemically as He^{++}, although it is more usually symbolized as the Greek α. Once outside the confines of the radioactive nucleus the alpha-particle will tend to capture two electrons from the surroundings and become gaseous helium. For this reason, ores containing such radioactive elements as uranium and thorium frequently contain occluded helium as well.

Another form of radioactive emission is the *beta-particle*. which is simply a high speed electron; and a third form of emission is the *gamma-ray*, which is not a particle in the ordinary sense but an electromagnetic radiation, just as ordinary light is, but much more energetic. The wave length of gamma-rays is much shorter than that of even the x-ray, and the radiation is correspondingly more energetic and penetrating. The emission of an electron from a nucleus containing only protons and neutrons represents the change of a neutron into a proton, the gain of a positive charge being equivalent to the loss or ejection of a negative charge in the form of an electron.

A nucleus which emits an alpha-particle loses four in mass number and two in atomic number. One which emits a beta-particle loses nothing in mass number but gains one in atomic number. One which emits a gamma-ray is unchanged in both respects. By using a combination of such changes successively, a nucleus such as that of uranium can change to a somewhat simpler nucleus, which will itself change to another and so on, until a stable nucleus such as one of lead or bismuth is reached and the process ceases.

These radioactive changes have two startling properties. In the first place, a tremendous amount of energy is released per mass of material as the changes take place, far more energy than can be released by even the most energetic chemical reaction known to man. Secondly, the rate at which the changes take place is unaffected by any change in temperature or pressure. Two questions may therefore be raised, one for each of these properties. First, where does the energy come from, and second, why are not all the radioactive nuclei long since decayed and gone?

Radioactive energy is derived from the supply within the atomic nucleus. Just as molecules contain a store of chemical energy which was utilized in the formation of various electronic bonds between atoms (see page 224), so nuclei contain a store of *nuclear energy* in the more powerful bonds between protons and neutrons. A rearrangement of nuclei into more stable configurations involves of necessity the release of some of this energy. Nuclear energy is of an order so great that it can be detected as mass. Einstein has shown that there is a definite mass-energy equivalence, and that one gram of mass represents the equivalent in energy of 9×10^{20}

ergs. The energy changes in ordinary chemical reactions represent changes in mass so small that our most delicate instruments can not detect them. In nuclear changes, however, gains and losses of energy are large enough to upset the "law" of conservation of mass if not allowed for.

The radioactive nuclei still exist simply because of the extremely low rate at which some of them change. Such elements as uranium and thorium are almost but not quite stable. U^{238} has a half life of nearly five billion years. That means that of a given mass of uranium, half will decay in five billion years, half of that remaining will decay in another five billion years, and so on. In the case of Th^{232}, the half life is even longer, nearly fourteen billion years. Although the elements into which they decay are much more unstable, with half lives that can be counted in centuries (radium is 1600 years), days (radon is four days), and split seconds (thorium C′ is one billionth of a second), uranium and thorium remain as unfailing sources of entire series of radioactive compounds. Since the age of the universe has been estimated as a mere three to four billion years, it can be seen that more than half of the original uranium and more than four-fifths of the original thorium exist. A third series of radioactive compounds has as its long-lived parent the now notorious U^{235}, whose half life—a mere nine hundred million years—is long enough to allow remnants of itself still to exist.

What happens when no sufficiently long-lived parent exists is indicated by the missing fourth series (there is room for four series all told, for reasons that we need not go into here) of naturally radioactive elements. The most long-lived member of that series is one of the isotopes of neptunium, Np^{237}, whose half life is a trifle of only two million years. During the lifetime of the earth, neptunium has decayed away into an indetectable remnant and is now known only because man has learned to create nuclei artificially (2).

Radioactivity is not confined to elements more complex than bismuth. Bismuth (Bi^{209}) itself is now thought to be an alpha-particle-emitter, with the extraordinarily long half life of 2×10^{17} years. Any element, if the neutrons in its nucleus are increased or decreased below certain narrow limits, becomes radioactive. Among the elements up to and including bismuth, however, radioactive isotopes usually have comparatively small half lives so that any which may have existed have long since vanished. A very few exceptions exist among naturally-occurring nuclei, which although relatively simple in structure, have long half lives. Among the elements of biochemical interest the best example is K^{40}, with a half life of about one and a fifth billion years. It emits beta rays. Tritium is also radioactive, emitting weak beta rays. It has a half life of only twelve years; thus it would not be expected to exist naturally but it is, however, continually recreated in trace quantities in natural transmutative processes.

A basic unit of radioactivity is the *curie*, which is defined as the quantity of a radioactive material in which 3.70×10^{10} disintegrations are taking place per second. In the case of Ra^{226}, 1 gram of the substance represents 1 curie of radioactivity. The mass of 1 curie of a given nuclide is in inverse proportion to its half life. Both the *millicurie* (1/1000 curie) and the *microcurie* (1/1000 millicurie) are frequently used. For example 2.8 grams of U^{238} represent 1 microcurie of radioactivity. A more recently introduced but less used unit of radioactivity is the *rutherford*, which represents a quantity of radioactive material in which 10^6 disintegrations are taking place per second. One millicurie is equal to 37 rutherfords. One rutherford of radioactivity is represented by 76 grams of U^{238} or by 27 micrograms of Ra^{226}.

NUCLEAR REACTIONS

Transmutation

The radiations of radioactive materials are so energetic that if allowed to impinge upon other atoms they do not merely disturb the outer electrons and thus effect chemical changes as do such less energetic forms of radiation as visible light, but actually disturb the nucleus, initiating changes in its structure and sometimes disrupting it. Such nuclear changes amount frequently to changes in the nature of the element and represent the first true success in the old alchemists' dream of transmutation. Devices such as the cyclotron and the atomic pile are used to produce high concentrations of nucleons. In addition, deuterons (the nuclei of deuterium, consisting of one neutron-proton pair) are used.

A typical nuclear reaction would be symbolized as follows

$$_{11}Na^{23} + {}_2He^4 \rightarrow {}_1H^1 + {}_{12}Mg^{26}$$

the significance of this is that an ordinary sodium atom when struck by alpha-particles ($_2He^4$) absorbs the impinging particle and emits a proton ($_1H^1$), leaving behind a magnesium isotope ($_{12}Mg^{26}$). Note that in such a reaction the atomic numbers and the mass numbers as well add up to the same values on each side of the equation.

Other symbols used in writing nuclear reactions are $_1H^2$ for the deuteron, $_0n^1$ for the neutron. Electrons may be represented as $_{-1}e^0$ and positrons (very short-lived "positive electrons" which are sometimes emitted by nuclei) as $_1e^0$. A gamma-ray is indicated by the Greek letter γ (just as electrons and positrons are sometimes symbolized by the Greek letter β, with a negative or positive superscript). The reader should be able to interpret without further explanation such nuclear reactions as

$$_{7}N^{14} + {}_{0}n^{1} \rightarrow {}_{6}C^{14} + {}_{1}H^{1}$$

$$_{1}H^{2} + {}_{1}H^{2} \rightarrow {}_{1}H^{3} + {}_{1}H^{1}$$

$$_{17}Cl^{37} + {}_{0}n^{1} \rightarrow {}_{17}Cl^{38} + \gamma$$

In general, the bombardment of atomic nuclei with these high-energy particles results in the emission by those nuclei of relatively simple particles and a simplified notation for such reactions has now arisen. The form of this notation may be described as

$$X^{x}(a, b)Y^{y}$$

which indicates that when the nucleus X with a mass number x is bombarded with a, it emits b, leaving behind the nucleus Y with the mass number y. In this notation, the proton and deuteron are represented by the symbols p and d, while the alpha-particle is represented as α. Other symbols are the same as in the case of the more extended notation shown.

In the condensed notation the four nuclear reactions given would be represented as

$$Na^{23}(\alpha, p)Mg^{26}$$

$$N^{14}(n, p)C^{14}$$

$$H^{2}(d, p)H^{3}$$

$$Cl^{37}(n, \gamma)Cl^{38}$$

In at least one variety of nuclear reaction, the nucleus bombarded does not remain largely intact while emitting small fragments of mass number 1 or 2, but splits into two or more reasonably equal portions. This is *nuclear fission*, and occurs when U^{235} is bombarded with slow neutrons. In splitting into such fragments as nuclei of barium, technetium, and so on, more neutrons are produced and if these are slowed by the inclusion of such "moderators" as heavy water in the system, more uranium atoms are split and a cyclic process is set up in which in extraordinarily short periods extraordinarily huge quantities of energy are released. The application of such chain reactions to the development of the atom bomb is known to all.

Artificial Radioactivity

In some nuclear reactions well known naturally occurring isotopes are created or "synthesized" in the process. In the first reaction listed Mg^{26} is formed, a respectable magnesium isotope occurring as 11.1 per cent of all magnesium atoms. In the second reaction, however, C^{14} is formed, and Cl^{38} is formed in the last. Many such "unnatural" isotopes have been artificially formed and in all cases they were found to be unstable, and to rearrange spontaneously, giving more stable nuclei.

Thus, C^{14} emits an electron to become the ordinary nitrogen isotope N^{14}. The half life of this reaction is 5570 years. The Cl^{38} emits an electron to become the stable argon isotope A^{38}. The half life of that reaction is 38.5 minutes. Such spontaneous decay is in every respect similar to that of the spontaneous decay of natural radioactive elements. The phenomenon is therefore known as artificial radioactivity. C^{14} has become of interest as a means of dating archaeological specimens. It is continually formed from the N^{14} of air by cosmic ray bombardment. As a result, all matter which incorporates atmospheric CO_2, directly or indirectly (i.e., all living matter) contains a fixed ratio of C^{14} to the stable carbon isotopes. Carbonaceous materials of organic origin (bones, wooden artifacts, etc.) which no longer exchange with atmospheric CO_2, slowly lose what C^{14} they contain. From the known half life of C^{14} and the change in C^{14}/C^{12} ratio, the time lapse since the object formed part of an active carbon cycle can be determined provided care is taken to make proper allowance for natural variations in C^{14}/C^{12} ratios (13).

Radioactive isotopes (or, as they are often called, *radioisotopes*), the vast majority artificially formed, are known for every chemical element. In most cases a considerable number are known. At least fifteen radioactive isotopes of tellurium, for instance, have been described. Artificial radioactivity generally involves the emission of beta-particles and gamma-rays. Alpha-particles are confined to those complex atoms with mass numbers of 190 or more, with the single exception of one naturally occurring samarium isotope, Sm^{147}, which emits the weakest known alpha-rays and has a half life of the order of two hundred billion years.

A great many radioactive isotopes have found use in one branch or another of biochemical research. Kamen (7) lists 201 of them, distributed among 46 different elements. Radioactive isotopes of elements significant in metabolism studies are listed in table APP-2, the data again being taken from Sullivan (16). The choice between several possible radioactive isotopes depends upon half life, characteristic radiation, the ease of manufacture, and so on. Details on the preparation of these isotopes and their comparative uses may be found in great detail in Kamen.

It will be noted that no radioactive isotopes of more than very fugitive life have been found for two important elements, nitrogen and oxygen, and it is not expected that any will be found in the future.

TABLE APP-2

Some radioactive isotopes of elements of biochemical significance

ELEMENT	MASS NUMBER	HALF LIFE
Hydrogen	3	12.5 years
Carbon	11	20 minutes
Carbon	14	5570 years
Fluorine	18	2 hours
Sodium	22	2.6 years
Sodium	24	15 hours
Phosphorus	32	14.3 days
Sulfur	35	87 days
Chlorine	34	33 minutes
Chlorine	36	440 thousand years
Chlorine	38	37.5 minutes
Potassium	40	1.2 billion years
Potassium	42	12.5 hours
Potassium	43	22.5 hours
Calcium	45	164 days
Calcium	47	5 days
Iron	55	3 years
Iron	59	46 days
Cobalt	55	18 hours
Cobalt	56	72 days
Cobalt	57	27 days
Cobalt	58	72 days
Cobalt	60	5.3 years
Copper	61	3.4 hours
Copper	64	12.8 hours
Copper	67	58 hours
Zinc	62	9.5 hours
Zinc	65	25 days
Zinc	69	13.8 hours
Zinc	72	2 days
Iodine	124	4 days
Iodine	125	56 days
Iodine	126	13 days
Iodine	129	17 million years
Iodine	130	13 hours
Iodine	131	8 days

METABOLIC STUDIES

In this book several applications of isotopes to biochemistry have been mentioned. Thus in Chapter 2 we spoke of the isotope dilution technique as applied to analytical biochemistry, where an isotopically "tagged" substance is added to a mixture containing the normal substance. Separation of a pure sample of the substance and the determination of the fraction which is isotopically tagged allows us to compute the original total quantity of the substance in the mixture. In Chapter 10 we referred to the application of radioactive isotopes in cancer therapy.

The most important application of isotope technique to the science of biochemistry, however, involves the elucidation of intermediary metabolic pathways. We know with considerable accuracy the nature of the substances ingested by man and how they are broken down in the alimentary canal prior to absorption. We know also the final materials which man biosynthesizes out of these absorbed foodstuff derivatives and the waste products excreted in the urine. But how, exactly, do the foodstuffs pass from glucose and amino acids to steroids and urea—by exactly what steps—through exactly which intermediate stages?

The simplest attempt at a solution has been to feed an animal (when possible, a human) measured quantities of a particular chemical, and observe any increase in some component of the urine. This would give a first stage and a last stage, with only conjecture as to the nature of the many possible intermediate steps. Even if it were possible to take animals which had fed on substance A, kill them at various intervals after ingestion, and analyze completely each tissue of each organ for all its components, such would be the multiplicity of the materials to be considered that it would be hopelessly impractical to attempt to assemble such a jigsaw puzzle.

It then occurred to biochemists to "label" an ingested compound, that is, add to its structure some chemically identified group by means of which its wanderings through the cell mechanisms of man might be traced. Before the days of widespread application of isotopic technique, such a label might include a particular metal or compound not ordinarily found in the body, such as lead or benzoic acid. These materials would be all "label". Such experiments sometimes had interesting results. Thus ingestion of benzoic acid resulted in the excretion in the urine of hippuric acid, a condensation product of benzoic acid and the amino acid, glycine. Such studies pointed up the use of glycine as a "detoxifying" substance.

Another example of such labeling is the study of Knoop of the fat metabolism of animals through their ingestion of various phenyl derivatives of fatty acids. It was possible to locate in the urine certain phenylated compounds such as hippuric acid or phenylacetylglycine, and from their occurrence advance a theory as to the beta-oxidation of fatty acids by the body.

The fault in all such studies is that the materials ingested are entirely or in part non-physiological. Since it is the essence of this labeling process that the label be a group which can be recognized by the chemist from among the myriad groups already present in the body, as a necessity it must be something the body does not ordinarily encounter. The result is that biochemists find themselves studying abnormal metabolism and the detoxication of poisons, not the normal processes they most wish to study.

The problem was, then, to find a substance which to the body was normal and indistinguishable from its usual raw materials, and yet which to the chemist was surely identifiable under all circumstances. This proposition, impossible on the face of it, was solved neatly and handily by the use of isotopes.

Earliest experiments were conducted with naturally occurring stable isotopes. In the early 1930's, ways were discovered for fractionating elements or simple compounds so as to obtain fractions richer than normal in one of the less common isotopes. Such concentrated sources of, say, deuterium or N^{15}, could be used in the synthesis of fatty acids or amino acids which, in turn, would be richer than normal in the rare isotope. If animals were fed these isotopically tagged substances and later killed and their organs analyzed, it would no longer be a question of trying to place all the compounds found, but only those which likewise were richer than normal in the ingested isotope. The jigsaw puzzle is tremendously simplified.

Except for those elements, notably nitrogen and oxygen, where usable radioactive isotopes do not exist, the use of stable isotopes has given way to that of the radioactive variety. This is so for one important reason—radioactive isotopes are far easier to estimate in far smaller quantities than are the stable isotopes. The stable isotopes can be estimated by means of the *mass spectrograph*, which operates on the principle that in an electric field and magnetic field of given intensities, two ions of similar charge will be deflected by an amount inversely proportional to their masses. Thus hydrogen ions travel in a path more curved than that of deuterium ions and therefore strike the photographic plate at a different point. Mass spectroanalysis is expensive. Radioactive isotopes, on the other hand, can be measured by means of their radiations, which can be made to actuate counting devices.

Most frequently, tracers, both stable and radioactive, are used to follow the pathways of intermediate metabolism. H^2 and H^3 can be used in studying fat and water metabolism. N^{15} is useful in the study of amino acids; S^{35} for cystine and methionine; P^{32} for nucleic acids; and Fe^{59} for iron. C^{14} can be applied most generally to the organic compounds of the body. In addition, tracer technique can be applied to a variety of other problems in biochemistry and physiology. K^{42} and Na^{24} have been employed in research on the permeability of cell membranes. I^{131} has been used in the study of thyroid function. Fe^{59} has been used to tag red cells in order to determine their half life, while P^{32}-tagged red cells have been employed in determining circulating red-cell volume and blood circulation time. Diiodofluorescein (containing I^{131}) has been found to localize in brain tumors, and radiation counters may be used to locate the exact position of the tumors prior to operation. For a discussion of some of these applications, see Popjak (12).

Radiation Sickness

The use of radioactive isotopes carries with it one important disadvantage. Because of the potential damage to living tissue by their radiations, unusual precautions must be taken which increase both the expense and the inconvenience of the procedure. Tissue changes may be induced by quite small doses of radiation (15). As little as 0.8 roentgen units per day may cause skin changes. (A roentgen unit, symbolized as r, is defined as the amount of x-rays or gamma-rays which, on passage through dry air under standard conditions, produces by ionization, one electrostatic unit of electricity of either sign per milliliter.) Some tissue changes may appear as long as 25 years after exposure to radiation ceases. This may be due to long-term retention of radioactive isotopes in bone, where atom exchange and consequent excretion is much slower than in the soft tissue. This is well known in the case of radium, which is chemically similar to calcium and is readily deposited in the bones, giving rise to the so-called radium poisoning. Similar bone retention may take place with an isotope as generally-used as C^{14} (14).

In studying the effects of radiation on man, alpha and beta particle emission must be considered as well as x-rays and gamma-rays. It is only for the electromagnetic radiations that the roentgen unit may be applied directly. To include the effects of particles, the *roentgen equivalent physical* or *rep* is defined as the quantity of ionizing radiation of any kind which, if absorbed by the soft tissues of the body, causes an energy gain per gram of tissue equivalent to that caused by 1 r of x-rays or gamma rays.

Since different types of radiation may contribute an equal amount of energy to the tissues yet give rise to differing degrees of damage because of differences in penetration or in the nature of their chemical effects, the *roentgen equivalent man* or *rem* is defined as the quantity of radiation of any type which produces the same amount of *damage* in man upon absorption as does 1 r of x-rays or gamma rays. For x-rays, gamma rays, and beta particles, 1 rep equals approximately 1 rem, but for alpha particles, 1 rep equals 10 to 20 rem. It would thus appear that alpha-particle emitters demand particular precautions in handling.

On the basis of current knowledge, the total permissible dose of all kinds of ionizing radiation has been set at 0.3 rem per week over the whole body. In defining this dose, it is understood that very concentrated whole-body exposure for short times is dangerous even within the limits of the permitted amount per week. Larger doses, up to 1.5 rem per week, may be permitted, if exposure is limited to hands and forearms. For permissible doses of individual radioisotopes, see (10).

While the physiological symptoms of *radiation sickness* are well known, the chemical changes involved are still a matter of dispute. It seems reasonable to suppose that the primary changes resulting from the impact of radiation on the body would occur in those molecules that form the largest part of the body and would, therefore, on a purely statistical basis, be most often hit. This would be the water molecule and the effect of radiation would be to produce highly reactive *free radicals* (11). A free radical is a chemical grouping distinguished by the absence of net charge and the presence of an unpaired electron. Two such free radicals that could be produced from water are *hydroxyl* $(\cdot\underset{\cdot\cdot}{\overset{\cdot\cdot}{O}}{:}H)$ and *peroxyl* $(\cdot\underset{\cdot\cdot}{\overset{\cdot\cdot}{O}}{:}\underset{\cdot\cdot}{\overset{\cdot\cdot}{O}}{:}H)$. These reactive free radicals are then able to produce the symptoms of radiation sickness by reacting with some of the key substances of the body. Minor contributions to the body's chemical disorder may result from the direct formation of NH_3 from amino acids, purines, or pyrimidines or of H_2S from cysteine or glutathione, as the result of exposure to radiation. Local pH changes along the track of an entering particle or gamma-ray may also be significant. The effect of radiation on individual compounds of biological importance is discussed in (6).

It would seem most likely that the substance which is most dangerously damaged by the free radicals is DNA. The nucleus is more radiosensitive than is the cytoplasm; mitotic cells are more radiosensitive than resting cells; and tissues, such as spleen, which are particularly rich in DNA, are also particularly sensitive. Epstein (5) reports data indicating that the radiosensitive portion of a virus is the nucleic acid and not the protein. The depolymerizing effect of x-rays on solutions of DNA *in vitro* is well known and probably takes place through the intermediate formation of free radicals derived from water. The key position of DNA as the chemical arbiter of the cell and its reproduction makes this theory seem an attractive one and would account moreover for the general increase in mutations occurring among the offspring of organisms exposed to radiation.

Certain chemicals have been found to increase radioresistance (9). The best-known are the —SH containing substances, such as cysteine and glutathione. These must be introduced into the laboratory animal either just before or just after exposure. The effect would seem to be that of an anti-oxidant. That is, the thiol group, being readily oxidized, reacts preferentially with the strongly oxidant free radicals, thus tending to decrease their concentration and protect the less reactive DNA. Induction of hypoxia during or just prior to irradiation also increases radioresistance, presumably through inhibition of oxidation effects owing to the lowered concentration of blood oxygen.

Precautions in Tracer Work

To insure validity of experimental results with tracers, either radioactive or stable, additional precautions must be taken.

(1) Sufficient isotope must be used to remain detectable after dilution in the body. Once introduced into the body, glycine containing C^{14}, for instance, becomes indistinguishable from all other glycine in the body and in not too long a time becomes evenly distributed throughout all the body glycine. Subsequent isolation of glycine from some organ of the animal will then contain only a fraction of the C^{14} contained by an equivalent mass of the starting material. This smaller amount must still be detectable or the initial concentration must be increased.

(2) The tracer used must be located on the molecule in a stable fashion and must not "exchange" with its environment. Thus, if ethyl alcohol is prepared with a deuterium atom bound to the oxygen in the place of the normal hydrogen, thus, C_2H_5OD, in the hope that ingestion of the compound will show the metabolic fate of the alcohol group, non-significant results would be obtained. The reason for this is that if C_2H_5OD were simply allowed to remain in water solution, it would be found that there was a rapid exchange of oxygen-bound hydrogen (or deuterium) between the alcohol and the water. In the body, therefore, the deuterium of the ethyl alcohol

$$C_2H_5OD + HOH \rightleftharpoons C_2H_5OH + HOD$$

would quickly be distributed throughout the body water and any organic compound in which it might later be identified might just as well have acquired it from water as from alcohol. In general, deuterium bound to oxygen, nitrogen, or to a carbon adjacent to a carbonyl group, suffers from this instability and can not be used in tracer experiments. Deuterium bound to carbons not adjacent to carbonyl groups stays put and may be used.

(3) Abnormalities in metabolism must not be brought about by the use of the isotopes. Radiation may alter the functioning of tissues and radioactive isotopes must therefore be used in the lowest concentrations consistent with their sure analysis. With stable isotopes, radiation damage does not occur, but with carbon and nitrogen, mass differences among isotopes result in partial fractionation during metabolism (1, 3). If allowance is not made for this, results of tracer studies may be misinterpreted. The effect is most extreme with hydrogen. Thus, rats fed on water containing deuterium and tritium are found to have incorporated deuterium preferentially to tritium in their liver glycogen to an extent of 8 per cent. The preference in the case of liver fatty acids was 18 per cent. It is surmised that the prefer-

ence for protium (H^1) over deuterium is of a similar order (4). Since in all cases, the heavier deuterium reacts more slowly than protium, high concentrations of H^2 inhibit respiration and fermentation.

(4) Allowance must be made for the loss in intensity of isotope radiation during the time of the experiment and the time required afterwards for analysis. Radioactive decay is a first-order phenomenon (see page 291) so that the rate at which the number of atoms, N, of the radioactive substance decreases with time is proportional to the number of atoms present at that time. Mathematically, this is presented as follows:

$$-\frac{dN}{dt} = \lambda N \qquad [1]$$

where λ is the *radioactive constant.* The radioactive constant is characteristic for a given species of radioactive atoms.

If the number of radioactive atoms present at the beginning of an experiment is symbolized as N_0 and the number at any subsequent time as N_t, then integration of equation [1] from 0 to t gives:

$$\ln N_t - \ln N_0 = -\lambda t \qquad [2]$$

where the symbol ln refers to logarithms to the base e. Converting these to logarithms to the base 10, setting N_0 arbitrarily equal to unity and remembering that the logarithm of one to any base is equal to zero, we have

$$\log N_t = -0.4343\lambda t \qquad [3]$$

Solving equation [3] for N_t will give us the fraction of the original number of atoms remaining after the lapse of time, t. It can be seen from equation [3] that no matter how large t may be made, N_t can never be zero. In other words, from a theoretical standpoint, a mass of radioactive material will never completely decay.

The decay of individual atoms is unpredictable. A given atom may decay after a second or not for a billion years. In dealing with huge numbers of atoms, however, we may make statistical predictions in the same way that insurance companies can predict how many Americans will die in a given year though the lifetime of any one individual is unpredictable.

Since masses of radioactive matter never decay completely regardless of the rate at which the constituent atoms break down, the "life" of a radioactive isotope is a meaningless term. Instead, we use the half life, which is defined as the time interval in which the number of radioactive atoms in a given mass has been reduced to half the original quantity. To determine the half life, set N_t equal to ½ in equation [3]. Since log 2 is equal to the negative of log ½, it follows that:

$$\log 2 = 0.4343\, \lambda t_{1/2} \qquad [4]$$

where $t_{1/2}$ is the half life. Log 2 is equal to 0.3010 and we have:

$$t_{1/2} = \frac{0.6391}{\lambda} \qquad [5]$$

Since λ is a constant, characteristic for each kind of radioactive isotope, so is the half life. We have already mentioned on page 882 the long half lives of some of the uranium and thorium isotopes and in table APP-2 we have listed the considerably shorter half lives of some of the radioactive isotopes used in biochemical research.

The practical significance of the rate of radioactive decay lies in the extent of decay over the actual time interval of a given experiment. Thus, in the case of C^{14}, with a half life in the neighborhood of 5,000 years, the loss of even 0.1 per cent of the radioactive atoms requires a time lapse of more than 7 years. In experiments using C^{14}, then, the progress of decay may be ignored.

In the case of most other radioactive isotopes used in research, the rate of decay has to be taken into account. The fraction, F, of any isotope remaining after a given time, t, is related to the half life, $t_{1/2}$, as follows:

$$F = 2^{-t/t_{1/2}} \qquad [6]$$

Utilizing equation [6], table APP-3 can be prepared, in which the extent of decay over a short period of time is given for some of the radioactive isotopes most frequently used in biochemical research. The case of C^{11}, sometimes used in short experiments, is more extreme than any of the cases listed in the table. With a half life of 20 minutes, 99.9 per cent of its activity is lost in three and a half hours.

Dynamic State of Body Constituents

The paths or partial paths of many biosyntheses have been determined by means of isotopic techniques. Thus it was found that glycine is used as

TABLE APP-3

Isotope decay over short intervals of time

RADIOACTIVE ISOTOPE	HALF LIFE	PER CENT DECAYED IN 6 HOURS
H^3	12.5 years	0.005
Na^{24}	15 hours	24.1
P^{32}	14.3 days	1.21
S^{35}	87 days	0.04
K^{42}	12.5 hours	28.4
Fe^{59}	46 days	0.37
Cu^{64}	12.8 hours	27.7
I^{131}	8 days	2.14

one of the building blocks for heme, and that acetic acid is an important precursor of the steroid nucleus. In Chapter 9 the precursors for each atom of uric acid were given as determined by isotope studies. Various carbons were found to arise from carbonate and formate ions; one of the nitrogens with two neighboring carbons arose from glycine, while the remaining two nitrogens were derived from aspartic acid and from glutamine.

It is the nature of the *metabolic pool* that was the most significant contribution of isotopic techniques to the understanding of intermediary metabolism. One of the earliest findings, once isotopes began to be used, was that the compounds of the body were anything but stable. The depot fat, for instance, had been thought to lie quiescent within its cells, never stirring except and until undue exertion or underfeeding required its mobilization. This is not the case. If an organism is fed on deuterium-containing fat, the deuterium is eventually found to occur uniformly through the depot fat. This is not because of isotope exchange as described earlier in the case of oxygen-bound deuterium. Here the deuterium was bound stably to carbon and the only hypothesis that fits the fact is that the molecules of fat within the stores of the body are constantly changing with incoming fat, in a continuous dynamic equilibrium.

This is even more dramatically demonstrated with experiments carried out with amino acids labeled with heavy nitrogen (N^{15}). Here the nitrogen isotope, fed as part of one amino acid, was quickly found to occur in greater or lesser degree among all the kinds of amino acids in the body (with the exception of threonine and lysine). The picture one perceives is of a ceaseless deamination and reamination of most amino acids.

It might seem that this dynamic equilibrium is a great waste of energy, that it would be more economical on the part of living matter to allow its protein, fat, and other constituents to remain fixed in position. Probably this idea of "rest" is a false one. Actually, the molecules with which the body deals are, of necessity, so complex that they lack sufficient inherent stability to "rest", just as the long rod balanced on the nose is too unstable to remain standing if the juggler stops moving his head in a calculated fashion. Their instability is such that they are forever falling apart and only a device whereby they are put together again at an equivalent rate can allow the body to make use of their tremendously complicated structure. Since the larger molecules which compose living tissue are constantly releasing smaller molecules and groups, such fragments are always available as raw materials for biosyntheses and are spoken of as the metabolic pool.

REFERENCES

1. Abbott, L. D., Jr., *et al.* Natural abundance of nitrogen[15] in hemin and plasma protein from normal blood. Proc. Soc. Exper. Biol. Med., **84:** 402–404, 1953.
2. Asimov, I. Naturally occurring radioisotopes. J. Chem. Ed., **30:** 398, 1953.
3. Buchanan, D. L., *et al.* Carbon-isotope effects on biological systems. Science, **117:** 541–545, 1953.
4. Eidenoff, M. L., *et al.* The fractionation of hydrogen isotopes in biological systems. J. Am. Chem. Soc., **75:** 248–249, 1953.
5. Epstein, H. T. Identification of radiosensitive volume with nucleic acid volume. Nature, **171:** 394–395, 1953.
6. Guzman Barron, E. S. The effect of ionizing radiations on systems of biological importance. Ann. N. Y. Acad. Sci., **59:** 574–593, 1955.
7. Kamen, M. D. *Radioactive Tracers in Biology.* 2nd edit. New York, Academic Press, 1951.
8. Kaplan, I. *Nuclear Physics.* Cambridge, Mass., Addison-Wesley Publishing Co., 1955.
9. Loutit, J. F. Drugs which may give protection against atomic radiation. Mfg. Chemist, **22:** 49–51, 1951.
10. Morgan, K. Z., and Ford, M. R. Developments in internal-dose determinations. Nucleonics, **12:** 32–39, 1954.
11. Patt, H. M. Protective mechanisms in ionizing radiation injury. Physiol. Rev., **33:** 35–76, 1953.
12. Popjak, G. Certain aspects of the medical application of isotopic tracers. A critical review. Quart. J. Med., **21:** 83–122, 1952.
13. Rafter, T. A. Carbon-14 variations in nature and the effect on radiocarbon dating. New Zealand J. Sci. Technol., **37B:** 20–38, 1955.
14. Skipper, H. E., *et al.* The hazard involved in the use of carbon[14]. III. Long-term retention in the bone. J. Biol. Chem., **189:** 159–166, 1951.
15. Stone, R. S. The concept of a maximum permissible exposure. Radiology, **58:** 639–661, 1952.
16. Sullivan, W. H. *Trilinear Chart of Nuclear Species.* New York, John Wiley and Sons, 1949.
17. Wickman, F. E. Variations in the relative abundance of the carbon isotopes in plants. Nature, **169:** 1051, 1952.

Index

Number in italics indicates formula on that page

C

D

F

G

J

K

W

X

Y

Z